Interactive Activities

Google Earth™ interactive satellite imagery gives you a geographic context for global places and topics discussed in the text. Globe icons indicate when to visit the *ARIS* site, where you will find links to locations mentioned in the text and corresponding exercises that will help you understand environmental topics.

Case Study — Family Planning in Thailand: A Success Story

Down a narrow lane off Bangkok's busy Sukhumvit Road, is a most unusual café. Called Cabbages and Condoms, it's not only highly rated for its spicy Thai food, but it's also the only restaurant in the world dedicated to birth control. In an adjoining gift shop, baskets of condoms stand next to decorative handicrafts of the northern hill tribes. Piles of T-shirts carry messages, such as, "A condom a day keeps the doctor away," and "Our food is guaranteed not to cause pregnancy." Both businesses are run by the Population and Community Development Association (PDA), Thailand's largest and most influential nongovernmental organization.

The PDA was founded in 1974 by Mechai Viravaidya, a genial and fun-loving former Thai Minister of Health, who is a genius at public relations and human motivation (fig. 7.1). While traveling around Thailand in the early 1970s, Mechai recognized that rapid population growth—particularly in poor rural areas—was an obstacle to community development. Rather than lecture people about their behavior, Mechai decided to use humor to promote family planning. PDA workers handed out condoms at theaters and traffic jams, anywhere a crowd gathered. They challenged governmental officials to condom balloon-blowing contests, and taught youngsters Mechai's condom song: "Too Many Children Make You Poor." The PDA even pays farmers to paint birth control ads on the sides of their water buffalo.

This campaign has been extremely successful at making birth control and family planning, which once had been taboo topics in polite society, into something familiar and unembarrassing. Although condoms—now commonly called "mechais" in Thailand—are the trademark of PDA, other contraceptives, such as pills, spermicidal foam, and IUDs, are promoted as well. Thailand was one of the first countries to allow the use of the injectable contraceptive DMPA, and remains a major user. Free non-scalpel vasectomies are available on the king's birthday. Sterilization has become the most widely used form of contraception in the

country. The campaign to encourage condom use has also been helpful in combating AIDS.

In 1974, when PDA started, Thailand's growth rate was 3.2 percent per year. In just fifteen years, contraceptive use among married couples increased from 15 to 70 percent, and the growth rate had dropped to 1.6 percent, one of the most dramatic birth rate declines ever recorded. Now Thailand's growth rate is 0.7 percent, or nearly the same as the United States. The fertility rate (or average number of children per woman) decreased from 7 in 1974 to 1.7 in 2006. The PDA is credited with the fact that Thailand's population is 20 million less than it would have been if it had followed its former trajectory.

In addition to Mechai's creative genius and flair for showmanship, there are several reasons for this success story. Thai people love humor and are more egalitarian than most developing countries. Thai spouses share in decisions regarding children, family life, and contraception. The government recognizes the need for family planning and is willing to work with volunteer organizations, such as the PDA. And Buddhism, the religion of 95 percent of Thais, promotes family planning.

The PDA hasn't limited itself to family planning and condom distribution. It has expanded into a variety of economic development projects. Microlending provides money for a couple of pigs, or a bicycle, or a small supply of goods to sell at the market. Thousands of water-storage jars and cement rainwater-catchment basins have been distributed. Larger scale community development grants include road building, rural electrification, and irrigation projects. Mechai believes that human development and economic security are keys to successful population programs.

This case study introduces several important themes of this chapter. What might be the effects of exponential growth in human populations? How might we manage fertility and population growth? And what are the links between poverty, birth rates, and our common environment? Keep in mind, as you read this chapter, that resource limits aren't simply a matter of total number of people on the planet, they also depend on consumption levels and the types of technology used to produce the things we use.

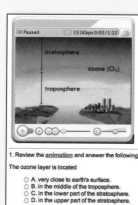

FIGURE 7.1 Mechai Viravaidya (right) is joined by Peter Piot, Executive Director of UNAIDS, in passing out free condoms on family planning and AIDS awareness day in Bangkok".

Case studies, flash cards, animation-based quizzes, and other activities are organized by chapter and designed to reinforce the concepts in the text.

DEFINITION

A REAL INCREASE IN WELL-BEING AND STANDARD OF LIVING FOR THE AVERAGE PERSON THAT CAN BE MAINTAINED OVER THE LONG-TERM WITHOUT DEGRADING THE ENVIRONMENT OR COMPROMISING THE ABILITY OF FUTURE GENERATIONS TO MEET THEIR OWN NEEDS.

TERM
NEXT
REMOVE CARD
RANDOMIZE DECK
CARDS REMAINING 14

CLICK ON THE "TERM" OR "DEFINITION" TO FLIP THE CARD OVER.

TERM

SUSTAINABLE DEVELOPMENT

DEFINITION
NEXT
REMOVE CARD
RANDOMIZE DECK
CARDS REMAINING 14

CLICK ON THE "TERM" OR "DEFINITION" TO FLIP THE CARD OVER.
PRESS "NEXT" TO MOVE ON TO THE NEXT TERM.
PRESS "REMOVE CARD" TO REMOVE THE CURRENT CARD FROM THE DECK.
PRESS "RANDOMIZE DECK" TO PUT ALL THE CARDS BACK IN THE DECK AND RESTART.

stratosphere
ozone (O₃)
troposphere

1. Review the animation and answer the following questions.

The ozone layer is located

- A. very close to earth's surface.
- B. in the middle of the troposphere.
- C. in the lower part of the stratosphere.
- D. in the upper part of the stratosphere.

2. One chlorofluorocarbon molecule has been estimated to be able to destroy

- A. one hundred ozone molecules.
- B. one thousand ozone molecules.
- C. one hundred thousand ozone molecules.
- D. one million ozone molecules.

3. Chlorofluorocarbons were used in all of the following EXCEPT

- A. refrigerators.
- B. the production of styrofoam.
- C. aerosol spray cans.
- D. the production of plastic.

Submit Test   Reset Test

TENTH EDITION

Environmental SCIENCE

A Global Concern

William P. Cunningham
University of Minnesota

Mary Ann Cunningham
Vassar College

McGraw-Hill Higher Education

Boston   Burr Ridge, IL   Dubuque, IA   New York   San Francisco   St. Louis
Bangkok   Bogotá   Caracas   Kuala Lumpur   Lisbon   London   Madrid   Mexico City
Milan   Montreal   New Delhi   Santiago   Seoul   Singapore   Sydney   Taipei   Toronto

McGraw-Hill
Higher Education

ENVIRONMENTAL SCIENCE: A GLOBAL CONCERN, TENTH EDITION

Published by McGraw-Hill, a business unit of The McGraw-Hill Companies, Inc., 1221 Avenue of the Americas, New York, NY 10020. Copyright © 2008 by The McGraw-Hill Companies, Inc. All rights reserved. Previous editions © 2007, 2005, 2003, 2001, 1999, and 1997. No part of this publication may be reproduced or distributed in any form or by any means, or stored in a database or retrieval system, without the prior written consent of The McGraw-Hill Companies, Inc., including, but not limited to, in any network or other electronic storage or transmission, or broadcast for distance learning.

Some ancillaries, including electronic and print components, may not be available to customers outside the United States.

⊕ This book is printed on recycled, acid-free paper containing 10% postconsumer waste.

1 2 3 4 5 6 7 8 9 0 QPD/QPD 0 9 8 7

ISBN 978–0–07–305138–3
MHID 0–07–305138–1

Publisher: *Janice Roerig-Blong*
Developmental Editor: *Rose M. Koos*
Editorial Coordinator: *Ashley A. Zellmer*
Senior Marketing Manager: *Tami Petsche*
Project Manager: *April R. Southwood*
Lead Production Supervisor: *Sandy Ludovissy*
Senior Media Project Manager: *Tammy Juran*
Senior Coordinator of Freelance Design: *Michelle D. Whitaker*
Cover/Interior Designer: *Jamie E. O'Neal*
 (USE) Cover Image: © *Keren Su / Corbis (Location Longji, China)*
Senior Photo Research Coordinator: *Lori Hancock*
Photo Research: *LouAnn K. Wilson*
Supplement Producer: *Melissa M. Leick*
Compositor: *Aptara, Inc.*
Typeface: *10/12 Times Roman*
Printer: *Quebecor World Dubuque, IA*

The credits section for this book begins on page 598 and is considered an extension of the copyright page.

Library of Congress Cataloging-in-Publication Data

Cunningham, William P.
  Environmental science : a global concern. — 10th ed. / William P. Cunningham, Mary Ann Cunningham.
    p. cm.
  Includes index.
  ISBN 978–0–07–305138–3 — ISBN 0–07–305138–1 (hard copy : alk. paper) 1. Environmental sciences—Textbooks. I. Cunningham, Mary Ann. II. Title.
GE105.C86 2008
363.7—dc22
                                                                    2007029516

www.mhhe.com

Periodic Table of the Elements

MAIN-GROUP ELEMENTS

Legend:
- Metals (main-group)
- Metals (transition)
- Metals (inner transition)
- Metalloids
- Nonmetals

As of late 2005, elements 112, 114, and 116 have not been named.

Main-Group Elements (left)

1A (1)	2A (2)
1 **H** 1.008	
3 **Li** 6.941	4 **Be** 9.012
11 **Na** 22.99	12 **Mg** 24.31
19 **K** 39.10	20 **Ca** 40.08
37 **Rb** 85.47	38 **Sr** 87.62
55 **Cs** 132.9	56 **Ba** 137.3
87 **Fr** (223)	88 **Ra** (226)

Transition Elements

3B (3)	4B (4)	5B (5)	6B (6)	7B (7)	8B (8)	8B (9)	8B (10)	1B (11)	2B (12)
21 **Sc** 44.96	22 **Ti** 47.88	23 **V** 50.94	24 **Cr** 52.00	25 **Mn** 54.94	26 **Fe** 55.85	27 **Co** 58.93	28 **Ni** 58.69	29 **Cu** 63.55	30 **Zn** 65.41
39 **Y** 88.91	40 **Zr** 91.22	41 **Nb** 92.91	42 **Mo** 95.94	43 **Tc** (98)	44 **Ru** 101.1	45 **Rh** 102.9	46 **Pd** 106.4	47 **Ag** 107.9	48 **Cd** 112.4
57 **La** 138.9	72 **Hf** 178.5	73 **Ta** 180.9	74 **W** 183.9	75 **Re** 186.2	76 **Os** 190.2	77 **Ir** 192.2	78 **Pt** 195.1	79 **Au** 197.0	80 **Hg** 200.6
89 **Ac** (227)	104 **Rf** (263)	105 **Db** (262)	106 **Sg** (266)	107 **Bh** (267)	108 **Hs** (277)	109 **Mt** (268)	110 **Ds** (281)	111 **Rg** (272)	112 (285)

Main-Group Elements (right)

3A (13)	4A (14)	5A (15)	6A (16)	7A (17)	8A (18)
					2 **He** 4.003
5 **B** 10.81	6 **C** 12.01	7 **N** 14.01	8 **O** 16.00	9 **F** 19.00	10 **Ne** 20.18
13 **Al** 26.98	14 **Si** 28.09	15 **P** 30.97	16 **S** 32.07	17 **Cl** 35.45	18 **Ar** 39.95
31 **Ga** 69.72	32 **Ge** 72.61	33 **As** 74.92	34 **Se** 78.96	35 **Br** 79.90	36 **Kr** 83.80
49 **In** 114.8	50 **Sn** 118.7	51 **Sb** 121.8	52 **Te** 127.6	53 **I** 126.9	54 **Xe** 131.3
81 **Tl** 204.4	82 **Pb** 207.2	83 **Bi** 209.0	84 **Po** (209)	85 **At** (210)	86 **Rn** (222)
	114 (289)		116 (292)		

Inner Transition Elements

Lanthanides (Period 6):

58 **Ce** 140.1	59 **Pr** 140.9	60 **Nd** 144.2	61 **Pm** (145)	62 **Sm** 150.4	63 **Eu** 152.0	64 **Gd** 157.3	65 **Tb** 158.9	66 **Dy** 162.5	67 **Ho** 164.9	68 **Er** 167.3	69 **Tm** 168.9	70 **Yb** 173.0	71 **Lu** 175.0

Actinides (Period 7):

90 **Th** 232.0	91 **Pa** (231)	92 **U** 238.0	93 **Np** (237)	94 **Pu** (242)	95 **Am** (243)	96 **Cm** (247)	97 **Bk** (247)	98 **Cf** (251)	99 **Es** (252)	100 **Fm** (257)	101 **Md** (258)	102 **No** (259)	103 **Lr** (260)

Period

WILLIAM P. CUNNINGHAM

William P. Cunningham is an emeritus professor at the University of Minnesota. In his 38-year career at the university, he taught a variety of biology courses, including Environmental Science, Conservation Biology, Environmental Health, Environmental Ethics, Plant Physiology, and Cell Biology. He is a member of the Academy of Distinguished Teachers, the highest teaching award granted at the University of Minnesota. He was a member of a number of interdisciplinary programs for international students, teachers, and nontraditional students. He also carried out research or taught in Sweden, Norway, Brazil, New Zealand, China, and Indonesia.

Professor Cunningham has participated in a number of governmental and nongovernmental organizations over the past 40 years. He was chair of the Minnesota chapter of the Sierra Club, a member of the Sierra Club national committee on energy policy, vice president of the Friends of the Boundary Waters Canoe Area, chair of the Minnesota governor's task force on energy policy, and a citizen member of the Minnesota Legislative Commission on Energy.

In addition to environmental science textbooks, Cunningham edited three editions of an *Environmental Encyclopedia* published by Thompson-Gale Press. He has also authored or coauthored about 50 scientific articles, mostly in the fields of cell biology and conservation biology as well as several invited chapters or reports in the areas of energy policy and environmental health. His Ph.D. from the University of Texas was in botany.

Professor Cunningham's hobbies include photography birding, hiking, gardening, and traveling. He lives in St. Paul, Minnesota with his wife, Mary. He has three children (one of whom is coauthor of this book) and six grandchildren.

Both authors have a long-standing interest in the topic in this book. Nearly half the photos in the book were taken on trips to the places we discuss.

MARY ANN CUNNINGHAM

Mary Ann Cunningham is an assistant professor of geography at Vassar College, where she holds the Mary Clark Rockefeller Chair in Geography. A biogeographer with interests in landscape ecology, geographic information systems (GIS), and remote sensing, she teaches environmental science, natural resource conservation, and land-use planning, as well as GIS and remote sensing. Field research methods, statistical methods, and scientific methods in data analysis are regular components of her teaching. As a scientist and educator, Mary Ann enjoys teaching and conducting research with both science students and nonscience liberal arts students. As a geographer, she likes to engage students with the ways their physical surroundings and social context shape their world experience. In addition to teaching at a liberal arts college, she has taught at community colleges and research universities.

Mary Ann has been writing in environmental science for over a decade, and she has been coauthor of this book since its seventh edition. She is also coauthor of *Principles of Environmental Science* (now in its fourth edition), and an editor of the *Environmental Encyclopedia* (third edition, Thompson-Gale Press). She has published work on pedagogy in cartography, as well as instructional and testing materials in environmental science. With colleagues at Vassar, she has published a GIS lab manual, *Exploring Environmental Science with GIS,* designed to provide students with an easy, inexpensive introduction to spatial and environmental analysis with GIS.

In addition to environmental science, Mary Ann's primary research activities focus on land-cover change, habitat fragmentation, and distributions of bird populations. This work allows her to conduct field studies in the grasslands of the Great Plains as well as in the woodlands of the Hudson Valley. In her spare time she loves to travel, hike, and watch birds.

Mary Ann holds a bachelor's degree from Carleton College, a master's degree from the University of Oregon, and a Ph.D. from the University of Minnesota.

Brief Contents

Contents

CHAPTER 3 Matter, Energy, and Life 51

CHAPTER 4 Evolution, Biological Communities,
        and Species Interactions 74

CHAPTER 5 Biomes: Global Patterns of Life 98

Preface

ENVIRONMENTAL SCIENCE HAS NEVER BEEN MORE IMPORTANT

We seem to have reached a turning point in environmental attitudes. A decade ago, few people took climate change seriously. Today, climate is a topic of headline news and political campaigns. Hundreds of colleges, communities, and local governments are working locally to reduce carbon emissions and improve efficiency. More than 400 bills have been passed in 40 states to require renewable energy. "Green" buildings are transforming architecture, and green business models are beginning to transform industry. The 2008 Summer Olympics in Beijing aims to be the greenest games ever with radical new strategies to conserve energy, water, and air quality. Most importantly, it's not just environmentalists who are involved in efforts to protect and improve our environment; it is business leaders finding ways to reduce costs by reducing waste, insurance companies concerned about rising sea levels, and inner-city communities who are trying to lower asthma rates in children. Environmental science is increasingly understood to be a pragmatic field that helps us understand issues that affect our lives.

In the twenty-five years since we began work on this book, the United States and the world have undergone several cycles of concern and neglect for our global environment. We have seen growth, declines, and a recent resurgence in public support for energy conservation, farmland protection, and environmental health. Global biodiversity, once a special interest of ecologists, is now an economic concern of drug companies and the global fishing industry. After years of purchasing ever-larger vehicles, Americans have recently found a renewed interest in automobile efficiency as we become increasingly aware of the costs of climate change—and of the political and economic costs of fighting to preserve our fossil fuel supplies. Major events like the Earth Summit in Rio de Janeiro in 1992 and the Kyoto conference on climate change in 1997 have raised awareness of how globally interconnected all of us are in terms of our environment, our economies, and our well-being. All these changes present new possibilities for building coalitions and finding new approaches to living sustainably. We hope this book will inform and inspire students as they consider their role in protecting our shared environment.

WHAT SETS THIS BOOK APART?

A positive viewpoint

We wrote this book because we think it's important for students to realize the difference they can make in their community. We believe a book focused on gloom and decay provides little inspiration to students, and in this time of exciting change, we think such a gloomy view is inaccurate. Many environmental problems remain severe, but there have been many improvements over past decades including cleaner water and cleaner air for most Americans. The Kyoto Protocol, despite its imperfections, is now pushing nations to reduce their climate impacts. The earth's population exceeds 6 billion people, but birth rates have plummeted as education and health care for women have improved. This book highlights these developments and presents positive steps that individuals can take, while acknowledging the many challenges we face. Case studies that show successful projects, a new chapter on Restoration Ecology, and "What Can You Do?" boxes are some of the features written to give students an applicable sense of direction. A number of other features also set this book apart.

An integrated, global perspective

Globalization spotlights the interconnectedness of environmental concerns, as well as economies. To remain competitive in a global economy, it is critical that we understand conditions in other countries and cultures. This book provides case studies and topics from regions around the world, as well as maps and data showing global issues. These examples also show the integration between environmental, social, and economic conditions at home and abroad.

A balanced presentation that encourages critical thinking

Environmental science often involves special interests, contradictory data, and conflicting interpretations of data. Throughout the text, one of the most important skills a student can learn is to think analytically and clearly about evidence, weigh the data, consider uncertainty, and skeptically evaluate the sources of information. We give students opportunities to practice critical thinking in brief "Think About It" boxes and in "What Do You Think?" readings. We present balanced evidence, while not suggesting that any opinion is on par with ideas accepted by the community of informed scientists, and we provide the tools for students to discuss and form their own opinions.

Emphasis on science

Science is critical for understanding environmental change. We emphasize principles and methods of science through coverage on uncertainty and probability, new graphing exercises, data analysis exercises, and "Exploring Science" readings that show how scientists observe the world and gather data.

OVERVIEW OF CHANGES TO *ENVIRONMENTAL SCIENCE,* TENTH EDITION

What's new to this edition?

The tenth edition includes marked changes in approach as well as a thorough update of topics and data. Among the important changes is an emphasis on positive lessons presented through recent events in environmental science. Also important are several new pedagogical features including data analysis exercises and web-based exercises.

New restoration ecology chapter

Chapter 13 is a brand new chapter on the important topic of ecological restoration. It explains how forests, grasslands, savannas, wetlands, and other ecosystems are being repaired and restored to ecological health, and provides many positive examples of restoration efforts.

New case studies and readings

Seventeen of the 25 chapter opening case studies are new, as are 10 of the boxed readings within chapters. A majority of the case studies and boxed readings in this edition are focused on current events and success stories that display the global progress being made in environmental protection.

Google Earth™ placemarks

An exciting new feature in this edition are the Google Earth™ placemarks in every chapter. Google Earth™ is an online program that provides interactive satellite imagery of the earth and will help students understand the geographic context of places and topics in the text. Wherever students see this icon 🌏 in the text, they can go to our website (http://www.mhhe.com/cunningham10e) to find a Google Earth™ placemark that will take them to the specific place being discussed. Students can zoom in to see amazing detail and zoom out to gain regional perspective.

Data Analysis exercises

A Data Analysis box has been added to the end of every chapter. These exercises ask students to graph and evaluate data, to practice looking at numbers and graphs, and to critically analyze what they see.

Learning Outcomes and other pedagogical tools

Chapter material now includes Learning Outcomes presented at the beginning of each chapter and a Reviewing Learning Outcomes section at the end of each chapter. These Learning Outcomes are connected to the major headings of each numbered section within the chapter to help students better organize the content and their study. Each chapter also includes a conclusion, which summarizes major points, a Practice Quiz to aid in understanding key concepts from the chapter, and Critical Thinking and Discussion Questions that challenge students to apply what they have learned.

Specific changes by chapter

- Chapter 1 includes a new case study on the Green Olympics 2008 that illustrates China's new concern for environmental quality and its importance in our global environment. The environmental history section has been expanded to include contemporary, diverse leaders. Chapter 1 also has a major revision of current environmental conditions using 2006 data. The Data Analysis box introduces graphing.

- Chapter 2 has a new case study on the ethics of climate change. It also has important new material on statistics, probability and uncertainty. The Data Analysis box discusses bar graphs, pie charts, and scatter plots.

- Chapter 3 adds a Data Analysis exercise on extracting data from a graph.

- Chapter 4 benefits from a strengthened section on evolution and revised material on competition, predation, symbiosis, keystone groups, and succession. The Data Analysis box presents Gause's historic experiments on interspecific competition with population growth graphs.

- Chapter 5 opens with a positive case study on saving the reefs of Apo Island (Philippines). It also has new material on coastal zones, coral reefs, estuaries, and shorelines. The Data Analysis box explains how to read climate graphs.

- Chapter 6 combines an extensive revision of population growth dynamics together with a new section on changes in human life expectancies. The Data Analysis box demonstrates graphing exponential and logistic growth.

- Chapter 7 opens with a case study on successful family planning in Thailand. Chapter 7 also presents a revised and strengthened discussion of factors influencing population, and different approaches to family planning in India are revealed in a What Do You Think box. The Data Analysis box displays how graphs can be used to explain, persuade, and inform.

- Chapter 8 starts with an encouraging case study on Guinea worm eradication. The discussion and data on infectious and emergent diseases is extensively revised and updated, and the Data Analysis box presents graphing multiple variables.

- Chapter 9 updates information on world food supply, farm policy, and soil conservation. A new section has been added on consumer choices and local food supplies. The Data Analysis box shows how to graph relative values on an index scale.

- Chapter 10 begins with a new case study on forgotten pollinators. The Data Analysis box assesses pesticide use in schools.

- Chapter 11 includes a significantly revised and updated opening case study on biodiversity and ecological resilience. The discussion of endangered and threatened species has been extended and strengthened. The Data Analysis box addresses confidence limits in data.

- Chapter 12 has undergone a major reorganization. It opens with a new case study on saving the Great Bear Rainforest and it has a new What Do You Think? box on forest thinning and salvage logging. A major new section on parks and preserves includes successes in preserving landscapes and relative percentages of different biomes protected, together with ecotourism and the role of indigenous people in biodiversity protection. The Data Analysis box is a practical exercise in detecting edge effects.

- Chapter 13 is a brand new chapter on the important topic of ecological restoration. It explains how forests, grasslands, savannas, wetlands, and other ecosystems are being repaired and restored to ecological health, and provides many positive examples of restoration efforts. Data Analysis box clarifies concept maps.

- Chapter 14 opens with a new case study on cyanide heap-leach gold mining. It continues with a rewritten section on earth structure and tectonic processes. A table of the world's worst polluted places shows the environmental effects of mining and smelting. The Data Analysis box explores mapping volcanoes and earthquakes.

- Chapter 15 contains a thoroughly updated discussion of critical issues of climate change, informed by the 2007 reports from the IPCC. It opens with an inspiring case study on California's new law to regulate greenhouse gases. It also includes a new Exploring Science box on free-air carbon enrichment studies and a Data Analysis box on graphing methane emissions.

- Chapter 16 opens with a new positive case study on controlling mercury pollution and market mechanisms for reducing greenhouse gases. It adds an encouraging case study on reducing air pollution in New Delhi, India. The Data Analysis box graphs air pollution abatement in Europe.

- Chapter 17 presents a new case study about China's gargantuan South-to-North Water Diversion project. It adds material on drying of the Aral Sea and Lake Chad, and it illustrates freshwater shortages, as well as the problems with dams and diversions. The Data Analysis box examines a water scarcity graph.

- Chapter 18 explains how low-cost natural systems can be used for wastewater treatment. The section on inorganic water pollutants has been extensively rewritten. The Data Analysis box elucidates water pollution graphs.

- Chapter 19 starts with a new positive case study on integrated gasification combined cycle (IGCC) "clean" coal power plants. It continues with information on carbon sequestration. The discussion on nuclear energy has been updated with the report of the first new reactor approval in 30 years. The Data Analysis box compares energy use by different countries.

- Chapter 20 presents an encouraging case study of Danish islands that depend entirely on renewable energy. It has a new discussion of hybrid gasoline-electric vehicles and adds an extensive new section on biofuels including the net energy balance of new energy sources. The Data Analysis box illustrates energy calculations.

- Chapter 21 has a new positive case study on waste recycling in New York City. The chapter includes new information on ocean pollution, waste export to developing countries, and electronic waste, and the Data Analysis box invites personal waste calculations.

- Chapter 22 revises and extends the heartening opening case study on Curitiba, Brazil. It retains the positive example of sustainable housing in London. The Data Analysis box elucidates graphing with a logarithmic scale.

- Chapter 23 begins with a new case study on Grameen bank microlending. It adds a new section on the Environmental Performance Index, and explores market mechanisms for pollution reduction. The Data Analysis box illustrates graphing the human development index.

- Chapter 24 retains its opening story on the snail darter and the Endangered Species Act. The Data Analysis box discusses scatter plots and regression analysis.

- Chapter 25 opens with a new case study on saving the gray whale nursery in Laguna San Ignacio. The chapter includes new sections on environmental leadership, campus greening, and the millennium development goals. The Data Analysis box shows how students can do a campus environmental audit.

Acknowledgements

We owe a great debt to the hardworking, professional team that has made this the best environmental science text possible. We express special thanks for editorial support to Marge Kemp, Janice Roerig-Blong, Joan Weber, Rose Koos, and Ashley Zellmer. We are grateful to the excellent production team led by April Southwood and marketing leadership by Tami Petsche. We also thank Kandis Elliot for her outstanding artwork, LouAnn Wilson for photo research, and Cathy Conroy for copyediting. We also thank Dr. Kim Chapman for essays that contributed to the text. Finally, we thank the many contributions of careful reviewers who shared their ideas with us during revisions.

Tenth Edition Reviewers

Professor Dwight W. Allen, Eminent Scholar of Educational Reform
Old Dominion University

Daphne Hall Babcock
Collin County Community College

David C. Belt
Johnson County Community College, Overland Park, Kansas

Donna H. Bivans
Pitt Community College

J. Christopher Brown
University of Kansas

Kelly S. Cartwright
College of Lake County

Robert S. Dill
Bergen Community College

Dr. Iver W. Duedall, Professor Emeritus
Florida Institute of Technology, Melbourne

Daniel Habib
Queens College of the City University of New York

Barbara Hollar
University of Detroit Mercy

Timothy V. Horger
Illinois Valley Community College

Charles Ide, Director, Environmental Institute
Western Michigan University

Richard R. Jurin
University of Northern Colorado

Dawn G. Keller
Hawkeye Community College

David Knowles
East Carolina University

Tim Marbach
California State University, Sacramento

Chris Migliaccio
Miami Dade College

Jay C. Odaffer
Manatee Community College

Julie Phillips
De Anza College

Michael Phillips
Illinois Valley Community College

Sarah Quast
Middlesex Community College

Dawn Ranish
Broward Community College

Michelle L. Stevens
Imperial Valley College

Julie Stoughton
University of Nevada, Reno

S. Kant Vajpayee
University of Southern Mississippi

Kelly Wessell
Grand Valley State University

Van Wheat
South Texas College

John J. Wielichowski
Milwaukee Area Technical College

Lorne Wolfe
Georgia Southern University

GUIDED TOUR

A Global Perspective Is Vital to Learning about Environmental Science

Case Studies

In the front of each chapter, case studies utilize stories to portray real-life global issues that affect our food, our quality of life, and our future. Seventeen new case studies have been added to further focus on current events and the success stories of environmental protection progress.

Google Earth™ Placemarks

This feature provides interactive satellite imagery of the earth to give students a geographic context of places and topics in the text. Students can zoom in for detail or they can zoom out for a more global perspective. Placemark links can be found on the website http://www.mhhe.com/cunningham10e.

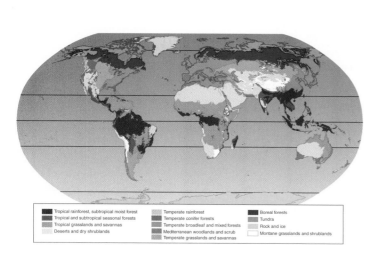

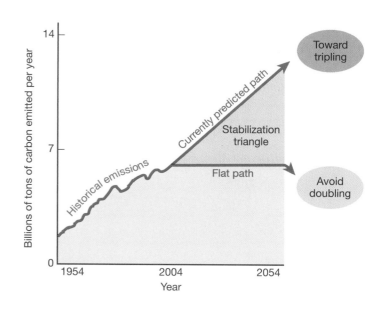

The Latest Global Data

Easy to follow graphs, charts, and maps display numerous examples from many regions of the world. Students are exposed to the fact that environmental issues cross borders and oceans.

xviii

Critical Thinking Skills Support Understanding of Environmental Change

Exploring Science

Real-life environmental issues drive these readings as students learn about the principles of scientific observation and proper data-gathering techniques.

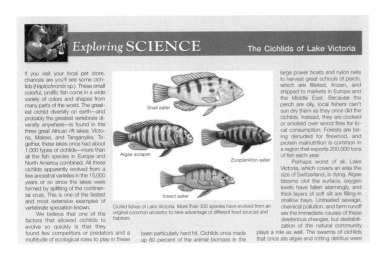

Exploring SCIENCE — The Cichlids of Lake Victoria

If you visit your local pet store, chances are you'll see some cichlids (*Haplochromis* sp.). These small colorful, prolific fish come in a wide variety of colors and shapes from many parts of the world. The greatest cichlid diversity on earth—and probably the greatest vertebrate diversity anywhere—is found in the three great African rift lakes: Victoria, Malawi, and Tanganyika. Together, these lakes once had about 1,000 types of cichlids—more than all the fish species in Europe and North America combined. All these cichlids apparently evolved from a few ancestral varieties in the 15,000 years or so since the lakes were formed by splitting of the continental crust. This is one of the fastest and most extensive examples of vertebrate speciation known.

We believe that one of the factors that allowed cichlids to evolve so quickly is that they found few competitors or predators and a multitude of ecological roles to play in these

large power boats and nylon nets to harvest great schools of perch, which are filleted, frozen, and shipped to markets in Europe and the Middle East. Because the perch are oily, local fishers can't sun dry them as they once did the cichlids. Instead, they are cooked or smoked over wood fires for local consumption. Forests are being denuded for firewood, and protein malnutrition is common in a region that exports 200,000 tons of fish each year.

Perhaps worst of all, Lake Victoria, which covers an area the size of Switzerland, is dying. Algae blooms clot the surface, oxygen levels have fallen alarmingly, and thick layers of soft silt are filling-in shallow bays. Untreated sewage, chemical pollution, and farm runoff are the immediate causes of these deleterious changes, but destabilization of the natural community plays a role as well. The swarms of cichlids that once ate algae and rotting detritus were

Snail eater

Algae scraper

Zooplankton eater

Insect eater

Cichlid fishes of Lake Victoria. More than 300 species have evolved from an original common ancestor to take advantage of different food sources and habitats.

been particularly hard hit. Cichlids once made up 80 percent of the animal biomass in the

What Do You Think?

This feature provides challenging environmental studies that offer an opportunity for students to consider contradictory data, special interests, and conflicting interpretations within a real scenario.

What Do You Think?

Too Many Deer?

A century ago, few Americans had ever seen a wild deer. Uncontrolled hunting and habitat destruction had reduced the deer population to about 500,000 animals nationwide. Some states had no deer at all. To protect the remaining deer, laws were passed in the 1920s and 1930s to restrict hunting, and the main deer predators—wolves and mountain lions—were exterminated throughout most of their former range.

As Americans have moved from rural areas to urban centers, forests have regrown, and deer populations have undergone explosive growth. Maturing at age two, a female deer can give birth to twin fawns every year for a decade or more. Increasing more than 20 percent annually, a deer population can double in just three years, an excellent example of irruptive, exponential growth.

Wildlife biologists estimate that the contiguous 48 states now have a population of more than 30 million white-tailed deer (*Odocoileus*

A white-tailed deer (*Odocoileus virginianus*)

virginianus), probably triple the number present in pre-Columbian times. Some areas have as many as 200 deer per square mile (77/km²). At this density, woodland plant diversity is generally reduced to a few species that deer won't eat. Most deer, in such conditions, suffer from malnourishment, and many die every year of disease and starvation. Other species are diminished as well. Many small mammals and ground-dwelling birds begin to disappear when deer populations reach 25 animals per square mile. At 50 deer per square mile, most ecosystems are seriously impoverished.

The social costs of large deer populations are high. In Pennsylvania alone, where deer numbers are now about 500 times greater than a century ago, deer destroy about $70 million worth of crops and $75 million worth of trees annually. Every year some 40,000 collisions with motor vehicles cause $80 million in property damage. Deer help spread Lyme disease, and, in many states, chronic wasting disease is found in wild deer herds. Some of the most heated criticisms of current deer management policies are in the suburbs. Deer love to browse on the flowers, young trees, and ornamental bushes in suburban yards. Heated disputes often arise between those who love to watch deer and their neighbors who want to exterminate them all.

In remote forest areas, many states have extended hunting seasons, increased the bag limit to four or more animals, and encouraged hunters to shoot does (females) as well as bucks (males). Some hunters criticize these changes because they believe that fewer deer will make it harder to hunt successfully and less likely that they'll find a trophy buck. Others, however, argue that a healthier herd and a more diverse ecosystem is better for all concerned.

In urban areas, increased sport hunting usually isn't acceptable. Wildlife biologists argue that the only practical way to reduce deer herds is culling by professional sharpshooters. Animal rights activists protest lethal control methods as cruel and inhumane. They call instead for fertility controls, reintroduction of predators, such as wolves and mountain lions, or trap and transfer programs. Birth control works in captive populations but is expensive and impractical with wild animals. Trapping, also, is expensive, and there's rarely anyplace willing to take surplus animals. Predators may kill domestic animals or even humans.

This case study shows that carrying capacity can be more complex than the maximum number of organisms an ecosystem can support. While it may be possible for 200 deer to survive in a square mile, there's an ecological carrying capacity lower than that if we consider the other species dependent on that same habitat. There's also an ethical carrying capacity if we don't want to see animals suffer from malnutrition and starve to death every winter. And there's a cultural carrying capacity if we consider the tolerable rate of depredation on crops and lawns or an acceptable number of motor vehicle collisions.

If you were a wildlife biologist charged with managing the deer herd in your state, how would you reconcile the different interests in this issue? How would you define the optimum deer population, and what methods would you suggest to reach this level? What social or ecological indicators would you look for to gauge whether deer populations are excessive or have reached an appropriate level?

Data Analysis

At the end of every chapter, these exercises ask students to graph and evaluate data while critically analyzing what they observe.

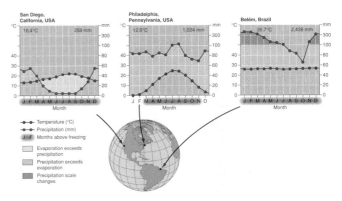

DATA analysis — Reading Climate Graphs

As you've learned in this chapter, temperature and moisture are critical factors in determining the distribution and health of ecosystems. But how do you read the climate and precipitation graphs that accompany the description of each biome? To begin, examine the three climate graphs in this box. These graphs show annual trends in temperature and precipitation (rainfall and snow). They also indicate the relationship between potential evaporation, which depends on temperature and precipitation. When evaporation exceeds precipitation, dry conditions result (yellow areas). Extremely wet months are shaded dark blue on the graphs. Moist climates may vary in precipitation rates, but evaporation rarely exceeds precipitation. Months above freezing temperature (shaded brown on the X-axis) have most evaporation. Comparing these climate graphs helps us understand the different seasonal conditions that control plant and animal lives in different biomes.

1. What are the maximum and minimum temperatures in each of the three locations shown?
2. What do these temperatures correspond to in Fahrenheit? (*Hint*: look at the table in the back of your book).
3. Which area has the wettest climate; which is driest?
4. How do the maximum and minimum monthly rainfalls in San Diego and Belém compare?
5. Describe these three climates.
6. What kinds of biomes would you expect to find in these areas?

San Diego, California, USA — 16.4°C — 259 mm

Philadelphia, Pennsylvania, USA — 12.5°C — 1,024 mm

Belém, Brazil — 26.7°C — 2,438 mm

Temperature (°C)
Precipitation (mm)
Months above freezing
Evaporation exceeds precipitation
Precipitation exceeds evaporation
Precipitation scale changes

Moisture availability depends on temperature as well as precipitation. The horizontal axis on these climate diagrams shows months of the year; vertical axes show temperature (*left side*) and precipitation (*right*). The number of dry months (*shaded yellow*) and wetter months (*blue*) varies with geographic location. Mean annual temperature (°C) and precipitation (mm) are shown at the top of each graph.

Sound Pedagogy Encourages Science Inquiry and Application

Learning Outcomes

Found at the beginning of each chapter, and organized by major headings, these outcomes give students an overview of the key concepts they will need to understand.

LEARNING OUTCOMES

After studying this chapter, you should be able to:

12.1 Discuss the types and uses of world forests.
12.2 Describe the location and state of grazing lands around the world.

12.3 Summarize the types and locations of nature preserves.

Conclusion

This section summarizes the chapter by highlighting key ideas and relating them to one another.

CONCLUSION

Forests and grasslands cover nearly 60 percent of global land area. The vast majority of humans live in these biomes, and we obtain many valuable materials from them. And yet, these biomes also are the source of much of the world's biodiversity on which we depend for life-supporting ecological services. How we can live sustainably on our natural resources while also preserving enough nature so those resources can be replenished represents one of the most important questions in environmental science.

There is some good news in our search for a balance between

Australia's great barrier reef shows that we can choose to protect some biodiverse areas in spite of forces that want to exploit them. Overall, nearly 12 percent of the earth's land area is now in some sort of protected status. While the level of protection in these preserves varies, the rapid recent increase in number and area in protected status exceeds the goals of the United Nations Millennium Project.

While we haven't settled the debate between focusing on individual endangered species versus setting aside representative samples of habitat, pursuing both strategies seems to be working.

Reviewing Learning Outcomes

Connected to the Learning Outcomes at the beginning of each chapter, this review clearly restates the important concepts associated with each outcome.

REVIEWING LEARNING OUTCOMES

By now you should be able to explain the following points:

12.1 Discuss the types and uses of world forests.
- Boreal and tropical forests are most abundant.
- Forests provide many valuable products.
- Tropical forests are being cleared rapidly.
- Temperate forests have competing uses.

12.2 Describe the location and state of grazing lands around the world.
- Grazing c
- Overgrazi
- Ranchers

12.3 Summarize the types and locations of nature preserves.
- Many countries have created nature preserves.
- Not all preserves are preserved.
- Marine ecosystems need greater protection.
- Conservation and economic development can work together.
- Native people can play important roles in nature protection.
- Species survival can depend on preserve size and shape.

Critical Thinking and Discussion Questions

Brief scenarios of everyday occurrences or ideas challenge students to apply what they have learned to their lives.

CRITICAL THINKING AND DISCUSSION QUESTIONS

1. Paper and pulp are the fastest growing sector of the wood products market, as emerging economies of China and India catch up with the growing consumption rates of North America, Europe, and Japan, What should be done to reduce paper use?

2. Conservationists argue that watershed protection and other ecological functions of forests are more economically valuable than timber. Timber companies argue that continued production supports stable jobs and local economies. If you were a judge attempting to decide which group was right, what evidence would you need on both sides? How would you gather this evidence?

3. Divide your class into a ranching group, a conservation grou
prot
natu
land
to m

4. Calculating forest area and forest losses is complicated by the difficulty of defining exactly what constitutes a forest. Outline a definition for what counts as forest in your area, in terms of size, density, height, or other characteristics. Compare your definition to those of your colleagues. Is it easy to agree? Would your definition change if you lived in a different region?

5. Why do you suppose dry topical forest and tundra are well represented in protected areas, while grasslands and wetlands are protected relatively rarely? Consider social, cultural, geographic, and economic reasons in your answer.

6. Oil and gas companies want to drill in several parks, monu-

Practice Quiz

Short-answer questions allow students to check their knowledge of chapter concepts.

PRACTICE QUIZ

1. What do we mean by *closed-canopy* forest and *old-growth* forest?

2. Which commodity is used most heavily in industrial economies: steel, plastic, or wood? What portion of the world's population depends on wood or charcoal as the main energy supply?

3. What is a *debt-for-nature swap*?

4. Why is fire suppression a controversial strategy? Why are forest thinning and salvage logging controversial?

5. Are pastures and rangelands always damaged by grazing animals? What are some results of overgrazing?

6. What is *rotational grazing*, and how does it mimic natural processes?

7. What was the first national park in the world, and when was it established? How have the purposes of this park and others changed?

8. How do the size and design of nature preserves influence their effectiveness? What do landscape ecologists mean by *interior habitat* and *edge effects*?

9. What is *ecotourism*, and why is it important?

10. What is a *biosphere reserve*, and how does it differ from a wilderness area or wildlife preserve?

What Can You Do?

This feature gives students realistic steps for applying their knowledge to make a positive difference in our environment.

What Can You Do?

Reducing Waste

1. Buy foods that come with less packaging; shop at farmers' markets or co-ops, using your own containers.

2. Take your own washable refillable beverage container to meetings or convenience stores.

3. When you have a choice at the grocery store between plastic, glass, or metal containers for the same food, buy the reusable or easier-to-recycle glass or metal.

Source: Minnesota Pollution Control Agency.

Think About It

These boxes provide several opportunities in each chapter for students to review material, practice critical thinking, and apply scientific principles.

Think About It

Why might you and your mother rank some risks differently? List some activities on which the two of you might disagree.

TEACHING AND LEARNING SUPPLEMENTS

McGraw-Hill offers various tools and technology products to support *Environmental Science: A Global Concern*. Students can order supplemental study materials by contacting their local bookstore or by calling 800-262-4729. Instructors can obtain teaching aids by calling the Customer Service Department at 800-338-3987, visiting the McGraw-Hill website at www.mhhe.com, or by contacting their local McGraw-Hill sales representative.

Teaching supplements for instructors

Presentation Center

ARIS Presentation Center is an online digital library containing assets such as photos, artwork, PowerPoints, animations, and other media types that can be used to create customized lectures, visually enhanced tests and quizzes, compelling course websites, and attractive printed support materials. The following digital assets are grouped by chapter:

- **Color Art** Full-color digital files of illustrations in the text can be readily incorporated into lecture presentations, exams, or custom-made classroom materials. These include all of the 3-D realistic art found in this edition, representing some of the most important concepts in environmental science.

- **Photos** Digital files of photographs from the text can be reproduced for multiple classroom uses.

- **Tables** Every table that appears in the text is provided in electronic format.

- **Videos** This special collection of 69 underwater video clips displays interesting habitats and behaviors for many animals in the ocean.

- **Animations** One hundred full-color animations that illustrate many different concepts covered in the study of environmental science are available for use in creating classroom lectures, testing materials, or online course communication. The visual impact of motion will enhance classroom presentations and increase comprehension.

- **Test Bank** A computerized test bank that uses testing software to quickly create customized exams is available on for this text. The user-friendly program allows instructors to search for questions by topic or format, edit existing questions or add new ones; and scramble questions for multiple versions of the same test. Word files of the test bank questions are provided for those instructors who prefer to work outside the test-generator software.

- **Global Base Maps** Eighty-eight base maps for all world regions and major subregions are offered in four versions: black-and-white and full-color, both with labels and without labels. These choices allow instructors the flexibility to plan class activities, quizzing opportunities, study tools, and PowerPoint enhancements.

- **PowerPoint Lecture Outlines** Ready-made presentations that combine art and photos and lecture notes are provided for each of the 25 chapters of the text. These outlines can be used as they are or tailored to reflect your preferred lecture topics and sequences.

- **PowerPoint Slides** For instructors who prefer to create their lectures from scratch, all illustrations, photos, and tables are preinserted by chapter into blank PowerPoint slides for convenience.

 McGraw-Hill's **ARIS—Assessment, Review, and Instruction System** (aris.mhhe.com) for *Environmental Science: A Global Concern* is a complete, online tutorial, electronic homework, and course management system, designed for greater ease of use than any other system available. Free upon adoption of *Environmental Science: A Global Concern,* instructors can create and share course materials and assignments with colleagues with a few clicks of the mouse. All PowerPoint lectures, assignments, quizzes, tutorials, and interactives are directly tied to text-specific materials in *Environmental Science,* but instructors can also edit questions, import their own content, and create announcements and due dates for assignments. ARIS has automatic grading and reporting of easy-to-assign homework, quizzing, and testing. All student activity within McGraw-Hill's ARIS website is automatically recorded and available to the instructor through a fully integrated grade book that can be downloaded to Excel.

eInstruction

This classroom performance system (CPS) utilizes wireless technology to bring interactivity into the classroom or lecture hall. Instructors and students receive immediate feedback through wireless response pads that are easy to use and engage students. eInstruction can assist instructors by:

- taking attendance
- administering quizzes and tests
- creating a lecture with intermittent questions
- using the CPS grade book to manage lectures and student comprehension
- integrating interactivity into PowerPoint presentations

Contact your local McGraw-Hill sales representative for more information.

Course Delivery Systems

With help from WebCT and Blackboard, professors can take complete control of their course content. Course cartridges containing website content, online testing, and powerful student tracking features are readily available for use within these platforms.

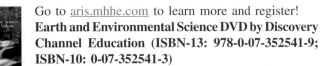

Go to aris.mhhe.com to learn more and register!
Earth and Environmental Science DVD by Discovery Channel Education (ISBN-13: 978-0-07-352541-9; ISBN-10: 0-07-352541-3)

Begin your class with a quick peek at science in action. The exciting NEW DVD by Discovery Channel Education offers 50 short (3–5 minute) videos on topics ranging from conservation to volcanoes. Search by topic and download into your PowerPoint lecture. Available to colleges and universities. See your McGraw-Hill sales representative for a detailed listing.

McGraw-Hill's Biology Digitized Videos (ISBN-13: 978-0-07-312155-0; ISBN-10: 0-07-312155-X)

Licensed from some of the highest quality life-science video producers in the world, these brief video clips on DVD range in length from 15 seconds to 2 minutes and cover all areas of general biology, from cells to ecosystems. Engaging and informative, McGraw-Hill's digitized biology videos will help capture students' interest while illustrating key biological concepts, applications, and processes.

Learning supplements for students

ARIS (aris.mhhe.com)

This site includes quizzes for each chapter, additional case studies, interactive base maps, and much more. Learn more about the exciting features provided for students through the enhanced *Environmental Science: A Global Concern* website.

Exploring Environmental Science with GIS by Stewart, Cunningham, Schneiderman, and Gold (ISBN-13: 978-0-07-297564-2; ISBN-10: 0-07-297564-4)

This short book provides exercises for students and instructors who are new to GIS, but are familiar with the Windows operating system. The exercises focus on improving analytical skills, understanding spatial relationships, and understanding the nature and structure of environmental data. Because the software used is distributed free of charge, this text is appropriate for courses and schools that are not yet ready to commit to the expense and time involved in acquiring other GIS packages.

Annual Editions: Environment 07/08 by Allen (ISBN-13: 978-0-07-351544-1; ISBN-10: 0-07-351544-2)

This twenty-sixth edition is a compilation of current articles from the best of the public press. The selections explore the global environment, the world's population, energy, the biosphere, natural resources, and pollutions.

Taking Sides: Clashing Views on Controversial Environmental Issues, Twelfth Edition by Easton (ISBN-13: 978-0-07-351443-7; ISBN-10: 0-07-351443-8)

This book represents the arguments of leading environmentalists, scientists, and policymakers. The issues reflect a variety of viewpoints and are staged as "pro" and "con" debates. Issues are organized around four core areas: general philosophical and political issues, the environment and technology, disposing of wastes, and the environment and the future.

Classic Edition Sources: Environmental Studies, Third Edition by Easton (ISBN-13: 978-0-07-352758-1; ISBN-10: 0-07-352758-0)

This volume brings together primary source selections of enduring intellectual value—classic articles, book excerpts, and research studies—that have shaped environmental studies and our contemporary understanding of it. The book includes carefully edited selections from the works of the most distinguished environmental observers, past and present. Selections are organized topically around the following major areas of study: energy, environmental degradation, population issues and the environment, human health and the environment, and environment and society.

Student Atlas of Environmental Issues, by Allen (ISBN-13: 978-0-69-736520-0; ISBN-10: 0-69-736520-4)

This atlas is an invaluable pedagogical tool for exploring the human impact on the air, waters, biosphere, and land in every major world region. This informative resource provides a unique combination of maps and data that help students understand the dimensions of the world's environmental problems and the geographical basis of these problems.

Environmental science students learn about wetland biology.

Learning to Learn

What kind of world do you want to live in? Demand that your teachers teach you what you need to know to build it.

—*Peter Kropotkin*—

LEARNING OUTCOMES

After studying this introduction, you should be able to:

L.1 Form a plan to organize your efforts and become a more effective and efficient student.

L.2 Make an honest assessment of the strengths and weaknesses of your current study skills.

L.3 Assess what you need to do to get the grade you want in this class.

L.4 Set goals, schedule your time, and evaluate your study space.

L.5 Use this textbook effectively, practice active reading, and prepare for exams.

L.6 Be prepared to apply critical and reflective thinking in environmental science.

L.7 Understand the advantages of concept mapping and use it in your studying.

Case Study Why Study Environmental Science?

Welcome to environmental science. We hope you'll enjoy learning about the material presented in this book, and that you'll find it both engaging and useful. There should be something here for just about everyone, whether your interests are in basic ecology, natural resources, or the broader human condition. You'll see, as you go through the book, that it covers a wide range of topics. It defines our environment, not only the natural world, but also the built world of technology, cities, and machines, as well as human social or cultural institutions. All of these interrelated aspects of our life affect us, and, in turn, are affected by what we do.

You'll find that many issues discussed here are part of current news stories on television or in newspapers. Becoming an educated environmental citizen will give you a toolkit of skills and attitudes that will help you understand current events and be a more interesting person. Because this book contains information from so many different disciplines, you will find connections here with many of your other classes. Seeing material in an environmental context may assist you in mastering subject matter in many courses, as well as in life after you leave school.

One of the most useful skills you can learn in any of your classes is critical thinking—a principal topic of this chapter. Much of the most important information in environmental science is highly contested. Facts vary depending on when and by whom they were gathered. For every opinion there is an equal and opposite opinion. How can you make sense out of this welter of ever-changing information? The answer is that you need to develop a capacity to think independently, systematically, and skillfully to form your own opinions (fig. L.1). These qualities and abilities can help you in many aspects of life. Throughout this book you will find "What Do You Think?" boxes that invite you to practice your critical and reflective thinking skills.

There is much to be worried about in our global environment. Evidence is growing relentlessly that we are degrading our environment and consuming resources at unsustainable rates.

FIGURE L.1 What does it all mean? Studying environmental science gives you an opportunity to develop creative, reflective, and critical thinking skills.

Biodiversity is disappearing at a pace unequaled since the end of the age of dinosaurs 65 million years ago. Irreplaceable topsoil erodes from farm fields, threatening global food supplies. Ancient forests are being destroyed to make newsprint and toilet paper. Rivers and lakes are polluted with untreated sewage, while soot and smoke obscure our skies. Even our global climate seems to be changing to a new regime that could have catastrophic consequences.

At the same time, we have better tools and knowledge than any previous generation to do something about these crises. Worldwide public awareness of—and support for—environmental protection is at an all-time high. Over the past 50 years, human ingenuity and enterprise have brought about a breathtaking pace of technological innovations and scientific breakthroughs. We have learned to produce more goods and services with less material. The breathtaking spread of communication technology makes it possible to share information worldwide nearly instantaneously. Since World War II, the average real income in developing countries has doubled; malnutrition has declined by almost one-third; child death rates have been halved; average life expectancy has increased by 30 percent; and the percentage of rural families with access to safe drinking water has risen from less than 10 percent to almost 75 percent.

The world's gross domestic product has increased more than tenfold over the past five decades, but the gap between the rich and poor has grown ever wider. More than a billion people now live in abject poverty without access to adequate food, shelter, medical care, education, and other resources required for a healthy, secure life. The challenge for us is to spread the benefits of our technological and economic progress more equably and to find ways to live sustainably over the long run without diminishing the natural resources and vital ecological services on which all life depends. We've tried to strike a balance in this book between enough doom and gloom to give you a realistic view of our problems, and enough positive examples to give hope that we can discover workable solutions.

What would it mean to become a responsible environmental citizen? What rights and privileges do you enjoy as a member of the global community? What duties and responsibilities go with that citizenship? In many chapters of this book you will find practical advice on things you can do to conserve resources and decrease adverse environmental impacts. Ethical perspectives are an important part of our relationship to the environment and the other people with whom we share it. The discussion of ethical principles and worldviews in chapter 2 is a key section of this book. We hope you'll think about the ethics of how we treat our common environment.

Clearly, to become responsible and productive environmental citizens each of us needs a basis in scientific principles, as well as some insights into the social, political, and economic systems that impact our global environment. We hope this book and the class you're taking will give you the information you need to reach those goals. As the noted Senegalese conservationist and educator, Baba Dioum, once said, "in the end, we will conserve only what we love, we will love only what we understand, and we will understand only what we are taught."

L.1 HOW CAN I GET AN A IN THIS CLASS?

"What have I gotten myself into?" you are probably wondering as you began to read this book. "Will environmental science be worth my while? Do I have a chance to get a good grade?" The answers to these questions depend, to a large extent, on you and how you decide to apply yourself. Expecting to be interested and to do either well or poorly in your classes often turns out to be a self-fulfilling prophecy. As Henry Ford once said, "If you think you can do a thing, or think you can't do a thing, you're right." Cultivating good study skills can help you to reach your goals and make your experience in environmental science a satisfying and rewarding one. The purpose of this introduction is to give you some tips to help you get off to a good start in studying. You'll find that many of these techniques are also useful in other courses and after you graduate, as well.

Environmental science, as you can see by skimming through the table of contents of this book, is a complex, transdisciplinary field that draws from many academic specialties. It is loaded with facts, ideas, theories, and confusing data. It is also a dynamic, highly contested subject. Topics such as environmental contributions to cancer rates, potential dangers of pesticides, or when and how much global warming may be caused by human activities are widely disputed. Often you will find distinguished and persuasive experts who take completely opposite positions on any particular question. It will take an active, organized approach on your part to make sense of the vast amount of information you'll encounter here. And it will take critical, thoughtful reasoning to formulate your own position on the many controversial theories and ideas in environmental science. Learning to learn will help you keep up-to-date on important issues after you leave this course. Becoming educated voters and consumers is essential for a sustainable future.

Develop good study habits

Many students find themselves unprepared for studying in college. In a survey released in 2003 by the Higher Education Research Institute, more than two-thirds of high school seniors nationwide reported studying outside of class less than one hour per day. Nevertheless, because of grade inflation, nearly half those students claim to have an A average. It comes as a rude shock to many to discover that the study habits they developed in high school won't allow them to do as well—or perhaps even to pass their classes—in college. Many will have to triple or even quadruple their study time. In addition, they need urgently to learn to study more efficiently and effectively.

What are your current study skills and habits? Making a frank and honest assessment of your strengths and weaknesses will help you set goals and make plans for achieving them during this class. Answer the questions in table L.1 as a way of assessing where you are as you begin to study environmental science and where you need to work to improve your study habits.

One of the first requirements for success is to set clear, honest, attainable goals for yourself. Are you willing to commit the time and effort necessary to do well in this class? Make goals for yourself in terms that you can measure and in time frames

TABLE L.1

Assess Your Study Skills

Rate yourself on each of the following study skills and habits on a scale of 1 (excellent) to 5 (needs improvement). If you rate yourself below 3 on any item, think about an action plan to improve that competence or behavior.

_____ How strong is your commitment to be successful in this class?

_____ How well do you manage your time (e.g., do you always run late or do you complete assignments on time)?

_____ Do you have a regular study environment that reduces distraction and encourages concentration?

_____ How effective are you at reading and note-taking (e.g., do you remember what you've read; can you decipher your notes after you've made them)?

_____ Do you attend class regularly and listen for instructions and important ideas? Do you participate actively in class discussions and ask meaningful questions?

_____ Do you generally read assigned chapters in the textbook before attending class or do you wait until the night before the exam?

_____ Are you usually prepared before class with questions about material that needs clarification or that expresses your own interest in the subject matter?

_____ How do you handle test anxiety (e.g., do you usually feel prepared for exams and quizzes or are you terrified of them? Do you have techniques to reduce anxiety or turn it into positive energy)?

_____ Do you actively evaluate how you are doing in a course based on feedback from your instructor and then make corrections to improve your effectiveness?

_____ Do you seek out advice and assistance outside of class from your instructors or teaching assistants?

within which you can see progress and adjust your approach if it isn't taking you where you want to go. Be positive but realistic. It's more effective to try to accomplish a positive action than to avoid a negative one. When you set your goals, use proactive language that states what you want rather than negative language about what you're trying to avoid. It's good to be optimistic, but setting impossibly high standards will only lead to disappointment. Be objective about the obstacles you face and be willing to modify your goals if necessary. As you gain more experience and information, you may need to adjust your expectations either up or down. Take stock from time to time to see whether you are on track to accomplish what you expect from your studies. In environmental planning, this is called adaptive management.

One of the most common mistakes many of us make is to procrastinate and waste time. Be honest, are you habitually late for meetings or in getting assignments done? Do you routinely leave your studying until the last minute and then frantically cram the night before your exams? If so, you need to organize your schedule so that you can get your work done and still have a life. Make a study schedule for yourself and stick to it. Allow enough time for sleep, regular meals, exercise, and recreation so that you will be rested, healthy, and efficient when you do study. Schedule regular study times between your classes and work. Plan some study times during the day when you are fresh; don't leave all your work until late night hours when you don't get much done. Divide your work into reasonable sized segments that you can accomplish on a daily basis. Plan to have all your reading and assignments completed several days before your exams so you will have adequate time to review and process information. Carry a calendar so you will remember appointments and assignments.

Establish a regular study space in which you can be effective and productive. It might be a desk in your room, a carrel in the library, or some other quiet, private environment. Find a place that works for you and be disciplined about sticking to what you need to do. If you get in the habit of studying in a particular place and time, you will find it easier to get started and to stick to your tasks. Many students make the mistake of thinking that they can study while talking to their friends or watching TV. They may put in many hours but not really accomplish much. On the other hand, some people think most clearly in the anonymity of a crowd. The famous philosopher, Immanuel Kant, found that he could think best while wandering through the noisy, crowded streets of Königsberg, his home town.

How you behave in class and interact with your instructor can have a big impact on how much you learn and what grade you get. Make an effort to get to know your instructor. She or he is probably not nearly as formidable as you might think. Sit near the front of the room where you can see and be seen. Pay attention and ask questions that show your interest in the subject matter. Practice the skills of good note-taking (table L.2). Attend every class and arrive on time. Don't fold up your papers and prepare to leave until after the class period is over. Arriving late and leaving early says to your instructor that you don't care much about either the class or your grade. If you think of yourself as a good student and act like one, you may well get the benefit of the doubt when your grade is assigned.

TABLE L.2

Learning Skills—Taking Notes

1. Identify the important points in a lecture and organize your notes in an outline form to show main topics and secondary or supporting points. This will help you follow the sense of the lecture.

2. Write down all you can. If you miss something, having part of the notes will help your instructor identify what you've missed.

3. Leave a wide margin in your notes in which you can generate questions to which your notes are the answers. If you can't write a question about the material, you probably don't understand it.

4. Study for your test under test conditions by answering your own questions without looking at your notes. Cover your notes with a sheet of paper on which you write your answers, then slide it to the side to check your accuracy.

5. Go all the way through your notes once in this test mode, then go back to review those questions you missed.

6. Compare your notes and the questions you generated with those of a study buddy. Did you get the same main points from the lecture? Can you answer the questions someone else has written?

7. Review your notes again just before test time, paying special attention to major topics and questions you missed during study time.

Source: Dr. Melvin Northrup, Grand Valley State University.

Practice active, purposeful learning. It isn't enough to passively absorb knowledge provided by your instructor and this textbook. You need to actively engage the material in order to really understand it. The more you invest yourself in the material, the easier it will be to comprehend and remember. It is very helpful to have a study buddy with whom you can compare notes and try out ideas (fig. L.2). You will get a valuable perspective

FIGURE L.2 Cooperative learning, in which you take turns explaining ideas and approaches with a friend, can be one of the best ways to comprehend material.

on whether you're getting the main points and understanding an adequate amount by comparing. It's an old adage that the best way to learn something is to teach it to someone else. Take turns with your study buddy explaining the material you're studying. You may think you've mastered a topic by quickly skimming the text but you're likely to find that you have to struggle to give a clear description in your own words. Anticipating possible exam questions and taking turns quizzing each other can be a very good way to prepare for tests.

Recognize and hone your learning styles

Each of us has ways that we learn most effectively. Discovering techniques that work for you and fit the material you need to learn is an important step in reaching your goals. Do any of the following fit your preferred ways of learning?

- **Visual Learner:** understands and remembers best by reading, looking at photographs, figures, and diagrams. Good with maps and picture puzzles. Visualizes image or spatial location for recall. Uses flash cards for memorization.
- **Verbal Learner:** understands and remembers best by listening to lectures, reading out loud, and talking things through with a study partner. May like poetry and word games. Memorizes by repeating item verbally.
- **Logical Learner:** understands and remembers best by thinking through a subject and finding reasons that make sense. Good at logical puzzles and mysteries. May prefer to find patterns and logical connections between items rather than memorize.
- **Active Learner:** understands and remembers best those ideas and skills linked to physical activity. Takes notes, makes lists, uses cognitive maps. Good at working with hands and learning by doing. Remembers best by writing, drawing, or physically manipulating items.

The list above represents only a few of the learning styles identified by educational psychologists. How can you determine which approaches are right for you? Think about the one thing in life that you most enjoy and in which you have the greatest skills. What hobbies or special interests do you have? How do you learn new material in that area? Do you read about a procedure in a book and then do it, or do you throw away the manual and use trial and error to figure out how things work? Do you need to see a diagram or a picture before things make sense, or are spoken directions most memorable and meaningful for you? Some people like to learn by themselves in a quiet place where there are no distractions, while others need to discuss ideas with another person to feel really comfortable about what they're learning.

Sometimes you have to adjust your preferred learning style to the specific material you're trying to master. You may be primarily an oral learner, but if what you need to remember for a particular subject is spatial or structural, you may need to try some visual learning techniques. Memorizing vocabulary items might be best accomplished by oral repetition, while developing your ability to work quantitative problems should be approached by practicing analytical or logical skills.

Use this textbook effectively

An important part of productive learning is to read assigned material in a purposeful, deliberate manner. Ask yourself questions as you read. What is the main point being made here? Does the evidence presented adequately support the assertions being made? What personal experience have you had or what prior knowledge can you bring to bear on this question? Can you suggest alternative explanations for the phenomena being discussed? What additional information would you need in order to make an informed judgment about this subject and how might you go about obtaining that information or making that judgment?

A study technique developed by Frances Robinson and called the SQ3R method (table L.3) can be a valuable aid in improving your reading comprehension. Start your study session with a *survey* of the entire chapter or section you are about to read so you'll have an idea of how the whole thing fits together. What are the major headings and subdivisions? Notice that there is usually a hierarchical organization that gives you clues about the relationship between the various parts. This survey will help you plan your strategy for approaching the material. Next, *question* what the main points are likely to be in each of the sections. Which parts look most important or interesting? Ask yourself where you should invest the most time and effort. Is one section or topic likely to be more relevant to your particular class? Has your instructor emphasized any of the topics you see? Being alert for important material can help you plan the most efficient way to study.

After developing a general plan, begin *active reading* of the text. Read in small segments and stop frequently for reflection and to make notes. Don't fall into a trance in which the words swim by without leaving any impression. Highlight or underline

TABLE L.3

The SQ3R Method for Studying Texts

Survey

Preview the information to be studied before reading.

Question

Ask yourself critical questions about the content of what you are reading.

Read

Conduct the actual reading in small segments.

Recite

Stop periodically to recite to yourself what you have just read.

Review

Once you have completed the section, review the main points to make sure you remember them clearly.

FIGURE L.3 Cooperative learning is an important part of mastering environmental science. You never grasp material as clearly as when you explain it to someone else.

the main points but be careful that you don't just paint the whole page yellow. If you highlight too much, nothing will stand out. Try to distinguish what is truly central to the argument being presented. Make brief notes in the margins that identify main points. This can be very helpful in finding important sections or ideas when you are reviewing. Check your comprehension at the end of each major section. Ask yourself: Did I understand what I just read? What are the main points being made here? Does this relate to my own personal experiences or previous knowledge? Are there details or ideas that need clarification or elaboration?

As you read, stop periodically to *recite* the information you've just acquired. Summarize the information in your own words to be sure that you really understand and are not just depending on rote memory. This is a good time to have a study group (fig. L.3). Taking turns to summarize and explain material really helps you internalize it. If you don't have a study group and you feel awkward talking to yourself, you can try writing your summary. Finally, *review* the section. Did you miss any important points? Do you understand things differently the second time through? This is a chance to think critically about the material. Do you agree with the conclusions suggested by the authors? Can you think of alternative explanations for the same evidence? As you review each section, think about how this may be covered on the test. Put yourself in the position of the instructor. What would be some good questions based on this material? Don't try to memorize everything but try to anticipate what might be the most important points.

After class, compare your lecture notes with your study notes. Do they agree? If not, where are the discrepancies? Is it possible that you misunderstood what was said in class, or does your instructor differ with what's printed in the textbook? Are there things that your instructor emphasized in lecture that you missed in your pre-class reading? This is a good time to go back over the readings to reinforce your understanding and memory of the material.

Will this be on the test?

Students often complain that test results don't adequately reflect what they know and how much they've learned in studying. It may well be that test questions won't cover what you think is important or use a style that appeals to you, but you'll probably be more successful if you adapt yourself to the realities of your instructor's test

methods rather than trying to force your instructor to accommodate to your preferences. One of your first priorities in studying, therefore, should be to learn your instructor's test style. Are you likely to have short-answer objective questions (multiple choice, true or false, fill in the blank) or does your instructor prefer essay questions? If you have an essay test, will the questions be broad and general or more analytical? You should develop a very different study strategy depending on whether you are expected to remember and choose between a multitude of facts and details, or whether you will be asked to write a paragraph summarizing some broad topic.

Organize the ideas you're reading and hearing in lecture. This course will probably include a great deal of information. Unless you have a photographic memory, you won't be able to remember every detail. What's most important? What's the big picture? If you see how pieces of the course fit together, it will all make more sense and be easier to remember. As you read and review, ask yourself what might be some possible test questions in each section. If you're likely to have factual questions, what are the most significant facts in the material you've read? Memorize some benchmark figures. Just a few will help a lot. Pay special attention to tables, graphs, and diagrams. They were chosen because they illustrate important points.

You probably won't be expected to remember all the specific numbers in this book but you probably should know orders of magnitude. The world population is about 6.5 *billion* people, not thousands, millions, or trillions. Highlight facts and figures in your lecture notes about which your instructor seemed especially interested. There is a good chance you'll see those topics again on a test. It often helps to remember facts and figures if you can relate them to some other familiar example. The United States, for instance, has about 295 million residents. The European Union is slightly larger, India is about three times and China is more than four times as large. Be sure you're familiar with the bold-face key terms in the textbook. Vocabulary terms make good objective questions. The Practice Quiz at the end of each chapter generally covers objective material that makes good short-answer questions.

A number of strategies can help you be successful in test-taking. Look over the whole test at the beginning and answer the questions you know well first, then tackle the harder ones. On multiple choice tests, find out whether there is a penalty for guessing. Use the process of elimination to narrow down the possible choices and improve the odds for guessing. Often you can get hints from the context of the question or from other similar questions. Notice that the longest or most specific answer often is right while those that are vague or general are more likely wrong. Be alert for absolutes (such as always, never, all) which could indicate wrong choices. Qualifiers (such as sometimes, may, or could) on the other hand, often point to correct answers. Exactly opposite answers may indicate that one of them is correct.

If you anticipate essay questions, practice writing one- or two-paragraph summaries of major points in each chapter. Develop your ability to generalize and to make connections between important facts and ideas. Notice that the Critical Thinking and Discussion Questions at the end of each chapter are open-ended topics that can work well either for discussion groups or as questions for an essay test. You'll have a big advantage on a test if you have some carefully thought out arguments for and against the major ideas presented in each chapter. If you don't have any idea what a particular essay question means, you often can make a transition to something you do understand. Look for a handle that links the question to a topic you are prepared to answer. Even if you have no idea what the question means, make an educated guess. You might get some credit. Anything is better than a zero. Sometimes if you explain your answer, you'll get at least some points. "If the question means such and such, then the answer would be _____" may get you partial credit.

Does your instructor like thought questions? Does she/he expect you to be able to interpret graphs or to draw inferences from a data table? Might you be asked to read a paragraph and describe what it means or relate it to other cases you've covered in the class? If so, you should practice these skills. Making up and sharing these types of questions with your study group can greatly increase your understanding of the material as well as improve your performance on exams. Writing a paragraph answer for each of the Critical Thinking and Discussion Questions could be a very good way to study for an essay test.

Concentrate on positive attitudes and building confidence before your tests. If you have fears and test anxiety, practice relaxation techniques and visualize success. Be sure you are rested and well prepared. You certainly won't do well if you're sleep-deprived and a bundle of nerves. Often the worst thing you can do is to stay up all night to cram your brain with a jumble of data. Being able to think clearly and express yourself well may count much more than knowing a pile of unrelated facts. Review your test when it is returned to learn what you did well and where you need to improve. Ask your instructor for pointers on how you might have answered the questions better. Carefully add your score to be sure you got all the points you deserve. Sometimes graders make simple mathematical errors in adding up points.

L.2 THINKING ABOUT THINKING

Perhaps the most valuable skill you can learn in any of your classes is the ability to think clearly, creatively, and purposefully. In a rapidly moving field such as environmental science, facts and explanations change constantly. It's often said that in six years approximately half the information you learn from this class will be obsolete. During your lifetime you will probably change careers four to six times. Unfortunately, we don't know which of the ideas we now hold will be outdated or what qualifications you will need for those future jobs. Developing the ability to learn new skills, examine new facts, evaluate new theories, and formulate your own interpretations is essential to keep up in a changing world. In other words, you need to learn how to learn on your own.

Even in our everyday lives most of us are inundated by a flood of information and misinformation. Competing claims and contradictory ideas battle for our attention. The rapidly growing complexity of our world and our lives intensifies the difficulties in knowing what to believe or how to act. Consider how the communications revolution has brought us computers, e-mail, cell phones, mobile faxes, pagers, the World Wide Web, hundreds of channels of satellite TV, and direct mail or electronic marketing that overwhelm us with conflicting information. We have more choices than we can possibly manage, and know more about the world around us than ever before but, perhaps, understand less. How can we deal with the barrage of often contradictory news and advice that inundates us?

To complicate our difficulty in knowing what to believe, distinguished authorities disagree vehemently about many important topics. A law of environmental science might be that for any expert there is always an equal and opposite expert. How can you decide what is true and meaningful in such a welter of confusing information? Is it simply a matter of what feels good at the moment or supports our preconceived notions? Or are there ways to use logical, orderly, creative thinking procedures to reach decisions?

By now, most of us know not to believe everything we read or hear (fig. L.4). "Tastes great . . . Low, low sale price . . . Vote for me . . . Lose 30 pounds in 3 weeks . . . You may already be a winner . . . Causes no environmental harm . . . I'll never lie to you . . . Two out of three doctors recommend . . . " More and more of the information we use to buy, elect, advise, judge, or heal has been created not to expand our knowledge but to sell a product or advance a cause. It would be unfortunate if we become cynical and apathetic due to information overload. It does make a difference what we think and how we act.

FIGURE L.4 "There is absolutely no cause for alarm at the nuclear plant!"
© Tribune Media Services. Reprinted with permission.

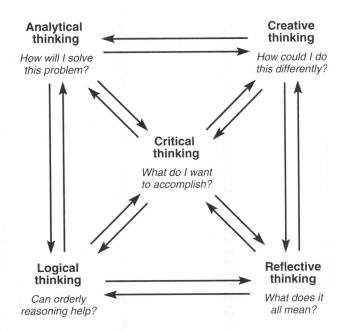

Analytical thinking — *How will I solve this problem?*

Creative thinking — *How could I do this differently?*

Critical thinking — *What do I want to accomplish?*

Logical thinking — *Can orderly reasoning help?*

Reflective thinking — *What does it all mean?*

FIGURE L.5 Different approaches to thinking are used to solve different kinds of problems or to study alternate aspects of a single issue.

TABLE L.4

Steps in Critical Thinking

1. What is the *purpose* of my thinking?
2. What precise *question* am I trying to answer?
3. Within what *point of view* am I thinking?
4. What *information* am I using?
5. How am I *interpreting* that information?
6. What *concepts* or ideas are central to my thinking?
7. What *conclusions* am I aiming toward?
8. What am I taking for granted; what *assumptions* am I making?
9. If I accept the conclusions, what are the *implications*?
10. What would the *consequences* be, if I put my thoughts into action?

Source: Courtesy of Karen Warren, Philosophy Department, Macalester College, St. Paul, MN.

Approaches to truth and knowledge

A number of skills, attitudes, and approaches can help us evaluate information and make decisions. **Analytical thinking** asks, "How can I break this problem down into its constituent parts?" **Creative thinking** asks, "How might I approach this problem in new and inventive ways?" **Logical thinking** asks, "How can orderly, deductive reasoning help me think clearly?" **Critical thinking** asks, "What am I trying to accomplish here and how will I know when I've succeeded?" **Reflective thinking** asks, "What does it all mean?" In this section, we'll look more closely at critical and reflective thinking as a foundation for your study of environmental science. We hope you will apply these ideas consistently as you read this book.

As figure L.5 suggests, critical thinking is central in the constellation of thinking skills. It challenges us to examine theories, facts, and options in a systematic, purposeful, and responsible manner. It shares many methods and approaches with other methods of reasoning but adds some important contextual skills, attitudes, and dispositions. Furthermore, it challenges us to plan methodically and to assess the process of thinking as well as the implications of our decisions. Thinking critically can help us discover hidden ideas and means, develop strategies for evaluating reasons and conclusions in arguments, recognize the differences between facts and values, and avoid jumping to conclusions. Professor Karen J. Warren of Macalester College identifies ten steps in critical thinking (table L.4).

Notice that many critical thinking processes are self-reflective and self-correcting. This form of thinking is sometimes called "thinking about thinking." It is an attempt to plan rationally how to analyze a problem, to monitor your progress while you are doing it, and to evaluate how your strategy worked and what you have learned when you are finished. It is not critical in the sense of finding fault, but it makes a conscious, active, disciplined effort to be aware of hidden motives and assumptions, to uncover bias, and to recognize the reliability or unreliability of sources (What Do You Think? p. 9).

What do I need to think critically?

Certain attitudes, tendencies, and dispositions are essential for well-reasoned analysis. Professor Karen Warren suggests the following list:

- *Skepticism and independence.* Question authority. Don't believe everything you hear or read—including this book; even experts sometimes are wrong.

- *Open-mindedness and flexibility.* Be willing to consider differing points of view and to entertain alternative explanations. Try arguing from a viewpoint different from your own. It will help you identify weaknesses and limitations in your own position.

- *Accuracy and orderliness.* Strive for as much precision as the subject permits or warrants. Deal systematically with parts of a complex whole. Be disciplined in the standards you apply.

- *Persistence and relevance.* Stick to the main point. Don't allow diversions or personal biases to lead you astray. Information may be interesting or even true, but is it relevant?

- *Contextual sensitivity and empathy.* Consider the total situation, relevant context, feelings, level of knowledge, and sophistication of others as you evaluate information. Imagine being in someone else's place to try to understand how they feel.

What Do You Think?

Don't Believe Everything You See or Hear on the News

For most of us, access to news is becoming ever more abundant and ubiquitous. Internet web logs comment on events even as they're happening. Cable television news is available around the clock. Live images are projected to our homes from all over the world. We watch video coverage of distant wars and disasters as if they are occurring in our living rooms, but how much do we really know about what's going on? At the same time that media is becoming more technically sophisticated, news providers are also becoming more adept at manipulating images and content to convey particular messages.

Many people watch TV news programs and read newspapers or web logs today not so much to be educated or to get new ideas as to reinforce their existing beliefs. A State of the Media study by the Center for Journalistic Excellence at Columbia University concluded that the news is becoming increasingly partisan and ideological.[1] The line between news and entertainment has become blurred in most media. Disputes and disasters are overdramatized, while too little attention is paid to complex issues. News reports are increasingly shallow and one-sided, with little editing or fact checking. On live media, such as television and radio, attack journalism is becoming ever more common. Participants try to ridicule and demean their opponents rather than listening respectfully and comparing facts and sources. Many shows simply become people shouting at each other. Print media also is moving toward tabloid journalism, featuring many photographs and sensationalist coverage of events.

According to the State of the Media report, most television stations have all but abandoned the traditional written and edited news story. Instead, more than two-third of all news segments now consist of on-site "stand-up" reports or live interviews in which a single viewpoint is presented as news without any background or perspective. Visual images seem more immediate and are regarded as more believable by most people: after all, pictures don't lie, but they can give a misleading impression of what's really important. Many topics, such as policy issues, don't make good visuals, and therefore never make it into TV coverage. Crime, accidents, disasters, lifestyle stories, sports, and weather make up more than 90 percent of the coverage on a typical television news program. If you watched cable TV news for an entire day, for instance, you'd see, on average, only 1 minute each about the environment and health care, 2 minutes each on science and education, and 4 minutes on art and culture. More than 70 percent of the segments are less than 1 minute long, meaning that they convey more emotion than substance. People who get their news primarily from TV are significantly more fearful and pessimistic than those who get news from print media.

Partisan journalism has become much more prevalent since the deregulation of public media. From the birth of the broadcasting industry, the airwaves were regarded and regulated as a public trust. Broadcasters, as a condition of their licenses, were required to operate in the "public interest" by covering important policy issues and providing equal time to both sides of contested issues. In 1988, however, the Federal Communications Commission ruled that the proliferation of mass media gives the public adequate access to diverse sources of information. Media outlets no longer are obliged to provide fair and balanced coverage of issues. Presenting a single perspective or even a deceptive version of events is no longer regarded as a betrayal of public trust.

A practice that further erodes the honesty and truthfulness of media coverage is the use of video news releases that masquerade as news stories. In these videos, actors, hired by public relations firms, pose as reporters or experts to promote a special interest. Businesses have long used this tactic to sell products, but a recent disturbing development is placement of news videos by governmental agencies. For example, in 2004, the federal Department of Health and Human Services sent video stories to TV stations promoting the benefits of the recently passed but controversial Medicare drug law. The actors in these videos appeared to be simply reporting news, but, in fact, were presenting a highly partisan viewpoint. Critics complained that these "stealth ads" undermine the credibility of both journalists and public officials. Kevin W. Keane, a Health Department spokesman, dismissed the criticism, saying this is "a common, routine practice in government and the private sector." In 2004, the federal government paid $88 million to public relations firms and news commentators to represent administration positions on policy issues.

How can you detect bias in a news report? Ask yourself the following questions:

1. What political positions are represented in the story?
2. What special interests might be involved here? Who stands to gain presenting a particular viewpoint? Who is paying for the message?
3. What sources are used as evidence in this story? How credible are they?
4. Are statistics cited in the presentation? Are citations provided so you can check the source?
5. Is the story one-sided, or are alternate viewpoints presented? Are both sides represented by credible spokespersons, or is one simply a patsy set up to make the other side look good?
6. Are the arguments presented based on facts and logic, or are they purely emotional appeals?

We need to practice critical thinking to detect bias and make sense out of what we see and hear. Although the immediacy and visual impact of television or the Internet may seem convincing, we have to use caution and judgment to interpret the information they present. Don't depend on a single source for news. Compare what different media outlets say about an issue before making up your mind.

[1] The State of the News Media 2004 available at http://www.journalism.org.

- *Decisiveness and courage.* Draw conclusions and take a stand when the evidence warrants doing so. Although we often wish for more definitive information, sometimes a well-reasoned but conditional position has to be the basis for action.

- *Humility.* Realize that you may be wrong and that reconsideration may be called for in the future. Be careful about making absolute declarations; you may need to change your mind someday.

While critical thinking shares many of the orderly, systematic approaches of formal logic, it also invokes traits like empathy, sensitivity, courage, and humility. Formulating intelligent opinions about some of the complex issues you'll encounter in environmental science requires more than simple logic. Developing these attitudes and skills is not easy or simple. It takes practice. You have to develop your mental faculties just as you need to train for a sport. Traits such as intellectual integrity, modesty, fairness, compassion, and fortitude are not things you can use only occasionally. They must be cultivated until they become your normal way of thinking.

Applying critical thinking

We all use critical or reflective thinking at times. Suppose a television commercial tells you that a new breakfast cereal is tasty and good for you. You may be suspicious and ask yourself a few questions. What do they mean by good? Good for whom or what? Does "tasty" simply mean more sugar and salt? Might the sources of this information have other motives in mind besides your health and happiness? Although you may not have been aware of it, you already have been using some of the techniques of critical analysis. Working to expand these skills helps you recognize the ways information and analysis can be distorted, misleading, prejudiced, superficial, unfair, or otherwise defective.

Here are some steps in critical thinking:

1. *Identify and evaluate premises and conclusions in an argument.* What is the basis for the claims made here? What evidence is presented to support these claims and what conclusions are drawn from this evidence? If the premises and evidence are correct, does it follow that the conclusions are necessarily true?

2. *Acknowledge and clarify uncertainties, vagueness, equivocation, and contradictions.* Do the terms used have more than one meaning? If so, are all participants in the argument using the same meanings? Are ambiguity or equivocation deliberate? Can all the claims be true simultaneously?

3. *Distinguish between facts and values.* Are claims made that can be tested? (If so, these are statements of fact and should be able to be verified by gathering evidence.) Are claims made about the worth or lack of worth of something? (If so, these are value statements or opinions and probably cannot be verified objectively.) For example, claims of what we *ought* to do to be moral or righteous or to respect nature are generally value statements.

4. *Recognize and assess assumptions.* Given the backgrounds and views of the protagonists in this argument, what underlying reasons might there be for the premises, evidence, or conclusions presented? Does anyone have an "axe to grind" or a personal agenda in this issue? What do they think you know, need, want, or believe? Is there a subtext based on race, gender, ethnicity, economics, or some belief system that distorts this discussion?

5. *Distinguish the reliability or unreliability of a source.* What makes the experts qualified in this issue? What special knowledge or information do they have? What evidence do they present? How can we determine whether the information offered is accurate, true, or even plausible?

6. *Recognize and understand conceptual frameworks.* What are the basic beliefs, attitudes, and values that this person, group, or society holds? What dominating philosophy or ethics control their outlook and actions? How do these beliefs and values affect the way people view themselves and the world around them? If there are conflicting or contradictory beliefs and values, how can these differences be resolved?

Some clues for unpacking an argument

In logic, an argument is made up of one or more introductory statements (called **premises**), and a **conclusion** that supposedly follows logically from the premises. Often in ordinary conversation, different kinds of statements are mixed together, so it is difficult to distinguish between them or to decipher hidden or implied meanings. Social theorists call the process of separating and analyzing textual components *unpacking*. Applying this type of analysis to an argument can be useful.

An argument's premises are usually claimed to be based on facts; conclusions are usually opinions and values drawn from, or used to interpret, those facts. Words that often introduce a premise include: *as, because, assume that, given that, since, whereas,* and *we all know that* . . . Words that generally indicate a conclusion or statement of opinion or values include: *and, so, thus, therefore, it follows that, consequently, the evidence shows,* and *we can conclude that.*

For instance, in the example we used earlier, the television ad might have said: "*Since* we all need vitamins, and *since* this cereal contains vitamins, *consequently* the cereal must be good for you." Which are the premises and which is the conclusion? Does one necessarily follow from the other? Remember that even if the facts in a premise are correct, the conclusions drawn from them may not be. Information may be withheld from the argument such as the fact that the cereal is also loaded with unhealthy amounts of sugar.

Avoiding logical errors and fallacies

Formal logic catalogs a large number of fallacies and errors that invalidate arguments. Although we don't have room here to include all of these fallacies and errors, it may be helpful to review a few of the more common ones.

- *Red herring:* Introducing extraneous information to divert attention from the important point.

- *Ad hominem attacks:* Criticizing the opponent rather than the logic of the argument.

- *Hasty generalization:* Drawing conclusions about all members of a group based on evidence that pertains only to a selected sample.

- *False cause:* Drawing a link between premises and conclusions that depends on some imagined causal connection that does not, in fact, exist.
- *Appeal to ignorance:* Because some facts are in doubt, therefore a conclusion is impossible.
- *Appeal to authority:* It's true because _____ says so.
- *Begging the question:* Using some trick to make a premise seem true when it is not.
- *Equivocation:* Using words with double meanings to mislead the listener.
- *Slippery slope:* A claim that some event or action will cause some subsequent action.
- *False dichotomy:* Giving either/or alternatives as if they are the only choices.

Avoiding these fallacies yourself or being aware of them in another's argument can help you be more logical and have more logical and reasonable discussions.

Using critical thinking in environmental science

As you go through this book, you will have many opportunities to practice critical thinking skills. Every chapter includes many facts, figures, opinions, and theories. Are all of them true? No, probably not. They were the best information available when this text was written, but much in environmental science is in a state of flux. Data change constantly as does our interpretation of them. Do the ideas presented here give a complete picture of the state of our environment? Unfortunately, they probably don't. No matter how comprehensive our discussion is of this complex, diverse subject, it can never capture everything worth knowing, nor can it reveal all possible points of view.

When reading this text, try to distinguish between statements of fact and opinion. Ask yourself if the premises support the conclusions drawn from them. Although we have tried to be fair and even-handed in presenting controversies, we, like everyone, have biases and values—some of which we may not even recognize—that affect how we see issues and present arguments. Watch for cases in which you need to think for yourself and utilize your critical and reflective thinking skills to find the truth.

L.3 CONCEPT MAPS

Concept mapping is a learning strategy that many students find useful in understanding complex ideas and clarifying ambiguous relationships. Creating a graphic representation of a topic often can help you visualize key concepts and organize your knowledge more clearly than will other methods of study. You may find that the physical process of drawing a map of a topic engages a different part of your brain than does ordinary reading or taking notes. Taking time to think carefully about what is most important about a particular topic will help you remember

it better at a later date. It can also point out weaknesses in your understanding as well as areas in which you need more study. Practicing this technique as you think about environmental science can help you comprehend difficult material and prepare for exams.

From ancient times, people have made maps to help them understand and remember important aspects of the world around them. Maps integrate and summarize knowledge. They suggest linkages that we may not have seen before, and they suggest routes for further exploration. But maps can never record every possible detail of the world or the ideas they represent. Only the most useful and meaningful information is put onto the map so that important points can easily be seen and remembered. No map is ever complete and finished. As we learn more, we revise our maps, correcting errors, adding new information, removing unnecessary features, and refining the presentation. The act of drawing a map exposes doubtful knowledge and shows us where we need more data and a clearer understanding.

If you ask different people to construct a map of the same area of a city, you would probably get very different results. A young child, for example, might draw in only the locations of her school, home, and playground. A commuter from the suburbs, on the other hand, might show the locations of the major office buildings, filling stations, freeway ramps, and the cheapest parking lots in that same city. Neither of these representations is wrong, they just emphasize the aspects of greatest importance to each person.

When we think about maps, we usually visualize physical features such as mountains, lakes, roads, buildings, and so forth. But we can also create maps of biodiversity, magnetic fields, economic flows, cultures, language families, or anything else that can be presented in graphical form. A **concept map** is a two-dimensional representation of the relationship between key ideas. It shows us how we think and suggests affinities and associations that might not otherwise be obvious. At first glance, a concept map looks like a flow chart in which key terms are placed in boxes connected by directional arrows. Based on educational psychology theories of how we organize information, concept maps are hierarchical, with broader, more general items at the top and more specific topics arranged in a cascade below them. They are metacognitive tools that empower the learner to take charge of his/her own learning in a highly organized and meaningful manner.

How do I create a concept map?

To create a concept map, start with what you already know. Build from what's familiar. What are the key components or ideas in the topic you're trying to understand? Place each concept in its own individual circle, box, or other geometrical shape. You might want to use different shapes to indicate relative levels or types of ideas. Connect concept boxes with directional arrows to show relationships. Label each arrow with descriptive terms so that your diagram can be read as a statement or proposition by following interconnections from the top down. In figure L.6, for example, you can read the proposition that "concept maps are used to develop study strategies that lead to learning that determines your grade" as one set of associations.

As you can see in this example, branches to one side or the other of the key concepts show related ideas. Where appropriate, cross links or bridges can connect branches of your map. Linking arrows can be bidirectional to indicate mutual interactions, but be careful not to make everything connected to everything else. Focus on the most important concepts and the most significant relationships. View this as an exercise in discrimination. Don't try to make your map perfect. Sometimes working briskly will help you cut away the superfluous while fostering creativity and synthesis. The point is not to create a work of art but to organize, discover, and understand central meanings. The map helps you learn how to learn.

A concept map can show just a small part or a subset of a broader field of knowledge. The top or key concept in one map may be subsumed into a lower position in a map with a different focus. A small branch in a general map could be expanded into a much more specific map of its own.

Remember that concept maps are works in progress. There are no right or wrong maps. Each one represents one possible way of understanding a particular set of ideas by an individual at the time the map was made. Expect to do several iterations, right from the outset. There are probably as many ways of representing a collection of concepts as there are concepts in the collection, but some—typically those discovered by a process of trial and error—are more elegant and easier to understand than others.

The benefits of mapping are mainly to the individual making the map. The process of simplifying concepts and arranging them on a page forces you to think about what's most important.

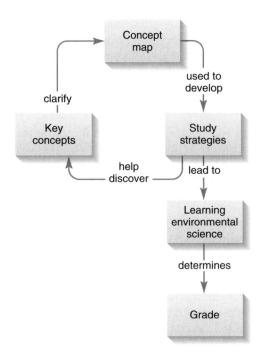

FIGURE L.6 A concept map can help you organize your thinking.

It helps you clarify your thoughts and understandings and makes learning more meaningful. A concept map can be a heuristic device; that is, a process in which you make discoveries and uncover meanings through trial and error. Although you benefit most directly from making your own map, it can be instructive to compare your map to that of a fellow student to see how her or his take on a topic compares to yours.

CONCLUSION

Whether you find environmental science interesting and useful depends largely on your own attitudes and efforts. Developing good study habits, setting realistic goals for yourself, taking the initiative to look for interesting topics, finding an appropriate study space, and working with a study partner can both make your study time more efficient and improve your final grade. Each of us has his or her own learning style. You may understand and remember things best if you see them in writing, hear them spoken by someone else, reason them out for yourself, or learn by doing. By determining your preferred style, you can study in the way that is most comfortable and effective for you.

PRACTICE QUIZ

1. Which study skills in table L.1 do you need to improve?
2. Describe some ways you can avoid procrastination and keep on schedule.
3. List four learning styles. Which fits you best?
4. Describe the SQ3R study techniques.
5. What are ten steps in critical thinking?
6. Name (and describe) seven attributes or dispositions essential for critical thinking.
7. List some adverbs or adverbial phrases that introduce premises and conclusions.
8. Distinguish between an *ad hominem* attack and an appeal to ignorance.
9. Describe three questions you'd ask to evaluate the reliability of Internet information.

CRITICAL THINKING AND DISCUSSION QUESTIONS

1. What is critical thinking? Why is it important?

2. Suppose your preferred learning style doesn't match the teaching style of your instructor. How might you find a common ground?

3. You may find that you fit two or more of the learning profiles described in this chapter. How would you decide which is most effective or appropriate for you?

4. Why is critical thinking important in environmental science?

5. What is the difference between critical and reflective thinking?

6. Suppose you see a claim on television or the Internet, how would you evaluate its reliability?

7. Why are empathy and contextual sensitivity important in critical thinking?

8. If some facts in this book are vague or doubtful, why mention them at all? How would you decide which "facts" to use and which to ignore?

9. Why are concept maps usually hierarchical and linked by active verbs?

As part of it's green Olympics effort, China is planting trees to stabilize soil, reduce erosion and flooding, and hold back the dust storms that plague northern China. In 2006, some 500 million Chinese planted more than 2 billion trees.

C H A P T E R

1

Understanding Our Environment

*Today we are faced with a challenge that calls for a shift
in our thinking, so that humanity stops threatening its
life-support system.*

—*Wangari Maathai*—
Winner of 2004 Nobel Peace Prize

LEARNING OUTCOMES

After studying this chapter, you should be able to:

1.1 Define environmental science and identify some important environmental concerns we face today.

1.2 Discuss the history of conservation and the different attitudes toward nature at various times in our past.

1.3 Think critically about the major environmental dilemmas and issues that shape our current environmental agenda.

1.4 Appreciate the human dimensions of environmental science, including the connection between poverty and environmental degradation.

1.5 Explain sustainable development and evaluate some of its requirements.

Case Study A Green Olympics?

The People's Republic of China has ambitious plans to make the 2008 summer Olympics the "greenest" ever. Plans for sustainable resource use and minimal environmental effects include buildings constructed with environmentally friendly materials and the latest energy-saving technology. Solar water heaters will provide hot showers for athletes, while windmills and photovoltaic cells will contribute 20 percent of the electricity used in the Olympic Village. Rain will be collected and used to water playing fields. Toilets will use recycled water. New wastewater treatment plants will reduce sewage effluents, and two new recycling plants will reduce solid waste disposal. Reforestation projects and fuel-switching are expected to offset carbon dioxide emissions and make the games climate neutral.

Although critics doubt how effective these measures will be, the fact that they're being mentioned at all is a hopeful sign. A generation ago, environmental conditions weren't even mentioned by Chinese officials. Mere survival dominated the political agenda. Providing enough food, jobs, and housing for a rapidly growing population and stabilizing a chaotic political system preoccupied public attention. Now, at least environmental quality and resource use is something that leaders feel they need to address.

China has many serious environmental challenges. Seventy percent of all Chinese cities don't meet national air quality standards. According to the United Nations Environment Programme, 16 of the 20 smoggiest cities in the world are in China, and one-third of the country is affected by acid precipitation (fig. 1.1). It's estimated that 400,000 people die each year from the effects of air pollution. China is now the world's largest producer of sulfur dioxide, chlorofluorocarbons, and carbon dioxide. Seventy percent of Chinese rivers and lakes fail to meet government water quality standards. About half the surface water is so contaminated that it isn't useful even for agriculture. Two-thirds of Chinese cities don't have enough water to meet demands, and at least 300 million people live in areas with severe water shortages. One-third of China's land has been degraded by unsustainable farming, grazing, and logging, and 400 million people are threatened by expanding deserts.

But there also is positive news among this litany of woes. In addition to planning for a green Olympics, Chinese leaders have pledged to take steps to improve environmental quality for the whole country. They plan to spend at least (U.S.)$125 billion over the next five years to reduce water pollution and bring clean drinking water to everyone in the country by 2015. Officials also promise that 10 percent of China's energy will come from renewable sources within a decade. Logging on steep hillsides has been officially banned, and more than 50 billion trees have been planted on 500,000 km² of marginal land to hold back deserts and reduce blowing dust. Construction has already begun on new cities designed to be largely self-sufficient in energy, water, and food and to be climate neutral with respect to greenhouse gases. Although there are doubts about whether the government can actually deliver on all these promises, it's encouraging that they're moving toward sustainability.

Some of the new-found governmental concern about the Chinese environment comes from public demand. In 2005 there were at least 85,000 public protests in China, many of which were about pollution, environmental health, land degradation, and similar issues. A campaign led by artists, students, and writers forced the government to cancel plans for a series of 13 large dams on the Nu River (the Chinese portion of the Salween) in an area of high biological and cultural diversity. There are now more than 2,000 nongovernmental organizations (NGOs) in China working on social and environmental issues, and for the first time, they have officially recognized status.

But why should you be concerned about what happens half a world away? Part of the answer is that we all share a single planet. What happens in one place affects all of us. You may have noticed that rising Chinese demand for commodities, such as oil, copper, steel, and seafood, are driving up the prices worldwide. Less well known is the fact that on some days, three-fourths of the air pollution in some North American west coast cities can be traced to China. And if China meets its rapidly growing power demand by burning more coal, the effects on our global climate could be disastrous.

Although China currently has the world's largest population, and, therefore, has a huge impact on our global resources and environment, its problems aren't unique. Many of the countries in which the poorer four-fifths of all humans live face similar environmental

FIGURE 1.1 Sixteen of the 20 smoggiest cities in the world are in China. Sulfur dioxide emissions are highest in the world, and the number of Chinese automobiles increases by about 5 million per year.

Case Study *continued*

challenges. Finding ways that all of us can live sustainably within the limits of our resource base and without damaging nature's life-support systems is the preeminent challenge of environmental science. And, as is the case with China, while the world faces many serious environmental problems, there are also signs of progress that give us hope for the future.

For more information, see

Liu, Jianguo, and Jared Diamond. 2005. China's environment in a globalizing world: How China and the rest of the world affect each other. *Nature* 435:1179–86.

Min, Shao, et al. 2006. City clusters in China: Air and surface water pollution. *Frontiers in Ecological Environment* 4 (7):353–61.

1.1 WHAT IS ENVIRONMENTAL SCIENCE?

Humans have always inhabited two worlds. One is the natural world of plants, animals, soils, air, and water that preceded us by billions of years and of which we are a part. The other is the world of social institutions and artifacts that we create for ourselves using science, technology, and political organization. Both worlds are essential to our lives, but integrating them successfully causes enduring tensions.

Where earlier people had limited ability to alter their surroundings, we now have power to extract and consume resources, produce wastes, and modify our world in ways that threaten both our continued existence and that of many organisms with which we share the planet. To ensure a sustainable future for ourselves and future generations, we need to understand something about how our world works, what we are doing to it, and what we can do to protect and improve it.

Environment (from the French *environner:* to encircle or surround) can be defined as (1) the circumstances or conditions that surround an organism or group of organisms, or (2) the complex of social or cultural conditions that affect an individual or community. Since humans inhabit the natural world as well as the "built" or technological, social, and cultural world, all constitute important parts of our environment (fig. 1.2).

Environmental science, then, is the systematic study of our environment and our proper place in it. A relatively new field, environmental science is highly interdisciplinary, integrating natural sciences, social sciences, and humanities in a broad, holistic study of the world around us. In contrast to more theoretical disciplines, environmental science is mission-oriented. That is, it seeks new, valid, contextual knowledge about the natural world and our impacts on it, but obtaining this information creates a responsibility to get involved in trying to do something about the problems we have created.

As distinguished economist Barbara Ward pointed out, for an increasing number of environmental issues, the difficulty is not to identify remedies. Remedies are now well understood. The problem is to make them socially, economically, and politically acceptable. Foresters know how to plant trees, but not how to establish conditions under which villagers in developing countries can manage plantations for themselves. Engineers know how to control pollution, but not how to persuade factories to install the necessary equipment. City planners know how to build housing and design safe drinking water systems, but not how to make them affordable for the poorest members of society. The solutions to these problems increasingly involve human social systems as well as natural science.

Criteria for environmental literacy suggested by the National Environmental Education Advancement Project in Wisconsin include: awareness and appreciation of the natural and built environment; knowledge of natural systems and ecological concepts; understanding of current environmental issues; and the ability to use critical-thinking and problem-solving skills on environmental issues. These are good overall goals to keep in mind as you study this book.

Chapter 2 looks more closely at science as a way of knowing, environmental ethics, and other tools that help us analyze and understand the world around us. For the remainder of this chapter, we'll complete our overview with a short history of environmental thought and a survey of some important current issues that face us.

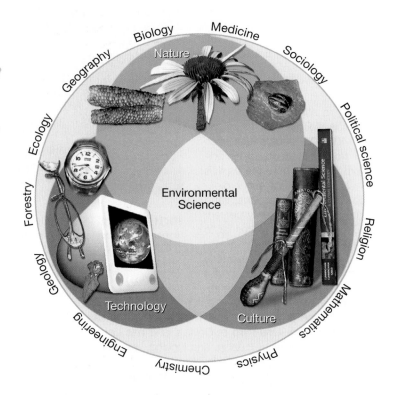

FIGURE 1.2 The intersections of the natural, cultural, and technological worlds outline the province of environmental science. Many disciplines contribute to our understanding and management of our environment.

1.2 A Brief History of Conservation and Environmentalism

Although many early societies had negative impacts on their surroundings, others lived in relative harmony with nature. In modern times, however, growing human populations and the power of our technology have increased our impacts on our environment. We can divide conservation history and environmental activism into at least four distinct stages: (1) pragmatic resource conservation, (2) moral and aesthetic nature preservation, (3) a growing concern about health and ecological damage caused by pollution, and (4) global environmental citizenship. Each era focused on different problems and each suggested a distinctive set of solutions. These stages are not necessarily mutually exclusive, however; parts of each persist today in the environmental movement and one person may embrace them all simultaneously.

Nature protection has historic roots

Recognizing human misuse of nature is not unique to modern times. Plato complained in the fourth century B.C. that Greece once was blessed with fertile soil and clothed with abundant forests of fine trees. After the trees were cut to build houses and ships, however, heavy rains washed the soil into the sea, leaving only a rocky "skeleton of a body wasted by disease." Springs and rivers dried up while farming became all but impossible. Many classical authors regarded Earth as a living being, vulnerable to aging, illness, and even mortality. Periodic threats about the impending death of nature as a result of human misuse have persisted into our own time. Many of these dire warnings have proven to be premature or greatly exaggerated, but others remain relevant to our own times. As Mostafa K. Tolba, former Executive Director of the United Nations Environment Programme has said, "The problems that overwhelm us today are precisely those we failed to solve decades ago."

Some of the earliest scientific studies of environmental damage were carried out in the eighteenth century by French and British colonial administrators who often were trained scientists and who considered responsible environmental stewardship as an aesthetic and moral priority, as well as an economic necessity. These early conservationists observed and understood the connection between deforestation, soil erosion, and local climate change. The pioneering British plant physiologist, Stephen Hales, for instance, suggested that conserving green plants preserved rainfall. His ideas were put into practice in 1764 on the Caribbean island of Tobago, where about 20 percent of the land was marked as "reserved in wood for rains."

Pierre Poivre, an early French governor of Mauritius, an island in the Indian Ocean, was appalled at the environmental and social devastation caused by destruction of wildlife (such as the flightless dodo) and the felling of ebony forests on the island by early European settlers. In 1769, Poivre ordered that one-quarter of the island was to be preserved in forests, particularly on steep mountain slopes and along waterways. Mauritius remains a model for balancing nature and human needs. Its forest reserves shelter a larger percentage of its original flora and fauna than most other human-occupied islands.

Resource waste inspired pragmatic, utilitarian conservation

Many historians consider the publication of *Man and Nature* in 1864 by geographer George Perkins Marsh as the wellspring of environmental protection in North America. Marsh, who also was a lawyer, politician, and diplomat, traveled widely around the Mediterranean as part of his diplomatic duties in Turkey and Italy. He read widely in the classics (including Plato) and personally observed the damage caused by the excessive grazing by goats and sheep and by the deforesting of steep hillsides. Alarmed by the wanton destruction and profligate waste of resources still occurring on the American frontier in his lifetime, he warned of its ecological consequences. Largely as a result of his book, national forest reserves were established in the United States in 1873 to protect dwindling timber supplies and endangered watersheds.

Among those influenced by Marsh's warnings were President Theodore Roosevelt (fig. 1.3a) and his chief conservation advisor, Gifford Pinchot (fig. 1.3b). In 1905, Roosevelt, who was the leader of the populist, progressive movement, moved

(a) President Teddy Roosevelt

(b) Gifford Pinchot

(c) John Muir

(d) Aldo Leopold

FIGURE 1.3 Some early pioneers of the American conservation movement. President Teddy Roosevelt (a) and his main advisor Gifford Pinchot (b) emphasized pragmatic resource conservation, while John Muir (c) and Aldo Leopold (d) focused on ethical and aesthetic relationships.

the Forest Service out of the corruption-filled Interior Department into the Department of Agriculture. Pinchot, who was the first native-born professional forester in North America, became the founding head of this new agency. He put resource management on an honest, rational, and scientific basis for the first time in our history. Together with naturalists and activists such as John Muir, William Brewster, and George Bird Grinnell, Roosevelt and Pinchot established the framework of our national forest, park, and wildlife refuge systems, passed game protection laws, and tried to stop some of the most flagrant abuses of the public domain. In 1908, Pinchot organized and chaired the White House Conference on Natural Resources, perhaps the most prestigious and influential environmental meeting ever held in the United States.

The basis of Roosevelt's and Pinchot's policies was pragmatic **utilitarian conservation.** They argued that the forests should be saved "not because they are beautiful or because they shelter wild creatures of the wilderness, but only to provide homes and jobs for people." Resources should be used "for the greatest good, for the greatest number for the longest time." "There has been a fundamental misconception," Pinchot said, "that conservation means nothing but husbanding of resources for future generations. Nothing could be further from the truth. The first principle of conservation is development and use of the natural resources now existing on this continent for the benefit of the people who live here now. There may be just as much waste in neglecting the development and use of certain natural resources as there is in their destruction." This pragmatic approach still can be seen in the multiple use policies of the Forest Service.

Ethical and aesthetic concerns inspired the preservation movement

John Muir (fig. 1.3c), geologist, author, and first president of the Sierra Club, strenuously opposed Pinchot's influence and policies. Muir argued that nature deserves to exist for its own sake, regardless of its usefulness to us. Aesthetic and spiritual values formed the core of his philosophy of nature protection. This outlook has been called **biocentric preservation** because it emphasizes the fundamental right of other organisms to exist and to pursue their own interests. Muir wrote: "The world, we are told, was made for man. A presumption that is totally unsupported by the facts. . . . Nature's object in making animals and plants might possibly be first of all the happiness of each one of them. . . . Why ought man to value himself as more than an infinitely small unit of the one great unit of creation?"

Muir, who was an early explorer and interpreter of the Sierra Nevada Mountains in California, fought long and hard for establishment of Yosemite and Kings Canyon National Parks. The National Park Service, established in 1916, was first headed by Muir's disciple, Stephen Mather, and has always been oriented toward preservation of nature in its purest state. It has often been at odds with Pinchot's utilitarian Forest Service. Environmental ethics is discussed further in chapter 2.

FIGURE 1.4 Aldo Leopold's Wisconsin shack, the main location for his *Sand County Almanac,* in which he wrote, "A thing is right when it tends to preserve the integrity, stability, and beauty of the biotic community. It is wrong when it tends otherwise." How might you apply this to your life?

In 1935, pioneering wildlife ecologist Aldo Leopold (fig. 1.3d) bought a small, worn-out farm in central Wisconsin. A dilapidated chicken shack, the only remaining building, was remodeled into a rustic cabin (fig. 1.4). Working together with his children, Leopold planted thousands of trees in a practical experiment in restoring the health and beauty of the land. "Conservation," he wrote, "is the positive exercise of skill and insight, not merely a negative exercise of abstinence or caution." The shack became a writing refuge and became the main focus of *A Sand County Almanac,* a much beloved collection of essays about our relation with nature. In it, Leopold wrote, "We abuse land because we regard it as a commodity belonging to us. When we see land as a community to which we belong, we may begin to use it with love and respect." Together with Bob Marshall and two others, Leopold was a founder of the Wilderness Society.

Rising pollution levels led to the modern environmental movement

The undesirable effects of pollution probably have been recognized at least as long as those of forest destruction. In 1273, King Edward I of England threatened to hang anyone burning coal in London because of the acrid smoke it produced. In 1661, the English diarist John Evelyn complained about the noxious air pollution caused by coal fires and factories and suggested that sweet-smelling trees be planted to purify city air. Increasingly dangerous smog attacks in Britain led, in 1880, to formation of a national Fog and Smoke Committee to combat this problem.

The tremendous industrial expansion during and after the Second World War added a new set of concerns to the environmental agenda. *Silent Spring,* written by Rachel Carson (fig. 1.5a) and published in 1962, awakened the public to the threats of pollution and toxic chemicals to humans as well as other species.

(a) Rachel Carson

(b) David Brower

(c) Barry Commoner

(d) Wangari Maathai

FIGURE 1.5 Among many distinguished environmental leaders in modern times, Rachel Carson (a), David Brower (b), Barry Commoner (c), and Wangari Maathai (d) stand out for their dedication, innovation, and bravery.

The movement she engendered might be called **environmentalism** because its concerns are extended to include both environmental resources and pollution. Among the pioneers of this movement were activist David Brower (fig. 1.5b) and scientist Barry Commoner (fig. 1.5c). Brower, while executive director of the Sierra Club, Friends of the Earth, and the Earth Island Institute, introduced many of the techniques of modern environmentalism, including litigation, intervention in regulatory hearings, book and calendar publishing, and using mass media for publicity campaigns. Commoner, who was trained as a molecular biologist, has been a leader in analyzing the links between science, technology, and society. Both activism and research remain hallmarks of the modern environmental movement.

In 1977, Professor Wangari Maathai (fig. 1.5d) founded the Green Belt Movement in her native Kenya as a way of both organizing poor rural women and restoring their environment. Beginning at a small, local scale, this organization has grown to more than 600 grassroots networks across Kenya. They have planted more than 30 million trees while mobilizing communities for self-determination, justice, equity, poverty reduction, and environmental conservation. Dr. Maathai was elected to the Kenyan Parliament and served as Assistant Minister for Environment and Natural Resources. Her leadership has helped bring democracy and good government to her country. In 2004, she received the Nobel Peace Prize for her work, the first time a Nobel has been awarded for environmental action. In her acceptance speech, she said, "Working together, we have proven that sustainable development is possible; that reforestation of degraded land is possible; and that exemplary governance is possible when ordinary citizens are informed, sensitized, mobilized and involved in direct action for their environment."

Under the leadership of a number of other brilliant and dedicated activists and scientists, the environmental agenda was expanded in the 1960s and 1970s to include issues such as human population growth, atomic weapons testing and atomic power, fossil fuel extraction and use, recycling, air and water pollution, wilderness protection, and a host of other pressing problems that are addressed in this textbook. Environmentalism has become well established on the public agenda since the first national Earth Day in 1970. A majority of Americans now consider themselves environmentalists, although there is considerable variation in what that term means.

Think About It

Suppose a beautiful grove of trees near your house is scheduled to be cut down for a civic project such as a swimming pool. Would you support this? Why or why not? Which of the philosophies described in this chapter best describes your attitude?

Global interconnections have expanded environmentalism

Increased opportunities to travel, as well as greatly expanded international communications, now enable us to know about daily events in places unknown to our parents or grandparents. We have become, as Marshal McLuhan announced in the 1960s, a global village. As in a village, we are all interconnected in various ways. Events that occur on the other side of the globe have profound and immediate effects on our lives.

Photographs of the earth from space (fig. 1.6) provide a powerful icon for the fourth wave of ecological concern that might be called **global environmentalism.** These photos remind us how small, fragile, beautiful, and rare our home planet is. We all share a common environment at this global scale. As our attention shifts from questions of preserving particular landscapes or preventing pollution of a specific watershed or airshed, we begin to worry about the life-support systems of the whole planet.

We now understand that we are changing planetary weather systems and atmospheric chemistry, reducing the natural variety of organisms, and degrading ecosystems in ways that could have devastating effects, both on humans and on all other life-forms. Protecting our environment has become an international cause and it will take international cooperation to bring about many necessary changes.

Some Chinese leaders are part of this global environmental movement. In 2006, Yu Xiaogang was awarded the Goldman Prize, the world's top honor for environmental protection. Yu was recognized for his work on Yunan's Lashi Lake where he brought

FIGURE 1.6 The life-sustaining ecosystems on which we all depend are unique in the universe, as far as we know.

FIGURE 1.7 Perhaps the most amazing feature of our planet is its rich diversity of life.

together residents, government officials, and entrepreneurs to protect wetlands, restore fisheries, and improve water quality. He also worked on sustainable development programs, such as women's schools and microcredit loans. His leadership was instrumental in stopping plans for dams on the Nu River, mentioned in the opening case study for this chapter. Another Goldman Prize winner is Dai Qing, who was jailed for her book that revealed the social and environmental costs of the Three Gorges Dam on the Yangtze River.

Other global environmental leaders include Professor Muhammad Yunus of Bangladesh, who won the Nobel Peace Prize in 2006 for his microcredit loan program at the Grameen Bank, and former Norwegian Prime Minister Gro Harlem Brundtland, who chaired the World Commission on Environment and Development, which coined the most widely accepted definition of sustainability.

1.3 CURRENT CONDITIONS

As you probably already know, many environmental problems now face us. Before surveying them in the following section, we should pause for a moment to consider the extraordinary natural world that we inherited and that we hope to pass on to future generations in as good—perhaps even better—a condition than when we arrived.

We live on a marvelous planet

Imagine that you are an astronaut returning to Earth after a long trip to the moon or Mars. What a relief it would be to come back to this beautiful, bountiful planet after experiencing the hostile, desolate environment of outer space. Although there are dangers and difficulties here, we live in a remarkably prolific and hospitable world that is, as far as we know, unique in the universe. Compared to the conditions on other planets in our solar system, temperatures on the earth are mild and relatively constant. Plentiful supplies of clean air, fresh water, and fertile soil are regenerated endlessly and spontaneously by geological and biological cycles (discussed in chapters 3 and 4).

Perhaps the most amazing feature of our planet is the rich diversity of life that exists here. Millions of beautiful and intriguing species populate the earth and help sustain a habitable environment (fig. 1.7). This vast multitude of life creates complex, interrelated communities where towering trees and huge animals live together with, and depend upon, tiny life-forms such as viruses, bacteria, and fungi. Together all these organisms make up delightfully diverse, self-sustaining communities, including dense, moist forests, vast sunny savannas, and richly colorful coral reefs. From time to time, we should pause to remember that, in spite of the challenges and complications of life on earth, we are incredibly lucky to be here. We should ask ourselves: what is our proper place in nature? What *ought* we do and what *can* we do to protect the irreplaceable habitat that produced and supports us?

But we also need to get outdoors and appreciate nature. As author Ed Abbey said, "It is not enough to fight for the land; it is even more important to enjoy it. While you can. While it is still there. So get out there and mess around with your friends, ramble out yonder and explore the forests, encounter the grizz, climb the mountains. Run the rivers, breathe deep of that yet sweet and lucid air, sit quietly for a while and contemplate the precious stillness, that lovely, mysterious and awesome space. Enjoy yourselves, keep your brain in your head and your head firmly attached to your body, the body active and alive."

We face many serious environmental problems

It's important for you to be aware of current environmental conditions. We'll cover all these issues in subsequent chapters of this book, but here's an overview to get you started. With more than 6.5 billion humans currently, we're adding about 75 million more to the world every year. While demographers report a transition to slower growth rates in most countries, present trends project a population between 8 and 10 billion by 2050. The impacts of that many people on our natural resources and ecological systems is a serious concern.

Water may well be the most critical resource in the twenty-first century. Already at least 1.1 billion people lack an adequate supply of safe drinking water, and more than twice that many don't have modern sanitation. Polluted water and lack of sanitation are estimated to contribute to the ill health of more than 1.2 billion people annually, including the death of 15 million children per year. About 40 percent of the world population lives in countries where water demands now exceed supplies, and by 2025 the UN projects that as many as three-fourths of us could live under similar conditions. Water wars may well become the major source of international conflict in coming decades.

Over the past century, global food production has more than kept pace with human population growth, but there are worries about whether we will be able to maintain this pace. Soil scientists report that about two-thirds of all agricultural lands show signs of degradation. Biotechnology and intensive farming techniques responsible for much of our recent production gains often are too expensive for poor farmers. Can we find ways to produce the food we need without further environmental degradation? And will that food be distributed equitably? In a world of food surpluses, the United Nations estimates that some 850 million people are now chronically undernourished, and at least 60 million face acute food shortages due to natural disasters or conflicts.

How we obtain and use energy is likely to play a crucial role in our environmental future. Fossil fuels (oil, coal, and natural gas) presently provide around 80 percent of the energy used in industrialized countries (fig. 1.8). Supplies of these fuels are diminishing, however, and problems associated with their acquisition and use—air and water pollution, mining damage, shipping accidents, and geopolitics—may limit what we do with remaining reserves. Cleaner renewable energy resources—solar power, wind, geothermal, and biomass—together with conservation, could give us cleaner, less destructive options if we invest in appropriate technology.

Burning fossil fuels, making cement, cultivating rice paddies, clearing forests, and other human activities release carbon dioxide and other so-called "greenhouse gases" that trap heat in the atmosphere. Over the past 200 years, atmospheric CO_2 concentrations have increased about 35 percent. By 2100, if current trends continue, climatologists warn that mean global temperatures will probably warm $1.5°$ to $6°C$ ($2.7°–11°F$). Although it's controversial whether specific recent storms were influenced by global warming, climate changes caused by greenhouse gases are

FIGURE 1.8 Fossil fuels supply about 80 percent of world commercial energy. They also produce a large percentage of all air pollutants and greenhouse gases, and contribute to economic and political instability.

very likely to cause increasingly severe weather events including droughts in some areas and floods in others. Melting alpine glaciers and snowfields could threaten water supplies on which millions of people depend.

Already, we are seeing dramatic climate changes in the Antarctic and Arctic where seasons are changing, sea ice is disappearing, and permafrost is melting (fig. 1.9). Rising sea levels

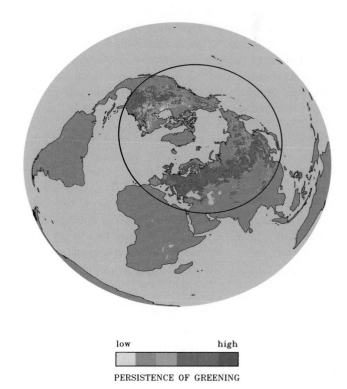

low   high

PERSISTENCE OF GREENING

FIGURE 1.9 Satellite images and surface temperature data show that polar regions, especially in Eurasia, are becoming green earlier and staying green longer than ever in recorded history. This appears to be evidence of a changing global climate.
Source: NASA, 2002.

FIGURE 1.10 At least half of all primates are considered threatened or endangered. Hunting and habitat destruction are the biggest problems.

Think About It

With your classmates or friends, list five important environmental issues in your area. What kinds of actions might you take to improve your local situation?

are flooding low-lying islands and coastal regions, while habitat losses and climatic changes are affecting many biological species. Canadian Environment Minister David Anderson has said that global climate change is a greater threat than terrorism because it could threaten the homes and livelihood of billions of people and trigger worldwide social and economic catastrophe.

Air quality has worsened dramatically in many areas. Over southern Asia, for example, satellite images recently revealed a 3-km (2-mile)-thick toxic haze of ash, acids, aerosols, dust, and photochemical products regularly covers the entire Indian subcontinent for much of the year. Nobel laureate Paul Crutzen estimates that at least 3 million people die each year from diseases triggered by air pollution. Worldwide, the United Nations estimates that more than 2 billion metric tons of air pollutants (not including carbon dioxide or wind-blown soil) are emitted each year. Air pollution no longer is merely a local problem. Mercury, polychlorinated biphenyls (PCB), DDT, and other long-lasting pollutants accumulate in arctic ecosystems and native people after being transported by air currents from industrial regions thousands of kilometers to the south. And during certain days, as much as 75 percent of the smog and particulate pollution recorded on the west coast of North America can be traced to Asia.

Biologists report that habitat destruction, overexploitation, pollution, and introduction of exotic organisms are eliminating species at a rate comparable to the great extinction that marked the end of the age of dinosaurs. The UN Environment Programme reports that over the past century, more than 800 species have disappeared and at least 10,000 species are now considered threatened. This includes about half of all primates and freshwater fish together with around 10 percent of all plant species (fig. 1.10). More than three-quarters of all global fisheries are overfished or harvested at their biological limit. At least half of the forests existing before the introduction of agriculture have been cleared, and much of the diverse "old growth" on which many species depend for habitat, is rapidly being cut and replaced by secondary growth or monoculture. All these biodiversity losses could threaten the ecological life-support systems on which we all depend.

Finding solutions to these problems requires good science as well as individual and collective actions. Becoming educated about our global environment is the first step in understanding how to control our impacts on it. We hope this book will help you in that quest.

There are many signs of hope

Is there hope that we can find solutions to these dilemmas? We think so. As the opening case study for this chapter shows, even countries, such as China, are making progress on social and environmental problems. China now has more than 200,000 wind generators and 10 million biogas generators (most in the world). Solar collectors on 35 million buildings furnish hot water. China could easily get all its energy from renewable sources, and it may be better able to provide advice and technology to other developing countries than can rich nations.

Many cities in Europe and North America are cleaner and much more livable now than they were a century ago. Population has stabilized in most industrialized countries and even in some very poor countries where social security and democracy have been established. Over the last 20 years, the average number of children born per woman worldwide has decreased from 6.1 to 2.7. By 2050, the UN Population Division predicts that all developed countries and 75 percent of the developing world will experience a below-replacement fertility rate of 2.1 children per woman. This prediction suggests that the world population will stabilize at about 8.9 billion rather than 9.3 billion, as previously estimated.

The incidence of life-threatening infectious diseases has been reduced sharply in most countries during the past century, while life expectancies have nearly doubled on average. Smallpox has been completely eradicated and polio has been vanquished except in a few countries. Since 1990, more than 800 million people have gained access to improved water supplies and modern sanitation. In spite of population growth that added nearly a billion people to the world during the 1990s, the number facing food insecurity and chronic hunger during this period actually declined by about 40 million.

Deforestation has slowed in Asia, from more than 8 percent during the 1980s to less than 1 percent in the 1990s. Nature preserves and protected areas have increased nearly fivefold over the past 20 years, from about 2.6 million km^2 to about 12.2 million km^2. This represents only 8.2 percent of all land area—less than the 12 percent thought necessary to protect a viable sample of the world's biodiversity—but is a dramatic expansion nonetheless.

Dramatic progress is being made in a transition to renewable energy sources. The European Union has pledged to get 20 percent of its energy from renewable sources (30 percent if other countries participate) by 2020. Former British Prime Minister Tony Blair laid out even more ambitious plans to fight global warming by cutting carbon dioxide emissions in his country by 60 percent through energy conservation and a switch to renewables. If nonpolluting, sustainable energy technology is made available to

What Do You Think?

Calculating Your Ecological Footprint

Can the earth sustain our current lifestyles? Will there be adequate natural resources for future generations? These questions are among the most important in environmental science today. We depend on nature for food, water, energy, oxygen, waste disposal, and other life-support systems. Sustainability implies that we cannot turn our resources into waste faster than nature can recycle that waste and replenish the supplies on which we depend. It also recognizes that degrading ecological systems ultimately threatens everyone's well-being. Although we may be able to overspend nature's budget temporarily, future generations will have to pay the debts we leave them. Living sustainably means meeting our own vital needs without compromising the ability of future generations to meet their own needs.

How can we evaluate and illustrate our ecological impacts? Redefining Progress, a nongovernmental environmental organization, has developed a measure called the **ecological footprint** to compute the demands placed on nature by individuals and nations. A simple questionnaire of 16 items gives a rough estimate of your personal footprint. A more complex assessment of 60 categories including primary commodities (such as milk, wood, or metal ores), as well as the manufactured products derived from them, gives a measure of national consumption patterns.

According to Redefining Progress, the average world citizen has an ecological footprint equivalent to 2.3 hectares (5.6 acres), while the biologically productive land available is only 1.9 hectares (ha) per person. How can this be? The answer is that we're using nonrenewable resources (such as fossil fuels) to support a lifestyle beyond the productive capacity of our environment. It's like living by borrowing on your credit cards. You can do it for a while, but eventually you have to pay off the deficit. The unbalance is far more pronounced in some of the richer countries. The average resident of the United States, for example, lives at a consumption level that requires 9.7 ha of bioproductive land. A dramatic comparison of consumption levels versus population size is shown in figure 1. If everyone in the world were to adopt a North American lifestyle, we'd need about four more planets to support us all. You can check out your own ecological footprint by going to www.redefiningprogress.org/.

Like any model, an ecological footprint gives a useful description of a system. Also like any model, it is built on a number of assumptions: (1) Various measures of resource consumption and waste flows can be converted into the biologically productive area required to maintain them; (2) different kinds of resource use and dissimilar types of productive land can be standardized into roughly equivalent areas; (3) because these areas stand for mutually exclusive uses, they can be added up to a total—a total representing humanity's demand—that can be compared to the total world area of bioproductive land. The model also implies that our world has a fixed supply of resources that can't be expanded. Part of the power of this metaphor is that we all can visualize a specific area of land and imagine it being divided into smaller and smaller parcels as our demands increase. But this perspective doesn't take into account technological progress. For example, since 1950, world food production has increased about fourfold. Some of this growth has come from expansion of croplands, but most has come from technological advances such as irrigation, fertilizer use, and higher-yielding crop varieties. Whether this level of production is sustainable is another question, but this progress shows that land area isn't always an absolute limit. Similarly, switching to renewable energy sources such as wind and solar power would make a huge impact on estimates of our ecological footprint. Notice that in figure 2 energy consumption makes up about half of the calculated footprint.

What do you think? Does analyzing our ecological footprint inspire you to correct our mistakes, or does it make sustainability seem an impossible goal? If we in the richer nations have the technology and political power to exploit a larger share of resources, do we have a right to do so, or do we have an ethical responsibility to restrain our consumption? And what about future generations? Do we have an obligation to leave resources for them, or can we assume they'll make technological discoveries to solve their own problems if resources become scarce? You'll find that many of the environmental issues we discuss in this book aren't simply a matter of needing more scientific data. Ethical considerations and intergenerational justice often are just as important as having more facts.

FIGURE 1 Ecological footprint by region in 2001. The height of each bar is proportional to each region's average footprint per person, the width of the bar is proportional to its population, and the area of the bar is proportional to the region's total ecological footprint.

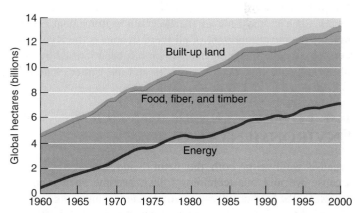

FIGURE 2 Humanity's ecological footprint grew by about 160 percent from 1961 to 2001, somewhat faster than population, which doubled over the same period.
WWF, 2004.

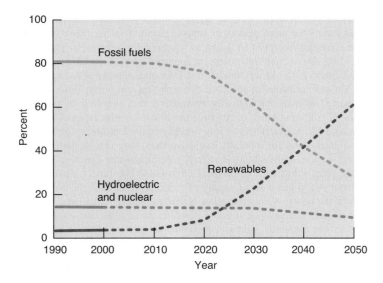

FIGURE 1.11 A possible energy future. Global warming and other environmental problems may require that we switch from our current dependence on fossil fuels to renewable sources such as wind and solar energy. **Source:** World Bank, 2000.

the world's poorer countries, it may be possible to promote human development while simultaneously reducing environmental damage (fig. 1.11).

Over the past two decades, the world has made dramatic progress in opening up political systems and expanding political freedoms. During this time, some 81 countries took significant steps toward democracy. Currently, nearly three-quarters of the world's 200 countries now hold multiparty elections. At least 60 developing countries claim to be transferring decision-making authority to local units of government. Of course, decentralization doesn't always guarantee better environmental stewardship, but it puts people with direct knowledge of local conditions in a position of power rather than distant elites or bureaucrats.

Currently, more than 500 international environmental protection agreements are now in force. Some, such as the Montreal Protocol on Stratospheric Ozone, have been highly successful. Others, such as the Law of the Sea, lack enforcement powers. Perhaps the most important of all these treaties is the Kyoto Protocol on global climate change, which has been ratified by every industrialized nation except Australia and the United States.

1.4 HUMAN DIMENSIONS OF ENVIRONMENTAL SCIENCE

Because we live in both the natural and social worlds, and because we and our technology have become such dominant forces on the planet, environmental science must take human institutions and the human condition into account. We live in a world of haves and have-nots; a few of us live in increasing luxury, while many others lack the basic necessities for a decent, healthy, productive life. The World Bank estimates that more than 1.4 billion people—almost one-fifth of the world's population—

FIGURE 1.12 Three-quarters of the world's poorest nations are in Africa. Millions of people lack adequate food, housing, medical care, clean water, and safety. The human suffering engendered by this poverty is tragic.

live in **extreme poverty** with an income of less than (U.S.)$1 per day (fig. 1.12). These poorest of the poor often lack access to an adequate diet, decent housing, basic sanitation, clean water, education, medical care, and other essentials for a humane existence. Seventy percent of those people are women and children. In fact, four out of five people in the world live in what would be considered poverty in the United States or Canada.

Policymakers are becoming aware that eliminating poverty and protecting our common environment are inextricably interlinked because the world's poorest people are both the victims and the agents of environmental degradation. The poorest people are often forced to meet short-term survival needs at the cost of long-term sustainability. Desperate for croplands to feed themselves and their families, many move into virgin forests or cultivate steep, erosion-prone hillsides, where soil nutrients are exhausted after only a few years. Others migrate to the grimy, crowded slums and ramshackle shantytowns that now surround most major cities in the developing world. With no way to dispose of wastes, the residents often foul their environment further and contaminate the air they breathe and the water on which they depend for washing and drinking.

The cycle of poverty, illness, and limited opportunities can become a self-sustaining process that passes from one generation to another. People who are malnourished and ill can't work productively to obtain food, shelter, or medicine for themselves or their children, who also are malnourished and ill. About 250 million children—mostly in Asia and Africa and some as young as 4 years old—are forced to work under appalling conditions weaving carpets, making ceramics and jewelry, or working in the sex trade. Growing up in these conditions leads to educational, psychological, and developmental deficits that condemn these children to perpetuate this cycle.

Faced with immediate survival needs and few options, these unfortunate people often have no choice but to overharvest resources; in doing so, however, they diminish not only their own options but also those of future generations. And in an increasingly interconnected world, the environments and resource bases damaged by poverty and ignorance are directly linked to those on which we depend.

The Worldwatch Institute warns that "poverty, disease and environmental decline are the true axis of evil." Terrorist attacks—and the responses they provoke—are the symptoms of the underlying sources of global instability, including the dangerous interplay among poverty, hunger, disease, environmental degradation, and rising resource competition. Failure to deal with these sources of insecurity could plunge the world into a dangerous downward spiral in which instability and radicalization grows. Unless the world takes action to promote sustainability and equity, Worldwatch suggests we will face an uphill battle to deal with the consequences of wars, terrorism, and natural disasters.

We live in an inequitable world

About one-fifth of the world's population lives in the 20 richest countries, where the average per capita income is above (U.S.) $25,000 per year. Most of these countries are in North America or Western Europe, but Japan, Singapore, and Australia also fall into this group. Almost every country, however, even the richest, such as the United States and Canada, has poor people. No doubt everyone reading this book knows about homeless people or other individuals who lack resources for a safe, productive life. According to the U.S. Census Bureau, 37 million Americans—one-third of them children—live in households below the poverty line.

The other four-fifths of the world's population lives in middle- or low-income countries, where nearly everyone is poor by North American standards. Nearly 3 billion people live in the poorest nations, where the average per capita income is below (U.S.)$620 per year. China and India are the largest of these countries, with a combined population of about 2.3 billion people. Among the 41 other nations in this category, 33 are in sub-Saharan Africa. All the other lowest-income nations, except Haiti, are in

TABLE 1.1

Quality of Life Indicators

	Least-Developed Countries	Most-Developed Countries
GDP/Person[1]	(U.S.)$329	(U.S.)$30,589
Poverty Index[2]	78.1%	~0
Life Expectancy	43.6 years	76.5 years
Adult Literacy	58%	99%
Female Secondary Education	11%	95%
Total Fertility[3]	5.0	1.7
Infant Mortality[4]	97	5
Improved Sanitation	23%	100%
Improved Water	61%	100%
CO_2/capita[5]	0.2 tons	13 tons

[1]Annual gross domestic product
[2]Percent living on less than (U.S.)$2/day
[3]Average births/woman
[4]Per 1,000 live births
[5]Metric tons/yr/person
Source: UNDP Human Development Index, 2006.

Asia. Although poverty levels in countries such as China and Indonesia have fallen in recent years, most countries in sub-Saharan Africa and much of Latin America have made little progress. The destabilizing and impoverishing effects of earlier colonialism continue to play important roles in the ongoing problems of these unfortunate countries. Meanwhile, the relative gap between rich and poor has increased dramatically.

As table 1.1 shows, the gulf between the richest and poorest nations affects many quality-of-life indicators. The average individual in the highest-income countries has an annual income nearly 100 times that of those in the lowest-income nations. Infant mortality in the least-developed countries is nearly 20 times as high as in the most-developed countries. Only 23 percent of residents in poorer countries have access to modern sanitation, while this ammenity is essentially universal in richer countries. Carbon dioxide emissions (a measure of both energy use and contributions to global warming) are 65 times greater in rich countries.

The gulf between rich and poor is even greater at the individual level. The richest 200 people in the world have a combined wealth of $1 trillion. This is more than the total owned by the 3 billion people who make up the poorest half of the world's population.

Is there enough for everyone?

Those of us in the richer nations now enjoy a level of affluence and comfort unprecedented in human history. But we consume an inordinate share of the world's resources, and produce an unsustainable

FIGURE 1.13 "And may we continue to be worthy of consuming a disproportionate share of this planet's resources."

FIGURE 1.14 A rapidly growing economy has brought increasing affluence to China that has improved standards of living for many Chinese people, but it also brings environmental and social problems associated with western life styles.

amount of pollution to support our lifestyle. What if everyone in the world tried to live at that same level of consumption? The United States, for example, with about 4.6 percent of the world's population, consumes about 25 percent of all oil while producing about 25 percent of all carbon dioxide and 50 percent of all toxic wastes in the world (fig. 1.13). What will the environmental effects be if other nations try to emulate our prosperity?

Take the example of China that we discussed in the opening case study for this chapter. In the early 1960s, it's estimated that 300 million Chinese suffered from chronic hunger, and at least 30 million starved to death in the worst famine in world history. Since then, however, China has experienced amazing economic growth. The national GDP has been growing at about 10 percent per year. If current trends continue, the Chinese economy will surpass the United States and become the world's largest by 2020. This rapid growth has brought many benefits. Hundreds of millions of people have been lifted out of extreme poverty. Chronic hunger has decreased from about 30 percent of the population 40 years ago to less than 10 percent today. Average life expectancy has increased from 42 to 72.5 years. And infant mortality dropped from 150 per 1,000 live births in 1960 to 24.5 today, while the annual per capita GDP has grown from less than (U.S.)$200 per year to more than $4,500.

Still, most Chinese live at a low level of material consumption by European or American standards. One way of measuring material consumption is by environmental footprint. It now takes about 9.7 global hectares to support the average American. By contrast, the average Chinese citizen has a footprint of only 1.6 global hectares. If all the 1.3 billion residents of China were to try to match the American level of consumption it would take about four extra planets using the same technology we now employ.

Many of the environmental problems mentioned in the opening case study for this chapter arise from poverty. China couldn't afford to worry (at least so they thought) about pollution and land

degradation in the past. Today, however, the greatest environmental worries are about the effects of rising affluence (fig. 1.14). In 1985, there were essentially no private automobiles in China. Bicycles and public transportation were how nearly everyone got around. Now, there are about 30 million automobiles in China, and by 2015, if current trends continue, there could be 150 million (fig. 1.15). Already, Chinese auto efficiency standards are higher than in the United States, but is there enough petroleum in the world to support all these vehicles? China is now the second largest source of CO_2

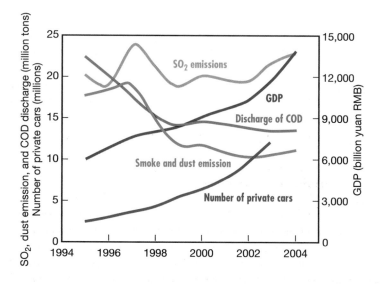

FIGURE 1.15 Between 1994 and 2004, Chinese GDP more than doubled, while the number of private automobiles grew about fourfold. Chemical oxygen demand (COD) fell by more than half as citizens demanded effluent controls on industry, but sulfur dioxide (SO_2) emissions increased as more coal was burned.
Source: Shao, M., et al., 2006.

(the United States is first) in the world. Both China and the United States depend on coal for about 75 percent of their electricity. Both have very large supplies of coal. There are many benefits of expanding China's electrical supply, but if they reach the same level of power consumption—which is now about one-tenth the amount per person as in the United States—by burning coal, the effects on our global climate will be disastrous.

On the other hand, China is now building a series of new cities that are expected to be self-sustaining in food production, water and energy supplies, and climate-neutral with respect to climate-changing gases. If more countries in both the developed and developing countries adopt these environmentally friendly technologies, the world could easily have enough resources for everyone.

Recent progress is encouraging

Over the past 50 years, human ingenuity and enterprise have brought about a breathtaking pace of technological innovations and scientific breakthroughs. The world's gross domestic product increased more than tenfold during that period, from $2 trillion to $22 trillion per year. While not all that increased wealth was applied to human development, there has been significant progress in increasing general standard of living nearly everywhere. In 1960, for instance, nearly three-quarters of the world's population lived in abject poverty. Now, less than one-third are still at this low level of development.

Since World War II, average real income in developing countries has doubled; malnutrition declined by almost one-third; child death rates have been reduced by two-thirds; average life expectancy increased by 30 percent. Overall, poverty rates have decreased more in the last 50 years than in the previous 500. Nonetheless, while general welfare has increased, so has the gap between rich and poor worldwide. In 1960, the income ratio between the richest 20 percent of the world and the poorest 20 percent was 30 to 1. In 2000, this ratio was 100 to 1. Because perceptions of poverty are relative, people may feel worse off compared to their rich neighbors than development indices suggest they are.

1.5 SUSTAINABLE DEVELOPMENT

Can we improve the lives of the world's poor without destroying our shared environment? A possible solution to this dilemma is **sustainable development,** a term popularized by *Our Common Future,* the 1987 report of the World Commission on Environment and Development, chaired by Norwegian Prime Minister Gro Harlem Brundtland (and consequently called the Brundtland Commission). In the words of this report, sustainable development means "meeting the needs of the present without compromising the ability of future generations to meet their own needs."

Another way of saying this is that we are dependent on nature for food, water, energy, fiber, waste disposal, and other life-support services. We can't deplete resources or create wastes faster than nature can recycle them if we hope to be here for the long term. Development means improving people's lives. Sustainable development, then, means progress in human well-being that can be extended or prolonged over many generations rather than just a few years. To be truly enduring, the benefits of sustainable development must be available to all humans rather than to just the members of a privileged group.

To many economists, it seems obvious that economic growth is the only way to bring about a long-range transformation to more advanced and productive societies and to provide resources to improve the lot of all people. As former President John F. Kennedy said, "A rising tide lifts all boats." But economic growth is not sufficient in itself to meet all essential needs. As the Brundtland Commission pointed out, political stability, democracy, and equitable economic distribution are needed to ensure that the poor will get a fair share of the benefits of greater wealth in a society. A study released in 2006 by researchers at Yale and Columbia Universities reported a significant correlation between environmental sustainability, open political systems, and good government. Of the 133 countries in this study, New Zealand, Sweden, Finland, Czech Republic, and the United Kingdom held the top five places (in that order). The United States ranked 28th, behind countries such as Japan, and most of Western Europe.

Can development be truly sustainable?

Many ecologists regard "sustainable" growth of any sort as impossible in the long run because of the limits imposed by nonrenewable resources and the capacity of the biosphere to absorb our wastes. Using ever-increasing amounts of goods and services to make human life more comfortable, pleasant, or agreeable must inevitably interfere with the survival of other species and, eventually, of humans themselves in a world of fixed resources. But, supporters of sustainable development assure us, both technology and social organization can be managed in ways that meet essential needs and provide long-term—but not infinite—growth within natural limits, if we use ecological knowledge in our planning.

While economic growth makes possible a more comfortable lifestyle, it doesn't automatically result in a cleaner environment. As figure 1.16 shows, people will purchase clean water and sanitation if they can afford to do so. For low-income people, however, more money tends to result in higher air pollution because they can afford to burn more fuel for transportation and heating. Given enough money, people will be able to afford both convenience *and* clean air. Some environmental problems, such as waste generation and carbon dioxide emissions, continue to rise sharply with increasing wealth because their effects are diffuse and delayed. If we are able to sustain economic growth, we will need to develop personal restraint or social institutions to deal with these problems.

Think About It

Examine figure 1.16. Describe in your own words how increasing wealth affects the three kinds of pollution shown. Why do the trends differ?

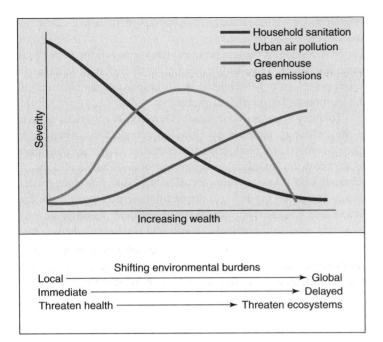

FIGURE 1.16 Environmental indicators show different patterns as incomes rise. Sanitation problems decrease when people can afford septic systems and clean water. Local air pollution, on the other hand, increases as more fuel is burned; eventually, however, development reaches a point at which people can afford both clean air and the benefits of technology. Delayed, distant problems, such as greenhouse gas emissions that lead to global climate change, tend to rise steadily with income because people make decisions based on immediate needs and wants rather than long-term consequences. Thus, we tend to shift environmental burdens from local and immediate to distant and delayed if we can afford to do so.
Graph from World Energy Assessment, UNDP 2000, Figure 3.10, p. 95.

FIGURE 1.17 A Mayan woman from Guatemala weaves on a backstrap loom. A member of a women's weaving cooperative, she sells her work to nonprofit organizations in the United States at much higher prices than she would get at the local market.

Some projects intended to foster development have been environmental, economic, and social disasters. Large-scale hydropower projects, like that in the James Bay region of Quebec or the Brazilian Amazon that were intended to generate valuable electrical power, also displaced indigenous people, destroyed wildlife, and poisoned local ecosystems with acids from decaying vegetation and heavy metals leached out of flooded soils. Similarly, introduction of "miracle" crop varieties in Asia and huge grazing projects in Africa financed by international lending agencies crowded out wildlife, diminished the diversity of traditional crops, and destroyed markets for small-scale farmers.

Other development projects, however, work more closely with both nature and local social systems. Socially conscious businesses and environmental, nongovernmental organizations sponsor ventures that allow people in developing countries to grow or make high-value products—often using traditional techniques and designs—that can be sold on world markets for good prices (fig. 1.17). Pueblo to People, for example, is a nonprofit organization that buys textiles and crafts directly from producers in Latin America. It sells goods in America, with the profits going to community development projects in Guatemala, El Salvador, and Peru. It also informs customers in wealthy countries about the conditions in the developing world.

As the economist John Stuart Mill wrote in 1857, "It is scarcely necessary to remark that a stationary condition of capital and population implies no stationary state of human improvement. There would be just as much scope as ever for all kinds of mental culture and moral and social progress; as much room for improving the art of living and much more likelihood of its being improved when minds cease to be engrossed by the art of getting on." Somehow, in our rush to exploit nature and consume resources, we have forgotten this sage advice.

What's the role of international aid?

Could we eliminate the most acute poverty and ensure basic human needs for everyone in the world? Many experts say this goal is eminently achievable. Economist Jeffery Sachs, director of the UN Millennium Development Project, says we could end extreme poverty worldwide by 2025 if the richer countries would donate just 0.7 percent of their national income for development aid in the poorest nations. These funds could be used for universal childhood vaccination against common infectious diseases, access to primary schools for everyone, family planning services

FIGURE 1.18 Every year, military spending equals the total income of half the world's people. The cost of a single large aircraft carrier equals ten years of human development aid given by all the world's industrialized countries.

for those who wish them, safe drinking water and sanitation for all, food supplements for the hungry, and strategic microcredit loans for self-employment.

How much would this cost? A rough estimate provided by the United Nations Development Agency is that it would take about (U.S.)$135 billion per year to abolish extreme poverty and the worst infectious diseases over the next 20 years. That's a lot of money—much more than we currently give—but it's not an impossible goal. Annual global military spending is now over $1 trillion (fig. 1.18). If we were to shift one-tenth of that to development aid, we'd not only reduce incalculable suffering but also be safer in the long run, according to many experts.

In 2005 the G-8, made up of the eight wealthiest nations, pledged (U.S.)$50 billion per year by 2010 to combat poverty. In addition, they promised to cancel at least $40 billion in debt of the 18 poorest nations. Meanwhile the United States, while the world's largest total donor, sets aside only 0.16 percent of its gross domestic product for development aid. Put another way, the United States currently donates about 18 cents per citizen per day for both private and government aid to foreign nations.

What do you think? Would you be willing to donate an extra dollar per day to reduce suffering and increase political stability? As former Canadian Prime Minister Jean Chrétien says, "Aid to developing countries isn't charity; it's an investment. It will make us safer, and when standards of living increase in those countries, they'll become customers who will buy tons of stuff from us."

Indigenous people are important guardians of nature

Often at the absolute bottom of the social strata, whether in rich or poor countries, are the indigenous or native peoples who are generally the least powerful, most neglected groups in the world. Typically descendants of the original inhabitants of an area taken over by more powerful outsiders, they often are distinct from their country's dominant language, culture, religion, and racial communities. Of the world's nearly 6,000 recognized cultures, 5,000 are indigenous ones that account for only about 10 percent of the total world population. In many countries, these indigenous people are repressed by traditional caste systems, discriminatory laws, economics, or prejudice. Unique cultures are disappearing, along with biological diversity, as natural habitats are destroyed to satisfy industrialized world appetites for resources. Traditional ways of life are disrupted further by dominant Western culture sweeping around the globe.

At least half of the world's 6,000 distinct languages are dying because they are no longer taught to children. When the last few elders who still speak the language die, so will the culture that was its origin. Lost with those cultures will be a rich repertoire of knowledge about nature and a keen understanding about a particular environment and a way of life (fig. 1.19).

Nonetheless, in many places, the 500 million indigenous people who remain in traditional homelands still possess valuable ecological wisdom and remain the guardians of little-disturbed habitats that are the refuge for rare and endangered species and relatively undamaged ecosystems. Author Alan Durning estimates that indigenous homelands harbor more biodiversity than all the

FIGURE 1.19 Do indigenous people have unique knowledge about nature and inalienable rights to traditional territories?

world's nature reserves and that greater understanding of nature is encoded in the languages, customs, and practices of native people than is stored in all the libraries of modern science. Interestingly, just 12 countries account for 60 percent of all human languages (fig. 1.20). Seven of those are also among the "megadiversity" countries that contain more than half of all unique plant and animal species. Conditions that support evolution of many unique species seem to favor development of equally diverse human cultures as well.

Recognizing native land rights and promoting political pluralism is often one of the best ways to safeguard ecological processes and endangered species. As the Kuna Indians of Panama say, "Where there are forests, there are native people, and where there are native people, there are forests." A few countries, such as Papua New Guinea, Fiji, Ecuador, Canada, and Australia acknowledge indigenous title to extensive land areas.

In other countries, unfortunately, the rights of native people are ignored. Indonesia, for instance, claims ownership of nearly three-quarters of its forest lands and all waters and offshore fishing rights, ignoring the interests of indigenous people who have lived in these areas for millennia. Similarly, the Philippine government claims

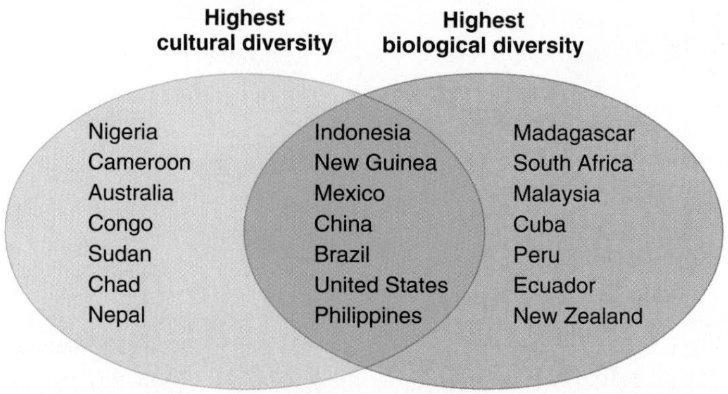

FIGURE 1.20 Cultural diversity and biodiversity often go hand in hand. Seven of the countries with the highest cultural diversity in the world are also on the list of "megadiversity" countries with the highest number of unique biological organisms. (Listed in decreasing order of importance.)
Source: Norman Myers, Conservation International and Cultural Survival Inc., 2002.

possession of all uncultivated land in its territory, while Cameroon and Tanzania recognize no rights at all for forest-dwelling pygmies who represent one of the world's oldest cultures.

CONCLUSION

We face many environmental dilemmas. As the case of China shows, air and water pollution, chronic hunger, water shortages, land degradation, and other environmental problems exact a terrible toll. China's problems aren't unique. About 3 billion people (nearly half the world's population) live on less than (U.S.)$2 per day, and face environmental hardships similar to those of the poorest Chinese. Finding ways that all of us can live sustainably within the limits of our resource base and without damaging nature's life-support systems is the preeminent challenge of environmental science.

Nature protection has deep roots reaching back into ancient history. Pragmatic resource conservation and moral or aesthetic concerns motivated early efforts at environmental defense. More recently, the health risks from pollution and the ecological dangers of habitat destruction and biodiversity losses have entered the dialog. Global environmentalism raises questions about sustainability. There are many signs of hope. Population growth is

slowing nearly everywhere, many terrible diseases have been conquered, progress is being made in a transition to renewable energy and pollution control. Some countries have made encouraging advances in reducing greenhouse gas emissions.

Understanding the links between poverty and environmental degradation, and recognizing the rights of indigenous people are essential if we are to protect natural resources and improve environmental quality. Former UN Ambassador Adlai Stevenson once said, "We travel together passengers on a little spaceship, dependent upon its vulnerable reserve of air and soil; all committed for our safety to its security and peace; preserved from annihilation only by the care, the work, and I will say, the love we give to our fragile craft. We cannot maintain it half fortunate, half miserable; half confident, half despairing; half slave to the ancient enemies of man; half free in a liberation of resources. No craft, no crew can travel safely with such vast contradictions. On their resolution, then, depends the survival of us all."

REVIEWING LEARNING OUTCOMES

By now you should be able to explain the following points:

1.1 Define environmental science and identify some important environmental concerns we face today.

- Environmental science is the systematic study of our environment and our proper place in it.
- China is a good case study of environmental concerns including population growth, poverty, food supplies, air and water pollution, energy choices, and the threats of global climate change.

1.2 Discuss the history of conservation and the different attitudes toward nature at various times in our past.

- Nature protection has historic roots.
- Resource waste inspired pragmatic, utilitarian conservation.
- Ethical and aesthetic concerns inspired the preservation movement.
- Rising pollution levels led to the modern environmental movement.
- Global interconnections have expanded environmentalism.

1.3 Think critically about the major environmental dilemmas and issues that shape our current environmental agenda.

- We live on a marvelous planet of rich biodiversity and complex ecological systems.
- We face many serious environmental problems including water supplies, safe drinking water, hunger, land degradation, energy, air quality, and biodiversity losses.
- There are many signs of hope in terms of social progress, environmental protection, energy choices, and the spread of democracy.

1.4 Appreciate the human dimensions of environmental science, including the connection between poverty and environmental degradation.

- We live in an inequitable world.
- Recent progress is encouraging.
- Is there enough for everyone?

1.5 Summarize sustainable development and evaluate some of its requirements.

- Can development be truly sustainable?
- What's the role of international aid?
- Indigenous people are important guardians of nature.

PRACTICE QUIZ

1. Define *environment* and *environmental science*.
2. Describe four stages in conservation history and identify one leader associated with each stage.
3. List six environmental dilemmas that we now face and summarize how each concerns us.
4. Identify some signs of hope for solving environmental problems.
5. What is extreme poverty, and why should we care?
6. How much difference is there in per capita income, infant mortality, and CO_2 production between the poorest and richest countries?
7. Why should we be worried about economic growth in China?
8. Define *sustainable development*.
9. How much would it cost to eliminate acute poverty and ensure basic human needs for everyone?
10. Why are indigenous people important as guardians of nature?

CRITICAL THINKING AND DISCUSSION QUESTIONS

1. Should environmental science include human dimensions? Explain.
2. Overall, do environmental and social conditions in China give you hope or fear about the future?
3. What are the underlying assumptions and values of utilitarian conservation and altruistic preservation? Which do you favor?
4. What resource uses are most strongly represented in the ecological footprint? What are the advantages and disadvantages of using this assessment?
5. Are there enough resources in the world for 8 or 10 billion people to live decent, secure, happy lives? What do these terms mean to you? Try to imagine what they mean to residents of other countries.
6. What would it take for human development to be truly sustainable?
7. Are you optimistic or pessimistic about our chances of achieving sustainability? Why?

DATA analysis Working with Graphs

Graphs are one of the most common and important ways that scientists communicate their results. To be a scientifically literate citizen, it's important for you to be familiar with graphing techniques.

Graphs are visual presentations of data that help us identify trends and understand relationships. A table of numbers is more precise, but most of us have difficulty visualizing patterns in a field of numbers. We can look at the patterns in a graphic representation and make comparisons much more easily than we could with raw data. The ability to read graphs and plot data are essential skills for

students of environmental science. You'll see graphs throughout this book. In this exercise, we'll explore simple line plots.

A line plot represents a data set that involves a sequence of some sort, or a change over time. Table 1.2 shows growth of a population of a single-cell protozoan called *Tetrahymena* in a laboratory culture. To start the experiment, 10 cells were placed in a growth medium (this is time 0 in the table). Each hour for the next 12 hours, the cell number was counted and recorded. How can you convert data such as these into a graph? First, you lay out the axes. By convention, the horizontal (X) axis shows

TABLE 1.2 Growth of *Tetrahymena*

Hours	Cells
0	10
1	15
2	20
3	40
4	80
5	160
6	320
7	450
8	550
9	600
10	620
11	630
12	640

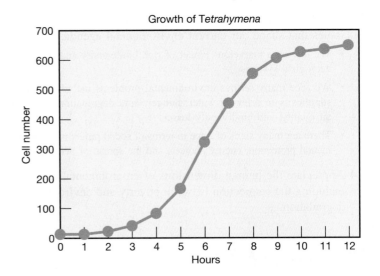

the **independent variable** (one whose values presumably don't depend on the other factor). The vertical axis (Y) records the **dependent variable** (which responds to changes in the other factor). Choose a range for each axis that approximately fits the range of values: this way your data will fill most of the graph, and the patterns should be big enough to see. Now plot the data: for hour 0, find the number of cells (10). Draw a dot *above the 0 on the X-axis* and *level with the 10 on the Y-axis*. Repeat these steps for hour 1, hour 2, and so on. After you've added all your dots, you can connect them with a line. The line makes it easier to see trends in the data. Your graph should look something like the one in this box.

Can you extract data from a line plot? Approximately. The process is simply the reverse of the way you just learned to create the graph. You draw a line from any point on the curve to the spot on the axes directly under and parallel to that point. For instance, the point on the curve for hour 6 corresponds to slightly more than 300 cells on the Y-axis.

To make comparisons between two different sets of dependent variables, it's often useful to plot them on the same graph. Look, for example, at figures 1.15 and 1.16 on pp. 26 and 27. You can see they have as many as five categories of independent variables in a single graph. To make the graphs simpler, the individual data points have been removed and each curve is a simple continuous line. You could still find the data for a specific point on either of these graphs with the process outlined above.

For Additional Help in Studying This Chapter, please visit our website at www.mhhe.com/cunningham10e. You will find additional practice quizzes and case studies, flashcards, regional examples, place markers for Google Earth™ mapping, and an extensive reading list, all of which will help you learn environmental science.

The Pacific Island nation of Tuvalu is being abandoned as global warming raises sea level. Do we have a moral obligation to poor nations or future generations for a livable environment?

Frameworks for Understanding Science, Systems, and Ethics

The ultimate test of a moral society is the kind of world that it leaves to its children.

—Dietrich Bonhoeffer—

LEARNING OUTCOMES

After studying this chapter, you should be able to:

2.1 Describe the scientific method and explain how it works.

2.2 Evaluate the role of scientific consensus and conflict.

2.3 Explain systems and how they're useful in science.

2.4 Discuss environmental ethics and worldviews.

2.5 Identify the roles of religious and cultural perspectives in conservation and environmental justice.

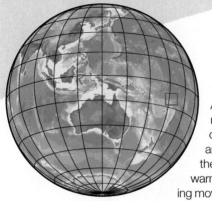

Case Study Is Climate Change a Moral Issue?

Across America, thousands of religious groups are meeting in churches, synagogues, mosques, and meeting houses to discuss the science and ethics of global warming. Congregations are watching movies about global climate change, sponsoring discussions about the science behind the issue, distributing educational kits, encouraging members to become ecologically-concerned citizens, and installing energy conservation measures. But why do all this in a religious context?

"At its core, global climate change is not about economic theory or political platforms," according to the United States Conference of Catholic Bishops. "It is about our human stewardship of God's creation and our responsibility to those who come after us." Ethical questions lie at the heart of most important environmental challenges, many religious leaders insist. "We share a deep conviction that global climate change presents an unprecedented threat to the integrity of life on Earth and a challenge to the universal values that bind us as human beings," says a statement written by a group of religious leaders from many faiths. "Global warming is harming God's creation: first the poor of the world, and eventually, all of us, and all life," says the Reverend Sally Bingham of the Grace Cathedral in San Francisco.

Although they have tended in the past to be politically, socially, and economically conservative, many evangelical Christians also have joined the crusade against global warming. In a widely circulated document entitled "An Evangelical Call to Action," the presidents of 39 evangelical colleges, leaders of religious social service agencies, and pastors of some of the largest evangelical congregations in the country wrote that (1) Human-induced climate change is real. (2) The consequences of climate change will be significant and will hit the poor the hardest. (3) Christian moral convictions demand our response to the climate change problem and (4) The time to act now is urgent. Governments, businesses, churches, and individuals all have a role to play in addressing climate change—starting now." Calling for "Creation Care," these leaders call for federal legislation that will require reductions in carbon dioxide emissions through "cost-effective, market-based mechanisms."

Of course, not all religious believers agree with their brethren. A group called the Interfaith Stewardship Alliance, for instance, which includes a number of high-profile evangelical leaders, argues that those who demand environmental stewardship are practicing a kind of neo-heathenism, placing nature, including the "creepy, crawly things," ahead of humans and the Creator. They claim that the science isn't settled about whether global warming is actually a problem or that human beings are causing it. Moreover, they assert, the solutions proposed by global warming opponents will harm business and cause energy costs to rise. In this view, free-market solutions to environmental problems are much better than government mandates, and individuals and private organizations should be trusted to care for their own property without outside intervention.

While there are debates about the timing and detailed effects of global warming, a vast majority of the world's climate scientists now agree that climate change is already occurring and that humans are the main cause. In its 2007 report on the consequences of global warming, the Intergovernmental Panel on Climate Change (IPCC) concluded that poorer countries (especially in the tropics, such as Tuvalu), which have contributed least to climate change and can least afford to accommodate to it, will likely suffer most while richer countries, which have produced the vast majority of greenhouse gases, will have much less trouble adapting. The burdens of environmental damage will also fall mostly on unborn generations, especially in developing countries. Is it fair for those of us who have benefited from burning fossil fuels leave the burden of dealing with this problem to our descendents?

This case study introduces two important aspects of environmental science. The first of these is ethics and values. What obligations do we have as world citizens? What rights and values do other people, other species, and future generations have? Should we be stewards of nature or have dominion over it? How you answer this question depends on your worldview, values, and religious beliefs. In this chapter, we'll examine some moral and ethical perspectives that shape different understandings of important environmental questions. An equally important topic is the nature of science. To be an informed citizen, you need to understand how science works and what questions it can—and can't—answer. We'll start this chapter by examining how science helps us understand our world.

2.1 What Is Science?

Science is a process for producing knowledge methodically and logically. Derived from *scire,* "to know" in Latin, science depends on making precise observations of natural phenomena. We develop or test theories (proposed explanations of how a process works) using these observations. "Science" also refers to the cumulative body of knowledge produced by many scientists. Science is valuable because it helps us understand the world and meet practical needs, such as new medicines, new energy sources, or new foods. In this section, we'll investigate how and why science follows standard methods.

Science rests on the assumption that the world is knowable and that we can learn about the world by careful observation (table 2.1). For early philosophers of science, this assumption was a radical departure from religious and philosophical approaches. In the Middle Ages, the ultimate sources of knowledge about matters, such as how crops grow, how diseases spread, or how the stars move, were religious authorities or cultural

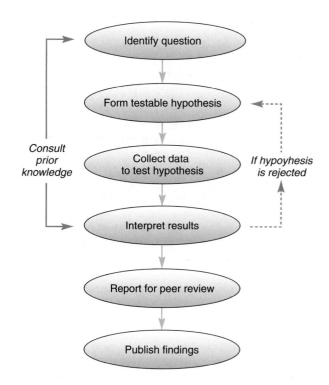

FIGURE 2.1 Ideally, scientific investigation follows a series of logical, orderly steps to formulate and test hypotheses.

traditions. While these sources provided many useful insights, there was no way to test their explanations independently and objectively. The benefit of scientific thinking is that it searches for testable evidence: If you suspect a disease spreads through contaminated water, you can close off access to the water source and see if the disease stops spreading.

Science depends on skepticism and accuracy

Ideally, scientists are skeptical. They are cautious about accepting proposed explanations until there is substantial evidence to support them. Even then, as we saw in the case study about global warming that opened this chapter, explanations are considered only provisionally true, because there is always a possibility that some additional evidence may appear to disprove them. Scientists also aim to be methodical and unbiased. Because bias and methodical errors are hard to avoid, scientific tests are subject to review by informed peers, who can evaluate results and conclusions (fig. 2.1). The peer review process is an essential part of ensuring that scientists maintain good standards in study design, data collection, and interpretation of results.

Scientists demand **reproducibility** because they are cautious about accepting conclusions. Making an observation or obtaining a result just once doesn't count for much. You have to produce the same result consistently to be sure that your first outcome wasn't a fluke. Even more important, you must be able to describe the conditions of your study so that someone else can reproduce your findings. Repeating studies or tests is known as **replication.**

Science also relies on accuracy and precision. Accuracy is correctness of measurements. Inaccurate data can produce sloppy and misleading conclusions (fig. 2.2). Precision means repeatability of results and level of detail. The classic analogy for repeatability is throwing darts at a dart board. You might throw ten darts and miss the center every time, but if all the darts hit nearly the same spot, they were very precise. Another way to think of precision is levels of detail. Suppose you want to measure how much snow fell last night, so you take out your ruler, which is marked in centimeters, and you find that the snow is just over 6 cm deep. You cannot tell if it is 6.3 cm or 6.4 cm because the ruler doesn't report that level of detail. If you average several measurements, you might find an average depth of 6.4333 cm. If you report all four decimal places, it will imply that you know more than you really do about the snow depth. If you had a ruler marked in millimeters (one-tenth of a centimeter), you could find a depth of 6.4 cm. Here, the one decimal place would be a **significant number,** or a level of detail you actually knew. Reporting 6.4333 cm would still involve three insignificant digits.

Deductive and inductive reasoning are both useful

Ideally, scientists deduce conclusions from general laws that they know to be true. For example, if we know that massive objects attract each other (because of gravity), then it follows

FIGURE 2.2 Making careful, accurate measurements and keeping good records are essential in scientific research.

that an apple will fall to the ground when it releases from the tree. This logical reasoning from general to specific is known as **deductive reasoning.** Often, however, we do not know general laws that guide natural systems. We observe, for example, that birds appear and disappear as a year goes by. Through many repeated observations in different places, we can infer that the birds move from place to place. We can develop a general rule that birds migrate seasonally. Reasoning from many observations to produce a general rule is **inductive reasoning.** Although deductive reasoning is more logically sound than inductive reasoning, it only works when our general laws are correct. We often rely on inductive reasoning to understand the world because we have few immutable laws.

Sometimes it is insight, as much as reasoning, that leads us to an answer. Many people fail to recognize the role that insight, creativity, aesthetics, and luck play in research. Some of our most important discoveries were made not because of superior scientific method and objective detachment, but because the investigators were passionately interested in their topics and pursued hunches that appeared unreasonable to fellow scientists. A good example is Barbara McClintock, the geneticist who discovered that genes in corn can move and recombine spontaneously. Where other corn geneticists saw random patterns of color and kernel size, McClintock's years of experience in corn breeding and an uncanny ability to recognize patterns, led her to guess that genes could recombine in ways that no one had yet imagined. Her intuitive understanding led to a theory that took other investigators years to accept.

Testable hypotheses and theories are essential tools

You may already be using the scientific method without being aware of it. Suppose you have a flashlight that doesn't work. The flashlight has several components (switch, bulb, batteries) that could be faulty. If you change all the components at once, your

flashlight might work, but a more methodical series of tests will tell you more about what was wrong with the system—knowledge that may be useful next time you have a faulty flashlight. So you decide to follow the standard scientific steps:

1. *Observe* that your flashlight doesn't light; also, there are three main components of the lighting system (batteries, bulb, and switch).

2. Propose a **hypothesis,** a testable explanation: "The flashlight doesn't work because the batteries are dead."

3. Develop a *test* of the hypothesis and *predict* the result that would indicate your hypothesis was correct: "I will replace the batteries; the light should then turn on."

4. Gather *data* from your test: After you replaced the batteries, did the light turn on?

5. *Interpret* your results: If the light works now, then your hypothesis was right; if not, then you should formulate a new hypothesis, perhaps that the bulb is faulty, and develop a new test for that hypothesis.

In systems more complex than a flashlight, it is almost always easier to prove a hypothesis wrong than to prove it unquestionably true. This is because we usually test our hypotheses with observations, but there is no way to make every possible observation. The philosopher Ludwig Wittgenstein illustrated this problem as follows: Suppose you saw hundreds of swans, and all were white. These observations might lead you to hypothesize that all swans were white. You could test your hypothesis by viewing thousands of swans, and each observation might support your hypothesis, but you could never be entirely sure that it was correct. On the other hand, if you saw just one black swan, you would know with certainty that your hypothesis was wrong.

As you'll read in later chapters, the elusiveness of absolute proof is a persistent problem in environmental policy and law. You can never absolutely prove that the toxic waste dump up the street is making you sick. The elusiveness of proof often decides environmental liability lawsuits.

When an explanation has been supported by a large number of tests, and when a majority of experts have reached a general consensus that it is a reliable description or explanation, we call it a **scientific theory.** Note that scientists' use of this term is very different from the way the public uses it. To many people, a theory is speculative and unsupported by facts. To a scientist, it means just the opposite: While all explanations are tentative and open to revision and correction, an explanation that counts as a scientific theory is supported by an overwhelming body of data and experience, and it is generally accepted by the scientific community, at least for the present (fig. 2.3).

Understanding probability helps reduce uncertainty

One strategy to improve confidence in the face of uncertainty is to focus on probability. Probability is a measure of how likely something is to occur. Usually, probability estimates are based

FIGURE 2.3 Data collection and repeatable tests support scientific theories. Here students use telemetry to monitor radio-tagged fish.

on a set of previous observations or on standard statistical measures. Probability does not tell you what *will* happen, but it tells you what *is likely* to happen. If you hear on the news that you have a 20 percent chance of catching a cold this winter, that means that 20 of every 100 people are likely to catch a cold. This doesn't mean that you will catch one. In fact, it's more likely that you won't catch a cold than that you will. If you hear that 80 out of every 100 people will catch a cold, you still don't know whether you'll get sick, but there's a much higher chance that you will.

Science often involves probability, so it is important to be familiar with the idea. Sometimes probability has to do with random chance: If you flip a coin, you have a random chance of getting heads or tails. Every time you flip, you have the same 50 percent probability of getting heads. The chance of getting ten heads in a row is small (in fact, the chance is 1 in 2^{10}, or 1 in 1,024), but on any individual flip, you have exactly the same 50 percent chance, since this is a random test. Sometimes probability is weighted by circumstances: Suppose that about 10 percent of the students in this class earn an A each semester. Your likelihood of being in that 10 percent depends a great deal on how much time you spend studying, how many questions you ask in class, and other factors. Sometimes there is a combination of chance and circumstances: The probability that you will catch a cold this winter depends partly on whether you encounter someone who is sick (largely random chance) and whether you take steps to stay healthy (get enough rest, wash your hands frequently, eat a healthy diet, and so on).

Scientists often increase their confidence in a study by comparing results to a random sample or a larger group. Suppose that 40 percent of the students in your class caught a cold last winter. This *seems* like a lot of colds, but is it? One way to decide is to compare to the cold rate in a larger group. You call your state epidemiologist, who took a random sample of the state population last year: She collected 200 names from the telephone book and called each to find out if each got a cold last year. A larger sample, say 2,000 people, would have been more likely to represent the actual statewide cold rate. But a sample of 200 is much better than a sample of 50 or 100. The epidemiologist tells you that in your state as a whole, only 20 percent of people caught a cold.

Now you know that the rate in your class was quite high, and you can investigate possible causes for the difference. Perhaps people in your class got sick because they were short on sleep, because they tended to stay up late studying. Could you test whether studying late was a contributing factor? One way to test the relationship is to separate the class into two groups: those who study long and late, and those who don't. Then compare the rate of colds in these groups. Suppose it turns out that among the 40 late-night studiers, 30 got colds (a rate of 75 percent). Among the 60 casual studiers, only 10 got colds (17 percent). This difference would give you a good deal of confidence that staying up late contributes to getting sick. (Note, however, that all 40 of the studying group got good grades!)

Statistics can calculate the probability that your results were random

Statistics can help in experimental design as well as in interpreting data (see Exploring Science, p. 38). Many statistical tests focus on calculating the probability that observed results could have occurred by chance. Often, the degree of confidence we can assign to results depends on sample size as well as the amount of variability between groups.

Ecological tests are often considered significant if there is less than 5 percent probability that the results were achieved by random chance. A probability of less than 1 percent gives still greater confidence in the results.

As you read this book, you will encounter many statistics, including many measures of probability. When you see these numbers, stop and think: Is the probability high enough to worry about? How high is it compared to other risks or chances you've read about? What are the conditions that make probability higher or lower? Science involves many other aspects of statistics.

Experimental design can reduce bias

The study of colds and sleep deprivation is an example of an observational experiment, one in which you observe natural events and interpret a causal relationship between the variables. This kind of study is also called a **natural experiment,** one that involves observation of events that have already happened. Many scientists depend on natural experiments: A geologist, for instance, might want to study mountain building, or an ecologist might want to learn about how species coevolve, but neither scientist can spend millions of years watching the process happen. Similarly, a toxicologist cannot give people a disease just to see how lethal it is.

Other scientists can use **manipulative experiments,** in which conditions are deliberately altered, and all other variables are held constant (fig. 2.4). In one famous manipulative study, ecologists Edward O. Wilson and Robert MacArthur were interested in how quickly species colonize small islands, depending on distance to the mainland. They fumigated several tiny islands in the Florida Keys, killing all resident insects, spiders, and other invertebrates. They then monitored the islands to learn how quickly ants and spiders recolonized them from the mainland or other islands.

Exploring SCIENCE

Statistics are numbers that let you evaluate and compare things. "Statistics" is also a field of study that has developed meaningful methods of comparing those numbers. By both definitions, statistics are widely used in environmental sciences, partly because they can give us a useful way to assess patterns in a large population, and partly because the numbers can give us a measure of confidence in our research or observations. Understanding the details of statistical tests can take years of study, but a few basic ideas will give you a good start toward interpreting statistics.

1. *Descriptive statistics help you assess the general state of a group.* In many towns and cities, the air contains a dust, or particulate matter, as well as other pollutants. From personal experience you might know your air isn't as clean as you'd like, but you may not know how clean or dirty it is. You could start by collecting daily particulate measurements to find average levels. An averaged value is more useful than a single day's values, because daily values may vary a great deal, but general, long-term conditions affect your general health. Collect a sample every day for a year; then divide the sum by the number of days, to get a **mean** (average) dust level. Suppose you found a mean particulate level of 30 micrograms per cubic meter ($\mu g/m^3$) of air. Is this level high or low? In 1997 the EPA set a standard of 50 $\mu g/m^3$ as a limit for allowable

levels of coarse particulates (2.5–10 micrometers in diameter). Higher levels tend to be associated with elevated rates of asthma and other respiratory diseases. Now you know that your town, with an annual average of 30 $\mu g/m^3$, has relatively clean air, after all.

2. *Statistical samples.* Although your town is clean by EPA standards, how does it compare with the rest of the cities in the country? Testing the air in *every* city is probably not possible. You could compare your town's air quality with a **sample,** or subset of cities, however. A large, random sample of cities should represent the general "population" of cities reasonably well. Taking a large sample reduces the effects of outliers (unusually high or low values) that might be included. A random sample minimizes the chance that you're getting only the worst sites, or only a collection of sites that are close together, which might all have similar conditions. Suppose you getaverage annual particulate levels from a sample of 50 randomly selected cities. You can draw a frequency distribution, or histogram, to display your results (fig. 1). The mean value of this group is 36.8 $\mu g/m^3$, so by comparison your town (at 30 $\mu g/m^3$) is relatively clean.

Many statistical tests assume that the sample has a normal, or Gaussian, frequency distribution, often described as a bell-shaped curve (fig. 2). In this distri-

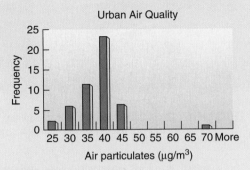

FIGURE 1 Average annual airborne dust levels for 50 cities in 2001.
Source: Data from U.S. Environmental Protection Agency.

bution, the mean is near the center of the range of values, and most values are fairly close to the mean. Large and random samples are more likely to fit this shape than are small and nonrandom samples.

3. *Confidence.* How do you know that the 50 cities you sampled really represent all the cities in the country? You can't ever be completely certain, but you can use estimates, such as confidence limits, to express the reliability of your mean statistic. Depending on the size of your sample (not 10, not 100, but 50) and the amount of variability in the sample data, you can calculate a confidence interval that the mean represents the whole population (all cities). Confidence levels, or confidence intervals,

Most manipulative experiments are done in the laboratory, where conditions can be carefully controlled. Suppose you were interested in studying whether lawn chemicals contributed to deformities in tadpoles. You might keep two groups of tadpoles in fish tanks, and expose one to chemicals. In the lab, you could ensure that both tanks had identical temperatures, light, food, and oxygen. By comparing a treatment (exposed) group and a control (unexposed) group, you have also made this a **controlled study.**

Often, there is a risk of experimenter bias. Suppose the researcher sees a tadpole with a small nub that looks like it might become an extra leg. Whether she calls this nub a deformity

FIGURE 2.4 Manipulative experiments attempt to control all variables except the tested variables. Here students control the number of species in a biodiversity experiment.

http://www.mhhe.com/cunningham10e

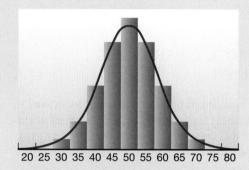

FIGURE 2 A normal distribution.

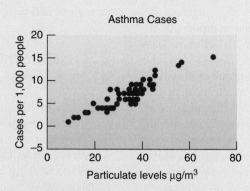

FIGURE 3 A plot showing relationships between variables.

represent the likelihood that your statistics represent the entire population correctly. For the mean of your sample, a confidence interval tells you the probability that your sample is similar to other random samples of the population. A common convention is to compare values with a 95 percent confidence level, or a probability of 5 percent or less that your conclusions are misleading. Using statistical software, we can calculate that, for our 50 cities, the mean is $36.8 \, \mu g/m^3$, and the confidence interval is 35.0 to 38.6. This suggests that, if you take 1,000 samples from the entire population of cities, 95 percent of those samples ought to be within $2 \, \mu g/m^3$ of your mean. This indicates that your mean is reliable and representative.

4. *Is your group unusual?* Once you have described your group of cities, you can compare it with other groups. For example, you might believe that Canadian cities have cleaner air than U.S. cities. You can compare mean air quality levels for the two groups. Then you can calculate confidence intervals for the difference between the means, to see if the difference is meaningful.

5. *Evaluating relationships between variables.* Are respiratory diseases correlated with air pollution? For each city in your sample, you could graph pollution and asthma rates (fig. 3). If the graph looks like a loose cloud of dots, there is no clear relationship. A tight, linear pattern of dots trending upward to the right indicates a strong and

positive relationship. You can also use a statistical package to calculate an equation to describe the relationship and, again, confidence intervals for the equation. This is known as a regression equation.

6. *Lies, damned lies, and statistics.* Can you trust a number to represent a complex or large phenomenon? One of the devilish details of representing the world with numbers is that those numbers can be tabulated in many ways. If we want to assess the greatest change in air quality statistics, do we report rates of change or the total amount of change? Do we look at change over five years? Twenty-five years? Do we accept numbers selected by the EPA, by the cities themselves, by industries, or by environmental groups? Do we trust that all the data were collected with a level of accuracy and precision that we would accept if we knew the hidden details in the data-gathering process? Like all information, statistics need to be interpreted in terms of who produced them, when, and why. Awareness of some of the standard assumptions behind statistics, such as sampling, confidence, and probability, will help you interpret statistics that you see and hear.

For more information about creating and interpreting graphs, see the Data Analysis exercise at the end of this chapter.

might depend on whether she knows that the tadpole is in the treatment group or the control group. To avoid this bias, **blind experiments** are often used, in which the researcher doesn't know which group is treated until after the data have been analyzed. In health studies, such as tests of new drugs, **double-blind experiments** are used, in which neither the subject (who receives a drug or a placebo) nor the researcher knows who is in the treatment group and who is in the control group.

In each of these studies there is one **dependent variable** and one, or perhaps more, **independent variables.** The dependent variable, also known as a response variable, is affected by the independent variables. In a graph, the dependent variable is on the vertical (Y) axis, by convention. Independent variables are rarely really independent (they are affected by the same environmental conditions as the dependent variable, for example). Many people prefer to call them explanatory variables, because we hope they will explain differences in the dependent variable.

Models are an important experimental strategy

Another way to gather information about environmental systems is to use **models.** A model is a simple representation of something. Perhaps you have built a model airplane. The model doesn't have all the elements of a real airplane, but it has the most important ones for your needs. A simple wood or plastic airplane has the proper shape, enough to allow a child to imagine it is flying (fig. 2.5). A more complicated model airplane might have a small gas engine, just enough to let a teenager fly it around for short distances.

Similarly, scientific models vary greatly in complexity, depending on their purposes. Some models are physical models: Engineers test new cars and airplanes in wind tunnels to see how they perform, and biologists often test theories about evolution and genetics using "model organisms" such as fruit flies or rats as a surrogate for humans.

FIGURE 2.5 A model uses just the essential elements to represent a complex system.

Most models are numeric, though. A model could be a mathematical equation, such as a simple population growth model ($N_t = rN_{(t-1)}$). Here the essential components are number (N) of individuals at time t (N_t), and the model proposes that N_t is equal to the growth rate (r) times the number in the previous time period ($N_{(t-1)}$). This model is a very simplistic representation of population change, but it is useful because it precisely describes a relationship between population size and growth rate. Also, by converting the symbols to numbers, we can predict populations over time. For example, if last year's rabbit population was 100, and the growth rate is 1.6 per year, then this year's population will be 160. Next year's population will be 160 × 1.6, or 460. This is a simple model, then, but it can be useful. A more complicated model might account for deaths, immigration, emigration, and other factors.

More complicated mathematical models can be used to describe and calculate more complex processes, such as climate change or economic growth (fig. 2.6). These models are also useful because they allow the researcher to manipulate variables without actually destroying anything. An economist can experiment with different interest rates to see how they affect economic growth. A climatologist can raise CO_2 levels and see how quickly temperatures respond. These models are often called simulation models, because they simulate a complex system. Of course, the results depend on the assumptions built into the models. One model might show temperature rising quickly in response to CO_2; another might show temperature rising more slowly, depending on how evaporation, cloud cover, and other variables are taken into account. Consequently, simulations can produce powerful but controversial results. If multiple models generally agree, though, as in the cases of climate models that agree on generally upward temperature trends, we can have confidence that the overall predictions are reliable. These models are also very useful in laying out and testing our ideas about how a system works.

2.2 SCIENTIFIC CONSENSUS AND CONFLICT

The scientific method outlined in figure 2.1 is the process used to carry out individual studies. Larger-scale accumulation of scientific knowledge involves cooperation and contributions from countless people. Good science is rarely carried out by a single individual working in isolation. Instead, a community of scientists collaborates in a cumulative, self-correcting process. You often hear about big breakthroughs and dramatic discoveries that change our understanding overnight, but in reality these changes are usually the culmination of the labor of many people, each working on different aspects of a common problem, each adding small insights to solve a problem. Ideas and information are exchanged, debated, tested, and re-tested to arrive at **scientific consensus,** or general agreement among informed scholars.

The idea of consensus is important. For those not deeply involved in a subject, the multitude of contradictory results can be bewildering: Are shark populations disappearing, and does it matter? Is climate changing, and how much? Among those who have performed and read many studies, there tends to emerge a general agreement about the state of a problem. Scientific consensus now holds that many shark populations are in danger, though opinions vary on how severe the problem is. Consensus is that global climates are changing, though models differ somewhat on how rapidly they will change under different policy scenarios.

Sometimes new ideas emerge that cause major shifts in scientific consensus. Two centuries ago, geologists explained many earth features in terms of Noah's flood. The best scientists held that the flood created beaches well above modern sea level, scattered boulders erratically across the landscape, and gouged enormous valleys where there is no water now (fig. 2.7). Then the Swiss glaciologist Louis Agassiz and others suggested that the earth had once been much colder and that glaciers had covered large areas. Periodic ice ages better explained changing sea levels, boulders transported far from their source rock, and the great, gouged valleys. This new idea completely altered the way geologists explained their subject. Similarly, the idea of tectonic plate movement, in which continents shift slowly around the earth's surface, revolutionized the ways geologists, biogeographers, ecologists, and others explained the development of the earth and its life-forms.

These great changes in explanatory frameworks were termed **paradigm shifts** by Thomas Kuhn (1967), who studied revolutions in scientific thought. According to Kuhn, paradigm shifts occur when a majority of scientists accept that the old explanation no longer explains new observations very well. The shift is often contentious and political, because whole careers and worldviews, based on one

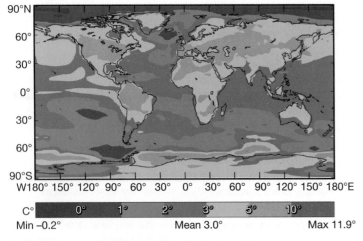

FIGURE 2.6 Numerical models, calculated from observed data, project scenarios such as global climate change. The Hadley model of summer temperature change for 70 to 100 years from now is shown.
Source: © Crown. Copyright 2005. Published by the Met Office.

FIGURE 2.7 Paradigm shifts change the ways we explain our world. Geologists now attribute Yosemite's valleys to glaciers, where once they believed Noah's flood carved its walls.

sort of research and explanation, can be undermined by a new model. Sometimes a revolution happens rather quickly. Quantum mechanics and Einstein's theory of relativity, for example, overturned classical physics in only about 30 years. Sometimes a whole generation of scholars has to retire before new paradigms can be accepted.

As you study this book, try to identify some of the paradigms that guide our investigations, explanations, and actions today. This is one of the skills involved in critical thinking, discussed in the introductory chapter of this book.

Detecting pseudoscience relies on independent, critical thinking

Ideally, science should serve the needs of society. Deciding what those needs are, however, is often a matter of politics and economics. Should water be taken from a river for irrigation or left in the river for wildlife habitat? Should we force coal-burning power plants to reduce air pollution in order to lower health costs and respiratory illnesses, or are society and our economy better served by having cheap but dirty energy? These thorny questions are decided by a combination of scientific evidence, economic priorities, political positions, and ethical viewpoints. Among these factors, science is most widely regarded as a source of truth: Scientific conclusions are based on observations, so science must be objective and rational.

On the other hand, in every political debate, lawyers and lobbyists can find scientists who will back either side. Politicians hold up favorable studies, proclaiming them "sound science," while they dismiss others as "junk science." Opposing sides dispute the scientific authority of the study they dislike. What is "sound" science, anyway? If science is often embroiled in politics, does this mean that science is always a political process?

The opening case study for this chapter illustrates some of the connections between science, politics, ethics, and values. If you judge only from reports in newspapers or on television about this issue, you'd probably conclude that scientific opinion is about equally divided on whether global warming is a threat or not. In fact, the vast majority of scientists working on this issue agree with the proposition that the earth's climate is being affected by human activities. Only a handful of maverick scientists disagree. In a study of 928 papers published in refereed scientific journals between 1993 and 2003, not one disagreed with the broad scientific consensus on global warming.

Why, then, is there so much confusion among the public about this issue? Why do politicians continue to assert that the dangers of climate change are uncertain at best, or "the greatest hoax ever perpetrated on the American people," as James Inhofe, former chair of the Senate Committee on Environment and Public Works, claims. A part of confusion lies in the fact that media often present the debate as if it's evenly balanced. The fact that an overwhelming majority of working scientists are mostly in agreement on this issue doesn't make good drama, so the media give equal time to minority viewpoints just to make an interesting fight.

Perhaps a more important source of misinformation comes from corporate funding for articles and reports denying climate change. The ExxonMobil corporation, for example, has donated at least $20 million over the past decade to more than 100 think tanks, media outlets, and consumer, religious, and civil rights groups that promote skepticism about global warming. Some of these organizations sound like legitimate science or grassroots groups but are really only public relations operations. Others are run by individuals who find it rewarding to offer contrarian views. This tactic of spreading doubt and disbelief through innocuous-sounding organizations or seemingly authentic experts isn't limited to the climate change debate. It was pioneered by the tobacco industry to mislead the public about the dangers of smoking. Interestingly, some of the same individuals, groups, and lobbying firms employed by tobacco companies are now working to spread confusion about climate change.

Given this highly sophisticated battle of "experts," how do you interpret these disputes, and how do you decide whom to trust? The most important strategy is to apply critical thinking as you watch or read the news. What is the position of the person making the report? What is the source of their expertise? What economic or political interests do they serve? Do they appeal to your reason or to your emotions? Do they use inflammatory words (such as "junk"), or do they claim that scientific uncertainty makes their opponents' study meaningless? If they use statistics, what is the context for their numbers?

It helps to seek further information as you answer some of these questions. When you watch or read the news, you can look for places where reporting looks incomplete, you can consider sources and ask yourself what unspoken interests might lie behind the story.

Another strategy for deciphering the rhetoric is to remember that there are established standards of scientific work, and to investigate whether an "expert" follows these standards: Is the report peer-reviewed? Do a majority of scholars agree? Are the methods used to produce statistics well documented?

TABLE 2.2

Questions for Baloney Detection

1. How reliable are the sources of this claim? Is there reason to believe that they might have an agenda to pursue in this case?

2. Have the claims been verified by other sources? What data are presented in support of this opinion?

3. What position does the majority of the scientific community hold in this issue?

4. How does this claim fit with what we know about how the world works? Is this a reasonable assertion or does it contradict established theories?

5. Are the arguments balanced and logical? Have proponents of a particular position considered alternate points of view or only selected supportive evidence for their particular beliefs?

6. What do you know about the sources of funding for a particular position? Are they financed by groups with partisan goals?

7. Where was evidence for competing theories published? Has it undergone impartial peer review or it is only in proprietary publication?

Harvard's Edward O. Wilson writes, "We will always have contrarians whose sallies are characterized by willful ignorance, selective quotations, disregard for communications with genuine experts, and destructive campaigns to attract the attention of the media rather than scientists. They are the parasite load on scholars who earn success through the slow process of peer review and approval." How can we identify misinformation and questionable claims? The astronomer Carl Sagan proposed a "Baloney Detection Kit" containing the questions in table 2.2.

Most scientists have an interest in providing knowledge that is useful, and our ideas of what is useful and important depend partly on our worldviews and priorities. Science is not necessarily political, but it is often used for political aims. The main task of educated citizens is to discern where it is being misused or disregarded for purposes that undermine public interests.

What's the relation between environmental science and environmentalism?

As you've learned already, environmental science uses scientific methods to study processes and systems in our environment. Environmentalism, on the other hand, works to influence attitudes and policies that affect our environment. Obviously, the former should be objective, while the latter has an subjective agenda. And yet, how can you gather information about what's happening to our environment without wanting to share that knowledge with policymakers? As the great conservationist Aldo Leopold wrote, "one of the penalties of an ecological education is that one lives alone in a world of wounds." He took the position that scientists have a duty not only to study the world, but to try to make it a better place.

This creates a dilemma for scientists. How can you gather information about habitat destruction, extinction of species, health risks of pollution, and other environmental threats without speaking out about the evidence you've collected? If you don't take a position, aren't you an accomplice in the damage done to our environment? Perhaps the best solution for this dilemma is to be very clear when you are acting as a scientist and when you are speaking as an advocate. When you're carrying out scientific studies, you need to be objective and impartial. When you're promoting solutions to the problems you've discovered, you need to make it clear if you're taking a partisan role. We'll try to separate these two functions as we discuss issues in this book, but you, also, need to be careful to distinguish between objective facts and opinions and interpretations as you read the text.

2.3 SYSTEMS

Environmental science seeks to understand **systems,** or networks of interactions among many interdependent factors. An ecosystem, for example, consists of living organisms and the soil, water, and air on which, or in which, these organisms live. Our global climate is an extremely complex system involving feedbacks between solar energy, atmosphere, oceans, fresh water, and living organisms. It's further complicated by our actions. You will encounter many systems in this book, such as ecological systems that include both living and nonliving components; geologic systems, which involve endless erosion and recycling of rocks in and on the earth's surface; and economic systems, which draw on natural resources and cultural information to circulate resources through societies and from one place to another.

A focus on systems is useful because it encourages us to inspect the relationships among components. Marine biologists worry about the disappearance of sharks because, as top predators, they affect population dynamics of many prey species. Loss of a top predator means more than just the loss of a species: it means possible disruption of the ecosystem as a whole. Biologists worry about sharks because we do not know how far cascading collapse could spread through marine ecosystems if sharks populations continue to fall.

Systems are composed of processes

We can think of systems as consisting of flows and storage compartments. A familiar example of a system is a fish tank (fig. 2.8a) in which plants grow (using sunlight). As the plants grow, they store solar energy and carbon (from carbon dioxide dissolved in the water) in their leaves. A fish grazes on the plant, taking energy and nutrients from the leaves. Waste from the fish fertilizes algae growing on the sides of the tank. A snail consumes the algae, and waste from the snail (and fish) fertilizes the plants, which feed the fish . . . In this balanced fish tank, nutrients and energy flow between the plants and animals. Plants and animals store nutrients temporarily. The fish tank stays healthy as long as the size of compartments and the rates of the flows remain in balance.

If the plants grow faster than the fish can eat them, and if there is plenty of sunlight, then the plants grow new stems, which grow new leaves, which provide energy to store more stems, which grow more leaves . . . until the tank becomes clogged with vegetation. This situation, when a flow leads to compartment

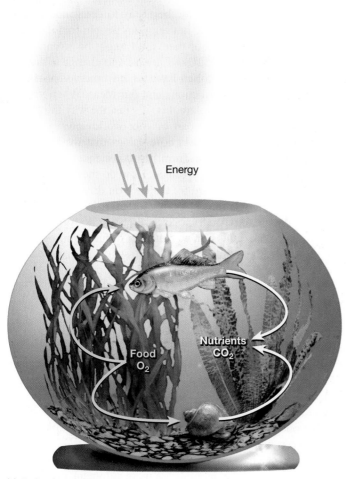

Energy

Food
O_2

Nutrients
CO_2

(a) A simple system

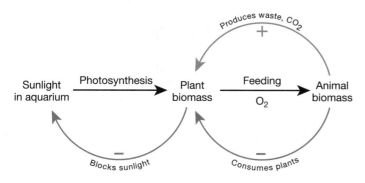

Produces waste, CO_2
+

Sunlight
in aquarium

Photosynthesis

Plant
biomass

Feeding
O_2

Animal
biomass

Blocks sunlight
−

Consumes plants
−

(b) Feedback loops

FIGURE 2.8 Systems consist of compartments (also known as state variables) such as fish and plants, and flows of resources, such as nutrients or O_2 (a). Feedback loops (b) enhance or inhibit flows and the growth of compartments.

changes that further enhance the flow, we call a **positive feedback loop** (fig. 2.8b). If unchecked, positive feedback loops can lead a system to become increasingly unbalanced. Systems also have **negative feedback loops,** which dampen flows. In the fish tank system, suppose fish reproduce after feeding on the plants. Reproduction leads to more fish, which graze more heavily. Once the

fish begin to overgraze the plants, there will be less food to go around, and reproduction will fall. Thus there is a negative feedback loop: reproduction → overgrazing → less reproduction.

When negative feedbacks prevent instability from positive feedback loops, we say the system is in homeostasis: It changes little from its stable condition. Your body is a system that is generally homeostatic. If your temperature changes more than slightly, you put on a sweater or take a cool drink; you tend to maintain fairly stable storage by eating when you are hungry and stopping when you are full.

If the system changes little over time, in the rate of flows or the relative size of storage compartments, we say the system is in **equilibrium.** Ecologists long considered that most forests normally existed in equilibrium: The number of trees, deer, birds, and other organisms should change little over the centuries if humans did not interfere. Increasingly, we see many ecosystems in terms of dynamic equilibrium: They undergo disturbance, then return to something like their previous state, then undergo disturbance again. Western lodgepole pine forests, for example, naturally experience periodic fires, which destroy expansive areas of pines. After the burn, sun-loving ground plants flourish for a while, and pine seeds released in the fire gradually grow to replace the dead pines. A century or so later, the mature pines begin to die, producing fuel that supports another stand-replacing fire.

Disturbances and emergent properties are important characteristics of many systems

Disturbances, periodic, destructive events such as fire or floods, are a natural part of many systems. Floodplains need periodic floods to replenish soil nutrients and maintain a diversity of understory vegetation; prairies depend on fires to recycle nutrients, slow tree growth, and reduce built-up, undecayed plant litter; ponderosa pine forests need periodic fires to clear understory vegetation below the mature pines. When systems recover quickly from disturbance, as in the case of a floodplain, prairie, or lodgepole pine ecosystem, we say the system has **resilience.**

We can think of systems as open or closed. A closed system, in theory, is entirely self-contained. It receives no inputs of energy or material from outside. There are few examples of closed systems, however. Your fish tank receives energy and oxygen from outside. Ecosystems receive energy, nutrients, and materials from their surroundings. Systems that take inputs from elsewhere are **open systems.**

Emergent properties are characteristics of a whole, functioning system that are quantitatively or qualitatively greater than the sum of the system's parts. For example, you are a system, consisting of flows (energy, nutrients, water, and information) and compartments (your body stores water, nutrients, and energy). But you are also much more than the matter of which you are made. You can sing, dance, talk, produce ideas and art, and share feelings with those around you. All of these are properties that emerge because you function well as a system. Similarly, a tree is more than a mass of stored carbon. It provides structure to a forest, habitat for other organisms, shades and cools the ground, and

FIGURE 2.9 Emergent properties of systems, including beautiful sights and sounds, make them exciting to study.

holds soil on a hillside with its roots. Often, we examine systems in terms of flows, compartments, and feedbacks, but it is the emergent properties that make systems exciting to study (fig. 2.9).

As you read this book, you will encounter many types of systems. Think about how flows, feedbacks, disturbance, and emergent properties occur in these systems.

2.4 ENVIRONMENTAL ETHICS AND WORLDVIEWS

As we study our environment, science informs many of our actions, but values and ethics can be just as important when we decide what to consume and how to vote. **Ethics** is a branch of philosophy concerned with what actions are right and wrong. The study of ethics helps us identify principles that guide what we should and should not do. Because most people, at some level, want to do the right thing, ethics are an important consideration in our dealings with environment, as well as with other people.

Environmental ethics has to do with our moral obligations to the world around us. Do we have duties, obligations, or responsibilities toward other species or to nature in general? Are there ethical principles that constrain how we use resources or modify our environment? How do we weigh obligations to nature against obligations to human interests? The way we answer these questions depends on our view of the world.

Worldviews express our deepest values

Worldviews are sets of basic beliefs, images, and understandings that shape how we see the world around us. They are very closely related to Thomas Kuhn's paradigms. Most of us learn these explanatory frameworks early in life, and we cannot change them easily. When we encounter evidence that doesn't fit our worldview, we often reject the evidence and cling to prior beliefs.

Our basic beliefs impact not only the way we think about ourselves and our place in the world; they also determine what questions are valid to ask. Slave owners in ancient Greece could

not question whether slaves had rights; modern Americans have difficulty asking if wealth is the best way to determine who has access to resources; many devout religious people find it unimaginable that the basic tenets of their faith might be wrong.

Part of the reason worldviews are hard to change is that they are hard to recognize. We often have a hard time verbalizing the underlying reasons for our actions and our ideas. When you find yourself saying, "I can't explain this, it's just the way things are," you are probably describing part of your worldview.

Each of us has deeply held core values that we learn, often without being aware of them, from life experiences. Try to identify two or three of your core values. How do these values influence your actions with regard to: the car you drive? relationships with your family? relationships with friends? the way you vote? where you donate, or spend, money? what you do for recreation? How do your core values or worldviews differ from, or overlap with, those of other students in your class? What can you learn from comparing lists of core values and worldviews?

Who (or what) has moral value?

Extending moral value to others is known as **moral extensionism.** Ancient Greeks granted moral value (the idea of worth, as well as responsibility) to adult male citizens. Slaves, women, and children had few rights and were essentially treated as property. Gradually, we have come to believe that all humans have certain inalienable rights, such as life, liberty, and the pursuit of happiness. No one can ethically treat another human as a mere object for their own pleasure or profit.

Do animals have rights? The philosopher René Descartes (1596–1650) argued that animals are mere automata, or machines, that can neither reason nor feel pain, so they also lack moral rights. In this view, other species are simply objects, and it is pointless to worry about their feelings. This debate is not just historical. In 2004, a report published in *Science* found that fish feel pain. While this may not seem surprising, it caused public outrage among recreational anglers, many of whom had, until then, managed to suppress worries about inflicting pain on fish.

For many people, moral extensionism now extends to granting some degree of moral value to animals, and even plants or inanimate objects (fig. 2.10).

Living things can have intrinsic or instrumental value

Rather than couch ethics strictly in terms of rights, some philosophers prefer to consider values. Value is a measure of the worth of something. But value can be either inherent or conferred. All humans, we believe, have **inherent value**—an intrinsic or innate worth—simply because they are human. They deserve moral consideration no matter who they are or what they do. Tools, on the other hand, have conferred, or **instrumental value.** They are worth something only because they are valued by someone who matters. If I hurt you without good reason, I owe you an apology. If I borrow your car and smash it into a tree, however, I don't

FIGURE 2.10 Moral extensionism describes an increasing consideration of moral value in other living things—or even nonliving things.

FIGURE 2.11 Mineral King Valley at the southern border of Sequoia National Park was the focus of an important environmental law case in 1969. The Disney Corporation wanted to build a ski resort here, but the Sierra Club sued to protect the valley on behalf of the trees, rocks, and native wildlife.

owe the car an apology. I owe *you* the apology—or reimbursement—for ruining your car. The car is valuable only because you want to use it. It doesn't have inherent values or rights of itself.

How does this apply to nonhumans? Domestic animals clearly have an instrumental value because they are useful to their owners. But some philosophers would say they also have inherent values and interests. By living, breathing, struggling to stay alive, the animal carries on its own life independent of its usefulness to someone else.

Some people believe that even nonliving things also have inherent worth. Rocks, rivers, mountains, landscapes, and certainly the earth itself, have value. These things were in existence before we came along and we couldn't re-create them if they are altered or destroyed. In a landmark 1969 court case, the Sierra Club sued the Disney Corporation on behalf of the trees, rocks, and wildlife of Mineral King Valley in the Sierra Nevada Mountains (fig. 2.11) where Disney wanted to build a ski resort. The Sierra Club argued that it represented the interests of beings that could not speak for themselves in court.

A legal brief entitled *Should Trees Have Standing?*, written for this case by Christopher D. Stone, proposed that organisms as well as ecological systems and processes should have standing (or rights) in court. After all, corporations—such as Disney—are treated as persons and given legal rights even though they are really only figments of our imagination. Why shouldn't nature have similar standing? The case went all the way to the Supreme Court but was overturned on a technicality. In the meantime, Disney lost interest in the project and the ski resort was never built. What do you think? Where would you draw the line of what deserves moral considerability? Are there ethical limits on what we can do to nature?

Is discrimination against other people related to our attitudes toward nature?

Many feminist philosophers argue that thoughtless mistreatment of nature is related to how we treat minorities, women, children, and others who lack power and prestige. **Ecofeminism,** the main framework for this perspective, holds that patriarchal attitudes in society are based on domination and ideas of superiority. If you believe that nature has only instrumental value, then you probably also believe you can do anything you please to it. In societies where some people are granted greater value than others, discrimination and exploitation of those with lower status also are likely to be regarded as acceptable. Ecofeminists argue that we need less domination and more cooperation with both nature and other people to achieve a peaceful, sustainable society.

2.5 FAITH-BASED CONSERVATION AND ENVIRONMENTAL JUSTICE

As the opening case study for this chapter shows, religious and ethical values can play important roles in environmental debates. Often people's most deeply held worldviews and values are expressed in their religious beliefs. Some of our most pressing environmental problems don't need technological or scientific solutions; they're not so much a question of what we're able to do, but what we're willing to do. Are we willing to take the steps necessary to stop global climate change? Do our values and ethics require us to do so? In this section, we'll look at some religious perspectives and how they influence our attitudes toward nature.

Environmental scientists have long been concerned about religious perspectives. In 1967, historian Lynn White, Jr., published a widely influential paper, "The Historic Roots of Our Ecological Crisis." He argued that Christian societies have often exploited natural resources carelessly because the Bible says that God commanded Adam and Eve to dominate nature: "Be fruitful, and multiply, and replenish the earth and subdue it: and have dominion over the fish of the sea, and over the fowl of the air, and over every living thing that moveth upon the earth" (Genesis 1:28). Since then, many religious scholars have pointed out that God also commanded Adam and Eve to care for the garden they were given, "to till it and keep it" (Genesis 2:15). Furthermore, Noah was commanded to preserve individuals of all living species, so that they would not perish in the great Flood. Passages such as these inspire many Christians to insist that it is our responsibility to act as stewards of nature, and to care for God's creations.

Both calls for environmental stewardship and anthropocentric domination over nature can be found in the writings of most major faiths. The Koran teaches that "each being exists by virtue of the truth and is also owed its due according to nature," a view that extends moral rights and value to all other creatures. Hinduism and Buddhism teach *ahimsa,* or the practice of not harming other living creatures, because all living beings are divinely connected (fig. 2.12).

Many faiths support environmental conservation

The idea of **stewardship,** or taking care of the resources we are given, inspires many religious leaders to promote conservation. "Creation care" is a term that has become prominent among evangelical Christians in the United States. In 1995, representatives of nine major religions met in Ohito, Japan, to discuss views of environmental stewardship in their various traditions. The resulting document, the Ohito Declaration, outlined common beliefs and responsibilities of these different faiths toward protecting the earth and its life (table 2.3). In recent years, religious organizations have played important roles in nature protection. A coalition of evangelical Christians has been instrumental in opposing Congressional attacks on the U.S. Endangered Species Act. Similarly, the Central Conference of American Rabbis has declared that desecration of the Headwaters Redwood Forest in California breaks our covenant with the Creator.

Religious concern extends beyond our treatment of plants and animals. Pope John Paul II and Orthodox Patriarch Bartholomew called on countries bordering the Black Sea to stop pollution, saying that "to commit a crime against nature is a sin." In addition to its campaign to combat global warming described at the beginning

FIGURE 2.12 Many religions emphasize the divine relationships among humans and the natural world. The Tibetan Buddhist goddess Tara represents compassion for all beings.

of this chapter, the Creation Care Network has also launched initiatives against energy inefficiency, mercury pollution, mountaintop removal mining, and endangered species destruction. For many people, religious beliefs provide the best justification for environmental protection.

Environmental justice combines civil rights and environmental protection

People of color in the United States and around the world are subjected to a disproportionately high level of environmental health risks in their neighborhoods and on their jobs. Minorities, who tend to be poorer and more disadvantaged than other residents, work in the dirtiest jobs where they are exposed to toxic chemicals and other hazards. More often than not they also live in urban ghettos, barrios, reservations, and rural poverty pockets that have shockingly high pollution levels and are increasingly the site of unpopular industrial facilities, such as toxic waste dumps, landfills, smelters, refineries, and incinerators. **Environmental justice** combines civil rights with environmental protection to demand a safe, healthy, life-giving environment for everyone.

TABLE 2.3

Principles and Actions in the Ohito Declaration

Spiritual Principles

1. Religious beliefs and traditions call us to care for the earth.
2. For people of faith, maintaining and sustaining environmental life systems is a religious responsibility.
3. Environmental understanding is enhanced when people learn from the example of prophets and of nature itself.
4. People of faith should give more emphasis to a higher quality of life, in preference to a higher standard of living, recognizing that greed and avarice are root causes of environmental degradation and human debasement.
5. People of faith should be involved in the conservation and development process.

Recommended Courses of Action

The Ohito Declaration calls upon religious leaders and communities to

1. emphasize environmental issues within religious teaching: faith should be taught and practiced as if nature mattered.
2. commit themselves to sustainable practices and encourage community use of their land.
3. promote environmental education, especially among youth and children.
4. pursue peacemaking as an essential component of conservation action.
5. take up the challenge of instituting fair trading practices devoid of financial, economic, and political exploitation.

FIGURE 2.13 Poor people and people of color often live in the most dangerous and least desirable places. Here children play next to a chemical refinery in Texas City, Texas.

Among the evidence of environmental injustice is the fact that three out of five African-Americans and Hispanics, and nearly half of all Native Americans, Asians, and Pacific Islanders live in communities with one or more uncontrolled toxic waste sites, incinerators, or major landfills, while fewer than 10 percent of all whites live in these areas. Using zip codes or census tracts as a unit of measurement, researchers found that minorities make up twice as large a population share in communities with these locally unwanted land uses (**LULUs**) as in communities without them. A 2006 study using "distance-based" methods found an even greater correlation between race and location of hazardous waste facilities.

Although it is difficult to distinguish between race, class, historical locations of ethnic groups, economic disparities, and other social factors in these disputes, racial origins often seem to play a role in exposure to environmental hazards. Simple correlation doesn't prove causation; still, while poor people in general are more likely to live in polluted neighborhoods than rich people, the discrepancy between the pollution exposure of middle class blacks and middle class whites is even greater than the difference between poorer whites and blacks. Where upper class whites can "vote with their feet" and move out of polluted and dangerous neighborhoods, blacks and other minorities are restricted by color barriers and prejudice (overt or covert) to the less desirable locations (fig. 2.13).

Environmental racism distributes hazards inequitably

Racial prejudice is a belief that someone is inferior merely because of their race. Racism is prejudice with power. **Environmental racism** is inequitable distribution of environmental hazards based on race. Evidence of environmental racism can be seen in lead poisoning in children. The Federal Agency for Toxic Substances and Disease Registry considers lead poisoning to be the number one environmental health problem for children in the United States. Some 4 million children—many of whom are African American, Latino, Native American, or Asian, and most of whom live in inner-city areas—have dangerously high lead levels in their bodies. This lead is absorbed from old lead-based house paint, contaminated drinking water from lead pipes or lead solder, and soil polluted by industrial effluents and automobile exhaust. The evidence of racism is that at every income level, whether rich or poor, black children are two to three times more likely than whites to suffer from lead poisoning.

Because of their quasi-independent status, most Native-American reservations are considered sovereign nations that are not covered by state environmental regulations. Court decisions holding that reservations are specifically exempt from hazardous waste storage and disposal regulations have resulted in a land rush of seductive offers from waste disposal companies to Native-American reservations for onsite waste dumps, incinerators, and landfills. The short-term economic incentives can be overwhelming for communities in which adult unemployment runs between 60 and 80 percent. Uneducated, powerless people often can be tricked or intimidated into signing environmentally and socially disastrous contracts. Nearly every tribe in America has been approached with proposals for some dangerous industry or waste facility.

The practice of targeting poor communities of color in the developing nations for waste disposal and/or experimentation with risky technologies has been described as **toxic colonialism.** Internationally, the trade in toxic waste has mushroomed in recent years as wealthy countries have become aware of the risks of industrial refuse. Poor, minority communities at home and abroad are being increasingly targeted as places to dump unwanted wastes. Although a treaty regulating international shipping of toxics was signed by 105 nations in 1989, millions of tons of toxic and hazardous materials continue to move—legally or illegally—from the richer countries to the poorer ones every year. This issue is discussed further in chapter 23.

Another form of toxic colonialism is the flight of polluting industries from developed nations and states where control requirements are stringent to less developed areas where regulations are lax and local politicians are easily co-opted. For example, more than 2,000 *maquiladoras,* or assembly plants operated by American, Japanese, or other foreign companies, are now located along the United States/Mexico border to take advantage of favorable import quotas, low wages, and weak pollution-control laws.

Mexican laborers work under appalling conditions in some of these factories, assembling imported components into consumer goods to be exported to the United States. Although the jobs are demeaning, dangerous, and pay far less than the minimum U.S. wage, large populations have been attracted to squalid shantytowns along the border where industrial effluents poison the air and water (fig. 2.14).

FIGURE 2.14 Living conditions in the *colonias* or unplanned settlements along the U.S./Mexican border often are substandard. Is this evidence of racism or simply a result of poverty and poor planning?

The U.S. Environmental Justice Act was established in 1992 to identify areas threatened by the highest levels of toxic chemicals, assess health effects caused by emissions of those chemicals, and ensure that groups or individuals residing within those areas have opportunities and resources to participate in public discussions concerning siting and cleanup of industrial facilities. Perhaps we need something similar worldwide.

CONCLUSION

Many of the most difficult environmental problems we face are moral or ethical dilemmas rather than technological crises. Are we willing to do what we need to leave a habitable world for our children and grandchildren? Do we have a moral obligation to the poor people of the world whose lives are harmed by the way we've chosen to use (or abuse) resources? Scientific uncertainties about what may happen in the future make many of these decisions even more difficult. However, those who oppose policies that may be expensive or personally inconvenient often claim more doubt or ambiguity than is scientifically justified.

Rigorous, objective, empirical science based on unbiased, systematic research and open, honest communication among a community of scientists is one of the most important features of the modern age. It's how we discover information about the world around us. It has given rise to most of the comforts and conveniences we enjoy today. Not everyone is able to—or wants to—be a practicing scientist, but because science is such a powerful force in modern society, all citizens should understand the basics about how science works, and what questions it is able, and not able, to answer.

"No man is an island," wrote John Donne; nor does any organism live in complete isolation. All of us are part of interlocking systems that involve interactions between many interdependent processes and factors. Often those systems have properties that are greater than or different from the individual components that produce them. Understanding how systems work is another essential element in an environmental education. Some systems exist in equilibrium, changing little over time. Others may be highly variable, responding to disturbances. Resilience, or the ability to recover from disturbance is an important quality of some systems.

As environmental citizens, science informs many of our decisions, but values and ethics can be just as important when we choose what to do or not do. Environmental ethics has to do with our moral obligations to the world around us. Do we have duties, obligations, or responsibilities toward nature? Do other species—or even inanimate objects or features in nature—have rights and values? How we answer these questions depends on our worldview, which generally expresses our most deeply held beliefs. While some people use religious beliefs to justify domination and exploitation of nature, others believe that God calls us to be stewards and caretakers of nature. Different religious interpretations have become critical aspects of debates about how to manage resources.

Environmental justice combines civil rights with environmental protection to demand a safe, healthy, life-giving environment for everyone. Too often, minorities and poor people are exposed unfairly to toxins, hazards, and degraded environments.

We're happy to have pollution and waste removed from our own neighborhood, but we often fail to recognize that there really is no "away" to throw our problems. We're simply dumping them on someone else, or leaving them to future generations. As the great ecologist Aldo Leopold wrote, "All history consists of successive excursions from a single starting point to which man returns again and again to organize yet another search for a durable scale of values."

REVIEWING LEARNING OUTCOMES

By now you should be able to explain the following points:

2.1 Describe the scientific method and explain how it works.
- Science depends on skepticism and accuracy.
- Deductive and inductive reasoning are both useful.
- Testable hypotheses and theories are essential tools.
- Understanding probability helps reduce uncertainty.
- Statistics can calculate the probability that your results were random.
- Experimental design can reduce bias.
- Models are an important experimental strategy.

2.2 Evaluate the role of scientific consensus and conflict.
- Detecting pseudoscience relies on independent, critical thinking.
- What's the relation between environmental science and environmentalism?

2.3 Explain systems and how they're useful in science.
- Systems are composed of processes.
- Disturbances and emergent properties are important characteristics of many systems.

2.4 Discuss environmental ethics and worldviews.
- Worldviews express our deepest values.
- Who (or what) has moral value?
- Living things can have intrinsic or instrumental value.
- Is discrimination against other people related to our attitudes toward nature?

2.5 Identify the roles of religious and cultural perspectives in conservation and environmental justice.
- Many faiths support environmental conservation.
- Environmental justice combines civil rights and environmental protection.
- Environmental racism distributes hazards inequitably.

PRACTICE QUIZ

1. What is science? What are some of its basic principles?
2. Why are widely accepted, well-defended scientific explanations called "theories"?
3. Explain the following terms: probability, dependent variable, independent variable, and model.
4. What are inductive and deductive reasoning? Describe an example in which you have used each.
5. Draw a diagram showing the steps of the scientific method, and explain why each is important.
6. What is scientific consensus and why is it important?
7. What is a positive feedback loop? What is a negative feedback loop? Give an example of each.
8. Describe anthropocentric, biocentric, utilitarian, and preservationist viewpoints.
9. What is stewardship, and why do many religious leaders feel it is important in current policy discussions?
10. What is environmental justice?

CRITICAL THINKING AND DISCUSSION QUESTIONS

1. Explain why scientific issues are or are not influenced by politics. Can scientific questions ever be entirely free of political interest? If you say no, does that mean that all science is merely politics? Why or why not?
2. Review the questions for "baloney detection" in table 2.2, and apply them to an ad on TV. How many of the critiques in this list are easily detected in the commercial?
3. How important is scientific thinking for you, personally? How important do you think it should be? How important is it for society to have thoughtful scientists? How would your life be different without the scientific method?
4. Try to put yourself in the place of a person from a minority community, an underdeveloped nation, or a developing country in discussing questions of environmental justice and

environmental quality. What preconceptions, values, beliefs, and contextual perspectives would you bring to the issue? What would you ask from the majority society?

5. Many scientific studies rely on models for experiments that cannot be done on real systems, such as climate, human health, or economic systems. If assumptions are built into models, then are model-based studies inherently weak? What would increase your confidence in a model-based study?

DATA analysis   More Graph Types

The simple line plots we discussed in chapter 1 are a good way to display data from a sequence that changes over time. But what if you want to compare groups of things or data from a single point in time? In this exercise, we'll explore some alternative graphing techniques.

Bar Graphs and Pie Charts

Bar graphs help you compare *values* of different *categories,* such as the number of apples and the number of oranges. Figure 1, for example, shows levels of urban air pollution. The particu-late quantities are classified by size rather than a sequence of

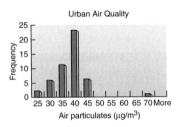

FIGURE 1   Bar graphs.

dates. Thus you can read from the graph that 40 μg/m^3 occurred about 25 times during the year studied, although you can't tell when that happened. As we discussed in the Exploring Science box on pp. 38–39, this type of graph allows you to see quickly which category is most abundant, where the approximate mean is, and to recognize *outliers* (unusually high or low values, such as 70 μg/m^3). Notice that the frequency distribution in this graph creates a roughly bell-shaped curve. A bell-shaped curve is often called a *normal distribution,* or a Gaussian distribution, because most large, randomly selected groups have approximately that distribution. More measurements would probably make this his-togram closer to a normal, or Gaussian, distribution.

Pie charts, like bar graphs, compare categories. The differ-ence is that categories in a pie chart add up to 100 percent of something. Figure 2, for example, shows major energy sources for the United States. To make comparison between these energy sources easier, each is expressed as a percentage of the total energy supply. They are grouped together in a circle, the circum-ference of which is made to be 100 percent. In effect, the outside

of the circle is the X-axis, and the size of each sector shows its value on the Y- (dependent variable) axis. Because neither of these axes have scales, it's customary to label the sectors with the name of the category and its value.

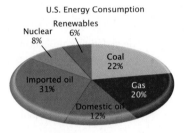

FIGURE 2 Pie chart

Scatter Plots

A scatter plot (fig. 3) shows the distribution of many observations that have no particular sequence. Trends in the scatter plot show a relationship between the two variables. A field of dots trending upward indicates that the depen-dent variable (Y-axis) increases as the independent variable (X-axis) increases. A tight field of dots shows that Y responds closely to X. A loose cloud shows that Y responds only generally to X. Often scatter plots show no trend, and no relationship between X and Y. Figure 3, for example, shows the number of asthma

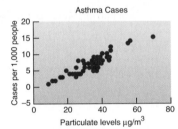

FIGURE 3 Scatter plot.

cases per 1,000 people in an urban area at different particulate levels. Although the data points don't fall on a single line, you can see a trend in which higher pollution levels are associated with higher numbers of asthma cases. To test whether this trend is reli-able or merely a chance occurrence, you could use statistical pro-cedures such as regression analysis. These are discussed in the Exploring Science box on statistics on pp. 38–39.

Think about these principles as you examine graphs through-out this book. In subsequent chapters, we'll have more discussion of data analysis techniques.

For Additional Help in Studying This Chapter, please visit our website at www.mhhe.com/cunningham10e. You will find additional practice quizzes and case studies, flashcards, regional examples, place markers for Google Earth™ mapping, and an extensive reading list, all of which will help you learn environmental science.

Together the fish, bears, birds, and other organisms transport and distribute nutrients in and around the McNeil River, Alaska.

C H A P T E R

3

Matter, Energy, and Life

*When one tugs at a single thing in nature, he finds it
attached to the rest of the world.*

—John Muir—

LEARNING OUTCOMES

After studying this introduction, you should be able to:

3.1 Describe matter, atoms, and molecules and give simple examples of the role of four major kinds of organic compounds in living cells.

3.2 Define energy and explain how thermodynamics regulates ecosystems.

3.3 Understand how living organisms capture energy and create organic compounds.

3.4 Define species, populations, communities, and ecosystems, and summarize the ecological significance of trophic levels.

3.5 Compare the ways that water, carbon, nitrogen, sulfur, and phosphorus cycle within ecosystems.

Case Study Why Trees Need Salmon

Ecologists have long known that salmon need clean, fast-moving streams to breed, and that clear streams need healthy forests. Surprising new evidence now indicates that some forests themselves need salmon to remain healthy, and that bears play an important intermediary role in this dynamic relationship.

The yearly return of salmon from the open Pacific Ocean to coastal waters of western North America is one of nature's grand displays. Salmon (*Onchorhyncus* sp.) are anadromous: They hatch in freshwater lakes and streams, spend most of their lives at sea, then return to the stream where they were born, to breed and die. To reproduce successfully, these fish require clear, cold, shaded streams and clean gravel riverbeds. If forests are stripped from riverbanks and surrounding hillsides, sediment washes down into streams, clogging gravel beds and suffocating eggs. Open to the sunlight, the water warms, lowering its oxygen levels, and reducing survival rates of eggs and young fish.

Every year, as millions of fish return to spawn and die in rivers of the Pacific Northwest, they provide a bonanza for bears, eagles, and other species (see photograph previous page). Ecologist Tom Reimchen estimates that each bear fishing in British Columbia's rivers catches about 700 fish during the 45-day spawn, and that 70 percent of the bear's annual protein comes from salmon. After a quick bite on the head to kill the fish, the bears drag their prey back into the forest, where they can feed undisturbed. Some bears have been observed carrying fish as much as 800 m (0.5 mi) from the river before feeding on them.

Bears don't eat everything they catch. They leave about half of each carcass to be scavenged by eagles, martens, crows, ravens, and gulls. A diversity of insects, including flies and beetles, also feed on the leftovers. Within a week, all the soft tissue is consumed, leaving only a bony skeleton. Reimchen calculates that between the nutrients leeching directly from decomposing carcasses and the excreta from bears and other scavengers, the fish provide about 120 kg of nitrogen per hectare of forest along salmon-spawning rivers. This is comparable to the rate of fertilizer applied by industry to commercial forest plantations. Altogether, British Columbia's 80,000 to 120,000 brown and black bears could be transferring 60 million kg of salmon tissue into the rainforest every year.

How do ecologists know that trees absorb nitrogen from salmon? Analyzing different kinds of nitrogen atoms, researchers can distinguish between marine-derived nitrogen (MDN) and that from terrestrial sources. Marine phytoplankton (tiny floating plant cells) have more of a rare, heavy form of nitrogen called ^{15}N compared to most terrestrial vegetation, in which ^{14}N, the more common, lighter form, predominates. Using a machine called a mass spectrometer, researchers can separate and measure the kinds and amounts of nitrogen in different tissues. We'll discuss different forms of atoms (called isotopes) later in this chapter. Because salmon spend most of their lives feeding on dense clouds of plankton far out to sea, they have higher ratio of $^{15}N/^{14}N$ in their bodies than do most freshwater or terrestrial organisms. When the fish die and decompose, they contribute their nitrogen to the ecosystem. Bears and other scavengers distribute this nitrogen throughout the forest where they drop fish carcasses or defecate in the woods.

Robert Naiman and James Helfield from the University of Washington found that foliage of spruce trees growing in bear-impacted areas is significantly enriched with MDN relative to similar trees growing at comparable distances from streams with and without spawning salmon (fig. 3.1). These results suggest that in feeding on salmon, bears play an important role in transferring MDN from the stream to the riparian (streamside) forest. Nitrogen is often a limiting nutrient for rainforest vegetation. Tree ring studies show that when salmon are abundant, trees grow up to three times as fast as when salmon are scarce. For some streamside trees, researchers estimate that between one-quarter to one-half of all their nitrogen is derived from salmon. Not only do salmon replenish the forest, but they also vitalize the streams and lakes with carbon, nitrogen, phosphorous, and micronutrients. Nearly 50 percent of the nutrients that juvenile salmon consume comes from dead parents.

This research is important because salmon stocks are dwindling throughout the Pacific Northwest. In Washington, Oregon, and California, most salmon populations have fallen by 90 percent from their historic numbers, and some stocks are now extinct. Because of the close relationship of salmon and the trees, biologists argue, forest, wildlife, and fish management need to be integrated. Each population—rainforest trees, bears, hatchlings, and ocean-going fish—affects the stability of the others. Salmon need healthy forests and streams to reproduce successfully, and forests and bears need abundant salmon. Stream ecosystems need standing trees to retain soil and provide shade. So healthy streams depend on fish, just as the fish depend on the streams. As this case shows, the flow of nutrients and energy between organisms can be intricate and complex. Relationships between apparently separate environments, such as rivers and forests, can be equally complex and important. In this chapter we'll explore some of these relationships among organisms and between organisms and their environment.

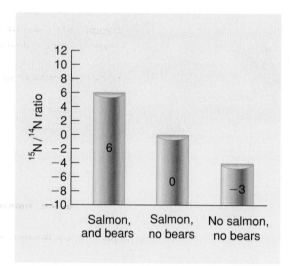

FIGURE 3.1 Nitrogen in trees near rivers with both salmon and bears have a significantly higher $^{15}N/^{14}N$ ratio in their foliage than do trees with salmon and no bears, or those without either salmon or bears. These findings suggest an important role for both fish and bears in distributing marine-derived nitrogen in riparian forests.

Case Study *continued*

For more information, see

Helfield, J. M., and R. J. Naiman. 2001. Effects of salmon-derived nitrogen on riparian forest growth and implications for stream productivity. *Ecology* 82(9):2403–9.

Reimchen, T., et al. 2003. Isotopic evidence for enrichment of salmon-derived nutrients in vegetation, soil and insects in riparian zones in coastal British Columbia. *American Fisheries Society Symposium* 34:59–69.

3.1 ELEMENTS OF LIFE

How are nutrients being exchanged between salmon, bears, and trees in the opening case study of this chapter? And what energy source keeps this whole system running? These questions are at the core of **ecology,** the scientific study of relationships between organisms and their environment. In this chapter we'll introduce a number of concepts that are essential to understanding how living things function in their environment.

As an introduction to principles of ecology, this chapter first reviews the nature of matter and energy, then explores the ways organisms acquire and use energy and chemical elements. Then we'll investigate feeding relationships among organisms—the ways that energy and nutrients are passed from one living thing to another—forming the basis of ecosystems. Finally, we'll review some of the key substances that cycle through organisms, ecosytems, and our environment.

In chapters 4 and 5 we'll continue investigating concepts of ecology: general organization of ecosystems by environmental types and landscapes, and principles of population growth.

In a sense, every organism is a chemical factory that captures matter and energy from its environment and transforms them into structures and processes that make life possible. To understand how these processes work, we will begin with some of the fundamental properties of matter and energy.

Matter is made of atoms, molecules, and compounds

Everything that takes up space and has mass is **matter.** Matter exists in three distinct states—solid, liquid, and gas—due to differences in the arrangement of its constitutive particles. Water, for example, can exist as ice (solid), as liquid water, or as water vapor (gas).

Under ordinary circumstances, matter is neither created nor destroyed but rather is recycled over and over again. Some of the molecules that make up your body probably contain atoms that once made up the body of a dinosaur and most certainly were part of many smaller prehistoric organisms, as chemical elements are used and reused by living organisms. Matter is transformed and combined in different ways, but it doesn't disappear; everything goes somewhere. These statements paraphrase the physical principle of **conservation of matter.**

How does this principle apply to human relationships with the biosphere? Particularly in affluent societies, we use natural resources to produce an incredible amount of "disposable" consumer goods.

If everything goes somewhere, where do the things we dispose of go after the garbage truck leaves? As the sheer amount of "disposed-of stuff" increases, we are having greater problems finding places to put it. Ultimately, there is no "away" where we can throw things we don't want any more.

Matter consists of **elements,** which are substances that cannot be broken down into simpler forms by ordinary chemical reactions. Each of the 118 known elements (92 natural, plus 26 created under special conditions) has distinct chemical characteristics. Just four elements—oxygen, carbon, hydrogen, and nitrogen—are responsible for more than 96 percent of the mass of most living organisms.

All elements are composed of **atoms,** which are the smallest particles that exhibit the characteristics of the element. Atoms are composed of positively charged protons, negatively charged electrons, and electrically neutral neutrons. Protons and neutrons, which have approximately the same mass, are clustered in the nucleus in the center of the atom (fig. 3.2). Electrons, which are tiny in comparison to the other particles, orbit the nucleus at the speed of light.

Each element has a characteristic number of protons per atom, called its **atomic number.** The number of neutrons in different atoms of the same element can vary slightly. Thus, the atomic mass, which is the sum of the protons and neutrons in each nucleus, also can vary. We call forms of a single element that differ in atomic mass **isotopes.** For example, hydrogen, the lightest element, normally has only one proton (and no neutrons) in its nucleus. A small percentage of hydrogen atoms have one proton and one neutron. We call this isotope deuterium (^{2}H). An even smaller percentage of natural hydrogen called tritium (^{3}H) has one proton plus two neutrons. The heavy form of nitrogen (^{15}N) mentioned in the opening story of this

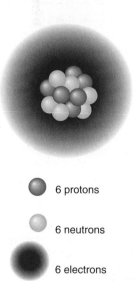

FIGURE 3.2 As difficult as it may be to imagine when you look at a solid object, all matter is composed of tiny, moving particles, separated by space and held together by energy. It is hard to capture these dynamic relationships in a drawing. This model represents carbon-12, with a nucleus of six protons and six neutrons; the six electrons are represented as a fuzzy cloud of potential locations rather than as individual particles.

- 6 protons
- 6 neutrons
- 6 electrons

chapter has one more neutron in its nucleus than does the more common ^{14}N. Both these nitrogen isotopes are stable but some isotopes are unstable—that is, they may spontaneously emit electromagnetic energy, or subatomic particles, or both. Radioactive waste, and nuclear energy, result from unstable isotopes of elements such as uranium and plutonium.

Chemical bonds hold molecules together

Atoms often join to form **compounds,** or substances composed of different kinds of atoms (fig. 3.3). A pair or group of atoms that can exist as a single unit is known as a **molecule.** Some elements commonly occur as molecules, such as molecular oxygen (O_2) or molecular nitrogen (N_2), and some compounds can exist as molecules, such as glucose ($C_6H_{12}O_6$). In contrast to these molecules, sodium chloride (NaCl, table salt) is a compound that cannot exist as a single pair of atoms. Instead it occurs in a large mass of Na and Cl atoms or as two ions, Na$^+$ and Cl$^-$, in solution. Most molecules consist of only a few atoms. Others, such as proteins and nucleic acids, can include millions or even billions of atoms.

When ions with opposite charges form a compound, the electrical attraction holding them together is an *ionic bond.* Sometimes atoms form bonds by *sharing* electrons. For example, two hydrogen atoms can bond by sharing a pair of electrons—they orbit the two hydrogen nuclei equally and hold the atoms

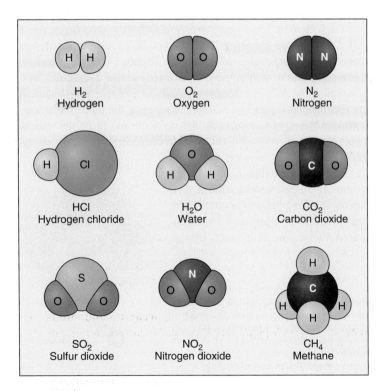

FIGURE 3.3 These common molecules, with atoms held together by covalent bonds, are important components of the atmosphere or important pollutants.

together. Such electron-sharing bonds are known as *covalent bonds.* Carbon (C) can form covalent bonds simultaneously with four other atoms, so carbon can create complex structures such as sugars and proteins. Atoms in covalent bonds do not always share electrons evenly. An important example in environmental science is the covalent bonds in water (H_2O). The oxygen atom attracts the shared electrons more strongly than do the two hydrogen atoms. Consequently, the hydrogen portion of the molecule has a slight positive charge, while the oxygen has a slight negative charge. These charges create a mild attraction between water molecules, so that water tends to be somewhat cohesive. This fact helps explain some of the remarkable properties of water (Exploring Science, p. 55).

When an atom gives up one or more electrons, we say it is *oxidized* (because it is very often oxygen that takes the electron, as bonds are formed with this very common and highly reactive element). When an atom gains electrons, we say it is *reduced.* Chemical reactions necessary for life involve oxidation and reduction: Oxidation of sugar and starch molecules, for example, is an important part of how you gain energy from food.

Forming bonds usually releases energy. Breaking bonds generally requires energy. For example, burning wood is exothermic (releases heat) because the energy released in the formation of carbon dioxide and water is greater than the energy required to break the bonds in cellulose (and other organic compounds in wood). Generally, some energy input (activation energy) is needed to initiate these reactions. In your fireplace, a match might provide the needed activation energy. In your car, a spark from the battery provides activation energy to initiate the oxidation (burning) of gasoline.

Electrical charge is an important chemical characteristic

Atoms frequently gain or lose electrons, acquiring a negative or positive electrical charge. Charged atoms (or combinations of atoms) are called **ions.** Negatively charged ions (with one or more extra electrons) are *anions.* Positively charged ions are *cations.* A hydrogen (H) atom, for example, can give up its sole electron to become a hydrogen ion (H$^+$). Chlorine (Cl) readily gains electrons, forming chlorine ions (Cl$^-$).

Substances that readily give up hydrogen ions in water are known as **acids.** Hydrochloric acid, for example, dissociates in water to form H$^+$ and Cl$^-$ ions. In later chapters, you may read about acid rain (which has an abundance of H$^+$ ions), acid mine drainage, and many other environmental problems involving acids. In general, acids cause environmental damage because the H$^+$ ions react readily with living tissues (such as your skin or tissues of fish larvae) and with nonliving substances (such as the limestone on buildings, which erodes under acid rain).

Substances that readily bond with H$^+$ ions are called **bases** or alkaline substances. Sodium hydroxide (NaOH), for example, releases hydroxide ions (OH$^-$) that bond with H$^+$ ions in water. Bases can be highly reactive, so they also cause significant

Exploring SCIENCE

If travelers from another solar system were to visit our lovely, cool, blue planet, they might call it Aqua rather than Terra because of its outstanding feature: the abundance of streams, rivers, lakes, and oceans of liquid water. Our planet is the only place we know where water exists as a liquid in any appreciable quantity. Water covers nearly three-fourths of the earth's surface and moves around constantly via the hydrologic cycle (discussed in chapter 15) that distributes nutrients, replenishes freshwater supplies, and shapes the land. Water makes up 60 to 70 percent of the weight of most living organisms. It fills cells, giving form and support to tissues. Among water's unique, almost magical qualities, are the following:

1. Water molecules are polar, that is, they have a slight positive charge on one side and a slight negative charge on the other side. Therefore, water readily dissolves polar or ionic substances, including sugars and nutrients, and carries materials to and from cells.

2. Water is the only inorganic liquid that exists under normal conditions at temperatures suitable for life. Most substances exist as either a solid or a gas, with only a very narrow liquid temperature range. Organisms synthesize organic compounds such as oils and alcohols that remain liquid at ambient temperatures and are therefore extremely valuable to life, but the original and predominant liquid in nature is water.

3. Water molecules are cohesive, tending to stick together tenaciously. You have experienced this property if you have ever done a belly flop off a diving board. Water has the highest surface tension of any common, natural liquid. Water also adheres to surfaces. As a result, water is subject to *capillary action:* it can be drawn into small channels. Without capillary action, movement of water and nutrients into groundwater reservoirs and through living organisms might not be possible.

4. Water is unique in that it expands when it crystallizes. Most substances shrink as they change from liquid to solid. Ice floats because it is less dense than liquid water. When temperatures fall below freezing, the surface layers of lakes, rivers, and oceans cool faster and freeze before deeper water. Floating ice then insulates underlying layers, keeping most water bodies liquid (and aquatic organisms alive) throughout the winter in most places. Without this feature, many aquatic systems would freeze solid in winter.

5. Water has a high heat of vaporization, using a great deal of heat to convert from liquid to vapor. Consequently, evaporating water is an effective way for organisms to shed excess heat. Many animals pant or sweat to moisten evaporative cooling surfaces. Why do you feel less comfortable on a hot, humid day than on a hot, dry day?

Surface tension is demonstrated by the resistance of a water surface to penetration, as when it is walked upon by a water strider.

Because the water vapor–laden air inhibits the rate of evaporation from your skin, thereby impairing your ability to shed heat.

6. Water also has a high specific heat; that is, a great deal of heat is absorbed before it changes temperature. The slow response of water to temperature change helps moderate global temperatures, keeping the environment warm in winter and cool in summer. This effect is especially noticeable near the ocean, but it is important globally.

All these properties make water a unique and vitally important component of the ecological cycles that move materials and energy and make life on earth possible.

---

environmental problems. Acids and bases can also be essential to living things: The acids in your stomach dissolve food, for example, and acids in soil help make nutrients available to growing plants.

We describe the strength of an acid and base by its **pH,** the negative logarithm of its concentration of H^+ ions (fig. 3.4). Acids have a pH below 7; bases have a pH greater than 7. A solution of exactly pH 7 is "neutral." Because the pH scale is logarithmic, pH 6 represents *ten times* more hydrogen ions in solution than pH 7.

A solution can be neutralized by adding buffers, or substances that accept or release hydrogen ions. In the environment, for example, alkaline rock can buffer acidic precipitation, decreasing its acidity. Lakes with acidic bedrock, such as granite, are especially vulnerable to acid rain because they have little buffering capacity.

Organic compounds have a carbon backbone

Organisms use some elements in abundance, others in trace amounts, and others not at all. Certain vital substances are concentrated within cells, while others are actively excluded. Carbon is a particularly important element because chains and rings of carbon atoms form the skeletons of **organic compounds,** the material of which biomolecules, and therefore living organisms, are made.

The four major categories of organic compounds in living things ("bio-organic compounds") are lipids, carbohydrates, proteins, and nucleic acids. Lipids (including fats and oils) store energy for cells, and they provide the core of cell membranes and other structures. Lipids do not readily dissolve in water, and their

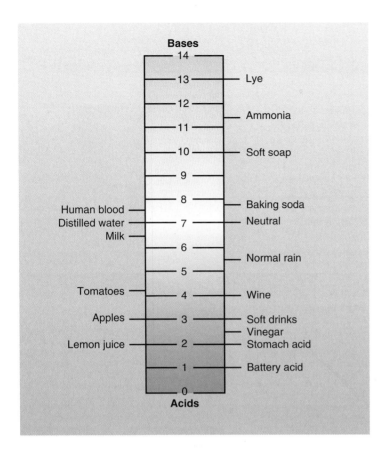

FIGURE 3.4 The pH scale. The numbers represent the negative logarithm of the hydrogen ion concentration in water.

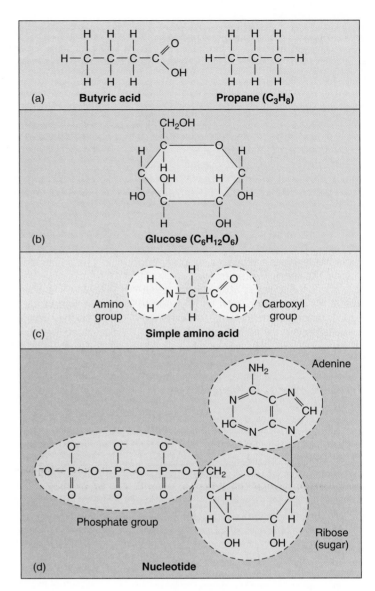

FIGURE 3.5 The four major groups of organic molecules are based on repeating subunits of these carbon-based structures. Basic structures are shown for (a) butyric acid (a building block of lipids) and a hydrocarbon, (b) a simple carbohydrate, (c) a protein, and (d) a nucleic acid.

basic structure is a chain of carbon atoms with attached hydrogen atoms. This structure makes them part of the family of hydrocarbons (fig. 3.5a). Carbohydrates (including sugars, starches, and cellulose) also store energy and provide structure to cells. Like lipids, carbohydrates have a basic structure of carbon atoms, but hydroxyl (OH) groups replace half the hydrogen atoms in their basic structure, and they usually consist of long chains of sugars. Glucose (fig. 3.5b) is an example of a very simple sugar.

Proteins are composed of chains of subunits called amino acids (fig. 3.5c). Folded into complex three-dimensional shapes, proteins provide structure to cells and are used for countless cell functions. Most enzymes, such as those that release energy from lipids and carbohydrates, are proteins. Proteins also help identify disease-causing microbes, make muscles move, transport oxygen to cells, and regulate cell activity.

Nucleotides are complex molecules made of a five-carbon sugar (ribose or deoxyribose), one or more phosphate groups, and an organic nitrogen-containing base called either a purine or pyrimidine (fig. 3.5d). Nucleotides are extremely important as signaling molecules (they carry information between cells, tissues, and organs) and as sources of intracellular energy. They also form long chains called *ribonucleic acid* **(RNA)** or **deoxyribonucleic acid (DNA)** that are essential for

storing and expressing genetic information. Only four kinds of nucleotides (adenine, guanine, cytosine, and thyamine) occur in DNA, but there can be billions of these molecules lined up in a very specific sequence. Groups of three nucleotides (called codons) act as the letters in messages that code for the amino acid sequences in proteins. Long chains of DNA bind together to form a stable double helix (fig. 3.6). These chains separate for replication in preparation for cell division or to express their genetic information during protein synthesis. Molecular biologists have developed techniques for extracting DNA from cells and reading its nucleotide sequence. These techniques are proving to be tremendously powerful in medical genetics and agriculture. They also are extremely

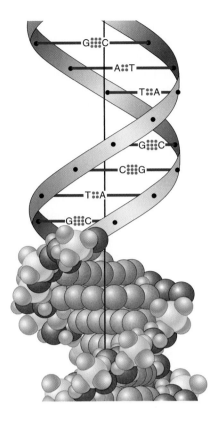

FIGURE 3.6 A composite molecular model of DNA. The lower part shows individual atoms, while the upper part has been simplified to show the strands of the double helix held together by hydrogen bonds (small dots) between matching nucleotides (A, T, G, and C). A complete DNA molecule contains millions of nucleotides and carries genetic information for many specific, inheritable traits.

useful in fields such as forensics and taxonomy. Because every individual has a unique set of DNA molecules, sequencing their nucleotide content can provide a distinctive individual identification. We'll discuss this technology further in chapters 9 and 11.

Cells are the fundamental units of life

All living organisms are composed of **cells,** minute compartments within which the processes of life are carried out (fig. 3.7). Microscopic organisms such as bacteria, some algae, and protozoa are composed of single cells. Most higher organisms are multicellular, usually with many different cell varieties. Your body, for instance, is composed of several trillion cells of about two hundred distinct types. Every cell is surrounded by a thin but dynamic membrane of lipid and protein that receives information about the exterior world and regulates the flow of materials between the cell and its environment. Inside, cells are subdivided into tiny organelles and subcellular particles that provide the machinery for life. Some of these organelles store and release energy. Others manage and distribute information. Still others create the internal structure that gives the cell its shape and allows it to fulfill its role.

All of the chemical reactions required to create these various structures, provide them with energy and materials to carry out their functions, dispose of wastes, and perform other functions of life at the cellular level are carried out by a special class of proteins called **enzymes.** Enzymes are molecular catalysts, that is, they regulate chemical reactions without being used up or inactivated in the process. Think of them as tools: Like hammers or wrenches, they do their jobs without being consumed or damaged as they work. There are generally thousands of different kinds of enzymes in every cell, all necessary to carry out the many processes on which life depends. Altogether, the multitude of enzymatic reactions performed by an organism is called its **metabolism.**

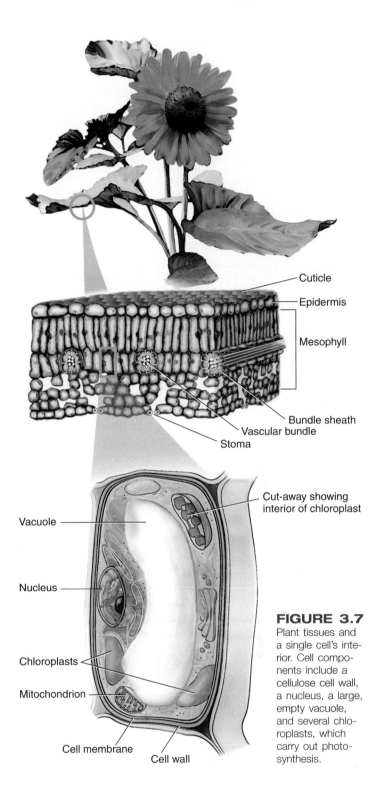

FIGURE 3.7 Plant tissues and a single cell's interior. Cell components include a cellulose cell wall, a nucleus, a large, empty vacuole, and several chloroplasts, which carry out photosynthesis.

3.2 ENERGY

If matter is the material of which things are made, energy provides the force to hold structures together, tear them apart, and move them from one place to another. In this section we will look at some fundamental characteristics of these components of our world.

FIGURE 3.8 Water stored behind this dam represents potential energy. Water flowing over the dam has kinetic energy, some of which is converted to heat.

Energy occurs in different types and qualities

Energy is the ability to do work such as moving matter over a distance or causing a heat transfer between two objects at different temperatures. Energy can take many different forms. Heat, light, electricity, and chemical energy are examples that we all experience. The energy contained in moving objects is called **kinetic energy.** A rock rolling down a hill, the wind blowing through the trees, water flowing over a dam (fig. 3.8), or electrons speeding around the nucleus of an atom are all examples of kinetic energy. **Potential energy** is stored energy that is latent but available for use. A rock poised at the top of a hill and water stored behind a dam are examples of potential energy. **Chemical energy** stored in the food that you eat and the gasoline that you put into your car are also examples of potential energy that can be released to do useful work. Energy is often measured in units of heat (calories) or work (joules). One joule (J) is the work done when one kilogram is accelerated at one meter per second per second. One calorie is the amount of energy needed to heat one gram of pure water one degree Celsius. A calorie can also be measured as 4.184 J.

Heat describes the energy that can be transferred between objects of different temperature. When a substance absorbs heat, the kinetic energy of its molecules increases, or it may change state: A solid may become a liquid, or a liquid may become a gas. We sense change in heat content as change in temperature (unless the substance changes state).

An object can have a high heat content but a low temperature, such as a lake that freezes slowly in the fall. Other objects, like a burning match, have a high temperature but little heat content. Heat storage in lakes and oceans is essential to moderating climates and maintaining biological communities. Heat

absorbed in changing states is also critical. As you will read in chapter 15, evaporation and condensation of water in the atmosphere helps distribute heat around the globe.

Energy that is diffused, dispersed, and low in temperature is considered low-quality energy because it is difficult to gather and use for productive purposes. The heat stored in the oceans, for instance, is immense but hard to capture and use, so it is low quality. Conversely, energy that is intense, concentrated, and high in temperature is high-quality energy because of its usefulness in carrying out work. The intense flames of a very hot fire or high-voltage electrical energy are examples of high-quality forms that are valuable to humans. Many of our alternative energy sources (such as wind) are diffuse compared to the higher-quality, more concentrated chemical energy in oil, coal, or gas.

Think About It

Can you describe one or two practical examples of the laws of physics and thermodynamics in your own life? Do they help explain why you can recycle cans and bottles but not energy? Which law is responsible for the fact that you get hot and sweaty when you exercise?

Thermodynamics regulates energy transfers

Atoms and molecules cycle endlessly through organisms and their environment, but energy flows in a one-way path. A constant supply of energy—nearly all of it from the sun—is needed to keep biological processes running. Energy can be used repeatedly as it flows through the system, and it can be stored temporarily in the chemical bonds of organic molecules, but eventually it is released and dissipated.

The study of thermodynamics deals with how energy is transferred in natural processes. More specifically, it deals with the rates of flow and the transformation of energy from one form or quality to another. Thermodynamics is a complex, quantitative discipline, but you don't need a great deal of math to understand some of the broad principles that shape our world and our lives.

The **first law of thermodynamics** states that energy is *conserved;* that is, it is neither created nor destroyed under normal conditions. Energy may be transformed, for example, from the energy in a chemical bond to heat energy, but the total amount does not change.

The **second law of thermodynamics** states that, with each successive energy transfer or transformation in a system, less energy is available to do work. That is, energy is degraded to lower-quality forms, or it dissipates and is lost, as it is used. When you drive a car, for example, the chemical energy of the gas is degraded to kinetic energy and heat, which dissipates, eventually, to space. The second law recognizes that disorder, or **entropy,** tends to increase in all natural systems. Consequently, there is always less *useful* energy available when you finish a process than there was before you started. Because of this loss, everything in the universe tends to fall apart, slow down, and get more disorganized.

How does the second law of thermodynamics apply to organisms and biological systems? Organisms are highly organized, both structurally and metabolically. Constant care and maintenance is required to keep up this organization, and a constant supply of energy is required to maintain these processes. Every time some energy is used by a cell to do work, some of that energy is dissipated or lost as heat. If cellular energy supplies are interrupted or depleted, the result—sooner or later—is death.

3.3 ENERGY FOR LIFE

Where does the energy needed by living organisms come from? How is it captured and used to do work? For nearly all plants and animals living on the earth's surface, the sun is the ultimate energy source, but for organisms living deep in the earth's crust or at the bottom of the oceans, where sunlight is unavailable, chemicals derived from rocks provide alternate energy sources. We'll consider this alternate energy pathway first because it seems to be more ancient. Before green plants existed, we believe that ancient bacteria-like cells probably lived by processing chemicals in hot springs.

Extremophiles live in severe conditions

Until recently, the deep ocean floor was believed to be a biological desert. Cold, dark, subject to crushing pressures, and without any known energy supply, it was a place where scientists thought nothing could survive. Undersea explorations in the 1970s, however, revealed dense colonies of animals—blind shrimp, giant tubeworms, strange crabs, and bizarre clams—clustered around vents called black chimneys, where boiling hot, mineral-laden water bubbles up through cracks in the earth's crust. But how were these organisms getting energy? The answer is **chemosynthesis,** the process in which inorganic chemicals, such as hydrogen sulfide (H_2S) or hydrogen gas (H_2), provide energy for synthesis of organic molecules.

Discovering organisms living under the severe conditions of deep-sea hydrothermal vents led to an interest in other sites that seem exceptionally harsh to us. A variety of interesting organisms have been discovered in hot springs in thermal areas such as Yellowstone National Park, in intensely salty lakes, and even in deep rock formations (up to 1,500 m, or nearly a mile deep) in Columbia River basalts, for example. Some species are amazingly hardy. The recently described *Pyrolobus fumarii* can withstand temperatures up to 113°C (235°F). Most of these extremophiles are archaea, single-celled organisms intermediate between bacteria or eukaryotic organisms (those with their genetic material enclosed in a nucleus (see fig. 3.7). Archaea are thought to be the most primitive of all living organisms, and the conditions under which they live are thought to be similar to those in which life first evolved.

Deep-sea explorations of areas without thermal vents also have found abundant life (fig. 3.9).We now know that archaea live in oceanic sediments in astonishing numbers. The deepest of these species (they can be 800 m or more below the ocean floor) make methane from gaseous hydrogen (H_2) and carbon dioxide

FIGURE 3.9 A colony of tube worms and mussels clusters over a cool, deep-sea methane seep in the Gulf of Mexico.

(CO_2), derived from rocks. Other species oxidize methane using sulfur to create hydrogen sulfide (H_2S), which is consumed by bacteria that serve as a food source for more complex organisms such as tubeworms. But why should we care about this exotic community? It's estimated that the total mass of microbes (microscopic organisms) living beneath the seafloor represents nearly one-third of all the biomass (organic material) on the planet. Furthermore, the vast supply of methane generated by this community could be either a great resource or a terrible threat to us.

The total amount of methane made by these microbes is probably greater than all the known reserves of coal, gas, and oil. If we could safely extract the huge supplies of methane hydrate in ocean sediments, it could supply our energy needs for hundreds of years. Of greater immediate importance is that if methane-eating microbes weren't intercepting the methane produced by their neighbors, more than 300 million tons per year of this potent greenhouse gas would probably be bubbling to the surface, and we'd have run-away global warming. Some geologists believe that sudden "burps" of methane from melting hydrate deposits may have been responsible for catastrophic landslides and tsunamis, sudden climatic shifts, and mass extinctions in the past. We'll look further at issues of global climate change in chapter 15.

Green plants get energy from the sun

Our sun is a star, a fiery ball of exploding hydrogen gas. Its thermonuclear reactions emit powerful forms of radiation, including potentially deadly ultraviolet and nuclear radiation (fig. 3.10), yet life here is nurtured by, and dependent upon, this searing, energy source. Solar energy is essential to life for two main reasons.

First, the sun provides warmth. Most organisms survive within a relatively narrow temperature range. In fact, each species has its own range of temperatures within which it can function normally. At high temperatures (above 40°C), biomolecules

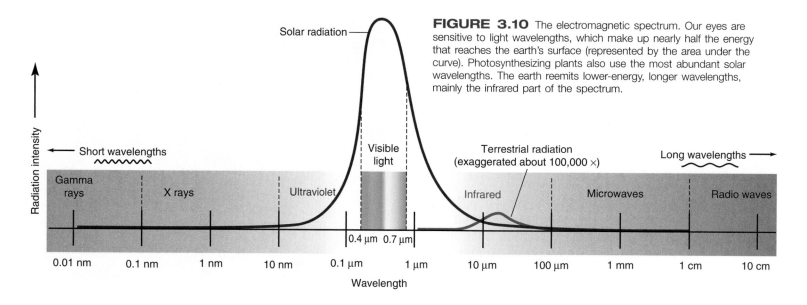

FIGURE 3.10 The electromagnetic spectrum. Our eyes are sensitive to light wavelengths, which make up nearly half the energy that reaches the earth's surface (represented by the area under the curve). Photosynthesizing plants also use the most abundant solar wavelengths. The earth reemits lower-energy, longer wavelengths, mainly the infrared part of the spectrum.

begin to break down or become distorted and nonfunctional. At low temperatures (near 0°C), some chemical reactions of metabolism occur too slowly to enable organisms to grow and reproduce. Other planets in our solar system are either too hot or too cold to support life as we know it. The earth's water and atmosphere help to moderate, maintain, and distribute the sun's heat.

Second, nearly all organisms on the earth's surface organisms depend on solar radiation for life-sustaining energy, which is captured by green plants, algae, and some bacteria in a process called **photosynthesis.** Photosynthesis converts radiant energy into useful, high-quality chemical energy in the bonds that hold together organic molecules.

How much of the available solar energy is actually used by organisms? The amount of incoming solar radiation is enormous, about 1,372 watts/m^2 at the top of the atmosphere (1 watt = 1 J per second). However, more than half of the incoming sunlight may be reflected or absorbed by atmospheric clouds, dust, and gases. In particular, harmful, short wavelengths are filtered out by gases (such as ozone) in the upper atmosphere; thus, the atmosphere is a valuable shield, protecting life-forms from harmful doses of ultraviolet and other forms of radiation. Even with these energy reductions, however, the sun provides much more energy than biological systems can harness, and more than enough for all our energy needs if technology could enable us to tap it efficiently.

Of the solar radiation that does reach the earth's surface, about 10 percent is ultraviolet, 45 percent is visible, and 45 percent is infrared. Most of that energy is absorbed by land or water or is reflected into space by water, snow, and land surfaces. (Seen from outer space, Earth shines about as brightly as Venus.)

Of the energy that reaches the earth's surface, photosynthesis uses only certain wavelengths, mainly red and blue light. (Most plants reflect green wavelengths, so that is the color they appear to us.) Half of the energy plants absorb is used in evaporating water. In the end, only 1 to 2 percent of the sunlight falling on plants is available for photosynthesis. This small percentage represents the energy base for virtually all life in the biosphere!

Photosynthesis captures energy while respiration releases that energy

Photosynthesis occurs in tiny membranous organelles called chloroplasts that reside within plant cells (see fig. 3.7). The most important key to this process is chlorophyll, a unique green molecule that can absorb light energy and use it to create high-energy chemical bonds in compounds that serve as the fuel for all subsequent cellular metabolism. Chlorophyll doesn't do this important job all alone, however. It is assisted by a large group of other lipid, sugar, protein, and nucleotide molecules. Together these components carry out two interconnected cyclic sets of reactions (fig. 3.11).

Photosynthesis begins with a series of steps called light-dependent reactions: These occur only while the chloroplast is receiving light. Enzymes split water molecules and release molecular oxygen (O_2). This is the source of all the oxygen in the atmosphere on which all animals, including you, depend for life. The light-dependent reactions also create mobile, high-energy molecules (adenosine triphosphate, or ATP, and nicotinamide adenine dinucleotide phosphate, or NADPH), which provide energy for the next set of processes, the light-independent reactions. As their name implies, these reactions do not use light directly. Here, enzymes extract energy from ATP and NADPH to add carbon atoms (from carbon dioxide) to simple sugar molecules, such as glucose. These molecules provide the building blocks for larger, more complex organic molecules.

In most temperate-zone plants, photosynthesis can be summarized in the following equation:

$$6H_2O + 6CO_2 + \text{solar energy} \xrightarrow{\text{chlorophyll}} C_6H_{12}O_6 \text{ (sugar)} + 6O_2$$

We read this equation as "water plus carbon dioxide plus energy produces sugar plus oxygen." The reason the equation uses six water and six carbon dioxide molecules is that it takes six carbon atoms to make the sugar product. If you look closely,

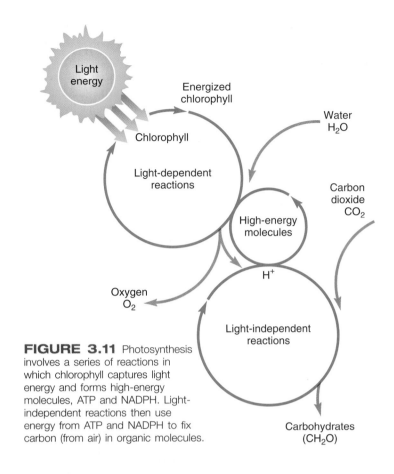

FIGURE 3.11 Photosynthesis involves a series of reactions in which chlorophyll captures light energy and forms high-energy molecules, ATP and NADPH. Light-independent reactions then use energy from ATP and NADPH to fix carbon (from air) in organic molecules.

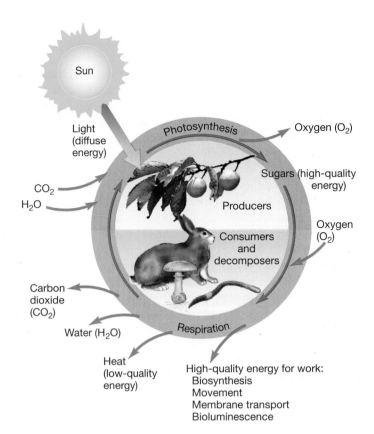

FIGURE 3.12 Energy exchange in ecosystems. Plants use sunlight, water, and carbon dioxide to produce sugars and other organic molecules. Consumers use oxygen and break down sugars during cellular respiration. Plants also carry out respiration, but during the day, if light, water, and CO_2 are available, they have a net production of O_2 and carbohydrates.

you will see that all the atoms in the reactants balance with those in the products. This is an example of conservation of matter.

You might wonder how making a simple sugar benefits the plant. The answer is that glucose is an energy-rich compound that serves as the central, primary fuel for all metabolic processes of cells. The energy in its chemical bonds—the ones created by photosynthesis—can be released by other enzymes and used to make other molecules (lipids, proteins, nucleic acids, or other carbohydrates), or it can drive kinetic processes such as movement of ions across membranes, transmission of messages, changes in cellular shape or structure, or movement of the cell itself in some cases. This process of releasing chemical energy, called **cellular respiration**, involves splitting carbon and hydrogen atoms from the sugar molecule and recombining them with oxygen to re-create carbon dioxide and water. The net chemical reaction, then, is the reverse of photosynthesis:

$$C_6H_{12}O_6 + 6O_2 \longrightarrow 6H_2O + 6CO_2 + \text{released energy}$$

Note that in photosynthesis, energy is *captured*, while in respiration, energy is *released*. Similarly, photosynthesis *consumes* water and carbon dioxide to *produce* sugar and oxygen, while respiration does just the opposite. In both sets of reactions, energy is stored temporarily in chemical bonds, which constitute a kind of energy currency for the cell. Plants carry out both photosynthesis and respiration, but during the day, if light, water, and CO_2 are available, they have a net production of O_2 and carbohydrates.

We animals don't have chlorophyll and can't carry out photosynthetic food production. We do have the components for cellular respiration, however. In fact, this is how we get all our energy for life. We eat plants—or other animals that have eaten plants—and break down the organic molecules in our food through cellular respiration to obtain energy (fig. 3.12). In the process, we also consume oxygen and release carbon dioxide, thus completing the cycle of photosynthesis and respiration. Later in this chapter we will see how these feeding relationships work.

3.4 FROM SPECIES TO ECOSYSTEMS

While cellular and molecular biologists study life processes at the microscopic level, ecologists study interactions at the species, population, biotic community, or ecosystem level. In Latin, *species* literally means *kind*. In biology, **species** refers to all organisms of the same kind that are genetically similar enough to breed in nature and produce live, fertile offspring. There are several qualifications and some important exceptions to this definition of species (especially among bacteria and plants), but for our purposes this is a useful working definition.

Organisms occur in populations, communities, and ecosystems

A **population** consists of all the members of a species living in a given area at the same time. Chapter 6 deals further with population growth and dynamics. All of the populations of organisms living and interacting in a particular area make up a **biological community.** What populations make up the biological community of which you are a part? The population sign marking your city limits announces only the number of humans who live there, disregarding the other populations of animals, plants, fungi, and microorganisms that are part of the biological community within the city's boundaries. Characteristics of biological communities are discussed in more detail in chapter 4.

An ecological system, or **ecosystem,** is composed of a biological community and its physical environment. The environment includes abiotic factors (nonliving components), such as climate, water, minerals, and sunlight, as well as biotic factors, such as organisms, their products (secretions, wastes, and remains), and effects in a given area. As you learned in chapter 2, systems often have properties above and beyond those of the individual components and processes that produce them. It is useful to think about the biological community and its environment together, because energy and matter flow through both. Understanding how those flows work is a major theme in ecology.

For simplicity's sake, we think of ecosystems as fixed ecological units with distinct boundaries. If you look at a patch of woods surrounded by farm fields, for instance, a relatively sharp line separates the two areas, and conditions such as light levels, wind, moisture, and shelter are quite different in the woods than in the fields around them. Because of these variations, distinct populations of plants and animals live in each place. By studying each of these areas, we can make important and interesting discoveries about who lives where and why and about how conditions are established and maintained there.

The division between the fields and woods is not always clear, however. Air, of course, moves freely from one to another, and the runoff after a rainfall may carry soil, leaf litter, and even live organisms between the areas. Birds may feed in the field during the day but roost in the woods at night, giving them roles in both places. Are they members of the woodland community or the field community? Is the edge of the woodland ecosystem where the last tree grows, or does it extend to every place that has an influence on the woods?

As you can see, it may be difficult to draw clear boundaries around communities and ecosystems. To some extent we define these units by what we want to study and how much information we can handle. Thus, an ecosystem might be as large as a whole watershed or as small as a pond or even the surface of your skin. Even though our choices may be somewhat arbitrary, we still can make useful discoveries about how organisms interact with each other and with their environment within these units. The woods are, after all, significantly different from the fields around them.

Like the woodland we just considered, most ecosystems are open, in the sense that they exchange materials and organisms with other ecosystems. A stream ecosystem is an extreme example. Water, nutrients, and organisms enter from upstream and are lost downstream. The species and numbers of organisms present may be relatively constant, but they are made up of continually changing individuals. Other ecosystems are relatively closed, in the sense that very little enters or leaves them. A balanced aquarium is a good example of a closed ecosystem. Aquatic plants, animals, and decomposers can balance material cycles in the aquarium if care is taken to balance their populations. Because of the second law of thermodynamics, however, every ecosystem must have a constant inflow of energy and a way to dispose of heat. Thus, at least with regard to energy flow, every ecosystem is open.

Many ecosystems have mechanisms that maintain composition and functions within certain limits. A forest tends to remain a forest, for the most part, and to have forestlike conditions if it isn't disturbed by outside forces. Some ecologists suggest that ecosystems—or perhaps all life on the earth—may function as superorganisms because they maintain stable conditions and can be resilient to change.

Food chains, food webs, and trophic levels link species

Photosynthesis (and rarely chemosynthesis) is the base of all ecosystems. Organisms that photosynthesize, mainly green plants and algae, are therefore known as **producers.** One of the major properties of an ecosystem is its **productivity,** the amount of **biomass** (biological material) produced in a given area during a given period of time. Photosynthesis is described as *primary productivity* because it is the basis for almost all other growth in an ecosystem. Manufacture of biomass by organisms that eat plants is termed *secondary productivity.* A given ecosystem may have very high total productivity, but if decomposers decompose organic material as rapidly as it is formed, the *net primary productivity* will be low.

Think about what you have eaten today and trace it back to its photosynthetic source. If you have eaten an egg, you can trace it back to a chicken, which ate corn. This is an example of a **food chain,** a linked feeding series. Now think about a more complex food chain involving you, a chicken, a corn plant, and a grasshopper. The chicken could eat grasshoppers that had eaten leaves of the corn plant. You also could eat the grasshopper directly—some humans do. Or you could eat corn yourself, making the shortest possible food chain. Humans have several options of where we fit into food chains.

In ecosystems, some consumers feed on a single species, but most consumers have multiple food sources. Similarly, some species are prey to a single kind of predator, but many species in an ecosystem are beset by several types of predators and parasites. In this way, individual food chains become interconnected to form a **food web.** Figure 3.13 shows feeding relationships among some of the larger organisms in a woodland and lake community. If we were to add all the insects, worms, and microscopic organisms that belong in this picture, however, we would have overwhelming

FIGURE 3.13 Each time an organism feeds, it becomes a link in a food chain. In an ecosystem, food chains become interconnected when predators feed on more than one kind of prey, thus forming a food web. The arrows in this diagram and in figure 3.14 indicate the direction in which matter and energy are transferred through feeding relationships. Only a few representative relationships are shown here. What others might you add?

complexity. Perhaps you can imagine the challenge ecologists face in trying to quantify and interpret the precise matter and energy transfers that occur in a natural ecosystem!

An organism's feeding status in an ecosystem can be expressed as its **trophic level** (from the Greek *trophe*, food). In our first example, the corn plant is at the producer level; it transforms solar energy into chemical energy, producing food molecules. Other organisms in the ecosystem are **consumers** of the chemical energy harnessed by the producers. An organism that eats producers is a primary consumer. An organism that eats primary consumers is a secondary consumer, which may, in turn, be eaten by a tertiary consumer, and so on. Most terrestrial food chains are relatively short (seeds → mouse → owl), but aquatic food chains may be quite long (microscopic algae → copepod → minnow → crayfish → bass → osprey). The length of a food chain also may reflect the physical characteristics of a particular ecosystem. A harsh arctic landscape has a much shorter food chain than a temperate or tropical one (fig. 3.14).

Organisms can be identified both by the trophic level at which they feed and by the *kinds* of food they eat (fig. 3.15). **Herbivores** are plant eaters, **carnivores** are flesh eaters, and **omnivores** eat both plant and animal matter. What are humans? We are natural omnivores, by history and by habit. Tooth structure is an important clue to understanding animal food preferences, and humans are no exception. Our teeth are suited for an omnivorous diet, with a combination of cutting and crushing surfaces that are not highly adapted for one specific kind of food, as are the teeth of a wolf (carnivore) or a horse (herbivore).

Think About It

What would have been the leading primary producers and top consumers in the native ecosystem where you now live? What are they now? Are fewer trophic levels now represented in your ecosystem than in the past?

One of the most important trophic levels is occupied by the many kinds of organisms that remove and recycle the dead bodies and waste products of others. **Scavengers** such as crows, jackals, and vultures clean up dead carcasses of larger animals. **Detritivores** such as ants and beetles consume litter, debris, and dung, while **decomposer** organisms such as fungi and bacteria complete the final breakdown and recycling of organic materials. It could be argued that these microorganisms are second in importance only to producers, because without their activity nutrients would remain locked up in the organic compounds of dead organisms and discarded body wastes, rather than being made available to successive generations of organisms.

FIGURE 3.14 Harsh environments tend to have shorter food chains than environments with more favorable physical conditions. Compare the arctic food chains depicted here with the longer food chains in the food web in figure 3.13.

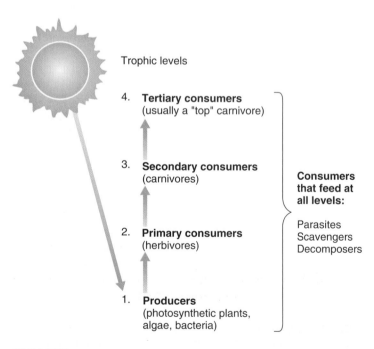

FIGURE 3.15 Organisms in an ecosystem may be identified by how they obtain food for their life processes (producer, herbivore, carnivore, omnivore, scavenger, decomposer, reducer) or by consumer level (producer; primary, secondary, or tertiary consumer) or by trophic level (1st, 2nd, 3rd, 4th).

Ecological pyramids describe trophic levels

If we arrange the organisms in a food chain according to trophic levels, they often form a pyramid with a broad base representing primary producers and only a few individuals in the highest trophic levels. This pyramid arrangement is especially true if we look at the energy content of an ecosystem (fig. 3.16). True to the second principle of thermodynamics, less food

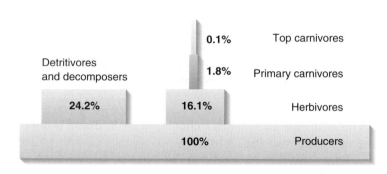

FIGURE 3.16 A classic example of an energy pyramid from Silver Springs, Florida. The numbers in each bar show the percentage of the energy captured in the primary producer level that is incorporated into the biomass of each succeeding level. Detritivores and decomposers feed at every level but are shown attached to the producer bar because this level provides most of their energy.

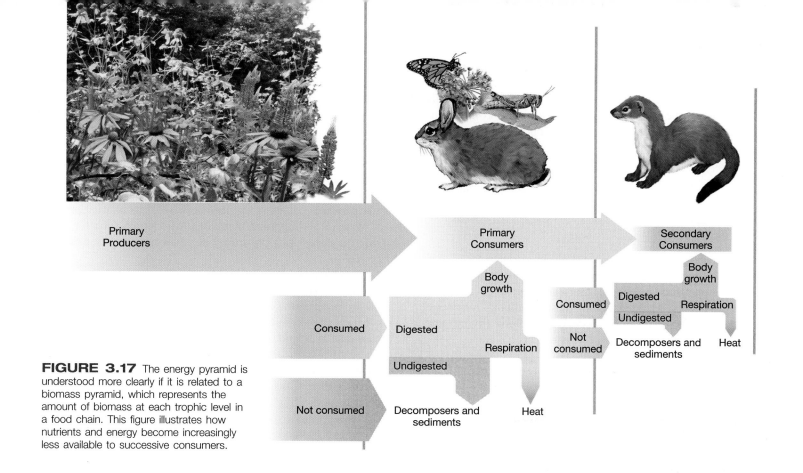

FIGURE 3.17 The energy pyramid is understood more clearly if it is related to a biomass pyramid, which represents the amount of biomass at each trophic level in a food chain. This figure illustrates how nutrients and energy become increasingly less available to successive consumers.

energy is available to the top trophic level than is available to preceding levels. For example, it takes a huge number of plants to support a modest colony of grazers such as prairie dogs. Several colonies of prairie dogs, in turn, might be required to feed a single coyote. And a very large top carnivore like a tiger may need a home range of hundreds of square kilometers to survive.

Why is there so much less energy in each successive level in figure 3.16? In the first place, some of the food that organisms eat is undigested and doesn't provide usable energy. Much of the energy that is absorbed is used in the daily processes of living or lost as heat when it is transformed from one form to another and thus isn't stored as biomass that can be eaten.

Furthermore, predators don't operate at 100 percent efficiency. If there were enough foxes to catch all the rabbits available in the summer when the supply is abundant, there would be too many foxes in the middle of the winter when rabbits are scarce. A general rule of thumb is that only about 10 percent of the energy in one consumer level is represented in the next higher level (fig. 3.17). The amount of energy available is often expressed in biomass. For example, it generally takes about 100 kg of clover to make 10 kg of rabbit and 10 kg of rabbit to make 1 kg of fox.

The total number of organisms and the total amount of biomass in each successive trophic level of an ecosystem also may form pyramids (fig. 3.18) similar to those describing energy content. The relationship between biomass and numbers is not as dependable as energy, however. The biomass pyramid, for

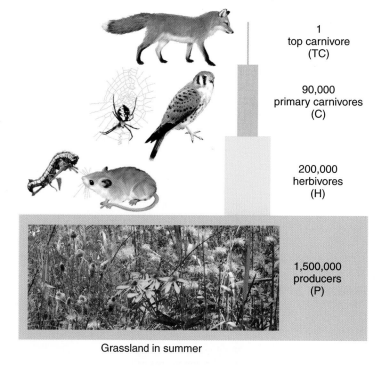

1 top carnivore (TC)

90,000 primary carnivores (C)

200,000 herbivores (H)

1,500,000 producers (P)

Grassland in summer

FIGURE 3.18 Usually, smaller organisms are eaten by larger organisms and it takes numerous small organisms to feed one large organism. The classic study represented in this pyramid shows numbers of individuals at each trophic level per 1,000 m² of grassland, and reads like this: to support one individual at the top carnivore level, there were 90,000 primary carnivores feeding upon 200,000 herbivores that in turn fed upon 1,500,000 producers.

instance, can be inverted by periodic fluctuations in producer populations (for example, low plant and algal biomass present during winter in temperate aquatic ecosystems). The numbers pyramid also can be inverted. One coyote can support numerous tapeworms, for example. Numbers inversion also occurs at the lower trophic levels (for example, one large tree can support thousands of caterpillars).

3.5 MATERIAL CYCLES AND LIFE PROCESSES

To our knowledge, Earth is the only planet in our solar system that provides a suitable environment for life as we know it. Even our nearest planetary neighbors, Mars and Venus, do not meet these requirements. Maintenance of these conditions requires a constant recycling of materials between the biotic (living) and abiotic (nonliving) components of ecosystems.

The hydrologic cycle moves water around the earth

The path of water through our environment, known as the **hydrologic cycle,** is perhaps the most familiar material cycle, and it is discussed in greater detail in chapter 17. Most of the earth's water is stored in the oceans, but solar energy continually evaporates this water, and winds distribute water vapor around the globe. Water that condenses over land surfaces, in the form of rain, snow, or fog, supports all terrestrial

(land-based) ecosystems (fig. 3.19). Living organisms emit the moisture they have consumed through respiration and perspiration. Eventually this moisture reenters the atmosphere or enters lakes and streams, from which it ultimately returns to the ocean again.

As it moves through living things and through the atmosphere, water is responsible for metabolic processes within cells, for maintaining the flows of key nutrients through ecosystems, and for global-scale distribution of heat and energy (chapter 15). Water performs countless services because of its unusual properties. Water is so important that when astronomers look for signs of life on distant planets, traces of water are the key evidence they seek.

Everything about global hydrological processes is awesome in scale. Each year, the sun evaporates approximately 496,000 km^3 of water from the earth's surface. More water evaporates in the tropics than at higher latitudes, and more water evaporates over the oceans than over land. Although the oceans cover about 70 percent of the earth's surface, they account for 86 percent of total evaporation. Ninety percent of the water evaporated from the ocean falls back on the ocean as rain. The remaining 10 percent is carried by prevailing winds over the continents where it combines with water evaporated from soil, plant surfaces, lakes, streams, and wetlands to provide a total continental precipitation of about 111,000 km^3.

What happens to the surplus water on land—the difference between what falls as precipitation and what evaporates? Some of it is incorporated by plants and animals into biological tissues. A large share of what falls on land seeps into the ground to be

FIGURE 3.19
The hydrologic cycle. Most exchange occurs with evaporation from oceans and precipitation back to oceans. About one-tenth of water evaporated from oceans falls over land, is recycled through terrestrial systems, and eventually drains back to oceans in rivers.

Movement of moist air from ocean to land 40,000 km^3

Precipitation over land 111,000 km^3

Transpiration from vegetation 41,000 km^3

Precipitation over ocean 385,000 km^3

Evaporation from soil, streams, rivers and lakes 30,000 km^3

Evaporation from ocean 425,000 km^3

Runoff 40,000 km^3

Percolation through porous rock and soil to groundwater

stored for a while (from a few days to many thousands of years) as soil moisture or groundwater. Eventually, all the water makes its way back downhill to the oceans. The 40,000 km^3 carried back to the ocean each year by surface runoff or underground flow represents the renewable supply available for human uses and sustaining freshwater-dependent ecosystems.

Carbon moves through the carbon cycle

Carbon serves a dual purpose for organisms: (1) it is a structural component of organic molecules, and (2) the energy-holding chemical bonds it forms represent energy "storage." The **carbon cycle** begins with the intake of carbon dioxide (CO_2) by photosynthetic organisms (fig. 3.20). Carbon (and hydrogen and oxygen) atoms are incorporated into sugar molecules during photosynthesis. Carbon dioxide is eventually released during respiration, closing the cycle. The carbon cycle is of special interest because biological accumulation and release of carbon is a major factor in climate regulation (Exploring Science, p. 68).

The path followed by an individual carbon atom in this cycle may be quite direct and rapid, depending on how it is used in an organism's body. Imagine for a moment what happens to a simple sugar molecule you swallow in a glass of fruit juice. The sugar molecule is absorbed into your bloodstream where it is made available to your cells for cellular respiration or for making more complex biomolecules. If it is used in respiration, you may exhale the same carbon atom as CO_2 the same day.

Can you think of examples where carbon may not be recycled for even longer periods of time, if ever? Coal and oil are the compressed, chemically altered remains of plants or microorganisms that lived millions of years ago. Their carbon atoms (and hydrogen, oxygen, nitrogen, sulfur, etc.) are not released until the coal and oil are burned. Enormous amounts of carbon also are locked up as calcium carbonate ($CaCO_3$), used to build shells and skeletons of marine organisms from tiny protozoans to corals. Most of these deposits are at the bottom of the oceans. The world's extensive surface limestone deposits are biologically formed calcium carbonate from ancient oceans, exposed by geological events. The carbon in limestone has been locked away for millennia, which is probably the fate of carbon currently being deposited in ocean sediments. Eventually, even the deep ocean deposits are recycled as they are drawn into deep molten layers and released via volcanic activity. Geologists estimate that every carbon atom on the earth has made about thirty such round trips over the last 4 billion years.

How does tying up so much carbon in the bodies and by-products of organisms affect the biosphere? Favorably. It helps balance CO_2 generation and utilization. Carbon dioxide is one of the so-called greenhouse gases because it absorbs heat radiated from the earth's surface, retaining it instead in the atmosphere. This phenomenon is discussed in more detail in chapter 15. Photosynthesis and deposition of $CaCO_3$ remove atmospheric carbon dioxide; therefore, vegetation (especially large forested areas such as the boreal forests) and the oceans are very important **carbon sinks** (storage deposits). Cellular respiration and combustion both release CO_2, so they are referred to as carbon sources of the cycle.

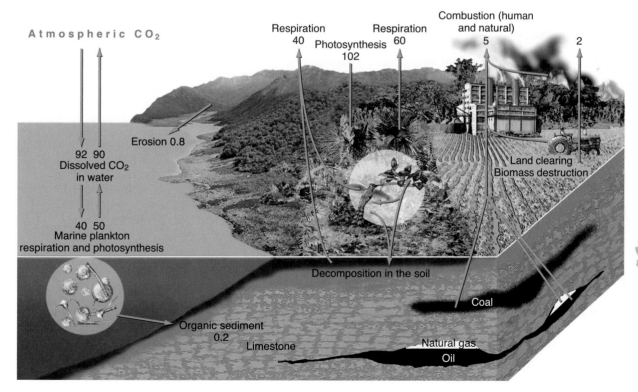

FIGURE 3.20
The carbon cycle. Numbers indicate approximate exchange of carbon in gigatons (Gt) per year. Natural exchanges are balanced, but human sources produce a net increase of CO_2 in the atmosphere.

Measuring primary productivity is important for understanding individual plants and local environments. Understanding rates of primary productivity is also key to understanding global processes, material cycling, and biological activity:

- In global carbon cycles, how much carbon is stored by plants, how quickly is it stored, and how does carbon storage compare in contrasting environments, such as the Arctic and the tropics?
- How does this carbon storage affect global climates (chapter 15)?
- In global nutrient cycles, how much nitrogen and phosphorus wash offshore, and where?

How can environmental scientists measure primary production (photosynthesis) at a global scale? In small-scale systems, you can simply collect all the biomass and weigh it. But that method is impossible for large ecosystems, especially for oceans, which cover 70 percent of the earth's surface. One of the newest methods of quantifying biological productivity involves remote sensing, or data collected from satellite sensors that observe the energy reflected from the earth's surface.

As you have read in this chapter, chlorophyll in green plants *absorbs* red and blue wavelengths of light and *reflects* green wavelengths. Your eye receives, or senses, these green wavelengths. A white-sand beach, on the other hand, reflects approximately equal amounts of all light wavelengths (see fig. 3.10) that reach it from the sun, so it looks white (and bright) to your eye. In a similar way, different surfaces of the earth reflect characteristic wavelengths. Snow-covered surfaces reflect light wavelengths; dark-green forests with abundant chlorophyll-rich leaves—and ocean surfaces rich in photosynthetic algae and plants—reflect greens and near-infrared wavelengths. Dry, brown forests with little active chlorophyll reflect more red

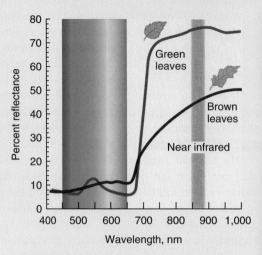

FIGURE 1 Energy wavelengths reflected by green and brown leaves.

and less infrared energy than do dark-green forests (fig. 1).

To detect land-cover patterns on the earth's surface, we can put a sensor on a satellite that orbits the earth. As the satellite travels, the sensor receives and transmits to earth a series of "snapshots." One of the best known earth- imaging satellites, *Landsat 7,* produces images that cover an area 185 km (115 mi) wide, and each pixel represents an area of just 30 × 30 m on the ground. *Landsat* orbits approximately from pole to pole, so as the earth spins below the satellite, it captures images of the entire surface every 16 days. Another satellite, *SeaWIFS,* was designed mainly for monitoring biological activity in oceans (fig. 2). *SeaWIFS* follows a path similar to *Landsat*'s but it revisits each point on the earth every day and produces images with a pixel resolution of just over 1 km.

Since satellites detect a much greater range of wavelengths than our eyes can, they are able to monitor and map chlorophyll abundance. In oceans, this is a useful measure of

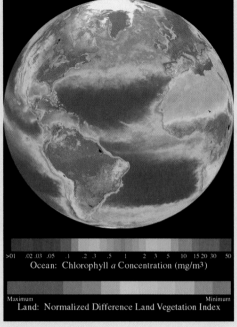

Ocean: Chlorophyll *a* Concentration (mg/m³)

Land: Normalized Difference Land Vegetation Index

FIGURE 2 *SeaWIFS* image showing chlorophyll abundance in oceans and plant growth on land (normalized difference vegetation index).

ecosystem health, as well as carbon dioxide uptake. By quantifying and mapping primary production in oceans, climatologists are working to estimate the role of ocean ecosystems in moderating climate change: for example, they can estimate the extent of biomass production in the cold, oxygen-rich waters of the North Atlantic (fig. 2). Oceanographers can also detect near-shore areas where nutrients washing off the land surface fertilize marine ecosystems and stimulate high productivity, such as near the mouth of the Amazon or Mississippi Rivers. Monitoring and mapping these patterns helps us estimate human impacts on nutrient flows (figs. 3.21, 3.23) from land to sea.

Presently, natural fires and human-created combustion of organic fuels (mainly wood, coal, and petroleum products) release huge quantities of CO_2 at rates that seem to be surpassing the pace of CO_2 removal. Scientific concerns over the linked problems of increased atmospheric CO_2 concentrations, massive deforestation, and reduced productivity of the oceans due to pollution are discussed in chapters 15 and 16.

Nitrogen moves via the nitrogen cycle

As the opening case study of this chapter shows, nitrogen often is one of the most important limiting factors in ecosystems. The complex interrelationships through which organisms exchange this vital element help shape these biological communities. Organisms cannot exist without amino acids, peptides, nucleic

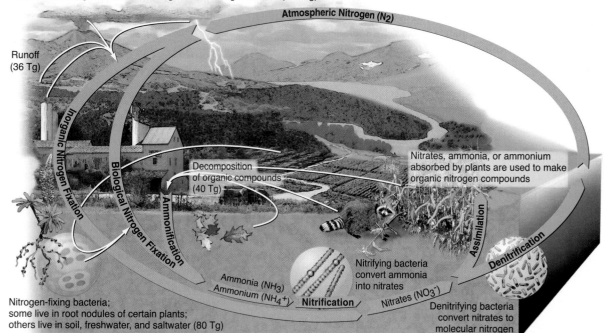

FIGURE 3.21
The nitrogen cycle. Human sources of nitrogen fixation (conversion of molecular nitrogen to ammonia or ammonium) are now about 50 percent greater than natural sources. Bacteria convert ammonia to nitrates, which plants use to create organic nitrogen. Eventually, nitrogen is stored in sediments or converted back to molecular nitrogen (1 Tg = 10^{12} g).

Lightning and volcanoes (10 Tg)
Fossil fuel burning, commercial and agricultural nitrogen fixation (140 Tg)

Atmospheric Nitrogen (N₂)

Runoff (36 Tg)

Inorganic Nitrogen Fixation

Biological Nitrogen Fixation

Ammonification

Decomposition of organic compounds (40 Tg)

Nitrates, ammonia, or ammonium absorbed by plants are used to make organic nitrogen compounds

Assimilation

Denitrification

Nitrifying bacteria convert ammonia into nitrates

Ammonia (NH₃)
Ammonium (NH₄⁺)

Nitrification

Nitrates (NO₃⁻)

Denitrifying bacteria convert nitrates to molecular nitrogen

Nitrogen-fixing bacteria; some live in root nodules of certain plants; others live in soil, freshwater, and saltwater (80 Tg)

acids, and proteins, all of which are organic molecules containing nitrogen. The nitrogen atoms that form these important molecules are provided by producer organisms. Plants assimilate (take up) inorganic nitrogen from the environment and use it to build their own protein molecules, which are eaten by consumer organisms, digested, and used to build their bodies. Even though nitrogen is the most abundant gas (about 78 percent of the atmosphere), however, plants cannot use N_2, the stable diatomic (2-atom) molecule in the air.

Where and how, then, do green plants get *their* nitrogen? The answer lies in the most complex of the gaseous cycles, the **nitrogen cycle.** Figure 3.21 summarizes the nitrogen cycle. The key natural processes that make nitrogen available are carried out by nitrogen-fixing bacteria (including some blue-green algae or cyanobacteria). These organisms have a highly specialized ability to "fix" nitrogen, meaning they change it to less mobile, more useful forms by combining it with hydrogen to make ammonia (NH_3).

Nitrite-forming bacteria combine the ammonia with oxygen, forming nitrites, which have the ionic form NO_2^-. Another group of bacteria then convert nitrites to nitrates, which have the ionic form NO_3^-, that can be absorbed and used by green plants. After nitrates have been absorbed into plant cells, they are reduced to ammonium (NH_4^+), which is used to build amino acids that become the building blocks for peptides and proteins.

Members of the bean family (legumes) and a few other kinds of plants are especially useful in agriculture because they have nitrogen-fixing bacteria actually living *in* their root tissues (fig. 3.22). Legumes and their associated bacteria enrich the soil, so interplanting and rotating legumes with

crops such as corn that use but cannot replace soil nitrates are beneficial farming practices that take practical advantage of this relationship.

Nitrogen reenters the environment in several ways. The most obvious path is through the death of organisms. Their bodies are decomposed by fungi and bacteria, releasing ammonia and

FIGURE 3.22 The roots of this adzuki bean plant are covered with bumps called nodules. Each nodule is a mass of root tissue containing many bacteria that help to convert nitrogen in the soil to a form the bean plants can assimilate and use to manufacture amino acids.

ammonium ions, which then are available for nitrate formation. Organisms don't have to die to donate proteins to the environment, however. Plants shed their leaves, needles, flowers, fruits, and cones; animals shed hair, feathers, skin, exoskeletons, pupal cases, and silk. Animals also produce excrement and urinary wastes that contain nitrogenous compounds. Urinary wastes are especially high in nitrogen because they contain the detoxified wastes of protein metabolism. All of these by-products of living organisms decompose, replenishing soil fertility.

How does nitrogen reenter the atmosphere, completing the cycle? Denitrifying bacteria break down nitrates into N_2 and nitrous oxide (N_2O), gases that return to the atmosphere; thus, denitrifying bacteria compete with plant roots for available nitrates. However, denitrification occurs mainly in waterlogged soils that have low oxygen availability and a high amount of decomposable organic matter. These are suitable growing conditions for many wild plant species in swamps and marshes, but not for most cultivated crop species, except for rice, a domesticated wetland grass.

In recent years, humans have profoundly altered the nitrogen cycle. By using synthetic fertilizers, cultivating nitrogen-fixing crops, and burning fossil fuels, we have more than doubled the amount of nitrogen cycled through our global environment. This excess nitrogen input is causing serious loss of soil nutrients such as calcium and potassium, acidification of rivers and lakes, and rising atmospheric concentrations of the greenhouse gas, nitrous oxide. It also encourages the spread of weeds into areas such as prairies occupied by native plants adapted to nitrogen-poor environments. In coastal areas, blooms of toxic algae and dinoflagellates result from excess nitrogen carried by rivers from farmlands and cities.

Phosphorus is an essential nutrient

Minerals become available to organisms after they are released from rocks. Two mineral cycles of particular significance to organisms are phosphorus and sulfur. Why do you suppose phosphorus is a primary ingredient in fertilizers? At the cellular level, energy-rich, phosphorus-containing compounds are primary participants in energy-transfer reactions, as we have discussed. The amount of available phosphorus in an environment can, therefore, have a dramatic effect on productivity. Abundant phosphorus stimulates lush plant and algal growth, making it a major contributor to water pollution.

The **phosphorus cycle** (fig. 3.23) begins when phosphorus compounds are leached from rocks and minerals over long periods of time. Because phosphorus has no atmospheric form, it is usually transported in aqueous form. Inorganic phosphorus is taken in by producer organisms, incorporated into organic molecules, and then passed on to consumers. It is returned to the environment by decomposition. An important aspect of the phosphorus cycle is the very long time it takes for phosphorus atoms to pass through it. Deep sediments of the oceans are significant phosphorus sinks of extreme longevity. Phosphate ores that now are mined to make detergents

FIGURE 3.23 The phosphorus cycle. Natural movement of phosphorus is slight, involving recycling within ecosytems and some erosion and sedimentation of phosphorus-bearing rock. Use of phosphate (PO_4^{-3}) fertilizers and cleaning agents increases phosphorus in aquatic systems, causing eutrophication. Units are teragrams (Tg) phosphorus per year.

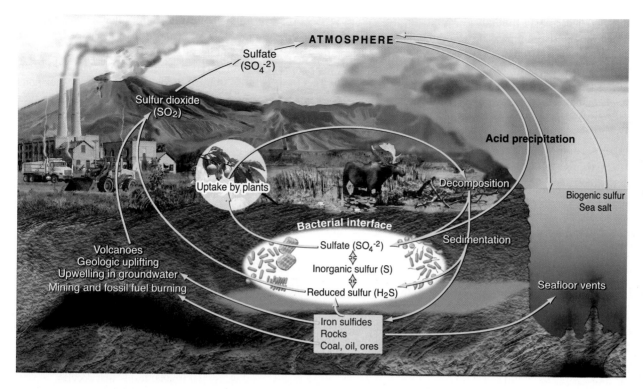

ATMOSPHERE

Sulfate (SO$_4^{-2}$)

Sulfur dioxide (SO$_2$)

Acid precipitation

Uptake by plants

Decomposition

Biogenic sulfur
Sea salt

Volcanoes
Geologic uplifting
Upwelling in groundwater
Mining and fossil fuel burning

Bacterial interface

Sulfate (SO$_4^{-2}$)

Inorganic sulfur (S)

Reduced sulfur (H$_2$S)

Sedimentation

Seafloor vents

Iron sulfides
Rocks
Coal, oil, ores

FIGURE 3.24
The sulfur cycle. Sulfur is present mainly in rocks, soil, and water. It cycles through ecosystems when it is taken in by organisms. Combustion of fossil fuels causes increased levels of atmospheric sulfur compounds, which create problems related to acid precipitation.

and inorganic fertilizers represent exposed ocean sediments that are millennia old. You could think of our present use of phosphates, which are washed out into the river systems and eventually the oceans, as an accelerated mobilization of phosphorus from source to sink. Aquatic ecosystems often are dramatically affected in the process because excess phosphates can stimulate explosive growth of algae and photosynthetic bacteria populations, upsetting ecosystem stability. Notice also that in this cycle, as in the others, the role of organisms is only one part of a larger picture.

Sulfur also cycles

Sulfur plays a vital role in organisms, especially as a minor but essential component of proteins. Sulfur compounds are important determinants of the acidity of rainfall, surface water, and soil. In addition, sulfur in particles and tiny air-borne droplets may act as critical regulators of global climate. Most of the earth's sulfur is tied up underground in rocks and minerals such as iron disulfide (pyrite) or calcium sulfate (gypsum). This inorganic sulfur

is released into air and water by weathering, emissions from deep seafloor vents, and by volcanic eruptions (fig. 3.24).

The **sulfur cycle** is complicated by the large number of oxidation states the element can assume, including hydrogen sulfide (H$_2$S), sulfur dioxide (SO$_2$), sulfate ion (SO$_4^{-2}$), and sulfur, among others. Inorganic processes are responsible for many of these transformations, but living organisms, especially bacteria, also sequester sulfur in biogenic deposits or release it into the environment. Which of the several kinds of sulfur bacteria prevail in any given situation depends on oxygen concentrations, pH, and light levels.

Human activities also release large quantities of sulfur, primarily through burning fossil fuels. Total yearly anthropogenic sulfur emissions rival those of natural processes, and acid rain caused by sulfuric acid produced as a result of fossil fuel use is a serious problem in many areas (see chapter 16). Sulfur dioxide and sulfate aerosols cause human health problems, damage buildings and vegetation, and reduce visibility. They also absorb UV radiation and create cloud cover that cools cities and may be offsetting greenhouse effects of rising CO$_2$ concentrations.

CONCLUSION

Matter is conserved as it cycles over and over through ecosystems, but energy is always degraded or dissipated as it is transformed or transferred from one place to another. These laws of physics and thermodynamics mean that elements are continuously recycled, but that living systems need a constant supply of external energy to replace that lost to entropy. Some extremophiles, living in harsh conditions, such as hot springs or the

bottom of the ocean, capture energy from chemical reactions. For most organisms, however, the ultimate source of energy is the sun. Plants capture sunlight through the process of photosynthesis, and use the captured energy for metabolic processes and to build biomass (organic material). Herbivores eat plants to obtain energy and nutrients, carnivores eat herbivores or each other, and decomposers eat the waste products of this food web.

This dependence on solar energy is a fundamental limit for most life on earth. It's estimated that humans now dominate roughly 40 percent of the potential terrestrial net productivity. We directly eat only about 10 percent of that total (mainly because of the thermodynamic limits on energy transfers in food webs), but the crops and livestock that feed, clothe, and house us represent the rest of that photosynthetic output. By dominating nature, as we do, we exclude other species.

While energy flows in a complex, but ultimately one-way path through nature, materials are endlessly recycled. Five of the major material cycles (water, carbon, nitrogen, phosphorus, and sulfur) are summarized in this chapter. Each of these materials is critically important to living organisms. As humans interfere with these material cycles, we make it easier for some organisms to survive and more difficult for others. Often, we're intent on manipulating material cycles for our own short-term gain, but we don't think about the consequences for other species or even for ourselves in the long-term. An example of that is the carbon cycle. Our lives are made easier and more comfortable by burning fossil fuels, but in doing so we release carbon dioxide into the atmosphere, causing global warming that could have disastrous results. Clearly, it's important to understand these environmental systems and to take them into account in our public policy.

Reviewing Learning Outcomes

By now you should be able to explain the following points:

3.1 Describe matter, atoms, and molecules and give simple examples of the role of four major kinds of organic compounds in living cells.

- Matter is made of atoms, molecules, and compounds.
- Chemical bonds hold molecules together.
- Electrical charge is an important characteristic.
- Organic compounds have a carbon backbone.
- Cells are the fundamental units of life.

3.2 Define energy and explain how thermodynamics regulates ecosystems.

- Energy occurs in different types and qualities.
- Thermodynamics regulates energy transfers.

3.3 Understand how living organisms capture energy and create organic compounds.

- Extremophiles live in severe conditions.

- Green plants get energy from the sun.
- Photosynthesis captures energy while respiration releases that energy.

3.4 Define species, populations, communities, and ecosystems, and summarize the ecological significance of trophic levels.

- Organisms occur in populations, communities, and ecosystems.
- Food chains, food webs, and trophic levels link species.
- Ecological pyramids describe trophic levels.

3.5 Compare the ways that water, carbon, nitrogen, sulfur, and phosphorus cycle within ecosystems.

- The hydrologic cycle moves water around the earth.
- Carbon moves through the carbon cycle.
- Nitrogen moves via the nitrogen cycle.
- Phosphorus is an essential nutrient.
- Remote sensing allows us to evaluate photosynthesis and material cycles.
- Sulfur also cycles.

Practice Quiz

1. Define *atom* and *element*. Are these terms interchangeable?

2. Your body contains vast numbers of carbon atoms. How is it possible that some of these carbon atoms may have been part of the body of a prehistoric creature?

3. What are six characteristics of water that make it so valuable for living organisms and their environment?

4. In the biosphere, matter follows a circular pathway while energy follows a linear pathway. Explain.

5. The oceans store a vast amount of heat, but (except for climate moderation) this huge reservoir of energy is of little use to humans. Explain the difference between high-quality and low-quality energy.

6. Ecosystems require energy to function. Where does this energy come from? Where does it go? How does the flow of energy conform to the laws of thermodynamics?

7. Heat is released during metabolism. How is this heat useful to a cell and to a multicellular organism? How might it be detrimental, especially in a large, complex organism?

8. Photosynthesis and cellular respiration are complementary processes. Explain how they exemplify the laws of conservation of matter and thermodynamics.

9. What do we mean by carbon-fixation or nitrogen-fixation? Why is it important to humans that carbon and nitrogen be "fixed"?

10. The population density of large carnivores is always very small compared to the population density of herbivores occupying the same ecosystem. Explain this in relation to the concept of an ecological pyramid.

11. A species is a specific kind of organism. What general characteristics do individuals of a particular species share? Why is it important for ecologists to differentiate among the various species in a biological community?

CRITICAL THINKING AND DISCUSSION QUESTIONS

1. A few years ago, laundry detergent makers were forced to reduce or eliminate phosphorus. Other cleaning agents (such as dishwasher detergents) still contain substantial amounts of phosphorus. What information would make you change your use of nitrogen, phosphorus, or other useful pollutants?

2. The first law of thermodynamics is sometimes summarized as "you can't get something for nothing." The second law is summarized as "you can't even break even." Explain what these phrases mean. Is it dangerous to oversimplify these important concepts?

3. The ecosystem concept revolutionized ecology by introducing holistic systems thinking as opposed to individualistic life history studies. Why was this a conceptual breakthrough?

4. If ecosystems are so difficult to delimit, why is this such a persistent concept? Can you imagine any other ways to define or delimit environmental investigation?

5. The properties of water are so unique and so essential for life as we know it that some people believe it proves that our planet was intentionally designed for our existence. What would an environmental scientist say about this belief?

6. Choose one of the material cycles (carbon, nitrogen, phosphorus, or sulfur) and identify the components of the cycle in which you participate. For which of these components would it be easiest to reduce your impacts?

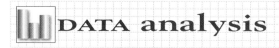 **DATA analysis** **Extracting Data from a Graph**

1. How can you extract data from a line plot? The process is simply the reverse of the way you just learned to create the graph. You draw lines from the axes to where they intersect on the graph at the point whose value you want to know. Let's look at figure 1, the line plot we examined in chapter 1.

- How many cells were there at the half point of the growth curve?

- How does that result compare to the cell population three hours earlier?

- When did the growth of this population start to slow?

  A curve, such as this, that increases slowly at first but then rapidly accelerates is called a logistic curve. We'll discuss the mathematics that create this pattern in chapter 6.

2. How do you extract data from a bar graph? Look again at the bar graph in figure 2. Using the same procedure described for a line plot, draw a horizontal line from the top of any bar straight to the Y-axis. Read the value for the frequency of that category on the Y-axis.

- Why don't you need to draw a vertical line with this type of graph?

- How many cities have 30 ug/m^3?

- Why aren't there any bars between 50 and 65 ug/m^3?

- Why are there so many more cities at 40 ug/m^3?

- Why is one city at 70 ug/m^3?

- How many more cities have 40 ug/m^3 than 30 ug/m^3?

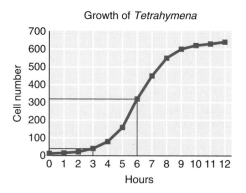

FIGURE 1 Growth of *Tetrahymena*.

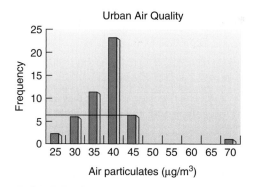

FIGURE 2 Urban air quality.

For Additional Help in Studying This Chapter, please visit our website at www.mhhe.com/cunningham10e. You will find additional practice quizzes and case studies, flashcards, regional examples, place markers for Google Earth™ mapping, and an extensive reading list, all of which will help you learn environmental science.

The relatively young and barren volcanic islands of the Galápagos, isolated from South America by strong, cold currents and high winds, have developed a remarkable community of unique plants and animals.

C H A P T E R

Evolution, Biological Communities, and Species Interactions

4

Any species of bug is an irreplaceable marvel, equal to the works of art which we religiously preserve in our museums.

—Claude Levi-Strauss—

LEARNING OUTCOMES

After studying this chapter, you should be able to:

4.1 Describe how evolution produces species diversity.

4.2 Discuss how species interactions shape biological communities.

4.3 Summarize how community properties affect species and populations.

4.4 Explain why communities are dynamic and change over time.

Case Study   Darwin's Voyage of Discovery

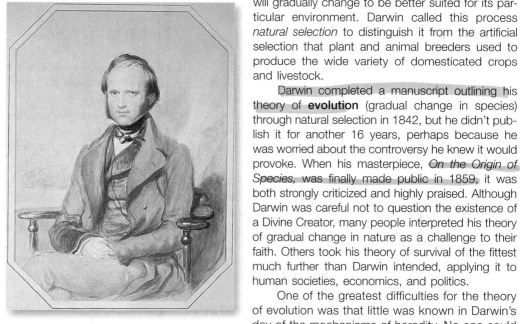

Charles Darwin was only 22 years old when he set out in 1831 on his epic five-year, around-the-world voyage aboard the H.M.S. *Beagle* (fig. 4.1). It was to be the adventure of a lifetime, and would lead to insights that would revolutionize the field of biology. Initially an indifferent student, Darwin had found inspiring professors in his last years of college. One of them helped him get a position as an unpaid naturalist on board the *Beagle.* Darwin turned out to be a perceptive observer, an avid collector of specimens, and an extraordinary scientist.

As the *Beagle* sailed slowly along the coast of South America, mapping coastlines and navigational routes, Darwin had time to go ashore on long field trips to explore natural history. He was amazed by the tropical forests of Brazil and the fossils of huge, extinct mammals in Patagonia. He puzzled over the fact that many fossils looked similar, but not quite identical, to contemporary animals. Could species change over time? In Darwin's day, most people believed that everything in the world was exactly as it had been created by God only a few thousand years earlier. But Darwin had read the work of Charles Lyell (1797–1875), who suggested that the world was much older than previously thought, and capable of undergoing gradual, but profound, change over time.

After four years of exploring and mapping, Darwin and the *Beagle* reached the Galápagos Islands, 900 km (540 mi) off the coast of Ecuador. The harsh, volcanic landscape of these remote islands (see opposite page) held an extraordinary assemblage of unique plants and animals. Giant land tortoises fed on tree-size cacti. Sea-going iguanas scraped algae off underwater shoals. Sea birds were so unafraid of humans that Darwin could pick them off their nests. The many finches were especially interesting: Every island had its own species, marked by distinct bill shapes, which graded from large and parrot-like to small and warbler-like. Each bird's anatomy and behavior was suited to exploit specific food sources available in its habitat. It seemed obvious that these birds were related, but somehow had been modified to survive under different conditions.

Darwin didn't immediately understand the significance of these observations. Upon returning to England, he began the long process of cataloging and describing the specimens he had collected. Over the next 40 years, he wrote important books on a variety of topics including the formation of oceanic islands from coral reefs, the geology of South America, and the classification and natural history of barnacles. Throughout this time, he puzzled about how organisms might adapt to specific environmental situations.

A key in his understanding was Thomas Malthus's *Essay on the Principle of Population* (1798). From Malthus, Darwin saw that most organisms have the potential to produce far more offspring than can actually survive. Those individuals with superior attributes are more likely to live and reproduce than those less well-endowed. Because the more fit individuals are especially successful in passing along their favorable traits to their offspring, the whole population will gradually change to be better suited for its particular environment. Darwin called this process *natural selection* to distinguish it from the artificial selection that plant and animal breeders used to produce the wide variety of domesticated crops and livestock.

Darwin completed a manuscript outlining his theory of **evolution** (gradual change in species) through natural selection in 1842, but he didn't publish it for another 16 years, perhaps because he was worried about the controversy he knew it would provoke. When his masterpiece, *On the Origin of Species,* was finally made public in 1859, it was both strongly criticized and highly praised. Although Darwin was careful not to question the existence of a Divine Creator, many people interpreted his theory of gradual change in nature as a challenge to their faith. Others took his theory of survival of the fittest much further than Darwin intended, applying it to human societies, economics, and politics.

One of the greatest difficulties for the theory of evolution was that little was known in Darwin's day of the mechanisms of heredity. No one could explain how genetic variation could arise in a natural population, or how inheritable traits could be sorted and recombined in offspring. It took nearly another century before biologists could use their understanding of molecular genetics to put together a modern synthesis of evolution that clarifies these details.

An overwhelming majority of biologists now consider the theory of evolution through natural selection to be the cornerstone of their science. The theory explains how the characteristics of organisms have arisen from individual molecules, to cellular structures, to tissues and organs, to complex behaviors and population traits. In this chapter, we'll look at the evidence for evolution and how it shapes species and biological communities. We'll examine the ways in which interactions between species and between organisms and their environment allow species to adapt to particular conditions as well as to modify both their habitat and their competitors.

FIGURE 4.1 Charles Darwin, in a portrait painted shortly after the voyage on the *Beagle.*

For more information, see

Darwin, Charles. *The Voyage of the* Beagle (1837) and *On the Origin of Species* (1859).

Quammen, David. *1996. The Song of the Dodo: Island Biogeography in an Age of Extinctions.* Scribners.

4.1 EVOLUTION PRODUCES SPECIES DIVERSITY

Why do some species live in one place but not another? A more important question to environmental scientists is, what are the mechanisms that promote the great variety of species on earth and that determine which species will survive in one environment but not another? In this section you will come to understand (1) concepts behind the theory of speciation by means of natural selection and adaptation (evolution); (2) the characteristics of species that make some of them weedy and others endangered; and (3) the limitations species face in their environments and implications for their survival. First we'll start with the basics: How do species arise?

Natural selection leads to evolution

How does a polar bear stand the long, sunless, super-cold arctic winter? How does the saguaro cactus survive blistering temperatures and extreme dryness of the desert? We commonly say that each species is *adapted* to the environment where it lives, but what does that mean? **Adaptation, the acquisition of traits that allow a species to survive in its environment,** is one of the most important concepts in biology.

We use the term *adapt* in two ways. An individual organism can respond immediately to a changing environment in a process called acclimation. If you keep a houseplant indoors all winter and then put it out in full sunlight in the spring, the leaves become damaged. If the damage isn't severe, your plant may grow new leaves with thicker cuticles and denser pigments that block the sun's rays. However, the change isn't permanent. After another winter inside, it will still get sun-scald in the following spring. The leaf changes are not permanent and cannot be passed on to offspring, or even carried over from the previous year. Although the capacity to acclimate is inherited, houseplants in each generation must develop their own protective leaf epidermis.

Another type of adaptation affects populations consisting of many individuals. Genetic traits are passed from generation to generation and allow a species to live more successfully in its environment. As the opening case study for this chapter shows, this process of adaptation to environment is explained by the theory of evolution. The basic idea of evolution is that species change over generations because individuals compete for scarce resources. Better competitors in a population survive—they have greater reproductive potential or fitness—and their offspring inherit the beneficial traits. Over generations, those traits become common in a population (fig. 4.2). The process of better-selected individuals passing their traits to the next generation is called **natural selection.** The traits are encoded in a species' DNA, but from where does the original DNA coding come, which then gives some individuals greater fitness? Every organism has a dizzying array of genetic diversity in its DNA. It has been demonstrated in experiments and by observing natural populations that changes to the DNA coding sequence of individuals occurs, and that the changed sequences are inherited by offspring. Exposure to ionizing radiation and toxic mate-

FIGURE 4.2 Giraffes don't have long necks because they stretch to reach tree-top leaves, but those giraffes that happened to have longer necks got more food and had more offspring, so the trait became fixed in the population.

rials, and random recombination and mistakes in replication of DNA strands during reproduction are the main causes of genetic mutations. Sometimes a single mutation has a large effect, but evolutionary change is mostly brought about by many mutations accumulating over time. Only mutations in reproductive cells (gametes) matter; body cell changes—cancers, for example— are not inherited. Most mutations have no effect on fitness, and many actually have a negative effect. During the course of a species' life span—a million or more years—some mutations are thought to have given those individuals an advantage under the **selection pressures** of their environment at that time. The result is a species population that differs from those of numerous preceding generations.

All species live within limits

Environmental factors exert selection pressure and influence the fitness of individuals and their offspring. For this reason, species are limited in where they can live. Limitations include the following: (1) physiological stress due to inappropriate levels of some critical environmental factor, such as moisture, light, temperature, pH, or specific nutrients; (2) competition with other species; (3) predation, including parasitism and disease; and (4) luck. In some cases, the individuals of a population that survive environmental catastrophes or find their way to a new habitat, where they start a new population, may simply be lucky rather than more fit than their contemporaries.

An organism's physiology and behavior allow it to survive only in certain environments. Temperature, moisture level, nutrient supply, soil and water chemistry, living space, and other environmental factors must be at appropriate levels for organisms to persist. In 1840, the chemist Justus von Liebig proposed that the single factor in shortest supply relative to demand is the **critical factor** determining where a species lives. The giant saguaro cactus (*Carnegiea gigantea*), which grows in the dry, hot Sonoran desert of southern Arizona and northern Mexico, offers an example (fig. 4.3). Saguaros are extremely sensitive to freezing temperatures. A single winter night with temperatures below freezing for 12 or more hours kills growing tips on the branches,

FIGURE 4.3 Saguaro cacti, symbolic of the Sonoran desert, are an excellent example of distribution controlled by a critical environmental factor. Extremely sensitive to low temperatures, saguaros are found only where minimum temperatures never dip below freezing for more than a few hours at a time.

preventing further development. Thus the northern edge of the saguaro's range corresponds to a zone where freezing temperatures last less than half a day at any time.

Ecologist Victor Shelford (1877–1968) later expanded Liebig's principle by stating that each environmental factor has both minimum and maximum levels, called **tolerance limits,** beyond which a particular species cannot survive or is unable to reproduce (fig. 4.4). The single factor closest to these survival limits, Shelford postulated, is the critical factor that limits where a particular organism can live. At one time, ecologists tried to identify unique factors limiting the growth of every plant and animal population. We now know that several factors working together, even in a clear-cut case like the saguaro, usually determine a species' distribution. If you have ever explored the rocky coasts of New England or the Pacific Northwest, you have probably noticed that mussels and barnacles

grow thickly in the intertidal zone, the place between high and low tides. No one factor decides this pattern. Instead, the distribution of these animals is determined by a combination of temperature extremes, drying time between tides, salt concentrations, competitors, and food availability.

In some species, tolerance limits affect the distribution of young differently than adults. The desert pupfish, for instance, lives in small, isolated populations in warm springs in the northern Sonoran desert. Adult pupfish can survive temperatures between 0° and 42°C (a remarkably high temperature for a fish) and tolerate an equally wide range of salt concentrations. Eggs and juvenile fish, however, can survive only between 20° and 36°C and are killed by high salt levels. Reproduction, therefore, is limited to a small part of the range of the adult fish.

Sometimes the requirements and tolerances of species are useful **indicators** of specific environmental characteristics. The presence or absence of such species indicates something about the community and the ecosystem as a whole. Lichens and eastern white pine, for example, are indicators of air pollution because they are extremely sensitive to sulfur dioxide and ozone, respectively. Bull thistle and many other plant weeds grow on disturbed soil but are not eaten by cattle; therefore, a vigorous population of bull thistle or certain other plants in a pasture indicates it is being overgrazed. Similarly, anglers know that trout species require cool, clean, well-oxygenated water; the presence or absence of trout is used as an indicator of good water quality.

The ecological niche is a species' role and environment

Habitat describes the place or set of environmental conditions in which a particular organism lives. A more functional term, **ecological niche,** describes either the role played by a species in a biological community or the total set of environmental factors that determine a species distribution. The concept of niche was

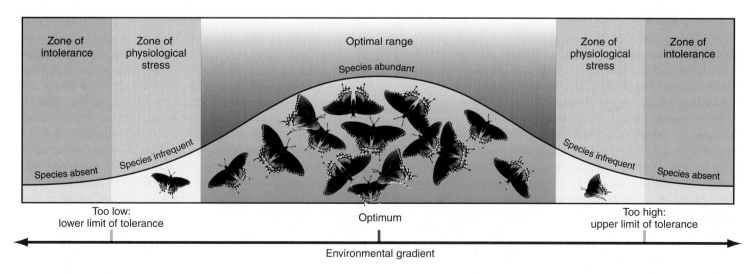

FIGURE 4.4 The principle of tolerance limits states that for every environmental factor, an organism has both maximum and minimum levels beyond which it cannot survive. The greatest abundance of any species along an environmental gradient is around the optimum level of the critical factor most important for that species. Near the tolerance limits, abundance decreases because fewer individuals are able to survive the stresses imposed by limiting factors.

FIGURE 4.5 Each of the animal and plant species living in this African savanna occupies an ecological niche. Some niches are broad and general, some are highly specialized.

FIGURE 4.6 The giant panda feeds exclusively on bamboo. Although its teeth and digestive system are those of a carnivore, it is not a good hunter, and has adapted to a vegetarian diet. In the 1970s, huge acreages of bamboo flowered and died, and many pandas starved.

first defined in 1927 by the British ecologist Charles Elton (1900–1991). To Elton, each species had a role in a community of species, and the niche defined its way of obtaining food, the relationships it had with other species, and the services it provided to its community. Thirty years later, the American limnologist G. E. Hutchinson (1903–1991) proposed a more biophysical definition of niche. Every species, he pointed out, exists within a range of physical and chemical conditions (temperature, light levels, acidity, humidity, salinity, etc.) and also biological interactions (predators and prey present, defenses, nutritional resources available, etc.). The niche is more complex than the idea of a critical factor (fig. 4.5). A graph of a species niche would be multidimensional, with many factors being simultaneously displayed, almost like an electron cloud.

For a generalist, like the brown rat, the ecological niche is broad. In other words, a generalist has a wide range of tolerance for many environmental factors. For others, such as the giant panda (*Ailuropoda melanoleuca*), only a narrow ecological niche exists (fig. 4.6). Bamboo is low in nutrients, but provides 95 percent of a panda's diet, requiring it to spend as much as 16 hours a day eating. There are virtually no competitors for bamboo, except other pandas, yet the species is endangered, primarily due to shrinking habitat. Giant pandas, like many species on earth, are habitat specialists. Specialists have more exacting habitat requirements, tend to have lower reproductive rates, and care for their young longer. They may be less resilient in response to environmental change. Plants can also be habitat specialists—for instance, **endemic** plant species exist on serpentine and other unusual types of rock outcrops, and nowhere else.

Over time, niches change as species develop new strategies to exploit resources. Species of greater intelligence or complex social structures, such as elephants, chimpanzees, and dolphins, learn from their social group how to behave and can invent new ways of doing things when presented with novel opportunities or challenges. In effect, they alter their ecological niche by passing on cultural behavior from one generation to the next. Most organisms, however, are restricted to their niche by their genetically determined bodies and instinctive behaviors. When two such spe-

cies compete for limited resources, one eventually gains the larger share, while the other finds different habitat, dies out, or experiences a change in its behavior or physiology so that competition is minimized. The idea that "complete competitors cannot coexist" was proposed by the Russian microbiologist G. F. Gause (1910–1986) to explain why mathematical models of species competition always ended with one species disappearing. The **competitive exclusion principle,** as it is called, states that no two species can occupy the same ecological niche for long. The one that is more efficient in using available resources will exclude the other (see Species Competition p. 97). We call this process of niche evolution **resource partitioning** (fig. 4.7). Partitioning can allow several species to utilize different parts of the same resource and coexist within a single habitat (fig. 4.8). Species can specialize in time, too. Swallows and insectivorous bats both catch insects, but some insect species are active during the day and others at night, providing noncompetitive feeding opportunities for day-active swallows and night-active bats. The competitive exclusion principle does not explain all situations, however. For example, many similar plant species coexist in some habitats. Do they avoid competition in ways we cannot observe, or are resources so plentiful that no competition need occur?

78   **PART 1**   Enviornmental Science A Global Concern

http://www.mhhe.com/cunningham10e

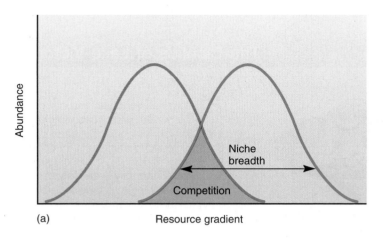

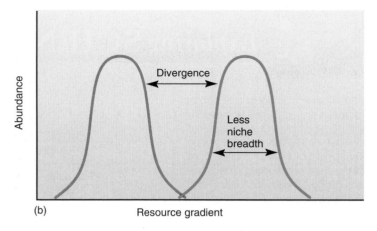

FIGURE 4.7 Competition causes resources partitioning and niche specialization. (a) Where niches of two species overlap along a resource gradient, competition occurs (*shaded area*). Individuals in this part of the niche have less success producing young. (b) Over time the traits of the populations diverge, leading to specialization, narrower niche breadth, and less competition between species.

Speciation maintains species diversity

As an interbreeding species population becomes better adapted to its ecological niche, its genetic heritage (including mutations passed from parents to offspring) gives it the potential to change further as circumstances, dictate. In the case of Galápagos finches studied a century and a half ago by Charles Darwin, evidence from body shape, behavior, and genetics leads to the idea that modern Galapagos finches look, behave, and bear DNA related to an original seed-eating finch species (fig. 4.9). It is proposed to have blown to the islands from the mainland where a similar species still exists. Today there are 13 distinct species on the islands that differ markedly in appearance, food preferences, and habitat. Fruit eaters have thick, parrot-like bills; seed eaters have heavy, crushing bills; insect eaters have thin, probing beaks to catch their prey. One of the most unusual species is the woodpecker finch, which pecks at tree bark for hidden insects. Lacking the woodpecker's long tongue, the finch uses a cactus spine as a tool to extract bugs.

The development of a new species is called **speciation.** No one has observed a new species springing into being, but in some organisms, especially plants, it is inferred to occur somewhat frequently. Nevertheless, given the evidence, speciation is a reasonable proposal for how species arise. Speciation may be relatively rapid on millennial timescales (punctuated equilibrium). For example, after a long period of stability, a new species may arise from parents following the appearance of a new food source, predator, or competitor, or a change in climate. One mechanism of speciation is **geographic isolation.** This is termed **allopatric speciation**—species arise in non-overlapping geographic locations. The original Galapagos finches were separated from the rest of the population on the mainland, could no longer share genetic material, and became reproductively isolated.

The barriers that divide subpopulations are not always physical. For example, two virtually identical tree frogs (*Hyla versicolor, H. chrysoscelis*) live in similar habitats of eastern North America but have different mating calls. This is an example of behavioral isolation. It also happens that one species has twice the chromosomes of

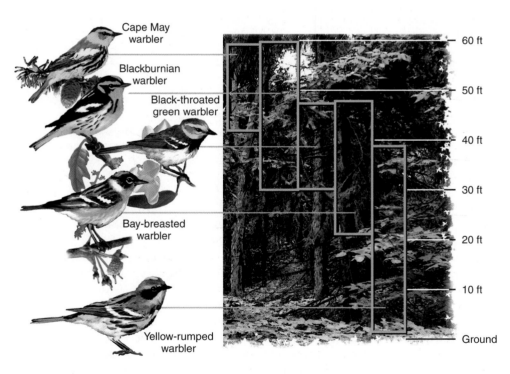

FIGURE 4.8 Several species of insect-eating wood warblers occupy the same forests in eastern North America. The competitive exclusion principle predicts that the warblers should partition the resource—insect food—in order to reduce competition. And in fact, the warblers feed in different parts of the forest.
Source: Original observations by R. H. MacArthur (1958).

If you visit your local pet store, chances are you'll see some cichlids (*Haplochromis* sp.). These small colorful, prolific fish come in a wide variety of colors and shapes from many parts of the world. The greatest cichlid diversity on earth—and probably the greatest vertebrate diversity anywhere—is found in the three great African rift lakes: Victoria, Malawi, and Tanganyika. Together, these lakes once had about 1,000 types of cichlids—more than all the fish species in Europe and North America combined. All these cichlids apparently evolved from a few ancestral varieties in the 15,000 years or so since the lakes were formed by splitting of the continental crust. This is one of the fastest and most extensive examples of vertebrate speciation known.

We believe that one of the factors that allowed cichlids to evolve so quickly is that they found few competitors or predators and a multitude of ecological roles to play in these new lakes. There are mud biters, algae scrapers, leaf chewers, snail crushers, zooplankton eaters, prawn predators, and fish feeders. Because they live in different habitats in the lakes, are active at different times of the day, and have developed different body sizes and shapes to feed on specialized prey, the cichlids have been ecologically isolated for long enough to evolve into an amazing variety of species. Cichlids are a good example of radiative evolution.

Unfortunately, a well-meaning but disastrous fish-stocking experiment has wiped out at least half the cichlid species in these lakes in just a few decades and set off a series of changes that are upsetting important ecological relationships. Lake Victoria, which lies between Kenya, Tanzania, and Uganda, has

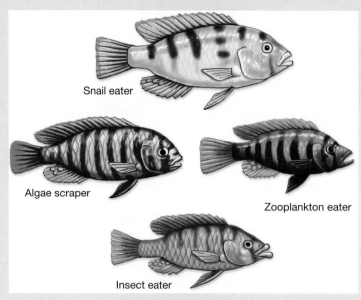

Cichlid fishes of Lake Victoria. More than 300 species have evolved from an original common ancestor to take advantage of different food sources and habitats.

been particularly hard hit. Cichlids once made up 80 percent of the animal biomass in the lake and were the base for a thriving local fishery, supplying much-needed protein for local people. Management agencies regarded the bony little cichlids as "trash fish," and decided in the 1960s to introduce Nile perch (*Lates niloticus*), a voracious, exotic predator that can weigh up to 100 kg (220 lbs) and grow up to 2 m long. The perch, they believed, would support a lucrative commercial export trade.

The perch gobbled up the cichlids so quickly that, by 1980, two-thirds of the haplochromine species in the lake were extinct. Although there are still lots of fish in the lake, 80 percent of the biomass is now made up of perch, which are too large and powerful for the small boats, papyrus nets, and woven baskets traditionally used to harvest cichlids. International fishing companies now use large power boats and nylon nets to harvest great schools of perch, which are filleted, frozen, and shipped to markets in Europe and the Middle East. Because the perch are oily, local fishers can't sun dry them as they once did the cichlids. Instead, they are cooked or smoked over wood fires for local consumption. Forests are being denuded for firewood, and protein malnutrition is common in a region that exports 200,000 tons of fish each year.

Perhaps worst of all, Lake Victoria, which covers an area the size of Switzerland, is dying. Algae blooms clot the surface, oxygen levels have fallen alarmingly, and thick layers of soft silt are filling-in shallow bays. Untreated sewage, chemical pollution, and farm runoff are the immediate causes of these deleterious changes, but destabilization of the natural community plays a role as well. The swarms of cichlids that once ate algae and rotting detritus were the lake's self-cleaning system. Eliminating them threatens the long-term ability of the lake to support any useful aquatic life.

As this example shows, species and ecological diversity are important. Misguided management and development schemes that destroyed native species in Lake Victoria have resulted in an ecosystem that no longer supports the natural community or the local people dependent on it. It's difficult to see how we could replace the variety of species and the ecological roles they played, which evolution provided for free.

For more information, see

Stiassny, M. L. J., and A. Meyer. 1999. Cichlids of the rift lakes. *Scientific American* 280(2):64–69.

the other. This example of **sympatric speciation** takes place in the same location as the ancestor species (Exploring Science above). Fern species and other plants seem prone to sympatric speciation by doubling or quadrupling the chromosome number of their ancestors.

Once isolation is imposed, the two populations begin to diverge in genetics and physical characteristics. Genetic drift ensures that DNA of two formerly joined populations eventually

diverges; in several generations, traits are lost from a population during the natural course of reproduction. Under more extreme circumstances, a die-off of most members of an isolated population strips much of the variation in traits from the survivors. The cheetah experienced a genetic bottleneck about 10,000 years ago and exists today as virtually identical individuals.

In isolation, selection pressures shape physical, behavioral, and genetic characteristics of individuals, causing population

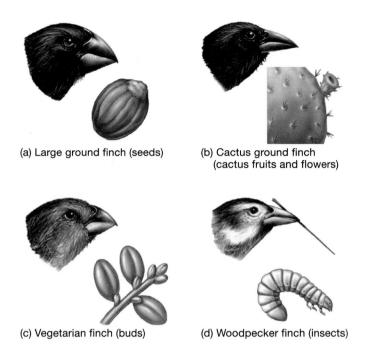

(a) Large ground finch (seeds)

(b) Cactus ground finch (cactus fruits and flowers)

(c) Vegetarian finch (buds)

(d) Woodpecker finch (insects)

FIGURE 4.9 Each of the 13 species of Galápagos finches, although originally derived from a common ancestor, has evolved distinctive anatomies and behaviors to exploit different food sources. The woodpecker finch (d) uses cactus thorns to probe for insects under tree bark.

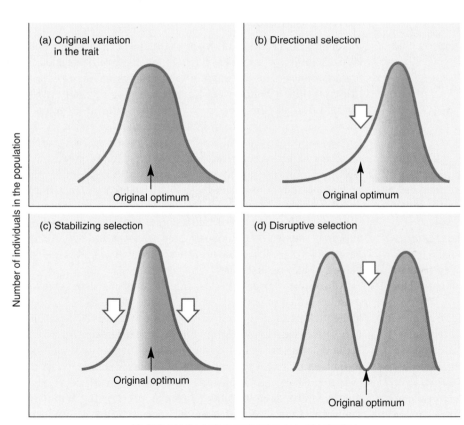

(a) Original variation in the trait

(b) Directional selection

(c) Stabilizing selection

(d) Disruptive selection

Number of individuals in the population

Variation in the trait experiencing natural selection

traits to shift over time (fig. 4.10). From an original range of characteristics, the shift can be toward an extreme of the trait (directional selection), it can narrow the range of a trait (stabilizing selection), or it can cause traits to diverge to the extremes (disruptive selection). Directional selection is implied by increased pesticide resistance in German cockroaches (*Blattella germanica*). Apparently some individuals can make an enzyme that detoxifies pesticides. Individual cockroaches that lack this characteristic are dying out, and as a result, populations of cockroaches with pesticide resistance are developing.

A small population in a new location—island, mountaintop, unique habitat—encounters new environmental conditions that favor some individuals over others (fig. 4.11). The physical and behavioral traits these individuals have are passed to the next generation, and the frequency of the trait shifts in the population. Where a species may have existed but has died out, others arise and contribute to the incredible variety of life-forms seen in nature. The fossil record is one of ever-increasing species diversity, despite several catastrophes, which were recorded in different geological strata and which wiped out a large proportion of the earth's species each time.

Evolution is still at work

You may think that evolution only occurred in the distant past, but it's an ongoing process. Ample evidence from both laboratory experiments and from nature shows evolution at work. Geneticists have modified many fruit fly properties—including body size, eye color, growth rate, life span, and feeding behavior—using artificial selection. In one experiment, researchers selected for flies with many bristles (stiff, hairlike structures) on their abdomen. In each generation, the flies with the most bristles were allowed to mate. After 86 generations, the number of bristles had quadrupled. In a similar experiment with corn, agronomists chose seeds with the highest oil content to plant and mate. After 90 generations, the average oil content had increased 450 percent.

Evolutionary change is also occurring in nature. A classic example is seen in some of the finches on the Galápagos Island of Daphne. Twenty years ago, a large-billed species (*Geospiza magnirostris*) settled on the island, which previously had only a medium-billed species (*Geospiza fortis*). The *G. magnirosris* were better at eating larger seeds and pushed *G. fortis* to depend more

FIGURE 4.10 A species trait, such as beak shape, changes in response to selection pressure. (a) The original variation is acted on by selection pressure (*arrows*) that (b) shifts the characteristics of that trait in one direction, (c) or to an intermediate condition. (d) Disruptive selection moves characteristics to the extremes of the trait. Which selection type plausibly resulted in two distinct beak shapes among Galápagos finches—narrow in tree finches versus stout in ground finches?

1. Single population

2. Geographically isolated populations

FIGURE 4.11 Geographic barriers can result in allopatric speciation. During cool, moist glacial periods, what is now Arizona was forest-covered and squirrels could travel and interbreed freely. As the climate warmed and dried, desert replaced forest on the plains. Squirrels were confined to cooler mountaintops, which acted as island refugia, where new, reproductively isolated species gradually evolved.

and more on smaller seeds. Gradually, birds with smaller bills suited to small seeds became more common in the *G. fortis* population. During a severe drought in 2003–2004, large seeds were scarce, and most birds with large beaks disappeared. This included almost all of the recently arrived *G. magnirosris* as well as the larger-beaked *G. fortis*. In just two generations, the *G. fortis* population changed to entirely small-beaked individuals. At first, this example of rapid evolution was thought to be a rarity, but subsequent research suggests that it may be more common than previously thought.

Similarly, the widespread application of pesticides in agriculture and urban settings has led to the rapid evolution of resistance in more than 500 insect species. Similarly, the extensive use of antibiotics in human medicine and livestock operations has led to antibiotic resistance in many microbes. The Centers for Disease Control estimates that 90,000 Americans die every year from hospital-acquired infections, most of which are resistant to one or more antibiotics. We're engaged in a kind of an arms race

with germs. As quickly as new drugs are invented, microbes become impervious to them. Currently, vancomycin is the drug of last resort. When resistance to it becomes widespread, we may have no protection from infections.

On the other hand, evolution sometimes works in our favor. We've spread a number of persistent organic pollutants (called POPs), such as pesticides and industrial solvents, throughout our environment. One of the best ways to get rid of them is with microbes that can destroy or convert them to a nontoxic form. It turns out that the best place to look for these species is in the most contaminated sites. The presence of a new food source has stimulated evolution of organisms that can metabolize it. A little artificial selection and genetic modification in the laboratory can turn these species into very useful bioremediation tools.

Taxonomy describes relationships among species

Taxonomy is the study of types of organisms and their relationships. With it you can trace how organisms have descended from common ancestors. Taxonomic relationships among species are displayed like a family tree. Botanists, ecologists, and other scientists often use the most specific levels of the tree, genus and species, to compose **binomials.** Also called scientific or Latin names, they identify and describe species using Latin, or Latinized nouns and adjectives, or names of people or places. Scientists communicate about species using these scientific names instead of common names (e.g., lion, dandelion, or ant lion), to avoid confusion. A common name can refer to any number of species in different places, and a single species might have many common names. The bionomial *Pinus resinosa,* on the other hand, always is the same tree, whether you call it a red pine, Norway pine, or just pine.

Taxonomy also helps organize specimens and subjects in museum collections and research. You are *Homo sapiens* (human) and eat chips made of *Zea mays* (corn or maize). Both are members of two well-known kingdoms. Scientists, however, recognize six kingdoms (fig. 4.12): animals, plants, fungi (molds and mushrooms), protists (algae, protozoans, slime molds), bacteria (or eubacteria), and archaebacteria (ancient, single-celled organisms that live in harsh environments, such as hot springs). Within these kingdoms are millions of different species, which you will learn more about in chapters 5 and 11.

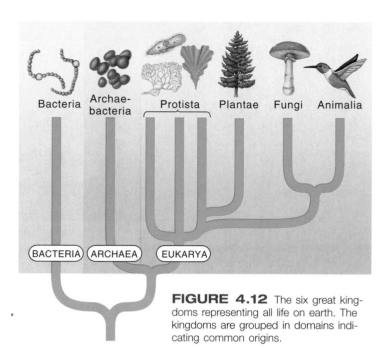

FIGURE 4.12 The six great kingdoms representing all life on earth. The kingdoms are grouped in domains indicating common origins.

FIGURE 4.13 In this tangled Indonesian rainforest, space and light are at a premium. Plants growing beneath the forest canopy have adaptations that help them secure these limited resources. The ferns and bromeliads seen here are epiphytes; they find space and get closer to the sun by perching on limbs and tree trunks. Strangler figs start out as epiphytes, but send roots down to the forest floor and, once contact is made, put on a growth spurt that kills the supporting tree. These are just some of the adaptations to life in the dark jungle.

4.2 SPECIES INTERACTIONS SHAPE BIOLOGICAL COMMUNITIES

We have learned that adaptation to one's environment, determination of ecological niche, and even speciation is affected not just by bodily limits and behavior, but also by competition and predation. Don't despair. Not all biological interactions are antagonistic, and many, in fact, involve cooperation or at least benign interactions and tolerance. In some cases, different organisms depend on each other to acquire resources. Now we will look at the interactions within and between species that affect their success and shape biological communities.

Competition leads to resource allocation

Competition is a type of antagonistic relationship within a biological community. Organisms compete for resources that are in limited supply: energy and matter in usable forms, living space, and specific sites to carry out life's, activities. Plants compete for growing space to develop root and shoot systems so that they can absorb and process sunlight, water, and nutrients (fig. 4.13). Animals compete for living, nesting, and feeding sites, and also for mates. Competition among members of the same species is called **intraspecific competition,** whereas competition between members of different species is called **interspecific competition.** Recall the competitive exclusion principle as it applies to interspecific competition. Competition shapes a species population and biological community by causing individuals and species to shift their focus from one segment of a resource type to another. Thus, warblers all competing with each other for insect food in New England tend to specialize on different areas of the forest's trees, reducing or avoiding competition. Since the 1950s there have been hundreds of interspecific competition studies in natural populations. In general,

scientists assume it does occur, but not always, and in some groups—carnivores and plants—it has little effect.

In intraspecific competition, members of the same species compete directly with each other for resource. Several avenues exist to reduce competition in a species population. First, the young of the year disperse. Even plants practice dispersal; seeds are carried by wind, water, and passing animals to less crowded conditions away from the parent plants. Second, by exhibiting strong territoriality, many animals force their offspring or trespassing adults out of their vicinity. In this way territorial species, which include bears, songbirds, ungulates, and even fish, minimize competition between individuals and generations. A third way to reduce intraspecific competition is resource partitioning between generations. The adults and juveniles of these species occupy different ecological niches. For instance, monarch caterpillars munch on milkweed leaves, while metamorphosed butterflies lap nectar. Crabs begin as floating larvae and do not compete with bottom-dwelling adult crabs.

We think of competition among animals as a battle for resources—"nature red in tooth and claw" is the phrase. In fact, many animals avoid fighting if possible, or confront one another with noise and predictable movements. Bighorn sheep and many other ungulates, for example, engage in ritualized combat, with the weaker animal knowing instinctively when to back off. It's worse to be injured than to lose. Instead, competition often is

FIGURE 4.14 Insect herbivores are predators as much as are lions and tigers. In fact, insects consume the vast majority of biomass in the world. Complex patterns of predation and defense have often evolved between insect predators and their plant prey.

FIGURE 4.15 Microscopic plants and animals form the basic levels of many aquatic food chains and account for a large percentage of total world biomass. Many oceanic plankton are larval forms that have habitats and feeding relationships very different from their adult forms.

simply about getting to food or habitat first, or being able to use it more efficiently. As we discussed, each species has tolerance limits for nonbiological (abiotic) factors. Studies often show that, when two species compete, the one living in the center of its tolerance limits for a range of resources has an advantage and, more often than not, prevails in competition with another species living outside its optimal environmental conditions.

Predation affects species relationships

All organisms need food to live. Producers make their own food, while consumers eat organic matter created by other organisms. As we saw in chapter 3, photosynthetic plants and algae are the producers in most communities. Consumers include herbivores, carnivores, omnivores, scavengers, detritivores, and decomposers. You may think only carnivores are predators, but ecologically a predator is any organism that feeds directly on another living organism, whether or not this kills the prey (fig. 4.14). Herbivores, carnivores, and omnivores, which feed on live prey, are predators, but scavengers, detritivores, and decomposers, which feed on dead things, are not. In this sense, parasites (organisms that feed on a host organism or steal resources from it without necessarily killing it) and even pathogens (disease-causing organisms) can be considered predator organisms. Herbivory is the type of predation practiced by grazing and browsing animals on plants.

Predation is a powerful but complex influence on species populations in communities. It affects (1) all stages in the life cycles of predator and prey species; (2) many specialized food-obtaining mechanisms; and (3) the evolutionary adjustments in behavior and body characteristics that help prey escape being eaten, and predators more efficiently catch their prey. Predation also interacts with competition. In **predator-mediated competition,** a superior competitor in a habitat builds up a larger population than its competing species; predators take note and increase their hunting pressure on the superior species, reducing its abundance and allowing the weaker competitor to increase its numbers. To test this idea, scientists remove predators from communities of competing species. Often the superior competitors eliminated other species from the habitat. In a classic example, the ochre starfish (*Pisaster ochraceus*) was removed from Pacific tidal zones and its main prey, the common mussel (*Mytilus californicus*), exploded in numbers and crowded out other intertidal species.

Knowing how predators affect prey populations has direct application to human needs, such as pest control in cropland. The cyclamen mite (*Phytonemus pallidus*), for example, is a pest of California strawberry crops. Its damage to strawberry leaves is reduced by predatory mites (*Typhlodromus* and *Neoseiulus*). which arrive naturally or are introduced into fields. Pesticide spraying to control the cyclamen mite can actually increase the infestation because it also kills the beneficial predatory mites.

Predatory relationships may change as the life stage of an organism changes. In marine ecosystems, crustaceans, mollusks, and worms release eggs directly into the water where they and hatchling larvae join the floating plankton community (fig. 4.15). Planktonic animals eat each other and are food for larger carnivores, including fish. As prey species mature, their predators change. Barnacle larvae are planktonic and are eaten by small fish, but as adults their hard shells protect them from fish, but not starfish and predatory snails. Predators often switch prey in the course of their lives. Carnivorous adult frogs usually begin their lives as herbivorous tadpoles. Predators also switch prey when it becomes rare, or something else becomes abundant. Many predators have morphologies and behaviors that make them highly adaptable to a changing prey base, but some, like the polar bear are highly specialized in their prey preferences.

Some adaptations help avoid predation

Predator-prey relationships exert selection pressures that favor evolutionary adaptation. In this world, predators become more efficient at searching and feeding, and prey become more effective at escape and avoidance. Toxic chemicals, body armor, extraordinary speed, and the ability to hide are a few strategies organisms use to protect themselves. Plants have thick bark,

FIGURE 4.16 Poison arrow frogs of the family Dendrobatidae use brilliant colors to warn potential predators of the extremely toxic secretions from their skin. Native people in Latin America use the toxin on blowgun darts.

(a)

(b)

FIGURE 4.17 An example of Batesian mimicry. The dangerous wasp (a) has bold yellow and black bands to warn predators away. The much rarer longhorn beetle (b) has no poisonous stinger, but looks and acts like a wasp and thus avoids predators as well.

spines, thorns, or distasteful and even harmful chemicals in tissues—poison ivy and stinging nettle are examples. Arthropods, amphibians, snakes, and some mammals produce noxious odors or poisonous secretions that cause other species to leave them alone. Animal prey are adept at hiding, fleeing, or fighting back. On the Serengeti Plain of East Africa, the swift Thomson's gazelle and even swifter cheetah are engaged in an arms race of speed, endurance, and quick reactions. The gazelle escapes often because the cheetah lacks stamina, but the cheetah accelerates from 0 to 72 kph in 2 seconds, giving it the edge in a surprise attack. The response of predator to prey and vice versa, over tens of thousands of years, produces physical and behavioral changes in a process known as **coevolution.** Coevolution can be mutually beneficial: many plants and pollinators have forms and behaviors that benefit each other. A classic case is that of fruit bats, which pollinate and disperse seeds of fruit-bearing tropical plants.

Often species with chemical defenses display distinct coloration and patterns to warn away enemies (fig. 4.16). In a neat evolutionary twist, certain species that are harmless resemble poisonous or distasteful ones, gaining protection against predators who remember a bad experience with the actual toxic organism. This is called **Batesian mimicry,** after the English naturalist H. W. Bates (1825–1892), a traveling companion of Alfred Wallace. Many wasps, for example, have bold patterns of black and yellow stripes to warn off potential predators (fig. 4.17a). The much rarer longhorn beetle has no stinger but looks and acts much like a wasp, tricking predators into avoiding it (fig. 4.17b). The distasteful monarch and benign viceroy butterflies are a classic case of Batesian mimicry. Another form of mimicry, **Müllerian mimicry** (after the biologist Fritz Müller) involves two unpalatable or dangerous species who look alike. When predators learn to avoid either species, both benefit. Species also display forms, colors, and patterns that help avoid being discovered. Insects that look like dead leaves or twigs are among the most remarkable examples (fig. 4.18). Unfortunately for prey, predators also often use camouflage to conceal themselves as they lie in wait for their next meal.

Symbiosis involves intimate relations among species

In contrast to predation and competition, some interactions between organisms can be nonantagonistic, even beneficial. In such relationships, called **symbiosis,** two or more species live intimately together, with their fates linked. Symbiotic relationships often enhance the survival of one or both partners. In lichens, a fungus and a photosynthetic partner (either an alga or a cyanobacterium) combine tissues to mutual benefit (fig. 4.19a). This association is called **mutualism.** Some ecologists believe that cooperative, mutualistic relationships may be more important in evolution than commonly thought (fig. 4.19b). Survival of the fittest may also mean survival of organisms that can live together.

Symbiotic relationships often entail some degree of coevolution of the partners, shaping—at least in part—their structural and behavioral characteristics. This mutualistic coadaptation is evident

FIGURE 4.18 This walking stick is highly camouflaged to blend in with the forest floor. Natural selection and evolution have created this remarkable shape and color.

between swollen thorn acacias (*Acacia collinsii*) and the ants (*Pseudomyrmex ferruginea*) that tend them in Central and South America. Acacia ant colonies live inside the swollen thorns on the acacia tree branches. Ants feed on nectar that is produced in glands at the leaf bases and also eat special protein-rich structures that are produced on leaflet tips. The acacias thus provide shelter and food for the ants. Although they spend energy to provide these services, the trees are not harmed by the ants. What do the acacias get in return? Ants aggressively defend their territories, driving away herbivorous insects that would feed on the acacias. Ants also trim away vegetation that grows around the tree, reducing competition by other plants for water and nutrients. You can see how mutualism is structuring the biological community in the vicinity of acacias harboring ants, just as competition or predation shapes communities.

Mutualistic relationships can develop quickly. In 2005 the Harvard entomologist E. O. Wilson pieced together evidence to explain a 500-year-old agricultural mystery in the oldest Spanish settlement in the New World, Hispaniola. Using historical accounts and modern research, Dr. Wilson reasoned that mutualism developed between the tropical fire ant (*Solenopsis geminata*), native to the Americas, and a sap-sucking insect that was probably introduced from the Canary Islands in 1516 on a shipment of plantains. The plantains were planted, the sap-suckers were distributed across Hispaniola, and in 1518 a great die-off of crops occurred. Apparently the native fire ants discovered the foreign sap-sucking insects, consumed their excretions of sugar and protein, and protected them from predators, thus allowing the introduced insect population to explode. The Spanish assumed the fire ants caused the agricultural blight, but a little ecological knowledge would have led them to the real culprit.

Commensalism is a type of symbiosis in which one member clearly benefits and the other apparently is neither benefited nor harmed. Many mosses, bromeliads, and other plants growing on trees in the moist tropics are considered commensals (fig. 4.19c). These epiphytes are watered by rain and obtain nutrients from leaf litter and falling dust, and often they neither help nor hurt the trees on which they grow. Robins and sparrows that inhabit suburban yards are commensals with humans. **Parasitism,** a form of predation, may also be considered symbiosis because of the dependency of the parasite on its host.

Keystone species have disproportionate influence

A **keystone species** plays a critical role in a biological community that is out of proportion to its abundance. Originally, keystone species were thought to be top predators—lions, wolves, tigers—which limited herbivore abundance and reduced the herbivory of plants. Scientists now recognize that less-conspicuous species also play keystone roles. Tropical figs, for example, bear fruit year-round at a low but steady rate. If figs are removed from a forest, many fruit-eating animals (frugivores) would starve in the dry season when fruit of other species is scarce. In turn, the disappearance of frugivores would affect plants that

(a) Lichen on a rock

(b) Oxpecker and impala

(c) Bromeliad

FIGURE 4.19 Symbiotic relationships. (a) Lichens represent an obligatory mutualism between a fungus and alga or cyanobacterium. (b) Mutualism between a parasite-eating red-billed oxpecker and parasite-infested impala. (c) Commensalism between a tropical tree and free-loading bromeliad.

FIGURE 4.20 Sea otters protect kelp forests in the northern Pacific Ocean by eating sea urchins that would otherwise destroy the kelp. But the otters are being eaten by killer whales. Which is the keystone in this community—or is there a keystone set of organisms?

depend on them for pollination and seed dispersal. It is clear that the effect of a keystone species on communities often ripples across trophic levels.

Keystone functions have been documented for vegetation-clearing elephants, the predatory ochre sea star, and frog-eating salamanders in coastal North Carolina. Even microorganisms can play keystone roles. In many temperate forest ecosystems, groups of fungi that are associated with tree roots (mycorrhizae) facilitate the uptake of essential minerals. When fungi are absent, trees grow poorly or not at all. Overall, keystone species seem to be more common in aquatic habitats than in terrestrial ones.

The role of keystone species can be difficult to untangle from other species interactions. Off the northern Pacific coast, a giant brown alga (*Macrocystis pyrifera*) forms dense "kelp forests," which shelter fish and shellfish species from predators, allowing them to become established in the community. It turns out, however, that sea otters eat sea urchins living in the kelp forests (fig. 4.20); when sea otters are absent, the urchins graze on and eliminate kelp forests. To complicate things, around 1990, killer whales began preying on otters because of the dwindling stocks of seals and sea lions, thereby creating a cascade of effects. Is the kelp, otter, or orca the keystone here? Whatever the case, keystone species exert their influence by changing competitive relationships. In some communities, perhaps we should call it a "keystone set" of organisms.

4.3 COMMUNITY PROPERTIES AFFECT SPECIES AND POPULATIONS

The processes and principles that we have studied thus far in this chapter—tolerance limits, species interactions, resource partitioning, evolution, and adaptation—play important roles in determining the characteristics of populations and species. In this section we will look at some fundamental properties of biological communities and ecosystems—productivity, diversity, complexity, resilience, stability, and structure—to learn how they are affected by these factors.

Productivity is a measure of biological activity

A community's **primary productivity** is the rate of biomass production, an indication of the rate of solar energy conversion to chemical energy. The energy left after respiration is net primary production. Photosynthetic rates are regulated by light levels, temperature, moisture, and nutrient availability. Figure 4.21 shows approximate productivity levels for some major ecosystems. As you can see, tropical forests, coral reefs, and estuaries (bays or inundated river valleys where rivers meet the ocean) have high levels of productivity because they have abundant supplies of all these resources. In deserts, lack of water limits photosynthesis. On the arctic tundra or in high mountains, low temperatures inhibit plant growth. In the open ocean, a lack of nutrients reduces the ability of algae to make use of plentiful sunshine and water.

Some agricultural crops such as corn (maize) and sugar cane grown under ideal conditions in the tropics approach the productivity levels of tropical forests. Because shallow water ecosystems such as coral reefs, salt marshes, tidal mud flats, and other highly productive aquatic communities are relatively rare compared to the vast extent of open oceans—which are effectively biological deserts—marine ecosystems are much less productive on average than terrestrial ecosystems.

Even in the most photosynthetically active ecosystems, only a small percentage of the available sunlight is captured and used to make energy-rich compounds. Between one-quarter and three-quarters of the light reaching plants is reflected by leaf surfaces. Most of the light absorbed by leaves is converted to heat that is either radiated away or dissipated by evaporation of water. Only 0.1 to 0.2 percent of the absorbed energy is used by chloroplasts to synthesize carbohydrates.

In a temperate-climate oak forest, only about half the incident light available on a midsummer day is absorbed by the leaves. Ninety-nine percent of this energy is used to evaporate water. A large oak tree can transpire (evaporate) several thousand liters of water on a warm, dry, sunny day while it makes only a few kilograms of sugars and other energy-rich organic compounds.

Abundance and diversity measure the number and variety of organisms

Abundance is an expression of the total number of organisms in a biological community, while **diversity** is a measure of the number of different species, ecological niches, or genetic variation present. The abundance of a particular species often is inversely related to the total diversity of the community. That is, communities with a very large number of species often have only a few members of any given species in a particular area. As a general rule, diversity decreases but abundance within species increases as we go from the equator toward the poles. The Arctic has vast

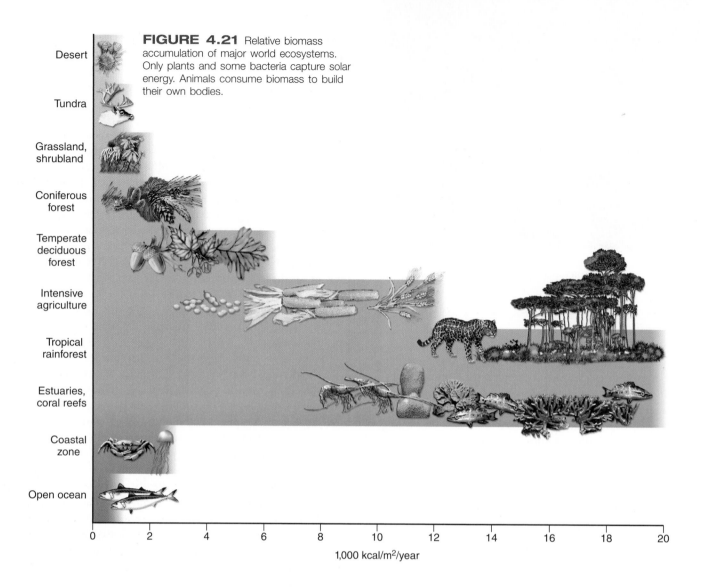

FIGURE 4.21 Relative biomass accumulation of major world ecosystems. Only plants and some bacteria capture solar energy. Animals consume biomass to build their own bodies.

Desert

Tundra

Grassland, shrubland

Coniferous forest

Temperate deciduous forest

Intensive agriculture

Tropical rainforest

Estuaries, coral reefs

Coastal zone

Open ocean

0   2   4   6   8   10   12   14   16   18   20

1,000 kcal/m²/year

numbers of insects such as mosquitoes, for example, but only a few species. The tropics, on the other hand, have vast numbers of species—some of which have incredibly bizarre forms and habits—but often only a few individuals of any particular species in a given area.

Consider bird populations. Greenland is home to 56 species of breeding birds, while Colombia, which is only one-fifth the size of Greenland, has 1,395. Why are there so many species in Colombia and so few in Greenland?

Climate and history are important factors. Greenland has such a harsh climate that the need to survive through the winter or escape to milder climates becomes the single most important critical factor that overwhelms all other considerations and severely limits the ability of species to specialize or differentiate into new forms. Furthermore, because Greenland was covered by glaciers until about 10,000 years ago, there has been little time for new species to develop.

Many areas in the tropics, by contrast, have relatively abundant rainfall and warm temperatures year-round so that

ecosystems there are highly productive. The year-round dependability of food, moisture, and warmth supports a great exuberance of life and allows a high degree of specialization in physical shape and behavior. Coral reefs are similarly stable, productive, and conducive to proliferation of diverse and amazing life-forms. The enormous abundance of brightly colored and fantastically shaped fish, corals, sponges, and arthropods in the reef community is one of the best examples we have of community diversity.

Productivity is related to abundance and diversity, both of which are dependent on the total resource availability in an ecosystem as well as the reliability of resources, the adaptations of the member species, and the interactions between species. You shouldn't assume that all communities are perfectly adapted to their environment. A relatively new community that hasn't had time for niche specialization, or a disturbed one where roles such as top predators are missing, may not achieve maximum efficiency of resource use or reach its maximum level of either abundance or diversity.

Working Locally for Ecological Diversity

You might think that diversity and complexity of ecological systems are too large or too abstract for you to have any influence. But you can contribute to a complex, resilient, and interesting ecosystem, whether you live in the inner city, a suburb, or a rural area.

- Keep your cat indoors. Our lovable domestic cats are also very successful predators. Migratory birds, especially those nesting on the ground, have not evolved defenses against these predators (Exploring Science p. 91).

- Plant a butterfly garden. Use native plants that support a diverse insect population. Native trees with berries or fruit also support birds. (Be sure to avoid non-native invasive species: see chapter 11.) Allow structural diversity (open areas, shrubs, and trees) to support a range of species.

- Join a local environmental organization. Often, the best way to be effective is to concentrate your efforts close to home. City parks and neighborhoods support ecological communities, as do farming and rural areas. Join an organization working to maintain ecosystem health; start by looking for environmental clubs at your school, park organizations, a local Audubon chapter, or a local Nature Conservancy branch.

- Take walks. The best way to learn about ecological systems in your area is to take walks and practice observing your environment. Go with friends and try to identify some of the species and trophic relationships in your area.

- Live in town. Suburban sprawl consumes wildlife habitat and reduces ecosystem complexity by removing many specialized plants and animals. Replacing forests and grasslands with lawns and streets is the surest way to simplify, or eliminate, ecosystems.

Community structure describes spatial distribution of organisms

Ecological structure refers to patterns of spatial distribution of individuals and populations within a community, as well as the relation of a particular community to its surroundings. At the local level, even in a relatively homogeneous environment, individuals in a single population can be distributed randomly, clumped together, or in highly regular patterns. In randomly arranged populations, individuals live wherever resources are available (fig. 4.22a). Ordered patterns may be determined by the physical environment but are more often the result of biological competition. For example, competition for nesting space in seabird colonies on the Falkland Islands is often fierce. Each nest tends to be just out of reach of the neighbors sitting on their own nests. Constant squabbling produces a highly regular pattern (fig. 4.22b). Similarly, sagebrush releases toxins from roots and fallen leaves, which inhibit the growth of competitors and create a circle of bare ground around each bush. As neighbors fill in empty spaces up to the limit of this chemical barrier, a regular spacing results.

Some other species cluster together for protection, mutual assistance, reproduction, or access to a particular environmental resource (fig. 4.22c). Dense schools of fish, for instance, cluster closely together in the ocean, increasing their chances of detecting and escaping predators. Similarly, predators, whether sharks, wolves, or humans, often hunt in packs to catch their prey. A flock of blackbirds descending on a cornfield or a troop of baboons traveling across the African savanna band together both to avoid predators and to find food more efficiently.

Plants can cluster for protection, as well. A grove of wind-sheared evergreen trees is often found packed tightly together at the crest of a high mountain or along the seashore. They offer mutual protection from the wind not only to each other but also to other creatures that find shelter in or under their branches.

Most environments are patchy at some scale. Organisms cluster or disperse according to patchy availability of water, nutrients, or other resources. Distribution in a community can be vertical as well as horizontal. The tropical forest, for instance, has many layers, each with different environmental conditions and combinations of species. Distinct communities of smaller plants, animals, and microbes live at different levels. Similarly, aquatic communities are often stratified into layers based on light penetration in the water, temperature, salinity, pressure, or other factors.

(a) Random

(b) Uniform

(c) Clustered

FIGURE 4.22 Distribution of members of a population in a given space can be (a) random, (b) uniform, or (c) clustered. The physical environment and biological interactions determine these patterns. The patterns may produce a graininess or patchiness in community structure.

Complexity and connectedness are important ecological indicators

Community complexity and connectedness generally are related to diversity and are important because they help us visualize and understand community functions. **Complexity** in ecological terms refers to the number of species at each trophic level and the number of trophic levels in a community. A diverse community may not be very complex if all its species are clustered in only a few trophic levels and form a relatively simple food chain.

By contrast, a complex, highly interconnected community (fig. 4.23) might have many trophic levels, some of which can be compartmentalized into subdivisions. In tropical rainforests, for instance, the herbivores can be grouped into "guilds" based on the specialized ways they feed on plants. There may be fruit eaters, leaf nibblers, root borers, seed gnawers, and sap suckers, each composed of species of very different size, shape, and even biological kingdom, but that feed in related ways. A highly interconnected community such as this can form a very elaborate food web.

Resilience and stability make communities resistant to disturbance

Many biological communities tend to remain relatively stable and constant over time. An oak forest tends to remain an oak forest, for example, because the species that make it up have self-perpetuating mechanisms. We can identify three kinds of stability or resiliency in ecosystems: *constancy* (lack of fluctuations in composition or functions), *inertia* (resistance to perturbations), and *renewal* (ability to repair damage after disturbance).

In 1955, Robert MacArthur, who was then a graduate student at Yale, proposed that the more complex and interconnected a community is, the more stable and resilient it will be in the face of disturbance. If many different species occupy each trophic level, some can fill in if others are stressed or eliminated by external forces, making the whole community resistant to perturbations and able to recover relatively easily from disruptions. This theory has been controversial, however. Some studies support it, while others do not. For example, Minnesota ecologist David Tilman, in studies of native prairie and recovering farm fields, found that plots with high diversity were better able to withstand and recover from drought than those with only a few species.

On the other hand, in a diverse and highly specialized ecosystem, removal of a few keystone members can eliminate many other associated species. Eliminating a major tree species from a tropical forest, for example, may destroy pollinators and fruit distributors as well. We might replant the trees, but could we replace the whole web of relationships on which they depend? In this case, diversity has made the forest less resilient rather than more.

Diversity is widely considered important and has received a great deal of attention. In particular, human impacts on diversity are a primary concern of many ecologists (Exploring Science p. 91).

Edges and boundaries are the interfaces between adjacent communities

An important aspect of community structure is the boundary between one habitat and its neighbors. We call these relationships **edge effects.** Sometimes, the edge of a patch of habitat is relatively sharp and distinct. In moving from a woodland patch into a grassland or cultivated field, you sense a dramatic change from the cool, dark, quiet forest interior to the windy, sunny, warmer,

FIGURE 4.23 Tropical rainforests are complex structurally and ecologically. Trees form layers, each with a different amount of light and a unique combination of flora and fauna. Many insects, arthropods, birds, and mammals spend their entire life in the canopy. In Brazil's Atlantic Rainforest, a single hectare had 450 tree species and many times that many insects. With so many species, the ecological relationships are complex and highly interconnected.

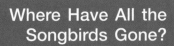

Every June, some 2,200 amateur ornithologists and bird watchers across the United States and Canada join in an annual bird count called the Breeding Bird Survey. Organized in 1966 by the U.S. Fish and Wildlife Service to follow bird population changes, this survey has discovered some shocking trends. While birds such as robins, starlings, and blackbirds that prosper around humans have increased their number and distribution over the past 30 years, many of our most colorful forest birds have declined severely. The greatest decreases have been among the true songbirds such as thrushes, orioles, tanagers, catbirds, vireos, buntings, and warblers. These long-distance migrants nest in northern forests but spend the winters in South or Central America or in the Caribbean Islands. Scientists call them neotropical migrants.

In many areas of the eastern United States and Canada, three-quarters or more of the neotropical migrants have declined significantly since the survey was started. Some that once were common have become locally extinct. Rock Creek Park in Washington, D.C., for instance, lost 75 percent of its songbird population and 90 percent of its long-distance migrant species in just 20 years. Nationwide, cerulean warblers, American redstarts, and ovenbirds declined about 50 percent in the single decade of the 1970s. Studies of radar images from National Weather Service stations in Texas and Louisiana suggest that only about half as many birds fly across the Gulf of Mexico each spring now compared to the 1960s. This could mean a loss of about half a billion birds in total.

What causes these devastating losses? Destruction of critical winter habitat is clearly a major issue. Birds often are much more densely crowded in the limited areas available to them during the winter than they are on their summer range. Unfortunately, forests throughout Latin America are being felled at an appalling rate. Central America, for instance, is losing about 1.4 million hectares (2 percent of its forests or an area about the size of Yellowstone National Park) each year. If this trend continues, there will be essentially no intact forest left in much of the region in 50 years.

But loss of tropical forests is not the only threat. Recent studies show that fragmentation of breeding habitat and nesting failures in the United States and Canada may be just as big a problem for woodland songbirds. Many of the most threatened species are adapted to deep woods and need an area of 10 hectares (25 acres) or more per pair to breed and raise their young. As our woodlands are broken up by roads, housing developments, and shopping centers, it becomes more and more difficult for these highly specialized birds to find enough contiguous woods to nest successfully.

This thrush has been equipped with a lightweight radio transmitter and antenna so that its movements can be followed by researchers

Predation and nest parasitism also present a growing threat to many bird species. In human-dominated landscapes, raccoons, opossums, crows, bluejays, squirrels, and house cats thrive. They are protected from larger predators like wolves or owls and find abundant supplies of food and places to hide. Cats are a particular problem. By some estimates, there are 100 million feral cats in the United States, and 73 million pet cats. A comparison of predation rates in the Great Smoky Mountain National Park and in small rural and suburban woodlands shows how devastating predators can be. In a 1,000-hectare study area of mature, unbroken forest in the national park, only one songbird nest in fifty was raided by predators. By contrast, in plots of 10 hectares or less near cities, up to 90 percent of the nests were raided.

Nest parasitism by brown-headed cowbirds is one of the worst threats for woodland songbirds. Rather than raise their young themselves, cowbirds lay their eggs in the nests of other species. The larger and more aggressive cowbird young either kick their foster siblings out of the nest, or claim so much food that the others starve. Well adapted to live around humans, there are now about 150 million cowbirds in the United States.

A study in southern Wisconsin found that 80 percent of the nests of woodland species were raided by predators and that three-quarters of those that survived were invaded by cowbirds. Another study in the Shawnee National Forest in southern Illinois found that 80 percent of the scarlet tanager nests contained cowbird eggs and that 90 percent of the wood thrush nests were taken over by these parasites. The sobering conclusion of this latter study is that there probably is no longer any place in Illinois where scarlet tanagers and wood thrushes can breed successfully.

What can we do about this situation? Elsewhere in this book, we discuss sustainable forestry and economic development projects that could preserve forests at home and abroad. Preserving corridors that tie together important areas also will help. In areas where people already live, clustering of houses protects remaining woods. Discouraging the clearing of underbrush and trees from yards and parks leaves shelter for the birds.

Could we reduce the number of predators or limit their access to critical breeding areas? Would you accept fencing or trapping of small predators in wildlife preserves? How would you feel about a campaign to keep house cats inside during the breeding season?

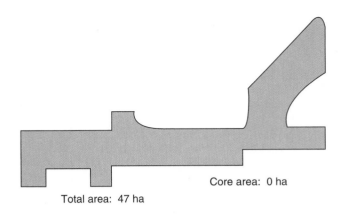

Total area: 47 ha   Core area: 0 ha

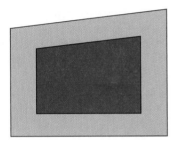

Total area: 47 ha   Core area: 20 ha

FIGURE 4.24 Ecological edges are known as ecotones. Temperature, wind, and humidity differ at the edges in a landscape. Edge conditions do extend into patches of habitat. Small or linear fragments may be mostly edge.

FIGURE 4.25 Shape can be as important as size in small preserves. While these areas are similar in size, no place in the top figure is far enough from the edge to have characteristics of core habitat, while the bottom patch has a significant core.

open space of the field or pasture (fig. 4.24). In other cases, one habitat type intergrades very gradually into another, so there is no distinct border.

Ecologists call the boundaries between adjacent communities **ecotones.** A community that is sharply divided from its neighbors is called a closed community. In contrast, communities with gradual or indistinct boundaries over which many species cross are called open communities. Often this distinction is a matter of degree or perception. As we saw earlier in this chapter, birds might feed in fields or grasslands but nest in the forest. As they fly back and forth, the birds interconnect the ecosystems by moving energy and material from one to the other, making both systems relatively open. Furthermore, the forest edge, while clearly different from the open field, may be sunnier and warmer than the forest interior, and may have a different combination of plant and animal species than either field or forest "core."

Depending on how far edge effects extend from the boundary, differently shaped habitat patches may have very dissimilar amounts of interior area (fig. 4.25). In Douglas fir forests of the Pacific Northwest, for example, increased rates of blowdown, decreased humidity, absence of shade-requiring ground cover, and other edge effects can extend as much as 200 m into a forest. A 40-acre block (about 400 m^2) surrounded by clear-cut would have essentially no true core habitat at all.

Many popular game animals, such as white-tailed deer and pheasants that are adapted to human disturbance, often are most plentiful in boundary zones between different types of habitat. Game managers once were urged to develop as much edge as possible to promote large game populations. Today, however, most wildlife conservationists recognize that the edge effects associated with habitat fragmentation are generally detrimental to biodiversity. Preserving large habitat blocks and linking smaller blocks with migration corridors may be the best ways to protect rare and endangered species (chapter 12).

4.4 COMMUNITIES ARE DYNAMIC AND CHANGE OVER TIME

If fire sweeps through a biological community, it's destroyed, right? Not so fast. Fire may be good for that community. Up until now, we've focused on the day-to-day interactions of organisms with their environments, set in a context of adaptation and selection. In this section, we'll step back and look at more dynamic aspects of communities and how they change over time.

The nature of communities is debated

For several decades starting in the early 1900s, ecologists in North America and Europe argued about the basic nature of communities. It doesn't make interesting party conversation, but those discussions affected how we study and understand communities, view the changes taking place within them, and ultimately use them. Both J. E. B. Warming (1841–1924) in Denmark and Henry Chandler Cowles (1869–1939) in the United States came up with the idea that communities develop in a sequence of stages, starting either from bare rock or after a severe disturbance. They worked in sand dunes and watched the changes as plants first took root in bare sand and, with further development, created forest. This example represents constant change, not stability. In sand dunes, the community that developed last and lasted the longest was called the **climax community.**

The idea of climax community was first championed by the biogeographer F. E. Clements (1874–1945). He viewed the process as a relay—species replace each other in predictable groups and in a fixed, regular order. He argued that every landscape has a characteristic climax community, determined mainly by climate. If left undisturbed, this community would mature to a characteristic set of organisms, each performing its optimal functions. A climax community to Clements represented the maximum complexity and stability that was possible. He and others made the analogy that the development of a climax community resembled the maturation of an organism. Both communities and organisms, they argued, began simply and primitively, maturing until a highly integrated, complex community developed.

This organismal theory of community was opposed by Clements' contemporary, H. A. Gleason (1882–1975), who saw community history as an unpredictable process. He argued that species are individualistic, each establishing in an environment according to its ability to colonize, tolerate the environmental conditions, and reproduce there. This idea allows for myriad temporary associations of plants and animals to form, fall apart, and reconstitute in slightly different forms, depending on environmental conditions and the species in the neighborhood. Imagine a time-lapse movie of a busy airport terminal. Passengers come and go; groups form and dissipate. Patterns and assemblages that seem significant may not mean much a year later. Gleason suggested that we think ecosystems are uniform and stable only because our lifetimes are too short and our geographic scope too limited to understand their actual dynamic nature.

Ecological succession describes a history of community development

In any landscape, you can read the history of biological communities. That history is revealed by the process of ecological succession. During succession, organisms occupy a site and change the environmental conditions. In **primary succession** land that is bare of soil—a sandbar, mudslide, rock face, volcanic flow—is colonized by living organisms where none lived before (fig. 4.26). When an existing community is disturbed, a new one develops from the biological legacy of the old in a process called **secondary succession.** In both kinds of succession, organisms change the environment by modifying soil, light levels, food supplies, and microclimate. This change permits new species to colonize and eventually replace the previous species, a process known as ecological development or facilitation.

In primary succession on land, the first colonists are hardy **pioneer species,** often microbes, mosses, and lichens that can withstand a harsh environment with few resources. When they die, the bodies of pioneer species create patches of organic matter. Organics and other debris accumulate in pockets and crevices, creating soil where seeds lodge and grow. As succession

proceeds, the community becomes more diverse and interspecies competition arises. Pioneers disappear as the environment favors new colonizers that have competitive abilities more suited to the new environment.

You can see secondary succession all around you, in abandoned farm fields, in clear-cut forests, and in disturbed suburbs and lots. Soil and possibly plant roots and seeds are present. Because soil lacks vegetation, plants that live one or two years (annuals and biennials) do well. Their light seeds travel far on the wind, and their seedlings tolerate full sun and extreme heat. When they die, they lay down organic material that improves the soil's fertility and shelters other seedlings. Soon long-lived and deep-rooted perennial grasses, herbs, shrubs, and trees take hold, building up the soil's organic matter and increasing its ability to store moisture. Forest species that cannot survive bare, dry, sunny ground eventually find ample food, a diverse community structure, and shelter from drying winds and low humidity.

Generalists figure prominently in early succession. Over thousands of years, however, competition should decrease as niches proliferate and specialists arise. In theory, long periods of community development lead to greater community complexity, high nutrient conservation and recycling, stable productivity, and great resistance to disturbance—an ideal state to be in when the slings and arrows of misfortune arrive.

Appropriate disturbances can benefit communities

Disturbances are plentiful on earth: landslides, mudslides, hailstorms, earthquakes, hurricanes, tornadoes, tidal waves, wildfires, and volcanoes, to name just the obvious. A **disturbance** is any force that disrupts the established patterns of species diversity and abundance, community structure, or community properties. Animals can cause disturbance. African elephants rip out small

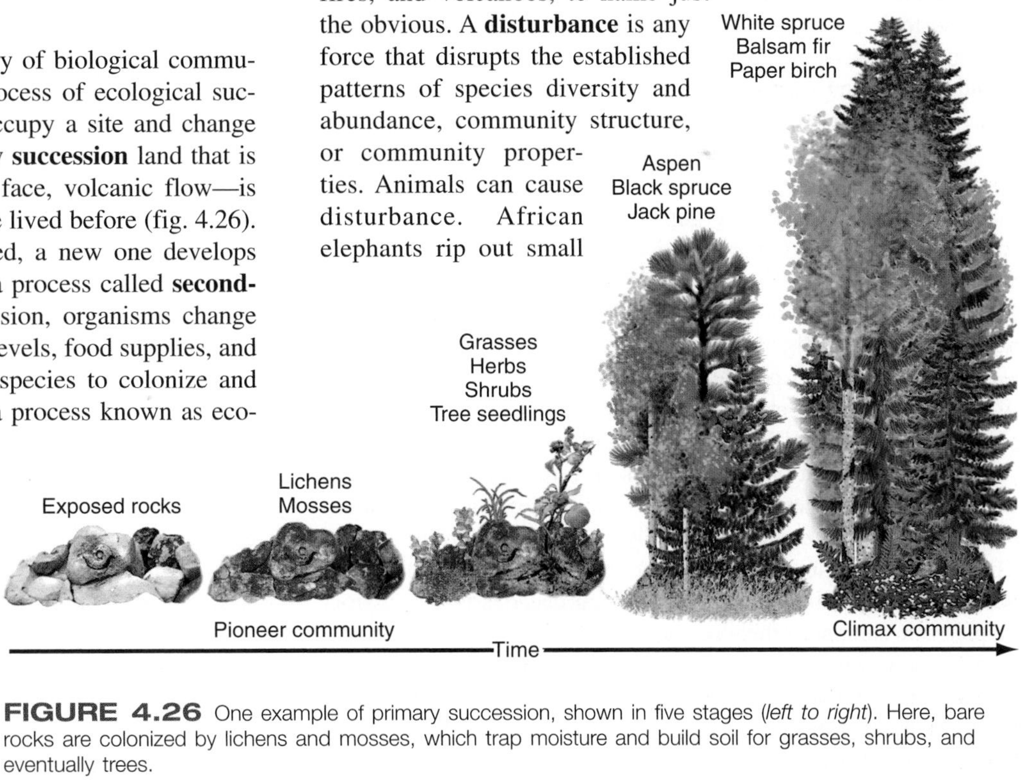

White spruce
Balsam fir
Paper birch

Aspen
Black spruce
Jack pine

Grasses
Herbs
Shrubs
Tree seedlings

Lichens
Mosses

Exposed rocks

Pioneer community ——Time—— Climax community

FIGURE 4.26 One example of primary succession, shown in five stages (*left to right*). Here, bare rocks are colonized by lichens and mosses, which trap moisture and build soil for grasses, shrubs, and eventually trees.

FIGURE 4.27 These "stump barrens" in Michigan's Upper Peninsula were created over a century ago when clear-cutting of dense white pine forest was followed by repeated burning. The stumps are left from the original forest, which has not grown back in more than 100 years.

trees, trample shrubs, and tear down tree limbs as they forage and move about, opening up forest communities and creating savannas. People also cause disturbances with agriculture, forestry, new roads and cities, and construction projects for dams and pipelines. It is customary in ecology to distinguish between natural disturbances and human-caused (or anthropogenic) disturbances, but a subtle point of clarification is needed.

Aboriginal populations have disturbed and continue to disturb communities around the world, setting fire to grasslands and savannas, practicing slash-and-burn agriculture in forests, and so on. Because their populations often are or were relatively small, the disturbances are patchy and limited in scale in forests, or restricted to quickly passing wildfires in grasslands, savannas, or woodland, which are comprised of species already adapted to fire.

The disturbances caused by technologically advanced and numerous people however, may be very different from the disturbances caused by small groups of aborigines. In the Kingston Plains of Michigan's Upper Peninsula, clear-cut logging followed by repeated human-set fires from 1880 to 1900 caused a change in basic ecological conditions such that the white pine forest has never regenerated (fig. 4.27). Given the right combination of disturbances by modern people, or by nature, it may take hundreds of years for a community to return to its predisturbance state.

Ecologists generally find that disturbance benefits most species, much as predation does, because it sets back supreme competitors and allows less-competitive species to persist. In northern temperate forests, maples (especially sugar maple) are more prolific seeders and more shade tolerant at different stages of growth than nearly any other tree species. Given decades of succession, maples out compete other trees for a place in the forest canopy. Most species of oak, hickory, and other light-requiring trees diminish in abundance, as do species of forest herbs. The dense shade of maples basically starves other species for light. When windstorms, tornadoes, wildfires, or ice storms hit a maple forest,

FIGURE 4.28 This lodgepole pine forest in Yellowstone National Park was once thought to be a climax forest, but we now know that this forest must be constantly renewed by periodic fire. It is an example of an equilibrium, or disclimax, community.

trees are toppled, branches broken, and light again reaches the forest floor and stimulates seedlings of oaks and hickories, as well as forest herbs. Breaking the grip of a supercompetitor is the helpful role disturbances often play.

Some landscapes never reach a stable climax in a traditional sense because they are characterized by periodic disturbance and are made up of **disturbance-adapted species** that survive fires underground, or resist the flames, and then reseed quickly after fires. Grasslands, the chaparral scrubland of California and the Mediterranean region, savannas, and some kinds of coniferous forests are shaped and maintained by periodic fires that have long been a part of their history (fig. 4.28). In fact, many of the dominant plant species in these communities need fire to suppress competitors, to prepare the ground for seeds to germinate, or to pop open cones or split thick seed coats and release seeds. Without fire, community structure would be quite different.

People taking an organismal view of such communities believe that disturbance is harmful. In the early 1900s this view merged with the desire to protect timber supplies from ubiquitous wildfires, and to store water behind dams while also controlling floods. Fire suppression and flood control became the central policies in American natural resource management (along with predator control) for most of the twentieth century. Recently, new concepts about natural disturbances are entering land management discussions and bringing change to land management policies. Grasslands and some forests are now considered "fire-adapted" and fires are allowed to burn in them if weather conditions are appropriate. Floods also are seen as crucial for maintaining floodplain and river health. Policymakers and managers increasingly consider ecological information when deciding on new dams and levee construction projects.

From another view, disturbance resets the successional clock that always operates in every community. Even though all seems chaotic after a disturbance, it may be that preserving species diversity by allowing in natural disturbances (or judiciously applied human disturbances) actually ensures stability over the long run, just as diverse prairies managed with fire recover after drought. In time, community structure and productivity get back to normal, species diversity is preserved, and nature seems to reach its dynamic balance.

FIGURE 4.29 Mongooses were released in Hawaii in an effort to control rats. The mongooses are active during the day, however, while the rats are night creatures, so they ignored each other. Instead, the mongooses attacked defenseless native birds and became as great a problem as the rats.

Introduced species can cause profound community change

Succession requires the continual introduction of new community members and the disappearance of previously existing species. New species move in as conditions become suitable; others die or move out as the community changes. New species also can be introduced after a stable community already has become established. Some cannot compete with existing species and fail to become established. Others are able to fit into and become part of the community, defining new ecological niches. If, however, an introduced species preys upon or competes more successfully with one or more populations that are native to the community, the entire nature of the community can be altered.

Human introductions of Eurasian plants and animals to non-Eurasian communities often have been disastrous to native species because of competition or overpredation. Oceanic islands offer classic examples of devastation caused by rats, goats, cats, and pigs liberated from sailing ships. All these animals are prolific, quickly developing large populations. Goats are efficient, nonspecific herbivores; they eat nearly everything vegetational, from grasses and herbs to seedlings and shrubs. In addition, their sharp hooves are

hard on plants rooted in thin island soils. Rats and pigs are opportunistic omnivores, eating the eggs and nestlings of seabirds that tend to nest in large, densely packed colonies, and digging up sea turtle eggs. Cats prey upon nestlings of both ground- and tree-nesting birds. Native island species are particularly vulnerable because they have not evolved under circumstances that required them to have defensive adaptations to these predators.

Sometimes we introduce new species in an attempt to solve problems created by previous introductions but end up making the situation worse. In Hawaii and on several Caribbean Islands, for instance, mongooses were imported to help control rats that had escaped from ships and were destroying indigenous birds and devastating plantations (fig. 4.29). Since the mongooses were diurnal (active in the day), however, and rats are nocturnal, they tended to ignore each other. Instead, the mongooses also killed native birds and further threatened endangered species. Our lessons from this and similar introductions have a new technological twist. Some of the ethical questions currently surrounding the release of genetically engineered organisms are based on concerns that they are novel organisms, and we might not be able to predict how they will interact with other species in natural ecosystems—let alone how they might respond to natural selective forces. It is argued that we can't predict either their behavior or their evolution.

CONCLUSION

Evolution is one of the key organizing principles of biology. It explains how species diversity originates, and how organisms are able to live in highly specialized ecological niches. Natural selection, in which beneficial traits are passed from survivors in one generation to their progeny, is the mechanism by which evolution occurs. Species interactions—competition, predation, symbiosis, and coevolution—are important factors in natural selection. The unique set of organisms and environmental conditions in an

ecological community give rise to important properties, such as productivity, abundance, diversity, structure, complexity, connectedness, resilience, and succession. Human introduction of new species as well as removal of existing ones can cause profound changes in biological communities and can compromise the life-supporting ecological services on which we all depend. Understanding these community ecology principles is a vital step in becoming an educated environmental citizen.

REVIEWING LEARNING OUTCOMES

By now you should be able to explain the following points:

4.1 Describe how evolution produces species diversity.

- Natural selection leads to evolution.
- All species live within limits.
- The ecological niche is a species' role and environment.
- Speciation maintains species diversity.
- Evolution is still at work.
- Taxonomy describes relationships among species.

4.2 Discuss how species interactions shape biological communities.

- Competition leads to resource allocation.
- Predation affects species relationships.
- Some adaptations help avoid predation.
- Symbiosis involves intimate relations among species.
- Keystone species have disproportionate influence.

4.3 Summarize how community properties affect species and populations.

- Productivity is a measure of biological activity.
- Abundance and diversity measure the number and variety of organisms.

- Community structure describes spatial distribution of organisms.
- Complexity and connectedness are important ecological indicators.
- Resilience and stability make communities resistant to disturbance.
- Edges and boundaries are the interfaces between adjacent communities.

4.4 Explain why communities are dynamic and change over time.

- The nature of communities is debated.
- Ecological succession describes a history of community development.
- Appropriate disturbances can benefit communities.
- Introduced species can cause profound community change.

PRACTICE QUIZ

1. Explain how tolerance limits to environmental factors determine distribution of a highly specialized species such as the saguaro cactus.

2. Productivity, diversity, complexity, resilience, and structure are exhibited to some extent by all communities and ecosystems. Describe how these characteristics apply to the ecosystem in which you live.

3. Define *selective pressure* and describe one example that has affected species where you live.

4. Define *keystone species* and explain their importance in community structure and function.

5. The most intense interactions often occur between individuals of the same species. What concept discussed in this chapter can be used to explain this phenomenon?

6. Explain how predators affect the adaptations of their prey.

7. Competition for a limited quantity of resources occurs in all ecosystems. This competition can be interspecific or intraspecific. Explain some of the ways an organism might deal with these different types of competition.

8. Describe the process of succession that occurs after a forest fire destroys an existing biological community. Why may periodic fire be beneficial to a community?

9. Which world ecosystems are most productive in terms of biomass (fig. 4.21)? Which are least productive? What units are used in this figure to quantify biomass accumulation?

10. Discuss the dangers posed to existing community members when new species are introduced into ecosystems.

CRITICAL THINKING AND DISCUSSION QUESTIONS

1. The concepts of natural selection and evolution are central to how most biologists understand and interpret the world, and yet the theory of evolution is contrary to the beliefs of many religious groups. Why do you think this theory is so important to science and so strongly opposed by others? What evidence would be required to convince opponents of evolution?

2. What is the difference between saying that a duck has webbed feet because it needs them to swim and saying that a duck is able to swim because it has webbed feet?

3. The concept of keystone species is controversial among ecologists because most organisms are highly interdependent. If each of the trophic levels is dependent on all the

others, how can we say one is most important? Choose an ecosystem with which you are familiar and decide whether it has a keystone species or keystone set.

4. Some scientists look at the boundary between two biological communities and see a sharp dividing line. Others looking at the same boundary see a gradual transition with much intermixing of species and many interactions between communities. Why are there such different interpretations of the same landscape?

5. The absence of certain lichens is used as an indicator of air pollution in remote areas such as national parks. How can we be sure that air pollution is really responsible? What evidence would be convincing?

6. We tend to regard generalists or "weedy" species as less interesting and less valuable than rare and highly specialized endemic species. What values or assumptions underlie this attitude?

DATA analysis   Species Competition

In a classic experiment on competition between species for a common food source, the Russian microbiologist G. F. Gause grew populations of different species of ciliated protozoans separately and together in an artificial culture medium. He counted the number of cells of each species and plotted the total volume of each population. The organisms were *Paramecium caudatum* and its close relative, *Paramecium aurelia*. He plotted the aggregate volume of cells rather than the total number in each population because *P. caudatum* is much larger than *P. aurelia* (this size difference allowed him to distinguish between them in a mixed culture). The graphs in this box show the experimental results. As we mentioned earlier in the text, this was one of the first experimental demonstrations of the principle of competitive exclusion. After studying these graphs, answer the following questions.

1. How do you read these graphs? What is shown in the top and bottom panels?

2. How did the total volume of the two species compare after 14 days of separate growth?

3. If *P. caudatum* is roughly twice as large as *P. aurelia*, how did the total number of cells compare after 14 days of separate growth?

4. How did the total volume of the two species compare after 24 days of growth in a mixed population?

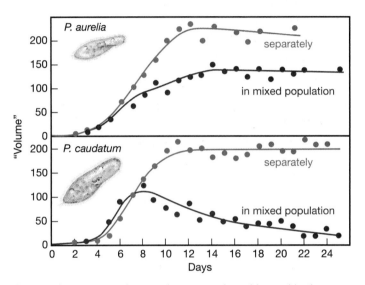

Growth of two *paramecium* species separately and in combination.
Source: Gause, Georgyi Frantsevitch. 1934 *The Struggle for Existence.* Dover Publications 1971 reprint of original text.

5. Which of the two species is the more successful competitor in this experiment?

6. Does the larger species always win in competition for food? Why not?

For Additional Help in Studying This Chapter, please visit our website at www.mhhe.com/cunningham10e. You will find additional practice quizzes and case studies, flashcards, regional examples, place markers for Google Earth™ mapping, and an extensive reading list, all of which will help you learn environmental science.

Traditional outrigger canoes and hand lines are still used by villagers on many islands in the southwestern Pacific, but these low-impact fishing methods are being threatened by trawlers, dynamite fishing, and other destructive techniques.

C H A P T E R

Biomes

Global Patterns of Life

What is the use of a house if you haven't got a
tolerable planet to put it on?

—Henry David Thoreau—

LEARNING OUTCOMES

After studying this chapter, you should be able to:

5.1 Recognize the characteristics of some major terrestrial biomes as well as the factors that determine their distribution.

5.2 Understand how and why marine environments vary with depth and distance from shore.

5.3 Compare the characteristics and biological importance of major freshwater ecosystems.

5.4 Summarize the overall patterns of human disturbance of world biomes.

Case Study   Saving the Reefs of Apo Island

As their outrigger canoes glide gracefully onto Apo Island's beach after an early morning fishing expedition, villagers call to each other to ask how fishing was. "Tunay mabuti!" (very good!) is the cheerful reply. Nearly every canoe has a basketful of fish; enough to feed a family for several days with a surplus to send to the market. Life hasn't always been so good on the island. Thirty years ago, this island, like many others in the Philippines, suffered a catastrophic decline in the seafood that was the mainstay of their diet and livelihood. Rapid population growth coupled with destructive fishing methods such as dynamite or cyanide fishing, small mesh gill nets, deep-sea trawling, and *Muro-ami* (a technique in which fish are chased into nets by pounding on coral with weighted lines) had damaged the reef habitat and exhausted fish stocks.

In 1979, scientists from Silliman University on nearby Negros Island visited Apo to explain how establishing a marine sanctuary could help reverse this decline. The coral reef fringing the island acts as a food source and nursery for many of the marine species sought by fishermen. Protecting that breeding ground, they explained, is the key to preserving a healthy fishery. The scientists took villagers from Apo to the uninhabited Sumilon Island, where a no-take reserve was teeming with fish.

After much discussion, several families decided to establish a marine sanctuary along a short section of Apo Island shoreline. Initially, the area had high-quality coral but few fish. The participating families took turns watching to make sure that no one trespassed in the no-fishing zone. Within a few years, fish numbers and sizes in the sanctuary increased dramatically, and "spillover" of surplus fish led to higher catches in surrounding areas. In 1985, Apo villagers voted to establish a 500 m (0.3 mi) wide marine sanctuary around the entire island.

Fishing is now allowed in this reserve, but only by low-impact methods such as hand-held lines, bamboo traps, large mesh nets, spearfishing without SCUBA gear, and hand netting. Coral-destroying techniques, such as dynamite, cyanide, trawling, and *Muro-ami* fishing are prohibited. By protecting the reef, villagers are guarding the nursery that forms the base for their entire marine ecosystem. Young fish growing up in the shelter of the coral move out as adults to populate the neighboring waters and yield abundant harvests. Fishermen report that they spend much less time traveling to distant fishing areas now that fish around the island are so much more abundant.

Apo Island's sanctuary is so successful that it has become the inspiration for more than 400 marine preserves throughout the Philippines and many others around the world. Not all are functioning as well as they might, but many, like Apo, have made dramatic progress in restoring abundant fish populations in nearby waters.

The rich marine life and beautiful coral formations in Apo Island's crystal clear water now attract international tourists (fig. 5.1). Two small hotels and a dive shop provide jobs for island residents. Other villagers take in tourists as boarders or sell food and T-shirts to visitors. The island government collects a diving/snorkeling fee, which has been used to build schools, improve island water supplies, and provide electricity to most of the island's 145 households. Almost all the island men still fish as their main occupation, but the fact that they don't have to go so far or work so hard for the fish they need means they have time for other activities, such as guiding diving tours or helping with household chores.

Higher family incomes now allow most island children to attend high school on Negros. Many continue their education with college or technical programs. Some find jobs elsewhere in the Philippines, and the money they send back home is a big economic boost for Apo families. Others return to their home island as teachers or to start businesses such as restaurants or dive shops. Seeing that they can do something positive to improve their environment and living conditions has empowered villagers to take on self-improvement projects that they may not otherwise have attempted.

Finding ways to live sustainably within the limits of the resource base available to us and without damaging the life-support systems provided by our ecosystem is a preeminent challenge of environmental science. Our answers to these challenges must be ecologically sound, economically sustainable, and socially acceptable if they are to succeed in the long term. Sometimes, as this case study shows, actions based on ecological knowledge and local action can spread to have positive effects on a global scale. Economics, policy, planning, and social organization all play vital roles in finding answers to human/environment problems. We'll discuss those disciplines later in this book, but first we'll look in this chapter, at some of the major terrestrial and aquatic biological communities as well as ways that humans are degrading them.

FIGURE 5.1 Coral reefs are among the most beautiful, species-rich, and productive biological communities on the planet. They serve as the nurseries for many open-water species. At least half the world's reefs are threatened by pollution, global climate change, destructive fishing methods, and other human activities, but they can be protected and restored if we care for them.

5.1 TERRESTRIAL BIOMES

Although all local environments are unique, it is helpful to understand them in terms of a few general groups with similar climate conditions, growth patterns, and vegetation types. We call these broad types of biological communities **biomes.** Understanding the global distribution of biomes, and knowing the differences in what grows where and why, is essential to the study of global environmental science. Biological productivity—and ecosystem resilience—varies greatly from one biome to another. Human use of biomes depends largely on those levels of productivity. Our ability to restore ecosystems and nature's ability to restore itself, depend largely on biome conditions. Clear-cut forests regrow relatively quickly in New England, but very slowly in Siberia, where current logging is expanding. Some grasslands rejuvenate quickly after grazing, and some are slower to recover. Why these differences? The sections that follow seek to answer this question.

Temperature and precipitation are among the most important determinants in biome distribution on land (fig. 5.2). If we know the general temperature range and precipitation level, we can predict what kind of biological community is likely to occur there, in the absence of human disturbance. Landforms, especially mountains, and prevailing winds also exert important influences on biological communities.

Because the earth is cooler at high latitudes (away from the equator), many temperature-controlled biomes occur in latitudinal bands. For example, a band of boreal (northern) forests crosses Canada and Siberia, tropical forests occur near the equator, and expansive grasslands lie near—or just beyond—the tropics (fig. 5.3). Many biomes are even named for their latitudes Tropical rainforests occur between the Tropic of Cancer (23° north) and the Tropic of Capricorn (23° south); arctic tundra lies near or above the Arctic Circle (66.6° north).

Temperature and precipitation change with elevation as well as with latitude. In mountainous regions, temperatures are cooler and precipitation is usually greater at high elevations. Communities can transition quickly from warm and dry to cold and wet as you go up a mountain. **Vertical zonation** is a term applied to vegetation zones defined by altitude. A 100 km transect from California's Central Valley up to Mt. Whitney, for example, crosses as many vegetation zones as you would find on a journey from southern California to northern Canada (fig. 5.4).

In this chapter, we'll examine the major terrestrial biomes, then we'll investigate ocean and freshwater communities and environments. Ocean environments are important because they cover

FIGURE 5.2 Biomes most likely to occur in the absence of human disturbance or other disruptions, according to average annual temperature and precipitation. *Note:* this diagram does not consider soil type, topography, wind speed, or other important environmental factors. Still, it is a useful general guideline for biome location.

two-thirds of the earth's surface, provide food for much of humanity, and help regulate our climate through photosynthesis. Wetlands are often small, but they have great influence on environmental health, biodiversity, and water quality. In chapter 12, we'll look at how we use these communities; and in chapter 13, we'll see how we preserve, manage, and restore them when they're degraded.

Think About It

As you look at the map in figure 5.3, which biomes do you think are most heavily populated by humans? Why? Which biomes are most altered by humans? (Check your answers by looking at figure 5.22.)

Tropical moist forests are warm and wet year-round

The humid tropical regions of the world support one of the most complex and biologically rich biome types in the world (fig. 5.5). Although there are several kinds of moist tropical forests, they share common attributes of ample rainfall and uniform temperatures. Cool **cloud forests** are found high in the mountains where fog and mist keep vegetation wet all the time. **Tropical rainforests** occur where rainfall is abundant—more than 200 cm (80 in.) per year—and temperatures are warm to hot year-round. For aid in reading the climate graphs in these figures, see the Data Analysis box at the end of this chapter.

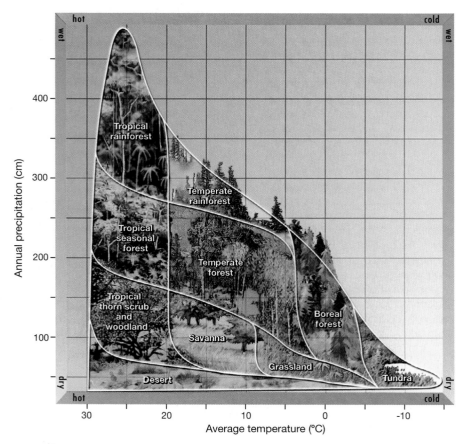

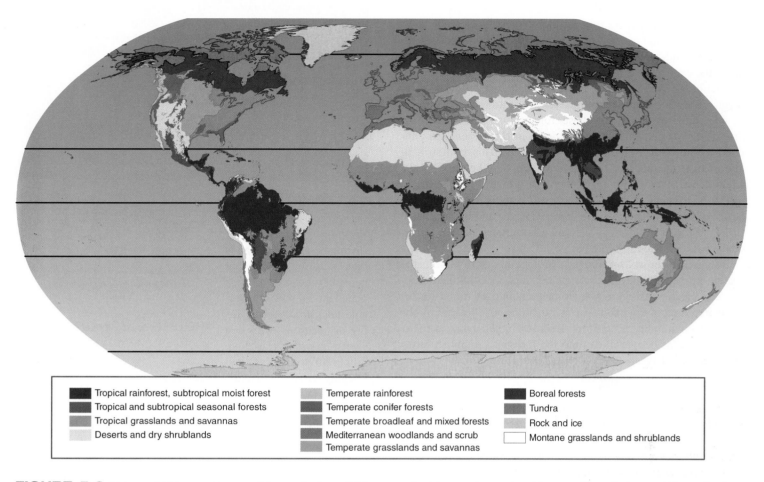

Tropical rainforest, subtropical moist forest

Tropical and subtropical seasonal forests

Tropical grasslands and savannas

Deserts and dry shrublands

Temperate rainforest

Temperate conifer forests

Temperate broadleaf and mixed forests

Mediterranean woodlands and scrub

Temperate grasslands and savannas

Boreal forests

Tundra

Rock and ice

Montane grasslands and shrublands

FIGURE 5.3 Major world biomes. Compare this map to figure 5.2 for generalized temperature and moisture conditions that control biome distribution. Also compare it to the satellite image of biological productivity (fig. 5.13).
Source: WWF Ecoregions.

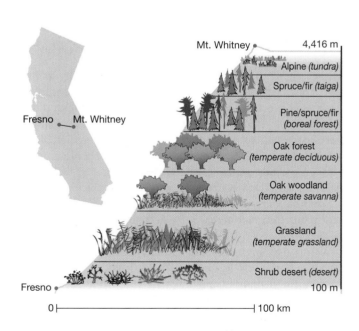

FIGURE 5.4 Vegetation changes with elevation because temperatures are lower and precipitation is greater high on a mountain side. A 100 km transect from Fresno, California, to Mt. Whitney (California's highest point) crosses vegetation zones similar to about seven different biome types.

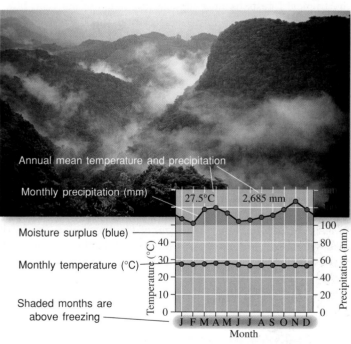

FIGURE 5.5 Tropical rainforests have luxuriant and diverse plant growth. Heavy rainfall in most months, shown in the climate graph, supports this growth.

The soil of both these tropical moist forest types tends to be old, thin, acidic, and nutrient-poor, yet the number of species present can be mind-boggling. For example, the number of insect species in the canopy of tropical rainforests has been estimated to be in the millions! It is estimated that one-half to two-thirds of all species of terrestrial plants and insects live in tropical forests.

The nutrient cycles of these forests also are distinctive. Almost all (90 percent) of the nutrients in the system are contained in the bodies of the living organisms. This is a striking contrast to temperate forests, where nutrients are held within the soil and made available for new plant growth. The luxuriant growth in tropical rainforests depends on rapid decomposition and recycling of dead organic material. Leaves and branches that fall to the forest floor decay and are incorporated almost immediately back into living biomass.

When the forest is removed for logging, agriculture, and mineral extraction, the thin soil cannot support continued cropping and cannot resist erosion from the abundant rains. And if the cleared area is too extensive, it cannot be repopulated by the rainforest community. Rapid deforestation is occurring in many tropical areas as people move into the forests to establish farms and ranches, but the land soon loses its fertility.

Tropical seasonal forests have annual dry seasons

Many tropical regions are characterized by distinct wet and dry seasons, although temperatures remain hot year-round. These areas support **tropical seasonal forests:** drought-tolerant forests that look brown and dormant in the dry season but burst into vivid green during rainy months. These forests are often called dry tropical forests because they are dry much of the year; however, there must be some periodic rain to support tree growth. Many of the trees and shrubs in a seasonal forest are drought-deciduous: They lose their leaves and cease growing when no water is available. Seasonal forests are often open woodlands that grade into savannas.

Tropical dry forests have typically been more attractive than wet forests for human habitation and have suffered greater degradation. Clearing a dry forest with fire is relatively easy during the dry season. Soils of dry forests often have higher nutrient levels and are more agriculturally productive than those of a rainforest. Finally, having fewer insects, parasites, and fungal diseases than a wet forest makes a dry or seasonal forest a healthier place for humans to live. Consequently, these forests are highly endangered in many places. Less than 1 percent of the dry tropical forests of the Pacific coast of Central America or the Atlantic coast of South America, for instance, remain in an undisturbed state.

Tropical savannas and grasslands are dry most of the year

Where there is too little rainfall to support forests, we find open **grasslands** or grasslands with sparse tree cover, which we call **savannas** (fig. 5.6). Like tropical seasonal forests, most tropical

FIGURE 5.6 Tropical savannas and grasslands experience annual drought and rainy seasons and year-round warm temperatures. Thorny acacias and abundant grazers thrive in this savanna. Yellow areas show moisture deficit.

savannas and grasslands have a rainy season, but generally the rains are less abundant or less dependable than in a forest. During dry seasons, fires can sweep across a grassland, killing off young trees and keeping the landscape open. Savanna and grassland plants have many adaptations to survive drought, heat, and fires. Many have deep, long-lived roots that seek groundwater and that persist when leaves and stems above the ground die back. After a fire, or after a drought, fresh green shoots grow quickly from the roots. Migratory grazers, such as wildebeest, antelope, or bison thrive on this new growth. Grazing pressure from domestic livestock is an important threat to both the plants and animals of tropical grasslands and savannas.

Deserts are hot or cold, but always dry

You may think of deserts as barren and biologically impoverished. Their vegetation is sparse, but it can be surprisingly diverse, and most desert plants and animals are highly adapted to survive long droughts, extreme heat, and often extreme cold. **Deserts** occur where precipitation is rare and unpredictable, usually with less than 30 cm of rain per year. Adaptations to these conditions include water-storing leaves and stems, thick epidermal layers to reduce water loss, and salt tolerance. As in other dry environments, many plants are drought-deciduous. Most desert plants also bloom and set seed quickly when a spring rain does fall.

Warm, dry, high-pressure climate conditions (chapter 15) create desert regions at about 30° north and south. Extensive deserts occur in continental interiors (far from oceans, which evaporate the moisture for most precipitation) of North America, Central Asia, Africa, and Australia (fig. 5.7). The rain shadow of the Andes produces the world's driest desert in coastal Chile. Deserts can also be cold. Antarctica is a desert. Some inland valleys apparently get almost no precipitation at all.

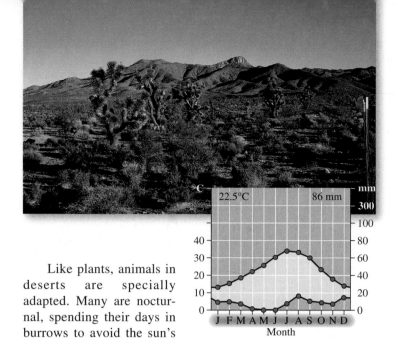

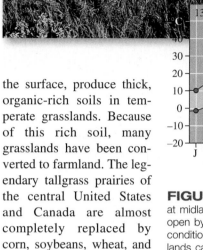

Like plants, animals in deserts are specially adapted. Many are nocturnal, spending their days in burrows to avoid the sun's heat and desiccation. Pocket mice, kangaroo rats, and gerbils can get most of their moisture from seeds and plants. Desert rodents also have highly concentrated urine and nearly dry feces that allow them to eliminate body waste without losing precious moisture.

FIGURE 5.7 Deserts generally receive less than 300 mm (30 cm) of precipitation per year. Hot deserts, as in the American Southwest, endure year-round drought and extreme heat in summer.

Deserts are more vulnerable than you might imagine. Sparse, slow-growing vegetation is quickly damaged by off-road vehicles. Desert soils recover slowly. Tracks left by army tanks practicing in California deserts during World War II can still be seen today.

Deserts are also vulnerable to overgrazing. In Africa's vast Sahel (the southern edge of the Sahara Desert), livestock are destroying much of the plant cover. Bare, dry soil becomes drifting sand, and restabilization is extremely difficult. Without plant roots and organic matter, the soil loses its ability to retain what rain does fall, and the land becomes progressively drier and more bare. Similar depletion of dryland vegetation is happening in many desert areas, including Central Asia, India, and the American Southwest and Plains states.

Temperate grasslands have rich soils

As in tropical latitudes, temperate (midlatitude) grasslands occur where there is enough rain to support abundant grass but not enough for forests (fig. 5.8). Usually grasslands are a complex, diverse mix of grasses and flowering herbaceous plants, generally known as forbs. Myriad flowering forbs make a grassland colorful and lovely in summer. In dry grasslands, vegetation may be less than a meter tall. In more humid areas, grasses can exceed 2 m. Where scattered trees occur in a grassland, we call it a savanna.

Deep roots help plants in temperate grasslands and savannas survive drought, fire, and extreme heat and cold. These roots, together with an annual winter accumulation of dead leaves on

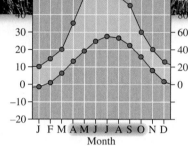

the surface, produce thick, organic-rich soils in temperate grasslands. Because of this rich soil, many grasslands have been converted to farmland. The legendary tallgrass prairies of the central United States and Canada are almost completely replaced by corn, soybeans, wheat, and other crops. Most remaining grasslands in this region

FIGURE 5.8 Grasslands occur at midlatitudes on all continents. Kept open by extreme temperatures, dry conditions, and periodic fires, grasslands can have surprisingly high plant and animal diversity.

are too dry to support agriculture, and their greatest threat is overgrazing. Excessive grazing eventually kills even deep-rooted plants. As ground cover dies off, soil erosion results, and unpalatable weeds, such as cheatgrass or leafy spurge, spread.

Temperate shrublands have summer drought

Often, dry environments support drought-adapted shrubs and trees, as well as grass. These mixed environments can be highly variable. They can also be very rich biologically. Such conditions are often described as Mediterranean (where the hot season coincides with the dry season producing hot, dry summers and cool, moist winters). Evergreen shrubs with small, leathery, sclerophyllous (hard, waxy) leaves form dense thickets. Scrub oaks, drought-resistant pines, or other small trees often cluster in sheltered valleys. Periodic fires burn fiercely in this fuel-rich plant assemblage and are a major factor in plant succession. Annual spring flowers often bloom profusely, especially after fires. In California, this landscape is called **chaparral,** Spanish for thicket. Resident animals are drought tolerant such as jackrabbits, kangaroo rats, mule deer, chipmunks, lizards, and many bird species. Very similar landscapes are found along the Mediterranean coast as well as southwestern Australia, central Chile, and South Africa. Although this biome doesn't cover a very large total area, it contains a high number of unique species and is often considered a "hot spot" for biodiversity. It also is highly desired for human habitation, often leading to conflicts with rare and endangered plant and animal species.

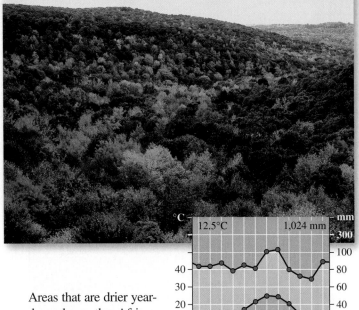

Areas that are drier year-round, such as the African Sahel (edge of the Sahara Desert), northern Mexico, or the American Intermountain West (or Great Basin) tend to have a more sparse, open shrubland, characterized by sagebrush (*Artemisia* sp.), chamiso (*Adenostoma* sp.), or saltbush (*Atriplex* sp.). Some typical animals of this biome in America are a wide variety of snakes and lizards, rodents, birds, antelope, and mountain sheep.

FIGURE 5.9 Temperate deciduous forests have year-round precipitation and winters near or below freezing.

Temperate forests can be evergreen or deciduous

Temperate, or midlatitudes, forests occupy a wide range of precipitation conditions but occur mainly between about 30 and 55 degrees latitude (see fig. 5.3). In general we can group these forests by tree type, which can be broad-leaf **deciduous** (losing leaves seasonally) or evergreen **coniferous** (cone-bearing).

Deciduous Forests

Broad-leaf forests occur throughout the world where rainfall is plentiful. In midlatitudes, these forests are deciduous and lose their leaves in winter. The loss of green chlorophyll pigments can produce brilliant colors in these forests in autumn (fig. 5.9). At lower latitudes, broad-leaf forest may be evergreen or drought-deciduous. Southern live oaks, for example, are broad-leaf evergreen trees.

Although these forests have a dense canopy in summer, they have a diverse understory that blooms in spring, before the trees leaf out. Spring ephemeral (short-lived) plants produce lovely flowers, and vernal (springtime) pools support amphibians and insects. These forests also shelter a great diversity of songbirds.

North American deciduous forests once covered most of what is now the eastern half of the United States and southern Canada. Most of western Europe was once deciduous forest but

was cleared a thousand years ago. When European settlers first came to North America, they quickly settled and cut most of the eastern deciduous forests for firewood, lumber, and industrial uses, as well as to clear farmland. Many of those regions have now returned to deciduous forest, though the dominant species have changed.

Deciduous forests can regrow quickly because they occupy moist, moderate climates. But most of these forests have been occupied so long that human impacts are extensive, and most native species are at least somewhat threatened. The greatest threat to broad-leaf deciduous forests is in eastern Siberia, where deforestation is proceeding rapidly. Siberia may have the highest deforestation rate in the world. As forests disappear, so do Siberian tigers, bears, cranes, and a host of other endangered species.

Coniferous Forests

Coniferous forests grow in a wide range of temperature and moisture conditions. Often they occur where moisture is limited: In cold climates, moisture is unavailable (frozen) in winter; hot climates may have seasonal drought; sandy soils hold little moisture, and they are often occupied by conifers. Thin, waxy leaves (needles) help these trees reduce moisture loss. Coniferous forests provide most wood products in North America. Dominant wood production regions include the southern Atlantic and Gulf coast states, the mountain West, and the Pacific Northwest (northern California to Alaska), but coniferous forests support forestry in many regions.

The coniferous forests of the Pacific coast grow in extremely wet conditions. The wettest coastal forests are known as **temperate rainforest,** a cool, rainy forest often enshrouded in fog (fig. 5.10). Condensation in the canopy (leaf drip) is a major form of precipitation in the understory. Mild year-round temperatures and abundant rainfall, up to 250 cm (100 in.) per year, result in luxuriant plant growth and giant trees such as the California redwoods, the largest trees in the world and the largest aboveground organism ever known to have existed. Redwoods once grew along the Pacific coast from California to Oregon, but logging has reduced them to a few small fragments.

Remaining fragments of ancient temperate rainforests are important areas of biodiversity. Recent battles over old-growth conservation (chapter 12) focus mainly on these areas. As with deciduous forests, Siberian forests are especially vulnerable to old-growth logging. The rate of this clearing, and its environmental effects, remain largely unknown.

Boreal forests occur at high latitudes

Because conifers can survive winter cold, they tend to dominate the **boreal forest,** or northern forests, that lie between about 50° and 60° north (fig. 5.11). Mountainous areas at lower latitudes may also have many characteristics and species of the boreal forest. Dominant trees are pines, hemlocks, spruce, cedar, and fir. Some deciduous trees are also present, such as maples, birch, aspen, and alder. These forests are slow-growing because of the cold temperatures and short frost-free growing

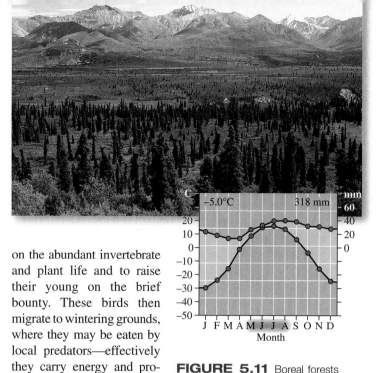

season, but they are still an expansive resource. In Siberia, Canada, and the western United States, large regional economies depend on boreal forests.

The extreme, ragged edge of the boreal forest, where forest gradually gives way to open tundra, is known by its Russian name, **taiga.** Here extreme cold and short summer limits the growth rate of trees. A 10 cm diameter tree may be over 200 years old in the far north.

FIGURE 5.10 Temperate rainforests have abundant but often seasonal precipitation that supports magnificent trees and luxuriant understory vegetation. Often these forests experience dry summers.

Tundra can freeze in any month

Where temperatures are below freezing most of the year, only small, hardy vegetation can survive. **Tundra,** a treeless landscape that occurs at high latitudes or on mountaintops, has a growing season of only two to three months, and it may have frost any month of the year. Some people consider tundra a variant of grasslands because it has no trees; others consider it a very cold desert because water is unavailable (frozen) most of the year.

Arctic tundra is an expansive biome that has low productivity because it has a short growing season (fig. 5.12). During midsummer, however, 24-hour sunshine supports a burst of plant growth and an explosion of insect life. Tens of millions of waterfowl, shorebirds, terns, and songbirds migrate to the Arctic every year to feast

on the abundant invertebrate and plant life and to raise their young on the brief bounty. These birds then migrate to wintering grounds, where they may be eaten by local predators—effectively they carry energy and protein from high latitudes to low latitudes. Arctic tundra is essential for global biodiversity, especially for birds.

Alpine tundra, occurring on or near mountaintops, has environmental conditions and vegetation similar to arctic tundra. These areas have a short, intense growing season. Often one sees a splendid profusion of flowers in alpine tundra; everything must flower at once in order to produce seeds in a few

FIGURE 5.11 Boreal forests have moderate precipitation but are often moist because temperatures are cold most of the year. Cold-tolerant and drought-tolerant conifers dominate boreal forests and taiga, the forest fringe.

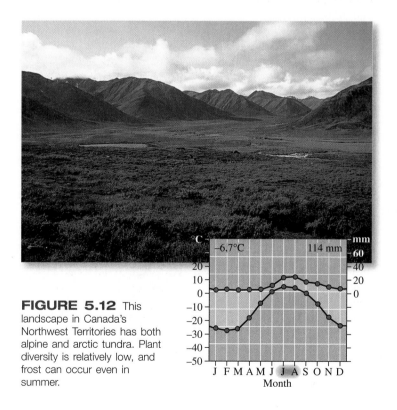

FIGURE 5.12 This landscape in Canada's Northwest Territories has both alpine and arctic tundra. Plant diversity is relatively low, and frost can occur even in summer.

weeks before the snow comes again. Many alpine tundra plants also have deep pigmentation and leathery leaves to protect against the strong ultraviolet light in the thin mountain atmosphere.

Compared to other biomes, tundra has relatively low diversity. Dwarf shrubs, such as willows, sedges, grasses, mosses, and lichens tend to dominate the vegetation. Migratory musk-ox, caribou, or alpine mountain sheep and mountain goats can live on the vegetation because they move frequently to new pastures.

Because these environments are too cold for most human activities, they are not as badly threatened as other biomes. There are important problems, however. Global climate change may be altering the balance of some tundra ecosystems, and air pollution from distant cities tends to accumulate at high latitudes (chapter 15). In eastern Canada, coastal tundra is being badly depleted by overabundant populations of snow geese, whose numbers have exploded due to winter grazing on the rice fields of Arkansas and Louisiana. Oil and gas drilling—and associated truck traffic—threatens tundra in Alaska and Siberia. Clearly, this remote biome is not independent of human activities at lower latitudes.

5.2 MARINE ECOSYSTEMS

The biological communities in oceans and seas are poorly understood, but they are probably as diverse and as complex as terrestrial biomes. In this section, we will explore a few facets of these fascinating environments. Oceans cover nearly three-fourths of the earth's surface, and they contribute in important, although often unrecognized, ways to terrestrial ecosystems. Like land-based systems, most marine communities depend on photosynthetic organisms. Often it is algae or tiny, free-floating photosynthetic plants (**phytoplankton**) that support a marine food web, rather than the trees and grasses we see on land. In oceans, photosynthetic activity tends to be greatest near coastlines, where nitrogen, phosphorus, and other nutrients wash offshore and fertilize primary producers. Ocean currents also contribute to the distribution of biological productivity, as they transport nutrients and phytoplankton far from shore (fig. 5.13).

As plankton, algae, fish, and other organisms die, they sink toward the ocean floor.

Deep-ocean ecosystems, consisting of crabs, filter-feeding organisms, strange phosphorescent fish, and many other life-forms, often rely on this "marine snow" as a primary nutrient source. Surface communities also depend on this material. Upwelling currents circulate nutrients from the ocean floor back to the surface. Along the coasts of South America, Africa, and Europe, these currents support rich fisheries.

Vertical stratification is a key feature of aquatic ecosystems. Light decreases rapidly with depth, and communities below the photic zone (light zone, often reaching about 20 m deep) must rely on energy sources other than photosynthesis to persist. Temperature also decreases with depth. Deep-ocean species often grow slowly in part because metabolism is reduced in cold conditions. In contrast, warm, bright, near-surface communities such as coral reefs and estuaries are among the world's most biologically productive environments. Temperature also affects the amount of oxygen and other elements that can be absorbed in water. Cold water holds abundant oxygen, so productivity is often high in cold oceans, as in the North Atlantic, North Pacific, and Antarctic.

Ocean systems can be described by depth and proximity to shore (fig. 5.14). In general, **benthic** communities occur on the bottom, and **pelagic** (from "sea" in Greek) zones are the water column. The epipelagic zone (*epi* = on top) has photosynthetic organisms. Below this are the mesopelagic (*meso* = medium), and bathypelagic (*bathos* = deep) zones. The deepest layers are the abyssal zone (to 4,000 m) and hadal zone (deeper than 6,000 m). Shorelines are known as littoral zones, and the area exposed by low tides is known as the intertidal zone. Often there is a broad, relatively shallow region along a continent's coast, which may reach a few kilometers or hundreds of kilometers from shore. This undersea area is the continental shelf.

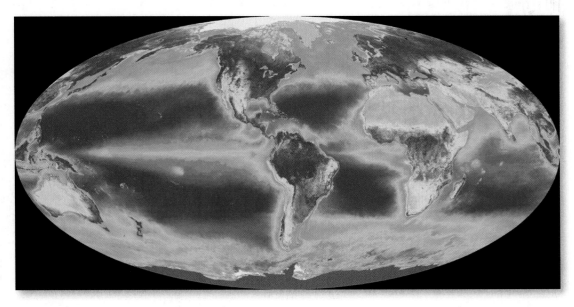

FIGURE 5.13 Satellite measurements of chlorophyll levels in the oceans and on land. Dark green to blue land areas have high biological productivity. Dark blue oceans have little chlorophyll and are biologically impoverished. Light green to yellow ocean zones are biologically rich. Courtesy Seawifs/NASA.

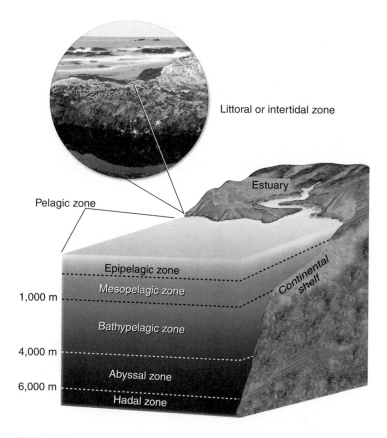

FIGURE 5.14 Light penetrates only the top 10–20 m of the ocean. Below this level, temperatures drop and pressure increases. Nearshore environments include the intertidal zone and estuaries.

FIGURE 5.15 Deep-ocean thermal vent communities were discovered only recently. They have great diversity and are unusual because they rely on chemosynthesis, not photosynthesis, for energy.

Open-ocean communities vary from surface to hadal zones

The open ocean is often referred to as a biological desert because it has relatively low productivity. But like terrestrial deserts, the open ocean has areas of rich productivity and diversity. Fish and plankton abound in regions such as the equatorial Pacific and Antarctic oceans, where nutrients are distributed by currents. Another notable exception, the Sargasso Sea in the western Atlantic, is known for its free-floating mats of brown algae. These algae mats support a phenomenal diversity of animals, including sea turtles, fish, and even eels that hatch amid the algae, then eventually migrate up rivers along the Atlantic coasts of North America and Europe.

Deep-sea thermal vent communities are another remarkable type of marine system (fig. 5.15) that was completely unknown until 1977 explorations with the deep-sea submarine *Alvin*. These communities are based on microbes that capture chemical energy, mainly from sulfur compounds released from thermal vents—jets of hot water and minerals on the ocean floor. Magma below the ocean crust heats these vents. Tube worms, mussels, and microbes on these vents are adapted to survive both extreme temperatures, often above 350°C (700°F), and the intense water pressure at depths of 7,000 m (20,000 ft)

or more. Oceanographers have discovered thousands of different types of organisms, most of them microscopic, in these communities (chapter 3).

Coastal zones support rich, diverse biological communities

As in the open ocean, shoreline communities vary with depth, light, nutrient concentrations, and temperature. Some shoreline communities, such as estuaries, have high biological productivity and diversity because they are enriched by nutrients washing from the land. But nutrient loading can be excessive. Around the world, more than 200 "dead zones" occur in coastal zones where excess nutrients stimulate bacterial growth that consumes almost all oxygen in the water and excludes most other life. We'll discuss this problem further in chapter 18.

Corals reefs are among the best-known marine ecosystems because of their extraordinary biological productivity and their diverse and beautiful organisms (see fig. 5.1). Reefs are aggregations of minute colonial animals (coral polyps) that live

(a) Coral reefs

(c) Estuary and salt marsh

(b) Mangroves

(d) Tide pool

FIGURE 5.16 Coastal environments support incredible diversity and help stabilize shorelines. Coral reefs (a), mangroves (b), and estuaries (c) also provide critical nurseries for marine ecosystems. Tide pools (d) also shelter highly specialized organisms.

symbiotically with photosynthetic algae. Calcium-rich coral skeletons build up to make reefs, atolls, and islands (fig. 5.16a). Reefs protect shorelines and shelter countless species of fish, worms, crustaceans, and other life-forms. Reef-building corals live where water is shallow and clear enough for sunlight to reach the photosynthetic algae. They need warm (but not too warm) water, and can't survive where high nutrient concentrations or runoff from the land create dense layers of algae, fungi, or sediment. Coral reefs also are among the most endangered biomes in the world. As the opening case study for this chapter shows, destructive fishing practices can damage or destroy coral communities. In addition, polluted urban runoff, trash, sewage and industrial effluent, sediment from agriculture, and unsustainable forestry are smothering coral reefs along coastlines that have high human populations. Introduced pathogens and predators also threaten many reefs. Perhaps the greatest threat to reefs is global warming. Elevated water temperatures cause **coral bleaching,** in which corals expel their algal partner and then die. The third UNESCO Conference on Oceans, Coasts, and Islands in 2006 reported that one-third of all coral reefs have been destroyed, and that 60 percent are now degraded and probably will be dead by 2030.

The value of an intact reef in a tourist economy, like that of Apo Island, can be upwards of (U.S.)$1 million per km^2. The costs of conserving these same reefs in a marine-protected area would be just (U.S.)$775 per km^2 per year, the UN Environment Programme estimates. Of the estimated 30 million small-scale fishers in the developing world, most are dependent to a greater or lesser extent on coral reefs. In the Philippines, the UN estimates that more than 1 million fishers depend directly on coral reefs for their livelihoods. We'll discuss reef restoration efforts further in chapter 13.

Sea-grass beds, or eel-grass beds, often occupy shallow, warm, sandy areas near coral reefs. Like reefs, these communities support a rich diversity of grazers, from snails and worms to turtles and manatees. Also like reefs, these environments are easily smothered by sediment originating from onshore agriculture and development.

Mangroves are trees that grow in salt water. They occur along calm, shallow, tropical coastlines around the world (fig. 5.16b). Mangrove forests or swamps help stabilize shorelines, and they are also critical nurseries for fish, shrimp, and other commercial species. Like coral reefs, mangroves line tropical and subtropical coastlines, where they are vulnerable to development, sedimentation, and overuse. Unlike reefs, mangroves provide commercial timber, and they can be clear-cut to make room for aquaculture (fish farming) and other activities. Ironically, mangroves provide the protected spawning beds for most of the fish and shrimp farmed in these ponds. As mangroves become increasingly threatened in tropical countries, villages relying on fishing for income and sustenance are seeing reduced catches and falling income.

Estuaries are bays where rivers empty into the sea, mixing fresh water with salt water. **Salt marshes,** shallow wetlands flooded regularly or occasionally with seawater, occur on shallow coastlines, including estuaries (fig. 5.16c). Usually calm, warm, and nutrient-rich, estuaries and salt marshes are biologically diverse and productive. Rivers provide nutrients and sediments, and a muddy bottom supports emergent plants (whose leaves emerge above the water surface), as well as the young forms of crustaceans, such as crabs and shrimp, and mollusks, such as clams and oysters. Nearly two-thirds of all marine fish and shellfish rely on estuaries and saline wetlands for spawning and juvenile development.

Estuaries near major American cities once supported an enormous wealth of seafood. Oyster beds and clam banks in the waters adjacent to New York, Boston, and Baltimore provided free and easy food to early residents. Sewage and other contaminants long ago eliminated most of these resources, however. Recently, major efforts have been made to revive Chesapeake Bay, America's largest and most productive estuary. These efforts have shown some success, but many challenges remain (see related story "Restoring the Chesapeake" at www.mhhe.com/cunningham10e).

In contrast to the shallow, calm conditions of estuaries, coral reefs, and mangroves, there are violent, wave-blasted shorelines that support fascinating life-forms in **tide pools.** Tide pools are depressions in a rocky shoreline that are flooded at high tide but retain some water at low tide. These areas remain rocky where wave action prevents most plant growth or sediment (mud) accumulation. Extreme conditions, with frigid flooding at high tide and hot, desiccating sunshine at low tide, make life impossible for most species. But the specialized animals and plants that do occur in this rocky intertidal zone are astonishingly diverse and beautiful (fig. 5.16d).

Barrier islands are low, narrow, sandy islands that form parallel to a coastline (fig. 5.17). They occur where the continental shelf is shallow and rivers or coastal currents provide a steady source of sediments. They protect brackish (moderately salty), inshore lagoons and salt marshes from storms, waves, and tides. One of the world's most extensive sets of barrier islands lines the Atlantic coast from New England to Florida, as well as along the Gulf coast of Texas. Composed of sand that is constantly reshaped by wind and waves, these islands can be formed or removed by a single violent storm. Because they are mostly beach, barrier islands are also popular places for real estate development. About 20 percent of the barrier island surface in the United States has been developed. Barrier islands are also critical to preserving coastal shorelines, settlements, estuaries, and wetlands.

Unfortunately, human occupation often destroys the value that attracts us there in the first place. Barrier islands and beaches are dynamic environments, and sand is hard to keep in place. Wind and wave erosion is a constant threat to beach develop-

FIGURE 5.17 A barrier island, Assateague, along the Maryland–Virginia coast. Grasses cover and protect dunes, which keep ocean waves from disturbing the bay, salt marshes, and coast at right. Roads cut through the dunes expose them to erosion.

ments. Walking or driving vehicles over dune grass destroys the stabilizing vegetative cover and accelerates, or triggers, erosion. Cutting roads through the dunes further destabilizes these islands, making them increasingly vulnerable to storm damage. When Hurricane Katrina hit the U.S. Gulf coast in 2005, it caused at least $200 billion in property damage and displaced 4 million people. Thousands of homes were destroyed (fig. 5.18), particularly on low-lying barrier islands.

Because of these problems, we spend billions of dollars each year building protective walls and barriers, pumping sand onto beaches from offshore, and moving sand from one beach area to

FIGURE 5.18 Winter storms have eroded the beach and undermined the foundations of homes on this barrier island. Breaking through protective dunes to build such houses damages sensitive plant communities and exposes the whole island to storm sand erosion. Coastal zone management attempts to limit development on fragile sites.

another. Much of this expense is borne by the public. Some planners question whether we should allow rebuilding on barrier islands, especially after they've been destroyed multiple times.

5.3 FRESHWATER ECOSYSTEMS

Freshwater environments are far less extensive than marine environments, but they are centers of biodiversity. Most terrestrial communities rely, to some extent, on freshwater environments. In deserts, isolated pools, streams, and even underground water systems, support astonishing biodiversity as well as provide water to land animals. In Arizona, for example, most birds gather in trees and bushes surrounding the few available rivers and streams.

Lakes have open water

Freshwater lakes, like marine environments, have distinct vertical zones (fig. 5.19). Near the surface a subcommunity of plankton, mainly microscopic plants, animals, and protists (single-celled organisms such as amoebae), float freely in the water column. Insects such as water striders and mosquitoes also live at the air-water interface. Fish move through the water column, sometimes near the surface, and sometimes at depth.

Finally, the bottom, or *benthos,* is occupied by a variety of snails, burrowing worms, fish, and other organisms. These make up the benthic community. Oxygen levels are lowest in the benthic environment, mainly because there is little mixing to introduce oxygen to this zone. Anaerobic bacteria (not using oxygen) may live in low-oxygen sediments. In the littoral zone, emergent plants such as cattails and rushes grow in the bottom sediment. These plants create important functional links between layers of an aquatic ecosystem, and they may provide the greatest primary productivity to the system.

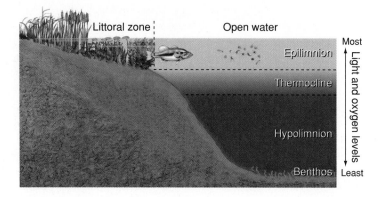

FIGURE 5.19 The layers of a deep lake are determined mainly by gradients of light, oxygen, and temperature. The epilimnion is affected by surface mixing from wind and thermal convections, while mixing between the hypolimnion and epilimnion is inhibited by a sharp temperature and density difference at the thermocline.

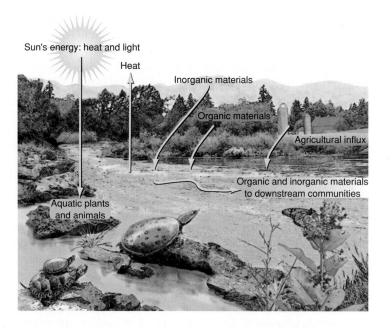

FIGURE 5.20 The character of freshwater ecosystems is greatly influenced by the immediately surrounding terrestrial ecosystems, and even by ecosystems far upstream or far uphill from a particular site.

Lakes, unless they are shallow, have a warmer upper layer that is mixed by wind and warmed by the sun. This layer is the *epilimnion.* Below the epilimnion is the hypolimnion (*hypo* = below), a colder, deeper layer that is not mixed. If you have gone swimming in a moderately deep lake, you may have discovered the sharp temperature boundary, known as the **thermocline,** between these layers. Below this boundary, the water is much colder. This boundary is also called the mesolimnion.

Local conditions that affect the characteristics of an aquatic community include (1) nutrient availability (or excess) such as nitrates and phosphates; (2) suspended matter, such as silt, that affects light penetration; (3) depth; (4) temperature; (5) currents; (6) bottom characteristics, such as muddy, sandy, or rocky floor; (7) internal currents; and (8) connections to, or isolation from, other aquatic and terrestrial systems (fig. 5.20).

Wetlands are shallow and productive

Wetlands are shallow ecosystems in which the land surface is saturated or submerged at least part of the year. Wetlands have vegetation that is adapted to grow under saturated conditions. These legal definitions are important because although wetlands make up only a small part of most countries, they are disproportionately important in conservation debates and are the focus of continual legal disputes around the world and in North America. Beyond these basic descriptions, defining wetlands is a matter of hot debate. How often must a wetland be saturated, and for how long? How large must it be to deserve legal protection? Answers can vary, depending on political, as well as ecological, concerns.

(a) Swamp, or wooded wetland

(b) Marsh

(c) Coastal saltmarsh

FIGURE 5.21 Wetlands provide irreplaceable ecological services, including water filtration, water storage and flood reduction, and habitat. Forested wetlands (a) are often called swamps; marshes (b) have no trees; coastal saltmarshes (c) are tidal and have rich diversity.

These relatively small systems support rich biodiversity, and they are essential for both breeding and migrating birds. Although wetlands occupy less than 5 percent of the land in the United States, the Fish and Wildlife Service estimates that one-third of all endangered species spend at least part of their lives in wetlands. Wetlands retain storm water and reduce flooding by slowing the rate at which rainfall reaches river systems. Floodwater storage is worth $3 billion to $4 billion per year in the United States. As water stands in wetlands, it also seeps into the ground, replenishing groundwater supplies. Wetlands filter, and even purify, urban and farm runoff, as bacteria and plants take up nutrients and contaminants in water. They are also in great demand for filling and development. They are often near cities or farms, where land is valuable, and once drained, wetlands are easily converted to more lucrative uses.

Wetlands are described by their vegetation. **Swamps** are wetlands with trees (fig. 5.21*a*). **Marshes** are wetlands without trees (fig. 5.21*b*). **Bogs** are areas of saturated ground, and usually the ground is composed of deep layers of accumulated, undecayed vegetation known as peat. **Fens** are similar to bogs except that they are mainly fed by groundwater, so that they have mineral-rich water and specially adapted plant species. Bogs are fed mainly by precipitation. Swamps and marshes have high biological productivity. Bogs and fens, which are often nutrient-poor, have low biological productivity. They may have unusual and interesting species, though, such as sundews and pitcher plants, which are adapted to capture nutrients from insects rather than from soil.

The water in marshes and swamps usually is shallow enough to allow full penetration of sunlight and seasonal warming (fig. 5.21*c*). These mild conditions favor great photosynthetic activity, resulting in high productivity at all trophic levels. In short, life is abundant and varied. Wetlands are major breeding, nesting, and migration staging areas for waterfowl and shorebirds.

Wetlands may gradually convert to terrestrial communities as they fill with sediment, and as vegetation gradually fills in toward the center. Often this process is accelerated by increased sediment loads from urban development, farms, and roads. Wetland losses are one of the areas of greatest concern among biologists.

5.4 HUMAN DISTURBANCE

Humans have become dominant organisms over most of the earth, damaging or disturbing more than half of the world's terrestrial ecosystems to some extent. By some estimates, humans preempt about 40 percent of the net terrestrial primary productivity of the biosphere either by consuming it directly, by interfering with its production or use, or by altering the species composition or physical processes of human-dominated ecosystems. Conversion of natural habitat to human uses is the largest single cause of biodiversity losses.

Researchers from the environmental group Conservation International have attempted to map the extent of human disturbance of the natural world (fig. 5.22). The greatest impacts have been in Europe, parts of Asia, North and Central America, and islands such as Madagascar, New Zealand, Java, Sumatra, and those in the Caribbean. Data from this study are shown in table 5.1.

Temperate broad-leaf forests are the most completely human-dominated of any major biome. The climate and soils that support such forests are especially congenial for human occupation. In eastern North America or most of Europe, for example, only remnants of the original forest still persist. Regions with a Mediterranean climate generally are highly desired for human habitation. Because these landscapes also have high levels of biodiversity, conflicts between human preferences and biological values frequently occur.

Temperate grasslands, temperate rainforests, tropical dry forests, and many islands also have been highly disturbed by human activities. If you have traveled through the American cornbelt states such as Iowa or Illinois, you have seen how thoroughly former prairies have been converted to farmlands. Intensive cultivation of this land exposes the soil to erosion and fertility losses (chapter 9). Islands, because of their isolation, often have high

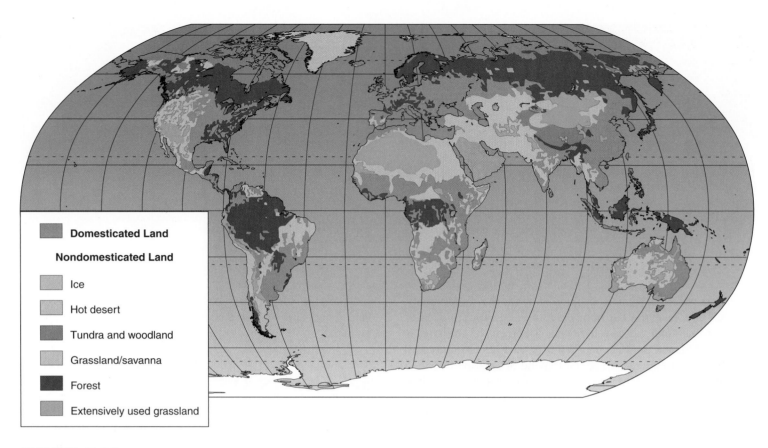

FIGURE 5.22 Domesticated land has replaced much of the earth's original land cover.
Source: United Nations Environment Programme, *Global Environment Outlook,* 1997.

TABLE 5.1

Human Disturbance

Biome	Total Area (10^6 km²)	% Undisturbed Habitat	% Human Dominated
Temperate broad-leaf forests	9.5	6.1	81.9
Chaparral	6.6	6.4	67.8
Temperate grasslands	12.1	27.6	40.4
Temperate rainforests	4.2	33.0	46.1
Tropical dry forests	19.5	30.5	45.9
Mixed mountain systems	12.1	29.3	25.6
Mixed island systems	3.2	46.6	41.8
Cold deserts/semideserts	10.9	45.4	8.5
Warm deserts/semideserts	29.2	55.8	12.2
Moist tropical forests	11.8	63.2	24.9
Tropical grasslands	4.8	74.0	4.7
Temperate coniferous forests	18.8	81.7	11.8
Tundra and arctic desert	20.6	99.3	0.3

Note: Where undisturbed and human-dominated areas do not add up to 100 percent, the difference represents partially disturbed lands.

Source: Hannah, Lee, et al., "Human Disturbance and Natural Habitat: A Biome Level Analysis of a Global Data Set," in *Biodiversity and Conservation,* 1995, Vol. 4:128–55.

numbers of endemic species. Many islands, such as Madagascar, Haiti, and Java have lost more than 99 percent of their original land cover.

Tundra and arctic deserts are the least disturbed biomes in the world. Harsh climates and unproductive soils make these areas unattractive places to live for most people. Temperate conifer forests also generally are lightly populated and large areas remain in a relatively natural state. However, recent expansion of forest harvesting in Canada and Siberia may threaten the integrity of this biome. Large expanses of tropical moist forests still remain in the Amazon and Congo basins but in other areas of the tropics such as West Africa, Madagascar, Southeast Asia, and the Indo-Malaysian peninsula and archipelago, these forests are disappearing at a rapid rate (chapter 12).

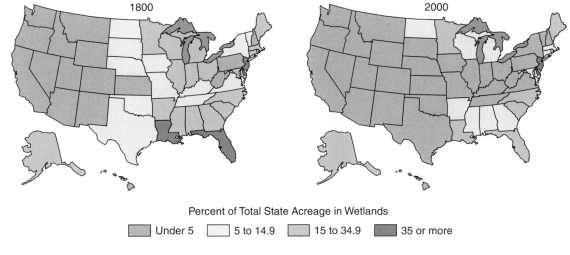

Percent of Total State Acreage in Wetlands

☐ Under 5   ☐ 5 to 14.9   ☐ 15 to 34.9   ■ 35 or more

FIGURE 5.23 Over the past two centuries, more than half of the original wetlands in the lower 48 states have been drained, filled, polluted, or otherwise degraded. Some of the greatest losses have been in midwestern farming states where up to 99 percent of all wetlands have been lost.

As mentioned earlier, wetlands have suffered severe losses in many parts of the world. About half of all original wetlands in the United States have been drained, filled, polluted, or otherwise degraded over the past 250 years. In the prairie states, small potholes and seasonally flooded marshes have been drained and converted to croplands on a wide scale. Iowa, for example, is estimated to have lost 99 percent of its presettlement wetlands (fig. 5.23). Similarly, California has lost 90 percent of the extensive marshes and deltas that once stretched across its central valley. Wooded swamps and floodplain forests in the southern United States have been widely disrupted by logging and conversion to farmland.

Similar wetland disturbances have occurred in other countries as well. In New Zealand, over 90 percent of natural wetlands have been destroyed since European settlement. In Portugal, some 70 percent of freshwater wetlands and 60 percent of estuarine habitats have been converted to agriculture and industrial areas. In Indonesia, almost all the mangrove swamps that once lined the coasts of Java have been destroyed, while in the Philippines and Thailand, more than two-thirds of coastal mangroves have been cut down for firewood or conversion to shrimp and fish ponds.

Slowing this destruction, or even reversing it, is a challenge that we will discuss in chapter 13.

CONCLUSION

The potential location of biological communities is determined in large part by climate, moisture availability, soil type, geomorphology, and other natural features. Understanding the global distribution of biomes, and knowing the differences in who lives where and why, are essential to the study of global environmental science. Human occupation and use of natural resources is strongly dependent on the biomes found in particular locations. We tend to prefer mild climates and the highly productive biological communities found in temperate zones. These biomes also suffer the highest rates of degradation and overuse.

Being aware of the unique conditions and the characteristics evolved by plants and animals to live in those circumstances can help you appreciate how and why certain species live in unique biomes, such as seasonal tropical forests, alpine tundra, or chaparral shrublands.

Oceans cover nearly three-fourths of the earth's surface, and yet we know relatively little about them. Some marine biomes, such as coral reefs, can be as biologically diverse and productive as any terrestrial biome. People have depended on rich, complex coastal ecosystems for eons, but in recent times rapidly growing human populations, coupled with more powerful ways to harvest resources, have led to damage—and, in some cases, irreversible destruction—of these irreplaceable treasures. Still, there is reason to hope that we'll find ways to live sustainably with nature. The opening case study of this chapter illustrates how, without expensive high technology, a group of local residents protected and restored their coral reef. It gives us optimism that we'll find similar solutions in other biologically rich but endangered biomes.

REVIEWING LEARNING OUTCOMES

By now you should be able to explain the following points:

5.1 Recognize the characteristics of some major terrestrial biomes as well as the factors that determine their distribution.

- Tropical moist forests are warm and wet year-round.
- Tropical seasonal forests have annual dry seasons.
- Tropical savannas and grasslands are dry most of the year.
- Deserts are hot or cold, but always dry.
- Temperate grasslands have rich soils.
- Temperate shrublands have summer drought.
- Temperate forests can be evergreen or deciduous.
- Boreal forests occur at high latitudes.
- Tundra can freeze in any month.

5.2 Understand how and why marine environments vary with depth and distance from shore.

- Open-ocean communities vary from surface to hadal zones.
- Coastal zones support rich, diverse biological communities.

5.3 Compare the characteristics and biological importance of major freshwater ecosystems.

- Lakes have open water.
- Wetlands are shallow and productive.

5.4 Summarize the overall patterns of human disturbance of world biomes.

- Biomes that humans find comfortable and profitable have high rates of disturbance, while those that are less attractive or have limited resources have large pristine areas.

PRACTICE QUIZ

1. Throughout the central portion of North America is a large biome once dominated by grasses. Describe how physical conditions and other factors control this biome.

2. What is taiga and where is it found? Why might logging in taiga be more disruptive than in southern coniferous forests?

3. Why are tropical moist forests often less suited for agriculture and human occupation than tropical deciduous forests?

4. Find out the annual temperature and precipitation conditions where you live (fig. 5.2). Which biome type do you occupy?

5. Describe four different kinds of wetlands and explain why they are important sites of biodiversity and biological productivity.

6. Forests differ according to both temperature and precipitation. Name and describe a biome that occurs in (a) hot, (b) cold, (c) wet, and (d) dry climates (one biome for each climate).

7. How do physical conditions change with depth in marine environments?

8. Describe four different coastal ecosystems.

CRITICAL THINKING AND DISCUSSION QUESTIONS

1. What physical and biological factors are most important in shaping your biological community? How do the present characteristics of your area differ from those 100 or 1,000 years ago?

2. Forest biomes frequently undergo disturbances such as fire or flooding. As more of us build homes in these areas, what factors should we consider in deciding how to protect people from natural disturbances?

3. Often humans work to preserve biomes that are visually attractive. What biomes might be lost this way? Is this a problem?

4. Disney World in Florida wants to expand onto a wetland. It has offered to buy and preserve a large nature preserve in a different area to make up for the wetland it is destroying. Is that reasonable? What conditions would make it reasonable or unreasonable?

5. Suppose further that the wetland being destroyed in question 4 and its replacement area both contain several endangered species (but different ones). How would you compare different species against each other? How many plant or insect species would one animal species be worth?

6. Historically, barrier islands have been hard to protect because links between them and inshore ecosystems are poorly recognized. What kinds of information would help a community distant from the coast commit to preserving a barrier island?

DATA analysis   Reading Climate Graphs

As you've learned in this chapter, temperature and moisture are critical factors in determining the distribution and health of ecosystems. But how do you read the climate and precipitation graphs that accompany the description of each biome? To begin, examine the three climate graphs in this box. These graphs show annual trends in temperature and precipitation (rainfall and snow). They also indicate the relationship between potential evaporation, which depends on temperature and precipitation. When evaporation exceeds precipitation, dry conditions result (yellow areas). Extremely wet months are shaded dark blue on the graphs. Moist climates may vary in precipitation rates, but evaporation rarely exceeds precipitation. Months above freezing temperature (shaded brown on the X-axis) have most evaporation. Comparing these climate graphs helps us understand the different seasonal conditions that control plant and animal lives in different biomes.

1. What are the maximum and minimum temperatures in each of the three locations shown?
2. What do these temperatures correspond to in Fahrenheit? (*Hint:* look at the table in the back of your book).
3. Which area has the wettest climate; which is driest?
4. How do the maximum and minimum monthly rainfalls in San Diego and Belém compare?
5. Describe these three climates.
6. What kinds of biomes would you expect to find in these areas?

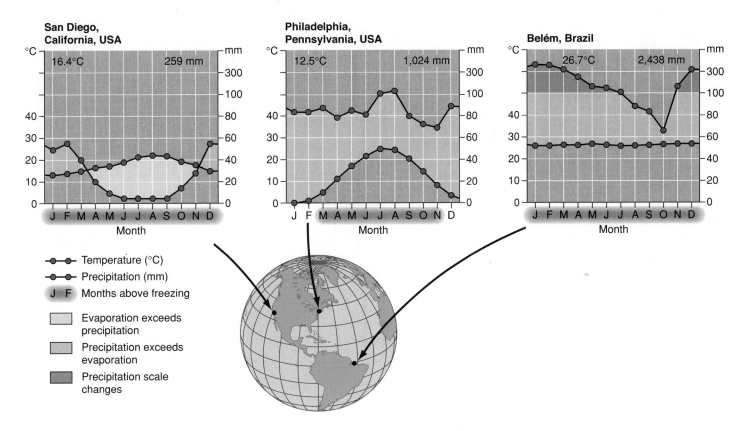

Moisture availability depends on temperature as well as precipitation. The horizontal axis on these climate diagrams shows months of the year; vertical axes show temperature (*left side*) and precipitation (*right*). The number of dry months (*shaded yellow*) and wetter months (*blue*) varies with geographic location. Mean annual temperature (°C) and precipitation (mm) are shown at the top of each graph.

For Additional Help in Studying This Chapter, please visit our website at www.mhhe.com/cunningham10e. You will find additional practice quizzes and case studies, flashcards, regional examples, place markers for Google Earth™ mapping, and an extensive reading list, all of which will help you learn environmental science.

Crew members sort fish on a trawler. As the large predators, such as cod, have been exhausted, we turn our attention to smaller prey. Some call this fishing down the food chain.

CHAPTER

Population Biology

6

Nature teaches more than she preaches.
—John Burroughs—

LEARNING OUTCOMES

After studying this chapter, you should be able to:

6.1 Describe the dynamics of population growth.

6.2 Summarize the factors that increase or decrease populations.

6.3 Compare and contrast the factors that regulate population growth.

6.4 Identify some applications of population dynamics in conservation biology.

Case Study How Many Fish in the Sea?

When John Cabot discovered Newfoundland in 1497, cod (*Gadus morhua*) were so abundant that sailors could simply scoop them up in baskets. Growing up to 2 meters (6 feet) long, weighing as much as 100 kg (220 lbs) and living up to 25 years, cod have been a major food resource for Europeans for more than 500 years. Because the firm, white flesh of the cod has little fat, it can be salted and dried to produce a long-lasting food that can be stored or shipped to distant markets.

No one knows how many cod there may have once been in the ocean. Coastal people recognized centuries ago that huge schools would gather to spawn on shoals and rocky reefs from Massachusetts around the North Atlantic to the British Isles. In 1990, Canadian researchers watched on sonar as a school estimated to contain several hundred million fish spawned on the Gorges Bank off Newfoundland. Because a single mature female cod can lay up to 10 million eggs in a spawning, a school like this—only one of many in the ocean—might have produced a quadrillion eggs.

It seems that such an abundant and fecund an animal could never be threatened by humans. In 1883, Thomas Huxley, the eminent biologist and friend of Charles Darwin, said, "I believe that the cod fishery . . . and probably all the great sea fisheries are inexhaustible . . . Nothing we do seriously affects the number of fish." But in Huxley's time, most cod were caught on handlines by fishermen in small wooden dories. He couldn't have imagined the size and efficiency of modern fishing fleets. Following World War II, fishing boats grew larger, more powerful, and more numerous, while their fish-finding and harvesting technology grew tremendously more effective.

Modern trawlers now pull nets with mouths large enough to engulf a dozen jumbo jets at a time. Heavy metal doors, connected by a thick metal chain, hold the net down on the ocean floor, where it crushes bottom-dwelling organisms and reduces habitat to rubble. A single pass of the trawler not only can scoop up millions of fish, it leaves a devastated community that may take decades to repair. Some environmental groups have called for a complete ban on trawling everywhere in the world.

It's difficult to know how many fish are in the ocean. We can't see them easily, and often we don't even know where they are. Our estimates of population size often are based on the harvest brought in by fishing boats. Biologists warn that many marine species are overfished and in danger of catastrophic population crashes. Research shows that 90 percent of large predators such as tuna, marlin, swordfish, sharks, cod, and halibut are gone from the ocean.

Fish and seafood (including freshwater species) contribute more than 140 million metric tons of highly valued food every year, and are the main animal protein source for about one-quarter of the world population. Marine biologists note, however, that we're "fishing down the food chain." First we pursued the top predators and ground fish until they were commercially extinct, then we went after smaller fish, such as pilchard, capelin, pollock, and eels. When they became scarce, we turned to squid, skates, and other species once discarded as unwanted by-catch. Finally, we've begun harvesting invertebrates, such as sea cucumbers and krill, that many people regard as inedible.

In 2006, an international team of researchers predicted that all the world's major fish and seafood populations will collapse by 2048 if current trends in overfishing and habitat destruction continue. Marine biodiversity, they found, has declined dramatically, particularly since the 1950s. Three-fourths of all major marine fisheries are reported to be fully exploited, overfished, or severely depleted. About one-third of those species are already in collapse—defined as having catches decline 90 percent from the maximum catch. Nevertheless, scientists say, it's not too late to turn this situation around. Many fish stocks can recover quickly if we change destructive fishing practices.

Some governments already have heeded warnings about declining marine fisheries. In 1972, Iceland unilaterally declared a 200 nautical mile (370 km) exclusive economic zone that excluded all foreign fishing boats. In 2003 the Canadian government, in response to declining populations of prized ground fish (fig. 6.1) banned all trawling in the Gulf of St. Lawrence and in the Atlantic Ocean northeast of Newfoundland and Labrador. More than 40,000 Canadians lost their jobs, and many fishing towns were decimated. Marine scientists have called for similar bans in European portions of the North Atlantic, but governments there have been reluctant to impose draconian regulations. They've closed specific fisheries, such as anchovy harvest in the Bay of Biscay and sand eel fishing off Scotland, but they've only gradually reduced quotas for fisheries, such as cod, despite growing evidence of population declines.

Industry trade groups deny that there's a problem with marine fish populations. If restrictions were lifted, they argue, they could catch plenty of fish. It's true that some cod stocks, including the Barents Sea and the Atlantic around Iceland, are stable or even increasing. Establishing marine preserves, like the one around Apo Island in the Philippines, described in chapter 5, can quickly replenish many species if enough fish are available for breeding.

It's questionable, however, if some areas will ever recover their former productivity. It appears that overharvesting may have irreversibly disrupted marine ecosystems and food webs. On the Gorges Bank off the coast of Newfoundland, trillions of tiny tentacled organisms called hydroids now prey on both the organisms that once fed young cod as well as the juvenile cod themselves. Although hydroids have probably always been present, they once were held in check by adult fish.

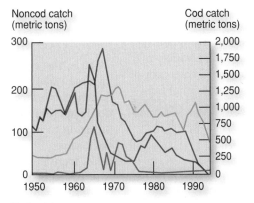

Noncod catch (metric tons)   **Cod catch (metric tons)**

Cod catch
— Atlantic cod

Noncod catch
— Haddock
— Flatfishes
— Red hake

FIGURE 6.1 Commercial harvests in the Northwest Atlantic of some important ground (bottom) fish, 1950–1995.
Source: World Resources Institute, 2000.

Case Study *continued*

Now not enough fish survive to regulate hydroid populations. Is this a shift to a permanent new state, or just a temporary situation?

This case study illustrates some of the complexities and importance of population biology. How can we predict the impacts of human actions and environmental change on different kinds organisms? What are acceptable harvest limits and minimum viable population sizes? In this chapter, we'll look at some of the factors that affect population dynamics of biological organisms.

For more information, see

Worm, B., et al. 2006. Impacts of biodiversity loss on ocean ecosystem services. *Science* 314 (5800):787–90.

6.1 DYNAMICS OF POPULATION GROWTH

Many biological organisms can produce unbelievable numbers of offspring if environmental conditions are right. Consider the common housefly (*Musca domestica*). Each female fly lays 120 eggs (assume half female) in a generation. In 56 days those eggs become mature adults, able to reproduce. In one year, with seven generations of flies being born and reproducing, that original fly would be the proud parent of 5.6 trillion offspring. If this rate of reproduction continued for ten years, the entire earth would be covered in several meters of housefly bodies. Luckily housefly reproduction, as for most organisms, is constrained in a variety of ways—scarcity of resources, competition, predation, disease, accident. The housefly merely demonstrates the remarkable amplification—the **biotic potential**—of unrestrained biological reproduction (fig. 6.2). Population dynamics describes these changes in the number of organisms in a population over time.

FIGURE 6.2 Reproduction gives many organisms the potential to expand populations explosively. The cockroaches in this kitchen could have been produced in only a few generations. A single female cockroach can produce up to 80 eggs every six months. This exhibit is in the Smithsonian Institute's National Museum of Natural History.

Growth without limits is exponential

As you learned in chapter 3, a population consists of all the members of a single species living in a specific area at the same time. The growth of the housefly population just described is **exponential,** having no limit and possessing a distinctive shape when graphed over time. An exponential growth rate (increase in numbers per unit of time) is expressed as a constant fraction, or exponent, which is used as a multiplier of the existing population. The mathematical formula for exponential growth is:

$$\frac{dN}{dt} = rN$$

That is, the change in numbers of individuals (*dN*) per change in time (*dt*) equals the rate of growth (*r*) times the number of individuals in the population (*N*). The *r* term (intrinsic capacity for increase) is a fraction representing the average individual contribution to population growth. If *r* is positive, the population is increasing. If *r* is negative, the population is shrinking. If *r* is zero, there is no change, and *dN/dt* = 0.

A graph of exponential population growth is described as a **J curve** (fig. 6.3) because of its shape. As you can see, the number of individuals added to a population at the beginning of an exponential growth curve can be rather small. But within a very short time, the numbers begin to increase quickly because a fixed percentage leads to a much larger increase as the population size grows.

The exponential growth equation is a very simple model; it is an idealized, simple description of the real world. The same equation is used to calculate growth in your bank account due to compounded interest rates; achieving the potential of your savings depends on you never making a withdrawal. Just as species populations lose individuals and experience reduced biotic potential, not all of your dollars will survive to a ripe old age and contribute fully to your future cash position.

Carrying capacity relates growth to its limits

In the real world there are limits to growth. Around 1970, ecologists developed the concept of **carrying capacity** to mean the number or biomass of animals that can be supported (without harvest) in a certain area of habitat. The concept is now used more generally to suggest a limit of sustainability that an environment has in relation to the size of a species population. Carrying capacity is helpful in understanding the population dynamics of some species, perhaps even humans.

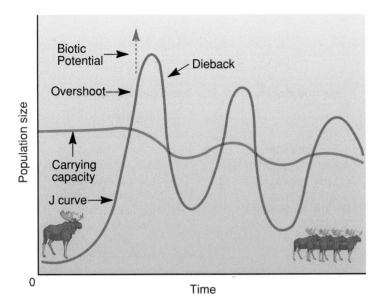

FIGURE 6.3 J curve, or exponential growth curve, with overshoot of carrying capacity. Exponential growth in an unrestrained population (*left side of curve*) leads to a population crash and oscillations below former levels. After the overshoot, carrying capacity may be reduced because of damage to the resources of the habitat. Moose on Isle Royale in Lake Superior may be exhibiting this growth pattern in response to their changing environment.

When a population **overshoots** or exceeds the carrying capacity of its environment, resources become limited and death rates rise. If deaths exceed births, the growth rate becomes negative and the population may suddenly decrease, a change called a **population crash** or dieback. (see fig. 6.3). Some populations can go through repeated **boom-and-bust cycles** in which they repeatedly overshoot the carrying capacity of their habitat and then crash catastrophically. This oscillation can eventually lower the environmental carrying capacity for an entire food web.

Moose and other browsers or grazers sometimes overgraze their food plants, for example, such that future populations of herbivores in the same habitat find less preferred food to sustain them, at least until the habitat recovers. Some species go through predictable cycles if simple factors are involved, such as the seasonal light- and temperature-dependent bloom of algae in a lake. Cycles can be irregular if complex environmental and biotic relationships exist. Irregular cycles include migratory locust outbreaks in the Sahara, or tent caterpillars in temperate forests—these represent irruptive population growth. Population dynamics are also affected by the emigration of organisms from an overcrowded habitat, or immigration of individuals into new habitat, as occurred in 2005 when owls suddenly invaded the northern United States due to a food shortage in their Canadian habitat.

Feedback produces logistic growth

Not all biological populations cycle through exponential overshoot and catastrophic dieback. Many species are regulated by both internal and external factors and come into equilibrium with their environmental resources while maintaining relatively stable population sizes. When resources are unlimited, they may even grow exponentially, but this growth slows as the carrying capacity of the environment is approached. This population dynamic is called **logistic growth** because of its constantly changing rate.

Mathematically, this growth pattern is described by the following equation, which adds a feedback term for carrying capacity (K) to the exponential growth equation:

$$\frac{dN}{dt} = rN \left(1 - \frac{N}{K}\right)$$

The logistic growth equation says that the change in numbers over time (dN/dt) equals the exponential growth rate (rN) times the portion of the carrying capacity (K) not already taken by the current population size (N). The term $(1 - N/K)$ establishes the relationship between population size at any given time step and the number of individuals the environment can support. Depending on whether N is less than or greater than K, the rate of growth will be positive or negative.

The logistic growth curve has a different shape than the exponential growth curve. It is a sigmoidal-shaped, or **S curve** (fig. 6.4). It describes a population that decreases if its numbers exceed the carrying capacity of the environment. In the Data Analysis exercise at the end of this chapter, you will learn how the terms in the exponential and logistic equations influence the size of a population at any time.

Population growth rates are affected by external and internal factors. External factors are habitat quality, food availability, and interactions with other organisms. As populations grow, food becomes scarcer and competition for resources more intense. With a larger population, there is an increased risk that disease or para-

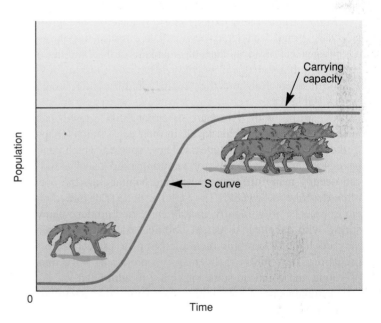

FIGURE 6.4 S curve, or logistic growth curve, describes a population's changing numbers over time in response to feedback from the environment or its own population density. Over the long run, a conservative and predictable population dynamic may win the race over an exponential population dynamic. Species with this growth pattern tend to be *K*-selected.

sites will spread, or that predators will be attracted to the area. Some organisms become physiologically stressed when conditions are crowded, but other internal factors of maturity, body size, and hormonal status may cause them to reduce their reproductive output. Overcrowded house mice ($>1,600/m^3$), for instance, average 5.1 baby mice per litter, while uncrowded house mice ($<34/m^3$) produce 6.2 babies per litter. All these factors are **density-dependent,** meaning as population size increases, the effect intensifies. With density-independent factors, a population is affected no matter what its size. Drought, an early killing frost, flooding, landslide, or habitat destruction by people—all increase mortality rates regardless of the population size. Density-independent limits to population are often nonbiological, capricious acts of nature.

Species respond to limits differently: *r*- and *K*-selected species

The story of the race between the hare and the tortoise has parallels to the way that species deal with limiting factors in their environment. Some organisms, such as dandelions and barnacles, depend on a high rate of reproduction and growth (rN) to secure a place in the environment. These organisms are called ***r*-selected species** because they employ a high reproductive rate (*r*) to overcome the high mortality of virtually ignored offspring. Without predators or diseases to control their population, those species can overshoot carrying capacity and experience population crashes, but as long as vast quantities of young are produced, a few will survive. Other organisms that reproduce more conservatively—longer generation times, late sexual maturity, fewer young—are referred to as ***K*-selected species,** because their growth slows as the carrying capacity (*K*) of their environment is approached.

Many species blend exponential (*r*-selected) and logistic (*K*-selected) growth characteristics. Still, it's useful to contrast the advantages and disadvantages of organisms at the extremes of the continuum. It also helps if we view differences in terms of "strategies" of adaptation and the "logic" of different reproductive modes (table 6.1).

Organisms with *r*-selected, or exponential, growth patterns tend to occupy low trophic levels in their ecosystems (chapter 3) or they are successional pioneers. These species, which generally have wide tolerance limits for environmental factors, and thus can occupy many different niches and habitats, are the ones we often describe as "weedy." They tend to occupy disturbed or new environments, grow rapidly, mature early, and produce many offspring with excellent dispersal abilities. As individual parents, they do little to care for their offspring or protect them from predation. They invest their energy in producing huge numbers of young and count on some surviving to adulthood.

A female clam, for example, can release up to 1 million eggs in her lifetime. The vast majority of young clams die before reaching maturity, but if even a few survive, the species will continue. Many marine invertebrates, parasites, insects, rodents, and annual plants follow this reproductive strategy. Also included in this group are most invasive and pioneer organisms, weeds, pests, and nuisance species.

TABLE 6.1

Reproductive Strategies

r-Selected Species	*K*-Selected Species
1. Short life	1. Long life
2. Rapid growth	2. Slower growth
3. Early maturity	3. Late maturity
4. Many, small offspring	4. Few, large offspring
5. Little parental care and protection	5. High parental care or protection
6. Little investment in individual offspring	6. High investment in individual offspring
7. Adapted to unstable environment	7. Adapted to stable environment
8. Pioneers, colonizers	8. Later stages of succession
9. Niche generalists	9. Niche specialists
10. Prey	10. Predators
11. Regulated mainly by intrinsic factors	11. Regulated mainly by extrinsic factors
12. Low trophic level	12. High trophic level

So-called *K*-selected organisms are usually larger, live long lives, mature slowly, produce few offspring in each generation, and have few natural predators. Elephants, for example, are not reproductively mature until they are 18 to 20 years old. In youth and adolescence, a young elephant belongs to an extended family that cares for it, protects it, and teaches it how to behave. A female elephant normally conceives only once every 4 or 5 years. The gestation period is about 18 months; thus, an elephant herd doesn't produce many babies in any year. Since elephants have few enemies and live a long life (60 or 70 years), this low reproductive rate produces enough elephants to keep the population stable, given good environmental conditions and no poachers.

When you consider the species you recognize from around the world, can you pigeonhole them into categories of *r*- or *K*-selected species? What strategies seem to be operating for ants, bald eagles, cheetahs, clams, dandelions, giraffes, or sharks?

Think About It

Which of the following strategies do humans follow: Do we more closely resemble wolves and elephants in our population growth, or does our population growth pattern more closely resemble that of moose and rabbits? Will we overshoot our environment's carrying capacity (or are we already doing so), or will our population growth come into balance with our resources?

6.2 FACTORS THAT INCREASE OR DECREASE POPULATIONS

Now that you have seen population dynamics in action, let's focus on what happens *within* populations, which are, after all, made up of individuals. In this section, we will discuss how new members

What Do You Think?

Too Many Deer?

A century ago, few Americans had ever seen a wild deer. Uncontrolled hunting and habitat destruction had reduced the deer population to about 500,000 animals nationwide. Some states had no deer at all. To protect the remaining deer, laws were passed in the 1920s and 1930s to restrict hunting, and the main deer predators—wolves and mountain lions—were exterminated throughout most of their former range.

As Americans have moved from rural areas to urban centers, forests have regrown, and deer populations have undergone explosive growth. Maturing at age two, a female deer can give birth to twin fawns every year for a decade or more. Increasing more than 20 percent annually, a deer population can double in just three years, an excellent example of irruptive, exponential growth.

Wildlife biologists estimate that the contiguous 48 states now have a population of more than 30 million white-tailed deer (*Odocoileus*

A white-tailed deer (*Odocoileus virginianus*)

virginianus), probably triple the number present in pre-Columbian times. Some areas have as many as 200 deer per square mile ($77/km^2$). At this density, woodland plant diversity is generally reduced to a few species that deer won't eat. Most deer, in such conditions, suffer from malnourishment, and many die every year of disease and starvation. Other species are diminished as well. Many small mammals and ground-dwelling birds begin to disappear when deer populations reach 25 animals per square mile. At 50 deer per square mile, most ecosystems are seriously impoverished.

The social costs of large deer populations are high. In Pennsylvania alone, where deer numbers are now about 500 times greater than a century ago, deer destroy about $70 million worth of crops and $75 million worth of trees annually. Every year some 40,000 collisions with motor vehicles cause $80 million in property damage. Deer help spread Lyme disease, and, in many states, chronic wasting disease is found in wild deer herds. Some of the most heated criticisms of current deer management policies are in the suburbs. Deer love to browse on the flowers, young trees, and ornamental bushes in suburban yards. Heated disputes often arise between those who love to watch deer and their neighbors who want to exterminate them all.

In remote forest areas, many states have extended hunting seasons, increased the bag limit to four or more animals, and encouraged hunters to shoot does (females) as well as bucks (males). Some hunters criticize these changes because they believe that fewer deer will make it harder to hunt successfully and less likely that they'll find a trophy buck. Others, however, argue that a healthier herd and a more diverse ecosystem is better for all concerned.

In urban areas, increased sport hunting usually isn't acceptable. Wildlife biologists argue that the only practical way to reduce deer herds is culling by professional sharpshooters. Animal rights activists protest lethal control methods as cruel and inhumane. They call instead for fertility controls, reintroduction of predators, such as wolves and mountain lions, or trap and transfer programs. Birth control works in captive populations but is expensive and impractical with wild animals. Trapping, also, is expensive, and there's rarely anyplace willing to take surplus animals. Predators may kill domestic animals or even humans.

This case study shows that carrying capacity can be more complex than the maximum number of organisms an ecosystem can support. While it may be possible for 200 deer to survive in a square mile, there's an ecological carrying capacity lower than that if we consider the other species dependent on that same habitat. There's also an ethical carrying capacity if we don't want to see animals suffer from malnutrition and starve to death every winter. And there's a cultural carrying capacity if we consider the tolerable rate of depredation on crops and lawns or an acceptable number of motor vehicle collisions.

If you were a wildlife biologist charged with managing the deer herd in your state, how would you reconcile the different interests in this issue? How would you define the optimum deer population, and what methods would you suggest to reach this level? What social or ecological indicators would you look for to gauge whether deer populations are excessive or have reached an appropriate level?

---

are added to and old members removed from populations. We also will examine the composition of populations in terms of age classes and introduce terminology that will apply in subsequent chapters.

Natality, fecundity, and fertility are measures of birth rates

Natality is the production of new individuals by birth, hatching, germination, or cloning, and is the main source of addition to most biological populations. Natality is usually sensitive to environmental conditions so that successful reproduction is tied strongly to nutritional levels, climate, soil or water conditions, and—in some species—social interactions between members of the species. The maximum rate of reproduction under ideal conditions varies widely among organisms and is a species-specific characteristic. We already have mentioned, for instance, the differences in natality between several different species.

Fecundity is the physical ability to reproduce, while **fertility** is a measure of the actual number of offspring produced. Because of lack of opportunity to mate and successfully produce offspring, many fecund individuals may not contribute to population growth. Human fertility often is determined by personal choice of fecund individuals.

Immigration adds to populations

Organisms are introduced into new ecosystems by a variety of methods. Seeds, spores, and small animals may be floated on winds or water currents over long distances. This is a major route of colonization for islands, mountain lakes, and other remote locations. Sometimes organisms are carried as hitchhikers in the fur, feathers, or intestines of animals traveling from one place to another. They also may ride on a raft of drifting vegetation. Some animals travel as adults—flying, swimming, or walking. In some ecosystems, a population is maintained only by a constant influx of immigrants. Salmon, for instance, must be important predators in some parts of the ocean, but their numbers are maintained only by recruitment from mountain streams thousands of kilometers away.

Mortality and survivorship measure longevity

An organism is born and eventually it dies; it is mortal. **Mortality,** or death rate, is determined by dividing the number of organisms that die in a certain time period by the number alive at the beginning of the period.

Since the number of survivors is more important to a population than is the number that died, mortality is often better expressed in terms of **survivorship** (the percentage of a cohort that survives to a certain age) or **life expectancy** (the probable number of years of survival for an individual of a given age). Aging has an interesting effect on life expectancy. For each year you survive, your life expectancy increases. If you were one year old in 2000, for example, you could expect to live, on average, 76.7 years more. But if you had

already reached age 75 that year, rather than having only 1.7 more years to live, you could expect to survive another 11.2 years, on average. How could this be? At age 1, many in your cohort were likely to die early for one reason or another. By age 75, most of those individuals are already dead, giving the rest of you a longer average life probability. Even at age 100, long past your starting life expectancy, you would still have 2.7 more years to live, on average.

Human life expectancies have risen dramatically nearly everywhere in the world over the past century. In 1900 the world average life expectancy was only about 30 years, which was not much higher than the average lifespan in the Roman Empire 2,000 years earlier. By 2006, the average was 64.3 years (fig. 6.5). Improved nutrition, sanitation, and medical care were responsible for most of that increase. Demographers wonder how much more life expectancies can increase. Notice the great discrepancy in life expectancies between rich and poor countries. Currently, microstates Andorra, San Marino, and Singapore have the world's highest life expectancies (83.5, 82.1, and 81.6 years, respectively). Japan is nearly as high with a countrywide average of 81.5 years. The lowest national life expectancies are in Africa, where diseases, warfare, poverty, and famine cause many early deaths. In Swaziland, Botswana, and Lesotho, for example, the average person lives only 32.6, 33.7, and 34.4 years, respectively. In many African countries AIDS has reduced life expectancies by about 25 percent in the past two decades. This can be seen in the lag in progress in life expectancies between 1980 and 2000 in these countries in fig. 6.5.

Large discrepancies also exist in the United States. While the nation-wide average life expectancy is 77.5 years, Asian American women in Bergen County, New Jersey, live 91 years on average, while Native American men on the Pine Ridge Reservation in South Dakota are reported to typically live only 48 years. Two-thirds of African countries have life expectancies greater than Pine Ridge. Women almost always have higher life expectancies than men. Worldwide, the average difference between sexes is 3 years, but in Russia the difference between men and women is 13 years.

Is this because women are biologically superior to men, and thus live longer? Or is it simply that men are generally employed in more hazardous occupations and often engage in more dangerous behaviors (drinking, smoking, reckless driving)?

Life span is the longest period of life reached by a given type of organism. The process of living entails wear and tear that eventually overwhelm every organism, but maximum age is dictated

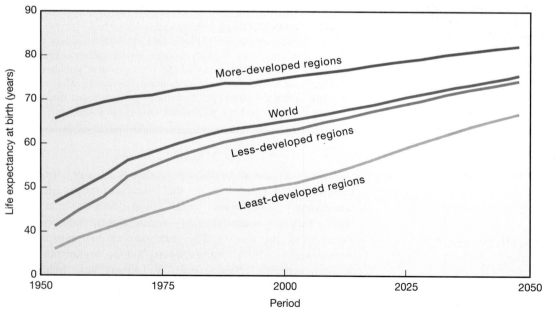

FIGURE 6.5 Life expectancy has increased nearly everywhere in the world, but the increase has lagged in the least-developed countries.

Data from the Population Division of the United Nations, 2006.

primarily by physiological aspects of the organism itself. There is an enormous difference in life span between different species. Some microorganisms live their whole life cycles in a matter of hours or minutes. Bristlecone pine trees in the mountains of California, on the other hand, have life spans up to 4,600 years. The maximum life span for humans appears to be about 120 years.

Most organisms do not live anywhere near the maximum life span for their species. The major factors in early mortality are predation, parasitism, disease, accidents, fighting, and environmental influences, such as climate and nutrition. Important differences in relative longevity among different species are reflected in the survivorship curves shown in figure 6.6.

Four general patterns of survivorship can be seen in this idealized figure. Curve (*a*) is the pattern of organisms that tend to live their full physiological life span if they reach maturity and then have a high mortality rate when they reach old age. This pattern is typical of many large mammals, such as whales, bears, and elephants (when not hunted by humans), as well as humans in developed countries. Interestingly, some very small organisms, including predatory protozoa and rotifers (small, multicellular, freshwater animals), have similar survivorship curves even though their maximum life spans may be hundreds or thousands of times shorter than those of large mammals. In general, curve (*a*) is the pattern for top consumers in an ecosystem, although many annual plants have a similar survivorship pattern.

Curve (*b*) represents the survivorship pattern for organisms for which the probability of death is generally unrelated to age. Sea gulls, for instance, die from accidents, poisoning, and other factors that act more or less randomly. Their mortality rate is generally constant with age, and their survivorship curve is a straight line.

Curve (*c*) is characteristic of many songbirds, rabbits, members of the deer family, and humans in less-developed countries (see chapter 7). They have a high mortality early in life when they are more susceptible to external factors, such as predation, disease, starvation, or accidents. Adults in the reproductive phase have a high level of survival. Once past reproductive age, they become susceptible again to external factors and the number of survivors falls quite rapidly.

Curve (*d*) is typical of organisms at the base of a food chain or those especially susceptible to mortality early in life. Many tree species, fish, clams, crabs, and other invertebrate species produce a very large number of highly vulnerable offspring, few of which survive to maturity. Those individuals that do survive to adulthood, however, have a very high chance of living most of the maximum life span for the species. Figure 6.7 shows some examples of organisms with each of these survivorship patterns.

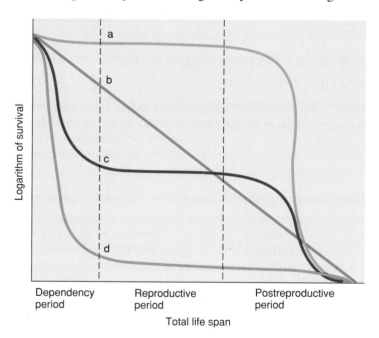

FIGURE 6.6 Four basic types of survivorship curves for organisms with different life histories. Curve (a) represents organisms such as humans or whales, which tend to live out the full physiological life span if they survive early growth. Curve (b) represents organisms such as sea gulls, which have a fairly constant mortality at all age levels. Curve (c) represents such organisms as white-tailed deer, which have high mortality rates in early and late life. Curve (d) represents such organisms as clams and redwood trees, which have a high mortality rate early in life but live a full life if they reach adulthood.

(a) Survive to old age

(c) High infant mortality

(b) Die randomly

(d) Long adult life span

FIGURE 6.7 Different survivorship patterns. (a) Most elephants survive to old age. (b) Seagulls die randomly at all ages from accidents. (c) Antelope have high infant mortality, but adults survive well. (d) Redwood trees have very high seedling losses, but mature trees live thousands of years.

Emigration removes members of a population

Emigration, the movement of members out of a population, is the second major factor that reduces population size. The dispersal factors that allow organisms to migrate into new areas are important in removing surplus members from the source population. Emigration can even help protect a species. For instance, if the original population is destroyed by some catastrophic change in their environment, their genes still are carried by descendants in other places. Many organisms have very specific mechanisms to facilitate migration of one or more of each generation of offspring.

6.3 FACTORS THAT REGULATE POPULATION GROWTH

So far, we have seen that differing patterns of natality, mortality, life span, and longevity can produce quite different rates of population growth. The patterns of survivorship and age structure created by these interacting factors not only show us how a population is growing but also can indicate what general role that species plays in its ecosystem. They also reveal a good deal about how that species is likely to respond to disasters or resource bonanzas in its environment. But what factors *regulate* natality, mortality, and the other components of population growth? In this section, we will look at some of the mechanisms that determine how a population grows.

Various factors regulate population growth, primarily by affecting natality or mortality, and can be classified in different ways. They can be *intrinsic* (operating within individual organisms or between organisms in the same species) or *extrinsic* (imposed from outside the population). Factors can also be either **biotic** (caused by living organisms) or **abiotic** (caused by nonliving components of the environment). Finally, the regulatory factors can act in a *density-dependent* manner (effects are stronger or a higher proportion of the population is affected as population density increases) or *density-independent* manner (the effect is the same or a constant proportion of the population is affected regardless of population density).

In general, biotic regulatory factors tend to be density-dependent, while abiotic factors tend to be density-independent. There has been much discussion about which of these factors is most important in regulating population dynamics. In fact, it probably depends on the particular species involved, its tolerance levels, the stage of growth and development of the organisms involved, the specific ecosystem in which they live, and the way combinations of factors interact. In most cases, density-dependent and density-independent factors probably exert simultaneous influences. Depending on whether regulatory factors are regular and predictable or irregular and unpredictable, species will develop different strategies for coping with them.

Population factors can be density-independent

In general, the factors that affect natality or mortality independently of population density tend to be abiotic components of the ecosystem. Often weather (conditions at a particular time) or climate (average weather conditions over a longer period) are among the most important of these factors. Extreme cold or even moderate cold at the wrong time of year, high heat, drought, excess rain, severe storms, and geologic hazards—such as volcanic eruptions, landslides, and floods—can have devastating impacts on particular populations.

Abiotic factors can have beneficial effects as well, as anyone who has seen the desert bloom after a rainfall can attest. Fire is a powerful shaper of many biomes. Grasslands, savannas, and some montane and boreal forests often are dominated—even created—by periodic fires. Some species, such as jack pine and Kirtland's warblers, are so adapted to periodic disturbances in the environment that they cannot survive without them.

In a sense, these density-independent factors don't really regulate population *per se,* since regulation implies a homeostatic feedback that increases or decreases as density fluctuates. By definition, these factors operate without regard to the number of organisms involved. They may have such a strong impact on a population, however, that they completely overwhelm the influence of any other factor and determine how many individuals make up a particular population at any given time.

Population factors also can be density-dependent

Density-dependent mechanisms tend to reduce population size by decreasing natality or increasing mortality as the population size increases. Most of them are the results of interactions *between* populations of a community (especially predation), but some of them are based on interactions *within* a population.

Interspecific Interactions Occur between Species

As we discussed in chapter 4, a predator feeds on—and usually kills—its prey species. While the relationship is one-sided with respect to a particular pair of organisms, the prey species as a whole may benefit from the predation. For instance, the moose that gets eaten by wolves doesn't benefit individually, but the moose *population* is strengthened because the wolves tend to kill old or sick members of the herd. Their predation helps prevent population overshoot, so the remaining moose are stronger and healthier.

Sometimes predator and prey populations oscillate in a sort of synchrony with each other as is shown in figure 6.8, which shows the number of furs brought into Hudson Bay Company trading posts in Canada between 1840 and 1930. As you can see, the numbers of Canada lynx fluctuate on about a ten-year cycle that is similar to, but slightly out of phase with, the population peaks of snowshoe hares. Although there are some doubts now about

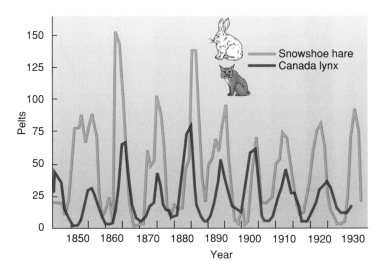

FIGURE 6.8 Ten-year oscillations in the populations of snowshoe hare and lynx in Canada suggest a close linkage of predator and prey, but may not tell the whole story. These data are based on the number of pelts received by the Hudson Bay Company each year, meaning fur-traders were unwitting accomplices in later scientific research.
Source: Data from D. A. MacLulich, *Fluctuations in the Numbers of the Varying Hare* (Lepus americus).Toronto: University of Toronto Press, 1937, reprinted 1974.

how and where these data were collected, this remains a classic example of population dynamics. When prey populations (hares) are abundant, predators (lynx) reproduce more successfully and their population grows. When hare populations crash, so do the lynx. This predator-prey oscillation is known as the Lotka-Volterra model after the scientists who first described it mathematically.

Not all interspecific interactions are harmful to one of the species involved. Mutualism and commensalism, for instance, are interspecific interactions that are beneficial or neutral in terms of population growth (chapter 4).

Intraspecific Interactions Occur within Species

Individuals within a population also compete for resources. When population density is low, resources are likely to be plentiful and the population growth rate will approach the maximum possible for the species, assuming that individuals are not so dispersed that they cannot find mates. As population density approaches the carrying capacity of the environment, however, one or more of the vital resources becomes limiting. The stronger, quicker, more aggressive, more clever, or luckier members get a larger share, while others get less and then are unable to reproduce successfully or survive.

Territoriality is one principal way many animal species control access to environmental resources. The individual, pair, or group that holds the territory will drive off rivals if possible, either by threats, displays of superior features (colors, size, dancing ability), or fighting equipment (teeth, claws, horns, antlers). Members of the opposite sex are attracted to individuals that are able to seize and defend the largest share of the resources. From a selective point of view, these successful individuals presumably represent superior members of the population and the ones best able to produce offspring that will survive.

FIGURE 6.9 Animals often battle over resources. This conflict can induce stress and affect reproductive success.

Stress and Crowding Can Affect Reproduction

Stress and crowding also are density-dependent population control factors. When population densities get very high, organisms often exhibit symptoms of what is called stress shock or **stress-related diseases.** These terms describe a loose set of physical, psychological, and/or behavioral changes that are thought to result from the stress of too much competition and too close proximity to other members of the same species. There is a considerable controversy about what causes such changes and how important they are in regulating natural populations. The strange behavior and high mortality of arctic lemmings or hares during periods of high population density may be a manifestation of stress shock (fig. 6.9). On the other hand, they could simply be the result of malnutrition, infectious disease, or some other more mundane mechanism at work.

Some of the best evidence for the existence of stress-related disease comes from experiments in which laboratory animals, usually rats or mice, are grown in very high densities with plenty of food and water but very little living space. A variety of symptoms are reported, including reduced fertility, low resistance to infectious diseases, and pathological behavior. Dominant animals seem to be affected least by crowding, while subordinate animals—the ones presumably subjected to the most stress in intraspecific interactions—seem to be the most severely affected.

Case Study A Plague of Locusts Schistocerca gregarius, the desert locust, has been called the world's most destructive insect. Throughout recorded human history, locust plagues have periodically swarmed out of deserts and into settled areas. Their impact on human lives has often been so disruptive that records of plagues have taken on religious significance and made their way into sacred and historical texts.

Locusts usually are solitary creatures resembling ordinary grasshoppers. Every few decades, however, when rain comes to the desert and vegetation flourishes, locusts reproduce rapidly until the ground is literally crawling with bugs. High population densities and stress bring ominous changes in these normally innocuous insects. They stop reproducing, grow longer wings, group together in enormous swarms, and begin to move across the desert. Dense clouds of insects darken the sky, moving as much as 100 km per day. Locusts may be small, but they can eat their own body weight of vegetation every day. A single swarm can cover 1,200 km² and contain 50 to 100 billion individuals. The swarm can strip pastures, denude trees, and destroy crops in a matter of hours, consuming as much food in a day as 500,000 people would need for a year. Eventually, having exhausted their food supply and migrated far from the desert where conditions favor reproduction, the locusts die and aren't seen again for decades.

Huge areas of crops and rangeland in northern Africa, the Middle East, and Asia are within the reach of the desert locust. This small insect, with its voracious appetite, can affect the livelihood of at least one-tenth of the world's population. During quiet periods, called recessions, African locusts are confined to the Sahara Desert, but when conditions are right, swarms invade countries as far away as Spain, Russia, and India. Swarms are even reported to have crossed the Atlantic Ocean from Africa to the Caribbean.

Unusually heavy rains in the Sahara in 2004 created the conditions for a locust explosion. Four generations bred in rapid succession, and swarms of insects moved out of the desert. Twenty-eight countries in Africa and the Mediterranean area were afflicted. Crop losses reached 100 percent in some places, and food supplies for millions of people were threatened. Officials at the United Nations warned that we could be headed toward another great plague. Hundreds of thousands of hectares of land were treated with pesticides, but millions of dollars of crop damage were reported anyway.

This case study illustrates the power of exponential growth and the disruptive potential of a boom-and-bust life cycle. Stress, population density, migration, and intraspecific interactions all play a role in this story. Although desert conditions usually keep locust numbers under control, their biotic potential for reproduction is a serious worry for residents of many countries.

6.4 CONSERVATION BIOLOGY

Small, isolated populations can undergo catastrophic declines due to environmental change, genetic problems, or stochastic (random or unpredictable) events. A critical question in conservation biology is the minimum population size of a rare and endangered species required for long-term viability. In this section, we'll look at some factors that limit species and genetic diversity. We'll also consider the interaction of collections of subpopulations of species in fragmented habitats.

Island biogeography describes isolated populations

In a classic 1967 study, R. H. MacArthur and E. O. Wilson asked why it is that small islands far from the mainland generally have far fewer species than larger or nearer islands. Their theory of **island biogeography** explains that diversity in isolated habitats is a balance between colonization and extinction rates. An island far from a population source has a lower colonization rate for terrestrial species because it is harder for organisms to reach (fig. 6.10). At

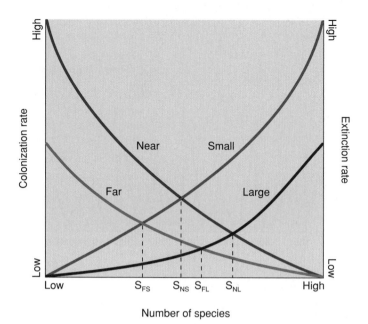

FIGURE 6.10 Predicted species richness on an island resulting from a balance between colonization (immigration) and extinction by natural causes. This island biogeography theory of MacArthur and Wilson (1967) is used to explain why large islands near a mainland (S_{NL}) tend to have more species than small, far islands (S_{FS}).

Source: Based on MacArthur and Wilson, *The Theory of Island Biogeography*, 1967, Princeton University Press.

the same time, the limited habitat on a small island can support fewer individuals of any given species. This creates a greater probability that a species could go extinct due to natural disasters, diseases, or demographic factors such as imbalance between sexes in a particular generation. Larger islands, or those closer to the mainland, on the other hand, are more likely to be colonized or to retain those species already successfully established. Thus they tend to have greater diversity than smaller, more remote places.

Island biogeographical effects have been observed in many places. Cuba, for instance, is 100 times as large and has about 10 times as many amphibian species as its Caribbean neighbor, Monserrat. Similarly, in a study of bird species on the California Channel Islands, Jared Diamond observed that on islands with fewer than 10 breeding pairs, 39 percent of the populations went extinct over an 80-year period. In contrast, only 10 percent of populations numbering between 10 and 100 pairs went extinct, and no species with over 1,000 pairs disappeared over this time (fig. 6.11). This theory of a balance between colonization and extinction is now seen to explain species dynamics in many small, isolated habitat fragments whether on islands or not.

Conservation genetics is important in survival of endangered species

Genetics plays an important role in the survival or extinction of small, isolated populations. In large populations, genetic variation

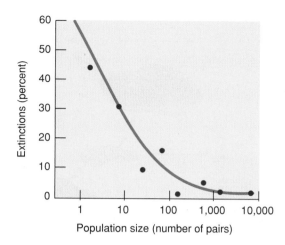

FIGURE 6.11 Extinction rates of bird species on the California Channel Islands as a function of population size over 80 years.
Source: H. L. Jones and J. Diamond, "Short-term-base studies of turnover in breeding bird populations on the California coast island," in *Condor,* vol. 78:526–49, 1976.

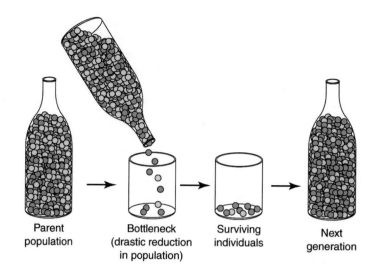

FIGURE 6.12 Genetic drift: the bottleneck effect. The parent population contains roughly equal numbers of blue and yellow individuals. By chance, the few remaining individuals that comprise the next generation are mostly blue. The bottleneck occurs because so few individuals form the next generation, as might happen after an epidemic or catastrophic storm.

tends to persist in what is called a Hardy-Weinberg equilibrium, named after the scientists who first described why this occurs. If mating is random, no mutations (changes in genetic material) occur, and there is no gene in-flow or selective pressure for or against particular traits, random distribution of gene types (alleles) will occur during gamete formation and sexual reproduction. That is, different alleles will be distributed in the offspring in the same ratio they occur in the parents, and genetic diversity is preserved.

In a large population, these conditions for maintaining genetic equilibrium are generally operative. The addition or loss of a few individuals or appearance of new genotypes makes little difference in the total gene pool, and genetic diversity is relatively constant. In small, isolated populations, however, immigration, mortality, mutations, or chance mating events involving only a few individuals can greatly alter the genetic makeup of the whole population. We call the gradual changes in gene frequencies due to random events **genetic drift.**

For many species, loss of genetic diversity causes a number of harmful effects that limit adaptability, reproduction, and species survival. A **founder effect** or **demographic bottleneck** occurs when just a few members of a species survive a catastrophic event or colonize new habitat geographically isolated from other members of the same species. Any deleterious genes present in the founders will be overrepresented in subsequent generations (fig. 6.12). Inbreeding, mating of closely related individuals, also makes expression of rare or recessive genes more likely.

Some species seem not to be harmed by inbreeding or lack of genetic diversity. The northern elephant seal, for example, was reduced by overharvesting a century ago, to fewer than 100 individuals. Today there are more than 150,000 of these enormous animals along the Pacific coast of Mexico and California. No marine mammal is known to have come closer to extinction and then made such a remarkable recovery. All northern elephant seals today appear to be essentially genetically identical and yet

they seem to have no apparent problems. Although interpretations of their situation are controversial, in highly selected populations, where only the most fit individuals reproduce, or in which there are few deleterious genes, inbreeding and a high degree of genetic identity may not be such a negative factor.

Cheetahs, also appear to have undergone a demographic bottleneck sometime in the not-too-distant past. All the male cheetahs throughout the world appear to be nearly genetically identical,

FIGURE 6.13 Sometime in the past, cheetahs underwent a severe population crash. Now all male cheetahs in the world are nearly genetically identical, and deformed sperm, low fertility levels, and low infant survival are common in the species.

suggesting that they all share a single male ancestor (fig. 6.13). This lack of diversity is thought to be responsible for an extremely low fertility rate, a high abundance of abnormal sperm, and low survival rate for offspring, all of which threatens the survival of the species.

Population viability analysis calculates chances of survival

Conservation biologists use the concepts of island biogeography, genetic drift, and founder effects to determine **minimum viable population size,** or number of individuals needed for long-term survival of rare and endangered species. A classic example is that of the grizzly bear (*Ursus arctos horribilis*) in North America. Before European settlement, grizzlies roamed from the Great Plains west to California and north to Alaska. Hunting and habitat destruction reduced the number of grizzlies from an estimated 100,000 in 1800 to less than 1,200 animals in six separate subpopulations that now occupy less than 1 percent of the historic range. Recovery target sizes—based on estimated environmental carrying capacities—call for fewer than 100 bears for some subpopulations. Conservation genetics predicts that a completely isolated population of 100 bears cannot be maintained for more than a few generations. Even the 600 bears now in Yellowstone National Park will be susceptible to genetic problems if completely isolated. Interestingly, computer models suggest that translocating only two unrelated bears into small populations every generation (about ten years) could greatly increase population viability.

Metapopulations are important interconnections

For mobile organisms, separated populations can have gene exchange if suitable corridors or migration routes exist. A **metapopulation** is a collection of populations that have regular or intermittent gene flow between geographically separate units (fig. 6.14). For example, the Bay checkerspot butterfly (*Euphydrays editha bayensis*) in California exists in several distinct habitat patches. Individuals occasionally move among these patches, mating with existing animals or recolonizing empty habitats. Thus, the apparently separate groups form a functional metapopulation.

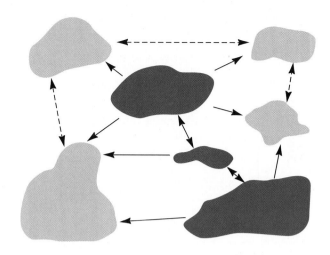

FIGURE 6.14 A metapopulation is composed of several local populations linked by regular (*solid arrows*) or occasional (*dashed lines*) gene flows. Source populations (*dark*) provide excess individuals, which emigrate to and colonize sink habitats (*light*).

A "source" habitat, where birth rates are higher than death rates, produces surplus individuals that can migrate to new locations within a metapopulation. "Sink" habitats, on the other hand, are places where mortality exceeds birth rates. Sinks may be spatially larger than sources but because of unfavorable conditions, the species would disappear in the sink habitat if it were not periodically replenished from a source population. In general, the larger a reserve is, the better it is for endangered species. Sometimes, however, adding to a reserve can be negative if the extra area is largely sink habitat. Individuals dispersing within the reserve may settle in unproductive areas if better habitat is hard to find. Recent studies using a metapopulation model for spotted owls predict just such a problem for this species in the Pacific Northwest.

Some conservation biologists argue that we ought to try to save every geographically distinct population or "evolutionarily significant unit (ESU)" possible in order to preserve maximum genetic diversity. Paul Ehrlich and Gretchen Daily estimate there may be an average of 220 ESU for every one of the 10 to 50 billion species in the world. Saving all of them would be a gargantuan task.

CONCLUSION

Given optimum conditions, populations of many organisms can grow exponentially; that is, they can expand at a constant *rate* per unit of time. This biotic potential can produce enormous populations that far surpass the carrying capacity of the environment if left unchecked. Obviously, no population grows at this rate forever. Sooner or later, predation, disease, starvation, or some other factor will cause the population to crash. Not all species follow this boom-and-bust pattern, however. Most top predators have intrinsic factors that limit their reproduction and prevent overpopulation.

Overharvesting of species, habitat destruction, predator elimination, introduction of exotic species, and other forms of human disruption can also drive populations to boom and/or crash. Population dynamics are an important part of conservation biology. Principles, such as island biogeography, genetic drift, demographic bottlenecks, and metapopulation interactions are critical in endangered species protection.

REVIEWING LEARNING OUTCOMES

By now you should be able to explain the following points:

6.1 Describe the dynamics of population growth.
- Growth without limits is exponential.
- Carrying capacity relates growth to its limits.
- Feedback produces logistic growth.
- Species respond to limits differently: *r*- and *K*-selected species.

6.2 Summarize the factors that increase or decrease populations.
- Natality, fecundity, and fertility are measures of birth rates.
- Immigration adds to populations.
- Mortality and survivorship measure longevity.
- Emigration removes members of a population.

6.3 Compare and contrast the factors that regulate population growth.
- Population factors can be density-independent.
- Population factors also can be density-dependent.

6.4 Identify some applications of population dynamics in conservation biology.
- Island biogeography describes isolated populations.
- Conservation genetics is important in survival of endangered species.
- Population viability analysis calculates chances of survival.
- Metapopulations are important interconnections.

PRACTICE QUIZ

1. What caused the sudden collapse of Atlantic cod populations?
2. Define *exponential growth*.
3. Given a growth rate of 3 percent per year, how long will it take for a population of 100,000 individuals to double? How long will it take to double when the population reaches 10 million?
4. What is environmental resistance? How does it affect populations?
5. What is the difference between fertility and fecundity?
6. Describe the four major types of survivorship patterns and explain what they show about the role of the species in an ecosystem.
7. What are the main interspecific population regulatory interactions? How do they work?
8. What is island biogeography and why is it important in conservation biology?
9. Why does genetic diversity tend to persist in large populations, but gradually drift or shift in small populations?
10. Draw a diagram showing gene flow between source and sink habitat in a metapopulation. Explain.

CRITICAL THINKING AND DISCUSSION QUESTIONS

1. Compare the advantages and disadvantages to a species that result from exponential or logistic growth. Why do you think hares have evolved to reproduce as rapidly as possible, while lynx appear to have intrinsic or social growth limits?
2. Are humans subject to environmental resistance in the same sense that other organisms are? How would you decide whether a particular factor that limits human population growth is ecological or social?
3. Species differ greatly in birth and death rates, survivorship, and life spans. There must be advantages and disadvantages in living longer or reproducing more quickly. Why hasn't evolution selected for the most advantageous combination of characteristics so that all organisms would be more or less alike?
4. Abiotic factors that influence population growth tend to be density-independent, while biotic factors that regulate population growth tend to be density-dependent. Explain.
5. Some people consider stress and crowding studies of laboratory animals highly applicable in understanding human behavior. Other people question the cross-species transferability of these results. What considerations would be important in interpreting these experiments?
6. What implications (if any) for human population control might we draw from our knowledge of basic biological population dynamics?

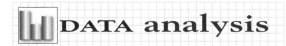

Exponential growth occurs in a series of time steps—days, months, years, or generations. Imagine cockroaches in a room multiplying (or some other species, if you must). Picture a population of ten cockroaches that together produce enough young to increase at a rate of 150 percent per month. What is r for this population?

Time Step (t)	Begin Step (N_b)	Intrinsic Growth Rate (r)	End Step (N_e)
0	10		15
1			
2			
3			
4			
5			
6			
7			

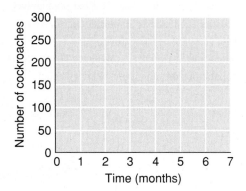

To find out how this population grows, fill out the table shown. (*Hint: r* remains constant.) Remember, for time step 0 (the first month), you begin with ten roaches, and end (N_e) with a larger number that depends on r, the intrinsic rate of growth. The beginning of the second time step (1) starts with the number at the end of step 0. Round N to the nearest whole number. When you are done, graph the results. At the end of 7 months, how large did this population become? What is the shape of the growth curve?

More than 1.1 billion people live in India, more than one-quarter of them in object poverty.

C H A P T E R

7
Human Populations

Live simply so that others may simply live.
—Mahatma Gandhi—

LEARNING OUTCOMES

After studying this chapter, you should be able to:

7.1 Trace the history of human population growth.

7.2 Summarize different perspectives on population growth.

7.3 Analyze some of the factors that determine population growth.

7.4 Explain how ideal family size is culturally and economically dependent.

7.5 Describe how a demographic transition can lead to stable population size.

7.6 Relate how family planning gives us choices.

7.7 Reflect on what kind of future we are creating.

Case Study Family Planning in Thailand: A Success Story

Down a narrow lane off Bangkok's busy Sukhumvit Road, is a most unusual café. Called Cabbages and Condoms, it's not only highly rated for its spicy Thai food, but it's also the only restaurant in the world dedicated to birth control. In an adjoining gift shop, baskets of condoms stand next to decorative handicrafts of the northern hill tribes. Piles of T-shirts carry messages, such as, "A condom a day keeps the doctor away," and "Our food is guaranteed not to cause pregnancy." Both businesses are run by the Population and Community Development Association (PDA), Thailand's largest and most influential nongovernmental organization.

The PDA was founded in 1974 by Mechai Viravaidya, a genial and fun-loving former Thai Minister of Health, who is a genius at public relations and human motivation (fig. 7.1). While traveling around Thailand in the early 1970s, Mechai recognized that rapid population growth—particularly in poor rural areas—was an obstacle to community development. Rather than lecture people about their behavior, Mechai decided to use humor to promote family planning. PDA workers handed out condoms at theaters and traffic jams, anywhere a crowd gathered. They challenged governmental officials to condom balloon-blowing contests, and taught youngsters Mechai's condom song: "Too Many Children Make You Poor." The PDA even pays farmers to paint birth control ads on the sides of their water buffalo.

This campaign has been extremely successful at making birth control and family planning, which once had been taboo topics in polite society, into something familiar and unembarrassing. Although condoms—now commonly called "mechais" in Thailand—are the trademark of PDA, other contraceptives, such as pills, spermicidal foam, and IUDs, are promoted as well. Thailand was one of the first countries to allow the use of the injectable contraceptive DMPA, and remains a major user. Free non-scalpel vasectomies are available on the king's birthday. Sterilization has become the most widely used form of contraception in the country. The campaign to encourage condom use has also been helpful in combating AIDS.

In 1974, when PDA started, Thailand's growth rate was 3.2 percent per year. In just fifteen years, contraceptive use among married couples increased from 15 to 70 percent, and the growth rate had dropped to 1.6 percent, one of the most dramatic birth rate declines ever recorded. Now Thailand's growth rate is 0.7 percent, or nearly the same as the United States. The fertility rate (or average number of children per woman) decreased from 7 in 1974 to 1.7 in 2006. The PDA is credited with the fact that Thailand's population is 20 million less than it would have been if it had followed its former trajectory.

FIGURE 7.1 Mechai Viravaidya (right) is joined by Peter Piot, Executive Director of UNAIDS, in passing out free condoms on family planning and AIDS awareness day in Bangkok.

In addition to Mechai's creative genius and flair for showmanship, there are several reasons for this success story. Thai people love humor and are more egalitarian than most developing countries. Thai spouses share in decisions regarding children, family life, and contraception. The government recognizes the need for family planning and is willing to work with volunteer organizations, such as the PDA. And Buddhism, the religion of 95 percent of Thais, promotes family planning.

The PDA hasn't limited itself to family planning and condom distribution. It has expanded into a variety of economic development projects. Microlending provides money for a couple of pigs, or a bicycle, or a small supply of goods to sell at the market. Thousands of water-storage jars and cement rainwater-catchment basins have been distributed. Larger scale community development grants include road building, rural electrification, and irrigation projects. Mechai believes that human development and economic security are keys to successful population programs.

This case study introduces several important themes of this chapter. What might be the effects of exponential growth in human populations? How might we manage fertility and population growth? And what are the links between poverty, birth rates, and our common environment? Keep in mind, as you read this chapter, that resource limits aren't simply a matter of total number of people on the planet, they also depend on consumption levels and the types of technology used to produce the things we use.

7.1 POPULATION GROWTH

Every second, on average, four or five children are born, somewhere on the earth. In that same second, two other people die. This difference between births and deaths means a net gain of roughly 2.3 more humans per second in the world's population. In mid-2007 the total world population stood at roughly 6.6 billion people and was growing at 1.17 percent per year. This means we are now adding nearly 79 million more people per year, and if this rate persists, our global population will double in about 58 years. Humans are now probably the most numerous vertebrate species on the earth. We also are more widely distributed and manifestly have a greater global environmental impact than any other species. For the families to whom these children are born, this may well be a joyous and long-awaited event (fig. 7.2). But is a continuing increase in humans good for the planet in the long run?

Many people worry that overpopulation will cause—or perhaps already is causing—resource depletion and environmental degradation that threaten the ecological life-support systems on which we all depend. These fears often lead to demands for immediate, worldwide birth control programs to reduce fertility rates and to eventually stabilize or even shrink the total number of humans.

Others believe that human ingenuity, technology, and enterprise can extend the world carrying capacity and allow us to overcome any problems we encounter. From this perspective, more people may be beneficial rather than disastrous. A larger population means a larger workforce, more geniuses, more ideas about what to do. Along with every new mouth comes a pair of hands. Proponents of this worldview—many of whom happen to be economists—argue that continued economic and technological growth can both feed the world's billions and enrich everyone enough to end the population explosion voluntarily. Not so, counter many ecologists. Growth is the problem; we must stop both population and economic growth.

Yet another perspective on this subject derives from social justice concerns. In this worldview, there are sufficient resources for everyone. Current shortages are only signs of greed, waste, and oppression. The root cause of environmental degradation, in this view, is inequitable distribution of wealth and power rather than population size. Fostering democracy, empowering women and minorities, and improving the standard of living of the world's poorest people are what are really needed. A narrow focus on population growth only fosters racism and an attitude that blames the poor for their problems while ignoring the deeper social and economic forces at work.

Whether human populations will continue to grow at present rates and what that growth would imply for environmental quality and human life are among the most central and pressing questions in environmental science. In this chapter, we will look at some causes of population growth as well as how populations are measured and described. Family planning and birth control are essential for stabilizing populations. The number of children a couple decides to have and the methods they use to regulate fertility, however, are strongly influenced by culture, religion,

FIGURE 7.2 A Mayan family in Guatemala with four of their six living children. Decisions on how many children to have are influenced by many factors, including culture, religion, need for old age security for parents, immediate family finances, household help, child survival rates, and power relationships within the family. Having many children may not be in the best interest of society at large, but may be the only rational choice for individual families.

politics, and economics, as well as basic biological and medical considerations. We will examine how some of these factors influence human demographics.

Human populations grew slowly until relatively recently

For most of our history, humans have not been very numerous compared to other species. Studies of hunting and gathering societies suggest that the total world population was probably only a few million people before the invention of agriculture and the domestication of animals around 10,000 years ago. The larger and more secure food supply made available by the agricultural revolution allowed the human population to grow, reaching perhaps 50 million people by 5000 B.C. For thousands of years, the number of humans increased very slowly. Archaeological evidence and historical descriptions suggest that only about 300 million people were living at the time of Christ (table 7.1).

Until the Middle Ages, human populations were held in check by diseases, famines, and wars that made life short and uncertain for most people (fig. 7.3). Furthermore, there is evidence that many early societies regulated their population size through cultural taboos and practices such as abstinence and infanticide. Among the most destructive of natural population controls were bubonic plagues (or Black Death) that periodically swept across Europe between 1348 and 1650. During the worst plague years (between 1348 and 1350), it is estimated that at least one-third of the European population perished. Notice, however, that this didn't retard population growth for very long. In 1650, at the end of the last great plague, there were about 600 million people in the world.

TABLE 7.1		
World Population Growth and Doubling Times		
Date	**Population**	**Doubling Time**
5000 B.C.	50 million	?
800 B.C.	100 million	4,200 years
200 B.C..	200 million	600 years
A.D. 1200	400 million	1,400 years
A.D. 1700	800 million	500 years
A.D. 1900	1,600 million	200 years
A.D. 1965	3,200 million	65 years
A.D. 2005	6,400 million	40 years
A.D. 2050 (estimate)	8,920 million	140 years

Source: United Nations Population Division.

As you can see in figure 7.3, human populations began to increase rapidly after A.D. 1600. Many factors contributed to this rapid growth. Increased sailing and navigating skills stimulated commerce and communication between nations. Agricultural developments, better sources of power, and better health care and hygiene also played a role. We are now in an exponential or J curve pattern of growth.

It took all of human history to reach 1 billion people in 1804, but little more than 150 years to reach 3 billion in 1960. To go from 5 to 6 billion took only 12 years. Another way to look at population growth is that the number of humans tripled during the twentieth

century. Will it do so again in the twenty-first century? If it does, will we overshoot the carrying capacity of our environment and experience a catastrophic dieback similar to those described in chapter 6? As you will see later in this chapter, there is evidence that population growth already is slowing, but whether we will reach equilibrium soon enough and at a size that can be sustained over the long term remains a difficult but important question.

7.2 PERSPECTIVES ON POPULATION GROWTH

As with many topics in environmental science, people have widely differing opinions about population and resources. Some believe that population growth is the ultimate cause of poverty and environmental degradation. Others argue that poverty, environmental degradation, and overpopulation are all merely symptoms of deeper social and political factors The worldview we choose to believe will profoundly affect our approach to population issues. In this section, we will examine some of the major figures and their arguments in this debate.

Does environment or culture control human populations?

Since the time of the Industrial Revolution, when the world population began growing rapidly, individuals have argued about the causes and consequences of population growth. In 1798 Thomas Malthus (1766–1834) wrote *An Essay on the Principle of Population,* changing the way European leaders thought about population growth. Malthus marshaled evidence to show that populations

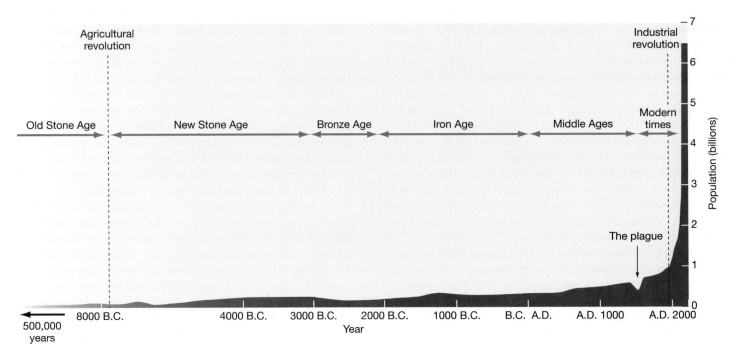

FIGURE 7.3 Human population levels through history. Since about A.D. 1000, our population curve has assumed a J shape. Are we on the upward slope of a population overshoot? Will we be able to adjust our population growth to an S curve? Or can we just continue the present trend indefinitely?

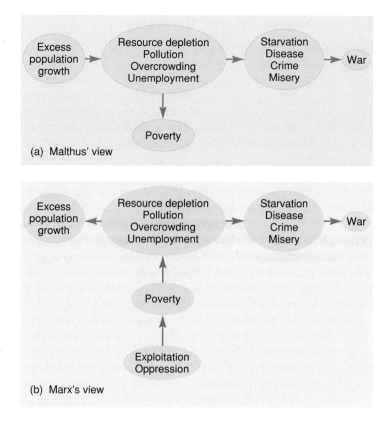

(a) Malthus' view

(b) Marx's view

FIGURE 7.4 (a) Thomas Malthus argued that excess population growth is the ultimate cause of many other social and environmental problems. (b) Karl Marx argued that oppression and exploitation are the real causes of poverty and environmental degradation. Population growth in this view is a symptom or result of other problems, not the source.

tended to increase at an exponential, or compound, rate while food production either remained stable or increased only slowly. Eventually human populations would outstrip their food supply and collapse into starvation, crime, and misery (fig. 7.4a). He converted most economists of the day from believing that high fertility increased gross domestic output to believing that per capita output actually fell with rapidly rising population.

In Malthusian terms, growing human populations stop growing when disease or famine kills many, or when constraining social conditions compel others to reduce their birth rates—late marriage, insufficient resources, celibacy, and "moral restraint." Several decades later, the economist Karl Marx (1818–1883) presented an opposing view, that population growth resulted from poverty, resource depletion, pollution, and other social ills. Slowing population growth, said Marx, required that people be treated justly, and that exploitation and oppression be eliminated from social arrangements (fig. 7.4b).

Both Marx and Malthus developed their theories about human population growth when understanding of the world, technology, and society were much different than they are today. But these different views of human population growth still inform competing approaches to family planning today. On the one hand, some believe that we are approaching, or may have surpassed, the earth's carrying capacity. Joel Cohen, a mathematical biologist at Rockefeller University, reviewed published estimates of the maximum human population size the planet can sustain. The estimates, spanning 300 years of thinking, converged on a median value of 10–12 billion. We are more than 6.5 billion strong today, and growing, an alarming prospect for some. Cornell University entomologist David Pimental, for example, has said: "By 2100, if current trends continue, twelve billion miserable humans will suffer a difficult life on Earth." In this view, birth control should be our top priority. On the other hand, many scholars agree with Marx that improved social conditions and educational levels can stabilize populations humanely. In this perspective, the earth is bountiful in its resource base, but poverty and high birth rates result from oppressive social relationships that unevenly distribute wealth and resources. Consequently, this position believes, technological development, education, and just social conditions are the means of achieving population control. Mohandas Gandhi stated it succinctly: "There is enough for everyone's need, but not enough for anyone's greed."

Technology can increase carrying capacity for humans

Optimists argue that Malthus was wrong in his predictions of famine and disaster 200 years ago because he failed to account for scientific and technical progress. In fact, food supplies have increased faster than population growth since Malthus' time. For example, according to the UN FAO Statistics Division, each person on the planet averaged 2,435 calories of food per day in 1970, while in 2000 the caloric intake reached 2,807 calories. Even poorer, developing countries saw a rise, from an average of 2,135 calories per day in 1970 to 2,679 in 2000. In that same period the world population went from 3.7 to more than 6 billion people. Certainly terrible famines have stricken different locations in the past 200 years, but they were caused more by politics and economics than by lack of resources or population size. Whether the world can continue to feed its growing population remains to be seen, but technological advances have vastly increased human carrying capacity so far.

The burst of world population growth that began 200 years ago was stimulated by scientific and industrial revolutions. Progress in agricultural productivity, engineering, information technology, commerce, medicine, sanitation, and other achievements of modern life have made it possible to support thousands of times as many people per unit area as was possible 10,000 years ago. Economist Stephen Moore of the Cato Institute in Washington, D.C., regards this achievement as "a real tribute to human ingenuity and our ability to innovate." There is no reason, he argues, to think that our ability to find technological solutions to our problems will diminish in the future.

Much of our growth and rising standard of living in the past 200 years, however, has been based on easily acquired natural resources, especially cheap, abundant fossil fuels (see chapter 19). Whether rising prices of fossil fuels will constrain that production and result in a crisis in food production and distribution, or in some other critical factor in human society, concerns many people.

However, technology can be a double-edged sword. Our environmental effects aren't just a matter of sheer population size; they also depend on what kinds of resources we use and how we use them. This concept is summarized as the **I = PAT** formula. It says that our environmental impacts (I) are the product of our population size (P) times affluence (A) and the technology (T) used to produce the goods and services we consume. A single American living an affluent lifestyle that depends on high levels of energy and material consumption, and that produces excessive amounts of pollution, probably has a greater environmental impact than a whole village of Asian or African farmers. Ideally, Americans will begin to use nonpolluting, renewable energy and material sources. Better yet, Americans will extend the benefits of environmentally friendly technology to those villages of Asians and Africans so everyone can enjoy the benefits of a better standard of living without degrading their environment.

Population growth could bring benefits

Think of the gigantic economic engine that China is becoming as it continues to industrialize and its population becomes more affluent. More people mean larger markets, more workers, and efficiencies of scale in mass production of goods. Moreover, adding people boosts human ingenuity and intelligence that will create new resources by finding new materials and discovering new ways of doing things. Economist Julian Simon (1932–1998), a champion of this rosy view of human history, believed that people are the "ultimate resource" and that no evidence suggests that pollution, crime, unemployment, crowding, the loss of species, or any other resource limitations will worsen with population growth. In a famous bet in 1980, Simon challenged Paul Ehrlich, author of *The Population Bomb,* to pick five commodities that would become more expensive by the end of the decade. Ehrlich chose metals that actually became cheaper, and he lost the bet. Leaders of many developing countries share this outlook and insist that, instead of being obsessed with population growth, we should focus on the inordinate consumption of the world's resources by people in richer countries.

> ### Think About It
>
> What larger worldviews are reflected in this population debate? What positions do you believe neo-Malthusians and neo-Marxists might take on questions of human rights, resource abundance, or human perfectability? Where do you stand on these issues?

7.3 MANY FACTORS DETERMINE POPULATION GROWTH

Demography is derived from the Greek words *demos* (people) and *graphos* (to write or to measure). It encompasses vital statistics about people, such as births, deaths, and where they live, as well as total population size. In this section, we will survey ways human populations are measured and described, and discuss demographic factors that contribute to population growth.

How many of us are there?

The estimate of 6.55 billion people in the world in 2006 quoted earlier in this chapter is only an educated guess. Even in this age of information technology and communication, counting the number of people in the world is like shooting at a moving target. People continue to be born and die. Furthermore, some countries have never even taken a census, and those that have been done may not be accurate. Governments may overstate or understate their populations to make their countries appear larger and more important or smaller and more stable than they really are. Individuals, especially if they are homeless, refugees, or illegal aliens, may not want to be counted or identified.

We really live in two very different demographic worlds. One is old, rich, and relatively stable. The other is young, poor, and growing rapidly. Most people in Asia, Africa, and Latin America inhabit the latter demographic world. These countries represent 80 percent of the world population but more than 90 percent of all projected growth (fig. 7.5).

The highest population growth rates occur in a few "hot spots," such as sub-Saharan Africa and the Middle East, where economics, politics, religion, and civil unrest keep birth rates high and contraceptive use low. In Niger, Yemen, and Palestine, for example, annual population growth is above 3.2 percent. Less than 10 percent of all couples use any form of birth control, women average more than seven children each, and nearly half the population is less than 15 years old. The world's highest current growth rate is in the United Arab Emirates, where births plus immigration are producing an annual increase of 6.8 percent (the highest immigration rate in the world is responsible for 80 percent of that growth). This means that the UAE is doubling its population size approximately every decade. Obviously, a small country with limited resources (except oil) and almost no fresh water or agriculture, can't sustain that high growth rate indefinitely.

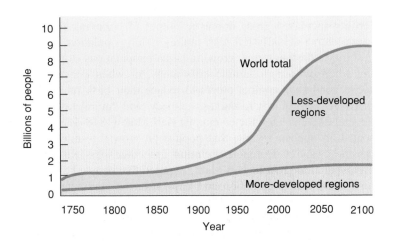

FIGURE 7.5 Estimated human population growth, 1750–2100, in less-developed and more-developed regions. Almost all growth projected for the twenty-first century is in the less-developed countries.
Source: UN Population Division, 2005.

TABLE 7.2

The World's Largest Countries

2006		2050*	
Country	Population (millions)	Country	Population (millions)
China	1,311	India	1,628
India	1,122	China	1,437
United States	299	United States	420
Indonesia	225	Nigeria	299
Brazil	187	Pakistan	295
Pakistan	166	Indonesia	285
Bangladesh	147	Brazil	260
Russia	142	Bangladesh	231
Nigeria	135	Dem. Rep. of Congo	183
Japan	128	Ethiopia	145

*Estimate.

Source: Population Reference Bureau, 2006.

Some countries in the developing world have experienced amazing growth rates and are expected to reach extraordinary population sizes by the middle of the twenty-first century. Table 7.2 shows the ten largest countries in the world, arranged by their estimated size in 2006 and projected size in 2050. Note that, while China was the most populous country throughout the twentieth century, India is expected to pass China in the twenty-first century. Nigeria, which had only 33 million residents in 1950, is forecast to have nearly 300 million in 2050. Ethiopia, with about 18 million people 50 years ago, is likely to grow nearly eight-fold over a century. In many of these countries, rapid population growth is a serious problem. Bangladesh, about the size of Iowa, is already overcrowded at 147 million people. Another 84 million people by 2050 will only add to current problems.

The other demographic world is made up of the richer countries of North America, western Europe, Japan, Australia, and New Zealand. This world is wealthy, old, and mostly shrinking. Italy, Germany, Hungary, and Japan, for example, all have negative growth rates. The average age in these countries is now 40, and life expectancy of their residents is expected to exceed 90 by 2050. With many couples choosing to have either one or no children, the populations of these countries are expected to decline significantly over the next century. Japan, which has 128 million residents now, is expected to shrink to about 100 million by 2050. Europe, which now makes up about 12 percent of the world population, will constitute less than 7 percent in 50 years, if current trends continue. Even the United States and Canada would have nearly stable populations if immigration were stopped.

It isn't only wealthy countries that have declining populations. Russia, for instance, is now declining by nearly 1 million people per year as death rates have soared and birth rates have plummeted. A collapsing economy, hyperinflation, crime, corruption, and despair have demoralized the population. Horrific pollution levels left from the Soviet era, coupled with poor nutrition and health care, have resulted in high levels of genetic abnormalities, infertility, and infant mortality. Abortions are twice as common as live births, and the average number of children per woman is now 1.3, one of the lowest in the world. Death rates, especially among adult men, have risen dramatically. Male life expectancy dropped from 68 years in 1990 to 59 years in 2006. Russia, which is the world's largest country geographically, is expected to decline from 142 million people in 2006 to less than 100 million in 2050. It will then have a smaller population than Vietnam, Egypt, or Uganda. Other former Soviet states are experiencing similar declines. Estonia, Bulgaria, Georgia, and Ukraine, for example, now have negative growth rates and are expected to lose about 40 percent of their population in the next 50 years.

The situation is even worse in many African countries, where AIDS and other communicable diseases are killing people at a terrible rate. In Zimbabwe, Botswana, Zambia, and Namibia, for example, up to 39 percent of the adult population have AIDS or are HIV positive. Health officials predict that more than two-thirds of the 15-year-olds now living in Botswana will die of AIDS before age 50. Without AIDS, the average life expectancy is estimated to be 69.7 years. Now, with AIDS, Botswana's life expectancy has dropped to only 31.6 years. The populations of many African countries are now falling because of this terrible disease (fig. 7.6). Altogether, Africa's population is expected to be nearly 200 million lower in 2050 than it would have been without AIDS.

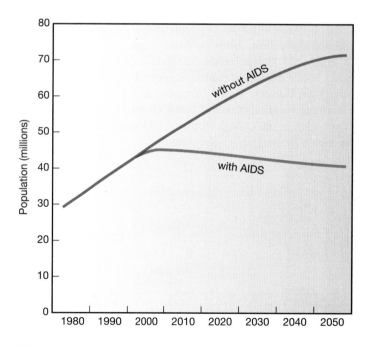

FIGURE 7.6 Projected population of south Africa with and without AIDS.
Data source: UN Population Division, 2006.

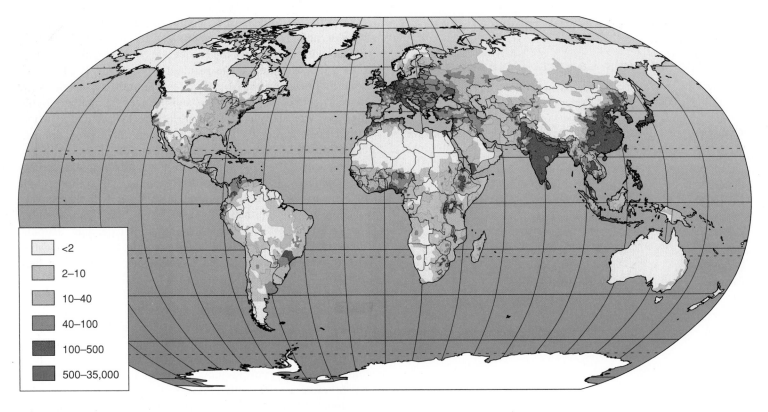

FIGURE 7.7 Population density in persons per square kilometer.
Source: World Bank, 2000.

AIDS is now spreading in Asia. Because of the large population there, Asia is expected to pass Africa in 2020 in total number of deaths. Although a terrible human tragedy, this probably won't affect total world population very much. Remember that the Black Death killed many people in the fourteenth century but had only a transitory effect on demography.

Figure 7.7 shows human population distribution around the world. Notice the high densities supported by fertile river valleys of the Nile, Ganges, Yellow, Yangtze, and Rhine Rivers and the well-watered coastal plains of India, China, and Europe. Historic factors, such as technology diffusion and geopolitical power, also play a role in geographic distribution.

Fertility measures the number of children born to each woman

As we pointed out in chapter 6, fecundity is the physical ability to reproduce, while fertility describes the actual production of offspring. Those without children may be fecund but not fertile. The most accessible demographic statistic of fertility is usually the **crude birth rate,** the number of births in a year per thousand persons. It is statistically "crude" in the sense that it is not adjusted for population characteristics such as the number of women in reproductive age. Table 7.3 shows birth rates more-developed and less-developed countries in 2006.

TABLE 7.3

Birth and Death Rates in 2006

	Population (Millions)	Births per 1,000	Deaths per 1,000	Total Fertility Rate	Natural Increase (Percent)
World	6,555	21	9	2.7	1.2
More-Developed Regions	1,216	11	10	1.6	0.1
Less-Developed Regions	5,339	23	8	2.9	1.5

Source: Population Reference Bureau, 2006.

The **total fertility rate** is the number of children born to an average woman in a population during her entire reproductive life. Upper-class women in seventeenth- and eighteenth-century England, whose babies were given to wet nurses immediately after birth and who were expected to produce as many children as possible, often had 25 or 30 pregnancies. The highest recorded total fertility rates for working-class people is among some Anabaptist agricultural groups in North America who have averaged up to 12 children per woman. In most tribal or traditional societies, food shortages, health problems, and cultural practices limit total fertility to about six or seven children per woman even without modern methods of birth control.

Zero population growth (ZPG) occurs when births plus immigration in a population just equal deaths plus emigration. It takes several generations of replacement level fertility (where people just replace themselves) to reach ZPG. Where infant mortality rates are high, the replacement level may be five or more children per couple. In the more highly developed countries, however, this rate is usually about 2.1 children per couple because some people are infertile, have children who do not survive, or choose not to have children.

Fertility rates have declined dramatically in every region of the world except Africa over the past 50 years (fig. 7.8). Only a few decades ago, total fertility rates above 6 were common in many countries. The average family in Mexico in 1975, for instance, had 7 children. By 2006, however, the average Mexican woman had only 2.4 children. According to the World Health Organization, 61 out of the world's 190 countries are now at or below a replacement rate of 2.1 children per couple, and by 2050, all but a few of the least-developed countries are expected to have reached that milestone. The greatest fertility reduction has been in Southeast Asia, where rates have fallen by more than half. Most

FIGURE 7.9 China's one-child-per-family policy, promoted in this billboard, has been remarkably successful in reducing birth rates. It may, however, have created a generation of "little emperors," since parents and grandparents focus all their attention on an only child.

of this decrease has occurred in just the past few decades and, contrary to what many demographers expected, some of the poorest countries in the world have been remarkably successful in lowering growth rates. As the opening case study for this chapter shows, Thailand reduced its total fertility rate from 7.0 in 1979 to 1.7 (lower than that in the United States) in 2006.

China's one-child-per-family policy decreased the fertility rate from 6 in 1970 to 1.8 in 1990 and 1.6 in 2006 (fig. 7.9). This policy, however, has sometimes resulted in abortions, forced sterilizations, and even infanticide. Another adverse result is that the only children (especially boys) allowed to families may grow up to be spoiled "little emperors" who have an inflated impression of their own importance. (fig. 7.9) Furthermore, there may not be enough workers to maintain the army, sustain the economy, or support retirees when their parents reach old age.

China reports that 119 boys are now being born for every 100 girls. Normal ratios would be about 105 boys to 100 girls. If this imbalance persists, there will be a shortage of brides in another generation. The government is considering easing the one-child policy. Macao, with a total average fertility rate of only 0.9, now has the lowest birth rate in the world.

Although the world as a whole still has an average fertility rate of 2.7, growth rates are now lower than at any time since World War II. If fertility declines like those in Thailand and China were to occur everywhere in the world, global population could begin to decline by 2050, and might be below 6 billion by 2150. Most of Eastern Europe now has fertility levels of 1.2 children per woman. Interestingly, Spain and Italy, although predominately Roman Catholic, have similar fertility rates. Several Indian states have reached zero population growth, but their means of doing so have been very different (What Do You Think? p. 140).

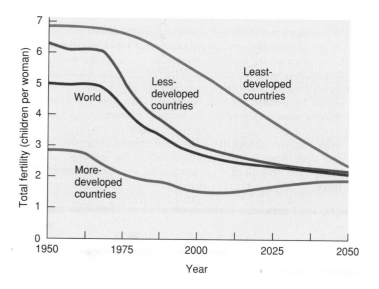

FIGURE 7.8 Average total fertility rates for less-developed countries fell by more than half over the past 50 years. Much of this dramatic change was due to China's one-child policy. Progress has been slower in the least-developed countries, but by 2050, they should be approaching the replacement rate of 2.1 children per woman of reproductive age.

How to Reduce Population Growth?

Every year, India adds more people to the world's population than any other country. In 2006, having added more than 185 million residents in the previous decade, India had more than 1.1 billion people. By 2050, if current growth rates persist, India will have increased its population by more than 50 percent over current levels, and will have around 1.63 billion residents, making it the most populous country in the world. How will the country, which already has more than a quarter of its population living in abject poverty, feed, house, educate, and employ all those being added each year? And what's the best way to slow this rapid growth? The fierce debate now taking place about how to control India's population has ramifications for the rest of the world as well.

Currently, fewer than half of all Indian women use any form of birth control. How can this percentage be raised? On one side of this issue are those who believe that the best way to reduce the number of children born is poverty eradication and progress for women. Drawing on social justice principles established at the 1994 UN Conference on Population and Development in Cairo, some argue that responsible economic development, a broad-based social welfare system, education and empowerment of women, and high-quality health care—including family planning services—are essential components of population control. Without progress in these areas, they believe, efforts to provide contraceptives or encourage sterilization are futile.

On the other side of this debate are those who contend that, while social progress is an admirable goal, India doesn't have the time or resources to wait for an indirect approach to population control. The government must push aggressively, they argue, to reduce births now or the population will be so huge and its use of resources so great that only mass starvation, class war, crime, and disease will be able to bring it down to a manageable size.

Unable to reach a consensus on population policy, the Indian government decided in 2000 to let each state approach the problem in its own way. Some states have chosen to focus on social justice, while others have adopted more direct, interventionist policies.

The model for the social justice approach is the southern state of Kerala, which achieved population stabilization in the mid-1980s, the first Indian state to do so. Although still one of the poorest places in the world, economically, Kerala's fertility rate is comparable to that of many industrialized nations, including the United States. Both women and men have a nearly 100 percent literacy rate and share affordable and accessible health care, family planning, and educational opportunities; therefore, women have only the number of children they want, usually two. The Kerala experience suggests that increased wealth isn't a prerequisite for zero population growth.

Taking a far different path to birth reduction is the nearby state of Andra Pradesh, which reached a stable growth rate in 2001. Boasting the most dramatic fertility decline of any large Indian state, Andra Pradesh has focused on targeted, strongly enforced sterilization programs. The poor are encouraged—some would say, compelled—to be sterilized after having only one or two children. The incentives include cash payments. You might receive 500 rupees—equivalent to (U.S.) $11 or a month's wages for an illiterate farm worker—if you agree to have "the operation." In addition, participants are eligible for better housing, land, wells, and subsidized loans.

The pressure to be sterilized is overwhelmingly directed at women, for whom the procedure is major abdominal surgery. Sterilizations often are done by animal husbandry staff and carried out in government sterilization camps. This practice raises troubling memories of the 1970s for many people, when then-Prime Minister Indira Gandhi suspended democracy and instituted a program of forced sterilization of poor people. There were reports at the time of people being rounded up like livestock and castrated or neutered against their will.

While many feminists and academics regard Andra Pradesh's policies as appallingly intrusive and coercive to women and the poor, the state has successfully reduced population growth. By contrast, the hugely populous northern states of Uttar Pradesh and Bihar have seen slightly increased growth rates over the past two decades to a current rate above 2.5 percent per year. How will they slow this exponential growth, and what might be the social and environmental costs of not doing so?

What do you think? How do birth control programs in India compare with those in Thailand describes in the opening case study for this chapter? Which of these models for population control would you favor?

Indian states have taken very different approaches to family planning and human development.

Mortality is the other half of the population equation

A traveler to a foreign country once asked a local resident, "What's the death rate around here?" "Oh, the same as anywhere," was the reply, "about one per person." In demographics, however, **crude death rates** (or crude mortality rates) are expressed in terms of the number of deaths per thousand persons in any given year. Countries in Africa where health care and sanitation are limited may have mortality rates of 20 or more per 1,000 people. Wealthier countries generally have mortality rates around 10 per 1,000. The number of deaths in a population is sensitive to the age structure of the population. Rapidly growing, developing countries such as Libya or Costa Rica have lower crude death rates (4 per 1,000) than do the more-developed, slowly growing countries, such as Denmark (12 per 1,000). This is because there are proportionately more youths and fewer elderly people in a rapidly growing country than in a more slowly growing one.

Crude death rate subtracted from crude birth rate gives the **natural increase** of a population. We distinguish natural increase from the **total growth rate,** which includes immigration and emigration, as well as births and deaths. Both of these growth rates are usually expressed as a percent (number per hundred people) rather than per thousand. A useful rule of thumb is that if you divide 70 by the annual percentage growth, you will get the approximate doubling time in years. Niger, for example, which is growing 3.4 percent per year, is doubling its population every 20 years. The United States, which has a natural increase rate of 0.6 percent per year, would double, without immigration, in 116.7 years. Belgium and Sweden, with natural increase rates of 0.1 percent, are doubling in about 700 years. Ukraine, on the other hand, with a growth rate of −0.8 percent, will lose about 40 percent of its population in the next 50 years. The world growth rate is now 1.2 percent, which means that the population will double in about 58 years if this rate persists.

Life span and life expectancy describe our potential longevity

Life span is the oldest age to which a species is known to survive. Although there are many claims in ancient literature of kings living a millennium or more, the oldest age that can be certified by written records was that of Jeanne Louise Calment of Arles, France, who was 122 years old at her death in 1997. The aging process is still a medical mystery, but it appears that cells in our bodies have a limited ability to repair damage and produce new components. At some point they simply wear out, and we fall victim to disease, degeneration, accidents, or senility.

Life expectancy is the average age that a newborn infant can expect to attain in any given society. It is another way of expressing the average age at death. For most of human history, we believe that average life expectancy in most societies has been about 30 years. This doesn't mean that no one lived past age 40, but rather that so many deaths at earlier ages (mostly early childhood) balanced out those who managed to live longer.

Declining mortality, not rising fertility, is the primary cause of most population growth in the past 300 years. Crude death rates began falling in western Europe during the late 1700s. Most of this advance in survivorship came long before the advent of modern medicine and is due primarily to better food and better sanitation.

The twentieth century has seen a global transformation in human health unmatched in history. This revolution can be seen in the dramatic increases in life expectancy in most places. Worldwide, the average life expectancy has risen from about 30 to 65 years over the past century (see fig. 6.5). Table 7.4 shows gains in some selected countries. Globally, the number of people

TABLE 7.4

Life Expectancy at Birth for Selected Countries in 1900 and 2006

Country	1900		2006	
	Males	Females	Males	Females
India	23	23	62	63
Japan	42	44	79	86
Russia	31	33	59	72
Sweden	57	60	78	83
United States	46	48	75	80

Source: Population Reference Bureau, 2006.

over 60 years old is expected to triple, increasing from 600 million today to nearly 2 billion in 2050. The oldest old (over 80 years) is projected to grow five-fold to about 400 million in that same period.

The greatest progress in life expectancy has been in developing countries. Take the case of Nicaragua, for example. In 1900, the average Nicaraguan man could expect to live only 29 years, while the average woman would reach just 33 years. By 2006, although Nicaragua had an annual per capita income of about (U.S.)$3,600, the average life expectancy for both men and women had more than doubled and was close to that of countries with ten times its income level. Longer lives were due primarily to better nutrition, improved sanitation, clean water, and education rather than miracle drugs or high-tech medicine. While the gains were not as great for the already industrialized countries, residents of the United States, Italy, and Japan, for example, now live about half-again as long as they did at the beginning of the twentieth century.

As figure 7.10 shows, there is a good correlation between annual income and life expectancy up to about (U.S.) $4,000 per person. Beyond that level—which is generally enough for adequate food, shelter, and sanitation for most people—life expectancies level out at about 75 years for men and 85 for women.

Large discrepancies in how benefits of modernization and social investment are distributed within countries are revealed in differential longevity of various groups. The greatest life expectancy reported anywhere in the United States is for Asian American women in New Jersey, who live to an average age of 91. By contrast, Native American men on Pine Ridge Indian Reservation in neighboring South Dakota, live, on average, only to age 48. Only a few countries in Africa have a lower life expectancy. The Pine Ridge Reservation is the poorest area in America with an unemployment rate near 75 percent and high rates of poverty, alcoholism, drug use, and cultural alienation. Similarly, African-American men in Washington, D.C., live, on average, only 57.9 years, or less than men in Lesotho or Swaziland.

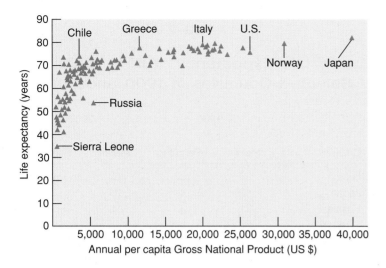

FIGURE 7.10 As incomes rise, so does life expectancy up to about (U.S.) $4,000. Russia is an exception, with a life expectancy nearly 20 years less than that of Chile, even though their GNP is about the same.
Source: The World Bank, *World Development Indicators 1997,* the World Bank, Washington, D.C., 1997.

Some demographers believe that life expectancy is approaching a plateau, while others predict that advances in biology and medicine might make it possible to live 150 years or more. If our average age at death approaches 100 years, as some expect, society will be profoundly affected. In 1970 the median age in the United States was 30. By 2100 the median age could be over 60.

If workers continue to retire at 65, half of the population could be unemployed, and retirees might be facing 35 or 40 years of retirement. We may need to find new ways to structure and finance our lives.

Living longer has demographic implications

A population that is growing rapidly by natural increase has more young people than does a stationary population. One way to show these differences is to graph age classes in a histogram as shown in figure 7.11. In Niger, which is growing at a rate of 3.4 percent per year, 49 percent of the population is in the prereproductive category (below age 15). Even if total fertility rates were to fall abruptly, the total number of births, and population size, would continue to grow for some years as these young people enter reproductive age. This phenomenon is called population momentum.

By contrast, a country with a stable population, like Sweden, has nearly the same number in each age cohort. A population that has recently entered a lower growth rate pattern, such as Singapore, has a bulge in the age classes for the last high-birth-rate generation. Notice that there are more females than males in the older age group in Sweden because of differences in longevity between the sexes.

Both rapidly growing countries and slowly growing countries can have a problem with their **dependency ratio,** or the number of nonworking compared to working individuals in a population. In Mexico, for example, each working person supports

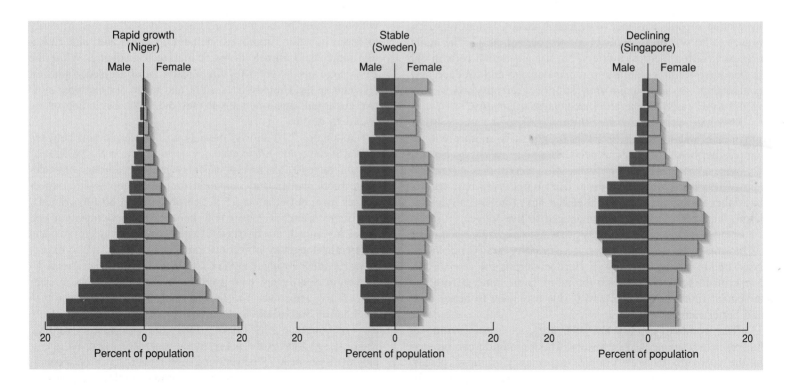

FIGURE 7.11 Age structure graphs for rapidly growing, stable, and declining populations.
Source: U.S. Census Bureau, 2006.

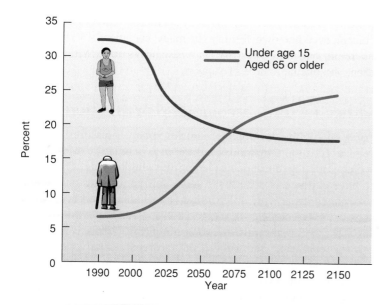

FIGURE 7.12 By the mid-twenty-first century, children under age 15 will make up a smaller percentage of world population, while people over age 65 will contribute a larger and larger share of the population.

a high number of children. In the United States, by contrast, a declining working population is now supporting an ever larger number of retired persons and there are dire predictions that the social security system will soon be bankrupt. This changing age structure and shifting dependency ratio are occurring worldwide (fig. 7.12). By 2050 the UN predicts there will be two older persons for every child in the world. Many countries are rethinking their population policies and beginning to offer incentives for marriage and child-rearing.

Emigration and immigration are important demographic factors

Humans are highly mobile, so emigration and immigration play a larger role in human population dynamics than they do in those of many species. Currently, about 800,000 people immigrate legally to the United States each year, but many more enter illegally. Western Europe receives about 1 million applications each year for asylum from economic chaos and wars in former socialist states and the Middle East. The United Nations High Commission on Refugees reported that in 2006 there were 20.8 million refugees who had left their countries for political or economic reasons, while about 25 million more were displaced persons in their own countries, and 175 million migrants had left their homes to look for work, greater freedom, or better opportunities.

The more-developed regions are expected to gain about 2 million immigrants per year for the next 50 years. Without migration, the population of the wealthiest countries would already be declining and would be more than 126 million less than the current 1.2 billion by 2050. In 2006, nearly 42 million

U.S. residents (14.5 percent of the total population) class themselves as Hispanic or Latino. They now constitute t largest U.S. minority.

Immigration is a controversial issue in many countries. "Guest workers" often perform heavy, dangerous, or disagreeable work that citizens are unwilling to do. Many migrants and alien workers are of a different racial or ethnic background than the majority in their new home. They generally are paid low wages and given substandard housing, poor working conditions, and few rights. Local residents often complain, however, that immigrants take away jobs, overload social services, and ignore established rules of behavior or social values. Anti-immigrant groups are springing up in many rich countries.

Some nations encourage, or even force, internal mass migrations as part of a geopolitical demographic policy. In the 1970s, Indonesia embarked on an ambitious "transmigration" plan to move 65 million people from the overcrowded islands of Java and Bali to relatively unpopulated regions of Sumatra, Borneo, and New Guinea. Attempts to turn rainforest into farmland had disastrous environmental and social effects, however, and this plan was greatly scaled back. China has announced a plan to move up to 100 million people to a sparsely populated region along the Amur River in Heilongjiang. By some estimates, more than 250 million internal migrants in China have moved from rural areas to the cities to look for work.

7.4 IDEAL FAMILY SIZE IS CULTURALLY AND ECONOMICALLY DEPENDENT

A number of social and economic pressures affect decisions about family size, which in turn affects the population at large. In this section we will examine both positive and negative pressures on reproduction.

Many factors increase our desire for children

Factors that increase people's desires to have babies are called **pronatalist pressures.** Raising a family may be the most enjoyable and rewarding part of many people's lives. Children can be a source of pleasure, pride, and comfort. They may be the only source of support for elderly parents in countries without a social security system. Where infant mortality rates are high, couples may need to have many children to be sure that at least a few will survive to take care of them when they are old. Where there is little opportunity for upward mobility, children give status in society, express parental creativity, and provide a sense of continuity and accomplishment otherwise missing from life. Often children are valuable to the family not only for future income, but even more as a source of current income and help with household chores. In much of the developing world, children as young as 6 years old tend domestic animals and younger siblings, fetch water, gather firewood, and help grow crops or sell things in the

(a)   (b)

FIGURE 7.13 In rural areas with little mechanized agriculture (a) children are needed to tend livestock, care for younger children, and help parents with household chores. Where agriculture is mechanized (b) rural families view children just as urban families do—helpful, but not critical to survival. This affects the decision about how many children to have.

marketplace (fig. 7.13) Parental desire for children rather than an unmet need for contraceptives may be the most important factor in population growth in many cases.

Society also has a need to replace members who die or become incapacitated. This need often is codified in cultural or religious values that encourage bearing and raising children. In some societies, families with few or no children are looked upon with pity or contempt. The idea of deliberately controlling fertility may be shocking, even taboo. Women who are pregnant or have small children are given special status and protection. Boys frequently are more valued than girls because they carry on the family name and are expected to support their parents in old age. Couples may have more children than they really want in an attempt to produce a son.

Male pride often is linked to having as many children as possible. In Niger and Cameroon, for example, men, on average, want 12.6 and 11.2 children, respectively. Women in these countries consider the ideal family size to be only about one-half that desired by their husbands. Even though a woman might desire

fewer children, however, she may have few choices and little control over her own fertility. In many societies, a woman has no status outside of her role as wife and mother. Without children, she has no source of support.

Birth reduction Pressures
Other factors discourage reproduction

In more highly developed countries, many pressures tend to reduce fertility. Higher education and personal freedom for women often result in decisions to limit childbearing. The desire to have children is offset by a desire for other goods and activities that compete with childbearing and childbearing for time and money. When women have opportunities to earn a salary, they are less likely to stay home and have many children. Not only are the challenge and variety of a career attractive to many women, but the money that they can earn outside the home becomes an important part of the family budget. Thus, education and socioeconomic status are usually inversely related to fertility in richer countries. In developing countries, however, fertility is likely to increase as educational levels and socioeconomic status rise. With higher income, families are better able to afford the children they want; more money means that women are likely to be healthier, and therefore better able to conceive and carry a child to term.

In less-developed countries where feeding and clothing children can be a minimal expense, adding one more child to a family usually doesn't cost much. By contrast, raising a child in the United States can cost hundreds of thousands of dollars by the time the child is through school and is independent. Under these circumstances, parents are more likely to choose to have one or two children on whom they can concentrate their time, energy, and financial resources.

Figure 7.14 shows U.S. birth rates between 1910 and 2000. As you can see, birth rates have fallen and risen in a complex pattern. The period between 1910 and 1930 was a time of industrialization and urbanization. Women were getting more education

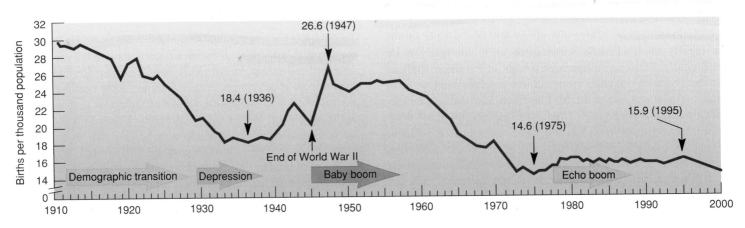

FIGURE 7.14 Birth rates in the United States, 1910–2000. The falling birth rate from 1910 to 1929 represents a demographic transition from an agricultural to an industrial society. The baby boom following World War II lasted from 1945 to 1965. A much smaller "echo boom" occurred around 1980 when the baby boomers started to reproduce.
Source: Data from Population Reference Bureau and U.S. Bureau of the Census.

than ever before and entering the workforce. The Great Depression in the 1930s made it economically difficult for families to have children, and birth rates were low. The birth rate increased at the beginning of World War II (as it often does in wartime). For reasons that are unclear, a higher percentage of boys are usually born during war years.

At the end of the war, there was a "baby boom" as couples were reunited and new families started. This high birth rate persisted through the times of prosperity and optimism of the 1950s, but began to fall in the 1960s. Part of this decline was caused by the small number of babies born in the 1930s. This meant fewer young adults to give birth in the 1960s. Part was due to changed perceptions of the ideal family size. Whereas in the 1950s women typically wanted four children or more, in the 1970s the norm dropped to one or two (or no) children. A small "echo boom" occurred in the 1980s as people born in the 1960s began to have babies, but changing economics and attitudes seem to have permanently altered our view of ideal family size in the United States.

> **Think About It**
>
> How many children (if any) do you want to have? Is this number different from that of your parents or grandparents? Why or why not?

Could we have a birth dearth?

Most European countries now have birth rates below replacement rates, and Italy, Russia, Austria, Germany, Greece, and Spain are experiencing negative rates of natural population increase. Asia, Japan, Singapore, and Taiwan are also facing a "child shock" as fertility rates have fallen well below the replacement level of 2.1 children per couple. There are concerns in all these countries about falling military strength (lack of soldiers), economic power (lack of workers), and declining social systems (not enough workers and taxpayers) if low birth rates persist or are not balanced by immigration. In a sense, the United States is fortunate to have a high influx of immigrants that provides youth and energy to its population.

Economist Ben Wattenberg warns that this "birth dearth" might seriously erode the powers of Western democracies in world affairs. He points out that Europe and North America accounted for 22 percent of the world's population in 1950. By the 1980s, this number had fallen to 15 percent, and by the year 2030, Europe and North America probably will make up only 9 percent of the world's population. Germany, Hungary. Denmark, and Russia now offer incentives to encourage women to bear children. Japan offers financial support to new parents, and Singapore provides a dating service to encourage marriages among the upper classes as a way of increasing population.

On the other hand, since Europeans and North Americans consume so many more resources per capita than most other people in the world, a reduction in the population of these countries will do more to spare the environment than would a reduction in population almost anywhere else.

One reason that birth rates have been falling in many industrialized countries may be that toxins and endocrine hormone disrupters in our environment interfere with sperm production. Sperm numbers and quality (fertilization ability) appear to have fallen by about half over the past 50 years in a number of countries. Widespread chemicals, such as phthalates—common ingredients in plastics—that disrupt sperm production may be responsible for this decline. We'll discuss this further in chapter 8.

7.5 A DEMOGRAPHIC TRANSITION CAN LEAD TO STABLE POPULATION SIZE

In 1945, demographer Frank Notestein pointed out that a typical pattern of falling death rates and birth rates due to improved living conditions usually accompanies economic development. He called this pattern the **demographic transition** from high birth and death rates to lower birth and death rates. Figure 7.15 shows an idealized model of a demographic transition. This model is often used to explain connections between population growth and economic development.

Economic and social development influence birth and death rates

Stage I in figure 7.15 represents the conditions in a premodern society. Food shortages, malnutrition, lack of sanitation and medicine, accidents, and other hazards generally keep death rates in such a society around 35 per 1,000 people. Birth rates are correspondingly high to keep population densities relatively constant. As economic development brings better jobs, medical care, sanitation, and a generally improved standard of living in Stage II, death rates often fall very rapidly. Birth rates may actually rise at first as more money and better nutrition allow people to have the children they always wanted. Eventually, in a mature industrial economy (Stage III), birth rates fall as people see that all their children are more likely to survive and that the whole family benefits from concentrating more resources on fewer children. Note that population continues to grow rapidly during this stage because of population momentum (baby boomers reaching reproductive age). Depending on how long it takes to complete the transition, the population may go through one or more rounds of doubling before coming into balance again.

Stage IV in figure 7.15 represents conditions in developed countries, where the transition is complete and both birth rates and death rates are low, often, a third or less than those in the predevelopment era. The population comes into a new equilibrium in this phase, but at a much larger size than before. Most of the countries of northern and western Europe went through a demographic transition in the nineteenth or early twentieth century similar to the curves shown in this figure.

Many of the most rapidly growing countries in the world, such as Kenya, Yemen, Libya, and Jordan, now are in the Stage I of this demographic transition. Their death rates have fallen

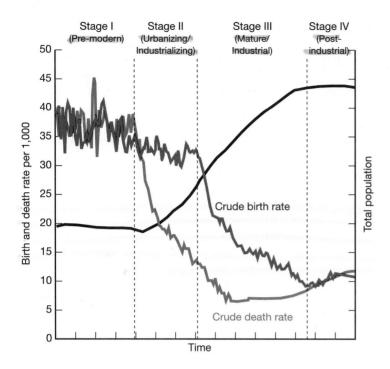

FIGURE 7.15 Theoretical birth, death, and population growth rates in a demographic transition accompanying economic and social development. In a predevelopment society, birth and death rates are both high, and total population remains relatively stable. During development, death rates tend to fall first, followed in a generation or two by falling birth rates. Total population grows rapidly until both birth and death rates stabilize in a fully developed society.

close to the rates of the fully developed countries, but birth rates have not fallen correspondingly. In fact, both their birth rates and total population are higher than those in most European countries when industrialization began 300 years ago. The large disparity between birth and death rates means that many developing countries now are growing at 3 to 4 percent per year. Such high growth rates in developing countries could boost total world population to 9 billion or more before the end of the twenty-first century. This raises what may be the two most important questions in this entire chapter: Why are birth rates not yet falling in these countries, and what can be done about it?

There are reasons to be optimistic about population

Four conditions are necessary for a demographic transition to occur: (1) improved standard of living, (2) increased confidence that children will survive to maturity, (3) improved social status of women, and (4) increased availability and use of birth control. As the example of Thailand in the opening case study for this chapter shows, these conditions can be met, even in relatively poor countries.

Some demographers claim that a demographic transition already is in progress in most developing nations. Problems in taking censuses and a normal lag between falling death and birth rates may hide this for a time, but the world population should stabilize sometime in the next century. Some evidence supports this view. As we mentioned earlier in this chapter, fertility rates have fallen dramatically nearly everywhere in the world over the past half century.

Some countries have had remarkable success in population control. In Thailand, Indonesia, Colombia, and Iran, for instance, total fertility dropped by more than half in 20 years. Morocco, Dominican Republic, Jamaica, Peru, and Mexico all have seen fertility rates fall between 30 percent and 40 percent in a single generation. The following factors could contribute to stabilizing populations:

- Growing prosperity and social reforms that accompany development reduce the need and desire for large families in most countries.

- Technology is available to bring advances to the developing world much more rapidly than was the case a century ago, and the rate of technology transfer is much faster than it was when Europe and North America were developing.

- Less-developed countries have historic patterns to follow. They can benefit from our mistakes and chart a course to stability more quickly than they might otherwise do.

- Modern communications (especially television) have caused a revolution of rising expectations that act as a stimulus to spur change and development.

Many people remain pessimistic about population growth

Economist Lester Brown takes a more pessimistic view. He warns that many of the poorer countries of the world appear to be caught in a "demographic trap" that prevents them from escaping from the middle phase of the demographic transition. Their populations are now growing so rapidly that human demands exceed the sustainable yield of local forests, grasslands, croplands, or water resources. The resulting resource shortages, environmental deterioration, economic decline, and political instability may prevent these countries from ever completing modernization. Their populations may continue to grow until catastrophe intervenes.

Many people argue that the only way to break out of the demographic trap is to immediately and drastically reduce population growth by whatever means are necessary. They argue strongly for birth control education and bold national policies to encourage lower birth rates. Some agree with Malthus that helping the poor will simply increase their reproductive success and further threaten the resources on which we all depend. Author Garret Hardin described this view as lifeboat ethics. "Each rich nation," he said, "amounts to a lifeboat full of comparatively rich people. The poor of the world are in other much more crowded lifeboats. Continuously, so to speak, the poor fall out of their lifeboats and swim for a while, hoping to be admitted to a rich lifeboat, or in some other way to benefit from the goodies on board. . . . We cannot risk the safety of all the passengers by helping others in need. What happens if you share space in a lifeboat? The boat is swamped and everyone drowns. Complete justice, complete catastrophe."

Social justice is an important consideration

A third view is that **social justice** (a fair share of social benefits for everyone) is the real key to successful demographic transitions. The world has enough resources for everyone, but inequitable social and economic systems cause maldistributions of those resources. Hunger, poverty, violence, environmental degradation, and overpopulation are symptoms of a lack of social justice rather than a lack of resources. Although overpopulation exacerbates other problems, a narrow focus on this factor alone encourages racism and hatred of the poor. A solution for all these problems is to establish fair systems, not to blame the victims. Small nations and minorities often regard calls for population control as a form of genocide. Figure 7.16 expresses the opinion of many people in less-developed countries about the relationship between resources and population.

An important part of this view is that many of the rich countries are, or were, colonial powers, while the poor, rapidly growing countries were colonies. The wealth that paid for progress and security for developed countries was often extracted from colonies, which now suffer from exhausted resources, exploding populations, and chaotic political systems. Some of the world's poorest countries such as India, Ethiopia, Mozambique, and Haiti had rich resources and adequate food supplies before they were impoverished by colonialism. Those of us who now enjoy abundance may need to help the poorer countries not only as a matter of justice but because we all share the same environment.

In addition to considering the rights of fellow humans, we should also consider those of other species. Rather than ask what is the maximum number of humans that the world can possibly support, perhaps we should think about the needs of other creatures. As we convert natural landscapes into agricultural or industrial areas, species are crowded out that may have just as much right to exist as we do. Perhaps we should seek the optimum number of people at which we can provide a fair and decent life for all humans while causing the minimum impact on nonhuman neighbors.

FIGURE 7.16 Controlling our population and resources—there may be more than one side to the issue.

Women's rights affect fertility

Opportunities for education and paying jobs are critical factors in fertility rates (fig. 7.17). Child survival also is crucial in stabilizing population. When Infant and child mortality rates are high, as they are in much of the developing world, parents tend to have high numbers of children to ensure that some will survive to adulthood. There has never been a sustained drop in birth rates that was not first preceded by a sustained drop in infant and child mortality. One of the most important distinctions in our demographically divided world is the high infant mortality rates in the less-developed countries. Better nutrition, improved health care, simple oral rehydration therapy, and immunization against infectious diseases (chapter 8) have brought about dramatic reductions in child mortality rates, which have been accompanied in most regions by falling birth rates. It has been estimated that saving 5 million children each year from easily preventable communicable diseases would avoid 20 or 30 million extra births.

Increasing family income does not always translate into better welfare for children since men in many cultures control most financial assets. Often the best way to improve child survival is to ensure the rights of mothers. Land reform, political rights, opportunities to earn an independent income, and improved health status of women often are better indicators of total fertility and family welfare than rising GNP.

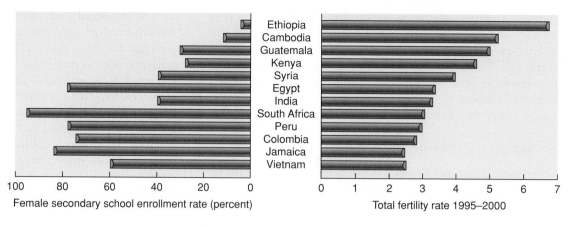

FIGURE 7.17 Total fertility declines as women's education increases.
Source: Worldwatch Institute, 2003.

7.6 FAMILY PLANNING GIVES US CHOICES

Family planning allows couples to determine the number and spacing of their children. It doesn't necessarily mean fewer children—people may use family planning to have the maximum number of children possible—but it does imply that the parents will control their reproductive lives and make rational, conscious decisions about how many children they will have and when those children will be born, rather than leaving it to chance. As the desire for smaller families becomes more common, birth control becomes an essential part of family planning in most cases. In this context, **birth control** usually means any method used to reduce births, including abstinence, delayed marriage, contraception, methods that prevent implantation of embryos, and induced abortions. As the opening case study in this chapter shows, there are many ways to encourage family planning.

Fertility control has existed throughout history

Evidence suggests that people in every culture and every historic period have used a variety of techniques to control population size. Studies of hunting and gathering people, such as the !Kung or San of the Kalahari Desert in southwest Africa, indicate that our early ancestors had stable population densities, not because they killed each other or starved to death regularly, but because they controlled fertility.

For instance, San women breast-feed children for three or four years. When calories are limited, lactation depletes body fat stores and suppresses ovulation. Coupled with taboos against intercourse while breast-feeding, this is an effective way of spacing children. Other ancient techniques to control population size include abstinence, folk medicines, abortion, and infanticide. We may find some or all of these techniques unpleasant or morally unacceptable, but we shouldn't assume that other people are too ignorant or too primitive to make decisions about fertility.

Today there are many options

Modern medicine gives us many more options for controlling fertility than were available to our ancestors. The major categories of birth control techniques include (1) avoidance of sex during fertile periods (for example, celibacy or the use of changes in body temperature or cervical mucus to judge when ovulation will occur), (2) mechanical barriers that prevent contact between sperm and egg (for example, condoms, spermicides, diaphragms, cervical caps, and vaginal sponges), (3) surgical methods that prevent release of sperm or egg (for example, tubal ligations in females and vasectomies in males), (4) hormone-like chemicals that prevent maturation or release of sperm or eggs or that prevent embryo implantation in the uterus (for example, estrogen plus progesterone, or progesterone alone, for females: gossypol for males), (5) physical barriers to implantation (for example, intrauterine devices), and (6) abortion.

TABLE 7.5

Some Birth Control Methods and Pregnancy Prevention Rates

Method	Number of Women in 100 Who Become Pregnant
Sterilization (male, female)	<1
IUD	<1
Oral contraceptive (the Pill)	1–2
Hormones (implant, patch, injection, etc.)	1–2
Male condom	11
Sponge and spermicide	14–28
Female condom (e.g., cervical cap)	15–23
Diaphragm together with spermicide	17
Abstinence during fertile periods	20
Morning-after-pill (e.g., Preven)	20
Spermicide alone	20–50
Actively seeking pregnancy	85

Source: U.S. Food and Drug Administration, *Birth Control Guide,* 2003 Revision.

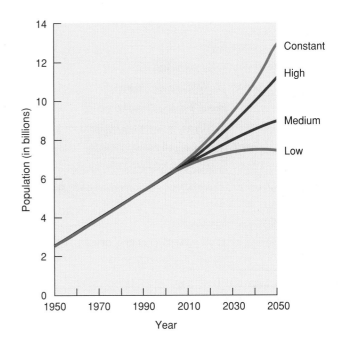

FIGURE 7.18 Population projections for different growth scenarios. Recent progress in family planning and economic development have led to significantly reduced estimates compared to a few years ago. The medium projection is 8.9 billion in 2050, compared to previous estimates of over 10 billion for that date.
Source: UN Population Division, 2004.

Not surprisingly, the most effective birth control methods are also the ones most commonly used (table 7.5). In the United States, the majority of women younger than 30 who eventually want to become pregnant use the Pill. Most women over 35, with their child-bearing years behind them, choose sterilization. Male condom use is more effective than the remaining techniques in the table, and increases in effectiveness when used with a spermicide, Only 2 to 6 women in a hundred become pregnant in a year using this combination method. Condoms have the added advantage of protecting partners against sexually transmitted diseases, including AIDS, if they are made of latex and used correctly. That may partly explain why their use in the United States went from 3.5 million users in 1980 to 8 million in 2000. Condoms are an ancient birth control method; the Egyptians used them some 3,000 years ago.

More than 100 new contraceptive methods are now being studied, and some appear to have great promise. Nearly all are biologically based (e.g., hormonal), rather than mechanical (e.g., condom, IUD). Recently, the U.S. Food and Drug Administration approved five new birth control products. Four of these use various methods to administer female hormones that prevent pregnancy. Other methods are years away from use, but take a new direction entirely. Vaccines for women are being developed that will prepare the immune system to reject the hormone choriononic gonadotropin, which maintains the uterine lining and allows egg implant, or that will cause an immune reaction against sperm. Injections for men are focused on reducing sperm production, and have proven effective in mice. Without a doubt, the contemporary couple has access to many more birth control options than their grandparents had.

7.7 WHAT KIND OF FUTURE ARE WE CREATING?

How many people will be in the world a century from now? Most demographers believe that world population will stabilize sometime during the next century. The total number of humans, when we reach that equilibrium, is likely to be somewhere around 8 to 10 billion people, depending on the success of family planning programs and the multitude of other factors affecting human populations. Figure 7.18 shows three scenarios projected by the UN Population Division in its 2004 revision. The optimistic (low) projection shows that world population might reach about 7 billion in 2050, and then fall back below 6 billion by 2150. The medium projection suggests that growth might continue to around 8.9 billion in 2050, and then stabilize. The most pessimistic projection assumes a constant rate of growth (no change from present) to 25 billion people by 2150.

Which of these scenarios will we follow? As you have seen in this chapter, population growth is a complex subject. To accomplish a stabilization or reduction of human populations will require substantial changes from business as usual.

An encouraging sign is that worldwide contraceptive use has increased sharply in recent years. About half of the world's married couples used some family planning techniques in 2000, compared to only 10 percent 30 years earlier, but another 100 million couples say they want, but do not have access to, family planning. Contraceptive use varies widely by region. More than

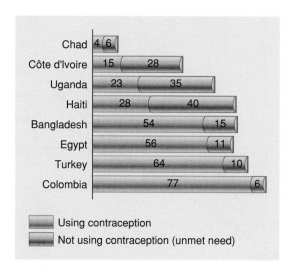

FIGURE 7.19 Unmet need for family planning in selected countries. Globally, more than 100 million women in developing countries would prefer to avoid pregnancy but do not have access to family planning. **Source:** Population Reference Bureau, 2003.

70 percent of women in Latin America use some form of birth control, compared to 51 percent in Asia (excluding China), and only 21 percent in Africa.

Figure 7.19 shows the unmet need for family planning among married women in some representative countries. When people in developing countries are asked what they want most, men say they want better jobs, but the first choice for a vast majority of women is family planning assistance. In general, a 15 percent increase in contraceptive use equates to about one fewer birth per woman per lifetime. In Chad, for example, where only 4 percent of all women use contraceptives, the average fertility is 6.6 children per woman. In Columbia, by contrast, where 77 percent of the women who would prefer not to be pregnant use contraceptives, the average fertility is 2.6.

Religion and politics complicate family planning

In 1994, the United Nations convened a historic meeting in Cairo, Egypt, to discuss women's rights and population. The United States played a lead role in the International Conference on Population and Development (ICPD), which identified links between population growth, economic development, environmental degradation, and the social status of women and girls. To address these issues, 179 countries, including the United States, endorsed the goal of universally available reproductive health services, including family planning, by 2015.

Since 2000, however, the United States has refused to reaffirm the ICPD because it maintained that the document could be interpreted as promoting abortion—even though the ICPD clearly states, "In no case should abortion be promoted as a method of family planning."

In particular, the United States has withheld funds from the United Nations Population Fund (UNFPA) due to claims that, by working in China, the fund tacitly supports the forced abortions reported to be part of that country's one-child policy. A fact-finding team sent to China in 2002 found "no evidence of UNFPA knowledge of or support for such measures," but funding was still halted. Ms. Thoraya Obaid, executive director of the UNFPA, said her organization "does not, and never will condone or support coercive activities of any kind, anywhere."

She also said, "The denial of these funds will, unfortunately, significantly affect millions of women and children worldwide for whom the life-saving services provided by the UNFPA will have to be discontinued." She estimated that the funds withheld by the United States could have prevented 2 million unwanted pregnancies, 800,000 abortions, 4,700 maternal deaths, 60,000 cases of serious maternal illness, and more than 77,000 infant and child deaths.

Many Muslim countries encourage couples to have as many children as possible. Access to birth control is difficult or forbidden outright. Still, some Islamic governments recognize the need for family planning. Iran, for example, decided, in the 1990s, to promote smaller families. It succeeded in cutting birth rates by more than half in ten years.

The World Health Organization estimates that nearly 1 million conceptions occur daily around the world as a result of some 100 million sex acts. At least half of those conceptions are unplanned or unwanted. But there are still places where people desire large families (fig. 7.20).

Deep societal changes are often required to make family planning programs successful. Among the most important of these are (1) improved social, educational, and economic status

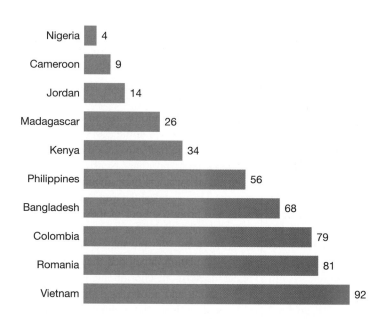

FIGURE 7.20 Percent of married reproductive-age women with two living children who do not want another child. **Data source:** UN Population Division, 2006.

for women (birth control and women's rights are often interdependent); (2) improved status for children (fewer children are born as parents come to regard them as valued individuals rather than possessions); (3) acceptance of calculated choice as a valid element in life in general and in fertility in particular (belief that we have no control over our lives discourages a sense of responsibility); (4) social security and political stability that give people the means and the confidence to plan for the future; (5) knowledge, availability, and use of effective and acceptable means of birth control.

CONCLUSION

A few decades ago, we were warned that a human population explosion was about to engulf the world. Exponential population growth was seen as a cause or corollary to nearly every important environmental problem. Some people still warn that the total number of humans might grow to 30 or 40 billion by the end of this century. Birth rates have fallen, however, almost everywhere, and most demographers now believe that we will reach an equilibrium around 9 billion people in about 2050. Some claim that if we promote equality, democracy, human development, and modern family planning techniques, population might even decline to below its current level of 6.5 billion in the next 50 years. How we should carry out family planning and birth control remains a controversial issue. Should we focus on political and economic reforms, and hope that a demographic transition will naturally follow; or should we take more direct action (or any action) to reduce births?

Whether our planet can support 9 billion—or even 6 billion—people on a long-term basis remains a vital question. If all those people try to live at a level of material comfort and affluence now enjoyed by residents of the wealthiest nations, using the old, polluting, inefficient technology that we now employ, the answer is almost certain that even 6 billion people is too many in the long run. If we find more sustainable ways to live, however, it may be that 9 billion people could live happy, comfortable, productive lives. If we don't find new ways to live, we probably face a crisis no matter what happens to our population size. We'll discuss pollution problems, energy sources, and sustainability in subsequent chapters of this book.

REVIEWING LEARNING OUTCOMES

By now you should be able to explain the following points:

7.1 Trace the history of human population growth.

- Human populations grew slowly until relatively recently.

7.2 Summarize different perspectives on population growth.

- Does environment or culture control human populations?
- Technology can increase carrying capacity for humans.
- Population growth could bring benefits.

7.3 Analyze some of the factors that determine population growth.

- How many of us are there?
- Fertility measures the number of children born to each woman.
- Mortality is the other half of the population equation.
- Life span and life expectancy describe our potential longevity.
- Living longer has demographic implications.
- Emigration and immigration are important demographic factors.

7.4 Explain how ideal family size is culturally and economically dependent.

- Many factors increase our desire for children.
- Other factors discourage reproduction.
- Could we have a birth dearth?

7.5 Describe how a demographic transition can lead to stable population size.

- Economic and social development influence birth and death rates.
- There are reasons to be optimistic about population.
- Many people remain pessimistic about population growth.
- Social justice is an important consideration.
- Women's rights affect fertility.

7.6 Relate how family planning gives us choices.

- Fertility control has existed throughout history.
- Today there are many options.

7.7 Reflect on what kind of future we are creating.

- Religion and politics complicate family planning.

PRACTICE QUIZ

1. At what point in history did the world population pass its *first* billion? What factors restricted population before that time, and what factors contributed to growth after that point?

2. How might growing populations be beneficial in solving development problems?

3. Why do some economists consider human resources more important than natural resources in determining the future of a country?

4. Where will most population growth occur in the next century? What conditions contribute to rapid population growth in some countries?

5. Define *crude birth rate, total fertility rate, crude death rate,* and *zero population growth.*

6. What is the difference between life expectancy and life span?

7. What is dependency ratio, and how might it affect the United States in the future?

8. What pressures or interests make people want or not want to have babies?

9. Describe the conditions that lead to a demographic transition.

10. Describe some choices in modern birth control.

CRITICAL THINKING AND DISCUSSION QUESTIONS

1. What do you think is the optimum human population? The maximum human population? Are the numbers different? If so, why?

2. Some people argue that technology can provide solutions for environmental problems; others believe that a "technological fix" will make our problems worse. What personal experiences or worldviews do you think might underlie these positions?

3. Karl Marx called Thomas Malthus a "shameless sycophant of the ruling classes." Why would the landed gentry of the eighteenth century be concerned about population growth of the lower classes? Are there comparable class struggles today?

4. Try to imagine yourself in the position of a person your age in a developing world country. What family planning choices and pressures would you face? How would you choose among your options?

5. Some demographers claim that population growth has already begun to slow; others dispute this claim. How would you evaluate the competing claims of these two camps? Is this an issue of uncertain facts or differing beliefs? What sources of evidence would you accept as valid?

6. What role do race, ethnicity, and culture play in our immigration and population policies? How can we distinguish between prejudice and selfishness on one hand and valid concerns about limits to growth on the other?

 DATA analysis   Telling a Story with Graphs

Graphs are pictorial representations of data designed to communicate information. They are particularly useful in conveying patterns and relationships. Most of us, when confronted by a page of numbers, tend to roll our eyes and turn to something else. A graph can simplify trends and highlight important information. If you read a newspaper or news magazine these days, you will undoubtedly see numerous graphs portraying a wide variety of statistics. Why are they used so widely and why do they make such powerful impressions on us?

Graphs can be used to explain, persuade, or inform. Like photographs or text, however, they also can be used to manipulate or even mislead the viewer. When you create a graph, you inevitably make decisions about what details to display, and which ones to omit. By selecting certain scales or methods of displaying data, you deliberately choose the message you intend to convey. It takes critical thinking skills to understand what a graph is (and isn't) showing.

Central questions addressed in this chapter are: How serious is the world population problem? Does it demand action? Many graphs present information about population growth, yet the impressions each conveys to the reader vary considerably, depending on the type of data used and the design of the graph. Reexamine figure 7.3, which presents a historical perspective on human population growth. What impressions do you draw from it? What conclusions do you suppose the person who originally created this graph intended for you to take away from this presentation? Does it give the impression that human numbers are growing explosively? Does choosing a 5,000-year time horizon emphasize this idea?

Now look at figure 1. It plots world population growth, as does figure 7.3, but gives a very different picture. By choosing only a 100-year time horizon, the author of figure 1 makes it much easier to see precisely what's happening in this century, but provides a very different emotional reaction than figure 7.3. Clearly,

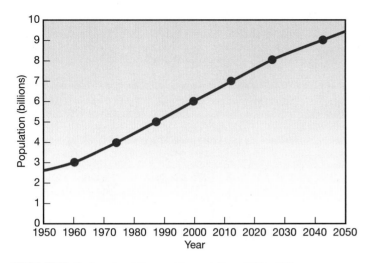

FIGURE 1 Growth of the world population, 1950–2050.
Source: U.S. Census Bureau, 2006.

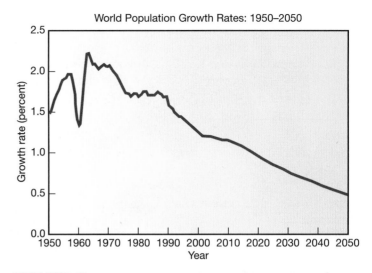

World Population Growth Rates: 1950–2050

FIGURE 2 World population growth rates, 1950–2050.
Source: U.S. Census Bureau, 2006.

these two graphs were intended to convey different messages. Is one more accurate than the other? Not necessarily; they're just showing different aspects of the same data set. But if you weren't aware that there's more than one way to display information, you might think that only one presentation is valid.

Now study figure 2. Why does it look so different from the other two? Notice that the Y-axis is rate of growth rather than total population. Again, this graph can be derived from the same data set, but it gives yet another message. The rate of population growth is slowing. Fifty years from now, we could be approaching a zero growth rate. This doesn't address what the total popu-

lation will be at that point. It could well be the 9 billion predicted by figure 1. In fact, if you look closely at the curve between 2040 and 2050 in figure 1, you might detect that the rate of growth is slowing, but when the Y-scale is in billions of people, the difference between a 0.7 and 0.5 percent growth rate is very small.

What do we learn from this exercise? All three graphs are probably correct, but each was designed to communicate different information. It's helpful to ask yourself, when you look at each graph, what did the author intend to say? Are there other ways to display these same data? Would it give a different impression, or tell a different story if you did so?

For Additional Help in Studying This Chapter, please visit our website at www.mhhe.com/cunningham10e. You will find additional practice quizzes and case studies, flashcards, regional examples, place markers for Google Earth™ mapping, and an extensive reading list, all of which will help you learn environmental science.

A public health worker explains to villagers that guinea worm is a parasitic disease, not the result of bad behavior or sorcery. Once the most guinea-worm-plagued region in the world, West Africa is now almost totally free of this terrible disease.

C H A P T E R

8

Environmental Health and Toxicology

To wish to become well is a part of becoming well.

—Seneca—

LEARNING OUTCOMES

After studying this chapter, you should be able to:

8.1 Describe health and disease and how global disease burden is now changing.

8.2 Summarize the principles of toxicology.

8.3 Discuss the movement, distribution, and fate of toxins in the environment.

8.4 Characterize mechanisms for minimizing toxic effects.

8.5 Explain ways we measure and describe toxicity.

8.6 Evaluate risk assessment and acceptance.

8.7 Relate how we establish health policy.

Case Study   Defeating the Fiery Serpent

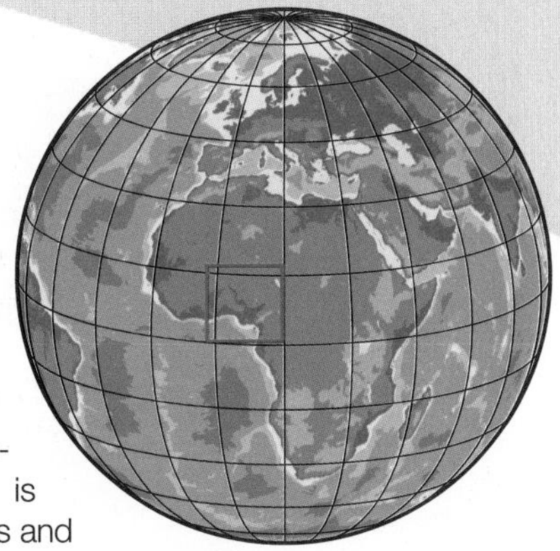

Fighting back tears, an African child cradles her swollen foot as a thin, white worm emerges from an oozing sore. The unfortunate youngster has been invaded by a guinea worm (*Dracunculus mediensis*). The pain is intense. Her whole leg feels like it's on fire. It's difficult to walk or work when you're afflicted with one of these terrible invaders. The disease is known in Africa as "empty granary," because the worms usually erupt during harvest season, making it impossible to work in the fields and harvest crops on which the whole family depends.

The infection cycle starts when someone suffering from the pain of an emerging worm (fig. 8.1) bathes their wound in a local lake or pond. The worm, sensing water, emerges to release thousands of larvae, which are ingested by freshwater copepods ("water fleas"). Inside the water fleas, the larvae develop into the infective stage in about two weeks. When villagers drink the contaminated water, the copepods are digested, but the worms survive and penetrate the wall of the intestine and move to the abdominal cavity. Over the next year, the worm grows to about a meter long (3 feet) and as thick as a spaghetti noodle. When fully grown, the worm migrates to the site where it will erupt, usually in the legs or feet—or even eye sockets—of victims. It takes several weeks for the worm to emerge completely. If you pull too hard on it, the worm breaks, and the part left in your body festers and decays. If the suffering host soaks the lesion in water to soothe the pain, the cycle begins again.

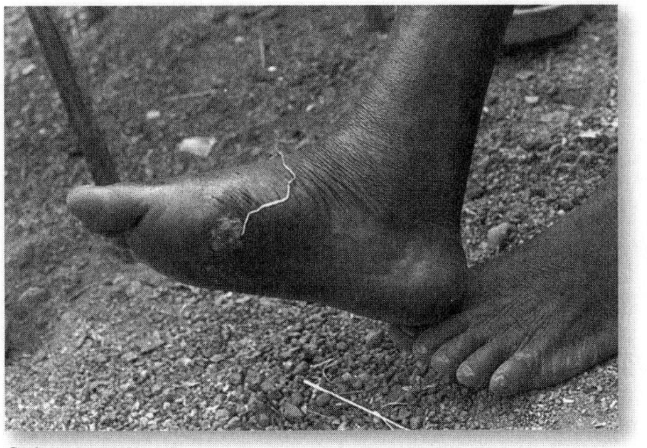

This terrible parasite has plagued tropical countries for thousands of years. It has been found in Egyptian mummies and is thought to be the "fiery serpent" described in the Old Testament as torturing the Israelites in the desert. As recently as 1986, at least 3 million people in 16 countries suffered from this affliction, and more than 100 million people were at risk worldwide. But there's a happy ending to this story. In 1988, the world health community, led by former U.S. President Jimmy Carter, began a campaign to eliminate guinea worms.

Eradicating the parasite is fairly simple and relatively inexpensive. There is no medicine to cure the infection once the worm is inside your body, but wells can be drilled to prevent patients from contaminating drinking-water sources. And local ponds can be treated with pesticides that are safe for human consumption but kill the worm larvae and copepods. Furthermore, families can be taught to pour their drinking water through a fine cloth filter to remove any remaining water fleas. The problem is simply to get the proper materials and information to remote villages. Having the prestige of a former American president has been a powerful tool in convincing local officials that the world cares, and that a cure is possible.

Progress has been spectacular. Only a few countries still are afflicted by this horrendous disease. Nigeria is an example of this remarkable success. In 1986, the country was the most Guinea-worm-plagued country in the world, with more than 650,000 cases in 36 states. By 2006, more than 99.9 percent of the infections had been eliminated, and only 120 people still suffered from infections. Worldwide, there now are fewer than 12,000 cases, mostly in Sudan, where civil war has made public health intervention difficult. When guinea worms are finally eradicated, it will be only the second disease ever completely eliminated (smallpox, which was abolished in 1977 was first), and the only human parasite totally exterminated worldwide.

Guinea worm

FIGURE 8.1 A guinea worm emerges from a patient's foot.

An encouraging outcome of this crusade is the demonstration that public health education and community organization can be effective, even in some of the poorest and most remote areas. Once people understand how the disease spreads and what they need to do to protect themselves and their families, they do change their behavior. And when the campaign is completed and guinea worms are completely vanquished, the health workers and volunteers will be available for further community development projects.

This case study reminds us of the importance of public health and how susceptible humans have always been to diseases and contaminants. In this chapter we'll look at the principles of environmental health to help you understand some of the risks we face and what we might do about them.

8.1 ENVIRONMENTAL HEALTH

What is health? The WHO defines **health** as a state of complete physical, mental, and social well-being, not merely the absence of disease or infirmity. By that definition, we all are ill to some extent. Likewise, we all can improve our health to live happier, longer, more productive, and more satisfying lives if we think about what we do.

What is disease? A **disease** is an abnormal change in the body's condition that impairs important physical or psychological functions. Diet and nutrition, infectious agents, toxic substances, genetics, trauma, and stress all play roles in **morbidity** (illness)

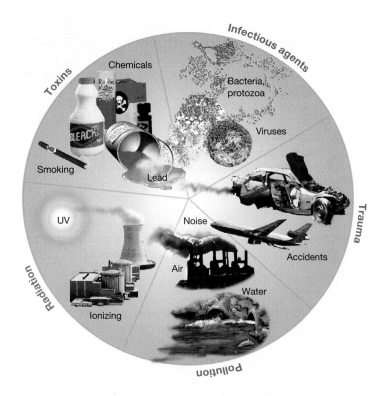

FIGURE 8.2 Major sources of environmental health risks.

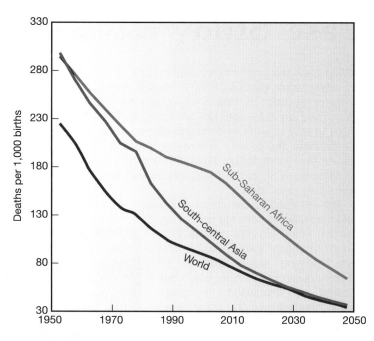

FIGURE 8.3 Child mortality has fallen dramatically over the past 50 years and is expected to continue this decline in the future. Note that south-central Asia saw a significant improvement in child survival in the 1970s and 80s, while sub-Saharan Africa lagged behind—mainly because of wars and AIDS. Data from the UN Population Division, 2006.

and **mortality** (death). **Environmental health** focuses on external factors that cause disease, including elements of the natural, social, cultural, and technological worlds in which we live. Figure 8.2 shows some major environmental disease agents as well as the media through which we encounter them. Ever since the publication of Rachel Carson's *Silent Spring* in 1962, the discharge, movement, fate, and effects of synthetic chemical toxins have been a special focus of environmental health. Later in this chapter, we'll study these topics in detail. First, however, let's look at some of the major causes of illness worldwide.

The global disease burden is changing

In addition to tremendous progress in eradicating guinea worms, there are many other public health successes. Smallpox was completely wiped out in 1977. Polio has been eliminated everywhere in the world except for a few remote villages in northern Nigeria. Epidemics of typhoid fever, cholera, and yellow fever that regularly killed thousands of people in North America a century ago are now rarely encountered. AIDS, which once was an immediate death sentence, has become a highly treatable disease. The average HIV-positive person in the United States now lives 24 years after diagnosis if treated faithfully with modern medicines.

A way of demonstrating these advances is to look at how much longer we're living. During the twentieth century, world average life expectancies more than doubled from 30 to 64.3 years. For richer countries, residents can expect to live, on average about three times as long as their great grandparents did a century earlier (see chapters 6 and 7).

A vital component of rising life expectancies is declining child mortality. In 1950, almost one-quarter (224 per 1,000) of all children born worldwide didn't live to their fifth birthday (fig. 8.3). By 2000, this rate had fallen to 86 per 1,000. By 2050, it's expected to be nearly 90 percent less than a century earlier. This is essential if we are to reach zero population growth. Sub-Saharan Africa and South-Central Asia had much higher child mortality than the world average in 1950. Asia has caught up with the rest of the world, but Africa, for a variety of reasons, lags behind.

In the past, health organizations have focused on the leading causes of death as the best summary of world health. Mortality data, however, fail to capture the impacts of nonfatal outcomes of disease and injury, such as dementia or blindness, on human well-being. When people are ill, work isn't done, crops aren't planted or harvested, meals aren't cooked, and children can't study and learn. Health agencies now calculate **disability-adjusted life years (DALYs)** as a measure of disease burden. DALYs combine premature deaths and loss of a healthy life resulting from illness or disability. This is an attempt to evaluate the total cost of disease, not simply how many people die. Clearly, many more years of expected life are lost when a child dies of neonatal tetanus than when an 80-year-old dies of pneumonia. Similarly, a teenager permanently paralyzed by a traffic accident will have many more years of suffering and lost potential than will a senior citizen who has a stroke. According to the WHO, chronic diseases now account for nearly 60 percent of the 56.5 million total deaths worldwide each year and about half of the global disease burden.

TABLE 8.1

Leading Causes of Global Disease Burden

Rank	1990	Rank	2020
1	Pneumonia	1	Heart disease
2	Diarrhea	2	Depression
3	Perinatal conditions	3	Traffic accidents
4	Depression	4	Stroke
5	Heart disease	5	Chronic lung disease
6	Stroke	6	Pneumonia
7	Tuberculosis	7	Tuberculosis
8	Measles	8	War
9	Traffic accidents	9	Diarrhea
10	Birth defects	10	HIV/AIDS
11	Chronic lung disease	11	Perinatal conditions
12	Malaria	12	Violence
13	Falls	13	Birth defects
14	Iron anemia	14	Self-inflicted injuries
15	Malnutrition	15	Respiratory cancer

Source: World Health Organization, 2002.

The world is now undergoing a dramatic epidemiological transition. Chronic conditions, such as cardiovascular disease and cancer, no longer afflict only wealthy people. Although the traditional killers in developing countries—infections, maternal and perinatal (birth) complications, and nutritional deficiencies—still take a terrible toll, diseases such as depression and heart attacks that once were thought to occur only in rich countries are rapidly becoming the leading causes of disability and premature death everywhere.

In 2020, the WHO predicts heart disease, which was fifth in the list of causes of global disease burden a decade ago, will be the leading source of disability and deaths worldwide (table 8.1). Most of that increase will be in the poorer parts of the world where people are rapidly adopting the lifestyles and diet of the richer countries. Similarly, global cancer rates will increase by 50 percent. By 2020, it's expected that 15 million people will have cancer and 9 million will die from it.

A silent epidemic of diabetes is now sweeping through our population. It's estimated that one-third of all children born today in North America will develop this disease in their lifetime. Obesity, diets high in sugar and fat, lack of exercise, and poverty (which encourages fast-food intake and makes health food unavailable) all play important roles in this disease. Blindness, circulatory problems, and kidney failure are common results of severe, uncontrolled diabetes. Seventy percent of all lower limb amputations are diabetes-related. In some Native American groups, more than half of all adults have this disease. It used to be thought that diabetes only affected rich people, but obesity, sedentary lifestyles, and diabetes are spreading around the world.

Taking disability as well as death into account in our assessment of disease burden reveals the increasing role of mental health as a worldwide problem. WHO projections suggest that psychiatric and neurological conditions could increase their share of the global burden from 10 percent currently to 15 percent of the total load by 2020. Again, this isn't just a problem of the developed world. Depression is expected to be the second largest cause of all years lived with disability worldwide, as well as the cause of 1.4 percent of all deaths. For women in both developing and developed regions, depression is the leading cause of disease burden, while suicide, which often is the result of untreated depression, is the fourth largest cause of female deaths.

Notice in table 8.1 that diarrhea, which was the second leading cause of disease burden in 1990, is expected to be ninth on the list in 2020, while measles and malaria are expected to drop out of the top 15 causes of disability. Tuberculosis, which is becoming resistant to antibiotics and is spreading rapidly in many areas (especially in the former Soviet Union), is the only infectious disease whose ranking is not expected to change over the next 20 years. Traffic accidents are now soaring as more people drive. War, violence, and self-inflicted injuries similarly are becoming much more important health risks than ever before.

Chronic obstructive lung diseases (e.g., emphysema, asthma, and lung cancer) are expected to increase from eleventh to fifth in disease burden by 2020. A large part of the increase is due to rising use of tobacco in developing countries, sometimes called "the tobacco epidemic." Every day about 100,000 young people—most of them in poorer countries—become addicted to tobacco. At least 1.1 billion people now smoke, and this number is expected to increase at least 50 percent by 2020. If current patterns persist, about 500 million people alive today will eventually be killed by tobacco. This is expected to be the biggest single cause of death worldwide (because illnesses such as heart attack and depression are triggered by multiple factors). In 2003, the World Health Assembly adopted a historic tobacco-control convention that requires countries to impose restrictions on tobacco advertising, establish clean indoor air controls, and clamp down on tobacco smuggling. Dr. Gro Harlem Brundtland, former director-general of the WHO, predicted that the convention, if implemented, could save billions of lives.

Think About It

What changes could you make in your lifestyle to lessen your risks from the diseases in table 8.1? What would have the greatest impact on your future well-being?

Infectious and emergent diseases still kill millions of people

Although the ills of modern life have become the leading killers almost everywhere in the world, communicable diseases still are responsible for about one-third of all disease-related mortality. Diarrhea, acute respiratory illnesses, malaria, measles, tetanus, and a few other infectious diseases kill about 11 million children

FIGURE 8.4 At least 3 million children die every year from easily preventable diseases. This billboard in Guatemala encourages parents to have their children vaccinated against polio, diphtheria, TB, tetanus, pertussis (whooping cough), and scarlet fever.

under age five every year in the developing world. Better nutrition, clean water, improved sanitation, and inexpensive inoculations could eliminate most of those deaths (fig. 8.4).

A wide variety of **pathogens** (disease-causing organisms) afflict humans, including viruses, bacteria, protozoans (single-celled animals), parasitic worms, and flukes (fig. 8.5). The greatest loss of life from an individual disease in a single year was the great influenza pandemic of 1918. Epidemiologists now estimate that at least one-third of all humans living at the time were infected, and that between 50 to 100 million died. Businesses, schools, churches, and sports or entertainment events were shut down for months. Public health authorities warn that if bird flu picks up the ability to spread through the human population as well as it moves from bird to bird, it could sicken 2 billion people, kill up to 150 million, and bring the world economy to a standstill. Influenza is caused by a family of viruses (fig. 8.5a) that mutate rapidly and move from wild and domestic animals to humans, making control of this disease very difficult.

Every year there are 76 million cases of foodborne illnesses in the United States, resulting in 300,000 hospitalizations and 5,000 deaths. Both bacteria and intestinal protozoa cause these illnesses (fig. 8.5b and c). They are spread from feces through food and water. In 2006, a deadly strain of E. coli was spread in lettuce that had been fertilized with cattle manure. At any given time, around 2 billion people—one-third of the world population—suffer from worms, flukes, and other internal parasites. Guinea worms (see opening case study) are one example, while people rarely die from parasites, they can be extremely debilitating, and can cause poverty that leads to other, more deadly, diseases.

Malaria is one of the most prevalent remaining infectious diseases. Every year about 500 million new cases of this disease occur, and about 2 million people die from it. The territory infected by this disease is expanding as global climate change allows mosquito vectors to move into new territory. Simply providing insecticide-treated bed nets and a few dollars worth of chloroquine pills could prevent tens of millions of cases of this debilitating disease every year. Tragically, some of the countries where malaria is most widespread tax both bed nets and medicine as luxuries, placing them out of reach for ordinary people.

Emergent diseases are those not previously known or that have been absent for at least 20 years. The new strain of bird flu now spreading around the world is a good example. There have been at least 39 outbreaks of emergent diseases over the past two decades, including the extremely deadly Ebola and Marburg fevers, which have afflicted Central Africa in at least six different locations in the past decade. Similarly, cholera, which had been absent from South America for more than a century, reemerged in Peru in 1992 (fig. 8.6). Some other examples include a new drug-resistant form of tuberculosis, now spreading in Russia; dengue fever, which is spreading through Southeast Asia and the Caribbean; malaria, which is reappearing in places where it had

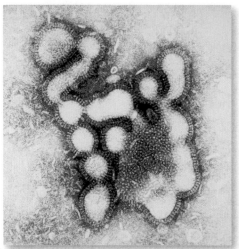

(a) Influenza virus

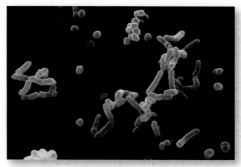

(b) Pathogenic bacteria

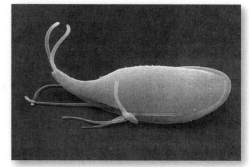

(c) *Giardia*

FIGURE 8.5 (a) A group of influenza viruses magnified about 300,000 times. (b) Pathogenic bacteria magnified about 50,000 times. (c) *Giardia*, a parasitic intestinal protozoan, magnified about 10,000 times.

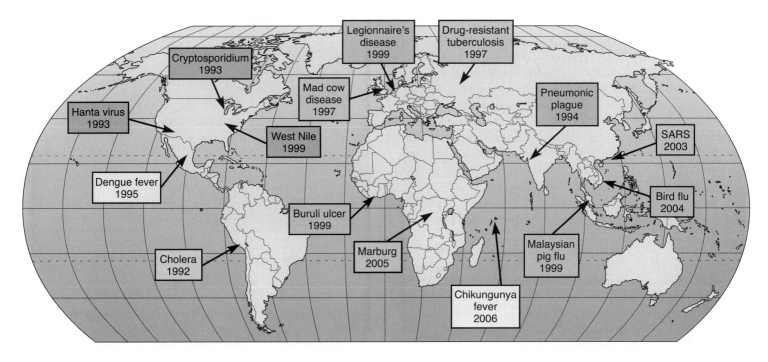

FIGURE 8.6 Some recent outbreaks of highly lethal infectious diseases. Why are supercontagious organisms emerging in so many different places? **Source:** Data from U.S. Centers for Disease Control and Prevention.

been nearly eradicated; and a new human lymphotropic virus (HTLV), which is thought to have jumped from monkeys into people in Cameroon who handled or ate bushmeat. Theses HTLV strains are now thought to infect 25 million people.

Growing human populations push people into remote areas where they encounter diseases that may have existed for a long time, but only now are exposed to humans. Rapid international travel makes it possible for these new diseases to spread around the world at jet speed. Epidemiologists warn that the next deadly epidemic is only a plane ride away.

West Nile virus shows how fast new diseases can travel. West Nile belongs to a family of mosquito-transmitted viruses that cause encephalitis (brain inflammation). Although recognized in Africa in 1937, the West Nile virus was absent from North America until 1999, when it apparently was introduced by a bird or mosquito from the Middle East. The disease spread rapidly from New York, where it was first reported, throughout the eastern United States in only two years (fig. 8.7). Now, it's found everywhere in the lower 48 states. The virus infects 230 species of animals, including 130 bird species. By 2006, more than 1,000 people had died from this disease.

The largest recent death toll from an emergent disease is HIV/AIDS. Although virtually unknown 15 years ago, acquired immune deficiency syndrome has now become the fifth greatest cause of contagious deaths. The WHO estimates that 60 million people are now infected with the human immune-deficiency virus, and that 3 million die every year from AIDS complications. Although two-thirds of all current HIV infections are now in sub-Saharan Africa, the disease is spreading rapidly in South and East Asia. Over the next 20 years, there could be an additional 65 million AIDS deaths.

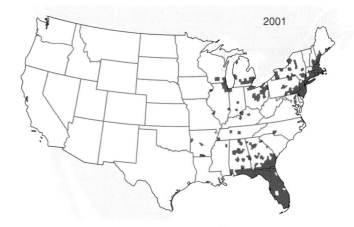

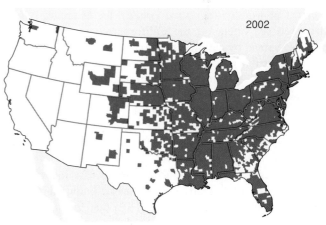

FIGURE 8.7 The spread of West Nile virus in birds, 2001–2002. By 2003, it occurred throughout the 48 contiguous states. **Source:** CDC and USGS.

In Swaziland, health officials estimate about one-third of all adults are HIV positive and that two-thirds of all current 15-year-olds will die of AIDS before age 50. As chapter 7 points out, without AIDS, the life expectancy in Swaziland would be expected to be about 65 years. With AIDS, the average life expectancy is now 32.6 years. Worldwide, more than 14 million children—the equivalent of every child under age five in America—have lost one or both parents to AIDS. The economic costs of treating patients and lost productivity from premature deaths resulting from this disease are estimated to be at least (U.S.) $35 billion per year or about one-tenth of the total GDP of sub-Saharan Africa.

Conservation medicine combines ecology and health care

Humans aren't the only ones to suffer from new and devastating diseases. Domestic animals and wildlife also experience sudden and widespread epidemics, which are sometimes called **ecological diseases.** Ebola hemorrhagic fever is one of the most virulent viruses ever seen, killing up to 90 percent of its victims. In 2002, an outbreak of Ebola fever began killing humans along the Gabon-Congo border. A few months later, researchers found that 221 of the 235 western lowland gorillas they had been studying in this area disappeared in just a few months. Many chimpanzees also died. Although the study team could only find a few of the dead gorillas, 75 percent of those tested positive for Ebola. Altogether, researchers estimate that 5,000 gorillas died in this small area of the Congo. Extrapolating to all of central Africa, it's possible that Ebola has killed one-quarter of all the gorillas in the world. It's thought that the spread of this disease in humans resulted from the practice of hunting and eating primates.

A viral hemorrhagic septicemia much like Ebola is spreading through fish in North America's Great Lakes. It is highly infectious and also has an extremely high fatality rate. The virus infects about 40 species of fish. Shipments of game fish, such as salmon, trout, walleye, and perch, have been banned across much of America in an effort to slow the spread of this disease, which probably was introduced in ballast water of ships from Europe.

A cat parasite (*Toxoplasma gondii*), for example, is killing endangered sea otters off the California coast. The disease is apparently spread by cat owners who empty kitty litter boxes into sewer systems that empty into the ocean. In another human/wildlife connection, 18 percent of all stranded California sea lions have tumors caused by the herpes I virus, which also causes Kaposi's sarcoma in AIDS patients. Presumably, this virus also is spread through sewer systems.

Chronic wasting disease (CWD) is spreading through deer and elk populations in North America (fig. 8.8). Caused by a strange protein called a prion, CWD is one of a family of irreversible, degenerative neurological diseases known as transmissible spongiform encephalopathies (TSE) that include mad cow disease in cattle, scrapie in sheep, and Creutzfelt-Jacob disease in humans. CWD probably started when elk ranchers fed contaminated animal by-products to their herds. Infected animals were sold to other ranches, and now the disease has spread to wild populations. First recognized in 1967 in Saskatchewan,

FIGURE 8.8 Wild elk and deer are widely affected by chronic wasting disease, which originated with domestic herds.

CWD has been identified in wild deer populations and ranch operations in at least eleven American states.

No humans are known to have contracted TSE from deer or elk, but there is a concern that we might see something like the mad cow disaster that inflicted Europe in the 1990s. At least 100 people died, and nearly 5 million European cattle and sheep were slaughtered in an effort to contain that disease.

In 1995, California residents noticed that the oak trees were dying from a fast-spreading new disease known as sudden oak death syndrome (SODS), The disease is caused by funguslike organism known as *Phytophthora ramorum,* part of a group known as water molds. Possibly imported with Asian rhododendrons shipped from European plant nurseries, this organism is a close relative of the pathogen that caused Ireland's potato blight and another species that has wiped out vast swaths of forest in western Australia. Recently, foresters reported that Douglas fir, one of the nation's most economically important timber species, and California coast redwood also are infected with SODS. In addition to being beautiful trees and dominant members of forest communities, each of these species provides commercial products worth more than a billion dollars per year in the United States (fig. 8.9).

One thing that emergent diseases in humans and ecological diseases in natural communities have in common is environmental change that stresses biological systems and upsets normal ecological relationships. We cut down forests and drain wetlands, destroying habitat for native species. Invasive organisms and diseases are accidentally or intentionally introduced into new areas where they can grow explosively. Increasing incursion into former wilderness is spurred by human population growth and ecotourism. In 1950, only about 3 million people a year flew on commercial jets; by 2000, more than 300 million did. Diseases can spread around the globe in mere days as people pass through international travel hubs.

We are coming to recognize that the delicate ecological balances that we value so highly—and disrupt so frequently—are important to our own health. **Conservation medicine** is an emerging discipline that attempts to understand how our environmental changes threaten our own health as well as that of the natural communities on which we depend for ecological services.

FIGURE 8.9 The pathogen that causes sudden oak death also infects redwoods, Douglas fir, and dozens of other species.

While still small, this new field is gaining recognition from mainstream funding sources such as the World Bank, the World Health Organization, and the U.S. National Institutes of Health.

Resistance to drugs, antibiotics, and pesticides is increasing

Malaria, the most deadly of all insect-borne illnesses, is an example of the return of a disease that once was thought to be nearly vanquished. Malaria now claims about a million lives every year—90 percent are in Africa, and most of them children. With the advent of modern medicines and pesticides, malaria had nearly been wiped out in many places but recently has come roaring back. The protozoan parasite that causes the disease is now resistant to most drugs, while the mosquitoes that transmit it have developed resistance to many insecticides. Spraying of DDT in India and Sri Lanka, for instance, reduced malaria from millions of infections per year to only a few thousand in the 1950s and 1960s. Now South Asia is back to its pre-DDT level of about half a million new cases of malaria every year. Other places that never had cases of malaria now have them as a result of climate change and habitat alteration.

Why have vectors such as mosquitoes, and pathogens such as the malaria parasite become resistant to pesticides and drugs? Part of the answer is natural selection and the ability of many organisms to evolve rapidly. Another factor is the human tendency to use control measures carelessly. When we discovered that DDT and other insecticides could control mosquito populations, we spread them everywhere. This not only harmed wildlife and beneficial insects, but it created perfect conditions for natural selection. Many pests and pathogens were exposed only minimally to control measures,

allowing those with natural resistance to survive and spread their genes through the population (fig. 8.10). After repeated cycles of exposure and selection, many microorganisms and their vectors are insensitive to almost all our weapons against them.

As chapter 9 discusses, raising huge numbers of cattle, hogs, and poultry in densely packed barns and feedlots helps spread antibiotic resistance in pathogens. Confined animals are dosed constantly with antibiotics and steroid hormones to keep them disease-free and to make them gain weight faster. More than half of all antibiotics used in the United States each year is fed to livestock. A significant amount of these antibiotics and hormones are excreted in urine and feces, which are spread, untreated, on the land or discharged into surface water where they contribute further to the evolution of supervirulent pathogens.

At least half of the 100 million antibiotic doses prescribed for humans every year in the United States are unnecessary or are the wrong ones. Furthermore, many people who start a course of antibiotic treatment fail to carry it out for the time prescribed. For your own health and that of the people around you, if you are taking an antibiotic, follow your doctor's orders and don't stop taking the medicine as soon as you start feeling better.

Who should pay for health care?

The heaviest burden of illness is borne by the poorest people who can afford neither a healthy environment nor adequate health care. Women in sub-Saharan Africa, for example, suffer six times the disease burden per 1,000 population as do women in most European countries. The WHO estimates that 90 percent of all disease burden occurs in developing countries where less than one-tenth of all health care dollars is spent. The group Medecins Sans Frontieres (MSF, or Doctors Without Borders) calls this the 10/90 gap. While wealthy nations pursue drugs to treat baldness and obesity,

(a) Mutation and selection create drug-resistant strains

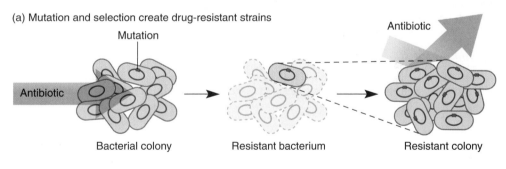

(b) Conjugation transfers drug resistance from one strain to another

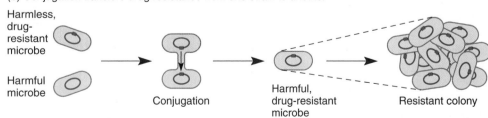

FIGURE 8.10 How microbes acquire antibiotic resistance. Random mutations make a few cells resistant. When challenged by antibiotics, only those cells survive to give rise to a resistant colony. Sexual reproduction (conjugation) or plasmid transfer moves genes from one strain or species to another.

depression in dogs, and erectile dysfunction, billions of people are sick or dying from treatable infections and parasitic diseases to which little attention is paid. Worldwide, only 2 percent of the people with AIDS have access to modern medicines. Every year, some 600,000 infants acquire HIV—almost all of them through mother-to-child transmission during birth or breast-feeding. Antiretroviral therapy costing only a few dollars can prevent most of this transmission. The Bill and Melinda Gates Foundation has pledged $200 million for medical aid to developing countries to help fight AIDS, TB, and malaria.

Dr. Jeffrey Sachs of the Columbia University Earth Institute says that disease is as much a cause as a consequence of poverty and political unrest, yet the world's richest countries now spend just $1 per person per year on global health. He predicts that raising our commitment to about $25 billion annually (about 0.1 percent of the annual GDP of the 20 richest countries) would not only save about 8 million lives each year, but would boost the world economy by billions of dollars. There also would be huge social benefits for the rich countries in not living in a world endangered by mass social instability, the spread of pathogens across borders, and the spread of other ills such as terrorism and drug trafficking caused by social problems. Sachs also argues that reducing disease burden would help reduce population growth. When parents believe their offspring will survive, they have fewer children and invest more in food, health, and education for smaller families.

The United States is the least generous of the world's rich countries, donating only about 12 cents per $100 of GDP on international development aid. Could the U.S. do better? During a time of fear of terrorism and rising anti-American feelings around the globe, it's difficult to interest legislators in international aid, and yet, helping to reduce disease might win the United States more friends and make the nation safer than buying more bombs and bullets. Improved health care in poorer countries may also help prevent the spread of emergent diseases in a globally interconnected world.

Think About It

If you were making a case for greater U.S. funding for international health care, what points would you stress? Do we have a moral obligation to help others?

8.2 TOXICOLOGY

Toxicology is the study of **toxins** (poisons) and their effects, particularly on living systems. Because many substances are known to be poisonous to life (whether plant, animal, or microbial), toxicology is a broad field, drawing from biochemistry, histology, pharmacology, pathology, and many other disciplines. Toxins damage or kill living organisms because they react with cellular components to disrupt metabolic functions. Because of this reactivity, toxins often are harmful even in extremely dilute concentrations. In some cases billionths, or even trillionths of a gram can cause irreversible damage.

All toxins are hazardous, but not all hazardous materials are toxic. Some substances, for example, are dangerous because they're flammable, explosive, acidic, caustic, irritants, or sensitizers. Many of these materials must be handled carefully in large doses or high concentrations, but can be rendered relatively innocuous by dilution, neutralization, or other physical treatment. They don't react with cellular components in ways that make them poisonous at low concentrations.

Environmental toxicology or ecotoxicology specifically deals with the interactions, transformation, fate, and effects of natural and synthetic chemicals in the biosphere including individual organisms, populations, and whole ecosystems. In aquatic systems the fate of the pollutants is primarily studied in relation to mechanisms and processes at interfaces of the ecosystem components. Special attention is devoted to the sediment/water, water/organisms, and water/air interfaces. In terrestrial environments, the emphasis tends to be on effects of metals on soil fauna community and population characteristics.

Table 8.2 lists the top 20 toxins compiled by the U.S. Environmental Protection Agency from the 275 substances regulated by the Comprehensive Environmental Response, Compensation, and

TABLE 8.2

Top 20 Toxic and Hazardous Substances

Material	Major Sources
1. Arsenic	Treated lumber
2. Lead	Paint, gasoline
3. Mercury	Coal combustion
4. Vinyl chloride	Plastics industrial uses
5. Polychlorinated biphenyls (PCBs)	Electric insulation
6. Benzene	Gasoline, industrial use
7. Cadmium	Batteries
8. Benzo(a)pyrene	Waste incineration
9. Polycyclic aromatic hydrocarbons	Combustion
10. Benzo(b)fluoranthene	Fuels
11. Chloroform	Water purification, industry
12. DDT	Pesticide use
13. Aroclor 1254	Plastics
14. Aroclor 1260	Plastics
15. Trichloroethylene	Solvents
16. Dibenz(a,h)anthracene	Incineration
17. Dieldrin	Pesticides
18. Chromium, hexavalent	Paints, coatings, welding, anticorrosion agents
19. Chlordane	Pesticides
20. Hexachlorobutadiene	Pesticides

Source: Data from U.S. Environmental Protection Agency, 2003.

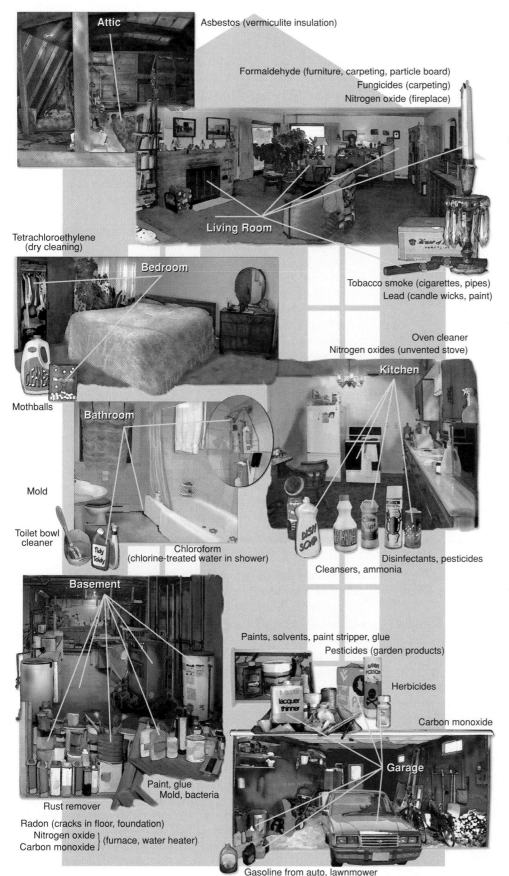

Attic Asbestos (vermiculite insulation)

Formaldehyde (furniture, carpeting, particle board)
Fungicides (carpeting)
Nitrogen oxide (fireplace)

Living Room

Tetrachloroethylene
(dry cleaning)

Bedroom

Tobacco smoke (cigarettes, pipes)
Lead (candle wicks, paint)

Mothballs

Oven cleaner
Nitrogen oxides (unvented stove)

Bathroom

Kitchen

Mold

Toilet bowl
cleaner

Chloroform
(chlorine-treated water in shower)

Disinfectants, pesticides
Cleansers, ammonia

Basement

Paints, solvents, paint stripper, glue
Pesticides (garden products)

Herbicides

Carbon monoxide

Paint, glue
Mold, bacteria

Garage

Rust remover

Radon (cracks in floor, foundation)
Nitrogen oxide } (furnace, water heater)
Carbon monoxide }

Gasoline from auto, lawnmower

FIGURE 8.11 Some sources of toxic and hazardous substances in a typical home.

Liability Act (CERCLA), commonly known as the Superfund Act. These materials are listed in order of assessed importance in terms of human and environmental health.

How do toxins affect us?

Allergens are substances that activate the immune system. Some allergens act directly as **antigens;** that is, they are recognized as foreign by white blood cells and stimulate the production of specific antibodies (proteins that recognize and bind to foreign cells or chemicals). Other allergens act indirectly by binding to and changing the chemistry of foreign materials so they become antigenic and cause an immune response.

Formaldehyde is a good example of a widely used chemical that is a powerful sensitizer of the immune system. It is directly allergenic and can also trigger reactions to other substances. Widely used in plastics, wood products, insulation, glue, and fabrics, formaldehyde concentrations in indoor air can be thousands of times higher than in normal outdoor air. Some people suffer from what is called **sick building syndrome:** headaches, allergies, chronic fatigue, and other symptoms caused by poorly vented indoor air contaminated by mold spores, carbon monoxide, nitrogen oxides, formaldehyde, and other toxins released from carpets, insulation, plastics, building materials, and other sources (fig. 8.11). The Environmental Protection Agency estimates that poor indoor air quality may cost the United States $60 billion a year in absenteeism and reduced productivity.

Immune system depressants are pollutants that suppress the immune system rather than activate it. Little is known about how this occurs or which chemicals are responsible. Immune system failure is thought to have played a role, however, in widespread deaths of seals in the North Atlantic and of dolphins in the Mediterranean. These dead animals generally contain high levels of pesticide residues, polychlorinated biphenyls (PCBs), and other contaminants that are suspected of disrupting the immune system and making it susceptible to a variety of opportunistic infections.

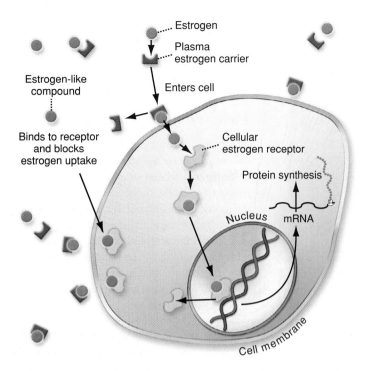

FIGURE 8.12 Steroid hormone action. Plasma hormone carriers deliver regulatory molecules to the cell surface, where they cross the cell membrane. Intracellular carriers deliver hormones to the nucleus, where they bind to and regulate expression of DNA. Estrogen-like compounds bind to receptors and either block uptake of endogenous hormone, or act as a substitute hormone to disrupt gene expression.

Endocrine disrupters are chemicals that disrupt normal hormone functions. Hormones are chemicals released into the blood stream by cells in one part of the body to regulate development and function of tissues and organs elsewhere in the body (fig. 8.12). You undoubtedly have heard about sex hormones and their powerful effects on how we look and behave, but these are only one example of the many regulatory hormones that rule our lives. Some other powerful hormones include thyroxin, insulin, adrenalin, and endorphins.

We now know that some of the most insidious effects of persistent chemicals such as dioxins and PCBs are that they interfere with normal growth, development, and physiology of a variety of animals—including humans—at very low doses. In some cases, picogram concentrations (trillionths of a gram per liter) may be enough to cause developmental abnormalities in sensitive organisms. Because these chemicals often cause sexual dysfunction (reproductive health problems in females or feminization of males, for example), these chemicals are sometimes called environmental estrogens or androgens. They are just as likely, however, to disrupt functions of other important regulatory molecules as they are to obstruct sex hormones.

Neurotoxins are a special class of metabolic poisons that specifically attack nerve cells (neurons). The nervous system is so important in regulating body activities that disruption of its activities is especially fast-acting and devastating. Different types of neurotoxins act in different ways. Heavy metals such as lead

and mercury kill nerve cells and cause permanent neurological damage. Anesthetics (ether, chloroform, halothane, etc.) and chlorinated hydrocarbons (DDT, Dieldrin, Aldrin) disrupt nerve cell membranes necessary for nerve action. Organophosphates (Malathion, Parathion) and carbamates (carbaryl, zeneb, maneb) inhibit acetylcholinesterase, an enzyme that regulates signal transmission between nerve cells and the tissues or organs they innervate (for example, muscle). Most neurotoxins are both extremely toxic and fast-acting.

Mutagens are agents, such as chemicals and radiation, that damage or alter genetic material (DNA) in cells. This can lead to birth defects if the damage occurs during embryonic or fetal growth. Later in life, genetic damage may trigger neoplastic (tumor) growth. When damage occurs in reproductive cells, the results can be passed on to future generations. Cells have repair mechanisms to detect and restore damaged genetic material, but some changes may be hidden, and the repair process itself can be flawed. It is generally accepted that there is no "safe" threshold for exposure to mutagens. Any exposure has some possibility of causing damage.

Teratogens are chemicals or other factors that specifically cause abnormalities during embryonic growth and development. Some compounds that are not otherwise harmful can cause tragic problems in these sensitive stages of life. Perhaps the most prevalent teratogen in the world is alcohol. Drinking during pregnancy can lead to **fetal alcohol syndrome**—a cluster of symptoms including craniofacial abnormalities, developmental delays, behavioral problems, and mental defects that last throughout a child's life. Even one alcoholic drink a day during pregnancy has been associated with decreased birth weight.

By some estimates, between 300,000 and 600,000 children born every year in the United States are exposed in the womb to unsafe levels of mercury. The effects are subtle, but include reduced intelligence, attention deficit, and behavioral problems. The total cost of these effects is estimated to be $8.7 billion per year.

Carcinogens are substances that cause **cancer,** invasive, out-of-control cell growth that results in malignant tumors. Cancer rates rose in most industrialized countries during the twentieth century, and cancer is now the second leading cause of death in the United States, killing more than half a million people in 2002. According to the American Cancer Society, 1 in 2 males and 1 in 3 females in the United States will have some form of cancer in their lifetime. Some authors blame this cancer increase on toxic synthetic chemicals in our environment and diet. Others argue that it is attributable mainly to lifestyle (smoking, sunbathing, alcohol) or simply living longer. The U.S. EPA estimates that 200 million U.S. residents live in areas where the combined lifetime cancer risk from environmental carcinogens exceeds 1 in 100,000, or ten times the risk normally considered acceptable.

How does diet influence health?

Diet also has an important effect on health. For instance, there is a strong correlation between cardiovascular disease and the amount of salt and animal fat in one's diet.

Fruits, vegetables, whole grains, complex carbohydrates, and dietary fiber (plant cell walls) often have beneficial health effects. Certain dietary components, such as pectins; vitamins A, C, and E; substances produced in cruciferous vegetables (cabbage, broccoli, cauliflower, brussels sprouts); and selenium, which we get from plants, seem to have anticancer effects.

Eating too much food is a significant dietary health factor in developed countries and among the well-to-do everywhere. Sixty percent of all U.S. adults are now considered overweight, and the worldwide total of obese or overweight people is estimated to be over 1 billion. Every year in the United States, 300,000 deaths are linked to obesity.

The U.S. Centers for Disease Control in Atlanta warn that one in three U.S. children will become diabetic unless many more people start eating less and exercising more. The odds are worse for Black and Hispanic children: nearly half of them are likely to develop the disease. And among the Pima tribe of Arizona, nearly 80 percent of all adults are diabetic. More information about food and its health effects is available in chapter 9.

8.3 MOVEMENT, DISTRIBUTION, AND FATE OF TOXINS

There are many sources of toxic and hazardous chemicals in the environment and many factors related to each chemical itself, its route or method of exposure, and its persistence in the environment, as well as characteristics of the target organism (table 8.3), that determine the danger of the chemical. We can think of both individuals and an ecosystem as sets of interacting compartments between which chemicals move, based on molecular size, solubility, stability, and reactivity (fig. 8.13). The dose (amount), route of entry, timing of exposure, and sensitivity of the organism all play important roles in determining toxicity. In this section, we will consider some of these characteristics and how they affect environmental health.

Solubility and mobility determine where and when chemicals move

Solubility is one of the most important characteristics in determining how, where, and when a toxic material will move through the

---

TABLE 8.3

Factors in Environmental Toxicity

Factors Related to the Toxic Agent

1. Chemical composition and reactivity
2. Physical characteristics (such as solubility, state)
3. Presence of impurities or contaminants
4. Stability and storage characteristics of toxic agent
5. Availability of vehicle (such as solvent) to carry agent
6. Movement of agent through environment and into cells

Factors Related to Exposure

1. Dose (concentration and volume of exposure)
2. Route, rate, and site of exposure
3. Duration and frequency of exposure
4. Time of exposure (time of day, season, year)

Factors Related to Organism

1. Resistance to uptake, storage, or cell permeability of agent
2. Ability to metabolize, inactivate, sequester, or eliminate agent
3. Tendency to activate or alter nontoxic substances so they become toxic
4. Concurrent infections or physical or chemical stress
5. Species and genetic characteristics of organism
6. Nutritional status of subject
7. Age, sex, body weight, immunological status, and maturity

Source: U.S. Department of Health and Human Services, 1995.

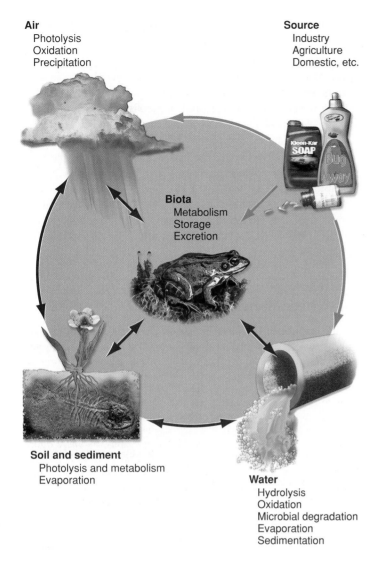

Air
Photolysis
Oxidation
Precipitation

Source
Industry
Agriculture
Domestic, etc.

Biota
Metabolism
Storage
Excretion

Soil and sediment
Photolysis and metabolism
Evaporation

Water
Hydrolysis
Oxidation
Microbial degradation
Evaporation
Sedimentation

FIGURE 8.13 Movement and fate of chemicals in the environment. Toxins also move directly from a source to soil and sediment.

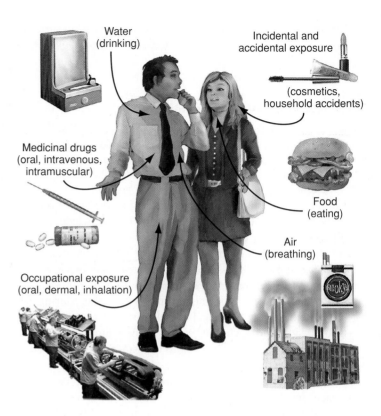

Water
(drinking)

Incidental and
accidental exposure

(cosmetics,
household accidents)

Medicinal drugs
(oral, intravenous,
intramuscular)

Food
(eating)

Air
(breathing)

Occupational exposure
(oral, dermal, inhalation)

FIGURE 8.14 Routes of exposure to toxic and hazardous environmental factors.

Exposure and susceptibility determine how we respond

Just as there are many sources of toxins in our environment, there are many routes for entry of dangerous substances into our bodies (fig. 8.14). Airborne toxins generally cause more ill health than any other exposure source. We breathe far more air every day than the volume of food we eat or water we drink. Furthermore, the lining of our lungs, which is designed to exchange gases very efficiently, also absorbs toxins very well. Epidemiologists estimate that some 3 million people—two-thirds of them children—die each year from diseases caused or exacerbated by air pollution. Still, food, water, and skin contact also can expose us to a wide variety of toxins. The largest exposures for many toxins are found in industrial settings, where workers may encounter doses thousands of times higher than would be found anywhere else. The European Agency for Safety and Health at Work warns that 32 million people (20 percent of all employees) in the European Union are exposed to unacceptable levels of carcinogens and other toxins in their workplace.

Condition of the organism and timing of exposure also have strong influences on toxicity. Healthy adults, for example, may be relatively insensitive to doses that would be very dangerous for young children or for someone already weakened by disease (What Do You Think? p. 167). Similarly, exposure to a toxin

environment or through the body to its site of action. Chemicals can be divided into two major groups: those that dissolve more readily in water and those that dissolve more readily in oil. Water-soluble compounds move rapidly and widely through the environment because water is ubiquitous. They also tend to have ready access to most cells in the body because aqueous solutions bathe all our cells. Molecules that are oil- or fat-soluble (usually organic molecules) generally need a carrier to move through the environment, into, and within, the body. Once inside the body, however, oil-soluble toxins penetrate readily into tissues and cells because the membranes that enclose cells are themselves made of similar oil-soluble chemicals. Once they get inside cells, oil-soluble materials are likely to be accumulated and stored in lipid deposits where they may be protected from metabolic breakdown and persist for many years.

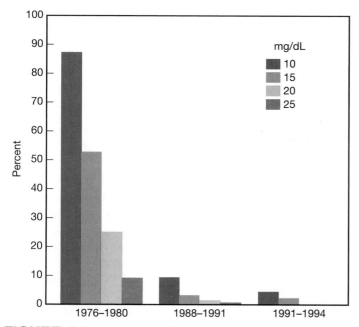

What Do You Think?

Protecting Children's Health

Increasing evidence shows that children are much more vulnerable than adults to environmental health hazards. Pound for pound, children drink more water, eat more food, and breathe more air than do adults. Putting fingers, toys, and other objects into their mouths increases children's exposure to toxins in dust or soil. Compared to adults, children generally have less-developed immune systems or processes to degrade or excrete toxins. The brain is especially sensitive to disruption. From before birth to adolescence, the brain undergoes a highly complex series of changes. Obviously, anything that interferes with brain development can have tragic long-term results.

In 2006, an international team of epidemiologists, led by Drs. Phillippe Grandjean, of Harvard, and Philip Landrigan, of the Mount Sinai School of Medicine, both of whom are pioneers in studying long-term effects of toxins on children, surveyed toxicity data on common industrial chemicals. They identified 202 toxic substances—about half of them used in very large volumes—known to damage human brains. They found that effects of these chemicals on children have generally been neglected.

"The human brain is a precious and vulnerable organ. And because optimal brain function depends on the integrity of the organ, even limited damage may have serious consequences," says Dr. Grandjean. He estimates that one out of every six children in America has a developmental disability, usually involving the nervous system. Among the neurodevelopmental disorders that may be associated with toxic chemical exposure are autism, mental retardation, attention deficit disorder, shortened attention spans, slowed motor coordination, increased aggressiveness, and diminished social skills.

Despite years of evidence that many industrial substances are dangerous, few have been studied thoroughly for their effects on children. Lead was the first toxicant identified as having effects at low levels on developing brains of children. Banning leaded gasoline and paint was one of the most successful steps ever taken to protect children's health. Between 1920 and 1970, virtually all children in industrialized countries were exposed to lead from house paint and leaded gasoline. Before these sources were banned in the 1970s, at least 4.4 million American children had more than 10 µg of lead per dl of blood, (the level considered safe by the U.S. EPA), and hundreds of children died each year from acute lead poisoning. Epidemiologists now estimate that for every 10 µg of lead per dl of blood, IQ scores drop 4.6 points. Grandjean and Landrigan suggest that high lead exposures in the United States reduced IQ scores above 130 (considered superior intelligence) by half, while the number of scores less than 70 increased comparably.

By 2006, the number of children with elevated blood lead levels had fallen more than 90 percent (fig. 1). This is a remarkable success story, but about 430,000 children—especially in poor urban areas—still have unsafe lead levels. Recent evidence suggests that even 10 µg/dl represents a risk of subtle but important effects in growing children's brains, including diminished learning ability, attention deficit disorder, and hyperactivity. Many public health experts argue that blood lead levels in all children should be below 2.5 µg/dl (fig. 2).

Grandjean and Landrigan point out that, in addition to lead, only four other toxins—methylmercury, arsenic, PDBs and toluene—are specifically regulated to protect children. Because industrialization has now become global, they warn that this represents a "silent pandemic" in which millions of children may be suffering reduced intelligence. They suggest that to protect children from industrial chemicals that cause mental retardation and developmental disabilities, we should adopt a precautionary approach to chemical testing and control. This philosophy is already being applied in Europe, where industry is required to show that a chemical is safe before it is introduced, rather than requiring regulatory agencies to prove that it is unsafe in order to restrict its use.

What do you think? How important is it to protect children's mental development? What does it cost society to have large numbers of slow learners and behavioral problems? Industry claims that more rigorous testing and stricter regulations on industrial chemicals will drive up consumer prices and limit consumer choices. Would you support stronger regulations now, which could be relaxed later if hazards are less than anticipated, or the current regulations that require a high level of proof before restricting use of toxic products?.

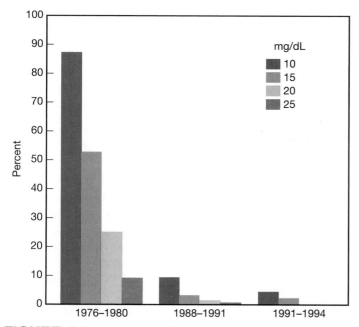

FIGURE 1 Blood lead levels in U.S. preschoolers fell 90 percent between the 1970s and 1990s. This is one of our greatest environmental health successes ever.
Source: J. Pickle et al. 1998.

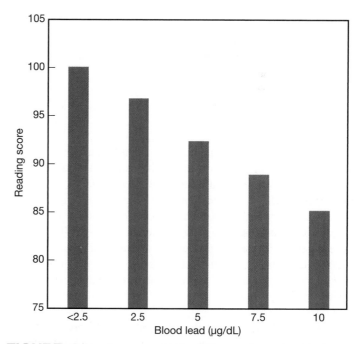

FIGURE 2 Reading score of U.S. children are inversely related to blood lead levels even below 10 µg/dL, the amount considered safe by the U.S. EPA.
Source: J. Pickle et al. 1998.

may be very dangerous at certain stages of developmental or metabolic cycles, but may be innocuous at other times. A single dose of the notorious teratogen thalidomide, for example, taken in the third week of pregnancy (a time when many women aren't aware they're pregnant) can cause severe abnormalities in fetal limb development. A complication in measuring toxicity is that great differences in sensitivity exist between species. Thalidomide was tested on a number of laboratory animals without any deler-ious effects. Unfortunately, however, it is a powerful teratogen in humans.

Bioaccumulation and biomagnification increase concentrations of chemicals

Cells have mechanisms for **bioaccumulation,** the selective absorption and storage of a great variety of molecules. This allows them to accumulate nutrients and essential minerals, but at the same time, they also may absorb and store harmful substances through these same mechanisms. Toxins that are rather dilute in the environment can reach dangerous levels inside cells and tissues through this process of bioaccumulation.

The effects of toxins also are magnified in the environment through food webs. **Biomagnification** occurs when the toxic burden of a large number of organisms at a lower trophic level is accumulated and concentrated by a predator in a higher trophic level. Phytoplankton and bacteria in aquatic ecosystems, for instance, take up heavy metals or toxic organic molecules from water or sediments (fig. 8.15). Their predators—zooplankton and small fish—collect and retain the toxins from many prey organisms, building up higher concentrations of toxins. The top carnivores in the food chain—game fish, fish-eating birds, and humans—can accumulate such high toxin levels that they suffer adverse health effects (chapter 18). One of the first known examples of bioaccumulation and biomagnification was DDT, which accumulated through food chains so that by the 1960s it was shown to be interfering with reproduction of peregrine falcons, brown pelicans, and other predatory birds at the top of their food chains (chapter 10).

Persistence makes some materials a greater threat

Some chemical compounds are very unstable and degrade rapidly under most environmental conditions so that their concentrations decline quickly after release. Most modern herbicides and pesticides, for instance, quickly lose their toxicity. Other substances are more persistent and last for years or even centuries in the environment. Metals—such as lead—PVC plastics, chlorinated hydrocarbon pesticides, and asbestos are valuable because they are resistant to degradation. This stability, however, also causes problems because these materials persist in the environment and have unexpected effects far from the sites of their original use.

Some **persistent organic pollutants (POPs)** have become extremely widespread, being found now from the tropics to the Arctic. They often accumulate in food webs and reach toxic

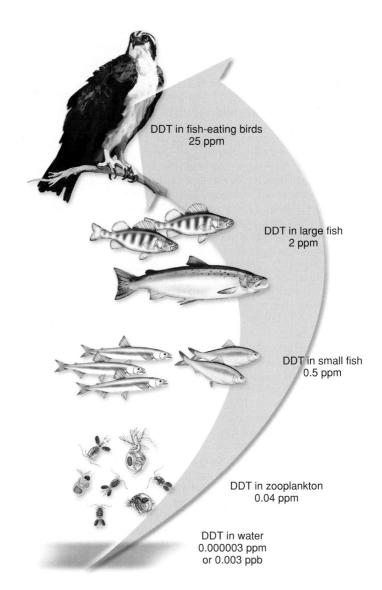

DDT in fish-eating birds
25 ppm

DDT in large fish
2 ppm

DDT in small fish
0.5 ppm

DDT in zooplankton
0.04 ppm

DDT in water
0.000003 ppm
or 0.003 ppb

FIGURE 8.15 Bioaccumulation and biomagnification. Organisms lower on the food chain take up and store toxins from the environment. They are eaten by larger predators, who are eaten, in turn, by even larger predators. The highest members of the food chain can accumulate very high levels of the toxin.

concentrations in long-living top predators such as humans, sharks, raptors, swordfish, and bears. Some of greatest current concerns are:

• Polybrominated diphenyl ethers (PBDE). Widely used as flame-retardants in textiles, foam in upholstery, and plastic in appliances and computers, these chemicals are now found in humans and other species everywhere in the world. Nearly 150 million metric tons (330 million lbs) of PBDEs are used every year worldwide. The toxicity and environmental persistence of PBDE is much like that of PCBs, to which it is closely related chemically. Relatively low exposures in the womb or shortly after birth can irreparably harm children's reproductive and nervous systems. The European Union has already banned this compound.

- Perfluorooctane sulfonate (PFOS) and perfluorooctanoic acid (PFOA, also known as C8) are members of a chemical family used to make nonstick, waterproof, and stain-resistant products such as Teflon, Gortex, Scotchguard, and Stainmaster. Industry makes use of their slippery, heat-stable properties to manufacture everything from airplanes and computers to cosmetics and household cleaners. Now these chemicals—which are reported to be infinitely persistent in the environment—are found throughout the world, even the most remote and seemingly pristine sites. Almost all Americans have one or more perfluorinated compounds in their blood. Heating some nonstick cooking pans above 500°F (260°C) can release enough PFOA to kill pet birds. This chemical family has been shown to cause liver damage as well as various cancers and reproductive and developmental problems in rats. Exposure may be especially dangerous to women and girls, who may be 100 times more sensitive than men to these chemicals. In 2005, the EPA announced the start of a study of human health effects of these chemicals.

- Phthalates (pronounced *thalates*) are found in cosmetics, deodorants, and many plastics (such as soft polyvinyl chloride or PVC) used for food packaging, children's toys, and medical devices. Some members of this chemical family are known to be toxic to laboratory animals, causing kidney and liver damage and possibly some cancers. In addition, many phthalates act as endocrine hormone disrupters, and have been linked to reproductive abnormalities and decreased fertility in humans. A correlation has been found between phthalate levels in urine and low sperm numbers and decreased sperm motility in men. Nearly everyone in the United States has phthalates in their body at levels reported to cause these problems. While not yet conclusive, these results could help explain a 50-year decline in semen quality in most industrialized countries.

- Perchlorate is a waterborne contaminant left over from propellants and rocket fuels. About 12,000 sites in the United States were used by the military for live munition testing and are contaminated with perchlorate. Polluted water used to irrigate crops such as alfalfa and lettuce has introduced the chemical into the human food chain. Tests of cow's milk and human breast milk detected perchlorate in nearly every sample from throughout the United States. Perchlorate can interfere with iodine uptake in the thyroid gland, disrupting adult metabolism and childhood development.

- Bisphenol A (BPA), a prime ingredient in polycarbonate plastic (commonly used for products ranging from water bottles to tooth-protecting sealants), has been widely found in humans, even those who have no known exposure to these chemicals. So far, there is little direct evidence linking BPA exposure to human health risks, but studies in animals have found that the chemical can cause abnormal chromosome numbers, a condition called aneuploidy, which is the leading cause of miscarriages and several forms of

mental retardation. It also is an environmental estrogen and may alter sexual development in both males and females.

- Atrazine is the most widely used herbicide in America. More than 60 million pounds of this compound are applied per year, mainly on corn and cereal grains, but also on golf courses, sugar cane, and Christmas trees. It has long been known to disrupt endocrine hormone functions in mammals, resulting in spontaneous abortions, low birth weights, and neurological disorders. Studies of families in corn-producing areas in the American Midwest have found higher rates of developmental defects among infants, and certain cancers in families with elevated atrazine levels in their drinking water. University of California professor Tyrone Hayes has shown that atrazine levels as low as 0.1 ppb (30 times less than the EPA maximum contaminant level) caused severe reproductive effects in amphibians, including abnormal gonadal development and hermaphroditism. Atrazine now is found in rain and surface waters nearly everywhere in the United States at levels that could cause abnormal development in frogs. In 2003, the European Union withdrew regulatory approval for this herbicide, and several countries banned its use altogether. Some toxicologists have suggested a similar rule in the United States.

Bills to ban Bisphenol A and certain phthalates—particularly in children's toys and feeding products—were introduced in several states in 2005. None of them passed, but the city of San Francisco did pass a similar measure. All these compounds already are regulated in Europe.

Everyone of us has dozens, if not hundreds, of persistent toxins in our body. This accumulation is called our **body burden.** We acquire it from our air, water, diet, and surroundings. Many of these toxins are present in parts per billion, or even parts per trillion. We don't know how dangerous this persistent burden is, but its presence is a matter of concern. If we're anything like the frogs that Tyrone Hayes studies, this accumulated dose of toxins may be a serious problem. Further discussion of POPs can be found in chapter 10.

Chemical interactions can increase toxicity

Some materials produce *antagonistic* reactions. That is, they interfere with the effects or stimulate the breakdown of other chemicals. For instance, vitamins E and A can reduce the response to some carcinogens. Other materials are *additive* when they occur together in exposures. Rats exposed to both lead and arsenic show twice the toxicity of only one of these elements. Perhaps the greatest concern is synergistic effects. **Synergism** is an interaction in which one substance exacerbates the effects of another. For example, occupational asbestos exposure increases lung cancer rates 20-fold. Smoking increases lung cancer rates by the same amount. Asbestos workers who also smoke, however, have a 400-fold increase in cancer rates. How many other toxic chemicals are we exposed to that are below threshold limits individually but combine to give toxic results?

8.4 MECHANISMS FOR MINIMIZING TOXIC EFFECTS

A fundamental concept in toxicology is that every material can be poisonous under some conditions, but most chemicals have some safe level or threshold below which their effects are undetectable or insignificant. Each of us consumes lethal doses of many chemicals over the course of a lifetime. One hundred cups of strong coffee, for instance, contain a lethal dose of caffeine. Similarly, one hundred aspirin tablets, or 10 kilograms (22 lbs) of spinach or rhubarb, or a liter of alcohol would be deadly if consumed all at once. Taken in small doses, however, most toxins can be broken down or excreted before they do much harm. Furthermore, damage they cause can be repaired. Sometimes, however, mechanisms that protect us from one type of toxin or at one stage in the life cycle become deleterious with another substance or in another stage of development. Let's look at how these processes help protect us from harmful substances as well as how they can go awry.

Metabolic degradation and excretion eliminate toxins

Most organisms have enzymes that process waste products and environmental poisons to reduce their toxicity. In mammals, most of these enzymes are located in the liver, the primary site of detoxification of both natural wastes and introduced poisons. Sometimes, however, these reactions work to our disadvantage. Compounds, such as benzopyrene, for example, that are not toxic in their original form are processed by these same enzymes into cancer-causing carcinogens. Why would we have a system that makes a chemical more dangerous? Evolution and natural selection are expressed through reproductive success or failure. Defense mechanisms that protect us from toxins and hazards early in life are "selected for" by evolution. Factors or conditions that affect postreproductive ages (like cancer or premature senility) usually don't affect reproductive success or exert "selective pressure."

We also reduce the effects of waste products and environmental toxins by eliminating them from our body through excretion. Volatile molecules, such as carbon dioxide, hydrogen cyanide, and ketones are excreted via breathing. Some excess salts and other substances are excreted in sweat. Primarily, however, excretion is a function of the kidneys, which can eliminate significant amounts of soluble materials through urine formation. Accumulation of toxins in the urine can damage this vital system, however, and the kidneys and bladder often are subjected to harmful levels of toxic compounds. In the same way, the stomach, intestine, and colon often suffer damage from materials concentrated in the digestive system and may be afflicted by diseases and tumors.

Repair mechanisms mend damage

In the same way that individual cells have enzymes to repair damage to DNA and protein at the molecular level, tissues and organs that are exposed regularly to physical wear-and-tear or to toxic or hazardous materials often have mechanisms for damage repair. Our skin and the epithelial linings of the gastrointestinal tract, blood vessels, lungs, and urogenital system have high cellular reproduction rates to replace injured cells. With each reproduction cycle, however, there is a chance that some cells will lose normal growth controls and run amok, creating a tumor. Thus any agent, such as smoking or drinking, that irritates tissues is likely to be carcinogenic. And tissues with high cell-replacement rates are among the most likely to develop cancers.

Think About It

Some of the mechanisms that help repair wounds or fight off infections can result in cancer when we're old. Why hasn't evolution eliminated these processes?

Hint: Do conditions of postreproductive age affect natural selection?

8.5 MEASURING TOXICITY

Almost 500 years ago, the Swiss scientist Paracelsus said "the dose makes the poison," by which he meant that almost everything is toxic at some level. This remains the most basic principle of toxicology. Sodium chloride (table salt), for instance, is essential for human life in small doses. If you were forced to eat a kilogram of salt all at once, however, it would make you very sick. A similar amount injected into your bloodstream would be lethal. How a material is delivered—at what rate, through which route of entry, and in what medium—plays a vitally important role in determining toxicity.

This does not mean that all toxins are identical, however. Some are so poisonous that a single drop on your skin can kill you. Others require massive amounts injected directly into the blood to be lethal. Measuring and comparing the toxicity of various materials is difficult because not only do species differ in sensitivity, but individuals within a species respond differently to a given exposure. In this section, we will look at methods of toxicity testing and at how results are analyzed and reported.

We usually test toxins on lab animals

The most commonly used and widely accepted toxicity test is to expose a population of laboratory animals to measured doses of a specific substance under controlled conditions. This procedure is expensive, time-consuming, and often painful and debilitating to the animals being tested. It commonly takes hundreds—or even thousands—of animals, several years of hard work, and hundreds of thousands of dollars to thoroughly test the effects of a toxin at very low doses. More humane toxicity tests using computer simulation of model reactions, cell cultures, and other substitutes for whole living animals are being developed. However, conventional large-scale animal testing is the method in which we have the most confidence and on which most public policies about pollution and environmental or occupational health hazards are based.

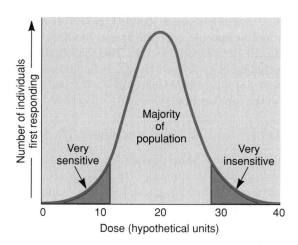

FIGURE 8.16 Probable variations in sensitivity to a toxin within a population. Some members of a population may be very sensitive to a given toxin, while others are much less sensitive. The majority of the population falls somewhere between the two extremes.

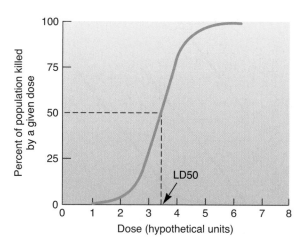

FIGURE 8.17 Cumulative population response to increasing doses of a toxin. The LD50 is the dose that is lethal to half the population.

In addition to humanitarian concerns, there are several problems in laboratory animal testing that trouble both toxicologists and policymakers. One problem is differences in sensitivity to a toxin of the members of a specific population. Figure 8.16 shows a typical dose/response curve for exposure to a hypothetical toxin. Some individuals are very sensitive to the toxin, while others are insensitive. Most, however, fall in a middle category forming a bell-shaped curve. The question for regulators and politicians is whether we should set pollution levels that will protect everyone, including the most sensitive people, or only aim to protect the average person. It might cost billions of extra dollars to protect a very small number of individuals at the extreme end of the curve. Is that a good use of resources?

Dose/response curves are not always symmetrical, making it difficult to compare toxicity of unlike chemicals or different species of organisms. A convenient way to describe toxicity of a chemical is to determine the dose to which 50 percent of the test population is sensitive. In the case of a lethal dose (LD), this is called the **LD50** (fig. 8.17).

Unrelated species can react very differently to the same toxin, not only because body sizes vary but also because of differences in physiology and metabolism. Even closely related species can have very dissimilar reactions to a particular toxin. Hamsters, for instance, are nearly 5,000 times less sensitive to some dioxins than are guinea pigs. Of 226 chemicals found to be carcinogenic in either rats or mice, 95 caused cancer in one species but not the other. These variations make it difficult to estimate the risks for humans since we don't consider it ethical to perform controlled experiments in which we deliberately expose people to toxins.

There is a wide range of toxicity

It is useful to group materials according to their relative toxicity. A moderate toxin takes about one gram per kilogram of body weight (about two ounces for an average human) to make a lethal dose. Very toxic materials take about one-tenth that amount, while extremely toxic substances take one-hundredth as much (only a few drops) to kill most people. Supertoxic chemicals are extremely potent; for some, a few micrograms (millionths of a gram—an amount invisible to the naked eye) make a lethal dose. These materials are not all synthetic. One of the most toxic chemicals known, for instance, is ricin, a protein found in castor bean seeds. It is so toxic that 0.3 billionths of a gram given intravenously will generally kill a mouse. If aspirin were this toxic, a single tablet, divided evenly, could kill 1 million people.

Many carcinogens, mutagens, and teratogens are dangerous at levels far below their direct toxic effect because abnormal cell growth exerts a kind of biological amplification. A single cell, perhaps altered by a single molecular event, can multiply into millions of tumor cells or an entire organism. Just as there are different levels of direct toxicity, however, there are different degrees of carcinogenicity, mutagenicity, and teratogenicity. Methanesulfonic acid, for instance, is highly carcinogenic, while the sweetener saccharin is a suspected carcinogen whose effects may be vanishingly small.

Acute and chronic doses and effects differ

Most of the toxic effects that we have discussed so far have been **acute effects.** That is, they are caused by a single exposure to the toxin and result in an immediate health crisis of some sort. Often, if the individual experiencing an acute reaction survives this immediate crisis, the effects are reversible. **Chronic effects,** on the other hand, are long lasting, perhaps even permanent. A chronic effect can result from a single dose of a very toxic substance, or it can be the result of a continuous or repeated sublethal exposure.

We also describe long-lasting *exposures* as chronic, although their effects may or may not persist after the toxin is removed. It usually is difficult to assess the specific health risks of chronic exposures because other factors, such as aging or normal diseases, act simultaneously with the factor under study. It often requires

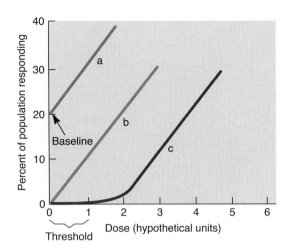

FIGURE 8.18 Three possible dose-response curves at low doses. (a) Some individuals respond, even at zero dose, indicating that some other factor must be involved. (b) Response is linear down to the lowest possible dose. (c) Threshold must be passed before any response is seen.

very large populations of experimental animals to obtain statistically significant results for low-level chronic exposures. Toxicologists talk about "megarat" experiments in which it might take a million rats to determine the health risks of some supertoxic chemicals at very low doses. Such an experiment would be terribly expensive for even a single chemical, let alone for the thousands of chemicals and factors suspected of being dangerous.

An alternative to enormous studies involving millions of animals is to give massive amounts—usually the maximum tolerable dose—of a toxin being studied to a smaller number of individuals and then to extrapolate what the effects of lower doses might have been. This is a controversial approach because it is not clear that responses to toxins are linear or uniform across a wide range of doses.

Figure 8.18 shows three possible results from low doses of a toxin. Curve (*a*) shows a baseline level of response in the population, even at zero dose of the toxin. This suggests that some other factor in the environment also causes this response. Curve (*b*) shows a straight-line relationship from the highest doses to zero exposure. Many carcinogens and mutagens show this kind of response. Any exposure to such agents, no matter how small, carries some risks. Curve (*c*) shows a threshold for the response where some minimal dose is necessary before any effect can be observed. This generally suggests the presence of some defense mechanism that prevents the toxin from reaching its target in an active form or repairs the damage that it causes. Low levels of exposure to the toxin in question may have no deleterious effects, and it might not be necessary to try to keep exposures to zero.

Which, if any, environmental health hazards have thresholds is an important but difficult question. The 1958 Delaney Clause to the U.S. Food and Drug Act forbids the addition of *any* amount of known carcinogens to food and drugs, based on the assumption that any exposure to these substances represents unacceptable risks. This standard was replaced in 1996 by a *no*

reasonable harm requirement, defined as less than one cancer for every million people exposed over a lifetime. This change was supported by a report from the National Academy of Sciences concluding that synthetic chemicals in our diet are unlikely to represent an appreciable cancer risk. We will discuss risk analysis in the next section.

Detectable levels aren't always dangerous

You may have seen or heard dire warnings about toxic materials detected in samples of air, water, or food. A typical headline announced recently that 23 pesticides were found in 16 food samples. What does that mean? The implication seems to be that any amount of dangerous materials is unacceptable and that counting the numbers of compounds detected is a reliable way to establish danger. We have seen, however, that the dose makes the poison. It matters not only what is there, but how much, where it is located, how accessible it is, and who is exposed. At some level, the mere presence of a substance is insignificant.

Toxins and pollutants may seem to be more widespread now than in the past, and this is surely a valid perception for many substances. The daily reports we hear of new materials found in new places, however, are also due, in part, to our more sensitive measuring techniques. Twenty years ago, parts per million were generally the limits of detection for most chemicals. Anything below that amount was often reported as zero or absent rather than more accurately as undetected. A decade ago, new machines and techniques were developed to measure parts per billion. Suddenly, chemicals were found where none had been suspected. Now we can detect parts per trillion or even parts per quadrillion in some cases. Increasingly sophisticated measuring capabilities may lead us to believe that toxic materials have become more prevalent. In fact, our environment may be no more dangerous; we are just better at finding trace amounts.

8.6 RISK ASSESSMENT AND ACCEPTANCE

Risk is the possibility of suffering harm or loss. **Risk assessment** is the scientific process of estimating the threat that particular hazards pose to human health. This process includes risk identification, dose response assessment, exposure appraisal, and risk characterization. In hazard identification, scientists evaluate all available information about the effects of a toxin to estimate the likelihood that a chemical will cause a certain effect in humans. The best evidence comes from human studies, such as physician case reports. Animal studies are also used to assess health risks. Risk assessment for identified toxicity hazards (for example, lead) includes collection and analysis of site data, development of exposure and risk calculations, and preparation of human health and ecological impact reports. Exposure assessment is the estimation or determination of the magnitude, frequency, duration, and route of exposure to a possible toxin. Toxicity assessment weighs all available evidence and estimates the potential for adverse health effects to occur.

Risk perception isn't always rational

A number of factors influence how we perceive relative risks associated with different situations.

- People with social, political, or economic interests—including environmentalists—tend to downplay certain risks and emphasize others that suit their own agendas. We also tend to tolerate risks that we choose—such as driving, smoking, or overeating—while objecting to risks we cannot control—such as potential exposure to slight amounts of toxic substances.

- Most people have difficulty understanding and believing probabilities. We feel that there must be patterns and connections in events, even though statistical theory says otherwise. If the coin turned up heads last time, we feel certain that it will turn up tails next time. In the same way, it is difficult to understand the meaning of a 1-in-10,000 risk of being poisoned by a chemical.

- Our personal experiences often are misleading. When we have not personally experienced a bad outcome, we feel it is more rare and unlikely to occur than it actually may be. Furthermore, the anxieties generated by life's gambles make us want to deny uncertainty and to misjudge many risks (fig. 8.19).

- We have an exaggerated view of our own abilities to control our fate. We generally consider ourselves above-average drivers, safer than most when using appliances or power tools, and less likely than others to suffer medical problems, such as heart attacks. People often feel they can avoid hazards because they are wiser or luckier than others.

- News media give us a biased perspective on the frequency of certain kinds of health hazards, overreporting some accidents or diseases while downplaying or underreporting others. Sensational, gory, or especially frightful causes of death like murders, plane crashes, fires, or terrible accidents occupy a disproportionate amount of attention in the public media. Heart diseases, cancer, and stroke kill nearly 15 times as many people in the United States as do accidents and 75 times as many people as do homicides, but the emphasis placed by the media on accidents and homicides is nearly inversely proportional to their relative frequency compared to either cardiovascular disease or cancer. This gives us an inaccurate picture of the real risks to which we are exposed.

- We tend to have an irrational fear or distrust of certain technologies or activities that leads us to overestimate their dangers. Nuclear power, for instance, is viewed as very risky, while coal-burning power plants seem to be familiar and relatively benign; in fact, coal mining, shipping, and combustion cause an estimated 10,000 deaths each year in the United States. An old, familiar technology seems safer and more acceptable than does a new, unknown one (see Data Analysis p. 177).

Think About It

Why might you and your mother rank some risks differently? List some activities on which the two of you might disagree.

Risk acceptance depends on many factors

How much risk is acceptable? How much is it worth to minimize and avoid exposure to certain risks? Most people will tolerate a higher probability of occurrence of an event if the harm caused by that event is low. Conversely, harm of greater severity is acceptable only at low levels of frequency. A 1-in-10,000 chance of being killed might be of more concern to you than a 1-in-100 chance of being injured. For most people, a 1-in-100,000 chance of dying from some event or some factor is a threshold for changing what we do. That is, if the chance of death is less than 1 in 100,000, we are not likely to be worried enough to change our ways. If

FIGURE 8.19 How dangerous are motorcycles? Many parents regard them as extremely risky, while many students—especially males—believe the risks (which are about the same as your chances of dying from surgery or other medical care) are acceptable. Perhaps the more important question is whether the benefits outweigh the risks.

the risk is greater, we will probably do something about it. The Environmental Protection Agency generally assumes that a risk of 1 in 1 million is acceptable for most environmental hazards. Critics of this policy ask, acceptable to whom?

For activities that we enjoy or find profitable, we are often willing to accept far greater risks than this general threshold. Conversely, for risks that benefit someone else we demand far higher protection. For instance, your chance of dying in a motor vehicle accident in any given year is about 1 in 5,000, but that doesn't deter many people from riding in automobiles. Your lifetime chance of dying from lung cancer if you smoke one pack of cigarettes per day is about 1 in 4. By comparison, the risk from drinking water with the EPA limit of trichloroethylene is about 1 in 10 million. Strangely, many people demand water with zero levels of trichloroethylene, while continuing to smoke cigarettes.

Table 8.4 lists lifetime odds of dying from a few leading diseases and accidents. These are statistical averages, of course, and there clearly are differences in where one lives or how one behaves that affect the danger level of these activities. Although the average lifetime chance of dying in an automobile accident is 1 in 100, there clearly are things you can do—like wearing a seatbelt, following safety rules, and avoiding risky situations—that improve your odds. Still, it is interesting how we readily accept some risks while shunning others.

Our perception of relative risks is strongly affected by whether risks are known or unknown, whether we feel in control of the outcome, and how dreadful the results are. Risks that are unknown or unpredictable and results that are particularly gruesome or disgusting seem far worse than those that are familiar and socially acceptable.

TABLE 8.4

Lifetime Chances of Dying in the U.S.

Source	Odds (1 in x)
Heart disease	2
Cancer	3
Smoking	4
Lung disease	15
Pneumonia	30
Automobile accident	100
Suicide	100
Falls	200
Firearms	200
Fires	1,000
Airplane accident	5,000
Jumping from high places	6,000
Drowning	10,000
Lightning	56,000
Hornets, wasps, bees	76,000
Dog bite	230,000
Poisonous snakes, spiders	700,000
Botulism	1 million
Falling space debris	5 million
Drinking water with EPA limit of trichloroethylene	10 million

Source: U.S. National Safety Council, 2003.

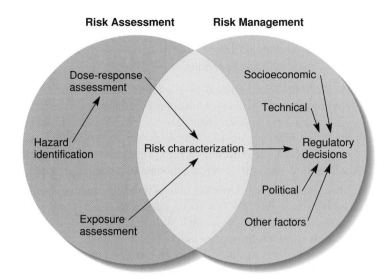

FIGURE 8.20 Risk assessment organizes and analyzes data to determine relative risk. Risk management sets priorities and evaluates relevant factors to make regulatory decisions.

Studies of public risk perception show that most people react more to emotion than statistics. We go to great lengths to avoid some dangers while gladly accepting others. Factors that are involuntary, unfamiliar, undetectable to those exposed, catastrophic, or that have delayed effects or are a threat to future generations are especially feared while those that are voluntary, familiar, detectable, or immediate cause less anxiety. Even though the actual number of deaths from automobile accidents, smoking, or alcohol, for instance, are thousands of times greater than those from pesticides, nuclear energy, or genetic engineering, the latter preoccupy us far more than the former.

8.7 ESTABLISHING HEALTH POLICY

Risk management combines principles of environmental health and toxicology together with regulatory decisions based on socio-economic, technical, and political considerations (fig. 8.20). The biggest problem in making regulatory decisions is that we are usually dealing with many sources of harm to which we are exposed, often without being aware of them. It is difficult to separate the effects of all these different hazards and to evaluate their risks accurately, especially when the exposures

are near the threshold of measurement and response. In spite of often vague and contradictory data, public policymakers must make decisions.

The case of the sweetener saccharin is a good example of the complexities and uncertainties of risk assessment in public health. Studies in the 1970s at the University of Wisconsin and the Canadian Health Protection Branch suggested a link between saccharin and bladder cancer in male rats. Critics of these studies pointed out that humans would have to drink 800 cans of diet soda *per day* to get a saccharin dose equivalent to that given to the rats. Furthermore, they argued this response may be unique to male rats. In 2000, the U.S. Department of Health concluded a study that found no association between saccharin and cancer in humans. Congress ordered that all warnings be removed from saccharin-containing products. Still, some groups like the Center for Science in the Public Interest consider this sweetener dangerous and urge us to avoid it if possible.

In setting standards for environmental toxins, we need to consider (1) combined effects of exposure to many different sources of damage, (2) different sensitivities of members of the population, and (3) effects of chronic as well as acute exposures. Some people argue that pollution levels should be set at the highest amount that does *not* cause measurable effects. Others demand that pollution be reduced to zero if possible, or as low as is technologically feasible. It may not be reasonable to demand that we be protected from every potentially harmful contaminant in our environment, no matter how small the risk. As we have seen, our bodies have mechanisms that enable us to avoid or repair many kinds of damage so that most of us can withstand some minimal level of exposure without harm (fig. 8.21).

On the other hand, each challenge to our cells by toxic substances represents stress on our bodies. Although each individual stress may not be life-threatening, the cumulative effects of all

FIGURE 8.21 "Do you want to stop reading those ingredients while we're trying to eat?"

TABLE 8.5

Relative Risks to Human Welfare

Relatively High-Risk Problems

Habitat alteration and destruction
Species extinction and loss of biological diversity
Stratospheric ozone depletion
Global climate change

Relatively Medium-Risk Problems

Herbicides/pesticides
Toxics and pollutants in surface waters
Acid deposition
Airborne toxics

Relatively Low-Risk Problems

Oil spills
Groundwater pollution
Radionuclides
Thermal pollution

Source: Environmental Protection Agency.

the environmental stresses, both natural and human-caused, to which we are exposed may seriously shorten or restrict our lives. Furthermore, some individuals in any population are more susceptible to those stresses than others. Should we set pollution standards so that no one is adversely affected, even the most sensitive individuals, or should the acceptable level of risk be based on the average member of the population?

Finally, policy decisions about hazardous and toxic materials also need to be based on information about how such materials affect the plants, animals, and other organisms that define and maintain our environment. In some cases, pollution can harm or destroy whole ecosystems with devastating effects on the life-supporting cycles on which we depend. In other cases, only the most sensitive species are threatened. Table 8.5 shows the Environmental Protection Agency's assessment of relative risks to human welfare. This ranking reflects a concern that our exclusive focus on reducing pollution to protect human health has neglected risks to natural ecological systems. While there have been many benefits from a case-by-case approach in which we evaluate the health risks of individual chemicals, we have often missed broader ecological problems that may be of greater ultimate importance.

CONCLUSION

We have made marvelous progress in reducing some of the worst diseases that have long plagued humans. Smallpox is the first major disease to be completely eliminated. Guinea worms and polio are nearly eradicated worldwide; typhoid fever, cholera, yellow fever, tuberculosis, mumps, and other highly communicable diseases are rarely encountered in advanced countries. Childhood mortality has decreased 90 percent globally, and people almost everywhere are living twice as long, on average, as they did a century ago.

But the technological innovations and affluence that have diminished many terrible diseases, have also introduced new risks. Chronic conditions, such as cardiovascular disease, cancer, depression, dementia, diabetes, and traffic accidents, that once were confined to richer countries, now have become leading health problems nearly everywhere. Part of this change is that we no longer die at an early age of infectious disease, so we live long enough to develop the infirmities of old age. Another factor is that affluent lifestyles, lack of exercise, and unhealthy diets aggravate these chronic conditions.

New, emergent diseases are appearing at an increasing rate. With increased international travel, diseases can spread around the globe in a few days. Epidemiologists warn that the next deadly epidemic may be only a plane ride away. In addition, modern industry is introducing thousands of new chemical substances every year, most of which aren't studied thoroughly for health effects. Endocrine disrupters, neurotoxins, carcinogens, mutagens, teratogens, and other toxins can have tragic outcomes. The effects of lead on children's mental development is an example of both how we have introduced materials with unintended consequences, and a success story of controlling a serious health risk. Many other industrial chemicals could be having similar harmful effects.

Determining what levels of environmental health risk are acceptable is difficult. We are exposed to many different health threats simultaneously. Furthermore, people consider some dangers tolerable, but dread others—especially those that are new, involuntary, difficult to detect, and whose effects are unknown to science. The situation is complicated by the fact that news media gives us a biased perspective on some hazards, while our personal experiences and our sense of our own abilities are often misleading.

There are many steps that each of us can take to protect our health. Eating a healthy diet, exercising regularly, drinking in moderation, driving prudently, and practicing safe sex are among the most important.

REVIEWING LEARNING OUTCOMES

By now you should be able to explain the following points:

8.1 Describe health and disease and how global disease burden is now changing.
- The global disease burden is changing.
- Infectious and emergent diseases still kill millions of people.
- Conservation medicine combines ecology and health care.
- Resistance to drugs, antibiotics, and pesticides is increasing.
- Who should pay for health care?

8.2 Summarize the principles of toxicology.
- How do toxins affect us?
- How does diet influence health?

8.3 Discuss the movement, distribution, and fate of toxins in the environment.
- Solubility and mobility determine where and when chemicals move.
- Exposure and susceptibility determine how we respond.

- Bioaccumulation and biomagnification increase concentrations of chemicals.
- Persistence makes some materials a greater threat.
- Chemical interactions can increase toxicity.

8.4 Characterize mechanisms for minimizing toxic effects.
- Metabolic degradation and excretion eliminate toxins.
- Repair mechanisms mend damage.

8.5 Explain ways we measure and describe toxicity.
- We usually test toxins on lab animals.
- There is a wide range of toxicity.
- Acute and chronic doses and effects differ.
- Detectable levels aren't always dangerous.

8.6 Evaluate risk assessment and acceptance.
- Risk perception isn't always rational.
- Risk acceptance depends on many factors.

8.7 Relate how we establish health policy.

PRACTICE QUIZ

1. What are guinea worms, and how are they acquired?
2. Define the terms *disease* and *health.*
3. What were some of the most serious diseases in the world in 1990? How is this list expected to change in the next 20 years?
4. What are emergent diseases? Give a few examples, and describe their cause and effects.
5. How do bacteria acquire antibiotic resistance? How might we prevent this?
6. What is the difference between toxic and hazardous? Give some examples of materials in each category.

7. How do the physical and chemical characteristics of materials affect their movement, persistence, distribution, and fate in the environment?
8. What is the difference between acute and chronic toxicity?
9. Define *carcinogenic, mutagenic, teratogenic,* and *neurotoxic.*
10. What are the relative risks of smoking, driving a car, and drinking water with the maximum permissible levels of trichloroethylene? Are these relatively equal risks?

CRITICAL THINKING AND DISCUSSION QUESTIONS

1. What consequences (positive or negative) do you think might result from defining health as a state of complete physical, mental, and social well-being? Who might favor or oppose such a definition?
2. Do rich countries bear any responsibilities if the developing world adopts unhealthy lifestyles or diets? What could (or should) we do about it?

3. Why do we spend more money on heart or cancer research than childhood illnesses?
4. What are the premises in the discussion of assessing risk? Could conflicting conclusions be drawn from the facts presented in this section? What is your perception of risk from your environment?

5. Should pollution levels be set to protect the average person in the population or the most sensitive? Why not have zero exposure to all hazards?

6. What level of risk is acceptable to you? Are there some things for which you would accept more risk than others?

DATA analysis   Graphing Multiple Variables

Is it possible to show relationships between two dependent variables on the same graph? Sometimes that's desirable when you want to make comparisons between them. The graph shown here does just that. It's a description of how people perceive different risks. We judge the severity of risks based on how familiar they are and how much control we have over our exposure. The Y-axis represents how mysterious, unknown or delayed the risk seems to be. Things that are unobservable, unknown to those exposed, delayed in their effects, and unfamiliar or unknown to science tend to be more greatly feared than those that are observable, known, immediate, familiar, and known to science. The X-axis represents a measure of dread, which combines how much control we feel we have over the risk, how terrible the results could potentially be, and how equitably the risks are distributed. The size of the symbol for each risk indicates the combined effect of these two variables. Notice that things such as DNA technology or nuclear waste, which have high levels of both mystery and dread, tend to be regarded with the greatest fear, while familiar, voluntary, personally rewarding behaviors such as riding in automobiles or on bicycles, or drinking alcohol are thought to be relatively minor risks. Actuarial experts (statisticians who gather mortality data) would tell you that automobiles, bicycles, and alcohol have killed far more people (so far) than DNA technology or radioactive waste. But this isn't just a question of data. It's a reflection of how much we fear various risks. Notice that this is a kind of scatter plot mapping categories of data that have no temporal sequence. Still, you can draw some useful inferences from this sort of graphic presentation.

Compare the graph with table 8.4.

1. What is the highest risk factor in table 8.4 that also appears on the perception of risk graph?

2. How do the lifetime risk for smoking and auto accidents calculated by the National Safety Council correspond to the perception of risk in this graph?

3. How do airplane accidents compare in these two assessments?

4. Why do DNA technology and radioactive waste appear to generate so much dread, but don't even appear in the lifetime chances of dying list?

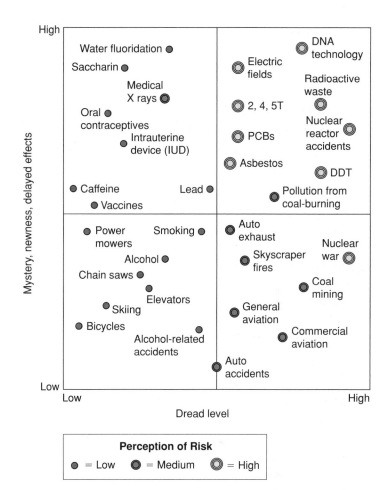

Public perception of risk depending on the familiarity, apparent potential for harm, and personal control over the risk.
Source: Data from Slovic, Paul, 1987. Perception of Risk, *Science*, 236 (4799): 286–90.

5. Thousands of people die every year from alcohol-related accidents, yet alcohol ranks very low in the two scales shown in this graph. On the other hand, no one has ever died—as far as we know—from DNA technology. Why the great discrepancy in rankings?

Enormous farms have been carved out of Brazil's Cerrado
(savanna), which once was the most biodiverse
grassland and open tropical forest complex
in the world.

Food and Agriculture

*We abuse the land because we regard it as a commodity
belonging to us. When we see land as a community to which
we belong, we may begin to use it with love and respect.*

—Aldo Leopold—

LEARNING OUTCOMES

After studying this chapter, you should be able to:

9.1 Describe world food supplies and nutritional requirements.

9.2 Compare key food sources.

9.3 Understand how farm policy can help or hurt soil conservation.

9.4 Illustrate why soil is a renewable resource.

9.5 Summarize the ways we use and abuse soil.

9.6 Identify other agricultural resources.

9.7 Explain new crops and genetic engineering.

9.8 Appraise sustainable agriculture.

Case Study Farming the Cerrado

A soybean boom is sweeping across South America. Inexpensive land, availability of new crop varieties, and government policies that favor agricultural expansion have made South America the fastest growing agricultural area in the world. The center of this rapid expansion is the Cerrado, a huge area of grassland and tropical forest stretching from Bolivia and Paraguay across the center of Brazil almost to the Atlantic Ocean (fig. 9.1). Biologically, this rolling expanse of grasslands and tropical woodland is the richest savanna in the world, with at least 130,000 different plant and animal species, many of which are threatened by agricultural expansion.

Until recently, the Cerrado, which is roughly equal in size to the American Midwest, was thought to be unsuitable for cultivation. Its red iron-rich soils are highly acidic and poor in essential plant nutrients. Furthermore, the warm, humid climate harbors many destructive pests and pathogens. For hundreds of years, the Cerrado was primarily cattle country with many poor-quality pastures producing low livestock yields.

In the past few decades, however, Brazilian farmers have learned that modest applications of lime and phosphorus can quadruple yields of soybeans, maize, cotton, and other valuable crops. Researchers have developed more than 40 varieties of soybeans—mostly through conventional breeding, but some created with molecular techniques—specially adapted for the soils and climate of the Cerrado. Until about 30 years ago, soybeans were a relatively minor crop in Brazil. Since 1975, however, the area planted with soy has doubled about every four years, reaching more than 25 million ha (60 million acres) in 2006. Although that's a large area, it represents only one-eighth of the Cerrado, more than half of which is still occupied by pasture.

Brazil is now the world's top soy exporter, shipping some 50 million metric tons per year, or about 10 percent more than the United States. With two crops per year, cheap land, low labor costs, favorable tax rates, and yields per hectare equal to those in the American Midwest, Brazilian farmers can produce soybeans for less than half the cost in America. Agricultural economists predict that, by 2020, the global soy crop will double from the current 160 million metric tons per year, and that South America could be responsible for most of that growth. In addition to soy, Brazil now leads the world in beef, corn

(maize), oranges, and coffee exports. This dramatic increase in South American agriculture helps answer the question of how the world may feed a growing human population.

A major factor in Brazil's current soy expansion is rising income in China. With more money to spend, the Chinese are consuming more soy, both directly as tofu and other soy products, and indirectly as animal feed. A decade ago, China was self-sufficient in soy production. Now, China imports about 30 million tons of soy annually. About half that amount comes from Brazil, which passed the United States in 2007 as the world's leading soy exporter. In 1997, Brazil shipped only 2 million tons of soy. A decade later, exports reached 28 million tons.

The outbreak of mad cow disease (or BSE, see chapter 8) in Europe, Canada, and Japan has fueled increased worldwide demand for Brazilian beef. With 175 million free-range, grass-fed (and presumably BSE-free) cattle, Brazil has become the world's largest beef exporter.

Increasing demand for both soybeans and beef create land conflicts in Brazil. The pressure for more cropland and pasture is a leading cause of deforestation and habitat loss, most of which is occurring in the "arc of destruction" between the Cerrado and the Amazon. Small family farms are being gobbled up, and farm workers, displaced by mechanization, often migrate either to the big cities or to frontier forest areas. Increasing conflicts between poor farmers and big landowners have led to violent confrontations. The Landless Workers Movement claims that 1,237 rural workers died in Brazil between 1985 and 2000 as a result of assassinations and clashes over land rights. In 2005, a 74-year-old Catholic nun, Sister Dorothy Stang, was shot by gunmen hired by ranchers who resented her advocacy for native people, workers, and environmental protection. Over the past 20 years, Brazil claims to have resettled 600,000 families from the Cerrado. Still, tens of thousands of landless farm workers and displaced families live in unauthorized squatter camps and shanty-towns across the country awaiting relocation.

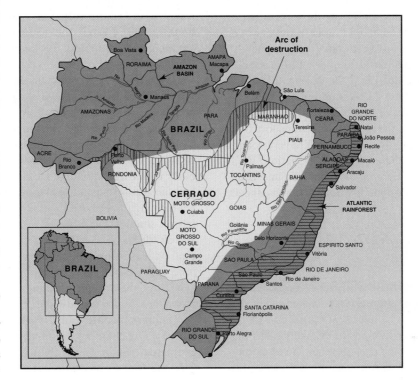

FIGURE 9.1 Brazil's Cerrado, 2 million ha of savanna (grassland) and open woodland, is the site of the world's fastest growing soybean production. Cattle ranchers and agricultural workers, displaced by mechanized crop production, are moving northward into the "arc of destruction" at the edge of the Amazon rainforest, where the continent's highest rate of forest clearing is occurring.

Case Study *continued*

As you can see, rapid growth of beef and soy production in Brazil have both positive and negative aspects. On one hand, more high-quality food is now available to feed the world. The 2 million km² of the Cerrado represents one of the world's last opportunities to open a large area of new, highly productive cropland. On the other hand, the rapid expansion and mechanization of agriculture in Brazil is destroying biodiversity and creating social conflicts as people move into formerly pristine lands. The issues raised in this case study illustrate many of the major themes in this chapter. Will there be enough food for everyone in the world? What will be the environmental and social consequences of producing the nutrition we need? In this chapter, we'll look at world food supplies, agricultural inputs, and sustainable approaches that can help solve some of the difficult questions we face.

9.1 WORLD FOOD AND NUTRITION

Despite dire predictions that runaway population growth would soon lead to terrible famines (chapter 7), world food supplies have more than kept up with increasing human numbers over the past two centuries. The past 40 years have seen especially encouraging strides in reducing world hunger. While population growth averaged 1.7 percent per year during that time, world food production increased an average of 2.2 percent. Increased use of irrigation, improved crop varieties, more readily available fertilizers, and distribution systems to transport food from regions with surpluses to those in need have brought improved nutrition to billions of people. In this section, we'll look at the causes and effects of remaining chronic and acute food shortages, as well as recommendations for a balanced, healthful diet.

Millions of people don't have enough to eat

In 1960, nearly 60 percent of the residents of developing countries were considered **chronically undernourished,** meaning their diet didn't provide the 2,200 kcal per day, on average, considered necessary for a healthy productive life. Today, despite the fact that their population has doubled over the past 46 years, the proportion in these countries suffering from chronic caloric deficiency has fallen to less than 14 percent.

The UN Food and Agriculture Organization (FAO) expects agricultural production to continue to grow over the next few decades. Where the current world food supply would be sufficient, if equitably distributed, to provide an average of 2,800 kcal per person per day, the FAO predicts that by 2030, there will be enough food available to supply 3,050 kcal per day to everyone, or about 30 percent more than most of us need. In countries such as the United States, the problem already has become what to do with surplus food. Farmers in these countries are paid billions of dollars per year not to grow crops.

Still, in a world of surplus food, at least 850 million people don't have enough to eat. Figure 9.2 shows countries with the highest risk of food shortages. As you can see, most of sub-Saharan Africa, South and Southeast Asia, and parts of Latin America fall in this category. Ninety-five percent of the chronically undernourished are in developing countries, but an increasing number are in transition countries (primarily states of the former Soviet Union and its allies undergoing a change from socialism to market economies) where bad weather, poor management, and social crises have resulted in sharply falling agricultural production (fig. 9.3). Even in the richest countries, where excess calories are the greatest problem for the majority, some 11 million people don't have enough to eat.

Poverty is the greatest threat to **food security,** or the ability to obtain sufficient food on a day-to-day basis. The 1.5 billion people in the world who live on less than $1 per day all too often can't buy the food they need and don't have access to resources to grow it for themselves. Food security occurs at multiple scales. In the poorest countries, hunger may affect nearly everyone. In other countries, although the average food availability may be satisfactory, some individual communities or families may not have enough to eat. And within families, males often get both the largest share as well as the most nutritious food, while women and children—who need food most—all too often get the poorest diet. At least 6 million children under 5 years old die every year (one every 5 seconds) from hunger and malnutrition. Providing a healthy diet might eliminate as much as 60 percent of all premature deaths worldwide.

Hungry people can't work their way out of poverty, the Nobel Prize-winning economist Robert Fogel points out. He estimates that in 1790, about 20 percent of the population of England and France was effectively excluded from the labor force because they were too weak and hungry to work. Improved nutrition, he calculates, accounted for about half of all European economic growth during the nineteenth century. Since many developing countries are as poor now (in relative terms) as Britain and France were in 1790, his analysis suggests that reducing hunger could yield more than (U.S.) $120 billion in economic growth produced by longer, healthier, more productive lives for several hundred million people.

The 2003 UN World Food Summit reaffirmed the goal set by previous conventions of reducing the number of chronically undernourished people from 850 million to 400 million by 2015. We aren't on track to meet that goal, but some countries have made impressive progress. China, alone, has reduced its number of undernourished people by 74 million over the past decade. Indonesia, Vietnam, Thailand, Nigeria, Ghana, and Peru each reduced chronic hunger by about 3 million people. In 47 other countries, however, the numbers of chronically underfed

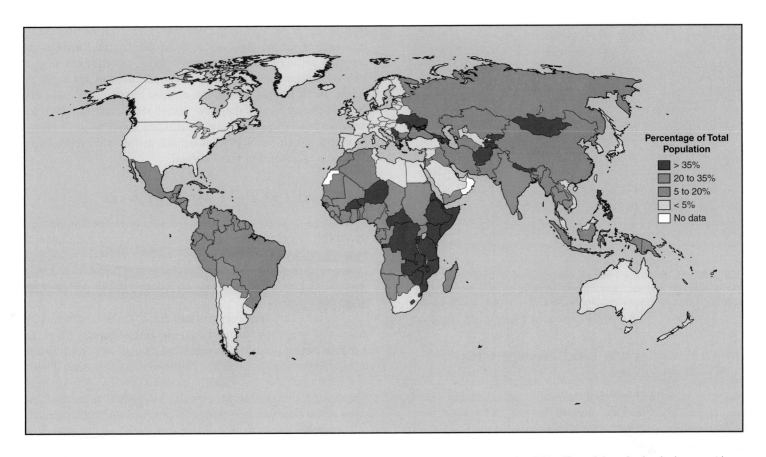

FIGURE 9.2 Hunger around the world. In 2005, the United Nations reported that 852 million people—815 million of them in developing countries— suffered from chronic hunger and malnutrition. Africa has the largest number of countries with food shortages.

people have increased. India now has by far the largest number of persistently hungry people in the world, while central Africa has the highest percentage (fig. 9.4).

Recognizing the role of women in food production is an important step toward food security for all. Throughout the developing world, women do 50 to 70 percent of all farm work but control only a tiny fraction of the land and rarely have access to capital or developmental aid. In Nigeria, for example, home gardens occupy only 2 percent of all cropland, but provide half the food families eat. Making land, credit, education, and access to markets available to women could contribute greatly to family nutrition.

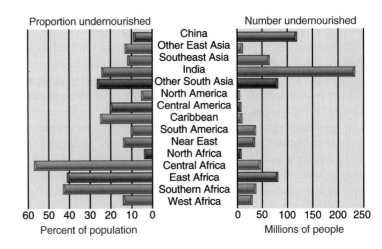

FIGURE 9.3 Changes in agricultural production in 2000. Transition nations are states from the former Soviet Union undergoing a change from socialism to capitalism.
Source: Food and Agriculture Organization (FAO), 2002.

FIGURE 9.4 Number and percent of population chronically undernourished by region in the developing world.
Source: Food and Agriculture Organization (FAO), 2002

FIGURE 9.5 Children wait for their daily ration of porridge at a feeding station in Somalia. When people are driven from their homes by hunger or war, social systems collapse, diseases spread rapidly, and the situation quickly becomes desperate.

Famines are acute food emergencies

The chronic hunger and malnutrition described in the previous section is silent and generally invisible, affecting individuals, families, and communities, often in the midst of general plenty, but **famines** are characterized by large-scale food shortages, massive starvation, social disruption, and economic chaos. Starving people eat their seed grain and slaughter their breeding stock in a desperate attempt to keep themselves and their families alive. Even if better conditions return, they have sacrificed their productive capacity and will take a long time to recover. Famines are characterized by mass migrations as starving people travel to refugee camps in search of food and medical care (fig. 9.5). Many die on the way or fall prey to robbers.

In 2006, the FAO reported that 58 million people in 36 countries (two-thirds of them in sub-Saharan Africa) needed emergency food aid. What causes these emergencies? Droughts, earthquakes, severe storms, and other natural disasters are usually the immediate trigger, but politics and economics are often equally important. Bad weather, insect outbreaks (see Case Study: A Plague of Locusts, p. 125), and other environmental factors cause crop failures and create food shortages. But the Nobel Prize-winning work of Harvard economist Amartya K. Sen shows that these factors have often been around for a long time, and local people usually have adaptations to get through hard times if they aren't thwarted by inept or corrupt governments and greedy elites. National politics, however, together with commodity hoarding, price gouging, poverty, wars, landlessness, and other external factors often conspire so that poor people can neither grow their own food nor find jobs to earn money to buy the food they need. Professor Sen points out that armed conflict and political oppression almost always are at the root of famine. No democratic country with a relatively free press, he says, has ever had a major famine.

The aid policies of rich countries often serve more to get rid of surplus commodities and make us feel good about our generosity than to get at the root causes of starvation. But herding people into feeding camps often is the worst thing to do for them. The stress of getting there kills many of them, and the crowding and lack of sanitation in the camps exposes them to epidemic diseases. There are no jobs in the refugee camps, so people can't support themselves if they try. Social chaos and family breakdown expose those who are weakest to robbery and violence. Having left their land and tools behind, people can't replant crops when the weather returns to normal.

We need the right kinds of food

In addition to energy (calories), we also need specific nutrients in our diet, such as proteins, vitamins, and certain trace minerals. You might have more than enough calories and still suffer from **malnourishment,** a nutritional imbalance caused by a lack of specific dietary components or an inability to absorb or utilize essential nutrients.

Many poor people can't afford meat, fruits, and vegetables that would provide a balanced diet. The FAO estimates that nearly 3 billion people (half the world population) suffer from vitamin, mineral, or protein deficiencies. This results in devastating illnesses and deaths as well as reduced mental capacity, developmental abnormalities, and stunted growth. Altogether, these problems bring an incalculable loss of human potential and social capital.

Anemia (low hemoglobin levels in the blood, usually caused by dietary iron deficiency) is the most common nutritional problem in the world. According to the FAO, more than 2 billion people (52 percent are pregnant women and 39 percent are children under age 5) suffer from iron deficiencies. The problem is most severe in India, where it's estimated that 80 percent of all pregnant women are anemic. Anemia increases the risk of maternal deaths from hemorrhage in childbirth and affects childhood development. Red meat, eggs, legumes, and green vegetables all are good sources of dietary iron.

Iodine is essential for synthesis of thyroxin, an endocrine hormone that regulates metabolism and brain development, among other things. Chronic iodine deficiency causes goiter (a swollen thyroid gland, fig. 9.6), stunted growth, and reduced mental ability. The FAO estimates that 740 million people—mainly in South and Southeast Asia—suffer from iodine

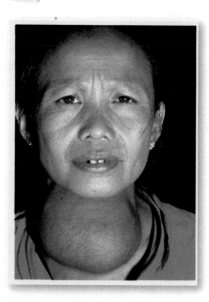

FIGURE 9.6 Goiter, a swelling of the thyroid gland at the base of the neck, is often caused by an iodine deficiency. It is a common problem in many parts of the world, particularly where soil iodine is low and seafood is unavailable.

deficiency and that 177 million children have stunted growth and development. Adding a few pennies worth of iodine to our salt has largely eliminated this problem in developed countries.

Starchy foods like maize (corn), polished rice, and manioc (tapioca), which form the bulk of the diet for many poor people, tend to be low in several essential vitamins as well as minerals. According to the FAO, vitamin A deficiencies affect 100–140 million children at any given time. At least 350,000 go blind every year from the effects of this vitamin shortage. Folic acid (found in dark green, leafy vegetables) is essential for early fetal development. Folic acid deficiencies have been linked to neurological problems in babies including microencephaly (an abnormally small head) or even anencephaly (lacking a brain).

Protein also is essential for normal growth and development. The two most widespread human protein deficiency diseases are kwashiorkor and marasmus. **Kwashiorkor** is a West African word meaning "displaced child." (A young child is displaced—and deprived of nutritious breast milk—when a new baby is born.) This condition most often occurs in young children who eat mainly cheap starchy food and don't get enough good-quality protein. Children with kwashiorkor often have reddish-orange hair, puffy, discolored skin, and a bloated belly. **Marasmus** (from the Greek "to waste away") is caused by a diet low in both calories and protein. A child suffering from severe marasmus is generally thin and shriveled, like a tiny, very old starving person (fig. 9.7). Children with both these deficiencies have low resistance to infections and are likely to suffer from stunted growth, mental retardation, and other developmental problems. Altogether, the FAO estimates that the annual losses due to deaths and diseases caused by calorie and nutrient deficiencies are equivalent to 46 million years of productive life (see more in chapter 8 on disability-adjusted life years).

Eating a balanced diet is essential for good health

Those of us in richer countries often eat too much meat, salt, sugar, and saturated fat, and too little fiber, vitamins, trace minerals, and other components lost from highly processed foods. On average, we consume about one-third more calories than we need and get too little exercise. Food provides solace in stressful lives, and we're constantly

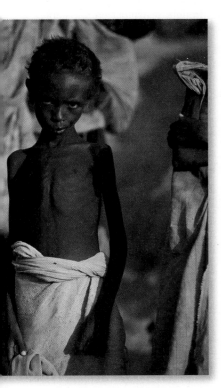

FIGURE 9.7 Marasmus is caused by combined energy (calorie) and protein deficiencies. Children with marasmus have the wizened look and dry, flaky skin of an old person.

bombarded with advertising tempting us to consume more. Ironically, there are now probably more people in the world (1 billion) who suffer from too many calories, than there are who suffer from too few. According to the U.S. Surgeon General, 62 percent of all adult Americans are overweight, up from 40 percent only a decade ago, and about one-third of us are seriously overweight or **obese**—generally considered to be a body mass greater than 30 kg/m^2, or roughly 30 pounds above normal for an average person.

Being overweight substantially raises your risk of hypertension, diabetes, heart attacks, stroke, gallbladder disease, osteoarthritis, respiratory problems, and certain cancers. Every year, about 300,000 in the United States die from illnesses related to obesity. This number is close to the 400,000 annual deaths from diseases related to smoking. Paradoxically, food insecurity and poverty can contribute to obesity. In one study, more than half of the women who reported not having enough to eat were overweight compared to one-third of food-secure women. Lack of time for cooking and access to health food choices along with ready availability of fast-food snacks and calorie-laden drinks lead to dangerous dietary imbalances for many people.

This trend isn't limited to richer countries. Obesity is spreading around the world (fig. 9.8). As chapter 8 points out, western diets and lifestyles are being adopted by many in the developing world, and diseases such as heart attack, stroke, diabetes, and depression that once were thought to inflict only wealthy nations are now becoming the most prevalent causes of death and disability everywhere.

What's the best way to be sure you're getting a healthy diet? Generally, it isn't necessary to take synthetic dietary supplements. Eating a good variety of foods should give you all the nutrients you need. For years, Americans were advised to eat daily servings of four major food groups: meat, dairy products, grains, and fruits and vegetables. These recommendations were revised in 1992 to emphasize only sparing servings of meat, dairy, fats, and sweets.

Some nutritionists believe that the 1992 recommendations for 6 to 11 servings a day of bread, cereal, rice, and pasta still provides too many simple sugars (or starches that are quickly converted to

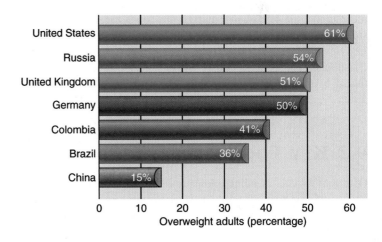

FIGURE 9.8 While nearly a billion people are chronically undernourished, people in wealthier countries are at risk from eating too much. **Source:** Worldwatch Institute, 2001.

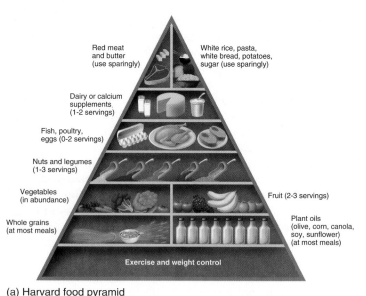

(a) Harvard food pyramid

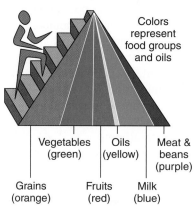

(b) USDA recommendations

Colors represent food groups and oils

Vegetables (green)   Oils (yellow)   Meat & beans (purple)

Grains (orange)   Fruits (red)   Milk (blue)

FIGURE 9.9 The Harvard food pyramid (a) emphasizes fruits, vegetables, and whole grains as the basis of a healthy diet. Red meat, white rice, pasta, and potatoes should be used sparingly. The new USDA food pyramid (b) is confusing to many people and seems to encourage more consumption of meat and milk.
Source: (a) Data from Willett and Stampfer, 2002 (b) USDA, 2005.

A few major crops supply most of our food

The three crops on which humanity depends for the majority of its nutrients and calories are wheat, rice, and maize (called corn in the United States). Together, more than 1,900 million metric tons of these three grains are grown each year. Wheat and rice are especially important since they are the staple foods for most of the 5.5 billion people in the developing countries of the world. These two grass species supply around 60 percent of the calories consumed directly by humans.

Potatoes, barley, oats, and rye are staples in mountainous regions and high latitudes (northern Europe, north Asia) because they grow well in cool, moist climates. Cassava, sweet potatoes, and other roots and tubers grow well in warm, wet areas and are staples in Amazonian, Africa, Melanesia, and the South Pacific. Barley, oats, and rye can grow in cool, short-season climates. Sorghum and millet are drought resistant and are staples in the dry regions of Africa.

Fruits, vegetables, and vegetable oils make a surprisingly large contribution to human diets. They are especially welcome because they typically contain high levels of vitamins, minerals, dietary fiber, and complex carbohydrates.

sugar). Based on observations of health effects of Mediterranean diets as well as a long-term study of 140,000 U.S. health professionals, Drs. Walter Willett and Meir Stampfer of Harvard University have recommended a new dietary pyramid (fig. 9.9a). Both red meat and starchy food such as white rice, white bread, potatoes, and pasta should be eaten sparingly. Nuts, legumes (beans, peas, and lentils), fruit, vegetables, and whole grain foods form the basis of this diet. Unsaturated plant oils should make up 30 to 40 percent of dietary calories, according to this view. Trans fat (the kind found in hydrogenated margarine), on the other hand, is not recommended at all. Combined with regular, moderate exercise, this food selection should provide most people with all the nutrition they need.

In 2005, the U.S. Department of Agriculture unveiled an alternative food pyramid (fig. 9.9b). Rather than give specific recommendations, the USDA pyramid groups foods together and shows some consumption of each food group at every level. Many people find this graphic confusing. Critics charge that it was developed by consultants who simultaneously represented the food industry, and that it overemphasizes meat and dairy while downplaying whole grains, vegetables, fruits, and unsaturated oils.

9.2 KEY FOOD SOURCES

Of the thousands of edible plants and animals in the world, only about a dozen types of seeds and grains, three root crops, twenty or so common fruits and vegetables, six mammals, two domestic fowl, and a few fish and other forms of marine life make up almost all of the food humans eat. Table 9.1 shows annual production of some important foods in human diets. In this section, we will highlight the characteristics of some important food sources.

TABLE 9.1

Some Important Food Sources

Crop	2006 Yield (Million Metric Tons)
Wheat	624
Rice (paddy)	635
Maize (corn)	683
Potatoes	320
Coarse grains*	1,009
Soybeans	222
Cassava and sweet potato	550
Sugar (cane and beet)	150
Pulses (beans, peas)	61
Oil seeds	397
Vegetables and fruits	1,220
Meat and milk	927
Fish and seafood	150

*Barley, oats, sorghum, rye, millet.
Source: Food and Agriculture Organization (FAO), 2006.

FIGURE 9.10 Meat and dairy consumption have quadrupled in the past 40 years, and China represents about 40 percent of that increased demand.

Meat and dairy are important protein sources

Protein-rich foods are prized by people nearly everywhere. In the past, the rich countries consumed a vast majority of the meat, dairy, and high-quality seafood traded internationally. The four-fifths of the world's people in less-developed countries generally raised 60 percent of the 3 billion domestic ruminants and 20 billion poultry in the world but consumed only one-fifth of all commercial animal products. The FAO reports, however, that as incomes rise in developing countries, food choices throughout the world are shifting toward higher-quality and more expensive foods.

Meat consumption in developing countries, for example, has risen from only 10 kg per person annually in the 1960s to 26 kg currently (fig. 9.10). By 2030, average annual meat consumption in those countries is expected to be nearly 40 kg per person.

Much of that increase in meat production could be provided by improved productivity in tropical areas. As the opening Case Study for this chapter shows, there has been a huge increase in beef and soy production in South America. Modern breeding techniques that combine the best qualities of Indian zebu (humped) cattle and beef-producing varieties from Europe and North America have provided animals that resist heat and tropical diseases but have meat characteristics preferred by consumers in developed countries (fig. 9.11a). Over the past 40 years, the area devoted to soybeans in Brazil has increased more than 20-fold (fig. 9.11b.), while the yield per hectare has doubled. Carving farmland and pasture out of tropical forest and savanna is helping to feed the world, but it is also reducing biological diversity and causing social conflict, as small subsistence farms are bought up by large-scale industrial agriculture, and landless farm workers are forced off the land into shantytowns (fig. 9.11c).

Soy farmers claim they aren't causing deforestation because they're only turning low-quality pasture, which was cleared years ago for ranching, into high-production row cropping. In 2006, Brazil's grain and processing industry declared a two-year moratorium on purchasing soybeans from farmers who deforest land in the Amazon. It's true that the land is more productive under intensive farming, but as grazing lands are turned into farms, ranchers and displaced rural families often move further into the forest in search of new land. In 2006, there were 1,690 violent clashes and 93 people were killed in disputes over landownership in Brazil, mostly involving indigenous or local people and large landowners.

Most of the livestock grown in North America are confined in **concentrated animal feeding operations (CAFOs)**

(a) Brazilian zebu cattle

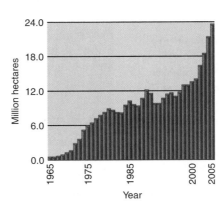

(b) Brazil soy area

(c) Favela

FIGURE 9.11 Beef and soy production in tropical countries, such as Brazil, could help satisfy the growing demand for meat in the developing world. Beef production and soy yields in Brazil have doubled in the past 40 years, while the area devoted to soybeans has increased more than 20-fold. This rapid expansion has had severe environmental costs, however. And thousands of rural families have been displaced to live in shanty towns called *favelas*.
Sources: (b) Data from USDA, 2005.

FIGURE 9.12 Concentrated feeding operations fatten animals quickly and efficiently, but create enormous amounts of waste and expose livestock to unhealthy living conditions.

for part or all of their lifetime. A diet rich in grain, oil, and protein fattens animals quickly and produces meat preferred by many consumers. Globally, some 680 million metric tons of cereals are used as livestock feed each year, representing more than one-third of the total maize, soy, and coarse grain production. It's often pointed out that we could feed about ten times as many people if we ate grain directly rather than feeding it to livestock. Surprisingly, the FAO claims that using cereals as animal feed does not contribute to hunger and under nutrition. Given current agricultural economics, if these cereals were not used as animal feed, they would probably not be produced at all, and thus would not be available as food. According to this view, if everyone became a vegetarian, the lack of demand for cereals for livestock production would simply lead to lower crop production, not more food for the hungry.

The rapid proliferation of CAFOs raises a number of social and environmental concerns. Housing up to a million animals in a single facility can cause serious local air and water pollution. Altogether, CAFOs in the United States are estimated to generate some 500 million metric tons of manure annually, or roughly 25 times all the human waste produced in the country each year (fig. 9.12). Storage and disposal of this waste is a serious problem. CAFOs are almost never served by sewage treatment plants. Land disposal of waste overloads soil with nutrients, and can contaminate food and groundwater supplies. In 2006, more than 200 people were sickened and four died from eating fresh spinach contaminated by run-off from a nearby cattle feed lot.

Local residents often find the stench of so many animals and so much waste overwhelming. A study by the University of North Carolina found that people living near large CAFOs suffer high levels of respiratory illness. Many communities have tried to restrict or regulate CAFOs, but find it difficult to do.

Animal wastes, particularly from hog farms, often are stored in huge open lagoons that can leak or rupture, poisoning local surface waters. In 1999, for example, Hurricane Floyd dumped torrential rains on North Carolina. An estimated 10 million m^3 (2.5 billion gal) of hog and poultry waste overflowed from storage lagoons into local rivers. Bacterial growth, which was stimulated by the manure, deoxygenated the water, killed millions of fish, and created a "dead zone" in Pamlico Sound that lasted for months.

The high density of animals in these facilities and the rich diet they're fed to speed weight gain requires a constant use of antibiotics and growth hormones. About 25 million pounds of antibiotics are added to animal feed annually in the United States, mostly in CAFOs. This is eight times as much as is used in human therapy. A big part of the rapid rise in antibiotic resistant pathogens is due to this massive and constant dosing of domestic livestock with drugs.

Seafood is another important protein source

We currently harvest about 95 million metric tons of wild fish and seafood every year, but we directly eat only about two-thirds of that amount. The rest is used as feed for captive-raised aquatic species, which now make up about half of the seafood we eat. Seafood is the main animal protein source for about 1.5 billion people in developing countries and is an important nutritional source for billions of others. Unfortunately, overharvesting and habitat destruction threaten most of the world's wild fisheries. Annual catches of ocean fish rose by about 4 percent annually between 1950 and 1988. Since 1989, however, 13 of 17 major marine fisheries have declined dramatically or become commercially unsustainable (see Data Analysis p. 205). An international team of marine biologists warn that if current trends continue, all the world's major fisheries will be exhausted by 2048.

The problem is too many boats using efficient but destructive technology to exploit a dwindling resource base. Boats as big as ocean liners travel thousands of kilometers and drag nets large enough to scoop up a dozen jumbo jets, sweeping a large patch of ocean clean of fish in a few hours. Long-line fishing boats set cables up to 10 km long with hooks every 2 meters that catch birds, turtles, and other unwanted "by-catch" along with targeted species. Trawlers drag heavy nets across the bottom, scooping up everything indiscriminately and reducing broad swaths of habitat to rubble. In some operations, up to 15 kg of dead and dying by-catch is dumped back into the ocean for every kilogram of marketable food. The FAO estimates that operating costs for the 4 million boats now harvesting wild fish exceed sales by (U.S.) $50 billion per year. Countries subsidize fishing fleets to preserve jobs and to ensure access to this valuable resource.

> ### *Think About It*
>
> Americans spend a smaller proportion of their income on food than any other industrial nation. A major reason for this is that we've industrialized our food supply in the same way that we produce cars, shoes, and television sets. Are there reasons that we should treat food differently from other commodities? What would be your ideal food production system? What could you do to attain that ideal?

Aquaculture (growing aquatic species in net pens or tanks) is of increasing importance in our diets. Fish can be grown in farm ponds that take relatively little space but are highly produc-

FIGURE 9.13 Pens for fish-rearing in Thailand.

tive. Cultivation of high-value carnivorous species, however, such as salmon, sea bass, and tuna threaten wild stocks exploited to stock captive operations or to provide fish food. Building coastal fish-rearing ponds causes destruction of hundreds of thousands of hectares of mangrove forests and wetlands that serve as irreplaceable nurseries for marine species. Net pens anchored in nearshore areas allow spread of diseases, escape of exotic species, and release of feces, uneaten food, antibiotics, and other pollutants into surrounding ecosystems (fig. 9.13).

Polyculture systems of mixed species of herbivores or filter feeders can alleviate many aquaculture problems. Raising species in enclosed, land-based ponds or warehouses can eliminate the pollution problems associated with net pens in lakes or oceans. In China, for example, most fish are raised in ponds or rice paddies. One ecologically balanced system uses four carp species that feed at different levels of the food chain. The grass carp, as its name implies, feeds largely on vegetation, while the common carp is a bottom feeder, living on detritus that settles to the bottom. Silver carp and bighead carp are filter feeders that feed on phytoplankton and zooplankton, respectively. Agricultural wastes such as manure, dead silkworms, and rice straw fertilize ponds and encourage phytoplankton growth. All these carp species are considered dangerous invasive species in North America. Still, these integrated polyculture systems typically boost fish yields per hectare by 50 percent or more compared with monoculture farming. Of the top ten seafood species recommended by the Monterey Aquarium as ecologically acceptable, six are farmed.

9.3 FARM POLICY

Much of the increase in food production over the past 50 years has been fueled by public support for agricultural education, research, and development projects such as irrigation systems, transportation networks, and crop insurance. While helping local farmers, agricultural subsidies also can distort markets and cripple production in developing countries. The World Bank estimates that rich countries pay their own farmers $350 billion per year, or nearly six times as much as all developmental aid to poor countries. A typical cow in Europe enjoys annual subsidies three times the average yearly income for most African farmers. In the past, European countries have had the highest aid per farm, but recent policy reforms have promised to decouple farm support from crop production.

Powerful political and economic interests protect agricultural assistance in many countries. Over the past decade, the United States, for example, has spent $143 billion in farm support. This aid is distributed very inequitably. According to the Environmental Working Group, 72 percent of all aid goes to the top 10 percent of recipients. One giant operation in Arkansas, for example, received $38 million over a five-year period. Aid also is concentrated geographically. Just 22 of the nation's 435 congressional districts (5 percent) collect more than half of all agricultural payments. Most of this aid goes to a few selected crops, such as corn, wheat, soybeans, rice, and cotton along with certain specially protected commodities including milk, sugar, and peanuts. Legislators claim that crop supports are intended to preserve "family farms," but critics claim this program really amounts to political payoff and corporate welfare. There have been repeated efforts to roll back agricultural payments, but Congress has been more inclined to slash conservation funds and food assistance for the poor rather than reduce rewards for their friends.

One unfortunate effect of these market interventions is that they tend to encourage unhealthy diets. High-calorie, fat- and sugar-rich, processed food is often cheaper and more readily available than fresh fruits and vegetables that would be better for us. Subsidies also allow American farmers to sell their products overseas at as much as 20 percent below the actual cost of production. When cheap commodities flood markets in developing countries, it drives local farmers out of business and destabilizes indigenous food production. The FAO argues that ending distorting financial support in the richer countries would have far more positive impact on local food supplies and livelihoods in the developing world than any aid program.

Every year millions of tons of topsoil and agricultural chemicals wash from U.S. farm fields into rivers, lakes, and, eventually, the ocean. Farmers know that erosion both impoverishes their land and pollutes water, but they're caught in a bind. For every $1 the U.S. government pays farmers to conserve soil and manage nutrients, it pays $7 to support commodities that promote soil loss and chemical runoff. The USDA estimates that farmers would shift 2.5 million ha (6 million acres) of row crops, which require intensive cultivation, into pasture grass and other perennial crops that minimize erosion if federal subsidies didn't favor certain row crops.

The United States already has an effective Conservation Reserve Program (CRP) that pays farmers to take highly erodible land out of production. Contracts expired in 2007 for more than 11 million ha (28 million acres) of this land. The USDA reports that CRP lands prevent the annual loss of 450 million tons of soil every year, protect 270,000 km (170,000 miles) of streams, and store 48 million tons of carbon per year. Just in the upper Midwest, CRP contracts will end on 6.5 million ha of critical wildlife habitat.

Many agronomists suggest that we should have more, not less CRP land. The United States could gradually shift payments from production subsidies to conservation programs that would truly support family farms while also protecting the environment. This could become a "whole farm" safety net that allows farmers to plant whatever crops are appropriate to market and environmental conditions rather than a few specific commodities. Farmers might find it more profitable to grow fruits and vegetables for local consumption rather than low-value grains. Nutrient management could be required as a condition of receiving crop aid.

We'll discuss soil conservation and low-input agriculture further in the next section of this chapter. In the meantime, please keep this question in mind: How can we support family farms and rural economies while also protecting our environment and helping our global neighbors feed themselves?

9.4 SOIL: A RENEWABLE RESOURCE

Growing the food and fiber needed to support human life is a complex enterprise that requires knowledge from many different fields and cooperation from many different groups of people. In this section, we'll survey some of the principles of soil science and look at some of the inputs necessary for continued agricultural production.

Of all the earth's crustal resources, the one we take most for granted is soil. We are terrestrial animals and depend on soil for life, yet most of us think of it only in negative terms. English is unique in using "soil" as an interchangeable word for earth and excrement. "Dirty" has a moral connotation of corruption and impurity. Perhaps these uses of the word enhance our tendency to abuse soil without scruples; after all, it's only dirt.

The truth is that **soil** is a marvelous substance, a living resource of astonishing beauty, complexity, and frailty. It is a complex mixture of weathered mineral materials from rocks, partially decomposed organic molecules, and a host of living organisms. It can be considered an ecosystem by itself. Soil is an essential component of the biosphere, and it can be used sustainably, or even enhanced, under careful management.

There are at least 15,000 different soil series or types in the United States and many thousands more worldwide. They vary because of the influences of parent material, time, topography, climate, and organisms on soil formation. There are young soils that, because they have not weathered much, are rich in soluble nutrients. There are old soils, like the red soils of the tropics, from which rainwater has washed away most of the soluble minerals and organic matter, leaving behind clay and rust-colored oxides.

To understand the potential for feeding the world on a sustainable basis we need to know how soil is formed, how it is being lost, and what can be done to protect and rebuild good agricultural soil. With careful husbandry, soil can be replenished and renewed indefinitely. Many farming techniques deplete soil nutrients, however, and expose the soil to the erosive forces of wind and moving water. As a result, in many places we are essentially mining this resource and using it much faster than it is being replaced.

Building good soil is a slow process. Under the best circumstances, good topsoil accumulates at a rate of about 10 tons per hectare (2.5 acres) per year—enough soil to make a layer about 1 mm deep when spread over a hectare. Under poor conditions, it can take thousands of years to build that much soil. Perhaps one-third to one-half of the world's current croplands are losing topsoil faster than it is being replaced. In some of the worst spots, erosion carries away about 2.5 cm (1 in.) of topsoil per year. With losses like that, agricultural production has already begun to fall in many areas.

Soil is a complex mixture

Most soil is about half mineral. The rest is plant and animal residue, air, water, and living organisms. The mineral particles are derived either from the underlying bedrock or from materials transported and deposited by glaciers, rivers, ocean currents, windstorms, or landslides. The weathering processes that break rocks down into soil particles are described in chapter 14.

Particle sizes affect the characteristics of the soil (fig. 9.14). The spaces between sand particles give sandy soil good drainage and usually allow it to be well aerated, but also cause it to dry out quickly when rains are infrequent. Tight packing of small particles in silty or clay soils makes them less permeable to air and water than sandy soils. Tiny capillary spaces between the particles, on the other hand, store water and mineral ions better than more porous soils. Because clay particles have a proportionately large surface area and a high ionic charge, they stick together tenaciously, giving clay its slippery plasticity, cohesiveness, and impermeability. Soils

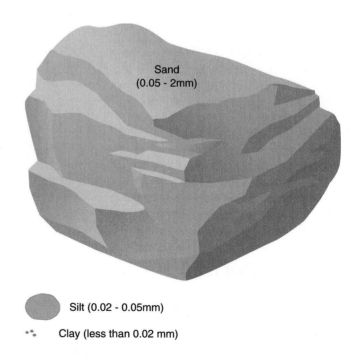

Sand (0.05 - 2mm)

Silt (0.02 - 0.05mm)

Clay (less than 0.02 mm)

FIGURE 9.14 Relative sizes of soil particles magnified about 100-fold.

with a high clay content are called "heavy soils," in contrast to easily worked "light soils" that are composed mostly of sand or silt. Varying proportions of these mineral particles occur in each soil type (fig. 9.15). Farmers usually consider sandy loam the best soil type for cultivating crops.

The organic content of soil can range from nearly zero for pure sand, silt, or clay, to nearly 100 percent for peat or muck, which is composed mainly of partly decomposed plant material. Much of the organic material in soil is **humus,** a sticky, brown, insoluble residue from the partially decomposed bodies of dead plants and animals. Humus is the most significant factor in the development of "structure," a description of how the soil particles clump together. Humus coats mineral particles and holds them together in loose crumbs, giving the soil a spongy texture that holds water and nutrients needed by plant roots, and maintains the spaces through which delicate root hairs grow.

Living organisms create unique properties of soil

Without soil organisms, the earth would be covered with sterile mineral particles far different from the rich, living soil ecosystems on which we depend for most of our food. The activity of the myriad organisms living in the soil help create structure, fertility, and tilth (condition suitable for tilling or cultivation) (fig. 9.16).

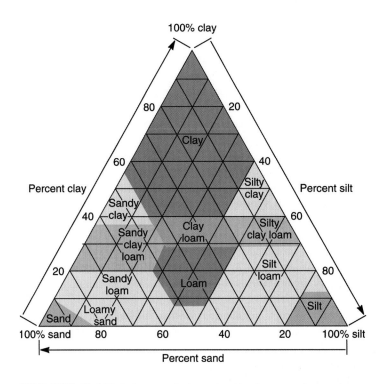

FIGURE 9.15 Soil texture is determined by the percentages of clay, silt, and sand particles in the soil. Soils with the best texture for most crops are loams, which have enough larger particles (sand) to be loose, yet enough smaller particles (silt and clay) to retain water and dissolved mineral nutrients.

FIGURE 9.16 Soil ecosystems include numerous consumer organisms, as depicted here: (1) snail, (2) termite, (3) nematodes and nematode-killing constricting fungus, (4) earthworm, (5) wood roach, (6) centipede, (7) carabid (ground) beetle, (8) slug, (9) soil fungus, (10) wireworm (click beetle larva), (11) soil protozoan, (12) sow bug, (13) ant, (14) mite, (15) springtail, (16) pseudoscorpion, and (17) cicada nymph.

Soil organisms usually stay close to the surface, but that thin living layer can contain thousands of species and billions of individual organisms per hectare. Algae live on the surface, while bacteria and fungi flourish in the top few centimeters of soil. A single gram of soil (about one-half teaspoon) can contain hundreds of millions of these microscopic cells. Algae and blue-green bacteria capture sunlight and make new organic compounds. Bacteria and fungi decompose organic detritus and recycle nutrients that plants can use for additional growth. The sweet aroma of freshly turned soil is caused by actinomycetes, bacteria that grow in funguslike strands and give us the antibiotics streptomycin and tetracyclines.

Micorrhizal symbiosis is an association between the roots of most plant species and certain fungi. The plant provides organic compounds to the fungus, while water and inorganic nutrients are absorbed from the soil by the fungus and transferred to the plant. As a result, micorrhizal plants often grow better—especially in infertile soil—than those lacking fungal partners. There can be 20 m of fungal strands in a gram of soil.

Roundworms, segmented worms, mites, and tiny insects swarm by the thousands in that same gram of soil from the surface. Some of them are herbivorous, but many of them prey upon one another. Soil roundworms (nematodes) attack plant rootlets and can cause serious crop damage. A carnivorous fungus snares nematodes with tiny loops of living cells that constrict like a noose when a worm blunders into it. Burrowing animals, such as gophers, moles, insect larvae, and worms, tunnel deeper in the soil, mixing and aerating it. Plant roots also penetrate lower soil levels, drawing up soluble minerals and secreting acids that decompose mineral particles. Fallen plant litter adds new organic material to the soil, returning nutrients to be recycled.

Soils are layered

Most soils are stratified into horizontal layers called **soil horizons** that reveal much about the history and usefulness of the soil. The thickness, color, texture, and composition of each horizon are used to classify the soil. Together these horizons make up a **soil profile** (fig. 9.17).

The soil surface is often covered with a layer of leaf litter, crop residues, or other fresh or partially decomposed organic material. This organic layer is known as the O horizon. Below this layer is the A horizon or **topsoil,** composed of mineral particles mixed with organic material. The A horizon can range from several meters thick under virgin prairie to almost nothing in dry deserts. The O and A horizons contain most of the living organisms and organic material in the soil, and it is in these layers that most plants spread their roots to absorb water and nutrients. An eluviated (leached) E horizon often lies below the A horizon. This layer is depleted of clays and soluble nutrients, which are removed by rainwater seeping down through the soil. These clays and nutrients generally accumulate in the B horizon, or **subsoil.** The B horizon may have a dense or clayey texture because of the accumulated clays. In desert regions, a dense, impermeable "hardpan" layer of accumulated minerals or salts may develop on the B horizon. This hardpan blocks plant root growth and prevents water from draining properly.

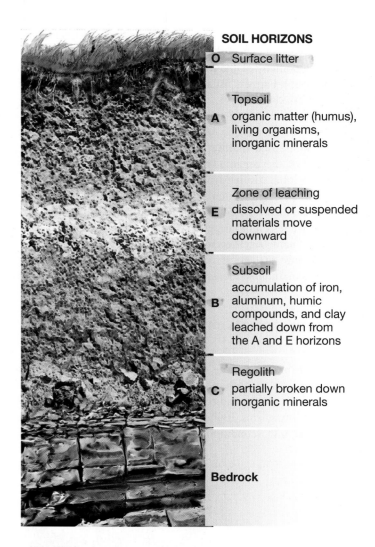

SOIL HORIZONS

O Surface litter

Topsoil
A organic matter (humus), living organisms, inorganic minerals

Zone of leaching
E dissolved or suspended materials move downward

Subsoil
B accumulation of iron, aluminum, humic compounds, and clay leached down from the A and E horizons

Regolith
C partially broken down inorganic minerals

Bedrock

FIGURE 9.17 Soil profile showing possible soil horizons. The actual number, composition, and thickness of these layers varies in different soil types.

Beneath the subsoil is parent material, called the C horizon or **regolith,** made of weathered rock fragments with very little organic material. Weathering in the C horizon, largely accomplished by rain water mixed with organic compounds from plants, produces new soil particles and allows downward expansion of the soil profile. Parent material can be sand, solid rock, or other material. About 70 percent of the parent material in the United States was transported to its present site by glaciers, wind, or water and is not directly related to the underlying bedrock.

Soils are classified according to their structure and composition

In the United States, soils are classified into 12 *soil orders.* The richest farming soils, with thick, organic-rich A horizons are *mollisols* (formed under grasslands) and *alfisols* (formed under deciduous forests). Both of these develop where rainfall and temperatures are moderate. *Spodosols* also develop in temperate climates, but usually under pine forests, where acidic needle litter causes a characteristic whitish, ashy-looking E horizon. In hot and rainy environments, *oxisols* and *ultisols* develop. These soils are severely depleted of nutrients and tend to be reddish because iron-rich minerals make up much of the remaining soil material. *Aridosols* form in arid environments, have little organic material, and often contain accumulated salt or hardpan layers. Some soil types are defined mainly by parent material: *andisols* develop from volcanic material, and *vertisols* develop from clay-rich material such as lake beds or shale bedrock. *Histosols* are formed of water-logged, incompletely decayed plant material. Other soils are largely defined by degree of development; *entisols* and *inceptisols* have little or no horizon development. This may be because these soils are on recently exposed parent material (such as glacial debris) or because they are in an environment where soil development is extremely slow because of slight rainfall or organic activity. A relatively new soil order is *gelisols*, soils in permafrost areas. These soils occur only at high latitudes or high altitudes.

FIGURE 9.18 In many areas, soil or climate constraints limit agricultural production. These hungry goats in Sudan feed on a solitary *Acacia* shrub.

9.5 WAYS WE USE AND ABUSE SOIL

Only about 12.5 percent of the earth's land area (16.6 million km² out of a total of 132.4 million km²) is currently in agricultural production. Perhaps four times as much land could potentially be converted to cropland, but much of this land serves as a refuge for cultural or biological diversity or suffers from constraints, such as steep slopes, shallow soils, poor drainage, tillage problems, low nutrient levels, metal toxicity, or excess soluble salts or acidity, that limit the types of crops that can be grown there (fig. 9.18). Agronomists worry that global climate change (chapter 15) could alter weather patterns and flood low-lying coastal areas so that world food production could be seriously reduced.

Arable land is unevenly distributed

In parts of Canada, the United States, and Europe, temperate climates, abundant water, and high soil fertility produce high crop yields that contribute to high standards of living. Other parts of the world, although rich in land area, lack suitable soil, topography, water, or climate to sustain our levels of productivity.

Worldwide, the cropland available for agriculture is shrinking. In 1970, there was a global average of 0.38 ha per person. By 2002, this had declined to 0.21 ha per person. The FAO forecasts that by 2030 the land supply will shrink to 0.16 ha per person. In Asia, cropland will be even more scarce—0.09 ha (0.22 acre) per person—30 years from now. If you live on a typical quarter-acre suburban lot, look at your yard and imagine feeding yourself for a year on what you could produce there.

In the developed countries, 95 percent of recent agricultural growth in the twentieth century has come from improved crop varieties or increased fertilization, irrigation, and pesticide use, rather than from bringing new land into production. In fact, less land is being cultivated now than 100 years ago in North America, or 600 years ago in Europe. As more effective use of labor, fertilizer, and water and improved seed varieties have increased in the more developed countries, productivity per unit of land has increased, and marginal land has been retired, mostly to forests and grazing lands. In many developing countries, land continues to be cheaper than other resources, and new land is still being brought under cultivation, mostly at the expense of forests and grazing lands. Still, at least two-thirds of recent production gains have come from new crop varieties and more intense cropping rather than expansion into new lands.

The largest increases in cropland over the last 30 years occurred in South America and Oceania where forests and grazing lands—such as Brazil's Cerrado, described in this chapter's opening case study—are rapidly being converted to farms. Many developing countries are reaching the limit of lands that can be exploited for agriculture without unacceptable social and environmental costs, but others still have considerable potential for opening new agricultural lands. East Asia, for instance, already uses about three-quarters of its potentially arable land. Most of its remaining land has severe restrictions for agricultural use. Further increases in crop production will probably have to come from higher yields per hectare. Latin America, by contrast, uses only about one-fifth of its potential land, and Africa uses only about one-fourth of the land that theoretically could grow crops. However, there would be serious ecological trade-offs in putting much of this land into agricultural production.

Land degradation reduces agricultural potential

Agriculture both causes and suffers from environmental degradation. The International Soil Reference and Information Centre in the Netherlands estimates that every year 3 million ha (7.4 million acres) of cropland are ruined by erosion, 4 million ha are turned into deserts, and 8 million ha are converted to nonagricultural uses such as homes, highways, shopping centers, factories, reservoirs, etc.

Over the past 50 years, some 1.9 *billion* ha of agricultural land (an area greater than that now in production) have been degraded to some extent. About 300 million ha of this land are strongly degraded (deep gullies, severe nutrient depletion, crops grow poorly, restoration is difficult and expensive), while 910 million ha—about the size of China—are moderately degraded. Nearly 9 million ha of former croplands are so degraded that they no longer support any crop growth at all. The causes of this extreme degradation vary: In Ethiopia it is water erosion, in Somalia it is wind, and in Uzbekistan salt and toxic chemicals are responsible.

Definitions of degradation are based on both biological productivity and our expectations about what the land should be like. Often this is a subjective judgment and it is difficult to distinguish between human-caused deterioration and natural processes like drought. We generally consider the land degraded when the soil is impoverished or eroded, water runs off or is contaminated more than is normal, vegetation is diminished, biomass production is decreased, or wildlife diversity diminishes. On farmlands this results in lower crop yields. On ranchlands it means fewer livestock can be supported per unit area. On nature reserves it means fewer species.

The amount and degree of land degradation varies by region and country. About 20 percent of land in Africa and Asia is degraded, but most is in either the light or moderate category. In Central America and Mexico, by contrast, 25 percent of all vegetated land suffers moderate to extreme degradation. Figure 9.19 shows some areas of greatest concern for soil degradation in the United States, as well as the prime mechanisms for this problem. Water and wind erosion provide the motive force for the vast majority of all soil degradation, worldwide. Chemical deterioration includes nutrient depletion, salinization (salt accumulation), acidification, and pollution. Physical deterioration includes compaction by heavy machinery or trampling by cattle, waterlogging—water accumulation—from excess irrigation and poor drainage, and laterization—solidification of iron and aluminum-rich tropical soil when exposed to sun and rain.

Soil erosion is widespread

Erosion is an important natural process, resulting in the redistribution of the products of geologic weathering, and is part of both soil formation and soil loss. The world's landscapes have been sculpted by erosion. When the results are spectacular enough, we enshrine them in national parks as we did with the Grand Canyon. Where erosion has worn down mountains and spread soil over the plains, or deposited rich alluvial silt in river bottoms, we gladly farm it. Erosion is a disaster only when it occurs in the wrong place at the wrong time.

In some places, erosion occurs so rapidly that anyone can see it happen. Deep gullies are created where water scours away the soil, leaving fenceposts and trees sitting on tall pedestals as the land erodes away around them. In most places, however, erosion is more subtle. It is a creeping disaster that occurs in small increments. A thin layer of topsoil is washed off fields

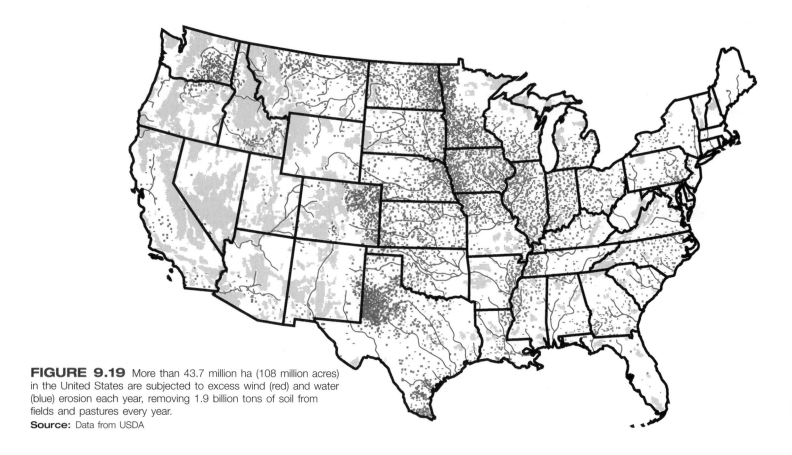

FIGURE 9.19 More than 43.7 million ha (108 million acres) in the United States are subjected to excess wind (red) and water (blue) erosion each year, removing 1.9 billion tons of soil from fields and pastures every year.
Source: Data from USDA

year after year until eventually nothing is left but poor-quality subsoil that requires more and more fertilizer and water to produce any crop at all.

The total annual soil loss from croplands is thought to be 25 billion metric tons. About twice that much soil is lost from rangelands, forests, and urban construction sites each year. The net effect, worldwide, of this general, widespread topsoil erosion is a reduction in crop production equivalent to removing about 1 percent of world cropland each year. Many farmers are able to compensate for this loss by applying more fertilizer and by bringing new land into cultivation.

In addition to reduced land fertility, this erosion results in sediment-loading of rivers and lakes, siltation of reservoirs, smothering of wetlands and coral reefs, and clogging of water intakes and waterpower turbines.

Wind and water are the main agents that move soil

A thin layer taken off the land surface is called **sheet erosion.** When little rivulets of running water gather together and cut small channels in the soil, the process is called **rill erosion** (fig. 9.20a). When rills enlarge to form bigger channels or ravines that are too large to be removed by normal tillage operations, we call the process **gully erosion** (fig. 9.20b). Streambank erosion refers to the washing away of soil from the banks of established streams, creeks, or rivers, often as a result of removing trees and brush along streambanks and by cattle damage to the banks.

Most soil loss on agricultural land is sheet or rill erosion. Large amounts of soil can be transported a little bit at a time without being very noticeable. A farm field can lose 20 metric tons of soil per hectare during winter and spring runoff in rills so small that they are erased by the first spring cultivation. That represents a loss of only a few millimeters of soil over the whole surface of the field,

hardly apparent to any but the most discerning eye. But it doesn't take much mathematical skill to see that if you lose soil twice as fast as it is being replaced, eventually it will run out.

Wind can equal or exceed water in erosive force, especially in a dry climate and on relatively flat land. When plant cover and surface litter are removed from the land by agriculture or grazing, wind lifts loose soil particles and sweeps them away. In extreme conditions, windblown dunes encroach on useful land and cover roads and buildings (fig. 9.20c). Over the past 30 years, China has lost 93,000 km² (about the size of Indiana) to **desertification, or conversion of productive land to desert.** Advancing dunes from the Gobi desert are now only 160 km (100 mi) from Beijing. Every year more than 1 million tons of sand and dust blow from Chinese drylands, often traveling across the Pacific Ocean to the West Coast of North America.

Some of the highest erosion rates in the world occur in the United States and Canada. The U.S. Department of Agriculture reports that 69 million hectares (170 million acres) of U.S. farmland and range are eroding at rates that reduce long-term productivity. Five tons per acre (11 metric tons per hectare) is generally considered the maximum tolerable rate of soil loss because that is generally the highest rate at which soil forms under optimum conditions. Some farms lose soil at twice that rate or more.

Intensive farming practices are largely responsible for this situation. Row crops, such as corn and soybeans, leave soil exposed for much of the growing season (fig. 9.21). Deep plowing and heavy herbicide applications create weed-free fields that look neat but are subject to erosion. Because big machines cannot easily follow contours, they often go straight up and down the hills, creating ready-made gullies for water to follow. Farmers sometimes plow through grass-lined watercourses and have pulled out windbreaks and fencerows to accommodate the large machines and to get every last square meter into production. Consequently, wind and water carry away the topsoil.

(a) Sheet and rill erosion

(b) Gullying

(c) Wind erosion and desertification

FIGURE 9.20 Land degradation affects more than 1 billion ha yearly, or about two-thirds of all global cropland. Water erosion (a) and (b) account for about half that total. Wind erosion affects a nearly equal area (c).

FIGURE 9.21 Annual row crops leave soil bare and exposed to erosion for most of the year, especially when fields are plowed immediately after harvest, as this one always is.

Pressed by economic conditions, many farmers have abandoned traditional crop rotation patterns and the custom of resting land as pasture or fallow every few years. Continuous monoculture cropping can increase soil loss tenfold over other farming patterns. A soil study in Iowa showed that a three-year rotation of corn, wheat, and clover lost an average of about 6 metric tons per hectare. By comparison, continuous wheat production on the same land caused nearly four times as much erosion, and continuous corn cropping resulted in seven times as much soil loss as the rotation with wheat and clover. The Mississippi River carries enough topsoil and fertilizer every year to create a "dead zone" in the Gulf of Mexico that can be as large as 57,000 km². Algal growth stimulated by high nitrogen in runoff from farms and cities depletes oxygen within this zone to levels that are lethal for most marine life. A task force recommended a 20 to 30 percent decrease in nitrogen loading to reduce the size and effects of this zone. Similar hypoxic zones occur near mouths of many other rivers that drain agricultural areas (chapter 18).

Deserts are spreading around the world

According to the United Nations, about one-third of the earth's surface and the livelihoods of at least one billion people are threatened by **desertification** (conversion of productive lands to desert), which contributes to food insecurity, famine, and poverty. UN Secretary General Kofi Annan calls this a "creeping catastrophe" that creates millions of environmental refugees every year. Forced by economic circumstances to overcultivate and overgraze their land, poor people often are both the agents and victims of desertification.

Rangelands and pastures, which generally are too dry for cultivation, are highly susceptible to desertification. According to the UN, 80 percent of the world's grasslands are suffering from overgrazing and soil degradation, and three-quarters of that area has undergone some degree of desertification. The world's 3 billion domestic grazing animals provide livelihood and food for many people, but can have severe environmental effects.

Two areas of particular concern are Africa and China. Arid lands, where rains are sporadic and infrequent and the economy is based mainly on crop and livestock raising, make up about two-thirds of Africa. Nearly 400 million people live around the edges of these deserts. Rapid population growth and poverty create unsustainable pressures on the fragile soils of these areas. Stripping trees and land cover for fodder and firewood exposes the soil to erosion and triggers climate changes that spread desertification, which now affects nearly three-quarters of the arable land in Africa to some extent. The fringes of the two great African deserts, the Sahara and the Kalahari, are particularly vulnerable. Nearly every year there is drought somewhere in Africa. About one-third of the 60 million people who required food aid in 2005 were victims of drought and desertification.

In China, large areas of natural woodland were cleared in the twentieth century to create farmland and provide timber for industry. Especially in the 1950s, trees were cut to fuel backyard iron smelters. This rampant deforestation has been blamed for disastrous summer floods on the Yangtze and Yellow Rivers and severe spring sandstorms that threaten the health of millions of people. Overgrazing and cultivation of fragile lands also contribute to desertification. The advancing Gobi Desert is now within 150 km (100 mi) from Beijing; and the Taklimakan, Asia's largest desert, is spreading eastward to unite with the Gobi. Sandstorms from these deserts have grown from about 5 per year in the 1960s to 24 per year in the 1990s.

China is trying to fight the spread of deserts with an ambitious ecological restoration program. Since 1985, more than 50 billion trees have been planted over an area the size of Germany. Every Chinese citizen between the ages of 11 and 60 is required to plant 3 to 5 trees per year. Many saplings aren't watered, however, and survival rates are reported to be low. Eventually this gargantuan program plans to create a "Green Wall" thousands of kilometers long that will stabilize soils, clear the air, and stop the advancing deserts. China is especially anxious to improve air quality in Beijing before the 2008 Olympics, as we discussed in chapter 1. We'll look at other methods to control soil erosion in our discussion of sustainable agriculture.

9.6 OTHER AGRICULTURAL RESOURCES

Soil is only part of the agricultural resource picture. Agriculture is also dependent upon water, nutrients, favorable climates to grow crops, productive crop varieties, and upon the mechanical energy to tend and harvest them.

All plants need water to grow

Agriculture accounts for the largest single share of global water use. About two-thirds of all fresh water withdrawn from rivers, lakes, and groundwater supplies is used for irrigation (chapter 17). Although estimates vary widely (as do definitions of irrigated land), about 15 percent of all cropland, worldwide, is irrigated.

Some countries are water rich and can readily afford to irrigate farmland, while other countries are water poor and must use water very carefully. The efficiency of irrigation water use is rather low in most countries. High evaporative and seepage losses from unlined and uncovered canals often mean that in some

FIGURE 9.22 Downward-facing sprinklers on this center-pivot irrigation system deliver water more efficiently than upward-facing sprinklers.

places, up to 80 percent of water withdrawn for irrigation never reaches its intended destination (chapter 17). Farmers often tend to over-irrigate because water prices are relatively low and because they lack the technology to meter water and distribute just the amount needed. In the United States and Canada, many farmers are adopting water-saving technologies such as drip irrigation or downward-facing sprinklers (fig. 9.22).

Excessive use not only wastes water; it often results in **waterlogging.** Waterlogged soil is saturated with water, and plant roots die from lack of oxygen. **Salinization,** in which mineral salts accumulate in the soil, occurs particularly when soils in dry climates are irrigated with saline water. As the water evaporates, it leaves behind a salty crust on the soil surface that is lethal to most plants. Flushing with excess water can wash away this salt accumulation but the result is even more saline water for downstream users.

Worldwide, irrigation problems are a major source of land degradation and crop losses. The Worldwatch Institute reports that 60 million ha (150 million acres) of cropland have been damaged by salinization and waterlogging. Water conservation techniques can greatly reduce problems arising from excess water use. Conservation also makes more water available for other uses or for expanded crop production where water is in short supply (chapter 17).

Many agronomists worry that global climate change (chapter 15) could reduce food production in many countries. It's expected that global warming will bring too little rain to some areas and too much to others. Warmer temperatures could extend growing seasons at high latitudes, but dry soils and stress crops elsewhere.

Plants need fertilizer

In addition to water, sunshine, and carbon dioxide, plants need small amounts of inorganic nutrients for growth. The major elements required by most plants are nitrogen, potassium, phosphorus, calcium, magnesium, and sulfur. Calcium and magnesium often are limited in areas of high rainfall and must be supplied in the form of lime. Lack of nitrogen, potassium, and phosphorus even more often limits plant growth. Adding these elements in fertilizer usually stimulates growth and greatly increases crop yields. A good deal of the doubling in worldwide crop production since 1950 has come from increased inorganic fertilizer use. In 1950, the average amount of fertilizer used was 20 kg per hectare. In 1990, this had increased to an average of 91 kg per hectare worldwide.

Farmers may overfertilize because they are unaware of the specific nutrient content of their soils or the needs of their crops. While European farmers use more than twice as much fertilizer per hectare as do North American farmers, their yields are not proportionally higher. Phosphates and nitrates from farm fields and cattle feedlots are a major cause of aquatic ecosystem pollution. Nitrate levels in groundwater have risen to dangerous levels in many areas where intensive farming is practiced. Young children are especially sensitive to the presence of nitrates. Using nitrate-contaminated water to mix infant formula can be fatal for newborns.

What are some alternative ways to fertilize crops? Manure and green manure (crops grown specifically to add nutrients to the soil) are important natural sources of soil nutrients. Nitrogen-fixing bacteria living symbiotically in root nodules of legumes are valuable for making nitrogen available as a plant nutrient (chapter 3). Interplanting or rotating beans or some other leguminous crop with such crops as corn and wheat are traditional ways of increasing nitrogen availability.

There is considerable potential for increasing world food supply by increasing fertilizer use in low-production countries if ways can be found to apply fertilizer more effectively and reduce pollution. Africa, for instance, uses an average of only 19 kg of fertilizer per hectare, or about one-fourth of the world average. It has been estimated that the developing world could at least triple its crop production by raising fertilizer use to the world average.

Farming consumes energy

Farming as it is generally practiced in the industrialized countries is highly energy-intensive. Fossil fuels supply almost all of this energy. Between 1920 and 1980, direct energy use on farms rose as gasoline and diesel fuels were consumed by increasing mechanization of agriculture. An even greater increase in indirect energy use, in the form of synthetic fertilizers, pesticides, and other agricultural chemicals, also occurred, especially after World War II.

After crops leave the farm, additional energy is used in food processing, distribution, storage, and cooking. It has been estimated that the average food item in the American diet travels 2,400 km between the farm that grew it and the person who consumes it. The energy required for this complex processing and distribution system may be five times as much as is used directly in farming. Altogether the food system in the United States consumes about 16 percent of the total energy we use. Most of our foods require more energy to produce, process, and get to market than they yield when we eat them. A British study concluded that eating locally grown food has less environmental impact—even if produced with conventional farming—than organic food from far away.

FIGURE 9.23 Biofuel crops, such as these sunflowers, can provide a good cash crop for farmers and contribute to national energy supplies.

Farmers could assist in moving to a renewable energy future by growing energy crops that can be converted into biofuels. Many Midwestern states have encouraged construction of corn- or soy-based ethanol factories in recent years. Mixing ethanol with gasoline helps reduce air pollution, and having a new market helps struggling farmers. There has been much debate about whether corn-based ethanol produces more energy than is required to grow and process the crop. State-of-the-art facilities do, in fact, appear to have a 150 percent net energy yield. Plant oils from sunflowers, soybeans, corn, and other oil seed crops can be burned directly in most diesel engines, and may be closer than ethanol to a net energy gain (fig. 9.23). Europe plans to produce 5.7 percent of its transportation energy from biofuels by 2010. Fast-growing trees, kenaf, switch grass, cattails, and other energy crops may be grown on marginal land and burned directly in power plants or used to create ethanol. Many American farmers are finding that their most lucrative crop may be the wind (chapter 20).

9.7 NEW CROPS AND GENETIC ENGINEERING

Although at least 3,000 species of plants have been used for food at one time or another, most of the world's food now comes from only 16 species. Many new or unconventional varieties might be valuable human food supplies, however, especially in areas where conventional crops are limited by climate, soil, pests, or other problems. The FAO predicts that 70 percent of future world production growth will come from higher yields and new crop varieties rather than expansion of arable lands. Among the plants now being investigated as potential additions to our crop roster is the winged bean (fig. 9.24), a perennial plant that grows well in hot climates where other legumes will not grow. It is totally edible (pods, mature seeds, shoots, flowers, leaves, and tuberous roots), resistant to diseases, and enriches the soil. Another promising crop is tricale, a hybrid between wheat (*Triticum*) and rye (*Secale*) that grows in light, sandy, infertile soil. It is drought resistant, has

FIGURE 9.24 Winged beans bear fruit year-round in tropical climates and are resistant to many diseases that prohibit growing other bean species. Whole pods can be eaten when they are green, or dried beans can be stored for later use. It is a good protein source in a vegetarian diet.

nutritious seeds, and is being tested for salt tolerance for growth in saline soils or irrigation with seawater. Some traditional crop varieties grown by Native Americans, such as tepary beans, amaranth, and Sonoran panicgrass are being collected by seed conservator Gary Nabhan both as a form of cultural revival for native people and as a possible food crop for harsh environments.

The "green revolution" produced dramatic increases in crop yields

So far, the major improvements in farm production have come from technological advances and modification of a few well-known species. Yield increases often have been spectacular. A century ago, when all maize (corn) in the United States was open-pollinated, average yields were about 25 bushels per acre. In 1999, average yields from hybrid maize were around 130 bushels per acre, and under optimum conditions, 250 bushels per acre are possible. Most of this gain was accomplished

FIGURE 9.25 Semi-dwarf wheat (*right*), bred by Norman Borlaug, has shorter, stiffer stems and is less likely to lodge (fall over) when wet than its conventional cousin (*left*). This "miracle" wheat responds better to water and fertilizer, and has played a vital role in feeding a growing human population.

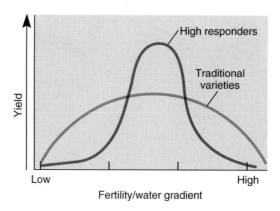

FIGURE 9.26 Green revolution miracle crops are really high responders, meaning that they have excellent yields under optimum conditions. For poor farmers who can't afford the fertilizer and water needed by high responders, traditional varieties may produce better yields.

by conventional plant breeding: Geneticists laboriously hand-pollinating plants, moving selected genes from one variety to another.

Starting about 50 years ago, agricultural research stations began to breed tropical wheat and rice varieties that would provide food for growing populations in developing countries. The first of the "miracle" varieties was a dwarf, high-yielding wheat developed by Norman Borlaug (who received a Nobel Peace Prize for his work) at a research center in Mexico (fig. 9.25). At about the same time, the International Rice Institute in the Philippines developed dwarf rice strains with three or four times the production of varieties in use at the time. The dramatic increases obtained as these new varieties spread around the world has been called the **green revolution.** It is one of the main reasons that world food supplies have more than kept pace with the growing human population over the past few decades.

Most green revolution breeds really are "high responders," meaning that they yield more than other varieties if given optimum levels of fertilizer, water, and protection from pests and diseases (fig. 9.26). Under suboptimum conditions, on the other hand, high responders may not produce as well as traditional varieties. Poor farmers who can't afford the expensive seed, fertilizer, and water required to become part of this movement, usually are left out of the green revolution. In fact, they may be driven out of farming altogether as rising land values and falling commodity prices squeeze them from both sides.

The first advances under the green revolution were for grains, such as wheat, corn, and rice, grown in temperate climates of wealthy countries. Little research was devoted to tropical crops. Recently, however, the Gates and Rockefeller Foundations announced a new initiative for research on uniquely African crops and agricultural technology. One aim of this research is to create hardy "climate-ready" crops in anticipation of global climate change.

Genetic engineering uses molecular techniques to produce new crop varieties

Genetic engineering involves removing genetic material from one organism and splicing it into the chromosomes of another (fig. 9.27). This new technology has the potential to greatly increase both the quantity and quality of our food supply. It is now possible to build entirely new genes, and even new organisms, taking a bit of DNA from here, a bit from there, even synthesizing artificial DNA sequences to create desired characteristics in **genetically modified organisms (GMOs).**

Proponents predict dramatic benefits from genetic engineering. Research is now underway to improve yields and create crops that resist drought, frost, or diseases. Other strains are being developed to tolerate salty, waterlogged, or low-nutrient soils, allowing degraded or marginal farmland to become productive. All of these could be important for reducing hunger in developing countries. Plants that produce their own pesticides might reduce the need for toxic chemicals; while engineering for improved protein or vitamin content could make our food more nutritious. Attempts to remove specific toxins or allergens from crops also could make our food safer. Crops such as bananas and potatoes have been altered to contain oral vaccines that can be grown in developing countries where refrigeration and sterile needles are unavailable. Plants have been engineered to make industrial oils and plastics. Animals, too, are being genetically modified to grow faster, gain weight on less food, and produce pharmaceuticals such as insulin in their milk. It may soon be possible to create animals with human cell-recognition factors that could serve as organ donors.

Opponents worry that moving genes willy-nilly could create a host of problems, some of which we can't even imagine. GMOs themselves might escape and become pests or they might interbreed with wild relatives. In either case, we may create superweeds or reduce native biodiversity. Constant presence of pesticides in plants could accelerate pesticide resistance in insects or leave toxic residues in soil or our food. Genes for toxicity or

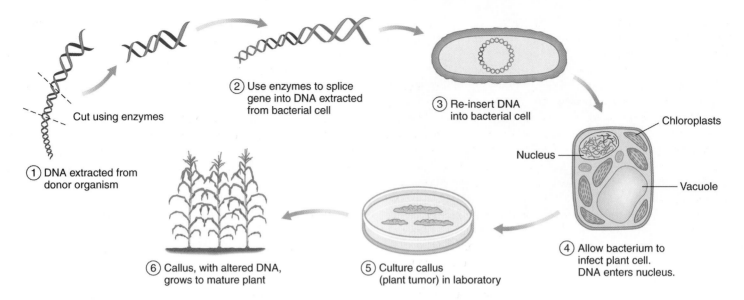

① DNA extracted from donor organism

Cut using enzymes

② Use enzymes to splice gene into DNA extracted from bacterial cell

③ Re-insert DNA into bacterial cell

Chloroplasts

Nucleus

Vacuole

④ Allow bacterium to infect plant cell. DNA enters nucleus.

⑤ Culture callus (plant tumor) in laboratory

⑥ Callus, with altered DNA, grows to mature plant

FIGURE 9.27 One method of gene transfer, using an infectious, tumor-forming bacterium such as agrobacterium. Genes with desired characteristics are cut out of donor DNA and spliced into bacterial DNA using special enzymes. The bacteria then infect plant cells and carry altered DNA into cells' nuclei. The cells multiply, forming a tumor, or callus, which can grow into a mature plant.

allergies could be transferred along with desirable genes, or novel toxins could be created as genes are mixed together. This technology may be available only to the richest countries or the wealthiest corporations, making family farms uncompetitive and driving developing countries even further into poverty.

Both the number of GMO crops and the acreage devoted to growing them is increasing rapidly. Between 1990 and 2006, more than 11,373 field tests for some 750 different crop varieties were carried out in the United States (fig. 9.28). Hawaii is the

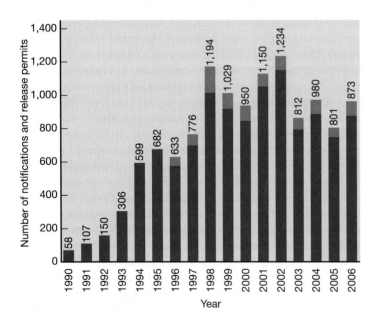

FIGURE 9.28 Transgenic crop field releases in the United States 1990 to 2006. Approved or acknowledged (*blue*) versus denied or withdrawn (*red*).
Source: Data from Information Systems for Biotechnology, Virginia Tech University.

most popular site for GMO testing, with about 1,500 field releases over the past 15 years. Illinois, Iowa, California, and Puerto Rico each have had more than 1,000 tests during that time. Worldwide, 90 million ha (about 25 percent of all cultivated land) were planted with GMO crops in 2006. The United States accounted for 56 percent of that acreage, followed by Argentina with 19 percent. Canada, Brazil, and China together make up 21 percent of all transgenic cropland. In 2005, China approved GMO rice for commercial production. This is the first GMO cereal grain approved for direct human consumption, and could move China into the forefront of GMO crop production.

Some transgenic crops have reached mainstream status. Currently about 82 percent of all soybeans, 71 percent of the cotton, and one-quarter of all maize (corn) grown in the United States are GMOs. You've already probably eaten some genetically modified food. Estimates are that at least 60 percent of all processed food in America contains GMO ingredients. Since the United States, Argentina, and Brazil account for 90 percent of international trade in maize and soybeans (in which GMOs often are mixed with other grains), a large fraction of the world most likely has been exposed to some GMO products.

Most GMOs have been engineered for pest resistance or weed control

Biotechnologists recently have created plants with genes for endogenous insecticides. *Bacillus thuringiensis* (Bt), a bacterium, makes toxins lethal to Lepidoptera (butterfly family) and Coleoptera (beetle family). The genes for some of these toxins have been transferred into crops such as maize (to protect against European cut worms), potatoes (to fight potato beetles), and cotton (for protection against boll weevils). This allows farmers to reduce insecticide spraying. Arizona cotton farmers, for example, report reducing their use of chemical insecticides by 75 percent. Small farms in India report an 80 percent yield increase with Bt cotton compared to neighboring plots growing conventional cotton.

Entomologists worry that Bt plants churn out toxin throughout the growing season, regardless of the level of infestation, creating perfect conditions for selection of Bt resistance in pests. The effectiveness of this natural pesticide—one of the few available to organic growers—is likely to be destroyed within a few years. One solution is to plant at least a part of every field in non-Bt crops that will act as a refuge for non-resistant pests. The hope is that interbreeding between these "wild-type" bugs and those exposed to Bt will dilute out recessive resistance genes. Deliberately harboring pests and letting them munch freely on crops is something that many farmers find hard to do. In addition, devoting a significant part of their land to nonproductive crops lowers the total yield and counteracts the profitability of engineered seed.

There also is a concern about the effects on nontarget species. In laboratory tests, about half of a group of monarch butterfly caterpillars died after being fed on plants dusted with pollen from Bt corn. Under field conditions, however, it was difficult to show any harm to butterflies. Critics still worry, however, that there may be unexpected problems associated with this new technology.

The other major group of transgenic crops are engineered to tolerate high doses of herbicides. Currently these crops occupy 20 million hectares worldwide, or about three-quarters of all genetically engineered acreage. The two main products in this category are Monsanto's "Roundup Ready" crops—so-called because they can withstand treatment with Monsanto's best-selling herbicide, Roundup (glyphosate)—and AgrEvo's "Liberty Link" crops, which resist that company's Liberty (glufosinate) herbicide. Because crops with these genes can grow in spite of high herbicide doses, farmers can spray fields heavily to exterminate weeds. This practice allows for conservation tillage and leaving more crop residue on fields to protect topsoil from erosion, which are both good ideas, but it also means using much more herbicide in higher doses than might otherwise be done.

Is genetic engineering safe?

Ever since scientists discovered how to move genes from one organism to another, critics have worried about irresponsible use of this technology or unforeseen consequences arising from novel combinations of genetic material. Environmental and consumer groups have campaigned against transgenic organisms, labeling them "Frankenfoods." Industry groups, on the other hand, describe their critics as Neo-Luddites, who blindly oppose any new technology. In 2002, while millions of its people faced famine, Zambia's government refused to accept thousands of tons of genetically modified maize from the United States, claiming that it might be unsafe for human consumption. Most European nations have bans on genetically engineered crops. The United States has filed a suit at the World Trade Organization claiming that European policies constitute an unwarranted restraint on trade. How can we decide what to believe in this welter of claims and counterclaims?

At the base of some people's unease with genetic modification is a feeling that it simply isn't right to mess with nature. It doesn't seem proper to recombine genes any way we want. Doing so raises specters of gruesome monsters stitched together from spare parts, like Mary Shelley's Frankenstein. Is this merely a fear of science, or is it a valid ethical issue?

The U.S. Food and Drug Administration declined to require labeling of foods containing GMOs, saying that these new varieties are "substantially equivalent" to related varieties bred via traditional practices. After all, proponents say, we have been moving genes around for centuries through plant and animal breeding. All domesticated organisms should be classified as genetically modified, some people argue. Biotechnology may be a more precise way of creating novel organisms than normal breeding procedures. We're moving only a few genes—or even part of one gene—at a time with biotechnology, rather than hundreds of unknown genes through classical techniques.

What if GMOs escape and interbreed with native species? In a ten-year study of genetically modified crops, a group from Imperial College, London, concluded that GMOs tested so far do not survive well in the wild, and are no more likely to invade other habitats than their unmodified counterparts. Other scientists counter that some GMO crops already have been shown to spread their genes to nearby fields. Normal rapeseed (canola oil) varieties in Canada were found to contain genes from genetically modified varieties in nearby fields. It isn't clear, however, if the genes involved will have any adverse effect or whether they will persist for many generations. In 2001, researchers reported finding traces of genetically modified corn in wild relatives in Mexico, which supposedly has banned planting of GMOs, especially in regions where species were thought to have originated. This report was later withdrawn due to criticism of the techniques used to detect genetic markers, but worries that GMO genes could spread to wild relatives continues to be among the greatest concerns about this technology.

The first genetically modified animal designed to be eaten by humans is an Atlantic salmon containing extra growth hormone genes from an oceanic pout. The greatest worry from this fish is not that it will introduce extra hormones into our diet—that's already being done by chickens and beef that get extra growth hormone via injections or their diet—but rather the ecological effects if the fish escape from captivity. The transgenic fish grow seven times faster and are more attractive to the opposite sex than a normal salmon. If they escape from captivity, they may outcompete already endangered wild relatives for food, mates, and habitat. Fish farmers say they will grow only sterile females and will keep them in secure net pens. Opponents point out that salmon frequently escape from aquaculture operations, and that if just a few fertile transgenic fish break out it could be catastrophic for wild stocks.

Many people argue that we should take a better-safe-than-sorry "precautionary approach" that errs—if at all—on the side of safety. The 5,000-member Society of Toxicology has concluded that, "Based on available tests, there's no reason to suspect that transgenic plants differ in any substantive way from traditional varieties." On the other hand, a panel of biologists and agricultural scientists convened by the U.S. National Academy of Sciences

What Do You Think?

Shade-Grown Coffee and Cocoa

Has it ever occurred to you that your purchases of coffee and chocolate may be contributing to the protection or destruction of tropical forests? Both coffee and cocoa are examples of food products grown exclusively in developing countries but consumed almost entirely in the wealthy nations (vanilla and bananas are some other examples). Coffee grows in cool, mountain areas of the tropics, while cocoa is native to the warm, moist lowlands. Both are small trees of the forest understory, adapted to low light levels.

Until a few decades ago, most of the world's coffee and cocoa were grown under a canopy of large forest trees. Recently, however, new varieties of both crops have been developed that can be grown in full sun. Yields for sun-grown crops are higher because more coffee or cocoa trees can be crowded into these fields, and they get more solar energy than in a shaded plantation.

There are costs, however, in this new technology. Sun-grown trees die earlier from the stress and diseases common in these fields. Furthermore, ornithologists have found that the number of bird species can be cut in half in full-sun plantations, and the number of individual birds may be reduced by 90 percent. Shade-grown coffee and cocoa generally require fewer pesticides (or sometimes none) because the birds and insects residing in the forest canopy eat many of the pests. Shade-grown plantations also need less chemical fertilizer because many of the plants in these complex forests add nutrients to the soil. In addition, shade-grown crops rarely need to be irrigated because heavy leaf fall protects the soil, while forest cover reduces evaporation.

Cocoa pods grow directly on the trunk and large branches of cocoa trees.

Currently, about 40 percent of the world's coffee and cocoa plantations have been converted to full-sun varieties and another 25 percent are in process. Traditional techniques for coffee and cocoa production are worth preserving. Thirteen of the world's 25 biodiversity hot spots occur in coffee or cocoa regions. If all the 20 million ha (49 million acres) of coffee and cocoa plantations in these areas are converted to monocultures, an incalculable number of species will be lost.

The Brazilian state of Bahia is a good example of both the ecological importance of these crops and how they might help preserve forest species. At one time, Brazil produced much of the world's cocoa, but in the early 1900s, the crop was introduced into West Africa. Now Côte d'Ivoire alone grows more than 40 percent of the world total, and the value of Brazil's harvest has dropped by 90 percent. Côte d'Ivoire is aided in this competition by a labor system that reportedly includes widespread child slavery. Even adult workers in Côte d'Ivoire get only about $165 per year (if they get paid at all) compared to a minimum wage of $850 per year in Brazil. As African cocoa production ratchets up, Brazilian landowners are converting their plantations to pastures or other crops.

The area of Bahia where cocoa was once king is part of Brazil's Atlantic forest, one of the most threatened forest biomes in the world. Only 8 percent of this forest remains undisturbed. Although cocoa plantations don't represent the full diversity of intact forests, they protect a surprisingly large sample of what once was there. And shade-grown cocoa can provide an economic rationale for preserving that biodiversity. Brazilian cocoa will probably never compete with that from other areas for lowest cost. There is room in the market, however, for specialty products. If consumers were willing to pay a small premium for organic, fair-trade, shade-grown chocolate and coffee it might provide the incentive needed to preserve biodiversity. Wouldn't you like to know that your chocolate or coffee wasn't grown with child slavery, and is helping protect plants and animal species that might otherwise go extinct?

urged the government to more carefully, and more publically, review the potential environmental impacts of genetically modified plants and animals before approving them for commercial use.

Finally, there are social and economic implications of GMOs. Will they help feed the world or will they lead to a greater consolidation of corporate power and economic disparity? Might higher yields and fewer losses to pests and diseases allow poor farmers in developing countries to stop using marginal land and avoid cutting down forests to create farmland? Is this simply a technological fix or could it help promote agricultural sustainability? Critics suggest that there are simpler and cheaper ways other than high-tech crop varieties to provide nutrients to people in developing countries or to increase the income of poor, rural families. Adding a cow or a fishpond or training people in water harvesting or regenerative farming techniques (as we'll discuss

in the next section) may have a longer-lasting impact than selling them expensive new seeds.

On the other hand, if we hope to reduce malnutrition and feed 9 billion people in 50 years, maybe we need all the tools we can get. Where do you stand in this debate? What additional information would you need to reach a reasoned judgment about the risks and benefits of this new technology?

Think About It

Suppose your grandmother asks you, "what's all this controversy about GMOs? Are they safe or not?" Could you summarize the arguments for and against genetic engineering in a few sentences? Which are the most important issues in this debate, in your opinion?

9.8 SUSTAINABLE AGRICULTURE

How, then, shall we feed the world? Can we make agriculture compatible with sustainable ecological and social systems? Having discussed some of the problems that beset modern agriculture, we will now consider some suggested ways to overcome problems and make farming and food production just and lasting enterprises. This goal is usually termed **sustainable agriculture,** regenerative farming, or agroecology, all of which aim to produce food and fiber on a sustainable basis and repair the damage caused by destructive practices. Some alternative methods are developed through scientific research; others are discovered in traditional cultures and practices nearly forgotten in our mechanization and industrialization of agriculture.

Soil conservation is essential

With careful husbandry, soil is a renewable resource that can be replenished and renewed indefinitely. Since agriculture is the area in which soil is most essential and also most often lost through erosion, agriculture offers the greatest potential for soil conservation and rebuilding. Some rice paddies in Southeast Asia, for instance, have been farmed continuously for a thousand years without any apparent loss of fertility. The rice-growing cultures that depend on these fields have developed management practices that return organic material to the paddy and carefully nurture the soil's ability to sustain life.

While American agriculture hasn't reached that level of sustainability, there is evidence that soil conservation programs are having a positive effect. In one Wisconsin study, erosion rates in one small watershed were 90 percent less in 1975–1993 than they were in the 1930s. Among the most important elements in soil conservation are land management, ground cover, climate, soil type, and tillage system.

Managing Topography

Water runs downhill. The faster it runs, the more soil it carries off the fields. Comparisons of erosion rates in Africa have shown that a 5 percent slope in a plowed field has three times the water runoff volume and eight times the soil erosion rate of a comparable field with a 1 percent slope. Water runoff can be reduced by leaving grass strips in waterways and by **contour plowing,** that is, plowing across the hill rather than up and down. Contour plowing is often combined with **strip farming,** the planting of different kinds of crops in alternating strips along the land contours (fig. 9.29). When one crop is harvested, the other is still present to protect the soil and keep water from running straight downhill. The ridges created by cultivation make little dams that trap water and allow it to seep into the soil rather than running off. In areas where rainfall is very heavy, intersecting, or "tied," ridges are often useful. This method involves a series of ridges running at right angles to each other, so that water runoff is blocked in all directions and is encouraged to soak into the soil.

Terracing involves shaping the land to create level shelves of earth to hold water and soil. The edges of the terrace are

FIGURE 9.29 Contour plowing and strip cropping help prevent soil erosion on hilly terrain as well as create a beautiful landscape.

planted with soil-anchoring plant species. This is an expensive procedure, requiring either much hand labor or expensive machinery, but makes it possible to farm very steep hillsides. Rice terraces in Asia create beautiful landscapes as well as highly productive and sustainable agroecosystems (fig. 9.30).

FIGURE 9.30 Terracing, as in these Balinese rice paddies, can control erosion and make steep hillsides productive.

Planting **perennial species** (plants that grow for more than two years) is the only suitable use for some lands and some soil types. Establishing forest, grassland, or crops such as tea, coffee, or other crops that do not have to be cultivated every year may be necessary to protect certain unstable soils on sloping sites or watercourses (low areas where water runs off after a rain).

Providing Ground Cover

Annual row crops such as corn or beans generally cause the highest erosion rates because they leave soil bare for much of the year (table 9.2). Often, the easiest way to provide cover that protects soil from erosion is to leave crop residues on the land after harvest. They not only cover the surface to break the erosive effects of wind and water, but they also reduce evaporation and soil temperature in hot climates and protect ground organisms that help aerate and rebuild soil. In some experiments, 1 ton of crop residue per acre (0.4 ha) increased water infiltration 99 percent, reduced runoff 99 percent, and reduced erosion 98 percent. Leaving crop residues on the field also can increase disease and pest problems, however, and may require increased use of pesticides and herbicides.

Where crop residues are not adequate to protect the soil or are inappropriate for subsequent crops or farming methods, such **cover crops** as rye, alfalfa, or clover can be planted immediately after harvest to hold and protect the soil. These cover crops can be plowed under at planting time to provide green manure. Another method is to flatten cover crops with a roller and drill seeds through the residue to provide a continuous protective cover during early stages of crop growth.

In some cases, interplanting of two different crops in the same field not only protects the soil but also is more efficient use of the land, providing double harvests. Native Americans and pioneer farmers, for instance, planted beans or pumpkins between the corn rows. The beans provided nitrogen needed by the corn, pumpkins crowded out weeds, and both crops provided foods that nutritionally balance corn. Traditional swidden (slash-and-burn) cultivators in Africa and South America often plant as many as 20 different crops together in small plots. The crops mature at different times so that there is always something to eat, and the soil is never exposed to erosion for very long.

Mulch is a general term for a protective ground cover that can include manure, wood chips, straw, seaweed, leaves, and other natural products. For some high-value crops, such as tomatoes, pineapples, and cucumbers, it is cost-effective to cover the ground with heavy paper or plastic sheets to protect the soil, save water, and prevent weed growth. Israel uses millions of square meters of plastic mulch to grow crops in the Negev Desert.

Reduced Tillage

Farmers have traditionally used a moldboard plow to till the soil, digging a deep trench and turning the topsoil upside down. In the 1800s, it was shown that tilling a field fully—until it was "clean"—increased crop production. It helped control weeds and pests, reducing competition; it brought fresh nutrients to the surface, providing a good seedbed; and it improved surface drainage and aerated the soil. This is still true for many crops and many soil types, but it is not always the best way to grow crops. We are finding that less plowing and cultivation often makes for better water management, preserves soil, saves energy, and increases crop yields.

There are several major **reduced tillage systems.** *Minimum till* involves reducing the number of times a farmer disturbs the soil by plowing, cultivating, etc. This often involves a disc or chisel plow rather than a traditional moldboard plow. A chisel plow is a curved chisel-like blade that doesn't turn the soil over but creates ridges on which seeds can be planted. It leaves up to 75 percent of plant debris on the surface between the rows, preventing erosion (fig. 9.31). *Conserv-till* farming uses a coulter, a

FIGURE 9.31 In ridge tilling, a chisel plow is used to create ridges on which crops are planted and shallow troughs filled with crop residue. Less energy is used in plowing and cultivation, weeds are suppressed, and moisture is retained by the ground cover, or crop residue, left on the field.

TABLE 9.2

Soil Cover and Soil Erosion

Cropping System	Average Annual Soil Loss (Tons/Hectare)	Percent Rainfall Runoff
Bare soil (no crop)	41.0	30
Continuous corn	19.7	29
Continuous wheat	10.1	23
Rotation: corn, wheat, clover	2.7	14
Continuous bluegrass	0.3	12

Source: Based on 14 years' data from Missouri Experiment Station, Columbia, MO.

sharp disc like a pizza cutter, which slices through the soil, opening up a furrow or slot just wide enough to insert seeds. This disturbs the soil very little and leaves almost all plant debris on the surface. *No-till* planting is accomplished by drilling seeds into the ground directly through mulch and ground cover. This allows a cover crop to be interseeded with a subsequent crop.

Farmers who use these conservation tillage techniques often must depend on pesticides (insecticides, fungicides, and herbicides) to control insects and weeds. Increased use of toxic agricultural chemicals is a matter of great concern. Massive use of pesticides is not, however, a necessary corollary of soil conservation. It is possible to combat pests and diseases with integrated pest management that combines crop rotation, trap crops, natural repellents, and biological controls (chapter 10).

Low-input agriculture can be good for farmers and their land

In contrast to the trend toward industrialization and dependence on chemical fertilizers, pesticides, antibiotics, and artificial growth factors common in conventional agriculture, some farmers are going back to a more natural, agroecological farming style. Finding that they can't—or don't want to—compete with factory farms, these folks are making money and staying in farming by returning to small-scale, low-input agriculture. The Minar family, for instance, operate a highly successful 150-cow dairy operation on 97 ha (240 acres) near New Prague, Minnesota. No synthetic chemicals are used on their farm. Cows are rotated every day between 45 pastures or paddocks to reduce erosion and maintain healthy grass. Even in the winter, livestock remain outdoors to avoid the spread of diseases common in confinement (fig. 9.32). Antibiotics are used only to fight diseases. Milk and meat from this operation are marketed through co-ops and a community-supported agriculture (CSA) program. Sand Creek, which flows

FIGURE 9.32 On the Minar family's 230-acre dairy farm near New Prague, Minnesota cows and calves spend the winter outdoors in the snow, bedding down on hay. Dave Minar is part of a growing counterculture that is seeking to keep farmers on the land and bring prosperity to rural areas.

across the Minar land, has been shown to be cleaner when leaving the farm than when it enters.

Research at Iowa State University has shown that raising animals on pasture grass rather than grain reduces nitrogen runoff by two-thirds while cutting erosion by more than half. If more Midwestern farmers followed the Minar's example, we could easily eliminate the Gulf of Mexico "Dead Zone" (chapter 18).

Similarly, the Franzen family, who raise livestock on their organic farm near Alta Vista, Iowa, allow their pigs to roam in lush pastures, where they can supplement their diet of corn and soybeans with grasses and legumes. Housing for these happy hogs is in spacious, open-ended hoop structures. As fresh layers of straw are added to the bedding, layers of manure beneath are composted, breaking down into odorless organic fertilizer.

Low-input farms such as these typically don't turn out the quantity of meat or milk that their intensive agriculture neighbors do, but their production costs are lower, and they get higher prices for their crops, so that the all-important net gain is often higher. The Franzens, for example, calculate that they pay 30 percent less for animal feed, 70 percent less for veterinary bills, and half as much for buildings and equipment as their neighboring confinement operations. And on the Minar's farm, erosion after an especially heavy rain was measured to be 400 times lower than a conventional farm nearby.

Preserving small-scale, family farms also helps preserve rural culture. As Marty Strange of the Center for Rural Affairs in Nebraska asks, "Which is better for the enrollment in rural schools, the membership of rural churches, and the fellowship of rural communities—two farms milking 1,000 cows each or twenty farms milking 100 cows each?" Family farms help keep rural towns alive by purchasing machinery at the local implement dealer, gasoline at the neighborhood filling station, and groceries at the mom-and-pop grocery store.

Consumers' choices play an important role

Adopting a vegetarian, organic diet can help reduce environmental impacts of the food you consume, but an even more environmentally friendly choice may be to become a **locavore** (that is, one who eats locally grown, seasonal food). Buying locally supports family farms and local economies. Where conventional foods were shipped an average of 2,400 km (1,500 mi) to markets in one frequently cited study, the average food item at a farmers' market traveled only 72 km (45 miles). Having any kind of food we want whenever we want it is gratifying, but the industrialized, global agriculture on which we have come to depend requires huge amounts of energy for fertilizer, fuel for farm equipment and shipping, plastics for food packaging, and climate control during shipping, storing, and ripening. While payments at a farmers' market go directly to many individual producers, most profits from conventional foods are retained by a tiny number of giant food corporations (the top three or four corporations in each commodity group typically control between 60 to 80 percent of the U.S. market—and often the same handful of companies, or their affiliates, dominate most food categories).

Co-ops often make an effort to carry locally grown and processed food. An even better way to know where your food comes from and how it's produced is to join a **community-supported agriculture (CSA)** program in which you make an annual contribution to a local farm in return for weekly deliveries of a "share" of whatever the farm produces. CSA farms generally practice organic or low-input agriculture, and they often invite members to visit and learn how their food is grown. A share often provides most fresh fruits and vegetables that an average family would consume during the growing season at a far lower price than similar food (especially if it's organic) would cost at a co-op or conventional grocery store. Much of America's most fertile land is around major cities. Farm failure fuels sprawl. Small-scale, artesinal farms are one of the best ways to preserve rural landscapes around metropolitan areas. There are now at least 1,500 CSA farms in the United States. You should be able to find one in your area.

Agroecology, or sustainable farming using ecological knowledge, also can be applied effectively in poorer countries. Brian Haliweil, of the Worldwatch Institute, reports that farmers in the Guatemalan Highlands raised their crop yields tenfold without the use of chemical fertilizers or pesticides. Encouraged by World Neighbors, a nongovernmental development organization, to find innovative, local solutions to their problems, the farmers came up with low-cost techniques such as planting cover crops and grass strips to reduce erosion, or rotating corn with beans, peas, and other legumes to add nitrogen to the soil. Corn harvests increased from 0.4 tons to 4.5 tons per hectare, yields comparable to many richer countries.

One of the largest experiments in low-input farming is currently taking place in Cuba, where loss of aid from the former Soviet Union coupled with a trade embargo imposed by the United States have forced farmers to turn to organic, nonmechanized agriculture (chapter 10).

CONCLUSION

World food supplies have increased dramatically over the past half century. Despite the fact that human populations have grown nearly threefold during that time, there's now more than enough food available for everyone. Nevertheless, at least 850 million people don't have enough to eat on a daily basis, and many more don't have the right kind of food. Whether we can meet the United Nations Millennium goal of cutting world hunger in half by 2020 remains to be seen. Declining fossil fuel supplies, soil erosion, and global climate change may make it difficult to continue to feed everyone. Ironically, there are now probably more obese people (1 billion) in the world than there are who have too few calories.

Billions of hectares of agricultural and pasture lands suffer soil degradation caused by erosion, nutrient depletion, salinization, and other forms of abuse. It's estimated that 25 billion metric tons of soil are lost from croplands every year. This not only reduces productivity, but also causes air and water pollution. Better farming techniques, including terracing, contour plowing, strip cropping, reduced tillage, providing ground cover, and planting perennial species could conserve our precious soil resources.

Some encouraging progress in low-input farming, community-supported agriculture, and agroecology show us how we can live more harmoniously with the land while still producing the food and fiber that we need. As pioneering ecologist Aldo Leopold wrote in *A Sand County Almanac,* "A thing is right only when it tends to preserve the integrity, stability and beauty of the community; and the community includes the soil, water, fauna and flora, as well as the people."

REVIEWING LEARNING OUTCOMES

By now you should be able to explain the following points:

9.1 Describe world food supplies and nutritional requirements.

- Millions of people don't have enough to eat.
- Famines are acute food emergencies.
- We need the right kinds of food.
- Eating a balanced diet is essential for good health.

9.2 Compare key food sources.

- A few major crops supply most of our food.
- Meat and dairy are important protein sources.
- Seafood is another important protein source.

9.3 Understand how farm policy can help or hurt soil conservation.

9.4 Illustrate why soil is a renewable resource.

- Soil is a complex mixture.
- Living organisms create unique properties of soil.
- Soils are layered.
- Soils are classified according to their structure and composition.

9.5 Summarize the ways we use and abuse soil.

- Arable land is unevenly distributed.
- Land degradation reduces agricultural potential.
- Soil erosion is widespread.
- Wind and water are the main agents that move soil.
- Deserts are spreading around the world.

9.6 Identify other agricultural resources.

- All plants need water to grow.
- Plants need fertilizer.
- Farming consumes energy.

9.7 Explain new crops and genetic engineering.

- The "green revolution" produced dramatic increases in crop yields.
- Genetic engineering uses molecular techniques to produce new crop varieties.

- Most GMOs have been engineered for pest resistance or weed control.
- Is genetic engineering safe?
- How could shade-grown coffee and cocoa save wildlife?

9.8 Appraise sustainable agriculture.

- Soil conservation is essential.
- Low-input agriculture can be good for farmers and their land.
- Consumers' choices play an important role.

PRACTICE QUIZ

1. How many people in the world are chronically undernourished and how many die each year from starvation and nutritionally related diseases?
2. What is the relation between poverty and food security?
3. Define *malnutrition* and *obesity*. How many Americans are considered overweight?
4. How have recommendations for a balanced diet changed over the years?
5. What do we mean by the "green revolution"?
6. Why are wild fish populations endangered? How might aquaculture contribute to our food supply?
7. What is the composition of soil? What is humus? Why are soil organisms so important?
8. What are four kinds of erosion? Why is erosion a problem?
9. What is genetic engineering, or biotechnology, and how might they help or hurt agriculture and the environment?
10. What is sustainable agriculture?

CRITICAL THINKING AND DISCUSSION QUESTIONS

1. Promoters claim that South American lands that are being converted to soybean farms had already been transformed from native savanna to pasture long ago. Critics argue that ranching is simply shifting to virgin lands, and that biodiversity will suffer. How would you weigh the need of growing populations for high-quality protein with the value of pristine nature?
2. Suppose that an agribusiness wants to start a CAFO near your home. What regulations or safeguards would you want to see imposed on its operation? How would you weigh the possible costs and benefits of this operation?
3. Debate the claim that famines are caused more by human actions (or inactions) than by environmental forces. What is the critical element or evidence in this debate?
4. Should farmers be forced to use ecologically sound techniques that serve farmers' best interests in the long run, regardless of

short-term consequences? How could we mitigate hardships brought about by such policies?
5. Should we try to increase food production on existing farmland, or should we sacrifice other lands to increase farming areas?
6. Some rice paddies in Southeast Asia have been cultivated continuously for a thousand years or more without losing fertility. Could we, and should we, adapt these techniques to our own country?
7. Outline the arguments you would make to your family and friends for why they should buy shade-grown, fair-trade coffee and cocoa. How much of a premium would you pay for these products?

 **DATA analysis**   **Using Relative Values**

The graph in this box shows changes in abundance of marine species in six oceanic regions between 1970 and 2000. Look at the scales on the axes. Do you see anything unusual about them? The X-axis is simply five-year intervals. Nothing unusual there. But what are the units for the Y-axis? What is a species index, and why

do all the curves start at 100? Essentially, the species index is a percent change from the abundance of species in each region in 1970. Why use an index rather than absolute numbers? We do this because the abundance values vary so much between regions. It's difficult to fit the full range of real numbers of species on the Y-axis scale, and if we did so, the curves would be so far apart that comparison would be difficult. What we're mostly interested in here are the comparative changes and the trends in different areas. A relative scale is useful because it makes it easy to see that species abundance in most areas has declined sharply over this 30-year period as well as where the losses have been greatest.

1. In which regions have species abundance remained relatively constant?

2. What do you think may be responsible for the sharp declines shown here?

3. Can you think of any reason that these regions have experienced less decline than other places?

4. How much has marine abundance declined in the South Atlantic during the period shown?

5. What is the significance of the thick gray line?

6. Why are some of the curves dotted at their ends?

7. How would you describe the marine biodiversity history of the North Atlantic over the time shown?

8. How do you think the agenda of the WWF, which created this graph, might be furthered by the way it's presented?

9. Might some other organization be likely to present the data in another way?

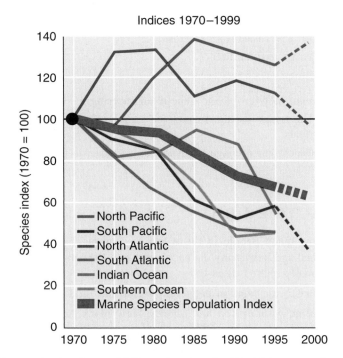

Indices 1970–1999

Legend:
- North Pacific
- South Pacific
- North Atlantic
- South Atlantic
- Indian Ocean
- Southern Ocean
- Marine Species Population Index

Species index (1970 = 100)

Changes in abundance of 217 species of marine fish, birds, mammals, and reptiles, from 1970 to 1999, by continent.
Source: World Wide Fund for Nature (WWF), 2000.

For Additional Help in Studying This Chapter, please visit our website at <u>www.mhhe.com/cunningham10e</u>. You will find additional practice quizzes and case studies, flashcards, regional examples, place markers for Google Earth™ mapping, and an extensive reading list, all of which will help you learn environmental science.

Pollinators, such as honeybees, provide services worth billions of dollars per year to farmers and orchardists. Pesticides can help protect pollinators from parasites, but can also have unexpected side effects.

C H A P T E R

10

Pest Control

It ain't the things we know that cause all the trouble; it's the things we think we know that ain't so.

—Will Rogers—

LEARNING OUTCOMES

After studying this chapter, you should be able to:

10.1 Define the major types of pesticides and describe the pest they are meant to manage.

10.2 Identify some important pesticide benefits.

10.3 Illustrate some important pesticide problems.

10.4 Summarize some alternatives to current pesticide uses.

10.5 Discuss how we might reduce pesticide exposure.

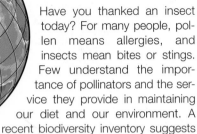

Case Study The Forgotten Pollinators

Have you thanked an insect today? For many people, pollen means allergies, and insects mean bites or stings. Few understand the importance of pollinators and the service they provide in maintaining our diet and our environment. A recent biodiversity inventory suggests that about 90 percent of all flowering plants depend on insects, birds, bats, and other animal species to assist in delivering the pollen that they need to create seeds. At least one-third of the crops we eat rely on animal pollinators for successful reproduction. The annual value of this ecological service is estimated to be (U.S.) $40 billion. At least 200,000 kinds of animal play a role in assisting plant reproduction, and the vast majority are wild species, many of which play roles unrecognized by science.

For decades, biologists have been worried about declining populations of both wild and domesticated pollinators. Many factors contribute to pollinator disappearance, including habitat loss, climate change, introduction of exotic species, and the spread of diseases, but one of the most important threats is the indiscriminate use of pesticides that eliminate beneficial species along with the pests they are intended to target. Few people realize that the United States now applies twice the amount of pesticides used when Rachel Carson wrote *Silent Spring* in 1962.

A great deal of attention has been focused recently on honeybees. These industrious little insects not only make honey, but they play a vital role in fertilizing crops. In fact, their importance in spreading pollen is worth at least 100 times the value of the honey they make. Bee-pollinated foods include squash, tomatoes, peppers, apples, and pears. Farmers pay commercial beekeepers to bring hives to their fields to pollinate crops. The $1.6 billion annual California almond crop, for example, is totally dependent on rented honeybee colonies to make a successful crop. Unfortunately, there's a shortage of honeybees. The United States has fewer than half as many honeybee colonies now as it did in the 1940s. In 2005, there was a severe shortage of honeybees in California. Beekeepers traveled from as far away as Florida and North Carolina to service California's almond groves. For the first time in 50 years, the U.S. borders were opened to honeybees from New Zealand and Australia.

Native pollinators are just as important for crops as domesticated ones. In the 1970s, the Canadian government began spraying forests in New Brunswick in an attempt to control spruce budworm outbreaks. Blueberry farmers noticed that their crop yields decreased about 1 million pounds per year compared to before spraying started. In the late 1970s, forest spraying was stopped, and blueberry harvests gradually returned to normal. At about the same time, ranchers in the western United States noticed that alfalfa crops (a staple food for dairy cows and other livestock) were failing. Research showed that alkali bees, a wild, solitary, ground-nesting species, were pollinating the alfalfa. Farmers believed that a shift in pesticide use on neighboring crops did the bees in, or perhaps the alfalfa farmer's own use of insecticides to combat pests was responsible. In any case, almost all the $5 billion annual U.S. alfalfa crop depends on leafcutter bees, which are imported from Canada.

This case study introduces some important themes for this chapter. Pesticides are an important tool in disease prevention, food preservation, and property protection. On the other hand, our battle to protect ourselves from biological enemies often has unintended consequences. In many cases, we need more thoughtful, sustainable methods of accomplishing our goals. In this chapter, we'll look at both some benefits and costs of using powerful pesticides, as well as some new ways to use them more effectively.

For more information see

Klein, Alexandra-Maria, et al. 2006. Importance of pollinators in changing landscapes for world crops. *Proceedings of the Royal Society B* 3721 27 Oct. 2006, Page FirstCite, DOI 10.1098/rspb. 2006.3721, URL http://dx.doi.org/10.1098/rspb.2006.3721.

10.1 PESTS AND PESTICIDES

A pest is something or someone that annoys us, detracts from some resource that we value, or interferes with a pursuit that we enjoy. In this chapter, we will concentrate on **biological pests, organisms that reduce the availability, quality, or value of resources useful to humans.** What's annoying or undesirable depends, of course, on your perspective. The mosquitoes that swarm in clouds over a marsh in the summer may be irritating to us, but they are an essential food source for birds and bats that feed on them. You may regard dandelions in your yard as tenacious and obnoxious weeds, but in some countries dandelions are cultivated as beautiful flowers and as a food source. Of the millions of species of organisms only about 100 plants, animals, fungi, and microbes cause 90 percent of all crop damage worldwide.

Insects tend to be the most frequent pests, in part, because they make up at least three-quarters of all species on the earth. Most pest organisms tend to be generalists, the opportunistic species that reproduce rapidly, migrate quickly into disturbed areas, and that are pioneers in ecological succession. They compete aggressively against more specialized endemic species and can often take over a biotic community, especially where humans have disrupted natural conditions and created an opening into which they can slip. Most Americans are familiar with dandelions, ragweed, English sparrows, starlings, European pigeons, and other "weedy" species that survive well in urban habitats. Chapter 11 describes how exotic aliens brought in by humans are crowding out native species in many places.

A **pesticide** is a chemical that kills pests. We generally think of toxic substances in this category, but chemicals that drive away pests or prevent their development are sometimes included as well.

FIGURE 10.1 Broad-spectrum toxins can eliminate pests quickly and efficiently, but what are the long-term costs to us and to our environment?

Pest control can also include activities such as killing pests by burning crop residues or draining wetlands to eliminate breeding sites. A broad-spectrum pesticide that kills a wide range of living organisms is called a **biocide** (fig. 10.1). Fumigants, such as ethylene dibromide or dibromochloropropane, used to protect stored grain or sterilize soil fall into this category. Generally, we prefer narrower spectrum agents that attack a specific type of pest: **herbicides** kill plants; **insecticides** kill insects; **fungicides** kill fungi; acaricides kill mites, ticks, and spiders; nematicides kill nematodes (microscopic roundworms); rodenticides kill rodents; and avicides kill birds. Pesticides can also be defined by their method of dispersal (fumigation, for example) or their mode of action, such as an ovicide, which kills the eggs of pests. In a sense, the antibiotics used in medicine to fight infections are pesticides as well.

People have always known of ways to control pests

Using chemicals to control pests may well have been among our earliest forms of technology. People in every culture have known that salt, smoke, and insect-repelling plants can keep away bothersome organisms and preserve food. The Sumerians controlled insects and mites with sulfur 5,000 years ago. Chinese texts 2,500 years old describe mercury and arsenic compounds used to control body lice and other pests. Greeks and Romans used oil sprays, ash and sulfur ointments, lime, and other natural materials to protect themselves, their livestock, and their crops from a variety of pests.

In addition to these metals and inorganic chemicals, people have used organic compounds, biological controls, and cultural practices for a long time. Alcohol from fermentation and acids in pickling solutions prevent growth of organisms that would otherwise ruin food. Spices were valued both for their flavors and because they deterred spoilage and pest infestations. Romans burned fields and rotated crops to reduce crop diseases. They also employed cover crops to reduce weeds. The Chinese developed plant-derived insecticides and introduced predatory ants in orchards to control caterpillars 1,200 years ago.

Modern pesticides provide benefits, but also create problems

The era of synthetic organic pesticides began in 1939 when Swiss chemist Paul Müller discovered the powerful insecticidal properties of dichloro-diphenyl-trichloroethane (DDT). Inexpensive, stable, easily applied, and highly effective, this compound seemed ideal for crop protection and disease prevention. DDT is remarkably lethal to a wide variety of insects but relatively nontoxic to

FIGURE 10.2 Before we realized the toxicity of DDT, it was sprayed freely on people to control insects as shown here at Jones Beach, New York, in 1948.

mammals. Mass production of DDT started during World War II, when Allied armies used it to protect troops from insect-borne diseases. Manufacture of the compound soared from a few kilograms to thousands of metric tons per year in less than a decade. It was sprayed on crops and houses, dusted on people and livestock, and used to combat insects nearly everywhere (fig. 10.2).

By the 1960s, however, evidence began to accumulate that indiscriminate use of DDT and other long-lasting industrial toxins was having unexpected effects on wildlife. Peregrine falcons, bald eagles, brown pelicans, and other carnivorous birds were disappearing from former territories in eastern North America. Studies revealed that egg shells were thinning in these species as DDT and its breakdown products were concentrated through food chains until it reached endocrine hormone-disrupting levels in top predators (see fig. 8.15). In 1962, biologist Rachel Carson published *Silent Spring,* warning that persistent organic pollutants, such as DDT, pose a threat to wildlife, and perhaps to humans. Most uses of DDT were banned in developed countries in the late 1960s, but it continues to be used in developing countries and remains the most prevalent contaminant on food imported to the United States.

Since the 1940s, many new synthetic pesticides have been invented. Many of them, like DDT, have proven to have unintended consequences on nontarget species. Assessing the relative costs and benefits of using these compounds continues to be a contentious topic, especially when unexpected complications arise, such as increasing pest resistance or damage to beneficial insects as described in the opening case study for this chapter.

According to the EPA, total pesticide use in the United States amounts to about 5.3 billion pounds (2.4 million metric tons) per year. Roughly half of that amount is chlorine and hypochlorites (bleach) used for water purification (fig. 10.3). Eliminating pathogens from drinking water prevents a huge number of infections and deaths, but there's concern that using so much chlorine and hypochlorite to do so may be creating other chronic health risks. The next largest category is conventional pesticides: primarily insecticides, herbicides, and fungicides.

Specialty biocides, such as preservatives used in adhesives and sealants, paints and coatings, leather, petroleum products, and

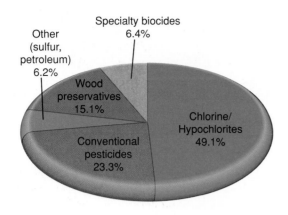

FIGURE 10.3 Of the 5.3 billion pounds of pesticides used in the United States each year, nearly half is chlorine/hypochlorite disinfectant. Specialty biocides include other antiseptics and sanitizers. **Source:** U.S. EPA, 2000.

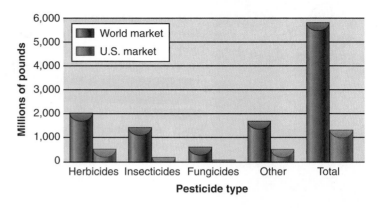

FIGURE 10.4 World and U.S. "conventional" pesticide use in 1999 in millions of pounds of active ingredients. U.S. consumption of these chemicals is about 20 percent of world total. **Source:** U.S. EPA, 2002.

plastics as well as recreational and industrial water treatment amount to some 300 million pounds per year, although they represent only about 6 percent of total pesticide use. The "other" category in figure 10.3 includes sulfur, oil, and chemicals used for insect repellents (such as DEET) and moth control. Wood preservatives represent just 15 percent of total pesticide consumption, but can be especially dangerous to our health because they tend to be both highly toxic and very long lasting. We'll discuss some of the concerns about these chemicals later in this chapter.

Roughly 80 percent of all conventional pesticides applied in the United States are used in agriculture or food storage and shipping. According to CropLife America, some 90 million ha of crops in the United States—including 96 percent of all corn (maize) and one-third of all soybeans—are treated with herbicides every year. In addition, 25 million ha of agricultural fields and 7 million ha of parks, lawns, golf courses, and other lands are treated with insecticides and fungicides. By some accounts, cotton has the highest rate of insecticide application of any crop, while golf courses often have higher rates of application of all conventional pesticides per unit area than any farm fields.

Homes and gardens account for only about 8 percent of total annual pesticide use in the United States, but three-quarters of all American homes use some type of pesticide, amounting to 20 million applications per year. Often people use much larger quantities of chemicals in their homes, yards, or gardens than farmers would to eradicate the same species in their fields. Storage and accessibility of toxins in homes also can be a problem. As we'll discuss later in this chapter, children's exposure to toxins in their home may be of greater concern than pesticide residues in food.

Information on global pesticide use is spotty and uncertain. The U.S. EPA estimates that world usage of conventional pesticides alone amounts to some 5.7 billion pounds (2.6 million metric tons) per year. In addition, millions of metric tons of so-called "inert" ingredients are added to pesticide formulations as carriers, stabilizers, emulsifiers, etc. Some of these materials are dangerous in their own right, and should be regulated more carefully than they often are. The 1.24 billion pounds of conventional pesticides used annually in the United States accounts for about 20 percent of total world consumption (fig. 10.4). Good data for

water purification compounds, wood preservatives, and specialty biocides are lacking for many countries.

There are many types of pesticides

One way to classify pesticides is by their chemical structure. This is useful because environmental properties—such as stability, solubility, and mobility—and toxicological characteristics of members of a particular chemical group are often similar.

Inorganic pesticides include compounds of arsenic, sulfur, copper, lead, and mercury. These broad-spectrum poisons are generally highly toxic and essentially indestructible, remaining in the environment forever. Seeds are sometimes coated with a mercury or arsenic powder to deter insects and rodents during storage or after planting. Handling such seeds with bare hands can be very dangerous for farmers or gardeners. They are generally neurotoxins and even a single dose can cause permanent damage.

Natural organic pesticides, or "botanicals," generally are extracted from plants. Some important examples are nicotine and nicotinoid alkaloids from tobacco; rotenone from the roots of derris and cubé plants; pyrethrum, a complex of chemicals extracted from the daisylike *Chrysanthemum cinerariaefolium* (fig. 10.5); and turpentine, phenols, and other aromatic oils from conifers. All are toxic to insects, but nicotine is also toxic to a broad spectrum of organisms including humans. Rotenone is commonly used to kill fish. Turpentine, phenols, and other natural hydrocarbons are effective pesticides, but synthetic forms such as pentachlorophenol are more stable and more toxic than natural forms. They penetrate surfaces well and are used to prevent wood decay, but they have been banned for most uses in many countries because of their persistence and toxicity.

Fumigants are generally small molecules such as carbon tetrachloride, carbon disulfide, ethylene dichloride, ethylene dibromide, methylene bromide, and dibromochloropropane that gasify easily and penetrate rapidly into a variety of materials. They are used to sterilize soil and prevent decay or rodent and insect infestation of stored grain. Because these compounds are extremely dangerous for workers who apply them, many have been restricted or banned altogether.

FIGURE 10.5 Chrysanthemum flowers are a source of pyrethrum, a natural insecticide.

Chlorinated hydrocarbons, or organochlorines such as DDT, chlordane, aldrin, dieldrin, toxaphene, paradichlorobenzene (mothballs), and lindane, are synthetic organic insecticides that inhibit nerve membrane ion transport and block nerve signal transmission. They are fast acting and highly toxic in sensitive organisms. Toxaphene, for instance, kills goldfish at 5 parts per billion (5 μg/liter), one of the highest toxicities for any compound in any organism. Chlorinated hydrocarbons may persist in soil for decades, become concentrated through food chains, and are stored in fatty tissues of a variety of organisms. The chloriphenoxy herbicides 2,4 D and 2,4,5 T have hormone-like growth-regulating properties and are selective for broad-leaf flowering plants. Because these compounds are so toxic and so long-lasting, many chlorinated hydrocarbons have been banned for most uses over much of the world.

Organophosphates, such as parathion, malathion, dichlorvos, chlorpyrifos, dimethyldichlorovinylphosphate (DDVP), and tetraethylpyrophosphate (TEPP), are an outgrowth of nerve gas research during World War II. They generally inhibit cholinesterase, an enzyme essential for removing excess neurotransmitter from synapses in the peripheral nervous system. Many organophosphates are extremely toxic to mammals, birds, and fish (generally 10 to 100 times more poisonous than most chlorinated hydrocarbons). A single drop of TEPP on your skin, for example, can be lethal. Because they are quickly degraded, they are much less persistent in the environment than organochlorines, generally lasting only a few hours or a few days. These compounds are very dangerous for workers who often are sent into fields too soon after they have been sprayed (fig. 10.6).

Glyphosate (N-phosphonomethylglycine) is a special organophosphate that does not react with the central nervous system as most others in this group do. Instead, it is a broad spectrum, nonselective herbicide used to kill weeds. Monsanto markets glyphosate as Roundup, and has genetically engineered soy beans, corn, wheat, and other crops to be Roundup resistant. This allows no-till agriculture, but encourages use of the herbicide.

Carbamates, or urethanes, such as carbaryl (Sevin), aldicarb (Temik), aminocarb (Zineb), carbofuran (Baygon), and Mirex share many organophosphate properties, including mode of action, toxicity, and lack of environmental persistence and low bioaccumulation. Car-

bamates generally are extremely toxic to bees and must be used carefully to prevent damage to these beneficial organisms.

Halogenated pyrroles are a new class of compounds based on a naturally occurring microbial toxin. The first of these compounds to be released commercially is chlorfenapyr, which is marketed as Pirate. It was hailed as a replacement for organophosphates like malathion, but has been shown to reduce egg laying and egg survival in ducks at very low concentrations. Opponents argue that it may be the new DDT and call for its withdrawal.

Microbial agents and **biological controls** are living organisms or toxins derived from them used in place of pesticides. Bacteria such as *Bacillus thuringiensis* or *Bacillus popilliae* kill caterpillars or beetles by producing a toxin that ruptures the digestive tract lining when eaten. Parasitic wasps such as the tiny *Trichogramma* genus attack moth caterpillars and eggs, while lacewings and ladybugs control aphids. Viral diseases also have been used against specific pests.

10.2 PESTICIDE BENEFITS

Like all organisms, humans compete with other species for food and shelter and struggle to protect ourselves from diseases and predators. Pesticides are important weapons in this fight for survival.

We have made progress in controlling many insect-borne diseases

Insects and ticks serve as vectors in the transmission of a number of disease-causing pathogens and parasites. Consider malaria, for example. About 300 million people suffer from this disease at any given time, and about 1 million die each year from *Plasmodium* protozoa

FIGURE 10.6 The United Farm Workers of America claims that 300,000 farmworkers in the United States suffer from pesticide-related illnesses each year. Worldwide, the WHO estimates that 25 million people suffer from pesticide poisoning and 20,000 die each year from improper use or storage of pesticides.

FIGURE 10.7 Malaria, spread by the *Anopheles* mosquito, is one of the largest causes of human disease and premature death in the world. By controlling mosquitoes, pesticides save millions of lives per year in tropical countries.

spread to humans by *Anopheles* mosquitoes (fig. 10.7). It is estimated that insecticidal mosquito control has prevented at least 50 million deaths from malaria over the past 50 years. Sri Lanka is a classic example of pesticide benefits. In the early 1950s, more than 2 million cases of malaria were reported in Sri Lanka each year. After DDT spraying began in 1954, new malaria cases almost completely disappeared. When DDT spraying was discontinued in 1964, however, malaria reappeared almost immediately.

Within three years the annual incidence was more than 1 million cases per year. The Sri Lankan government resumed DDT spraying in 1968 and continues limited use of this insecticide despite environmental concerns. They concluded that the reductions in medical expenses, social disruption, pain and suffering, and lost work resulting from malaria outweighed the direct costs of insecticide spraying by a thousand to one. Many people advocate resumed use of DDT in other malaria-prone countries as well.

Some other diseases spread by biting insects include yellow fever and related viral diseases such as encephalitis and West Nile, also carried by mosquitoes, and trypanosomiasis or sleeping sickness caused by protozoa transmitted by the tsetse fly. Onchocerciasis (river blindness) and filariasis (one form of which is elephantiasis, fig. 10.8) are caused by tiny worms spread by biting flies that afflict hundreds of millions of people in tropical countries.

FIGURE 10.8 Elephantiasis is caused by parasitic worms (filaria) that block lymph vessels and cause fluid accumulation in various parts of the body.

All of these terrible diseases can be reduced by judicious use of pesticides. If you were faced with a choice of going blind before age 30 because of masses of worms accumulating in your eyeballs or a small chance of cancer due to pesticide exposures if you live to age 50 or 60, which would you choose?

Without pesticides, we might lose two-thirds of conventional crops

Although reliable data on crop losses are difficult to obtain, it is thought that plant diseases, insect and bird predation, and competition by weeds reduce crop yields worldwide by at least one-third. Postharvest losses to rodents, insects, and fungi may be as high as another 20 to 30 percent. Without modern chemical pesticides, these losses might be much higher. Although we said in chapter 9 that there is more than enough food in the world to adequately feed everyone now living—if food were equitably distributed—this most certainly would not be true, given current intensive farming practices, if modern chemical pesticides were unavailable.

A commonly quoted estimate is that farmers save $3 to $5 for every $1 spent on pesticides. This means lower costs and generally more reliable quality for consumers. In some cases, insects and fungal diseases cause only small losses in terms of the total crop quantity, but the cosmetic damage they cause greatly reduces the economic value of crops. For example, although codling moth larvae consume very little of the apples they infest, the brown trails they leave as they crawl through the fruit and the possibility of biting into a living worm often make an unsprayed crop unsalable. Which is the more worrisome risk for you—consuming a bug or consuming pesticides?

CropLife America calculates that prohibiting pesticides in the United States would result in a $21 billion per year loss in food and fiber production. Without herbicides, they conclude, crop yields would be reduced up to 67 percent, soil erosion would increase by at least 1 billion metric tons per year (because of more cultivation to remove weeds), and it would take up to 70 million additional farm laborers to remove weeds by hand. This assumes, however, that we would continue growing the same single-species crops year after year without crop rotation or other sustainable agriculture practices.

Think About It

Which of the benefits of pesticides described in this section is most persuasive for you? Do these benefits change your overall attitude toward pesticides? You should be aware that the claims quoted here come from pesticide proponents. Compare them to problems described in the next section from the perspective of pesticide opponents.

10.3 PESTICIDE PROBLEMS

While synthetic chemical pesticides have brought us valuable economic and social benefits, they also cause a number of serious problems. In this section we will examine some of the worst of those problems.

FIGURE 10.9 This machine sprays insecticide on orchard trees—and everything nearby. Up to 90 percent of pesticides applied in this fashion never reach target organisms.

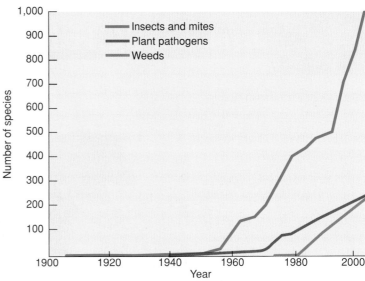

FIGURE 10.10 Many pests have developed resistance to pesticides. Because insecticides were the first class of pesticide to be used widely, selection pressures led insects to show resistance early. More recently, plant pathogens and weeds are also becoming insensitive to pesticides. **Source:** Worldwatch Institute, 2003.

Pesticides often poison nontarget species

It is estimated that in many cases, up to 90 percent of the pesticides we use never reach their intended targets (fig. 10.9). Many beneficial organisms are poisoned unintentionally as a result, as described in the opening case study for this chapter.

In some cases, the effects of poisoning nontarget species are immediate and unmistakable. In one episode in 1972, a single application of the insecticide Azodrin to combat potato aphids on a farm in Dade County, Florida, killed 10,000 migrating robins in three days. Similarly, a 1991 derailment of a Southern Pacific tanker car on a tricky canyon bridge just north of Dunsmuir, California, dumped 75,000 liters (20,000 gallons) of highly toxic metam sodium herbicide into the Sacramento River. The entire river ecosystem—including aquatic plants, insects, amphibians, and at least 100,000 trout—was completely wiped out for 45 kilometers (27 miles) downstream.

In other cases, the effects are more difficult to pin down, although not less serious. Insecticide spraying in Canadian forests has been linked to dramatic declines (up to 77 percent) in Atlantic salmon. The chemical suspected in this case is 4-nonylphenol, a powerful endocrine-hormone disrupter (Exploring Science, p. 214).

Pesticide resistance is often rapid and widespread

The resistance evolved by Varroa mites is a common phenomenon. Pesticides almost never kill 100 percent of a target species even under the most ideal conditions. As we discussed in chapter 4, every population contains some diversity in tolerance to adverse environmental factors. The most resistant members of a population survive pesticide treatment and produce more offspring like themselves with genes that enable them to withstand further chemical treatment. Because most pests propagate rapidly and produce many offspring, the population quickly rebuilds with pesticide-resistant individuals. We call this phenomenon **pest resurgence** or rebound. The Worldwatch Institute reports that at least 1,000 insect pest species and another 550 or so weeds and plant pathogens worldwide have developed chemical resistance (fig. 10.10). Of the 25 most serious insect pests in California, three-quarters are now resistant to one or more insecticides. This resistance means that it takes constantly increasing doses to get the same effect or that farmers who are caught on a **pesticide treadmill** must constantly try newer and more toxic chemicals in an attempt to stay ahead of the pests. Cornell University entomologist David Pimentel reports that a larger percentage of crops are lost now to insects, diseases, and weeds than in 1944 despite the use of millions of tons more pesticides. Some people worry that we are losing the battle to control pests.

One of the most ominous developments in this race to find effective pesticides is that we increasingly find pests that are resistant to chemicals to which they have never been exposed. Apparently, genes for pesticide resistance are being transferred from one species to another by means of vectors such as viruses and plasmids (naked pieces of virus-like DNA). Often a whole cluster of genes jump between species so that multiple chemical-tolerance is inherited before a particular pest is ever exposed to any of the chemicals. Widespread use of crops genetically engineered to produce pesticides endogenously (chapter 9) is likely to cause even more pesticide resistance.

In a way, every pesticide has a very limited useful life span before target species become resistant to it or the pesticide builds up intolerable environmental concentrations. We may have made a big mistake in broadcasting pesticides recklessly and extravagantly. DDT, for instance, is such a helpful insecticide that we

What might alligators in Florida, seals in the North Sea, salmon in the Great Lakes, and you have in common? All are at the top of their respective food chains and all appear to be accumulating threatening levels of toxic environmental chemicals in their body tissues. One of the most serious effects of those chemicals is that many disrupt endocrine hormones that regulate many important bodily functions, including reproductive and immune systems. Evidence for this seems quite convincing in some wildlife populations, but whether it also is true for humans is one of the most contentious and important questions in environmental toxicology today.

A shocking decrease in fertility and hatchling survival of alligators in Lake Apopka has been linked to pesticide pollution. Are humans affected by similar toxins?

One of the first examples of hormone-disrupting chemicals in the environment was a dramatic decline in alligators two decades ago in Florida's Lake Apopka just north of Orlando. Surveys showed that 90 percent of the alligator eggs laid each year were infertile and that of the few that hatched, only about half survived more than two weeks. Male hatchlings had shrunken penises and unusually low levels of the male hormone testosterone. Female alligators, meanwhile, had highly elevated estrogen levels and abnormal ovaries. The explanation seems to be that a DDT spill in the lake in the 1980s, along with pesticide-laden runoff from adjacent farm fields, has led to high levels of DDE (a persistent breakdown product of DDT) in the reptiles' tissues and eggs. Because of a similarity in chemical structure, DDE appears to interfere with the action of androgens and estrogens, the normal sex hormones (see fig. 8.12).

Researchers have begun to suspect that mysterious outbreaks of health and reproductive problems and immune system failures in other wildlife populations may have similar origins. A USGS survey of 139 U.S. rivers, for instance, found that 40 percent had hormonally active industrial chemicals. In many rivers, fish that appeared superficially to be females were actually genetically males.

In some rivers and lakes, all male fish are feminized. Recent evidence suggests that atrazine converts male frogs into hermaphrodites and makes both sexes more susceptible to parasites and limb deformities.

Humans may be affected as well. DDE and other compounds have been correlated with elevated cases of estrogen-sensitive breast and vaginal cancers in women, with developmental disorders in children, and possibly with falling sperm counts and low fertility in men.

Good evidence exists from controlled laboratory experiments that rats and mice exposed *in utero* or through mother's milk to very low levels of estrogen-like compounds develop physical, reproductive, and behavioral problems. Among some pesticides shown to have hormone-disrupting effects are the herbicides alachlor and atrazine, the fungicides mancozeb and benomyl, and the insecticides carbaryl, clicofol, endosulfan, methomyl, methoxychlor, parathion, linuron, chlorfenapyr, and sythetic pyrethroids.

We know that some of these chemicals act as synthetic hormones, others are antagonists that block normal hormone function. Furthermore, there can be striking synergy between some compounds. When endosulfan and DDT or chlordane are applied

together, for example, the combination is 1,600 times more estrogenic than either chemical alone.

The question is whether specific chemical exposures are linked to particular human health problems. Many of these compounds are hundreds or thousands of times less active than normal hormones, leading skeptics to doubt that they have any noticeable effects except in people exposed to extremely high levels. Furthermore, we may have protective mechanisms that are lacking in highly inbred laboratory rodents, and we can eat a highly varied diet that includes protective factors as well as toxins.

A wide variety of industrial products—including phthalates in plastics and cosmetics, polychlorinated biphenyls (PCBs), dioxins, chlorinated solvents, and metals such as lead, mercury, arsenic, and cadmium—also can disrupt hormone actions. Every human in the world now carries dozens of these chemicals in his or her body from before birth to death. Determining which of these toxins—or which combination—might be responsible for a particular disease or developmental abnormality may never be possible.

The bottom line is that we don't know (and we may never know for sure) whether falling sperm counts, increasing cancers, birth defects, immune diseases, and behavioral disorders in humans are caused by hormonally active, bioaccumulative, environmental chemicals. In 1996, the EPA ordered pesticide manufacturers to begin testing for disrupting effects. Given the continuing uncertainty about the dangers we face, what more do you think we should do? Is this threat serious enough to warrant drastic steps to reduce our risk? If you were head of the Environmental Protection Agency or the Food and Drug Administration, how much certainty would you demand before acting to protect our environment and ourselves from this ominous threat?

should have used it sparingly and carefully, so that it would still be effective against the worst insect pests. It has been spread so widely that 50 of the 60 malaria-carrying mosquitoes are now resistant to it, and the environmental side effects outweigh its benefits. You can think of the useful life of DDT as a nonrenewable resource that we squandered.

Pesticide misuse can create new pests

Often the worst side effect of broadcast spraying a pesticide is that we kill beneficial predators that previously kept a number of pests under control. Many of the agricultural pests that we want to eliminate, for instance, are herbivores such as aphids, grasshoppers, or moth larvae that eat crop plants. Under natural conditions, their populations are kept under control by predators such as wasps, ladybugs, and praying mantises. As we discussed in chapter 3, there are generally far fewer predators in a food web than the species they prey upon. When we use a broad-spectrum pesticide, higher trophic levels are more likely to be knocked out than lower ones. This means that species that previously were insignificant can be released from natural controls and suddenly become major pests.

Consider the case of the Canete Valley in Peru. Before DDT was introduced in 1949, cotton yields were about 500 kg per hectare. By 1952 yields had risen to nearly 750 kg per hectare, but DDT-resistant boll weevils also had appeared (fig. 10.11). Toxaphene replaced DDT, but within two years it also became ineffective against boll weevils, which rebounded to higher levels than ever. Even worse, *Heliothis worms*—which had not previously been a problem—began increasing rapidly. The wasps that earlier had kept both organisms in check were poisoned by the increasing pesticide doses. By 1955, cotton yields were down to 330 kg per hectare, one-third less than before any pesticides were used.

Similarly, in California in the 1990s, cotton yields plunged by 20 percent despite a doubling of the number of insecticide applications and a 43 percent increase in pesticide costs. With the advent of chemical pest controls, farmers tend to abandon mixed crops, rotation regimes, and other traditional methods of management. This creates a greater potential for pest damage and pest resistance and thus makes it economical to use even higher levels of pesticide treatment. The application of large amounts of pesticides over wide expanses of land, together with changes in the ways the land is managed, have undermined the biological and ecological forces and interactions that previously governed population dynamics among species. A lack of understanding of the ways pesticides adversely affect beneficial organisms is one reason pest managers stick with chemical-based systems despite declining pesticide efficiency.

Some persistent pesticides can move long distances in the environment

The qualities that make DDT and other chlorinated hydrocarbons so effective—stability, high solubility, and high toxicity—also make them environmental nightmares. Because they persist for

FIGURE 10.11 The cotton boll weevil caused terrible devastation in southern U.S. cotton fields in the 1930s and 1940s. DDT and other insecticides controlled this pest briefly, but the beetle quickly became resistant.

years, even decades in some cases, and move freely through air, water, and soil, they often show up far from the point of original application. Some of these compounds have been discovered far from any possible source and long after they most likely were used. Because they have an affinity for fat, many chlorinated hydrocarbons are bio-concentrated and stored in the bodies of predators—such as porpoises, whales, polar bears, trout, eagles, ospreys, and humans—that feed at the top of food webs. In a study of human pesticide uptake and storage, Canadian researchers found the levels of chlorinated hydrocarbons in the breast milk of Inuit mothers living in remote arctic villages were five times that of women from Canada's industrial region some 2,500 km (1,600 mi) to the south. Inuit people have the highest levels of these persistent pollutants of any human population except those contaminated by industrial accidents.

These compounds accumulate in polar regions by what has been called the "grasshopper effect," in which they evaporate from water and soil in warm areas and then condense and precipitate in colder regions. In a series of long-distance hops, they eventually collect in polar regions, where they accumulate in top predators. Polar bears, for instance, have been shown to have concentrations of certain chlorinated compounds 3 billion times greater than the seawater around them. In Canada's St. Lawrence estuary, beluga (white whales), which suffer from a wide range of infectious diseases and tumors thought to be related to environmental toxins, have such high levels of chlorinated hydrocarbons that their carcasses must be treated as toxic waste.

Although DDT hasn't been used legally in the United States or Canada for more than 20 years, in 1999 researchers reported discovery of p, p′ -DDE, a DDT breakdown by-product, in the amniotic fluid of 30 percent of a sample of pregnant women in Los Angeles, California. A growing baby is surrounded by amniotic fluid during its nine months in the mother's womb. Retrospective studies have shown that mothers with high DDT or DDE levels in their blood are more likely to have premature or low-birthweight babies, and that these children have lower fertility rates as adults than those not exposed to pesticides in the womb.

Atrazine and alochlor are the most widely used herbicides in North America. More than 96 percent of the maize (corn) and soybean crops in the United States are treated with these compounds. The 31 million kg (70 million lbs) of atrazine used every year in the United States makes it the biggest commercial pesticide. Surveys of wells have found that up to 30 percent of all community wells and as much as 60 percent of all private wells in the Midwestern U.S. corn belt are contaminated with atrazine and alochlor. University of California researcher Tyrone Hayes has shown that extremely low levels of atrazine cause sexual abnormalities in frogs. Although the chemical industry has tried to discredit his research and suppress its publication, others have confirmed the estrogenic effects of this herbicide. In 2003, the European Union withdrew regulatory approval for atrazine, citing its human and environmental health effects.

Because **persistent organic pollutants (POPs)** are so long-lasting and so dangerous, 127 countries agreed in 2001 to a global ban on the worst of them, including aldrin, clordane, dieldrin, DDT, endrin, hexachlorobenzene, neptachlor, mirex, toxaphene, polychlorinated biphenyls (PCBs), dioxins, and furans. Most of this "dirty dozen" had been banned or severely restricted in developed countries for years. However, their production continued. Between 1994 and 1996, U.S. ports shipped more than 100,000 tons of POPs each year. Most of this was sent to developing countries where regulations were lax. Ironically, many of these pesticides returned to the United States on bananas and other imported crops. According to the 2001 POPs treaty, eight of the dirty dozen were banned immediately; PCBs, dioxins, and furans will be phased out; and use of DDT, still allowed for limited uses such as controlling malaria, must be publically registered in order to permit monitoring. The POPs treaty has been hailed as a triumph for environmental health and international cooperation. Unfortunately, other compounds—perhaps just as toxic—have been introduced to replace POPs.

Many pesticides cause human health problems

Pesticide effects on human health can be divided into two categories: (1) acute effects, including poisoning and illnesses caused by relatively high doses and accidental exposures, and (2) chronic effects suspected to include cancer, birth defects, immunological problems, endometriosis, neurological problems, Parkinson's disease, and other chronic degenerative diseases.

The World Health Organization (WHO) estimates that 25 million people suffer pesticide poisoning and at least 20,000 die each year (fig. 10.12). At least two-thirds of this illness and death results from occupational exposures in developing countries where people use pesticides without proper warnings or protective clothing. A tragic example of occupational pesticide exposure is found among workers in the Latin American flower industry. Fueled by the year-round demand in North

FIGURE 10.12 Handling pesticides requires protective clothing and an effective respirator. Pesticide applicators in tropical countries, however, often can't afford these safeguards or can't bear to wear them because of the heat.

America for fresh vegetables, fruits, and flowers, a booming export trade has developed in countries such as Guatemala, Colombia, Chile, and Ecuador. To meet demands in North American markets for perfect flowers, table grapes and other produce, growers use high levels of pesticides, often spraying daily with fungicides, insecticides, nematicides, and herbicides. Working in warm, poorly ventilated greenhouses with little protective clothing, the workers—70 to 80 percent of whom are women—find it hard to avoid pesticide contact. Nearly two-thirds of nearly 9,000 workers surveyed in Colombia experienced blurred vision, nausea, headaches, conjunctivitis, rashes, and asthma. Although harder to document, they also reported serious chronic effects such as stillbirths, miscarriages, and neurological problems.

Long-term health effects caused by chronic pesticide exposures are difficult to document conclusively. As we discussed in chapter 8, isolating a specific pesticide-related disease from the myriad of other risks we encounter is complex. For example, numerous studies have shown that farmers who use pesticides have elevated rates of prostate cancer, non-Hodgkin's lymphoma, Parkinson's, and several other chronic diseases, but distinguishing between pesticide effects, smoking, exposure to gasoline and solvents, and other health risks is complex. Similarly, children of farmers who apply herbicides and fungicides have higher rates of birth defects, but it's difficult to know which of the mixture of chemicals used by these farmers may be responsible.

In an ongoing study of prenatal exposure to PCBs and other persistent contaminants, researchers from Wayne State University in Detroit, Michigan, have documented significant learning and attention problems in children whose mothers ate Lake Michigan fish regularly in the years prior to pregnancy. At age 11, the most highly exposed children had difficulty paying attention, suffered from poorer short- and long-term memory, were twice as likely to be below average in reading comprehension, and were three times as likely to have low IQ scores as their age-matched peers.

Researchers in a similar study in Mexico showed striking differences in behavior and motor skills in children exposed to pesticides compared to peers with minimal pesticide risk. Two groups of Yaqui children from northwestern Mexico were tested in this study. The children were similar in every respect except for their pesticide exposure. Children who lived in the foothills come from ranches where pesticides are used sparingly, if at all. Children from the Yaqui valley, live in a farming area where pesticide use has been heavy for the past 50 years. Samples of mother's milk and umbilical cord blood taken from valley families contained high amounts of several persistent pesticides including aldrin, endrin, dieldrin, heptachlor, and DDE. In tests designed to measure growth and development, the valley children fell far behind their foothill-dwelling peers (fig. 10.13). Farm children also exhibited diminished memory, decreased physical stamina, higher irritability, a greater tendency to fight, and poorer eye-hand coordination.

Farm families aren't the only ones to suffer from pesticide exposure. The 63 million pounds of pesticides used annually in American homes and gardens pose a threat to families, especially those with young children. In California, for example, children from homes fumigated by professional bug exterminators have been found to be three times as likely to have acute lymphocytic

FIGURE 10.14 Our approach to nature is to beat it into submission, but at what cost?

leukemia as those from homes where no pesticides were used (fig. 10.14). One recent study concluded that more than 1 million children under age five in the United States may exceed the safe daily dose of organophosphate insecticides. Children are much more susceptible to toxic chemicals than adults because—pound for pound—they eat more food, drink more water, and breathe more air. Children spend more time playing on carpets, grass, or in dirt where pesticides may accumulate. They put their fingers in their mouths more often, and children's rapid growth and development make them vulnerable to toxic effects (see What Do You Think? p. 221). Many health professionals believe that the sudden rise in autism and hyperactivity/attention deficit disorder in America may be linked to environmental toxins.

All of us now have traces of persistent pollutants in our bodies. In its "Body Burden Study," the Environmental Working Group found detectable levels of 91 different industrial chemicals per person on average (161 different chemicals in all) in a sample of people none of whom had any known occupational exposure. Of these chemicals, 76 are suspected to be cancer-causing, 94 are known to be neurotoxins in laboratory animals, and 79 are suspected of causing birth defects (the total adds up to more than 161 because some have multiple effects). Of course, some materials can be detected at levels far below those known to have toxic effects, but the presence of so many different compounds in ordinary people suggests the ubiquity of these chemicals in our environment. Still, there are steps you can take to reduce your exposure to pesticides (What Can You Do? p. 218).

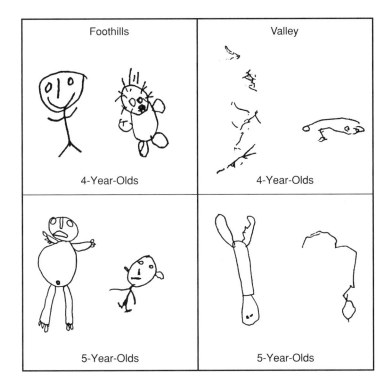

FIGURE 10.13 Representative examples of drawings of people by 4- and 5-year-old Yaqui children relatively unexposed to pesticides (foothills) and those heavily exposed (valley).

Controlling Pests

Based on the principles of Integrated Pest Management, the U.S. EPA has released a helpful and informative guide to pest control. Among their recommendations:

1. *Identify the pest problem.* What kinds of pests and how many do you have? Many free resources are available from your library or County Cooperative Extension Service to help you understand what you face and how best to deal with it.

2. *Decide how much pest control is necessary.* Does your lawn really need to be totally weed free? Could you tolerate some blemished fruits and vegetables, or could you replace some of the plants you now grow with ones less sensitive to pests?

3. *Eliminate pest sources.* Remove food, water, and habitat that encourage pest growth from your house or yard. Block off hiding places or routes of entry into your home. Don't plant the same crops in the same spot year after year.

4. *Develop a healthy, weed-resistant yard.* Make sure your soil has the right pH balance, key nutrients, and good texture. Grow a type of grass or cover crop that does well in your climate. Set realistic weed and pest controls.

5. *Use biological controls.* Beneficial predators such as purple martins and bats eat insects; lady beetles (ladybugs) and their larvae eat aphids, mealybugs, whiteflies, and mites. Other beneficial predators include spiders, centipedes, dragonflies, wasps, and ants. Planting a border of marigolds may keep insects or herbivores away from your crops or flowers.

6. *Use simple manual methods.* Cultivate your garden and hand-pick weeds and pests from your garden. Use a flyswatter. Set traps to control rats, mice, and some insects. Mulch to reduce weed growth.

7. *Use chemical pesticides carefully.* If you decide that the best solution is chemical, choose the right pesticide product, read safety warnings and handling instructions, buy the amount you need, store the product safely, and dispose of excess properly.

8. *Evaluate the results.* Compare pretreatment and posttreatment conditions. Is there clear evidence of pest reduction? Are the benefits of short-term chemical pesticide control worth the risks? Would other treatments, including prevention of pest buildup work as well?

Source: Citizen's Guide to Pest Control and Pesticide Safety: EPA 730-K-95-001.

10.4 ALTERNATIVES TO CURRENT PESTICIDE USES

Can we avoid using toxic pesticides or lessen their environmental and human-health impacts? In many cases, improved management programs can cut pesticide use between 50 and 90 percent without reducing crop production or creating new diseases. Some of these techniques are relatively simple and save money while maintaining disease control and yielding crops with just as high quality and quantity as we get with current methods. In this section, we will examine behavioral changes, biological controls, and integrated pest management systems that could substitute for current pest-control methods.

FIGURE 10.15 The praying mantis looks ferocious and is an effective predator against garden pests, but it is harmless to humans. They can even make interesting and useful pets.

We can change our behavior

Crop rotation (growing a different crop in a field each year in a two- to six-year cycle) keeps pest populations from building up. For instance, a soybean/corn/hay rotation is effective and economical against white-fringed weevils. Mechanical cultivation can substitute for herbicides but increases erosion. Flooding fields before planting or burning crop residues and replanting with a cover crop can suppress both weeds and insect pests. Habitat diversification, such as restoring windbreaks, hedgerows, and ground cover on watercourses, not only prevents soil erosion but also provides perch areas and nesting space for birds and other predators that eat insect pests. Growing crops in areas where pests are absent makes good sense. Adjusting planting times can avoid pest outbreaks, while switching from huge monoculture fields to mixed polyculture (many crops grown together) makes it more difficult for pests to find the crops they like. Tillage at the right time can greatly reduce pest populations. For instance, spring or fall plowing can help control overwintering corn earworms.

Useful organisms can help us control pests

Biological controls such as predators (wasps, ladybugs, praying mantises; fig. 10.15) or pathogens (viruses, bacteria, fungi) can control many pests more cheaply and safely than broad-spectrum, synthetic chemicals. *Bacillus thuringiensis* or Bt, for example, is a naturally occurring bacterium that kills the larvae of lepidopteran (butterfly and moth) species but is generally harmless to mammals. A number of important insect pests such as tomato hornworm, corn rootworm, cabbage loopers, and others can be controlled by spraying bacteria on crops. Larger species are effective as well. Ducks, chickens, and geese, among other species, are used to rid fields of both insect pests and weeds. These biological organisms are self-reproducing and often have wide prey tolerance. A few mantises or ladybugs released in your garden in the spring will keep producing

FIGURE 10.16 A Nigerian woman examines a neem tree, the leaves, seeds, and bark of which provide a natural insecticide.

offspring and protect your fruits and vegetables against a multitude of pests for the whole growing season.

Herbivorous insects also have been used to control weeds. For example, the prickly pear cactus was introduced to Australia about 150 years ago as an ornamental plant. This hardy cactus escaped from gardens and found an ideal home in the dry soils of the outback. It quickly established huge, dense stands that dominated 25 million ha (more than 60 million acres) of grazing land. A natural predator from South America, the cactoblastis moth, was introduced into Australia in 1935 to combat the prickly pear. Within a few years, cactoblastis larvae had eaten so much prickly pear that the cactus has become rare and is no longer economically significant.

Some plants make natural pesticides and insect repellents. The neem tree (*Azadirachta indica*) is native to India, but is now grown in many tropical countries (fig. 10.16). The leaves, bark, roots and flowers all contain compounds that repel insects and can be used to combat a number of crop pests and diseases.

Genetics and bioengineering can help in our war against pests. Traditional farmers have long known to save seeds of disease-resistant crop plants or to breed livestock that tolerate pests well. Modern science can speed up this process through selection regimes or by using biotechnology to transfer genes between closely related or even totally unrelated species. (Often we don't know, however what unintended consequences, such as super weeds or new pests, might result from these activities.) Insect pest reproduction has sometimes been reduced by releasing sterile males. Screwworms, for example, are the flesh-eating larvae of flies that lay their eggs in scratches or skin wounds of livestock. They were a terrible problem for ranchers in Texas and Florida in the 1950s, but release of massive numbers of radiation-sterilized males disrupted reproduction and eliminated this pest in Florida. In Texas, where flies continue to cross the border from Mexico, control has been more difficult, but continual vigilance keeps the problem manageable.

Sometimes we can change the biology or behavior of species that are being attacked by pests. Researchers at the University of Minnesota, for example, are fighting Varroa mites by raising honeybees with an unusual behavior. These bees recognize mites and remove them from the hive. When bee-keepers obtain enough honeybees with this behavior, they can save their hives while also reducing reliance on pesticides.

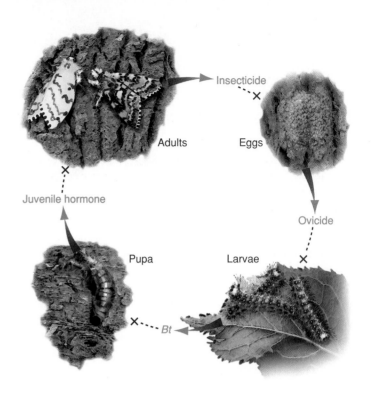

FIGURE 10.17 Different strategies can be used to control pests at various stages of their life cycles. *Bacillus thuringiensis* (*Bt*) kills caterpillars when they eat leaves with these bacteria on the surface. Releasing juvenile hormone in the environment prevents maturation of pupae. Predators attack at all stages.

Other promising approaches are to use hormones that upset development or sex attractants to bait traps containing toxic pesticides. Many municipalities control mosquitoes with these techniques rather than aerial spraying of insecticides because of worries about effects on human health. Briquettes saturated with insect juvenile hormone are scattered in wetlands where mosquitoes breed. The presence of even minute amounts of this hormone prevent larvae from ever turning into biting adults (fig. 10.17). Unfortunately, many beneficial insects as well as noxious ones are affected by the hormone, and birds or amphibians that eat insects may be adversely affected when their food supply is reduced. Some communities that formerly controlled mosquitoes have abandoned these programs, believing that having naturally healthy wetlands is worth getting a few bites in the summer.

Integrated pest management uses a combination of techniques to fight pests

Integrated pest management (IPM) is a flexible, ecologically based strategy that is applied at specific times and aimed at specific crops and pests. It often uses mechanical cultivation and techniques such as vacuuming bugs off crops as an alternative to chemical application (fig. 10.18). IPM doesn't give up chemical pest controls entirely but rather tries to use the minimum amount necessary only as a last resort and avoids broad-spectrum, ecologically disruptive products. IPM relies on preventive practices that encourage growth

FIGURE 10.18 This machine, nicknamed the "salad vac," vacuums bugs off crops as an alternative to treating them with toxic chemicals.

and diversity of beneficial organisms and enhance plant defenses and vigor. Careful, scientific monitoring of pest populations to determine **economic thresholds,** the point at which potential economic damage justifies pest control expenditures, and the precise time, type, and method of pesticide application is critical in IPM.

Trap crops, small areas planted a week or two earlier than the main crop, are also useful. This plot matures before the rest of the field and attracts pests away from other plants. The trap crop then is sprayed heavily with enough pesticides so that no pests are likely to escape. The trap crop is destroyed so that workers will not be exposed to the pesticide and consumers will not be at risk. The rest of the field should be mostly free of both pests and pesticides.

IPM programs are already in use all over the United States on a variety of crops. Massachusetts apple growers who use IPM have cut pesticide use by 43 percent in the past ten years while maintaining per-acre yields of marketable fruit equal to that of farmers who use conventional techniques. Some of the most dramatic IPM success stories come from the developing world. Cuba, for example (What Do You Think? p. 221), has turned almost entirely to organic farming. In Brazil, pesticide use on soybeans has been reduced up to 90 percent with IPM. In Costa Rica, use of IPM on banana plantations has eliminated pesticides altogether in one region. In Africa, mealybugs

were destroying up to 60 percent of the cassava crop (the staple food for 200 million people) before IPM was introduced in 1982. A tiny wasp that destroys mealybug eggs was discovered and now controls this pest in over 65 million ha (160 million acres) in 13 countries.

A successful IPM program that could serve as a model for other countries is found in Indonesia, where brown planthoppers developed resistance to virtually every insecticide and threatened the country's hard-won self-sufficiency in rice. In 1986, President Suharto banned 56 of 57 pesticides previously used in Indonesia and declared a crash program to educate farmers about IPM and the dangers of pesticide use. Researchers found that farmers were spraying their fields habitually—sometimes up to three times a week—regardless of whether fields were infested. By allowing natural predators to combat pests and spraying only when absolutely necessary with chemicals specific for planthoppers, Indonesian farmers using IPM have had higher yields than their neighbors using normal practices and they cut pesticide costs by 75 percent. In 1988, only two years after its initiation, the program was declared a success. It has been extended throughout the whole country. Since nearly half the people in the world depend on rice as their staple crop, this example could have important implications elsewhere (fig. 10.19).

Alternative Pest Control Strategies

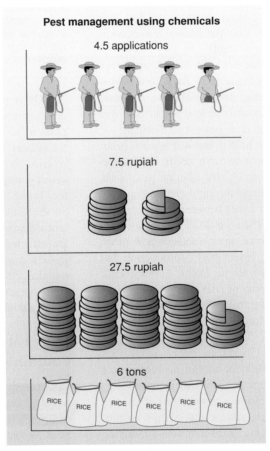
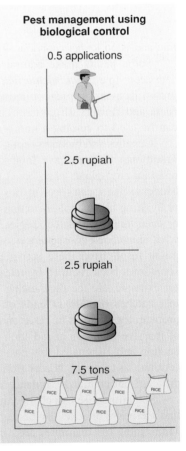

	Pest management using chemicals	Pest management using biological control
Number of times insecticide used in rice season	4.5 applications	0.5 applications
Cost to farmers per hectare	7.5 rupiah	2.5 rupiah
Cost to government per hectare	27.5 rupiah	2.5 rupiah
Rice yield per hectare	6 tons	7.5 tons

FIGURE 10.19 Indonesia has one of the world's most successful integrated pest management (IPM) programs. Switching from toxic chemicals to natural pest predators has saved money while also increasing rice production.
Source: Tolba, et al., *World Environment,* 1972–1992, p. 307, Chapman & Hall, 1992 United Nations Environment Programme.

What Do You Think?

Organic Farming in Cuba

The biggest experiment in low-input, sustainable agriculture in world history is occurring now in Cuba. The sudden collapse of the socialist bloc, upon which Cuba had been highly dependent for trade and aid, has forced an abrupt and difficult conversion from conventional agriculture to organic farming on a nationwide scale.

Between the Cuban revolution in 1959 and the breakdown of trading relations with the Soviet Union in 1989, Cuba experienced rapid modernization, a high degree of social equity and welfare, and a strong dependence on external aid. Cuba's economy was supported during this period by the most modern agricultural system in Latin America. Farming techniques, levels of mechanization, and output often rivaled those in the United States. The main crop was sugarcane, almost all of which was grown on huge state farms and sold to the former Soviet Union at premium prices. More than half of all food eaten by Cubans came from abroad, as did most fertilizers, pesticides, fuel, and other farm inputs on which agricultural production depended.

Under the theory of comparative advantage, it seemed reasonable for Cuba to rely on international trade. With the collapse of the socialist bloc, however, Cuba's economy also fell apart. In 1990, wheat and grain imports decreased by half and other foodstuffs declined even more. At the same time, fertilizer, pesticide, and petroleum imports were down 60 to 80 percent. Farmers faced a dual challenge: how to produce twice as much food using half the normal inputs.

The crisis prompted a sudden turn to a new model of agriculture. Cuba was forced to adopt sustainable, organic farming practices based on indigenous, renewable resources. Typically, it takes three to five years for a farmer in the United States to make the change from conventional to organic farming profitable. Cuba, however, didn't have that long; it needed food immediately.

This organic, communal garden in Havana, Cuba, provides most of the fresh produce for the housing complex surrounding it.

Cuba's agricultural system is based on a combination of old and new ideas. Broad community participation and use of local knowledge is essential. Scientific, adaptive management is another key. Diverse crops suitable to local microclimates, soil types, and human nutritional needs have been adopted. Natural, renewable energy sources such as wind, solar, and biomass fuels are being substituted for fossil fuels. Oxen and mules have replaced some 500,000 tractors idled by lack of fuel.

Soil management is vital for sustainable agriculture. Organic fertilizers substitute for synthetic chemicals. Livestock manure, green manure crops, composted municipal garbage, and industrial-scale cultivation of high-quality humus in earthworm farms all replenish soil fertility. In 1995 more than 100,000 metric tons of worm compost were produced and spread on fields.

Pests are suppressed by crop rotation and biological controls rather than chemical pesticides. For example, the parasitic fly (*Lixophaga diatraeae*) controls sugarcane borers; wasps in the genus *Trichogramma* feed on the eggs of grain weevils; while the predatory ant (*Pheidole megacephala*) attacks sweet potato weevils. Pest control also involves innovative use of biopesticides, such as *Bacillis thuringiensis,* that are poisonous or repellent to crop pests. Finally, integrated pest management includes careful monitoring of crops and measures to build populations of native beneficial organisms and to enhance the vigor and defenses of crop species.

Worker brigades from schools and factories help provide farm labor during harvest season. In addition to state farms and rural communes, urban gardening provides a much-needed supplement to city diets. Individual gardens are encouraged, but community or institutional gardens—schools, factories, and mass organizations—also produce large amounts of food.

Although food supplies in Cuba still are limited and diets are austere, the crisis wasn't as bad as many feared. In some ways, this draconian transition is fortunate. Cuba is now a world leader in sustainable agriculture. What do you think; is Cuba's experiment with organic agriculture a model that could be exported to other countries? Does the centrally planned economy of Cuba create special conditions for this project? Which of the aspects of Cuban farming might work in your country?

While IPM can be a good alternative to chemical pesticides, it also presents environmental risks in the form of exotic organisms. Wildlife biologist George Boettner of the University of Massachusetts reported in 2000 that biological controls of gypsy moths, which attack fruit trees and ornamental plants, have also decimated populations of native North American moths. *Compsilura* flies, introduced in 1905 to control the gypsy moths, have a voracious appetite for other moth caterpillars as well. One of the largest North American moths, the Cecropia moth (*Hyalophora cecropia*), with a 15 cm wingspan, was once ubiquitous in the eastern United States, but it is now rare in regions where *Compsilura* flies were released.

10.5 REDUCING PESTICIDE EXPOSURE

The total numbers and amounts of different chemicals to which we are potentially exposed is overwhelming. The 2.4 million metric tons of pesticides used in the United States every year contain about 600 active ingredients combined with more than 1,200 presumably inactive carriers, solvents, preservatives, and other ingredients in about 25,000 commercial products. Less than 10 percent of the active pesticide ingredients have been subjected to a full battery of ten chronic health-effect tests, and studies of the inactive ingredients have started only recently (fig. 10.20). Of the 321 pesticides screened so far, the EPA reports that 146 are probable human carcinogens. Since 1972, only 40 pesticides have been banned. Critics charge that this slow pace puts all of us at risk and that assessment and regulation must proceed more quickly.

Who regulates pesticides?

Three federal agencies share responsibility for regulating pesticides used in food production in the United States: the Environmental Protection Agency (EPA), the Food and Drug Administration (FDA), and the Department of Agriculture (USDA). The EPA regulates the sale and use of pesticides under the Federal Insecticide, Fungicide, and Rodenticide Act (FIFRA), which mandates the "registration" (licensing) of all pesticide products. Based on scientific studies, the EPA determines which pesticides will not pose significant risks to human health or the environment when used according to label directions. Under the Federal Food, Drug, and Cosmetic Act (FFDCA), the EPA also sets "tolerance levels" or limits for the amount of pesticide residues that lawfully may remain in or on foods marketed in the United States, whether grown domestically or imported from abroad. The FDA and USDA enforce pesticide use and tolerance levels set by the EPA. These agencies can seize and destroy food shipments found to contain pesticide residues in violation of limits set by the EPA.

Think About It

Pesticide residues in food are a major concern for many people, but most of us also use toxic chemicals in other aspects of our life as well. Look around your home, how many different toxic products can you find? Are they all necessary? Would you have alternatives to these products?

In Canada, the Pest Control Products Act and Regulations, administered by Agriculture Canada, performs similar functions. Individual Canadian provinces also set local standards. The Ontario Ministry of Agriculture and Food, for instance, introduced a comprehensive program in 1988 to reduce pesticide use by 50 percent. Nonchemical pest control methods, on-farm education, and biotechnology and the development of pest-resistant crop varieties all played a role in this program. Savings on chemical costs are about (C)$100 million per year.

In 1996, as a result of growing evidence of pesticide risks to children, Congress passed the Food Quality Protection Act. Key provisions of this act direct the EPA to set pesticide tolerance levels based on potential aggregate pesticide exposures. Benefits of pesticides can be considered when evaluating tolerances for carcinogens in adults, but not in calculating risks on reproduction and prenatal development or on exposure levels for infants and children. This act requires that by 2006 the EPA reassess the risks of over 9,700 pesticides, reevaluate the allowable levels of pesticides that can remain in or on food, and apply a safety factor of 10 where comprehensive and complete information is not available regarding the cumulative impacts on children.

For the first time, under this act, the EPA is required to examine exemptions given to "inert" pesticide ingredients used to dilute or carry the active agents. Of the 2,500 substances added to pesticides but not usually named on product labels, more than 650 have been identified as hazardous by federal, state, or international agencies. Nearly half of these "inert" ingredients have been classified as carcinogens, occupation hazards, or air and water pollutants. Naphthalene, for instance, which is commonly described as an inert ingredient, is designated a hazardous air pollutant under the Clean Air Act and a priority pollutant under the Clean Water Act.

Based on these new rules, the EPA recently banned the use of methyl parathion on all fruits and many vegetables. It also limited the quantity of azinphos methyl that can be used on foods common in children's diets, such as apples, peaches, and pears. Similarly, the EPA has prohibited most uses of chlorpyrifos, commonly known by its trade name Dursban. Found in more than 1,000 commercial products, this organophosphate represented 20 percent of all residential insecticide applications. It was shown, however, to cause brain damage in developing rat pups, and was judged an unacceptable risk to children. Some 36 other organophosphates are still being reviewed. The EPA is also studying hormone-disrupting chemicals and will probably regulate products on this basis in the near future.

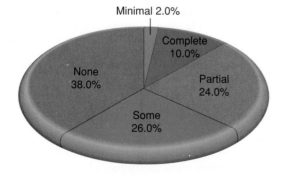

Minimal 2.0%
Complete 10.0%
None 38.0%
Partial 24.0%
Some 26.0%

FIGURE 10.20 Level of testing for health effects of 3,350 active and inert pesticide ingredients. Complete testing includes a battery of ten tests for both chronic and acute toxicity and genetic effects.
Source: National Academy of Sciences.

Wood preservatives are another example of pesticide use especially dangerous to children. Thirty years ago, the principal wood preservatives were petroleum-based compounds like tar and creosote. These highly toxic biocides prevented growth of molds, invasion by termites, and just about all other living organisms, but also were dangerous to humans. Many Superfund sites are contaminated with these wood preservatives. As understanding of the lingering toxicity of creosote and its chemical cousins emerged, consumers switched to CCA (chromated copper arsenate) pressure-treated lumber. We now know that the arsenic in CCA-treated wood is a serious hazard, especially to workers exposed to sawdust or children who come in contact with the lumber in play equipment, decks, sidewalks, and other outdoor settings (see What Do You Think? p. 221). CCA-treated wood was banned for virtually all residential uses in the United States in 2003, but are the substitutes safer?

With globalization, our food comes increasingly from other countries where rules governing food safety may be less strict than we would like. The U.S. General Accounting Office reports that the volume of imported food has doubled over the last five years, while food inspections have decreased during that same time. Currently, about 38 percent of the fruits and 12 percent of the vegetables consumed by Americans are imported. The Food and Drug Administration (FDA) inspects only about 2 percent of all food shipments. Less than 0.2 percent of imported fruits and vegetables are examined for microbial contamination, and even a smaller fraction is tested for pesticides. In studies of a wide range of foods collected by the USDA, the State of California, and the Consumers Union between 1994 and 2000, 73 percent of conventionally grown food had residue from at least one pesticide and were six times as likely as organic foods to contain multiple pesticide residues. Only 23 percent of the organic samples of the same groups had any residues. Using these data, the Environmental Working Group has assembled a list of the fruits and vegetables most commonly contaminated with pesticides (table 10.1).

In 2001, the Canadian Supreme Court ruled that pesticides can be excluded from residential areas. In response, the province of Quebec announced that it would immediately ban the use of 30 "cosmetic" pesticides on public lands including parks, schools, day-care centers and hospitals. By 2005, the province will prohibit the use of most nonfarm pesticides. "People's health is more important than a perfect lawn," said Quebec Environment Minister Andre Boisclair. "I enjoin Quebecers to no longer use pesticides."

Is organic the answer?

Many farmers and consumers are turning to organic agriculture as a way to reduce pesticide exposures. Numerous studies have shown that organic, sustainable agriculture is more eco-friendly and leaves soils healthier than intensive, chemical-based monoculture cropping. A Swiss study spanning two decades found that average yields on organic plots were 20 percent less than adjacent fields farmed by conventional methods. Costs also were lower and prices paid for organic produce were higher, so that net returns were

TABLE 10.1	
The Twelve Most Contaminated Foods	
Rank	**Food**
1.	Strawberries
2.	Bell peppers
3.	Spinach
4.	Cherries (U.S.)
5.	Peaches
6.	Cantaloupe (Mexican)
7.	Celery
8.	Apples
9.	Apricots
10.	Green beans
11.	Grapes (Chilean)
12.	Cucumbers

Source: Environmental Working Group, 2002.

actually higher with organic crops. Energy use, for example, was 56 percent less per unit of yield in organic farming than conventional approaches. In addition, root fungi were 40 percent higher, earthworms were three times as abundant, and spiders and other pest-eating predators were doubled in the organic plots. Both the organic farmers and their families reported better health and greater satisfaction than their conventional neighbors. A study of food quality in Sweden reported that organic food contained more cancer-fighting polyphenolics and antioxidants than pesticide-treated produce. As mentioned in chapter 9, farms using sustainable techniques can have up to 400 times less erosion after heavy rains than monoculture row crops.

Currently, less than 1 percent of all American farmland is devoted to organic growing, but the market for organic products may stimulate more conversion to this approach in the future. Organic food is much more popular in Europe than in North America. Tiny Liechtenstein is probably the leader among industrialized nations with 18 percent of its land in certified organic agriculture. Sweden is second with 11 percent of its land in organic production.

Considerable controversy and confusion have existed in the United States over what constitutes organic agriculture. How do you know if you can believe claims that food has been grown safely (fig. 10.21)? Walmart, for example, has vowed to become the top seller of organic products in the United States. Will this mean that much of our organic food will come from overseas? Already, more than 2,000 farms in China and India are certified "organic," but what does that really mean?

With the market for organic food generating $11 billion per year, some farmers and marketers are likely to try to pass off foods grown inexpensively with pesticides as more valuable organic produce.

FIGURE 10.21 These strawberries were grown organically, but how could you tell? The USDA reports more pesticides in commercial strawberries than any other fruit.

FIGURE 10.22 Your local farmers' market is a good source of locally grown, organic produce.

According to U.S. Department of Agriculture rules, products labeled "100 percent organic" must be produced without hormones, antibiotics, pesticides, synthetic fertilizers, or genetic modification. "Organic" means that at least 95 percent of the ingredients must be organic. "Made with organic ingredients" must contain at least 70 percent organic contents. Products containing less than 70 percent organic ingredients can list them individually. Organic animals must be raised on organic feed, given access to the outdoors, given no steroidal growth hormones, and treated with antibiotics only to treat diseases.

Many who endorse the concept of organic food are disappointed by the limited scope of these definitions. They point out that there are three goals for organic, sustainable agriculture. One is growing food in harmony with nature—a nonindustrial approach to food production that treats animals humanely and avoids the use of chemical pesticides. Another is a food distribution system based on co-ops, farmers' markets, community supported agriculture, and local production rather than crops flown and trucked from far-away places where consumers have no connection with producers (fig. 10.22). The third goal is simple, wholesome, nutritious food. It's true that the USDA rules eliminate many of the pesticides, but a myriad of highly processed ingredients are allowed in organic food. You might find yourself someday buying organic Coca-Cola to go with your organic TV dinner. You can already buy organic, intercontinental grapes in which thousands of calories of jet and diesel fuel were consumed to transport every calorie of food energy from Chile to your supermarket. Some of the best farmers are refusing to apply for organic certification, regarding it as a meaningless term under these rules.

Other people doubt that organic growers can produce enough—even within these limited definitions—to feed everyone. We need the efficiency of chemical pesticides and largescale, chemical-intensive farming to provide food for the 8 to 9 billion people expected by the middle of this century. According to Vaclav Smil, a Canadian geographer, without synthetic fertilizer and pesticides we could feed only 2 to 3 billion people. Dennis Avery, director of global food issues for the conservative Hudson Institute calculates that if we were to depend entirely on animal manure to fertilize crops, America would need to increase its cattleherd ninefold and convert almost the entire U.S. landmass to pastureland to support all those animals.

Furthermore, eating organic food doesn't protect you from all risks. Dr. Elsa Murano, undersecretary for food safety in the U.S. Department of Agriculture warns that consumers should be wary of organic food. "We must remember that bacteria and parasites are also all-natural," she said. "As a microbiologist, I know that preservatives are used in foods for a reason . . . to preserve food against the growth of microorganisms." Supporting her point, a Danish research group found more campylobacteria—one of the most common causes of diarrhea—in organic chickens than in standard chickens. Some natural pesticides allowed in organic farming have been linked to health problems in both producers or consumers. Rotenone, for example, has been associated with Parkinson's disease in farmers, while *Bacillus thuringiensis* (Bt) toxin has been found to cause allergies and asthma in crop pickers and handlers.

Biochemist Bruce Ames, who is famous for developing the most widely used cancer screening test, argues that pesticides confer a public-health advantage that far outweighs any risks posed by their toxicity or hormone-disrupting effects. "If the EPA eliminates all pesticides," he says, "all it's going to do is increase the price of fruits and vegetables, and there will be more cancer." It's not that he argues that pesticides are safe, but rather that we're ignoring the real risks we face like smoking, poor nutrition, lack of health care, and accidents. We could save more lives, he claims, by focusing on the factors that kill the vast majority of us than by spending so much time and money worrying about relatively minor hazards.

You can reduce your own risks

Even though pesticide exposure may not be the greatest threat that most of us face, why not take reasonable steps to reduce your risks? There are many things each of us can do to minimize pesticides in our food and to encourage sustainable agriculture.

- Wash and scrub all fresh fruits and vegetables thoroughly under running water.
- Peel fruits and vegetables when possible. Throw away the outer leaves of leafy vegetables such as lettuce and cabbage.
- Store food carefully so it doesn't get moldy or pick up contaminants from other foods. Use food as soon as possible to ensure freshness.

- Cook or bake foods that you suspect have been treated with pesticides to break down chemical residues.
- Trim the fat from meat, chicken, and fish. Eat lower on the food chain where possible to reduce bioaccumulated chemicals.
- Don't pick and eat berries or other wild foods that grow on the edges of roadsides where pesticides may have been sprayed.
- Grow your own fruits and vegetables without using pesticides or with minimal use of dangerous chemicals.
- Ask for organically grown food at your local grocery store or shop at a farmers' market or co-op where you can get such food.

CONCLUSION

Pesticides bring us many benefits, but they have costs as well. Indiscriminate and profligate use of a variety of powerful chemicals kills beneficial organisms, such as honeybees, and often does as much harm as good. Often highly persistent and mobile in the environment, many pesticides move through air, water, and soil to bioaccumulate in food chains and cause serious ecological and human health problems.

Several good alternatives can reduce our dependence on the most dangerous pesticides. Among these are behavioral changes, such as crop rotation, use of biological controls, mechanical cultivation, and planting mixed polycultures rather than vast monocultures. Integrated pest management is a flexible, ecologically based strategy that carefully monitors pest populations to apply just the right kind and amount of pesticide at just the right time to do the most good. It can result in dramatic reductions in pesticide use and unintended consequences while minimizing economic losses. Consumers may have to learn to accept less than perfect fruits and vegetables if they want to reduce their pesticide exposure. In the end, both food producers and consumers will need to be wiser, more perceptive, and, perhaps, more cautious about how we try to defend ourselves against pests.

REVIEWING LEARNING OUTCOMES

By now you should be able to explain the following points:

10.1 Define the major types of pesticides and describe the pest they are meant to manage.
- People have always known of ways to control pests.
- Modern pesticides provide benefits, but also create problems.
- There are many types of pesticides.

10.2 Identify some important pesticide benefits.
- We have made progress in controlling many insect-borne diseases.
- Without pesticides, we might lose two-thirds of conventional crops.

10.3 Illustrate some important pesticide problems.
- Pesticides often poison nontarget species.
- Pesticide resistance is often rapid and widespread.
- Pesticide misuse can create new pests.
- Some persistent pesticides can move long distances in the environment.
- Many pesticides cause human health problems.

10.4 Summarize some alternatives to current pesticide uses.
- We can change our behavior.
- Useful organisms can help us control pests.
- Integrated pest management uses a combination of techniques to fight pests.

10.5 Discuss how we might reduce pesticide exposure.
- Who regulates pesticides?
- Is organic the answer?
- You can reduce your own risks.

PRACTICE QUIZ

1. What is a pest and what are pesticides? What is the difference between a biocide, a herbicide, an insecticide, and a fungicide?

2. What is DDT and why was it considered a "magic bullet"? Why was it listed among the "dirty dozen" persistent organic pollutants (POPs)?

3. How much pesticide is used worldwide and which of the general categories accounts for the greatest use? Why are "inert" ingredients of concern?

4. Describe fumigants, botanicals, chlorinated hydrocarbons, organophosphates, carbamates, and microbial pesticides.

5. What are endocrine disrupters and why are they dangerous?

6. Explain why pests often resurge or rebound after treatment with pesticides and how they become pesticide resistant. What is a pesticide treadmill?

7. Identify three major categories of alternatives to synthetic pesticides and describe, briefly, how each one works.

8. What is IPM, and how is it used in pest control?

9. Why are children more susceptible to pesticides than adults are? What is being done to protect children?

10. List eight things you could do to reduce your dietary exposure to pesticides.

CRITICAL THINKING AND DISCUSSION QUESTIONS

1. Paul Müller, the discoverer of DDT, received a Nobel Prize for his work. What do you think; was it a gift or curse for the world?

2. If you were a public health official in a country in which malaria, filariasis, or onchocerciasis were rampant, would you spray DDT to eradicate vector organisms? Would you spray it in your own house?

3. Pesticide treadmill, "dirty dozen" pesticides, and environmental estrogens are all highly emotional terms. Why would some people choose to use or not use these terms? Can you suggest alternative terms for the same phenomena that convey different values?

4. Many farmworkers who suffer from pesticide poisoning are migrants or minorities. Is this evidence of environmental

racism? What evidence would you look for to determine whether environmental justice is being served?

5. Suppose that a developing country believes that it needs a pesticide banned in the United States or Canada to feed or protect the health of its people. Are we right to refuse to sell that pesticide?

6. Do you believe we could grow enough food for everyone if we switched to organic, low-input, sustainable agriculture? How would you design a research program to test this question?

7. What would you personally consider a "negligible" risk? Would you eat grapes if you knew they had a measurable amount of some pesticide? How small would the amount have to be?

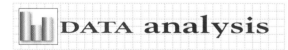 Assessing Health Risks

As we discussed in chapter 8, children are much more vulnerable than adults to environmental health hazards. Pound for pound, children drink more water, eat more food, and breathe more air than do adults. Putting fingers, toys, and other objects into their mouths increases children's exposure to toxins in dust or soil. And compared to adults, children generally have less-developed immune systems or processes to degrade or excrete toxins. Because children spend a significant portion of time at school, pesticides used in schools may be an important contributor to overall exposure. There is no federal mandate to collect data on pesticide use in schools, and few states require reporting of this

information, so we know little about how pesticides are used in the nation's 110,000 schools.

In 1999 the state of Minnesota conducted a survey on pesticide use in schools. This study focused on indoor uses of pesticides, because pesticide residues can be persistent indoors, and because children spend most of their time indoors when they are at school. The graph below presents results from this study. It presents information about the frequency of spraying and where these pesticides were used. These measures are only a surrogate for exposure. They don't provide information about the toxicity of pesticides used or details about how they were applied and thus cannot provide a complete representation of the risk of adverse effects following exposure. Nevertheless, the frequency of exposure is valuable information about the extent of risk.

After studying the graph, answer the following questions:

1. Do the numbers for all categories add up to 100 percent?

2. Why does this matter?

3. Which areas are sprayed most "as needed"?

4. Why might this be important?

5. Which areas are sprayed most often overall?

6. How might you find an answer for question 5 other than adding up the first three groups of bars?

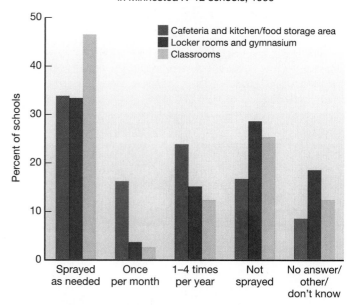

Frequency of application of pesticides in Minnestoa K–12 schools, 1999

Pesticide use in Minnesota schools.
Data from U.S. EPA, 2003.

Student interns weed experimental biodiversity plots at Cedar Creek Natural History Area in Minnesota.

Biodiversity

Preserving Species

The first rule of intelligent tinkering is to save all the pieces.

—Aldo Leopold—

LEARNING OUTCOMES

After studying this chapter, you should be able to:

11.1 Discuss biodiversity and the species concept.

11.2 Summarize some of the ways we benefit from biodiversity.

11.3 Characterize the threats to biodiversity.

11.4 Evaluate endangered species management.

11.5 Scrutinize captive breeding and species survival plans.

Case Study Species Diversity Promotes Ecological Resilience

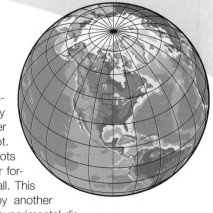

Does diversity matter? Ecologists long believed that a diversity of species in a community is not just beautiful and fascinating, but also functionally important. If disturbed, do communities recover quickly when they contain many species? Are they more stable, in other words? This question, raised by ecologists in the 1950s, suggests a strong reason to preserve the earth's biodiversity, but not until 1994 did someone present data that helped answer it. In a long-term study of prairie plant communities, ecologist Dave Tilman and his colleagues planted dozens of experimental plots at Cedar Creek Natural History Area, each plot containing different numbers of species (fig. 11.1). In 1988, central Minnesota experienced its hottest, driest summer in 50 years. As Tilman's team watched their plots recover from the drought, they measured growth in the plots by carefully clipping, counting, and weighing every plan (see opposite page). By 1992, it was clear that plots with five or fewer species were recovering slowly, while higher-diversity plots had reached or exceeded their pre-drought productivity (fig. 11.2).

Because the question of diversity and recovery was controversial, the strength of Tilman's results depended on duplicated plots (replicates) and a long period of study. If he had used only one plot for each level of species diversity, chance alone might have depressed productivity. Another question in any experiment is how well experimental results represent other situations and other samples Tilman's team approximated an answer to this question by reporting a range of values around the average recovery rates, in this case the "standard error." This standard statistical measure and its related value, confidence interval, show the range within which nearly all means should fall if someone else did the same experiment. In this way we describe how confident we are that the results can be applied to other situations (For more about confidence limits and uncertainty, see the Data Analysis exercise at the end of this chapter.)

Later experiments at Cedar Creek showed that species-rich plots recovered more fully after drought because some species were severely harmed by drought, others went dormant but didn't die, and still others continued growing but at a slow rate. When rain returned, surviving plants revived filling the plots with green leaves and stems.

The chance that a species-poor plot contained quickly recovering species was lower than for a species-rich plot. Consequently, low-diversity plots were more likely to attain their former productivity slowly, if at all. This explanation was bolstered by another experimental discovery that nitrogen—a limiting factor in the sterile, sandy soils of the plots—was used up more completely in species-rich plots than in species-poor plots.

The experimental data from Cedar Creek prairies stimulated tests in other ecosystems. Is resilience after disturbance always evident in species-rich plant communities, such as rainforests of Central and South America? (Preliminary results show that it depends on how big and severe the disturbance is.) Do species-poor plant communities, such as dune grasslands and salt marshes, recover only slowly following disturbance? Actually, some simple plant communities appear quite stable where disturbances like wind, waves, and storms are frequent.

What are the implications of Tilman's research for agricultural fields populated by a single crop plant? One important application may be in biofuels production. Ethanol (a biomass fuel) made from starch or cellulose may be an attractive alternative to petroleum-based fuels (chapter 20). Tilman's study suggests that, given variable weather over several years, a diverse biomass crop can achieve an economical net energy yield while providing greater environmental benefits than does a single-species crop. The general lesson is that keeping a diversity of species as a backdrop to human existence may serve us best in the long run; and if species diversity is maintained, nature may be more resilient than we imagine.

One of the most important debates in conservation biology is whether the best way to preserve biodiversity and ecological resilience is through protecting rare and endangered species or by conserving whole communities and landscapes. In this chapter, we'll look at the benefits from, and threats to, species diversity. In the next chapter, we'll look at preservation of broader ecological units.

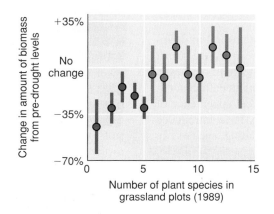

FIGURE 11.1 Experimental grassland plots at Cedar Creek.

FIGURE 11.2 By 1992, plots with five or fewer species were still below pre-drought productivity, while higher-diversity plots had recovered, (Bars show standard error, or variability, around the means.)
Source: Tilman & Dowling (1994) *Nature 367*: 363–365.

11.1 BIODIVERSITY AND THE SPECIES CONCEPT

From the driest desert to the dripping rainforests, from the highest mountain peaks to the deepest ocean trenches, life on earth occurs in a marvelous spectrum of sizes, colors, shapes, life cycles, and interrelationships. Think for a moment how remarkable, varied, abundant, and important the other living creatures are with whom we share this planet (fig. 11.3). How will our lives be impoverished if this biological diversity diminishes?

What is biodiversity?

Previous chapters of this book have described some of the fascinating varieties of organisms and complex ecological relationships that give the biosphere its unique, productive characteristics. Three kinds of **biodiversity** are essential to preserve these ecological systems: (1) *genetic diversity* is a measure of the variety of different versions of the same genes within individual species; (2) *species diversity* describes the number of different kinds of organisms within individual communities or ecosystems; and (3) *ecological diversity* assesses the richness and complexity of a biological community, including the number of niches, trophic levels, and ecological processes that capture energy, sustain food webs, and recycle materials within this system.

Within species diversity, we can distinguish between *species richness* (the total number of species in a community) and *species evenness* (the relative abundance of individuals within each species). To illustrate this difference, imagine two ecological communities, each with 10 species and 100 individual plants or animals. Suppose that one community has 82 individuals of one species and two each of nine other species. In the other community, all 10 species are equally abundant, meaning they have 10 individuals each. Although the species richness is the same, if you were to walk through these communities, you'd have the impression that the second is much more diverse because you'd be much more likely to encounter a greater variety of organisms.

What are species?

As you can see, the concept of species is fundamental in defining biodiversity, but what, exactly, do we mean by the term? When Carolus Linnaeus, the great Swedish taxonomist, began our system of scientific nomenclature in the eighteenth century, classification was based entirely on the physical appearance of adult organisms. In recent years, taxonomists have introduced other characteristics as means of differentiating species. In chapter 3, we defined species in terms of *reproductive isolation;* that is, all the organisms potentially able to breed in nature and produce fertile offspring. As we pointed out, this definition has some serious problems, especially among plants and protists, many of which either reproduce asexually or regularly make fertile hybrids.

Another definition favored by many evolutionary biologists is the *phylogenetic species concept* (PSC), which emphasizes the

FIGURE 11.3 This coral reef has both high abundance of some species and high diversity of different genera. What will be lost if this biologically rich community is destroyed?

branching (or cladistic) relationships among species or higher taxa regardless of whether organisms can breed successfully.

A third definition, favored by some conservation biologists, is the *evolutionary species concept* (ESC), which defines species in evolutionary and historic terms rather than reproductive potential. The advantage of this definition is that it recognizes that there can be several "evolutionarily significant" populations within a genetically related group of organisms. Unfortunately, we rarely have enough information about a population to judge what its evolutionary importance or fate may be. Paul Ehrlich and Gretchen Daily calculate that each species averages 220 evolutionarily significant populations. This calculation could mean that there are up to 10 billion different populations in total. Deciding which ones we should protect becomes an even more daunting prospect.

Molecular techniques are revolutionizing taxonomy

Increasingly, DNA sequencing and other molecular techniques are giving us insights into taxonomic and evolutionary relationships. As we described in chapter 3, each individual has a unique hereditary complement called the *genome.* The genome is made up of the millions or billions of nucleotides in DNA arranged in a very specific sequence that spells out the structure of all the proteins that make up cellular structure and machinery of every organism. As you know from modern court cases and paternity suits, we can use that DNA sequence to identify individuals with a very high degree of certainty. Now, this very precise technology is being applied to identify species in nature.

Because only a small amount of tissue is needed for DNA analysis, species classification—or even the identity of individual animals—can be made on samples such as feathers, fur, or feces when it's impossible to capture living creatures. For example, whale meat for sale in Japanese markets was shown to be from protected species using DNA analysis. Samples of hair from

FIGURE 11.4 DNA analysis revealed a new tiger subspecies (*T. panthera jacksoni*) in Malaysia. This technology has become essential in conservation biology.

indistinguishable from the black-tailed flycatcher (*Polioptila californica pontilis*), which is abundant in adjacent areas of Mexico.

In some cases, molecular taxonomy is causing a revision of the basic phylogenetic ideas of how we think evolution proceeded. Studies of corals and other cnidarians (jellyfish and sea anemones), for example, show that they share more genes with primates than do worms and insects. This evidence suggests a branching of the family tree very early in evolution rather than a single sequence from lower to higher animals.

How many species are there?

At the end of the great exploration era of the nineteenth century, some scientists confidently declared that every important kind of living thing on earth would soon be found and named. Most of those explorations focused on charismatic species such as birds and mammals. Recent studies of less conspicuous organisms such as insects and fungi suggest that millions of new species and varieties remain to be studied scientifically.

Think About It

Compare the estimates of known and threatened species in table 11.1. Are some groups over-represented? Are we simply more interested in some organisms or are we really a greater threat to some species?

scratching pads has allowed genetic analysis of lynx and bears in North America without the trauma of capturing them. Similarly, a new tiger subspecies (*Tigris panthera jacksoni*) was detected in Southeast Asia based on blood, skin, and fur samples from zoo and museum specimens (fig. 11.4).

This new technology can help resolve taxonomic uncertainties in conservation. In some cases, an apparently widespread and low-risk species may, in reality, comprise a complex of distinct species, some rare or endangered. Such is the case for a unique New Zealand reptile, the tuatara. Genetic marker studies revealed two distinct species, one of which needed additional protection. Similar studies have shown that the northern spotted owl (*Strix occidentalis caurina*) is a genetically distinct subspecies from its close relatives the California spotted owl (*S. occidentalis occidentalis*) and the Mexican spotted owl (*S. occidentalis lucida*), and therefore deserves continued protection.

On the other hand, in some cases, genetic analysis shows that a protected population is closely related to another much more abundant one. For example, the colonial pocket gopher from Georgia is genetically identical to the common pocket gopher and probably doesn't deserve endangered status. The California gnatcatcher (*Polioptila californica californica*), which lives in the coastal sage scrub between Los Angeles and the Mexican border, was listed as a threatened species in 1993, and thousands of hectares of land worth billions of dollars were put off-limits for development. Genetic studies showed, however, that this population is

TABLE 11.1

Current Estimates of Known and Threatened Living Species by Taxonomic Group

	Known	Threatened
Mammals	5,416	1,093
Birds	9,934	1,206
Reptiles	8,240	341
Amphibians	5,918	1,811
Fishes	29,300	1,173
Insects	950,000	623
Molluscs	70,000	975
Crustaceans	40,000	459
Other animals	130,000	44
Mosses	15,000	80
Ferns and allies	13,025	139
Gymnosperms	980	306
Dicotyledons	199,350	7,086
Monocotyledons	59,300	770
Lichens	10,000	2
Mushrooms	16,000	1
Total	**1,562,663**	**16,118**

Source: Data from IUCN Red List, 2006.

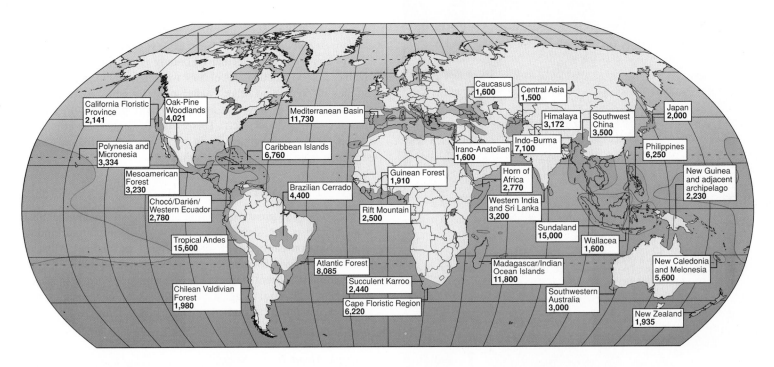

FIGURE 11.5 Biodiversity "hot spots," identified by Conservation International, tend to be in tropical or Mediterranean climates and on islands, coastlines, or mountains where many habitats exist and physical barriers encourage speciation. Numbers indicate endemic species.
Source: Conservation International, 2005.

The 1.5 million species presently known (table 11.1) probably represent only a small fraction of the total number that exist. Based on the rate of new discoveries by research expeditions—especially in the tropics—taxonomists estimate that there may be somewhere between 3 million and 50 million different species alive today. In fact, some taxonomists estimate that there are 30 million species of tropical insects alone. The upper limits for these estimates assume a high degree of ecological specialization among tropical insects. A recent study in New Guinea, however, found that 51 plant species were host to 900 species of herbivorous insects. This evidence would suggest no more than 4 to 6 million insect species worldwide.

About 76 percent of all known species are invertebrates (animals without backbones, such as insects, sponges, clams, worms, etc.). This group probably makes up the vast majority of organisms yet to be discovered and may constitute 95 percent of all species. What constitutes a species in bacteria and viruses is even less certain than for other organisms, but there are large numbers of physiologically or genetically distinct varieties of these organisms.

The numbers of threatened species shown in table 11.1 are those officially listed as endangered by the IUCN. This represents only a small fraction of those actually at risk. It's estimated that one-third of all amphibians, for example, are declining and threatened with extinction. We'll discuss this issue later in this chapter.

Hot spots have exceptionally high biodiversity

Of all the world's currently identified species, only 10 to 15 percent live in North America and Europe. The greatest concentration of different organisms tends to be in the tropics, especially in tropical rainforests and coral reefs. Norman Myers, Russell Mittermeier, and others have identified **biodiversity hot spots** that have at least 1,500 endemics (species that occur nowhere else) and have lost at least 70 percent of their habitat owing to, for example, deforestation or invasive species. Using plants and land-based vertebrates as indicators, they have proposed 34 hot spots that represent a high-priority for conservation because they have both high biodiversity and a high risk of disruption by human activities (fig. 11.5). Although they occupy only 1.4 percent of the world's land area, these hot spots house three-quarters of the worlds most threatened mammals, birds, and amphibians. The hot spots also account for about half of all known higher plant species and 42 percent of all terrestrial vertebrate species. The hottest of these hot spots tend to be tropical islands such as Madagascar, Indonesia, and the Philippines, where geographic isolation has resulted in large numbers of unique plants and animals. Special climatic conditions such as those found in South Africa, California, and the Mediterranean Basin also produce highly distinctive flora and fauna.

Some areas with high biodiversity—such as Amazonia, New Guinea, and the Congo basin, for example—aren't included in this hot spot map because most of their land area is relatively undisturbed. Other groups prefer different criteria for identifying important conservation areas. Aquatic biologists, for example, point out that coral reefs, estuaries, and marine shoals host some of the most diverse wildlife communities in the world, and warn that freshwater species are more highly endangered than terrestrial ones. Other scientists worry that the hot spot approach neglects many rare species and major groups that live in less biologically rich areas (cold spots). Nearly half of all terrestrial vertebrates, after all, aren't represented in Myers's hot spots. Focusing on a few hot spots also doesn't recognize the importance of certain species and ecosystems to human beings. Wetlands, for instance, may contain just a few, common plant species but perform valuable ecological services, such as filtering water, regulating floods, and serving as nurseries for fish. Some conservationists argue that we should concentrate on saving important biological communities or landscapes rather than rare species.

Anthropologists point out that regions with high biodiversity are also often home to high cultural diversity as well (see fig. 1.20). It isn't a precise correlation; some countries, like Madagascar, New Zealand, and Cuba, with a high percentage of endemic species, have only a few cultural groups. Often, however, the varied habitat and high biological productivity of places like Indonesia, New Guinea, and the Philippines that allow extensive species specialization also have fostered great cultural variety. By preserving some of the 7,200 recognized language groups in the world—more than half of which are projected to disappear in this century—we might also protect some of the natural setting in which those cultures evolved.

11.2 HOW DO WE BENEFIT FROM BIODIVERSITY?

We benefit from other organisms in many ways, some of which we don't appreciate until a particular species or community disappears. Even seemingly obscure and insignificant organisms can play irreplaceable roles in ecological systems or be the source of genes or drugs that someday may be indispensable.

All of our food comes from other organisms

Many wild plant species could make important contributions to human food supplies either as new crops or as a source of genetic material to provide disease resistance or other desirable traits to current domestic crops. Norman Myers estimates that as many as 80,000 edible wild plant species could be utilized by humans. Villagers in Indonesia, for instance, are thought to use some 4,000 native plant and animal species for food, medicine, and other valuable products. Few of these species have been explored for possible domestication or more widespread cultivation. A 1975 study by the National Academy of Science (U.S.) found that Indonesia has 250 edible fruits, only 43 of which have been cultivated widely (fig. 11.6).

Living organisms provide us with many useful drugs and medicines

More than half of all modern medicines are either derived from or modeled on natural compounds from wild species (table 11.2) The United Nations Development Programme estimates the value of pharmaceutical products derived from developing world plants, animals, and microbes to be more than $30 billion per year. Indigenous communities that have protected and nurtured the biodiversity on which these products are based are rarely acknowledged—much less compensated—for the resources extracted from them. Many

FIGURE 11.6 Mangosteens from Indonesia have been called the world's best-tasting fruit, but they are practically unknown beyond the tropical countries where they grow naturally. There may be thousands of other traditional crops and wild food resources that could be equally valuable but are threatened by extinction.

consider this expropriation "biopiracy" and call for royalties to be paid for folk knowledge and natural assets.

Consider the success story of vinblastine and vincristine. These anticancer alkaloids are derived from the Madagascar periwinkle (*Catharanthus roseus*) (fig. 11.7). They inhibit the growth of cancer cells and are very effective in treating certain kinds of cancer. Before these drugs were introduced, childhood leukemias were invariably fatal. Now the remission rate for some childhood leukemias is 99 percent. Hodgkin's disease was 98 percent fatal a few years ago, but is now only 40 percent fatal, thanks to these compounds. The total value of the periwinkle crop is roughly $15 million per year, although Madagascar gets little of those profits.

TABLE 11.2

Some Natural Medicinal Products

Product	Source	Use
Penicillin	Fungus	Antibiotic
Bacitracin	Bacterium	Antibiotic
Tetracycline	Bacterium	Antibiotic
Erythromycin	Bacterium	Antibiotic
Digitalis	Foxglove	Heart stimulant
Quinine	Chincona bark	Malaria treatment
Diosgenin	Mexican yam	Birth-control drug
Cortisone	Mexican yam	Anti-inflammation treatment
Cytarabine	Sponge	Leukemia cure
Vinblastine, vincristine	Periwinkle plant	Anticancer drugs
Reserpine	Rauwolfia	Hypertension drug
Bee venom	Bee	Arthritis relief
Allantoin	Blowfly larva	Wound healer
Morphine	Poppy	Analgesic

FIGURE 11.7 The rosy periwinkle from Madagascar provides anticancer drugs that now make childhood leukemias and Hodgkin's disease highly remissible.

FIGURE 11.8 Birdwatching and other wildlife observation contribute more than $29 million each year to the U.S. economy.

Pharmaceutical companies are actively prospecting for useful products in many tropical countries. Merck, the world's largest biomedical company, paid (U.S.) $1.4 million to the Instituto Nacional de Biodiversidad (INBIO) of Costa Rica for plant, insect, and microbe samples to be screened for medicinal applications. INBIO, a public/private collaboration, trained native people as practical "parataxonimists" to locate and catalog all the native flora and fauna—between 500,000 and 1 million species—in Costa Rica. This effort may be a good model both for scientific information gathering and as a way for developing countries to share in the profits from their native resources.

The UN Convention on Biodiversity calls for a more equitable sharing of the gains from exploiting nature between rich and poor nations. Bioprospectors who discover useful genes or biomolecules in native species will be required to share profits with the countries where those species originate. This is not only a question of fairness; it also provides an incentive to poor nations to protect their natural heritage.

Biodiversity provides ecological services

Human life is inextricably linked to ecological services provided by other organisms. Soil formation, waste disposal, air and water purification, nutrient cycling, solar energy absorption, and management of biogeochemical and hydrological cycles all depend on the biodiversity of life (chapter 3). Total value of these ecological services is at least $33 trillion per year, or more than double total world GNP.

There has been a great deal of controversy about the role of biodiversity in ecosystem stability. It seems intuitively obvious that having more kinds of organisms would make a community better able to withstand or recover from disturbance, but few empirical studies show an unequivocal relationship. The opening case study for this chapter describes one of the most famous studies of the stability/diversity relationship.

Because we don't fully understand the complex interrelationships between organisms, we often are surprised and dismayed at the effects of removing seemingly insignificant members of biological communities. For instance, wild insects provide a valuable but often unrecognized service in suppressing pests and disease-carrying organisms. It is estimated that 95 percent of the potential pests and disease-carrying organisms in the world are controlled by other species that prey upon them or compete with them in some way. Many unsuccessful efforts to control pests with synthetic chemicals (chapter 10) have shown that biodiversity provides essential pest control services.

Biodiversity also brings us many aesthetic and cultural benefits

Millions of people enjoy hunting, fishing, camping, hiking, wildlife watching, and other outdoor activities based on nature. These activities keep us healthy by providing invigorating physical exercise. Contact with nature also can be psychologically and emotionally restorative. In some cultures, nature carries spiritual connotations, and a particular species or landscape may be inextricably linked to a sense of identity and meaning. Many moral philosophies and religious traditions hold that we have an ethical responsibility to care for creation and to save "all the pieces" as far as we are able (chapter 2).

Nature appreciation is economically important. The U.S. Fish and Wildlife Service estimates that Americans spend $104 billion every year on wildlife-related recreation (fig. 11.8). This compares to $81 billion spent each year on new automobiles. Forty percent of all adults enjoy wildlife, including 39 million who hunt or fish and 76 million who watch, feed, or photograph wildlife. Ecotourism can be a good form of sustainable economic development, although we have to be careful that we don't abuse the places and cultures we visit.

For many people, the value of wildlife goes beyond the opportunity to shoot or photograph, or even see, a particular species. They argue that **existence value,** based on simply knowing that a species exists, is reason enough to protect and preserve it. We contribute to programs to save bald eagles, redwood trees, whooping cranes, whales, and a host of other rare and endangered organisms because we like to know they still exist somewhere, even if we may never have an opportunity to see them.

11.3 WHAT THREATENS BIODIVERSITY?

Extinction, the elimination of a species, is a normal process of the natural world. Species die out and are replaced by others, often their own descendants, as part of evolutionary change. In undisturbed ecosystems, the rate of extinction appears to be about one species lost every decade. In this century, however, human impacts on populations and ecosystems have accelerated that rate, causing hundreds or perhaps even thousands of species, subspecies, and varieties to become extinct every year. If present trends continue, we may destroy *millions* of kinds of plants, animals, and microbes in the next few decades. In this section, we will look at some ways we threaten biodiversity.

Extinction is a natural process

Studies of the fossil record suggest that more than 99 percent of all species that ever existed are now extinct. Most of those species were gone long before humans came on the scene. Species arise through processes of mutation and natural selection and disappear the same way (chapter 4). Often, new forms replace their own parents. The tiny *Hypohippus,* for instance, has been replaced by the much larger modern horse, but most of its genes probably still survive in its distant offspring.

Periodically, mass extinctions have wiped out vast numbers of species and even whole families (table 11.3). The best studied of these events occurred at the end of the Cretaceous

TABLE 11.3

Mass Extinctions

Historic Period	Time (Before Present)	Percent of Species Extinct
Ordovician	444 million	85
Devonian	370 million	83
Permian	250 million	95
Triassic	210 million	80
Cretaceous	65 million	76
Quaternary	Present	33–66

Source: W. W. Gibbs, 2001.

period when dinosaurs disappeared, along with at least 50 percent of existing genera and 15 percent of marine animal families. An even greater disaster occurred at the end of the Permian period about 250 million years ago when 95 percent of all marine species and nearly half of all plant and animal families died out over a period of about 10,000 years—a short time by geological standards. Current theories suggest that these catastrophes were caused by climate changes, perhaps triggered when large asteroids struck the earth. Many ecologists worry that global climate change caused by our release of "greenhouse" gases in the atmosphere could have similarly catastrophic effects (chapter 15).

We are accelerating extinction rates

The rate at which species are disappearing appears to have increased dramatically over the last 150 years. Between A.D. 1600 and 1850, human activities appear to have been responsible for the extermination of two or three species per decade. By some estimates, we are now losing species at hundreds or even thousands of times natural rates. If present trends continue, the United Nations Environment Program warns, between 22 and 47 percent of all known plant species—and the animals dependent on them—could be extinct in the next 50 years. The eminent biologist E. O. Wilson says the impending biodiversity crash could be more abrupt than any previous mass extinction. Some biologists call this the sixth mass extinction, but note that this time it's not asteroids or volcanoes, but human impacts that are responsible.

Accurate predictions of biodiversity losses are difficult when many species probably haven't yet been identified. Most predictions of anthropogenic mass extinction are based on an assumption that habitat area and species abundance are tightly correlated. E. O. Wilson calculates, for example, that if you cut down 90 percent of a forest, you'll eliminate at least half of the species originally present. In some of the best studied biological communities, however, this seems not to be true. More than 90 percent of Costa Rica's dry seasonal forest, for instance, has been converted to pasture land. Yet entomologist Dan Janzen reports that no more than 10 percent of the original flora and fauna appear to have been permanently lost. Wilson and others respond that remnants of the native species may be hanging on temporarily, but that in the long run they're doomed without sufficient habitat.

Still, it's clear that habitat is being destroyed in many places, and that numerous species are less abundant than they once were. Shouldn't we try to protect and preserve as much as we can? E. O. Wilson summarizes human threats to biodiversity with the acronym **HIPPO,** which stands for Habitat destruction, Invasive species, Pollution, Population (human), and Overharvesting. Let's look in more detail at each of these issues.

Habitat Destruction

The most important extinction threat for most species—especially terrestrial ones—is habitat loss. Perhaps the most obvious example of habitat destruction is clear-cutting of forests and conversion of

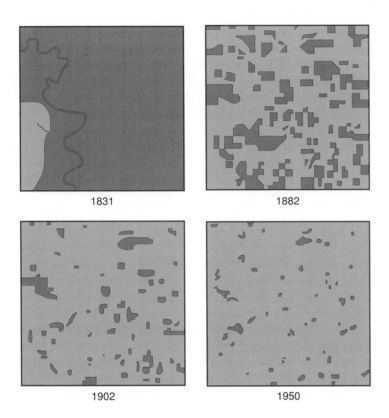

FIGURE 11.9 Decrease in wooded area of Cadiz Township in southern Wisconsin during European settlement. Green areas represent the amount of land in forest each year.

1831

1882

1902

1950

grasslands to crop fields (fig. 11.9). Over the past 10,000 years, humans have transformed billions of hectares of former forests and grasslands to croplands, cities, roads, and other uses. These human-dominated spaces aren't devoid of wild organisms, but they generally favor weedy species adapted to coexist with us.

Today, forests cover less than half the area they once did, and only around one-fifth of the original forest retains its old-growth characteristics. Species like the northern spotted owl (*Strix occidentalis caurina*) that depend on the varied structure and resources of old-growth forest vanish as their habitat disappears (chapter 12). Grasslands currently occupy about 4 billion ha (roughly equal to the area of closed-canopy forests). Much of the most highly productive and species-rich grasslands—for example, the tallgrass prairie that once covered the United States corn belt—has been converted to cropland. Much more may need to be used as farmland or pasture if human populations continue to expand.

Sometimes we destroy habitat as side effects of resource extraction such as mining, dam-building, and indiscriminate fishing methods. Surface mining, for example, strips off the land covering along with everything growing on it. Waste from mining operations can bury valleys and streams with toxic material (see mountaintop removal, chapter 14). Dam-building floods vital stream habitat under deep reservoirs and eliminates food sources and breeding habitat for some aquatic species. Our current fishing methods are highly unsustainable. One of the most destructive fishing techniques is bottom trawling, in which heavy nets are dragged across the ocean floor, scooping up every living thing and crushing the bottom structure to lifeless rubble (chapter 9).

Preserving small, scattered areas of habitat often isn't sufficient to maintain a complete species collection. Large mammals, like tigers or wolves, need large expanses of contiguous range relatively free of human incursion. Even species that occupy less space individually suffer when habitat is fragmented into small, isolated pieces. If the intervening areas create a barrier to migration, isolated populations become susceptible to environmental catastrophes such as bad weather or disease epidemics. They also can become inbred and vulnerable to genetic flaws (chapter 6).

Invasive Species

A major threat to native biodiversity in many places is from accidentally or deliberately introduced species. Called a variety of names—alien, exotic, non-native, non-indigenous, unwanted, disruptive, or invaders—**invasive species** are organisms that move into new territory. These migrants often flourish where they are free of predators, diseases, or resource limitations that may have controlled their population in their native habitat. Although humans have probably transported organisms into new habitats for thousands of years, the rate of movement has increased sharply in recent years with the huge increase in speed and volume of travel by air, water, and land. We move species around the world in a variety of ways. Some are deliberately released because people believe they will be aesthetically pleasing or economically beneficial. Others hitch a ride in ship ballast water, in the wood of packing crates, inside suitcases or shipping containers, in the soil of potted plants, even on people's shoes.

Over the past 300 years, approximately 50,000 non-native species have become established in the United States. Many of these introductions such as corn, wheat, rice, soybeans, cattle, poultry, and honeybees have proved to be both socially and economically beneficial. At least 4,500 of these species have established free-living populations, of which 15 percent cause environmental or economic damage (fig. 11.10). Invasive species are estimated to cost the United States some $138 billion annually and are forever changing a variety of many ecosystems.

A few important examples of invasive species include the following:

- Eurasian milfoil (*Myriophyllum spicatum* L.) is an exotic aquatic plant native to Europe, Asia, and Africa. Scientists believe that milfoil arrived in North America during the late nineteenth century in shipping ballast. It grows rapidly and tends to form a dense canopy on the water surface, which displaces native vegetation, inhibits water flow, and obstructs boating, swimmings, and fishing. Humans spread the plant between water body systems from boats and boat trailers carrying the plant fragments. Herbicides and mechanical harvesting are effective in milfoil control but can be expensive (up to $5,000 per hectare per year). There is also concern that the methods may harm nontarget organisms. A native milfoil weevil, *Euhrychiopsis lecontei,* is being studied as an agent for milfoil biocontrol.

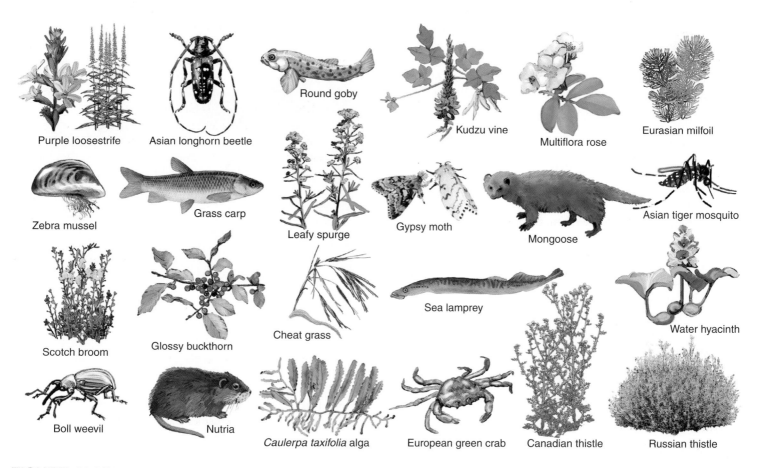

FIGURE 11.10 A few of the approximately 50,000 invasive species in North America. Do you recognize any that occur where you live? What others can you think of?

Labels in figure:
Purple loosestrife, Asian longhorn beetle, Round goby, Kudzu vine, Multiflora rose, Eurasian milfoil, Zebra mussel, Grass carp, Leafy spurge, Gypsy moth, Mongoose, Asian tiger mosquito, Scotch broom, Glossy buckthorn, Cheat grass, Sea lamprey, Water hyacinth, Boll weevil, Nutria, *Caulerpa taxifolia* alga, European green crab, Canadian thistle, Russian thistle

- European green crab (*Carcinils maenav*), a native of the Atlantic coasts of Europe and Africa, was introduced into the waters of Massachusetts in the mid-1800s and has spread both north and south. A voracious predator of mussels, oysters, clams, scallops, and juvenile crabs, this exotic invader is believed to be a major contributor to the demise of the New England soft clam fishery. In the 1990s, this species was found on the Pacific coast, posing a potential threat to the region's clam- and oyster-growing industries and to the Dungeness crab fishery. Despite its name, the crab isn't always green; it can change color to orange and red with yellow patches on its abdomen during its mating cycle.

- Kudzu vine (*Pueraria lobata*) has blanketed large areas of the southeastern United States. Long cultivated in Japan for edible roots, medicines, and fibrous leaves and stems used for paper production, kudzu was introduced by the U.S. Soil Conservation Service in the 1930s to control erosion. Unfortunately, it succeeded too well. In the ideal conditions of its new home, kudzu can grow 18 to 30 m in a single season. Smothering everything in its path, it kills trees, pulls down utility lines, and causes millions of dollars in damage every year.

- Purple loosestrife (*Lythrum salicaria*) grows in wet soil. Originally cultivated by gardeners for its bright purple flower spikes, this tall wetland plant escaped into New England marshes about a century ago. Spreading rapidly across the Great Lakes, it now fills wetlands across much of the northern United States and southern Canada. Because it crowds out indigenous vegetation and has few native predators or symbionts, it tends to reduce biodiversity wherever it takes hold.

- Zebra mussels (*Dreissena polymorpha*) probably made their way from their home in the Caspian Sea to the Great Lakes in ballast water of transatlantic cargo ships, arriving sometime around 1985. Attaching themselves to any solid surface, zebra mussels reach enormous densities—up to 70,000 animals per square meter—covering fish spawning beds, smothering native mollusks, and clogging utility intake pipes. Found in all the Great Lakes, zebra mussels have recently moved into the Mississippi River and its tributaries. Public and private costs for zebra mussel removal now amount to some $400 million per year. On the good side, mussels have improved water clarity in Lake Erie at least fourfold by filtering out algae and particulates.

Disease organisms, or pathogens, could also be considered predators. To be successful over the long term, a pathogen must establish a balance in which it is vigorous enough to reproduce, but not so lethal that it completely destroys its host. When a disease is introduced into a new environment, however, this balance may be lacking and an epidemic may sweep through the area.

The American chestnut was once the heart of many Eastern hardwood forests. In the Appalachian Mountains, at least one of every four trees was a chestnut. Often over 45 m (150 ft) tall, 3 m (10 ft) in diameter, fast growing, and able to sprout quickly from a cut stump, it was a forester's dream. Its nutritious nuts were important for birds (like the passenger pigeon), forest mammals, and humans. The wood was straight grained, light, rot-resistant and used for everything from fence posts to fine furniture, and its bark was used to tan leather. In 1904, a shipment of nursery stock from China brought a fungal blight to the United States, and within 40 years, the American chestnut had all but disappeared from its native range. Efforts are now underway to transfer blight-resistant genes into the few remaining American chestnuts that weren't reached by the fungus or to find biological controls for the fungus that causes the disease.

Of course, the most ubiquitous, ecosystem-changing, invasive species is us. We and our domesticated companions have occupied and altered the whole planet. One study calculated that the familiar and docile cow (*Bos tarus*), through grazing and trampling, endangers three times as many rare plant and animal species as any nondomesticated invader.

Island ecosystems are particularly susceptible to invasive species

New Zealand is a prime example of the damage that can be done by invasive species in island ecosystems. Having evolved for thousands of years without predators, New Zealand's flora and fauna are particularly susceptible to the introduction of alien organisms. Originally home to more than 3,000 endemic species, including flightless birds such as the kiwi and giant moas, New Zealand has lost at least 40 percent of its native flora and fauna since humans first landed there 1,000 years ago. More than 20,000 plant species have been introduced to New Zealand, and at least 200 have become pests that can create major ecological and economic problems. Many animal introductions (both intentional and accidental) also have become major threats to native species. Cats, rats, mice, deer, dogs, goats, pigs, and cattle accompanying human settlers consume native vegetation and eat or displace native wildlife.

One of the most notorious invasive species is the Australian brush-tailed possum *Trichosurus vulpecula*. This small, furry marsupial was introduced to New Zealand in 1837 to establish a fur trade. In Australia, where their population is held in check by dingoes, fires, diseases, and inhospitable vegetation, possums are rare and endangered. Freed from these constraints in New Zealand, however, possum populations exploded. Now at least 70 million possums chomp their way through at least 7 million tons of vegetation per year in their new home. They destroy habitat needed by indigenous New Zealand species, and also eat eggs, nestlings, and even adult birds of species that lack instincts for protection.

Several dozen of New Zealand's offshore islands have been declared nature sanctuaries. Efforts are being made to eliminate invasive pests and to restore endangered species and native ecosystems. One of the most successful examples is Kapiti Island, off the southwest coast of the North Island (fig. 11.11). In

FIGURE 11.11 New Zealand has been highly successful in eradicating invasive species from island nature reserves. Native vegetation has reemerged, and ground-dwelling birds, such as the kiwi, kakapo, and weka, breed once again in the absence of predators. Landings by visitors are strictly controlled. Note the rat trap at the base of the sign board.

the 1980s, the Department of Conservation eradicated 22,500 brush-tailed possums—along with all the feral cats, ferrets, stoats, weasels, dogs, pigs, goats, cattle, and rats—on the 10 km long by 2 km wide island. The ecological benefits were immediately apparent. Native vegetation reappeared as seeds left in the soil sprouted and germinated without being eaten by foreign herbivores. Many native birds, such as the little brown kiwi, saddleback, stitchbird, kokako, and takahe, that are rare and endangered on the main islands now breed successfully in the predator-free environment.

Not everyone agrees with attempts to protect native flora and fauna. A major controversy in New Zealand is over programs to eradicate the Himalayan thar (*Hemitragus jemlahicus*), a goat-like animal from Asia that was introduced to New Zealand's southern Alps to provide sport hunting. Living in high alpine areas, the thar do extensive damage to fragile alpine vegetation, such as tundra and rare tussock grasslands. The Department of Conservation has hired professional hunters to eradicate thar. Sport hunters are angry about not having an opportunity to kill thar themselves. In 2004, the Department of Conservation received several anonymous letters claiming that brush-tailed possums had been released on Kapiti Island in retaliation for the thar eradication program, and that more would be released unless thar control stopped.

Completely eradicating all invasive species is impossible in most places, but the benefits of controlling their populations is demonstrated by the example of New Zealand's island nature sanctuaries.

Think About It

Domestic and feral house cats are estimated to kill 1 billion birds and small mammals in the United States annually. In 2005, a bill was introduced in the Wisconsin legislature to declare an open hunting season year-round on cats that roam out of their owner's yard. Would you support such a measure? Why or why not? What other measures (if any) would you propose to control feline predation?

FIGURE 11.12 Brown pelicans, and other bird species at the top of the food chain, were decimated by DDT in the 1960s. Pelicans and other species have largely recovered since DDT was banned in the United States.

Pollution

We have known for a long time that toxic pollutants can have disastrous effects on local populations of organisms. Pesticide-linked declines of top predators, such as eagles, osprey, falcons, and pelicans, was well documented in the 1970s (fig. 11.12). Declining populations of marine mammals, alligators, fish, and other wildlife alert us to the connection between pollution and health. This connection has led to a new discipline of conservation medicine (chapter 8). Mysterious, widespread deaths of thousands of seals on both sides of the Atlantic in recent years are thought to be linked to an accumulation of persistent chlorinated hydrocarbons, such as DDT, PCBs, and dioxins, in fat, causing weakened immune systems that make animals vulnerable to infections. Similarly, mortality of Pacific sea lions, beluga whales in the St. Lawrence estuary, and striped dolphins in the Mediterranean are thought to be caused by accumulation of toxic pollutants.

Lead poisoning is another major cause of mortality for many species of wildlife. Bottom-feeding waterfowl, such as ducks, swans, and cranes, ingest spent shotgun pellets that fall into lakes and marshes. They store the pellets, instead of stones, in their gizzards and the lead slowly accumulates in their blood and other tissues. The U.S. Fish and Wildlife Service (USFWS) estimates that 3,000 metric tons of lead shot are deposited annually in wetlands and that between 2 and 3 million waterfowl die each year from lead poisoning.

Population

Human population growth represents a threat to biodiversity in several ways. If our consumption patterns remain constant, with more people, we will need to harvest more timber, catch more fish, plow more land for agriculture, dig up more fossil fuels and minerals, build more houses, and use more water. All of these demands impact wild species. Unless we find ways to dramatically increase the crop yield per unit area, it will take much more land than is currently domesticated to feed everyone if our population grows to 8 to 10 billion as current projections predict. This will be especially true if we abandon intensive (but highly productive) agriculture and introduce more sustainable practices. The human population growth curve is leveling off (chapter 7), but it remains unclear whether we can reduce global inequality and provide a tolerable life for all humans while also preserving healthy natural ecosystems and a high level of biodiversity.

Overharvesting

Overharvesting is responsible for depletion or extinction of many species. A classic example is the extermination of the American passenger pigeon (*Ectopistes migratorius*). Even though it inhabited only eastern North America, 200 years ago this was the world's most abundant bird with a population of between 3 and 5 billion animals (fig. 11.13). It once accounted for about one-quarter of all birds in North America. In 1830, John James Audubon saw a single flock of birds estimated to be ten miles wide, hundreds of miles long, and thought to contain perhaps a billion birds. In spite of this vast abundance, market hunting and habitat destruction caused the entire population to crash in only about 20 years between 1870 and 1890. The last known wild bird was shot in 1900 and the last existing passenger pigeon, a female named Martha, died in 1914 in the Cincinnati Zoo.

At about the same time that passenger pigeons were being extirpated, the American bison or buffalo (*Bison bison*) was being hunted to near extinction on the Great Plains. In 1850, some 60 million bison roamed the western plains. Many were killed only for their hides or tongues, leaving millions of carcasses to rot. Some of the bison's destruction was carried out by the U.S. Army to deprive native peoples who depended on bison for food, clothing, and shelter of these resources, thereby forcing them onto reservations. By 1900, there were only about 150 wild bison left and another 250 in captivity.

Fish stocks have been seriously depleted by overharvesting in many parts of the world. A huge increase in fishing fleet size and efficiency in recent years has led to a crash of many oceanic populations. Worldwide, 13 of 17 principal fishing zones are now reported to be commercially exhausted or in steep decline. At least three-quarters of all commercial oceanic species are overharvested. Canadian fisheries biologists estimate that only 10 percent of the top predators such as swordfish, marlin, tuna, and shark remain in the Atlantic Ocean. If

FIGURE 11.13 A pair of stuffed passenger pigeons (*Ectopistes migratorius*). The last member of this species died in the Cincinnati Zoo in 1914.

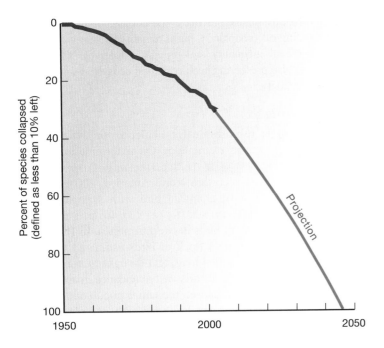

FIGURE 11.14 About one-third of all marine fish species are already in a state of population collapse. If current tends continue, all salt water fish may reach this state by 2050.
Source: SeaWeb.

the only source of animal protein in their diet. If we hope to protect the animals targeted by bushmeat hunters, we will need to help them find alternative livelihoods and replacement sources of high-quality protein. The emergence of SARS in 2003 resulted from the wild food trade in China and Southeast Asia, where millions of civets, monkeys, snakes, turtles, and other animals are consumed each year as luxury foods.

Commercial Products and Live Specimens

In addition to harvesting wild species for food, we also obtain a variety of valuable commercial products from nature. Much of this represents sustainable harvest, but some forms of commercial exploitation are highly destructive and represent a serious threat to certain rare species. Despite international bans on trade in products from endangered species, smuggling of furs, hides, horns, live specimens, and folk medicines amounts to millions of dollars each year.

Developing countries in Asia, Africa, and Latin America with the richest biodiversity in the world are the main sources of wild animals and animal products, while Europe, North America, and some of the wealthy Asian countries are the principal importers. Japan, Taiwan, and Hong Kong buy three-quarters of all cat and snake skins, for instance, while European countries buy a similar percentage of live birds (fig. 11.15b). The United States imports 99 percent of all live cacti and 75 percent of all orchids sold each year.

The profits to be made in wildlife smuggling are enormous. Tiger or leopard fur coats can bring $100,000 in Japan or Europe. The population of African black rhinos dropped from approximately 100,000 in the 1960s to about 3,000 in the 1980s because of a demand for their horns. In Asia, where it is prized for its supposed medicinal properties, powdered rhino horn fetches (U.S.) $28,000 per kg.

Plants also are threatened by overharvesting. Wild ginseng has been nearly eliminated in many areas because of the Asian demand for roots that are used as an aphrodisiac and folk medicine. Cactus "rustlers" steal cacti by the ton from the American southwest and Mexico. With prices ranging as high as $1,000 for rare specimens, it's not surprising that many are now endangered.

The trade in wild species for pets is an enormous business. Worldwide, some 5 million live birds are sold each year for pets.

current trends continue, researchers warn, all major fish stocks could be in collapse—defined as 90 percent depleted—within 50 years (fig. 11.14). You can avoid adding to this overharvest by eating only abundant, sustainably harvested varieties (What Can You Do? p. 241). At the UN Conference on Biodiversity in Paris in 2005, more than 700 scientists from 83 nations called for a ban on longline fishing that threatens sea birds, turtles, and marine mammals.

Perhaps the most destructive example of harvesting terrestrial wild animal species today is the African bushmeat trade. Wildlife biologists estimate that 1 million tons of bushmeat, including antelope, elephants, primates, and other animals, are sold in African markets every year (fig. 11.15a). For many poor Africans, this is

(a) Bush meat market

(b) Hyacinth macaws

(c) Cyanide fishing

FIGURE 11.15 Threats to wildlife. (a) More than 1 million tons of wild animals are sold each year for human consumption. (b) Wild birds, like these Brazilian hyacinth macaws, are endangered by the pet trade. (c) Cyanide fishing not only kills fish, it also destroys the entire reef community.

What Can You Do?

Don't Buy Endangered Species Products

You probably are not shopping for a fur coat from an endangered tiger, but there might be other ways you are supporting unsustainable harvest and trade in wildlife species. To be a sustainable consumer, you need to learn about the source of what you buy. Often plant and animal products are farm-raised, not taken from wild populations. But some commercial products are harvested in unsustainable ways. Here are a few products about which you should inquire before you buy:

Seafood includes many top predators that grow slowly and reproduce only when many years old. Despite efforts to manage many fisheries, the following have been severely, sometimes tragically, depleted:

- Top predators: swordfish, marlin, shark, bluefin tuna, albacore ("white") tuna.
- Groundfish and deepwater fish: orange roughy, Atlantic cod, haddock, pollack (source of most fish sticks, artificial crab, generic fish products), yellowtail flounder, monkfish.
- Other species, especially shrimp, yellowfin tuna, and wild sea scallops, are often harvested with methods that destroy other species or habitats.
- Farm-raised species such as shrimp and salmon can be contaminated with PCBs, pesticides, and antibiotics used in their

rearing. In addition, aquaculture operations often destroy coastal habitat, pollute surface waters, and deplete wild fish stocks to stock ponds and provide fish meal.

Pets and plants are often collected from wild populations, some sustainably and others not:

- Aquarium fish (often harvested by stunning with dynamite and squirts of cyanide, which destroy tropical reefs and many fish).
- Reptiles: snakes and turtles, especially, are often collected in the wild.
- Plants: orchids and cacti are the best-known, but not the only, group collected in the wild.

Herbal products such as wild ginseng, wild echinacea (purple coneflower), should be investigated before purchasing.

Do buy some of these sustainably harvested products:

- Shade-grown (or organic) coffee, nuts, and other sustainably harvested forest products.
- Pets from the Humane Society, which works to protect stray animals.
- Organic cotton, linen, and other fabrics.
- Fish products that have relatively little environmental impact or fairly stable populations: farm-raised catfish or tilapia, wild-caught salmon, mackerel, Pacific pollack, dolphinfish (mahimahi), squids, crabs, and crayfish.
- Wild freshwater fish like bass, sunfish, pike, catfish, and carp, which are usually better managed than most ocean fish.

This trade endangers many rare species. It also is highly wasteful. Up to 60 percent of the birds die before reaching market. After the United States banned sale of wild birds in 1992, imports declined 88 percent. Still, pet traders import (often illegally) some 2 million reptiles, 1 million amphibians and mammals, and 128 million tropical fish into the United States each year. About 75 percent of all saltwater tropical aquarium fish sold come from coral reefs of the Philippines and Indonesia.

Many of these fish are caught by divers using plastic squeeze bottles of cyanide to stun their prey (fig. 11.15c). Far more fish die with this technique than are caught. Worst of all, it kills the coral animals that create the reef. A single diver can destroy all of the life on 200 m² of reef in a day. Altogether, thousands of divers currently destroy about 50 km² of reefs each year. Net fishing would prevent this destruction, and it could be enforced if pet owners would insist on net-caught fish. More than half the world's coral reefs are potentially threatened by human activities, with up to 80 percent at risk in the most populated areas.

11.4 ENDANGERED SPECIES MANAGEMENT

Over the years, we have gradually become aware of the harm we have done—and continue to do—to wildlife and biological resources. Slowly, we are adopting national legislation and international

treaties to protect these irreplaceable assets. Parks, wildlife refuges, nature preserves, zoos, and restoration programs have been established to protect nature and rebuild depleted populations. There has been encouraging progress in this area, but much remains to be done. While most people favor pollution control or protection of favored species such as whales or gorillas, surveys show that few understand what biological diversity is or why it is important.

Hunting and fishing laws have been effective

In 1874, a bill was introduced in the United States Congress to protect the American bison, whose numbers were already falling to dangerous levels. This initiative failed, however, because most legislators believed that all wildlife—and nature in general—was so abundant and prolific that it could never be depleted by human activity. As we discussed earlier in this chapter, however, by the end of the nineteenth century, bison had plunged from some 60 million to only a few hundred animals. By the 1890s most states had enacted some hunting and fishing restrictions. The general idea behind these laws was to conserve the resource for future human use rather than to preserve wildlife for its own sake. The wildlife regulations and refuges established since that time have been remarkably successful for many species. At the turn of the century, there were an estimated half a million white-tailed deer in the United States; now there are some 14 million—more in some places than the environment can support. Wild turkeys and wood ducks were nearly all gone 50 years ago. By restoring habitat, planting food crops, transplanting breeding stock, building shelters or houses, protecting these birds during breeding season, and other conservation measures, populations of these beautiful and interesting birds have been restored to several million each. Snowy egrets, which were almost wiped out by plume hunters 80 years ago, are now common again.

Legislation is key to biodiversity protection

The U.S. Endangered Species Act (ESA) and the Canadian Species at Risk law are powerful tools for wildlife protection. Where earlier regulations had been focused almost exclusively on "game" animals, these programs seek to identify all endangered species and populations and to save as much biodiversity as possible, regardless of its usefulness to humans. As defined by the ESA, **endangered species** are those considered in imminent danger of extinction, while **threatened species** are those that are likely to become endangered—at least locally—within the foreseeable future. Bald eagles, gray wolves, brown (or grizzly) bears, and sea otters, for instance, together with a number of native orchids and other rare plants are considered to be locally threatened even though they remain abundant in other parts of their former range (fig. 11.16). **Vulnerable species** are naturally rare or have been locally depleted by human activities to a level that puts them at risk. They often are candidates for future listing. For vertebrates, protected categories include species, subspecies, and local races or ecotypes.

The ESA regulates a wide range of activities involving endangered species including "taking" (harassing, harming, pursuing, hunting, shooting, trapping, killing, capturing, or collecting) either accidentally or on purpose; importing into or exporting out of the United States; possessing, selling, transporting, or shipping; and selling or offering for sale any endangered species. Prohibitions apply to live organisms, body parts, and products made from endangered species. Violators of the ESA are subject to fines up to $50,000 and one year imprisonment. Vehicles and equipment used in violations may be subject to forfeiture. In 1995 the Supreme Court ruled that critical habitat—habitat essential for a species' survival—must be protected whether on public or private land.

Currently, the United States has 1,264 species on its endangered and threatened species lists and about 386 candidate species waiting to be considered. The number of listed species in different taxonomic groups reflects much more about the kinds of organisms that humans consider interesting and desirable than the actual number in each group. In the United States, invertebrates make up about three-quarters of all known species but only 9 percent of those deemed worthy of protection. Worldwide, the International Union for Conservation of Nature and Natural Resources (IUCN) lists a total of 16,118 endangered and threatened species (see table 11.1).

Listing of endangered species is highly selective. We tend to be concerned about the species that we find interesting or useful rather than strive for equal representation from every phylum. Notice that 20 percent of all known mammals on the IUCN red

FIGURE 11.16 Sea otters (*Enhydra lutris*) are considered threatened throughout most of their former range in the north Pacific. Overharvesting, pollution, competition for food, and predation by orca whales are among the causes of their decline.

list are described as threatened or endangered, for example, while only 0.06 percent of insects are listed as threatened. This is inequitable in two ways. First, there are probably far more endangered insect species than this, even among those we have identified. Furthermore, it's extremely rare to find a new mammal species, while the million known insect species may represent only one-thirtieth of the total on earth.

Listing of new species in the United States has been very slow, generally taking several years from the first petition to final determination. Limited funding, political pressures, listing moratoria, and changing administrative policies have created long delays. In December 2000, flooded by lawsuits over species protection and badly underfunded, the USFWS stopped reviewing new cases altogether. Hundreds of species are classified as "warranted (deserving of protection) but precluded" for lack of funds or local support. At least 18 species have gone extinct since being nominated for protection.

When Congress passed the original ESA, it probably intended to protect only a few charismatic species like birds and big game animals. Sheltering obscure species such as the Delhi Sands flower-loving fly, the Coachella Valley fringe-toed lizard, Mrs. Furbisher's lousewort, or the orange-footed pimple-back mussel most likely never occurred to those who voted for the bill. This raises some interesting ethical questions about the rights and values of seemingly minor species. Although uncelebrated, these species may be indicators of environmental health. Protecting them usually preserves habitat and a host of unlisted species.

Recovery plans rebuild populations of endangered species

Once a species is officially listed as endangered, the Fish and Wildlife Service is required to prepare a recovery plan detailing how populations will be rebuilt to sustainable levels. It usually takes years to reach agreement on specific recovery plans. Among the difficulties are costs, politics, interference on local economic

FIGURE 11.17 The California condor is recovering from near extinction. This is one of a few species that have received the bulk of restoration dollars.

interests, and the fact that once a species is endangered, much of its habitat and ability to survive is likely compromised. The total cost of recovery plans for all currently listed species is estimated to be nearly $5 billion.

The United States currently spends about $150 million per year on endangered species protection and recovery. About half that amount is spent on a dozen charismatic species like the California condor, and the Florida panther and grizzly bear, which receive around $13 million per year (fig. 11.17). By contrast, the 137 endangered invertebrates and 532 endangered plants get less than $5 million per year altogether. Our funding priorities often are based more on emotion and politics than biology. A variety of terms are used for rare or endangered species thought to merit special attention:

- *Keystone species* are those with major effects on ecological functions and whose elimination would affect many other members of the biological community; examples are prairie dogs (*Cynomys ludovicianus*) or bison (*Bison bison*).
- *Indicator species* are those tied to specific biotic communities or successional stages or set of environmental conditions. They can be reliably found under certain conditions but not others; an example is brook trout (*Salvelinus fontinalis*).
- *Umbrella species* require large blocks of relatively undisturbed habitat to maintain viable populations. Saving this habitat also benefits other species. Examples of umbrella species are the northern spotted owl (*Strix occidentalis caurina*) and elephant (*Loxodonta africana*).
- *Flagship species* are especially interesting or attractive organisms to which people react emotionally. These species can motivate the pubic to preserve biodiversity and contribute to conservation; an example is the giant panda (*Ailuropoda melanoleuca*).

Some recovery plans have been gratifyingly successful. The American alligator was listed as endangered in 1967 because hunting (for meat, skins, and sport) and habitat destruction had reduced populations to precarious levels. Protection has been so effective that the species is now plentiful throughout its entire southern range. Florida alone estimates that it has at least 1 million alligators.

Sometimes, restoring a single species can bring benefits to an entire ecosystem, especially when that species plays a keystone role in the community. Alligators, for example, dig out swimming holes, or wallows, that become dry season refuges for fish and other aquatic species. Another notable example is restoration of top predators. Returning wolves to the Greater Yellowstone ecosystem is credited by some ecologists with having brought elk populations under control and restoring the health and diversity of the entire ecosystem (Exploring Science p. 244). See further discussion in chapter 13.

Some other successful recovery programs include bald eagles, peregrine falcons, and whooping cranes. Forty years ago, due mainly to DDT poisoning, there were only 417 nesting pairs of bald eagles (*Haliaeetus leucocephalis*) remained in the contiguous United States. By 2007, the population had rebounded to more than 9,800 nesting pairs, and the birds were removed from the endangered species list. This doesn't mean that eagles are unprotected. Killing, selling or otherwise harming eagles, their nests or eggs is still prohibited. Similarly, peregrine falcons, which had been down to only 39 breeding pairs in the 1970s, had rebounded to 1,650 pairs in 1999 and also were taken off the endangered species list. In addition to eagles and falcons, 29 other species, including mammals, fish, reptiles, birds, plants, and even one insect (the Tinian monarch) have been removed or downgraded from the endangered species list.

Opponents of the ESA have repeatedly tried to require that economic costs and benefits be incorporated into endangered species planning. An important test of the ESA occurred in 1978 in Tennessee where construction of the Tellico Dam threatened a tiny fish called the snail darter. As a result of this case, a federal committee (the so-called "God Squad") was given power to override the ESA for economic reasons. Another important example is the case of the northern spotted owl (chapter 12), whose protection depends on preserving old-growth forest in the Pacific Northwest. Timber-industry economists estimate that saving a population of 1,600 to 2,400 owls will cost $33 billion, eliminate 40,000 jobs, and devastate local economies. Conservationists dispute these numbers and claim the owl is an umbrella species whose protection would aid many other organisms and resources (fig. 11.18). They point out that since logging was reduced in the 1990s, jobs and the economy in the Pacific Northwest have actually increased.

An even more costly recovery program may be required for Columbia River salmon and steelhead endangered by hydropower dams and water storage reservoirs that block their migration to the sea. Opening floodgates to allow young fish to run downriver and adults to return to spawning grounds would have high economic costs to barge traffic, farmers, and electric rate payers who have come to depend on abundant water and cheap electricity. On the other hand, commercial and sport fishing for salmon is worth $1 billion per year and employs about 60,000 people directly or indirectly.

Predators Help Restore Biodiversity in Yellowstone

In the past dozen years, Yellowstone National Park has been the site of a grand experiment in managing biodiversity. The park is renowned for its wildlife and wilderness. But between about 1930 an 1990, park managers and ecologists observed that growing elk populations were overbrowsing willow and aspen. Plant health and diversity were declining, and populations of smaller mammals, such as beaver, were gradually dwindling, Most ecologists blamed these changes on the eradication of one of the region's main predators, the gray wolf (*Canis lupus*). Now, a project to restore wolves has provided a rare opportunity to watch environmental change unfold.

Wolves reintroduced to Yellowstone National Park help to restore biodiversity and ecosystem health.

Wolves were once abundant in the Greater Yellowstone ecosystem. In the 1890s, perhaps 100,000 wolves roamed the western United States. Farmers and ranchers, aided by government programs to eradicate predators, gradually poisoned, shot, and trapped nearly all the country's wolves. In the northern Rocky Mountains, elk and deer populations expanded rapidly without predators to restrain their numbers. Yellowstone's northern range elk herd grew to about 20,000 animals. Some ecologists warned that without a complete guild of predators, the elk, bison, and deer could damage one of our best-loved parks. After much controversy, 31 wolves were captured in the Canadian Rockies and relocated to Yellowstone National Park in 1995 and 1996.

Wolves expanded quickly in their new home. With abundant elk for prey, as well as occasional moose and bison, the wolves tripled their population in just three years. The population now exceeds 100 animals in over 20 packs, well beyond the recovery goal of 10 packs.

Following the wolf reintroductions, elk numbers have fallen, and regenerating willow and aspen show signs of increased stature and abundance. Elk carcasses left by wolves provide a feast for scavengers, and wolves appear to have driven down numbers of coyotes, which prey on small mammals. Many ecologists see evidence of more songbirds, hawks, foxes, voles, and ground squirrels,

which they attribute to wolf reintroductions. Other ecologists caution that Yellowstone is a complex system, and wolves are one of many factors that may influence regeneration. For example, recent mild winters may have allowed elk to find alternate food sources, reducing browsing pressure on willows and aspen, and drought may have contributed to declining elk numbers. Debate on the exact impact of wolves underscores the difficulty of interpreting change in complex ecosystems. Whether or not wolves are driving ecological change, most ecologists agree that wolves contribute to a diverse and healthy ecosystem.

New Technology

How do ecologists study these changes? For years, ecologists have skiied and snowshoed out to remote observation posts to record wolf kills and observe wolf behavior. But the difficulty of this field work has limited the amount of data that could be collected. One of the exciting new alternatives involves a remotely controlled video camera that allows researchers to monitor wolf ecology and behavior without intruding on the wolves.

Ecologist Dan MacNulty, of the University of Minnesota, together with Glenn Plumb of the National Park Service, has installed a pair of computer-controlled video cameras in an open meadow near hot springs where

bison gather to seek exposed grass in the winter. Wolves, hoping to prey on the bison, frequent the area as well. Instead of skiing long, cold miles to the site, MacNulty can now turn on a camera from his office desktop The camera scans an area of 17 km^2, and it can zoom in to identify individual animals up to 9.5 km away. In the past, his team was lucky to get a few weeks of winter wolf observation in a year, but now MacNulty and his 12 undergraduate research assistants can watch for wolves year-round, any time it's light enough to see them. Working in shifts, the group records sightings and behavior of wolves, bears, elk, bison, coyotes, and fox.

Because they can observe the animals at length without disturbing them, MacNulty's group has made some intriguing observations. Wolves are masterful at detecting when a bison is vulnerable. They almost never attack a large bull, unless it has wandered from the group or gotten into deep snow, where bison quickly get bogged down. Bison will also work together to fend off wolves—if they can avoid leaving the snow-free areas. Wolves can also be surprisingly persistent: in one case McNulty says wolves harassed a vulnerable bison for 36 hours before finally killing it. All these observations help ecologists understand how wolves hunt, select their prey, and avoid injury in the chase.

Dan MacNulty observes wolves from his office.

FIGURE 11.18 Endangered species often serve as a barometer for the health of an entire ecosystem and as surrogate protector for a myriad of less well-known creatures.

Source: Copyright 1990 by Herb Block in *The Washington Post.* Reprinted by permission of The Herb Block Foundation.

Private land is vital in endangered species protection

Eighty percent of the habitat for more than half of all listed species is on nonpublic property. The Supreme Court has ruled that destroying habitat is as harmful to endangered species as directly taking (killing) them. Many people, however, resist restrictions on how they use their own property to protect what they perceive to be insignificant or worthless organisms. This is especially true when the land has potential for economic development. If property is worth millions of dollars as the site of a housing development or shopping center, most owners don't want to be told they have to leave it undisturbed to protect some rare organism. Landowners may be tempted to "shoot, shovel, and shut up," if they discover endangered species on their property. Many feel they should be compensated for lost value caused by ESA regulations.

Recently, to avoid crises like the northern spotted owl, the Fish and Wildlife Service has been negotiating agreements called **habitat conservation plans (HCP)** with private landowners. Under these plans, landowners are allowed to harvest resources or build on part of their land as long as the species benefits overall. By improving habitat in some areas, funding conservation research, removing predators and competitors, or other steps that benefit the endangered species, developers are allowed to destroy habitat or even "take" endangered organisms.

Scientists and environmentalists often are critical of HCPs, claiming these plans often are based more on politics than biology, and that the potential benefits are frequently overstated. Defenders argue that by making the ESA more landowner-friendly, HCPs benefit wildlife in the long run.

Among the more controversial proposals for HCPs are the so-called Safe Harbor and No-Surprises Policies. Under the Safe Harbor clause, any increase in an animal's population resulting from a property owner's voluntary good stewardship would not increase their responsibility or affect future land-use decisions. As long as the property owner complies with the terms of the agreement, he or she can make any use of the property. The No-Surprises provision says that the property owner won't be faced with new requirements or regulations after entering into an HCP. Scientists warn that change, uncertainty, dynamics, and flux are characteristic of all ecosystems. We can't say that natural catastrophes or environmental events won't make it necessary to modify conservation plans in the future.

Endangered species protection is controversial

The U.S. ESA officially expired in 1992. Since then, Congress has debated many alternative proposals ranging from outright elimination to substantial strengthening of the act. Perhaps no other environmental issue divides Americans more strongly than the ESA. In the western United States, where traditions of individual liberty and freedom are strong and the federal government is viewed with considerable suspicion and hostility, the ESA seems to many to be a diabolical plot to take away private property and trample on individual rights. Many people believe that the law puts the welfare of plants and animals above that of humans. Farmers, loggers, miners, ranchers, developers, and other ESA opponents repeatedly have tried to scuttle the law or greatly reduce its power. Environmentalists, on the other hand, see the ESA as essential to protecting nature and maintaining the viability of the planet. They regard it as the single most effective law in their arsenal and want it enhanced and improved.

Critical habitat protection is especially onerous to local residents because it often involves protecting lands that don't now have the endangered species. Conservationists view this as absolutely necessary. How can we hope to restore species if there's no place for them to live? Locals, on the other hand, resent having to curtail their activities for some animal they don't want living in their neighborhood, and which isn't there anyway.

You Can Help Preserve Biodiversity

If you live in an urban area, as most Americans do, you may not think that you have much influence on wildlife, but there are important ways that you can help conserve biodiversity.

- Protect or restore native biomes. If you inquire about environmental organizations or nature preserves near where you live, you'll find opportunities to remove invasive species, gather native seeds, replant native vegetation or find other ways to preserve or improve habitat.
- Plant local, native species in your garden. Exotic nursery plants often escape and threaten native ecosystems.
- Don't transport firewood from one region to another. It may carry diseases and insects.
- Follow legislation and management plans for natural areas you value. Lobby or write letters supporting funding and biodiversity-friendly policies.
- Help control invasive species. Never release non-native animals (fish, leaches, turtles, etc) or vegetation into waterways or sewers. If you boat, wash your boat and trailer when moving from one lake or river to another.
- Don't discard worms in the woods. You probably think of earthworms as beneficial for soils—and they are in the proper place—but many northern deciduous biomes evolved without them. Worms discarded by anglers are now causing severe habitat destruction in many places.
- Keep your cat indoors. House cats are major predators of woodland birds and other native animals. It's estimated that house cats in the United States kill 1 billion birds and small mammals every year.

Conservationists, too, have criticisms of our current endangered species protection. Perhaps chief of these is the focus on individual organisms. As we pointed out earlier in this chapter, protecting a keystone or umbrella species, such as wolves or elephants, can benefit entire ecological communities, but often we spend millions of dollars attempting to save a single kind of organism when those funds might have done more good, ecologically, by protecting a functional—if less unique—community. Take the case of the California condor, for example. They flourished during the time when woolly mammoths and other Pleistocene megafauna roamed the western plains and there was lots of carrion to eat. Now that only smaller animals populate the landscape, turkey vultures, magpies, crows, and other smaller scavengers apparently forage more efficiently than do condors. We're spending a lot of money to breed and release condors without much success in restoring a self-reproducing population. Losing condors may not have any ecological effect. Perhaps it would be better to try to preserve representative samples of many different kinds of biological communities and ecological services (even if those communities are missing a few of their historic members) than to save a few rare species that may be at the end of their evolutionary life cycle anyway.

Large-scale, regional planning is needed

Over the past decade, growing numbers of scientists, land managers, policymakers, and developers have been making the case that it is time to focus on a rational, continent-wide preservation of ecosystems that support maximum biological diversity rather than a species-by-species battle for the rarest or most popular organisms. By focusing on populations already reduced to only a few individuals, we spend most of our conservation funds on species that may be genetically doomed no matter what we do. Furthermore, by concentrating on individual species we spend millions of dollars to breed plants or animals in captivity that have no natural habitat where they can be released. While flagship species such as mountain gorillas or Indian tigers are reproducing well in zoos and wild animal parks, the ecosystems that they formerly inhabited have largely disappeared.

A leader of this new form of conservation is J. Michael Scott, who was project leader of the California condor recovery program in the mid-1980s and had previously spent ten years working on endangered species in Hawaii. In making maps of endangered species, Scott discovered that even Hawaii, where more than 50 percent of the land is federally owned, has many vegetation types completely outside of natural preserves (fig. 11.19). The gaps between protected areas may contain more endangered species than are preserved within them.

This observation has led to an approach called **gap analysis** in which conservationists and wildlife managers look for unprotected landscapes that are rich in species. Computers and geographical information systems (GIS) make it possible to store, manage, retrieve, and analyze vast amounts of data and create detailed, high-resolution

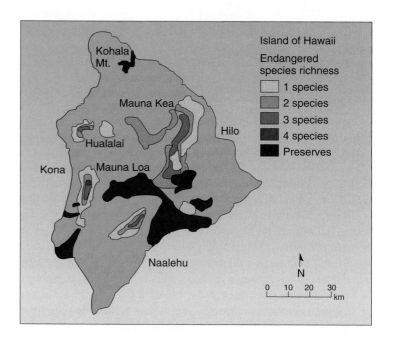

FIGURE 11.19 An example of the biodiversity maps produced by J. Michael Scott and the U.S. Fish and Wildlife Service. Notice that few of the areas of endangered species richness are protected in preserves, which were selected more for scenery or recreation than for biology.

maps relatively easily. This broad-scale, holistic approach seems likely to save more species than a piecemeal approach.

Conservation biologist R. E. Grumbine suggests four remanagement principles for protecting biodiversity in a large-scale, long-range approach:

1. Protect enough habitat for viable populations of all native species in a given region.

2. Manage at regional scales large enough to accommodate natural disturbances (fire, wind, climate change, etc.).

3. Plan over a period of centuries so that species and ecosystems may continue to evolve.

4. Allow for human use and occupancy at levels that do not result in significant ecological degradation.

International wildlife treaties are important

The 1975 Convention on International Trade in Endangered Species (CITES) was a significant step toward worldwide protection of endangered flora and fauna. It regulated trade in living specimens and products derived from listed species, but has not been foolproof. Species are smuggled out of countries where they are threatened or endangered, and documents are falsified to make it appear they have come from areas where the species are still common. Investigations and enforcement are especially difficult in developing countries where wildlife is disappearing most rapidly. Still, eliminating markets for endangered wildlife is an effective way of stopping poaching. Appendix I of CITES lists 700 species threatened with extinction by international trade.

11.5 CAPTIVE BREEDING AND SPECIES SURVIVAL PLANS

Breeding programs in zoos and botanical gardens are one way to attempt to save severely threatened species. Institutions like the Missouri Botanical Garden and the Bronx Zoo's Wildlife Conservation Society sponsor conservation and research programs. Botanical gardens, such as the Kew Gardens in England, and research stations, such as the International Rice Institute in the Philippines, are repositories for rare and endangered plant species that sometimes have ceased to exist in the wild. Valuable genetic traits are preserved in these collections, and in some cases, plants with unique cultural or ecological significance may be reintroduced into native habitats after being cultivated for decades or even centuries in these gardens and seed banks (fig. 11.20).

Zoos can help preserve wildlife

Until fairly recently, zoos depended on primarily wild-caught animals for most of their collections. This was a serious drain on wild populations, because up to 80 percent of the animals caught died from the trauma of capture and shipping. With better understanding of reproductive biology and better breeding facilities, most mammals in North American zoos now are produced by captive breeding programs.

FIGURE 11.20 Rare plants can be preserved and studied in botanical gardens.

Some zoos now participate in programs that reintroduce endangered species to the wild. The California condor (see fig. 11.17) is one of the best known cases of successful captive breeding. In 1986, only nine of these birds existed in their native habitat. Fearing the loss of these last condors, biologists captured them and brought them to the San Diego and Los Angeles zoos, which had begun breeding programs in the 1970s. By 2005 the population had reached 242 birds, with 110 reintroduced to the wild. The endemic nene of Hawaii (*Nesochen sandvicensis*) also has been successfully bred in captivity and reintroduced into the wild. When Captain Cook arrived in the Hawaiian Islands in 1778, there were probably 25,000 of these land-dwelling geese. By the 1950s, however, habitat destruction and invasive predators had reduced the population to fewer than 30 birds. Today there are about 500 wild nene, and more fledglings are introduced every year (fig. 11.21).

Such breeding programs have limitations, however. Bats, whales, and many reptiles rarely reproduce in captivity and still come mainly from the wild. Furthermore, we will never be able to protect the complete spectrum of biological variety in zoos. According to one estimate, if all the space in U.S. zoos were used for captive breeding, only about 100 species of large mammals could be maintained on a long-term basis.

FIGURE 11.21 Nearly extirpated in the 1950s, the land-dwelling nene of Hawaii has been successfully restored by captive breeding programs. From fewer than 30 birds half a century ago, the wild population has grown to more than 500 birds.

FIGURE 11.22 The *KM Minnesota* anchored in Tamanjaya Bay in west Java. Funds raised by the Minnesota Zoo paid for local construction of this boat, which allows wardens to patrol Ujung Kulon National Park and protect rare Javanese rhinos from poachers.

These limitations lead to what is sometimes called the "Noah question": how many species can or should we save? How much are we willing to invest to protect the slimy, smelly, crawly things? Would you favor preserving disease organisms, parasites, and vermin or should we use our limited resources to protect only beautiful, interesting, or seemingly useful organisms?

Even given adequate area and habitat conditions to perpetuate a given species, continued inbreeding of a small population in captivity can lead to the same kinds of fertility and infant survival problems described earlier for wild populations. To reduce genetic problems, zoos often exchange animals or ship individuals long distances to be bred. It sometimes turns out, however, that zoos far distant from each other unknowingly obtained their animals from the same source. Computer databases operated by the International Species Information System located at the Minnesota Zoo, now keep track of the genealogy of many species. This system can tell the complete reproductive history of every animal in every zoo in the world for some species. Comprehensive species survival plans based on this genealogy help match breeding pairs and project resource needs.

The ultimate problem with captive breeding, however, is that natural habitat may disappear while we are busy conserving the species itself. Large species such as tigers or apes are sometimes called "umbrella species." As long as they persist in their native habitat, many other species survive as well.

We need to save rare species in the wild

Renowned zoologist George Schaller says that ultimately "zoos need to get out of their own walls and put more effort into saving the animals in the wild." An interesting application of this principle is a partnership between the Minnesota Zoo and the Ujung Kulon National Park in Indonesia, home to the world's few remaining Javanese rhinos. Rather than try to capture rhinos and move them to Minnesota, the zoo is helping to protect them in their native habitat by providing patrol boats, radios, housing, training, and salaries for Indonesian guards (fig. 11.22). There are no plans to bring any rhinos to Minnesota and chances are very slight that any of us will ever see one, but we can gain satisfaction that, at least for now, a few Javanese rhinos still exist in the wild.

CONCLUSION

Biodiversity provides food, fiber, medicines, clean water, and many other products and services we depend upon every day. Yet nearly one-third of native species in the United States are at risk of disappearing. The Endangered Species Act has proven to be one of the most powerful tools we have for environmental protection. Because of its effectiveness, the act, itself, is endangered; opponents have succeeded in limiting its scope, and have threatened to eliminate it altogether. Still, the act remains a cornerstone of our most basic environmental protections. It has given new hope for survival to numerous species that were on the brink of extinction—less than 1 percent of species listed under the ESA have gone extinct since 1973, while 10 percent of candidate species still waiting to be listed have suffered that fate.

Biodiversity protection has gone far beyond the intent of the original framers of this act thirty years ago. In light of the serious threats facing our environment today—including pollution, habitat destruction, invasive species, and global climate change—we probably need to reevaluate which species we will protect, and how we will protect them. It's clear that we need to be concerned about the other organisms on which we depend for a host of ecological services, and with which we share this planet. In the next two chapters, we'll look at programs that work to protect and restore whole communities and landscapes.

REVIEWING LEARNING OUTCOMES

By now you should be able to explain the following points:

11.1 Discuss biodiversity and the species concept.
- What is biodiversity?
- What are species?
- Molecular techniques are revolutionizing taxonomy.
- How many species are there?
- Hot spots have exceptionally high biodiversity.

11.2 Summarize some of the ways we benefit from biodiversity.
- All of our food comes from other organisms.
- Living organisms provide us with many useful drugs and medicines.
- Biodiversity provides ecological services.
- Biodiversity also brings us many aesthetic and cultural benefits.

11.3 Characterize the threats to biodiversity.
- Extinction is a natural process.
- We are accelerating extinction rates.

- Island ecosystems are particularly susceptible to invasive species.

11.4 Evaluate endangered species management.
- Hunting and fishing laws have been effective.
- Legislation is key to biodiversity protection.
- Recovery plans rebuild populations of endangered species.
- Private land is vital in endangered species protection.
- Endangered species protection is controversial.
- Large-scale, regional planning is needed.
- International wildlife treaties are important.

11.5 Scrutinize captive breeding and species survival plans.
- Zoos can help preserve wildlife.
- We need to save rare species in the wild.

PRACTICE QUIZ

1. What is the range of estimates of the total number of species on the earth? Why is the range so great?

2. What group of organisms has the largest number of species?

3. Define *extinction*. What is the natural rate of extinction in an undisturbed ecosystem?

4. What are rosy periwinkles and what products do we derive from them?

5. Describe some foods we obtain from wild plant species.

6. Define *HIPPO* and describe what it means for biodiversity conservation.

7. What is the current rate of extinction and how does this compare to historic rates?

8. Compare the scope and effects of the Endangered Species Act and CITES.

9. Define *endangered* and *threatened*. Give an example of each.

10. What is gap analysis and how is it related to ecosystem management and design of nature preserves?

CRITICAL THINKING AND DISCUSSION QUESTIONS

1. Many ecologists would like to move away from protecting individual endangered species to concentrate on protecting whole communities or ecosystems. Others fear that the public will only respond to and support glamorous "flagship" species such as gorillas, tigers, or otters. If you were designing conservation strategy, where would you put your emphasis?

2. Put yourself in the place of a fishing industry worker. If you continue to catch many species they will quickly become economically extinct if not completely exterminated. On the other hand, there are few jobs in your village and welfare will barely keep you alive. What would you do?

3. Only a few hundred grizzly bears remain in the contiguous United States, but populations are healthy in Canada and Alaska. Should we spend millions of dollars for grizzly recovery and management programs in Yellowstone National Park and adjacent wilderness areas?

4. How could people have believed a century ago that nature is so vast and fertile that human actions could never have a lasting impact on wildlife populations? Are there similar examples of denial or misjudgment occurring now?

5. In the past, mass extinction has allowed for new growth, including the evolution of our own species. Should we assume that another mass extinction would be a bad thing? Could it possibly be beneficial to us? To the world?

6. Some captive breeding programs in zoos are so successful that they often produce surplus animals that cannot be released into the wild because no native habitat remains. Plans to euthanize surplus animals raise storms of protests from animal lovers. What would you do if you were in charge of the zoo?

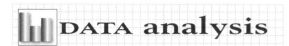 **DATA analysis**   Confidence Limits in the Breeding Bird Survey

At the beginning of this chapter (fig. 11.2), you saw a graph with vertical lines on each point. What did those lines mean? They represent standard error, a measure of how much variation there is in a group of observations. This is one way scientists show uncertainty, or their level of confidence in their results.

A central principle of science is the recognition that all knowledge involves uncertainty. No study can observe every possible event in the universe, so there is always missing information. Scientists try to define the limits of their uncertainty, in order to allow a realistic assessment of their results. A corollary of this principle is that the more data we have, the less uncertainty we have. More data increase our confidence that our observations represent the range of possible observations.

One of the most detailed records of wildlife population trends in North America is the Breeding Bird Survey (BBS). Every June, more than a thousand volunteers drive established routes and count every bird they see or hear. The accumulated data from thousands of routes, over many years, indicates population *trends,* telling which populations are increasing, decreasing, or expanding into new territory.

Because many scientists use BBS data, it is essential to communicate how much confidence there is in the data. The online BBS database reports measures of data quality, including:

- **N:** the number of survey routes from which population trends are calculated.

- **Confidence limits:** because the reported trend is an average of a small *sample* of year-to-year changes on routes, confidence limits tell us how close the sample's average probably is to the average for the *entire population* of that species. Statistically, 95 percent of all samples should fall in-between the confidence limits. In effect, we can be 95 percent sure that the entire population's actual trend falls between the upper and lower confidence limits. (These limits are different from the standard error bars in figure 11.2, but they serve a comparable purpose.)

1. Examine the table at right, which shows 10 species taken from the online BBS database. How many

species have a positive population trend (>0)? If a species had a trend of 0, how much would it change from year to year?

Species	Trend 1966–2004	N	Lower limit	Upper limit
Ring-necked pheasant	−0.5	397	−1.2	0.2
Ruffed grouse	0.7	26	−19.7	21.1
Sage grouse	−2.4	23	−5.6	0.9
Sharp-tailed grouse	0	104	−3.4	3.5
Greater prairie-chicken	−6.8	41	−12.8	−0.8
Wild turkey	8.2	308	5.1	11.2
Northern bobwhite	−2.2	541	−2.7	−1.8
Clapper rail	−0.7	8	−7.3	5.8
King rail	−6.8	22	−11.7	−1.8
Virginia rail	4.7	21	1.4	7.9

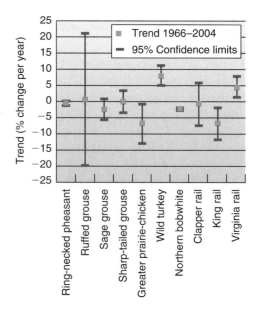

2. Which species has the greatest decline per year? For every 100 birds this year, how many fewer will there be next year?

3. If the distance between upper and lower confidence limits (the *confidence interval*) is narrow, then we can be reasonably sure the trend in our *sample* is close to the trend for the *total population* of that species. What is the reported trend for ring-necked pheasant? What is the number of routes (N) on which this trend is based? What is the range in which the pheasant's true population trend probably falls? Is there a reasonable chance that the pheasant population's average annual change is 0? That the population trend is actually −5?

   Now look at the ruffed grouse, on either the table or the graph. Does the trend show that the population is increasing or decreasing? Can you be certain that the actual trend is not 7, or 17? On how many routes is this trend based?

4. In general, confidence limits depend on the number of observations (N), and how much all the observed values (trends on routes, in this case) differ from the average value. If the values vary greatly, the confidence interval will be wide. Examine the table and graph. Does a *large* N tend to widen or narrow the confidence interval?

5. A trend of 0 would mean no change at all. When 0 falls within the confidence interval, we have little certainty that the trend is *not* 0. In this case, we say that the trend is not significant. How many species have trends that are not significant (at 95 percent certainty)?

   Can we be certain that the sharp-tailed grouse and greater prairie-chicken are changing at different rates? How about the sharp-tailed grouse and wild turkey?

6. Does uncertainty in the data mean results are useless? Does reporting of confidence limits increase or decrease your confidence in the results?

For further information on the Breeding Bird Survey, see http://www.mbr-pwrc.usgs.gov/bbs/.

For Additional Help in Studying This Chapter, please visit our website at www.mhhe.com/cunningham10e. You will find additional practice quizzes and case studies, flashcards, regional examples, place markers for Google Earth™ mapping, and an extensive reading list, all of which will help you learn environmental science.

British Columbia's Great Bear Rainforest will preserve the home for
rare, white-phase black (or spirit) bears along with
salmon streams, misty fjords, rich tidal
estuaries, and the largest remaining
area of old-growth, coastal,
temperate rainforest in
the world.

C H A P T E R

Biodiversity

Preserving Landscapes

*What a country chooses to save is what a country
chooses to say about itself.*

—*Mollie Beatty, former director, U.S. Fish and Wildlife Service*—

LEARNING OUTCOMES

After studying this chapter, you should be able to:

12.1 Discuss the types and uses of world forests.

12.2 Describe the location and state of grazing lands around
the world.

12.3 Summarize the types and locations of nature preserves.

Case Study Saving the Great Bear Rainforest

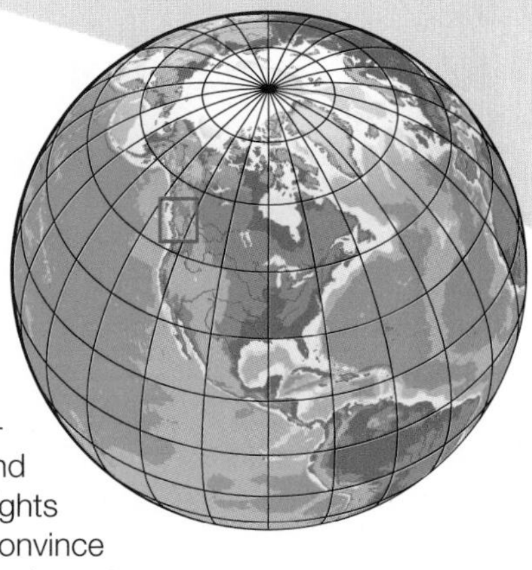

The wild, rugged coast of British Columbia is home to one of the world's most productive natural communities: the temperate rainforest. Nurtured by abundant rainfall and mild year-round temperatures, forests in the deep, misty fjords shelter giant cedar, spruce, and fir trees. Since this cool, moist forest rarely burns, trees often live for 1,000 years or more, and can be 5 m (16 ft) in diameter and 70 m tall. In addition to huge, moss-draped trees, the forest is home to an abundance of wildlife. One animal, in particular, has come to symbolize this beautiful landscape: it's a rare, white or cream-colored black bear. Called a Kermode bear by scientists, these animals are more popularly known as "spirit bears," the name given to them by native Gitga'at people.

The wetlands and adjacent coastal areas also are biologically rich. Whales and dolphins feed in the sheltered fjords and interisland channels. Sea otters float on the rich offshore kelp forests. It's estimated that 20 percent of the world's remaining wild salmon migrate up the wild rivers of this coastline.

In 2006, officials from the provincial government, Native Canadian nations, logging companies, and environmental groups announced a historic agreement for managing the world's largest remaining intact temperate coastal rainforest. This Great Bear Rainforest encompasses about 6 million ha (15.5 million acres) or about the size of Switzerland (fig. 12.1). One-third of the area will be entirely protected from logging. In the rest of the land, only selective, sustainable logging will be allowed rather than the more destructive clear-cutting that has devastated surrounding forests. At least $120 million will be provided for conservation projects and ecologically sustainable business ventures, such as ecotourism lodges and an oyster farm.

A series of factors contributed to preserving this unique area. The largest environmental protest in Canadian history took place at Clayoquot Sound on nearby Vancouver Island in the 1980s, when logging companies attempted to clear-cut land claimed by First Nations people. This educated the public about the values of and threats to the coastal temperate rainforest. As a result of the lawsuits and publicity generated by this controversy, most of the largest logging companies have agreed to stop clear-cutting in the remaining virgin forest. The rarity of the spirit bears also caught the public imagination. Tens of thousands of schoolchildren from across Canada wrote to the provincial government begging them to set aside a sanctuary for this unique animal. And a growing recognition of the rights of native people also helped convince public officials that traditional lands and ways of living need to be preserved.

FIGURE 12.1 The Great Bear Rainforest stretches along the British Columbia coast (including the Queen Charlotte Islands) from Victoria Island to the Alaska border. The area will be managed as a unit with some pristine wilderness, some First Nations lands, and some commercial production.

More than 60 percent of the world's temperate rainforest has already been logged or developed. The Great Bear Rainforest contains one-quarter of what's left. It also contains about half the estuaries, coastal wetlands, and healthy, salmon-bearing streams in British Columbia.

How did planners choose the areas to be within the protected area? One of the first steps was a biological survey. Where were the biggest and oldest trees? Which areas are especially valuable for wildlife? Protecting water quality in streams and coastal regions was also a high priority. Keeping logging and roads out of riparian habitats is particularly important. Interestingly, native knowledge of the area was also consulted in drawing boundaries. Which places are mentioned in oral histories? What are the traditional uses of the forest? While commercial logging is prohibited in the protected areas, First Nations people will be allowed to continue their customary harvest of selected logs for totem poles, longhouses, and canoes. They also will be allowed to harvest berries, catch fish, and hunt wildlife for their own consumption.

Because it's so remote, few people will ever visit the Great Bear Rainforest, yet many of us like knowing that some special places like this continue to exist. Although we depend on wildlands for many products and services, perhaps we don't need to exploit every place on the planet. Which areas we choose to set aside, and how we protect and manage those special places says a lot about who we are. In chapter 11, we looked at efforts to save individual endangered species. Many biologists believe that we should focus instead on saving habitat and representative biological communities. In this chapter, we'll look at how we use and preserve landscapes.

12.1 WORLD FORESTS

Forests and grasslands together occupy almost 60 percent of global land cover (fig. 12.2). These ecosystems provide many of our essential resources, such as lumber, paper pulp, and grazing lands for livestock. They also provide essential ecological services, including regulating climate, controlling water runoff, providing wildlife habitat, purifying air and water, and supporting rainfall. Forests and grasslands also have scenic, cultural, and historic values that deserve protection. These biomes are also among the most heavily disturbed ecosystems (chapter 5).

In many cases, these competing land uses and needs are incompatible. Most political debates over conservation have concerned protection or use of forests, prairies, and rangelands. This chapter examines the ways we use and abuse these biological communities, as well as some of the ways we can protect them and conserve their resources. We discuss forests first, followed by grasslands and then strategies for conservation, restoration, and preservation.

Boreal and tropical forests are most abundant

Forests are widely distributed, but most remaining forests are in the humid tropics and the cold boreal ("northern") or taiga regions (fig. 12.3). Assessing forest distribution is tricky, because forests vary in density and height, and many are inaccessible. The UN Food and Agriculture Organization (FAO) defines "forest" as any area where trees cover more than 10 percent of the land. This definition includes woodlands ranging from open **savannas,** whose trees cover less than 20 percent of the ground, to **closed-canopy forests,** in which tree crowns cover most of the ground. The largest tropical forest is in the Amazon River basin. The highest rates of forest loss are in Africa, especially tropical West Africa, and in South America (fig. 12.4). Some of the world's most biologically diverse regions are undergoing rapid deforestation, including Southeast Asia and Central America (see related story "Protecting Forests to Preserve Rain" at www.mhhe.com/cunningham4e).

Forests are a huge carbon sink, storing some 422 billion metric tons of carbon in standing biomass. Clearing and burning of forests

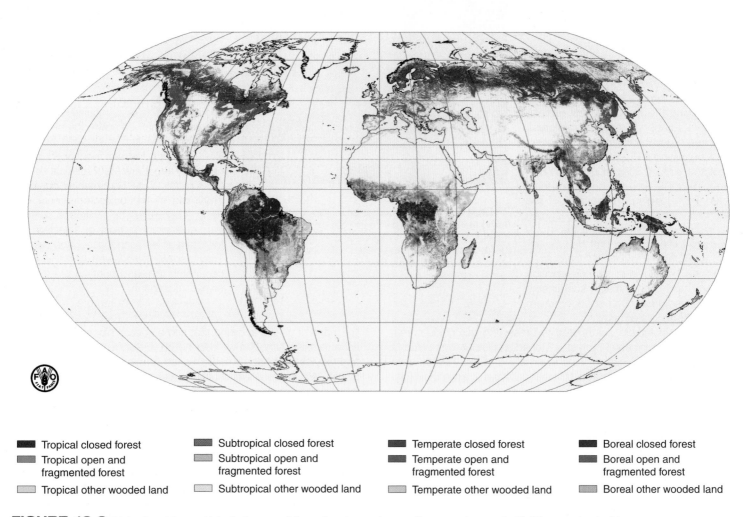

■ Tropical closed forest	■ Subtropical closed forest	■ Temperate closed forest	■ Boreal closed forest
■ Tropical open and fragmented forest	■ Subtropical open and fragmented forest	■ Temperate open and fragmented forest	■ Boreal open and fragmented forest
■ Tropical other wooded land	■ Subtropical other wooded land	■ Temperate other wooded land	■ Boreal other wooded land

FIGURE 12.2 Major forest types. Note that some of these forests are dense; others may have only 10–20 percent actual tree cover.
Source: United Nations Food and Agriculture Organization, 2002.

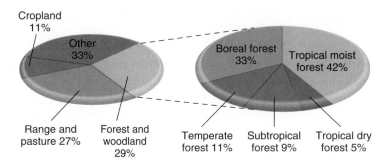

FIGURE 12.3 World land use and forest types. The "other" category includes tundra, desert, wetlands, and urban areas.
Source: UN Food and Agriculture Organization (FAO).

releases much of this carbon into the atmosphere (chapter 9) and may contribute substantially to global climate change. Moisture released from forests also contributes to rainfall. For example, recent climate studies suggest that deforestation of the Amazon could reduce precipitation in the American Midwest.

Among the forests of greatest ecological importance are the remnants of primeval forests that are home to much of the world's biodiversity, endangered species, and indigenous human cultures. Sometimes called frontier forests, **old-growth forests** are those that cover a large enough area and have been undisturbed by human activities long enough that trees can live out a natural life cycle and ecological processes can occur in relatively normal

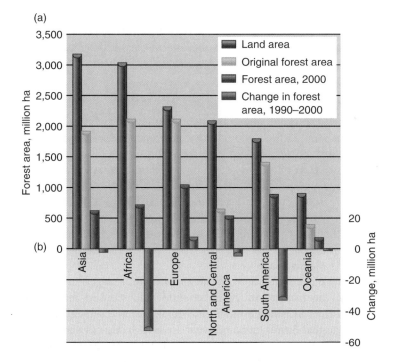

FIGURE 12.4 Regional distribution of original and current forests (a) and regional forest losses, 1990 to 2000 (b). Europe includes Russia's Siberian forests.
Source: UN Food and Agriculture Organization.

fashion. That doesn't mean that all trees need be enormous or thousands of years old. In some old-growth forests, most trees live less than a century before being killed by disease or some natural disturbance, such as a fire. Nor does it mean that humans have never been present. Where human occupation entails relatively little impact, an old-growth forest may have been inhabited by people for a very long time. Even forests that have been logged or converted to cropland often can revert to old-growth characteristics if left alone long enough.

While forests still cover about half the area they once did worldwide, only one-quarter of those forests retain old-growth features. The largest remaining areas of old-growth forest are in Russia, Canada, Brazil, Indonesia, and Papua New Guinea. Together, these five countries account for more than three-quarters of all relatively undisturbed forests in the world. In general, remoteness rather than laws protect those forests. Although official data describe only about one-fifth of Russian old-growth forest as threatened, rapid deforestation—both legal and illegal—especially in the Russian Far East, probably put a much greater area at risk.

Forests provide many valuable products

Wood plays a part in more activities of the modern economy than does any other commodity. There is hardly any industry that does not use wood or wood products somewhere in its manufacturing and marketing processes. Think about the amount of junk mail, newspapers, photocopies, and other paper products that each of us in developed countries handles, stores, and disposes of in a single day. Total annual world wood consumption is about 4 billion m^3. This is more than steel and plastic consumption combined. International trade in wood and wood products amounts to more than $100 billion each year. Developed countries produce less than half of all industrial wood but account for about 80 percent of its consumption. Less-developed countries, mainly in the tropics, produce more than half of all industrial wood but use only 20 percent.

Paper pulp, the fastest growing type of forest product, accounts for nearly a fifth of all wood consumption. Most of the world's paper is used in the wealthier countries of North America, Europe, and Asia. Global demand for paper is increasing rapidly, however, as other countries develop. The United States, Russia, and Canada are the largest producers of both paper pulp and industrial wood (lumber and panels). Much industrial logging in Europe and North America occurs on managed plantations, rather than in untouched old-growth forest However, paper production is increasingly blamed for deforestation in Southeast Asia, West Africa, and other regions.

Fuelwood accounts for nearly half of global wood use. More than half of the people in the world depend on firewood or charcoal as their principal source of heating and cooking fuel (fig. 12.5). The average amount of fuelwood used in less-developed countries is about 1 m^3 per person per year, roughly equal to the amount that each American consumes each year as paper products alone. Demand for fuelwood, which is increasing at slightly less than the global

FIGURE 12.5 Firewood accounts for almost half of all wood harvested worldwide and is the main energy source for one-third of all humans.

FIGURE 12.6 A tropical rainforest in Queensland, Australia. Primary, or old-growth forests, such as this, aren't necessarily composed entirely of huge, old trees. Instead, they have trees of many sizes and species that contribute to complex ecological cycles and relationships.

population growth rate, is causing severe fuelwood shortages and depleting forests in some developing areas, especially around growing cities. About 1.5 billion people have less fuelwood than they need, and many experts expect shortages to worsen as poor urban areas grow. Because fuelwood is rarely taken from closed-canopy forest, however, it does not appear to be a major cause of deforestation. Some analysts argue that fuelwood could be produced sustainably in most developing countries, with careful management.

> ### Think About It
>
> How could modern technology be used to reduce people's dependence on firewood? How might we distribute that technology to the people who need it?

Approximately one-quarter of the world's forests are managed for wood production. Ideally, forest management involves scientific planning for sustainable harvests, with particular attention paid to forest regeneration. In temperate regions, according to the UN Food and Agriculture Organization, more land is being replanted or allowed in regenerate naturally than is being permanently deforested. Much of this reforestation, however, is in large plantations of single-species, single-use, intensive cropping called **monoculture forestry.** Although this produces rapid growth and easier harvesting than a more diverse forest, a dense, single-species stand

often supports little biodiversity and does poorly in providing the ecological services, such as soil erosion control and clean water production, that may be the greatest value of native forests.

Some of the countries with the most successful reforestation programs are in Asia. China, for instance, cut down most of its forests 1,000 years ago and has suffered centuries of erosion and terrible floods as a consequence. Recently, however, timber cutting in the headwaters of major rivers has been outlawed, and a massive reforestation project has begun. Since 1990, China has planted 50 billion trees, mainly in Xinjiang Province, to stop the spread of deserts. Korea and Japan also have had very successful forest restoration programs. After being almost totally denuded during World War II. both countries are now about 70 percent forested.

Tropical forests are being cleared rapidly

Tropical forests are among the richest and most diverse terrestrial systems (fig. 12.6). Although they now occupy less than 10 percent of the earth's land surface, these forests are thought to contain more than two-thirds of all higher plant biomass and at least half of all the plant, animal, and microbial species in the world.

A century ago, an estimated 12.5 million km^2 of tropical lands were covered with closed-canopy forest. This was an area larger than the entire United States. The FAO estimates that about 9.2 million ha, or about 0.6 percent, of the remaining tropical forest is cleared each year.

There is considerable debate about current rates of deforestation in the tropics. In 2003, satellite data showed more than 30,000 fires in a single month in Brazil. Remote sensing experts calculate that 3 million ha per year are now being cut and burned in the Amazon basin alone. However, there are different definitions of **deforestation.** Some scientists insist that it means a complete change from forest to agriculture, urban areas, or desert. Others include any area that has been logged, even if the cut was selective and regrowth will be rapid. Furthermore, savannas, open woodlands, and succession following natural disturbance are hard to distinguish from logged areas. Consequently, estimates for total tropical forest losses range from about 5 million to more than 20 million ha per year. The FAO estimates of 12.3 million ha deforested per year are generally the most widely accepted. To put that figure in perspective, it means that about 1 acre—or the area of a football field—is cleared every second, on average, around the clock.

By most accounts, Brazil has the highest total deforestation rate in the world, but it also has by far the largest tropical forests. Indonesia and Malaysia together may be losing as much primary forest each year as Brazil, even though their original amount was far lower. In 1997 forest fires on Borneo and Sumatra, exacerbated by severe drought, burned 20,000 km^2 (8,000 mi^2). The fires reportedly were set both to clear land for agriculture and to hide illegal logging.

In Africa, the coastal forests of Senegal, Sierra Leone, Ghana, Madagascar. Cameroon, and Liberia already have been mostly destroyed. Haiti was once 80 percent forested; today, essentially all that forest has been destroyed, and the land lies barren and eroded. India, Burma, Kampuchea (Cambodia). Thailand, and Vietnam all have little old-growth lowland forest left. In Central America, nearly two-thirds of the original moist tropical forest has been destroyed, mostly within the past 30 years and primarily due to conversion of forest to cattle range. (See related story on "Disappearing Butterfly Forests" at www.mhhe.com/cunningham4e.)

Causes of Deforestation

A variety of factors contribute to deforestation, and, as figure 12.7 shows, different forces predominate in various parts of the world. Logging for valuable tropical hardwoods, such as teak and mahogany, is generally the first step. Although loggers may take only one or two of the largest trees per hectare, the canopy of tropical forests

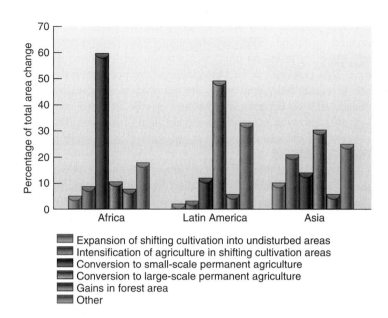

FIGURE 12.7 Tropical forest losses by different processes.
Source: Data from Food and Agriculture Organization of the United Nations, 2002.

is usually so strongly linked by vines and interlocking branches that felling one tree can bring down a dozen others. Building roads to remove logs kills more trees, but even more important, it allows entry to the forest by farmers, miners, hunters, and others who cause further damage (fig. 12.8).

In Africa, conversion of forest into small-scale agriculture accounts for nearly two-thirds of all tropical forest destruction. In Latin America, poor, landless farmers often start the deforestation but are bought out—or driven out—after a few years by large-scale farmers or ranchers. The thin, nutrient-poor tropical forest soils frequently are worn out after a few years of cropping,

1975   1989   2001

FIGURE 12.8 Forest destruction in Rondonia, Brazil, between 1975 and 2001. Construction of logging roads creates a feather-like pattern that opens forests to settlement by farmers.

and much of this land becomes unproductive pasture or impoverished scrubland. Shifting cultivation (sometimes called "slash and burn" or milpa farming) is often blamed for forest destruction. This practice can be sustainable where population densities are low and individual plots are allowed to regenerate for a decade or two between cultivation periods. In some countries, however, growing populations and shrinking forests lead to short rotation cycles, and cropping intensification can lead to permanent deforestation.

As forests are cleared, rainfall patterns can change. A computer model created by Pennsylvania State University scientists suggests this phenomenon might create a kind of chain reaction. As forests are cut down, plant transpiration and rainfall decrease. Drought kills more vegetation, and fires become more numerous and extensive. In a worst-case scenario, an area as large as the entire Amazonian forest might be permanently damaged in just a few decades.

Forest Protection

What can be done to stop this destruction and encourage forest protection? While much of the news is discouraging, there are some hopeful signs for forest conservation in the tropics. Many countries now recognize that forests are valuable resources. Intensive scientific investigations are underway to identify the best remaining natural areas (Exploring Science, p. 259).

A recent survey of world forests found that many countries are more thickly forested now than they were 200 years ago. The United States and China have had the greatest gains in recent years, while Brazil and Indonesia have lost the most. In looking at number and size of trees, rather than merely the area covered, researchers found that 22 of the 50 countries with the most forest had increased their total forest biomass in the past two decades. The transition from deforestation to forest protection is correlated with prosperity. Almost every country with a per capita GDP over (U.S.)$4,600 has moved to reforestation in the past few decades. With industrialization, people move to cities where jobs are available, and farm technology improves so that agriculture concentrates on fertile lands, and fewer people are pushed into marginal land at the edge of forests. With more money and leisure time, populations also value forest recreation and are willing to pay taxes to protect forest landscapes.

About 12 percent of all world forests are now in some form of protected status, but the effectiveness of that protection varies greatly. Costa Rica has one of the best plans for forest guardianship in the world. Attempts are being made there not only to rehabilitate the land (make an area useful to humans) but also to restore the ecosystems to naturally occurring associations. One of the best-known of these projects is Dan Janzen's work in Guanacaste National Park. Like many dry, tropical forests, the northwestern part of Costa Rica had been almost completely converted to ranchland. By controlling fires, however, Janzen and his coworkers are bringing back the forest. One of the keys to this success is involving local people in the project. Janzen also permits grazing in the park. The original forest evolved, he reasons, together with ancient grazing animals that are now extinct. Horses and cows can play a valuable role as seed dispersers.

People also are working on the grassroots level to protect and restore forests in other countries. India, for instance, has a long history of nonviolent, passive resistance movements—called *satyagrahas*—to protest unfair government policies. These protests go back to the beginning of Indian culture and often have been associated with forest preservation, Gandhi drew on this tradition in his protests of British colonial rule in the 1930s and 1940s During the 1970s, commercial loggers began large-scale tree felling in the Garhwal region in the state of Uttar Pradesh in northern India. Landslides and floods resulted from stripping the forest cover from the hills. The firewood on which local people depended was destroyed, and the way of life of the traditional forest culture was threatened. In a remarkable display of courage and determination, the village women wrapped their arms around the trees to protect them, sparking the Chipko Andolan movement (literally, movement to hug trees). They prevented logging on 12,000 km^2 of sensitive watersheds in the Alakanada basin. Today the Chipko Andolan movement has grown to more than 4,000 groups working to save India's forests.

Debt-for-Nature Swaps

Those of us in developed countries also can contribute toward saving tropical forests. Financing nature protection is often a problem in developing countries, where the need is greatest. One promising approach is called **debt-for-nature swaps.** Banks, governments, and lending institutions now hold nearly $1 trillion in loans to developing countries. There is little prospect of ever collecting much of this debt, and banks are often willing to sell bonds at a steep discount—perhaps as little as 10 cents on the dollar. Conservation organizations buy debt obligations on the secondary market at a discount and then offer to cancel the debt if the debtor country agrees to protect or restore an area of biological importance.

There have been many such swaps. Conservation International, for instance, bought $650,000 of Bolivia's debt for $100,000—an 85 percent discount. In exchange for canceling this debt, Bolivia agreed to protect nearly 1 million ha (2.47 million acres) around the Beni Biosphere Reserve in the Andean foothills. Ecuador and Costa Rica have had a different kind of debt-for-nature swap. They have exchanged debt for local currency bonds that fund activities of local private conservation organizations in the country. This has the dual advantage of building and supporting indigenous environmental groups while protecting the land. Critics, however, charge that these swaps compromise national sovereignty and do little to reduce the developing world's debt or to change the situations that led to environmental destruction in the first place.

Temperate forests have competing uses

Tropical countries aren't unique in harvesting forests at an unsustainable rate. Northern countries, such as the United States and Canada, also have allowed controversial forest management practices in many areas. For many years, the official policy of the U.S. Forest Service was "multiple use," which implied that the forests could be used for everything that we might want to do there simultaneously. Some uses are

Protecting areas in remote places, such as Central Africa, is hard because information is so elusive. Deep, remote, swampy, tropical jungles are difficult to enter, map, and assess for their ecological value. Yet without information about their ecological importance, most people have little reason to care about these remote, trackless forests. How can you conserve ecosystems if you don't know what's there?

For most of history, understanding the extent and conditions of a remote area required an arduous trek to see the place in person. Even on publicly owned lands, only those who could afford the time, or who could afford to pay surveyors, might understand the resources. Over time, maps improved, but maps usually show only a few features, such as roads, rivers, and some boundaries.

In recent years, details about public lands and resources have suddenly burst into public view through the use of geographic information systems **(GIS).** A GIS consists of spatial data, such as boundaries or road networks, and software to display and analyze the data. Spatial data can include variables that are hard to see on the ground—watershed boundaries, annual rainfall, landownership, or historical land use. Data can also represent phenomena much larger than we can readily see—land surface slopes and elevation, forested regions, river networks, and so on. By overlaying these layers, GIS analysts can investigate completely new questions about conservation, planning, and restoration.

You have probably used a GIS. Online mapping programs such as MapQuest or Google Earth organize and display spatial data. They let you turn layers on and off, or zoom in and out to display different scales. You can also use an online mapping program to calculate distances and driving directions between places. An ecologist, meanwhile, might use a GIS to calculate the extent of habitat areas, to monitor changes in area, to calculate the size of habitat fragments, or to calculate the length of waterways in a watershed.

Identifying Priority Areas

Recently, a joint effort of several conservation organizations has used GIS and spatial data to identify priority areas for conservation in Central Africa. The project was initiated because new data, including emerging GIS data, were showing dramatic increases in planned logging, in a region that contains the world's second greatest extent of tropical forest (fig. 1).

Researchers from the Wildlife Conservation Society, Worldwide Fund for Nature,

World Resources Institute, USGS, and other agencies and groups, began collecting GIS data on a variety of variables. They identified the range of great apes and other rare or threatened species. They identified areas of extreme plant diversity. They calculated the sizes of forest frag-

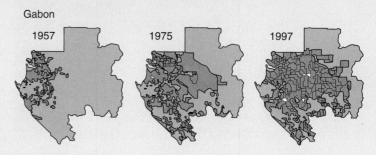

FIGURE 1 Gabon, Central Africa, has seen a steady increase in logging concessions.
Source: Wildlife Conservation Society.

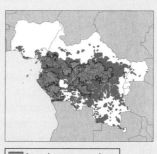 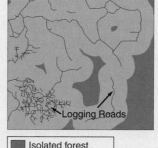 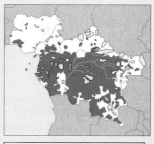

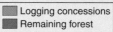 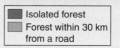

Logging concessions
Remaining forest

Isolated forest
Forest within 30 km from a road

Existing conservation areas
Proposed

FIGURE 2 A few GIS layers used to identify priority conservation areas.
Source: Wildlife Conservation Society.

ments to identify concentrations of intact, ancient forests. Using maps of logging roads, they calculated the area within a 30 km "buffer" around roads, since loggers, settlers, and hunters usually threaten biodiversity near roads. They also mapped existing and planned conservation areas (fig. 2).

By overlaying these and other layers, analysts identified priority conservation areas of extensive original forest, which have high biodiversity and rare species. Overlaying these priority areas with a map of protected lands and a map of timber concessions, they identified *threatened* priority areas (fig. 3).

Most of the unprotected priority areas may never be protected, but having this map provides two important guides for future conservation. First, it assesses the state of the problem. With this map, we know that most of the forest is unprotected but also that the region's primary forest is extensive. Second, this map provides priorities for conservation planning. In addition, maps are very effective tools for publicizing an issue. When a map like this is published, more people become enthusiastic about joining the conservation effort.

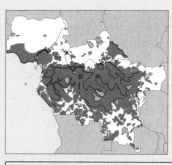

National parks
Original forest
Priority areas

FIGURE 3 Priority areas outside of national parks.
Source: Wildlife Conservation Society.

GIS has become an essential tool for conserving forests, grasslands, ecosystems, and nature preserves. GIS has revolutionized the science of planning and conservation—examining problems using quantitative data—just as it may have revolutionized the way you plan a driving trip.

FIGURE 12.9 The temperate Great Bear Rainforest has the highest biomass per hectare of any landscape in the world.

FIGURE 12.10 Only about 2,000 pairs of northern spotted owls remain in the old-growth forests of the Pacific Northwest. Cutting old-growth forests threatens the endangered species, but reduced logging threatens the jobs of many timber workers.

incompatible, however. Bird-watching, for example, isn't very enjoyable in a dirt bike racing course. And protecting species that need unbroken old-growth forest isn't feasible when you cut down the forest.

Old-Growth Forests

The most contentious forestry issues in the United States and Canada in recent years have centered on logging in old-growth forests in the Pacific Northwest. As the opening case study for this chapter shows, these forests have incredibly high levels of biodiversity, and they accumulate more total biomass in standing vegetation per unit area than any other ecosystem on earth (fig. 12.9). Many endemic species, such as the northern spotted owl (fig. 12.10), Vaux's swift, and the marbled murrelet, are so highly adapted to the unique conditions of these ancient forests that they live nowhere else.

Less than 10 percent of old-growth forest in the United States remains intact, and 80 percent of what remains is scheduled to be cut in the near future. Just a century ago, most of the coastal ranges of Washington, Oregon, northern California, British Columbia, and southeastern Alaska were clothed in a lush forest of huge trees, many a thousand years old or more. British Columbia has felled at least 60 percent of its richest and most productive ancient forests. Protection of the Great Bear Rainforest is a dramatic victory in the fight to save these magnificent landscapes. Many conservationists hope that Canada's example will encourage the United States to preserve similar areas in the adjacent Tougass National Forest of southeast Alaska.

In 1989 wildlife conservationists sued the U.S. Forest Service over logging rates in Washington and Oregon, arguing that northern spotted owls are endangered and must be protected under the Endangered Species Act. Federal courts agreed and ordered some 1 million ha (2.5 million acres) of ancient forest set aside to preserve the last 2,000 pairs of owls. This is about half the remaining

virgin forest in Washington and Oregon. The timber industry claimed that 40,000 jobs would be lost, although economic studies show that most logging jobs were lost as a result of log exports and mechanization. Still, outrage in the logging communities was loud and clear. Convoys of logging trucks converged on protest sites, while angry crowds burned conservationists in effigy. Bumper stickers urged, "Save a logger; eat an owl."

A compromise forest management plan was finally agreed upon that allows some continued logging but protects the most valuable forests, including selected old-growth preserves, riparian (streamside) buffer strips, and prime wildlife habitat. This plan may not be enough protection, however, to ensure the survival of endangered salmon and trout populations in northwestern rivers. An even bigger battle may be shaping up as specific fish populations are considered for listing as endangered species. Several salmon and steelhead trout populations were listed as endangered under the Clinton administration. These listings have been challenged by the timber industry, but the $1 billion-per-year fishing industry in the Pacific Northwest may be at least as powerful a constituency as were those concerned about spotted owls.

Harvest Methods

Most lumber and pulpwood in the United States and Canada currently are harvested by **clear-cutting,** in which every tree in a given area is cut, regardless of size (fig. 12.11). This method is effective for producing even-age stands of sun-loving species, such as aspen or pines, but often increases soil erosion and eliminates habitat for many forest species when carried out on large blocks. It was once thought that good forest management required immediate removal of all dead trees and logging residue. Research has shown, however, that standing snags and coarse woody debris play important ecological roles, including soil protection, habitat for a variety of organisms, and nutrient recycling.

FIGURE 12.11 This huge clear-cut in Washington's Gifford Pinchot National Forest threatens species dependent on old-growth forest and exposes steep slopes to soil erosion. Restoring something like the original forest will take hundreds of years.

FIGURE 12.12 Selective harvesting in this forest on the Menominee Reservation has left the most valuable red and white pines. A century of careful forest management has doubled the volume of standing trees while also providing a steady sustainable yield.

Some alternatives to clear-cutting include **shelterwood harvesting,** in which mature trees are removed in a series of two or more cuts, and **strip-cutting,** in which all the trees in a narrow corridor are harvested. For many forest types, the least disruptive harvest method is **selective cutting,** in which only a small percentage of the mature trees are taken in each 10- or 20-year rotation. Figure 12.12 shows a small selective cut on the Menominee Indian Reservation in Wisconsin. After more than a century of sustainable harvesting, the tribe's forests now have twice as much standing biomass (harvestable board feet of timber) than they did when the reservation was established. Rather than manage for short-term yields, the tribe aims to increase the quality and quantity of timber on their lands. "Instead of cutting the best, we cut the worst first," they say. Notice in figure 12.12, that most logging slash has been removed to prevent crown fires, but enough debris is left to suppress weeds and prevent erosion. With competition reduced, the seed trees left behind will grow to be more valuable lumber and will supply seedlings to regenerate a new pine forest. (See related story "Forestry for the Seventh Generation" at www.mhhe.com/cunningham10e.)

Should We Subsidize Logging on Public Lands?

An increasing number of people in the United States are calling for an end to all logging on federal lands. Many argue that ecological services, from maintaining river levels for fish and irrigation to recreation, generate more revenue at lower costs. Many remote communities depend on logging jobs, but these jobs depend on subsidies. The federal government builds roads, manages forests, fights fires, and sells timber for less than the administrative costs of the sales. What is the best policy?

Some argue that logging should be restricted to privately owned lands. Just 4 percent of the nation's timber comes from national forests, and this harvest adds about $4 billion to the American economy per year. In contrast, recreation, fish and wildlife, clean water, and other ecological services provided by the forest, by their calculations, are worth at least $224 billion each year. Timber industry officials, on the other hand, dispute these claims, arguing that logging not only provides jobs and supports rural communities but also keeps forests healthy. What do you think? Could we make up for decreased timber production from public lands by more intensive management of private holdings and by substitution or recycling of wood products? Are there alternative ways you could suggest to support communities now dependent on timber harvesting?

Roads on public lands are another controversy. Over the past 40 years, the Forest Service has expanded its system of logging roads more than ten-fold, to a current total of nearly 550,000 km (343,000 mi), or more than ten times the length of the interstate highway system. Government economists regard road building as a benefit because it opens up the country to motorized recreation and industrial uses. Wilderness enthusiasts and wildlife supporters, however, see this as an expensive and disruptive program. In 2001 the Clinton administration announced a plan to protect 23.7 million ha (58.5 million acres) of de facto wilderness from roads. Timber, mining, and oil companies protested this rule. In 2002 a federal judge in Idaho ruled that blocking road construction would do "irreparable harm" to the forest. He also ruled that the 600 public meetings held over a 12-month period to discuss this plan, and the record 1.6 million written comments gathered (90 percent of which supported increased protection), did not represent adequate public input. The Bush administration overturned the "roadless rule" and ordered resource managers to expedite logging, mining, and motorized recreation. What do you think? How much of the remaining old growth should be protected as ecological reserves?

What Can You Do?

Lowering Your Forest Impacts

For most urban residents, forests—especially tropical forests—seem far away and disconnected from everyday life. There are things that each of us can do, however, to protect forests.

- Reuse and recycle paper. Make double-sided copies. Save office paper, and use the back for scratch paper.

- Use email. Store information in digital form, rather than making hard copies of everything.

- If you build, conserve wood. Use wafer board, particle board, laminated beams, or other composites, rather than plywood and timbers made from old-growth trees.

- Buy products made from "good wood" or other certified sustainably harvested wood.

- Don't buy products made from tropical hardwoods, such as ebony, mahogany, rosewood, or teak, unless the manufacturer can guarantee that the hardwoods were harvested from agroforestry plantations or sustainable-harvest programs.

- Don't patronize fast-food restaurants that purchase beef from cattle grazing on deforested rainforest land. Don't buy coffee, bananas, pineapples, or other cash crops if their production contributes to forest destruction.

- Do buy Brazil nuts, cashews, mushrooms, rattan furniture, and other nontimber forest products harvested sustainably by local people from intact forests. Remember that tropical rainforest is not the only biome under attack. Contact the Taiga Rescue Network (www.taigarescue.org) for information about boreal forests.

- If you hike or camp in forested areas, practice minimum-impact camping, Stay on existing trails, and don't build more or bigger fires than you absolutely need. Use only downed wood for fires. Don't carve on trees or drive nails into them.

- Write to your congressional representatives, and ask them to support forest protection and environmentally responsible government policies. Contact the U.S. Forest Service, and voice your support for recreation and nontimber forest values.

Fire Management

Following a series of disastrous fire years in the 1930s, in which hundreds of millions of hectares of forest were destroyed, whole towns burned to the ground, and several hundred people died, the U.S. Forest Service adopted a policy of aggressive fire control in which every blaze on public land was to be out before 10 A.M. Smokey Bear was adopted as the forest mascot and warned us that "only you can prevent forest fires." Recent studies, however, of fire's ecological role suggest that our attempts to suppress all fires may have been misguided. Many biological communities are fire-adapted and require periodic burning for regeneration. Furthermore, eliminating fire from these forests has allowed woody debris to accumulate, greatly increasing the chances of a very big fire (fig. 12.13).

FIGURE 12.13 By suppressing fires and allowing fuel to accumulate, we make major fires such as this more likely. The safest and most ecologically sound management policy for some forests may be to allow natural or prescribed fires, that don't threaten property or human life, to burn periodically.

Forests that once were characterized by 50 to 100 mature, fire-resistant trees per hectare and an open understory now have a thick tangle of up to 2,000 small, spindly, mostly dead saplings in the same area. The U.S. Forest Service estimates that 33 million ha (73 million acres), or about 40 percent of all federal forestlands, are at risk of severe fires. To make matters worse, Americans increasingly live in remote areas where wildfires are highly likely. Because there haven't been fires in many of these places in living memory, many people assume there is no danger, but by some estimates, 40 million U.S. residents now live in areas with high wildfire risk.

Much of the federal and state firefighting efforts are controlled, in effect, by these homeowners who build in fire-prone areas. A government audit found that 90 percent of the Forest Service firefighting outlays go to save private property. If people who build in forested areas would take some reasonable precautions to protect themselves, we could let fires play their normal ecological role in forests in many cases. For example, you shouldn't build a log cabin with a wood shake roof surrounded by dense forest at the end of a long, narrow, winding drive that a fire truck can't safely navigate. If you're going to have a home in the forest, you should use fire proof materials, such as a metal roof and rock or brick walls, and clear all trees and brush from at least 60 m (200 ft) around any buildings.

A recent prolonged drought in the western United States has heightened fire danger. In 2006 more than 96,000 wildfires burned 4 million ha (10 million acres) of forests and grasslands in the United States. Federal agencies spent about $1.6 billion to fight these fires, nearly four times the previous ten-year average.

The dilemma is how to undo years of fire suppression and fuel buildup. Fire ecologists favor small, prescribed burns to clean out debris. Loggers decry this approach as a waste of valuable timber, and local residents of fire-prone areas fear that prescribed fires will escape and threaten them. Recently the Forest Service proposed a massive new program of forest

What Do You Think?

Forest Thinning and Salvage Logging

For more than 70 years, firefighting has been a high priority for forest managers. Unfortunately, our efforts have been so successful that dead wood and brush have now built up to dangerous levels in many forests. Lands that would once have been cleaned out by frequent low-temperature ground fires now have so much accumulated fuel that a catastrophic wildfire is all but inevitable, To make matters worse, increasing numbers of people are building cabins and homes in remote, fire-prone areas, where they expect to be protected from unavoidable risks.

People living in or near national forests demand protection from wildfires. In response, federal agencies are starting thinning programs to remove excess fuel. To make it profitable for loggers to remove fire-prone dead wood, small trees, and brush, the government is allowing them to harvest large, valuable, and fire-resistant trees located in the backcountry, often miles away from the nearest communities. And to avoid what many loggers claim is "red tape and litigation" that have tied up forest managers in "analysis paralysis," most thinning projects are exempt from public comment and administrative appeals, as well as from environmental reviews under the National Environmental Policy Act (NEPA).

Environmental groups denounce these projects as merely logging without laws. Thinning, they argue, looks very much like clear-cutting. Forest ecologists argue that thinning, unless it is repeated every few years, actually makes the forests more, rather than less, fire-prone. Removing big, old trees opens up the forest canopy and encourages growth of brush and new tree seedlings. Furthermore, logging compacts soil and introduces invasive species. Many forest ecologists maintain that small,

prescribed burns can reduce fuel and produce healthier forests than commercial logging.

A 2005 study of an Oregon fire found that salvage logging actually increased fire susceptibility and decreased the ability of the forest to regenerate naturally. Post-fire logging harms the soil, introduces invasive species, and makes future fire much more likely (see figure 1). Dry branches and debris left on the ground provided much better fuel than did standing dead trees. In addition, seedling growth among standing dead trees was robust, with three times the density of seedlings in salvage areas.

Other research calls into question a main justification for thinning; protecting housing that has proliferated around the edges of federal forests. The only thinning needed to protect houses, according to fire experts, is within about 60 m around the building. Regardless of how intense the fire is, building damage can be minimized by installing a metal roof, clearing pine needles and brush around the house, and removing trees immediately around the building.

After a fire has burned a forest, local residents often call for emergency salvage logging to utilize dead trees before they fall and become fuel for future fires. Foes say that salvage sales allow timber companies to cut many trees that survive the flames and, like thinning, leave roads and scars from heavy equipment that last far longer than any effects of the fires themselves. This kind of logging costs far more than it yields. And, critics contend, the salvage contracts rarely require timber companies to remove the very fuel that stokes wildfires—the underbrush and downed timber for which there is little or no market. The result, some conservationists say, is the eco-logical equivalent of mugging a fire victim.

What do you think? How can we get out of the crisis we've created by preventing fires and letting dead wood accumulate? If you were a forest manager, would you authorize thinning and salvage logging, or can you think of other ways to remove forest fuels in an ecologically and economically sustainable manner? What research projects would you direct your staff to undertake in order to inform your decision in this dilemma?

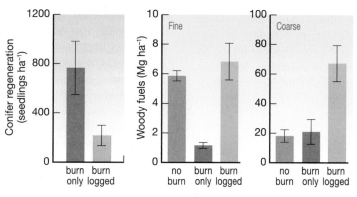

FIGURE 1 Research results from the 2005 Bisent fire in Oregon. (a) Post-fire logging reduces regeneration by about half. (b) Far more fine and coarse debris (fuelwood) are present in burned and logged forests than in burned only.
Data from Donato, et al., 2006.

thinning and emergency salvage operations (removing trees and flammable material from mature or recently burned forests) on 16 million ha (40 million acres) of national forest. Carried out over a 20-year period, this program could cost as much as $12 billion and would open up much roadless, de facto wilderness to large-scale logging. Critics complain that this program is ill advised and environmentally destructive. Proponents argue that the only way to save the forest is to log it (see What Do You Think, p. 263).

Ecosystem Management

In the 1990s the U.S. Forest Service began to shift its policies from a timber production focus to **ecosystem management,** which attempts to integrate sustainable ecological, economic, and social goals in a unified, systems approach. Some of the principles of this new philosophy include

- Managing across whole landscapes, watersheds, or regions over ecological time scales

- Considering human needs and promoting sustainable economic development and communities

- Maintaining biological diversity and essential ecosystem processes

- Utilizing cooperative institutional arrangements

- Generating meaningful stakeholder and public involvement and facilitating collective decision making

- Adapting management over time, based on conscious experimentation and routine monitoring

TABLE 12.1

Draft Criteria for Sustainable Forestry

1. Conservation of biological diversity
2. Maintenance of productive capacity of forest ecosystems
3. Maintenance of forest ecosystem health and vitality
4. Maintenance of soil and water resources
5. Maintenance of forest contribution to global carbon cycles
6. Maintenance and enhancement of long-term socioeconomic benefits to meet the needs of legal, institutional, and economic framework for forest conservation and sustainable management

Source: Data from USFS, 2002.

FIGURE 12.14 Grasslands are expansive, open environments that can support surprising biodiversity.

Some critics argue that we don't understand ecosystems well enough to make practical decisions in forest management on this basis. They argue we should simply set aside large blocks of untrammeled nature to allow for chaotic, catastrophic, and unpredictable events. Others see this new approach as a threat to industry and customary ways of doing things. Still, elements of ecosystem management appear in the *National Report on Sustainable Forests* prepared by the U.S. Forest Service. Based on the Montreal Working Group criteria and indicators for forest health, this report suggests goals for sustainable forest management (table 12.1).

12.2 GRASSLANDS

After forests, grasslands are among the biomes most heavily used by humans. Prairies, savannas, steppes, open woodlands, and other grasslands occupy about one-quarter of the world's land surface. Much of the U.S. Great Plains and the Prairie Provinces of Canada fall in this category (fig. 12.14). The 3.8 billion ha (12 million mi^2) of pastures and grazing lands in this biome make up about twice the area of all agricultural crops. When you add to this about 4 billion ha of other lands (forest, desert, tundra, marsh, and thorn scrub) used for raising livestock, more than half of all land is used at least occasionally for grazing. More than 3 billion cattle, sheep, goats, camels, buffalo, and other domestic animals on these lands make a valuable contribution to human nutrition. Sustainable pastoralism can increase productivity while maintaining biodiversity in a grassland ecosystem.

Because grasslands, chaparral, and open woodlands are attractive for human occupation, they frequently are converted to cropland, urban areas, or other human-dominated landscapes. Worldwide the rate of grassland disturbance each year is three times that of tropical forest. Although they may appear to be uniform and monotonous to the untrained eye, native prairies can be highly productive and species-rich. According to the U.S. Department of Agriculture, more threatened plant species occur in rangelands than in any other major American biome.

Grazing can be sustainable or damaging

By carefully monitoring the numbers of animals and the condition of the range, ranchers and **pastoralists** (people who live by herding animals) can adjust to variations in rainfall, seasonal plant conditions, and the nutritional quality of forage to keep livestock healthy and avoid overusing any particular area. Conscientious management can actually improve the quality of the range.

When grazing lands are abused by overgrazing—especially in arid areas—rain runs off quickly before it can soak into the soil to nourish plants or replenish groundwater. Springs and wells dry up. Seeds can't germinate in the dry, overheated soil. The barren ground reflects more of the sun's heat, changing wind patterns, driving away moisture-laden clouds, and leading to further desiccation. This process of conversion of once fertile land to desert is called **desertification.**

This process is ancient, but in recent years it has been accelerated by expanding populations and the political conditions that force people to overuse fragile lands. According to the International Soil Reference and Information Centre in the Netherlands, nearly three-quarters of all rangelands in the world show signs of either degraded vegetation or soil erosion. Overgrazing is responsible for about one-third of that degradation (fig. 12.15). The highest percentage of moderate, severe, and extreme land degradation is in Mexico and Central America, while the largest total area is in Asia, where the world's most extensive grasslands occur. Can we reverse this process? In some places, people are reclaiming deserts and repairing the effects of neglect and misuse.

Overgrazing threatens many U.S. rangelands

The health of most public grazing lands in the United States is not good. Political and economic pressures encourage managers to increase grazing allotments beyond the carrying capacity of

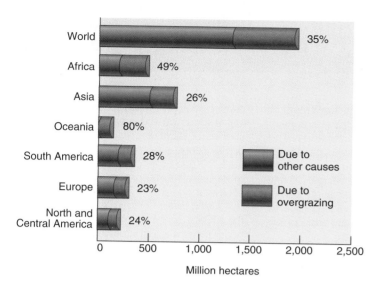

FIGURE 12.15 Rangeland soil degradation due to overgrazing and other causes. Notice that in Europe, Asia, and the Americas, farming, logging, mining, urbanization, etc., are responsible for about three-quarters of all soil degradation. In Africa and Oceania, where more grazing occurs and desert or semiarid scrub make up much of the range, grazing damage is higher.
Source: World Resources Institute, 2004.

FIGURE 12.16 More than half of all publicly owned grazing land in the United States is in poor or very poor condition. Overgrazing and invasive weeds are the biggest problems.

the range. Lack of enforcement of existing regulations and limited funds for range improvement have resulted in **overgrazing,** damage to vegetation and soil including loss of native forage species and erosion. The Natural Resources Defense Council claims that only 30 percent of public range-lands are in fair condition, and 55 percent are poor or very poor (fig. 12.16).

Overgrazing has allowed populations of unpalatable or inedible species, such as sage, mesquite, cheatgrass, and cactus, to build up on both public and private rangelands. Wildlife conservation groups regard cattle grazing as the most ubiquitous form of ecosystem degradation and the greatest threat to endangered species in the southwestern United States. They call for a ban on cattle and sheep grazing on all public lands, noting that it provides only 2 percent of the total forage consumed by beef cattle and supports only 2 percent of all livestock producers.

Like federal timber management policy, grazing fees charged for use of public lands often are far below market value and represent an enormous hidden subsidy to western ranchers. Holders of grazing permits generally pay the government less than 25 percent the amount of leasing comparable private land. The 31,000 permits on federal range bring in only $11 million in grazing fees but cost $47 million per year for administration and maintenance. The $36 million difference amounts to a massive "cow welfare" system of which few people are aware.

On the other hand, ranchers defend their way of life as an important part of western culture and history. Although few cattle go directly to market from their ranches, they produce almost all the beef calves subsequently shipped to feedlots. And without a viable ranch economy, they claim, even more of the western landscape would be subdivided into small ranchettes to the detriment of both wildlife and environmental quality. Many conservation groups are recognizing that preserving ranches may be the best way to protect wildlife habitat. What do you think? How can we best protect traditional lifestyles and rural communities while also preserving natural resources?

Ranchers are experimenting with new methods

Where a small number of livestock are free to roam a large area, they generally eat the tender, best-tasting grasses and forbs first, leaving the tough, unpalatable species to flourish and gradually dominate the vegetation. In some places, farmers and ranchers find that short-term, intensive grazing helps maintain forage quality. As South African range specialist Allan Savory observed, wild ungulates (hoofed animals), such as gnus or zebras in Africa or bison (buffalo) in America, often tend to form dense herds that graze briefly but intensively in a particular location before moving on to the next area. Rest alone doesn't necessarily improve pastures and rangelands. Short-duration, **rotational grazing**—confining animals to a small area for a short time (often only a day or two) before shifting them to a new location—simulates the effects of wild herds (fig. 12.17). Forcing livestock to eat everything equally, to trample the ground thoroughly, and to fertilize heavily with manure before moving on helps keep weeds in check and encourages the growth of more desirable forage species. This approach doesn't work everywhere, however. Many plant communities in the U.S. desert Southwest, for example, apparently evolved in the absence of large, hoofed animals and can't withstand intensive grazing.

Restoring fire and managing grasslands as regional units can have many benefits for both ranchers and wildlife. The Nature Conservancy has participated with private landowners in a number of innovative experiments in range restoration (Exploring Science p. 266).

For decades, environmentalists have tried to limit grazing on public lands, where ranchers lease pastures from the government. Now some scientists and conservationists are saying that cattle ranches may be the last best hope for preserving habitat for many native species. Maintaining ranches may also be the only way to restore the periodic fires that keep brush and cactus from taking over western grasslands. In a number of places in the United States, ranchers are forming alliances with environmental organizations and government officials to find new ways to protect and manage rangelands.

One of the pressures driving this new model of collaboration is the growing popularity of western hobby ranches and rural homesteads for city folk. Falling commodity prices, drought, taxes, and other forces are causing many ranchers to consider selling their land. Why continue to struggle to make a living with ranching when you can make millions by cutting up your land into 40-acre ranchettes? But when the land is subdivided, invasive species move in along with people and their pets, and fewer native species can survive. Furthermore, it becomes much harder, if not impossible, to let fires burn across the land periodically, a process that is now thought to be essential in the southwestern landscape.

Some grazing practices clearly have been detrimental. Studies have found extensive damage from grazing in and around streams in the desert West, for instance. But few studies have compared the alternatives to ranching on these lands, which are home not only to ranchers but to many native animal and plant species. Recent research has found that ranches have at least as many species of birds, carnivores, and plants as similar areas protected as wildlife refuges. Ranches also have fewer invasive weeds.

An outstanding example of new cooperative relationships between ranchers, conservationists, and government agencies is the Malpai Borderlands Group (MBG). This community-based ecosystem management effort was created by landowners in a region

called the "boot heel," where New Mexico, Arizona, and Mexico meet. Malpai is derived from the Spanish word for badlands. The craggy mountains, grassy plains, and scrub-covered desert hills of this region are home to more than 20 threatened species. Nearly 400,000 ha (about 1 million acres) are part of the collaboration, including private property, state trust lands, national forest, and Bureau of Land Management acreage.

This pioneering collaboration began in 1993 in an effort to address threats to ranch-

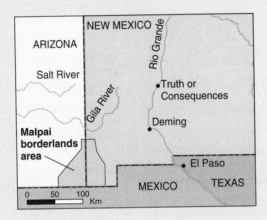

The Malpai borderlands are in the boot heel of New Mexico, where it meets Arizona and Mexico. Ranchers in this area have joined with government agents and conservation organizations to restore fire, protect wildlife habitat, and regenerate grasslands.

ing. Thirty-five neighbors got together to discuss common problems. They agreed that excluding wildfire from the range was contributing to increasing brush and declining grass cover, resulting in the loss of watershed stability, wildlife habitat, and livestock forage.

Early on, community leaders approached The Nature Conservancy (TNC) for assistance. TNC, in turn, brought in ecologists familiar with the borderlands ecosystems to aid in organizing a science program. This input from scientists was crucial in giving the MBG efforts a systems-based approach to conservation. This approach emphasized

conservation of natural processes rather than just a focus on the management of single species or particular resources as is typical of many conservation programs. Since 1993, the MBG and collaborators have established more than 200 monitoring plots to assess ecosystem health.

Using range science as a starting point, the MBG's goal has evolved to a comprehensive natural resource management and rural development agenda. Their stated goal is, "To preserve and maintain the natural processes that create and protect a healthy, unfragmented landscape to support a diverse, flourishing community of human, plant, and animal life in the borderlands region." One of the key features of this plan is to restore fire as a management tool.

With a large, contiguous area under common management, it's now possible to set prescribed fires that have a significant effect on vegetation. Before formation of the MBG coalition, the patchwork of ownership and management in the area made large-scale operations all but impossible. A key to MBG success was purchase of the 120,000-hectare Gray Ranch by TNC. This large landholding in the heart of the Malpai borderlands makes it possible to carry out an innovative program called "grass banking." If a neighbor rancher has a bad season (perhaps due to prolonged drought), he can move his cattle onto Gray Ranch until his own ranchland is able to recover. The rancher grants a conservation easement of equal value over to the MBG that prohibits future subdivision.

The MBG isn't the only innovative initiative in ranching country. The Quivira Coalition, also based in New Mexico, brings together ranchers, conservationists, and land managers from throughout the West to foster scientifically guided ranch management and riparian restoration. TNC-owned Matador Ranch in Montana also has a grass-banking program. Perhaps the success of these programs will inspire similar cooperation rather than confrontation and litigation, not only on rangeland problems, but also on other contentious environmental issues.

FIGURE 12.17 Intensive, rotational grazing encloses livestock in a small area for a short time (often only one day) within a movable electric fence to force them to eat vegetation evenly and fertilize the area heavily.

Another approach to ranching in some areas is to raise wild species, such as red deer, impala, wildebeest, or oryx (fig. 12.18). These animals forage more efficiently, resist harsh climates, often are more pest- and disease-resistant, and fend off predators better than usual domestic livestock. Native species also may have different feeding preferences and needs for water and shelter than cows, goats, or sheep. The African Sahel, for instance, can provide only enough grass to raise about 20 to 30 kg (44 to 66 lbs) of beef per hectare. Ranchers can produce three times as much

meat with wild native species in the same area because these animals browse on a wider variety of plant materials.

In the United States, ranchers find that elk, American bison, and a variety of African species take less care and supplemental feeding than cattle or sheep and result in a better financial return because their lean meat can bring a better market price than beef or mutton. Media mogul Ted Turner has become both the biggest private landholder in the United States and the owner of more American bison than anyone other than the government.

12.3 PARKS AND PRESERVES

While most forests and grasslands serve utilitarian purposes, many nations have set aside some natural areas for ecological, cultural, or recreational purposes. Some of these preserves have existed for thousands of years. Ancient Greeks and Druids, for example, protected sacred groves for religious purposes. Royal hunting grounds preserved wild areas in many countries. Although these areas were usually reserved for elite classes in society, they maintained biodiversity and natural landscapes in regions where most lands were heavily used.

The first public parks open to ordinary citizens may have been the tree-sheltered agoras in planned Greek cities. But the idea of providing natural space for recreation or to preserve natural environments, has really developed in the past half century (fig. 12.19). Currently, nearly 12 percent of the land area of the earth is protected in some sort of park, preserve, or wildlife management area. This represents about 19.6 million ha (7.6 million mi^2) in 107,000 different preserves.

Many countries have created nature preserves

Different levels of protection are found in nature preserves. The World Conservation Union divides protected areas into five categories depending on the intended level of allowed human use

FIGURE 12.18 Red deer (*Cervus elaphus*) are raised in New Zealand for antlers and venison.

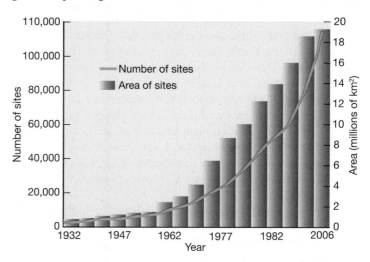

FIGURE 12.19 Growth of protected areas worldwide, 1932–2003.
Source: UN World Commission on Protected Areas.

TABLE 12.2

IUCN Categories of Protected Areas

Category	Allowed Human Impact or Intervention
1. Ecological reserves and wilderness areas	Little or none
2. National parks	Low
3. Natural monuments and archaeological sites	Low to medium
4. Habitat and wildlife management areas	Medium
5. Cultural or scenic landscapes, recreation areas	Medium to high
6. Managed resource area	High

Source: Data from World Conservation Union, 1990.

FIGURE 12.20 Canada's Quttinirpaaq National Park at the north end of Ellesmere Island plenty of solitude and pristine landscapes, but little biodiversity.

(table 12.2). In the most stringent category (ecological reserves and wilderness areas) little or no human impacts are allowed. In some strict nature preserves, where particularly sensitive wildlife or natural features are located, human entry may be limited only to scientific research groups that visit on rare occasions. In some wildlife sanctuaries, for example, only a few people per year are allowed to visit to avoid introducing invasive species or disrupting native species. In the least restrictive categories (national forests and other natural resource management areas), on the other hand, there may be a high level of human use.

Venezuela claims to have the highest proportion of its land area protected (70 percent) of any country in the world. About half this land is designated as preserves for indigenous people or for sustainable resource harvesting. With little formal management, protection from poaching by hunters, loggers, and illegal gold hunters is minimal. Unfortunately, it's not uncommon in the developing world to have "paper parks" that exist only as a line drawn on a map with no budget for staff, management, or infrastructure. The United States, by contrast, has only about 15.8 percent of its land area in protected status, and less than one-third of that amount is in IUCN categories I or II (nature reserves, wilderness areas, national parks). The rest is in national forests or wildlife management zones that are designated for sustainable use. With hundreds of thousands of state and federal employees, billions of dollars in public funding, and a high level of public interest and visibility, U.S. public lands are generally well managed.

Brazil, with more than one-quarter of all the world's tropical rainforest, is especially important in biodiversity protection. Currently, Brazil has the largest total area in protected status of any country. More than 1.6 million km² or 18.7 percent of the nation's land—mostly in the Amazon basin—is in some protected status. In 2006, the northern Brazilian state of Para, in collabora-

tion with Conservation International (CI) and other nongovernmental organizations, announced the establishment of nine new protected areas along the border with Suriname and Guyana. These new areas, about half of which will be strictly protected nature preserves, will link together several existing indigenous areas and nature preserves to create the largest tropical forest reserve in the world. More than 90 percent of the new 15 million ha (58,000 mi², or about the size of Illinois) Guyana Shield Corridor is in pristine natural state. CI president Russ Mittermeir says, "If any tropical rainforest on earth remains intact a century from now, it will be this portion of northern Amazonia." In contrast to this dramatic success, the Pantanal, the world's largest wetland/savanna complex, which lies in southern Brazil and is richer in some biodiversity categories than the Amazon, is almost entirely privately owned. There are efforts to set aside some of this important wetland, but so far, little is in protected status.

Some other countries with very large reserved areas include Greenland (with a 972,000 km² national park that covers most of the northern part of the island), and Saudi Arabia (with a 640,000 km² wildlife management area in its Empty Quarter). These areas are relatively easy to set aside, however, being mostly ice covered (Greenland) or desert (Saudi Arabia). Canada's Quttinirpaaq National Park on Ellesmere Island is an example of a preserve with high wilderness values but little biodiversity. Only 800 km (500 miles) from the North Pole, this remote park gets fewer than 100 human visitors per year during its brief, three-week summer season (fig. 12.20). With little evidence of human occupation, it has abundant solitude and stark beauty, but very little wildlife and almost no vegetation. By contrast, the Great Bear Rainforest management area described in the opening case study for this chapter has a rich diversity of both marine and terrestrial life, but the valuable timber, mineral, and wildlife resources in the area make protecting it expensive and controversial.

TABLE 12.3

Region	Total Area Protected (km²)	Protected Percent	Number of Areas
North America	4,459,305	16.2%	13,447
South America	1,955,420	19.3%	1,456
North Eurasia	1,816,987	7.7%	17,724
East Asia	1,764,648	14.0%	3,265
Eastern and Southern Africa	1,696,304	14.1%	4,060
Brazil	1,638,867	18.7%	1,287
Australia/ New Zealand	1,511,992	16.9%	9,549
North Africa and Middle East	1,320,411	9.8%	1,325
Western and Central Africa	1,131,153	8.7%	2,604
Southeast Asia	867,186	9.6%	2,689
Europe	785,012	12.4%	46,194
South Asia	320,635	6.5%	1,216
Central America	158,193	22.5%	781
Antarctic	70,323	0.5%	122
Pacific	67,502	1.9%	430
Caribbean	66,210	8.2%	958
Total	**19,630,149**	**11.6%**	**107,107**

Source: World Commission on Protected Areas, 2007.

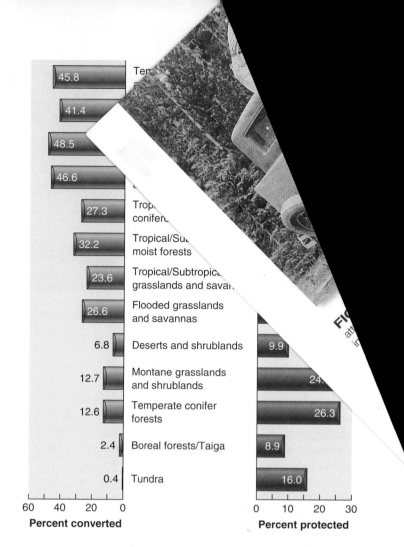

FIGURE 12.21 With a few exceptions, the percent of each biome converted to human use is roughly inverse to the percent protected in parks and preserves.
Data from J. Hoekstra, et al., 2005.

Collectively, according to the World Commission on Protected Areas, Central America has 22.5 percent of its land area in some protected status (table 12.3). The Pacific region, at 1.9 percent in nature reserves, has both the lowest percentage and the lowest total area. With land scarce on small islands, it's hard to find space to set aside for nature sanctuaries. Some biomes are well represented in nature preserves, while others are relatively underprotected. Figure 12.21 shows a comparison between the percent of each major biome in protected status. Not surprisingly, there's an inverse relationship between the percentage converted to human use (and where people live) and the percentage protected. Temperate grasslands and savannas (such as the American Midwest) and Mediterranean woodlands and scrub (such as the French Riviera) are highly domesticated, and, therefore, expensive to set aside in large areas. Temperate conifer forests (think of Siberia, or Canada's vast expanse of boreal forest) are relatively uninhabited, and therefore easy to put into some protected category.

Not all preserves are preserved

Even parks and preserves designated with a high level of protection aren't always safe from exploitation or changes in political priorities. Serious problems threaten natural resources and environmental quality in many countries. In Greece, the Pindus National Park is threatened by plans to build a hydroelectric dam in the center of the park. Furthermore, excessive stock grazing and forestry exploitation in the peripheral zone are causing erosion and loss of wildlife habitat. In Columbia, dam building also threatens the Paramillo National Park. Ecuador's largest park, Yasuni National Park, which contains one of the world's most megadiverse regions of lowland Amazonian forest, has been opened to oil drilling, while miners and loggers in Peru have invaded portions of Huascaran National Park. In Palau, coral reefs identified as a potential biosphere reserve are damaged by dynamite fishing, while on some beaches in Indonesia, every egg laid by endangered sea turtles is taken by egg hunters. These are just a few of the many problems faced by parks and preserves around the world. Often countries with the most important biomes lack funds, trained personnel, and experience to manage the areas under their control.

Even in rich countries, such as the United States, some of the "crown jewels" of the National Park System suffer from overuse and degradation. Yellowstone and Grand Canyon National Parks, for example have large budgets and are highly regulated, but are being "loved to death" because they are so popular. When the U.S. National Park Service was established in 1916, Stephen Mather, the first director, reasoned that he needed to make

FIGURE 12.22 Wild animals have always been one of the main attractions in national parks. Many people lose all common sense when interacting with big, dangerous animals. This is not a petting zoo.

FIGURE 12.23 Thousands of people wait for an eruption of Old Faithful geyser in Yellowstone National Park. Can you find the ranger who's giving a geology lecture?

the parks comfortable and entertaining for tourists as a way of building public support. He created an extensive network of roads in the largest parks so that visitors could view famous sights from the windows of their automobiles, and he encouraged construction of grand lodges in which guests could stay in luxury.

His plan was successful; the National Park System is cherished and supported by many American citizens. But sometimes entertainment seems to have trumped over nature protection. Visitors were allowed—in some cases even encouraged—to feed wildlife. Bears lost their fear of humans and became dependent on an unhealthy diet of garbage and handouts (fig. 12.22). In Yellowstone and the Grand Teton National Parks, the elk herd was allowed to grow to 25,000 animals, or about twice the carrying capacity of the habitat. The excess population overgrazed the vegetation to the detriment of many smaller species and the biological community in general. As we discussed earlier in this chapter, 70 years of fire suppression resulted in changes of forest composition and fuel buildup that made huge fires all but inevitable. In Yosemite, you can stay in a world-class hotel, buy a pizza, play video games, do laundry, play golf or tennis, and shop for curios, but you may find it difficult to experience the solitude or enjoy the natural beauty extolled by John Muir as a prime reason for creating the park.

Think About It

If you were superintendent of a major national park, how would you reconcile the demand for comfort and recreation with the need to protect nature? If no one comes to your park, you will probably lose public support. But if the landscape is trashed, what's the purpose of having a park?

In many of the most famous parks, traffic congestion and crowds of people stress park resources and detract from the experience of unspoiled nature (fig. 12.23). Some parks, such as Yosemite, and Zion National Park, have banned private automobiles from the most congested areas. Visitors must park in remote lots and take clean, quiet electric or natural gas-burning buses to popular sites. Other parks are considering limits on the number of visitors admitted each day. How would you feel about a lottery system that might allow you to visit some famous parks only once in your lifetime, but to have an uncrowded, peaceful experience on your one allowed visit? Or would you prefer to be able to visit whenever you wish even if it means fighting crowds and congestion?

Originally, the great wilderness parks of Canada and the United States were distant from development and isolated from most human impacts. This has changed in many cases. Forests are clear-cut right up to some park boundaries. Mine drainage contaminates streams and groundwater. At least 13 U.S. National Monuments are open to oil and gas drilling, including Texas's Padre Island, the only breeding ground for endangered Kemps Ridley sea turtles. Even in the dry desert air of the Grand Canyon, where visibility was once up to 150 km, it's often too smoggy now to see across the canyon due to air pollution from power plants just outside the park. Snowmobiles and off-road vehicles (ORV) create pollution and noise and cause erosion while disrupting wildlife in many parks (fig. 12.24).

Chronically underfunded, the U.S. National Park System now has a maintenance backlog estimated to be at least $5 billion. Politicians from both major political parties vow to repair park facilities during election campaigns, but then find other uses for public funds once in office. Ironically, a recent study found that, on average, parks generate $4 in user fees for every $1 they receive in federal subsidies. In other words, they more than pay their own way, and should have a healthy surplus if they were allowed to retain all the money they generate.

In recent years, the U.S. National Park System has begun to emphasize nature protection and environmental education over entertainment. This new agenda is being adopted by other countries as well. The IUCN has developed a **world conservation**

FIGURE 12.24 Off-road vehicles cause severe, long-lasting environmental damage when driven through wetlands.

FIGURE 12.25 Australia's Great Barrier Reef is the world's largest marine reserve. Stretching for nearly 2,000 km (1,200 mi) along Australia's northeast coast, this reef complex is one of the biological wonders of the world.

strategy for protecting natural resources that includes the following three objectives: (1) to maintain essential ecological processes and life-support systems (such as soil regeneration and protection, nutrient recycling, and water purification) on which human survival and development depend; (2) to preserve genetic diversity essential for breeding programs to improve cultivated plants and domestic animals; and (3) to ensure that any utilization of wild species and ecosystems is sustainable.

Marine ecosystems need greater protection

As ocean fish stocks become increasingly depleted globally (chapter 11), biologists are calling for protected areas where marine organisms are sheltered from destructive harvest methods. As the opening case study for chapter 1 describes, limiting the amount and kind of fishing in marine reserves can quickly replenish fish stocks in surrounding areas. In a study of 100 marine refuges around the world, researchers found that, on average, the number of organisms inside no-take preserves was twice as high as surrounding areas where fishing was allowed. In addition, the biomass of organisms was three times as great and individuals animals were, on average, 30 percent larger inside the refuge compared to outside. Recent research has shown that closing reserves to fishing even for a few months can have beneficial results in restoring marine populations. The size necessary for a safe haven to protect flora and fauna depends on the species involved, but some marine biologists call on nations to protect at least 20 percent of their nearshore territory as marine refuges.

Coral reefs are among the most threatened marine ecosystems in the world. Remote sensing surveys show that living coral covers only about 285,000 km^2 (110,000 mi^2), or an area about the size of Nevada. This is less than half of previous estimates, and 90 percent of all reefs face threats from rising sea temperatures, destructive fishing methods, coral mining, sediment runoff, and other human disturbance. In many ways, coral reefs are the old-growth rainforests of the ocean. Biologically rich, these sensitive communities can take a century or more to recover from damage.

If current trends continue, some researchers predict that in 50 years there will be no viable coral reefs anywhere in the world.

What can be done to reverse this trend? Some countries are establishing large marine reserves specifically to protect coral reefs (fig. 12.25). Australia has the largest marine reserve in the world in its 344,000 km^2 Great Barrier Reef (a large portion of which is open ocean). The recently established Northwest Hawaiian Islands National Monument is nearly as large. And Canada is considering closing much of the Grand Banks off its eastern shore to all fish harvesting. Altogether, however, aquatic reserves make up less than 10 percent of all the world's protected areas despite the fact that 70 percent of the earth's surface is water. A survey of marine biological resources identified the ten richest and most threatened "hot spots," including the Philippines, the Gulf of Guinea and Cape Verde Islands (off the west coast of Africa), Indonesia's Sunda Islands, the Mascarene Islands in the Indian Ocean, South Africa's coast, southern Japan and the east China Sea, the western Caribbean, and the Red Sea and Gulf of Aden. We urgently need more no-take preserves to protect marine resources.

Conservation and economic development can work together

Many of the most biologically rich communities in the world are in developing countries, especially in the tropics. These countires are the guardians of biological resources important to all of us. Unfortunately, where political and economic systems fail to provide residents with land, jobs, food, and other necessities of life, people do whatever necessary to meet their own needs. Immediate survival takes precedence over long-term environmental goals. Clearly the struggle to save species and ecosystems can't be divorced from the broader struggle to meet human needs.

People in some developing countries are beginning to realize that their biological resources may be their most valuable assets, and that their preservation is vital for sustainable development. **Ecotourism** (tourism that is ecologically and socially sustainable) can be more beneficial in many places over the long term

What Can You Do?

Being a Responsible Ecotourist

1. *Pretrip preparation.* Learn about the history, geography, ecology, and culture of the area you will visit. Understand the do's and don'ts that will keep you from violating local customs and sensibilities.

2. *Enviromental impact.* Stay on designated trails and camp in established sites. If available. Take only photographs and memories and leave only goodwill wherever you go.

3. *Resource impact.* Minimize your use of scarce fuels, food, and water resources. Do you know where your wastes and garbage go?

4. *Cultural impact.* Respect the privacy and dignity of those you meet and try to understand how you would feel in their place. Don't take photos without asking first. Be considerate of religious and cultural sites and practices. Be as aware of cultural pollution as you are of environmental pollution.

5. *Wildlife impact.* Don't harass wildlife or disturb plant life. Modern cameras make it possible to get good photos from a respectful, safe distance. Don't buy ivory, tortoise shell, animal skins, feathers, or other products taken from endangered species.

6. *Environmental benefit.* Is your trip strictly for pleasure, or will it contribute to protecting the local environment? Can you combine ecotourism with work on cleanup campaigns or delivery of educational materials or equipment to local schools or nature clubs?

7. *Advocacy and education.* Get involed in letter writing, lobbying, or educational campaigns to help protect the lands and cultures you have visited. Give talks at schools or to local clubs after you get home to inform your friends and neighbors about what you have learned.

than extractive industries, such as logging and mining. The What Can You Do? box (p. 272) suggests some ways to ensure that your vacations are ecologically responsible.

Native people can play important roles in nature protection

The American ideal of wilderness parks untouched by humans is unrealistic in many parts of the world. As we mentioned earlier, some biological communities are so fragile that human intrusions have to be strictly limited to protect delicate natural features or particularly sensitive wildlife. In many important biomes, however, indigenous people have been present for thousands of years and have a legitimate right to pursue traditional ways of life. Furthermore, many of the approximately 5,000 indigenous or native people that remain today possess ecological knowledge about their ancestral homelands that can be valuable in ecosystem management. According to author Alan Durning, "encoded in indigenous languages, customs, and practices may be as much understanding of nature as is stored in the libraries of modern science."

FIGURE 12.26 Some parks take draconian measures to expel residents and prohibit trespassing. How can we reconcile the rights of local or indigenous people with the need to protect nature?

Some countries have adopted draconian policies to remove native people from parks (fig. 12.26). In South Africa's Kruger National Park, for example, heavily armed solders keep intruders out with orders to shoot to kill. This is very effective in protecting wildlife. In all fairness, before this policy was instituted there was a great deal of poaching by mercenaries armed with automatic weapons. But it also means that people who were forcibly displaced from the park could be killed on sight merely for returning to their former homes to collect firewood or to hunt for rabbits and other small game. Similarly, in 2006, thousands of peasant farmers on the edge of the vast Mau Forest in Kenya's Rift Valley were forced from their homes at gun point by police who claimed that the land needed to be cleared to protect the country's natural resources. Critics claimed that the forced removal amounted to "ethnic cleansing" and was based on tribal politics rather than nature protection.

Other countries recognize that finding ways to integrate local human needs with those of nature is essential for successful conservation. In 1986, UNESCO (United Nations Educational, Scientific, and Cultural Organization) initiated its **Man and Biosphere (MAB) program,** which encourages the designation of **biosphere reserves,** protected areas divided into zones with different purposes. Critical ecosystem functions and endangered wildlife are protected in a central core region, where limited scientific study is the only human access allowed. Ecotourism and research facilities are located in a relatively pristine buffer zone around the core, while sustainable resource harvesting and permanent habitation are allowed in multiple-use peripheral regions (fig. 12.27).

While not yet given a formal MAB designation, the Great Bear Rainforest described in the opening case study for this chapter is organized along this general plan. A well established example of a biosphere reserve is Mexico's 545,000 ha (2,100 mi²) Sian Ka'an Reserve on the Tulum Coast of the Yucatán. The core area includes 528,000 ha (1.3 million acres) of coral reefs, bays, wetlands, and lowland tropical forest. More than 335 bird species have

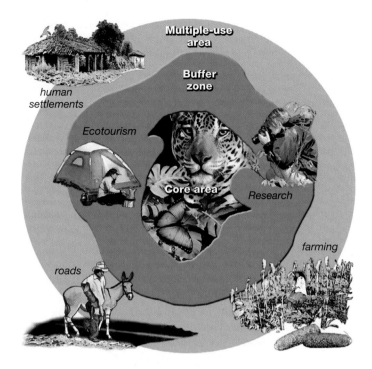

FIGURE 12.27 A model biosphere reserve. Traditional parks and wildlife refuges have well-defined boundaries to keep wildlife in and people out. Biosphere reserves, by contrast, recognize the need for people to have access to resources. Critical ecosystem is preserved in the core. Research and tourism are allowed in the buffer zone, while sustainable resource harvesting and permanent habitations are situated in the multiple-use area around the perimeter.

been observed within the reserve, along with endangered manatees, five types of jungle cats, spider and howler monkeys, and four species of increasingly rare sea turtles. Approximately 25,000 people (about the same number who live in the Great Bear Rainforest) reside in communities and the countryside around Sian Ka'an. In addition to tourism, the economic base of the area includes lobster fishing, small-scale farming, and coconut cultivation.

The Amigos de Sian Ka'an, a local community organization, played a central role in establishing the reserve and is working to protect natural resources while also improving living standards for local people. New intensive farming techniques and sustainable harvesting of forest products enable residents to make a living without harming their ecological base. Better lobster harvesting techniques developed at the reserve have improved the catch without depleting native stocks. Local people now see the reserve as a benefit rather than an imposition from the outside. Similar success stories from many parts of the world show how we can support local people and recognize indigenous rights while still protecting important environmental features.

Species survival can depend on preserve size and shape

Many natural parks and preserves are increasingly isolated, remnant fragments of ecosystems that once extended over large areas. As park ecosystems are shrinking, however, they are also becoming

more and more important for maintaining biological diversity. Principles of landscape design and landscape structure become important in managing and restoring these shrinking islands of habitat.

For years, conservation biologists have disputed whether it is better to have a *single large or several small* reserves (the SLOSS debate). Ideally, a reserve should be large enough to support viable populations of endangered species, keep ecosystems intact, and isolate critical core areas from damaging external forces. For some species with small territories, several small, isolated refuges can support viables populations, and having several small reserves provides insurance against a disease, habitat destruction, or other calamities that might wipe out a single population. But small preserves can't support species such as elephants or tigers, which need large amounts of space. Given human needs and pressures, however, big preserves aren't always possible. One proposed solution has been to create **corridors** of natural habitat that can connect to smaller habitat areas (fig. 12.28). Corridors could effectively create a large preserve from several small ones. Corridors could also allow populations to maintain genetic diversity or expand into new breeding territory. The effectiveness of corridors probably depends on how long and wide they are, and on how readily a species will use them.

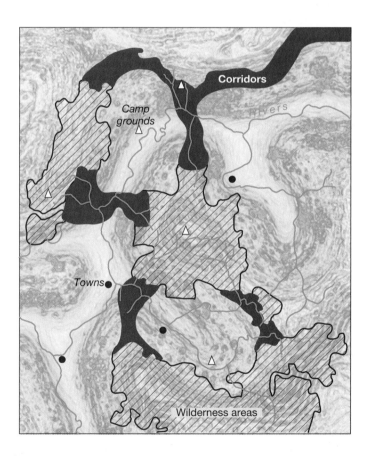

FIGURE 12.28 Corridors serve as routes of migration, linking isolated populations of plants and animals in scattered nature preserves. Although individual preserves may be too small to sustain viable populations, connecting them through river valleys and coastal corridors can facilitate interbreeding and provide an escape route if local conditions become unfavorable.

Perhaps the most ambitious corridor project in the world today is the Yellowstone to Yukon (Y2Y) proposal. Linking more than two dozen existing parks, preserves, and wilderness areas, this corridor would stretch 3,200 cm (2,000 cm) from the Wind River Range in Wyoming to northern Alaska. More than half this corridor is already forested. Some 31 different Fist Nations and American Indian tribes occupy parts of this land, and are being consulted in ecosystem management.

One of the reasons large preserves are considered better than small preserves is that they have more **core habitat,** areas deep in the interior of a habitat area, and that core habitat has better conditions for specialized species than do edges. **Edge effects** is a term generally used to describe habitat edges: for example, a forest edge is usually more open, bright, and windy than a forest interior, and temperatures and humidity are more varied. For a grassland, on the other hand, edges may be wooded, with more shade, and perhaps more predators, than in the core of the grassland area. As human disturbance fragments an ecosystem, habitat is broken into increasingly isolated islands, with less core and more edge. Small, isolated fragments of habitat often support fewer species, especially fewer rare species, than do extensive, uninterrupted ecosystems. The size and isolation of a wildlife preserve, then, may be critical to the survival of rare species.

(a) Patch size and shape: these influence core/edge ratio

  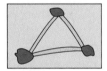

(b) Number of habitat patches, and patch proximity and connectivity

(c) Contrast between habitat patches and surrounding landscape

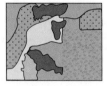 

(d) Complexity of the landscape mosaic

FIGURE 12.29 Some spatial variables examined in landscape ecology. Size, arrangement, context, and other factors often vary simultaneously. Together, they can strongly influence the ability of a wildlife preserve to support rare species.

Landscape ecology is a science that examines the relationship between these spatial patterns and ecological processes, such as species movement or survival. Landscape ecologists measure variables like habitat size, shape, and the relative amount of core habitat and edge (fig. 12.29). Landscape ecologists also examine the kind of land cover that surrounds habitat areas—is it a city, farm fields, or clear-cut forest? And they include utilitarian landscapes, such as farm fields, in their analysis, as well as pristine wilderness. By quantifying factors such as habitat shape and landscape complexity, and by monitoring the number of species or the size of popuations, landscape ecologists try to guide more effective design of nature preserves and parks.

A dramatic experiment in reserve size, shape, and isolation is being carried out in the Brazilian rainforest. In a project funded by the World Wildlife Fund and the Smithsonian Institution, loggers left 23 test sites when they clear-cut a forest. Test sites range from 1 ha (2.47 acres) to 10,000 ha. Clear-cuts surround some, and newly-created pasture surrounds others (fig. 12.30); others remain connected to the surrounding forest. Selected species are regularly inventoried to monitor their survival after disturbance. As expected some species disappear very quickly, especially from small areas. Sun-loving species flourish inn the newly-created forest edges, but deep-forest, shade-loving species disappear, particularly when the size or shape of a reserve reduces availability of core habitat. This experiment demonstrates the importance of maintaining core habitat in preserves.

FIGURE 12.30 How small can a nature preserve be? In an ambitious research project, scientists in the Brazilian rainforest are carefully tracking wildlife in plots of various sizes, either connected to existing forests or surrounded by clear-cuts. As you might expect, the largest and most highly specialized species are the first to disappear.

CONCLUSION

Forests and grasslands cover nearly 60 percent of global land area. The vast majority of humans live in these biomes, and we obtain many valuable materials from them. And yet, these biomes also are the source of much of the world's biodiversity on which we depend for life-supporting ecological services. How we can live sustainably on our natural resources while also preserving enough nature so those resources can be replenished represents one of the most important questions in environmental science.

There is some good news in our search for a balance between exploitation and preservation. Although deforestation and land degradation are continuing at unacceptable rates—particularly in some developing countries—many countries are more thickly forested now than they were two centuries ago. Protection of the Great Bear Rainforest in Canada and

Australia's great barrier reef shows that we can choose to protect some biodiverse areas in spite of forces that want to exploit them. Overall, nearly 12 percent of the earth's land area is now in some sort of protected status. While the level of protection in these preserves varies, the rapid recent increase in number and area in protected status exceeds the goals of the United Nations Millennium Project.

While we haven't settled the debate between focusing on individual endangered species versus setting aside representative samples of habitat, pursuing both strategies seems to be working. Protecting charismatic umbrella organisms, such as the "spirit bears" of the Great Bear Rainforest can result in preservation of innumerable unseen species. At the same time, protecting whole landscapes for aesthetic or recreational purposes can also achieve the same end.

REVIEWING LEARNING OUTCOMES

By now you should be able to explain the following points:

12.1 Discuss the types and uses of world forests.

- Boreal and tropical forests are most abundant.
- Forests provide many valuable products.
- Tropical forests are being cleared rapidly.
- Temperate forests have competing uses.

12.2 Describe the location and state of grazing lands around the world.

- Grazing can be sustainable or damaging.
- Overgrazing threatens many rangelands.
- Ranchers are experimenting with new methods.

12.3 Summarize the types and locations of nature preserves.

- Many countries have created nature preserves.
- Not all preserves are preserved.
- Marine ecosystems need greater protection.
- Conservation and economic development can work together.
- Native people can play important roles in nature protection.
- Species survival can depend on preserve size and shape.

PRACTICE QUIZ

1. What do we mean by *closed-canopy* forest and *old-growth* forest?

2. Which commodity is used most heavily in industrial economies: steel, plastic, or wood? What portion of the world's population depends on wood or charcoal as the main energy supply?

3. What is a *debt-for-nature swap*?

4. Why is fire suppression a controversial strategy? Why are forest thinning and salvage logging controversial?

5. Are pastures and rangelands always damaged by grazing animals? What are some results of overgrazing?

6. What is *rotational grazing,* and how does it mimic natural processes?

7. What was the first national park in the world, and when was it established? How have the purposes of this park and others changed?

8. How do the size and design of nature preserves influence their effectiveness? What do landscape ecologists mean by *interior habitat* and *edge effects*?

9. What is *ecotourism,* and why is it important?

10. What is a *biosphere reserve,* and how does it differ from a wilderness area or wildlife preserve?

CRITICAL THINKING AND DISCUSSION QUESTIONS

1. Paper and pulp are the fastest growing sector of the wood products market, as emerging economies of China and India catch up with the growing consumption rates of North America, Europe, and Japan, What should be done to reduce paper use?

2. Conservationists argue that watershed protection and other ecological functions of forests are more economically valuable than timber. Timber companies argue that continued production supports stable jobs and local economies. If you were a judge attempting to decide which group was right, what evidence would you need on both sides? How would you gather this evidence?

3. Divide your class into a ranching group, a conservation group, and a suburban home-builders group, and debate the protection of working ranches versus the establishment of nature preserves. What is the best use of the land? What landscapes are most desirable? Why? How do you propose to maintain these landscapes?

4. Calculating forest area and forest losses is complicated by the difficulty of defining exactly what constitutes a forest. Outline a definition for what counts as forest in your area, in terms of size, density, height, or other characteristics. Compare your definition to those of your colleagues. Is it easy to agree? Would your definition change if you lived in a different region?

5. Why do you suppose dry topical forest and tundra are well represented in protected areas, while grasslands and wetlands are protected relatively rarely? Consider social, cultural, geographic, and economic reasons in your answer.

6. Oil and gas companies want to drill in several parks, monuments, and wildlife refuges. Do you think this should be allowed? Why or why not? Under what conditions would drilling be allowable?

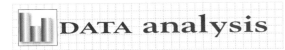 **DATA analysis**   **Detecting Edge Effects**

Edge effects are a fundamental consideration in nature preserves. We usually expect to find dramatic edge effects in pristine habitat with many specialized species. But you may be able to find interior-edge differences on your own college campus, or in a park or other unbuilt area near you. Here are three testable questions you can examine using your own local patch of habitat: (1) Can an edge effect be detected or not? (2) Which species will indicate the difference between edge and interior conditions? (3) At what distance can you detect a difference between edge and interior conditions? To answer these questions, you can form a hypothesis and test it as follows:

1. Choose a study area. Find a distinct patch of habitat, such as woods, unmowed grass, or marshy but walkable wetland, about 50 m wide or larger. With other students, list some local, familiar plant species that you would expect to find in your study area. If possible, make this list on a visit to your site.

2. Form a hypothesis. Examine your list, and predict which species will occur most on edges, and (if you can) which you think will occur more in the interior. Form a hypothesis, or a testable statement, based on one of the three questions above. For example, "I will be able to detect an edge effect in my patch," or "I think an edge effect will be indicated by these species: _____," or "I think changes in

species abundance will indicate an edge-interior change at _____ m from the edge of the patch."

3. Gather data. Get a meter tape and lay it along the ground from the edge of your habitat patch toward the interior. (You can also use a string and pace distances; treat one pace as a meter.) This line is your *transect*. At the edge end of the tape (or string), count the number of different species you can see within 1 m^2 on either side of your line. Repeat this count at each 5 m interval, up to 25 m. Thus you will create a list of species at 0, 5, 10, 15, 20, and 25 m in from the edge.

4. Examine your lists, and determine whether your hypothesis was correct. Can you see a change in species presence/absence from 0 to 25 m? Were you correct in your prediction of which species disappeared with distance from the edge? If you can identify an edge effect, at what distance did it occur?

5. Consider ways that your test could be improved. Should you take more frequent samples? Larger samples? Should you compare abundance rather than presence/absence? Might you have gotten different results if you had chosen a different study site? Or if your class had examined many sites and averaged the results? How else might you modify your test to improve the quality of your results?

Student researchers sample aquatic fauna in a Louisiana coastal wetland. Ecological restoration requires careful monitoring such as this.

C H A P T E R

Restoration Ecology

13

When we heal the earth, we heal ourselves.
—David Orr—

LEARNING OUTCOMES

After studying this chapter, you should be able to:

13.1 Illustrate ways that we can help nature heal.

13.2 Show how nature is resilient.

13.3 Explain how restoring forests has benefits.

13.4 Summarize plans to restore prairies.

13.5 Compare approaches to restoring wetlands and streams.

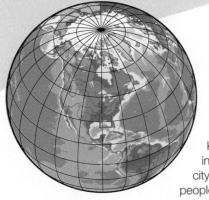

Case Study Restoring Louisiana's Coastal Defenses

Wetland restoration on the Louisiana coast has been in the news ever since Hurricane Katrina devastated New Orleans in September 2005. Most of the city flooded, and at least 1,500 people died in the worst natural catastrophe to hit the United States in a generation. Many failures were blamed for the disaster, including weak flood walls and inaction by disaster relief agencies. But one factor that caught the public eye was erosion of Louisiana's vast coastal marshes.

Historically, coastal wetlands helped protect New Orleans and the region's other towns, farms, and forests from storm surges in the Gulf of Mexico. In the past 60 years, these wetlands have shrunk by 4,000 km² (fig. 13.1) By some estimates, each kilometer of coastal marsh reduces the height of storm surges by 5 cm. Wetland losses have heightened storm floods in some parts of the state by up to 3 m. New Orleans lies mostly below sea level and is highly vulnerable to Gulf Coast storms. Restoring these coastal marshes has become a priority in efforts to defend New Orleans and other coastal cities from future hurricanes.

Why are these wetlands shrinking and dying? Sediment loss, salinization, and physical degradation are three main factors that have deteriorated this dynamic system. The whole northern Gulf Coast is built of sediment dumped by the Mississippi River, which is thick with mud and clay drained from half the continental United States. Over thousands of years, this meandering, sediment-rich river has deposited deltas—expanses of sand and silt—all along the Gulf Coast. Salt marsh grasses grow in these shifting, soggy sediments, creating root mats that stabilize the ground and provide nurseries for fish, birds, and the shrimp that make Cajun cuisine

legendary. As wetlands expanded outward, the inland areas became less and less salty—saturated increasingly by rainwater and less by seawater. These freshwater marshes support even more plant growth and biodiversity than do the coastal salt marshes.

This coastal wetland system has been radically altered since the 1930s. First the Army Corps of Engineers straightened and contained the Mississippi outlet. Levees now guide the river 50 km out into the Gulf. Mud and silt that once built the coastal wetlands is now dumped off the coastal shelf. Furthermore, dredging and filling has destroyed about 400,000 ha (1 million acres) per year. Oil and gas companies have dug canals deep into the wetlands, allowing boat access to oil and gas wells. Thousands of kilometers of access canals now perforate the region's wetlands, and each one allows salt water and storm surges, both lethal to the freshwater marshes, deep into the interior of the coastal marsh system. Freshwater marshes now cover less than one-fourth as much area as 60 years ago. This loss represents 80 percent of all coastal wetland loss in the United States and causes an estimated one-half billion dollars per year in economic losses.

Flooding in New Orleans has added urgency to efforts to reverse, or at least slow, these losses. A first step in restoration is to reduce the cause of the problem. For example, the biggest, most expensive navigation canal, the Mississippi River Gulf Outlet (MRGO), may be closed. The MRGO helped guide storm surges into New Orleans, and it has destroyed thousands of hectares of valuable cypress swamp by flooding them with salt water. A more difficult proposal has been to close and restore some of the web of oil and gas access canals. Who should pay for this is disputed: companies propose that the public is responsible for restoration, but some taxpayers think the oil and gas corporations should pay to clean up the damage.

An additional step in restoration would be to return some of the Mississippi River sediment and fresh water to the marshes. In theory, gaps could be cut in the levees below New Orleans, allowing fresh, muddy river water to reenter the wetlands. A portion of the sediment would return, but just as important, fresh water could help the wetlands rebuild themselves. In a few experimental restoration areas, such as the Carnaervon diversion east of New Orleans, simply replenishing freshwater flow has already helped expand the area of wetland vegetation. In addition to these strategies, revegetation is going on in some areas, with volunteers replanting wetland grasses by hand to speed the restoration of critical areas, or monitoring the vigor of plants in remaining wetlands.

Restoration is a new, exciting, and experimental field that applies ecological principles to healing nature. Full restoration of a vast ecosystem is a staggering task, but even small steps can make a difference. In coastal Louisiana and elsewhere, ecological restoration can be essential to human populations and economies, not just natural systems. In this chapter, we'll examine these and other aspects of restoration ecology.

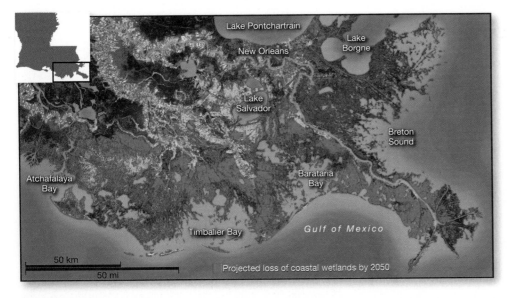

FIGURE 13.1 Historic and projected loss of coastal marshes in the Mississippi delta 1932–2050, given a "business-as-usual" scenario.
Source: USGS.

13.1 HELPING NATURE HEAL

Humans have disturbed and degraded nature as long as we've existed. With the availability of industrial technology, however, our impacts have increased dramatically in both scope and severity. We've chopped down forests, plowed the prairies, slaughtered wildlife, filled in wetlands, and polluted air and water. Our greatly enhanced power also makes it possible, however, to repair some of the damage we've caused. The relatively new field of **ecological restoration** attempts to do this based both on good science and pragmatic approaches. Some see this as a new era in conservation history. As we discussed in chapter 1, the earliest phases in conservation and nature protection consisted in efforts to use resources sustainably and to protect special places from degradation.

Now, thousands of projects are underway to restore or rehabilitate nature. These range from individual efforts to plant native vegetation on small urban yards to huge efforts to restore millions of hectares of prairie or continental-sized forests. We'll look in this chapter, for instance, at proposals to recreate a vast Buffalo Commons something like Lewis and Clark saw when they crossed the North American Great Plains in 1804. Undoubtedly, the biggest reforestation project in history is the Chinese effort to create a "green wall" to hold back the encroaching Gobi and Taklamakan Deserts. More than a billion people are reported to have planted 50 billion trees in China over the past 30 years.

The success of these projects varies. In some cases, the land and biota are so degraded that restoration is impossible. In other situations, some sort of ecosystem can be recreated, but soil, water, nutrients, topography, or genetic diversity available limit what can be done on a particular site. Figure 13.2 presents a schematic overview of restoration options. Note that restoration to an original pristine condition is rarely possible. More often, the best choices are some alternative condition that has desirable characteristics or even an entirely new use for the site. A variety of descriptions are used to describe different management goals.

Restoration projects range from modest to ambitious

A variety of descriptions are used to describe different management goals. Table 13.1 summarizes some of the most common terms employed in ecological restoration. A strict definition of restoration would be to return a biological community as nearly as possible to a pristine, predisturbance condition. Often, this isn't possible. A broader definition of restoration has a more pragmatic goal simply to develop a self-sustaining, useful ecosystem with as many of its original elements as possible. *Rehabilitation* may seek only to repair ecosystem functions. This may take the form of a community generally similar to the original one on a site, or it may air for an entirely different community that can carry out the desired functions. Sometimes it's enough to leave nature alone to heal itself, but often we need to intervene in some way to remove or discourage unwanted organisms while also promoting the growth of more desirable species. *Reintroduction* generally implies transplanting organisms from some external

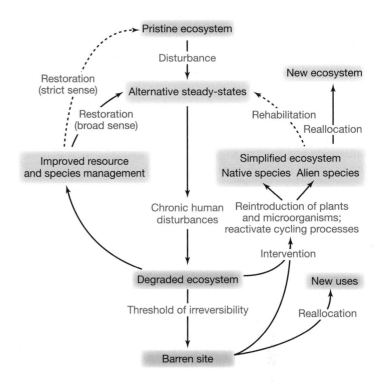

FIGURE 13.2 A model of ecosystem degradation and potential management options.
Data source: Walker and Moral, 2003.

TABLE 13.1

A Restoration Glossary

Some commonly used terms in restoration ecology:

- **Restoration** (strict sense) to return a biological community to its predisturbance structure and function.
- **Restoration** (broad sense) to reverse degradation and reestablish some aspects of an ecosystem that previously existed on a site.
- **Rehabilitation** to rebuild a community to a useful, functioning state but not necessarily its original condition.
- **Intervention** to apply techniques to discourage or reduce undesired organisms and favor or promote desired species.
- **Reallocation** to use a site (and its resources) to create a new and different kind of biological community rather than the existing one.
- **Remediation** to clean chemical contaminants from a polluted area using relatively mild or nondestructive methods.
- **Reclamation** to use powerful chemical or physical methods to clean and repair severely degraded or even barren sites.
- **Re-creation** to construct an entirely new ecosystem on a severely degraded site.
- **Mitigation** to replace a degraded site with one of more or less equal ecological value somewhere else.

source (often a nursery or hatchery where native species are grown under controlled conditions) to a site where they have been reduced or eliminated.

Remediation uses chemical, physical, or biological methods to remove pollution, generally with the intention of causing as little disruption as possible. *Reclamation* employs stronger, more extreme techniques to clean up severe pollution or create a newly functioning ecosystem on a seriously degraded or barren site. *Mitigation* implies compensation for destroying a site by purchasing or creating one of more or less equal ecological value somewhere else.

Restoration ecologists tend to be idealistic but pragmatic

Restoration ecologists work in the real world and, while they may dream of returning a disturbed site to its untouched situation, they have to deal with the actual constraints of a specific place as they find it. Recovery is linked to classical succession theory that suggests each site has a predictable, stable, climax community that will eventually result if nature is left to its own devices. This view is congruent with an engineering approach: all actions have predictable consequences. Modern ecology has shown, however, that primary succession is a stochastic (nondeterministic or random) process. Individual species occupy ecological niches according to the arbitrary and unpredictable historical trajectory of a particular place (see discussion of theories of Clements and Gleason in chapter 4).

Restoration ecologists often find it useful to express their goals as ecosystem health or integrity, but research ecologists protest that these concepts can be meaningfully applied only to entities that have been directly shaped by evolution, such as individual organisms. Organisms normally have clearly defined boundaries as well as homeostatic mechanisms that maintain internal conditions and perpetuate those boundaries as the organism grows, develops, matures, and reproduces. Today, most research ecologists don't accept that ecosystems and communities, which aren't shaped by evolution, behave as tightly coordinated entities. They point out that the aims of restoration often are more driven by human values, such as beauty, recreation, utility, and other social or philosophical issues rather than pure science.

Sometimes there are debates about what the outcome of ecosystem restoration should be. Suppose natural forces, such as wind storms or fires, disrupt a wilderness area. Should we use the techniques of ecological restoration to tidy up or improve an area, or leave it to natural processes? While we're improving an area, would it be all right to make it more scenic or attractive? "If we just cut down a few trees or move these rocks a little to the left, the view would be much better. Do you have a problem with that?" What do you think, are there limits to our reinvention of nature?

There may be more than one historic state to which we could restore a landscape. In the Nature Conservancy's Hassayampa River Preserve near Phoenix, for instance, pollen grains preserved in sediments reveal that 1,000 years ago the area was a grassy marsh that was unique in the surrounding desert landscape. Some ecologists would try to rebuild a similar marsh. Corn pollen in the same sediments, however, shows that 500 years ago Native Americans began farming the marsh. Which is more important, the natural or early agricultural landscape?

Unfortunately, it may be impossible to return to conditions of either 500 or 1,000 years ago since climate change and evolution may have made the communities existing at that time incompatible with current conditions. This creates a dilemma for restoration ecologists. Should we be attempting to restore areas to what they used to be, or creating a community that will be more compatible with future conditions? If change is natural and inevitable, who is to say that present conditions—whatever they are or however they have come about—are bad? How should we distinguish between desirable and undesirable change? If it's simply a matter of human preference, some people might prefer a golf course or a shopping mall on the site. What would you say to someone who claims that humans are a part of nature, and that therefore any changes we make to the landscape also are natural?

Restoration projects have common elements

General principles in restoration are drawn from a variety of sciences: an understanding of the importance of biodiversity comes from ecology; principles of groundwater movement comes from hydrology; soil science provides insights into soil health; and so on. Different types of restoration involve separate challenges. Wetland restoration, for example, involves a variety of efforts at reestablishing hydrologic connections, plant and animal diversity, weed removal, and sometimes salinity control. Forest restoration, on the other hand, requires a different set of steps that may include controlled burning, selective cutting, weed removal, control of deer and other herbivores, and so on. Even so, there are at least five main components of restoration, and most efforts share a majority of these activities.

1. *Removing physical stressors.* The first step in most restoration efforts is to remove the cause of degradation or habitat loss. Physical stressors such as pollutants, inadequate moisture, or vehicle traffic may need to be corrected. In wetland restoration, water flow and storage usually must be restored before other steps, such as replanting, can proceed. In forest restoration, clear-cut practices might be replaced with selective logging. In prairie restoration, cultivation might be ended so that land can be replanted with grassland plants, which then provide habitat for grassland butterflies, beetles, and birds.

2. *Controlling invasive species.* Often a few aggressive, "weedy" species suppress the growth of other plants or animals. These invasives may be considered biotic stressors. Removing invasive species can be extremely difficult, but without removal, subsequent steps often fail. Some invasives can be controlled by introducing pests that only eat those species: for example, purple loosestrife (*Lythrum salicaria*) is a wetland invader

that has succeeded in North America partly because it lacks predators. Loosestrife beetles (*Galerucella* sp.), introduced after careful testing to ensure that their introduction wouldn't lead to further invasions, have successfully set back loosestrife populations in many areas. Leafy spurge (*Euphorbia esula*), similarly, has invaded and blanketed vast areas of the Great Plains. Insects, including flea beetles (*Aphthona* sp.) have been a successful strategy to reduce the spread of spurge in many areas, allowing native grassland plants to compete again.

3. *Replanting.* Restoring a site or ecosystem usually involves some replanting of native plant species. Often restorationists try to collect seeds from nearby sources (so that the plants will be genetically similar to the original plants of the area), then grow them in a greenhouse before transplanting to the restoration site. In some cases, restoration ecologists can encourage existing plants to grow by removing other plants that outcompete the target plants for space, nutrients, sunshine, or moisture.

4. *Captive breeding and reestablishing fauna.* In some cases, restoration involves reintroducing animals. Peregrine falcon restoration involved releasing captive-bred birds, which then managed to survive and nest on their own. In some cases invertebrates, such as butterflies or beetles, may be released. Sometimes a top predator is reintroduced, or allowed to reinvade. Yellowstone National Park has had a 20-year experimental restoration of wolves. Evidence indicates that these predators are reducing excessive deer and elk populations, thus helping to restore vegetation, as well as reducing mid-level carnivores such as coyotes.

5. *Monitoring.* Without before-and-after monitoring, restoration ecologists cannot know if their efforts are working as hoped. Therefore, a central aspect of this science is planned, detailed, ongoing studies of key factors. Repeated counts of species diversity and abundance can tell whether biodiversity is improving. Repeated measures of water quality, salinity, temperature, or other factors can indicate whether suitable conditions have been established for target species.

Early conservationists showed the promise of restoration

As settlers spread across North America in the eighteenth and nineteenth centuries, the woods, prairies and wildlife populations seemed vast, much too large for anything that humans could do to affect them. As we discussed in chapter 1, however, a few pioneers recognized that the rapid destruction of natural communities was unsustainable. The most influential American forester was Gifford Pinchot. His first job after graduating from college was to manage the Vanderbilt's Biltmore estate in North Carolina. Pinchot introduced a system of selective harvest and replanting of choice tree species that increased the value of the forest while also producing a sustainable harvest. Pinchot went on to become the first head of the U.S. Forest Service, where he

FIGURE 13.3 Aldo Leopold's Sand County farm in central Wisconsin served as a refuge from the city and as a laboratory to test theories about land conservation, environmental ethics, and ecologically based land management.

put resource management on an honest, utilitarian, and scientific basis for the first time in U.S. history.

Another pioneer in restoration ecology was Aldo Leopold. In 1935, Leopold bought a small, worn-out farm on the banks of the Wisconsin River, not far from his home in Madison. Originally intended to be merely a hunting camp, the farm quickly became a year-round retreat from the city, as well as a laboratory in which Leopold could test his theories about conservation, game management, and land restoration (fig. 13.3). The whole Leopold family participated in planting as many as 6,000 trees each spring. "I have read many definitions of what is a conservationist, and written not a few myself," Leopold wrote, "but I suspect that the best one is written not with a pen, but with an axe. It is a matter of what a man thinks about while chopping, or while deciding what to chop. A conservationist is one who is humbly aware that with each stroke he is writing his signature on the face of his land. . . A land ethic then, reflects the existence of an ecological conscience, and this in turn reflects a conviction of individual responsibility for the health of the land. . . Health is the capacity of the land for self-renewal. Conservation is our effort to understand and preserve this capacity."

As mentioned earlier, many modern ecologists regard goals of restoring the health, beauty, stability, or integrity of nature as unscientific, but they generally view Aldo Leopold as a visionary pioneer and an important figure in conservation history.

13.2 NATURE CAN BE REMARKABLY RESILIENT

As is the case in rebuilding Gulf Coast marshes, the first step in conservation and ecological restoration is generally to stop whatever is causing damage. Sometimes, that's all that's necessary.

Nature has amazing regenerative power. If the damage hasn't passed a threshold of irreversibility, natural successional processes can often rebuild a diverse, stable, interconnected biological community, given enough time.

Protection is the first step in restoration

The first official wildlife refuge established in the United States was Pelican Island, a small sand spit in the Indian River estuary, not far from the present-day Cape Canaveral. The island was recognized by ornithologists in the mid-1800s as being especially rich in bird life. Pelicans, terns, egrets, and other wading birds nested in huge, noisy colonies. Others discovered the abundant birds as well. Boat loads of tourists slaughtered birds just for fun, while professional plume hunters shot thousands of adults during the breeding season, leaving fledglings to starve to death. In 1900 the American Ornithological Union, worried about the wanton destruction of colonies, raised private funds to hire wardens to protect the birds during the breeding season. And in 1903, President Theodore Roosevelt signed an executive order establishing Pelican Island as America's first National Bird Reservation. This set an important precedent that the government could set aside public land for conservation. It also was the first of 51 wildlife refuges created by President Roosevelt that may have saved species such as roseate spoonbills and snowy egrets from extinction.

Many of the forests and nature preserves described in chapter 12 suffered some degree of degradation before being granted protected status. Often, simply prohibiting logging, mining, or excessive burning is enough to allow nature to heal itself. Consider the forests of New England, for example. When the first Europeans arrived in America, New England was mostly densely forested. As settlers spread across the land, they felled the forest to create pastures and farm fields. In 1811, sheep were introduced in New England, and sheep farming expanded rapidly to provide wool to the mills in New Hampshire and Massachusetts. By 1840, Vermont had nearly 2 million sheep on its 9,600 mi^2, and only 20 percent of the landscape remained forested. Opening of the Erie Canal brought competition from western farmers, however, and Eli Whitney's invention of the cotton gin made wool production less profitable. Within a few decades, most Vermont farmers had abandoned large-scale sheep farming, and the land was allowed to revert to forest. Today, 80 percent of the land in Vermont is once again forested, and less than 20 percent is farmland (fig. 13.4). Given a century or more to undergo natural successional processes, much of this forest has reacquired many characteristics of old-growth forest. A mixture of native species of different sizes and ages gives the forest diversity and complexity. Many of the original animal species—moose, bear, bobcats, pine martins—have become reestablished. There are reports of lynx, mountain lions, and even wolves migrating south from Canada. Now, before woodlot owners can log their land, Vermont law requires that they consult with a professional forester to develop a plan to sustain the biodiversity and quality of their forest.

FIGURE 13.4 A mosaic of cropland, pasture, and sugar bush clothes the hills of Vermont. Two hundred years ago, this area was 80 percent cropland and only 20 percent forest. Today that ratio is reversed as the forests have invaded abandoned fields. Many of these forests are reaching late successional stages and are reestablishing ecological associations characteristic of old-growth forests.

Native species often need help to become reestablished

Sometimes rebuilding populations of native plants and animals is a simple process of restocking breeding individuals. In other cases, however, it's more difficult. Recovery of a unique indigenous seabird in Bermuda is an inspiring conservation story that gives us hope for other threatened and endangered species.

When Spanish explorers discovered Bermuda early in the fifteenth century, they were frightened by the eerie nocturnal screeching they thought came from ghosts or devils. In fact, the cries were those of extremely abundant, ground-nesting seabirds now known as the hook-billed petrel, or Bermuda cahow (*Pterodroma cahow*), endemic to the island archipelago (fig. 13.5*a*). It's thought that there may originally have been half a million of these small, agile, gadfly petrels. Colonists soon found that the birds were easy to catch and good to eat. Those overlooked by humans were quickly devoured by the hogs, rats, and cats that accompanied settlers. Although Bermuda holds the distinction of having passed the first conservation laws in the New World, protecting native birds as early as 1616, the cahow was thought to be extinct by the mid-1600s.

In 1951, however, scientists found 18 nesting pairs of cahows on several small islands in Bermuda's main harbor. A protection and recovery program was begun immediately, including establishment of a sanctuary on 6-hectare (15-acre) Nonsuch Island, which has become an excellent example of ecological restoration.

Nonsuch was a near desert after centuries of abuse, neglect, and habitat destruction. All the native flora and faun were gone, along with most of the soil in which the cahows once dug nesting burrows. This was a case of re-creating nature rather than merely protecting what was left. Sanctuary superintendent David Wingate, who devoted his entire professional career to this project, brought about a remarkable transformation of the island (fig. 13.5*b*).

(a) Bermuda cahow

(b) David Wingate examines a cahow nest

(c) Long-tail excluder at mouth of burrow

FIGURE 13.5 For more than three centuries, the Bermuda cahow, or hook-billed petrel (a), was thought to be extinct until a few birds were discovered nesting on small islets in the Bermuda harbor. David Wingate (b) devoted his entire career to restoring cahows and their habitat. A key step in this project was to build artificial burrows (c). A long-tail excluder device (a board with a hole just the size of the cahow) keeps other birds out of the burrow. A round cement lump at the back end of the burrow can be removed to view the nesting cahows.

The first step in restoration was to remove invasive species and reintroduce native vegetation. Wingate and many volunteers trapped and poisoned pigs and rats and other predators that threatened both wildlife and native vegetation. They uprooted millions of exotic plants and replanted native species including mangroves and Bermuda cedars (*Juniperus bermudiana*). Initial progress was slow as trees struggled to get a foothold; once the forest knit itself into a dense thicket that deflected salt spray and ocean winds, however, the natural community began to reestablish itself. As was the case in New Zealand's Kapiti Island Nature Reserve (chapter 11) native plants that hadn't been seen for decades began to reappear. Once the rats and pigs that ate seedlings were removed, and competition from weedy invasives was eliminated, native seeds that had lain dormant in the soil began to germinate again.

Still, there wasn't enough soil for cahows to dig the underground burrows they need for nesting. Wingate's crews built artificial cement burrows for the birds. Each pair of cahows lays only one egg per year and only about half survive under ideal conditions. It takes eight to ten years for fledglings to mature, giving the species a low reproductive potential. They also compete poorly against the more common long-tailed tropic birds that steal nesting sites and destroy cahow eggs and fledglings. Wingate designed wooden baffles for the burrow entrances, with holes just large enough for cahows but too small for the larger tropic birds (fig. 13.5*c*). The round cement cap at the back of the burrow can be removed to monitor the nesting cahows.

It takes constant surveillance to eradicate exotic plant species that continue to invade the sanctuary. Rats, cats, and toxic toads also swim from the mainland and must be removed regularly. By 2002, however, the cahow population had rebounded to about 200 individuals with 60 breeding pairs. Hurricane Fabian destroyed many nesting burrows on smaller islets in 2003. Fortunately, the restored native forest on Nonsuch Island withstood the winds and preserved the rebuilt soil. The larger island is now being repopulated

with chicks, their translocation timed so they will imprint on their new home. Reestablishing a viable population of cahows (which are now Bermuda's national bird) has had the added benefit of rebuilding an entire biological community (fig. 13.6).

It's too early to know if the population is large enough to be stable over the long term, but the progress to date is encouraging. Perhaps more important than rebuilding this single species is that Nonsuch has become a living museum of precolonial Bermuda that benefits many species besides its most famous resident. It is a heartening example of what can be done with vision, patience, and a great deal of hard work.

FIGURE 13.6 This brackish pond on Nonsuch Island is a reconstructed wetland. Note the Bermuda cedars on the shore and the mangroves planted in the center of the pond by David Wingate.

Some other notable reintroduction programs include restoration of peregrine falcons in the eastern U.S. and California condors in the American west. Both of these species were locally extinct. Successful captive breeding programs produced enough birds to repopulate much of their former range. Similarly, Arabian oryx have been successfully reinstated in the deserts of Saudi Arabia, and endemic Nene geese, which were nearly exterminated from Hawaii, were raised in captivity and reintroduced to the Volcano National Park, where they exist in a small, but self-reproducing population (see fig. 11.21).

13.3 RESTORING FORESTS HAS MANY BENEFITS

Restoration can be a cornerstone of managing economic resources and a source of cultural pride, not just an altruistic ecological activity. In the United States, the largest restoration projects ever attempted have been reforestation of cut-over or degraded forest lands. Building on the policies of Gifford Pinchot, lumber companies routinely replant forests they have harvested to prepare a future crop. Seedlings are grown in huge nurseries, and tractor-drawn planters allow a team of workers to plant thousands of new trees per day (fig. 13.7). Usually, this mechanical reforestation results in a monoculture of uniformly spaced trees (fig. 13.8). These plantings are designed to produce wood quickly, but they have little resemblance to diverse, complex native forests. Still, these commercial forests supply ground cover, provide habitat for some wildlife species, and grow valuable lumber for paper pulp. As we saw in chapter 12, a recent United Nations survey of world forests found that many countries are more thickly forested now than they were 200 years ago. Both the total biomass and the quality of the forests in most of these countries has increased as forests have been protected and replanted over the past two centuries. Japan, for example, was almost completely deforested at

FIGURE 13.8 Monoculture forests, such as this tree plantation in Austria, often have far less biodiversity than natural forests.

the end of World War II. In the 60 years since then, Japan has carried out a massive reforestation program. Now, more than 60 percent of the country is forest-covered. Tight restrictions on logging help preserve this forest, which has great cultural value for the Japanese people. Rather than cut their own forest, Japan buys wood from its neighbors (a policy that has drawn some criticism from ecologists and human rights groups).

What must be the biggest reforestation project in history is now taking place in China, where at least a billion people have planted 50 billion trees over the past 30 years to create a "green wall" to hold back the encroaching Gobi and Taklamakan Deserts. This effort has been intensified recently in an effort to improve air quality in Beijing for the 2008 Olympics. Dust blowing from these expanding deserts has increasingly caused choking dust storms in eastern China, and these conditions could be devastating for the thousands of athletes and tourists visiting for the Olympics. The success of this restoration project is unclear, however: many of the trees ordered to be planted by the Chinese government haven't been watered or cared for, and the mortality rate is said to be very high.

Tree planting can improve our quality of life

Planting trees within cities can be effective in improving air quality, providing shade, and making urban environments more pleasant. Figure 13.9 shows student volunteers planting native

FIGURE 13.7 Mechanical planters can plant thousands of trees per day.

FIGURE 13.9 Student volunteers plant native trees and shrubs to create an urban forest in the Mississippi River corridor within Minneapolis and St. Paul, Minnesota. This provides wildlife habitat as well as beautifying the urban landscape.

trees and bushes in a project called Greening the Great River. Over the past decade, this nonprofit organization has mobilized more than 10,000 volunteers to plant 160,000 native trees and shrubs on vacant land within the Mississippi River corridor as it winds through Minneapolis and St. Paul, Minnesota. This project helps beautify the cities, reduces global warming, and provides habitat for wildlife. Deer, fox, raccoons, coyotes, bobcats, and even an occasional wolf are now seen in the newly revegetated area.

In 2007, the United Nations announced a "billion tree campaign" that hopes to gather pledges to plant one billion new trees around the world. Everyone can participate in this global reforestation effort. If you look around, there's sure to be some space in your yard (or that of your friends or relatives), or an abandoned piece of land in your neighborhood that could house a tree. Most states have tree farms that will provide seedlings at little or no cost. The American Forestry Association has done an extensive remote sensing survey of the United States. They estimate the national urban tree deficit now stands at more than 634 million trees. They suggest that everyone has a duty to plant at least one tree every year to help restore our environment (What Can You Do? p. 285).

The billion tree campaign is inspired by the work of Nobel Peace Prize winner Wangari Maathai. As chapter 1 mentions, Dr. Maathai founded the Kenyan Green Belt Movement in 1976 as a way of controlling erosion, providing fodder and food, and empowering women. This network of more than 600 local women's groups from throughout Kenya has planted more than 30 million trees while mobilizing communities for self-determination, justice, equality, poverty reduction, and environmental conservation.

What Can You Do?

Ecological Restoration in Your Own Neighborhood

Everyone can participate in restoring and improving the quality of their local environment.

1. *Pick up litter*
- With your friends or fellow students, designate one day to carry a bag and pick up litter as you go. A small amount of collective effort will make your surroundings cleaner and more attractive, but also draw attention to the local environment. Try establishing a competition to see who can pick up the most. You may be surprised how a simple act can influence others. Working together can also be fun!
- Join a group in your community or on your campus that conducts cleanup projects. If you look around, you'll likely find groups doing stream cleanups, park cleanups, and other group projects. If you can't find such a group, you can volunteer to organize an event for a local park board, campus organization, fraternity, or other group.

2. *Remove invasive species*
- Educate yourself on what exotic species are a problem in your area: In your environmental science class, assign an invasive species to each student and do short class presentations to educate each other on why these species are a problem and how they can be controlled.
- Do your own invasive species removal: In your yard, your parents' yard, or on campus, you can do restoration by pulling weeds. (Make sure you know what you're pulling!) Find a local group that organizes invasive species removal projects. Park boards, wildlife refuges, nature preserves, and organizations such as The Nature Conservancy frequently have volunteer opportunities to help eradicate invasive species.

3. *Replant native species*
- Once the invasives are eliminated, volunteers are needed to replant native species. Many parks and clubs use volunteers to gather seeds from existing native prairie or wetlands, and to help plant seeds or seedlings. Many of the parks and natural areas you enjoy have benefited from such volunteer labor in the past.
- Create your own native prairie, wetland, or forest restoration project in your own yard, if you have access to available space. You can learn a great deal and have fun with such a project.

4. *Plant a tree or a garden*
- Everyone ought to plant at least one tree in their life. It's both repaying a debt to nature and a gratifying experience to see a seedling you've planted grow to a mature tree. If you don't have a yard of your own, ask your neighbors and friends if they would like a tree planted.
- Plant a garden, or join a community garden. Gardening is good for your mind, your body, and your environment. Flowering or fruiting plants provide habitat for butterflies and pollinating insects. Especially if you live in a city, small fragments of garden can provide critical patches of green space and habitat—even a windowsill garden or potted tomatoes are a good start. You might find your neighbors appreciate the effort, too!

FIGURE 13.10 Oak savannas are parklike forests where the tree canopy covers 10 to 50 percent of the area, and the ground is carpeted with prairie grasses and flowers. This biome once covered a broad swath between the open prairies of the Great Plains and the dense deciduous forest of eastern North America.

Fire is essential for savannas

Oak savannas once covered a broad band at the border between the prairies of the American Great Plains and the eastern deciduous forest. Millions of hectares in what is now Minnesota, Wisconsin, Iowa, Illinois, Michigan, Indiana, Ohio, Missouri, Arkansas, Oklahoma, and Texas had parklike savannas, oak openings, or oak barrens. Although definitions vary, an oak savanna is a forest with scattered "open-grown" trees where the canopy covers between 10 to 50 percent of the area and the dappled sunlight reaching the ground supports a variety of grasses and flowering plants (fig. 13.10). The most common tree species is bur oak (*Quercus macrocarpa*), but other species, including white and red oak, also occur in savannas. Due to reduced sunlight, however, some dominant prairie species such as big and little blue stem grasses, most goldenrods, and asters are generally missing from savannas.

Because people found them attractive places to live, most oak savannas were converted to agriculture or degraded by logging, overgrazing, and fire suppression. Wisconsin, for example, which probably had the greatest amount of savanna of any Midwestern state (at least 2 million ha) before European colonization, now has less than 200 ha (less than 0.01 percent of its original area) of high-quality savanna. Throughout North America, oak savanna rivals the tallgrass prairie as one of the rarest and most endangered plant communities.

It's difficult to restore, or even maintain, authentic oak savannas. Fire was historically important in controlling vegetation. Before settlement, periodic fires swept in from the prairies and removed shrubs and most trees. Mature oaks, however, have thick bark that allow them to survive low-intensity fires. Grazing by bison and elk may also have helped keep the savanna open. When settlers eliminated fire and grazing by native animals, savannas were invaded by a jumble of shrub and tree growth. Unfortunately, simply resuming occasional burning doesn't result in a high-quality savanna. If fires burn too frequently, they kill oak saplings—which

are much more vulnerable than mature trees—and prevent forest regeneration. Excessive grazing can do the same thing. If fires are too infrequent, on the other hand, shrubs, dense tree cover, and dead wood accumulate so that when fire does occur, it will be so hot that it kills mature oaks as well as invading species. It's often necessary to clear most of the accumulated vegetation and fuel before starting fires if you want to maintain a savanna. Herbicide treatment may be necessary to prevent regrowth of invasive species until native vegetation can become established.

The Somme Prairie Grove in Cook County, Illinois, is one of the largest and oldest efforts to restore a native oak savanna. The land was purchased by the city of Chicago in the 1920s, and restoration activities began about 50 years later. A complex of wetland, prairie, and forest, much of the upland area was an oak savanna. Although the area was used as pasture for more than a century, most of it was never plowed. Limited grazing probably helped keep brush at bay and preserved the parklike character of the remaining forest. Spring burning was started in 1981 in an effort to preserve and restore the savanna. Burning alone, it was discovered, wasn't enough to eliminate aggressive invasive species, such as glossy buckthorn (*Rhamnus frangula*).

After several years of burning, the ground was mostly bare except for weedy species. In 1985 a more intensive management program was begun. Invasive trees were removed. Seeds of native species were collected from nearby areas and broadcast under the oaks. Weeds thought to be a threat to the restoration were reduced by pulling and scything; these included garlic mustard (*Alliaria officinalis*), burdock (*Arctium minus*), briers (*Rubus* sp.), and tall goldenrod (*Solidago altissima*). A research program monitored the results of this restoration effort. Four transects were established and repeatedly sampled for a wide variety of biota (trees, shrubs, herbs, cryptogams, invertebrates, birds, and small mammals) over a six-year period. Results from this survey are shown in fig. 13.11.

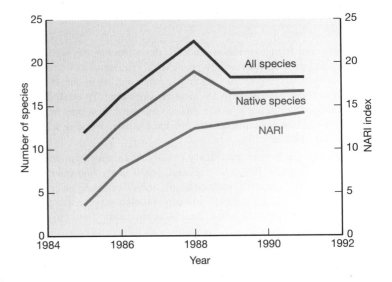

FIGURE 13.11 Biodiversity survey of the Somme Savanna restoration program. NARI, the Natural Area Rating Index, measures the frequency of native species associated with a high-quality community. Data from S. Packard and J. Balaban, 1994.

FIGURE 13.12 The conifer-hardwood mixture and complex matrix of ages and species in the Great Lakes Forest is dependent on regular but random fire. Maintaining this forest requires prescribed burning.

As you can see in this graph, the biodiversity of the forest increased significantly over the six years of sampling. Native species increased more than total biodiversity. NARI, the Natural Area Rating Index, is a measure of the relative abundance of native species characteristic of high-quality natural communities. Notice that this index shows the greatest increase of any of the measurements in the study. It also continues to increase after 1988, when native species began to replace invasive pioneer species. This suggests that the restoration efforts have been successful, resulting in an increasingly high-quality oak savanna.

Land managers now recognize the role of fire in preserving and restoring many forest types. The Superior National Forest, which manages the Boundary Waters Canoe Area Wilderness in northern Minnesota, for example, has started an ambitious program of prescribed fires to maintain the complex mosaic of mixed conifers and hardwoods in the forest. Pioneering work in the 1960s by forest ecologist M. L. Heinselman showed that the mixture of ages and species in this forest was maintained primarily by burning. Any given area in the forest experiences a low-intensity ground fire about once per decade, on average, and an intense, crown fire about once per century. As these fires burned randomly across the forest, they produced the complex forest we see today (fig. 13.12). But to maintain this forest, we need to allow natural fires to burn once again and to reintroduce prescribed fires where necessary to control fuel buildup.

Similarly, many national parks now recognize the necessity for fire to maintain forests. In Sequoia National Park, for example, rangers recognized that giant sequoias have survived for thousands of years because their thick bark is highly fire resistant, and they shed lower branches so that fire can't get up into their crown. Seventy years of fire suppression, however, allowed a dense undergrowth to crowd around the base of the sequoias. These smaller trees provide a "ladder" for fire

to climb up into the sequoia and kill them. Fuel removal and periodic prescribed fires are now a regular management tool for protecting the giant trees.

13.4 RESTORING PRAIRIES

Before European settlement, prairies covered most of the middle third of what is now the United States (fig. 13.13). The eastern edge of the Great Plains was covered by tallgrass prairies where big bluestem (*Andropogon gerardii*) reached heights of 2 m (6 ft). Their roots could extend more than 4 m into the soil and formed a dense, carbon-rich sod. This prairie has almost entirely disappeared, having been plowed and converted to corn and soybean fields. Less than 2 percent of the original 1 million km^2 (400,000 mi^2) of tall-grass prairie remains in its original condition. In Iowa, for example, which once was almost entirely covered by this biome, the largest sample of unplowed tallgrass prairie is only about 80 ha (200 acres). The middle of the Great Plains contained a mixed prairie with both bunch and sod-forming grasses. The drier climate west of the 100th meridian meant that few grasses grew to heights of more than 1 m. The westernmost band of the Great Plains, in the rain shadow of the Rocky Mountains, received only 10 to 12 inches (25 to 30 cm) of rain per year. In this shortgrass region, the sparse bunch grasses rarely grew to more than 30 or 40 cm tall.

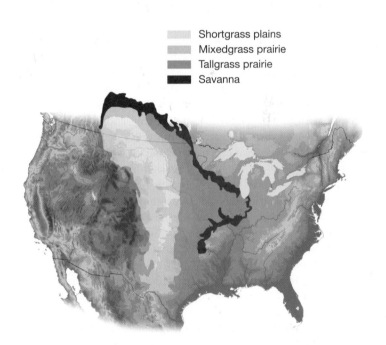

Shortgrass plains
Mixedgrass prairie
Tallgrass prairie
Savanna

FIGURE 13.13 The eastern edge of the Great Plains was covered by tallgrass prairie, where some grasses reached heights of 2 m (6 ft) and had roots more than 4 m long that formed a dense, carbon-rich sod. Less than 2 percent of the original 10 million km^2 (400,000 mi^2) of tall-grass prairie remains in its original condition. The middle of the Great Plains contained a mixed prairie with both bunch and sod-forming grasses. Few grasses in this region grew to heights of more than 1 m. The westernmost region of the Great Plains, in the rain shadow of the Rocky Mountains, had a shortgrass prairie where sparse bunch grasses rarely grew to more than 30 or 40 cm tall.

(a)

(b)

FIGURE 13.14 In 1934, workers from the Civilian Conservation Corps dug up old farm fields and planted native prairie seeds for the University of Wisconsin's Curtis Prairie (a). The restored prairie (b) has taught ecologists much about ecological succession and restoration.

Like the oak savanna immediately to the east, these prairies were maintained by frequent fires and grazing by bison, elk, and other native wildlife. Native American people understood the role of fire in regenerating the prairie. They frequently set fires to provide good hunting grounds and to ease travel.

Fire is also crucial for prairie restoration

One of the earliest attempts to restore native prairie occurred at the University of Wisconsin. Starting in 1934, Aldo Leopold and others worked to re-create a tallgrass prairie on an abandoned farm field at the University Arboretum in Madison. Student volunteers and workers from the Civilian Conservation Corps gathered seed from remnant prairies along railroad rights-of-way and in pioneer cemeteries, and then hand-planted and cultivated them (fig. 13.14a). Prairie plants initially had difficulty getting established and competing against exotics until it was recognized that fire is an essential part of this ecosystem. Fire not only kills many weedy species, but it also removes nutrients (especially nitrogen). This gives native species, which are adapted to low-nitrogen soils, an advantage. The Curtis Prairie is now an outstanding example of the tallgrass prairie community and a valuable research site (fig. 13.14b).

The Nature Conservancy (TNC) has established many preserves throughout the eastern Great Plains to protect fragments of tallgrass prairie. The biggest of these (and the largest remaining fragment of this once widespread biome) is in Oklahoma, where TNC has purchased or obtained conservation easements on about 18,000 ha (45,000 acres) of land that was grazed but never plowed just northwest of Tulsa. In cooperation with the University of Tulsa, TNC has established the Tallgrass Prairie Ecological Research Station, which is carrying out a number of experimental ecological restoration projects on both public and private land. Approximately three dozen prescribed burns are conducted each year, totaling 15,000 to 20,000 acres (fig. 13.15). Since 1991, over 350 randomly selected prescribed burns have been conducted, totaling 210,000 acres. About one-third of each pasture is burned each year. About half of the burns are done in spring and the rest in late summer.

In addition, Bison were reintroduced to the preserve in the early 1990s. By summer 2005, the herd numbered about 2,400 head. The long-term goal is to have 2,700 bison on about 23,000 acres. Patch burning and bison grazing create a habitat that can support the diverse group of plants and animals that make up the tallgrass prairie ecosystem. So far, more than three dozen research projects are active on the preserve, and 78 reports from these studies have been published in scientific journals.

The second largest example of this biome in the United States is the Tallgrass Prairie National Preserve in the Flint Hills region of Kansas. Because the land in this area is too rocky for agriculture, the land was grazed but never plowed. Most of this preserve was purchased originally by The Nature Conservancy, but it is now being managed by the National Park Service. Like its neighbor to the south, a part of this preserve—known as the Konza Prairie—is reserved for scientific research. A long-term ecological research (LTER) program, funded by the National

FIGURE 13.15 A burn-crew technician sets a back fire to control the prescribed fire that will restore a native prairie.

FIGURE 13.16 Much of the Great Plains is being depopulated as farms and ranches are abandoned and small towns disappear. This provides an opportunity to restore a buffalo commons much like that discovered two centuries ago by Lewis and Clark.

FIGURE 13.17 Millions of hectares of shortgrass prairie are being converted to nature preserves. Some may be a buffalo commons devoted to bison and other wildlife rather than cattle, as envisioned by artist George Catlin in 1832

Science Foundation and managed by Kansas State University, sponsors a wide variety of basic scientific research and applied ecological restoration experiments. Bison reintroduction and varied fire regimes are being studied at this prairie. The role of grasslands in carbon sequestration and responses of this biological community to climate change are of special interest.

Huge areas of shortgrass prairie are being preserved

Much of the middle, or mixed grass section, of the Great Plains has been converted to crop fields irrigated by water from the Ogallala Aquifer (chapter 17). As this fossil water is being used up, it may become impossible to continue farming using current techniques over much of the area. A great deal of worry and debate exist about the future of this region of the country. Interestingly, John Wesley Powell, the Colorado River explorer and first head of the U.S. Geological Survey, warned in 1878 that there wasn't enough water to settle this region. He opposed opening the Great Plains to homesteaders, and said, "I tell you gentlemen, you are piling up a heritage of conflict and litigation of water rights, for there is not sufficient water to supply the land."

The exodus of people from the land predicted by Powell is already occurring in the more arid regions of the Great Plains. In the 1990s, more than half the counties between the 100th meridian and the Rocky Mountains experienced declining populations (fig. 13.16). As farms and ranches fade away, the small towns that once supplied them also dry up. Of the people who remain, about twice as many are above 65 years as the national average. Ironically, several hundred thousand square miles now have less than six persons per square mile—the threshold Frederick Jackson Turner used to declare the frontier "closed" in 1890. Large areas have less than two persons per square mile.

A number of people are now suggesting that we should have followed Powell's recommendations and never tried to settle the arid regions of the American west. Some believe that the best use of much of the empty land is to return it to a **buffalo commons** where bison and other native wildlife could roam freely. This idea was publicized in 1987 when Drs. Frank and Debora Popper, geographers from New Jersey, proposed a grand ecological and social reorganization of the Great Plains in an article in the journal *Planning*. The Poppers were fiercely criticized by many residents of Plains states, but today steps are being taken to accomplish something much like they proposed.

Interestingly, the idea of a huge park or preserve didn't originate with the Poppers. In 1832, the artist George Catlin, who was deeply concerned by disappearance of both bison and the Native American cultures that depended on them, proposed that most of the Great Plains should be set aside as a national park populated by great herds of buffalo and other wildlife and home to native people living a traditional lifestyle. Native people may no longer be interested in living in tepees and depending entirely on buffalo for their sustenance, but some groups are working to create something like the huge park Catlin proposed.

There are two competing approaches to saving shortgrass prairie (fig. 13.17). The Nature Conservancy is cooperating with ranchers on joint conservation and restoration programs. On the 24,000 ha (60,000 acre) Matador ranch, which TNC bought in 2000, 13 ranchers graze cattle in exchange for specific conservation measures on their home range. The ranchers have agreed to protect about 900 ha of prairie dog colonies and sage grouse leks. All the ranchers have also agreed to control weeds, resulting in almost 120,000 ha of weed-free range. Together with the Conservancy, ranchers are experimenting with fire to improve the prairie, and they plan and manage a grass-banking arrangement in which ranchers access more Conservancy land in drought emergencies. In one case, TNC has given the deed for a ranch it owns to a young couple who promise to manage it sustainably and to protect some rare wetlands on the property. The Conservancy believes that keeping ranch families on the land is the best way to preserve both the social fabric and the biological resources of the Plains.

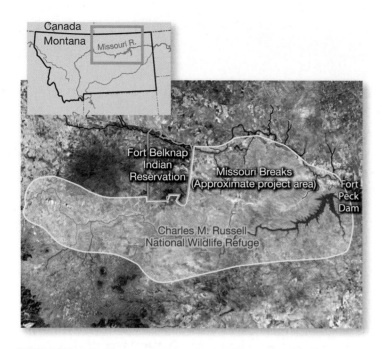

FIGURE 13.18 Both The Nature Conservancy and the American Prairie Foundation have bought large tracts of land in the Missouri Breaks between the Charles M. Russell National Wildlife Refuge and the Fort Belknap Indian Reservation. Ultimately, conservationists hope to protect and restore as much as 2 million ha of shortgrass prairie.

FIGURE 13.19 Bison can be an important tool in prairie restoration. Their trampling and intense grazing disturb the ground and provide an opening for pioneer species. The buffalo chips they leave behind fertilize the soil and help the successional process. Where there were only about two dozen wild bison a century ago, there are now more than 400,000 on ranches and preserves across the Great Plains.

Near the Matador land, another group is pursuing a very different strategy for preserving the buffalo commons. The American Prairie Foundation (APF), which is closely linked to the World Wildlife Fund, has also bought about 24,000 ha of former ranchland. Rather than keep it in cattle production, however, this group intends to pull out fences, eliminate all the ranch buildings, and turn the land back into wilderness. Ultimately, the APF hopes to create a reserve of at least 1.5 million ha in the Missouri Breaks region between the Charles M. Russell National Wildlife Refuge and the Fort Belknap Indian Reservation (fig. 13.18) The APF plans to reintroduce native wildlife including elk, bison, wolves, and grizzly bears to its lands. Neighboring ranchers don't mind having elk or bison nearby, but they object to reintroducing wolves and grizzlies, which their parents and grandparents exterminated a century ago. They also bristle at the source of the funding for the APF project. Donations, many of them very large, are coming from Wall Street or California's Silicon Valley. Wealthy individuals, such as media mogul Ted Turner, are using their money to make striking changes in how western land is used. Often locals, who struggle just to stay on the land, resent having rich outsiders change the range.

With donations of more than $12 million so far, and fundraising goals of at least $100 million, the APF points out that they aren't forcing anyone from the land. They're only buying from ranchers who want to sell. Locals worry, nonetheless, that the land will be restricted for hunting and other uses. The APF says it will allow tourism, bird-watching, and hunting on nearly all its land. Small towns also are anxious about the effect of taking land out of production on the local economy. The APF points out that without federal subsidies, the average return on ranchland in the Missouri Breaks is less than $5 per acre. Tourism and hunting already bring in more income per acre than does raising cows.

Bison help maintain prairies

So far, the APF has only 19 bison on its lands in Montana, but it intends to expand its herd to several thousand animals eventually. Others also are returning bison to the shortgrass prairie (fig. 13.19). The biggest of TNC's herds are in southern plains, but they have herds of several hundred animals in the Niobara River Valley in Nebraska and on the Ordway Prairie in north-central South Dakota. The Fort Belknap Indian reservation currently has a herd of 700 animals on a 5,000 ha tribal buffalo reserve. They also have a large prairie dog town that is a refuge for the highly endangered black-footed ferret. Although the native people don't live the lifestyle envisioned by George Catlin, they do welcome tourists for bird-watching, cultural tours, picnicking, sightseeing, wildlife viewing, and other forms of ecotourism.

Together with fire, bison are considered important tools in prairie restoration. Cattle graze selectively, giving noxious weeds a competitive advantage over many native species. Bison, on the other hand, tend to move in dense herds eating almost everything in their path. Their trampling and intense grazing disturb the ground and provide an opening for pioneer species, many of which disappear when bison are removed. The buffalo chips they leave behind fertilize the soil and help maintain plants that need nutrient-rich conditions. Bison also dig out wallows, in which they take dust baths, creating even more disturbed surface for primary succession. Having grazed an area, bison will tend to move on, and, if they have enough space in which to roam, won't come back for several years. This pattern of intense, short-duration grazing by wild ungulates is the origin of the idea of rotational grazing described in chapter 9.

Many ranchers are coming to recognize the benefits of raising native animals. Where cows require shelter from harsh weather and lots of water to drink, bison are well adapted to the harsh conditions of the prairie. Bison meat is lean and flavorful. It brings a higher price than beef in the marketplace. It's estimated that there are now about 400,000 bison on ranches and farms in America. This is probably less than 1 percent of the original population, but it's 10,000 times more than the tiny remnant left after the wanton slaughter in the nineteenth century.

13.5 RESTORING WETLANDS AND STREAMS

Wetlands and streams provide important ecological services. They play irreplaceable roles in the hydrologic cycle. They also are often highly productive, and provide food and habitat for a wide variety of species. As the opening case study of this chapter points out, the freshwater marshes and swamps of coastal Louisiana (fig. 13.20) play important roles in both biological and human communities. Although wetlands currently occupy less than 5 percent of the land in the United States, the Fish and Wildlife Service estimates that one-third of all endangered species spend at least part of their lives in wetlands. As the opening case study for this chapter shows, coastal wetlands are vital for absorbing storm surges. Storage of flood waters in wetlands is worth an estimated $3 billion to $4 billion per year. Wetlands also improve water quality by acting as natural water purification systems, removing silt and absorbing nutrients and toxins.

Throughout much of our history, however, wetlands have been considered disagreeable, dangerous, and useless. This attitude was reflected in public policies, such as the U.S. Swamp Lands Act of 1850, which allowed individuals to buy swamps and marshes for as little as 10 cents per acre. Until recently, federal, state, and local

FIGURE 13.21 Louisiana has nearly 40 percent of the remaining coastal wetlands in the United States, but diversion of river sediments that once replenished these marshes and swamps, together with channels and boat wakes, are causing the Gulf shoreline to retreat about 4 m (13.8 ft) per year.

governments encouraged wetland drainage and filling to create land for development. In an effort to boost crop production, the government paid farmers to ditch and tile millions of acres of wet meadows and potholes. Cutting channels for oil and gas exploration and shortcuts for navigation through Gulf Coast marshes is one of the main reasons for destruction of the barrier that once protected New Orleans from hurricanes (fig. 13.21).

The result of these policies was that from the 1950s to the mid-1970s, the United States lost about 200,000 ha (nearly 500,000 acres) of wetlands per year. The 1972 Clean Water Act began protecting streams and wetlands by requiring discharge permits (called Section 404 permits) for dumping waste into surface waters. In 1977, federal courts interpreted this rule to prohibit both pollution and filling of wetlands (but not drainage). The 1985 Farm Bill went further with a "swamp buster" provision that blocked agricultural subsidies to farmers who drain, fill, or damage wetlands. Many states now have "no net loss" wetlands policies. Between 1998 and 2004, America had a net gain of about 80,000 ha (197,000 acres) of wetlands. This total area concealed an imbalance, however. Continued losses of 210,000 ha of swamp and marsh wetlands were offset by a net gain of 290,000 ha of small ponds and shallow-water wetlands (which are easy to construct and good for duck production). Since 2004, when President Bush promised to restore, improve, and protect wetlands, another 700,000 ha have been added to the national total. This is a good start, but the U.S. needs much more wetland restoration to undo decades of damage. In this section, we'll look at some efforts to accomplish this goal in the U.S. and elsewhere.

Reinstating water supplies helps wetlands heal

As is the case with other biological communities, sometimes all that's needed is to stop the destructive forces. For wetlands, this often means simply to restore water supplies that have been diverted elsewhere.

FIGURE 13.20 A Louisiana cypress swamp of the sort being threatened by coastal erosion and salt infiltration. Wetlands provide habitat for a wide variety of species, and play irreplaceable ecological roles. **Source:** U.S. Army Corps.

For millennia, the delta where the Tigris and Euphrates empty into the Persian Gulf created a vast wetland with unique biological and human importance. Covering an area the size of Wales, these marshes provided a resting spot for millions of wild-fowl migrating between Eurasia and Africa. The marshes also were the home of a unique group of people, the Marsh Arabs, who built their homes on floating platforms and depended on the wetland for most of their sustenance. Some people believed the verdant marshes to be the biblical Garden of Eden.

Saddam Hussein accused the Marsh Arabs of supporting his enemies during the 1980–1988 war with Iran. In retaliation, he ordered the marshes dammed and drained. About 90 percent of the marshes were destroyed and more than 140,000 people were forced from their homes. As the marsh dried out, much of it also was burned. After Saddam was toppled in 2003, the Marsh Arabs cut through Saddam's dikes and plugged diversion canals, allowing water to flow once again into the marshes. Aided by a UN-led restoration project, the wetlands have rebounded to about half their former size. Flora and fauna have repopulated the area and people also have begun to return. The marsh hasn't returned to its original state. Given the political situation in Iraq it isn't clear how much restoration will be accomplished, but the situation is much better than it was a few decades ago.

Another simple physical solution to wetland degradation can be seen in the American Midwest. When the U.S. Army Corps of Engineers built 26 locks and dams on the upper Mississippi River to facilitate barge traffic in the 1950s, it created a series of large impoundments. Sediment from the surrounding farm fields began to fill these pools. This might have created a valuable network of wetlands, but waves and currents created by wind, floods, and river traffic keep the sediment constantly roiled up so that wetland plants can't take root. The result is a series of wide, shallow, semisolid mud puddles that are too thick for fish but too liquid for vegetation. To remedy this situation, the Army Corps is experimenting with wing dams (see thin white lines connecting islands in fig. 13.22) that separate the backwaters from the main river channel. It's hoped that these dams will allow the mud to solidify and turn into marshlands.

Much of the restoration of the Louisiana wetlands described in the opening story of this chapter depends on an assumption that controlling water flow and sediment deposition will result in restoration of a healthy biological community. This may turn out to be true, but monitoring is needed to evaluate results. How do we know whether restoration is working or not? (See Exploring Science p. 293.)

The Everglades are being replumbed

A huge, expensive restoration is now underway in the Florida Everglades. Famously described as a "river of grass" the Everglades are created by a broad, shallow sheet of water that starts in springs near Orlando, in the center of the state, then moves through Lake Okeechobee and flows southward to the Gulf. Spreading over most of the southern tip of the peninsula, and

FIGURE 13.22 Dams on the upper Mississippi River have created a series of large lakes. Currents created by wind, floods, and river traffic keep the soft sediment stirred up so that wetland plants can't take root. To restore these backwaters, the Army Corps is experimenting with wing dams (see thin white lines between islands) that will allow the mud to solidify and turn into marshlands. Photo courtesy of Army Corps of Engineers.

moving imperceptibly across the flat landscape, the fresh water nourishes a vast marshland (fig. 13.23) that supports myriad fish, invertebrates, birds, alligators, and the rare Florida panther.

Farmers found the rich, black muckland of the Everglades could grow fantastic crops if it was drained. Ditching and diverting the water started more than a century ago. A series of floods that threatened the wealthy coastal cities also triggered a demand for

FIGURE 13.23 The Florida Everglades, often described as a "river of grass," is threatened by water pollution and diversion projects.

The science of restoration is complex, because there are rarely simple answers. Restoration is also highly optimistic, because it often deals with environmental problems that are huge and persistent. But increasingly we recognize that ecological restoration is a necessity if we are to preserve economies, cultures, and ways of life.

As the opening case study of this chapter reveals, restoring Louisiana's coastal wetlands has been discussed for half a century, but the projects have gained a new urgency since 2005, when Hurricane Katrina flooded New Orleans. Restoring this vast system is almost an inconceivably large project, but without restoration, Louisiana will continue to lose communities, roads, and economic activity. Hundreds of small projects have been planned, and some are in progress. A prominent project is the Caernarvon Diversion, a series of structures built on the south bank of the Mississippi east of New Orleans. At a cost of $4.5 million, the project is using culverts (1.25 m diameter pipes) to divert water from the Mississippi, together with plugs that block abandoned gas-field canals.

Engineering this project involved a decade of planning, but monitoring is the most extensive part of the project. Monitoring is a key aspect of all restoration: otherwise how do we know if the project has been successful—and if it's worth doing again? In addition to the project area, two "reference areas" are also being monitored. Ideally, the project area will improve considerably over the reference areas, and over the baseline conditions before the project started. The project plan outlines a 20-year monitoring, with three central concerns:

1. *Ratio of land to open water.* Land and water are being mapped using aerial photos.

Using a GIS, the monitoring team can calculate the amount of land and water in 2000, then again in 2006 and in 2018. In theory, the amount of land will increase from one time period to the next. With a GIS, they can also overlay one year's map on top of another year's map. By subtracting one layer from the other, analysts identify not only the amount of change, but which areas have changed.

2. *Plant species composition and relative abundance.* Before water diversions began, ecologists designated a series of square plots of wetland, 2 m on a side. These 4 m² plots were placed in reference areas, as well as in the project area. Plant ecologists then visited each plot to list all species in each one. They also estimated the relative abundance of the different species. By leaving permanent markers at the plots, they can revisit the same location years later to monitor change. By sampling a number of replicate plots, they can get a sense of aggregate change in the area.

3. *Salinity.* Salt concentrations will be continuously recorded with a salinity sensor—an electrode that measures the concentration of salt ions in water—that is hooked up to a small computer that automatically records data once an hour. By monitoring hourly, hydrologists and ecologists can observe changes in salinity during storms, spring floods, and other events that might cause rapid variations.

To more clearly organize and structure the data collection and interpretation, scientists framed specific hypotheses (testable statements). For example, one goal is to increase abundance and diversity of plants. A testable hypothesis is: "After the project implementation, diversity will be significantly greater than before project implementation." By gathering data before and after, the restoration team can test this hypothesis by comparing the average number of plant species, or the average abundance of each species, in the different plots. A simple "yes" or "no" answer will diagnose whether the project is working as planned.

Similarly, reducing salinity is a goal. A testable hypothesis is: "After project implementation, salinity will be significantly less than before project implementation." Again, this hypothesis can be tested by comparing before-project samples and after-project samples for mean salinity value and variation from the average.

Restoration often relies on inputs from a variety of sciences, including hydrology, ecology, geology, and other fields. Restoration is also experimental, partly because it is a new science, and much remains uncertain in restoration projects. But restoration is an extremely important—and exciting—field within environmental science.

Student volunteers sample plant biomass and species composition to evaluate the health of a coastal wetland.

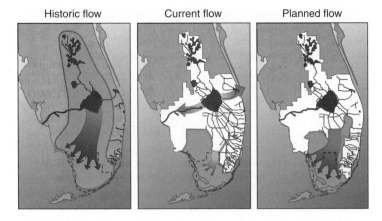

Historic flow   Current flow   Planned flow

FIGURE 13.24 Planned results of the Comprehensive Everglades Restoration Plan. Red dashed outline shows the national park boundary.

FIGURE 13.25 The naturally meandering Kissimmee River (*right channel*) was straightened by the Army Corps of Engineers (*left*) for flood control 30 years ago. Now the Corps is attempting to reverse its actions and restore the Kissimmee and its associated wetlands to their original state.

more water management. Once-meandering rivers were straightened to shunt surplus water out to sea. Altogether, the Army Corps of Engineers has built more than 1,600 km of canals, 1,000 km of levees, and 200 water control structures to intercept normal water flow, drain farmlands, and divert flood water (fig. 13.24). Ironically, many of the cities that demanded the water be diverted are experiencing water shortages during the dry season. Water that might have been stored in the natural wetlands is no longer available. Nature, also, is suffering from water shortages. The Everglades National Park has lost 90 percent of its wading birds and is worried that the entire aquatic ecosystem may be collapsing.

After years of debate and acrimony, the various stakeholders have finally agreed on a massive reengineering of the south Florida water system. The plan aims to return some water to the Everglades, yet retain control to prevent flooding. More than 400 km of levees and canals will be removed. New reservoirs will store water currently lost to the ocean, and 500 million liters of water per day will be pumped into underground aquifers for later release. Rivers are being dechannelized to restore natural meanders that store storm water and provide wildlife habitat (fig. 13.25). It's hoped that simply restoring some of the former flow to the Everglades will allow the biological community to recover, although whether it will be that simple remains to be seen. Altogether, this project is expected to cost at least $8 billion, making it one of the largest restoration projects ever undertaken.

Although announced with great fanfare in 2000, the restoration project, so far, is behind schedule, over budget, and at risk of losing congressional support. A blunt internal memo written by a top manager in the Army Corps of Engineers was leaked to the press in 2005. It said, "We haven't built a single project during the first five years, and we've missed almost every milestone." The memo echoed concerns of conservationists, the Miccosukee Tribe, and others who have been complaining about the lack of progress for years. The state of Florida has put up more than $1 billion, nearly five times more

than federal agencies to date. These problems show some of the difficulty in carrying out such a huge, complex, and highly political project.

The Chesapeake Bay is being rehabilitated

Largest and richest of all U.S. estuaries, the Chesapeake Bay is a drowned river valley. Its drainage basin covers more than 166,000 km^2 (64,000 mi^2) in six states, while its main stem stretches some 321 km (200 mi) from the tidal flats in Maryland to its mouth in the Tidewater area of Virginia (fig. 13.26). More than 150 tributary streams mix with saltwater in its broad, shallow basin, creating complex ecological gradients. Salinity ranges from near zero in the Susquehanna River at the bay's upper end, to thirty parts per thousand (near that of seawater) in its lower reaches. Fresh water generally flows south in the bay's upper layers, while saltier water flows north (up the bay) in the lower layers. Tides and

FIGURE 13.26 Chesapeake Bay is America's largest and richest estuary. Pollution has reduced fish and shellfish harvest, degraded water quality, and threatened biological diversity. Efforts are now being made to restore this complex ecosystem.

currents mix the waters, distributing nutrients and flushing out wastes. Shoals and salt marshes provide food, shelter, and living space for myriad plants and animals. Some 2,700 different species spend all or part of their lives in, on, or beside the bay.

Oysters, blue crabs, rockfish, white perch, shad, sturgeon, flounder, eel, menhaden, alewives, and soft-shell clams once created bountiful harvests for Tidewater residents. Even with over-harvesting of some species, the annual shellfish and finfish catch once averaged over 20,000 metric tons per year. But some important species have declined severely. Shad and striped bass, which once came into the bay to spawn in astronomical numbers, now are much less abundant. Sturgeon are rarely caught anymore. The oyster harvest, which was 15 to 20 million bushels per year in the 1890s, has now declined to less than 1 percent of that amount.

The reasons for loss of this productive fishery are many, including overfishing, sewage discharge, silt from erosion, heavy metals, toxic chemicals and heat from industry, pesticides and herbicides from agricultural runoff, oil spills, and habitat destruction. Filter feeders, such as shellfish, concentrate pathogenic viruses and bacteria in sewage. Contaminated seafoods transmit hepatitis and other dangerous diseases. Excess nitrogen and phosphorus from sewage and agricultural runoff effluent stimulate growth of phytoplankton in the bay's broad, shallow waters. As these innumerable tiny creatures die and settle to the bottom, they decompose, using up all the oxygen in the bottom layers of the bay, killing oysters and other bottom dwellers. It's common to describe these oxygen-depleted waters as a "dead zone" because few organisms can survive there.

Dredging, filling, and pollution of coastal areas have damaged or destroyed vast salt marshes that once filtered runoff and reduced silt and nutrient runoff. Increased sediment and nutrients has smothered beds of submerged aquatic vegetation that once served as a nursery for numerous marine animals. In particular, one especially valuable species, eelgrass (*Zostera marina*), which once covered large areas of shallow waters in the bay, and is considered a keystone species in its ecosystem, almost completely disappeared in the 1970s. Restoration programs are now replanting both dune grasses and eelgrass (fig. 13.27). The goal is to revegetate at least 200,000 ha of coastline, salt marsh, and shallow water areas of the bay.

The problems facing the bay are huge. Some 15 million people now live within the Chesapeake watershed. Major metropolitan areas include Baltimore, Maryland; Washington, D.C; and Scranton/Wilkes Barre, Pennsylvania. More than 66 major industrial facilities—including power plants, oil refineries, chemical factories, steel mills, and shipyards discharge wastes into the bay. More than 420,000 septic systems—many of them old and leaky—are located in the watershed. And thousands of farms, suburban yards, urban streets, and construction sites discharge nutrients, toxins, and sediment to streams that empty into the bay.

Just getting the federal government, four states, and thousands of towns and counties within the watershed to cooperate on a restoration program has taken several decades. In 2005, the Chesapeake Bay Commission, which coordinates efforts to protect and rehabilitate the bay, celebrated its 25th anniversary. In a

FIGURE 13.27 Volunteers plant dune grasses as part of Chesapeake Bay restoration. Ultimately, this effort hopes to revegetate at least 200,000 ha of coastline, salt marsh, and shallow water areas of the bay. The large cylinder is a "bio log" made of biodegradable material that will help stabilize the shore.
Source: NOAA.

summary of current conditions, the Commission admitted, "Today, despite 25 years of effort, poor water quality and loss or degradation of habitat threatens a majority of the Bay's living resources." Nevertheless there has been progress. New water quality laws aim to remove 3,400 metric tons of nitrogen and 120 metric tons of phosphorus from water entering the bay each year. The state of Maryland plans to spend $12.5 million per year to reduce runoff to the bay. Sixty percent of this money will go to upgrading septic systems, and 40 percent will be spent on cover crops to reduce erosion. Maryland has also spent between $1 million and $40 million annually over the past 40 years to dig up old oystershell deposits and spread them in appropriate areas in an effort to re-create productive oyster beds.

So far, most of these restoration efforts have had minimal success, but the increasing public awareness of the value of and threats to this magnificent biological resource is encouraging. And this public concern is reflected in the willingness of officials to take action. It may take a long time to repair the damage, but there's hope that the Chesapeake Bay may once again be the healthy, productive ecosystem it once was.

Wetland mitigation can replace damaged areas

The slow progress in improving water quality and re-creating biological communities in the Everglades and Chesapeake Bay illustrate the difficulties in working in large, complex, highly political ecosystems which have been damaged by a large variety of human actions. Smaller ecosystems can be much easier to restore or replace. An encouraging example is the re-creation of prairie potholes in the Great Plains. Before European settlement, millions of these shallow ponds and grassy

FIGURE 13.28 Millions of prairie potholes once covered the Great Plains from Iowa and South Dakota, north to Canada's Prairie Provinces. More than half of these shallow ponds have been drained for agricultural crop production, but now many are being replaced.

FIGURE 13.29 An example of wetland mitigation. In many states, developers aren't required to plant new vegetation. They simply dig a hole and wait for it to fill with rainwater. This is a poor replacement for a natural wetland.

wetlands once spread across the prairies from Alberta, Saskatchewan, and Manitoba, to northeastern Kansas and western Iowa (fig. 13.28). They produced more than half of all North American migratory waterfowl, and mediated flooding by storing enormous amounts of water. As agriculture expanded across the Plains, however, at least half of all prairie wetlands were drained and filled. In the 1900s, Canada's Prairie Provinces had about 10 million potholes; by 1964, an estimated 7 million were gone. The United States had similar losses.

Passage of the Migratory Bird Hunting Stamp Act (popularly known as the Duck Stamp Act) in 1934 marked a turning point in wetland conservation. In the past 70 years, this program has collected $700 million that has been used to acquire more than 5.2 million acres of habitat for the National Wildlife Refuge System. In 1934, when the Duck Stamp Act was passed, drought, predators, and habitat destruction had reduced migratory duck populations in the U.S. to about 26 million birds. John Phillips and Frederick Lincoln, in a 1930 report on the state of American waterfowl wrote, "we believe it soon will be too late to save [wild-fowl] in numbers sufficient to be of any real importance for recreation in the future." By 2006, however, conservation efforts, a wetter climate, and habitat restoration had increased duck production to about 44 million birds. Duck stamps now provide about $25 million annually for wildlife conservation.

While it's relatively easy to dig new ponds, it's much more difficult to replace other wetland types. As we mentioned earlier in this chapter, between 1998 and 2004, the United States lost 210,000 ha of swamp and marshes. This was offset by repair or construction of 290,000 ha of small ponds and shallow-water wetlands, such as restored prairie potholes. Replacing a damaged wetland with a substitute is called **wetland mitigation.** It's required whenever development destroys a natural wetland. Often, this process doesn't really replace the native species and ecological functions represented by the original biological

community. Figure 13.29, for example, shows a replacement wetland created by a housing developer in Minnesota. In building a housing project in the Minnesota River valley, the developer destroyed about 10 ha of a complex, native wetland that contained rare native orchids and several scarce sedge species. To compensate for this loss, the developer simply dug a hole and waited for it to fill with rainwater. He wasn't required to replant wetland species. The law assumes that natural succession will revegetate the disturbed area. In fact, it was soon revegetated, but entirely with exotic invasive species.

Constructed wetlands can filter water

Many cities are finding that artificial wetlands provide a low-cost way to filter and treat sewage effluent. Arcata, California, for instance, needed an expensive sewer plant upgrade. Instead, the city transformed a 65 ha garbage dump into a series of ponds and marshes that serve as a simple, low-cost, waste treatment facility. Arcata saved millions of dollars and improved its environment simultaneously. The marsh is a haven for wildlife and has become a prized recreation area for the city. Eventually, the purified water flows into Humbolt Bay, where marine life flourishes. Similarly, small pools and rain gardens—shallow pits lined with porous surface material and planted with water-tolerant vegetation—are used to collect storm runoff and allow it to seep into the ground rather than run into rivers or lakes. And constructed marshes allow industrial cooling water to equilibrate before entering streams or other surface water bodies. All these created wetlands can be both useful to humans and wildlife. For more on this topic, see chapter 18.

Many streams need rebuilding

Streams and rivers are damaged and degraded by many of the same factors that threaten other ecosystems. Pollution, pathogens and diseases, industrial toxins, invasive organisms, erosion, and

FIGURE 13.30 Many former streams have been turned into concrete-lined ditches to control erosion and speed runoff. The result is an artificial system with little resemblance to the living biological community that once made up the stream.
Source: Federal Interagency Stream Restoration Working Group.

FIGURE 13.31 A lunker structure is a multilevel wooden framework that rests on the stream bottom and can be anchored to the shore. The top of the box is covered with rock, soil, and vegetation. Openings in the structure provide hiding places for fish.

a host of other insults harm streams. The United States has more than 5.6 million km of rivers and streams. Together with their upland watersheds, these waterways are complex ecosystems that perform a number of important ecological functions and play valuable economic roles for society. Over the years, human activities have caused changes in the dynamic equilibrium of stream systems across the nation.

In 1994, of the nearly 1 million km of rivers and streams that were monitored by the Environmental Protection Agency, only 56 percent fully supported multiple uses, including drinking-water supply, fish and wildlife habitat, recreation, and agriculture, as well as flood prevention and erosion control. Sedimentation and excess nutrients were the most significant causes of degradation in the remaining 44 percent. Presumably these results could be extrapolated to the rest of the nation's waterways. Given these statistics, the need for stream restoration is obvious.

One response to erosion and flooding in urban streams is to turn them into cement channels that rush rainwater off into some larger body of water (fig. 13.30). The result is an artificial system with little resemblance to the living biological community that once made up the stream. In many places, streams have been buried in culverts and turned into underground sewers. Some cities have come to recognize, however, that natural streams can increase property values and improve the livability of the urban environment. Buried streams are being "daylighted," and channelized ditches are being turned back into living biological communities. Other streams clogged with silt or degraded by erosion and pollution also are being reconstructed.

A variety of restoration techniques have been developed for streams. This field has become an important job opportunity for environmental science majors. A simple approach in which everyone can participate is to reduce sediment influx by planting ground cover on uplands and filling gullies with rocks or brush. Sometimes for small streams, the quickest way to rebuild a channel is

to use heavy earthmoving equipment to simply dig a new one. This can be very disruptive, however, stirring up sediment that can be harmful for fish and other aquatic organisms. A number of alternative, less intrusive steam improvement methods are available. Most of these methods involve placing barriers (weirs, vanes, dams, log barriers, brush bundles, root wads, or other obstructions) in streams to deflect current away from the banks or trap sediment. Often, these barriers will cause currents to scour out deep pools in the stream bottom that provide places for fish to hide and rest.

Other techniques also create fish habitat. Logs, root wads, brush bundles, and boulders can shelter fish. An expensive but effective way to create fish hiding places is the so-called "lunker" structure. This is a wood framework that rests on the stream bottom and is anchored securely to the shore. The top of the box is covered with rock, soil, and vegetation. Openings in the structure provide hiding places for fish (fig. 13.31).

Sometimes what's needed is to speed up the current rather than slow it down. Figure 13.32a shows a spring-fed Minnesota trout stream that was degraded by crop production in its uplands, and grazing that broke down the banks and filled the stream with sediment. The stream, which had been about 1 m wide and 1 m deep, spread out to be 5 m wide and only about 10 cm deep. The formerly cold, swiftly moving water was warmed by the sun as it passed through these shallows so that it became uninhabitable for trout. Furthermore, there was no place to hide from predators. Because this was the last remaining trout stream in the Minneapolis/St. Paul metropolitan area, a decision was made to restore it. Two very different restoration approaches are shown in figure 13.32b and c. Trout Unlimited, an angler's group, offered to bring in a backhoe and completely rebuild the stream. A demonstration section they reconstructed is shown in figure 13.32b. They narrowed the channel with large stone blocks, which also provide a good surface for anglers to walk along the stream. There isn't much biological production in this stream model, but that isn't important to many anglers. They expected the Department of Natural Resources to stock the stream with hatchery-raised fish, which usually are caught before they have time to learn to forage for natural food.

Other groups involved in stream restoration objected to the artificial nature and lack of a native biological community in this design. Figure 13.32c, shows an alternative approach that was ultimately adopted for most of the stream. This is the same section shown in figure 13.32a. Straw bales were placed in the deepest part of the stream. This narrowed the stream and increased the speed of the current, which then cut down into the soft sediment. Within three months, the stream had deepened about 50 cm. The shallow area between the straw bales and the original shore quickly filled with sediment and became a cattail marsh. The bales were anchored in place with green willow stakes, which rooted in the moist soil and sprouted to make saplings that arched over the stream and shaded it from the sun.

Seeds of native wetland vegetation were scattered on top of the straw bales. They sprouted, and after the first summer, the stream was surrounded by a dense growth of vegetation that keeps the water cool, provides fish shelter, and supports a rich community of invertebrates that feed the native trout. An aquatic invertebrate survey found that while the stream was wide and shallow, 58 percent of the aquatic species were snails and copepods, while stone flies, caddis flies, and other preferred trout food made up only 20 percent of all invertebrates. After the straw bale restoration, snails and copepods made up only 15 percent of the invertebrate population, while stone flies and caddis flies had increased to 47 percent.

Stabilizing banks is an important step in stream restoration. Where banks have been undercut by erosion and stream action, they will continue to be unstable and cave into the stream. Ideally, the bank should be recontoured to a slope of no more than 45 degrees (fig. 13.33). Soil can then be held in place by rocks, planted vegetation, or other ground cover. It may be necessary to install erosion control fabric or mulch to hold soil until vegetation is established. If space isn't available for recontouring, steep banks may have to be supported by rock walls, riprap, or embedded tree trunks.

Severely degraded or polluted sites can be repaired or reconstructed

Remediation means finding remedies for problems. In restoration science, it usually involves relatively noninvasive techniques for removing or stabilizing pollutants or repairing damaged landscapes. A number of plant species can selectively eliminate toxins from the soil. Some types of mustard, for example, can extract lead, arsenic, zinc, and other metals from contaminated soil. Radioactive strontium and cesium have been removed from soil near the Chernobyl nuclear power plant using common sunflowers. Poplar trees can absorb and break down toxic organic chemicals. In a relatively small area—say an old industrial site—it may be economical to simply excavate and replace contaminated soil. If the pollutants are organic, it may be possible to pass soil through an incinerator to eliminate contaminants. After this treatment, the soil won't be worth much for growing vegetation, however.

If the problem is polluted surface or groundwater, bioremediation (treatment with living organisms) can be effective. Naturally occurring bacteria in groundwater, when provided with oxygen and nutrients, can decontaminate many kinds of toxins. Experiments have shown that pumping air *into* aquifers can be more effective

(a)

(b)

(c)

FIGURE 13.32 Different visions for restoring a trout stream. (a) Degraded by a century of agriculture and grazing, the stream had become too wide, shallow, and warm for native trout. (b) Trout Unlimited rebuilt a section of the stream to show their preferred option, which featured banks made of large stone blocks. (c) Other environmental organizations preferred a more organic approach. They used straw bales to narrow the stream and increase its current. This washed away the soft sediment and re-created the deep, narrow, cool, deeply shaded channel that favored native trout.

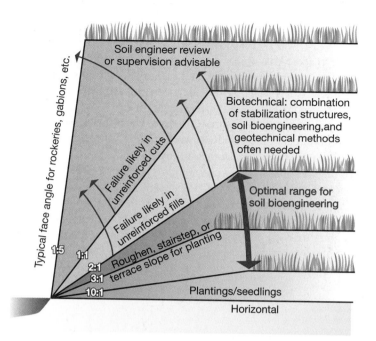

FIGURE 13.33 Steep streambanks need to be reinforced or recontoured to avoid erosion. Shallow slopes can be stabilized with vegetation or mulch. Steeper slopes need stabilization structures or reinforcement to hold the soil.

FIGURE 13.34 The Berkely mine pit in Butte, Montana, may be the most toxic water body in the United States. Water entering the pit is now being treated and eventually there may be an effort to pump water out of the pit and decontaminate it, but the pit itself will probably never be filled in.

than pumping water *out* for treatment. For hostile environments or exotic, human-made chemicals that can't be metabolized by normal organisms, it's sometimes possible to genetically engineer new varieties of bacteria that can survive in extreme conditions and consume materials that would kill ordinary species.

Many cities are finding that decontaminating urban "brown fields" (abandoned, contaminated industrial sites) can turn unusable inner-city property into valuable assets. This is a good way to control urban sprawl and make use of existing infrastructure. Cleaning up hazardous and toxic wastes is now a big business in America, and probably will continue to be so for a long time in the future. This is a growth industry in places where most other industry is disappearing.

Reclamation implies using intense physical or chemical methods to clean and repair severely degraded or even totally barren sites. Historically, reclamation meant irrigation projects that brought wetlands and deserts (considered useless wastelands) into agricultural production. Thus, the Bureau of Reclamation and the Army Corps of Engineers, in the early part of this century dredged, diked, drained, and provided irrigation water to convert millions of acres of wild lands into farm fields. Many of those projects were highly destructive to natural ecosystems. Ironically, we are now using ecological restoration to restore some of these "reclaimed" lands to a more natural state.

Today, reclamation means the repairing of human-damaged land. The Surface Mining Control and Reclamation Act (SMCRA), for example, requires mine operators to restore the shape of the land to its original contour and revegetate it to minimize impacts on local surface and groundwaters. According to the U.S. Office of Surface Mining, more than 8,000 km^2 (3,000 mi^2) of former strip mines have been reclaimed and returned to beneficial uses, such as recreation areas, farming and rangeland, wetlands, wildlife refuges, and sites for facilities such as hospitals, shopping centers, schools, and office and industrial parks.

Ideally, if topsoil is set aside during surface mining, overburden and tailings (waste rock discarded during mining and ore-enrichment) could be returned to the pit, smoothed out, and recovered with good soil that will support healthy vegetation. Unfortunately, topsoil is often buried deeply during mining, and what ends up on the surface is crushed rock that won't revegetate very well without a great deal of fertilizer and water.

The largest mine pits will never be returned to their original contour. Figure 13.34 shows a view of the Berkely mine pit in Butte, Montana. From the 1860s to the 1980s, this area was one of the world's richest sources of metals, including copper, silver, lead, zinc, manganese, and gold. In the early days all mining was in deep shafts. In 1955, the Anaconda Mining Company switched to open-pit mining and dug the hole you see now. After mining ended in 1981, groundwater, previously controlled with pumps, began to fill the pit. It has now accumulated to make a lake 1.6 km wide and 300 m deep.

The water has a pH of 2.5 (about the same as vinegar) and is laden with heavy metals and toxic chemicals, such as arsenic, cadmium, zinc, and sulfuric acid. In 1995, a whole flock of migrating snow geese were found dead after landing by mistake in the pit. The pit is now a Superfund site. A water treatment plant has been built on the far shore of the pit (you can see a thin, white stream of treated water from the plant cascading into the lake). By 2018, or whenever the water level hits the critical elevation of 1,649 m above sea level (the height at which it will threaten groundwater used by the city of Butte), the plant will begin to treat the contents of the pit. This is considered a reclamation project, but it's highly unlikely that the pit will ever be filled in.

CONCLUSION

Humans have caused massive damage and degradation to a wide variety of biological communities, but there are many ways to repair this damage and to restore or rehabilitate nature. Ideally, we might prefer to return a site to its pristine, predisturbance condition, but that often isn't possible. A more pragmatic goal is simply to develop a useful, stable, self-sustaining ecosystem with as many of its original ecological elements as possible. Sometimes it's enough to leave nature alone to heal itself, but often we need to intervene in some way to remove or discourage unwanted organisms while also promoting the growth of more desirable species.

Restoration pioneer Aldo Leopold wrote, "A thing is right when it tends to preserve the integrity, stability, and beauty of the biotic community. It is wrong when it tends otherwise." Some modern ecologists object that ecosystems are highly dynamic and species appear and disappear stochastically and individually. Characteristics such as integrity, health, stability, beauty, and moral responsibility tend to be human interpretations rather than scientific facts. Still, we need to have goals for restoration that the public can understand and accept.

Many ecological restoration projects are now underway. Some are huge efforts, such as rehabilitating Louisiana coastal wetlands, Florida's everglades, or the Chesapeake Bay, and returning a huge swath of shortgrass prairie to a buffalo commons. Others are much more modest: building a rain garden to trap polluted storm runoff, planting trees on empty urban land, or turning a part of your yard into a native prairie or woodland. Within this wide range, there are lots of opportunities for all of us to get involved.

REVIEWING LEARNING OUTCOMES

By now you should be able to explain the following points:

13.1 Illustrate ways that we can help nature heal.
- Restoration projects range from modest to ambitious.
- Restoration ecologists tend to be idealistic but pragmatic.
- Restoration projects have common elements.
- Early conservationists showed the promise of restoration

13.2 Show how nature is resilient.
- Protection is the first step in restoration.
- Native species often need help to become reestablished.

13.3 Explain how restoring forests has benefits.
- Tree planting can improve our quality of life.
- Fire is essential for savannas.

13.4 Summarize plans to restore prairies.
- Fire is also crucial for prairie restoration.
- Huge areas of shortgrass prairie are being preserved.
- Bison help maintain prairies.

13.5 Compare approaches to restoring wetlands and streams.
- Reinstating water supplies helps wetlands heal.
- The Everglades are being replumbed.
- The Chesapeake Bay is being rehabilitated.
- Wetland mitigation can replace damaged areas.
- Constructed wetlands can filter water.
- Many streams need rebuilding.
- Severely degraded or polluted sites can be repaired or reconstructed.

PRACTICE QUIZ

1. Why have levees on the lower Mississippi River starved coastal wetlands of sediments?
2. Why does coastal wetland loss matter to New Orleans?
3. Define *ecological restoration.*
4. What's the difference between strict and broad restoration?
5. Give an example of letting nature heal itself.
6. Why is restoring savannas difficult?
7. Why are fires essential for prairies?
8. What is the buffalo commons?
9. Why does the Everglades need restoration?
10. What is wetland mitigation?

CRITICAL THINKING AND DISCUSSION QUESTIONS

1. Should we be trying to restore biological communities to what they were in the past, or modify them to be more compatible to what we anticipate future conditions might be? What future conditions would you consider most likely to be problematic?

2. How would you balance human preferences (aesthetics, utility, cost) with biological considerations (biodiversity, ecological authenticity, evolutionary potential)?

3. The case of The Nature Conservancy's Hassayampa River Preserve near Phoenix illustrates a situation in which there may be more than one historic condition to which we may wish to restore a landscape. How would you reconcile these different values and goals?

4. Restoring savannas often requires the use of herbicides to remove invasive species. Some people regard this as dangerous and unnatural; how would you respond?

5. The Nature Conservancy believes it's essential to keep productive ranches on the land both to sustain rural society and for effective protection of the range. The American Prairie Foundation (APF), is buying up ranches and converting them to wilderness where wild animals can roam freely. Which of these approaches would you favor?

6. Sometimes the quickest and easiest way to restore a stream is to simply to reconstruct it with heavy equipment. You might even create something more interesting and useful (at least from a human perspective) than the original. Is it okay to replace real nature with something synthetic?

DATA analysis   Concept Maps

Figure 13.2 on p. 279 is a graphic representation we haven't used very often in this book. It's a concept map, or a two-dimensional representation of the relationship between key ideas. It could also be considered a decision flow chart because it represents an organized presentation of different policy options. This kind of chart shows how we might think about a situation, and suggests affinities and associations that might not otherwise be obvious. You might like to look at the introductory chapter of this book for more information about concept maps.

1. Using one of the examples presented in this chapter—or another familiar example—replace the descriptions in the boxes with brief descriptions of an actual ecosystem. Does this help you see the relationship between different states for the community you've chosen?

2. Now replace the verbs associated with the arrows between boxes with actions that cause changes in your particular biological community as well as restoration treatments that could accomplish the proposed restoration outcomes.

3. What do you suppose the authors meant by a "threshold of irreversibility"? If the system is irreversible, why are there arrows for reallocation or intervention?

4. The box for simplified ecosystems has two subcategories for native and alien species. What does this mean? What would be some examples for the ecological community you've chosen?

5. There are two arrows labeled "reallocation" in this diagram. One leads to "new uses," while the other leads to "new ecosystem." What's the difference? Why are the two boxes separated in the diagram?

6. The arrow labeled "restoration (strict sense)" is a dotted line. What do you think the authors meant by this detail?

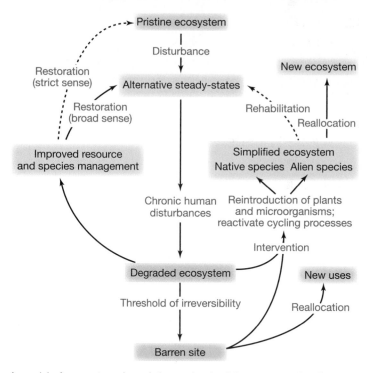

A model of ecosystem degradation and potential management options.
Data source: Walker and Moral, 2003.

Now closed and undergoing reclamation, the Beartrack Mine was an open-pit, cyanide heap-leach gold mine.

14
Geology and Earth Resources

Learn geology or die.

—Louis Agassiz—

LEARNING OUTCOMES

After studying this chapter, you should be able to:

14.1 Summarize the processes that shape the earth and its resources.

14.2 Explain how rocks and minerals are formed.

14.3 Think critically about economic geology and mineralogy.

14.4 Critique the environmental effects of resource extraction.

14.5 Discuss ways we could conserve geological resources.

14.6 Describe geological hazards.

Case Study Leaching gold

The Beartrack Mine is an open-pit, cyanide heap-leach gold mine (see photo opposite page) located within the Salmon National Forest of Idaho (fig. 14.1). The mine is owned and operated by the Meridian Gold Company. Mine construction began in 1994, and the first gold was produced in 1995. At its peak, Beartrack was the single largest gold mine in Idaho's history. In five years of operation, it produced 550,000 ounces of gold worth about $220 million at prevailing prices.

The ore in this deposit contained only about 1 g of gold per ton of rock. This wouldn't have been economical to extract using conventional underground mining and smelting, but with surface mining techniques and the cyanide leaching technology that became popular in the 1970s, gold was extracted from the Beartrack at a cost of about $190 per ounce. The mine consisted of two large open pits (upper right in photo), a 40-million-ton waste rock disposal site in a nearby valley, and heap-leaching facilities (center of photo). During operations, ore excavated from the pits was crushed and piled in 20 ft thick layers on the heap-leach pad, (large tan square), which is underlain by a thick plastic liner. The heap covers about 40 ha to a maximum depth of about 30 m, and contains about 15 million metric tons of ore.

Gold was separated from the ore by spraying a sodium cyanide solution over the top of the heap. This solution bonds with the gold in the ore, percolates through the heap, and drains through a ditch (lower right) into catch basins (large blue squares). The gold-bearing cyanide solution is pumped from the catch basins to a processing plant (white buildings) where the gold is absorbed by activated carbon in large tanks. When the carbon is saturated, the gold is removed by zinc precipitation. The "used" sodium cyanide is sprayed again on the pile (see fig. 14.14).

The mine closed in 2000 when falling gold prices made production uneconomical. Restoration began shortly thereafter and continues today. The reclamation goal is to restore the site to allow essentially the same land uses as existed prior to mining. Ultimately, this will require removal of buildings, impoundments, leach pads, and other mining structures, and the contouring, covering, and revegetation of waste rock piles and other disturbed areas. The photo on the opposite page was taken six years after reclamation began, and shows few changes from when mining ceased. One reason that reclamation is proceeding slowly is that about 4,000 ounces of gold per year continue to be rinsed out of the heap-leach pile.

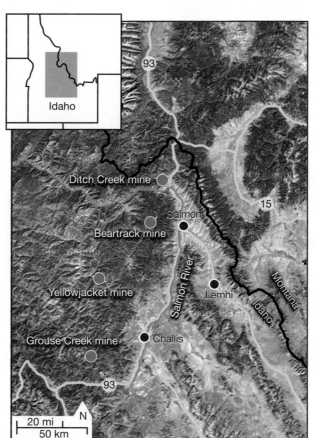

FIGURE 14.1 The Beartrack Mine is near the historic mining town of Leesburg in Lemhi County of east-central Idaho. The wild and scenic Salmon River runs nearby.

Nevertheless, in 2006, the Bureau of Land Management awarded Meridian Gold the Hardrock Mineral Environmental Award for being an "outstanding example of a mining project with strong community relations, recognizable achievements in reclamation of the site, and use of cutting-edge technology to bring an environmentally sound closure to a mining project." At the same time the Environmental Protection Agency fined Meridian $30,000 for violating its wastewater discharge permit at Beartrack.

Some Idaho residents are understandably concerned about what will happen to the Beartrack Mine, because wastewater from the facility is discharged into Napias Creek, which empties into the Salmon River. Known as the "River of No Return" from the adventures of Lewis and Clark, the Salmon is one of the premier trout streams—and one of the initial seven Wild and Scenic Rivers—of the United States. Every year, thousands of anglers, backpackers, and white water rafters visit the area and make up an important part of Idaho's $2 billion annual tourism industry.

Cyanide heap-leach operations in other parts of the world have had poor environmental records. At about the same time that Beartrack closed, a cyanide leak at a similar facility not far away spilled about 20,000 gallons of cyanide into the Yankee Fork of the Salmon River. The mine owner charged that vandals must have cut the plastic liner. Others believe that ice tore the plastic. Some engineers believe all heap-leach liners eventually tear and leak. Despite industry attempts to block it, a citizen's initiative to phase out open-pit, cyanide-leach mining passed in 1998 in Montana and is now state law.

This case study illustrates some of the dilemmas we face in using geological resources. We need materials from the earth to sustain our modern lifestyle, but often the methods we use to get those materials have severe environmental consequences. How can we supply the resources we need without harming our environment? In this chapter, we'll look at the processes that shape the earth and how rocks and minerals are formed, as well as how we obtain minerals. We'll also look at geological hazards that threaten us and what we might do to reduce our risks.

14.1 EARTH PROCESSES SHAPE OUR RESOURCES

Every one of us shares the benefits of geological resources. Right now you are probably wearing several geologic products; plastics, including glasses and synthetic fabric, are made from oil; iron, copper, and aluminum mines produced your snaps, zippers, and the screws in your glasses; silver, gold, and diamond mines may have produced your jewelry. All of us also share responsibility for the environmental and social devastation that often results from mining and drilling. Fortunately, there are many promising solutions to reduce these costs, including recycling and alternative materials. The question is whether voters will demand that we use these technologies—and whether consumers will share the costs of responsible production.

Why are these resources distributed as they are? To understand how and where earth resources are created, we must examine the earth's structure and the processes that shape it.

Earth is a dynamic planet

Although we think of the ground under our feet as solid and stable, the earth is a dynamic and constantly changing structure. Titanic forces inside the earth cause continents to split, move apart, and then crash into each other in slow but inexorable collisions.

The earth is a layered sphere. The **core,** or interior, is composed of a dense, intensely hot mass of metal—mostly iron—thousands of kilometers in diameter (fig. 14.2). Solid in the center but more fluid in the outer core, this immense mass generates the magnetic field that envelops the earth.

Surrounding the molten outer core is a hot, pliable layer of rock called the **mantle.** The mantle is much less dense than the core because it contains a high concentration of lighter elements, such as oxygen, silicon, and magnesium.

The outermost layer of the earth is the cool, lightweight, brittle rock **crust.** The crust below oceans is relatively thin (8–15 km), dense, and young (less than 200 million years old) because of constant recycling. Crust under continents is relatively thick (25–75 km), light, and as old as 3.8 billion years, with new material being added continually. It also is predominantly granitic, while oceanic crust is mainly dense basaltic rock. Table 14.1 compares the composition of the whole earth (dominated by the dense core) and the crust.

Tectonic processes reshape continents and cause earthquakes

The huge convection currents in the mantle are thought to break the overlying crust into a mosaic of huge blocks called **tectonic plates** (fig. 14.3). These plates slide slowly across the earth's surface like wind-driven ice sheets on water, in some places breaking up into smaller pieces, in other places crashing ponderously into each other to create new, larger landmasses. Ocean basins form where continents crack and pull apart. The Atlantic Ocean, for example, is growing slowly as Europe and Africa move away from the Americas. **Magma** (molten rock) forced

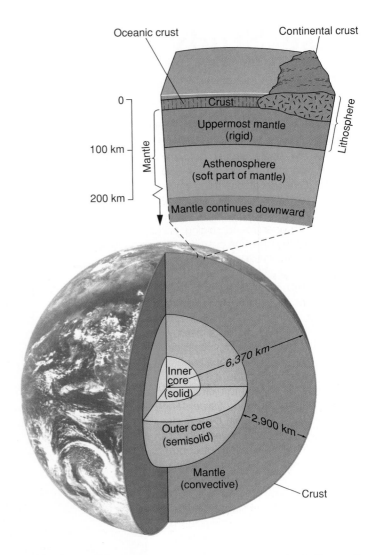

FIGURE 14.2 Earth's cross-section. Slow convection in the mantle causes the thin, brittle crust to move.

TABLE 14.1

Eight Most Common Chemical Elements (Percent)

Whole Earth		Crust	
Iron	33.3	Oxygen	45.2
Oxygen	29.8	Silicon	27.2
Silicon	15.6	Aluminum	8.2
Magnesium	13.9	Iron	5.8
Nickel	2.0	Calcium	5.1
Calcium	1.8	Magnesium	2.8
Aluminum	1.5	Sodium	2.3
Sodium	0.2	Potassium	1.7

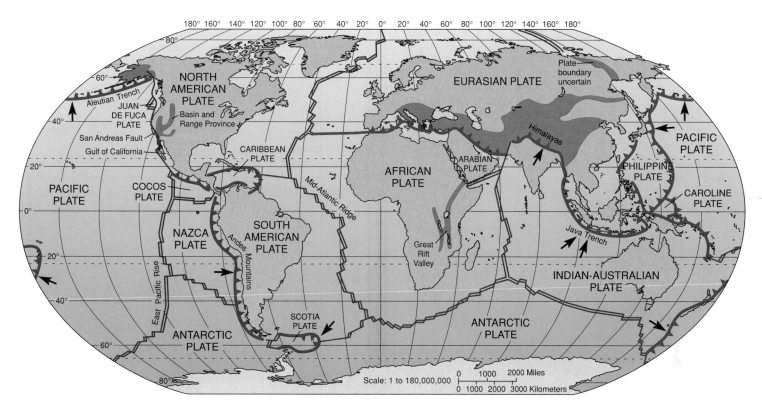

FIGURE 14.3 Map of tectonic plates. Plate boundaries are dynamic zones, characterized by earthquakes and volcanism and the formation of great rifts and mountain ranges. Arrows indicate direction of subduction where one plate is diving beneath another. These zones are sites of deep trenches in the ocean floor and high levels of seismic and volcanic activity.

up through the cracks forms new oceanic crust that piles up underwater in **mid-ocean ridges.** Creating the largest mountain range in the world, these ridges wind around the earth for 74,000 km (46,000 mi) (see fig. 14.3). Although concealed from our view, this jagged range boasts higher peaks, deeper canyons, and sheerer cliffs than any continental mountains. Slowly spreading from these fracture zones, ocean plates push against continental plates.

Earthquakes are caused by grinding and jerking as plates slide past each other. Mountain ranges like those on the west coast of North America and in Japan are pushed up at the margins of colliding continental plates. The Himalayas are still rising as the Indian subcontinent grinds slowly into Asia. Southern California is slowly sailing north toward Alaska. In about 30 million years, Los Angeles will pass San Francisco, if both still exist by then.

When an oceanic plate collides with a continental landmass, the continental plate usually rides up over the seafloor, while the oceanic plate is **subducted,** or pushed down into the mantle, where it

melts and rises back to the surface as magma (fig. 14.4). Deep ocean trenches mark these subduction zones, and volcanoes form where the magma erupts through vents and fissures in the

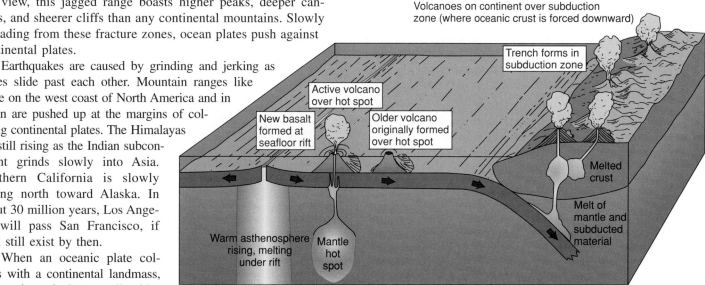

FIGURE 14.4 Tectonic plate movement. Where thin, oceanic plates diverge, upwelling magma forms mid-ocean ridges. A chain of volcanoes, like the Hawaiian Islands, may form as plates pass over a "hot spot." Where plates converge, melting can cause volcanoes, such as the Cascades.

FIGURE 14.5 Pangaea, an ancient supercontinent of 200 million years ago, combined all the world's continents in a single landmass. Continents have combined and separated repeatedly.

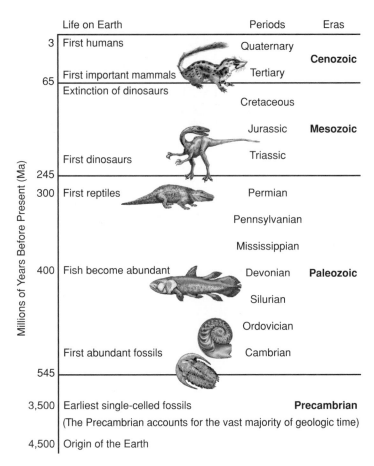

Life on Earth		Periods	Eras
3	First humans	Quaternary	**Cenozoic**
65	First important mammals	Tertiary	
	Extinction of dinosaurs	Cretaceous	
		Jurassic	**Mesozoic**
245	First dinosaurs	Triassic	
300	First reptiles	Permian	**Paleozoic**
		Pennsylvanian	
		Mississippian	
400	Fish become abundant	Devonian	
		Silurian	
		Ordovician	
	First abundant fossils	Cambrian	
545			
3,500	Earliest single-celled fossils		**Precambrian**
	(The Precambrian accounts for the vast majority of geologic time)		
4,500	Origin of the Earth		

Millions of Years Before Present (Ma)

FIGURE 14.6 Periods and eras in geological time, and major life-forms that mark some periods.

overlying crust. Trenches and volcanic mountains ring the Pacific Ocean rim from Indonesia to Japan to Alaska and down the west coast of the Americas, forming a so-called ring of fire where oceanic plates are being subducted under the continental plates. This ring is the source of more earthquakes and volcanic activity than any other region on the earth.

Over millions of years, continents can drift long distances. Antarctica and Australia once were connected to Africa, for instance, somewhere near the equator and supported luxuriant forests. Geologists suggest that several times in the earth's history most or all of the continents have gathered to form supercontinents, which have ruptured and re-formed over hundreds of millions of years (fig. 14.5). The redistribution of continents has profound effects on the earth's climate and may help explain the periodic mass extinctions of organisms marking the divisions between many major geologic periods (fig. 14.6).

14.2 ROCKS AND MINERALS

A **mineral** is a naturally occurring, inorganic, solid element or compound with a definite chemical composition and a regular internal crystal structure. Naturally occurring means not created by humans (or synthetic). Organic materials, such as coal, produced by living organisms or biological processes are generally not minerals. The two fundamental characteristics of a mineral that distinguish it from all other minerals are its chemical composition and its crystal structure. No two minerals are identical in both respects. Once purified, metals such as iron, aluminum,

or copper lack a crystal structure, and thus are not minerals. The ores from which they are extracted, however, are minerals and make up an important part of economic mineralogy.

A **rock** is a solid, cohesive, aggregate of one or more minerals. Within the rock, individual mineral crystals (or grains) are mixed together and held firmly in a solid mass (fig. 14.7). The grains may be large or small, depending on how the rock was

FIGURE 14.7 Crystals of different minerals create beautifully colored patterns in a rock sample seen in a polarizing microscope.

formed, but each grain retains its own unique mineral qualities. Each rock type has a characteristic mixture of minerals (and therefore of different chemical elements), grain sizes, and ways in which the grains are mixed and held together. Granite, for example, is a mixture of quartz, feldspar, and mica crystals. Different kinds of granite have distinct percentages of these minerals and particular grain sizes depending on how quickly the rock solidified. These minerals, in turn, are made up of a few elements such as silicon, oxygen, potassium, and aluminum.

The rock cycle creates and recycles rocks

What could be harder and more permanent than rocks? Like the continents they create, rocks are also part of a relentless cycle of formation and destruction. They are made and then torn apart, cemented together by chemical and physical forces, crushed, folded, melted, and recrystallized by dynamic processes related to those that shape the large-scale features of the crust. We call this cycle of creation, destruction, and metamorphosis the **rock cycle** (fig. 14.8). Understanding something of how this cycle works helps explain the origin and characteristics of different types of rocks, as well as how they are shaped, worn away, transported, deposited, and altered by geological forces.

There are three major rock classifications: igneous, sedimentary, and metamorphic. In this section, we will look at how they are made and some of their properties.

Igneous Rocks

The most common rock-type in the earth's crust is solidified from magma, welling up from the earth's interior. These rocks are classed as **igneous rocks** (from *igni,* the Latin word for fire). Magma

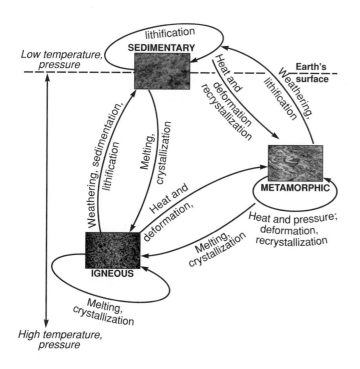

FIGURE 14.8 The rock cycle includes a variety of geological processes that can transform any rock.

FIGURE 14.9 Weathering slowly reduces an igneous rock to loose sediment. Here, exposure to moisture expands minerals in the rock, and frost may also force the rock apart.

extruded to the surface from volcanic vents cools quickly to make basalt, rhyolite, andesite, and other fine-grained rocks. Magma that cools slowly in subsurface chambers or is intruded between overlying layers makes granite, gabbro, or other coarse-grained crystalline rocks, depending on its specific chemical composition.

Metamorphic Rocks

Preexisting rocks can be modified by heat, pressure, and chemical agents to create new forms called **metamorphic rock.** Deeply buried strata of igneous, sedimentary, and metamorphic rocks are subjected to great heat and pressure by deposition of overlying sediments or while they are being squeezed and folded by tectonic processes. Chemical reactions can alter both the composition and structure of the rocks as they are metamorphosed. Some common metamorphic rocks are marble (from limestone), quartzite (from sandstone), and slate (from mudstone and shale). Metamorphic rocks are often the host rock for economically important minerals such as talc, graphite, and gemstones.

Weathering and sedimentation wear down rocks

Most of these crystalline rocks are extremely hard and durable, but exposure to air, water, changing temperatures, and reactive chemical agents slowly breaks them down in a process called **weathering** (fig. 14.9). *Mechanical weathering* is the physical breakup of rocks into smaller particles without a change in chemical composition of the constituent minerals. You have probably seen mountain valleys scraped by glaciers or river and shoreline pebbles that are rounded from being rubbed against one another as they are tumbled by waves and currents. *Chemical weathering* is the selective removal or alteration of specific components that leads to weakening and disintegration of rock. Among the more important chemical weathering processes are oxidation (combination of oxygen with an element to form an oxide or hydroxide mineral) and hydrolysis (hydrogen atoms from water molecules combine with other chemicals to form acids). The products of these reactions are more susceptible to both mechanical weathering and to dissolving in water. For instance, when carbonic acid

FIGURE 14.10 Different colors of soft sedimentary rocks deposited in ancient seas during the Tertiary period 63 to 40 million years ago have been carved by erosion into the fluted spires and hoodoos of the Pink Cliffs of Bryce Canyon National Park.

(formed when CO_2 and H_2O combine) percolates through porous limestone layers in the ground, it dissolves the calcium carbonate (limestone) and creates caves.

Particles of rock are transported by wind, water, ice, and gravity until they come to rest again in a new location. The deposition of these materials is called **sedimentation.** Waterborne particles from sediments cover ocean continental shelves and fill valleys and plains. Most of the American Midwest, for instance, is covered with a layer of sedimentary material hundreds of meters thick in the form of glacierborne till (rock debris deposited by glacial ice), windborne loess (fine dust deposits), riverborne sand and gravel, and ocean deposits of sand, silt, and clay. Deposited material that remains in place long enough, or is covered with enough material to compact it, may once again become rock. Some examples of **sedimentary rock** are shale (compacted mud), sandstone (cemented sand), tuff (volcanic ash), and conglomerates (aggregates of gravel, sand, silt, and clay).

Sedimentary rocks are also formed from crystals that precipitate out of, or grow from a solution. An example is rock salt, made of the mineral halite, which is the name for ordinary table salt (sodium chloride). Salt deposits often form when a body of salt water dries up and salt crystals are left behind.

Sedimentary formations often have distinctive layers that show different conditions when they were laid down. Sedimentary rocks such as sandstone and limestone can be shaped by erosion into striking features (fig. 14.10). Geomorphology is the study of the processes that shape the earth's surface and the structures they create.

Humans have become a major force in shaping landscapes. Geomorphologist Roger Hooke, of the University of Maine, looking only at housing excavations, road building, and mineral production, estimates that we move somewhere around 30 to 35 gigatons (billion tons or Gt) per year worldwide. When combined with the 10 Gt each year that we add to river sediments through erosion, our earth-moving prowess is comparable to, or greater than, any other single geomorphic agent except plate tectonics.

14.3 ECONOMIC GEOLOGY AND MINERALOGY

Economic mineralogy is the study of minerals that are valuable for manufacturing and are, therefore, an important part of domestic and international commerce. Most economic minerals are metal-bearing ores, such as the gold-bearing deposit at the Beartrack Mine (table 14.2). Nonmetallic economic minerals are mostly graphite, some feldspars, quartz crystals, diamonds, and many other crystals that are valued for their usefulness, beauty, and/or rarity. Metals have been so important in human affairs that major epochs of human history are commonly known by the dominant materials and the technology to use them (Stone Age, Bronze Age, Iron Age, etc.). The mining, processing, and distribution of these materials have broad and varied implications both for culture and our environment. Most economically valuable crustal resources exist everywhere in small amounts; the important thing is to find them concentrated in economically recoverable levels.

Public policy in the United States has encouraged mining on public lands as a way of boosting the economy and utilizing natural resources. Today many people think these laws seem outmoded and in need of reform (What Do You Think? p. 309).

Metals are essential to our economy

How has the quest for minerals and metals affected global development? We will focus first on world use of metals, earth resources that always have received a great deal of human attention. The availability of metals and the methods to extract and use them have determined technological developments, as

TABLE 14.2

Primary Uses of Some Major Metals Consumed in the United States

Metal	Use
Aluminum	Packaging foods and beverages (38%), transportation, electronics
Chromium	High-strength steel alloys
Copper	Building construction, electric and electronic industries
Iron	Heavy machinery, steel production
Lead	Leaded gasoline, car batteries, paints, ammunition
Manganese	High-strength, heat-resistant steel alloys
Nickel	Chemical industry, steel alloys
Platinum-group	Automobile catalytic converters, electronics, medical uses
Gold	Medical, aerospace, electronic uses; accumulation as monetary standard
Silver	Photography, electronics, jewelry

What Do You Think?

Should We Revise Mining Laws?

In 1872, the U.S. Congress passed the General Mining Law intended to encourage prospectors to open up the public domain and promote commerce. This law, which has been in effect more than a century, allows miners to stake an exclusive claim anywhere on public lands and to take—for free—any minerals they find. Three-forths of the Beartrack Mine, for example, is on public land. Claim holders can "patent" (buy) the land for $2.50 to $5 per acre (0.4 hectares) depending on the type of claim. Once the patent fee is paid, the owners can do anything they want with the land, just like any other private property. Although $2.50 per acre may have been a fair market value in 1872, many people regard it as ridiculously low today, amounting to a scandalous give-away of public property.

In Nevada, for example, a mining company paid $9,000 for federal land that contains an estimated $20 billion worth of precious metals. Similarly, Colorado investors bought about 7,000 ha (17,000 acres) of rich oil-shale land in 1986 for $42,000 and sold it a month later for $37 million. You don't actually have to find any minerals to patent a claim. A Colorado company paid a total of $400 for 65 ha (160 acres) it claimed would be a gold mine. Almost 20 years later, no mining has been done, but the property—which just happens to border the Keystone Ski Area—is being subdivided for condos and vacation homes.

According to the Bureau of Land Management (BLM), some $4 billion in minerals are mined each year on U.S. public lands. Under the 1872 law, mining companies don't pay a penny for the ores they take. Furthermore, they can deduct a depletion allowance from taxes on mineral profits. Former Senator Dale Bumpers of Arkansas, who

Thousands of abandoned mines on public lands poison streams and groundwater with acid, metal-laced drainage. This old mine in Montana drains into the Blackfoot River, the setting of Norman Maclean's book. *A River Runs Through It.*

calls the 1872 mining law "a license to steal," has estimated that the government could derive $320 million per year by charging an 8 percent royalty on all minerals and probably could save an equal amount by requiring larger bonds to be posted to clean up after mining is finished. The Meridian Gold Company, for example, has posted a $2 million bond for cleaning up the Beartrack Mine (a larger than normal amount). Reclamation, however, is expected to cost 15 times that amount. Chapter 13 has more information on how reclamation and restoration can return damaged sites to beneficial uses.

On the other hand, mining companies argue they would be forced to close down if they had to pay royalties or post larger bonds. Many people would lose jobs and the economies of western mining towns would collapse if mining becomes uneconomic. We provide subsidies and economic incentives to many industries to stimulate economic growth. Why not mining for metals essential for our industrial economy?

Mining is a risky and expensive business. Without subsidies, mines would close down and we would be completely dependent on unstable foreign supplies.

Mining critics respond that other resource-based industries have been forced to pay royalties on materials they extract from public lands. Coal, oil, and gas companies pay 12.5 percent royalties on fossil fuels obtained from public lands. Timber companies—although they don't pay the full costs of the trees they take—have to bid on logging sales and clean up when they are finished. Even gravel companies pay for digging up the public domain. Ironically, we charge for digging up gravel, but give gold away free.

Over the past decade, numerous mining bills have been introduced in Congress. Those supported by environmental groups generally would require companies mining on federal lands to pay a higher royalty on their production. They also would eliminate the patenting process, impose stricter reclamation requirements, and give federal managers authority to deny inappropriate permits. In contrast, bills offered by Western legislators, and enthusiastically backed by mining supporters, tend to leave most provisions of the 1872 bill in place. They would charge a 2 percent royalty, but only after exploration, production, and other costs were deducted. Permit processes would consider local economic needs before environmental issues in this version. What do you think we should do about this mining law? How could we separate legitimate public-interest land use from private speculation and profiteering? Are current subsidies necessary and justifiable or are they just a form of corporate welfare?

well as economic and political power for individuals and nations. We still are strongly dependent on the unique lightness, strength, and malleability of metals.

The metals consumed in greatest quantity by world industry include iron (740 million metric tons annually), aluminum (40 million metric tons), manganese (22.4 million metric tons), copper and chromium (8 million metric tons each), and nickel (0.7 million metric tons). Most of these metals are consumed in the United States, Japan, and Europe, in that order. They are produced in many countries, often in mountainous areas, where heat and

pressure of mountain building tend to concentrate ores (fig. 14.11). The worldwide mineral trade has become crucially important to the economic and social stability of all nations involved.

Nonmetal minerals are a broad class that covers resources from silicate minerals (gemstones, mica, talc, and asbestos) to sand, gravel, salts, limestone, and soils. Durable, highly valuable, and easily portable, gem stones and precious metals have long been a way to store and transport wealth. Unfortunately, these valuable materials also have bankrolled despots, criminal gangs, and terrorism in many countries. In recent years, brutal civil wars

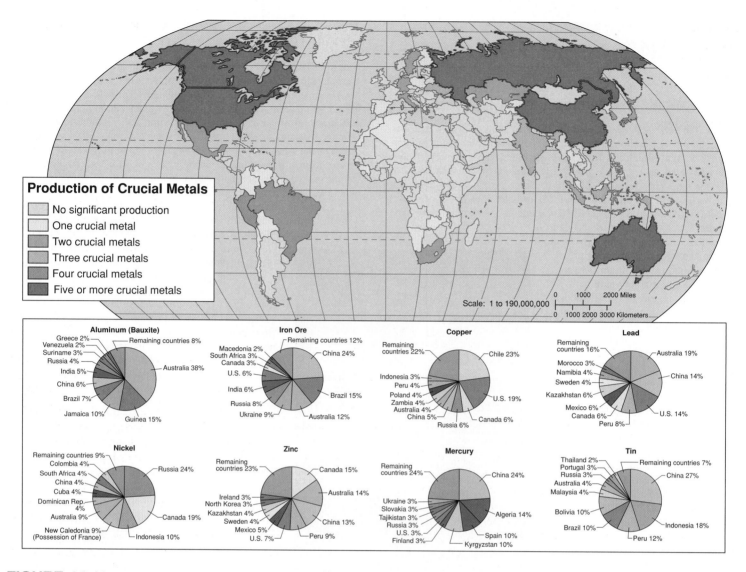

FIGURE 14.11 World production of metals most essential for an industrial economy. Principal consumers are the United States, Western Europe, Japan, and China.

in Africa have been financed—and often motivated by—gold, diamonds, tantalum ore, and other high-priced commodities. Much of this illegal trade ends up in the $100-billion-per-year global jewelry trade, two-thirds of which sells in the United States. Many people who treasure a diamond ring or a gold wedding band as a symbol of love and devotion are unaware that it may have been obtained through inhumane labor conditions and environmentally destructive mining and processing methods. Civil rights organizations are campaigning to require better documentation of the origins of gems and precious metals to prevent their use as financing for crimes against humanity (see "Conflict Diamonds" at www.mhhe.com/environmentalscience).

In 2004, a group of Nobel Peace Laureates called on the World Bank to overhaul its policies on lending for resource extractive industries. "War, poverty, climate change, and ongoing violations of human rights—all of these scourges are all too often linked to the oil and mining industries" wrote Archbishop Desmond Tutu,

winner of the 1984 Nobel Peace Prize for helping to eliminate apartheid in South Africa. In response, the World Bank appointed an Extractive Industries Review headed by former Indonesian environment minister, Emil Salim. In its final report, the committee recommended that some areas of exceptionally high biodiversity value should be "no-go" zones for extractive industries, and that the rights of those affected by extractive projects need better protection.

Nonmetal minerals include gravel, clay, sand, and salts

You might be surprised to learn that sand and gravel production comprise by far the greatest volume and dollar value of all nonmetal mineral resources and a far greater volume than all metal ores. Sand and gravel are used mainly in brick and concrete construction, paving, as loose road filler, and for sandblasting. High-purity silica sand is our source of glass. These materials usually are retrieved from surface pit mines and quarries, where they have been deposited by glaciers, winds, or ancient oceans.

Limestone, like sand and gravel, is mined and quarried for concrete and crushed for road rock. It also is cut for building stone, pulverized for use as an agricultural soil additive that neutralizes acidic soil, and roasted in lime kilns and cement plants to make plaster (hydrated lime) and cement.

Evaporites (materials deposited by evaporation of chemical solutions) are mined for halite, gypsum, and potash. These are often found at or above 97 percent purity. Halite, or rock salt, is used for water softening and melting ice on winter roads in some northern areas. Refined, it is a source of table salt. Gypsum (calcium sulfate) now makes our plaster wallboard, but it has been used for plaster ever since the Egyptians plastered the walls of their frescoed tombs along the Nile River some 5,000 years ago. Potash is an evaporite composed of a variety of potassium chlorides and potassium sulfates. These highly soluble potassium salts have long been used as a soil fertilizer.

Sulfur deposits are mined mainly for sulfuric acid production. In the United States, sulfuric acid use amounts to more than 200 lbs per person per year, mostly because of its use in industry, car batteries, and some medicinal products.

14.4 ENVIRONMENTAL EFFECTS OF RESOURCE EXTRACTION

Each of us depends daily on geologic resources mined from sites around the world. We use scores of metals and minerals, many of which we've never even heard of, in our lights, computers, watches, fertilizers, and cars. Mining and purifying all these resources can have severe environmental and social consequences. The most obvious effect of mining is often the disturbance or removal of the land surface. Farther-reaching effects, though, include air and water pollution. The EPA lists more than 100 toxic air pollutants, from acetone to xylene, released from U.S. mines every year. Nearly 80,000 metric tons of particulate matter (dust) and 11,000 tons of sulfur dioxide are released from nonmetal mining alone. Chemical- and sediment-runoff pollution is a major problem in many local watersheds.

Resource extraction can affect water quality for several reasons. Gold and other metals are often found in sulfide ores, as is the case at the Beartrack Mine. Sulfide minerals often occur in hydrothermal (hot water) deposits, where sulfur and other elements are mobilized, then deposited as they are heated by deep magma chambers. When sulfur-bearing minerals are exposed to air and water, they produce sulfuric acid, which is highly mobile and strongly acidic. In addition, metal elements often occur in very low concentrations—10 to 20 parts per billion may be economically extractable for gold, platinum, and other metals. Consequently, vast quantities of ore must be crushed and washed to extract metals, as the opening case study shows. A great deal of water is used in cyanide heap-leaching and other washing techniques. In arid Nevada, the USGS estimates that mining consumes about 230,000 m^3 (60 million gal) of fresh water per day. After use in ore processing, much of this water contains sulfuric acid, arsenic, heavy metals, and other contaminants. Mine runoff leaking into lakes and streams damages or destroys aquatic ecosystems.

Mining can have serious environmental impacts

There are many techniques for extracting geological materials. The most common methods are open-pit mining, strip mining, and underground mining. An ancient method of accumulating gold, diamonds, and coal is placer mining, in which pure nuggets are washed from stream sediments. Since the California gold rush of 1849, placer miners have used water cannons to blast away hillsides. This method, which chokes stream ecosystems with sediment, is still used in Alaska, Canada, and many other regions. Another ancient, and much more dangerous, method is underground mining. Ancient Roman, European, and Chinese miners tunneled deep into tin, lead, copper, coal, and other mineral seams. Mine tunnels occasionally collapse, and natural gas in coal mines can explode. Water seeping into mine shafts also dissolves toxic minerals. Contaminated water seeps into groundwater; it is also pumped to the surface, where it enters streams and lakes.

In underground coal mines, another major environmental risk is fires. Hundreds of coal mines smolder in the United States, China, Russia, India, South Africa, and Europe. The inaccessibility and size of these fires make many impossible to extinguish or control. One mine fire in Centralia, Pennsylvania, has been burning since 1962; control efforts have cost at least $40 million, but the fire continues to expand. China, which depends on coal for much of its heating and electricity, has hundreds of smoldering mine fires; one has been burning for 400 years. According to a recent study from the International Institute for Aerospace Survey in the Netherlands, these fires consume up to 200 million tons of coal every year and emit as much carbon dioxide as all the cars in the United States. Toxic fumes, explosive methane, and other hazardous emissions are also released from these fires.

Open-pit mines are used to extract massive beds of metal ores and other minerals. The size of modern open pits can be hard to comprehend. The Bingham Canyon mine, near Salt Lake City, Utah, is 800 m (2,640 ft) deep and nearly 4 km (2.5 mi) wide at the top. More than 5 billion tons of copper ore and waste material have been removed from the hole since 1906. A chief environmental challenge of open-pit mining is that groundwater accumulates in the pit. In metal mines, a toxic soup results. No one yet knows how to detoxify these lakes, which endanger wildlife and nearby watersheds.

Half the coal used in the United States comes from surface or strip mines (fig. 14.12). Since coal is often found in expansive, horizontal beds, the entire land surface can be stripped away to cheaply and quickly expose the coal. The overburden, or surface material, is placed back into the mine, but usually in long ridges called spoil banks. Spoil banks are very susceptible to erosion and chemical weathering. Since the spoil banks have no topsoil (the complex organic mixture that supports vegetation—see chapter 9), revegetation occurs very slowly.

The 1977 federal Surface Mining Control and Reclamation Act (SMCRA) requires better restoration of strip-mined lands,

FIGURE 14.12 Some giant mining machines stand as tall as a 20-story building and can scoop up thousands of cubic meters of rock per hour.

FIGURE 14.13 Mountaintop removal mining is a relatively new, and deeply controversial, method of extracting Appalachian coal.

especially where mines replaced prime farmland. Since then, the record of strip-mine reclamation has improved substantially. Complete mine restoration is expensive, often more than $10,000 per hectare. Restoration is also difficult because the developing soil is usually acidic and compacted by the heavy machinery used to reshape the land surface.

Bitter controversy has grown recently over mountaintop removal, a coal mining method mainly practiced in Appalachia. Long, sinuous ridge-tops are removed by giant, 20-story-tall shovels to expose horizontal beds of coal (fig. 14.13). Up to 215 m (700 ft) of ridge-top is pulverized and dumped into adjacent river valleys. The debris can be laden with selenium, arsenic, coal, and other toxic substances. At least 900 km (560 mi) of streams have been buried in West Virginia alone. Environmental lawyers have sued to stop the destruction of streams, arguing that it violates the Clean Water Act (chapter 24). In response, the Bush administration issued a "clarification" of the Clean Water Act, rewriting the law to allow stream filling by any sort of mining or industrial debris, rather than forbidding any stream destruction. Environmentalists charge that the new wording subverts the Clean Water Act, giving a green light to all sorts of stream destruction. West Virginians are deeply divided about mountaintop removal because they live in affected stream valleys but also depend on a coal economy.

The Mineral Policy Center in Washington, D.C., estimates that 19,000 km (12,000 mi) of rivers and streams in the United States are contaminated by mine drainage. The EPA estimates that cleaning up impaired streams, along with 550,000 abandoned mines in the U.S., may cost $70 billion. Worldwide, mine closing and rehabilitation costs are estimated in the trillions of dollars. Because of the volatile prices of metals and coal, many mining companies have gone bankrupt before restoring mine sites, leaving the public responsible for cleanup. In 2002, more than 500 leading mine executives and their critics convened at the Global Mining Initiative, a meeting in Toronto aimed at improving the sustainability of mining. Executives acknowledged that in the future they will

increasingly be held liable for environmental damages, and they said they were seeking ways to improve the industry's social and environmental record. Jay Hair, secretary general of the International Council on Mining and Metals, stated that "environmental protection and social responsibility are important," and that mining companies were interested in participating in sustainable development. Mine executives also recognized that, increasingly, big cleanup bills will cut into company values and stock prices. Finding creative ways to keep mines cleaner from the start will make good economic sense, even if it's not easy to do.

Processing ores also has negative consequences

Metals are extracted from ores by heating or with chemical solvents. Both processes release large quantities of toxic materials that can be even more environmentally hazardous than mining. **Smelting**—roasting ore to release metals—is a major source of air pollution. One of the most notorious examples of ecological devastation from smelting is a wasteland near Ducktown, Tennessee. In the mid-1800s, mining companies began excavating the rich copper deposits in the area. To extract copper from the ore, they built huge, open-air, wood fires, using timber from the surrounding forest. Dense clouds of sulfur dioxide released from sulfide ores poisoned the vegetation and acidified the soil over a 50 mi^2 (13,000 ha) area. Rains washed the soil off the denuded land, creating a barren moonscape.

Sulfur emissions from Ducktown smelters were reduced in 1907 after Georgia sued Tennessee over air pollution. In the 1930s the Tennessee Valley Authority (TVA) began treating the soil and replanting trees to cut down on erosion. Recently, upwards of $250,000 per year has been spent on this effort. While the trees and other plants are still spindly and feeble, more than two-thirds of the area is considered "adequately" covered with vegetation. Similarly, smelting of copper-nickel ore in Sudbury, Ontario, a century ago

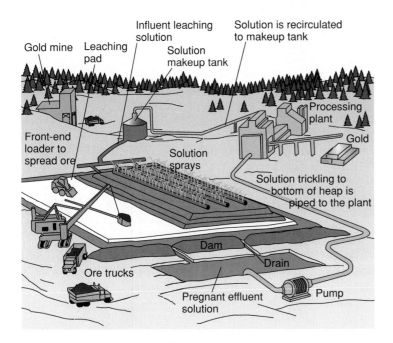

Gold mine
Leaching pad
Influent leaching solution
Solution makeup tank
Solution is recirculated to makeup tank
Processing plant
Gold
Front-end loader to spread ore
Solution sprays
Solution trickling to bottom of heap is piped to the plant
Dam
Drain
Ore trucks
Pregnant effluent solution
Pump

FIGURE 14.14 In a heap-leach operation, huge piles of low-grade ore are heaped on an impervious pad and sprayed with a cyanide solution. As the leaching solution trickles through the crushed ore, it extracts gold and other precious metals. This technique is highly profitable but carries large environmental risks.

TABLE 14.3

The World's Worst Polluted Places

Place	Pollution Source
Chernobyl, Ukraine	Nuclear power plant
Dzerzhinsk, Russia	Weapons production
Haina, Dominican Republic	Lead smelter
Kabwe, Zambia	Lead mine and smelter
La Oroya, Peru	Mine and smelter
Lifen, China	Coal mine and power plants
Maiuu Suu, Kyrgyzstan	Uranium mine
Norlisk, Russia	Copper-nickel mine and smelter
Ranipet, India	Tannery
Rudnava Pristan/Dalnegorsk, Russia	Lead mine

Source: Blacksmith Institute, 2006.

caused widespread ecological destruction that is slowly being repaired following pollution-control measures (see fig. 16.23).

As discussed in the opening case study for this chapter, chemical extraction methods have a high potential for environmental contamination. The cyanide solution sprayed on a heap-leach pile (fig. 14.14) can leak into surface or groundwater. A case in point is the Summitville Mine near Alamosa, Colorado. After extracting $98 million in gold, the absentee owners declared bankruptcy in 1992, abandoning millions of tons of mine waste and huge leaking ponds of cyanide. The Environmental Protection Agency may spend more than $100 million trying to clean up the mess and keep the cyanide pool from spilling into the Alamosa River.

Every year, the Blacksmith Institute compiles a list of the world's worst polluted places (table 14.3). For 2006, seven of the top ten worst places were mines and/or smelter facilities. These problems are especially disastrous in the developing world or in the former Soviet Union, where funds and political will aren't available to deal with pollution or help people suffering from terrible health effects of pollution.

14.5 CONSERVING GEOLOGICAL RESOURCES

Conservation offers great potential for extending our supplies of economic minerals and reducing the effects of mining and processing. The advantages of conservation are also significant: less waste to dispose of, less land lost to mining, and less consumption of money, energy, and water resources.

Recycling saves energy as well as materials

Some waste products already are being exploited, especially for scarce or valuable metals. Aluminum, for instance, must be extracted from bauxite ore by electrolysis, an expensive, energy-intensive process. Recycling waste aluminum, such as beverage cans, on the other hand, consumes one-twentieth of the energy of extracting new aluminum. Today, nearly two-thirds of all aluminum beverage cans in the United States are recycled, up from only 15 percent 20 years ago. The high value of aluminum scrap ($650 a ton versus $60 for steel, $200 for plastic, $50 for glass, and $30 for paperboard) gives consumers plenty of incentive to deliver their cans for collection. Recycling is so rapid and effective that half of all the aluminum cans now on a grocer's shelf will be made into another can within two months. The energy cost of extracting other materials is shown in table 14.4.

Platinum, the catalyst in automobile catalytic exhaust converters, is valuable enough to be regularly retrieved and recycled from used cars (fig. 14.15). Other metals commonly recycled are gold, silver, copper, lead, iron, and steel. The latter four are readily available in a pure and massive form, including copper pipes, lead batteries, and steel and iron auto parts. Gold and silver are valuable enough to warrant recovery, even through more difficult means. See chapter 21 for further discussion of this topic.

While total U.S. steel production has fallen in recent decades—largely because of inexpensive supplies from new and efficient Japanese steel mills—a new type of mill subsisting entirely on a readily available supply of scrap/waste steel and iron is a growing industry. Minimills, which remelt and reshape scrap iron and steel, are smaller and cheaper to operate than traditional integrated mills that perform every process from preparing raw ore to finishing iron and steel products. Minimills produce steel at between $225 and $480 per metric ton, while steel from integrated mills costs $1,425 to $2,250 per metric

TABLE 14.4

Energy Requirements in Producing Various Materials from Ore and Raw Source Materials

Product	Energy Requirement (Mj/Kg)	
	New	From Scrap
Glass	25	25
Steel	50	26
Plastics	162	n.a.
Aluminum	250	8
Titanium	400	n.a.
Copper	60	7
Paper	24	15

Source: E. T. Hayes, *Implications of Materials Processing,* 1997.

ton on average. The energy cost is likewise lower in minimills: 5.3 million BTU/ton of steel compared to 16.08 million BTU/ton in integrated mill furnaces. Minimills now produce about half of all of U.S. steel production. Recycling is slowly increasing as raw materials become more scarce and wastes become more plentiful.

New materials can replace mined resources

Mineral and metal consumption can be reduced by new materials or new technologies developed to replace traditional uses. This is a long-standing tradition; for example, bronze replaced stone technology and iron replaced bronze. More recently, the introduction of plastic pipe has decreased our consumption of copper, lead, and steel pipes. In the same way, the development of fiber-optic technology and satellite communication reduces the need for copper telephone wires.

Iron and steel have been the backbone of heavy industry, but we are now moving toward other materials. One of our primary uses for iron and steel has been machinery and vehicle parts. In automobile production, steel is being replaced by polymers (long-chain organic molecules similar to plastics), aluminum, ceramics, and new, high-technology alloys. All of these reduce vehicle weight and cost, while increasing fuel efficiency. Some of the newer alloys that combine steel with titanium, vanadium, or other metals wear much better than traditional steel. Ceramic engine parts provide heat insulation around pistons, bearings, and cylinders, keeping the rest of the engine cool and operating efficiently. Plastics and glass fiber-reinforced polymers are used in body parts and some engine components.

Electronics and communications (telephone) technology, once major consumers of copper and aluminum, now use ultra-high-purity glass cables to transmit pulses of light, instead of metal wires carrying electron pulses. Once again, this technology has been developed for its greater efficiency and lower cost, but it also affects consumption of our most basic metals.

FIGURE 14.15 The richest metal source we have—our mountains of scrapped cars—offers a rich, inexpensive, and ecologically beneficial resource that can be "mined" for a number of metals.

14.6 GEOLOGICAL HAZARDS

Earthquakes, volcanic eruptions, floods, and landslides are among the geological forces that have shaped the world around us (fig. 14.16). Catastrophic events, such as the impact of a giant asteroid 65 million years ago off the coast of what is now Yucatan, or volcanic eruptions, such as those that covered 2 million km^2 of Siberia with basalt up to 2 km deep 250 million years ago, are thought to have triggered the mass extinctions that mark transitions between major historic eras. The asteroid impact 65 million years ago that ended the age of the dinosaurs is calculated to have created a tsunami hundreds of meters high that could have swept around the world several times before subsiding. This impact also ejected so much dust into the air that sunlight was blocked for years and a global winter decimated much of the life on the earth. A similar impact 250 million years ago off the coast of Australia is thought to be related to the end of the Permian era, when 90 to 95 percent of all marine species perished. This asteroid impact may also have caused fissures in the earth's crust that allowed the vast outpouring of lava in Siberia. Together, these cataclysmic events probably filled the atmosphere with ash and sulfuric acid aerosols that plunged the earth into a massive deep freeze.

Fortunately, such massive events are rare. Still, as the devastating 2004 Sumatran tsunami shows, geological hazards represent a

(a) A meteor impact crater in Arizona

(b) Volcanic eruption

(c) Glacier in Alaska

FIGURE 14.16 Geologic events such as meteor or asteroid impacts (a), massive volcanic eruptions (b), or climate change (c), are thought to trigger mass extinctions that mark major eras in the earth's history.

huge threat. Diseases kill more people than any other cause. Some pandemics have been truly enormous. In the 1350s, for example, bubonic plague (the Black Death) is thought to have killed at least 135 million people in Europe and Asia. A similar pandemic in the sixth century (the Justinian plague) probably killed at least 100 million and uprooted nearly every existing major culture. Incidentally, the Justinian plague may have been triggered by an explosion of the predecessor of Indonesia's Krakatoa volcano, which exploded again in 1883. Drought-caused famines are probably the second biggest

killers. A severe El Niño (chapter 15) from 1876 to 1879, for example, is reported to have caused 50 million deaths around the world. Bad weather and extremely bad governance are thought to have caused a famine in China from 1959 to 1961 that killed at least 30 million people.

Among direct natural disasters, floods take the largest number of human lives, while wind storms (hurricanes, cyclones, tornadoes) cause the greatest property damage. Table 14.5 shows the estimated loss of life in the 20 worst natural disasters in the past 250 years.

TABLE 14.5

Worst Natural Disasters, 1755–2005

Date	Type	Deaths	Location
1931	Flood	3.7 million	Yangtze River, China
1959	Flood	2 million	Hwang Ho (Yellow) River, China
1938	Flood	900,000	Yangtze River, China
1877	Flood	900,000	Hwang Ho (Yellow) River, China
1939	Flood	500,000	Honan province, China
1970	Cyclone/Flood	500,000	Bangladesh, India
2004	Tsunami	300,000	Indian Ocean (mag. 9.2)
1850	Earthquake	300,000	Sichuan, China (mag. 8.0)
1976	Earthquake	275,000	Tianjin, China (mag. 8.2)
1920	Earthquake	200,000	Gansu, China (mag. 8.6)
1927	Earthquake	200,000	Xining, China (mag. 8.3)
1923	Earthquake	140,000	Tokyo, Japan (mag. 8.2)
1935	Flood	142,000	Changiyang River, China
1991	Cyclone/Flood	138,000	Bangladesh
1948	Earthquake	110,000	Turkmenistan, USSR (mag. 7.3)
1883	Tsunami	100,000	Java (Krakatoa volcano)
1911	Flood	100,000	Yangtze River, China
1815	Volcano	92,000	Tambora Sumbawa, Indonesia
1908	Earthquake	83,000	Messina, Italy (mag. 7.5)
1755	Earthquake	77,000	Lisbon, Portugal (mag. 8.7)

Source: The Disaster Center, 2005.

As you can see, floods have been the most lethal and, because of both its geology and large population, China has suffered a majority of these disasters. The 2004 earthquake/tsunami is seventh on this list, but it was the greatest death toll of any earthquake in recorded history. The only other earthquake/tsunami on this list was the explosion of Krakatoa, at the opposite end of Sumatra from the epicenter of the 2004 quake.

Think About It

What are the most dangerous geological hazards where you live? Are you aware of emergency preparedness plans? How would you evacuate your area if it becomes necessary?

Earthquakes can be very destructive

Earthquakes are sudden movements in the earth's crust that occur along faults (planes of weakness) where one rock mass slides past another one. When movement along faults occurs gradually and relatively smoothly, it is called creep or seismic slip and may be undetectable to the casual observer. When friction prevents rocks from slipping easily, stress builds up until it is finally released with a sudden jerk, as was the case in the 2004 Sumatran earthquake. The point on a fault at which the first movement occurs during an earthquake is called the epicenter.

Earthquakes have always seemed mysterious, sudden, and violent, coming without warning and leaving in their wake ruined cities and dislocated landscapes (table 14.6). Cities such as Tokyo, Japan, or Mexico City, parts of which are built on soft landfill or poorly consolidated soil, usually suffer the greatest damage from earthquakes. Water-saturated soil can liquify when shaken. Buildings sometimes sink out of sight or fall down like a row of dominoes under these conditions.

FIGURE 14.17 An earthquake near the town of Izmit in western Turkey in 1999 killed at least 50,000 people and injured an equal number. Poorly constructed buildings collapsed and trapped residents, while stronger structures nearby remained standing.

The worst death toll usually occurs in cities with poorly constructed buildings (fig. 14.17). Today contractors in earthquake zones are attempting to prevent damage and casualties by constructing buildings that can withstand tremors. The primary methods used are heavily reinforced structures, strategically placed weak spots in the building that can absorb vibration from the rest of the building, and pads or floats beneath the building on which it can shift harmlessly with ground motion. The problem of planning and building for earthquakes is at the center of a persistent dispute about nuclear waste storage at Yucca Mountain, Nevada (Exploring Science, p. 317).

As you can see in figure 14.18, the most seismically active region in the lower 48 U.S. states is along the west coast where tectonic plates are colliding. Nearly every state has had significant

TABLE 14.6

Worldwide Frequency of Earthquakes

Richter Scale Magnitude*	Description	Average Number Per Year
2–2.9	Unnoticeable	300,000
3–3.9	Smallest felt	49,000
4–4.9	Minor earthquake	6,200
5–5.9	Damaging earthquake	800
6–6.9	Destructive earthquake	120
7–7.9	Major earthquake	18
>8	Great earthquake	1 or 2

*For every unit increase in the Richter scale, ground displacement increases by a factor of 10, while energy release increases by a factor of 30. There is no upper limit to the scale, but the largest earthquake recorded was 9.5, in Chile in 1960.

Source: B. Gutenberg in *Earth* by F. Press and R. Siever, 1978, W. H. Freeman & Company.

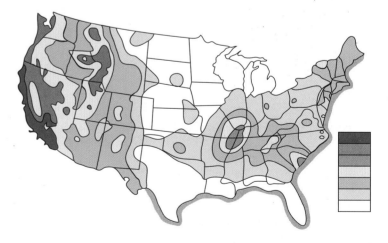

FIGURE 14.18 A seismic map of the lower 48 states shows the risk of earthquakes. Although the highest risk is along the Pacific coast and the Rocky Mountains, you may be surprised to see that an area along the Mississippi River around New Madrid, Missouri, also has a high potential for seismic activity.

Exploring SCIENCE

In July 2002 the U.S. Senate voted to designate Yucca Mountain, Nevada, as the permanent resting place for 77,000 metric tons of high-level nuclear waste, most of it spent fuel from the nation's 107 nuclear power plants (see fig. 19.25). Nuclear power proponents celebrated: After twenty years of research and lobbying, the site was finally approved. With a repository on the horizon, plans for new nuclear plants could proceed, and existing plants could look forward to clearing out 50,000 tons of spent nuclear fuel currently in temporary, sometimes inadequate, storage at reactor sites. The state of Nevada, meanwhile, immediately filed a lawsuit to stop the plan. Nevada Governor Kenny Guinn charged that the Department of Energy (DOE) had lowered its scientific standards for evaluating the geological integrity of the site. Opponents charged that heavy-handed eastern politics, not sound geology, was behind the decision to put the entire nation's nuclear waste in Nevada.

The stakes at Yucca Mountain are high. The DOE and the nuclear power industry have invested more than $4 billion in research, testing, and promotion of the site. The lack of permanent storage has been an obstacle to construction of nuclear power plants since the 1980s. Further, cleanup is beginning at old military installations, such as Hanford, Washington; and Rocky Flats, Colorado. These are some of the nation's worst toxic waste sites, and permanent storage is needed for their radioactive materials.

Geology is the central problem in siting a waste repository of any kind. For high-level radioactive wastes, the DOE needed to find a

Yucca Mountain, Nevada (long ridge in center of photo), has been chosen as the United States' first permanent high-level nuclear waste storage site.
Source: Department of Energy.

site where a labyrinth of deep tunnels could keep extremely dangerous materials isolated and secure for 10,000 years, more than twice the length of recorded human history. (The waste will remain highly radioactive for more than 500,000 years, but the DOE considers itself unlikely to be responsible for the site for that long.) This time span requires a geologically stable area—no active faults like those in California, no volcanoes like those in Washington and Oregon. Bedrock needs to be relatively impermeable, not riddled with underground channels and sinkholes as in Florida or parts of Texas. There must be no groundwater, which would soak storage casks and mobilize radioactive materials, possibly allowing tainted water to reach the surface. This rules out the central Great Plains and most of the eastern United States, where plentiful rainfall keeps water tables high and aquifers full.

Yucca Mountain fits these requirements better than most places in the United States. But there is conflicting evidence, and conflicting interpretations of evidence, about whether even Yucca Mountain is good enough. DOE geologists insist that the only aquifers in the area are 300 m below the site; other geologists say there is evidence in the rocks that aquifers have risen in the past, suggesting that they might rise again to the level of the repository tunnels. Revelation that some of the hydrologic data were fake cast even more doubt about the safety of the site.

Critics of the site point out that the area has more than 30 known faults. Geologists disagree on how long these faults have been dormant, and on how long it will be before they shift again. In addition, there are seven dormant volcanoes in the area. While these are likely to remain dormant for 10,000 years, their activity in several hundred thousand years is hard to predict. Opponents of the site also worry about the potential dangers of shipping waste, potentially 28,000 truckloads and 10,000 rail cars over the site's 30-year lifespan. Defenders point out that high-level waste is currently stored at more than 130 temporary sites, and those sites are clearly unsafe for the long term.

Do you think Nevada should be the repository for the nation's nuclear waste? Should we build more nuclear plants even if Yucca Mountain turns out to be unacceptable, given that we will have to find something to do with existing waste? Would our economy be different today if we had waited to build nuclear power plants until a permanent storage site was available?

---

tremors, however. The largest North American earthquake in recorded history occurred, surprisingly, in New Madrid, Missouri, in 1811. The tremor, which had an estimated magnitude of 8.8, made bells ring as far away as Boston, and caused the Mississippi River to run backwards for several hours. The region was sparsely inhabited, so there were few deaths. If a similar quake occurred today, however, a large part of Memphis would be destroyed.

Earthquakes are almost always followed by a series of aftershocks that can continue long after the initial shock. In the case of the 2004 Sumatran quake, hundreds of aftershocks followed in succeeding months. One in April was magnitude 8.7 (about half the energy of the original 9.2 quake). It didn't cause a

tsunami, but it did kill about 2,000 people on the nearby island of Nias, which is about 35 km from the epicenter of the earlier quake. No tsunami occurred—apparently because there was less displacement of the ocean floor.

Tsunamis can be generated by earthquakes, as the 2004 one was, or by landslides or underwater volcanic eruptions. The largest wave ever recorded was caused by a landslide. In July 1958, a massive rock avalanche fell into Lituya Fjord in Alaska's Glacier Bay National Park. The rockfall created a wave 525 m (over 1,700 ft) high. Several eyewitnesses lived to tell about it, including the crew of a fishing boat that was picked up from its anchorage in the fjord and thrown out into the ocean. A major concern in the Atlantic Ocean is a huge bulge forming on the Cumbre Vieja volcano on La Palma in the Canary Islands. This volcano erupts every 25 to 200 years. The last eruption was in 1971, but indications are that another eruption could be imminent. If this crater were to collapse all at once, it could dump half a trillion tons of rock into the ocean, which would send a tsunami 50 m high to the east coast of North America.

The ring of seismic activity and active volcanoes (often called the "ring of fire") around the edge of the Pacific Ocean makes it the most likely place in the world for tsunami formation. The United States already has six tsunami warning buoys in the Pacific Ocean. In January 2005, President Bush announced a $37.5 million plan to deploy 32 new Deep-ocean Assessment and Reporting of Tsunami (DART) buoys by mid-2007 in the Pacific, Atlantic, and Caribbean to protect U.S. coastal areas (fig. 14.19). Each buoy has an anchored seafloor bottom pressure recorder and a companion moored surface buoy for real-time communications. An acoustic link transmits data from the pressure recorder on the seafloor to the surface buoy. The data are then relayed by a satellite link to ground stations. The Indian Ocean has had few tsunamis in recent years, and had no such warning system in 2004. In the wake of the disaster, however, 53 nations met in Brussels and pledged to create a worldwide system, including the Indian Ocean, by 2015.

Volcanoes eject gas and ash, as well as lava

Volcanoes and undersea magma vents produce much of the earth's crust. Over hundreds of millions of years, gaseous emissions from these sources formed the earth's earliest oceans and atmosphere. Many of the world's fertile soils are weathered volcanic materials. Volcanoes have also been an ever-present threat to human populations. One of the most famous historic volcanic eruptions was that of Mount Vesuvius in southern Italy, which buried the cities of Herculaneum and Pompeii in A.D. 79. The

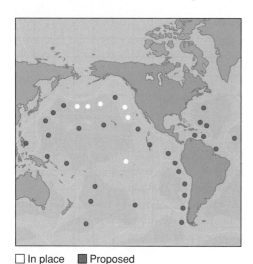

FIGURE 14.19 Proposed location of buoys for the Deep-ocean Assessment and Reporting of Tsunami (DART) warning system maintained by the United States.

☐ In place ■ Proposed

FIGURE 14.20 Pyroclastic flows spill down the slopes of Mayon volcano in the Philippines in this September 23, 1984, image. This eruption caused no casualties because more than 73,000 people evacuated danger zones.

mountain had been giving signs of activity before it erupted, but many citizens chose to stay and take a chance on survival. On August 24, the mountain buried the two towns in ash. Thousands were killed by the dense, hot, toxic gases that accompanied the ash flowing down from the volcano.

Today, more than 500 million people live in the danger zone around volcanoes. Many assume that because the volcano hasn't erupted for years it never will again. They don't realize that the time between eruptions can be centuries, but the next big one could come any time without warning.

Nuees ardentes (French for "glowing clouds") are deadly, denser-than-air mixtures of hot gases and ash like those that inundated Pompeii and Herculaneum. Temperatures in these clouds may exceed 1,000°C, and they move at more than 100 km/hour (60 mph). Nuees ardentes destroyed the town of St. Pierre on the Caribbean island of Martinique on May 8, 1902. Mount Pelee released a cloud of nuees ardentes that rolled down through the town, killing somewhere between 25,000 and 40,000 people within a few minutes. All the town's residents died except a single prisoner being held in the town dungeon.

Mudslides are also disasters sometimes associated with volcanoes. The 1985 eruption of Nevado del Ruiz, 130 km (85 mi) northwest of Bogotá, Colombia, caused mudslides that buried most of the town of Armero and devastated the town of Chinchina. An estimated 25,000 people were killed. Heavy mudslides also accompanied the eruption of Mount St. Helens in Washington in 1980. Sediments mixed with melted snow and the waters of Spirit Lake at the mountain's base and flowed many kilometers from their source. Extensive damage was done to roads, bridges, and property, but because of sufficient advance warning, there were few casualties.

Volcanic eruptions often release large volumes of ash and dust into the air (fig. 14.20). Mount St. Helens expelled 3 km^3 of dust and ash, causing ash fall across much of North America. This was only a minor eruption. An eruption in a bigger class of volcanoes was that of Tambora, Indonesia, in 1815, which expelled 175 km^2 of dust and ash, more than 58 times that of

Mount St. Helens. These dust clouds circled the globe and reduced sunlight and air temperatures enough so that 1815 was known as the year without a summer.

It is not just a volcano's dust that blocks sunlight. Sulfur emissions from volcanic eruptions combine with rain and atmospheric moisture to produce sulfuric acid (H_2SO_4). The resulting droplets of H_2SO_4 interfere with solar radiation and can significantly cool the world climate. In 1991, Mt. Pinatubo in the Philippines emitted 20 million tons of sulfur dioxide that combined with water to form tiny droplets of sulfuric acid. This acid aerosol reached the stratosphere where it circled the globe for two years. This thin haze cooled the entire earth by 1°C and postponed global warming for several years. It also caused a 10 to 15 percent reduction in stratospheric ozone, allowing increased ultraviolet light to reach the earth's surface.

Landslides are examples of mass wasting

Gravity constantly pulls downward on every material everywhere on earth, causing a variety of phenomena collectively termed **mass wasting** or mass movement, in which geologic materials are moved downslope from one place to another. The resulting movement is often slow and subtle, but some slope processes such as rockslides, avalanches, and land slumping can be swift, dangerous, and very obvious. Landslide is a general term for rapid downslope movement of soil or rock. In the United States alone, over $1 billion in property damage is done every year by landslides and related mass wasting.

In some areas, active steps are taken to control landslides or to limit the damage they cause. On the other hand, many human activities such as road construction, forest clearing, agricultural cultivation, and building houses on steep, unstable slopes increase

FIGURE 14.21 Parts of an expensive house slide down the hillside in Pacific Palisades, California. Building at the edge of steep slopes made of unconsolidated sediment in an earthquake-prone region is a risky venture.

both the frequency and the damage done by landslides. In some cases, people are unaware of the risks they face by locating on or under unstable hillsides. In other cases, they simply deny clear and obvious danger. Southern California, where people build expensive houses on steep hills of relatively unconsolidated soil, is often the site of large economic losses from landslides. Chapparal fires expose the soil to heavy winter rains. Resulting mudslides carry away whole neighborhoods and bury downslope areas in debris flows (fig. 14.21). Generally these processes are slow enough that few lives are lost, but the property damage can be high.

CONCLUSION

We need materials from the earth to sustain our modern lifestyle, but often the methods we use to get those materials have severe environmental consequences. Still, there are ways that we can extend resources through recycling and the development of new materials and more efficient ways of using them. We can do much to repair the damage caused by resource extraction, although operations, such as that at the Beartrack Mine, will probably never be returned to their original, pristine condition. We will need to be eternally vigilant to ensure that mine pits full of toxic water or the nuclear waste disposal site at Yucca Mountain, Nevada (assuming it ever overcomes critics objections and is actually used to store nuclear waste), don't leak.

Water contamination is one of the main environmental costs of mining. Pollutants include acids, cyanide, mercury, heavy metals, and sediment. Air pollution from smelting also can cause widespread damage. We'll discuss this risk further in chapter 16.

We also should be aware of geological hazards, such as floods, earthquakes, volcanoes, and landslides. Because these hazards often occur on a geological time scale, residents who haven't experienced one of these catastrophic events assume that they never will. People move into highly risky places without considering what the consequences may be of ignoring nature. The earth may seem extremely stable to us, but it's really a highly dynamic system. We remain ignorant of these forces at our own peril.

REVIEWING LEARNING OUTCOMES

By now you should be able to explain the following points:

14.1 Summarize the processes that shape the earth and its resources.
- Earth is a dynamic planet.
- Tectonic processes reshape continents and cause earthquakes.

14.2 Explain how rocks and minerals are formed.
- The rock cycle creates and recycles rocks.
- Weathering and sedimentation wear down rocks.

14.3 Think critically about economic geology and mineralogy.

- Metals are essential to our economy.
- Nonmetal minerals include gravel, clay, sand, and salts.

14.4 Critique the environmental effects of resource extraction.

- Mining can have serious environmental impacts.
- Processing ores also has negative consequences.

14.5 Discuss ways we could conserve geological resources.

- Recycling saves energy as well as materials.
- New materials can replace mined resources.

14.6 Describe geological hazards.

- Earthquakes can be very destructive.
- Volcanoes eject gas and ash, as well as lava.
- Landslides are examples of mass wasting.

PRACTICE QUIZ

1. Describe the layered structure of the earth.
2. Define *mineral* and *rock*.
3. What are tectonic plates and why are they important to us?
4. Why are there so many volcanoes, earthquakes, and tsunamis along the "ring of fire" that rims the Pacific Ocean?
5. Describe the rock cycle and name the three main rock types that it produces.
6. Figure 14.11 maps sources of some of the most important metals. What are they, and where do they come from?

7. Give some examples of nonmetal mineral resources and describe how they are used.
8. Describe some ways metals and other mineral resources can be recycled.
9. What are some environmental hazards associated with mineral extraction?
10. Describe some of the leading geological hazards and their effects.

CRITICAL THINKING AND DISCUSSION QUESTIONS

1. Look at the walls, floors, appliances, interior, and exterior of the building around you. How many earth materials were used in their construction?
2. Is your local bedrock igneous, metamorphic, or sedimentary? If you don't know, who might be able to tell you?
3. Suppose you live in a small, mineral-rich country, and a large, foreign mining company proposes to mine and market your minerals. How should revenue be divided between the company, your government, and citizens of the country? Who bears the greatest costs? How should displaced people be compensated? Who will make sure that compensation is fairly distributed?
4. Geological hazards affect a minority of the population who build houses on unstable hillsides, in flood-prone areas, or

on faults. What should society do to ensure safety for these people and their property?
5. A persistent question in this chapter is how to reconcile our responsibility as consumers with the damage from mineral extraction and processing. How responsible are you? What are some steps you could take to reduce this damage?
6. If gold jewelry is responsible for environmental and social devastation, should we stop wearing it? Should we worry about the economy in producing areas if we stop buying gold and diamonds? What further information would you need to answer these questions?

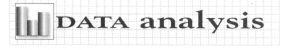 **DATA analysis** Mapping

Volcanoes

Look at the animated map of volcanoes in the animated National Atlas. http:nationalatlas.gov/dynamic.html. (You will need Shockwave to display this movie; you can download Shockwave at the link on this page) Where are most of the volcanoes in the lower 48 states? Roll over the red dots to see photos and elevations of the volcanoes. As you move down the chain of mountains from north to south, is there a trend in elevations? Look at the Alaska map. How are the volcanoes arranged here? Based on your readings in this chapter, can you explain the pattern?

Where Are Recent Earthquakes?

You can find out about recent earthquakes, where they were and how big they were, by looking at the USGS earthquake information page, http://earthquake.usgs.gov/. Click on the map to investigate recent events.

1. Describe where most recent earthquakes have occurred, either by dominant states or by mountain ranges/coastlines or other geographic features. Using what you know from this chapter about causes of earth movement and earthquakes, can you

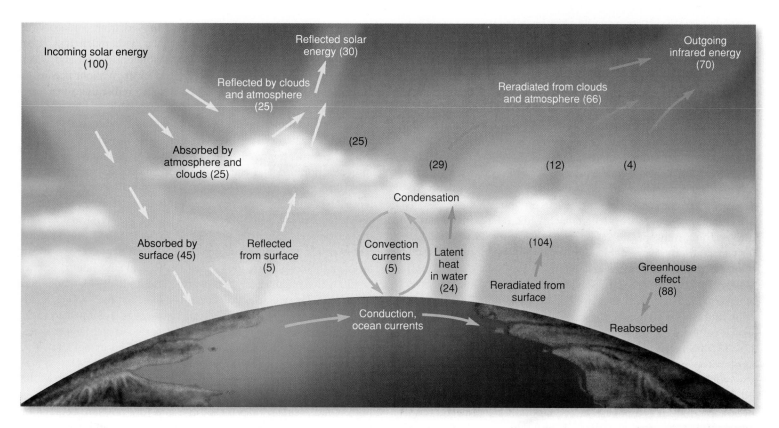

FIGURE 15.3 Energy balance between incoming and outgoing radiation. The atmosphere absorbs or reflects about half of the solar energy reaching the earth. Most of the energy reemitted from the earth's surface is long-wave, infrared energy. Most of this infrared energy is absorbed by aerosols and gases in the atmosphere and is re-radiated toward the planet, keeping the surface much warmer than it would otherwise be. This is known as the greenhouse effect.

Unlike the troposphere, the stratosphere is relatively calm. There is so little mixing in the stratosphere that volcanic ash or human-caused contaminants can remain in suspension there for many years.

Above the stratosphere, the temperature diminishes again, creating the mesosphere, or middle layer. The thermosphere (heated layer) begins at about 80 km. This is a region of highly ionized (electrically charged) gases, heated by a steady flow of high-energy solar and cosmic radiation. In the lower part of the thermosphere, intense pulses of high-energy radiation cause electrically charged particles (ions) to glow. This phenomenon is what we know as the *aurora borealis* and *aurora australis,* or northern and southern lights.

No sharp boundary marks the end of the atmosphere. Pressure and density decrease with distance from the earth until they become indistinguishable from the near-vacuum of interstellar space.

The sun warms our world

The sun supplies the earth with an enormous amount of energy, but that energy is not evenly distributed over the globe. Incoming solar radiation (insolation) is much stronger near the equator than at high latitudes. Of the solar energy that reaches the outer atmosphere, about one-quarter is reflected by clouds and atmospheric gases, and another quarter is absorbed by carbon dioxide, water vapor, ozone, methane, and a few other gases (fig. 15.3). This energy absorption

warms the atmosphere slightly. About half of insolation reaches the earth's surface. Some of this energy is reflected by bright surfaces, such as snow, ice, and sand. The rest is absorbed by the earth's surface and by water. Surfaces that *reflect* energy have a high **albedo** (reflectivity). Most of these surfaces appear bright to us because they reflect light as well as other forms of radiative energy. Surfaces that *absorb* energy have a low albedo and generally appear dark. Black soil, pavement, dark green vegetation, and open water, for example, have low albedos (table 15.2).

TABLE 15.2

Albedo (Reflectivity) of Surfaces

Surface	Albedo (%)
Fresh snow	80–85
Dense clouds	70–90
Water (low sun)	50–80
Sand	20–30
Water (sun overhead)	5
Forest	5–10
Black soil	3
Earth/atmosphere average	30

Absorbed energy heats the absorbing surface (such as an asphalt parking lot in summer), evaporates water, or provides the energy for photosynthesis in plants. Following the second law of thermodynamics, absorbed energy is gradually reemitted as lower-quality heat energy. A brick building, for example, absorbs energy in the form of light and reemits that energy in the form of heat.

The change in energy quality is very important because the atmosphere selectively absorbs longer wavelengths. Most solar energy comes in the form of intense, high-energy light or near-infrared wavelengths. This short-wavelength energy passes relatively easily through the atmosphere to reach the earth's surface. Energy re-released from the earth's warmed surface ("terrestrial energy") is lower-intensity, longer-wavelength energy in the far-infrared part of the spectrum. Atmospheric gases, especially carbon dioxide and water vapor, absorb much of this long-wavelength energy, re-releasing it in the lower atmosphere and letting it leak out to space only slowly. This terrestrial energy provides most of the heat in the lower atmosphere. If the atmosphere were as transparent to infrared radiation as it is to visible light, the earth's average surface temperature would be about 18°C (33°F) colder than it is now.

This phenomenon is called the **greenhouse effect** because the atmosphere, loosely comparable to the glass of a greenhouse, transmits sunlight while trapping heat inside. The greenhouse effect is a natural atmospheric process that is necessary for life as we know it. However, too strong a greenhouse effect, caused by burning of fossil fuels and deforestation, may cause harmful environmental change.

The current rapid melting of Arctic sea ice is a good example both of the importance of albedo and a **positive feedback loop.** When the Arctic Ocean was entirely covered with ice, as it was most of the time previously, it acted as a giant mirror reflecting solar radiation back to space. Now that the ice is melting, open water is exposed to sunlight. The much lower albedo of water, compared to ice, means that much more solar energy is absorbed by the ocean. This leads to warmer water, which melts the ice further, leading to even more warming. This ever accelerating system could bring huge changes to our world.

Water stores energy, and winds redistribute it

Much of the incoming solar energy is used to evaporate water. Every gram of evaporating water absorbs 580 calories of energy as it transforms from liquid to gas. Globally, water vapor contains a huge amount of stored energy, known as **latent heat.** When water vapor condenses, returning from a gas to a liquid form, the 580 calories of heat energy are released. Imagine the sun shining on the Gulf of Mexico in the winter. Warm sunshine and plenty of water allow continuous evaporation that converts an immense amount of solar (light) energy into latent heat stored in evaporated water. Now imagine a wind blowing the humid air north from the Gulf toward Canada. The air cools as it moves north (especially if it encounters cold air moving south). Cooling causes the water vapor to condense. Rain (or snow) falls as a consequence.

Note that not only water has moved from the Gulf to the Midwest: 580 calories of heat have also moved with every gram

of moisture. The heat and water have moved from a place with strong incoming solar energy to a place with much less solar energy and much less water. The redistribution of heat and water around the globe are essential to life on earth. Without oceans to absorb and store heat, and wind currents to redistribute that heat in the latent energy of water vapor, the earth would undergo extreme temperature fluctuations like those of the moon, where it is 100°C (212°F) during the day and −130°C (−200°F) at night. Water performs this vital function because of its unique properties in heat absorption and energy of vaporization (chapter 3).

Uneven heating, with warm air close to the equator and colder air at high latitudes, also produces pressure differences that cause wind, rain, storms, and everything else we know as weather. As the sun warms the earth's surface, the air nearest the surface warms and expands, becoming less dense than the air above it. The warm air must then rise above the denser air. Vertical convection currents result, which circulate air from warm latitudes to cool latitudes and vice versa. These convection currents can be as small and as localized as a narrow column of hot air rising over a sun-heated rock, or they can cover huge regions of the earth. At the largest scale, the convection cells are described by a simplified model known as Hadley cells, which redistribute heat globally (fig. 15.4).

Where air rises in convection currents, air pressure at the surface is low. Where air is sinking, or subsiding, air pressure is high. On a weather map these high and low pressure centers, or rising and sinking currents of air, move across continents. In most of North America, they generally move from west to east. Rising air tends to cool with altitude, releasing latent heat that causes further rising. Very warm and humid air can rise very vigorously,

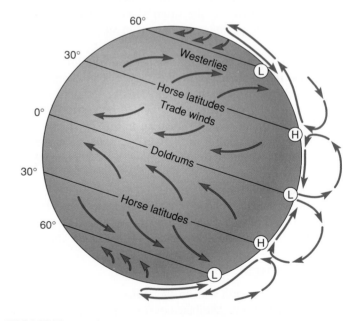

FIGURE 15.4 General circulation patterns. The actual boundaries of the circulating convection cells vary from day to day and season to season, as do the local directions of surface winds. Surface topography also complicates circulation patterns, but within these broad regions, winds usually have a predominant and predictable direction. L and H represent low and high pressure.

especially if it is rising over a mass of very cold air. As water vapor carried aloft cools and condenses, it releases energy that fuels violent storms, which we will discuss later.

Pressure differences are an important cause of wind. There is always someplace with sinking (high pressure) air and someplace with low pressure (rising) air. Air moves from high-pressure centers toward low-pressure areas, and we call this movement wind.

15.2 WEATHER HAPPENS

Weather is a description of the physical conditions in the atmosphere (humidity, temperature, air pressure, wind, and precipitation) over short time scales. Although most of us live in controlled environments, it's useful to know how and why weather happens.

Why does it rain?

To understand why it rains, remember two things: Water condenses as air cools, and air cools as it rises. Any time air is rising, clouds, rain, or snow might form. Cooling occurs because of changes in pressure with altitude: Air cools as it rises (as pressure decreases); air warms as it sinks (as pressure increases). Air rises in convection currents where solar heating is intense, such as over the equator. Moving masses of air also rise over each other and cool. Air also rises when it encounters mountains. If the air is moist (if it has recently come from over an ocean or an evaporating forest region, for example), condensation and rainfall are likely as the air is lifted (fig. 15.5). Regions with intense solar

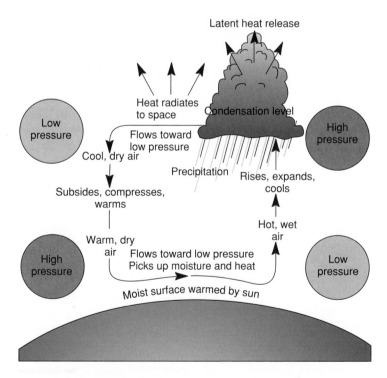

FIGURE 15.5 Convection currents and latent energy cause atmospheric circulation and redistribute heat and water around the globe.

heating, frequent colliding air masses, or mountains tend to receive a great deal of precipitation.

Where air is sinking, on the other hand, it tends to warm because of increasing pressure. As it warms, available moisture evaporates. Rainfall occurs relatively rarely in areas of high pressure. High pressure and clear, dry conditions occur where convection currents are sinking. High pressure also occurs where air sinks after flowing over mountains. Figure 15.4 shows sinking, dry air at about 30° north and south latitudes. If you look at a world map, you will see a band of deserts at approximately these latitudes.

Another ingredient is usually necessary to initiate condensation of water vapor: condensation nuclei. Tiny particles of smoke, dust, sea salts, spores, and volcanic ash all act as condensation nuclei. These particles form a surface on which water molecules can begin to coalesce. Without them even supercooled vapor can remain in gaseous form. Even apparently clear air can contain large numbers of these particles, which are generally too small to be seen by the naked eye.

Large-scale winds don't move in a straight line

In the Northern Hemisphere, winds generally bend clockwise (right), and in the Southern Hemisphere, they bend counterclockwise (left). This curving pattern results from the fact that the earth rotates in an eastward direction as the winds move above the surface. The apparent curvature of the winds is known as the **Coriolis effect.** On a global scale, this effect produces steady, reliable wind patterns, such as the trade winds and the midlatitude Westerlies (see fig. 15.4). Ocean currents similarly curve clockwise in the Northern Hemisphere and counterclockwise in the south. On a regional scale, the Coriolis effect produces cyclonic winds, or wind movements controlled by the earth's spin. Cyclonic winds spiral clockwise out of an area of high pressure in the Northern Hemisphere and counterclockwise into a low-pressure zone. If you look at a weather map in the newspaper, can you find this counterclockwise spiral pattern?

Why does this curving or spiraling motion occur? Imagine you were looking down on the North Pole of the rotating earth. Now imagine that the earth was a merry-go-round in a playground, with the North Pole at its center and the equator around the edge. As it spins counterclockwise (eastward), the spinning edge moves very fast (a full rotation, 39,800 km, every 24 hours for the real earth, or more than 1,600 km/hour). Near the center, though, there is very little eastward velocity. If you threw a ball from the edge toward the center, it would be traveling faster (edge speed) than the middle. It would appear, to someone standing on the merry-go-round, to curve toward the right. If you threw the ball from the center toward the edge, it would start out with no eastward velocity, but the surface below it would spin eastward, making the ball end up, to a person on the merry-go-round, west of its starting point.

Winds move above the earth's surface much as the ball does. If you were looking down at the South Pole, you would see the earth spinning clockwise, and winds—or thrown balls—would appear to bend left. Incidentally, this effect does not apply to the

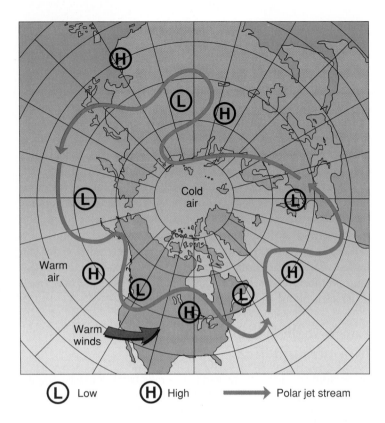

| Ⓛ Low | Ⓗ High | ⟶ Polar jet stream |

FIGURE 15.6 A typical pattern of the arctic circumpolar vortex. This large, circulating mass of cold air sends "fingers," or lobes, across North America and Eurasia, spreading storms in their path. If the vortex becomes stalled, weather patterns stabilize, causing droughts in some areas and excess rain elsewhere.

way a sink drains. It's a myth that bathtubs and sinks spiral in opposite directions in the Northern and Southern Hemispheres. Those movements are far too small to be affected by the spinning of the earth.

At the top of the troposphere are **jet streams,** hurricane-force winds that circle the earth. These powerful winds follow an undulating path approximately where the vertical convection currents known as the Hadley and Ferrell cells meet. The approximate path of one jet stream over the Northern Hemisphere is shown in figure 15.6. Although we can't perceive jet streams on the ground, they are important to us because they greatly affect weather patterns. Sometimes jet streams dip down near the top of the world's highest mountains, exposing mountain climbers to violent, brutally cold winds.

Ocean currents modify our weather

Warm and cold ocean currents strongly influence climate conditions on land. Surface ocean currents result from wind pushing on the ocean surface, as well as from the Coriolis effect. As surface water moves, deep water wells up to replace it, creating deeper ocean currents. Differences in water density—depending on the temperature and saltiness of the water—also drive ocean

circulation. Huge cycling currents called gyres carry water north and south, redistributing heat from low latitudes to high latitudes (see fig. 17.6). The Alaska current, flowing from Alaska southward to California, keeps San Francisco cool and foggy during the summer.

The Gulf Stream, one of the best known currents, carries warm Caribbean water north past Canada's maritime provinces to northern Europe. This current is immense, some 800 times the volume of the Amazon, the world's largest river. The heat transported from the Gulf keeps Europe much warmer than it should be for its latitude. As the warm Gulf Stream passes Scandinavia and swirls around Iceland, the water cools and evaporates, becomes dense and salty, and plunges downward, creating a strong, deep, southward current. Oceanographer Wallace Broecker calls this the ocean conveyor system (see fig. 17.6).

Ocean circulation patterns were long thought to be unchanging, but now oceanographers believe that currents can shift abruptly. About 11,000 years ago, for example, as the earth was gradually warming at the end of the Pleistocene ice age, a huge body of meltwater, called Lake Agassiz, collected along the south margin of the North American ice sheet. At its peak, it contained more water than all the current freshwater lakes in the world. Drainage of this lake to the east was blocked by ice covering what is now the Great Lakes. When that ice dam suddenly gave way, it's estimated that some 163,000 km^3 of fresh water roared down the St. Lawrence Seaway and out into the North Atlantic, where it layered on top of the ocean and prevented the sinking of deep, cold, dense seawater. This, in turn, stopped the oceanic conveyor and plunged the whole planet back into an ice age (called the Younger Dryas after a small tundra flower that became more common in colder conditions) that lasted for another 1,300 years.

Could this happen again? Meltwater from Greenland glaciers is now flooding into the North Atlantic just where the Gulf Stream sinks and creates the deep south-flowing current. Already, evidence shows that the deep return flow has weakened by about 30 percent. Even minor changes in the strength or path of the Gulf Stream might give northern Europe a climate more like that of Siberia—not a pleasant prospect for inhabitants of the area.

> ### Think About It
>
> Find London and Stockholm on a globe. Then find cities in North America at a similar latitude. Temperatures in London and Stockholm rarely get much below freezing. How do you think their climate compares with the cities you've identified in North America? Why are they so different?

Seasonal winds and monsoons have powerful effects

While figure 15.4 shows regular global wind patterns, large parts of the world, especially the tropics, receive seasonal winds and

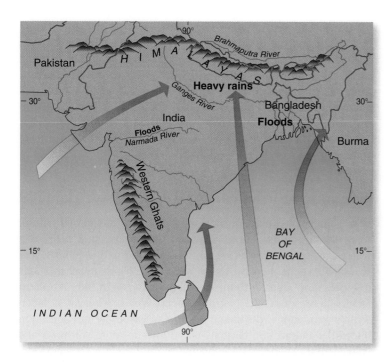

FIGURE 15.7 Summer monsoon air flows over the Indian subcontinent. Warming air rises over the plains of central India in the summer, creating a low-pressure cell that draws in warm, wet oceanic air. As this moist air rises over the Western Ghats or the Himalayas, it cools and heavy rains result. These monsoon rains flood the great rivers, bringing water for agriculture, but also causing much suffering.

FIGURE 15.8 Failure of monsoon rains brings drought, starvation, and death to both livestock and people in the Sahel desert margin of Africa. Although drought is a fact of life in Africa, many governments fail to plan for it, and human suffering is much worse than it needs to be.

rainy seasons that are essential for sustaining both ecosystems and human life. Sometimes these seasonal rains are extreme, causing disastrous flooding. In 2003, immense floods in China forced 100 million people from their homes. Many more millions elsewhere in Asia were similarly affected.

Sometimes the rains fail, causing crop failures and famine. The most regular seasonal winds and rains are known as **monsoons.** In India and Bangladesh, monsoon rains come when seasonal winds blow hot, humid air from the Indian Ocean (fig. 15.7). Strong convection currents lift this air, causing heavy rain across the subcontinent. When the rising air reaches the Himalayas, it rises even further, creating some of the heaviest rainfall in the world. During the 5-month rainy season of 1970, a weather station in the foothills of the Himalayas recorded 25 m (82 ft) of rain!

Tropical and subtropical regions around the world have seasonal rainy and dry seasons (see the discussion of tropical biomes, chapter 5). The main reason for this variable climate is that the region of most intense solar heating and evaporation shifts through the year. Remember that the earth's axis of rotation is at an angle. In December and January the sun is most intense just south of the equator; in June and July the sun is most intense just north of the equator. Wherever the sun shines most directly, evaporation and convection currents—and rainfall and thunderstorms—are very strong. As the earth orbits the sun, the tilt of its axis creates seasons with varying amounts of wind, rain, and heat or cold.

Seasonal rains support seasonal tropical forests, and they fill some of the world's greatest rivers, including the Ganges and the Amazon. As the year shifts from summer to winter, solar heating weakens, the rainy season ends, and little rain may fall for months.

During the 1970s and 1980s, rains failed repeatedly in parts of the Sahel in northern Africa. Although rains have often been irregular in this region, increasing populations and civil disorder made these droughts result in widespread starvation and death (fig. 15.8). Climate scientists predict that droughts in some areas, and severe storms in others, will become more common with global climate change.

Frontal systems create local weather

The boundary between two air masses of different temperature and density is called a front. When cooler air displaces warmer air, we call the moving boundary a **cold front.** Since cold air tends to be more dense than warm air, a cold front will hug the ground and push under warmer air as it advances. As warm air is forced upward, it cools adiabatically (without loss or gain of energy), and its cargo of water vapor condenses and precipitates. Upper layers of a moving cold air mass move faster than those in contact with the ground because of surface friction or drag, so the boundary profile assumes a curving, "bull-nose" appearance (fig. 15.9, *left*). Notice that the region of cloud formation and precipitation is relatively narrow. Cold fronts generate strong convective currents and often are accompanied by violent surface winds and destructive storms. An approaching cold front generates towering clouds called thunderheads that reach into the stratosphere where the jet stream pushes the cloud tops into a

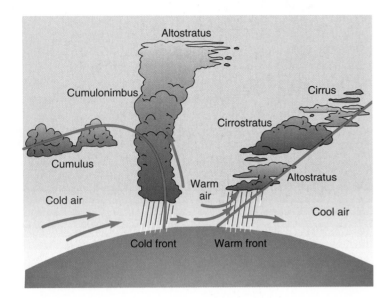

FIGURE 15.9 A cold front assumes a bulbous, "bull-nose" appearance because ground drag retards forward movement of surface air. As warm air is lifted up over the advancing cold front, it cools, producing precipitation. When warm air advances, it slides up over cooler air in front and produces a long, wedge-shaped zone of clouds and precipitation. The high cirrus clouds that mark the advancing edge of the warm air mass may be 1,000 km and 48 hours ahead of the front at ground level.

characteristic anvil shape. The weather after the cold front passes is usually clear, dry, and invigorating.

If the advancing air mass is warmer than local air, a **warm front** results. Since warm air is less dense than cool air, an advancing warm front will slide up over cool, neighboring air parcels, creating a long, wedge-shaped profile with a broad band of clouds and precipitation (fig. 15.9, *right*). Gradual uplifting and cooling of air in the warm front avoids the violent updrafts and strong convection currents that accompany a cold front. A warm front will have many layers of clouds at different levels. The highest layers are often wispy cirrus (mare's tail) clouds that are composed mainly of ice crystals. They may extend 1,000 km (621 mi) ahead of the contact zone with the ground and appear as much as 48 hours before any precipitation. A moist warm front can bring days of drizzle and cloudy skies.

Cyclonic storms can cause extensive damage

Few people experience a more powerful and dangerous natural force than cyclonic storms spawned by low-pressure cells over warm tropical oceans. As we discussed earlier in this chapter, low pressure is generated by rising warm air. Winds swirl into this low-pressure area, turning counterclockwise in the Northern Hemisphere due to the Coriolis effect. When rising air is laden with water vapor, the latent energy released by condensation intensifies convection currents and draws up more warm air and water vapor. As long as a temperature differential exists between air and ground and a supply of water vapor is available, the storm cell will continue to pump energy into the atmosphere.

Called **hurricanes** in the Atlantic and eastern Pacific, typhoons in the western Pacific, or cyclones in the Indian Ocean, winds near the center of these swirling air masses can reach hundreds of kilometers per hour and cause tremendous suffering and destruction. Often hundreds of kilometers across, these giant storms can generate winds as high as 320 km/hr (200 mph) and push walls of water called storm surges far inland (fig. 15.10*a*). In July 1931, torrential rains spawned by a typhoon caused massive flooding on China's Yangtze River that killed an

(a) Hurricane Floyd 1999

(b) Gulf Shores, Alabama 2005

(c) A tornado touches down

FIGURE 15.10 (a) Hurricane Floyd was hundreds of kilometers wide as it approached Florida in 1999. Note the hole, or eye, in the center of the storm. (b) Destruction caused by Hurricane Katrina in 2005. More than 230,000 km² (90,000 mi²) of coastal areas were devastated by this massive storm and many cities were almost completely demolished. (c) Tornadoes are much smaller than hurricanes, but can have stronger local winds.

estimated 3.7 million people, the largest known storm death toll in human history. A similar storm in July 1959 caused flooding of the Huang Ho (or Yellow River) that killed 2 million people.

On August 29, 2005, Hurricane Katrina slammed into the Gulf of Mexico shoreline. The category 4 storm, with 232 km/ hr winds (145 mph), pushed a storm surge up to 9 m (29 ft) high onto coastal areas and caused the greatest natural disaster in North American history. Coastal areas from Florida to Louisiana were damaged and many ocean-front cities were almost completely destroyed (fig. 15.10*b*).

Tornadoes, swirling funnel clouds that form over land, also are considered to be cyclonic storms, although their rotation isn't generated by Coriolis forces. While never as large or powerful as hurricanes, tornadoes can be just as destructive in the limited areas where they touch down (fig. 15.10*c*). Some meteorologists consider smaller, less destructive phenomena such as waterspouts, dust devils, and smoke funnels over forest fires to be miniature tornadoes, but others think they have very different origins.

Tornadoes are generated on the American Great Plains by giant "supercell" frontal systems where strong, dry air cold fronts from Canada collide with warm humid air moving north from the Gulf of Mexico. Greater air temperature differences cause more powerful storms, which is why most tornadoes occur in the spring, when arctic cold fronts penetrate far south over the warming plains. As warm air rises rapidly over dense, cold air, intense vertical convection currents generate towering thunderheads with anvil-shaped leading edges and domed tops up to 20,000 m (65,000 ft) high. Water vapor cools and condenses as it rises, releasing latent heat and accelerating updrafts within the supercell. Penetrating into the stratosphere, the tops of these clouds can encounter jet streams, which help create even stronger convection currents.

Sometimes a supercell isn't organized enough to spawn tornadoes, but strong downdrafts, called **downbursts,** can generate straight-line winds well over 160 km/hr (100 mph). In January 2007, straight-line winds in Europe reached hurricane strength, downing trees, grounding airplanes, derailing trains, and killing dozens of people.

The same strong updrafts that create tornadoes can also generate hail. Falling water droplets rise on powerful updrafts, freeze, fall again, and rise again. Droplets accumulate water, or ice, as they rise and fall, until they become heavy enough to resist being lofted by updrafts. If they remain frozen all the way to the ground, we call them hail. If you cut open large hail stones, you often can see accretion rings where successive layers of ice were added.

There seems to be a connection between strong monsoon winds over West Africa and the frequency of killer hurricanes in the Caribbean and along the Atlantic coast of North and Central America. In years when the Atlantic surface temperatures are high, a stronger than average monsoon trough forms over Africa. This pulls in moist maritime air that brings rain (and life) to the Sahel (the margin of the Sahara). This trough gives rise to tropical depressions (low-pressure storms) that follow one another in regular waves across the Atlantic. The weak trade winds associated with "wet" years allow these storms to organize and gain strength. Evaporation from the warm surface water provides energy.

Think About It

How would you describe the weather patterns where you live? Which of the factors described so far in this chapter are most important in determining your local weather?

During the 1970s and 1980s when the Sahel had devastating droughts, the weather was relatively quiet in North America. Only one killer hurricane (winds over 70 km/hr or 112 mph) reached the United States in two decades. By contrast, in 1995 when ocean surface temperatures were high and rain returned to the Sahel, seven hurricanes and six named tropical storms were spawned in the Atlantic. During the summer of 1996, satellite images of the Atlantic showed five active hurricanes roaring across the Caribbean at the same time. Are we witnessing a long-term climatic change or merely a temporary aberration?

15.3 CLIMATE CAN BE AN ANGRY BEAST

If weather is a description of physical conditions in the atmosphere, then climate is the *pattern* of weather in a region over long time periods. The interactions of atmospheric systems are so complex that climatic conditions are never exactly the same at any given location from one time to the next. While it is possible to discern patterns of average conditions over a season, year, decade, or century, complex fluctuations and cycles within cycles make generalizations doubtful and forecasting difficult. We always wonder whether anomalies in local weather patterns represent normal variations, a unique abnormality, or the beginnings of a shift to a new regime.

Climates have changed dramatically throughout history

Ice cores from glaciers have revolutionized our understanding of climate history. In this research, a hollow tube is drilled down through the ice. Every 30 m, or so, the tube is pulled up and an ice cylinder is pushed out of the center (fig. 15.11). Back illumination reveals a pattern

FIGURE 15.11 Dr. Mark Twickler, of the University of New Hampshire, holds a section of the 3,000 m Greenland ice sheet core, which records 250,000 years of climate history.

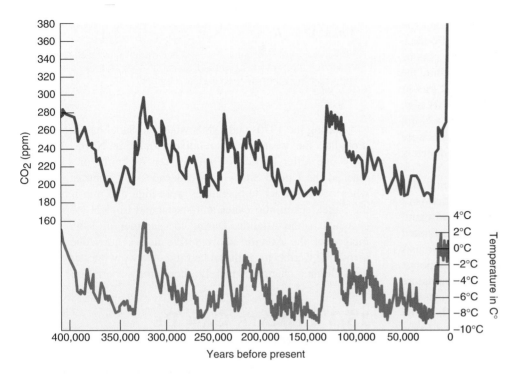

FIGURE 15.12 Atmospheric CO_2 concentrations and global mean temperatures estimated from the antarctic Vostok ice core. Note the relatively rapid changes and close correlation in both climate and atmospheric chemistry.
Source: Data from United Nations Environment Programme.

of lighter and darker bands caused by annual snow accumulation on top of the glacier (which gives us an estimate of climate at the time). Gas bubbles trapped in the ice give us data on atmospheric composition when the snow was deposited. Ash layers and sulfate concentrations in the ice can be correlated with volcanic eruptions. The longest ice record ever collected is the Vostok ice core, which is 3,100 m long, and gives us a record of both global temperatures and atmospheric CO_2 over the past 420,000 years (fig. 15.12). A team of Russian scientists worked for 37 years at a site about 1,000 km from the South Pole to extract this ice core. A similar core, nearly as long as the Vostok, has been drilled from the Greenland ice sheet. Other glaciers throughout the world have also now been cored. All these ice samples show that climate has varied dramatically over time, but that there is a close correlation between atmospheric temperatures and CO_2 concentrations.

Major climatic changes, such as those of the Ice Ages, can have catastrophic effects on living organisms. If climatic change is gradual, species may have time to adapt or migrate to more suitable locations. Where climatic change is relatively abrupt, many organisms are unable to respond before conditions exceed their tolerance limits. Whole communities may be destroyed, and if the climatic change is widespread, many species may become extinct.

Perhaps the most well-studied example of this phenomenon is the great die-off that occurred about 65 million years ago at the end of the Cretaceous period. Most dinosaurs—along with 75 percent

of all previously existing plant and animal species—became extinct, apparently as a result of sudden cooling of the earth's climate. Geologic evidence suggests that this catastrophe was not an isolated event. There appear to have been several great climatic changes, perhaps as many as a dozen, in which large numbers of species were exterminated (see table 11.3, p. 235).

What causes catastrophic climatic swings?

There are many explanations for climatic catastrophes. One of these is that changes in solar energy associated with 11-year sunspot cycles or 22-year solar magnetic cycles might play a role. Another theory is that a regular 18.6-year cycle of shifts in the angle at which our moon orbits the earth alters tides and atmospheric circulation in a way that affects climate. A theory that has received attention in recent years suggests that orbital variations as the earth rotates around the sun might be responsible for cyclic weather changes.

Milankovitch cycles, named after Serbian scientist Milutin Milankovitch, who first described them in the 1920s, are periodic shifts in the earth's orbit and tilt (fig. 15.13). The earth's elliptical orbit stretches and shortens

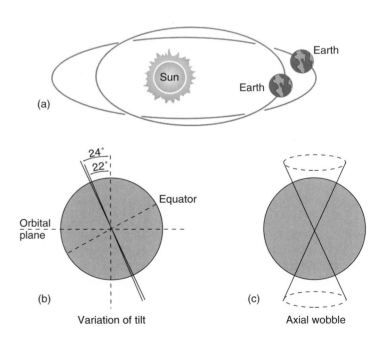

FIGURE 15.13 Milankovitch cycles, which may affect long-term climate conditions: (a) changes in the elliptical shape of the earth's orbit, (b) shifting tilt of the axis, and (c) wobble of the earth.

in a 100,000-year cycle, while the axis of rotation changes its angle of tilt in a 40,000-year cycle. Furthermore, over a 26,000-year period, the axis wobbles like an out-of-balance spinning top. These variations change the distribution and intensity of sunlight reaching the earth's surface and, consequently, global climate. Bands of sedimentary rock laid in the oceans seem to match both these Milankovitch cycles and the periodic cold spells associated with worldwide expansion of glaciers every 100,000 years or so.

A historical precedent for global climate change that had disastrous effects on humans was the "little ice age" that began in the 1400s. Temperatures dropped so that crops failed repeatedly in parts of northern Europe that once were good farmland. Scandinavian settlements in Greenland founded during the warmer period around A.D. 1000 lost contact with Iceland and Europe as ice blocked shipping lanes. It became too cold to grow crops, and fish that once migrated along the coast stayed farther south. The settlers slowly died out, perhaps in battles with Inuit people who were driven south from the high Arctic by colder weather.

Evidence from ice cores drilled in the Greenland ice cap suggests that world climate may change much more rapidly than previously thought. During the last major interglacial period 135,000 to 115,000 years ago, it appears that temperatures flipped suddenly from warm to cold or vice versa over a period of years or decades rather than centuries.

Volcanic eruptions may have been responsible for some of these climate flips. When Mt. Toba exploded in western Sumatra about 73,000 years ago, for example, it was the largest volcanic cataclysm in the past 28 million years. It's estimated that Mt. Toba ejected at least 2,800 km^3 of material compared to only 1 km^3 emitted by Mt. St. Helens in Washington State in 1980. Mt. Toba produced enough ash to cover the entire earth an average of 10 cm deep. The signal from this eruption can be detected in the Greenland ice core. The sulfuric acid and particulate material ejected into the atmosphere from Mt. Toba are estimated to have dimmed incoming sunlight by 75 percent and to have cooled the whole planet by as much as 16°C for more than 160 years. The ensuing volcanic winter is calculated to have reduced terrestrial biomass by 90 percent, and to have created a genetic bottleneck still evident in many species, including humans.

El Niño/Southern Oscillations are powerful cycles

What do sea surface temperatures near Indonesia, anchovy fishing off the coast of Peru, and the direction of tropical trade winds have to do with rainfall and temperatures over the United States? Strangely enough, they all may be interrelated. If the terms El Niño, La Niña, and Southern Oscillation are not yet a part of your vocabulary, perhaps they should be. They describe a connection between the ocean and atmosphere that appears to affect these factors as well as weather patterns throughout the world.

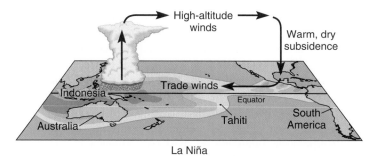

FIGURE 15.14 The El Niño/La Niña Southern Oscillation Cycle. During El Niño years, surface trade winds that normally push warm water westward toward Indonesia weaken and allow this pool of water to flow eastward, bringing storms to the Americas.

The core of this climatic system is a huge pool of warm surface water in the Pacific Ocean that sloshes slowly back and forth between Indonesia and South America like water in a giant bathtub. Most years, this pool is held in the western Pacific by steady equatorial trade winds pushing ocean surface currents westward (fig. 15.14). These surface winds are generated by a huge low-pressure cell formed by upwelling convection currents of moist air warmed by the ocean. Towering thunderheads created by rising air bring torrential summer rains to the tropical rainforests of Northern Australia and Southeast Asia. Winds high in the troposphere carry a return flow back to the eastern Pacific where dry subsiding currents create deserts from Chile to southern California. Surface waters driven westward by the trade winds are replaced by upwelling of cold, nutrient-rich, deep waters off the west coast of South America that support dense schools of anchovies and other finfish.

Every three to five years, for reasons that we don't fully understand, the Indonesian low collapses and the mass of warm surface water surges back east across the Pacific. One theory is that the high cirrus clouds atop the cloud columns block the sun and cool the ocean surface enough to reverse trade winds and ocean surface currents so they flow eastward rather than westward. Another theory is that eastward-flowing deep currents called baroclinic waves periodically interfere with coastal upwelling, warming the sea surface off South America and eliminating the temperature gradient across the Pacific. At any rate, the shift in position of the tropical depression sets off a chain of events lasting a year or more with repercussions in weather systems across North and South America and perhaps around the world.

Peruvian fishermen were the first to notice irregular cycles of rising ocean temperatures that resulted in disappearance of the anchovy schools on which they depended. They named these events **El Niño** (Spanish for the Christ child) because they often occur around Christmas time. We have come to call the intervening years **La Niña** (or little girl). Together, this cycle is called the El Niño Southern Oscillation (ENSO).

Actually, sea surface temperature measurements in the central Pacific near Tahiti are the best measurement of this giant oscillation. In May 1998, as an unusually strong El Niño was ending, surface temperatures in this region plunged 7°C (12.5°F) in a little over a week, a dramatic event considering the enormous heat-storing capacity of ocean waters.

How does the ENSO cycle affect you? During an El Niño year, the northern jet stream—which normally is over Canada—splits and is drawn south over the United States. This pulls moist air from the Pacific and Gulf of Mexico inland, bringing intense storms and heavy rains from California across the midwestern states. The intervening La Niña years bring hot, dry weather to these same areas. An exceptionally long El Niño event from 1991 to 1995 broke a seven-year drought over the western United States and resulted in floods of the century in the Mississippi Valley. Oregon, Washington, and British Columbia, on the other hand, tend to have warm, sunny weather in El Niño years rather than their usual rain. Droughts in Australia and Indonesia during El Niño episodes cause disastrous crop failures and forest fires, including one in Borneo in 1983 that burned 3.3 million ha (8 million acres). An even stronger El Niño in 1997 spread health-threatening forest fire smoke over much of Southeast Asia.

Are ENSO events becoming stronger or more irregular because of global climate change? Studies of corals up to 130,000 years old suggest this is true. Furthermore, there are signs that warm ocean surface temperatures are spreading. In addition to the pool of warm water in the western Pacific associated with La Niña years, oceanographers recently discovered a similar warm region in the Indian Ocean. High sea surface temperatures spawn larger and more violent storms such as hurricanes and typhoons. On the other hand, increased cloud cover would raise the albedo while upwelling convection currents generated by these storms could pump heat into the stratosphere. This might have an overall cooling effect and act as a safety valve for global warming.

A longer-scale ocean/weather connection called the **Pacific Decadal Oscillation (PDO)** involves a very large pool of warm water that moves back and forth across the North Pacific every 30 years or so. From about 1977 to 1997, surface water temperatures in the middle and western part of the North Pacific Ocean were cooler than average, while waters off the western United States were warmer. During this time, salmon runs in Alaska were bountiful, while those in Washington and Oregon were greatly diminished. In 1997, however, ocean surface temperatures along the coast of western North America turned significantly cooler, perhaps marking a return to conditions that prevailed between 1947 and 1977. Under this cooler regime, Alaskan salmon runs declined while those in Washington and Oregon improved somewhat. A

similar oscillation occurs in the North Atlantic. Together these decades-long climate patterns can be seen in tree-ring growth throughout North America.

15.4 GLOBAL WARMING IS HAPPENING

As the opening case study for this chapter shows, many scientists regard anthropogenic (human-caused) global climate change to be the most important environmental issue of our times. The possibility that humans might alter world climate is not a new idea. In 1895, Svante Arrhenius, who subsequently received a Nobel Prize for his work in chemistry, predicted that CO_2 released by coal burning could cause global warming. Most of Arrhenius' contemporaries dismissed his calculations as impossible, but his work seems remarkably prescient now.

A scientific consensus is emerging

The first evidence that human activities are increasing atmospheric CO_2 came from an observatory on top of the Mauna Loa volcano in Hawaii. The observatory was established in 1957 as part of an International Geophysical Year, and was intended to provide data on air chemistry in a remote, pristine environment. Surprisingly, measurements showed CO_2 levels increasing about 0.5 percent per year, rising from 315 ppm in 1958 to 383 ppm in 2006 (fig. 15.15). This increase isn't a straight line, however. Because a majority of the world's land and vegetation are in the Northern Hemisphere, northern seasons dominate the signal. Every May, CO_2 levels drop slightly as plant growth on northern continents use CO_2 in photosynthesis. During the northern winter, levels rise again as respiration releases CO_2.

The **Intergovernmental Panel on Climate Change (IPCC)** brings together scientists and government representatives from 130 countries to review scientific evidence on the causes and likely effects of human-caused climate change. In 2007, the IPCC issued its fourth report. The result of 6 years of work by 2,500 scientists, the four volumes of the report represent a consensus by more than 90 percent of all the scientists working on climate

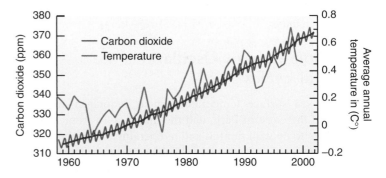

FIGURE 15.15 Carbon dioxide concentrations on Mauna Loa (monthly and annual averages) and global annual average temperature variations from 1960 to 2000.
Data source: NASA Earth Observatory, 2002.

change. This report has the strongest expression of certainty to date from the climate community. It says that global warming is "very likely" caused by humans. This implies a 90 percent probability that changes we are now seeing are caused by human actions. It's reported, however, that most delegates wanted the statement to say that human causation is "virtually certain," meaning a 99 percent probability. To achieve a consensus, the lead authors accepted the less definite wording. Most scientists, however, regard the evidence as "indisputable."

The estimated range of global temperature increase by the end of this century also has been increased in this report. The third IPCC report, published in 2001, calculated that average global surface temperatures would increase 1.4–5.8°C by 2100. The fourth report gives a variety of scenarios depending on population and economic growth, energy conservation and efficiency, and adoption (or lack thereof) of greenhouse gas controls. This gives a range of increases from 1.1–6.4°C (2–11.5°F) higher than now by 2100. According to the IPCC, the "best estimate," for temperature rise is now 1.8–4.0°C (3.2–7.8°F). To put that in perspective, the change in average global temperature between now and the middle of the last glacial period is estimated to be about 5°C.

The worst outcome from global warming could mean more than 1 million human deaths and hundreds of billions of dollars in costs by 2100. For most people, the harshest effects of global warming will be from extreme weather, including droughts, floods, heat waves, and hurricanes. These extremes have already increased significantly in the past decade, and are expected to get even worse in the future.

Sea levels are projected to rise 17–57 cm (7–23 in.) by the end of this century. If recent rapid melting of polar ice sheets and Greenland glaciers continues, it could add another 15 cm (6 in.) to that total. If nothing is done to halt greenhouse emissions, complete melting of Greenland's ice sheet would raise sea level by more than 60 m (nearly 20 ft). This would flood most of Florida, a broad swath of the Gulf Coast, most of Manhattan Island, Shanghai, Hong Kong, Tokyo, Kolkata, Mumbai, and many of the other largest cities in the world.

Following the release of the IPCC report, 45 nations called for a new global environmental body to regulate greenhouse emissions, perhaps with policing powers to punish violators. Conspicuously absent from this initiative were the world's biggest emitters of these gases, including the United States, China, and India. The Bush administration praised the IPCC report but said it still opposes mandatory cuts in greenhouse gas emissions. Voluntary programs, the president maintains, will be enough.

Greenhouse gases have many sources

Since preindustrial times atmospheric concentrations of CO_2, CH_4, and N_2O have climbed by over 31 percent, 151 percent, and 17 percent, respectively. Carbon dioxide is by far the most important cause of anthropogenic climate change (fig. 15.16a). Burning fossil fuels, making cement, burning forests and grasslands, and other human activities release nearly 30 billion tons of CO_2 every year, on average, containing some 8 billion tons of carbon (fig. 15.16b).

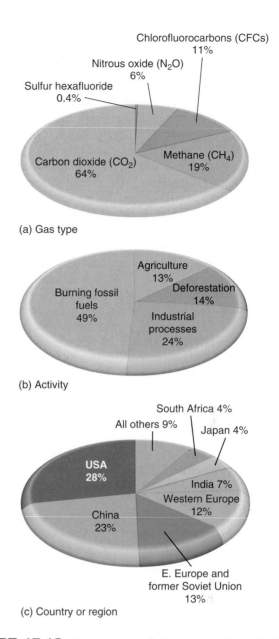

FIGURE 15.16 Percentage contributions to global warming by (a) gas type, (b) source activity, and (c) country or region. Data from world Resources Institute.

About 3 billion tons of this excess carbon is taken up by terrestrial ecosystems, and around 2 billion tons are absorbed by the oceans, leaving an annual atmospheric increase of some 3 billion tons per year. If current trends continue, CO_2 concentrations could reach about 500 ppm (approaching twice the preindustrial level of 280 ppm) by the end of the twenty-first century.

Although rarer than CO_2, methane absorbs 23 times as much infrared and is accumulating in the atmosphere about twice as fast as CO_2. Methane is released by ruminant animals, wet-rice paddies, coal mines, landfills, wetlands, and pipeline leaks. Hydroelectricity is usually promoted as a clean power source that offers an alternative to fossil fuels, but this may not always be true.

Philip Fearnside, an ecologist at Brazil's National Institute for Amazon Research, calculates that decaying vegetation in the reservoir behind the Cura-Una dam in Para Province emits so much carbon dioxide and methane every year that it causes three and a half times as much global warming as would generating the same amount of energy by burning fossil fuels. Chlorofluorocarbons (CFCs) also are powerful infrared absorbers. CFC releases in developed countries have declined since many of their uses were banned, but increasing production in developing countries, such as China and India, remains a problem. Nitrous oxide is produced by burning organic material and by soil denitrification. As fig. 15.16a shows, CFCs and N_2O together are thought to account for only about 17 percent of human-caused global warming.

The United States, with less than 5 percent of the world's population, produces 28 percent of all anthropogenic CO_2 (fig. 15.16c). China, with 1.3 billion people, is second in total CO_2 emissions, but fourteenth in per capita production. Japan and western Europe, which by most measures have standards of living at least equal to the United States, produce only about half as much CO_2 per person. Eastern Europe and the former Soviet Union have many old, inefficient factories and coal-buring power plants that emit high CO_2 levels with relatively little economic return.

Evidence of climate change is overwhelming

There is little doubt that our planet is warming. The American Geophysical Union, one of the nation's largest and most respected scientific organizations, says, "As best as can be determined, the world is now warmer than it has been at any point in the last two millennia, and, if current trends continue, by the end of the century it will likely be hotter than at any point in the last two million years." Some of the evidence for this warming includes the following:

- Over the last century the average global temperature has climbed about 0.6°C (1°F). Nineteen of the 20 warmest years in the past 150 have occurred since 1980. The hottest year since the instrumental temperature record began was 2005, and 2007 is expected to be even hotter.

- Polar regions have warmed much faster than the rest of the world. In Alaska, western Canada, and eastern Russia, average temperatures have increased as much as 4°C (7°F) over the past 50 years. Permafrost is melting; houses, roads, pipelines, sewage systems, and transmission lines are being damaged as the ground sinks beneath them. Trees are tipping over, and a beetle infestation in Alaska (made possible by warmer winters) killed 250 million spruce trees on the Kenai Peninsula. Coastal villages, such as Shishmaref, Point Hope, and Barrow, are having to relocate inland as ice breakup and increasingly severe storms erode the arctic shoreline.

- Arctic sea ice is only half as thick now as it was 30 years ago and the area covered by sea ice has decreased by more than 1 million km^2 (an area larger than Texas and Oklahoma combined) in just three decades. By 2040, the Arctic Ocean could be totally ice-free in the summer, if present trends continue.

FIGURE 15.17 Diminishing arctic sea ice prevents polar bears from hunting seals, their main food source.

This is bad news for polar bears, which depend on the ice to hunt seals. An aerial survey in 2005 found bears swimming across as much as 260 km (160 mi) of open water to reach the pack ice. When the survey was repeated after a major storm, dozens of bears were missing or spotted dead in the water. In 2007 the United States considered adding polar bears to the endangered species list because of loss of arctic ice (fig. 15.17). Loss of sea ice is also devastating for Inuit people whose traditional lifestyle depends on ice for travel and hunting.

- Ice shelves on the Antarctic Peninsula are breaking up and disappearing rapidly. A recent survey found that 90 percent of the glaciers on the peninsula are now retreating an average of 50 m per year. The Greenland ice cap also is reported to be melting twice as fast as it did a few years ago. Because ice shelves are floating, they don't affect sea level when they melt. Greenland's massive ice cap, however, holds enough water to raise sea level by about 7 m (about 23 ft) if it all melts.

- Alpine glaciers everywhere are retreating rapidly (fig. 15.18). Mt. Kilimanjaro has lost 85 percent of its famous ice cap since 1915. By 2015, all the permanent ice on the mountaintop is expected to be gone. In 1972, Venezuela had six glaciers; now it has only two. When Montana's Glacier National Park was created in 1910, it held some 150 glaciers. Now, fewer than 30, greatly shrunken glaciers remain (fig. 15.18). If current trends continue, all will have melted by 2030.

- So far, the oceans have been buffering the effects of our greenhouse emissions both by absorbing CO_2 directly and by storing heat. Deep-diving sensors show that the oceans are absorbing 0.85 watts per m^2 more than is radiated back to space. This absorption slows current warming, but also means that even if we reduce our greenhouse gas emissions today, it will take centuries to dissipate that stored heat. The higher levels of CO_2 being absorbed are acidifying

FIGURE 15.18 Alpine glaciers everywhere are retreating rapidly. These images show the Grinnell Glacier in 1914 and 1998. By 2030, if present melting continues, there will be no glaciers in Glacier National Park.

the oceans, and could have adverse effects on sea life. Mollusks and corals, for example, have a more difficult time making calcium carbonate shells and skeletons at the lower pH.

- Sea level has risen worldwide approximately 15–20 cm (6–8 in.) in the past century. About one-quarter of this increase is ascribed to melting glaciers; roughly half is due to thermal expansion of seawater. If all of Antarctica were to melt, the resulting rise in sea level could be several hundred meters.

- Satellite images and surface measurements show that growing seasons are now as much as three weeks longer in a band across northern Eurasia and North America than they were 30 years ago (see fig. 1.9). New plants and animals, never before seen in the far north, are now extending their territories, while native species, such as musk ox, caribou, walrus, and seal populations are declining as climate changes.

- Droughts are becoming more frequent and widespread. In Africa, for example, droughts have increased about 30 percent since 1970. In 2005, the United Nations reported that 60 million people in 36 countries needed emergency food aid. Drought was the single largest cause of famine, although wars, economics, and internal politics always exacerbate these emergencies.

- Biologists report that many animals are breeding earlier or extending their range into new territory as the climate changes. In Europe and North America, for example, 57 butterfly species have either died out at the southern end of their range, or extended the northern limits, or both. Plants, also, are moving into new territories. Given enough time and a route for migration, many species may adapt to new conditions, but we now are forcing them to move much faster than many achieved at the end of the last ice age (fig. 15.19).

- The disappearance of amphibians, such as the beautiful golden toads from the cloud forests of Costa Rica or western toads from Oregon's Cascade Range, are thought to have

been caused at least in part by changing weather patterns. Emperor and Adélie penguin populations have declined by half over the past 50 years as the ice shelves on which they depend for feeding and breeding disappear.

- Coral reefs worldwide are "bleaching," losing key algae and resident organisms, as water temperatures rise above 30°C (85°F). With reefs nearly everywhere threatened by pollution, overfishing, and other stressors, scientists worry that rapid climate change could be the final blow for many species in these complex, biologically rich ecosystems.

- Storms are becoming stronger and more damaging. The 2005 Atlantic storm season was the most severe on record, with 26 named tropical storms, twice as many as the average

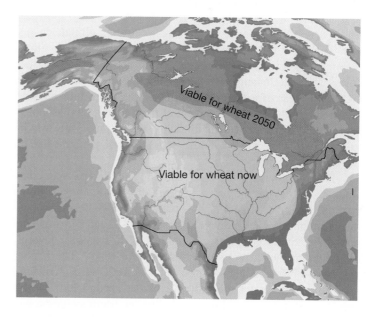

FIGURE 15.19 Most of the central United States is suitable for growing wheat now, but if current trends continue, the climatic conditions for wheat could be in central Canada in 2050.

FIGURE 15.20 More than 1 million people had to be evacuated from New Orleans in 2005 when a storm surge created by Hurricane Katrina breached the levees and allowed Lake Pontchartrain to flood into the city.

over the past 30 years. Some of this increased frequency could be a natural cycle, but the greater strength of these storms (an unprecedented 7 hurricanes reached level 5, the highest category) undoubtedly reflects higher sea surface temperatures. Hurricane Katrina, which demolished a 230,000 km^2 (90,000 mi^2) area of the U.S. Gulf Coast (an area about the size of Great Britain) and flooded New Orleans (fig. 15.20) caused at least $200 billion in damage, the greatest storm loss in American history.

Global warming will be expensive

In 2006, Sir Nicholas Stern, former chief economist of the World Bank, issued a study on behalf of the British government on the costs of global climate change. It was one of the strongest worded warnings to date from a government report. He said, "Scientific evidence is now overwhelming: climate change is a serious global threat, and it demands an urgent global response." If we don't act soon, Stern estimates, immediate costs of climate change will be at least 5 percent of the global GDP each year. If a wider range of risks is taken into account, the damage could equal 20 percent of the annual global economy. That would disrupt our economy and society on a scale similar to the great wars and economic depression of the first half of the twentieth century.

The report estimates that reducing greenhouse gas emissions now to avoid the worst impacts of climate change would cost only about 1 percent of the annual global GDP. That means that a dollar invested now could save us $20 later in this century. "We can't wait the five years it took to negotiate Kyoto," Sir Nicholas says, "We simply don't have time." The actions we take—or fail to take—in the next 10 to 20 years will have a profound effect on those living in the second half of this century and in the next. Energy production, Stern suggests, will have to be at least 80 percent decarbonized by 2050 to stabilize our global climate.

As chapter 2 points out, global climate change is a moral and ethical issue as well as a practical one. Many religious leaders are joining with scientists and business leaders to campaign for measures to reduce greenhouse gas emissions. Ultimately, the people likely to suffer most from global warming are the poorest in Africa, Asia, and Latin America who have contributed least to the problem. Those of us in the richer countries will likely have resources to blunt problems caused by climate change, but residents of poorer countries will have fewer options. The Stern report says that without action, at least 200 million people could become refugees as their homes are hit by drought or floods. Furthermore, there's a question of intergenerational equity. What kind of world are we leaving to our children and grandchildren? What price will they pay if we fail to act?

The Stern review recommends four key elements for combating climate change. They are: (1) *emissions trading* to promote cost-effective emissions reductions, (2) *technology sharing* that would double research investment in clean energy technology and accelerate spread of that technology to developing countries, (3) *reduce deforestation,* which is a quick and highly cost-effective way to reduce emissions, and (4) *help poorer countries* by honoring pledges for development assistance to adapt to climate change.

The fourth IPCC report—issued in 2007—estimates the cost of preventing CO_2 doubling to be even less than Sir Nicholas calculates. The IPCC says it would cost only 0.12 percent of annual global GDP to reduce carbon emissions the 2 percent per year necessary to stabilize world climate. The Bush administration continues to claim this to be too expensive even though reports, such as that described in the opening case study for this chapter, predict that regulating carbon could stimulate the global economy and create millions of jobs.

A vigorous market for emissions trading already exists. In 2006, about 700 million tons of carbon equivalent credits were exchanged for some (U.S.) $3.5 billion. In 2008, trade is expected to grow to the equivalent of 4.8 billion tons of carbon reduction. Ultimately, this market may grow to $500 billion per year by 2050, and could represent an important source of development assistance for poorer countries. Market mechanisms for controlling climate change are discussed further in chapter 23.

In many cases, both organisms and human infrastructure will not be able to move or adapt quickly enough to accommodate the rates at which climate is now changing. Tropical areas will probably not get much warmer than they are now, but some middle- and high-latitude areas could experience unaccustomed heat (fig. 15.21). Of course there could also be some good results from climate change. Residents of northern Canada, Siberia, and Alaska may enjoy warmer temperatures, longer growing seasons, and longer ice-free shipping seasons from their ports. The opening of the fabled Northwest Passage through the Arctic Ocean could cut thousands of kilometers off a trip from London to Vladivostok. Stronger monsoons might

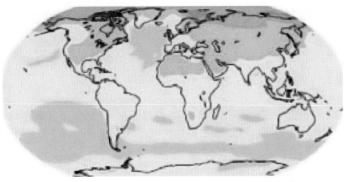

2020–2029

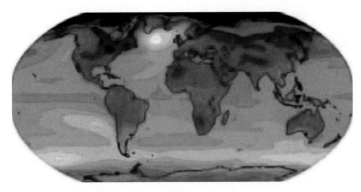

2090–2099

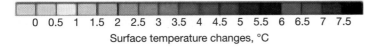

0 0.5 1 1.5 2 2.5 3 3.5 4 4.5 5 5.5 6 6.5 7 7.5

Surface temperature changes, °C

FIGURE 15.21 Surface temperature projections from IPCC scenario B1. This storyline assumes that global population peaks in mid-century and declines thereafter. It also infers rapid introduction of new, cleaner, and more-efficient technologies, but without additional climate initiatives. **Source:** IPCC, 2007.

bring welcome rains to Africa's Sahel, but we're already seeing extreme droughts in other parts of Africa that put millions of people at risk of food shortages.

Think About It

Looking at the predictions for the greatest warming by the end of this century shown in figure 15.21, which of the signs of warming discussed in this chapter can you locate on the map?

Carbon dioxide is a fertilizer for plants, and higher CO_2 levels bring faster growth for many plant species. Sherwood Idso of the U.S. Department of Agriculture claims that increased CO_2 concentrations could trigger lush plant growth that would make crops abundant. But other researchers point out that the effects of elevated CO_2 in real-world situations may be very different from conditions in a greenhouse (Exploring Science, p. 340).

Enough water is stored in glacial ice caps in Greenland and Antarctica to raise sea levels around 100 m (300 ft) if they all melt. About one-third of the world's population now live in areas that would be flooded if that happens. Even the 75 cm (30 in.) sea-level rise expected by 2050 will flood much of south Florida, Bangladesh, Pakistan, and many other low-lying coastal areas. Most of the world's largest urban areas are on coastlines. Wealthy cities such as New York or London can probably afford to build dikes to keep out rising seas, but poorer cities such as Jakarta, Kolkata, or Manila might simply be abandoned as residents flee to higher ground. Several small island countries such as the Maldives, the Bahamas, Kiribati, and the Marshall Islands could become uninhabitable if sea levels rise a meter or more. The South Pacific nation of Tuvalu has already announced that it is abandoning its island homeland. All 11,000 residents will move to New Zealand, perhaps the first of many climate change refugees.

Insurance companies worry that the $2 trillion in insured property along U.S. coastlines is at risk increased from a combination of high seas and catastrophic storms. At least 87,000 homes in the United States within 150 m (500 ft) of a shoreline are in danger of coastal erosion or flooding in the next 50 years. Accountants warn that loss of land and structures to flooding and coastal erosion together with damage to fishing stocks, agriculture, and water supplies could raise worldwide insurance claims from about $50 billion, which they were in the 1970s, to more than $150 billion per year in 2010. Some of this increase in insurance claims is that more people are living in dangerous places, but extra-severe storms only exacerbate this problem.

Infectious diseases are likely to increase as the insects and rodents that carry them spread to new areas. Already we have seen diseases such as malaria, dengue fever, and West Nile virus appear in parts of North America where they had never before been reported. Coupled with the movement of hundreds of millions of environmental refugees and greater crowding as people are forced out of areas made uninhabitable by rising seas and changing climates, the spread of epidemic diseases will also increase.

An ominous consequence of arctic permafrost melting might be the release of vast stores of methane hydrate now locked in frozen ground and in sediments in the ocean floor. Together with the increased oxidation of high-latitude peat lands potentially caused by warmer and dryer conditions, release of these carbon stores could add as much CO_2 to the atmosphere as all the fossil fuels ever burned. We could trigger a disastrous positive feedback loop in which the effects of warming cause even more warming. On the other hand, increased ocean evaporation might intensify snowfall at high latitudes so that arctic glaciers and snow pack would increase rather than decrease. Ironically, the increased

Among the greenhouse gases, none has been studied more than carbon dioxide (CO_2). As you've learned already in this chapter, CO_2 has been analyzed in 400,000-year-old air bubbles taken from glacial ice, measured atop Hawaii's Mauna Loa volcano since 1957, and implicated in a whole host of current climate changes. No one disputes that CO_2 levels in the troposphere increased from 280 to 383 ppm between 1750 and 2006. Few argue that it may exceed 500 ppm—nearly twice its pre-industrial concentration—before the twenty-first century ends, but we're not sure what the effects of these high CO_2 levels may be on ecological processes.

Plants need CO_2 to carry out photosynthesis. We know that many plant species flourish in elevated CO_2 levels. That's why greenhouses often pipe in extra CO_2 to accelerate growth. Some horticulturists predict that doubling CO_2 atmospheric concentrations will produce such lush vegetation that the earth will return to a verdant Eden. Others contend that how plants behave in the ideal conditions of a greenhouse or growth chamber is very different from what happens in the real world. How can we know what to expect if current trends continue?

In 1997, Peter Reich and his colleagues began experiments at the University of Minnesota's Cedar Creek Natural History Area that attempt to mimic natural conditions more faithfully than is possible in a growth chamber. At Cedar Creek, the researchers erected several large rings of towers from which CO_2 is released (fig. 1). Sensors track wind speeds and CO_2 concentrations within the ring, while a computer regulates gas flows to maintain 560 ppm throughout the experimental plot. Even when winds are swirling around the research site, the computer can adjust gas flows to keep CO_2 levels constant. Because the CO_2 isn't contained, these experiments are called Free Air Carbon Enrichment (FACE). Control plots have exactly the same design as the experimental ones except that the control towers blow plain air rather than CO_2-enriched air. The Cedar Creek Long-Term Ecological Research is an important center for the study of the relationship between biodiversity and prairie ecosystem stability, so the FACE research also included several different levels of native prairie species

FIGURE 1 Student researchers collect samples for analysis at Cedar Creek's Free Air Carbon Enrichment study. The ring of striped poles blow CO_2 over the prairie plants growing in experimental plots.

richness, as well as different fertilizer treatments, in experimental plots.

Dr. Reich and colleagues discovered that the response of plants to elevated CO_2 and nitrogen fertilization depended on how many species were in a test plot. In a paper published in 2001 in *Nature,* they showed that the amount of plant material (biomass) produced each year rose when CO_2 and nitrogen were added to the test plots. But test plots containing 9 or 16 plant species benefited much more than test plots with a single plant species (fig. 2). In a later paper in the *Proceedings of the National Academy of Sciences,* Dr. Reich and his team also demonstrated that, if test plots contained ecologically different plants (e.g., a legume and a summer-blooming grass), more biomass developed than in test plots with a single one of those functional plant groups. In other words, vegetation that contained a single plant species responded positively to rising carbon dioxide, but several plant species with different ecological functions responded much better, even without fertilization. The different species and functional groups compensated for each other in the face of a varying climate and environment.

Similar research is being done on forest species and agricultural crops. Peter Curtis, an Ohio State University researcher, writing in the journal *Phytologist,* summarized the results of dozens of FACE experiments on a number of species. In grain crop plants the total seed weight—and therefore food yield—increased

by up to 25 percent with a doubling of CO_2 concentration, but in all species except legumes the nutritional value of those plants fell. In one study of tall fescue (*Festuca alatior*), one of the most important pasture grasses in United States, the digestible protein that grazers could extract from the CO_2-stimulated grass actually fell below levels in existing forage, despite fertilization. So, while we may have luxuriant plant growth under elevated CO_2, it may be less valuable for certain uses than what we have today.

In the end, raising CO_2 through human activities represents a huge global experiment not unlike the FACE experiments, but with an important difference. If CO_2 reaches 500 ppm by 2100, we'll be stuck with those conditions—and their effects—for many years. What basic research tells us in the meantime about plant responses, including our important crop plants, will be important in understanding how to adapt to the challenges we face.

Dr. Reich says, "The idea that carbon dioxide with uniformly increase plant growth and therefore provide a large and predictable sink for carbon dioxide is naive at best. This type of model fails to consider the real complexity of the carbon dioxide fertilization effect. Species composition, community diversity, and the amount of nitrogen in the system interact to produce varying responses. In particular, reduced biodiversity and the widespread limited amount of nitrogen in the soil will reduce the CO_2 fertilization effect in terrestrial plant biomass."

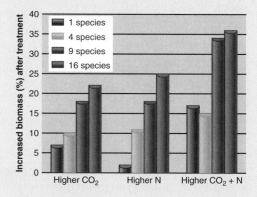

FIGURE 2 The Cedar Creek BioCON study demonstrates strength in numbers. Single species plots did not respond as well to CO_2 increases as multi-species plots.

Data From Peter Reich.

albedo of colder, snow-covered surfaces might then trigger a new ice age. Clearly, we're disturbing a complex system and facing consequences that we don't fully understand.

Think About It

Most of us find it difficult to contemplate global climate change. The potential effects are so disastrous and the time scale is so long that we simply put it out of our minds. How can we frame the issue in positive, practical terms that people can do something about immediately?

15.5 THE KYOTO PROTOCOL ATTEMPTS TO SLOW CLIMATE CHANGE

One of the centerpieces of the 1992 United Nations Earth Summit meeting in Rio de Janeiro was the Framework Convention on Climate Change, which set an objective of stabilizing greenhouse gas emissions to reduce the threats of global warming. At a follow-up conference in Kyoto, Japan, in 1997, 160 nations agreed to roll back CO_2, methane, and nitrous oxide emissions about 5 percent below 1990 levels by 2012. Three other greenhouse gases, hydrofluorocarbons, perfluorocarbons, and sulfur hexafluoride, would also be reduced, although from what level was not decided. Known as the **Kyoto Protocol,** this treaty sets different limits for individual nations, depending on their output before 1990. Poorer nations, such as China and India, were exempted from emission limits to allow development to increase their standard of living. Wealthy countries created the problem, the poorer nations argue, and the wealthy should deal with it.

Although the United States took a lead role in negotiating a compromise at Kyoto that other countries could accept, President George W. Bush refused to honor U.S. commitments. Claiming that reducing carbon emissions would be too costly for the U.S. economy, he said, "We're going to put the interests of our own country first and foremost." The United States, which depends solely on voluntary efforts to limit greenhouse gas emissions, had a 2 percent rise in those releases in 2007. At this rate, the U.S. will be 25 percent above 1990 emissions by 2012.

Meanwhile, 126 countries have ratified the Kyoto Protocol. If the U.S. continues to refuse to comply, it could have severe ramifications on U.S. corporations engaged in international business. Thousands of American businesses—including most of the largest ones—fall into this category. Having to modify their products and practices for overseas markets while not doing so for domestic sales would be expensive and would put them at a disadvantage with competitors not subject to these costs. Moreover, because Kyoto is based on a global cap-and-trade program, companies that get in on it early will have an advantage—they can buy cheap emissions credits before the price gets bid up.

We'll discuss market mechanisms for controlling climate change further in chapter 23. In 2007, the heads of ten of the largest business conglomerates in America joined four environmental groups to call for strong national legislation to achieve significant reductions of greenhouse gas emissions. The corporations included Alcoa, BP America, Caterpillar, Duke Energy, DuPont, FPL Group, General Electric, Lehman Brothers, Pacific Gas & Electric, and PNM Resources. The nongovernmental organizations were Environmental Defense, Pew Center, Natural Resources Defense Council, and World Resources Institute. Many companies would prefer a single national standard rather than a jumble of conflicting local and state rules. Knowing that climate controls are inevitable, they'd rather know now how they'll have to adapt rather than wait until a crisis causes us to demand sudden, radical changes.

Even if the United States joins the rest of the world in accepting the Kyoto provisions, however, many climatologists argue that much more will have to be done to combat global warming. According to some calculations, we will have to reduce CO_2 emissions by 60 percent or more to stabilize atmospheric concentrations. In the next section, we'll look at some of the ways we might accomplish this.

There are many ways we can control greenhouse emissions

Is the situation hopeless? Can we do anything to reverse the path we're on? No single step can reverse our release of greenhouse gases, but many simple policy changes could help stabilize our climate. One of the most encouraging analyses in this area is that of Robert Socolow and his colleagues at Princeton University. They say we already have the scientific, technical, and industrial know-how to stabilize our CO_2 emissions over the next half century, while still meeting the world's energy needs. In several recent publications, this group examines a portfolio of 14 policy options for reducing our CO_2 output. No single step can do the entire job by itself, they conclude, but the potential benefits of the entire portfolio are large enough that not every option must be used.

To avoid a doubling of atmospheric CO_2 we need to reduce our annual carbon emissions by about 7 billion tons by 2054. To make this easier to understand, the Princeton group divides the "stabilization triangle" between our current path and the flat emissions that we need in order to avoid doubling CO_2 levels into seven "wedges" (fig. 15.22). Each wedge represents 1 GT (1 billion tons) of carbon emissions avoided in 2054, or a cumulative total of 25 billion tons over the next 50 years, compared to a "business as usual" scenario.

Because most of our CO_2 emissions come from fossil fuel combustion, energy conservation and a switch to renewable fuels probably are the first places we should look to reduce our climate problem. For example, 1.5 wedges (1.5 GT reduction in carbon emissions) could be accomplished by doubling our average fuel economy from the expected 30 miles per gallon in 2054 to 60 mpg, and by halving the number of miles driven each year by switching to walking, biking, or using mass transit. Simply installing the most efficient lighting and appliances available, along with improved insulation in buildings, could save another 2 GT of carbon emissions. Similarly, improving power plant efficiency and reducing the energy consumption of industrial processes could reduce carbon output one wedge (1 GT) by 2054. Capturing and storing CO_2

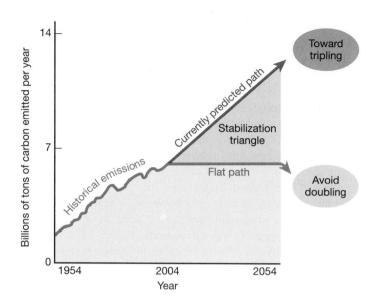

FIGURE 15.22 What would it take to stabilize our carbon emissions? Some authors argue that we already have the science and technology to do so. All we need is the political will.
Data from Socolow, et al. Environment vol. 46, no. 10 p. 11, 2004.

released by power plants, methane wells, and other large sources could save another billion tons of carbon. Much of this CO_2 could be injected into oil wells to improve crude oil production. Building 2 million wind turbines (of the current one-megawatt size) or doubling our current global nuclear generating capacity would each reduce carbon emissions by about 1 GT in 50 years. This would, of course, require solving the energy storage problems with wind power as well as public worries about the safety of nuclear reactors and storing radioactive waste. Increasing production of biofuels, such as ethanol and vegetable oil, 50-fold, could replace 1 GT of carbon from fossil fuels. Halting tropical forest destruction and replanting 250 million ha of forest (or 400 million ha in temperate regions) would avoid almost 1 GT of carbon emissions, while practices such as conservation tillage and planting of cover crops could reduce carbon releases by another GT.

As you can see, altogether these policy changes add up to considerably more than the 7 GT per year reduction we need to make by 2054 to avoid doubling atmospheric CO_2 concentrations. At least seven other immediately available options exist for reducing greenhouse gases by the equivalent of 1 GT of carbon. If we did all of them, we could reverse the present trajectory and move toward zero greenhouse emissions.

There also are many ways individuals can contribute to this effort (What Can You Do? p. 342). Perhaps the easiest way to become carbon neutral is to simply pay someone to do it for you. Many organizations will now plant a tree for you or arrange some other carbon reduction to offset the costs of your lifestyle. British Petroleum, for example, will compensate for the CO_2 emitted by a year's driving of the average car for about $40. You could make a guilt-free round-trip airplane flight from New York to Los Angeles for about $6.25. Or Carbonfund.org will

What Can You Do?

Reducing Carbon Dioxide Emissions

Individuals can help reduce global warming. Although some actions may cause only a small impact, collectively they add up. Many of these options save money in the long run and have other environmental benefits such as reducing air pollution and resource consumption.

1. Drive less, walk, bike, take public transportation, carpool, or buy a vehicle that gets at least 30 mpg (12.6 km/l). Average annual CO_2 reduction: about 20 lbs for each gallon of gasoline saved. Increasing your mileage by 10 mpg, for example, eliminates about 2,500 lbs (1.1 metric tons) of CO_2 per year. Becoming entirely self-propelled would save, on average, about 12,000 lbs (5.45 metric tons) per year.

2. Plant trees to shade your house during the summer, and paint your house a light color if you live in a warm climate or a dark color in a cold climate. Average annual CO_2 reduction: about 5,000 lbs (2.2 metric tons).

3. Insulate your house and seal all drafts. Average annual CO_2 reduction for the highest efficiency insulation, weatherstripping, and windows: about 5,000 lbs (2.2 metric tons).

4. Replace old appliances with new, energy-efficient models. Average annual CO_2 reduction for the most efficient refrigerator, for instance: about 3,000 lbs (1.4 metric tons).

5. Produce less waste. Buy minimally packaged goods and reusable products. Recycle. Average annual CO_2 reduction: about 1,000 lbs (0.45 metric tons) for 25 percent less garbage.

6. Turn down your thermostat in the winter and turn it up in the summer. Average annual CO_2 reduction: about 500 lbs (0.23 metric tons) for every 1°C (1.8°F) change.

7. Replace standard light bulbs with long-lasting compact fluorescent ones. Average annual CO_2 reduction: about 500 lbs (0.23 metric tons) for every bulb.

8. Wash laundry in warm or cold water, not hot. Average annual CO_2 reduction: about 500 lbs (0.23 metric tons) for two loads per week.

9. Set your water heater thermostat no higher than 48.9°C (120°F). Average annual CO_2 reduction: about 500 lbs (0.23 metric tons) for each 5°C (9°F) temperature change.

10. Buy renewable energy from your local utility if possible. Potential annual CO_2 reduction: about 30,000 lbs (13.5 metric tons).

offset for all the greenhouse gas emissions (both direct and indirect) for the average American for $99 per year.

There are some worries about these programs, however. In some places, planting trees is inappropriate. Fast-growing trees, such as the eucalyptus or pines often used in reforestation programs, can dry streams and springs and deplete soil nutrients. Furthermore, if the trees are grown for a few years and then harvested and burned, you haven't done anything for the planet. If you're going to finance tree planting, ask where the trees will be planted and how they'll be cared for.

Progress is being made

Many countries are working to reduce greenhouse emissions. The United Kingdom, for example, had already rolled CO_2 emissions back to 1990 levels by 2000 and vowed to reduce them 60 percent by 2050. Britain already has started to substitute natural gas for coal, promote energy efficiency in homes and industry, and raise its already high gasoline tax. Plans are to "decarbonize" British society and to decouple GNP growth from CO_2 emissions. A revenue-neutral carbon levy is expected to lower CO_2 releases and trigger a transition to renewable energy over the next five decades. New Zealand Prime Minister Helen Clark has pledged that her country will be the first to be "**carbon neutral,**" that is, to reduce net greenhouse gas emissions to zero, although she didn't say when this will occur.

Germany, also, has reduced its CO_2 emissions at least 10 percent by switching from coal to gas and by encouraging energy efficiency throughout society. Atmospheric scientist Steve Schneider calls this a "no regrets" policy; even if we don't need to stabilize our climate, many of these steps save money, conserve resources, and have other environmental benefits. Nuclear power also is being promoted as an alternative to fossil fuels. It's true that nuclear reactions don't produce greenhouse gases, but security worries and unresolved problems of how to store wastes safely make this option unacceptable to many people.

Many people believe renewable energy sources offer the best solution to climate problems. Chapter 20 discusses options for conserving energy and switching to renewable sources, such as solar, wind, geothermal, biomass, and fuel cells. Denmark, the world's leader in wind power, now gets 20 percent of its electricity from windmills. Plans are to generate half of the nation's electricity from offshore wind farms by 2030. Even China reduced its CO_2 emissions 20 percent between 1997 and 2005 through greater efficiency in coal burning and industrial energy use.

In addition to reducing its output, there are options for capturing and storing CO_2. Dr. Klaus S. Lackner of Columbia University has proposed building synthetic "trees" to capture CO_2 from the air (fig. 15.23). A calcium hydroxide solution trickling across the "leaves" would absorb CO_2. A structure with 2,500 m^2 of collecting area could remove 90,000 tons of CO_2 per year from the air (1,000 times more than a natural tree), an amount equal to the CO_2 produced by 15,000 cars. About 250,000 synthetic trees worldwide would be needed to soak up the 22 billion tons of CO_2 produced annually by burning fossil fuels. Critics claim that synthesizing the hydroxide needed for this reaction and disposing of the vast amount of calcium carbonate it would create may take more energy than the fuel that produced the CO_2 released. The first of these "trees" is now being built in Arizona. Planting real trees can also be effective if they're allowed to mature to old-growth status or are made into products, such as window frames and doors, that will last for many years before they're burned or recycled. Farmland also can serve as a carbon sink if farmers change their crop mixture and practice minimum till cultivation that keeps carbon in the soil.

FIGURE 15.23 Synthetic "trees," could capture CO_2 from the air. Each of these structures would be about 100 m (300 ft) tall and would have an array of metal slats 50 × 50 m across. Each of these trees could remove 90,000 tons of CO_2 per year from the air, an amount equal to the CO_2 produced by 15,000 cars. It would take a quarter of a million of them to remove all the CO_2 produced annually by humans.

In 1988 John H. Martin suggested that phytoplankton growth in the ocean may be limited by iron deficiency. This theory was tested in 2000, when researchers spread about 3.5 tons of dissolved iron over 75 km^2 of the South Pacific. Monitored by satellite, a ten-fold rise in chlorophyll concentration eventually spread over about 1,700 km^2 and was calculated to have pulled several thousand tons of CO_2 out of the air. Still, some ecologists warn about such large-scale tinkering. We don't know whether the carbon fixed by phytoplankton growth will be stored in sediments or simply eaten by predators and returned immediately to the atmosphere. It's possible that dying algae might create an anoxic zone that would devastate oceanic food webs.

Another way to store CO_2 is to inject it into underground strata or deep ocean waters. Since 1996, Norway's Statoil has been pumping more than 1 million metric tons of CO_2 per year into an aquifer 1,000 m below the seafloor at a North Sea gas well. This enhances oil recovery. It also saves money because otherwise the company would have to pay a $50 per ton carbon tax on its emissions. Around the world, deep, briny aquifers could store a century worth of CO_2 output at current fossil fuel consumption rates.

It might also be possible to pump liquid CO_2 into deep ocean trenches, where it would form lakes contained by enormous water pressures. There are worries, however, about what this might do to deep ocean fauna and what might happen if earthquakes or landslides caused a sudden release of this CO_2. This suggestion got a boost, however, with the recent discovery of a natural lake of liquid CO_2 off the coast of Japan. Microbial life around the edge of the lake was flourishing.

In Canada, an Alberta power plant is injecting CO_2 into a coal seam too deep to be mined. The CO_2 releases natural gas, which is burned to produce electricity and more CO_2. Proponents

FIGURE 15.24 Burning fossil fuels produces about half our greenhouse gas emissions, and transportation accounts for about half our fossil fuel consumption. Driving less, choosing efficient vehicles, carpooling, and other conservation measures are among our most important personal choices in the effort to control global warming.

of **carbon management,** as these various projects are called, argue that it may be cheaper to clean up fossil fuel effluents than to switch to renewable energy sources.

Most attention is focused on CO_2 because it lasts in the atmosphere, on average, for about 120 years. Methane and other greenhouse gases are much more powerful infrared absorbers but remain in the air for a much shorter time. NASA's Dr. James Hansen made a controversial suggestion, however, that the best short-term attack on warming might come by focusing not on carbon dioxide but on methane. Reducing gas pipeline leaks would conserve this valuable resource as well as help the environment.

Methane from landfills, oil wells, and coal mines that once would have been simply vented into the air is now being collected and used to generate electricity. Rice paddies are a major methane source. Changing flooding schedules and fertilization techniques can reduce anoxia (oxygen depletion) that produces marsh gas. Finally, ruminant animals (such as cows, camels, and buffalo) create large amounts of digestive system gas. Modified diets can reduce methane releases significantly.

CONCLUSION

We appear to be at a tipping point in our attitudes toward climate change. Where, a few years ago, most people in America thought that this was just a wild theory of a small group of crazy scientists, we've gone to a widespread acceptance that the science of global warming is indisputable. The surprising success of former Vice President Al Gore's documentary, *An Inconvenient Truth,* has been a powerful force in this social transformation. Perhaps more important is that many people can actually see tangible proof in their own lives that something strange is happening to our climate.

But is it too late to do anything? Most scientists believe there is time, if we act quickly and effectively, to avoid the worst

Soot, while not mentioned in the Kyoto Protocol, might also be important in global warming. Dark, airborne particles absorb both ultraviolet and visible light, converting them to heat energy. According to some calculations, reducing soot from diesel engines, coal-fired generators, forest fires, and wood stoves might reduce net global warming by 40 percent within three to five years. Curbing soot emissions would also have beneficial health effects.

Some individual cities and states have announced their own plans to combat global warming. Among the first of these were Toronto, Copenhagen, and Helsinki, which pledged to reduce CO_2 emissions 20 percent from 1990 levels by 2010. Some corporations are following suit. British Petroleum has set a goal of cutting CO_2 releases from all its facilities by 10 percent before 2010. Each of us can make a contribution in this effort. As Professor Socolow and his colleagues point out, simply driving less and buying high-mileage vehicles could save about 1.5 billion tons of carbon emissions by 2054 (fig. 15.24).

In the midst of all the debate about how serious the consequences of global climate change may or may not be, we need to remember that many of the proposed solutions are advantageous in their own right. Even if climate change turns out not to be as much of a threat as we think now, they have other positive benefits. Moving from fossil fuels to renewable energy sources such as solar or wind power, for example, would free us from dependence on foreign oil and improve air quality. Planting trees makes cities more pleasant places to live and provides habitat for wildlife. Making buildings more energy efficient and buying high-mileage vehicles saves money in the long run. Walking, biking, and climbing stairs are good for your health as well as reducing traffic congestion and energy consumption. Reducing waste, recycling, and other forms of sustainable living improve our environment in many ways in addition to helping fight climate change. It's important to focus on these positive effects rather than to look only at the gloom-and-doom scenarios for global climate catastrophes. As the Irish statesman and philosopher Edmund Burke said, "Nobody made a greater mistake than he who did nothing because he could do only a little."

consequences of global climate change. It's encouraging to know that we don't need a technological miracle to save us from this impending catastrophe. We have the ability to dramatically reduce our greenhouse gas emissions right now. And it will probably save money and create jobs to do so.

It's said that Napoleon Bonaparte ordered his government to plant trees along all French roads to provide shade and fodder for local people. "But sir," the bureaucrats protested, "it will take a generation for the trees to grow." "All the more reason to start immediately," Bonaparte responded.

REVIEWING LEARNING OUTCOMES

By now you should be able to explain the following points:

15.1 Illustrate how the atmosphere is a complex machine.
- The sun warms our world.
- Water stores energy, and winds redistribute it.

15.2 Explain how weather happens.
- Why does it rain?
- Large-scale winds don't move in a straight line.
- Ocean currents modify our weather.
- Seasonal winds and monsoons have powerful effects.
- Frontal systems create local weather.
- Cyclonic storms can cause extensive damage.

15.3 Clarify how climate can be an angry beast.
- Climates have changed dramatically throughout history.
- What causes catastrophic climate swings?
- El Niño/Southern Oscillations are powerful cycles.

15.4 Understand why global warming is happening.
- A scientific consensus is emerging.
- Greenhouse gases have many sources.
- Evidence of climate change is overwhelming.
- Global warming will be expensive.

15.5 Evaluate how the Kyoto Protocol attempts to slow climate change.
- There are many ways we can control greenhouse emissions.
- Progress is being made.

PRACTICE QUIZ

1. What are weather and climate? How do they differ?
2. Name and describe the four layers of air in the atmosphere.
3. What is the greenhouse effect?
4. What are the ENSO cycle and the PDO?
5. What are the jet streams? How do they influence weather patterns?
6. How do tornadoes form and why are they so destructive?
7. Describe the Coriolis effect. What causes it?
8. Explain some of the observed or expected effects of climate change.
9. Summarize some actions we could take, collectively and as individuals, to combat global climate change.
10. Why has the United States not ratified the Kyoto Protocol? What effects may this have for the U.S. economy?

CRITICAL THINKING AND DISCUSSION QUESTIONS

1. In most places, weather changes in highly variable, seemingly chaotic ways. How would you distinguish between meaningful patterns and random variations?
2. Most scientists say global warming is real; politicians, however, dispute this claim. How can we decide what to do when the science is uncertain and the risk is great?
3. Has there been a major change recently in the weather where you live? Can you propose any reasons for such changes?
4. What forces determine the climate in your locality? Are they the same for neighboring states?
5. Which of the seven options recommended by Professor Socolow do you think has the greatest chance of success?
6. Would you favor building nuclear power plants to reduce CO_2 emissions? Why or why not?

Methane has about 23 times the climate change potential of CO_2. Although methane is a much more minor component of the atmosphere than CO_2, it has been accumulating at a faster rate. In 2007 the IPCC reported that atmospheric methane concentrations have increased about 2.5 times from preindustrial levels (measured from ice core data). The IPCC concluded it is *"very likely"* that the observed increase is do to anthropogenic activities, predominantly agriculture and fossil fuel use. Look at the graphs on this page and answer the following questions:

1. What do the curves look different on these two graphs? (Hint: look at the x axis)

2. Is one graph more accurate than the other?

3. Do these graphs give different impressions about methane trends?

4. What are the units of measurement in these two graphs?

5. Assuming that the industrial revolution started in about 1800, what was the preindustrial methane concentration?

6. In which graph is it easier to determine the 1800 concentration?

7. What was the atmospheric methane concentration in 2000?

8. How does the current world methane concentration differ from 10,000 years ago?

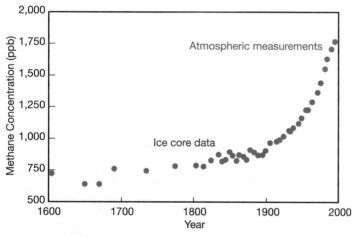

Changes in atmospheric methane measured from ice core data.

For Additional Help in Studying This Chapter, please visit our website at www.mhhe.com/cunningham10e. You will find additional practice quizzes and case studies, flashcards, regional examples, place markers for Google Earth™ mapping, and an extensive reading list, all of which will help you learn environmental science.

Coal-burning power plants produce about two-thirds of the sulfur oxides, one-third of all nitrogen oxides, and half of the mercury emitted in the United States each year.

16

Air Pollution

The only thing we have to fear is fear itself.
—Franklin D. Roosevelt—

LEARNING OUTCOMES

After studying this chapter, you should be able to:

16.1 Describe the air around us.

16.2 Identify natural sources of air pollution.

16.3 Discuss human-caused air pollution.

16.4 Explain how climate topography and atmospheric processes affect air quality.

16.5 Compare the effects of air pollution.

16.6 Evaluate air pollution control.

16.7 Summarize current conditions and future prospects.

Case Study Controlling Mercury Pollution

Often referred to as quicksilver, mercury is the only common metal that is liquid at room temperature. It's useful in a wide variety of industrial processes, and found in a host of commercial products including paints, batteries, fluorescent light bulbs, electrical switches, pesticides, skin creams, antifungal agents, and old thermometers.

Mercury also is a powerful neurotoxin that destroys the brain and central nervous system at high doses. Even tiny amounts of this toxic metal can cause neurological and developmental defects in young children. In a survey of freshwater fish from 260 lakes across the United States, the EPA found that every fish sampled had at least low levels of mercury. More than half the fish contained mercury levels unsafe for women of childbearing age, and three-quarters exceed the safe limit for young children. Fifty states have issued warnings about eating freshwater or ocean fish; mercury contamination is by far the most common reason for these advisories. In 1994, the EPA declared mercury a hazardous pollutant regulated under the Clean Air Act. Municipal and medical incinerators were ordered to reduce their mercury emissions by 90 percent. Chlorine-caustic soda plants and gold mining operations, the largest commercial mercury consumers, also agreed to reduce mercury releases by about 50 percent. Left unregulated, however, were the 1,032 coal-burning power plants, which are estimated to emit some 48 tons of mercury per year, or nearly half of total annual U.S. emissions.

Finally, in 2000, the EPA declared that mercury from power plants also is a risk to public health. Had the agency applied existing air-toxin regulations, utilities would have required installation of maximum achievable control technology, which would reduce power plant emissions by about 90 percent (similar to that achieved by incinerators) within three to five years. Rather than impose mandatory rules on industry, the EPA chose to regulate mercury through "cap and trade" market mechanisms that allow utilities to buy and sell pollution rights rather than have government specified pollution controls. This plan aims to reduce mercury releases by 70 percent by 2018, but the Congressional Research Service predicts that this reduction will be unlikely before 2030.

Critics contend that while cap and trade systems work well for some pollutants, they are an inappropriate form of regulation for a toxic substance, such as mercury. They say it will allow utilities to continue to emit mercury for years longer than necessary. It also will create mercury "hot spots," often in poor communities, where utilities find it cheaper to purchase pollution credits from distant facilities rather than clean up their own plants. Many eastern states oppose this approach because they suffer from high mercury pollution generated far away and blown in by prevailing winds (fig. 16.1).

Officially, the EPA estimates the benefits from reducing mercury emissions from power plants to be only $50 million a year, while the cost to industry would be $750 million a year. An independent study, on the other hand, concluded that mercury controls could save nearly $5 billion a year through reduced neurological and cardiac harm. Part of the discrepancy is that the EPA estimates that three-quarters of the mercury deposited in U.S. waterways comes from abroad and that 70 percent of the U.S. mercury emissions fell on other countries or the oceans. Thus, changes in mercury emissions from U.S. power plants would benefit the global environment more than U.S. residents.

Part of the debate here is whether we have an ethical responsibility to the rest of the world for the pollutants we emit.

Meanwhile, in the Allegheny Mountains of West Virginia, a huge coal-fired power plant is adding fuel to the mercury debate. Aptly named Mount Storm, it's a 1,600-megawatt Goliath that just a few years ago ranked second in the nation in mercury emissions. Since new pollution controls were installed to scrub sulfur and nitrogen oxides from its smokestack, however, 95 percent of its mercury also is being captured at no extra cost. This success story casts an odd light on the United States' mercury rules. If existing technology can remove so much mercury economically, why wait until 2018 to impose similarly stringent limits on other power plants? Utility executives, however, protest that technology that works for a specific plant may not be suitable everywhere.

This case study illustrates both the importance and the difficulty of regulating air pollution. Highly mobile, widely dispersed, produced by a variety of sources, and having diverse impacts, air pollutants can be challenging to regulate. Often air quality controversies—such as mercury control—pit a diffuse public interest (improving general health levels or IQ a few points) versus a very specific private interest (making utilities pay millions of dollars per year to control pollutants). In this chapter, we'll look at the major types and sources of air pollution, as well as more on the controversy about how best to ensure a healthy environment.

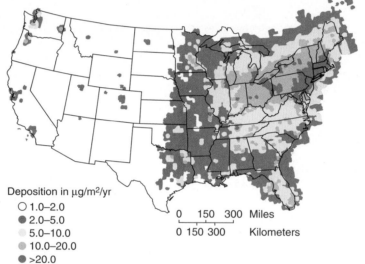

Deposition in μg/m²/yr
○ 1.0–2.0
● 2.0–5.0
○ 5.0–10.0
○ 10.0–20.0
● >20.0

0   150   300   Miles
0 150 300   Kilometers

FIGURE 16.1 Atmospheric mercury deposition in the United States. Due to prevailing westerly winds, and high levels of industrialization, eastern states have high mercury deposition.
Source: EPA, 1998.

FIGURE 16.2 On a smoggy day in Shanghai (*left*) visibility is less than 1 km. Twenty-four hours later, after a rainfall (*right*), the air has cleared dramatically.

16.1 THE AIR AROUND US

How does the air taste, feel, smell, and look in your neighborhood? Chances are that wherever you live, the air is contaminated to some degree. Smoke, haze, dust, odors, corrosive gases, noise, and toxic compounds are present nearly everywhere, even in the most remote, pristine wilderness. Air pollution is generally the most widespread and obvious kind of environmental damage. According to the Environmental Protection Agency (EPA), some 147 million metric tons of air pollution (not counting carbon dioxide or wind-blown soil) are released into the atmosphere each year in the United States by human activities. Total worldwide emissions of these pollutants are around 2 billion metric tons per year.

Over the past 30 years, however, air quality has improved appreciably in most cities in Europe, North America, and Japan. Many young people might be surprised to learn that a generation ago most American cities were much dirtier than they are today. This is an encouraging example of improvement in environmental conditions. Our success in controlling some of the most serious air pollutants gives us hope for similar progress in other environmental problems.

While developed countries have been making progress, however, air quality in the developing world has been getting much worse. Especially in the burgeoning megacities of rapidly industrializing countries (chapter 22), air pollution often exceeds World Health Organization standards most of the time. In many Chinese cities, for example, airborne dust, smoke, and soot often are ten times higher than levels considered safe for human health (fig. 16.2). Currently, seven of the ten smoggiest cities in the world are in China.

16.2 NATURAL SOURCES OF AIR POLLUTION

It is difficult to give a simple, comprehensive definition of pollution. The word comes from the Latin *pollutus,* which means made foul, unclean, or dirty. Some authors limit the use of the term to damaging materials that are released into the environment by human activities. There are, however, many natural sources of air quality degradation. Volcanoes spew out ash, acid mists, hydrogen sulfide, and other toxic gases (fig. 16.3). Sea spray and decaying vegetation are major sources of reactive sulfur compounds in the air. Trees and bushes emit millions of tons of volatile organic compounds (terpenes and isoprenes), creating, for example, the blue haze that gave the Blue Ridge Mountains their name. Pollen, spores, viruses, bacteria, and other small bits of organic material in the air cause widespread suffering from allergies and airborne infections. Storms in arid regions raise dust clouds that transport millions of tons of soil and can be detected half a world away. Bacterial metabolism of decaying vegetation in swamps and of cellulose in the guts of termites and ruminant animals is responsible for as much as two-thirds of the methane (natural gas) in the air.

Does it make a difference whether smoke comes from a natural forest fire or one started by humans? In many cases, the chemical compositions of pollutants from natural and human-related sources are identical, and their effects are inseparable. Sometimes, however, materials in the atmosphere are considered innocuous at naturally occurring levels, but when humans add to these levels, overloading of natural cycles or disruption of essential processes can occur. While the natural sources of suspended

FIGURE 16.3 Natural pollution sources, such as volcanoes, can be important health hazards.

FIGURE 16.4 Primary pollutants are released directly from a source into the air. A point source is a specific location of highly concentrated discharge, such as this smokestack.

particulate material in the air outweigh human sources at least tenfold worldwide, in many cities more than 90 percent of the airborne particulate matter is anthropogenic (human-caused).

16.3 HUMAN-CAUSED AIR POLLUTION

What are the major types of anthropogenic air pollutants and where do they come from? In this section, we will define some general categories and sources of air pollution.

We Categorize pollutants according to their source

Primary pollutants are those released directly from the source into the air in a harmful form (fig. 16.4). **Secondary pollutants,** by contrast, are modified to a hazardous form after they enter the air or are formed by chemical reactions as components of the air mix and interact. Solar radiation often provides the energy for these reactions. Photochemical oxidants and atmospheric acids formed by these mechanisms are probably the most important secondary pollutants in terms of human health and ecosystem damage. We will discuss several important examples of such pollutants in this chapter.

Fugitive emissions are those that do not go through a smokestack. By far the most massive example of this category is dust from soil erosion, strip mining, rock crushing, and building construction (and destruction). In the United States, natural and

anthropogenic sources of fugitive dust add up to some 100 million metric tons per year. The amount of CO_2 released by burning fossil fuels and biomass is nearly equal in mass to fugitive dust. Fugitive industrial emissions are also an important source of air pollution. Leaks around valves and pipe joints contribute as much as 90 percent of the hydrocarbons and volatile organic chemicals emitted from oil refineries and chemical plants.

The U.S. Clean Air Act of 1970 designated seven major pollutants (sulfur dioxide, carbon monoxide, particulates, hydrocarbons, nitrogen oxides, photochemical oxidants, and lead) for which maximum **ambient air** (air around us) levels are mandated. These seven **conventional** or **criteria pollutants** contribute the largest volume of air-quality degradation and also are considered the most serious threat of all air pollutants to human health and welfare. Figure 16.5 shows the major sources of the first six criteria pollutants. Table 16.1 shows an estimate of the total annual worldwide emissions of some important air pollutants. Now let's look more closely at the sources and characteristics of each of these major pollutants.

We also categorize pollutants according to their content

Seven "conventional" pollutants were regulated by the original Clean Air Act.

Sulfur Compounds

Natural sources of sulfur in the atmosphere include evaporation of sea spray, erosion of sulfate-containing dust from arid soils, fumes from volcanoes and fumaroles, and biogenic emissions of hydrogen sulfide (H_2S) and organic sulfur-containing compounds, such as dimethylsulfide, methyl mercaptan, carbon disulfide, and carbonyl

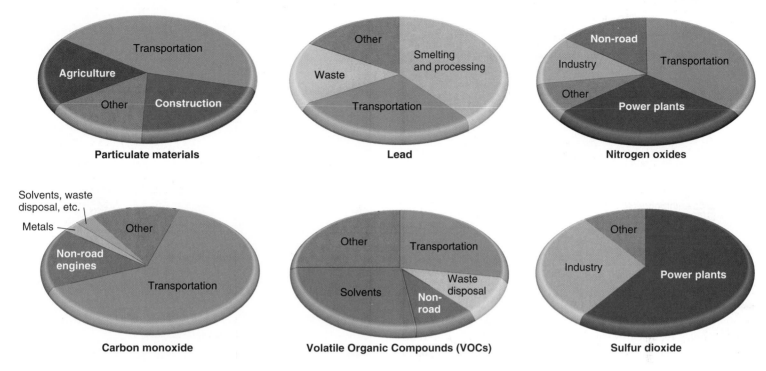

FIGURE 16.5 Anthropogenic sources of six of the primary "criteria" air pollutants in the United States.
Source: UNEP, 1999.

TABLE 16.1

Estimated Fluxes of Pollutants and Trace Gases to the Atmosphere

Species	Sources	Approximate Annual Flux (Millions of Metric Tons/Yr) Natural	Anthropogenic
CO_2 (carbon dioxide)	Respiration, fossil fuel burning, land clearing, industrial processes	370,000	29,600*
CH_4 (methane)	Rice paddies and wetlands, gas drilling, landfills, animals, termites	155	350
CO (carbon monoxide)	Incomplete combustion, CH_4 oxidation, biomass burning, plant metabolism	1,580	930
NMHC (nonmethane hydrocarbons)	Fossil fuels, industrial uses, plant isoprenes and other biogenics	860	92
NO_x (nitrogen oxides)	Fossil fuel burning, lightning, biomass burning, soil microbes	90	140
SO_x (sulfur oxides)	Fossil fuel burning, industry, biomass burning, volcanoes, oceans	35	79
SPM (suspended particulate materials)	Biomass burning, dust, sea salt, biogenic aerosols, gas-to-particle conversion	583	362

*Only 27.3 percent of this amount—or 8 billion tons—is carbon.
Source: UNEP, 1999.

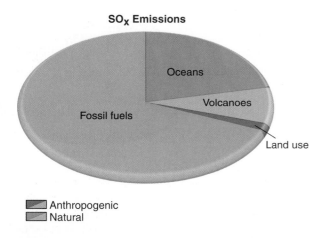

SOₓ Emissions

Oceans

Volcanoes

Fossil fuels

Land use

■ Anthropogenic
■ Natural

FIGURE 16.6 Sulfur fluxes into the atmosphere.
Source: UNEP, 1999.

sulfide. Total yearly emissions of sulfur from all sources amount to some 114 million metric tons (fig. 16.6). Worldwide, anthropogenic sources represent about two-thirds of the total sulfur flux, but in most urban areas they contribute as much as 90 percent of the sulfur in the air. The predominant form of anthropogenic sulfur is sulfur dioxide (SO_2) from combustion of sulfur-containing fuel (coal and oil), purification of sour (sulfur-containing) natural gas or oil, and industrial processes, such as smelting of sulfide ores. China and the United States are the largest sources of anthropogenic sulfur, primarily from coal burning.

Sulfur dioxide is a colorless corrosive gas that is directly damaging to both plants and animals. Once in the atmosphere, it can be further oxidized to sulfur trioxide (SO_3), which reacts with water vapor or dissolves in water droplets to form sulfuric acid (H_2SO_4), a major component of acid rain. Very small solid particles or liquid droplets can transport the acidic sulfate ion (SO_4^{-2}) long distances through the air or deep into the lungs where it is very damaging. Sulfur dioxide and sulfate ions are probably second only to smoking as causes of air pollution-related health damage. Sulfate particles and droplets reduce visibility in the United States as much as 80 percent. Some of the smelliest and most obnoxious air pollutants are sulfur compounds, such as hydrogen sulfide from pig manure lagoons or mercaptans (organosulfur thiols) from papermills (fig. 16.7).

Nitrogen Compounds

Nitrogen oxides are highly reactive gases formed when nitrogen in fuel or combustion air is heated to temperatures above 650°C (1,200°F) in the presence of oxygen, or when bacteria in soil or water oxidize nitrogen-containing compounds. The initial product, nitric oxide (NO), oxidizes further in the atmosphere to nitrogen dioxide (NO_2), a reddish brown gas that gives photochemical smog its distinctive color. Because of their interconvertibility, the general term NO_x is used to describe these gases. Nitrogen oxides combine with water to make nitric acid (HNO_3), which is also a major component of atmospheric acidification.

The total annual emissions of reactive nitrogen compounds into the air are about 230 million metric tons worldwide (see table 16.1).

FIGURE 16.7 The most annoying pollutants from this paper mill are pungent organosulfur thiols and sulfides. Chlorine bleaching can also produce extremely dangerous organochlorines, such as dioxins.

Anthropogenic sources account for 60 percent of these emissions (fig. 16.8). About 95 percent of all human-caused NO_x in the United States is produced by fuel combustion in transportation and electric power generation. Nitrous oxide (N_2O) is an intermediate in soil denitrification that absorbs ultraviolet light and plays an important role in climate modification (chapter 15). Excess nitrogen is causing fertilization and eutrophication of inland waters and coastal seas. It also may be adversely affecting terrestrial plants both by excess fertilization and by encouraging growth of weedy species that crowd out native varieties.

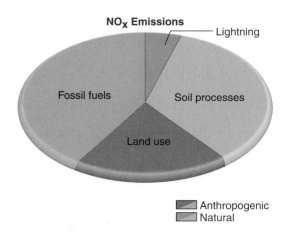

NOₓ Emissions

Lightning

Fossil fuels

Soil processes

Land use

■ Anthropogenic
■ Natural

FIGURE 16.8 Worldwide sources of reactive nitrogen gases in the atmosphere.
Source: UNEP, 1999.

Carbon Oxides

The predominant form of carbon in the air is carbon dioxide (CO_2). It is usually considered nontoxic and innocuous, but increasing atmospheric levels (about 0.5 percent per year) due to human activities is now causing global climate change that may have disastrous effects on both human and natural communities. As table 16.1 shows, more than 90 percent of the CO_2 emitted each year is from respiration (oxidation of organic compounds by plant and animal cells). These releases are usually balanced, however, by an equal uptake by photosynthesis in green plants.

Anthropogenic (human-caused) CO_2 releases are difficult to quantify because they spread across global scales. The best current estimate from the Intergovernmental Panel on Climate Change (IPCC) is that between 7 and 8 billion tons of carbon (in the form of CO_2) are released each year by fossil fuel combustion and that another 1 to 2 billion tons are released by forest and grass fires, cement manufacturing, and other human activities. Typically, terrestrial ecosystems take up about 3 billion tons of this excess carbon every year, while oceanic processes take up another 2 billion tons. This leaves an average of at least 3 billion tons to accumulate in the atmosphere. The actual releases and uptakes vary greatly, however, from year to year. Some years almost all anthropogenic CO_2 is reabsorbed; in other years, almost none of it is. The ecological processes that sequester CO_2 depend strongly on temperature, nutrient availability, and other environmental factors.

United States negotiators at the Global Climate meetings claim that forests and soils in North America act as carbon sinks—that is, they take up more carbon than is released by other activities. Over the past decade, CO_2 levels in air coming ashore on the U.S. West Coast have averaged about 2 ppm higher than air leaving from the East Coast. If we assume that there isn't a major inflow of CO_2-depleted air entering from Canada or Mexico, this would mean that somewhere between 1.6 and 2.2 billion tons of CO_2 are being taken up every year than are being released in the United States. Other countries doubt these measurements, however, and refuse to give the United States credit for this large carbon sequestration.

Carbon monoxide (CO) is a colorless, odorless, nonirritating but highly toxic gas produced by incomplete combustion of fuel (coal, oil, charcoal, or gas), incineration of biomass or solid waste, or partially anaerobic decomposition of organic material. CO inhibits respiration in animals by binding irreversibly to hemoglobin. About 1 billion metric tons of CO are released to the atmosphere each year, half of that from human activities. In the United States, two-thirds of the CO emissions are created by internal combustion engines in transportation. Land-clearing fires and cooking fires also are major sources. About 90 percent of the CO in the air is consumed in photochemical reactions that produce ozone.

Particulate Material

An **aerosol** is any system of solid particles or liquid droplets suspended in a gaseous medium. For convenience, we generally describe all atmospheric aerosols, whether solid or liquid, as **particulate material.** This includes dust, ash, soot, lint, smoke, pollen, spores, algal cells, and many other suspended materials. Anthropogenic particulate emissions amount to about 362 million metric tons per year worldwide.

Wind-blown dust, volcanic ash, and other natural materials may contribute considerably more suspended particulate material.

Particulates often are the most apparent form of air pollution since they reduce visibility and leave dirty deposits on windows, painted surfaces, and textiles. Respirable particles smaller than 2.5 micrometers are among the most dangerous of this group because they can be drawn into the lungs, where they damage respiratory tissues. Asbestos fibers and cigarette smoke are among the most dangerous respirable particles in urban and indoor air because they are carcinogenic.

Diesel fumes also are highly toxic because they contain both fine particulates and chemicals such as benzene, dioxins, and mercury. The EPA has proposed new rules to require low-sulfur fuel and antipollution devices, particularly for off-road engines such as bulldozers, tractors, pumps, and generators. Epidemiologists estimate that these new standards will prevent more than 360,000 asthma attacks and 8,300 premature deaths annually. Diesel owners protest that expenses will be exorbitant, but Europe has had standards ten times more stringent than the EPA proposes for several years.

In the 1930s, America experienced terrible soil erosion known as the dust bowl. Poor agricultural practices and policies, coupled with years of drought, left soil on 5 million ha of the southern plains exposed to the wind. Billowing clouds of dust darkened the skies for days and reached as far as Washington, D.C. In 1935, at the peak of the drought, an estimated 850 million tons of topsoil blew away in "black blizzards."

Soil conservation techniques have reduced dust storms in North America, but deserts and dust storms have increased elsewhere. Soil scientists report that 3 billion tons of sand and soil blow from drylands around the world every year. Although these storms start out gritty, coarse sediments soon fall out, leaving finer silts and clays to rise up to 4,500 m and can travel thousands of kilometers before settling out. Dust from Africa's Sahara desert regularly crosses the Atlantic and raises particulate levels above federal health standards in Miami and San Juan, Puerto Rico (fig. 16.9).

There are some benefits from these storms. Research has shown that existence of the Amazon rainforest depends on mineral nutrients carried in dust from Africa. Remarkably, more than half the 50 million tons of dust transported to South America each year has been traced to the bed of the former Lake Chad in Africa (see Chapter 17). There also are adverse effects. Huge storms blow out of China's Gobi desert. Every spring, dust clouds from China shut down airports and close schools in Japan and Korea. The dust plume follows the jet stream across the Pacific to Hawaii and then to the west coast of North America, where it sometimes makes up as much as half the particulate air pollution in Seattle, Washington. Some Asian dust storms have polluted the skies as far east as Savannah, Georgia, and Portland, Maine.

As we discussed in chapter 9, as much as one-third of the earth's surface is threatened by desertification. Population growth and poverty drive people into fragile, marginal lands that blow away when rains fail. Poor farming practices expose soils to wind erosion. The result is an escalating crisis that not only threatens food production but also pollutes air around the globe. The resulting haze reduces visibility in remote locations such as California's Sequoia National Park or Big Bend National Park in Texas.

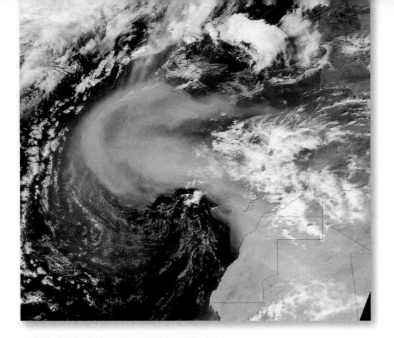

FIGURE 16.9 A massive dust storm extends more than 1,600 km (1,000 mi) from the coast of western Sahara and Morocco. Storms such as this can easily reach the Americas, and they have been linked both to the decline of coral reefs in the Caribbean and to the frequency and intensity of hurricanes formed in the eastern Atlantic Ocean.

Human health also suffers when dust fills the air. Epidemiological studies have found that cities with chronically high levels of particulates have higher death rates, mostly from heart and lung disease. Emergency-room visits and death rates rise in days following a dust storm. Some of this health risk comes from the particles themselves, which clog tiny airways and make breathing difficult. The dust also carries pollen, bacteria, viruses, fungi, herbicides, acids, radioactive isotopes, and heavy metals between continents. As we mentioned in the opening case study for this chapter, roughly half the mercury contamination falling on North America is thought to come from Asia; much of it may arrive attached to airborne particulates. Scientists once thought that living organisms couldn't survive the trip across oceans, but thick dust helps shield cells from sunlight and desiccation.

Airborne dust is considered the primary source of allergies worldwide. Saharan dust storms are suspected of raising asthma rates in Trinidad and Barbados, where cases have increased 17-fold in 30 years. *Aspergillus sydowii,* a soil fungus from Africa, has been shown to be causing death of corals and sea fans in remote reefs in the Caribbean. Europe also receives airborne pathogens via dust storms. Outbreaks of foot-and-mouth disease in Britain have been traced to dust storms from North Africa. The recent discovery of nanobacteria—the smallest known self-replicating organisms, about 100 times smaller than regular bacteria—in dust clouds suggests an even wider role of airborne pathogens in global disease. Although the effect of these tiny cells is controversial, they have been implicated in diseases as different as heart disease, kidney stone formation, and HIV.

Metals and Halogens

Many toxic metals are mined and used in manufacturing processes or occur as trace elements in fuels, especially coal. These metals are released to the air in the form of metal fumes or suspended particulates by fuel combustion, ore smelting, and disposal of wastes. Worldwide atmospheric lead emissions amount to about 2 million metric tons per year, or two-thirds of all metallic air pollution. Most of this lead is from leaded gasoline. Lead is a metabolic poison and a neurotoxin that binds to essential enzymes and cellular components and inactivates them.

Banning leaded gasoline is one of the most successful pollution-control measures in American history. Since 1986, when the ban was enforced, children's average blood lead levels have dropped 90 percent and average IQs have risen three points. Now, 50 nations have renounced leaded gasoline. The global economic benefit of this step is estimated to be more than $200 billion per year.

Mercury, described in the opening case study of this chapter, also is an extremely toxic substance. Young children and developing fetuses are particularly vulnerable. The most famous case of mercury poisoning occurred in Minamata, Japan, in the 1950s. A chemical factory dumped mercury-contaminated waste into the ocean, where it concentrated in fish. Babies whose mothers ate mercury-contaminated fish suffered profound neurological disabilities including deafness, blindness, mental retardation, and cerebral palsy. In adults, mercury poisoning can cause numbness, loss of muscle control, dementia, and death. The most dangerous form, dimethyl mercury, is so toxic that a single drop on your skin can kill you.

Mercury contamination is the most common cause of impairment of U.S. rivers and lakes. In 2007, researchers sampled more than 2,700 fish from 626 rivers and streams in 12 western states and found mercury in every one of them. Forty-five states have issued warnings about eating locally caught freshwater fish (fig. 16.10). Pregnant women and young children also are advised to limit their consumption of certain seafood, including swordfish, marlin, tuna,

FIGURE 16.10 Mercury contamination is the most common cause of impairment of U.S. rivers and lakes. Forty-five states have issued warnings about eating locally caught freshwater fish. Long-lived, top predators are especially likely to bioaccumulate toxic concentrations of mercury.

shark, and lobster. The U.S. National Institutes of Health (NIH) estimates that one woman in 12 in the United States has more mercury in her blood than 5.8 ug/l considered safe by the EPA. Between 300,000 and 600,000 of the 4 million children born each year in the United States are exposed in the womb to mercury levels that could cause diminished intelligence or developmental problems. According to the NIH, elevated mercury levels cost the U.S. economy $8.7 billion a year in higher medical and educational costs and lost productivity in the workforce.

Mercury comes from many sources. It's released by volcanoes, weathering of rocks, and evaporation from oceans and wetlands. Eighty-five percent of anthropogenic emissions come from smokestacks, primarily power plants and municipal and medical waste incinerators. Most mercury in the air is elemental metallic vapor. In this relatively insoluble and unreactive form, it can circumnavigate the globe for up to two years. When mercury gets into wetlands, however, it can be converted by microbes into an organic form that is absorbed by other organisms and concentrated as it passes through food webs. Top predators such as game fish can accumulate dangerously high levels of mercury.

Environment Canada's models indicate that 38 percent of the mercury pollution in the Great Lakes region, home to more than 9 million Canadians, comes from U.S. sources. They also note that airborne mercury from industrial regions, including the United States, is having "a serious and disproportionate impact on Canada's Northern and Arctic communities." Meanwhile, federal, provincial, and territorial governments in Canada are working under nationwide standards to reduce up to 90 percent of mercury emissions by 2010. Some states in America have similar programs. Massachusetts, for example, which has long had a serious mercury contamination problem, is requiring all its power plants to install scrubbing equipment that will remove 95 percent of mercury from smokestacks. Some observers wonder why the federal government can't do the same.

Think About It

Industry has a good idea of how much it will cost to install mercury scrubbers on power plants. But how much is it worth to add a point or two to a child's IQ? Would the answer be different if the child were yours?

Other toxic metals of concern are nickel, beryllium, cadmium, thallium, uranium, cesium, and plutonium. Some 780,000 tons of arsenic, a highly toxic metalloid, are released from metal smelters, coal combustion, and pesticide use each year. Halogens (fluorine, chlorine, bromine, and iodine) are highly reactive and generally toxic in their elemental form. Chlorofluorocarbons (CFCs) have been banned for most uses in industrialized countries, but about 600 million tons of these highly persistent chemical compounds are used annually worldwide in spray propellants, refrigeration compressors, and for foam blowing. They diffuse into the stratosphere where they release chlorine and fluorine atoms that destroy the ozone shield that protects the earth from ultraviolet radiation. We'll return to this topic later in this chapter.

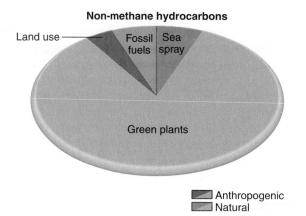

FIGURE 16.11 Sources of non-methane hydrocarbons in the atmosphere.
Source: UNEP, 1999.

Volatile Organic Compounds

Volatile organic compounds (VOCs) are organic chemicals that exist as gases in the air. Plants are the largest source of VOCs, releasing an estimated 350 million tons of isoprene (C_5H_8) and 450 million tons of terpenes ($C_{10}H_{15}$) each year (fig. 16.11). About 400 million tons of methane (CH_4) are produced by natural wetlands and rice paddies and by bacteria in the guts of termites and ruminant animals. These volatile hydrocarbons are generally oxidized to CO and CO_2 in the atmosphere.

In addition to these natural VOCs, a large number of other synthetic organic chemicals, such as benzene, toluene, formaldehyde, vinyl chloride, phenols, chloroform, and trichloroethylene, are released into the air by human activities. About 28 million tons of these compounds are emitted each year in the United States, mainly unburned or partially burned hydrocarbons from transportation, power plants, chemical plants, and petroleum refineries. These chemicals play an important role in the formation of photochemical oxidants.

Photochemical Oxidants

Photochemical oxidants are products of secondary atmospheric reactions driven by solar energy (fig. 16.12). One of the most important of these reactions involves formation of singlet (atomic)

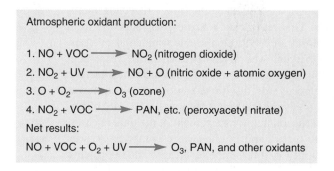

Atmospheric oxidant production:

1. NO + VOC ⟶ NO_2 (nitrogen dioxide)
2. NO_2 + UV ⟶ NO + O (nitric oxide + atomic oxygen)
3. O + O_2 ⟶ O_3 (ozone)
4. NO_2 + VOC ⟶ PAN, etc. (peroxyacetyl nitrate)

Net results:

NO + VOC + O_2 + UV ⟶ O_3, PAN, and other oxidants

FIGURE 16.12 Secondary production of urban smog oxidants by photochemical reactions in the atmosphere.

oxygen by splitting nitrogen dioxide (NO_2). This atomic oxygen then reacts with another molecule of O_2 to make **ozone** (O_3). Ozone formed in the stratosphere provides a valuable shield for the biosphere by absorbing incoming ultraviolet radiation. In ambient air, however, O_3 is a strong oxidizing reagent and damages vegetation, building materials (such as paint, rubber, and plastics), and sensitive tissues (such as eyes and lungs). Ozone has an acrid, biting odor that is a distinctive characteristic of photochemical smog. Hydrocarbons in the air contribute to accumulation of ozone by removing NO in the formation of compounds, such as peroxyacetyl nitrate (PAN), which is another damaging photochemical oxidant.

Air Toxins

Although most air contaminants are regulated because of their potential adverse effects on human health or environmental quality, a special category of toxins is monitored by the U.S. EPA because they are particularly dangerous. Called **hazardous air pollutants (HAPs),** these chemicals include carcinogens, neurotoxins, mutagens, teratogens, endocrine system disrupters, and other highly toxic compounds (chapter 8). Twenty of the most "persistent bioaccumulative toxic chemicals" (see table 8.2) require special reporting and management because they remain in ecosystems for long periods of time, and accumulate in animal and human tissues. Most of these chemicals are either metal compounds, chlorinated hydrocarbons, or volatile organic compounds.

Only about 50 locations in the United States regularly measure concentrations of HAPs in ambient air. Often the best source of information about these chemicals is the **Toxic Release Inventory (TRI)** collected by the EPA as part of the community right-to-know program. Established by Congress in 1986, the TRI requires 23,000 factories, refineries, hard rock mines, power plants, and chemical manufacturers to report on toxin releases (above certain minimum amounts) and waste management methods for 667 toxic chemicals. Although this total is less than 1 percent of all chemicals registered for use, and represents a limited range of sources, the TRI is widely considered the most comprehensive source of information about toxic pollution in the United States.

In 2005, U.S. industries released 4 billion pounds (1.8 million metric tons) of toxic chemicals into the environment from 24,000 facilities and disposed of roughly four times that much through waste management and recycling methods. This represents a 52 percent reduction from releases in 1999. Of the environmental releases, 65 percent was discharged on land, 27 percent (486,000 metric tons) was released into the air, 4 percent was discharged into water, and 4 percent was injected into deep wells. Twenty chemicals accounted for 88 percent of the total releases, with metals and mining waste comprising a vast majority of that amount. Nearly 200,000 metric tons of persistent bioaccumulative chemicals are emitted annually, with mercury and lead compounds comprising 97 percent of that total.

While most HAP releases are decreasing, discharges of mercury and dioxins—both of which are bioaccumulative and toxic at extremely low levels—have increased in recent years. Dioxins are created mainly by burning plastics and medical waste containing chlorine. The EPA reports that 100 million Americans live in areas where the cancer rate from HAPs exceeds 10 in 1 million or ten times the normally accepted standard for action (fig. 16.13). Benzene, formaldehyde, acetaldehyde, and 1,3 butadiene are responsible for most of this HAP cancer risk. Furthermore, twice that many (70 percent of the U.S. population) live in areas where the non-cancer risk of death exceeds 1 in 1 million. To help residents track local air quality levels, the EPA recently estimated the concentration of HAPs in localities across the continental United States (over 60,000 census tracts). You can access this information on the Environmental Defense Fund web page at www.scorecard.org/env-releases/hap/.

Unconventional pollutants also are important

In addition to toxic air pollutants, some other unconventional forms of air pollution deserve mention. **Aesthetic degradation** includes any undesirable changes in the physical characteristics or chemistry of the atmosphere. Noise, odors, and light pollution are examples of atmospheric degradation that may not be life-threatening but reduce the quality of our lives. This is a very subjective category. Odors and noise (such as loud music) that are offensive to some may be attractive to others. Often the most sensitive device for odor detection is the human nose. We can smell styrene, for example, at 44 parts per billion (ppb). Trained panels of odor testers often are used to evaluate air samples. Factories that emit noxious chemicals sometimes spray "odor maskants" or perfumes into smokestacks to cover up objectionable odors.

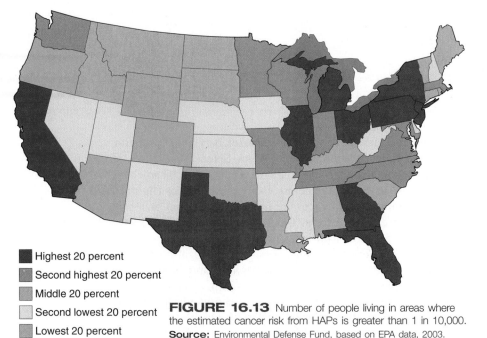

Highest 20 percent

Second highest 20 percent

Middle 20 percent

Second lowest 20 percent

Lowest 20 percent

FIGURE 16.13 Number of people living in areas where the estimated cancer risk from HAPs is greater than 1 in 10,000. **Source:** Environmental Defense Fund, based on EPA data, 2003.

In most urban areas, it is difficult or impossible to see stars in the sky at night because of dust in the air and stray light from buildings, outdoor advertising, and streetlights. This light pollution has become a serious problem for astronomers.

Indoor air is more dangerous for most of us than outdoor air

We have spent a considerable amount of effort and money to control the major outdoor air pollutants, but we have only recently become aware of the dangers of indoor air pollutants. The EPA has found that indoor concentrations of toxic air pollutants are often higher than outdoors. Furthermore, people generally spend more time inside than out and therefore are exposed to higher doses of these pollutants.

Smoking is without doubt the most important air pollutant in the United States in terms of human health. The Surgeon General estimates that more than 400,000 people die each year in the United States from emphysema, heart attacks, strokes, lung cancer, or other diseases caused by smoking. These diseases are responsible for 20 percent of all mortality in the United States, or four times as much as infectious agents. Lung cancer has now surpassed breast cancer as the leading cause of cancer deaths for U.S. women. Advertising aimed at making smoking appear stylish and liberating has resulted in a 600 percent increase in lung cancer among women since 1950.

Total costs for early deaths and smoking-related illnesses in the United States are estimated to be $100 billion per year. Eliminating smoking probably would save more lives than any other pollution-control measure. Smoking restrictions in many places have resulted in dramatic declines in second-hand smoke exposure to nonsmokers, EPA data show. In just a decade after indoor smoking bans were passed, levels of tobacco by-products in nonsmokers' blood dropped 75 percent. With increasing restrictions on smoking in Western countries, tobacco companies are now turning their attention to developing countries. Persuading consumers (especially women, who traditionally don't smoke) that American cigarettes are modern and stylish could recruit billions of new customers—and cause hundreds of millions of cancer deaths.

In some cases, indoor air in homes has concentrations of chemicals that would be illegal outside or in the workplace. The EPA has found that concentrations of such compounds as chloroform, benzene, carbon tetrachloride, formaldehyde, and styrene can be seventy times higher in indoor air than in outdoor air. "Green design" principles can make indoor spaces both healthier and more pleasant (Exploring Science, p. 358).

In the less-developed countries of Africa, Asia, and Latin America where such organic fuels as firewood, charcoal, dried dung, and agricultural wastes make up the majority of household energy, smoky, poorly ventilated heating and cooking fires represent the greatest source of indoor air pollution (fig. 16.14). The World Health Organization (WHO) estimates that 2.5 billion people—nearly half the world's population—are adversely affected by pollution from this source. Women and small children spend long hours each day around open fires or unventilated stoves in enclosed

FIGURE 16.14 Smoky cooking and heating fires may cause more ill health effects than any other source of indoor air pollution except tobacco smoking. Some 2.5 billion people, mainly women and children, spend hours each day in poorly ventilated kitchens and living spaces where carbon monoxide, particulates, and cancer-causing hydrocarbons often reach dangerous levels.

spaces. The levels of carbon monoxide, particulates, aldehydes, and other toxic chemicals can be 100 times higher than would be legal for outdoor ambient concentrations in the United States. Designing and building cheap, efficient, nonpolluting energy sources for the developing countries would not only save shrinking forests but would make a major impact on health as well.

16.4 CLIMATE, TOPOGRAPHY, AND ATMOSPHERIC PROCESSES

Topography, climate, and physical processes in the atmosphere play an important role in transport, concentration, dispersal, and removal of many air pollutants. Wind speed, mixing between air layers, precipitation, and atmospheric chemistry all determine whether pollutants will remain in the locality where they are produced or go elsewhere. In this next section, we will survey some environmental factors that affect air pollution levels.

Temperature inversions trap pollutants

Temperature inversions occur when a stable layer of warmer air overlays cooler air, reversing the normal temperature decline with increasing height and preventing convection currents from dispersing pollutants. Several mechanisms create inversions. When a cold front slides under an adjacent warmer air mass or when cool air subsides down a mountain slope to displace warmer air in the valley below, an inverted temperature gradient is established. These inversions are usually not stable, however, because winds

How safe is the air in your home, office, or school room? As we decrease air-infiltration into buildings to conserve energy, we often trap indoor air pollutants within spaces where most of us spend the vast majority of our time. In what has come to be known as "sick building syndrome," people complain of headaches, fatigue, nausea, upper-respiratory problems, and a wide variety of allergies from workplace or home exposure to airborne toxins. While these symptoms often are vague and difficult to verify scientifically, the U.S. Environmental Protection Agency estimates that sick building syndrome may cost $60 billion a year in medical expenses, absenteeism, and reduced productivity.

What might be making us sick? Mold spores are probably the greatest single cause of allergic reactions in indoor air. Moisture trapped in air-tight houses often accumulates in walls where molds flourish. Air ducts provide both a good environment for growth of pathogens such as Legionnaire's disease bacteria as well as a path for their dispersal. Legionnaire's pneumonia is much more prevalent than most people realize in places like California and Australia where air-conditioning is common. Uranium-bearing rocks and sediment are widespread across North America. When uranium decays, it produces carcinogenic radon gas that can seep into buildings. The EPA warns that one home in ten in the United States may exceed the recommended maximum radon concentration of 4 picocuries per liter.

In addition, we are exposed to a variety of synthetic chemicals emitted from carpets, wall coverings, building materials, and combustion gases (see fig. 8.11). You might be surprised to learn how many toxic, synthetic compounds are used to construct buildings and make furniture. Formaldehyde, for instance is a component of more than 3,000 products, including building materials such as particle board, waferboard, and urea-formaldehyde foam insulation. Vinyl chloride is used in plastic plumbing pipe, floor and wall coverings, and countertops. Volatile organic solvents make up as much as half the volume of some paint. New carpets and drapes typically contain up to two dozen chemical compounds designed to kill bacteria and molds, resist stains, bind fibers, and retain colors.

What can you do if you suspect that your living spaces are exposing you to materials that make you sick? Probably few students will be in a position anytime soon to build a new house with nontoxic materials, but there are some principles from the emerging field of "green design" that you might apply if you're house hunting, redecorating your apartment, or interviewing for a job. Low-volatile paint is now available for indoor use. Nontoxic, formaldehyde-free plywood, particle board, and insulation can be used in new construction. Nonallergenic carpets, drapes, and wall coverings are available, but some architects recommend natural wood, stone, and plaster surfaces that are easier to clean and less allergenic than any fabric.

High rates of air exchange can help rid indoor air of moisture, odors, mold spores, radon, and toxins. Does that mean energy inefficiency? Not necessarily. Air-to-air heat exchangers keep heat in during the winter and out during the summer, while still providing a healthy rate of fresh-air flow. Bathrooms and kitchens should have outdoor vents. Gas or oil furnaces should be checked for carbon monoxide production. Although many cooks prefer gas stoves because they heat quickly, they can produce toxic carbon monoxide and nitrogen oxides. Contact your city housing authority or county extension service for further tips on how to make your home, work, or study environment healthier. No matter what your situation, there are things that each of us can do to make our indoor air cleaner and safer.

---

accompanying these air exchanges tend to break up the temperature gradient fairly quickly and mix air layers.

The most stable inversion conditions are usually created by rapid nighttime cooling in a valley or basin where air movement is restricted. Los Angeles is a classic example of the conditions that create temperature inversions and photochemical smog (fig. 16.15). The city is surrounded by mountains on three sides and the climate is dry and sunny. Millions of automobiles and trucks create high pollution levels. Skies are generally clear at night, allowing rapid radiant heat loss, and the ground cools quickly. Surface air layers are cooled by conduction, while upper layers remain relatively warm. Density differences retard vertical mixing. During the night, cool, humid, onshore breezes slide in under the contaminated air, squeezing it up against the cap of warmer air above and concentrating the pollutants accumulated during the day.

Morning sunlight is absorbed by the concentrated aerosols and gaseous chemicals of the inversion layer. This complex mixture quickly cooks up a toxic brew of hazardous compounds. As the ground warms later in the day, convection currents break up the temperature gradient and pollutants are carried back down to the surface where more contaminants are added. Nitric oxide (NO) from automobile exhaust is oxidized to nitrogen dioxide. As nitrogen oxides are used up in reactions with unburned hydrocarbons, the ozone levels begin to rise. By early afternoon, an acrid brown haze fills the air, making eyes water and throats burn. In the 1970s, before pollution controls were enforced, the Los Angeles basin often would reach 0.34 ppm or more by late afternoon and the pollution index could be 300, the stage considered a health hazard.

Cities create dust domes and heat islands

Even without mountains to block winds and stabilize air layers, many large cities create an atmospheric environment quite different from the surrounding conditions. Sparse vegetation and high levels of concrete and glass in urban areas allow rainfall to run off quickly and create high rates of heat absorption during the day and radiation at night. Tall buildings create convective updrafts that sweep pollutants into the air. Temperatures in the center of large cities are frequently 3–5°C (5–9°F) higher than the surrounding countryside. Stable air masses created by this "heat island" over the city concentrate pollutants in a "dust dome." Rural areas downwind from major industrial areas often have significantly decreased visibility and increased rainfall (due to increased condensation nuclei in the dust plume) compared to neighboring areas with cleaner air. In the late 1960s,

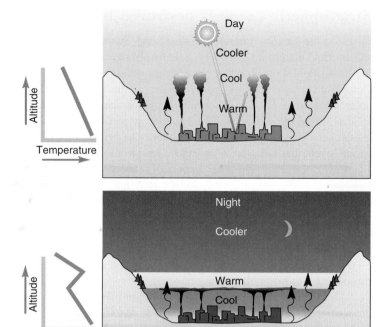

FIGURE 16.15 Atmospheric temperature inversions occur where ground level air cools more quickly than upper levels. This temperature differential prevents mixing and traps pollutants close to the ground.

for instance, areas downwind from Chicago and St. Louis reported up to 30 percent more rainfall than upwind regions.

Aerosols and dust in urban air seem to trigger increased cloud-to-ground lightning strikes. Houston and Lake Charles, Louisiana, for instance, which have many petroleum refineries, have among the highest number of lightning strikes in the United States and twice as many as nearby areas with similar climate but cleaner air.

Wind currents carry pollutants intercontinentally

Dust and contaminants can be carried great distances by the wind. Areas downwind from industrial complexes often suffer serious contamination, even if they have no pollution sources of their own (fig. 16.16). Pollution from the industrial belt between the Great Lakes and the Ohio River Valley, for example, regularly contaminates the Canadian Maritime Provinces, and sometimes can be traced as far as Ireland. As we've already seen in this chapter, as much as 70 percent of the mercury that falls on North America may come from abroad, much of it from Asia.

Studies of air pollutants over southern Asia reveal a 3 km thick toxic cloud of ash, acids, aerosols, dust, and photochemical reactants regularly covers the entire Indian subcontinent for much of the year. Nobel laureate Paul Crutzen estimates that up to 2 million people in India alone die each year from atmospheric pollution. Produced by forest fires, the burning of agricultural wastes, and dramatic increases in the use of fossil fuels, the Asian smog layer cuts the amount of solar energy reaching the earth's surface beneath it by up to 15 percent. Meteorologists suggest

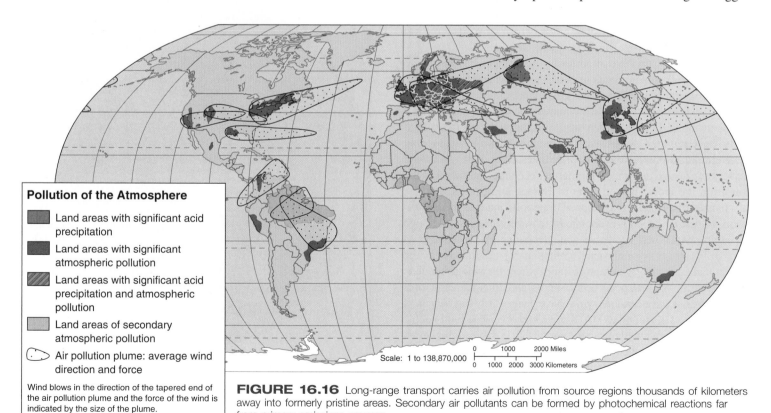

Pollution of the Atmosphere

- Land areas with significant acid precipitation
- Land areas with significant atmospheric pollution
- Land areas with significant acid precipitation and atmospheric pollution
- Land areas of secondary atmospheric pollution
- Air pollution plume: average wind direction and force

Wind blows in the direction of the tapered end of the air pollution plume and the force of the wind is indicated by the size of the plume.

Scale: 1 to 138,870,000

0   1000   2000 Miles
0   1000   2000   3000 Kilometers

FIGURE 16.16 Long-range transport carries air pollution from source regions thousands of kilometers away into formerly pristine areas. Secondary air pollutants can be formed by photochemical reactions far from primary emissions sources.

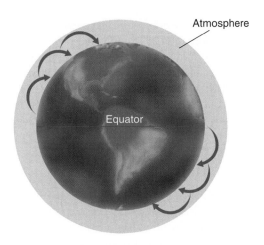

FIGURE 16.17 Air pollutants evaporate from warmer areas and then condense and precipitate in cooler regions. Eventually, this "grasshopper" redistribution leads to accumulation in the Arctic and Antarctic.

that the cloud—80 percent of which is human-made—could disrupt monsoon weather patterns and may be disturbing rainfall and reducing rice harvests over much of South Asia. Shifting monsoon flows may also have contributed to catastrophic floods in Nepal, Bangladesh, and eastern India that killed at least 1,000 people in 2002, and left more than 25 million homeless.

When this "Asian Brown Cloud" drifts out over the Indian Ocean at the end of the monsoon season, it cools sea temperatures and may be changing El Niño/Southern Oscillation patterns in the Pacific Ocean as well (chapter 15). As UN Environment Programme executive director, Klaus Töpfer, said, "There are global implications because a pollution parcel like this, which stretches three km high, can travel half way round the globe in a week."

Increasingly sensitive monitoring equipment has begun to reveal industrial contaminants in places usually considered among the cleanest in the world. Samoa, Greenland, and even Antarctica and the North Pole, all have heavy metals, pesticides, and radioactive elements in their air. Since the 1950s, pilots flying in the high Arctic have reported dense layers of reddish-brown haze clouding the arctic atmosphere. Aerosols of sulfates, soot, dust, and toxic heavy metals such as vanadium, manganese, and lead travel to the pole from the industrialized parts of Europe and Russia.

In a process called "grasshopper" transport, or atmosphere distillation, volatile compounds evaporate from warm areas, travel through the atmosphere, then condense and precipitate in cooler regions (fig. 16.17). Over several years, contaminants accumulate in the coldest places, generally at high latitudes where they bioaccumulate in food chains. Whales, polar bears, sharks, and other top carnivores in polar regions have been shown to have dangerously high levels of pesticides, metals, and other HAPs in their bodies. The Inuit people of Broughton Island, well above the Arctic Circle, have higher levels of polychlorinated biphenyls (PCBs) in their blood than any other known population, except victims of industrial accidents. Far from any source of this industrial by-product, these people accumulate PCBs from the flesh of fish, caribou, and other animals they eat. This exacerbates the cultural crisis caused by climate change.

Stratospheric ozone is destroyed by chlorine

In 1985, the British Antarctic Atmospheric Survey announced a startling and disturbing discovery: **Stratospheric ozone** levels over the South Pole were dropping precipitously during September and October every year as the sun reappears at the end of the long polar winter (fig. 16.18). This ozone depletion has been occurring at least since the 1960s but was not recognized because earlier researchers programmed their instruments to ignore changes in ozone levels that were presumed to be erroneous.

Chlorine-based aerosols, especially **chlorofluorocarbons (CFCs)** and other halon gases, are the principal agents of ozone depletion. Nontoxic, nonflammable, chemically inert, and cheaply produced, CFCs were extremely useful as industrial gases and in refrigerators, air conditioners, styrofoam inflation, and aerosol spray cans for many years. From the 1930s until the 1980s, CFCs were used all over the world and widely dispersed through the atmosphere.

Although ozone is a pollutant in the ambient air, ozone in the stratosphere is important because it absorbs much of the ultraviolet (UV) radiation entering the atmosphere. UV radiation harms plant and animal tissues, including the eyes and the skin. A 1 percent loss of ozone could result in about a million extra human skin cancers per year worldwide if no protective measures are taken. Excessive UV exposure could reduce agricultural production and disrupt ecosystems. Scientists worry, for example, that high UV levels in Antarctica could reduce populations of

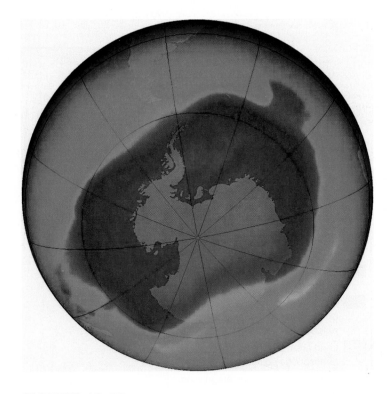

FIGURE 16.18 In 2006, stratospheric ozone was depleted over an area (*dark, irregular circle*) that covered 29.5 million km², or more than the entire Antarctic continent. Although CFC production is declining, this was the largest area ever recorded.

TABLE 16.2

Stratospheric Ozone Destruction by Chlorine Atoms and UV Radiation

Step	Products
1. $CFCl_3$ (chlorofluorocarbon) + UV energy	$CFCl_2$ + Cl
2. $Cl + O_3$	$ClO + O_2$
3. O_2 + UV energy	2O
4. $ClO + 2O$	O_2 + Cl
5. Return to step 2	

plankton, the tiny floating organisms that form the base of a food chain that includes fish, seals, penguins, and whales in Antarctic seas.

Antarctica's exceptionally cold winter temperatures (-85–$-90°C$) help break down ozone. During the long, dark, winter months, strong winds known as the circumpolar vortex isolate Antarctic air and allow stratospheric temperatures to drop low enough to create ice crystals at high altitudes—something that rarely happens elsewhere in the world. Ozone and chlorine-containing molecules are absorbed on the surfaces of these ice particles. When the sun returns in the spring, it provides energy to liberate chlorine ions, which readily bond with ozone, breaking it down to molecular oxygen (table 16.2). It is only during the Antarctic spring (September through December) that conditions are ideal for rapid ozone destruction. During that season, temperatures are still cold enough for high-altitude ice crystals, but the sun gradually becomes strong enough to drive photochemical reactions.

As the Antarctic summer arrives, temperatures moderate somewhat, the circumpolar vortex breaks down, and air from warmer latitudes mixes with Antarctic air, replenishing ozone concentrations in the ozone hole. Slight decreases worldwide result from this mixing, however. Ozone re-forms naturally, but not nearly as fast as it is destroyed. Since the chlorine atoms are not themselves consumed in reactions with ozone, they continue to destroy ozone for years, until they finally precipitate or are washed out of the air. Almost every year since it was discovered, the Antarctic ozone hole has grown. In 2006 the region of ozone depletion covered 29.5 million km^2 (larger than North America).

Although not as pronounced, about 10 percent of all stratospheric ozone worldwide has been destroyed in recent years, and levels over the Arctic have averaged 40 percent below normal. Ozone depletion has been observed over the North Pole as well, although it is not as concentrated as that in the south.

The Montreal Protocol is a resounding success

The discovery of stratospheric ozone losses brought about a remarkably quick international response. In 1987 an international meeting in Montreal, Canada, produced the Montreal Protocol, the first of

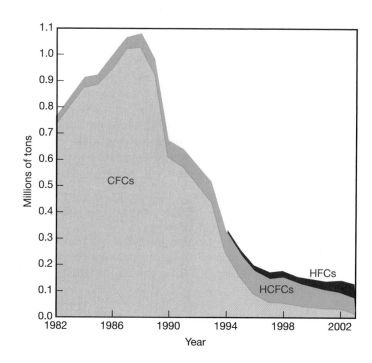

FIGURE 16.19 The Montreal Protocol has been remarkably successful in eliminating CFC production. The remaining HFC and HCFC use in primarily in developing countries, such as China and India.

several major international agreements on phasing out most use of CFCs by 2000. As evidence accumulated, showing that losses were larger and more widespread than previously thought, the deadline for the elimination of all CFCs (halons, carbon tetrachloride, and methyl chloroform) was moved up to 1996, and a $500 million fund was established to assist poorer countries in switching to non-CFC technologies. Fortunately, alternatives to CFCs for most uses already exist. The first substitutes are hydrochlorofluorocarbons (HCFCs), which release much less chlorine per molecule. Eventually, scientists hope to develop halogen-free molecules that work just as well and are no more expensive than CFCs.

The Montreal Protocol is often cited as the most effective international environmental agreement ever established. Global CFC production has been cut by more than 95 percent since 1988 (fig. 16.19). Some of that has been replaced by hydrochlorofluorocarbons (HCFCs), which release chlorine, but not as much as CFCs. The amount of chlorine entering the atmosphere already has begun to decrease, suggesting that stratospheric O_3 levels should be back to normal by about 2049. You might wonder, then, why the 2006 O_3 hole was the largest ever. The answer is global warming (see chapter 15). Greenhouse gases are warming the troposphere, which is causing the stratosphere to cool. This increases ice crystal formation over the Antarctic, and results in more O_3 depletion.

There's another interesting connection to climate change. Under the Montreal Protocol, China, India, Korea, and Argentina were allowed to continue to produce 72,000 tons (combined) of CFCs per year until 2010. Most of the funds appropriated through the Montreal Protocol are going to these countries to help them phase out CFC production and to destroy their existing stocks. Because CFCs are potent greenhouse gases, this phase-out also makes these countries eligible for credits in the climate trading market. In 2006, nearly two-thirds of the greenhouse gas emissions credits traded internationally were for HFC-23 elimination, and almost half of all payments went to China. Some critics think this is double-dipping, but if it eliminates a dangerous risk to all of us, isn't it worth it?

In 1995, chemists Sherwood Rowland, Mario Molina, and Paul Crutzen shared the Nobel Prize for their work on atmospheric chemistry and stratospheric ozone. This was the first Nobel Prize for an environmental issue.

16.5 EFFECTS OF AIR POLLUTION

So far we have looked primarily at the major types and sources of air pollutants. Now we will focus more closely on the effects of those pollutants on human health, physical materials, ecosystems, and global climate.

Polluted air is dangerous

The World Health Organization estimates that some 5 to 6 million people die prematurely every year from illnesses related to air pollution. Heart attacks, respiratory diseases, and lung cancer all are significantly higher in people who breathe dirty air, compared to matching groups in cleaner environments. Residents of the most polluted cities in the United States, for example, are 15 to 17 percent more likely to die of these illnesses than those in cities with the cleanest air. This can mean as much as a 5- to 10-year decrease in life expectancy if you live in the worst parts of Los Angeles or Baltimore, compared to a place with clean air. Of course your likelihood of suffering ill health from air pollutants depends on the intensity and duration of exposure as well as your age and prior health status. You are much more likely to be at risk if you are very young, very old, or already suffering from some respiratory or cardiovascular disease. Some people are super-sensitive because of genetics or prior exposure. And those doing vigorous physical work or exercise are more likely to succumb than more sedentary folks.

Conditions are often much worse in other countries than Canada or the United States. The United Nations estimates that at least 1.3 billion people around the world live in areas where outdoor air is dangerously polluted. In Madrid, Spain, smog is estimated to shave one-half year off the life of each resident. This adds up to more than 50,000 years lost annually for the whole city. In China, city dwellers are four to six times more likely than country folk to die of lung cancer. As mentioned earlier, the greatest air quality problem is often in poorly ventilated homes in poorer countries where smoky fires are used for cooking and heating. Billions of women and children spend hours each day in these unhealthy conditions. The World Health Organization estimates that 2 million children under age 5 die each year from acute respiratory diseases exacerbated by air pollution.

In industrialized countries, one of the biggest health threats from air pollution is from soot or fine particulate material. We once thought that particles smaller than 10 μm (10 millionths of a meter) were too small to be trapped in the lungs. Now we know

FIGURE 16.20 Soot and fine particulate material from diesel engines, wood stoves, power plants, and other combustion sources have been linked to asthma, heart attacks, and a variety of other diseases.

that small particles (less than 2.5 μm diameter) called PM2.5 are an even greater risk than larger ones. They have been linked with heart attacks, asthma, bronchitis, lung cancer, immune suppression, and abnormal fetal development, among other health problems. Fine particulates have many sources. Until recently, power plants were the largest source, but with recent clean air rules, they will be required to install filters and precipitators to remove at least 70 percent of their particulate emissions.

Diesel engines have long been a major source of both soot and SO_2 in the United States (fig. 16.20). Under a new rule announced in 2006, new engines in trucks and buses, in combination with low-sulfur diesel fuel that is now required nationwide, will reduce particulate emissions by up to 98 percent when the rule is fully implemented in 2012. These standards will also be applied to off-road vehicles, such as tractors, bulldozers, locomotives, and barges, whose engines previously emitted more soot than all the nation's cars, trucks, and buses together. The sulfur content of diesel fuel is now 500 parts per million (ppm) compared to an average of 3,400 ppm before the regulations were imposed. By 2012, only 15 ppm of sulfur will be allowed in diesel fuel. Europe has had low-sulfur fuel and clean diesel engines since the early 1990s, but in spite of this, diesel buses and trucks are banned from the centers of many European cities.

In some rural areas, smoke from wood stoves or burning crops remain an important soot source. Resort towns, such as Telluride and Aspen, Colorado, are beginning to limit or ban wood stoves and open fireplaces because of the pollution they produce. The U.S. EPA estimates that at least 160 million Americans—more than half the population—live in areas with unhealthy concentrations of fine particulate matter. More than 450 counties in 32 states are considered in nonattainment of clean air rules. The EPA reports that PM2.5 levels have decreased about 30 percent over the past 25 years, but health officials argue that the remaining pollution should also be cleaned up.

How does pollution harm us?

The most common route of exposure to air pollutants is by inhalation, but direct absorption through the skin or contamination of food and water also are important pathways. Because they are strong oxidizing agents, sulfates, SO_2, NO_x, and O_3 act as irritants that damage delicate tissues in the eyes and respiratory passages. Fine particulates penetrate deep into the lungs and are irritants in their own right, as well as carrying metals and other HAPs on their surfaces. Inflammatory responses set in motion by these irritants impair lung function and trigger cardiovascular problems as the heart tries to compensate for lack of oxygen by pumping faster and harder. If the irritation is really severe, so much fluid seeps into lungs through damaged tissues that the victim actually drowns.

Carbon monoxide binds to hemoglobin and decreases the ability of red blood cells to carry oxygen. Asphyxiants such as this cause headaches, dizziness, heart stress, and can even be lethal if concentrations are high enough. Lead also binds to hemoglobin and reduces oxygen-carrying capacity at high levels. At lower levels, lead causes long-term damage to critical neurons in the brain that results in mental and physical impairment and developmental retardation.

Some important chronic health effects of air pollutants include bronchitis and emphysema.

Bronchitis is a persistent inflammation of bronchi and bronchioles (large and small airways in the lung) that causes mucus buildup, a painful cough, and involuntary muscle spasms that constrict airways. Severe bronchitis can lead to emphysema, an irreversible **chronic obstructive lung disease** in which airways become permanently constricted and alveoli are damaged or even destroyed. Stagnant air trapped in blocked airways swells the tiny air sacs in the lung (alveoli), blocking blood circulation. As cells die from lack of oxygen and nutrients, the walls of the alveoli break down, creating large empty spaces incapable of gas exchange (fig. 16.21). Thickened walls of the bronchioles lose elasticity and breathing becomes more difficult. Victims of emphysema make a characteristic whistling sound when they breathe. Often they need supplementary oxygen to make up for reduced respiratory capacity.

Irritants in the air are so widespread that about half of all lungs examined at autopsy in the United States have some degree of alveolar deterioration. The Office of Technology Assessment (OTA) estimates that 250,000 people suffer from pollution-related bronchitis and emphysema in the United States, and some 50,000 excess deaths each year are attributable to complications of these diseases, which are probably second only to heart attack as a cause of death.

Smoking is undoubtedly the largest cause of obstructive lung disease and preventable death in the world. The World Health Organization says that tobacco kills some 3 million people each year. This makes it rank with AIDS as one of the world's leading killers. Because of cardiovascular stress caused by carbon monoxide in smoke and chronic bronchitis and emphysema, about twice as many people die of heart failure as die from lung cancer associated with smoking.

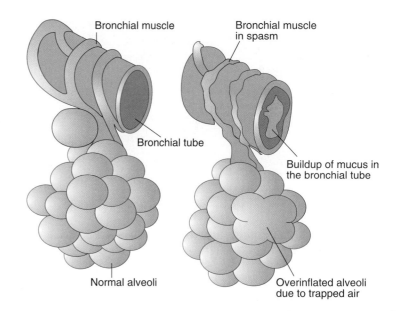

FIGURE 16.21 Bronchitis and emphysema can result in constriction of airways and permanent damage to tiny, sensitive air sacs called alveoli, where oxygen diffuses into blood vessels.

Plants are susceptible to pollution damage

In the early days of industrialization, fumes from furnaces, smelters, refineries, and chemical plants often destroyed vegetation and created desolate, barren landscapes around mining and manufacturing centers. The copper-nickel smelter at Sudbury, Ontario, is a spectacular and notorious example of air pollution effects on vegetation and ecosystems. In 1886, the corporate ancestors of the International Nickel Company (INCO) began open-bed roasting of sulfide ores at Sudbury. Sulfur dioxide and sulfuric acid released by this process caused massive destruction of the plant community within about 30 km of the smelter. Rains washed away the exposed soil, leaving a barren moonscape of blackened bedrock (fig. 16.22). Super-tall, 400 m smokestacks were installed in the 1950s and sulfur scrubbers were added 20 years later. Emissions were reduced by 90 percent and the surrounding ecosystem is beginning to recover (fig. 16.23). Similar destruction occurred at many other sites during the nineteenth century. Copperhill, Tennessee; Butte, Montana; and the Ruhr Valley in Germany are some well-known examples, but these areas also are showing signs of recovery since corrective measures were taken.

There are two probable ways that air pollutants damage plants. They can be directly toxic, damaging sensitive cell membranes much as irritants do in human lungs. Within a few days of exposure to toxic levels of oxidants, mottling (discoloration) occurs in leaves due to chlorosis (bleaching of chlorophyll), and then necrotic (dead) spots develop (fig. 16.24). If injury is severe, the whole plant may be killed. Sometimes these symptoms are so distinctive that positive identification of the source of damage is possible. Often, however, the symptoms are vague and difficult to separate from diseases or insect damage.

FIGURE 16.22 In 1975, acid precipitation from the copper-nickel smelters (tall stacks in background) had killed all the vegetation and charred the pink granite bedrock black for a large area around Sudbury, Ontario.

FIGURE 16.23 By 2005, a scrubby forest was growing again around Sudbury, but the rock surfaces remain burned black.

Certain combinations of environmental factors have **synergistic effects** in which the injury caused by exposure to two factors together is more than the sum of exposure to each factor individually. For instance, when white pine seedlings are exposed to subthreshold concentrations of ozone and sulfur dioxide individually, no visible injury occurs. If the same concentrations of pollutants are given together, however, visible damage occurs. In alfalfa, however, SO_2 and O_3 together cause less damage than either one alone. These complex interactions point out the unpredictability of future effects of pollutants. Outcomes might be either more or less severe than previous experience indicates.

Pollutant levels too low to produce visible symptoms of damage may still have important effects. Field studies using open-top chambers (fig. 16.25) and charcoal-filtered air show that yields in some sensitive crops, such as soybeans, may be reduced as much as 50 percent by currently existing levels of oxidants in ambient air. Some plant pathologists suggest that ozone and photochemical oxidants are responsible for as much as 90 percent of agricultural, ornamental, and forest losses from air pollution. The total costs of this damage may be as much as $10 billion per year in North America alone.

Acid deposition has many negative effects

Most people in the United States became aware of problems associated with **acid precipitation** (the deposition of wet acidic solutions or dry acidic particles from the air) within the last decade or so, but English scientist Robert Angus Smith coined the term "acid rain" in his studies of air chemistry in Manchester, England, in the 1850s. By the 1940s, it was known that pollutants, including atmospheric acids, could be transported long distances by wind currents. This was thought to be only an academic curiosity until it was shown that precipitation of these acids can have far-reaching ecological effects.

We describe acidity in terms of pH (chapter 3). Values below 7 are acidic, while those above 7 are alkaline. Normal, unpolluted rain generally has a pH of about 5.6 due to carbonic acid created by CO_2 in air. Volcanic emissions, biological decomposition, and chlorine and sulfates from ocean spray can drop the pH of rain well below 5.6, while alkaline dust can raise it above 7. In indus-

trialized areas, anthropogenic acids in the air usually far outweigh those from natural sources. Acid rain is only one form in which acid deposition occurs. Fog, snow, mist, and dew also trap and deposit atmospheric contaminants. Furthermore, fallout of dry sulfate, nitrate, and chloride particles can account for as much as half of the acidic deposition in some areas.

Aquatic Effects

It has been known for about 30 years that acids—principally H_2SO_4 and HNO_3—generated by industrial and automobile emissions in northwestern Europe are carried by prevailing winds to Scandinavia where they are deposited in rain, snow, and dry precipitation. The thin, acidic soils and oligotrophic lakes and streams in the mountains of southern Norway and Sweden have been severely affected by this acid deposition. Some 18,000 lakes

FIGURE 16.24 Soybean leaves exposed to 0.8 parts per million sulfur dioxide for 24 hours show extensive chlorosis (chlorophyll destruction) in white areas between leaf veins.

FIGURE 16.25 An open-top chamber tests air pollution effects on plants under normal conditions for rain, sun, field soil, and pest exposure.

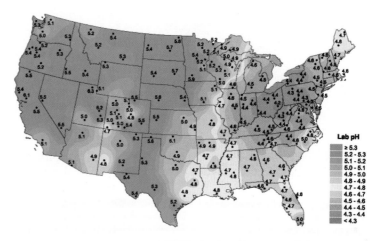

FIGURE 16.26 Acid precipitation over the United States.
Source: National Atmospheric Deposition Program/National Trends Network, 2000. http://nadp.sws.uiuc.edu.

in Sweden are now so acidic that they will no longer support game fish or other sensitive aquatic organisms.

Generally, reproduction is the most sensitive stage in fish life cycles. Eggs and fry of many species are killed when the pH drops to about 5.0. This level of acidification also can disrupt the food chain by killing aquatic plants, insects, and invertebrates on which fish depend for food. At pH levels below 5.0, adult fish die as well. Trout, salmon, and other game fish are usually the most sensitive. Carp, gar, suckers, and other less desirable fish are more resistant.

In the early 1970s, evidence began to accumulate suggesting that air pollutants are acidifying many lakes in North America. Studies in the Adirondack Mountains of New York revealed that about half of the high-altitude lakes (above 1,000 m or 3,300 ft) are acidified and have no fish. Areas showing lake damage correlate closely with average pH levels in precipitation (fig. 16.26). Some 48,000 lakes in Ontario are endangered and nearly all of Quebec's surface waters, including about 1 million lakes, are believed to be highly sensitive to acid deposition.

Sulfates account for about two-thirds of the acid deposition in eastern North America and most of Europe, while nitrates contribute most of the remaining one-third. In urban areas, where transportation is the major source of pollution, nitric acid is equal to or slightly greater than sulfuric acids in the air. A vigorous program of pollution control has been undertaken by both Canada and the United States, and SO_2 and NO_x emissions have decreased dramatically over the past three decades over much of North America.

Forest Damage

In the early 1980s, disturbing reports appeared of rapid forest declines in both Europe and North America. One of the earliest was a detailed ecosystem inventory on Camel's Hump Mountain in Vermont. A 1980 survey showed that seedling production, tree density, and viability of spruce-fir forests at high elevations had declined about 50 percent in 15 years. A similar situation was found on Mount Mitchell in North Carolina where almost all red spruce and Fraser fir above 2,000 m (6,000 ft) are in a severe decline. Nearly all the trees are losing needles and about half of them are dead (fig. 16.27). The stress of acid rain and fog, other air pollutants, and attacks by an invasive insect called the woody aldegid are killing the trees.

Many European countries reported catastrophic forest destruction in the 1980s. It still isn't clear what caused this injury. In the longest-running forest-ecosystem monitoring record in North America, researchers at the Hubbard Brook Experimental Forest in New Hampshire have shown that forest soils have become depleted of natural buffering reserves of basic cations such as calcium and magnesium through years of exposure to acid rain. Replacement of these cations by hydrogen and aluminum ions seems to be one of the main causes of plant mortality.

FIGURE 16.27 A Fraser fir forest on Mount Mitchell, North Carolina, killed by acid rain, insect pests, and other stressors.

FIGURE 16.28 Atmospheric acids, especially sulfuric and nitric acids, have almost completely eaten away the face of this medieval statue. Each year, the total losses from air pollution damage to buildings and materials amounts to billions of dollars.

Buildings and Monuments

In cities throughout the world, some of the oldest and most glorious buildings and works of art are being destroyed by air pollution. Smoke and soot coat buildings, paintings, and textiles. Limestone and marble are destroyed by atmospheric acids at an alarming rate. The Parthenon in Athens, the Taj Mahal in Agra, the Colosseum in Rome, frescoes and statues in Florence, medieval cathedrals in Europe (fig. 16.28). and the Lincoln Memorial and Washington Monument in Washington, D.C., are slowly dissolving and flaking away because of acidic fumes in the air. Medieval stained glass windows in Cologne's gothic cathedral are so porous from etching by atmospheric acids that pigments disappear and the glass literally crumbles away. Restoration costs for this one building alone are estimated at three to 1.5 billion Euros (U.S. $1.8 billion).

On a more mundane level, air pollution also damages ordinary buildings and structures. Corroding steel in reinforced concrete weakens buildings, roads, and bridges. Paint and rubber deteriorate due to oxidization. Limestone, marble, and some kinds of sandstone flake and crumble. The Council on Environmental Quality estimates that U.S. economic losses from architectural damage caused by air pollution amount to about $4.8 billion in direct costs and $5.2 billion in property value losses each year.

Smog and haze reduce visibility

We have realized only recently that pollution affects rural areas as well as cities. Even supposedly pristine places like our national parks are suffering from air pollution. Grand Canyon National Park, where maximum visibility used to be 300 km, is now so smoggy on some winter days that visitors can't see the opposite rim only 20 km across the canyon. Mining operations, smelters, and power plants (some of which were moved to the desert to improve air quality in cities like Los Angeles) are the main culprits. Similarly, the vistas from Shenandoah National Park just outside Washington, D.C., are so hazy that summer visibility is often less than 1.6 km because of smog drifting in from nearby urban areas.

Historical records show that over the past four or five decades human-caused air pollution has spread over much of the United States. Researchers report that a gigantic "haze blob" as much as 3,000 km across covers much of the eastern United States in the summer, cutting visibility as much as 80 percent. Smog and haze are so prevalent that it's hard for people to believe that the air once was clear. Studies indicate, however, that if all human-made sources of air pollution were shut down, the air would clear up in a few days and there would be about 150 km visibility nearly everywhere rather than the 15 km to which we have become accustomed.

16.6 AIR POLLUTION CONTROL

"Dilution is the solution to pollution" was one of the early approaches to air pollution control. Tall smokestacks were built to send emissions far from the source, where they became unidentifiable and largely untraceable. But dispersed and diluted pollutants are now the source of some of our most serious pollution problems. We are finding that there is no "away" to which we can throw our waste products. While most of the discussion in this section focuses on industrial solutions, each of us can make important personal contributions to this effort (What Can You Do? p. 366).

What Can You Do?

Saving Energy and Reducing Pollution

- Conserve energy: carpool, bike, walk, use public transport, buy compact fluorescent bulbs, and energy-efficient appliances (see chapter 20 for other suggestions).
- Don't use polluting two-cycle gasoline engines if cleaner four-cycle models are available for lawn mowers, boat motors, etc.
- Buy refrigerators and air conditioners designed for CFC alternatives. If you have old appliances or other CFC sources, dispose of them responsibly.
- Plant a tree and care for it (every year).
- Write to your Congressional representatives and support a transition to an energy-efficient economy.
- If green-pricing options are available in your area, buy renewable energy.
- If your home has a fireplace, install a high-efficiency, clean-burning, two-stage insert that conserves energy and reduces pollution up to 90 percent.
- Have your car tuned every 10,000 miles (16,000 km) and make sure that its antismog equipment is working properly. Turn off your engine when waiting longer than one minute. Start trips a little earlier and drive slower—it not only saves fuel but it's safer, too.
- Use latex-based, low-volatile paint rather than oil-based (alkyd) paint.
- Avoid spray can products. Light charcoal fires with electric starters rather than petroleum products.
- Don't top off your fuel tank when you buy gasoline; stop when the automatic mechanism turns off the pump. Don't dump gasoline or used oil on the ground or down the drain.
- Buy clothes that can be washed rather than dry-cleaned.

The most effective strategy for controlling pollution is to minimize production

Since most air pollution in the developed world is associated with transportation and energy production, the most effective strategy would be conservation: Reducing electricity consumption, insulating homes and offices, and developing better public transportation could all greatly reduce air pollution in the United States, Canada, and Europe. Alternative energy sources, such as wind and solar power, produce energy with little or no pollution, and these and other technologies are becoming economically competitive (chapter 20). In addition to conservation, pollution can be controlled by technological innovation.

Particulate removal involves filtering air emissions. Filters trap particulates in a mesh of cotton cloth, spun glass fibers, or asbestos-cellulose. Industrial air filters are generally giant bags 10 to 15 m long and 2 to 3 m wide. Effluent gas is blown through the bag, much like the bag on a vacuum cleaner. Every few days or weeks, the bags are opened to remove the dust cake. Electrostatic precipitators are the most common particulate controls in power plants. Ash particles pick up an electrostatic surface charge as they pass between large electrodes in the effluent stream (fig. 16.29). Charged particles then collect on an oppositely charged collecting plate. These precipitators consume a large amount of electricity, but maintenance is relatively simple, and collection efficiency can be as high as 99 percent. The ash collected by both of these techniques is a solid waste (often hazardous due to the heavy metals and other trace components of coal or other ash source) and must be buried in landfills or other solid-waste disposal sites.

Sulfur removal is important because sulfur oxides are among the most damaging of all air pollutants in terms of human health and ecosystem viability. Switching from soft coal with a high sulfur content to low-sulfur coal is the surest way to reduce sulfur

emissions. High-sulfur coal is frequently politically or economically expedient, however. In the United States, Appalachia, a region of chronic economic depression, produces most high-sulfur coal. In China, much domestic coal is rich in sulfur. Switching to cleaner oil or gas would eliminate metal effluents as well as sulfur. Cleaning fuels is an alternative to switching. Coal can be crushed, washed, and gasified to remove sulfur and metals before combustion. This improves heat content and firing properties, but may replace air pollution with solid-waste and water pollution problems; furthermore, these steps are expensive.

Sulfur can also be removed to yield a usable product instead of simply a waste disposal problem. Elemental sulfur, sulfuric acid, or ammonium sulfate can all be produced using catalytic converters to oxidize or reduce sulfur. Markets have to be reasonably close and fly ash contamination must be reduced as much as possible for this procedure to be economically feasible.

Nitrogen oxides (NO_x) can be reduced in both internal combustion engines and industrial boilers by as much as 50 percent by carefully controlling the flow of air and fuel. Staged burners, for example, control burning temperatures and oxygen flow to prevent formation of NO_x. The catalytic converter on your car uses platinum-palladium and rhodium catalysts to remove up to 90 percent of NO_x, hydrocarbons, and carbon monoxide at the same time.

Hydrocarbon controls mainly involve complete combustion or controlling evaporation. Hydrocarbons and volatile organic compounds are produced by incomplete combustion of fuels or by solvent evaporation from chemical factories, paints, dry cleaning, plastic manufacturing, printing, and other industrial processes. Closed systems that prevent escape of fugitive gases can reduce many of these emissions. In automobiles, for instance, positive crankcase ventilation (PCV) systems collect oil that escapes from around the pistons and unburned fuel and channels them back to the engine for combustion. Controls on fugitive losses from industrial valves, pipes, and storage tanks can have a significant impact on air quality. Afterburners are often the best method for destroying volatile organic chemicals in industrial exhaust stacks.

Fuel switching and fuel cleaning also are effective

Switching from soft coal with a high sulfur content to low-sulfur coal can greatly reduce sulfur emissions. This may eliminate jobs, however, in such areas as Appalachia that are already economically depressed. Changing to another fuel, such as natural gas or nuclear energy, can eliminate all sulfur emissions as well as those of particulates and heavy metals. Natural gas is more expensive and more difficult to ship and store than coal, however, and many people prefer the sure dangers of coal pollution to the uncertain dangers of nuclear power (chapter 19). Alternative energy sources, such as wind and solar power, are preferable to either fossil fuel or nuclear power, and are becoming economically competitive (chapter 20) in many areas. In the interim, coal can be crushed, washed, and gassified to remove sulfur and metals before combustion. This improves heat content and firing properties but may replace air pollution with solid waste and water pollution problems.

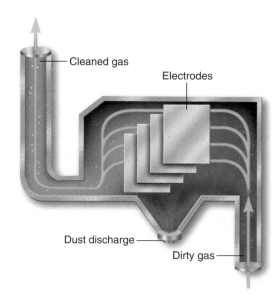

FIGURE 16.29 An electrostatic precipitator traps particulate material on electrically charged plates as effluent makes its way to the smokestack.

Ironically, reducing air pollution could increase global warming. The amount of sunlight reaching the earth's surface is reported to have been declining over the past few decades, especially over large cities. This "global dimming" has been ascribed to greater reflection of sunlight by atmospheric aerosols such as sulfate droplets and fine particulates. If they are removed, dimming may reverse and the earth may warm more.

Clean air legislation is controversial

Throughout history, countless ordinances have prohibited emission of objectionable smoke, odors, and noise. Air pollution traditionally has been treated as a local problem, however. The Clean Air Act of 1963 was the first national legislation in the United States aimed at air pollution control. The act provided federal grants to states to combat pollution but was careful to preserve states' rights to set and enforce air quality regulations. It soon became obvious that some pollution problems cannot be solved on a local basis.

In 1970, an extensive set of amendments essentially rewrote the Clean Air Act. These amendments identified the "criteria" pollutants discussed earlier in this chapter, and established primary and secondary standards for ambient air quality. Primary standards (table 16.3) are intended to protect human health, while secondary standards are set to protect materials, crops, climate, visibility, and personal comfort.

Since 1970 the Clean Air Act has been modified, updated, and amended many times. The most significant amendments were in the 1990 update. Amendments have involved acrimonious debate, with bills sometimes languishing in Congress from one session to the next because of disputes over burdens of responsibility

FIGURE 16.30 Should old power plants be required to install costly pollution-control equipment? This is the critical issue in the "new source review" under the Clean Air Act.

and cost and definitions of risk. A 2002 report concluded that simply by enforcing existing clean air legislation, the United States could save at least another 6,000 lives per year and prevent 140,000 asthma attacks.

Throughout its history the Clean Air Act has been controversial. Victims of air pollution demand more protection; industry and special interest groups complain that controls are too expensive.

One of the most contested aspects of the act is the "new source review," which was established in 1977. This provision was originally adopted because industry argued that it would be intolerably expensive to install new pollution-control equipment on old power plants and factories that were about to close down anyway. Congress agreed to "grandfather" or exempt existing equipment from new pollution limits with the stipulation that when they were upgraded or replaced, more stringent rules would apply (fig. 16.30). The result was that owners kept old facilities operating precisely because they were exempted from pollution control. In fact, corporations poured millions into aging power plants and factories, expanding their capacity rather than build new ones. Thirty years later, most of those grandfathered plants are still going strong, and continue to be among the biggest contributors to smog and acid rain.

The Clinton administration attempted to force utilities to install modern pollution control on old power plants when they replaced or repaired equipment. President Bush, however, said that determining which facilities are new, and which are not, represented a cumbersome and unreasonable imposition on industries. The EPA

TABLE 16.3

National Ambient Air Quality Standards (NAAQS)

Pollutant	Primary (Health-Based) Averaging Time	Standard Concentration
TSP[a]	Annual geometric mean[b]	50 µg/m³
	24 hours	150 µg/m³
SO$_2$	Annual arithmetic mean[c]	80 µg/m³ (0.03 ppm)
	24 hours	120 µg/m³ (0.14 ppm)
CO	8 hours	10 mg/m³ (9 ppm)
	1 hour	40 mg/m³ (35 ppm)
NO$_2$	Annual arithmetic mean	80 µg/m³ (0.05 ppm)
O$_3$	Daily max 8 hour avg.	157 µg/m³ (0.08 ppm)
Lead	Maximum quarterly avg.	1.5 µg/m³

[a]Total suspended particulate material.

[b]The geometric mean is obtained by taking the nth root of the product of n numbers. This tends to reduce the impact of a few very large numbers in a set.

[c]An arithmetic mean is the average determined by dividing the sum of a group of data points by the number of points.

subsequently announced it would abandon new source reviews, depending instead on voluntary emissions controls and a trading program for air pollution allowances.

Environmental groups generally agree that cap-and-trade (which sets maximum amounts for pollutants, and then lets facilities facing costly cleanup bills to pay others with lower costs to reduce emissions on their behalf) has worked well for sulfur dioxide. When trading began in 1990, economists estimated that eliminating 10 million tons of sulfur dioxide would cost $15 billion per year. Left to find the most economical ways to reduce emissions, however, utilities have been able to reach clean air goals for one-tenth that price. A serious shortcoming of this approach is that while trading has resulted in overall pollution reduction, some local "hot spots" remain where owners have found it cheaper to pay someone else to reduce pollution than to do it themselves. Knowing that the average person is enjoying cleaner air isn't much comfort if you're living in one of the persistently dirty areas.

Many environmentalists argue that carbon dioxide should be classified as a pollutant because of its role in global warming. Some also complain that market mechanisms allow industry to postpone installing pollution controls and forces residents of many states to continue to breathe dirty air for far longer than is necessary. And industry contends that "command and control" mechanisms aren't effective because they don't provide an incentive to continue to search for new, more efficient means of pollution control. What do you think? Which of these regulatory approaches would you favor? Does the kind of pollutant or its effects influence your answer?

16.7 CURRENT CONDITIONS AND FUTURE PROSPECTS

Although the United States has not yet achieved the Clean Air Act goals in many parts of the country, air quality has improved dramatically in the last decade in terms of the major large-volume pollutants. For 23 of the largest U.S. cities, the number of days each year in which air quality reached the hazardous level is down 93 percent from a decade ago. Of 97 metropolitan areas that failed to meet clean air standards in the 1980s, 41 are now in compliance. For many cities, this is the first time they met air quality goals in 20 years.

There have been some notable successes and some failures. The EPA estimates that between 1970 and 1998, lead fell 98 percent, SO_2 declined 35 percent, and CO shrank 32 percent (fig. 16.31). Filters, scrubbers, and precipitators on power plants and other large stationary sources are responsible for most of the particulate and SO_2 reductions. Catalytic converters on automobiles are responsible for most of the CO and O_3 reductions.

The only conventional "criteria" pollutants that have not dropped significantly are particulates and NO_x. Because automobiles are the main source of NO_x, cities, such as Nashville, Tennessee, and Atlanta, Georgia, where pollution comes largely from traffic, still have serious air quality problems. Rigorous pollution controls are having a positive effect on Southern California air quality. Los Angeles, which had the dirtiest air in the nation for decades, wasn't even in the top 20 polluted cities in 2005.

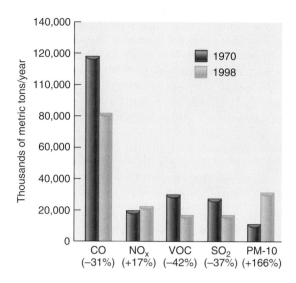

FIGURE 16.31 Air pollution trends in the United States, 1970 to 1998. Although population and economic activity increased during this period, emissions of all "criteria" air pollutants, except for nitrogen oxides and particulate matter, decreased significantly.
Source: Environmental Protection Agency, 2002.

Particulate matter (mostly dust and soot) is produced by agriculture, fuel combustion, metal smelting, concrete manufacturing, and other activities. Industrial cities, such as Baltimore, Maryland, and Baton Rouge, Louisiana, also have continuing problems. Eighty-five other urban areas are still considered nonattainment regions. In spite of these local failures, however, 80 percent of the United States now meets the National Ambient Air Quality Standards. This improvement in air quality is perhaps the greatest environmental success story in our history.

Air pollution remains a problem in many places

The outlook is not so encouraging in other parts of the world. The major metropolitan areas of many developing countries are growing at explosive rates to incredible sizes (chapter 22), and environmental quality is abysmal in many of them. Mexico City remains notorious for bad air. Pollution levels exceed WHO health standards 350 days per year, and more than half of all city children have lead levels in their blood high enough to lower intelligence and retard development. Mexico City's 131,000 industries and 2.5 million vehicles spew out more than 5,500 tons of air pollutants daily. Santiago, Chile, averages 299 days per year on which suspended particulates exceed WHO standards of 90 mg/m³.

While China is making efforts to control air and water pollution (see chapter 1), many of China's 400,000 factories have no air pollution controls. Experts estimate that home coal burners and factories emit 10 million tons of soot and 15 million tons of sulfur dioxide annually and that emissions have increased rapidly over the past 20 years. Seven of the ten cities in the world with the worst air quality are in China. Sheyang, an industrial city in northern China, is thought to have the world's worst continuing particulate

FIGURE 16.32 Air quality in Delhi, India, has improved dramatically since buses, auto-rickshaws, and taxis were required to switch from liquid fuels to compressed natural gas. This is one of the most encouraging success stories in controlling pollution in the developing world.

FIGURE 16.33 Cubatao, Brazil, was once considered one of the most polluted cities in the world. Better environmental regulations and enforcement along with massive investments in pollution-control equipment have improved air quality significantly.

problem, with peak winter concentrations over 700 mg/m³ (nine times U.S. maximum standards). Airborne particulates in Sheyang exceed WHO standards on 347 days per year. It's estimated that air pollution is responsible for 400,000 premature deaths every year in China. Beijing, Xi'an, and Guangzhou also have severe air pollution problems. The high incidence of cancer in Shanghai is thought to be linked to air pollution (see fig. 16.2).

Every year, the Blacksmith Institute compiles a list of the world's worst polluted places (see table 14.3). For 2006, air pollution was the main problem in nine of the top ten worst places, and all but two of those were mines and/or smelter complexes. These problems are especially disastrous in the developing world and the former Soviet Union, where funds and political will aren't available to deal with pollution or help people suffering from terrible health effects of pollution. You can learn more about these places at www.blacksmithinstitute.org.

Norilsk, Russia (one Blacksmith's pick) is a notorious example of toxic air pollution. Founded in 1935 as a slave labor camp, this Siberian city is considered one of the most polluted places on earth. Norilsk houses the world's largest nickel mine and heavy metals smelting complex, which discharge over 4 million tons of cadmium, copper, lead, nickel, arsenic, selenium, and zinc into the air every year. The snow turns black as quickly as it falls, the air tastes of sulfur, and the average life expectancy for factory workers is ten years below the Russian average (which already is lowest of any industrialized country). Difficult pregnancies and premature births are much more common in Norilsk than elsewhere in Russia. Children living near the nickel plant are ill twice as much as Russia's average, and birth defects are reported to affect as much as 10 percent of the population. Why do people stay in such a place? Many were attracted by high wages and hardship pay, and now that they're sick, they can't afford to move.

There are signs of hope

Not all is pessimistic, however. There have been some spectacular successes in air pollution control. Sweden and West Germany (countries affected by forest losses due to acid precipitation) cut their sulfur emissions by two-thirds between 1970 and 1985. Austria and Switzerland have gone even further, regulating even motorcycle emissions. The Global Environmental Monitoring System (GEMS) reports declines in particulate levels in 26 of 37 cities worldwide. Sulfur dioxide and sulfate particles, which cause acid rain and respiratory disease, have declined in 20 of these cities.

Even poor countries can control air pollution. Delhi, India, for example was once considered one of the world's ten most polluted cities. Visibility often was less than 1 km on smoggy days. Health experts warned that breathing Delhi's air was equivalent to smoking two packs of cigarettes per day. Pollution levels exceeded World Health Organization standards by nearly five times. Respiratory diseases were widespread, and the cancer rate was significantly higher than surrounding rural areas. The biggest problem was vehicle emissions, which contributed about 70 percent of air pollutants (industrial emissions made up 20 percent, while burning of garbage and firewood made up most of the rest).

In the 1990s, catalytic converters were required for automobiles, and unleaded gasoline and low-sulfur diesel fuel were introduced. In 2000, more than private automobiles were required to meet European standards, and in 2002, more than 80,000 buses, auto-rickshaws, and taxis were required to switch from liquid fuels to compressed natural gas (fig. 16.32). Sulfur dioxide and carbon monoxide levels have dropped 80 percent and 70 percent, respectively, since 1997. Particulate emissions dropped by about 50 percent. Residents report that the air is dramatically clearer and more healthy. Unfortunately, rising prosperity, driven by globalization of information management, has doubled the number of vehicles on the roads, threatening this progress. Still, the gains made in New Delhi are encouraging for people everywhere.

Twenty years ago, Cubatao, Brazil, was described as the "Valley of Death," one of the most dangerously polluted places in the world. A steel plant, a huge oil refinery, and fertilizer and chemical factories churned out thousands of tons of air pollutants every year that were trapped between onshore winds and the uplifted plateau on which São Paulo sits (fig. 16.33). Trees died on the surrounding hills. Birth defects and respiratory diseases were alarmingly high. Since then, however, the citizens of Cubatao have made remarkable progress in cleaning up their environment. The end of military rule and restoration of democracy allowed residents to publicize their complaints. The environment became an important political issue. The state of São Paulo invested about $100 million and the private sector spent twice as much to clean up most pollution sources in the valley. Particulate pollution was reduced 75 percent, ammonia emissions were reduced 97 percent, hydrocarbons that cause ozone and smog were cut 86 percent, and sulfur dioxide production fell 84 percent. Fish are returning to the rivers, and forests are regrowing on the mountains. Progress is possible! We hope that similar success stories will be obtainable elsewhere.

CONCLUSION

Air pollution is often the most obvious and widespread type of pollution. It can spread from a single source over the entire earth. No matter where you live, from the most remote island in the Pacific, to the highest peak in the Himalayas, to the frigid ice cap over the North Pole, there are traces of human–made contaminants, remnants of the 2 billion metric tons of pollutants released into the air worldwide every year by human activities.

There are many adverse effects of air pollution, from destroying the protective ozone layer in the stratosphere, poisoning whole forests with acid rain, and corroding building materials, to causing respiratory diseases, birth defects, heart attacks, or cancer in individual humans. We have made encouraging progress in controlling air pollution in many places. Many students aren't aware of how much worse air quality was in the industrial centers of North America and Europe a century or two ago than they are now. Cities such as London, Pittsburg, Chicago, Baltimore, and New York had air quality as bad or worse than most megacities of the developing world now. The progress in reducing air pollution in these cities gives us hope that residents can do so elsewhere as well.

The success of the Montreal Protocol in eliminating CFCs is a landmark in international cooperation on an environmental problem. While the stratospheric ozone hole continues to grow because of global warming effects and the residual chlorine in the air released decades ago, we expect the ozone depletion to end in about 50 years. This is one of the few global environmental threats that has had such a rapid and successful resolution. Let's hope that others will follow.

Progress in reducing local pollution in developing countries, such as Brazil and India, also is encouraging. Problems that once seemed overwhelming can be overcome. In some cases, it requires lifestyle changes or different ways of doing things to bring about progress, but as the Chinese philosopher Lao Tsu wrote, "A journey of a thousand miles must begin with a single step."

REVIEWING LEARNING OUTCOMES

By now you should be able to explain the following points:

16.1 Describe the air around us.

16.2 Identify natural sources of air pollution.

16.3 Discuss human-caused air pollution.

- We categorize pollutants according to their source.
- We also categorize pollutants according to their content.
- Unconventional pollutants also are important.
- Indoor air is more dangerous for most of us than outdoor air.

16.4 Explain how climate topography and atmospheric processes affect air quality.

- Temperature inversions trap pollutants.
- Cities create dust domes and heat islands.
- Wind currents carry pollutants intercontinentally.
- Stratospheric ozone is destroyed by chlorine.
- The Montreal Protocol is a resounding success.

16.5 Compare the effects of air pollution.

- Polluted air is dangerous.
- How does pollution harm us?.
- Plants are susceptible to pollution damage.
- Acid deposition has many negative effects.
- Smog and haze reduce visibility.

16.6 Evaluate air pollution control.

- The most effective strategy for controlling pollution is to minimize production.
- Fuel switching and fuel cleaning also are effective.
- Clean air legislation is controversial.

16.7 Summarize current conditions and future prospects.

- Air pollution remains a problem in many places.
- There are signs of hope.

PRACTICE QUIZ

1. Define *primary* and *secondary air pollutants*.
2. What are the seven "criteria" pollutants in the original Clean Air Act? Why were they chosen? How many more hazardous air toxins have been added?
3. What pollutants in indoor air may be hazardous to your health? What is the greatest indoor air problem globally?
4. What is acid deposition? What causes it?

5. What is an atmospheric inversion and how does it trap air pollutants?

6. What is the difference between ambient and stratospheric ozone? What is destroying stratospheric ozone?

7. What is long-range air pollution transport? Give two examples.

8. What is "new source review," and why is it controversial?

9. Which of the conventional pollutants has decreased most in the recent past and which has decreased least?

10. Give one example of current air quality problems and one success in controlling pollution in a developing country.

CRITICAL THINKING AND DISCUSSION QUESTIONS

1. What might be done to improve indoor air quality? Should the government mandate such changes? What values or world-views are represented by different sides of this debate?

2. Debate the following proposition: Our air pollution blows onto someone else; therefore, installing pollution controls will not bring any direct economic benefit to those of us who have to pay for them.

3. Utility managers once claimed that it would cost $1,000 per fish to control acid precipitation in the Adirondack lakes and that it would be cheaper to buy fish for anglers than to put scrubbers on power plants. Suppose that is true. Does it justify continuing pollution?

4. Developing nations claim that richer countries created global warming and stratospheric ozone depletion, and therefore should bear responsibility for fixing these problems. How would you respond?

5. If there are thresholds for pollution effects, is it reasonable or wise to depend on environmental processes to disperse, assimilate, or inactivate waste products?

6. How would you choose between government "command and control" regulations versus market-based trading programs for air pollution control? Are there situations where one approach would work better than the other?

 **DATA analysis**   **Graphing Air Pollution Control**

Reduction of acid-forming air pollutants in Europe is an inspiring success story. The first evidence of ecological damage from acid rain came from disappearance of fish from Scandinavian lakes and rivers in the 1960s. By the 1970s, evidence of air pollution damage to forests in northern and central Europe alarmed many people. International agreements, reached since the mid-1980s have been highly successful in reducing emissions of SO_2 and NO_x as well as photochemical oxidants, such as O_3. The graph on this page shows reductions in SO_2 emissions in Europe between 1990 and 2002. The light blue area shows actual SO_2 emissions. Blue represents changes due to increased nuclear and renewable energy. Orange shows reductions due to energy conservation. Green shows improvement from switching to low-sulfur fuels. Purple shows declines due to increased abatement measures (flue gas scrubbers). The upper boundary of each area indicates what emissions would have been without pollution control.

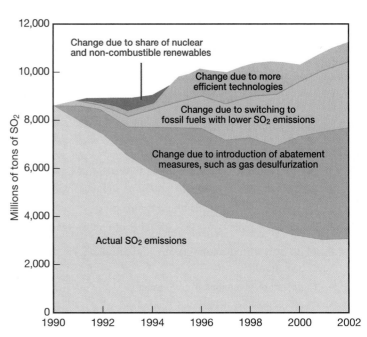

Sulfur dioxide emission reductions in Europe, 1990–2002.

1. How much have actual SO_2 emissions declined since 1990?

2. How much lower were SO_2 emissions in 2002 than they would have been without pollution control (either in percentage or actual amount)?

3. What percentage of this reduction was due to abatement measures, such as flue gas scrubbers?

4. What percent was gained by switching to low-sulfur fuels?

5. How much did energy conservation contribute?

6. What happened to nuclear power?

For Additional Help in Studying This Chapter, please visit our website at www.mhhe.com/cunningham10e. You will find additional practice quizzes and case studies, flashcards, regional examples, place markers for Google Earth™ mapping, and an extensive reading list, all of which will help you learn environmental science.

Water is a precious and beautiful resource.

C H A P T E R

17

Water Use and Management

Of all the natural resources available to human beings, water
is the most essential for virtually every human activity.

—Anna Kajumulo Tibaijuka, UN Undersecretary General—

LEARNING OUTCOMES

After studying this chapter, you should be able to:

17.1 Summarize why water is a precious resource and why shortages occur.

17.2 Compare major water compartments.

17.3 Summarize water availability and use.

17.4 Investigate freshwater shortages.

17.5 Illustrate the benefits and problems of dams and diversions.

17.6 Understand how we might increase water supplies.

Case Study China's South-to-North Water Diversion

Water is inequitably distributed in China. In the south, torrential monsoon rains cause terrible floods. A 1931 flood on the Yangtze displaced 56 million people and killed 3.7 million (the worst natural disaster in recorded history). Northern and western China, on the other hand, are too dry, and getting drier. At least 200 million Chinese live in areas without sufficient fresh water. The government has warned that unless new water sources are found soon, many of those people (including the capital Beijing, with 14 million residents) will have to be moved. But where could they go? Southern China has water, but doesn't need more people.

The solution, according to the government, is to transfer some of the extra water from south to north (fig. 17.1). A gargantuan project is now underway to do just that. Work has begun to build three major canals to carry water from the Yangtze River to northern China. Ultimately, it's planned to move 45 billion m³ per year (more than twice the flow of the Colorado River through the U.S. Grand Canyon) up to 1,600 km (1,000 mi) north. The initial cost estimate of this scheme is about 400 billion yuan (roughly U.S.$62 billion), but it could easily be twice that much.

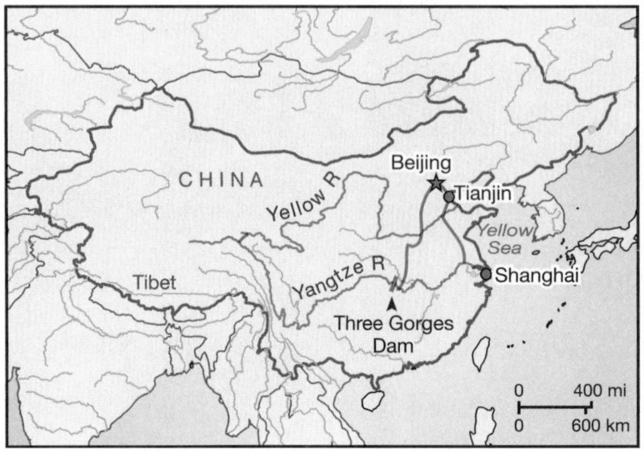

The eastern route (fig. 17.2) uses the Grand Canal, built by Zhou and Sui emperors 1,500 years ago across the coastal plain between Shanghai and Beijing. This project is already operational. It's relatively easy to pump water through the existing waterways, but they're so polluted by sewage and industrial waste that northern cities—even though they're desperately dry—are reluctant to accept this water.

FIGURE 17.1 With the Yellow River nearly depleted by overuse, northern China now plans canals (*red*) to deliver Yangtze water to Beijing.

The central route will draw water from the reservoir behind the nearly completed Three Gorges Dam on the Yangtze. Part of the motivation for building this controversial dam and flooding the historic Three Gorges (fig. 17.3) was to provide energy and raise the river level for the South-to-North project. This middle canal will cross several major mountain ranges and dozens of rivers, including the Han and the Yellow River. Currently, work is in progress on raising the Danjiangkou Dam and enlarging its reservoir as part of this route. This will displace some 200,000 people, but planners say it's worthwhile to benefit a thousand times as many. It's hoped this segment will be finished by 2020.

The western route is the most difficult and expensive. It would tunnel through rugged mountains, across aqueducts, and over deep canyons for more than 250 km (160 mi), from the upper Yangtze to the Yellow River, where they both spill off the Tibetan Plateau. This phase won't be finished until at least 2050. If global warming melts all Tibet's glaciers, however, it may not be feasible anyway.

Planners have waited a lifetime to see this project move forward. Revolutionary leader Mao Zedong proposed it 50 years ago. Environmental scientists worry, however, that drawing down the Yangtze will worsen pollution problems (already exacerbated by the Three Gorges Dam), dry up downstream wetlands, and possibly even alter ocean circulation and climate along China's eastern coast. Although southern China has too much water during the rainy season, even there cities face water shortages because of rapidly growing populations and severe pollution problems. At least half of all major Chinese rivers are too polluted for human consumption. Drawing water away from the rivers on which millions rely only makes pollution problems worse.

FIGURE 17.2 The 1,500-year-old Grand Canal is being cleaned and remodeled to carry water from Shanghai to Tianjin as part of the most ambitious water diversion plan in human history.

FIGURE 17.3 The famous Three Gorges region of the Yangtze River is being flooded in part so water can be pumped 1,600 km north to Beijing.

Case Study *continued*

China isn't unique in its water woes. As you'll read in this chapter, water shortages increasingly threaten economies, societies, and the environment in many places. Growing populations, changing climate, and demands on agriculture and industry stress supplies. Conflicts have already arisen within and between nations.

Other countries have dreamed about—and some have started—massive water relocation schemes similar to China's. How to conserve and allocate water resources is an important theme in environmental science. In this chapter, we'll survey water supplies, water budgets, and strategies for using water more efficiently.

17.1 WATER RESOURCES

Water is a marvelous substance—flowing, rippling, swirling around obstacles in its path, seeping, dripping, trickling, constantly moving from sea to land and back again. Water can be clear, crystalline, icy green in a mountain stream, or black and opaque in a cypress swamp. Water bugs skitter across the surface of a quiet lake; a stream cascades down a stairstep ledge of rock; waves roll endlessly up a sand beach, crash in a welter of foam, and recede. Rain falls in a gentle mist, refreshing plants and animals. A violent thunderstorm floods a meadow, washing away stream banks. Water is a most beautiful and precious resource.

Water is also a great source of conflict. Some 2 billion people, a third of the world's population, live in countries with insufficient fresh water. Some experts estimate this number could double in 25 years. To understand this resource, let's first ask, where does our water come from, and why is it so unevenly distributed?

The hydrologic cycle distributes water in our environment

The hydrologic cycle (water cycle) describes the circulation of water as it evaporates from land, water, and organisms; enters the atmosphere; condenses and is precipitated to the earth's surfaces; and moves underground by infiltration or overland by runoff into rivers, lakes, and seas (see fig. 3.19). This cycle supplies fresh water to the landmasses, maintains a habitable climate, and moderates world temperatures. Movement of water back to the sea in rivers and glaciers is a major geological force that shapes the land and redistributes material. Plants play an important role in the hydrologic cycle, absorbing groundwater and pumping it into the atmosphere by **transpiration** (transport plus evaporation). In tropical forests, as much as 75 percent of annual precipitation is returned to the atmosphere by plants. Much of this moisture falls again, keeping the rainforest humid.

Solar energy drives the hydrologic cycle by evaporating surface water. **Evaporation** is the process in which a liquid is changed to vapor (gas phase) at temperatures well below its boiling point. Water also can move between solid and gaseous states without ever becoming liquid in a process called **sublimation.** On bright, cold, windy winter days, when the air is very dry, snowbanks disappear by sublimation, even though the temperature never gets above freezing. This is the same process that causes "freezer burn" of frozen foods.

In both evaporation and sublimation, molecules of water vapor enter the atmosphere, leaving behind salts and other contaminants and thus creating purified fresh water. This is essentially distillation on a grand scale.

The amount of water vapor in the air is called humidity. Warm air can hold more water than cold air. When a volume of air contains as much water vapor as it can at a given temperature, we say that it has reached its **saturation point. Relative humidity** is the amount of water vapor in the air expressed as a percentage of the maximum amount (saturation point) that could be held at that particular temperature.

When the saturation concentration is exceeded, water molecules begin to aggregate in the process of **condensation.** If the temperature at which this occurs is above 0°C, tiny liquid droplets result. If the temperature is below freezing, ice forms. For a given amount of water vapor, the temperature at which condensation occurs is the **dew point.** Tiny particles, called **condensation nuclei,** float in the air and facilitate this process. Smoke, dust, sea salts, spores, and volcanic ash all provide such particles. Even apparently clear air can contain large numbers of these particles, which are generally too small to be seen by the naked eye. Sea salt is an excellent source of such nuclei, and heavy, low clouds frequently form in the humid air over the ocean.

A cloud, then, is an accumulation of condensed water vapor in droplets or ice crystals. Normally, cloud particles are small enough to remain suspended in the air, but when cloud droplets and ice crystals become large enough, gravity overcomes uplifting air currents, and precipitation occurs.

Water supplies are unevenly distributed

China isn't alone in suffering from droughts in some places and excess water elsewhere. Rain falls unevenly over the planet (fig. 17.4). Some places get almost no precipitation, while others receive heavy rain almost daily. At Iquique, in Chile's Atacama Desert, no rain has fallen in recorded history. At the other end of the scale, Cherrapunji, in northeastern India, received nearly 23 m (897 in.) of rain in 1861.

Three principal factors control these global water deficits and surpluses. First, global atmospheric circulation creates regions of persistent high air pressure and low rainfall about 20° to 40° north and south of the equator (chapter 15). These same circulation patterns produce frequent rainfall near the equator and between about 40° and 60° north and south latitude. Second, proximity to water sources influences precipitation. Where prevailing winds come over oceans, they bring moisture to land. Areas far from oceans—in a windward direction—are usually relatively dry.

A third factor in water distribution is topography. Mountains act as both cloud formers and rain catchers. As air sweeps up the windward side of a mountain, air pressure decreases and air cools. As the air cools, it reaches the saturation point, and moisture condenses as either rain or snow. Thus the windward side of a

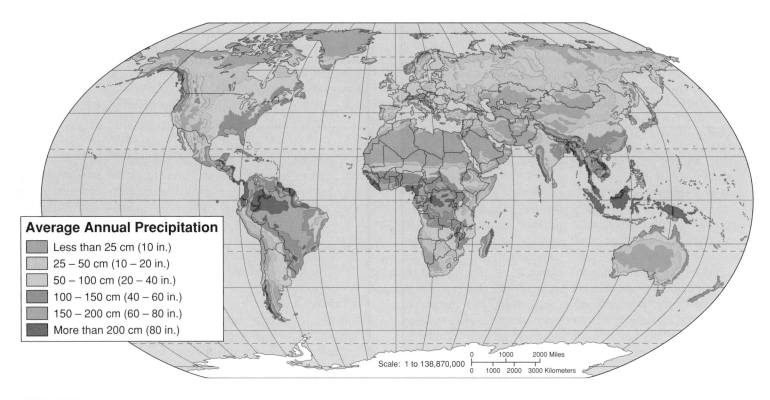

Average Annual Precipitation

- Less than 25 cm (10 in.)
- 25 – 50 cm (10 – 20 in.)
- 50 – 100 cm (20 – 40 in.)
- 100 – 150 cm (40 – 60 in.)
- 150 – 200 cm (60 – 80 in.)
- More than 200 cm (80 in.)

Scale: 1 to 138,870,000

0   1000   2000 Miles

0   1000 2000 3000 Kilometers

FIGURE 17.4 Average annual precipitation. Note wet areas that support tropical rainforests occur along the equator, while the major world deserts occur in zones of dry, descending air between 20° and 40° north and south.

mountain range, as in the Pacific Northwest, is usually wet much of the year. Precipitation leaves the air drier than it was on its way up the mountain. As the air passes the mountaintop and descends the other side, air pressure rises, and the already-dry air warms, increasing its ability to hold moisture. Descending, warming air rarely produces any rain or snow. Places in the **rain shadow,** the dry, leeward side of a mountain range, receive little precipitation. A striking example of the rain shadow effect is that of Mount Waialeale, on the island of Kauai, Hawaii (fig. 17.5). The windward side of the island receives nearly 12 m of rain per year, while the leeward side, just a few kilometers away, receives just 46 cm.

Usually a combination of factors affects precipitation. In Cherrapunji, India, atmospheric circulation sweeps moisture from the warm Indian Ocean toward the high ridges of the Himalayas. Iquique, Chile, lies in the rain shadow of the Andes and in a high-pressure desert zone. Prevailing winds are from the east, so even though Iquique lies near the ocean, it is far from the winds' moisture source—the Atlantic. In the American Southwest, Australia, and the Sahara, high-pressure atmospheric conditions tend to keep the air and land dry. The global map of precipitation represents a complex combination of these forces of atmospheric circulation, prevailing winds, and topography.

Human activity also explains some regions of water deficit. As noted earlier, plant transpiration recycles moisture and produces rain. When forests are cleared, falling rain quickly enters streams and returns to the ocean. In Greece, Lebanon, parts of Africa, the Caribbean, South Asia, and elsewhere, desert-like conditions have developed since the original forests were destroyed.

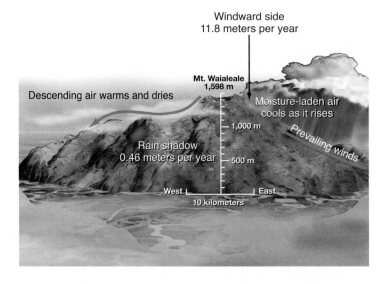

FIGURE 17.5 Rainfall on the east side of Mount Waialeale in Hawaii is more than 20 times as much as on the west side. Prevailing trade winds bring moisture-laden sea air onshore. The air cools as it rises up the flanks of the mountain and the water it carries precipitates as rain—11.8 m (38 ft) per year!

TABLE 17.1

Earth's Water Compartments—Estimated Volume of Water in Storage, Percent of Total, and Average Residence Time

	Volume (Thousands of km^3)	% Total Water	Average Residence Time
Total	1,403,377	100	2,800 years
Ocean	1,370,000	97.6	3,000 years to 30,000 years*
Ice and snow	29,000	2.07	1 to 16,000 years*
Groundwater down to 1 km	4,000	0.28	From days to thousands of years*
Lakes and reservoirs	125	0.009	1 to 100 years*
Saline lakes	104	0.007	10 to 1,000 years*
Soil moisture	65	0.005	2 weeks to a year
Biological moisture in plants and animals	65	0.005	1 week
Atmosphere	13	0.001	8 to 10 days
Swamps and marshes	3.6	0.003	From months to years
Rivers and streams	1.7	0.0001	10 to 30 days

*Depends on depth and other factors.
Source: Data from U.S. Geological Survey.

17.2 MAJOR WATER COMPARTMENTS

The distribution of water often is described in terms of interacting compartments in which water resides for short or long times. Table 17.1 shows the major water compartments in the world. Because these compartments are very large, we use special units, such as acre-feet, to describe their volume (table 17.2).

Oceans hold 97 percent of all water on earth

Together, the oceans contain more than 97 percent of all the *liquid* water in the world. (The water of crystallization in rocks is far larger than the amount of liquid water.) Oceans are too salty for most human uses, but they contain 90 percent of the world's living biomass. While the ocean basins really form a continuous reservoir, shallows and narrows between them reduce water exchange, so they have different compositions, climatic effects, and even different surface elevations.

Oceans play a crucial role in moderating the earth's temperature (fig. 17.6). Vast river-like currents transport warm water from the equator to higher latitudes, and cold water flows from the poles to the tropics (fig. 17.7). The Gulf Stream, which flows northeast from the coast of North America toward northern Europe, flows at a steady rate of 10–12 km per hour (6–7.5 mph) and carries more than 100 times more water than all rivers on earth put together.

In tropical seas, surface waters are warmed by the sun, diluted by rainwater and runoff from the land, and aerated by wave action. In higher latitudes, surface waters are cold and much more dense. This dense water subsides or sinks to the bottom of deep ocean basins and flows toward the equator. Warm surface water of the tropics stratifies or floats on top of this cold, dense water as currents carry warm water to high latitudes. Sharp boundaries form between different water densities, different salinities, and different temperatures, retarding mixing between these layers.

While parts of the hydrologic cycle occur on a time scale of hours or days, other parts take centuries. The average **residence time** of water in the ocean (the length of time that an individual molecule spends circulating in the ocean before it evaporates and starts through the hydrologic cycle again) is about 3,000 years. In the deepest ocean trenches, movement is almost nonexistent and water may remain undisturbed for tens of thousands of years.

TABLE 17.2

Some Units of Water Measurement

One cubic kilometer (km^3) equals 1 billion cubic meters (m^3), 1 trillion liters, or 264 billion gallons.

One acre-foot is the amount of water required to cover an acre of ground 1 foot deep. This is equivalent to 325,851 gallons, or 1.2 million liters, or 1,234 m^3, about the amount consumed annually by a family of four in the United States.

One cubic foot per second of river flow equals 28.3 liters per second or 449 gallons per minute.

See table at end of back for conversion factors.

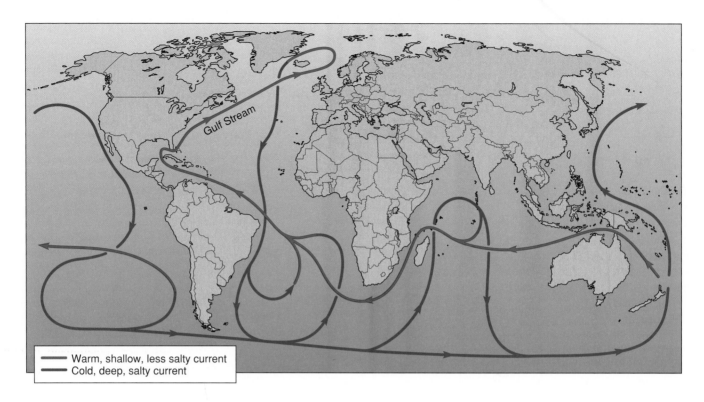

FIGURE 17.6 Ocean currents act as a global conveyor system, redistributing warm and cold water around the globe. These currents moderate our climate. For example, the Gulf Stream keeps northern Europe much warmer than northern Canada.

Glaciers, ice, and snow contain most surface fresh water

Of the 2.4 percent of all water that is fresh, nearly 90 percent is tied up in glaciers, ice caps, and snowfields (fig. 17.8). Glaciers are really rivers of ice flowing downhill very slowly

FIGURE 17.7 Ocean currents, such as the warm Gulf Stream, redistribute heat as they flow around the globe. Here, orange and yellow indicate warm water temperatures (25–30°C); blue and green are cold (0–5°C).

(fig. 17.9). They now occur only at high altitudes or high latitudes, but as recently as 18,000 years ago about one-third of the continental landmass was covered by glacial ice sheets. Most of this ice has now melted and the largest remnant is in Antarctica. As much as 2 km (1.25 mi) thick, the Antarctic glaciers cover all but the highest mountain peaks and contain nearly 85 percent of all ice in the world.

A smaller ice sheet on Greenland, together with floating sea ice around the North Pole, makes up another 10 percent of the world's frozen water reservoirs. Mountain snow pack and ice constitute the remaining 5 percent.

Groundwater stores large resources

After glaciers, the next largest reservoir of fresh water is held in the ground as **groundwater.** Precipitation that does not evaporate back into the air or run off over the surface percolates through the soil and into fractures and spaces of permeable rocks in a process called **infiltration** (fig. 17.10). Upper soil layers that hold both air and water make up the **zone of aeration.** Moisture for plant growth comes primarily from these layers. Depending on rainfall amount, soil type, and surface topography, the zone of aeration may be very shallow or quite deep. Lower soil layers where all spaces are filled with water make up the **zone of saturation.** The top of this zone is the **water table.** The water table is not flat, but undulates according to the surface topography and subsurface structure. Water tables also rise and fall seasonally, depending on precipitation and infiltration rates.

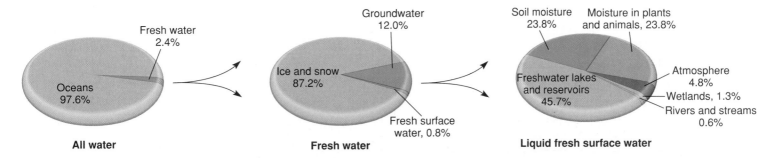

FIGURE 17.8 Less than 1 percent of fresh water, and less than 0.02 percent of all water, is fresh, liquid surface water on which terrestrial life depends. **Source:** U.S. Geological Survey.

Porous layers of sand, gravel, or rock lying below the water table are called **aquifers.** Aquifers are always underlain by relatively impermeable layers of rock or clay that keep water from seeping out at the bottom (fig. 17.11).

Folding and tilting of the earth's crust by geologic processes can create shapes that generate water pressure in confined aquifers (those trapped between two impervious, confining rock layers). When a pressurized aquifer intersects the surface, or if it is penetrated by a pipe or conduit, an **artesian** well or spring results from which water gushes without being pumped.

Areas in which infiltration of water into an aquifer occurs are called **recharge zones.** The rate at which most aquifers are refilled is very slow, however, and groundwater presently is being removed faster than it can be replenished in many areas. Urbanization, road building, and other development often block recharge zones and prevent replenishment of important aquifers. Contamination of surface water in recharge zones and seepage of pollutants into abandoned wells have polluted aquifers in many places, making them unfit for most uses (chapter 18). Many cities protect aquifer recharge zones from pollution or development, both as a way to drain off rainwater and as a way to replenish the aquifer with pure water.

Some aquifers contain very large volumes of water. The groundwater within 1 km of the surface in the United States is more than 30 times the volume of all the freshwater lakes, rivers, and reservoirs on the surface. While water can flow through limestone caverns in underground rivers, most movement in aquifers is a dispersed and almost imperceptible trickle through tiny fractures and spaces. Depending on geology, it can take anywhere from a few hours to several years for contaminants to move a few hundred meters through an aquifer.

Rivers, lakes, and wetlands cycle quickly

Precipitation that does not evaporate or infiltrate into the ground runs off over the surface, drawn by the force of gravity back toward the sea. Rivulets accumulate to form streams, and streams join to form rivers. Although the total amount of water contained

FIGURE 17.9 Glaciers are rivers of ice sliding very slowly downhill. Together, polar ice sheets and alpine glaciers contain more than three times as much fresh water as all the lakes, ponds, streams, and rivers in the world. The dark streaks on the surface of this Alaskan glacier are dirt and rocks marking the edges of tributary glaciers that have combined to make this huge flow.

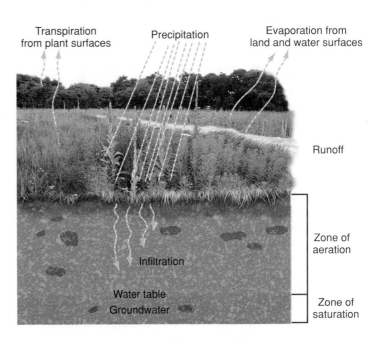

FIGURE 17.10 Precipitation that does not evaporate or run off over the surface percolates through the soil in a process called infiltration. The upper layers of soil hold droplets of moisture between air-filled spaces. Lower layers, where all spaces are filled with water, make up the zone of saturation, or groundwater.

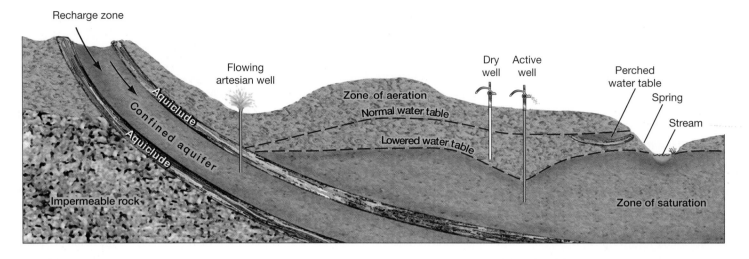

FIGURE 17.11 An aquifer is a porous or cracked layer of rock. Impervious rock layers (aquicludes) keep water within a confined aquifer. Pressure from uphill makes an artesian well flow freely. Pumping can create a cone of depression, which leaves shallower wells dry.

at any one time in rivers and streams is small compared to the other water reservoirs of the world (see table 17.1), these surface waters are vitally important to humans and most other organisms. Most rivers, if they were not constantly replenished by precipitation, meltwater from snow and ice, or seepage from groundwater, would begin to diminish in a few weeks.

We measure the size of a river in terms of its **discharge,** the amount of water that passes a fixed point in a given amount of time. This is usually expressed as liters or cubic feet of water per second. The 16 largest rivers in the world carry nearly half of all surface runoff on earth. The Amazon is by far the largest river in the world (table 17.3), carrying roughly ten times the volume of the Mississippi. Several Amazonian tributaries such as the Maderia, Rio Negro, and Ucayali would be among the world's top rivers in their own right.

Ponds are generally considered to be small temporary or permanent bodies of water shallow enough for rooted plants to grow over most of the bottom. Lakes are inland depressions that hold standing fresh water year-round. Maximum lake depths range from a few meters to over 1,600 m (1 mi) in Lake Baikal in Siberia. Surface areas vary in size from less than one-half hectare (one acre) to large inland seas, such as Lake Superior or the Caspian Sea, covering hundreds of thousands of square kilometers. Both ponds and lakes are relatively temporary features on the landscape because they eventually fill with silt or are emptied by cutting of an outlet stream through the barrier that creates them.

While lakes contain nearly 100 times as much water as all rivers and streams combined, they are still a minor component of total world water supply. Their water is much more accessible than groundwater or glaciers, however, and they are important in many ways for humans and other organisms.

Wetlands play a vital and often unappreciated role in the hydrologic cycle. Their lush plant growth stabilizes soil and holds back surface runoff, allowing time for infiltration into aquifers

and producing even, year-long stream flow. In the United States, about 20 percent of the 1 billion ha of land area was once wetland. In the past 200 years, more than one-half of those wetlands have been drained, filled, or degraded. Agricultural drainage accounts for the bulk of the losses.

When wetlands are disturbed, their natural water-absorbing capacity is reduced and surface waters run off quickly, resulting in floods and erosion during the rainy season and dry, or nearly dry, stream beds the rest of the year. This has a disastrous effect on biological diversity and productivity, as well as on human affairs.

TABLE 17.3

Major Rivers of the World

River	Countries in River Basin	Average Annual Discharge at (m^3/sec)
Amazon	Brazil, Peru	175,000
Orinoco	Venezula, Colombia	45,300
Congo	Congo	39,200
Yangtze	Tibet, China	28,000
Bramaputra	Tibet, India, Bangladesh	19,000
Mississippi	United States	18,400
Mekong	China, Laos, Burma, Thailand, Cambodia, Vietnam	18,300
Parana	Paraguay, Argentina	18,000
Yenisey	Russia	17,200
Lena	Russia	16,000

$1 \; m^3 = 264$ gallons.
Source: World Resources Institute.

The atmosphere is among the smallest of compartments

The atmosphere is among the smallest of the major water reservoirs of the earth in terms of water volume, containing less than 0.001 percent of the total water supply. It also has the most rapid turnover rate. An individual water molecule resides in the atmosphere for about ten days, on average. While water vapor makes up only a small amount (4 percent maximum at normal temperatures) of the total volume of the air, movement of water through the atmosphere provides the mechanism for distributing fresh water over the landmasses and replenishing terrestrial reservoirs.

Think About It

Locate the ten rivers in table 17.3 on the physiographic map in the back of your book. Also, check their approximate locations in figure 17.4. How many of these rivers are tropical? In rainy regions? In populous regions? How might some of these rivers affect their surrounding environment or populations?

17.3 WATER AVAILABILITY AND USE

Clean, fresh water is essential for nearly every human endeavor. Perhaps more than any other environmental factor, the availability of water determines the location and activities of humans on earth (fig. 17.12). **Renewable water supplies** are made up, in general, of surface runoff plus the infiltration into accessible freshwater aquifers. About two-thirds of the water carried in rivers and streams every year occurs in seasonal floods that are too large or violent to be stored or trapped effectively for human uses. Stable runoff is the dependable, renewable, year-round supply of surface water. Much of this occurs, however, in sparsely inhabited regions or where technology, finances, or other factors make it difficult to use it productively. Still, the readily accessible, renewable water supplies are very large, amounting to some 1,500 km³ (about 400,000 gal) per person per year worldwide.

When discussing water use, you will encounter several different units of volume (table 17.2). Becoming familiar with these terms will help you evaluate the statistics that you read.

Many people lack access to clean water

As you can see in figure 17.4. South America, West Central Africa, and South and Southeast Asia all have areas of very high rainfall. Brazil and the Democratic Republic of Congo, because they have high precipitation levels and large land areas, are among the most water-rich countries on earth. Canada and Russia, which are both very large, also have large annual water supplies. The highest per capita water supplies generally occur in countries with wet climates and low population densities. Iceland, for example, has about 160 million gallons per person per year. In contrast, Bahrain, where temperatures are extremely high and rain almost never falls, has essentially no natural fresh water. Almost all of Bahrain's water comes from imports and desalinized seawater. Egypt, in spite of the fact that the Nile River flows through it, has only about 11,000 gallons of water annually per capita, or about 15,000 times less than Iceland.

Periodic droughts create severe regional water shortages. Droughts are most common and often most severe in semiarid zones where moisture availability is the critical factor in determining plant and animal distribution. Undisturbed ecosystems often survive extended droughts with little damage, but introduction of domestic animals and agriculture disrupts native vegetation and undermines natural adaptations to low moisture levels.

Droughts are often cyclic, and land-use practices exacerbate their effects. In the United States, the cycle of drought seems to be about 30 years. There were severe dry years in the 1870s, 1900s, 1930s, 1950s, and 1970s. The worst of these in economic and social terms were the 1930s. Poor soil conservation practices and a series of dry years in the Great Plains combined to create the "dust bowl." Wind stripped topsoil from millions of hectares of land, and billowing dust clouds turned day into night. Thousands of families were forced to leave farms and migrate to cities.

Much of the western United States continues to be plagued by drought and overexploitation of limited water supplies (fig.17.13). The El Niño, Southern Oscillation (ENSO) system plays an important role in droughts in North America and elsewhere. There now is a great worry that global warming (see chapter 15) will bring about major climatic changes and make droughts both more frequent and more severe than in the past in some places.

FIGURE 17.12 Water has always been the key to survival. Who has access to this precious resource and who doesn't has long been a source of tension and conflict.

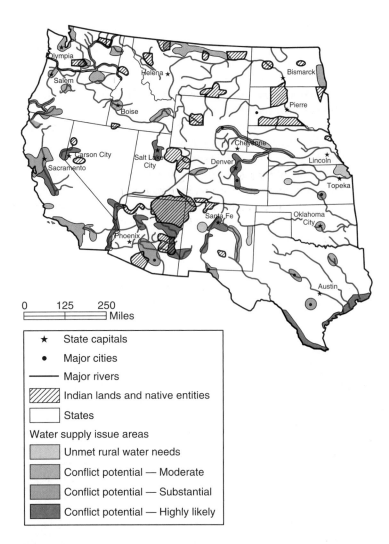

FIGURE 17.13 Rapidly growing populations in arid regions are straining available water supplies. By 2025, the Department of the Interior warns, shortages could cause conflicts in many areas.
Source: Data from U.S. Department of Interior.

Map legend:
★ State capitals
• Major cities
— Major rivers
Indian lands and native entities
States
Water supply issue areas
- Unmet rural water needs
- Conflict potential — Moderate
- Conflict potential — Substantial
- Conflict potential — Highly likely

Water consumption is less than withdrawal

Most water we use eventually returns to rivers and streams. Therefore, it is important to distinguish between withdrawal and consumption. **Withdrawal** is the total amount of water taken from a lake, river, or aquifer for any purpose. Much of this water is employed in nondestructive ways and is returned to circulation in a form that can be used again. **Consumption** is the fraction of withdrawn water that is lost in transmission, evaporation, absorption, chemical transformation, or otherwise made unavailable for other purposes as a result of human use. Note that much water that is withdrawn but not consumed may be **degraded**—polluted or heated so that it is unsuitable for other uses.

Many societies have always treated water as if there is an inexhaustible supply. It has been cheaper and more convenient for most people to dump all used water and get a new supply than to determine what is contaminated and what is not. The natural cleansing and renewing functions of the hydrologic cycle do replace the water we need if natural systems are not overloaded or damaged. Water is a renewable resource, but renewal takes time. The rate at which many of us are using water now may make it necessary to conscientiously protect, conserve, and replenish our water supply.

Water use is increasing

Human water use has been increasing about twice as fast as population growth over the past century (fig. 17.14). Water use is stabilizing in industrialized countries, but demand will increase in developing countries where supplies are available. The average amount of water withdrawn worldwide is about 646 m^3 (170,544 gal) per person per year. This overall average hides great discrepancies in the proportion of annual runoff withdrawn in different areas. Some countries with a plentiful water supply withdraw a very small percentage of the water available to them. Canada, Brazil, and the Congo, for instance, withdraw less than 1 percent of their annual renewable supply.

By contrast, in countries such as Libya and Israel, where water is one of the most crucial environmental resources, groundwater and surface water withdrawal together amount to more than 100 percent of their renewable supply. They are essentially "mining" water—extracting groundwater faster than it is being replenished. Obviously, this is not sustainable in the long run.

The total annual renewable water supply in the United States amounts to an average of about 9,000 m^3 (nearly 2.4 million gal) per person per

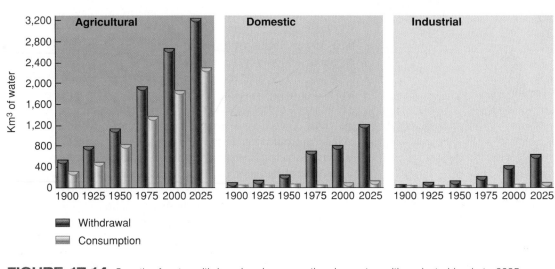

FIGURE 17.14 Growth of water withdrawal and consumption, by sector, with projected levels to 2025.
Source: UNEP, 2002.

Chart legend:
- Withdrawal
- Consumption

year. We now withdraw about one-fifth of that amount, or some 5,000 l (1,300 gal) per person per day, including industrial and agricultural water. By comparison, the average water use in Haiti is less than 30 l (8 gal) per person per day.

Agriculture is the greatest water consumer worldwide

We can divide water use into three major sectors: agricultural, domestic, and industrial. Of these, agriculture accounts for by far the greatest use and consumption. Worldwide, crop irrigation is responsible for two-thirds of water withdrawal and 85 percent of consumption. Evaporation and seepage from unlined irrigation canals are the principal consumptive water losses. Agricultural water use varies greatly, of course. Over 90 percent of water used in India is agricultural; in Kuwait, where water is especially precious, only 4 percent is used for crops. In the United States, which has both a large industrial sector and a highly urbanized population, about half of all water withdrawal, and about 80 percent of consumption, is agricultural.

A tragic case of water overconsumption is the Aral Sea, which lies in Kazakhstan and Uzbekistan (see map at the end of this book). Once the fourth largest inland water body in the world, this giant saline lake lost 75 percent of its surface area and 80 percent of its volume between 1975 and 2004 (fig. 17.15) when, under the former Soviet Union, 90 percent of the natural flow of the Amu Dar'ya and Syr Dar'ya Rivers was diverted to irrigate rice and cotton. Towns that once were prosperous fish processing and shipping ports now lie 100 km from the lake shore.

Vozrojdenie Island, which was used for biological weapons productions in the Soviet era, has become connected to the mainland. The salt concentration in the remaining water doubled, and fishing, which once produced 20,000 tons per year, ceased completely. Today, more than 200,000 tons of salt, sand, and toxic chemicals are blown every day from the dried lake bottom. This polluted cloud is destroying pastures, poisoning farm fields, and damaging the health of residents who remain in the area.

As water levels dropped, the lake split into two lobes. The "Small Aral" in Kazakhstan is now being reclaimed. Some of the river flow has been restored (mainly because Soviet-style rice and cotton farming have been abandoned), and a dam has been built to separate this small lobe from the larger one in Uzbekistan. Water levels in the small, northern lake have risen more than 8 m and surface area has expanded by 30 percent. With cleaner water pouring into the Small Aral, native fish are being reintroduced, and it's hoped that commercial fishing might one day be resumed. The fate of the larger lake remains clouded. There may never be enough water to refill it, and if there were, the toxins left in the lake bed could make it unusable anyway.

An similar catastrophe has befallen Lake Chad in northern Africa. Sixty thousand years ago, during the last ice age, this area was a verdant savanna sprinkled with freshwater lakes and occupied by crocodiles, hippopotamuses, elephants, and gazelles. At that time, Lake Chad was about the present size of the Caspian Sea (400,000 km^2). Climate change has turned the Sahara into a desert, and by the mid-1960s, Lake Chad had shrunk to 25,000 km^2 (as large as the United States' Lake Erie). With a maximum depth of 7 m, the lake is highly sensitive to climate, and it expands and contracts dramatically. Persistent drought coupled with increased demand by massive irrigation projects in the

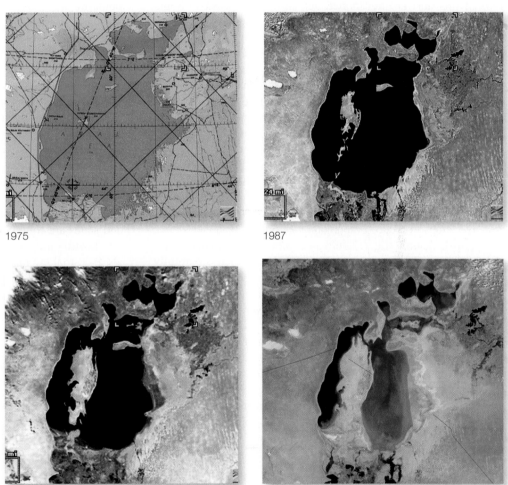

1975

1987

1997

2005

FIGURE 17.15 For 30 years, rivers feeding the Aral Sea have been diverted to irrigate cotton and rice fields. The Aral Sea has lost more than 80 percent of its water. The "Small Aral" (*upper right lobe*) has separated from the main lake, and is now being refilled.

(a) Flood irrigation

(b) Rolling sprinklers

(c) Drip irrigation

FIGURE 17.16 Agricultural irrigation consumes more water than any other use. Methods vary from flood and furrow (a), which use extravagant amounts of water but also flush salts from soils, to sprinklers (b), to highly efficient drip systems (c).

1970s and 1980s has reduced Lake Chad to less than 1,000 km². The silty sand left on the dry lake bed is whipped aloft by strong winds funneled between adjacent mountain ranges. In the winter, the former lake bed, known as the Bodélé Depression, produces an average of 700,000 tons of dust every day. About 40 million tons of this dust are transported annually from Africa to South America, where it is thought to be the main source of mineral nutrients for the Amazon rainforest (chapter 16).

Irrigation can be very inefficient. Traditionally, the main method has been flood or furrow irrigation, in which water- floods a field (fig. 17.16a). As much as half of this water can be lost directly through evaporation. Much of the rest runs off before it is used by plants. In arid lands, flood irrigation is needed to help remove toxic salts from soil, but these salts contaminate streams, lakes, and wetlands downstream. Repeated flood irrigation also waterlogs the soil, reducing crop growth. Sprinkler systems can also be inefficient (fig. 17.16b). Water spraying high in the air quickly evaporates, rather than watering crops. In recent years, growing pressure on water resources has led to more efficient sprinkler systems that hang low over crops to reduce evaporation (see fig. 9.22).

Drip irrigation (fig. 17.16c) is a promising technology for reducing irrigation water use. These systems release carefully regulated amounts of water just above plant roots, so that nearly all water is used by plants. Only about 1 percent of the world's croplands currently use these systems, however.

Irrigation infrastructure, such as dams, canals, pumps, and reservoirs, is expensive. Irrigation is also the economic foundation of many regions. In the United States, the federal government has taken responsibility for providing irrigation for nearly a century. The argument for doing so is that irrigated agriculture is a public good that cannot be provided by individual farmers. A consequence of this policy has frequently been heavily subsidized crops whose costs, in water and in dollars, far outweigh their value.

Domestic and industrial water use are greatest in wealthy countries

Worldwide, domestic water use accounts for about one-fifth of water withdrawals. Because little of this water evaporates or seeps into the ground, consumptive water use is slight, about 10 percent on average. Where sewage treatment is unavailable, water can be badly degraded by urban uses, however. In wealthy countries, each person uses about 500 to 800 l per day (180,000 to 280,000 l per year), far more than in developing countries (30 to 150 l per day). In North America, the largest single user of domestic water is toilet flushing (fig. 17.17). On average, each person in the United States uses about 50,000 l (13,000 gal) of drinking-quality water annually to flush toilets. Bathing accounts for nearly a third of water use, followed by laundry and washing. In western cities such as Palm Desert and Phoenix, lawn watering is also a major water user.

Urban and domestic water use have grown approximately in proportion with urban populations, about 50 percent between 1960 and 2000. Although individual water use seems slight on the scale of world water withdrawals, the cumulative effect of inefficient appliances, long showers, liberal lawn-watering, and other uses is enormous. California has established increasingly stringent standards for washing machines, toilets, and other appliances, in order to reduce urban water demands. Many other cities and states are following this lead to reduce domestic water use.

Industry accounts for 20 percent of global freshwater withdrawals. Industrial use rates range from 70 percent in industrialized parts of Europe to less than 5 percent in countries with little industry.

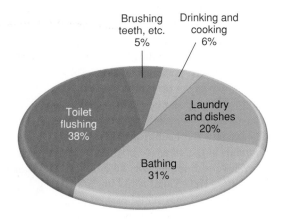

FIGURE 17.17 Typical household water use in the United States. Where could you save the most water?
Source: EPA, 2004.

Power production, including hydropower, nuclear, and thermoelectric power, make up 50 to 70 percent of industrial uses, and industrial processes make up the remainder. As with domestic water, little of this water is made unavailable after use, but it is often degraded by defouling agents, chlorine, or heat when it is released to the environment. The greatest industrial producer of degraded water is mining. Ores must be washed and treated with chemicals such as mercury and cyanide (chapter 14). As much as 80 percent of water used in mining and processing is released with only minimal treatment. In developed countries, industries have greatly improved their performance in recent decades, however. Water withdrawal and consumption have both fallen relative to industrial production.

FIGURE 17.18 Village water supplies in Ghana.

17.4 FRESHWATER SHORTAGES

As the case study for this chapter shows, parts of China are desperately dry. The Gobi Desert is expanding rapidly, and its leading edge is now only 100 km (60 mi) from Beijing. Of 600 major Chinese cities, more than two-thirds have water shortages, and 100 of those cities have severe water scarcity. Pollution exacerbates this situation. Nearly half the population is reported to drink unsafe water.

In addition to the South-to-North water diversion project, the Chinese government has spent nearly 900 billion yuan (about U.S.$125 billion) over the past five years on sewage treatment, desalinization plants, and distribution systems. It's reported that 91 percent of all industrial waste water is now treated, although illegal dumping of toxic chemicals is much too common—particularly from factories shielded by local or regional governments. So far, only 39 percent of domestic sewage is treated in China, but that's probably ten times the amount treated a few decades ago. In 2006, 20 million rural Chinese got safe drinking water. The government has promised to provide clean water to everyone in the country by 2020.

Many countries experience water scarcity and stress

The United Nations estimates that at least a billion people worldwide lack access to safe drinking water, and 2.6 billion don't have adequate sanitation. Depending on what happens to our global climate and human population growth (see the Data Analysis box at the end of this chapter) those numbers could be much higher in 50 years than they are now.

At least 45 countries—most of them in Africa or the Middle East—are considered to have serious water stress. They can't provide the 1,000 m³ of water per person annually to meet the essential needs of their citizens. As is the case in some parts of China, the problem often isn't so much the total water supply as it is lack of access to clean water. In Mali, for example, 88 percent of the population lacks clean water; in Ethiopia it is 94 percent. Rural people often have less access to safe water than do city dwellers. In 33 of the most water-stressed countries, 60 percent of urban residents can get clean water as opposed to only 20 percent of those in the country.

Many developing countries have adequate water, but lack delivery systems. More than two-thirds of the world's households have to fetch water from outside the home. Women and children do most of this work, which takes several hours per day of heavy carrying. Improved public systems bring many benefits to poor families. In Mozambique, for example, the World Bank reports that the average time women spent carrying water decreased from 2 hours per day to only 25 minutes when village wells were installed. The time saved could be used to garden, tend livestock, trade in the market, care for children, or even rest!

Clean water is available in most countries—for those who can pay for it. Water from vendors (fig. 17.18) often is the only source for many in the crowded slums and shantytowns around the major cities of the developing world. Although the quality is often questionable, this water generally costs about ten times more than a piped city supply. Naturally, sanitation levels decline when water is so expensive. A typical poor family in Lima, Peru, for instance, uses one-sixth as much water as a middle-class American family but pays three times as much for it. Following government recommendations that all water be boiled to prevent cholera would take up to one-third of the total income for such a poor family.

Would you fight for water?

Many environmental scientists have warned that water shortages could lead to wars between nations. *Fortune* magazine has written that "water will be to the 21st century what oil was to the 20th." With one-third of all humans living in areas with water stress now, the situation could become much worse as population grows and climate change dries up some areas and brings more severe storms to others. Already we've seen skirmishes—if not outright warfare—over insufficient water. In Kenya, for instance, nomadic tribes have fought over dwindling water and grazing. An underlying cause of the genocide now occurring in the Darfur region of Sudan is water scarcity. When rain was plentiful, Arab pastoralists and African farmers coexisted peacefully. Drought—perhaps caused by global warming—has upset that truce. The hundreds of thousands who have fled to Chad could be considered climate refugees as well as war victims.

Although they haven't usually risen to the level of war, there have been at least 37 military confrontations in the past 50 years in which water has been at least one of the motivating factors. Thirty of those conflicts have been between Israel and its neighbors. India, Pakistan, and Bangladesh also have confronted each other over water rights, and Turkey and Iraq threatened to send their armies to protect access to the water in the Tigris and Euphrates Rivers. Water can even be used as a weapon. As chapter 13 reports, Saddam Hussein cut off water flow into the massive Iraq marshes as a way of punishing his enemies among the Marsh Arabs. Drying of the marshes drove 140,000 people from their homes and destroyed a unique way of life. It also caused severe ecological damage to what is regarded by many as the original Garden of Eden.

Public anger over privatization of the public water supply in Bolivia sparked a revolution that overthrew the government in 2000. Water sales are already a $400-billion-a-year business. Multinational corporations are moving to take control of water systems in many countries. Who owns water and how much they are able to charge for it could become the question of the century. Investors are now betting on scarce water resources by buying future water rights. One Canadian water company, Global Water Corporation, puts it best: "Water has moved from being an endless commodity that may be taken for granted to a rationed necessity that may be taken by force."

Freshwater shortages may become much worse in many places because of global climate change. In 2007, the Intergovernmental Panel on Climate Change (IPCC) issued its fourth report on climate change. Figure 17.19 shows a summary of predictions from several models on likely changes in global precipitation for the period 2090–2099 compared to 1980–1999. White areas are where less than two-thirds of the models agree on likely outcomes; stippled areas are where more than 90 percent of the models agree. How does this map compare to figure 17.4? Which areas do you think are most likely to suffer from water shortages by the end of this century? If you lived in one of those areas, would you fight for water?

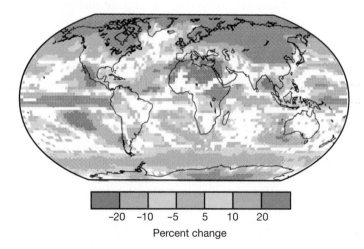

-20 -10 -5   5   10   20

Percent change

FIGURE 17.19 Relative changes in precipitation (in percentage) for the period 2090–2099 compared to 1980–1999, predicted by the Intergovernmental Panel on Climate Change.
Source: IPCC, 2007.

17.5 DAMS AND DIVERSIONS

One way to make more water available is to store runoff in lakes or reservoirs and to ship it to places that need it. People have been moving water around for thousands of years, although none of these projects has ever been as big as the Chinese South-to-North plan. Some of the great civilizations (Mesopotamia, Egypt, China, Harrapa, and the Inca, for example) were based on large-scale water systems that controlled floods and brought irrigation water to farm fields. In fact, some historians argue that organizing people to build and operate these systems was the catalyst for emergence of civilization. Some of these systems are still in use. Roman aqueducts built 2,000 years ago are still in use. Those early water engineers probably never dreamed of moving water on the scale now being proposed.

According to the World Dam Commission, there were only about 250 high dams (more than 15 m tall) in the world before 1900. In the twentieth century, however, at least 45,000 dams were built, about half of them in China. Other countries with many dams include Turkey, Japan, Iran, India, Russia, Brazil, Canada, and the United States. The total cost of this building boom is estimated to have been $2 trillion. At least one-third aren't justified on economic grounds, and less than half have planned for social or environmental impacts.

The Army Corps of Engineers and the Bureau of Reclamation are the primary agencies responsible for building federal dams and water diversion projects in the United States. Building dams provided cheap, renewable power, created jobs for workers, stimulated regional economic development, stored water to reduce flooding, and allowed farming on lands that would otherwise be too dry (fig. 17.20). But not everyone agrees that these dams are an unmitigated benefit. Their storage reservoirs drown free-flowing rivers and often submerge towns and valuable riparian farmlands. They block fish migrations and change aquatic habitats essential for endemic species. In this section, we'll look at some of the advantages and disadvantages of dams, as well as what can be done to mitigate their effects.

Dam failure can be disastrous

When a dam collapses, it can send a wall of water roaring down the river valley. One of the most infamous catastrophes in American history was the Johnstown flood of 1889. The city was built in the Little Conemaugh River valley just east of Pittsburg, Pennsylvania. Twenty kilometers upstream, and 150 m above Johnstown was a 5 km long lake created by an old, poorly maintained private dam. On May 31, 1889, the dam failed, sending 20 million tons of water crashing down the narrow valley. At times the wall of flood water grew to 60 ft in height. Moving at 60 km/hr, it swept away everything in its path, including much of Johnstown. More than 2,200 people died.

A much larger disaster occurred in China in 1975. Heavy monsoon rains caused 62 modern dams in China's Henan Province to fall like dominoes. At least 230,000 people were killed directly or died in subsequent famine and epidemics.

FIGURE 17.20 Hoover Dam powers Las Vegas, Nevada. Lake Mead, behind the dam, loses about 1.3 billion m³ per year to evaporation.

FIGURE 17.21 A Yangtze River town. This photo was taken just before the three Gorges Dam was finished and the lower, older section of the town was flooded. Most residents moved up to the large, new apartment buildings on the hills above the reservoir.

If the Three Gorges Dam on the Yangtze were to fail similarly, it could cause a flood of biblical proportions. More than 100 million people live downstream. The dam is built on an active seismic fault. If it were ruptured by an earthquake or landslide upstream, it could send a wall of water 200 m high racing downstream. What would be your chances of survival if you were in its path? More than 3,200 Chinese dams have failed since 1949.

The Rogun Dam on the Vakhsh River in Tajikistan also is regarded as highly risky. Construction started in 1987, but was abandoned in 1993 after the collapse of the Soviet Union. In 2006, the Tajik government decided to build the dam by itself. Designed to be 335 m high (1,100 ft), it would be the tallest in the world. This dam also is being built in an active seismic zone at the edge of a tectonic plate. If the reservoir behind the dam is ever fully filled, calculations are that its weight will likely trigger earthquakes. The dam, built of rock and earth fill, probably will be highly vulnerable to seismic action.

Dams often displace people and damage ecosystems

The 600 km long reservoir created by China's Three Gorges Dam flooded 1,500 towns and displaced 1.4 million people. Not all those people were unhappy about being resettled. Many merely moved uphill to newly constructed towns that were much better than the ancient river towns where they formerly lived (fig. 17.21). But many farmers, who were promised equal land to that of what they lost, have been disappointed by what they were offered.

Similarly, a series of dams on India's Narmada River has been the focus of decades of protest. Many of the 1 million villagers and tribal people being displaced by this project have engaged in mass resistance and civil disobedience when police try to remove them forcibly. Some have vowed to drown rather than leave their homes. They regard the river as sacred, and don't believe it should be shackled with dams. Furthermore, they don't trust government promises of resettlement.

Canada, also, has been the site of decades of protest by First Nations people over flooding of ancestral lands for hydroelectric projects. The James Bay project, built by Hydro-Quebec between 1971 and 2004, diverted three major rivers flowing west into Hudson Bay and created huge lakes that flooded more than 10,000 km² (4,000 mi²) of forest and tundra, to generate 26,000 megawatts of electrical power. In 1984, shortly after Phase I of this project was completed, 10,000 caribou drowned trying to follow their usual migration route across the newly flooded land. The loss of traditional hunting and fishing sites has been culturally devastating for native Cree people. In addition, mercury, leeched out of rocks in recently submerged land, has entered the food chain, and many residents of the area suffer from mercury poisoning.

In a similar, but less well-known case, Manitoba Hydro has diverted most water in the Churchill River on the west shore of Hudson Bay into the Nelson River to generate hydroelectricity. More than 125,000 km² (50,000 mi²) of tundra and boreal forest have been flooded. This area was traditional hunting land of First Nations people. Much of this land is underlain by permafrost, which is slowly melted by the impounded water, causing the lake shore to continually cave in and the broad, shallow lakes to expand ever further. Trees that fall into the lakes when the shore collapses become a navigational hazard for native people, who depend on boats for travel. As is the case in Quebec, most of the electricity generated by these dams is exported to the United States, where it's marketed as renewable energy. It's true that the hydropower is renewable, but the land that's being destroyed isn't.

Dams kill fish

Dams are especially lethal for migratory fish, such as salmon. Adult fish are blocked from migrating to upstream spawning areas. And juvenile fish die if they go through hydroelectric turbines. The slack water in reservoirs behind dams is also a serious problem. Juvenile salmon evolved to ride the surge of spring runoff downstream to the ocean in two or three weeks. Reservoirs slow this journey to as much as three months, throwing off the time-sensitive physiological changes that allow the fish to survive in salt water when they reach the ocean. Reservoirs expose young salmon to predators, and warm water in reservoirs increases disease in both young and older fish.

Some dams have fish ladders—a cascading series of pools and troughs—that allow fish to bypass the dam. Another option is to move both adults and juveniles by barge. This can result in the strange prospect of barges of wheat moving downstream while passing barges of fish moving the opposite direction. Both these options are expensive and only partially effective in restoring blocked salmon runs.

The tide may be turning against dams. In 1998, the Army Corps announced that it would no longer be building large dams and diversion projects. In the few remaining sites where dams might be built, public opposition is so great that getting approval for projects is unlikely. Instead, the new focus may be on removing existing dams and restoring natural habitats. Former Interior Secretary Bruce Babbitt said, "Of the 75,000 large dams in the United States, most were built a long time ago and are now obsolete, expensive, and unsafe. They were built with no consideration of the environmental costs. As operating licenses come up for renewal, removal and restoration to original stream flows will be one of the options." (See What Do You Think? p. 389).

An example of this option is the removal of four aging dams on California's Klamath River. For years, Indian tribes, commercial fishermen, and conservation groups have urged the government to remove these dams because of their adverse impacts on salmon. PacifiCorp, which owns and operates the dams, offered to trap fish and truck them around the dams, but in a move that surprised everyone, the National Marine Fisheries Service ordered that fish ladders, that would cost about $300 million, be built at the dams. Because removing the dams would cost about $100 million less than modifying them, it seems possible that they may soon be demolished.

Sedimentation limits reservoir life

Rivers with high sediment loads can fill reservoirs quickly (fig. 17.22). In 1957 the Chinese government began building the Sanmenxia Dam on the Huang He (Yellow River) in Shaanxi Province. From the beginning, engineers warned that the river carried so much sediment that the reservoir would have a very limited useful life. Dissent was crushed, however, and by 1960, the dam began filling the river valley and inundating fertile riparian fields that once had been part of China's traditional granaries.

Within two years, sediment accumulation behind the dam had become a serious problem. It blocked the confluence of the Wei and Yellow Rivers and backed up the Wei so it threatened to flood the historic city of Xi'an. By 1962, the reservoir was almost completely filled with sediment and hydropower production dropped by 80 percent. The increased elevation of the riverbed raised the underground water table and caused salinization of wells and farm fields. By 1991, the riverbed was 4.6 m above the surrounding

FIGURE 17.22 This dam is now useless because its reservoir has filled with silt and sediment.

landscape. The river is only kept in check by earthen dams that frequently fail and flood the surrounding countryside. By the time the project was complete, more than 400,000 people had been relocated, far more than planners expected.

Problems are similar, although not so severe, in some American rivers. As the muddy Colorado River slows behind the Glen Canyon and Boulder Dams, it drops its load of suspended sand and silt. More than 10 million metric tons of sediment collect every year behind these dams. Imagine a line of 20,000 dump trucks backed up to Lake Mead and Lake Powell every day, dumping dirt into the water. Within about a century, these reservoirs could be full of mud and useless for either water storage or hydroelectric generation. Elimination of normal spring floods—and the sediment they would usually drop to replenish beaches—has changed the riverside environment in the Grand Canyon. Invasive species crowd out native riparian plants. Beaches that campers use have disappeared. Boulders dumped in the canyon by side streams fill the riverbed. On several occasions, dam managers have released large surges of water from the Glen Canyon Dam to try to replicate normal spring floods. The results have been gratifying, but they don't last long. The canyon needs regular floods to maintain its character.

These lakes also lose huge amounts of water through evaporation in the dry desert air and seepage into porous rock. Together, these two lakes lose more than 2 billion m^3 of water every year. During a severe drought between 1995 and 2004, Lake Powell lost more than 60 percent of its volume and the lake surface dropped more than 50 m (150 ft), leaving a giant bathtub ring of precipitated salts on canyon walls (fig. 17.23). At its lowest point, the reservoir was only a few meters above the minimum level for hydroelectric generation. River runners and canyon aficionados hoped this would convince Congress to breach the dam and let the river run free, but better rains and snows in 2006 began to refill the lake.

The accumulating sediments that clog reservoirs also represents a loss of valuable nutrients. The Aswan High Dam in Egypt, for example, was built to supply irrigation water to make

What Do You Think?

Should We Remove Dams?

The first active hydroelectric dam in the United States to be breached against the wishes of its owners was the 162-year-old Edwards Dam, on the Kennebec River in Augusta, Maine. For years, the U.S. Fish and Wildlife Service advocated removal of this dam, which blocked migration of the endangered Atlantic salmon. After the dam was destroyed in 1999, anglers reported seeing salmon, striped bass, shad, alewive, and sturgeon upstream of the dam. This was the first of at least 140 dams removed over the next five years.

A much larger project is the removal of the Elwah and Glines Dams in Olympic National Park in Washington. Before these dams were built a century ago to provide power to lumber and paper mills in the town of Port Angeles, the Elwah River was one of the most productive salmon rivers in the world. Fifty kilogram (110 lb) King salmon once migrated upstream to spawn. Destruction of these dams is scheduled for 2008. Simply breaching the dams won't restore the salmon, however. Deep sediment beds deposited behind the dams will have to be removed or stabilized to restore the streambed (chapter 13). The costs for dam removal and stream restoration are expected to be at least $200 million.

Another challenging dam removal project is now underway in Montana. The Milltown Dam on the Clark Fork River just east of Missoula is a toxic nightmare. For more than a century, mining, milling, and smelting upstream at Anaconda and Butte (see fig 13.34) dumped millions of tons of waste into the river. In 2004, the Environmental Protection Agency declared the river from the Milltown Dam upstream for 200 km (120 mi) the largest Superfund site in the United States.

The greatest concentration of contaminated sediment is behind the dam, which is now old and unstable. The high levels of arsenic, copper, lead, zinc, nickel, and other toxins that would be released if the dam fails threaten water supplies of Missoula and the water quality in the Columbia, one of the most important rivers in the American west. But you can't just blow out the dam and hope for the best. In 2006, work started to build a bypass channel around the dam and to remove 2.6 million m^3 of toxic sediment. When this is completed, restoration work will reconstruct the streambed and revegetate the surrounding land.

A more controversial topic is the proposed removal of four high dams on the Snake River. These dams were authorized by Congress in 1945 to generate electricity and make Lewistown, Idaho—750 km inland—a seaport. At the time, Idaho produced half of the chinook salmon in the Columbia watershed. Salmon and steelhead trout runs almost immediately plummeted when the dams closed off the river. In 1991, Snake River sockeye became the first of 13 salmon and steelhead stocks to be declared endangered species. The government was forced to propose a plan for species recovery.

Over the next 15 years, federal judges have repeatedly rejected proposed plans said to be inadequate for salmon protection. The major issue is what to do about the dams. On one side, irrigators and businesses that benefit from cheap water, power, and shipping maintain that the dams are indispensable. On the other side are Native American tribes, conservation groups, and commercial and sport fishermen. They point out that the dams produce a very small fraction of the electricity consumed in the Pacific Northwest, benefit very few farmers, and that other shipping options exist. The gains from tourist revenue and commercial salmon fishing could more than offset the potential losses from removing the dams.

Pressure for salmon recovery intensified in 2006, when only three sockeye—all hatchery fish—made the 1,600 km journey from the Pacific to the headwaters of the Salmon River in central Idaho. U.S. District Judge James Redden, who has rejected federal plans for salmon recovery three times, warned recalcitrant agencies that time has run out. Either they submit a viable restoration plan, or he will order breaching of the four Snake River dams.

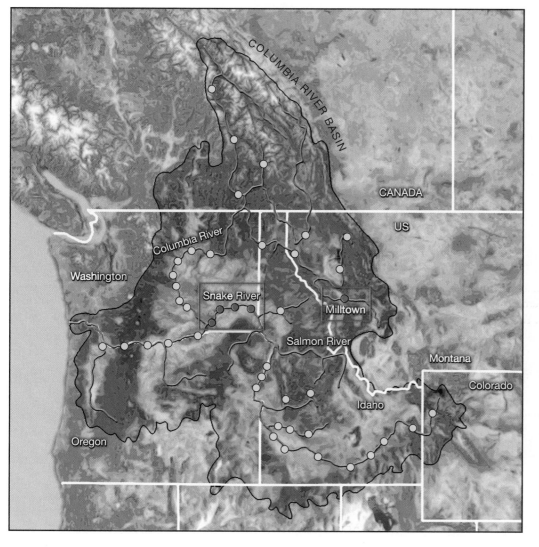

What do you think? How would you weigh the different ecological, economic, cultural, and health impacts of outdated dams? Is it simply an economic calculation of relative costs and benefits, or should ethical considerations be included? How would you compare the value of endangered salmon survival with the ability of farmers to grow food? These questions don't have simple answers. As you read subsequent chapters on water pollution, public policy, and environmental economics, keep in mind the practical questions about removing dams.

agriculture more productive. Although thousands of hectares are being irrigated, the water available is only about half that anticipated because of evaporation in Lake Nasser behind the dam, and seepage losses in unlined canals that deliver the water. Controlling the annual floods of the Nile also has stopped the deposition of nutrient-rich mud on which farmers depended for fertilizing their fields. Commercial fertilizer used to replace naturally provided nutrients costs more than $100 million annually. Furthermore, the nutrients carried by the river once supported a rich fishery in the Mediterranean that was a valuable food source for Egypt. After the dam was built, sardine fishing declined 97 percent. To make matters worse, snails living in shallow irrigation canals has led to an epidemic of schistosomiasis, a debilitating disease caused by parasitic flatworms, for which the snails are an alternate host. In some areas, 80 percent of the residents are infected.

Diversion projects sometimes dry up rivers

While China's South-to-North project probably won't deplete the Yangtze, in many places diversion projects take so much water out of smaller rivers that they are completely dried up for much of the year. China's Huang He (Yellow River), for example, loses so much water to evaporation and diversions that it's reduced to a muddy trickle in much of its lower section, and is completely dry 226 days per year. Similarly, the Colorado River, which once carved the United States' Grand Canyon, is so salty and depleted by the time it reaches the Mexican border that it's useless for agriculture. Most of the year, no water at all reaches the Sea of Cortez.

Environmental scientists point out the ecological disruption caused by overdraft of water from streams. Obviously, fish and other aquatic organisms suffer when the medium in which they live disappears. Anglers, whitewater boaters and others who enjoy the beauty of free-flowing rivers mourn the loss of canyons drowned in reservoirs, and streams dried up by diversion projects (fig. 17.24). Many battles have been fought over the relative value of natural aquatic ecosystems versus the economic gains from appropriating the water for agriculture, industry, or domestic use. Management agencies charged with regulating resources as a public good have to find a balance between competing demands.

California's Owens Valley and Mono Lake, just east of Yosemite National Park, are historic examples of the conflict and controversy over water diversion. In 1905, Frederic Eaton, mayor of Los Angeles, and his friend, William Mulholland, the superintendent of the L.A. Department of Water and Power, surreptitiously bought farms in the Owens Valley. Eventually, Los Angeles became the largest landowner in the valley, controlling more than 90 percent of the water rights. The city built a huge aqueduct to transport the water about 400 km over

FIGURE 17.23 Lake Powell, on the Colorado River, loses more than 1 billion m³ of water to evaporation and seepage every year. During a severe drought between 1995 and 2004, the lake lost more than 60 percent of its volume and its surface dropped more than 50 m (150 ft).

FIGURE 17.24 The recreational and aesthetic values of free-flowing wild rivers and wilderness lakes may be their greatest assets. Competition between *in situ* values and extractive uses can lead to bitter fights and difficult decisions.

FIGURE 17.25 Mono Lake in eastern California. Diversion of tributary rivers to provide water for Los Angeles has shrunk the lake's surface area by one-third, threatening migratory bird flocks that feed here. These formations were created underwater where calcium-rich springs entered the brine-laden lake.

the mountains. The movie *Chinatown* is a fictional account of this project. So much water was diverted that the Owens River was completely dry for most of its 100 km course, and Owens Lake, which was fed by the river, disappeared.

In 1941, Los Angeles needed even more water and began tapping streams that fed Mono Lake just to the north of the Owens Valley. Deprived of its freshwater sources, the volume of the lake was cut in half. The lake surface fell by about 20 m, exposing strange tufa towers created where mineral-rich underwater springs had once entered the salty lake (fig. 17.25). Salinity of the lake water doubled, killing the prolific brine shrimp that nourished huge flocks of migratory waterfowl that once stopped here to feed.

After years of legal wrangling, the California Water Resources Board ruled in 1994 that Los Angeles must allow some water to replenish Mono Lake. The recovery plan set 2014 as the target date to return the lake to its 1964 level. So far, the surface of the lake has risen about 3 m. Annual turnover of the lake, which brought nutrients to the surface where the brine shrimp could access them, has resumed again. The ecology of tributary streams is also recovering, although it will take several decades for complete revival. Similarly, in 2006, Los Angeles began to allow a small amount of water to flow, once again, down the Owens River.

In another inspiring river restoration story, the Deschutes River, a tributary of the Columbia, also is being revived. Much of the river flow was diverted a century ago to irrigate farms in central Oregon. Native American tribes on the Warm Springs Reservation, downstream from this diversion, sued over the destruction of their fishing rights. As part of the settlement, irrigation districts upstream are implementing conservation measures that will return some water to the river. By lining canals and switching from flood irrigation to more efficient sprinkler systems, they can save water but get the same crop yields. Both native people and farmers, seeing the disastrous confrontation over water rights in the Klamath River just to their south, decided it's better to work cooperatively rather than risk solutions that no one likes.

Groundwater is depleted when withdrawals exceed recharge

Groundwater is the source of nearly 40 percent of the fresh water for agricultural and domestic use in the United States. Nearly half of all Americans and about 95 percent of the rural population depend on groundwater for drinking and other domestic purposes. Overuse of these supplies causes several kinds of problems, including drying of wells, natural springs, and disappearance of surface water features such as wetlands, rivers, and lakes.

In many areas of the United States, groundwater is being withdrawn from aquifers faster than natural recharge can replace it. On a local level, this causes a cone of depression in the water table, as is shown in figure 17.11. A heavily pumped well can lower the local water table so that shallower wells go dry. On a broader scale, heavy pumping can deplete a whole aquifer. The Ogallala Aquifer, for example, underlies eight states in the arid high plains between Texas and North Dakota (fig. 17.26). As deep as 400 m (1,200 ft) in its center, this porous bed of sand, gravel, and sandstone once held more water than all the freshwater lakes, streams, and rivers on earth. Excessive pumping for

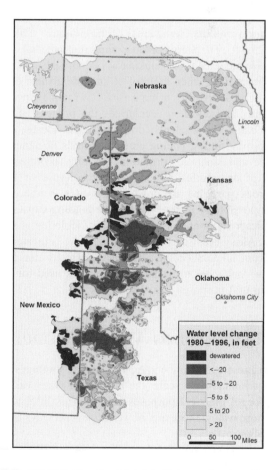

FIGURE 17.26 The Ogallala/High Plains regional aquifer supports a multimillion-dollar agricultural economy, but withdrawal far exceeds recharge. Some countries are down to less than 3 m of saturated thickness.

irrigation and other uses has removed so much water that wells have dried up in many places, and farms, ranches, even whole towns are being abandoned.

Many aquifers have slow recharge rates, so it will take thousands of years to refill them once they are emptied. Much of the groundwater we now are using probably was left there by the glaciers thousands of years ago. It is fossil water, in a sense. It will never be replaced in our lifetimes, and is, essentially, a non-renewable resource. Covering aquifer recharge zones with urban development or diverting runoff that once replenished reservoirs ensures that they will not refill.

Withdrawal of large amounts of groundwater causes porous formations to collapse, resulting in **subsidence** or settling of the surface above. The U.S. Geological Survey estimates that the San Joaquin Valley in California, for example, has sunk more than 10 m in the last 50 years because of excessive groundwater pumping. Around the world, many cities are experiencing subsidence. Many are coastal cities, built on river deltas or other unconsolidated sediments. Flooding is frequently a problem as these coastal areas sink below sea level (chapter 13). Some inland areas also are affected by severe subsidence. Mexico City is one of the worst examples. Built on an old lake bed, it has probably been sinking since Aztec times. In recent years, however, rapid population growth and urbanization (chapter 22) have caused groundwater overdrafts. Some areas of the city have sunk as much as 8.5 m (25.5 ft). The Shrine of Guadalupe, the cathedral, and many other historic monuments are sinking at odd and perilous angles.

Sinkholes form when the roof of an underground channel or cavern collapses, creating a large surface crater. Drawing water from caverns and aquifers accelerates the process of collapse. Sinkholes can form suddenly, dropping cars, houses, and trees without warning into a gaping crater hundreds of meters across. Subsidence and sinkhole formation generally represent permanent loss of an aquifer. When caverns collapse or the pores between rock particles are crushed as water is removed, it is usually impossible to restore their water-holding capacity.

A widespread consequence of aquifer depletion is **saltwater intrusion.** Along coastlines and in areas where saltwater deposits are left from ancient oceans, overuse of freshwater reservoirs often allows saltwater to intrude into aquifers used for domestic and agricultural purposes (fig. 17.27).

17.6 INCREASING WATER SUPPLIES

Where do present and impending freshwater shortages leave us now? On a human time scale, the amount of water on the earth is fixed, for all practical purposes, and there is little we can do to make more water. There are, however, several ways to increase local supplies.

In the dry prairie states of the 1800s and early 1900s, desperate farmers paid self-proclaimed "rainmakers" in efforts to save their withering crops. Centuries earlier, Native Americans danced and prayed to rain gods. We still pursue ways to make rain. Seeding clouds with dry ice or potassium iodide particles has been tested for many years with mixed results. Recently, researchers have been

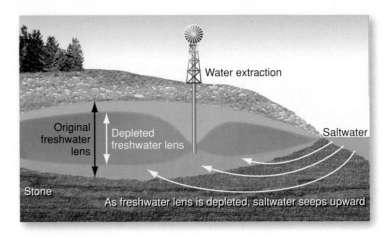

FIGURE 17.27 Saltwater intrusion into a coastal aquifer as the result of groundwater depletion. Many coastal regions of the United States are losing freshwater sources due to saltwater intrusion.

having more success using hygroscopic salts that seem to significantly increase rainfall amounts. This technique is being tested in Mexico, South Africa, and the western United States. There is a concern, however, that rain induced to fall in one area decreases the precipitation somewhere else. Furthermore, there are worries about possible contamination from the salts used to seed clouds.

Desalination provides expensive water

A technology that might have great potential for increasing freshwater supplies is **desalination** of ocean water or brackish saline lakes and lagoons. The most common methods of desalination are distillation (evaporation and recondensation) or reverse osmosis (forcing water under pressure through a semipermeable membrane whose tiny pores allow water to pass but exclude most salts and minerals). In 2007, the global capacity was about 40 million m^3 per day, less than 0.2 percent of all freshwater withdrawals worldwide. This is expected to grow to about 100 million m^3 per day by 2015. Middle Eastern oil-rich states produce about 60 percent of desalinated water. Saudi Arabia is the largest single producer, at about 34 percent of world total. The United States is second, at 20 percent. Although desalination is still three to four times more expensive than most other sources of fresh water, it provides a welcome water supply in such places as Oman and Bahrain where there is no other access to fresh water. If a cheap, inexhaustible source of energy were available, however, the oceans could supply all the water we would ever need.

Domestic conservation can save water

We could probably save as much as half of the water we now use for domestic purposes without great sacrifice or serious changes in our lifestyles. Simple steps, such as taking shorter showers, stopping leaks, and washing cars, dishes, and clothes as efficiently as possible, can go a long way toward forestalling the water shortages that many authorities predict. Isn't it better to adapt to more conservative uses now when we have a choice than to be forced to do it by scarcity in the future?

FIGURE 17.28 By using native plants in a natural setting, residents of Phoenix save water and fit into the surrounding landscape.

The use of conserving appliances, such as low-volume shower heads and efficient dishwashers and washing machines, can reduce water consumption greatly (What Can You Do? p. 394). If you live in an arid part of the country, you might consider whether you really need a lush green lawn that requires constant watering, feeding, and care. Planting native ground cover in a "natural lawn" or developing a rock garden or landscape in harmony with the surrounding ecosystem can be both ecologically sound and aesthetically pleasing (fig. 17.28). There are about 30 million ha (75 million acres) of cultivated lawns, golf courses, and parks in the United States. They receive more water, fertilizer, and pesticides per hectare than any other kind of land.

Our largest domestic water use is toilet flushing (see fig. 17.17). There are now several types of waterless or low-volume toilets. Waterless composting systems can digest both human and kitchen wastes by aerobic bacterial action, producing a rich, nonoffensive compost that can be used as garden fertilizer. There are also low-volume toilets that use recirculating oil or aqueous chemicals to carry wastes to a holding tank, from which they are periodically taken to a treatment plant. Anaerobic digesters use bacterial or chemical processes to produce usable methane gas from domestic wastes. These systems provide valuable energy and save water but are more difficult to operate than conventional toilets. Few cities are ready to mandate waterless toilets, but a number of cities (including Los Angeles, California; Orlando, Florida; Austin, Texas; and Phoenix, Arizona) have ordered that water-saving toilets, showers, and faucets be installed in all new buildings. The motivation was twofold: to relieve overburdened sewer systems and to conserve water.

Significant amounts of water can be reclaimed and recycled. In California, water recovered from treated sewage constitutes the fastest growing water supply, growing about 30 percent per year. Despite public squeamishness, purified sewage effluent is being used for everything from agricultural irrigation to flushing toilets (fig. 17.29). In a statewide first, San Diego is currently piping water from the local sewage plant directly into a drinking-water reservoir. Residents of Singapore and Queensland, Australia, also are now drinking purified sewage effluent. "Don't rule out desalination because it's expensive, or recycling because it sounds yucky," says Morris Iemma, premier of New South Wales. "We're not getting rain; we have no choice."

FIGURE 17.29 Recycled water is being used in California and Arizona for everything from agriculture, to landscaping, to industry. Some cities even use treated sewage effluent for human drinking-water supplies.

Recycling can reduce consumption

In many developing countries as much as 70 percent of all the agricultural water used is lost to leaks in irrigation canals, application to areas where plants don't grow, runoff, and evaporation. Better farming techniques, such as leaving crop residue on fields and ground cover on drainage ways, intercropping, use of mulches, and low-volume irrigation, could reduce these water losses dramatically.

Nearly half of all industrial water use is for cooling of electric power plants and other industrial facilities. Some of this water use could be avoided by installing dry cooling systems similar to the radiator of your car. In many cases, cooling water could be reused for irrigation or other purposes in which water does not have to be drinking quality. The waste heat carried by this water could be a valuable resource if techniques were developed for using it.

Prices and policies have often discouraged conservation

Through most of U.S. history, water policies have generally worked against conservation. In the well-watered eastern United States, water policy was based on riparian usufructuary (use) rights—those who lived along a river bank had the right to use as much water as they liked as long as they didn't interfere with its quality or availability to neighbors downstream. It was assumed that the supply would always be endless and that water had no value until it was used. In the drier western regions where water often is a limiting

What Can You Do?

Saving Water and Preventing Pollution

Each of us can conserve much of the water we use and avoid water pollution in many simple ways.

- Don't flush every time you use the toilet. Take shorter showers; don't wash your car so often.

- Don't let the faucet run while washing hands, dishes, food, or brushing your teeth. Draw a basin of water for washing and another for rinsing dishes. Don't run the dishwasher when half full.

- Dispose of used motor oil, household hazardous waste, batteries, etc., responsibly. Don't dump anything down a storm sewer that you wouldn't want to drink.

- Avoid using toxic or hazardous chemicals for simple cleaning or plumbing jobs. A plunger or plumber's snake will often unclog a drain just as well as caustic acids or lye. Hot water and soap will clean brushes more safely than organic solvents.

- If you have a lawn, use water sparingly. Water your grass and garden at night, not in the middle of the day. Consider planting native plants, low-maintenance ground cover, a rock garden, or some other xeriphytic landscaping.

- Use water-conserving appliances: low-flow showers, low-flush toilets, and aerated faucets.

- Use recycled (gray) water for lawns, house plants, car washing.

- Check your toilet for leaks. A leaky toilet can waste 50 gallons per day. Add a few drops of dark food coloring to the tank and wait 15 minutes. If the tank is leaking, the water in the bowl will change color.

---

resource, water law is based primarily on the Spanish system of prior appropriation rights, or "first in time are first in right." Even if the prior appropriators are downstream, they can legally block upstream users from taking or using water flowing over their property. But the appropriated water had to be put to "beneficial" use by being consumed. This creates a policy of "use it or lose it." Water left in a stream, even if essential for recreation, aesthetic enjoyment, or to sustain ecological communities, is not being appropriated or put to "beneficial" (that is, economic) use. Under this system, water rights can be bought and sold, but water owners frequently are reluctant to conserve water for fear of losing their rights.

In most federal "reclamation" projects, customers were charged only for the immediate costs of water delivery. The costs of building dams and distribution systems was subsidized, and the potential value of competing uses was routinely ignored. Farmers in California's Central Valley, for instance, for many years paid only about one-tenth of what it cost the government to supply water to them. This didn't encourage conservation. Subsidies created by underpriced water amounted to as much as $500,000 per farm per year in some areas.

Growing recognition that water is a precious and finite resource has changed policies and encouraged conservation across the United States. Despite a growing population, the United States is now saving some 144 million liters (38 million gal) per day—or enough water to fill Lake Erie in a decade—compared to per capita consumption rates of 20 years ago. With 37 million more people in the United Sates now than there were in 1980, we get by with 10 percent less water. New requirements for water-efficient fixtures and low-flush toilets in many cities help to conserve water on the home front. More efficient irrigation methods on farms also are a major reason for the downward trend.

Charging a higher proportion of real costs to users of public water projects has helped encourage conservation, and so have water marketing policies that allow prospective users to bid on water rights. Both the United States and Australia have had effective water pricing and allocation policies that encourage the most socially beneficial uses and discourage wasteful water uses. Market mechanisms for water allotment can be sensitive, however, in developing countries where farmers and low-income urban residents could be outbid for irreplaceable water supplies.

It will be important, as water markets develop, to be sure that environmental, recreational, and wildlife values are not sacrificed to the lure of high-bidding industrial and domestic uses. Given prices based on real costs of using water and reasonable investments in public water supplies, pollution control, and sanitation, the World Bank estimates that everyone in the world could have an adequate supply of clean water by the year 2030 (fig. 17.30). We will discuss the causes, effects, and solutions for water pollution in chapter 18.

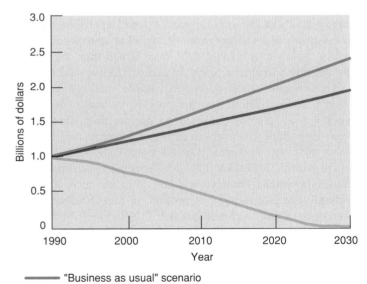

FIGURE 17.30 Three scenarios for government investments on clean water and sanitation services, 1990 to 2030.
Source: World Bank estimates based on research paper by Dennis Anderson and William Cavendish, "Efficiency and Substitution in Pollution Abatement: Simulation Studies in Three Sectors."

CONCLUSION

Water is a precious resource. As human populations grow and climate change affects rainfall patterns, water is likely to become even more scarce in the future. Already, about 2 billion people live in water-stressed countries (where there are inadequate supplies to meet all demands), and at least half those people don't have access to clean drinking water. Depending on population growth rates and climate change, it's possible that by 2050 there could be 7 billion people (about 60 percent of the world population) living in areas with water stress or scarcity. Conflicts over water rights are becoming more common between groups within countries and between neighboring countries that share water resources. This is made more likely by the fact that most major rivers cross two or more countries before reaching the sea. Many experts agree with *Fortune* magazine that "water will be to the 21st century what oil was to the 20th."

There are many ways to make more water available. Huge diversion projects, such as the Chinese South-to-North diversion, are already underway. Building dams and shipping water between watersheds, however, can have severe ecological and social effects. Perhaps a better way is to practice conservation and water recycling. These efforts, also, are underway in many places, and show great promise for meeting our needs for this irreplaceable resource. There are things you can do as an individual to save water and prevent pollution. Even if you don't have water shortages now where you live, it may be wise to learn how to live in a water-limited world.

REVIEWING LEARNING OUTCOMES

By now you should be able to explain the following points:

17.1 Summarize why water is a precious resource and why shortages occur.
- The hydrologic cycle distributes water in our environment.
- Water supplies are unevenly distributed.

17.2 Compare major water compartments.
- Oceans hold 97 percent of all water on earth.
- Glaciers, ice, and snow contain most surface fresh water.
- Groundwater stores large resources.
- Rivers, lakes, and wetlands cycle quickly.
- The atmosphere is among the smallest of compartments.

17.3 Summarize water availability and use.
- Many people lack access to clean water.
- Water consumption is less than withdrawal.
- Water use is increasing.
- Agriculture is the greatest water consumer worldwide.
- Domestic and industrial water use are greatest in wealthy countries.

17.4 Investigate freshwater shortages.
- Many countries experience water scarcity and stress.
- Would you fight for water?

17.5 Illustrate the benefits and problems of dams and diversions.
- Dam failure can be disastrous.
- Dams often displace people and damage ecosystems.
- Dams kill fish.
- Sedimentation limits reservoir life.
- Diversion projects sometimes dry up rivers.
- Groundwater is depleted when withdrawals exceed recharge.

17.6 Understand how we might increase water supplies.
- Desalination provides expensive water.
- Domestic conservation can save water.
- Recycling can reduce consumption.
- Prices and policies have often discouraged conservation.

PRACTICE QUIZ

1. What is the difference between withdrawal, consumption, and degradation of water?
2. Explain how water can enter and leave an aquifer (see fig. 17.11).
3. Describe the changes in water withdrawal and consumption by sector shown in figure 17.14.
4. Describe some problems associated with dam building and water diversion projects.
5. Describe the path a molecule of water might follow through the hydrologic cycle from the ocean to land and back again.
6. Where are the five largest rivers in the world (table 17.3)?
7. How do mountains affect rainfall distribution? Does this affect your part of the country?
8. Identify and explain three consequences of overpumping aquifers.
9. How much water is fresh (as opposed to saline) and where is it?
10 Explain how saltwater intrusion happens (fig. 17.27).

CRITICAL THINKING AND DISCUSSION QUESTIONS

1. What changes might occur in the hydrologic cycle if our climate were to warm or cool significantly?

2. Why does it take so long for the deep ocean waters to circulate through the hydrologic cycle? What happens to substances that contaminate deep ocean water or deep aquifers in the ground?

3. Are there ways you could use less water in your own personal life? What obstacles prevent you from taking these steps?

4. Should we use up underground water supplies now or save them for some future time?

5. How should we compare the values of free-flowing rivers and natural ecosystems with the benefits of flood control, water diversion projects, hydroelectric power, and dammed reservoirs?

6. Would it be feasible to change from flush toilets and using water as a medium for waste disposal to some other system? What might be the best way to accomplish this?

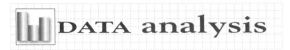 **DATA analysis** Graphing Global Water Stress and Scarcity

According to the United Nations, **water stress** is when annual water supplies drop below 1,700 m^3 per person. **Water scarcity** is defined as annual water supplies below 1,000 m^3 per person. More than 2.8 billion people in 48 countries will face either water stress or scarcity conditions by 2025. Of these countries, 40 are expected to be in West Asia or Africa. By 2050, far more people could be facing water shortages, depending both on population projections and scenarios for water supplies based on global warming and consumption patterns. The graph in this box shows an estimate for water stress and scarcity in 1995 together with three possible scenarios (high, medium, and low population projections) for 2050. You'll remember from chapter 7 that according to the 2004 UN population revision, the low projection for 2050 is about 7.6 billion, the medium projection is 8.9 billion, and the high projection is 10.6 billion.

1. What are the combined numbers of people could experience water stress and scarcity under the low, medium, and high scenarios in 2050?

2. What proportion (percentage) of 7.6 billion, 8.9 billion, and 10.6 billion would this be?

3. How does the percentage of the population in these two categories vary in the three estimates?

4. Why is the proportion of people in the scarce category so much larger in the high projection?

5. How many liters are in 1,000 m^3? How many gallons?

6. How does 1,000 m^3 compare to the annual consumption by the average family of four in the United States? (Hint: Look at table 17.2 and the table of units of measurement conversions at the end of this book).

7. Why isn't the United States (as a whole) considered to be water stressed?

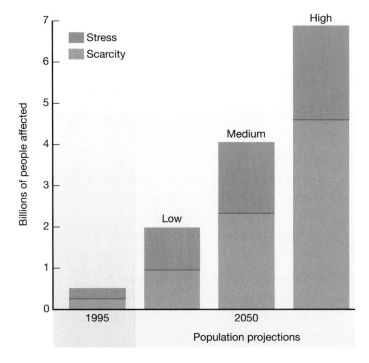

Global water stress and scarcity.

For Additional Help in Studying This Chapter, please visit our website at www.mhhe.com/cunningham10e. You will find additional practice quizzes and case studies, flashcards, regional examples, place markers for Google Earth™ mapping, and an extensive reading list, all of which will help you learn environmental science.

Arcata, California, built an artificial marsh as a low-cost, ecologically based treatment system for sewage effluent.

CHAPTER

18

Water Pollution

Water, water everywhere; nor any drop to drink.
—Samuel Taylor Coleridge—

LEARNING OUTCOMES

After studying this chapter, you should be able to:

18.1 Define water pollution.

18.2 Describe the types and effects of water pollutants.

18.3 Investigate water quality today.

18.4 Explain water pollution control.

18.5 Summarize water legislation.

Case Study A Natural System for Wastewater Treatment

Arcata, a small town in north-western California's redwood country, sits at the north end of Humbolt Bay, about 450 km (280 mi) north of San Francisco. Logging, fishing, and farming, long the economic engines of the area, are now giving way to tourism, education, and diversified industries. Many of the 17,000 residents of Arcata choose to live there because of the outstanding environmental quality and recreational opportunities of the area. Until the early 1970s, however, Arcata's sewage system discharged unchlorinated sewage effluent directly into the nearly enclosed bay. Algal blooms discolored the water, while fishing and swimming appeal declined.

In 1974 California enacted a policy prohibiting discharge of untreated wastewater into bays and estuaries. State planners proposed a large and expensive regional sewage treatment plant for the entire Humbolt Bay area. The plan was for large interceptor sewers to encircle the bay with a major underwater pipeline crossing the main navigation channel. Effluent from the proposed plant was to be released offshore into an area of shifting sea bottom in heavy winter storms. The cost and disruption required by this project prompted city officials to look for alternatives.

Ecologists from Humbolt State University pointed out that natural processes could be harnessed to solve Arcata's wastewater problems at a fraction of the cost and disturbance of a conventional treatment plant. After several years of study and experimentation, the city received approval to build a constructed wetland for wastewater treatment. This solved two problems at the same time. Arcata's waterfront was blighted by an abandoned lumbermill pond, channelized sloughs, marginal pasture land, and an abandoned sanitary landfill. Building a new wetland on this degraded area would beautify it while also solving the sewage treatment problem.

Today, Arcata's waterfront has been transformed into 40 ha (about 100 acres) of freshwater and saltwater marshes, brackish ponds, tidal sloughs, and estuaries (fig. 18.1). This diverse habitat supports a wide variety of plants and animals, and is now a wildlife sanctuary and a major tourist attraction. As a home or rest stop for over 200 bird species, the wetland is regarded as one of the best birding sites along the Pacific North Coast. In 1987, the city was awarded a prize by the Ford Foundation for innovative local government projects that included $100,000 to build an interpretive center in the marsh, The wetland is now an outstanding place for outdoor education, scientific study, and recreation.

Arcata's constructed wetland is also an unqualified success in wastewater treatment. After primary clarification to remove grit and sediment, wastewater passes through several oxidation ponds that remove about half the organic material (measured by biological oxygen demand) and suspended solids. Sludge captured by the clarifiers goes to digesters that generate methane gas, which is burned to provide heat that speeds the digestion process. Effluent from the oxidation ponds passes through treatment marshes, where sunlight and oxygen kill pathogens while aquatic plants and animals remove remaining organic material along with nitrogen, phosphorus, and other plant nutrients. The effluent is treated with chlorine gas to kill any residual pathogens. Finally, the treated water trickles through several large enhancement marshes that allow chlorine to evaporate and complete pollutant removal.

After 20 years of successful operation, Arcata's constructed wetland has inspired many other communities to find ecological solutions to their problems. Arcata Bay now produces more than half the oysters grown in California. The city also operates a wastewater aquaculture project where salmon, trout, and other fish species are raised in a mixture of wastewater effluent and seawater. Economically, the project is a success, providing excellent water quality treatment far below the cost of conventional systems. And the innovative approach of working with nature rather than against it has made Arcata an important ecotourist destination. Around the world, hundreds of communities have followed Arcata's model and now use constructed wetlands to eliminate up to 98 percent of the pollutants from their wastewater.

In this chapter, we'll look at both the causes and effects of water pollution as well as our options for controlling or treating water contaminants.

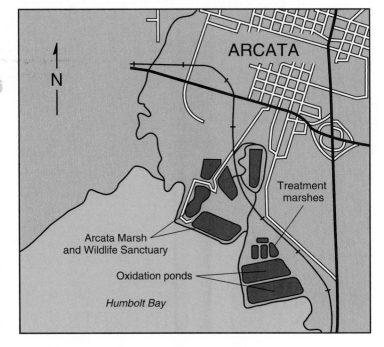

FIGURE 18.1 Arcata's constructed wetlands remove pathogens, pollutants, and nutrients from municipal wastewater using native biological communities and ecological processes in a setting that also serves as a wildlife sanctuary and tourist attraction.

18.1 WATER POLLUTION

Most students today are too young to appreciate that water in most industrialized countries was once far more polluted and dangerous than it is now. Thirty years ago, Lake Erie was on the brink of ecological collapse. The Cuyahoga River, choked with oil and floating debris, burned regularly. Factories and cities routinely dumped untreated chemicals, metals, oil, solvents, and sewage into rivers and lakes. Toxic solvents and organic chemicals were commonly dumped or buried in the ground, poisoning groundwater that we're now paying billions to clean up. In 1972, President Nixon signed the Clean Water Act, which has been called the United States' most successful and popular environmental legislation. This act established a goal that all the nation's waters should be "fishable and swimmable." While this goal is far from being achieved, the Clean Water Act remains popular because it protects public health (thus saving taxpayer dollars), as well as reducing environmental damage. In addition, water has an aesthetic appeal: The view of a clean lake, river, or seashore makes people happy, and water provides for recreation, so many people feel their quality of life has improved as water quality has been restored.

Clean water is a national, as well as global, priority. Recent polls have found repeatedly that 90 percent of Americans believe we should invest more in clean water and 70 percent would support establishing a trust fund to help communities repair water facilities.

We still have a long way to go in improving water quality. While pollution from factory pipes has been vastly reduced in the past 30 years, erosion from farm fields, construction sites, and streets has, in many areas, gotten worse since 1972. Airborne mercury, sulfur, and other substances are increasingly contaminating lakes and wetlands. Concentrated livestock production and agricultural runoff, threaten underground water as well as surface water systems. Increasing industrialization in developing countries has led to widespread water pollution in impoverished regions with little environmental regulation.

Water pollution is anything that degrades water quality

Any physical, biological, or chemical change in water quality that adversely affects living organisms or makes water unsuitable for desired uses can be considered pollution. There are natural sources of water contamination, such as poison springs, oil seeps, and sedimentation from erosion, but in this chapter we will focus primarily on human-caused changes that affect water quality or usability.

Pollution-control standards and regulations usually distinguish between point and nonpoint pollution sources. Factories, power plants, sewage treatment plants, underground coal mines, and oil wells are classified as **point sources** because they discharge pollution from specific locations, such as drain pipes, ditches, or sewer outfalls (fig. 18.2). These sources are discrete and identifiable, so they are relatively easy to monitor and

FIGURE 18.2 Sewer outfalls, industrial effluent pipes, acid draining out of abandoned mines, and other point sources of pollution are generally easy to recognize.

regulate. It is generally possible to divert effluent from the waste streams of these sources and treat it before it enters the environment.

In contrast, **nonpoint sources** of water pollution are scattered or diffuse, having no specific location where they discharge into a particular body of water. Nonpoint sources include runoff from farm fields and feedlots (fig. 18.3), golf courses, lawns and gardens, construction sites, logging areas, roads, streets, and parking lots. Whereas point sources may be fairly uniform and predictable throughout the year, nonpoint sources are often highly episodic. The first heavy rainfall after a dry period may flush high concentrations of gasoline, lead, oil, and rubber residues off city streets, for instance, while subsequent runoff may have lower levels of these pollutants. Spring snowmelt carries high levels of atmospheric acid deposition into streams and lakes in some areas. The irregular timing of these

FIGURE 18.3 This bucolic scene looks peaceful and idyllic, but allowing cows to trample stream banks is a major cause of bank erosion and water pollution. Nonpoint sources such as this have become the leading unresolved cause of stream and lake pollution in the United States.

events, as well as their multiple sources and scattered location, makes them much more difficult to monitor, regulate, and treat than point sources.

Perhaps the ultimate in diffuse, nonpoint pollution is **atmospheric deposition** of contaminants carried by air currents and precipitated into watersheds or directly onto surface waters as rain, snow, or dry particles. The Great Lakes, for example, have been found to be accumulating industrial chemicals such as PCBs and dioxins, as well as agricultural toxins such as the insecticide toxaphene that cannot be accounted for by local sources alone. The nearest sources for many of these chemicals are sometimes thousands of kilometers away (chapter 16).

Amounts of these pollutants can be quite large. It is estimated that there are 600,000 kg of the herbicide atrazine in the Great Lakes, most of which is thought to have been deposited from the atmosphere. Concentration of persistent chemicals up the food chain can produce high levels in top predators. Several studies have indicated health problems among people who regularly eat fish from the Great Lakes.

Ironically, lakes can be pollution sources as well as recipients. In the past 12 years, about 26,000 metric tons of PCBs have "disappeared" from Lake Superior. Apparently, these compounds evaporate from the lake surface and are carried by air currents to other areas where they are redeposited.

18.2 TYPES AND EFFECTS OF WATER POLLUTANTS

Although the types, sources, and effects of water pollutants are often interrelated, it is convenient to divide them into major categories for discussion (table 18.1). Let's look more closely at some of the important sources and effects of each type of pollutant.

Infectious agents remain an important threat to human health

The most serious water pollutants in terms of human health worldwide are pathogenic organisms (chapter 8). Among the most important waterborne diseases are typhoid, cholera, bacterial and amoebic dysentery, enteritis, polio, infectious hepatitis, and schistosomiasis. Malaria, yellow fever, and filariasis are transmitted by insects that have aquatic larvae. Altogether, at least 25 million deaths each year are blamed on these water-related diseases. Nearly two-thirds of the mortalities of children under 5 years old are associated with waterborne diseases.

The main source of these pathogens is from untreated or improperly treated human wastes. Animal wastes from feedlots or fields near waterways and food processing factories with inadequate waste treatment facilities also are sources of disease-causing organisms.

In developed countries, sewage treatment plants and other pollution-control techniques have reduced or eliminated most of the worst sources of pathogens in inland surface waters. Furthermore, drinking water is generally disinfected by chlorination so epidemics of waterborne diseases are rare in these countries. The United Nations estimates that 90 percent of the people in developed countries have adequate (safe) sewage disposal, and 95 percent have clean drinking water.

The situation is quite different in less-developed countries (fig. 18.4). The United Nations estimates that at least 2.5 billion people in these countries lack adequate sanitation, and that about half these people also lack access to clean drinking water. Conditions are especially bad in remote, rural areas where sewage treatment is usually primitive or nonexistent, and purified water is either unavailable or too expensive to obtain (fig. 18.5). The World Health Organization estimates

TABLE 18.1

Major Categories of Water Pollutants

Category	Examples	Sources
A. Causes Health Problems		
1. Infectious agents	Bacteria, viruses, parasites	Human and animal excreta
2. Organic chemicals	Pesticides, plastics, detergents, oil, and gasoline	Industrial, household, and farm use
3. Inorganic chemicals	Acids, caustics, salts, metals	Industrial effluents, household cleansers, surface runoff
4. Radioactive materials	Uranium, thorium, cesium, iodine, radon	Mining and processing of ores, power plants, weapons production, natural sources
B. Causes Ecosystem Disruption		
1. Sediment	Soil, silt	Land erosion
2. Plant nutrients	Nitrates, phosphates, ammonium	Agricultural and urban fertilizers, sewage, manure
3. Oxygen-demanding wastes	Animal manure and plant residues	Sewage, agricultural runoff, paper mills, food processing
4. Thermal	Heat	Power plants, industrial cooling

FIGURE 18.4 New Delhi residents wash food in polluted water. At least a billion people lack access to clean water.

FIGURE 18.6 Our national goal of making all surface waters in the United States "fishable and swimmable" has not been fully met, but scenes like this have been reduced by pollution-control efforts.

that 80 percent of all sickness and disease in less-developed countries can be attributed to waterborne infectious agents and inadequate sanitation.

If everyone had pure water and satisfactory sanitation, the World Bank estimates that 200 million fewer episodes of diarrheal illness would occur each year, and 2 million childhood deaths would be avoided. Furthermore, 450 million people would be spared debilitating roundworm or fluke infections. Surely these are goals worth pursuing.

Detecting specific pathogens in water is difficult, time-consuming, and costly; thus, water quality control personnel usually analyze water for the presence of **coliform bacteria,** any of the

many types that live in the colon or intestines of humans and other animals. The most common of these is *Eschericha coli* (or *E. coli*). Many strains of bacteria are normal symbionts in mammals, but some, such as *Shigella, Salmonella,* or *Lysteria* can cause fatal diseases. It is usually assumed that if any coliform bacteria are present in a water sample, infectious pathogens are present also.

To test for coliform bacteria, a water sample (or a filter through which a measured water sample has passed) is placed in a dish containing a nutrient medium that supports bacterial growth. After 24 hours in an incubator, living cells will have produced small colonies. If *any* colonies are found in drinking water samples, the U.S. Environmental Protection Agency considers the water unsafe and requiring disinfection. The EPA-recommended maximum coliform count for swimming water is 200 colonies per 100 ml, but some cities and states allow higher levels. If the limit is exceeded, the contaminated pool, river, or lake usually is closed to swimming (fig. 18.6).

Bacteria are detected by measuring oxygen levels

The amount of oxygen dissolved in water is a good indicator of water quality and of the kinds of life it will support. Water with an oxygen content above 6 parts per million (ppm) will support game fish and other desirable forms of aquatic life. Water with less than 2 ppm oxygen will support mainly worms, bacteria, fungi, and other detritus feeders and decomposers. Oxygen is added to water by diffusion from the air, especially when turbulence and mixing rates are high, and by photosynthesis of green plants, algae, and cyanobacteria. Oxygen is removed from water by respiration and chemical processes that consume oxygen.

Organic waste such as sewage, paper pulp, or food waste is rich in nutrients, especially nitrogen and phosphorus. These nutrients stimulate the growth of oxygen-demanding decomposing

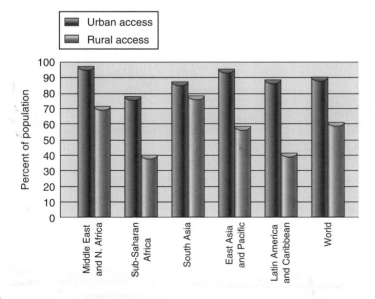

FIGURE 18.5 Proportion of people in developing regions with access to safe drinking water.
Source: UNESCO, 2002.

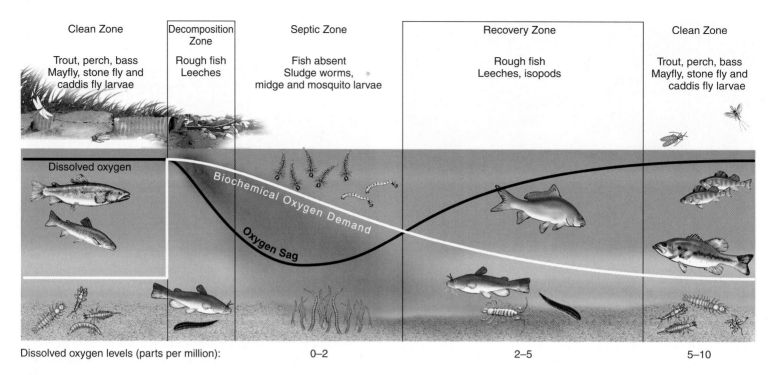

Clean Zone	Decomposition Zone	Septic Zone	Recovery Zone	Clean Zone
Trout, perch, bass Mayfly, stone fly and caddis fly larvae	Rough fish Leeches	Fish absent Sludge worms, midge and mosquito larvae	Rough fish Leeches, isopods	Trout, perch, bass Mayfly, stone fly and caddis fly larvae

Dissolved oxygen

Biochemical Oxygen Demand

Oxygen Sag

Dissolved oxygen levels (parts per million):   0–2   2–5   5–10

FIGURE 18.7 Oxygen sag downstream of an organic source. A great deal of time and distance may be required for the stream and its inhabitants to recover.

bacteria. **Biochemical oxygen demand (BOD)** is thus a useful test for the presence of organic waste in water. Usually, BOD tests involve incubating a water sample for five days, then comparing oxygen levels in the water before and after incubation. An alternative method, called the chemical oxygen demand (COD), uses a strong oxidizing agent (dichromate ion in 50 percent sulfuric acid) to completely break down all organic matter in a water sample. This method is much faster than the BOD test, but it records inactive organic matter as well as bacteria, so it is less useful. A third method of assaying pollution levels is to measure **dissolved oxygen (DO) content** directly, using an oxygen electrode. The DO content of water depends on factors other than pollution (for example, temperature and aeration), so it is best for indicating the health of the aquatic system.

The effects of oxygen-demanding wastes on rivers depends to a great extent on the volume, flow, and temperature of the river water. Aeration occurs readily in a turbulent, rapidly flowing river, which is, therefore, often able to recover quickly from oxygen-depleting processes. Downstream from a point source, such as a municipal sewage plant discharge, a characteristic decline and restoration of water quality can be detected either by measuring dissolved oxygen content or by observing the flora and fauna that live in successive sections of the river.

The oxygen decline downstream is called the **oxygen sag** (fig. 18.7). Upstream from the pollution source, oxygen levels support normal populations of clean-water organisms. Immediately below the source of pollution, oxygen levels begin to fall as decomposers metabolize waste materials. Rough fish, such as carp, bullheads, and gar, are able to survive in this oxygen-poor environment where they eat both decomposer organisms and the

waste itself. Further downstream, the water may become so oxygen-depleted that only the most resistant microorganisms and invertebrates can survive. Eventually, most of the nutrients are used up, decomposer populations are smaller, and the water becomes oxygenated once again. Depending on the volumes and flow rates of the effluent plume and the river receiving it, normal communities may not appear for several miles downstream.

Nutrient enrichment leads to cultural eutrophication

Water clarity (transparency) is affected by sediments, chemicals, and the abundance of plankton organisms, and is a useful measure of water quality and water pollution. Rivers and lakes that have clear water and low biological productivity are said to be **oligotrophic** (*oligo* = little + *trophic* = nutrition). By contrast, **eutrophic** (*eu* + *trophic* = truly nourished) waters are rich in organisms and organic materials. Eutrophication is an increase in nutrient levels and biological productivity. Some amount of eutrophication is a normal part of successional changes in most lakes. Tributary streams bring in sediments and nutrients that stimulate plant growth. Over time, ponds or lakes may fill in, eventually becoming marshes. The rate of eutrophication and succession depends on water chemistry and depth, volume of inflow, mineral content of the surrounding watershed, and the biota of the lake itself.

As with BOD, eutrophication often results from nutrient enrichment sewage, fertilizer runoff, even decomposing leaves in street gutters can produce a human-caused increase in biological productivity called **cultural eutrophication.** Cultural eutrophication can also result from higher temperatures, more sunlight

FIGURE 18.8 Eutrophic lake. Nutrients from agriculture and domestic sources have stimulated growth of algae and aquatic plants. This reduces water quality, alters species composition, and lowers the lake's recreational and aesthetic values.

reaching the water surface, or a number of other changes. Increased productivity in an aquatic system sometimes can be beneficial. Fish and other desirable species may grow faster, providing a welcome food source.

Often, however, eutrophication has undesirable results. Elevated phosphorus and nitrogen levels stimulate "blooms" of algae or thick growths of aquatic plants (fig. 18.8). Bacterial populations also increase, fed by larger amounts of organic matter. The water often becomes cloudy or turbid and has unpleasant tastes and odors. In extreme cases, plants and algae die and decomposers deplete oxygen in the water. Collapse of the aquatic ecosystem can result.

Eutrophication can cause toxic tides and "dead zones"

According to the Bible, the first plague to afflict the Egyptians when they wouldn't free Moses and the Israelites was that the water in the Nile turned into blood. All the fish died and the people were unable to drink the water, a terrible calamity in a desert country. Some modern scientists believe this may be the first recorded history of a **red tide** or a bloom of deadly aquatic microorganisms. Red tides—and other colors, depending on the species involved—have become increasingly common in slow-moving rivers, brackish lagoons, estuaries, and bays, as well as nearshore ocean waters where nutrients and wastes wash down our rivers.

Eutrophication in marine ecosystems occurs in nearshore waters and partially enclosed bays or estuaries. Some areas such as the Gulf of Mexico, the Caspian Sea, the Baltic, and Bohai Bay in the Yellow Sea tend to be in especially critical condition. During the tourist season, the coastal population of the Mediterranean, for example, swells to 200 million people. Eighty-five percent of the effluents from large cities go untreated into the sea.

Beach pollution, fish kills, and contaminated shellfish result. Extensive "dead zones" often form where rivers dump nutrients into estuaries and shallow seas. The second largest in the world occurs during summer months in the Gulf of Mexico at the mouth of the Mississippi River. (Exploring Science, p. 404). Studies indicate that as human populations, cities, and agriculture expand, these hypoxic zones will be come increasingly common.

It appears that fish and other marine species die in these hypoxic zones not only because oxygen is depleted, but also because of high concentrations of harmful organisms including toxic algae, pathogenic fungi, parasitic protists, and other predators. One of the most notorious and controversial of these is *Pfiesteria piscicida,* a dinoflagellate (a single-celled organism that swims with two whip-like flagella). Researchers at North Carolina State University reported in 1997 that they found this new species in Pamlico Sound, where thousands of fish died during a toxic tide. Initial reports described this organism as having a complex life cycle including dormant cysts, free-floating amoebae, and swimming flagellates. *Pfiesteria* cells were reported to secrete nerve-damaging toxins, allowing amoeba forms to attack wounded fish. *Pfiesteria* toxins have also been blamed for human health problems including nerve damage. Other studies, however, have raised doubts over the complexity of *Pfiesteria's* life cycle and its role in both fish mortality and human illnesses. In 2007, however, researchers succeeded in identifying the *Pfiesteria* toxin. It's an organic molecule containing a copper atom linked to two thiol (sulfur-containing) ligands. This toxin creates free radicals that destroy cells and tissues. As the compound generates free radicals, it decomposes. The fact that it remains active for only a few days is part of the reason it has been so elusive.

Inorganic pollutants include metals, salts, acids, and bases

Some toxic inorganic chemicals are released from rocks by weathering, are carried by runoff into lakes or rivers, or percolate into groundwater aquifers. This pattern is part of natural mineral cycles (chapter 3). Humans often accelerate the transfer rates in these cycles thousands of times above natural background levels through the mining, processing, using, and discarding of minerals.

In many areas, toxic, inorganic chemicals introduced into water as a result of human activities have become the most serious form of water pollution. Among the chemicals of greatest concern are heavy metals, such as mercury, lead, tin, and cadmium. Supertoxic elements, such as selenium and arsenic, also have reached hazardous levels in some waters. Other inorganic materials, such as acids, salts, nitrates, and chlorine, that normally are not toxic at low concentrations may become concentrated enough to lower water quality or adversely affect biological communities.

Metals

Many metals, such as mercury, lead, cadmium, tin, and nickel, are highly toxic in minute concentrations. Because metals are highly persistent, they can accumulate in food webs and have a cumulative effect in top predators—including humans.

In the 1980s shrimp boat crews noticed that certain locations off the Gulf Coast of Louisiana were emptied of all aquatic life. Since the region supports shrimp, fish, and oyster fisheries worth $250 to $450 million per year, these "dead zones" were important to the economy as well as to the Gulf's ecological systems. In 1985, Nancy Rabelais, a scientist working with Louisiana Universities Marine Consortium, began mapping areas of low oxygen concentrations in the Gulf waters. Her results, published in 1991, showed that vast areas, just above the floor of the Gulf, had oxygen concentration less than 2 parts per million (ppm), a level that eliminated all animal life except primitive worms. Healthy aquatic systems usually have about 10 ppm dissolved oxygen. What caused this hypoxic (oxygen-starved) area to develop?

Rabelais and her team tracked the phenomenon for several years, and it became clear that the dead zone was growing larger over time, that poor shrimp harvests coincided with years when the zone was large, and that the size of the dead zone, which ranges from 5,000 to 20,000 km^2 (about the size of New Jersey), depended on rainfall and runoff rates from the Mississippi River. Excessive nutrients, mainly nitrogen, from farms and cities far upstream on the Mississippi River, were the suspected culprit.

How did Rabelais and her team know that nutrients were the problem? They noticed that each year, 7–10 days after large spring rains in the agricultural parts of the upper Mississippi watershed, oxygen concentrations in the Gulf drop from 5 ppm to below 2 ppm. These rains are known to wash soil, organic debris, and last year's nitrogen-rich fertilizers from farm fields. The scientists also knew that saltwater ecosystems normally have little available nitrogen, a key nutrient for algae and plant growth. Pulses of agricultural runoff were followed by a profuse growth of algae and phytoplankton (tiny floating plants). Such a burst of biological activity produces an excess of dead plant cells and fecal matter that drifts to the seafloor. Shrimp, clams, oysters, and other filter feeders normally consume this debris, but they can't keep up with the sudden flood of material. Instead, decomposing bacteria in the sediment break down the debris, and they consume most of the available dissolved oxygen as well. Putrefying sediments also produce hydrogen sulfide, which further poisons the water near the seafloor.

In well-mixed water bodies, as in the open ocean, oxygen from upper layers of water is frequently mixed into lower water layers. Warm, protected water bodies are often stratified, however, as abundant sunlight keeps the upper layers warmer, and less dense, than lower layers. Denser lower layers cannot mix with upper layers unless strong currents or winds stir the water.

Many enclosed coastal waters, including Chesapeake Bay, Long Island Sound, the Mediterranean Sea, and the Black Sea, tend to be stratified and suffer hypoxic conditions that destroy bottom and near-bottom communities. There are about 200 dead zones around the world, and the number has doubled each decade since dead zones were first observed in the 1970s. The Gulf of Mexico is second in size behind a 100,000 km^2 dead zone in the Baltic Sea.

Can dead zones recover? Yes. Water is a forgiving medium, and organisms use nitrogen quickly. In 1996 in the Black Sea region, farmers in collapsing communist economies cut their nitrogen applications by half out of economic necessity; the Black Sea dead zone disappeared, while farmers saw no drop in their crop yields. In the Mississippi watershed, farmers can afford abundant fertilizer, and they fear they can't afford to risk underfertilizing. Because of the great geographic distance between the farm states and the Gulf, Midwestern states have been slow to develop an interest in the dead zone. At the same time, concentrated feedlot production of beef and pork is rapidly increasing, and feedlot runoff is the fastest growing, and least regulated, source of nutrient enrichment in rivers.

In 2001, federal, state, and tribal governments forged an agreement to cut nitrogen inputs by 30 percent and reduce the size of the dead zone to 5,000 km^2. This agreement represented astonishingly quick research and political response to scientific results, but it doesn't appear to be enough. Computer models suggest that it would take a 40–45 percent reduction in nitrogen to achieve the 5,000 km^2 goal.

Human activities have increased the flow of nitrogen reaching U.S. coastal waters by four to eight times since the 1950s. Phosphorus, another key nutrient, has tripled. This case study shows how water pollution can connect far-distant places, such as Midwestern farmers and Louisiana shrimpers.

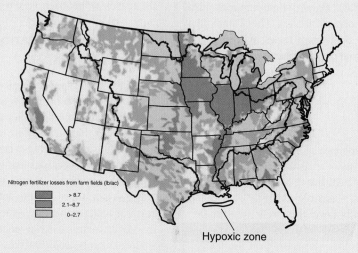

Nitrogen fertilizer losses from farm fields (lb/ac)
- > 8.7
- 2.1–8.7
- 0–2.7

Hypoxic zone

The Mississippi River drains 40 percent of the conterminous United States, including the most heavily farmed states. Nitrogen fertilizer produces a summer "dead zone" in the Gulf of Mexico.

Currently, the most widespread toxic metal contamination problem in North America is mercury released from coal-burning power plants. As chapter 16 mentions, an EPA survey of 2,500 fish from 260 lakes across the United States found at least low levels of mercury in every fish sampled. More than half the fish contained mercury levels unsafe for women of childbearing age, and three-quarters exceed the safe limit for young children.

Fifty states have issued warnings about eating freshwater or ocean fish; mercury contamination is by far the most common reason for these advisories. Top marine predators, such as shark, swordfish, marlin, king mackerel, and blue-fin tuna, should be avoided completely. You should check local advisories about the safety of fish caught in your local lakes, rivers, and coastal areas. If no advice is available, eat no more than one meal of such fish per week. Public health officials estimate that 600,000 American children now have mercury levels in their bodies high enough to cause mental and developmental problems, while one woman in six in the United States has blood-mercury concentrations that would endanger a fetus.

Mine drainage and leaching of mining wastes are serious sources of metal pollution in water. A survey of water quality in eastern Tennessee—where there has been a great deal of surface mining—found that 43 percent of all surface streams and lakes and more than half of all groundwater used for drinking supplies were contaminated by acids and metals from mine drainage. In some cases, metal levels were 200 times higher than what is considered safe for drinking water.

Nonmetallic Salts

Some soils contain high concentrations of soluble salts, including toxic selenium and arsenic. You have probably heard of poison springs and seeps in the desert, where percolating groundwater brings these materials to the surface. Irrigation and drainage of desert soils can mobilize these materials on a larger scale and result in serious pollution problems, as in Kesterson Marsh in California, where selenium poisoning killed thousands of migratory birds in the 1980s.

Salts, such as sodium chloride (table salt), that are nontoxic at low concentrations also can be mobilized by irrigation and concentrated by evaporation, reaching levels that are toxic for many plants and animals. Globally, 20 percent of the world's irrigated farmland is estimated to be affected by salinization, and half that land has enough salt buildup to decrease yields significantly. In the northern United States, millions of tons of sodium chloride and calcium chloride are used every year to melt road ice. Leaching of these road salts into surface waters is having adverse effects on some aquatic ecosystems.

Perhaps the largest human population threatened by naturally occurring arsenic in groundwater is in West Bengal, India, and adjacent areas of Bangladesh (fig. 18.9). Arsenic occurs naturally in the sediments that make up the Ganges River delta. Rapid population growth, industrialization, and intensification of irrigated agriculture have consumed or polluted limited surface water supplies. In an effort to provide clean drinking water for local residents, thousands of deep tube wells were sunk in the 1960s throughout the area. Much of this humanitarian effort was financed by loans from the World Bank.

By the 1980s, health workers became aware of widespread signs of chronic arsenic poisoning among Bengali villagers. Symptoms include watery and inflamed eyes, gastrointestinal cramps, gradual loss of strength, scaly skin and skin tumors, anemia, confusion, and, eventually, death. Some villages have had wells for centuries without a problem; why is arsenic poisoning appearing now? One theory is that excessive withdrawals now lower the water table during the dry season, exposing arsenic-bearing minerals

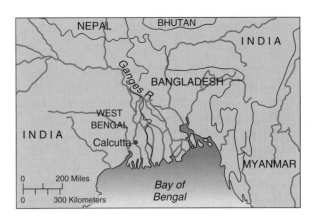

FIGURE 18.9 West Bengal and adjoining areas of Bangladesh have hundreds of millions of people who may be exposed to dangerous arsenic levels in well water.

to oxidation, which converts normally insoluble salts to soluble oxides. When aquifers refill during the next monsoon season, dissolved arsenic can be pumped out. Health workers estimate that the total number of potential victims in India and Bangladesh may exceed 200 million people.

Acids and Bases

Acids are released as by-products of industrial processes, such as leather tanning, metal smelting and plating, petroleum distillation, and organic chemical synthesis. Coal mining is an especially important source of acid water pollution. Sulfur compounds in coal react with oxygen and water to make sulfuric acid. Thousands of kilometers of streams in the United States have been acidified by acid mine drainage, some so severely that they are essentially lifeless.

Coal and oil combustion also leads to formation of atmospheric sulfuric and nitric acids (chapter 16), which are disseminated by long-range transport processes and deposited via precipitation (acidic rain, snow, fog, or dry deposition) in surface waters. Where soils are rich in such alkaline material as limestone, these atmospheric acids have little effect because they are neutralized. In high mountain areas or recently glaciated regions where crystalline bedrock is close to the surface and lakes are oligotrophic, however, there is little buffering capacity (ability to neutralize acids) and aquatic ecosystems can be severely disrupted. These effects were first recognized in the mountains of northern England and Scandinavia about 30 years ago.

Aquatic damage due to acid precipitation has been reported in about 200 lakes in the Adirondack Mountains of New York State and in several thousand lakes in eastern Quebec, Canada. Game fish, amphibians, and sensitive aquatic insects are generally the first to be killed by increased acid levels in the water. If acidification is severe enough, aquatic life is limited to a few resistant species of mosses and fungi. Increased acidity may result in leaching of toxic metals, especially aluminum, from soil and rocks, making water unfit for drinking or irrigation, as well.

Organic pollutants include pesticides and other industrial substances

Thousands of different natural and synthetic organic chemicals are used in the chemical industry to make pesticides, plastics, pharmaceuticals, pigments, and other products that we use in everyday life. Many of these chemicals are highly toxic (chapter 8). Exposure to very low concentrations (perhaps even parts per quadrillion in the case of dioxins) can cause birth defects, genetic disorders, and cancer. Some can persist in the environment because they are resistant to degradation and toxic to organisms that ingest them. Contamination of surface waters and groundwater by these chemicals is a serious threat to human health.

The two most important sources of toxic organic chemicals in water are improper disposal of industrial and household wastes and runoff of pesticides from farm fields, forests, roadsides, golf courses, and other places where they are used in large quantities. The U.S. EPA estimates that about 500,000 metric tons of pesticides are used in the United States each year. Much of this material washes into the nearest waterway, where it passes through ecosystems and may accumulate in high levels in nontarget organisms. The bioaccumulation of DDT in aquatic ecosystems was one of the first of these pathways to be understood. Dioxins, and other chlorinated hydrocarbons (hydrocarbon molecules that contain chlorine atoms) have been shown to accumulate to dangerous levels in the fat of salmon, fish-eating birds, and humans and to cause health problems similar to those resulting from toxic metal compounds (fig. 18.10).

As chapter 8 reports, atrazine, the most widely used herbicide in America, has been shown to disrupt normal sexual development in frogs at concentrations as low as 0.1 ppb. This level is found regularly wherever farming occurs. Could this be a problem for us as well?

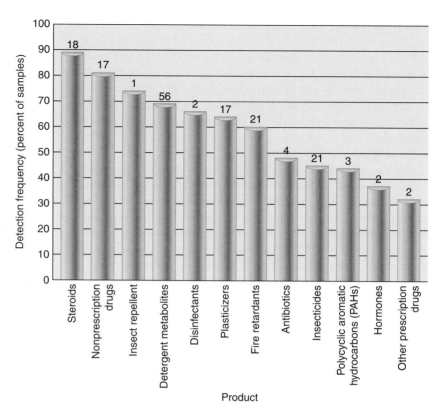

FIGURE 18.11 Detection frequency of organic, wastewater contaminants in a recent USGS survey. Maximum concentrations in water samples are shown above the bars in micrograms per liter. Dominant substances included DEET insect repellent, caffeine, and triclosan, which comes from antibacterial soaps.

FIGURE 18.10 The deformed beak of this young robin is thought to be due to dioxins, DDT, and other toxins in its mother's diet.

Hundreds of millions of tons of hazardous organic wastes are thought to be stored in dumps, landfills, lagoons, and underground tanks in the United States (chapter 21). Many, perhaps most, of these sites have leaked toxic chemicals into surface waters or groundwater or both. The EPA estimates that about 26,000 hazardous waste sites will require cleanup because they pose an imminent threat to public health, mostly through water pollution.

Countless additional organic compounds enter our water unmonitored. In 2002, the USGS released the first-ever study of pharmaceuticals and hormones in streams. Scientists sampled 130 streams, looking for 95 contaminants, including antibiotics, natural and synthetic hormones, detergents, plasticizers, insecticides, and fire retardants (fig. 18.11). All these substances were found, usually in low concentrations. One stream had 38 of the compounds tested. Drinking water standards exist for only 14 of the 95 substances. A similar study found the same substances in groundwater, which is much harder to clean than surface waters. What are the effects of these widely used chemicals, on our environment or on people consuming the water? Nobody knows. This study is a first step toward filling huge gaps in our knowledge about their distribution, though.

FIGURE 18.12 A plume of sediment and industrial waste flows from this drainage canal into Lake Erie.

Sediment also degrades water quality

Rivers have always carried sediment to the oceans, but erosion rates in many areas have been greatly accelerated by human activities. Some rivers carry astounding loads of sediment. Erosion and runoff from croplands contribute about 25 billion metric tons of soil, sediment, and suspended solids to world surface waters each year. Forests, grazing lands, urban construction sites, and other sources of erosion and runoff add at least 50 billion additional tons. This sediment fills lakes and reservoirs, obstructs shipping channels, clogs hydroelectric turbines, and makes purification of drinking water more costly. Sediments smother gravel beds in which insects take refuge and fish lay their eggs. Sunlight is blocked so that plants cannot carry out photosynthesis, and oxygen levels decline. Murky, cloudy water also is less attractive for swimming, boating, fishing, and other recreational uses (fig. 18.12).

Sediment also can be beneficial. Mud carried by rivers nourishes floodplain farm fields. Sediment deposited in the ocean at river mouths creates valuable deltas and islands. The Ganges River, for instance, builds up islands in the Bay of Bengal that are eagerly colonized by land-hungry people of Bangladesh. Louisiana's coastal wetlands require constant additions of sediment from the muddy Mississippi to counteract coastal erosion. These wetlands are now disappearing at a disastrous rate: Levees now channel the river and its load straight out to the Gulf of Mexico, where sediments are dumped beyond the continental shelf.

Thermal pollution is dangerous for organisms

Raising or lowering water temperatures from normal levels can adversely affect water quality and aquatic life. Water temperatures are usually much more stable than air temperatures, so aquatic organisms tend to be poorly adapted to rapid temperature changes. Lowering the temperature of tropical oceans by even

one degree can be lethal to some corals and other reef species. Raising water temperatures can have similar devastating effects on sensitive organisms. Oxygen solubility in water decreases as temperatures increase, so species requiring high oxygen levels are adversely affected by warming water.

Humans cause thermal pollution by altering vegetation cover and runoff patterns, as well as by discharging heated water directly into rivers and lakes.

The cheapest way to remove heat from an industrial facility is to draw cool water from an ocean, river, lake, or aquifer, run it through a heat-exchanger to extract excess heat, and then dump the heated water back into the original source. **A thermal plume** of heated water is often discharged into rivers and lakes, where raised temperatures can disrupt many processes in natural ecosystems and drive out sensitive organisms. Nearly half the water we withdraw is used for industrial cooling. Electric power plants, metal smelters, petroleum refineries, paper mills, food-processing factories, and chemical manufacturing plants all use and release large amounts of cooling water.

To minimize thermal pollution, power plants frequently are required to construct artificial cooling ponds or cooling towers in which heat is released into the atmosphere and water is cooled before being released into natural water bodies.

Some species find thermal pollution attractive. Warm water plumes from power plants often attract fish, birds, and marine mammals that find food and refuge there, especially in cold weather. This artificial environment can be a fatal trap, however. Organisms dependent on the warmth may die if they leave the plume or if the flow of warm water is interrupted by a plant shutdown. Endangered manatees in Florida, for example, are attracted to the abundant food and warm water in power plant thermal plumes and are enticed into spending the winter much farther north than they normally would. On several occasions, a midwinter power plant breakdown has exposed a dozen or more of these rare animals to a sudden thermal shock that they could not survive.

18.3 WATER QUALITY TODAY

Surface-water pollution is often both highly visible and one of the most common threats to environmental quality. In more developed countries, reducing water pollution has been a high priority over the past few decades. Billions of dollars have been spent on control programs and considerable progress has been made. Still much remains to be done. In developed countries, poor water quality often remains a serious problem. In this section, we will look at progress as well as continuing obstacles in this important area.

The Clean Water Act protects our water

Like most developed countries, the United States and Canada have made encouraging progress in protecting and restoring water quality in rivers and lakes over the past 40 years. In 1948, only about one-third of Americans were served by municipal sewage systems, and most of those systems discharged sewage without any treatment or with only primary treatment (the bigger lumps of waste are removed). Most people depended on cesspools and septic systems to dispose of domestic wastes.

The 1972 Clean Water Act established a National Pollution Discharge Elimination System (NPDES), which requires an easily revoked permit for any industry, municipality or other entity dumping wastes in surface waters. The permit requires disclosure of what is being dumped and gives regulators valuable data and evidence for litigation. As a consequence, only about 10 percent of our water pollution now comes from industrial or municipal point sources. One of the biggest improvements has been in sewage treatment.

Since the Clean Water Act was passed in 1972, the United States has spent more than $180 billion in public funds and perhaps ten times as much in private investments on water pollution control. Most of that effort has been aimed at point sources, especially to build or upgrade thousands of municipal sewage treatment plants. As a result, nearly everyone in urban areas is now served by municipal sewage systems and no major city discharges raw sewage into a river or lake except as overflow during heavy rainstorms.

This campaign has led to significant improvements in surface-water quality in many places. Fish and aquatic insects have returned to waters that formerly were depleted of life-giving oxygen. Swimming and other water-contact sports are again permitted in rivers, lakes, and at ocean beaches that once were closed by health officials.

The Clean Water Act goal of making all U.S. surface waters "fishable and swimmable" has not been fully met, but in 1999 the EPA reported that 91.4 percent of all monitored river miles and 87.5 percent of all assessed lake acres are suitable for their designated uses. This sounds good, but you have to remember that not all water bodies are monitored. Furthermore, the designated goal for some rivers and lakes is merely to be "boatable." Water quality doesn't have to be very high to be able to put a boat in it. Even in "fishable" rivers and lakes, there isn't a guarantee that you can catch anything other than rough fish like carp or bullheads, nor can you be sure that what you catch is safe to eat. Even with billions of dollars of investment in sewage treatment plants, elimination of much of the industrial dumping and other gross sources of pollutants, and a general improvement in water quality, the EPA reports that 21,000 water bodies still do not meet their designated uses. According to the EPA, an overwhelming majority of the American people—almost 218 million—live within 16 km (10 mi) of an impaired water body.

In 1998, a new regulatory approach to water quality assurance was instituted by the EPA. Rather than issue standards on a river by river approach or factory by factory permit discharge, the focus is being changed to watershed-level monitoring and protection. Some 4,000 watersheds are monitored for water quality (fig. 18.13). You can find information about your watershed at www.epa.gov/owow/tmdl/. The intention of this program is to give the public more and better information about the health of their watersheds. In addition, states will have greater flexibility as they identify impaired water bodies and set priorities, and new tools will be used to achieve goals. States are required to identify waters not meeting water quality goals and to develop **total**

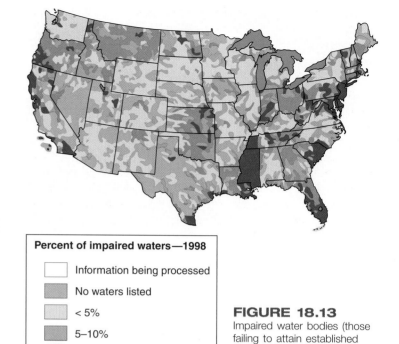

Percent of impaired waters—1998

- Information being processed
- No waters listed
- < 5%
- 5–10%
- 10–25%
- > 25%

FIGURE 18.13
Impaired water bodies (those failing to attain established water quality standards). As of 2005, the EPA had not released the 2000 impairment data.

maximum daily loads (TMDL) for each pollutant and each listed water body. A TMDL is the amount of a particular pollutant that a water body can receive from both point and nonpoint sources. It considers seasonal variation and includes a margin of safety.

By 1999, all 56 states and territories had submitted TMDL lists, and the EPA had approved most of them. Of the 3.5 million mi (5.6 million km) of rivers monitored, only 300,000 mi (480,000 km) fail to meet their clean water goals. Similarly, of 40 million lake acres (99 million ha), only 12.5 percent (in about 20,000 lakes) failed to meet their goal. To give states more flexibility in planning, the EPA has proposed new rules that include allowances for reasonably foreseeable increases in pollutant loadings to encourage "Smart Growth." In the future, TMDLs also will include load allocations from all nonpoint sources, including air deposition and natural background levels.

An encouraging example of improved water quality is seen in Lake Erie. Although widely regarded as "dead" in the 1960s, the lake today is promoted as the "walleye capital of the world." Bacteria counts and algae blooms have decreased more than 90 percent since 1962. Water that once was murky brown is now clear. Interestingly, part of the improved water quality is due to immense numbers of invasive zebra mussels, which filter the lake water very efficiently. Swimming is now officially safe along 96 percent of the lake's shoreline. Nearly 40,000 nesting pairs of double-crested cormorants nest in the Great Lakes region, up from only about 100 in the 1970s. Anglers now complain that the cormorants eat too many fish. In 1998 wildlife agents found 800 cormorants shot to death in a rookery on Galloo Island at the east end of Lake Ontario.

Canada's 1970 Water Act has produced comparable results. Seventy percent of all Canadians in towns over 1,000 population are now served by some form of municipal sewage treatment. In Ontario, the vast majority of those systems include tertiary treatment. After ten years of controls, phosphorus levels in the Bay of Quinte in the northeast corner of Lake Ontario have dropped nearly by half, and algal blooms that once turned waters green are less frequent and less intense than they once were. Elimination of mercury discharges from a pulp and paper mill on the Wabigoon-English River system in western Ontario has resulted in a dramatic decrease in mercury contamination. Twenty years ago this mercury contamination was causing developmental retardation in local residents. Extensive flooding associated with hydropower projects has raised mercury levels in fish to dangerous levels elsewhere, however.

Water quality problems remain

The greatest impediments to achieving national goals in water quality in both the United States and Canada are sediment, nutrients, and pathogens, especially from nonpoint discharges of pollutants. These sources are harder to identify and to reduce or treat than are specific point sources. About three-fourths of the water pollution in the United States comes from soil erosion, fallout of air pollutants, and surface runoff from urban areas, farm fields, and feedlots. In the United States, as much as 25 percent of the 46,800,000 metric tons (52 million tons) of fertilizer spread on farmland each year is carried away by runoff (fig. 18.14).

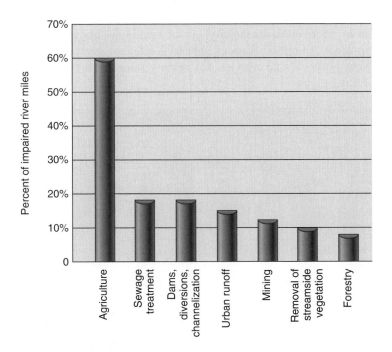

FIGURE 18.14 Percentage of impaired river miles in the United States by source of damage. Totals add up to more than 100 percent because one river can be affected by many sources.
Source: USDA and Natural Resources Conservation Service.

Cattle in feedlots produce some 129,600,000 metric tons (144 million tons) of manure each year, and the runoff from these sites is rich in viruses, bacteria, nitrates, phosphates, and other contaminants. A single cow produces about 30 kg (66 lb) of manure per day, or about as much as that produced by ten people. Some feedlots have 100,000 animals with little provision for capturing or treating runoff water. Imagine drawing your drinking water downstream from such a facility. Pets also can be a problem. It is estimated that the wastes from about a half million dogs in New York City are disposed of primarily through storm sewers, and therefore do not go through sewage treatment.

Loading of both nitrates and phosphates in surface water have decreased from point sources but have increased about fourfold since 1972 from nonpoint sources. Fossil fuel combustion has become a major source of nitrates, sulfates, arsenic, cadmium, mercury, and other toxic pollutants that find their way into water. Carried to remote areas by atmospheric transport, these combustion products now are found nearly everywhere in the world. Toxic organic compounds, such as DDT, PCBs, and dioxins, also are transported long distances by wind currents.

Developing countries often have serious water pollution

Japan, Australia, and most of western Europe also have improved surface-water quality in recent years. Sewage treatment in the wealthier countries of Europe generally equals or surpasses that in the United States. Sweden, for instance, serves 98 percent of its population with at least secondary sewage treatment (compared with 70 percent in the United States), and the other 2 percent have primary treatment. Poorer countries have much less to spend on sanitation. Spain serves only 18 percent of its population with even primary sewage treatment. In Ireland, it is only 11 percent, and in Greece, less than 1 percent of the people have even primary treatment. Most of the sewage, both domestic and industrial, is dumped directly into the ocean.

The fall of the "iron curtain" in 1989 revealed appalling environmental conditions in much of the former Soviet Union and its satellite states in eastern and central Europe. The countries closest geographically and socially to western Europe, the Czech Republic, Hungary, East Germany, and Poland, have made massive investments and encouraging progress toward cleaning up environmental problems. Parts of Russia itself, however, along with former socialist states in the Balkans and Central Asia, remain some of the most polluted places on earth. In Russia, for example, only about half the tap water is fit to drink. In cities like St. Petersburg, even boiling and filtering isn't enough to make municipal water safe. As we saw in chapter 17, at least 200 million Chinese live in areas without sufficient fresh water. Sadly, pollution makes much of the limited water unusable (fig. 18.15). It's estimated that 70 percent of China's surface water is unsafe for human consumption, and that the water in half the country's major rivers is so contaminated that it's unsuited for any use, even agriculture. The situation in Shanxi Province exemplifies the problems of water pollution in China. An industrial powerhouse,

FIGURE 18.15 Half the water in China's major rivers is too polluted to be suitable for any human use. Although the government has spent billions of yuan in recent years, dumping of industrial and domestic waste continues at dangerous levels.

in the north-central part of the country, Shanxi has about one-third of China's known coal resources and currently produces about two-thirds of the country's energy. In addition to power plants, major industries include steel mills, tar factories, and chemical plants.

Economic growth has been pursued in recent decades at the expense of environmental quality. According to the Chinese Environmental Protection Agency, the country's ten worst polluted cities are all in Shanxi. Factories have been allowed to exceed pollution discharges with impunity. For example, 3 million tons of wastewater is produced every day in the province with two-thirds of it discharged directly into local rivers without any treatment. Locals complain that the rivers, which once were clean and fresh, now run black with industrial waste. Among the 26 rivers in the province, 80 percent were rated Grade V (unfit for any human use) or higher in 2006. More than half the wells in Shanxi are reported to have dangerously high arsenic levels. Many of the 85,000 reported public protests in China in 2006 involved complaints about air and water pollution.

There are also some encouraging pollution-control stories. In 1997, Minamata Bay in Japan, long synonymous with mercury poisoning, was declared officially clean again. Another important success is found in Europe, where one of its most important rivers has been cleaned up significantly through international cooperation. The Rhine, which starts in the rugged Swiss Alps and winds 1,320 km through five countries before emptying through a Dutch delta into the North Sea, has long been a major commercial artery into the heart of Europe. More than 50 million people live in its catchment basin and nearly 20 million get their drinking water from the river or its tributaries. By the 1970s, the Rhine had become so polluted that dozens of fish species disappeared and swimming was discouraged along most of its length.

Efforts to clean up this historic and economically important waterway began in the 1950s, but a disastrous fire at a chemical warehouse near Basel, Switzerland, in 1986 provided the impetus for major changes. Through a long and sometimes painful series of international conventions and compromises, land-use practices, waste disposal, urban runoff, and industrial dumping have been changed and water quality has significantly improved. Oxygen concentrations have gone up fivefold since 1970 (from less than 2 mg/l to nearly 10 mg/l or about 90 percent of saturation) in long stretches of the river. Chemical oxygen demand has fallen fivefold during this same period, and organochlorine levels have decreased as much as tenfold. Many species of fish and aquatic invertebrates have returned to the river. In 1992, for the first time in decades, mature salmon were caught in the Rhine.

The less-developed countries of South America, Africa, and Asia have even worse water quality than do the poorer countries of Europe. Sewage treatment is usually either totally lacking or woefully inadequate. In some urban areas, 95 percent of all sewage is discharged untreated into rivers, lakes, or the ocean. Low technological capabilities and little money for pollution control are made even worse by burgeoning populations, rapid urbanization, and the shift of much heavy industry (especially the dirtier ones) from developed countries where pollution laws are strict to less-developed countries where regulations are more lenient.

Appalling environmental conditions often result from these combined factors (fig. 18.16). Two-thirds of India's surface waters are contaminated sufficiently to be considered dangerous to human health. The Yamuna River in New Delhi has 7,500 coliform bacteria per 100 ml (37 times the level considered safe for swimming in the United States) *before* entering the city. The coliform count increases to an incredible 24 *million* cells per 100 ml as the river leaves the city! At the same time, the river picks up some 20 million liters of industrial effluents every day from New Delhi. It's no wonder that disease rates are high and life

FIGURE 18.16 Ditches in this Haitian slum serve as open sewers into which all manner of refuse and waste are dumped. The health risks of living under these conditions are severe.

expectancy is low in this area. Only 1 percent of India's towns and cities have any sewage treatment, and only eight cities have anything beyond primary treatment.

In Malaysia, 42 of 50 major rivers are reported to be "ecological disasters." Residues from palm oil and rubber manufacturing, along with heavy erosion from logging of tropical rainforests, have destroyed all higher forms of life in most of these rivers. In the Philippines, domestic sewage makes up 60 to 70 percent of the total volume of Manila's Pasig River. Thousands of people use the river not only for bathing and washing clothes but also as their source of drinking and cooking water. China treats only 2 percent of its sewage. Of 78 monitored rivers in China, 54 are reported to be seriously polluted. Of 44 major cities in China, 41 use "contaminated" water supplies, and few do more than rudimentary treatment before it is delivered to the public.

Groundwater is hard to monitor and clean

About half the people in the United States, including 95 percent of those in rural areas, depend on underground aquifers for their drinking water. This vital resource is threatened in many areas by overuse and pollution and by a wide variety of industrial, agricultural, and domestic contaminants. For decades it was widely assumed that groundwater was impervious to pollution because soil would bind chemicals and cleanse water as it percolated through. Springwater or artesian well water was considered to be the definitive standard of water purity, but that is no longer true in many areas.

One of the serious sources of groundwater pollution throughout the United States is MTBE (methyl tertiary butyl ether), a suspected carcinogen. MTBE is a gasoline additive that has been used since the 1970s to reduce the amount of carbon monoxide and unburned hydrocarbons in vehicle exhaust. By the time the health dangers of MTBE were confirmed in the late 1990s, aquifers across the country had been contaminated—mainly from leaking underground storage tanks at gas stations. About 250,000 of these tanks are leaking MTBE into groundwater nationwide. In one U.S. Geological Survey (USGS) study, 27 percent of shallow urban wells tested contained MTBE. The additive is being phased out, but plumes of tainted water will continue to move through aquifers for decades to come. (Surface waters have also been contaminated, especially by two-stroke engines, such as those on personal watercraft.)

Treating MTBE-laced aquifers is expensive but not impossible. Douglas MacKay of the University of Waterloo in Ontario suggests that if oxygen could be pumped into aquifers, then naturally occurring bacteria could metabolize (digest) the compound. It could take decades or even centuries for natural bacteria to eliminate MTBE from a water supply, however. Water can also be pumped out of aquifers, reducing the flow and spread of contamination. Thus far, little funding has been invested in finding cost-effective remedies, however.

The U.S. EPA estimates that every day some 4.5 trillion l (1.2 trillion gal) of contaminated water seep into the ground in the United States from septic tanks, cesspools, municipal and industrial landfills and waste disposal sites, surface impoundments, agricultural fields, forests, and wells (fig. 18.17). The most toxic of these are probably waste disposal sites. Agricultural chemicals and wastes are responsible for the largest total volume of pollutants and area affected. Because deep underground aquifers often have residence times of thousands of years, many contaminants are extremely stable once underground. It is possible,

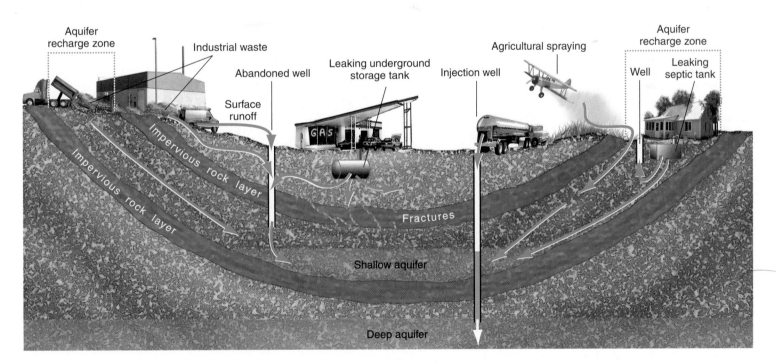

FIGURE 18.17 Sources of groundwater pollution. Septic systems, landfills, and industrial activities on aquifer recharge zones leach contaminants into aquifers. Wells provide a direct route for injection of pollutants into aquifers.

but expensive, to pump water out of aquifers, clean it, and then pump it back.

In farm country, especially in the Midwest's corn belt, fertilizers and pesticides commonly contaminate aquifers and wells. Herbicides such as atrazine and alachlor are widely used on corn and soybeans and show up in about half of all wells in Iowa, for example. Nitrates from fertilizers often exceed safety standards in rural drinking water. These high nitrate levels are dangerous to infants (nitrate combines with hemoglobin in the blood and results in "blue-baby" syndrome). They also are transformed into cancer-causing nitrosamines in the human gut. In Florida, 1,000 drinking water wells were shut down by state authorities because of excessive levels of toxic chemicals, mostly ethylene dibromide (EDB), a pesticide used to kill nematodes (roundworms) that damage plant roots.

Although most of the leaky, single-walled underground storage tanks once common at filling stations and factories have now been removed and replaced by more modern ones, a great deal of soil in American cities remains contaminated by previous careless storage and disposal of petroleum products. Considering that a single gallon (3.8 l) of gasoline can make a million gallons of water undrinkable, soil contamination remains a serious problem.

In addition to groundwater pollution problems, contaminated surface waters and inadequate treatment make drinking water unsafe in many areas. A 1996 survey concluded that nearly 20,000 public drinking water systems in the United States expose consumers to contaminants such as lead, pesticides, and pathogens at levels that violate EPA rules (fig. 18.18). A vast majority of these systems are small, serving fewer than 3,000 people, but altogether some 50 million people are sometimes at risk. Problems often occur because small systems can't afford modern purification and distribution equipment, regular testing, and trained operators to bring water quality up to acceptable standards.

Every year epidemiologists estimate that around 1.5 million Americans fall ill from infections caused by fecal contamination. In 1993, for instance, a pathogen called cryptosporidium got into the Milwaukee public water system, making 400,000 people sick and killing at least 100 people. The total costs of these diseases amount to billions of dollars per year. Preventive measures such as protecting water sources and aquifer recharge zones, providing basic treatment for all systems, installing modern technology and distribution networks, consolidating small systems, and strengthening the Clean Water Act and the Safe Drinking Water Act would cost far less. Unfortunately, in the present climate of budget-cutting and anti-regulation, these steps seem unlikely.

There are few controls on ocean pollution

Coastal zones, especially bays, estuaries, shoals, and reefs near large cities or the mouths of major rivers, often are overwhelmed by human-caused contamination. Suffocating and sometimes poisonous blooms of algae regularly deplete ocean waters of oxygen and kill enormous numbers of fish and other marine life. High levels of toxic chemicals, heavy metals, disease-causing organisms, oil, sediment, and plastic refuse are adversely affecting some of the most attractive and productive ocean regions. The potential losses caused by this pollution amount to billions of dollars each year.

Discarded plastic flotsam and jetsam are lightweight and nonbiodegradable. They are carried thousands of miles on ocean currents and last for years (fig. 18.19). Even the most remote beaches of distant islands are likely to have bits of polystyrene foam containers or polyethylene packing material that were discarded half a world away. It has been estimated that some 6 million metric tons of plastic bottles, packaging material, and other litter are tossed from ships every year into the ocean where they ensnare

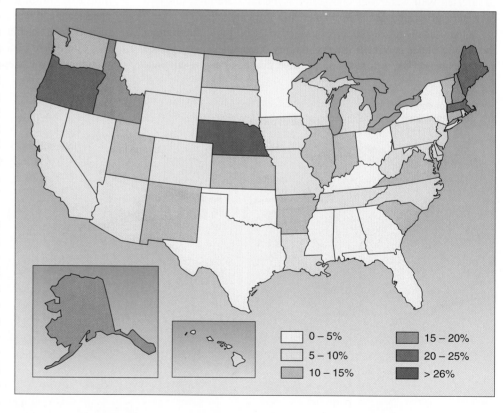

FIGURE 18.18 Percentage of drinking-water systems within states with violations of EPA health standards in 2000.
Source: EPA Safe Drinking Water Information System, 2001.

FIGURE 18.19 Beach pollution, including garbage, sewage, and contaminated runoff, is a growing problem associated with ocean pollution.

and choke seabirds, mammals (fig. 18.20), and even fish. Sixteen states now require that six-pack yokes be made of biodegradable or photodegradable plastic, limiting their longevity as potential killers. In one day, volunteers in Texas gathered more than 300 tons of plastic refuse from Gulf Coast beaches.

FIGURE 18.20 A deadly necklace. Marine biologists estimate that castoff nets, plastic beverage yokes, and other packing residue kill hundreds of thousands of birds, mammals, and fish each year.

Few coastlines in the world remain uncontaminated by oil or oil products. Oceanographers estimate that somewhere between 3 million and 6 million metric tons of oil are discharged into the world's oceans each year from both land- and sea-based operations. About half of this amount is due to maritime transport. Most oil spills result not from catastrophic, headliner accidents, but from routine open-sea bilge pumping and tank cleaning. These procedures are illegal but are easily carried out once ships are beyond sight of land. Much of the rest comes from land-based municipal and industrial runoff or from atmospheric deposition of residues from refining and combustion of fuels.

The transport of huge quantities of oil creates opportunities for major oil spills through a combination of human and natural hazards. Military conflict in the Middle East and oil drilling in risky locations, such as the notoriously rough North Sea and the Arctic Ocean, make it likely that more oil spills will occur. Plans to drill for oil along the seismically active California and Alaska coasts have been controversial because of the damage that oil spills could cause to these biologically rich coastal ecosystems.

18.4 WATER POLLUTION CONTROL

Appropriate land-use practices and careful disposal of industrial, domestic, and agricultural wastes are essential for control of water pollution.

Source reduction is often the cheapest way to reduce pollution

The cheapest and most effective way to reduce pollution is usually to avoid producing it or releasing it to the environment in the first place. Elimination of lead from gasoline has resulted in a widespread and significant decrease in the amount of lead in surface waters in the United States. Studies have shown that as much as 90 percent less road deicing salt can be used in many areas without significantly affecting the safety of winter roads. Careful handling of oil and petroleum products can greatly reduce the amount of water pollution caused by these materials. Although we still have problems with persistent chlorinated hydrocarbons spread widely in the environment, the banning of DDT and PCBs in the 1970s has resulted in significant reductions in levels in wildlife.

Modifying agricultural practices in headwater streams in the Chesapeake Bay watershed and the Catskill Mountains of New York have had positive and cost-effective impacts on downstream water quality (What Do You Think? pg. 414).

Industry can reduce pollution by recycling or reclaiming materials that otherwise might be discarded in the waste stream. Both of these approaches usually have economic as well as environmental benefits. It turns out that a variety of valuable metals can be recovered from industrial wastes and reused or sold for other purposes. The company benefits by having a product to sell, and the municipal sewage treatment plant benefits by not having to deal with highly toxic materials mixed in with millions of gallons of other types of wastes.

What Do You Think?

Watershed Protection in the Catskills

New York City has long been proud of its excellent municipal drinking water. Drawn from the rugged Catskill Mountains 100 km (60 mi) north of the city, stored in hard-rock reservoirs, and transported through underground tunnels, the city water is outstanding for so large an urban area. Yielding 450,000 m^3 (1.2 billion gal) per day, and serving more than 9 million people, this is the largest surface-water storage and supply complex in the world (see figure). As the metropolitan agglomeration has expanded, however, people have moved into the area around the Catskill Forest Preserve, and water quality is not as high as it was a century ago.

When the 1986 U.S. Safe Drinking Water Act mandated filtration of all public surface-water systems, the city was faced with building an $8 billion water treatment plant that would cost up to $500 million per year to operate. In 1989, however, the EPA ruled that the city could avoid filtration if it could meet certain minimum standards for microbial contaminants such as bacteria, viruses, and protozoan parasites. In an attempt to avoid the enormous cost of filtration, the city proposed land-use regulations for the five counties (Green, Ulster, Sullivan, Schoharie, and Delaware) in the Catskill/Delaware watershed from which it draws most of its water.

With a population of 50,000 people, the private land within the 520 km^2 (200 mi^2) watershed is mostly devoted to forestry, small farms, housing, and recreational activities. Among the changes the city called for was elimination of storm water runoff from barnyards, feedlots, or grazing areas into watersheds. In addition, farmers would be required to reduce erosion and surface runoff from crop fields and logging operations. Property owners objected strenuously to what they regarded as onerous burdens that would cost enough to put many

of them out of business. They also bristled at having the huge megalopolis impose rules on them. It looked like a long and bitter battle would be fought through the courts and the state legislature.

To avoid confrontation, a joint urban/rural task force was set up to see if a compromise could be reached, and to propose alternative solutions to protect both the water supply and the long-term viability of agriculture in the region. The task force agreed that agriculture is the "preferred land use" on private land, and that agriculture has "significant present and future environmental benefits." In addition, the task force proposed a voluntary, locally developed and administered program of "whole farm planning and best management approaches" very similar to ecosystem-based, adaptive management (chapter 9).

This grass-roots program, financed mainly by the city, but administered by local farmers themselves, attempts to educate landowners, and provides alternative marketing opportunities that help protect the watershed. Economic incentives are offered to encourage farmers and foresters to protect the water supply. Collecting feedlot and barnyard runoff in infiltration ponds together with solid conservation practices such as terracing, contour plowing, strip farming, leaving crop residue on fields, ground cover on waterways, and cultivation of perennial crops such as orchards and sugarbush have significantly improved watershed water quality. As of 1999, about 400 farmers—close to the 85 percent participation goal—have signed up for the program. The cost, so far, to the city has been about $50 million—or less than 1 percent of constructing a treatment plant.

Although landowners often object to any restrictions on development, many in the Catskills have found that land-use rules also protect rural lifestyles. Protection of the forests and waters has also helped the area retain in recreational economy and regional identity. Watershed management saved New York billions of dollars; it can also save traditional land uses and livelihoods.

What do you think? Are land-use restrictions a reasonable approach for saving on water treatment? How much should cities pay for watershed protection?

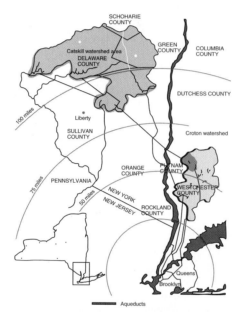

Investment in pollution prevention efforts in the Catskills has saved New York City billions of dollars in water filtration costs.

Controlling nonpoint sources requires land management

Among the greatest remaining challenges in water pollution control are diffuse, nonpoint pollution sources. Unlike point sources, such as sewer outfalls or industrial discharge pipes, which represent both specific locations and relatively continuous emissions, nonpoint sources have many origins and numerous routes by which contaminants enter ground and surface waters. It is difficult to identify—let alone monitor and control—all these sources and routes. Some main causes of nonpoint pollution are:

- *Agriculture:* The EPA estimates that 60 percent of all impaired or threatened surface waters are affected by sediment from eroded fields and overgrazed pastures; fertilizers, pesticides, and nutrients from croplands; and animal wastes from feedlots.

- *Urban runoff:* Pollutants carried by runoff from streets, parking lots, and industrial sites contain salts, oily residues, rubber, metals, and many industrial toxins. Yards, golf courses, parklands, and urban gardens often are treated with far more fertilizers and pesticides per unit area than farmlands. Excess chemicals are carried by storm runoff into waterways.

- *Construction sites:* New buildings and land development projects such as highway construction affect relatively small areas but produce vast amounts of sediment, typically ten to twenty times as much per unit area as farming (fig. 18.21).

- *Land disposal:* When done carefully, land disposal of certain kinds of industrial waste, sewage sludge, and biodegradable garbage can be a good way to dispose of unwanted materials. Some poorly run land disposal sites, abandoned dumps, and leaking septic systems, however, contaminate local waters.

FIGURE 18.21 Erosion on construction sites produces a great deal of sediment and is a major cause of nonpoint water pollution. Builders are generally required to install barriers to contain sediments, but these measures are often ineffectual.

Generally, soil conservation methods (chapter 9) also help protect water quality. Applying precisely determined amounts of fertilizer, irrigation water, and pesticides saves money and reduces contaminants entering the water. Preserving wetlands that act as natural processing facilities for removing sediment and contaminants helps protect surface and groundwaters.

In urban areas, reducing materials carried away by storm runoff is helpful. Citizens can be encouraged to recycle waste oil and to minimize use of fertilizers and pesticides. Regular street sweeping greatly reduces contaminants. Runoff can be diverted away from streams and lakes. Many cities are separating storm sewers and municipal sewage lines to avoid overflow during storms.

A good example of watershed management is seen in Chesapeake Bay, the United States' largest estuary. Once fabled for its abundant oysters, crabs, shad, striped bass, and other valuable fisheries, the Bay had deteriorated seriously by the early 1970s. Citizens' groups, local communities, state legislatures, and the federal government together established an innovative pollution-control program that made the bay the first estuary in America targeted for protection and restoration.

Among the principal objectives of this plan is reducing nutrient loading through land-use regulations in the six watershed states to control agricultural and urban runoff. Pollution prevention measures such as banning phosphate detergents also are important, as are upgrading wastewater treatment plants and improving compliance with discharge and filling permits. Efforts are underway to replant thousands of hectares of seagrasses and to restore wetlands that filter out pollutants. Since the 1980s, annual phosphorous discharges into the bay dropped 40 percent. Nitrogen levels, however, have remained constant or have even risen in some tributaries. Although progress has been made, the goals of reducing both nitrogen and phosphate levels by 40 percent and restoring viable fish and shellfish populations are still decades away. Still, as former EPA Administrator Carol Browner says, it demonstrates the "power of cooperation" in environmental protection.

Human waste disposal occurs naturally when concentrations are low

As we have already seen, human and animal wastes usually create the most serious health-related water pollution problems. More than 500 types of disease-causing (pathogenic) bacteria, viruses, and parasites can travel from human or animal excrement through water. In this section, we will look at how to prevent the spread of these diseases.

Natural Processes

In the poorer countries of the world, most rural people simply go out into the fields and forests to relieve themselves as they have always done. Where population densities are low, natural processes eliminate wastes quickly, making this a feasible method of sanitation. The high population densities of cities make this practice unworkable, however. Even major cities of many less-developed countries are often littered with human waste which has been left for rains to wash away or for pigs, dogs, flies, beetles, or other scavengers to consume. This is a major cause of disease, as well as being extremely unpleasant. Studies have shown that a significant portion of the airborne dust in Mexico City is actually dried, pulverized human feces.

Where intensive agriculture is practiced—especially in wet rice paddy farming in Asia—it has long been customary to collect "night soil" (human and animal waste) to be spread on the fields as fertilizer. This waste is a valuable source of plant nutrients, but it is also a source of disease-causing pathogens in the food supply. It is the main reason that travelers in less-developed countries must be careful to surface sterilize or cook any fruits and vegetables they eat. Collecting night soil for use on farm fields was common in Europe and America until about 100 years ago when the association between pathogens and disease was recognized.

Until about 50 years ago, most rural American families and quite a few residents of towns and small cities depended on a pit toilet or "outhouse" for waste disposal. Untreated wastes tended to seep into the ground, however, and pathogens sometimes contaminated drinking water supplies. The development of septic tanks and properly constructed drain fields represented a considerable improvement in public health (fig. 18.22). In a typical septic system, wastewater is first drained into a septic tank. Grease and oils rise to the top and solids settle to the bottom, where they are subject to bacterial decomposition. The clarified effluent from the septic tank is channeled out through a drainfield of small perforated pipes embedded in gravel just below the surface of the soil. The rate of aeration is high in this drainfield so that pathogens (most of which are anaerobic) will be killed, and soil microorganisms can metabolize any nutrients carried by the water. Excess water percolates up through the gravel and evaporates. Periodically, the solids in the septic tank are pumped out into a tank truck and taken to a treatment plant for disposal.

Where land is available and population densities are not too high, this can be an effective method of waste disposal. It is widely used in rural areas, but aging, leaky septic systems can be a huge cumulative problem. As chapter 13 points out, the Chesapeake Bay

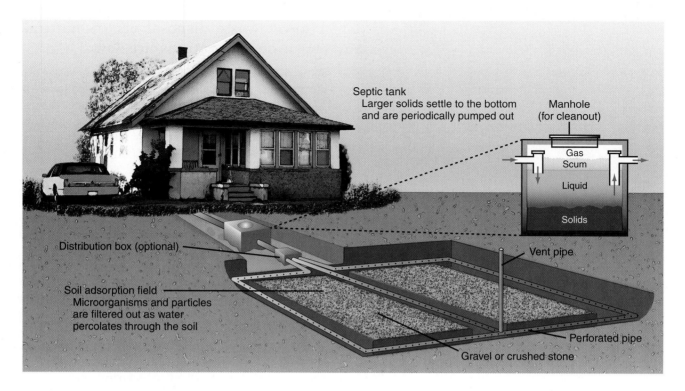

Septic tank
Larger solids settle to the bottom
and are periodically pumped out

Manhole
(for cleanout)

Gas
Scum

Liquid

Solids

Distribution box (optional)

Vent pipe

Soil adsorption field
Microorganisms and particles
are filtered out as water
percolates through the soil

Perforated pipe

Gravel or crushed stone

FIGURE 18.22 A domestic septic tank and drain field system for sewage and wastewater disposal. To work properly, a septic tank must have healthy microorganisms, which digest toilet paper and feces. For this reason, antimicrobial cleaners and chlorine bleach should never be allowed down the drain.

watershed has 420,000 individual septic systems, which constitute a major source of nutrients. Maryland, alone, plans to spend $7.5 million annually to upgrade failing septic systems.

Municipal Sewage Treatment

Over the past 100 years, sanitary engineers have developed ingenious and effective municipal wastewater treatment systems to protect human health, ecosystem stability, and water quality. This topic is an important part of pollution control, and is a central focus of every municipal government; therefore, let's look more closely at how a typical municipal sewage treatment facility works.

Primary treatment is the first step in municipal waste treatment. It physically separates large solids from the waste stream. As raw sewage enters the treatment plant, it passes through a metal grating that removes large debris (fig. 18.23a). A moving screen then filters out smaller items. Brief residence in a grit tank allows sand and gravel to settle. The waste stream then moves to the primary sedimentation tank where about half the suspended, organic solids settle to the bottom as sludge. Many pathogens remain in the effluent and it is not yet safe to discharge into waterways or onto the ground.

Secondary treatment consists of biological degradation of the dissolved organic compounds. The effluent from primary treatment flows into a trickling filter bed, an aeration tank, or a sewage lagoon. The trickling filter is simply a bed of stones or corrugated plastic sheets through which water drips from a sys-

tem of perforated pipes or a sweeping overhead sprayer. Bacteria and other microorganisms in the bed catch organic material as it trickles past and aerobically decompose it.

Aeration tank digestion is also called the activated sludge process. Effluent from primary treatment is pumped into the tank and mixed with a bacteria-rich slurry (fig. 18.23b). Air pumped through the mixture encourages bacterial growth and decomposition of the organic material. Water flows from the top of the tank and sludge is removed from the bottom. Some of the sludge is used as an inoculum for incoming primary effluent. The remainder would be valuable fertilizer if it were not contaminated by metals, toxic chemicals, and pathogenic organisms. The toxic content of most sewer sludge necessitates disposal by burial in a landfill or incineration. Sludge disposal is a major cost in most municipal sewer budgets (fig. 18.24). In some communities this is accomplished by land farming, composting, or anaerobic digestion, but these methods don't inactivate metals and some other toxic materials.

Where space is available for sewage lagoons, the exposure to sunlight, algae, aquatic organisms, and air does the same job more slowly but with less energy costs. Effluent from secondary treatment processes is usually disinfected with chlorine, UV light, or ozone to kill harmful bacteria before it is released to a nearby waterway.

Tertiary treatment removes plant nutrients, especially nitrates and phosphates, from the secondary effluent. Although wastewater is usually free of pathogens and organic material after secondary treatment, it still contains high levels of inorganic nutrients, such as nitrates and phosphates. When discharged into

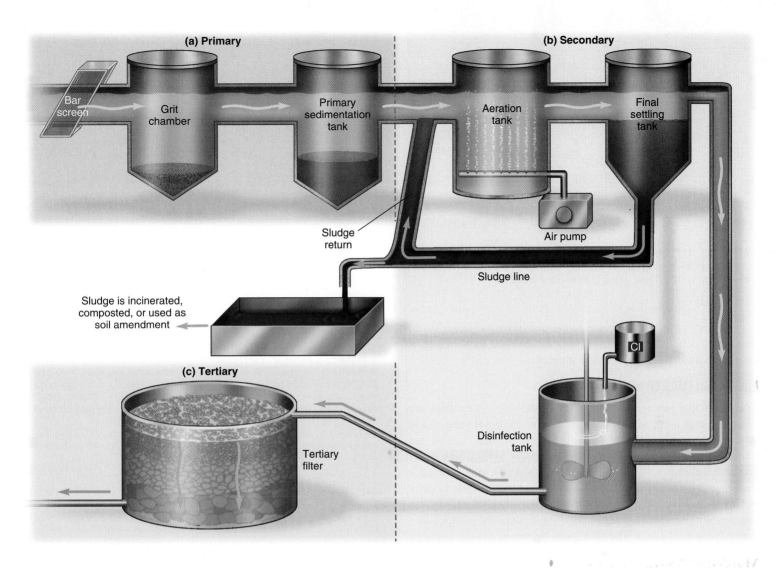

FIGURE 18.23 (a) Primary sewage treatment removes only solids and suspended sediment. (b) Secondary treatment, through aeration of activated sludge (or biosolids), followed by sludge removal and chlorination of effluent, kills pathogens and removes most organic material. (c) During tertiary treatment, passage through a trickling bed evaporator and/or a lagoon or marsh further removes inorganic nutrients, oxidizes any remaining organics, and reduces effluent volume.

surface waters, these nutrients stimulate algal blooms and eutrophication. To preserve water quality, these nutrients also must be removed. Passage through a wetland or lagoon can accomplish this. Alternatively, chemicals often are used to bind and precipitate nutrients (see fig. 18.23c).

In many American cities, sanitary sewers are connected to storm sewers, which carry runoff from streets and parking lots. Storm sewers are routed to the treatment plant rather than discharged into surface waters because runoff from streets, yards, and industrial sites generally contains a variety of refuse, fertilizers, pesticides, oils, rubber, tars, lead (from gasoline), and other undesirable chemicals. During dry weather, this plan works well. Heavy storms often overload the system, however, causing bypass dumping of large volumes of raw sewage and toxic surface runoff directly into receiving waters. To prevent this overflow, cities are spending hundreds of millions of dollars to separate storm

FIGURE 18.24 "Well, if *you* can't use it, do you know anyone who *can* use 3,000 tons of sludge every day?"

FIGURE 18.25 In India, a poplar plantation thrives on raw sewage water piped directly from nearby homes. Innovative solutions like this can make use of nutrients that would pollute water systems.

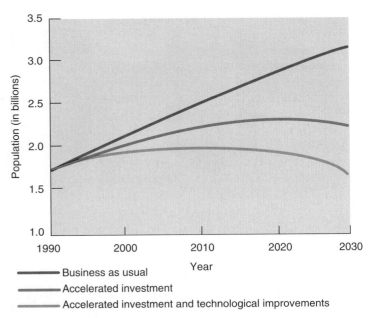

FIGURE 18.26 World population without adequate sanitation—three scenarios in the year 2030. If business as usual continues, more than 3 billion people will lack safe sanitation. Accelerated investment in sanitation services could lower this number. Higher investment, coupled with technological development, could keep the number of people without adequate sanitation from growing even though the total population increases.

Source: World Bank estimates based on research paper by Dennis Anderson and William Cavendish, "Efficiency and Substitution in Pollution Abatement: Simulation Studies in Three Sectors."

and sanitary sewers. These are huge, disruptive projects. When they are finished, surface runoff will be diverted into a river or lake and cause another pollution problem.

Low-Cost Waste Treatment

The municipal sewage systems used in developed countries are often too expensive to build and operate in the developing world where low-cost, low-tech alternatives for treating wastes are needed. One option is **effluent sewerage,** a hybrid between a traditional septic tank and a full sewer system. A tank near each dwelling collects and digests solid waste just like a septic system. Rather than using a drainfield, however, to dispose of liquids—an impossibility in crowded urban areas—effluents are pumped to a central treatment plant. The tank must be emptied once a year or so, but because only liquids are treated by the central facility, pipes, pumps, and treatment beds can be downsized and the whole system is much cheaper to build and run than a conventional operation.

Another alternative is to use natural or artificial wetlands to dispose of wastes. Constructed wetlands can cut secondary treatment costs to one-third of mechanical treatment costs, or less. As the opening case study for this chapter shows, this can be a critical savings for small municipalities.

Constructed wetland waste treatment systems are now operating in many American cities and many developing countries. Effluent from these operations can be used to irrigate crops or raise fish for human consumption if care is taken to first destroy pathogens (fig. 18.25). Usually 20 to 30 days of exposure to sun, air, and aquatic plants is enough to make the water safe. These systems make an important contribution to human food supplies. A 2,500 ha (6,000-acre) waste-fed aquaculture facility in Calcutta, for example, supplies about 7,000 metric tons of fish annually to local markets. The World Bank estimates that some 3 billion people will be without sanitation services by the middle of the next century under a business-as-usual scenario (fig. 18.26). With investments in innovative programs, however, sanitation could be

provided to about half those people and a great deal of misery and suffering could be avoided.

Water remediation may involve containment, extraction, or phytoremediation

Remediation means finding remedies for problems. Just as there are many sources for water contamination, there are many ways to clean it up. New developments in environmental engineering are providing promising solutions to many water pollution problems.

Containment methods confine or restrain dirty water or liquid wastes *in situ* (in place) or cap the surface with an impermeable layer to divert surface water or groundwater away from the site and to prevent further pollution. Where pollutants are buried too deeply to be contained mechanically, materials sometimes can be injected to precipitate, immobilize, chelate, or solidify them. Bentonite slurries, for instance, can effectively stabilize liquids in porous substrates. Similarly, straw or other absorbent material is spread on surface spills to soak up contaminants.

Extraction techniques pump out polluted water so it can be treated. Many pollutants can be destroyed or detoxified by chemical reactions that oxidize, reduce, neutralize, hydrolyze, precipitate, or otherwise change their chemical composition.

Where chemical techniques are ineffective, physical methods may work. Solvents and other volatile organic compounds, for instance, can be stripped from solution by aeration and then burned in an incinerator. Some contaminants can be removed by semipermeable membranes or resin filter beds that bind selectively to specific materials. Some of the same techniques used to stabilize liquids *in situ* can also be used *in vitro* (in a reaction vessel). Metals, for instance, can be chelated or precipitated in insoluble, inactive forms.

Often, living organisms can be used effectively and inexpensively to clean contaminated water. We call this bioremediation (chapter 21). Restored wetlands, for instance, along stream banks or lake margins can be very effective in filtering out sediment and removing pollutants. They generally cost far less than mechanical water treatment facilities and provide wildlife habitat as well.

Lowly duckweed (*Lemna* sp.), the green scum you often see covering the surface of eutrophic ponds, grows fast and can remove large amounts of organic nutrients from water. Under optimal conditions, a few square centimeters of these tiny plants can grow to cover nearly a hectare (about 2.5 acres) in four months. Large duckweed lagoons are being used as inexpensive, low-tech sewage treatment plants in developing countries. Where conventional wastewater purification typically costs $300 to $600 per person served, a duckweed system can cost one-tenth as much. The duckweed can be harvested and used as feed, fuel, or fertilizer. Up to 35 percent of its dry mass is protein—about twice as much as alfalfa, a popular animal feed.

Where space for open lagoons is unavailable, bioremediation can be carried out in reaction vessels. This has the advantage of controlling conditions more precisely and doesn't release organisms into the environment. Some of the most complex, holistic systems for water purification are designed by Ocean Arks International (OAI) in Falmouth, Massachusetts. Their "living machines" combine living organisms—chosen to perform specific functions—in contained environments. In a typical living machine, water flows through a series of containers, each with a distinct ecological community designed for a particular function. Wastes generated by the inhabitants of one vessel become the food for inhabitants of another. Sunlight provides the primary source of energy.

OAI has created or is in the process of building water treatment plants in a dozen states and foreign countries. Designs range from remediating toxic wastes from Superfund sites to simply treating domestic wastes. Starting with microorganisms in aerobic and anaerobic environments where different kinds of wastes are metabolized or broken down, water moves through a series of containers containing hundreds of different kinds of plants and animals, including algae, rooted aquatic plants, clams, snails, and fish, each chosen to provide a particular service. Technically, the finished water is drinkable, although few people feel comfortable doing so. More often, the final effluent is used to flush toilets or for irrigation. Called ecological engineering, this novel approach can save resources and money as well as clean up our environment and serve as a valuable educational tool (fig. 18.27).

FIGURE 18.27 In-house wastewater treatment in Oberlin College's Environmental Studies building. Constructed wetlands outside, and tanks inside, allow water plants to filter water and use nutrients.

18.5 WATER LEGISLATION

Water pollution control has been among the most broadly popular and effective of all environmental legislation in the United States. It has not been without controversy, however. In this section, we will look at some of the major issues concerning water quality laws and their provisions (table 18.2).

The Clean Water Act was ambitious, bipartisan, and largely successful

Passage of the Clean Water Act of 1972 was a bold, bipartisan step determined to "restore and maintain the chemical, physical, and biological integrity of the Nation's waters" that made clean water a national priority. Along with the Endangered Species Act and the Clean Air Act, this is one of the most significant and effective pieces of environmental legislation ever passed by the U.S. Congress. It also is an immense and complex law, with more than 500 sections regulating everything from urban runoff, industrial discharges, and municipal sewage treatment to land-use practices and wetland drainage.

The ambitious goal of the Clean Water Act was to return all U.S. surface waters to "fishable and swimmable" conditions. For specific "point" sources of pollution such as industrial discharge pipes or sewage outfalls, the act requires discharge permits and **best practicable control technology (BPT).** It sets national

TABLE 18.2

Some Important U.S. and International Water Quality Legislation

1. *Federal Water Pollution Control Act* (1972). Established uniform nationwide controls for each category of major polluting industries.
2. *Marine Protection Research and Sanctuaries Act* (1972). Regulates ocean dumping and established sanctuaries for protection of endangered marine species.
3. *Ports and Waterways Safety Act* (1972). Regulates oil transport and the operation of oil handling facilities.
4. *Safe Drinking Water Act* (1974). Requires minimum safety standards for every community water supply. Among the contaminants regulated are bacteria, nitrates, arsenic, barium, cadmium, chromium, fluoride, lead, mercury, silver, pesticides; radioactivity and turbidity also regulated. This act also contains provisions to protect groundwater aquifers.
5. *Resource Conservation and Recovery Act* (RCRA) (1976). Regulates the storage, shipping, processing, and disposal of hazardous wastes and sets limits on the sewering of toxic chemicals.
6. *Toxic Substances Control Act* (TOSCA) (1976). Categorizes toxic and hazardous substances, establishes a research program, and regulates the use and disposal of poisonous chemicals.
7. *Comprehensive Environmental Response, Compensation, and Liability Act* (CERCLA) (1980) and *Superfund Amendments and Reauthorization Act* (SARA) (1984). Provide for sealing, excavation, or remediation of toxic and hazardous waste dumps.
8. *Clean Water Act* (1985) (amending the 1972 Water Pollution Control Act). Sets as a national goal the attainment of "fishable and swimmable" quality for all surface waters in the United States.
9. *London Dumping Convention* (1990). Calls for an end to all ocean dumping of industrial wastes, tank washing effluents, and plastic trash. The United States is a signatory to this international convention.

---

goals of **best available, economically achievable technology (BAT),** for toxic substances and zero discharge for 126 priority toxic pollutants. As we discussed earlier in this chapter, these regulations have had a positive effect on water quality. While

What Can You Do?

Steps You Can Take to Improve Water Quality

Individual actions have important effects on water quality. Here are some steps you can take to make a difference.

- Compost your yard waste and pet waste. Nutrients from decayed leaves, grass, and waste are a major urban water pollutant. Many communities have public compost sites available.
- Don't fertilize your lawn or apply lawn chemicals. Untreated grass can be just as healthy, and it won't poison your pets or children.
- Make sure your car doesn't leak fluids, oil, or solvents on streets and parking lots, from which contaminants wash straight into rivers and lakes. Recycle motor oil at a gas station or oil change shop.
- Don't buy lawn mowers, personal watercraft, or other vehicles with two-cycle engines, which release abundant fuel and oil into air and water. Instead, buy more efficient, four-stroke engines.
- Visit your local sewage treatment plant. Often public tours are available or group tours can be arranged, and these sites can be fascinating.
- Keep informed about water policy debates at local and federal levels. Policies change often, and public input is important.

not yet swimmable or fishable everywhere, surface-water quality in the United States has significantly improved on average over the past quarter century. Perhaps the most important result of the act has been investment of $54 billion in federal funds and more than $128 billion in state and local funds for municipal sewage treatment facilities.

Not everyone, however, is completely happy with the Clean Water Act. Industries, state and local governments, farmers, land developers, and others who have been forced to change their operations or spend money on water protection often feel imposed upon. One of the most controversial provisions of the act has been Section 404, which regulates draining or filling of wetlands. Although the original bill only mentions wetlands briefly, this section has evolved through judicial interpretation and regulatory policy to become one of the principal federal tools for wetland protection. Many people applaud the protection granted to these ecologically important areas that were being filled in or drained at a rate of about half a million hectares per year before the passage of the Clean Water Act. Farmers, land developers, and others who are prevented from converting wetlands to other uses often are outraged by what they consider "taking" of private lands.

Another sore point for opponents of the Clean Water Act are what are called "unfunded mandates," or requirements for state or local governments to spend money that is not repaid by Congress. You will notice that the $128 billion already spent by cities to install sewage treatment and stormwater diversion to meet federal standards far exceeds the $54 billion in congressional assistance for these projects. Estimates are that local units of government could be required to spend another $130 billion to finish the job without any further federal funding. Small cities that couldn't afford or chose not to participate in earlier water quality programs, in which the federal government paid up to 90 percent of the costs, are especially hard hit by requirements that they upgrade municipal sewer and water systems. They now are faced with carrying out those same projects entirely on their own funds.

Clean water reauthorization remains contentious

Opponents of federal regulation have tried repeatedly to weaken or eliminate the Clean Water Act. They regard restriction of their "right" to dump toxic chemicals and waste into wetlands and waterways to be an undue loss of freedom. They resent being forced to clean up municipal water supplies, and call for cost/benefit analysis that places greater weight on economic interests in all environmental planning. Most of all, they view any limitation on use of private property to be a "taking" for which they should be fully compensated.

Even those who support the Clean Water Act in principle would like to see it changed and strengthened. Among these proposals are a shift from "end-of-the-pipe" focus on removing specific pollutants from effluents to more attention to changing industrial processes so toxic substances won't be produced in the first place. Another important issue is nonpoint pollution from agricultural runoff and urban areas, which has become the largest source of surface-water degradation in the United States. Regulating these sources remains a difficult problem.

Environmentalists also would like to see stricter enforcement of existing regulations, mandatory minimum penalties for violations, more effective community right-to-know provisions, and increased powers for citizen lawsuits against polluters. Studies have found that, in practice, polluters are given infrequent and light fines for polluting. Under the current law, using data that polluters themselves are required to submit, groups such as the Natural Resources Defense Council and the Citizens for a Better Environment have won million-dollar settlements in civil lawsuits (the proceeds generally are applied to clean-up projects) and some transgressors have even been sent to jail. Not surprisingly, environmentalists want these powers expanded, while polluters find them very disagreeable.

Other important legislation also protects water quality

In addition to the Clean Water Act, several other laws help to regulate water quality in the United States and abroad. Among these is the Safe Drinking Water Act, which regulates water

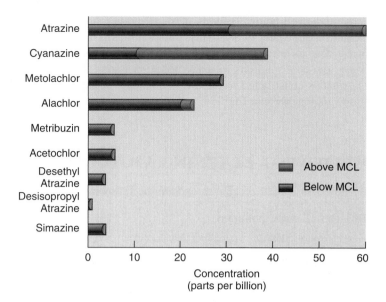

FIGURE 18.28 Nine herbicide active ingredients and metabolites found in drinking water samples in Fort Wayne, Indiana, in spring of 1995, compared to federal Maximum Concentration Levels (MCL). The use of these chemicals, and their presence in water sources, has increased in subsequent years.
Source: B. Cohen and E. Bondoc, "Weed Killers by the Glass," Environmental Working Group, 1995.

quality in commercial and municipal systems. Critics complain that standards and enforcement policies are too lax, especially for rural water districts and small towns. Some researchers report pesticides, herbicides, and lead in drinking water at levels they say should be of concern (fig. 18.28). Atrazine, for instance, was detected in 96 percent of all surface-water samples in one study of 374 communities across 12 states. Remember, however, that simply detecting a toxic compound is not the same as showing dangerous levels.

The Superfund program for remediation of toxic waste sites was created in 1980 by the Comprehensive Environmental Response, Compensation, and Liability Act (CERCLA) and was amended by the Superfund Amendments and Reauthorization Act (SARA) of 1984. This program is designed to provide immediate response to emergency situations and to provide permanent remedies for abandoned or inactive sites. These programs provide many jobs for environmental science majors in monitoring and removal of toxic wastes and landscape restoration. A variety of methods have been developed for remediation of problem sites.

CONCLUSION

Forty years ago, rivers in the United States were so polluted that some caught fire while others ran red, black, orange, or other unnatural colors with toxic industrial wastes. Many cities still dumped raw sewage into local rivers and lakes, so that warnings had to be posted to avoid any bodily contact. We've made huge progress since that time. Not all rivers and lakes are "fishable or swimmable," but federal, state, and local pollution controls have greatly improved our water quality in most places.

In rapidly developing countries, such as China and India, water pollution remains a serious threat to human health and ecosystem well-being. Billions of people don't have access to clean drinking water or adequate sanitation. It will take a massive investment to correct this growing problem. But there are relatively low-cost solutions to many pollution issues. The example of Arcata's constructed wetland for ecological sewage treatment shows us that we can find low-tech, inexpensive ways to reduce pollution. Living machines for water treatment in individual buildings or communities also offer hope for better ways to treat our wastes. Perhaps you can use the information you've learned by studying environmental science to make constructive suggestions for your own community.

REVIEWING LEARNING OUTCOMES

By now you should be able to explain the following points:

18.1 Define water pollution.
- Water pollution is anything that degrades water quality.

18.2 Describe the types and effects of water pollutants.
- Infectious agents remain an important threat to human health.
- Bacteria are detected by measuring oxygen levels.
- Nutrient enrichment leads to cultural eutrophication.
- Eutrophication can cause toxic tides and "dead zones."
- Inorganic pollutants include metals, salts, acids, and bases.
- Organic pollutants include pesticides and other industrial substances.
- Sediment also degrades water quality.
- Thermal pollution is dangerous for organisms.

18.3 Investigate water quality today.
- The Clean Water Act protects our water.
- Water quality problems remain.

- Developing countries often have serious water pollution.
- Groundwater is hard to monitor and clean.
- There are few controls on ocean pollution.

18.4 Explain water pollution control.
- Source reduction is often the cheapest way to reduce pollution.
- Controlling nonpoint sources requires land management.
- Human waste disposal occurs naturally when concentrations are low.
- Water remediation may involve containment, extraction, or phytoremediation.

18.5 Summarize water legislation.
- The Clean Water Act was ambitious, bipartisan, and largely successful.
- Clean water reauthorization remains contentious.
- Other important legislation also protects water quality.

PRACTICE QUIZ

1. Define *water pollution.*
2. List eight major categories of water pollutants and give an example for each category.
3. Describe eight major sources of water pollution in the United States. What pollution problems are associated with each source?
4. What are red tides, and why are they dangerous?
5. What is eutrophication? What causes it?
6. What is an oxygen sag? How much dissolved oxygen, in ppm, is present at each stage?
7. What are the origins and effects of siltation?

8. Describe primary, secondary, and tertiary processes for sewage treatment. What is the quality of the effluent from each of these processes?
9. Why do combined storm and sanitary sewers cause water quality problems? Why does separating them also cause problems?
10. What pollutants are regulated by the Clean Water Act? What goals does this act set for abatement technology?
11. What is MTBE? Why is it so widespread and hard to control?
12. Describe remediation techniques and how they work.

CRITICAL THINKING AND DISCUSSION QUESTIONS

1. Cost is the greatest obstacle to improving water quality. How would you decide how much of the cost of pollution control should go to private companies, government, or individuals?

2. How would you define *adequate* sanitation? Think of some situations in which people might have different definitions for this term.

3. What sorts of information would you need to make a judgment about whether water quality in your area is getting better or worse? How would you weigh different sources, types, and effects of water pollution?

4. Imagine yourself in a developing country with a severe shortage of clean water. What would you miss most if your water supply were suddenly cut by 90 percent?

5. Proponents of deep well injection of hazardous wastes argue that it will probably never be economically feasible to pump water out of aquifers more than 1 kilometer below the surface. Therefore, they say, we might as well use those aquifers for hazardous waste storage. Do you agree? Why or why not?

6. Arsenic contamination in Bangladesh results from geological conditions, World Bank and U.S. aid, poverty, government failures, and other causes. Who do you think is responsible for finding a solution? Why? Would you answer differently if you were a poor villager in Bangladesh?

 **DATA analysis**　　**Examining Pollution Sources**

Understanding the origins of pollution is the first step toward considering policies for reducing it. The chapter you have just read includes several graphs displaying pollution sources. The following questions ask you to think more about these sources:

1. In figure 18.14 (p. 409), which sources do you think are mainly point sources? Which are mainly nonpoint sources? How might you consider urban runoff to be either point or nonpoint? (If you need ideas to start, see figure 18.17, p. 411)

2. Explain the units on the Y-axis in figure 18.14. Does the graph show you the amount of actual pollution, by weight or volume, from different sources? Why might it be useful to show the extent of impacts?

3. Based on your knowledge of water contaminants, describe the kinds of pollutants you would expect from each of the sources shown:

   Agriculture _____

   Sewage treatment _____

   Dams, diversions, channelization _____

   Urban runoff _____

   Mining _____

   Removal of streamside vegetation _____

   Forestry _____

4. How would you design a sampling strategy to assess water pollution on your school campus?

5. Figure 18.11 (p. 406) shows some of the new developments in water pollution assessment. Conventional treatment systems were not designed to remove thousands of newly invented chemical compounds, or increasingly widespread compounds, including those shown in the figure. Explain the units used on the Y-axis. Also explain why the numbers (maximum concentrations found) are shown.

6. How many of the pollutants shown do *you* use? Don't forget to include caffeine (which is classed as a nonprescription drug) and antibacterial soaps (disinfectants). Steroids include cholesterol, which occurs naturally in foods.

7. Try to think of additional substances that you might contribute to wastewater.

8. This graph results from a reconnaissance study done by the U.S. Geological Survey. The researchers wanted to assess whether a list of 95 contaminants could be detected at all in public waterways. If you wanted to design a study like this, what sorts of sites would you select for sampling? How might your results and your conclusions differ if you did a random sample?

Mountaintop removal is an extremely destructive method of coal mining.

CHAPTER

19

Conventional Energy

America is addicted to oil.

—George W. Bush—

LEARNING OUTCOMES

After studying this chapter, you should be able to:

19.1 Define energy, work, and how our energy use has varied over time.

19.2 Describe the benefits and disadvantages of using coal.

19.3 Explain the consequences and rewards of exploiting oil.

19.4 Illustrate the advantages and disadvantages of natural gas.

19.5 Summarize the potential and risk of nuclear power.

19.6 Evaluate the problems of radioactive wastes.

19.7 Discuss the changing fortunes of nuclear power.

19.8 Identify the promise and peril of nuclear fusion.

Case Study Clean Coal?

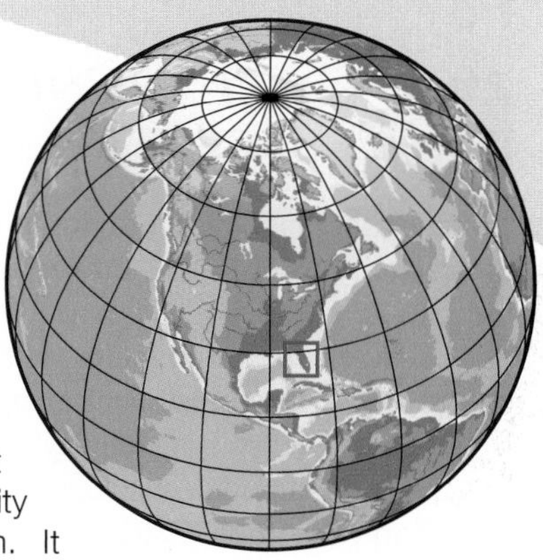

Most Americans are becoming aware of the dangers of our dependence on fossil fuels. The carbon dioxide released when we burn carbon-based fuels is altering our global climate. Furthermore, Americans spend about $250 billion annually to import 4 billion barrels of oil per day, much of it from politically unstable countries. What can we do about this problem? The coal industry points out that we have a huge supply of coal. If we could find ways to use it in environmentally benign ways, it could go a long way toward energy independence and global climate control.

There's a technology that offers hope for coal-based, zero-emissions electricity and hydrogen fuel. It's called **integrated gasification combined cycle (IGCC).** Power plants using this system could generate electricity while capturing and permanently storing carbon dioxide and other pollutants. An IGCC plant has been operating successfully for the past decade just outside of Tampa, Florida. Every day, the Polk power plant converts 2,400 tons of coal into 250 megawatts (MW) of electricity, or enough power for about 100,000 homes. Unlike conventional coal-fired power plants, an IGCC doesn't actually burn the coal. It converts the coal into gas and then burns the gas in a turbine (fig. 19.1). To do this, the coal is first ground into a fine powder and mixed with water to create a slurry. The slurry is pumped at high pressure into a gasification chamber, where it mixes with 96 percent pure oxygen, and is heated to 1,370°C (2,500°F). The coal doesn't burn; instead it reacts with the oxygen and breaks down into a variety of gases, mostly hydrogen and carbon dioxide. The gases are cooled, separated, and converted into easily managed forms.

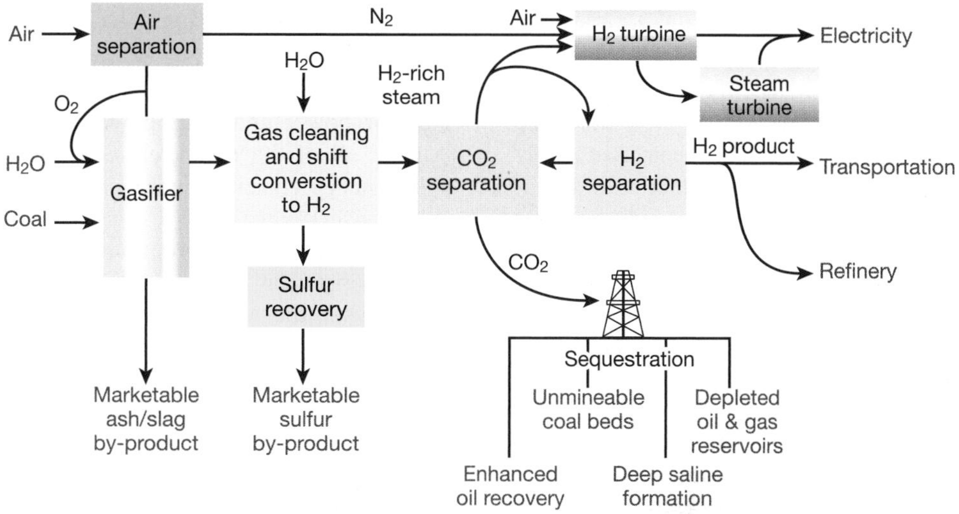

After purification, the synthetic hydrogen gas (or syngas) is pumped to the combustion turbine, which spins a huge magnet to produce electricity. Superheated gases from the turbine are fed into a steam generator that drives another turbine to produce more electrical current. Combining these two turbines makes an IGCC about 15 percent more efficient than a normal coal-fired power plant. Perhaps even better is that the hydrogen gas could power fuel cells if they become commercially feasible.

Contaminants, such as sulfur dioxide (SO_2), ash, and mercury, that often go up the smokestack in a normal coal-burning plant, are captured and sold to make the IGCC cleaner and more economical. Sulfur is marketed as fertilizer; ash and slag are sold to cement

companies. Mercury removal is an important public health benefit. All the slurry water is recycled to the gasifier; there is no waste water and very little solid waste. Because of these efficiencies, the Polk plant produces the cheapest electricity in the whole Tampa system. It doesn't now capture carbon dioxide, because it isn't required to, but it could easily do so. If we had CO_2 emission limits, IGCC plants could either pump it into deep wells for storage, or use it to enhance oil and natural gas recovery.

Because the Polk plant is so successful, Tampa Electric now plans to build another, even larger IGCC plant. Unfortunately this progressive outlook isn't typical of the whole power industry. Of the 80 or so new coal-fired power plants planned for construction over the next decade, only ten are slated to be IGCC, largely because of construction costs. While an IGCC can be very economical to operate, it costs 15 to 20 percent more to build than a conventional design. If industries were required to either sequester their CO_2 or pay a tax for not doing so, clean coal technology would be much more attractive, and our contributions to global warming would be far lower.

FIGURE 19.1 Clean coal technology could contribute to energy independence, while also reducing our greenhouse gas emissions.

Although the Tampa plant is the only one in the United States, Japan has about 18 IGCC plants.

While electrical production is far cleaner in an IGCC than in a conventional power plant, there's still a serious environmental problem in digging the coal out of the ground (see photo opposite page and discussion later in this chapter). Achieving energy independence is a wonderful goal, but the social and environmental costs of extracting coal are very high. This case study illustrates the dilemma in which we find ourselves. We have become dependent—some would say addicted—to the energy we now get from fossil fuels, such as coal, oil, and natural gas. Can we find alternative energy sources, or ways to obtain and use fossil fuels in more environmentally benign ways? In this chapter, we'll look more closely at conventional energy supplies and the problems associated with their use. In chapter 20, we'll look at some renewable alternatives for providing energy.

19.1 What is Energy and Where do We Get it?

Work is the application of force through a distance. **Energy** is the capacity to do work. **Power** is the rate of flow of energy, or the rate at which work is done. The energy that we use to move our muscles, to think, and to carry out metabolic functions comes from stored chemical energy (or potential energy) in our food. Food energy is generally measured in calories (cal). One calorie is the amount of energy to heat 1 gram of water 1°C. A kilocalorie (or food Calorie) is 1,000 calories. In physics, the basic metric unit of force is a newton, which is the force necessary to accelerate 1 kilogram 1 meter per second. A **joule (J)** is the amount of work done when a force of 1 newton is exerted over 1 meter or 1 amp per second flows through 1 ohm. One J equals 0.238 cal. Some other common energy units are presented in table 19.1.

Energy use is changing

Fire was probably the first human energy technology. Charcoal from fires has been found at sites occupied by our early ancestors 1 million years ago. Muscle power provided by domestic animals has been important at least since the dawn of agriculture some 10,000 years ago. Wind and water power have been used nearly as long. The invention of the steam engine, together with diminishing supplies of wood in industrializing countries, caused a switch to coal as the major energy source at the beginning of the nineteenth century. Coal, in turn, was replaced by oil in the twentieth century due to the ease of shipping, storing, and burning of liquid fuels. As easily accessible petroleum supplies have been depleted, we look increasingly to remote places, such as deep-ocean formations and the Arctic, for the fossil fuels on which we have become dependent. Operations in these locations is frequently difficult and dangerous, and the environmental, social, and economic costs of relying on them are high (fig. 19.2).

FIGURE 19.2 In our search for continuing supplies of fossil fuels, we increasingly turn to places like deep oceans or the high Arctic, but the social, environmental, and economic costs of our dependence on these energy sources can be high.

Currently, **fossil fuels** (petroleum, natural gas, and coal) supply about 87 percent of world commercial energy needs (fig. 19.3). Oil makes up at least 37 percent of that total. Nuclear power and hydropower supply about 6 percent each. Almost all of the renewable energy is hydropower. Wind and solar energy currently make only about 1 percent of our total energy use.

World energy consumption rose slightly more than 1 percent annually between 1970 and 2000, but that growth rate is expected to rise over the next few decades. Rapid economic growth in developing countries—especially in China—is responsible for much of the expected increase in energy use. For many years, the richer countries with about 20 percent of the world population consumed roughly 80 percent of all commercial energy, while the other 80 percent of the world had only 20 percent of the total supply. That situation is changing now. By 2025, energy experts expect that emerging economies, such as China and India, will be consuming nearly as much energy as Europe and North America.

TABLE 19.1

Some Energy Units

1 joule (J) = the force exerted by a current of 1 amp per second flowing through a resistance of 1 ohm

1 watt (W) = 1 joule (J) per second

1 kilowatt-hour (kWh) = 1 thousand (10^3) watts exerted for 1 hour

1 megawatt (MW) = 1 million (10^6) watts

1 gigawatt (GW) = 1 billion (10^9) watts

1 petajoule (PJ) = 1 quadrillion (10^{15}) joules

1 PJ = 947 billion BTU, or 0.278 billion kWh

1 British thermal unit (BTU) = energy to heat 1 lb of water 1°F

1 standard barrel (bbl) of oil = 42 gal (160 l) or 5.8 million BTU

1 metric ton of standard coal = 27.8 million BTU or 4.8 bbl oil

---

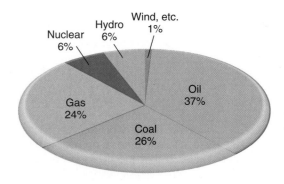

FIGURE 19.3 Worldwide commercial energy consumption. This does not include energy collected for personal use or traded in informal markets.
Source: Data from British Petroleum, 2006.

India's oil consumption has doubled since 1992, while China went from near self-sufficiency in the mid-1990s to the world's second largest importer in 2004. Aware that a reliable energy supply is essential for their economy, China and India have been quietly, but aggressively, pursuing political alliances and economic ties with oil-producing countries. Competition for limited oil supplies already plays a large role in geopolitics—witness recent events in the Middle East—but it may do so even more in the future.

Oil prices have fluctuated rapidly over the past 40 years. In the 1970s, shortages caused by the 1973 Arab oil embargo and the 1979 Iranian revolution triggered a tenfold price surge. These price shocks caused skyrocketing inflation and economic stagnation that lasted for a decade. They also were a major source of the crushing debt burdens that still hold back many developing countries. The results had less devastating effects in richer countries, but Americans waiting in long lines at gas stations became aware for the first time that much of the lifestyle they enjoy was dependent on a limited, unstable energy supply (fig. 19.4). Gasoline shortages brought an increased concern about conservation and renewable energy resources in the early 1980s, but this commitment didn't last very long. Reduced demand and increased production soon made prices fall almost as suddenly as they had risen, and Americans quickly went back to big cars and pickup trucks.

In 2006, oil prices shot up again to $78 per barrel. Americans saw gasoline prices over $3 per gallon. This brought record profits to oil companies (Exxon Mobile reported nearly $40 billion for the year), but pained many consumers. Montana governor, Brian Schweitzer, summarizes the situation succinctly, "We Americans use 6.5 billion bbl of oil a year. We produce 2.5 billion ourselves. We import 4 billion from the world's worst dictators. We need to stop doing that. We can save 1 billion bbl through conservation. We can produce another 1 billion bbl of biofuels from agricultural crops like corn, soybeans and canola.

We can produce 2 billion bbl a year turning our enormous coal reserves to clean-burning gas. We can achieve energy independence in 10 years, create a whole new industry with tens of thousands of high-paying jobs, and you'll never have to send your grandchildren to war in the Middle East." Sounds good, doesn't it? We'll look further, in this chapter, at our fossil fuel supplies and our options for using them.

How much energy do you use every year? Most of us don't think about it much, but maintaining the luxuries we enjoy usually requires an enormous energy input. On average, each person in the United States or Canada uses more than 300 gigajoules (GJ) (the equivalent of about 60 standard barrels or 8 metric tons of oil) per year. By contrast, in the poorest countries of the world, such as Bangladesh, Yemen, and Ethiopia, each person, on average, consumes less than one GJ per year. Put another way, each of us in the richer countries consumes nearly as much energy in a single day as the poorest people in the world consume in a year. In general, income and standards of living rise with increasing energy availability, but the correlation isn't absolute (see Data Analysis, p. 447). Some energy-rich countries, such as Qatar, use vast amounts of energy, although their level of human development isn't correspondingly high. Perhaps more important is that some countries, such as Norway, Denmark, and Japan, have a much higher standard of living by almost any measure than the United States, while using about half as much energy. This suggests abundant opportunities for energy conservation without great sacrifices.

> **Think About It**
>
> Years ago, Europe decided to discourage private automobiles and encourage mass transit by making gasoline expensive (about $5 per gal, on average). What changes would America have to make to achieve the same result?

Where do we get energy currently?

Similar to the rest of the world, fossil fuels supply about 85 percent of the energy used in the United States, and oil makes up 43 percent of that amount (fig. 19.5a). Because North America has large coal deposits, coal currently provides slightly more energy than natural gas, but that balance is shifting as we substitute cleaner burning gas for highly polluting coal. Nuclear reactors provide about 8 percent of U.S. commercial energy (but about 20 percent of all electricity). Renewable sources (mostly hydropower, which some people would argue isn't truly sustainable because reservoirs eventually silt up) make up about 6 percent of the energy mix.

Interestingly, until 1947, the United States was the largest exporter of oil in the world. Those easily accessible reserves have been depleted, however, and the United States is now the world's largest oil importer, dependent on foreign sources for nearly three-quarters of its supply. Contrary to what you might think, middle Eastern countries don't provide most of that oil. Canada

FIGURE 19.4 Sudden price shocks in the 1970s caused by anticipated oil shortages showed Americans that our energy-intensive lifestyles may not continue forever.

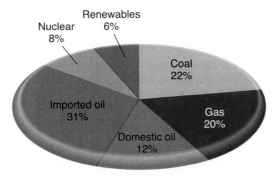

(a) U.S. energy consumption

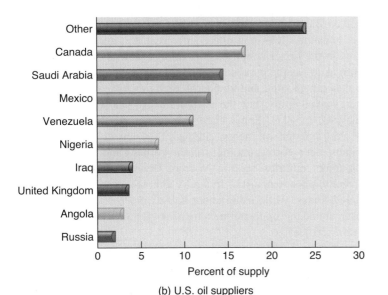

(b) U.S. oil suppliers

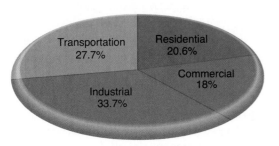

(c) Energy use sectors

FIGURE 19.5 Fossil fuels supply 85 percent of the energy used in the United States, and oil (three-quarters of it imported) makes up half that amount. Canada is now the largest oil supplier for the United States, and industry is the largest energy-use sector. Transportation, however, uses the vast bulk of all oil.

is now the largest single source of oil for the United States (fig. 19.5b). Saudi Arabia is the second largest, but, as you can see, a variety of other countries make up significant portions of U.S. oil imports.

The largest share (roughly one-third) of the energy used in the United States is consumed by industry (fig. 19.5c). Mining, milling, smelting, and forging of primary metals consume about

one-quarter of the industrial energy share. The chemical industry is the second largest industrial user of fossil fuels, but only half of its use is for energy generation. The remainder is raw material for plastics, fertilizers, solvents, lubricants, and hundreds of thousands of organic chemicals in commercial use. Residential and commercial buildings use about 20 percent of the primary energy consumed in the United States, mostly for space heating, air conditioning, lighting, and water heating.

Transportation consumes nearly 28 percent of all energy used in the United States each year. About 98 percent of that energy comes from petroleum products refined into liquid fuels. Almost three-quarters of all transport energy is used by motor vehicles. Nearly 3 trillion passenger miles and 600 billion ton miles of freight are carried annually by cars and trucks. About 75 percent of all freight traffic in the United States is carried by trains, barges, ships, and pipelines, but because they are very efficient, they use only 12 percent of all transportation fuel.

Finally, analysis of how energy is used has to take into account waste and loss of potential energy. About *half* of all the energy in primary fuels is lost during conversion to more useful forms, while it is being shipped to the site of end use, or during its use. Electricity, for instance, is generally promoted as a clean, efficient source of energy because when it is used to run a resistance heater or an electrical appliance almost 100 percent of its energy is converted to useful work and no pollution is given off.

What happens before then, however? We often forget that huge amounts of pollution are released during mining and burning of the coal that fires power plants. Nearly two-thirds of the energy in the coal that generated that electricity was lost in thermal conversion in the power plant. About 10 percent more is lost during transmission and stepping down to household voltages. Similarly, about 75 percent of the original energy in crude oil is lost during distillation into liquid fuels, transportation of that fuel to market, storage, marketing, and combustion in vehicles.

Natural gas is our most efficient fuel. Only 10 percent of its energy content is lost in shipping and processing since it moves by pipelines and usually needs very little refining. Ordinary gas-burning furnaces are about 75 percent efficient, and high-economy furnaces can be as much as 95 percent efficient. Because natural gas has more hydrogen per carbon atom than oil or coal, it produces about half as much carbon dioxide—and therefore half as much contribution to global warming—per unit of energy.

19.2 COAL

Coal is fossilized plant material preserved by burial in sediments and altered by geological forces that compact and condense it into a carbon-rich fuel. Coal is found in every geologic system since the Silurian Age 400 million years ago, but graphite deposits in very old rocks suggest that coal formation may date back to Precambrian times. Most coal was laid down during the Carboniferous period (286 million to 360 million years ago)

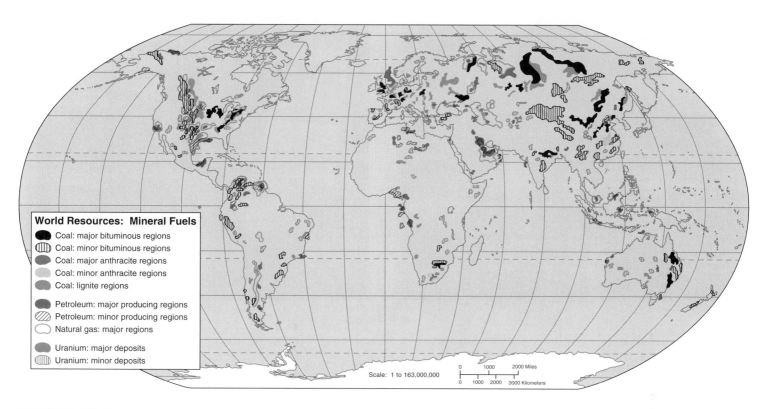

FIGURE 19.6 Where are fossil fuels and uranium located? North America, Europe, the Middle East, and parts of Asia are richly endowed. Africa, most of South America, and island states, like Japan, generally lack these fuels, greatly limiting their economic development.

when the earth's climate was warmer and wetter than it is now. Because coal takes so long to form, it is essentially a nonrenewable resource.

Coal resources are vast

World coal deposits are ten times greater than conventional oil and gas resources combined. Coal seams can be 100 m thick and can extend across tens of thousands of square kilometers that were vast swampy forests in prehistoric times. The total resource is estimated to be 10 trillion metric tons. If all this coal could be extracted, and if coal consumption continued at present levels, this would amount to several thousand years' supply. At present rates of consumption, these proven-in-place reserves—those explored and mapped but not necessarily economically recoverable—will last about 200 years. Note that "known reserves" have been identified but not thoroughly mapped. **"Proven reserves"** have been mapped, measured, and shown to be economically recoverable. Ultimate reserves include unknown as well as known resources.

Where are these coal deposits located? They are not evenly distributed throughout the world (fig. 19.6). North America, Europe, and Asia contain more than 90 percent of the world's coal, and five nations (United States, Russia, China, India, and Australia) account for three-quarters of that amount. In part, countries with large land areas are more likely to have coal deposits, but this resource is very rare in Africa, the Middle East, or Central and South America (fig. 19.7). Both China and India plan to

greatly increase coal consumption to raise standards of living. If they do so, CO_2 released will exacerbate global warming. Antarctica is thought to have large coal deposits, but they would be difficult, expensive, and ecologically damaging to mine.

It would seem that the abundance of coal deposits is a favorable situation. But do we really want to use all of the coal? In the next section, we will look at some of the disadvantages and dangers of mining and burning coal using conventional techniques.

Coal mining is a dirty, dangerous business

Underground mines are subject to cave-ins, fires, accidents, and accumulation of poisonous or explosive gases (carbon monoxide, carbon dioxide, methane, hydrogen sulfide). Between 1870 and 1950, more than 30,000 coal miners died of accidents and injuries

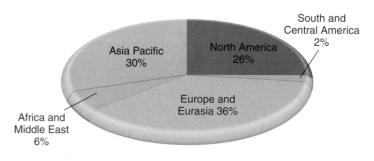

FIGURE 19.7 Proven-in-place coal reserves by region, 2004.

FIGURE 19.8 One of the most environmentally destructive methods of coal mining is mountaintop removal. Up to 100 m of a ridge line is scraped off and pushed into the valley below, burying forests, streams, farms, cemeteries, and sometimes houses.

FIGURE 19.9 Coal delivery in China. In parts of Guizhou Province, coal can have 35,000 ppm arsenic, and peppers dried over smokey coal fires can have 50,000 times as much arsenic as is considered safe.

in Pennsylvania alone, equivalent to one man per day for 80 years. Untold thousands have died of respiratory diseases. In some mines, nearly every miner who did not die early from some other cause was eventually disabled by **black lung disease,** inflammation and fibrosis caused by accumulation of coal dust in the lungs or airways (chapter 8). Few of these miners or their families were compensated for their illnesses by the companies for which they worked.

China is reported to have the world's most dangerous coal mines currently. In 2006, at least 6,000 workers were killed in mine accidents. This is more than ten times the mortality rate during the worst years of underground mining in the United States.

Strip mining or surface mining is cheaper and safer than underground mining but often makes the land unfit for any other use. Mine reclamation is now mandated in the United States, but land is rarely restored to its original contour or biological community. Coal mining also contributes to water pollution. Sulfur and other water soluble minerals make mine drainage and runoff from coal piles and mine tailings acidic and highly toxic. Thousands of miles of streams in the United States have been poisoned by coal-mining operations.

Perhaps the most egregious type of strip mining is "mountaintop removal," practiced mainly in Appalachia, where the tops of mountain ridges are scraped off and dumped into valleys below to get at coal seams (fig. 19.8). Streams, farms, even whole towns are buried by this practice under hundreds of meters of toxic rubble.

Burning coal releases many pollutants

Many people aren't aware that coal burning releases radioactivity and many toxic metals. Uranium, arsenic, lead, cadmium,

mercury, rubidium, thallium, and zinc—along with a number of other elements—are absorbed by plants and concentrated in the process of coal formation. These elements are not destroyed when the coal is burned, instead they are released as gases or concentrated in fly ash and bottom slag. You are likely to get a higher dose of radiation living next door to a coal-burning power plant than a nuclear plant under normal (nonaccident) conditions. Coal combustion is responsible for about 25 percent of all atmospheric mercury pollution in the United States.

In China alone, several hundred million people commonly burn coal in unvented stoves that permeate homes with high levels of toxic metals including arsenic, mercury, selenium, and fluorine (fig. 19.9). Coal samples from Guizhou Province in south central China, for example, have been found to have as much as 35,000 ppm arsenic. Chili peppers dried over coal fires (as they typically are in this area) can have 500 ppm arsenic, or 50,000 times the acceptable level for drinking water in the United States. At least 3,000 people from Guizhou were found in one survey to be suffering from severe arsenic poisoning, as much as 80 percent of which is believed to come from contaminated food.

Coal also contains up to 10 percent sulfur (by weight). Unless this sulfur is removed by washing or flue-gas scrubbing, it is released during burning and oxidizes to sulfur dioxide (SO_2) or sulfate (SO_4) in the atmosphere. The high temperatures and rich air mixtures ordinarily used in coal-fired burners also oxidize nitrogen compounds (mostly from the air) to nitrogen monoxide, dioxide, and trioxide. Every year the 900 million tons of coal burned in the United States (83 percent for electric power generation) releases 18 million metric tons of SO_2, 5 million metric tons of nitrogen oxides (NO_x), 4 million metric tons of airborne particulates, 600,000 metric tons of hydrocarbons and carbon monoxide, and about 2 trillion metric tons of CO_2. This is about three-quarters of the

SO_2, one-third of the NO_x, and about half of the industrial CO_2 released in the United States each year. Coal burning is the largest single source of greenhouse gases and acid rain in many areas.

These air pollutants have many deleterious effects, including human health costs, injury to domestic and wild plants and animals, and damage to buildings and property (chapter 16). Total losses from air pollution are estimated to be between $5 billion and $10 billion per year in the United States alone. By some accounts, at least 5,000 excess human deaths per year can be attributed to coal production and burning.

Sulfur can be removed from coal before it is burned, or sulfur compounds can be removed from the flue gas after combustion. Formation of nitrogen oxides during combustion also can be minimized. Perhaps the ultimate limit to our use of coal as a fuel will be the release of carbon dioxide into the atmosphere. As we discussed in chapter 15, carbon traps heat in the atmosphere and is a major contributor to global warming.

Clean coal technology could be helpful

As the opening case study for this chapter shows, new technologies, such as IGCC, could solve many of the problems currently caused by coal combustion. While sulfur-removal from flue gases in conventional power plants has been effective in reducing acid rain in much of the United States, it might be done much better in an IGCC before the coal is burned. Similarly, mercury can be removed from flue gases, but it is captured more cheaply in an IGCC. And NO_x formation is said to be much lower in an IGCC than in most coal-fired boilers.

Perhaps the ultimate limit to our use of coal in conventional boilers is CO_2 emissions. As we discussed in chapter 15, greenhouse gases are now changing our global climate in ways that could have catastrophic consequences. We need urgently to reduce these emissions. **Carbon sequestration** appears to be a good option. The United Nations estimates that at least half the CO_2 we release every year could be pumped into deep geologic formations. This can enhance gas and oil recovery. Norway's Statoil already is doing this. Since 1996, the company has injected more than 1 million tons of CO_2 into an oil reservoir beneath the North Sea because otherwise it would have to pay a (U.S.)$50 per ton carbon tax on its emissions. Alternatively, CO_2 can be stored in depleted oil or gas wells; forced into tight sandstone formations; injected into deep, briny aquifers; or compressed and pumped to the bottom of the ocean.

Many of the synthetic fuel (synfuel) projects currently underway, however, are merely get-rich-quick schemes. United States rules say that any treatment that chemically transforms coal qualifies for synfuel tax breaks. This means that companies can merely spray diesel fuel (or almost anything else) on a coal pile, then burn it in any old boiler. One North Carolina company reported $400 million in losses in a synfuel operation, but then claimed $850 million in tax credits.

In his 2007 State of the Union speech, President Bush announced a "20 in 10" initiative to reduce gasoline use 20 percent in 10 years. Ultimately, he promised to replace 35 billion gallons of oil consumption per year with "renewable and alternative" fuels. Most people assume this means a huge increase in ethanol and biodiesel production. Agricultural economists point out, however, that even if we convert the entire U.S. corn and soybean crop to ethanol (and doing so would drive food prices up astronomically), we couldn't produce 35 billion gallons of ethanol.

The key to this proposal is *alternative* fuels. The most likely option is **coal-to-liquid (CTL)** technology. This has been possible for a long time. Nazis did it in Germany during the World War II when their oil supplies were cut off. Sasol, a South African company, is currently negotiating to build a CTL factory in India. Sasol developed its expertise during apartheid days when oil sanctions were imposed on South Africa. While using IGCC in a stationary power plant has some attractive features, CTL is one of the worst possible alternatives we could adopt. It isn't clean by any stretch of the imagination. There are massive CO_2 emissions and waste production during the solid-to-liquid conversion, and then even more CO_2 releases when the fuel is burned in vehicles, from which it would be impossible to capture emissions. Furthermore, if CTL were cost effective, it could kill incentives to build efficient vehicles or develop cleaner alternatives, such as fuel cells or electric vehicles.

19.3 OIL

Like coal, petroleum is derived from organic molecules created by living organisms millions of years ago and buried in sediments where high pressures and temperatures concentrated and transformed them into energy-rich compounds. Depending on its age and history, a petroleum deposit will have varying mixtures of oil, gas, and solid tarlike materials. Some very large deposits of heavy oils and tars are trapped in porous shales, sandstone, and sand deposits in the western areas of Canada and the United States.

Liquid and gaseous hydrocarbons can migrate out of the sediments in which they formed through cracks and pores in surrounding rock layers. Oil and gas deposits often accumulate under layers of shale or other impermeable sediments, especially where folding and deformation of systems create pockets that will trap upward-moving hydrocarbons. Contrary to the image implied by its name, an oil pool is not usually a reservoir of liquid in an open cavern but rather individual droplets or a thin film of liquid permeating spaces in a porous sandstone or limestone, much like water saturating a sponge.

As oil exploration techniques improve, we are finding deposits more effectively and in places once thought to be devoid of oil. Ultra deep wells have been drilled in the ocean under 3,000 m (10,000 ft) of water, and some land-based rigs are claimed to be capable of drilling to depths of 12,400 m (40,000 ft). Sometimes it isn't practical or cost-effective to drill directly over an oil deposit. Directional drilling allows well heads to be as much as 6 km (3.75 mi) horizontally away from their intended target. Up to 20 wells from a single rig can access multiple

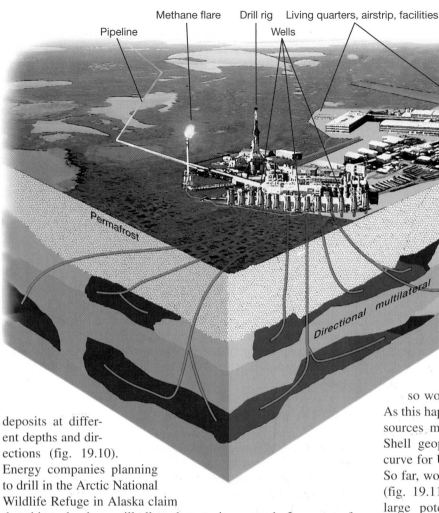

Pipeline Methane flare Drill rig Living quarters, airstrip, facilities
                                Wells   Waste well

Permafrost

Directional multilateral drilling

FIGURE 19.10 Horizontal, or directional drilling allows wells to extend up to 6 km horizontally from the well head. Up to 50 individual wells can be drilled from a single rig to reach isolated oil pockets.

deposits at different depths and directions (fig. 19.10). Energy companies planning to drill in the Arctic National Wildlife Refuge in Alaska claim that this technology will allow them to impact only 2 percent of the land surface while drilling down to oil-bearing strata. Critics doubt that damage to the land will be so limited.

Pumping oil out of a reservoir is much like sucking liquid out of a sponge. The first fraction comes out easily, but removing subsequent fractions requires increasing effort. We never recover all the oil in a formation; in fact, a 30 to 40 percent yield is about average. There are ways of forcing steam or CO_2 into the oil-bearing formations to "strip" out more of the oil, but at least half the total deposit usually remains in the ground at the point at which it is uneconomical to continue pumping. Methods for squeezing more oil from a reservoir are called secondary recovery techniques.

Oil resources aren't evenly distributed

The total amount of oil in the world is estimated to be about 4 trillion barrels (600 billion metric tons), half of which is thought to be ultimately recoverable. Some 465 billion barrels of oil already have been consumed. In 2006, the proven reserves were roughly 1.15 trillion bbls, enough to last only 40 years at the current consumption rate of 28.5 billion barrels per year. It is estimated that another 800 billion barrels either remain to be discovered or are not recoverable at current prices with present technology. As oil

resources become depleted and prices rise, it probably will become economical to find and bring this oil to the market unless alternative energy sources are developed. This estimate of the resource does not take into account the very large potential from unconventional liquid hydrocarbon resources, such as shale oil and tar sands, which might double the total reserve if they can be mined with acceptable social, economic, and environmental impacts.

Many geologists expect that within a decade or so world oil production will peak and then begin to decline. As this happens, oil prices will rise, which will make other energy sources more economically attractive. Dr. M. King Hubbert, a Shell geophysicist, first predicted in the 1940s a bell-shaped curve for U.S. oil production, peaking in the 1970s, which it did. So far, world oil production seems to be following a similar path (fig. 19.11). These estimates don't take into account the very large potential from unconventional semisolid hydrocarbon resources, such as shale oil and tar sands, which might double the total reserve if they can be mined with acceptable social, economic, and environmental impacts.

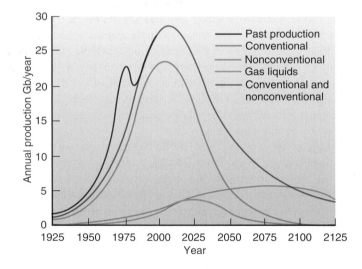

FIGURE 19.11 Worldwide production of crude oil with predicted Hubbert production. Gb = billion barrels.
Source: Jean Laherrère, www.hubbertpeak.org.

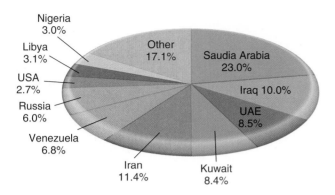

FIGURE 19.12 Proven oil reserves. Ten countries account for nearly 84 percent of all known recoverable oil.
Source: Data from British Petroleum, 2006.

FIGURE 19.13 The *Amoco Cadiz* ran aground off the coast of Brittany, France, on March 16, 1978, spilling 1.6 million barrels of oil and contaminating more than 350 km of coastline. The risk of similar spills is one cost of depending on imported oil.

Rapidly growing economies in China, India, South Korea, and other developing countries coupled with rising standards of living are already causing fierce competition for limited oil supplies. High gasoline prices may stimulate a new interest in conservation, as they did in the 1970s.

By far the largest supply of proven-in-place oil is in Saudi Arabia, which claims 262.7 billion barrels, almost one-fourth of the total proven world reserve (fig. 19.12). Kuwait had more than 10 percent of the proven world oil reserves before Iraq invaded in 1990. Some 600 wells were blown up and set on fire by the retreating Iraqis and at least 5 billion barrels of Kuwait's oil was burned, spilled, or otherwise lost. Together, the Persian Gulf countries in the Middle East contain nearly two-thirds of the world's proven petroleum supplies. With our insatiable appetite for oil (some would say addiction), it is not difficult to see why this volatile region plays such an important role in world affairs.

Note that this discussion has been of *proven* reserves. Oil companies estimate that reservoirs around and under the Caspian Sea may hold 200 billion barrels. If true, this would make Azerbaijan and Kazakhstan second only to Saudi Arabia in oil riches.

Altogether, the United States has already used more than half of its original recoverable petroleum resource. Of the 120 billion barrels thought to remain, about 31 billion barrels are proven-in-place. If we stopped importing oil and depended exclusively on indigenous supplies, our proven reserves would last only four years at current rates of consumption.

Like other fossil fuels, oil has negative impacts

Oil extraction isn't as destructive to landscapes as strip-mining coal, but oil wells can be dirty and disruptive, especially in pristine landscapes. The largest remaining untapped oil field in the United States is thought to be in the Arctic National Wildlife Refuge (ANWR) in northeastern Alaska. Drilling on the coastal plain could disrupt wildlife and wilderness in what has been called "North America's Serengeti" (What Do You Think? p. 434).

Like other fossil fuels, burning petroleum produces CO_2 emissions and contributes to global climate change. Sulfur is generally removed from gasoline, so that automobiles don't contribute very much to SO_2 emissions. Until recently, however, diesel fuel in the United States had high sulfur levels and also produced high amounts of very unhealthy particulate emissions. Internal combustion engines also produce large amounts of nitrogen oxides (NO_x). In many cities, vehicles are the largest source of degraded air quality, including secondary air pollutants, such as ambient ozone (chapter 16).

Our dependence on imported oil costs the United States about $250 billion per year in direct payments to oil-producing countries. In addition, a good share of our $800 billion annual military budget goes to protecting our access to oil in unfriendly or politically unstable countries. To support our oil habit, we tolerate—or even sponsor—some of the world's worst dictators and least democratic countries.

One of the costs of importing oil is the environmental damage from oil spills. Every year about 1.5 billion tons (90 billion barrels) of oil are shipped in ocean tankers. On average, about 1 percent of the cargo (1.5 million tons) is spilled or discharged annually. Most of the oil enters the ocean from bilge washing; accidents represent only about one-quarter of the oil lost from tankers. Still, the effects can be catastrophic. On March 16, 1978, for example, the *Amoco Cadiz* ran aground 2 km off the coast of Brittany, France, when its steering failed during a storm (fig. 19.13). The entire cargo of 1.6 million barrels (68.7 million gallons) spilled into the sea. The oil contaminated approximately 350 km of the Brittany coastline, including the beaches of 76 different communities. Fishing and tourism were devastated. Total economic losses were claimed to be about (U.S.) $1 billion. This was the largest tanker spill in history, more than six times as large as the wreck of the *Exxon Valdez* in Alaska in 1989. International tankers are supposed to be double hulled now to reduce the chances of such a disastrous spill, but many ship owners have ignored this requirement.

What Do You Think?

Oil Drilling in ANWR

A narrow strip of coastal plain in Alaska's Arctic National Wildlife Refuge (ANWR) presents one of nature's grandest spectacles as well as one of the most contentious environmental battles in recent years (fig. 1). For a few months during the brief arctic summer, the tundra teems with wildlife. It is the calving ground of the 130,000 caribou of the Porcupine herd; a nesting area for millions of snow geese, tundra swans, shorebirds, and other migratory waterfowl; a denning area for polar bears, arctic foxes, and arctic wolves; and a year-round home to about 350 shaggy musk ox.

The coastal plain may also be the site of the last big, onshore, liquid petroleum field in North America. Although only limited exploratory drilling has been permitted in the refuge, geologists estimate that it may contain as much as 12 billion barrels of oil and several trillion cubic feet of natural gas, mostly in small, isolated deposits. Can we extract these valuable fossil fuels without driving away the wildlife and polluting the pristine landscape? The oil industry believes it can access resources without doing lasting environmental harm; biologists and environmentalists doubt this is so.

Conservationists argue that oil drilling would do irreparable harm in this fragile arctic environment. They claim that excavating gravel for drill pads and pumping millions of gallons of water for ice roads would destroy wetlands on which wildlife depend for summer habitat. The noise and odors of up to 700 workers and their vehicles, construction equipment, drill rigs, helipads, and giant power-generating turbines could drive away wildlife. Pointing to the problems of other arctic oil drilling operations where drilling crews dumped garbage, sewage, and toxic drilling waste into surface pits, environmentalists predict disaster if drilling is allowed in the refuge. Pipeline and drilling spills at Prudhoe Bay have contaminated the tundra and seeped into waterways. Scars from bulldozer tracks made 50 years ago can still be seen clearly today.

Oil company engineers, on the other hand, claim that old, careless ways are no longer permitted in their operations. Wastes now are collected and either burned or injected into deep wells. Heavy equipment is hauled to the sites during the winter when most wildlife is either absent or hibernating. Modern directional drilling will allow up to 50 wells to be placed on a single pad, greatly reducing land impacts (see fig. 19.10). The native people of Alaska are divided on this issue. Inupiat people, many of whom work in the oil fields, and who will benefit from oil royalties, generally favor oil development. The Gwich'in people, most of whom live south of the Brooks Range (and therefore don't work in the oil fields or get oil royalties), and who still depend on caribou for much of their diet, oppose drilling.

In total, ANWR contains about 7.9 million ha (19.6 million acres) of land, but it's the

FIGURE 2 Alaska's Arctic National Wildlife Refuge is home to one of the world's largest caribou herds as well as 200 other wildlife and plant species

600,00 ha of coastal plain where all the oil is thought to be and where the most crucial caribou calving grounds are located (fig. 2). When the wildlife refuge was expanded to its present size in 1980, Congress exempted this "1002 area" (called that because of the section of the bill designating it) and reserved it for possible future oil exploration. Although the coastal plain represents less than 10 percent of the refuge, biologists worry that industrial activity there may frighten away the caribou and jeopardize calving success. It may not take very many years of calving failure to doom the whole herd.

The amount of economically recoverable oil in ANWR depends not only on geology but also on market prices and shipping costs. At current wholesale prices of about (U.S.) $60 per barrel, the U.S. Geological Survey estimates that about 7 billion barrels could be extracted profitably from ANWR. If prices drop back below $40 per barrel, where they were in the early 1990s, the economic resource might be less than a billion barrels, or only about 2 months supply at current consumption rates. Under the most optimistic scenarios, it will take at least a decade to begin to get this oil to market, and the peak production rate will probably be about one million barrels of oil per day in 2030. Conservationists point out that improving the average fuel efficiency of American cars and light trucks by just one mile per gallon would save more oil than is ever likely to be recovered from ANWR, and it would do so far faster and cheaper than extracting and transporting arctic oil. Furthermore, because there are few ports on the U.S. West Coast to receive ANWR oil, it will probably be sent to China or Japan, helping meet their demands, but doing little for us.

Oil companies have been pressing to drill in ANWR for 25 years. As oil supplies have dwindled in adjacent Prudhoe Bay and the oil revenues on which the state of Alaska has come to depend have shrunk, the pressure for drilling has mounted. Opening of the National Petroleum Reserve east of Prudhoe Bay to oil leases in 1998 didn't eliminate pressures on ANWR.

Citing rising gasoline prices and our growing dependence on foreign oil, supporters of ANWR drilling claim this oil is essential for maintaining a strong economy. Conservationists agree that we need independence from foreign oil, although they'd generally prefer that our energy come from sustainable sources such as solar and wind power. What do you think? Can we have all the energy we want and still have a tolerable environment? Will drilling in Alaska give us a stopgap while we develop other resources, or will it just postpone that development?

FIGURE 1 The Trans-Alaska pipeline is conveniently located to transport oil from the Arctic National Wildlife Refuge.

Oil shales and tar sands contain huge amounts of petroleum

Estimates of our recoverable oil supplies usually don't account for the very large potential from unconventional resources. The World Energy Council estimates that oil shales, tar sands, and other unconventional deposits contain ten times as much oil as liquid petroleum reserves. **Tar sands** are composed of sand and shale particles coated with bitumen, a viscous mixture of long-chain hydrocarbons. Shallow tar sands are excavated and mixed with hot water and steam to extract the bitumen, then fractionated to make useful products. For deeper deposits, superheated steam is injected to melt the bitumen, which can then be pumped to the surface, like liquid crude. Once the oil has been retrieved, it still must be cleaned and refined to be useful.

Canada and Venezuela have the world's largest and most accessible tar sand resources. Canadian deposits in northern Alberta are estimated to be equivalent to 1.7 trillion bbl of oil, and Venezuela has nearly as much. Together, these deposits are three times as large as all conventional liquid oil reserves. By 2010 Alberta is expected to increase its flow to 2 million bbl per day, or twice the maximum projected output of the Arctic National Wildlife Refuge (ANWR). Furthermore, because Athabascan tar sand beds are 40 times larger and much closer to the surface than ANWR oil, the Canadian resource will last longer and may be cheaper to extract. Canada is already the largest supplier of oil to the United States, having surpassed Saudi Arabia in 2000.

There are severe environmental costs, however, in producing this oil. A typical plant producing 125,000 bbl of oil per day creates about 15 million m^3 of toxic sludge, releases 5,000 tons of greenhouse gases, and consumes or contaminates billions of liters of water each year. Surface mining in Canada could destroy millions of hectares of boreal forest. Native Cree, Chipewyan, and Metis people worry about effects on traditional ways of life if forests are destroyed and wildlife and water are contaminated. Many Canadians dislike becoming an energy colony for the United States, and environmentalists argue that investing billions of dollars to extract this resource simply makes us more dependent on fossil fuels.

Similarly, vast deposits of oil shale occur in the western United States. Actually, **oil shale** is neither oil nor shale but a fine-grained sedimentary rock rich in solid organic material called kerogen (fig. 19.14). When heated to about 480°C (900°F), the kerogen liquefies and can be extracted from the stone. Oil shale beds up to 600 m (1,800 ft) thick occur in the Green River Formation in Colorado, Utah, and Wyoming, and lower-grade deposits are found over large areas of the eastern United States. If these deposits could be extracted at a reasonable price and with acceptable environmental impacts, they might yield the equivalent of several trillion barrels of oil.

Mining and extracting shale oil also creates many problems. It is expensive; it uses vast quantities of water, a scarce resource in the arid west; it has a high potential for air and water pollution; and it produces enormous quantities of waste. In the early 1980s,

FIGURE 19.14 Oil shale, a sedimentary rock that contains a solid organic material called kerogen, underlies large areas in Colorado and Utah. It could contain several trillion barrels of oil, more than the entire world supply of liquid oil, but extracting that resource could cause severe environmental and social problems.

when the search for domestic oil supplies was at fever pitch, serious discussions occurred about filling whole canyons, rim to rim, with oil shale waste. One experimental mine used a nuclear explosion to break up the oil shale. All the oil shale projects dried up when oil prices fell in the mid-1980s, however. What do you think? Should we declare some western states sacrifice areas so we can have cheap gasoline?

19.4 Natural Gas

Natural gas is the world's third largest commercial fuel (after oil, and coal), making up 24 percent of global energy consumption. Composed primarily of methane (90 percent), natural gas also contains propane, ethane, and small amounts of heavier hydrocarbons. Some gas (called "sour" gas) is contaminated with sulfur, but this is relatively easily removed before shipping. Gas is the most rapidly growing energy source because it is convenient, cheap, and clean burning. Because natural gas produces only half as much CO_2 as an equivalent amount of coal, substitution could help reduce global warming. World consumption of natural gas is growing by about 2.2 percent per year, considerably faster than either coal or oil. Much of this increase is in the developing world, where concerns about urban air pollution encourage the switch to cleaner fuel.

Gas can be shipped easily and economically through buried pipelines. The United States has been fortunate to have abundant gas resources accessible by an extensive pipeline system. It is difficult and dangerous, however, to ship and store gas between continents. To make the process economic, gas is compressed and liquefied. At −160°C (−260°F) the liquid takes up about one-six-hundredth the volume of gas.

FIGURE 19.15 As domestic supplies of natural gas dwindle, the United States is turning increasingly to shipments of liquefied gas in specialized ships, such as this one at an Australian terminal. An explosion of one of these ships would release about as much energy as a small atomic bomb.

Special refrigerated ships transport liquefied natural gas (LNG) (fig. 19.15). LNG weighs only about half as much as water, so the ships are very buoyant. Finding sites for terminals to load and unload these ships is difficult. Many cities are unwilling to accept the risk of an explosion of the volatile cargo. A fully loaded LNG ship contains about as much energy as a medium-size atomic bomb. Furthermore, huge amounts of seawater are used to warm and re-gasify the LNG. This can have deleterious effects on coastal ecology. In 2005, the U.S. Senate voted to eliminate state or local authority to expedite siting of LNG terminals.

In many places, gas and oil are found together in sediments, and both can be recovered at the same time. In remote areas, where no shipping facilities exist for the gas, it is simply flared (burned) off—a terrible waste of a valuable resource (see fig. 19.2). The World Bank estimates that 100 billion m^3 of gas are flared every year, or 1.5 times the amount used annually in Africa. Increasingly, however, these "stranded" gas deposits are being captured and shipped to market. Until 2001, Canada was the primary source of natural gas for the United States, providing about 105 billion m^3 per year. Over the next 20 years, Canadian exports are expected to decrease as more of its gas supply is used to process tar sands. LNG imports, on the other hand, are expected to increase to about one-fourth of the 600 billion m^3 of natural gas consumed in the United States each year.

Most of the world's known natural gas is in a few countries

Two-thirds of all proven natural gas reserves are in the Middle East and the former Soviet Union (fig. 19.16). The republics of the former Soviet Union have nearly 31 percent of known natural gas reserves (mostly in Siberia and the Central Asian republics) and account for about 40 percent of all production. Both eastern and western Europe buy substantial quantities of gas from these wells.

The total ultimately recoverable natural gas resources are estimated to be 10,000 trillion ft^3, corresponding to about 80 percent as much energy as the recoverable reserves of crude oil. The

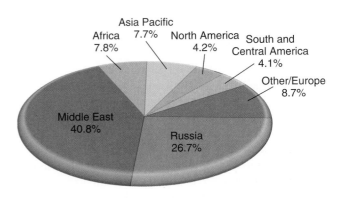

FIGURE 19.16 Proven natural gas reserves by region, 2002.
Source: Data from British Petroleum, 2006.

proven world reserves of natural gas are 6,200 trillion ft^3 (176 million metric tons). Because gas consumption rates are only about half of those for oil, current gas reserves represent roughly a 60-year supply at present usage rates. Proven reserves in North America are about 185 trillion ft^3, or 3 percent of the world total. This is a ten-year supply at current rates of consumption. Known reserves are more than twice as large.

As it breaks down, coal is slowly transformed into methane. Accumulation of this explosive gas is one of the things that makes coal mining so dangerous. In many places, where mining coal seams isn't economically feasible, it is relatively cheap and easy to extract the methane. This has led to a huge battle in the American West (What Do You Think? p. 437).

There may be vast unconventional gas sources

Natural gas resources have been less extensively investigated than petroleum reserves. There may be extensive "unconventional" sources of gas in unexpected places. Prime examples are recently discovered methane hydrate deposits in arctic permafrost and beneath deep ocean sediments. **Methane hydrate** is composed of small bubbles or individual molecules of natural gas trapped in a crystalline matrix of frozen water. At least 50 oceanic deposits and a dozen land deposits are known. Altogether, they are thought to hold some 10,000 gigatons (10^{13} tons) of carbon or twice as much as the combined amount of all coal, oil, and conventional natural gas. This could be a valuable energy source but would be difficult to extract, store, and ship. If climate change causes melting of these deposits, it could trigger a catastrophic spiral of global warming because methane is 20 times as powerful a greenhouse gas as CO_2. Japan plans exploratory extraction of methane hydrate in the next few years, first on land near Prudhoe Bay, Alaska, and then in Japanese waters.

Methane also can be produced by digesting garbage or manure. Some U.S. cities collect methane from landfills and sewage sludge digestion. Because methane is so much more potent than CO_2 as a greenhouse gas, stopping leaks from pipelines and other sources is important in preventing global warming. In developing countries, small-scale manure digesters provide a valuable, renewable source of gas for heating, lighting, and cooking (chapter 20).

Coal-Bed Methane

Vast deposits of coal, oil, and gas lie under the sage scrub and arid steppe of North America's intermountain West. Geologists estimate that at least 346 trillion ft^3 of "technically recoverable" natural gas and 62 billion barrels of petroleum liquids occur in five intermountain basins stretching from Montana to New Mexico. These deposits would provide a 15-year supply of gas at present usage rates, and at least four times as much oil as the most optimistic estimates for the Arctic National Wildlife Refuge. About half of that gas and oil is in or around relatively shallow coal seams, which makes it vastly cheaper to extract than most other gas supplies. Drilling a typical offshore gas well costs tens of millions of dollars, while a deep conventional gas well costs several million dollars, but a coal-bed methane well is generally less than $100,000. The total value of the methane and petroleum liquids from the Rocky Mountains could be as much as $200 billion over the next decade.

Most coal-bed methane is held in place by pressure from overlying aquifers. Pumping the water out these aquifers releases the gas, but creates phenomenal quantities of effluent that often is contaminated with salt and other minerals. A typical coal-bed well produces 75,000 liters of water per day. Dumping it on the surface can poison fields and pastures, erode stream banks, contaminate rivers, and harm fish and wildlife. Drawing down aquifers depletes the wells on which many ranches depend, and also dries up natural springs and wetlands essential for wildlife. Ranchers complain that livestock and wildlife are killed by traffic and poisoned by discarded toxic waste around well sites. "It may be a clean fuel," says one rancher, "but it's a dirty business."

Another objection to coal-bed methane extraction is simply the enormous scope of the enterprise. In Wyoming's Powder River Basin, energy companies have already installed 12,000 wells and have proposed 39,000 more. Eventually, this area could contain as many as 140,000 wells, together with the sprawling network of roads, pipelines, compressor stations, and waste water pits necessary for such a gargantuan undertaking. The Green River Basin and the San Juan Basin, with three to five times as much potential gas and oil as Powder River, have even greater probability for environmental damage.

An aerial view of the Jonah Field in the Upper Green River Basin.

An unlikely coalition of ranchers, hunters, anglers, conservationists, water users, and renewable energy activists have banded together to fight against coal-bed gas extraction, calling on Congress to protect private property rights, preserve water quality, and conserve sensitive public lands.

An example of the wildlife conflicts occurring in the gas fields can be seen in the Upper Green River Basin (see figure). Every year, 50,000 pronghorn antelope and 10,000 elk migrate through a narrow corridor on their way between summer and winter ranges. The gas fields lie across this migration route, and biologists worry that the noise, traffic, polluted waste water pits, and activity around the wells may interrupt the migration and doom the animals. Much of this land is also habitat for the greater sage-grouse. The Bush administration rejected endangered species protection for these birds even though populations have declined by 90 percent in recent decades, because doing so might interfere with gas and oil drilling.

What do you think? Does having access to cleaner fuels justify the social and environmental costs of their extraction? If you had a vote in this issue, what restrictions would you impose on the companies carrying out these projects? Could renewable energy sources, such as wind or solar, substitute for coal-bed methane (chapter 20)?

19.5 NUCLEAR POWER

In 1953, President Dwight Eisenhower presented his "Atoms for Peace" speech to the United Nations. He announced that the United States would build nuclear-powered electrical generators to provide clean, abundant energy. He predicted that nuclear energy would fill the deficit caused by predicted shortages of oil and natural gas. It would provide power "too cheap to meter" for continued industrial expansion of both the developed and the developing world. It would be a supreme example of "beating swords into plowshares." Technology and engineering would tame the evil genie of atomic energy and use its enormous power to do useful work.

Glowing predictions about the future of nuclear energy continued into the early 1970s. Between 1970 and 1974, American utilities ordered 140 new reactors for power plants (fig. 19.17). Some advocates predicted that by the end of the century there would be 1,500 reactors in the United States alone. In 1970, the International Atomic Energy Agency (IAEA) projected worldwide nuclear power generation of at least 4.5 million megawatts (MW) by the year 2000, 18 times more than our current nuclear capacity and twice as much as present world electrical capacity from all sources.

Rapidly increasing construction costs, declining demand for electric power, and safety fears have made nuclear energy much less attractive than promoters expected. Electricity from nuclear plants was about half the price of coal in 1970 but twice as much by 1990. Wind energy is already cheaper than nuclear power in many areas and solar power or hydropower are becoming cheaper as well (chapter 20).

FIGURE 19.17 Two nuclear reactors (domes) at the San Onofre Nuclear Generating Station sit between the beach and Interstate 5, the major route between Los Angeles and San Diego.

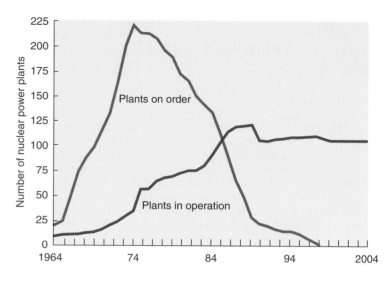

FIGURE 19.18 The changing fortunes of nuclear power in the United States are evident in this graph showing the number of nuclear plants on order and plants in operation.

After 1975, only 13 orders were placed for new nuclear reactors in the United States, and all of those orders subsequently were canceled (fig. 19.18). In fact, 100 of the 140 reactors on order in 1975 were canceled. It began to look as if the much-acclaimed nuclear power industry might have been a very expensive wild goose chase that would never produce enough energy to compensate for the amount invested in research, development, mining, fuel preparation, and waste storage.

How do nuclear reactors work?

The most commonly used fuel in nuclear power plants is U^{235}, a naturally occurring radioactive isotope of uranium. Ordinarily, U^{235} makes up only about 0.7 percent of uranium ore, too little to sustain a chain reaction in most reactors. It must be purified and concentrated by mechanical or chemical procedures (fig. 19.19). Mining and processing uranium to create nuclear fuel is even more dirty and dangerous than coal mining. In some uranium mines 70 percent of the workers—most of whom were Native Americans—have died from lung cancer caused by high radon and dust levels. In addition, mountains of

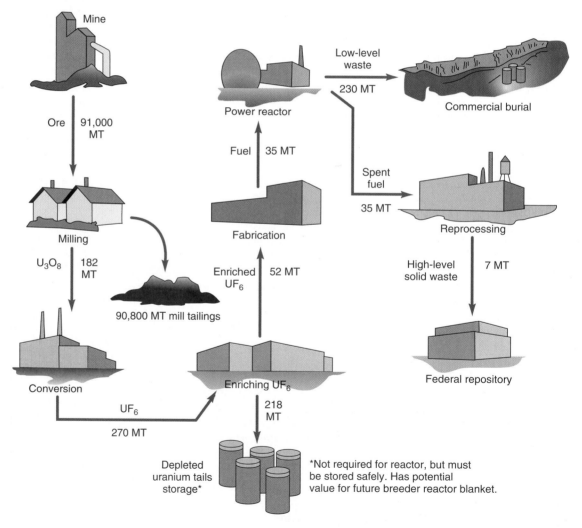

FIGURE 19.19 The nuclear fuel cycle. Quantities represent the average annual fuel requirements for a typical 1,000 MW light water reactor (MT = metric tons). About 35 MT or one-third of the reactor fuel is replaced every year. Reprocessing is not currently done in the United States.

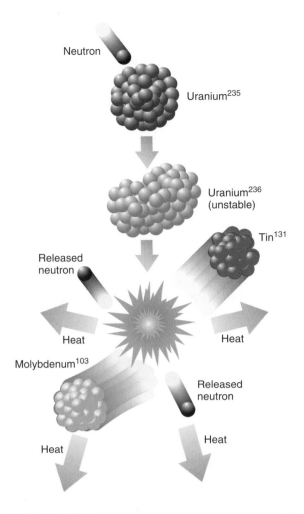

FIGURE 19.20 The process of nuclear fission is carried out in the core of a nuclear reactor. In the sequence shown here, the unstable isotope, uranium-235, absorbs a neutron and splits to form tin-131 and molybdenum-103. Two or three neutrons are released per fission event and continue the chain reaction. The total mass of the reaction product is slightly less than the starting material. The residual mass is converted to energy (mostly heat).

radioactive tailings and debris have been left around fuel preparation plants.

When the U^{235} concentration reaches about 3 percent, the uranium is formed into cylindrical pellets slightly thicker than a pencil and about 1.5 cm long. Although small, these pellets pack an amazing amount of energy. Each 8.5-gram pellet is equivalent to a ton of coal or four barrels of crude oil.

The pellets are stacked in hollow metal rods approximately 4 m long. About 100 of these rods are bundled together to make a **fuel assembly.** Thousands of fuel assemblies containing 100 tons of uranium are bundled together in a heavy steel vessel called the reactor core. Radioactive uranium atoms are unstable—that is, when struck by a high-energy subatomic particle called a neutron, they undergo **nuclear fission** (splitting), releasing energy and more neutrons. When uranium is packed tightly in the reactor core, the neutrons released by one atom will trigger the fission of another uranium atom and the release of still more neutrons (fig. 19.20). Thus a self-sustaining **chain reaction** is set in motion and vast amounts of energy are released.

The chain reaction is moderated (slowed) in a power plant by a neutron-absorbing cooling solution that circulates between the fuel rods. In addition, **control rods** of neutron-absorbing material, such as cadmium or boron, are inserted into spaces between fuel assemblies to shut down the fission reaction or are withdrawn to allow it to proceed. Water or some other coolant is circulated between the fuel rods to remove excess heat.

The greatest danger in one of these complex machines is a cooling system failure. If the pumps fail or pipes break during operation, the nuclear fuel quickly overheats and a "meltdown" can result that releases deadly radioactive material. Although nuclear power plants cannot explode like a nuclear bomb, the radioactive releases from a worst-case disaster like the 1986 fire at Chernobyl in the Ukraine are just as devastating as a bomb.

There are many different reactor designs

Seventy percent of the nuclear plants in the United States and in the world are pressurized water reactors (PWR) (fig. 19.21). Water is circulated through the core, absorbing heat as it cools the fuel rods. This primary cooling water is heated to 317°C (600°F) and reaches a pressure of 2,235 psi. It then is pumped to a steam generator where it heats a secondary water-cooling loop. Steam from the secondary loop drives a high-speed turbine generator that produces electricity. Both the reactor vessel and the steam generator are contained in a thick-walled concrete and steel containment building that prevents radiation from escaping and is designed to withstand high pressures and temperatures in

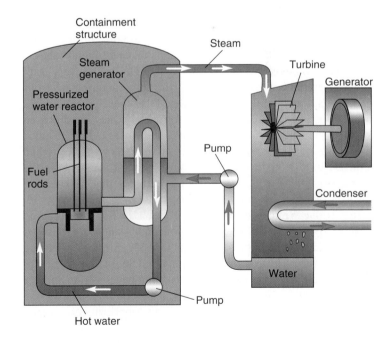

FIGURE 19.21 Pressurized water nuclear reactor. Water is superheated and pressurized as it flows through the reactor core. Heat is transferred to nonpressurized water in the steam generator. The steam drives the turbogenerator to produce electricity.

case of accidents. Engineers operate the plant from a complex, sophisticated control room containing many gauges and meters to tell them how the plant is running.

Overlapping layers of safety mechanisms are designed to prevent accidents, but these fail-safe controls make reactors very expensive and very complex. A typical nuclear power plant has 40,000 valves compared to only 4,000 in a fossil fuel-fired plant of similar size. In some cases, the controls are so complex that they confuse operators and cause accidents rather than prevent them. Under normal operating conditions, a PWR releases very little radioactivity and is probably less dangerous for nearby residents than a coal-fired power plant.

A simpler, but dirtier and more dangerous reactor design is the boiling water reactor (BWR). In this model, water from the reactor core boils to make steam, which directly drives the turbine-generators. This means that highly radioactive water and steam leave the containment structure. Controlling leaks is difficult, and the chances of releasing radiation in an accident are very high.

Canadian nuclear reactors use heavy water containing deuterium (H^2 or ^2H, the heavy, stable isotope of hydrogen) as both a cooling agent and a moderator. These *Cana*dian *deu*terium (CANDU) reactors operate with natural, unconcentrated uranium (0.7 percent U^{235}) for fuel, eliminating expensive enrichment processes. "Heavy water," deuterium-containing water (^2H_2O), is expensive, however, and these reactors are susceptible to overheating and meltdown if cooling pumps fail.

In Britain, France, and the former Soviet Union, a common reactor design uses graphite, both as a moderator and as the structural material for the reactor core. In the British MAGNOX design (named for the magnesium alloy used for its fuel rods), gaseous carbon dioxide is blown through the core to cool the fuel assemblies and carry heat to the steam generators. In the Soviet design, called RBMK (the Russian initials for a graphite-moderated, water-cooled reactor), low-pressure cooling water circulates through the core in thousands of small metal tubes.

These designs were originally thought to be very safe because graphite has high capacity for both capturing neutrons and dissipating heat. Designers claimed that these reactors could not possibly run out of control; unfortunately, they were proven wrong. The small cooling tubes are quickly blocked by steam if the cooling system fails and the graphite core burns when exposed to air. The two most disastrous reactor accidents in the world, so far, involved fires in graphite cores that allowed the nuclear fuel to melt and escape into the environment. In 1956, a fire at the Windscale Plutonium Reactor in England released roughly 100 million curies of radionuclides and contaminated hundreds of square kilometers of countryside. Similarly, burning graphite in the Chernobyl nuclear plant in Ukraine made the fire much more difficult to control than it might have been in another reactor design.

The most serious accident at a North American commercial reactor occurred in 1979 when the Three Mile Island nuclear plant near Harrisburg, Pennsylvania, suffered a partial meltdown of the reactor core. The containment vessel held in most radioactive material. No deaths or serious injuries were verified, but the accident was a serious blow of future nuclear development.

Concerns about nuclear plant safety were further heightened in 2002 when inspectors found that leaking boric acid had eaten nearly all the way through the reactor vessel lid of First Energy's Davis-Besse plant near Toledo, Ohio. The terrorist attacks of September 11, 2001, also reminds those living near nuclear plants of their risks. Utility engineers claim that reactor containment structures (the large cement domes over the reactor) can withstand a direct hit by a loaded jumbo jet. Critics doubt this claim, which has not been tested empirically.

Some alternative reactor designs may be safer

Several other reactor designs are inherently safer than the ones we now use. Among these are the modular High-Temperature, Gas-Cooled Reactor (HTGCR) and the Process-Inherent Ultimate-Safety (PIUS) reactor.

In HTGCR, which is sometimes called a "pebble-bed reactor," uranium is encased in tiny ceramic-coated pellets; gaseous helium blown around these pellets is the coolant. If the reactor core is kept small, it cannot generate enough heat to melt the ceramic coating, even if all coolant is lost; thus, a meltdown is impossible and operators could walk away during an accident without risk of a fire or radioactive release. Fuel pellets are loaded into the reactor from the top, shuffle through the core as the uranium is consumed, and emerge from the bottom as spent fuel. This type of reactor can be reloaded during operation. Since the reactors are small, they can be added to a system a few at a time, avoiding the costs, construction time, and long-range commitment of large reactors. Only two of these reactors have been tried in the United States: the Brown's Ferry reactor in Alabama and the Fort St. Vrain reactor near Loveland, Colorado. Both were continually plagued with problems (including fires in control buildings and turbine-generators), and both were closed without producing much power.

A much more successful design has been built in Europe by General Atomic. In West German tests, a HTGCR was subjected to total coolant loss while running at full power. Temperatures remained well below the melting point of fuel pellets and no damage or radiation releases occurred. These reactors might be built without expensive containment buildings, emergency cooling systems, or complex controls. They would be both cheaper and safer than current designs.

The PIUS design features a massive, 60-meters-high pressure vessel of concrete and steel, within which the reactor core is submerged in a very large pool of boron-containing water (fig. 19.22). As long as the primary cooling water is flowing, it keeps the borated water away from the core. If the primary coolant pressure is lost, however, the surrounding water floods the core, and the boron poisons the fission reaction. There is enough secondary water in the pool to keep the core cool for at least a week without any external power or cooling. This should be enough time to resolve the problem. If not, operators can add more water and evaluate conditions further.

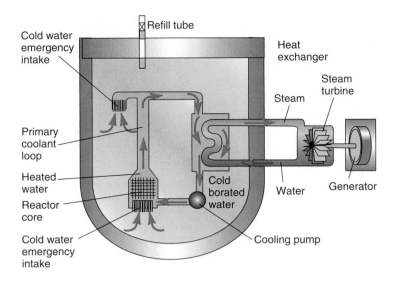

FIGURE 19.22 A PIUS reactor consists of a core and primary cooling system immersed in a very large pool of cold, borated water. As long as the reactor is operating, the cool water is excluded from the cooling circuit by the temperature and pressure differential. Any failure of the primary cooling system would allow cold water to flood the core and shut down the nuclear reaction. The large volume of the pool would cool the reactor for several days without any replenishment.

The Canadian "slow-poke" is a small-scale version of the PIUS design that doesn't produce electricity but might be a useful substitute for coal, oil, or gas burners in district heating plants. The core of this "mini-nuke" is only about half the size of a kitchen stove and generates only one-tenth to one-one-hundredth as much power as a conventional reactor. The fuel sits in a large pool of ordinary water, which it heats to just below boiling and sends to a heat exchanger. A secondary flow of hot water from the exchanger is pumped to nearby buildings for space heating. Promoters claim that a runaway reaction is impossible in this design and that it makes an attractive and cost-efficient alternative to fossil fuels. Despite a widespread aversion to anything nuclear, Switzerland and Germany are developing similar small nuclear heating plants.

If these reactors had been developed initially, the history of nuclear power might have been very different. Neither reactor type is suitable, however, for mobile power plants, such as nuclear submarines, which tells you something about the history of nuclear power and the motivation for its development. Aside from being inherently safer in case of coolant pressure loss, neither of these alternative reactor designs eliminates other problems, such as waste disposal, that we will discuss later in this chapter. Still, one of these alternate forms of nuclear power might make it an attractive energy source some day.

Breeder reactors could extend the life of our nuclear fuel

For more than 30 years, nuclear engineers have been proposing high-density, high-pressure, **breeder reactors** that produce fuel rather than consume it. These reactors create fissionable pluto-

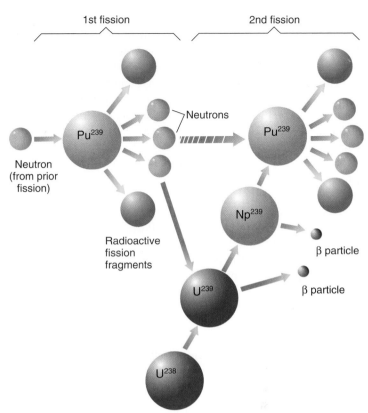

FIGURE 19.23 Reactions in a "breeder" fission process. Neutrons from a plutonium fission change U^{238} to U^{239} and then to Pu^{239} so that the reactor creates more fuel than it uses.

nium and thorium isotopes from the abundant, but stable, forms of uranium (fig. 19.23). The starting material for this reaction is plutonium reclaimed from spent fuel from conventional fission reactors. After about ten years of operation, a breeder reactor would produce enough plutonium to start another reactor. Sufficient uranium currently is stockpiled in the United States to produce electricity for 100 years at present rates of consumption, if breeder reactors can be made to work safely and dependably.

Several problems have held back the breeder reactor program in the United States. One problem is the concern about safety. The reactor core of the breeder must be at a very high density for the breeding reaction to occur. Water does not have enough heat capacity to carry away the high heat flux in the core, so liquid sodium generally is used as a coolant. Liquid sodium is very corrosive and difficult to handle. It burns with an intense flame if exposed to oxygen, and it explodes if it comes into contact with water. Because of its intense heat, a breeder reactor will melt down and self-destruct within a few seconds if the primary coolant is lost, as opposed to a few minutes for a normal fission reactor.

Another very serious concern about breeder reactors is that they produce excess plutonium that can be used for bombs. It is essential to have a spent-fuel reprocessing industry if breeders are used, but the existence of large amounts of weapons-grade plutonium in the world would surely be a dangerous and destabilizing development. The chances of some of that material falling into the hands of terrorists or other troublemakers are very high. Japan planned to purchase 30 tons of this dangerous material from France and ship it half way around the world through some of the most dangerous and congested shipping lanes on the planet to fuel a breeder program. In 1995, a serious accident at Japan's Moju breeder reactor caused reevaluation of the whole program.

A proposed $1.7 billion breeder-demonstration project in Clinch River, Tennessee, was on and off for 15 years. At last estimate, it would cost up to five times the original price if it is ever completed. In 1986, France put into operation a full-sized commercial breeder reactor, the SuperPhenix, near Lyons. It cost three times the original estimate to build and produces electricity at twice the cost per kilowatt of conventional nuclear power. After only a year of operation, a large crack was discovered in the inner containment vessel of the SuperPhenix, and in 1997 it was shut down permanently.

FIGURE 19.24 Spent fuel is being stored temporarily in large, aboveground "dry casks" at many nuclear power plants.

19.6 RADIOACTIVE WASTE MANAGEMENT

One of the most difficult problems associated with nuclear power is the disposal of wastes produced during mining, fuel production, and reactor operation. How these wastes are managed may ultimately be the overriding obstacle to nuclear power.

What will we do with radioactive wastes?

Until 1970, the United States, Britain, France, and Japan disposed of radioactive wastes in the ocean. Dwarfing all these dumps, however, are those of the former Soviet Union, which has seriously and permanently contaminated the Arctic Ocean. Rumors of Soviet nuclear waste dumping had circulated for years, but it was not until after the collapse of the Soviet Union that the world learned the true extent of what happened. Starting in 1965, the Soviets disposed of eighteen nuclear reactors—seven loaded with nuclear fuel—in the Kara Sea off the eastern coast of Novaya Zemlya island, and millions of liters of liquid waste in the nearby Barents Sea. Two other reactors were sunk in the Sea of Japan.

Altogether, the former Soviet Union dumped 2.5 million curies of radioactive waste into the oceans, more than twice as much as the combined amounts that 12 other nuclear nations have reported dumping over the past 45 years. In 1993, despite protests from Japan, Russia dumped 900 tons of additional radioactive waste into the Sea of Japan.

Enormous piles of mine wastes and abandoned mill tailings in all uranium-producing countries represent serious waste disposal problems. Production of 1,000 tons of uranium fuel typically generates 100,000 tons of tailings and 3.5 million liters of liquid waste. There now are approximately 200 million tons of radioactive waste in piles around mines and processing plants in the United States. This material is carried by the wind and washes into streams, contaminating areas far from its original source. Canada has even more radioactive mine waste on the surface than does the United States.

In addition to the leftovers from fuel production, there are about 100,000 tons of low-level waste (contaminated tools, clothing, building materials, etc.) and about 15,000 tons of high-level (very radioactive) wastes in the United States. The high-level wastes consist mainly of spent fuel rods from commercial nuclear power plants and assorted wastes from nuclear weapons production. For the past 40 years, spent fuel assemblies from commercial reactors have been stored in deep water-filled pools at the power plants. These pools were originally intended only as temporary storage until the wastes were shipped to reprocessing centers or permanent disposal sites.

With internal waste storage pools now full but neither reprocessing nor permanent storage available, a number of utility companies are beginning to store nuclear waste in large metal dry casks placed outside power plants (fig. 19.24). These projects are meeting with fierce opposition from local residents who fear the casks will leak. Most nuclear power plants are built near rivers, lakes, or seacoasts. Extremely toxic radioactive materials could spread quickly over large areas if leaks occur. A hydrogen gas explosion and fire in a dry storage cask at Wisconsin's Point Beach nuclear plant intensified opponents' suspicions about this form of waste storage.

In 1987, the U.S. Department of Energy announced plans to build the first **high-level waste repository** on a barren desert ridge near Yucca Mountain, Nevada. Intensely radioactive wastes are to be buried deep in the ground where it is hoped that they will remain unexposed to groundwater and earthquakes for the thousands of years required for the radioactive materials to decay to a safe level (fig. 19.25). Although the area is very dry now, we can't be sure that it will always remain that way (see Exploring Science, in ch. 14, p. 317). A billion dollars already has been spent

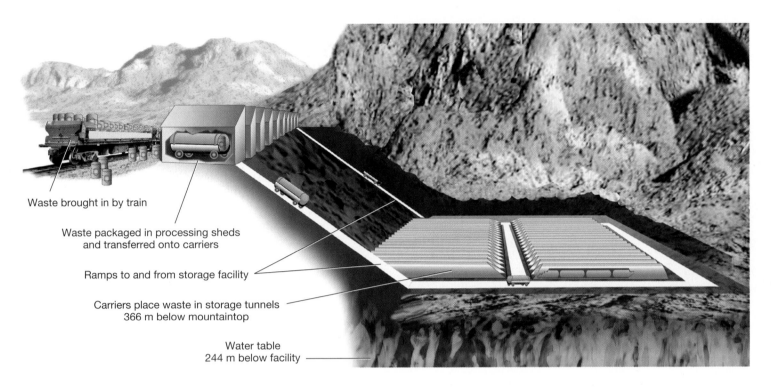

Waste brought in by train

Waste packaged in processing sheds and transferred onto carriers

Ramps to and from storage facility

Carriers place waste in storage tunnels 366 m below mountaintop

Water table 244 m below facility

FIGURE 19.25 The U.S. Department of Energy is planning to build the nation's first high-level underground waste repository at Yucca Mountain, Nevada.

at Yucca Mountain and total costs now are expected to be \$10 billion to \$35 billion. Although the facility was supposed to open in 1998, the earliest possible date is now 2010.

Several states have nuclear power plants that will be forced to close if waste storage isn't approved soon. They have been pushing for temporary storage in Nevada until Yucca Mountain is ready. Nevada has threatened to close its borders if such a plan is approved.

Russia has offered to store nuclear waste from other countries. Plans are to transport wastes to the Mayak in the Ural mountains. The storage site is near Chelyabinsk, where an explosion at a waste facility in 1957 contaminated about 24,000 km^2 (9,200 mi^2). The region is now considered the most radioactive place on earth, so the Russians feel it can't get much worse. They expect that storing 20,000 tons of nuclear waste should pay about \$20 billion.

Think About It

Several Native American tribes have offered to store nuclear waste if the price is right. The Skull Valley Band of the Goshutes, for example, whose barren desert land west of the Great Salt Lake in Utah, already is surrounded by hazardous waste dumps, believe their lives could improve with millions of dollars in revenue from nuclear waste storage. Would it be safe—or ethical—to let them do so? Do we have a right to interfere if they decide to proceed?

Some nuclear experts believe that **monitored, retrievable storage** would be a much better way to handle wastes. This method involves holding wastes in underground mines or secure surface facilities where they can be watched. If canisters begin to leak, they could be removed for repacking. Safeguarding the wastes would be expensive and the sites might be susceptible to wars or terrorist attacks. We might need a perpetual priesthood of nuclear guardians to ensure that the wastes are never released into the environment.

Shipping nuclear waste to a storage site worries many people—especially those whose cities will be on the shipping route. The Energy Department has performed crash tests on the shipping containers and assures us they are safe (fig. 19.26). Some people still worry about accidents or terrorist attacks. How would you feel if these trains were coming through your city? Would it be better to keep the waste where it is at 100 separate power plants?

Decommissioning old nuclear plants is expensive

Old power plants themselves eventually become waste when they have outlived their useful lives. Most plants are designed for a life of only 30 years. After that, pipes become brittle and untrustworthy because of the corrosive materials and high radioactivity to which they are subjected. Plants built in the 1950s and early 1960s already are reaching the ends of their lives. You don't just lock the door and walk away from a nuclear power plant; it is

FIGURE 19.26 A test railcar carrying a spent nuclear fuel shipping cask slams into a concrete wall at 130 km/hr (81 mph.). The cask survived without injury. Even so, many people don't want nuclear waste shipped through their city. Would you?

much too dangerous. It must be taken apart, and the most radioactive pieces have to be stored just like other wastes. This includes not only the reactor and pipes but also the meter-thick, steel-reinforced concrete containment building. The pieces have to be cut apart by remote-control robots because they are too dangerous to be worked on directly.

Only a few plants have been decommissioned so far, but it has generally cost two to ten times as much to tear them down as to build them in the first place. One example is Unit 1 of California's San Onofre Nuclear Plant (see fig. 19.17). After having operated for 24 years, the reactor was shut down in 1992. It took a decade to dismantle the reactor and shell and to pack them in a 950-ton shipping assembly, which is scheduled to be shipped 18,000 km around the tip of South America to Barnwell, South Carolina, for permanent storage. The Panama Canal refused passage to this cargo, and Chile may prohibit it as well. Another option is to go west from California, around the world to whichever East Coast port chooses to accept it. At the time of this writing, none has.

Altogether, the 103 reactors now in operation might cost somewhere between $200 billion and $1 trillion to decommission. No one knows how much it will cost to store the debris for thousands of years or how it will be done. However, we would face this problem, to some degree, even without nuclear electric power plants. Plutonium production plants and nuclear submarines also have to be decommissioned. Originally, the Navy proposed to just tow old submarines out to sea and sink them. The risk that the nuclear reactors would corrode away and release their radioactivity into the ocean makes this method of disposal unacceptable, however.

19.7 CHANGING FORTUNES OF NUCLEAR POWER

Although promoted originally as a new wonder of technology that could open the door to wealth and abundance, nuclear power has long been highly controversial. Around the world, many countries have decided to forego this power source. Germany, for example, has decided to close all its nuclear power plants and to depend, instead, on wind energy. In 2003, the first of 19 German nuclear power plants was shut down. About 500 million euros (U.S. $590 million) will be spent to dismantle this facility, four times as much as to build it. What to do with the contaminated waste, however, remains contentious.

For many U.S. environmental organizations, opposition to nuclear power is a perennial priority. Antinuclear groups have organized mass protest rallies featuring popular actors, musicians, and celebrities who help raise funds and attract attention to their cause. Protests and civil disobedience at some sites have gone on for decades, and plants that were planned—or in some cases, already built—have been abandoned or modified to burn fossil fuels because of public opposition.

On the other side, workers, consumers, utility officials, and others who stand to benefit from this new technology rally in support of nuclear power. They argue that abandoning this energy source is foolish since a great deal of money already has been invested and the risks may be lower than is commonly believed.

Public opinion about nuclear power has fluctuated over the years. Before the Three Mile Island accident in 1978, two-thirds of Americans supported nuclear power. By the time Chernobyl exploded in 1985, however, less than one-third of Americans favored this power source. More recently, however, memories of these earlier incidents have faded. Now about half of all Americans support atomic energy, and about one-quarter say they wouldn't mind having a nuclear plant within 16 km (10 mi) from their home.

The 103 nuclear reactors now operating in 31 states produce about 20 percent of all electricity consumed in the United States. Vermont, which gets 85 percent of its electricity from nuclear power, leads the nation in this category. With oil and natural gas prices soaring and worries about global warming causing concern about coal usage, advocates—and even some prominent conservationists—are promoting nuclear reactors as clean and environmentally friendly because they don't emit greenhouse gases.

Critics, on the other hand, point out that huge amounts of fossil fuels are used to mine, process, and ship fuels as well as to dismantle reactors and store wastes. They also regard the problems of reactor safety and waste disposal reasons to abandon this technology as quickly as possible. We've entered into a Faustian bargain that our descendants will regret for thousands of years to come, they claim. Over the past 50 years the U.S. government has supported nuclear power with more than $150 billion in subsidies. During that same time, less than $5 billion was spent on renewable energy research and development. Where might we be now if that ratio had been reversed?

In 2007, for the first time in 35 years, the U.S. Nuclear Regulatory Commission approved a site for a new nuclear power plant near Clinton, Illinois. An energy bill passed by the U.S. Senate provides $50 billion in subsidies for nuclear power. The nuclear industry hopes to build 28 new plants at 19 sites, but insists that the government must provide insurance and construction loans. Opponents argue that safety and waste disposal questions must be solved first.

> **Think About It**
>
> Is the energy from nuclear power worth the costs? Should we build new reactors and allow existing ones to continue to operate in order to reduce our dependence on fossil fuels? How would you evaluate the risks and benefits of this technology?

19.8 NUCLEAR FUSION

Fusion energy is an alternative to nuclear fission that could have virtually limitless potential. **Nuclear fusion** energy is released when two smaller atomic nuclei fuse into one larger nucleus. Nuclear fusion reactions, the energy source for the sun and for hydrogen bombs, have not yet been harnessed by humans to produce useful net energy. The fuels for these reactions are deuterium and tritium, two heavy isotopes of hydrogen.

It has been known for 50 years that if temperatures in an appropriate fuel mixture are raised to 100 million degrees Celsius and pressures of several billion atmospheres are obtained, fusion of deuterium and tritium will occur. Under these conditions, the electrons are stripped away from atoms and the forces that normally keep nuclei apart are overcome. As nuclei fuse, some of their mass is converted into energy, some of which is in the form

of heat. There are two main schemes for creating these conditions: magnetic confinement and inertial confinement.

Inertial confinement involves a small pellet (or a series of small pellets) bombarded from all sides at once with extremely high-intensity laser light. The sudden absorption of energy causes an implosion (an inward collapse of the material) that will increase densities by 1,000 to 2,000 times and raise temperatures above the critical minimum. So far, no lasers powerful enough to create fusion conditions have been built.

Magnetic confinement involves the containment and condensation of plasma, a hot, electrically neutral gas of ions and free electrons in a powerful magnetic field inside a vacuum chamber. Compression of the plasma by the magnetic field should raise temperatures and pressures enough for fusion to occur. The most promising example of this approach, so far, has been a Russian design called *tokomak* (after the Russian initials for "torodial magnetic chamber"), in which the vacuum chamber is shaped like a large donut (fig. 19.27b).

In both of these cases, high-energy neutrons escape from the reaction and are absorbed by molten lithium circulating in the walls of the reactor vessel. The lithium absorbs the neutrons and transfers heat to water via a heat exchanger, making steam that drives a turbine generator, as in any steam power plant. The advantages of fusion reactions, if they are ever feasible, include production of fewer radioactive wastes, the elimination of fissionable products that could be made into bombs, and a fuel supply that is much larger and less hazardous than uranium.

Despite 50 years of research and a $25 billion investment, fusion reactors never have reached the break-even point at which they produce more energy than they consume. A major setback occurred in 1997, when Princeton University's Tokomak Fusion Test Reactor was shut down. Three years earlier, this reactor had

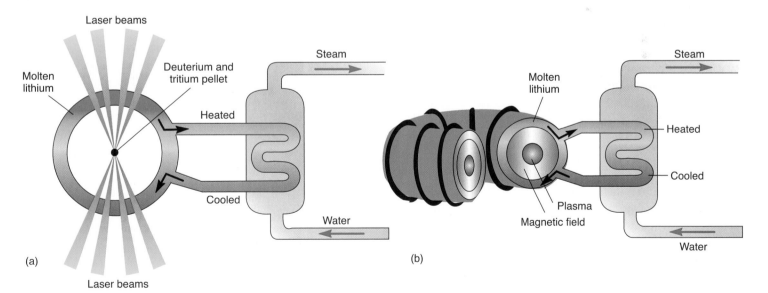

FIGURE 19.27 Nuclear fusion devices. (a) Inertial confinement is created by laser beams that bombard and ignite fuel pellets. Molten lithium transfers heat to a steam generator. (b) In the tokomak design, a powerful magnetic field confines the plasma and compresses it so that critical temperatures and pressures are reached.

set a world's record by generating 10.7 million watts for one second, but researchers conceded that the technology was still decades away from self-sustaining power generation. In 2006, China, South Korea, Russia, Japan, the United States, and the European Union announced a new (U.S.) $13 billion fusion reactor to be built jointly near Marseilles, France. Opponents view this project as just another expensive wild-goose chase and predict that it will never generate enough energy to pay back the fortune spent on its development.

CONCLUSION

Our energy future is far from certain. We have probably used half of the easily accessible liquid petroleum reserves in the world. This provided a lifestyle of luxury and convenience for those of us lucky enough to live in the industrialized countries of the world, but it has created titanic environmental problems—including acid rain, strip-mined landscapes, huge payments to unstable countries, and—perhaps most importantly—global climate change. There are still very large supplies of unconventional fossil fuels, including tar sands, oil shale, coal-bed methane, and methane hydrates, but the environmental costs of extracting those resources may preclude their use.

What, then, should we do? Some people hold out the promise of technological solutions to this dilemma. They point to IGCC, nuclear power, and, possibly, fusion reactors as ways that we may be able to get clean, affordable energy. Others argue that we ought to move immediately toward conservation and renewable energy, such as solar, wind, biofuels, small-scale hydro, and geothermal power. Even if we do this, however, it will probably take decades to replace our dependence on fossil fuels. Therefore, it's important to understand the relative benefits and disadvantages of each of our conventional energy sources.

As consumers, each of us needs to examine our energy use and its environmental impacts. In chapter 20, we'll investigate conservation and renewable energy options.

REVIEWING LEARNING OUTCOMES

By now you should be able to explain the following points:

19.1 Define energy, work, and how our energy use has varied over time.
- Energy use is changing.
- Where do we get energy currently?

19.2 Describe the benefits and disadvantages of using coal.
- Coal resources are vast.
- Coal mining is a dirty, dangerous business.
- Burning coal releases many pollutants.
- Clean coal technology could be helpful.

19.3 Explain the consequences and rewards of exploiting oil.
- Oil resources aren't evenly distributed.
- Like other fossil fuels, oil has negative impacts.
- Oil shales and tar sands contain huge amounts of petroleum.

19.4 Illustrate the advantages and disadvantages of natural gas.
- Most of the world's known natural gas is in a few countries.
- There may be vast unconventional gas sources.

19.5 Summarize the potential and risk of nuclear power.
- How do nuclear reactors work?
- There are many different reactor designs.
- Some alternative reactor designs may be safer.
- Breeder reactors could extend the life of our nuclear fuel.

19.6 Evaluate the problems of radioactive wastes.
- What will we do with radioactive wastes?
- Decommissioning old nuclear plants is expensive.

19.7 Discuss the changing fortunes of nuclear power.

19.8 Identify the promise and peril of nuclear fusion.

PRACTICE QUIZ

1. What is energy? What is power?
2. What are the major sources of commercial energy worldwide and in the United States? Why are data usually presented in terms of commercial energy?
3. How does energy use in the United States compare with that in other countries?
4. How much coal, oil, and natural gas are in proven reserves worldwide? Where are those reserves located?
5. What is coal-bed methane, and why is it controversial?
6. What are the most important health and environmental consequences of our use of fossil fuels?
7. Describe how a nuclear reactor works and why reactors can be dangerous.
8. What are the four most common reactor designs? How do they differ from each other?
9. What are the advantages and disadvantages of the breeder reactor?
10. Describe methods proposed for storing and disposing of nuclear wastes.

CRITICAL THINKING AND DISCUSSION QUESTIONS

1. We have discussed a number of different energy sources and energy technologies in this chapter. Each has advantages and disadvantages. If you were an energy policy analyst, how would you compare such different problems as the risk of a nuclear accident versus air pollution effects from burning coal?

2. If your local utility company were going to build a new power plant in your community, what kind would you prefer? Why?

3. The nuclear industry is placing ads in popular magazines and newspapers claiming that nuclear power is environmentally friendly since it doesn't contribute to the greenhouse effect. How do you respond to that claim?

4. Our energy policy effectively treats some strip-mine and well-drilling areas as national sacrifice areas knowing they will never be restored to their original state when extraction is finished. How do we decide who wins and who loses in this transaction?

5. Storing nuclear wastes in dry casks outside nuclear power plants is highly controversial. Opponents claim the casks will inevitably leak. Proponents claim they can be designed to be safe. What evidence would you consider adequate or necessary to choose between these two positions?

6. The policy of the United States has always been to make energy as cheap and freely available as possible. Most European countries charge three to four times as much for gasoline as we do. Who benefits and who or what loses in these different approaches? How have our policies shaped our lives? What does existing policy tell you about how governments work?

DATA analysis   Comparing Energy Use and Standards of Living

In general, income and standard of living increase with energy availability. This makes sense because cheap energy makes it possible to heat and air condition our homes, travel easily and frequently, obtain fresh foods out of season, have a wide variety of entertainment, work, and educational opportunities, and use machines to extend our productivity. However, energy use per capita isn't strictly tied to quality of life. Some countries use energy extravagantly without corresponding increases in income or standard of living. Look at the graph below and answer the following questions:

1. What country in this graph has the highest energy use?

2. How much energy does it use, and how much per capita income does it have?

3. What do you think might explain these values?

4. What do you know about the standard of living in this country?

5. How much energy per person do the United States and Denmark use annually?

6. How do you think the standard of living in the U.S. and Denmark compare?

7. How would you characterize energy use and income in Malaysia and Poland compared to Luxembourg?

8. In which of these countries would you rather live?.

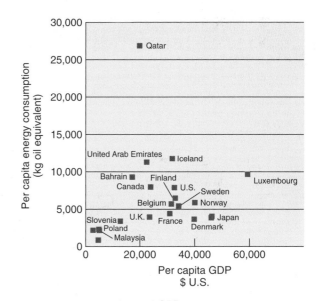

Per capita energy consumption and GDP.

Windmills supply all the electricity used on Denmark's Ærø Island, while solar and biomass energy provide space heating and vehicle fuel. Altogether, 100 percent of the island's energy comes from renewable sources.

C H A P T E R

20

Sustainable Energy

Two roads diverged in a wood, and I—I took the one less traveled by, And that has made all the difference.

—Robert Frost—

LEARNING OUTCOMES

After studying this chapter, you should be able to:

20.1 Remember that conservation can help us meet our energy needs.

20.2 Explain how we could tap solar energy.

20.3 Discuss high-temperature solar energy.

20.4 Grasp the potential of fuel cells.

20.5 Explain how we get energy from biomass.

20.6 Investigate energy from the earth's forces.

Case Study Renewable Energy Islands

Denmark has substantial oil and gas supplies under the North Sea, but the Danes have chosen to wean themselves away from dependence on fossil fuels. Currently the world leader in renewable energy, Denmark now gets 20 percent of its power from solar, wind, and biomass. Some parts of this small, progressive country have moved even further toward sustainability. One of the most inspiring examples of these efforts are the small islands of Samsø and Ærø, which now get 100 percent of their energy from renewable sources.

Samsø and Ærø lie between the larger island of Zealand (home to Copenhagen) and the Jutland Peninsula. The islands are mostly agricultural. Together, they have an area of about 200 km² (77 mi²) and a population of about 12,000 people. In 1997, Samsø and Ærø were chosen in a national competition to be renewable energy demonstration projects. The first step in energy independence is conservation, As you'll learn in this chapter, Denmark uses roughly half as much energy per person as the United States, although by most measures the Danes have a higher standard of living than most Americans. Danish energy conservation is achieved with high-efficiency appliances, superior building insulation, high-mileage vehicles, and other energy-saving measures. Most homes are clustered in small villages, both to save agricultural land and to facilitate district heating. Living closely together also makes having a private automobile less necessary.

Some 30 large wind generators provide 100 percent of Samsø and Ærø's electricity. Two-thirds of these windmills are located offshore, and are publicly owned. The 11 onshore wind turbines are mostly privately owned, but a share of the profits is used to finance other community energy projects. Space heating accounts for about one-third of the energy consumption on the islands. District heating systems provide most of this energy. Several large solar collector arrays supply about half the hot water for space heating and household use (fig. 20.1). Biomass-based (straw, wood chips, manure) systems supply the remainder of the island's heating needs. Some of this biomass comes from energy crops (fast-growing elephant grass and hybrid poplars, for example, are grown on marginal farmland), while the rest comes from agricultural waste.

Biodiesel (primarily from rapeseed oil) fuels farm tractors and ferries, while most passenger vehicles are electric.

Geothermal pumps supplement the solar water heaters, and in one village a recently closed landfill produces methane that is used to run a small electric generator. Nuclear power is considered an unacceptable option in Denmark and doesn't feature in current energy plans. Samsø and Ærø have won numerous prizes and awards for their pioneering conversion to renewable energy, Over the past 20 years, as a result of other projects like those on Samsø and Ærø, both Denmark's fossil fuel consumption and their greenhouse gas emissions have remained constant. All of us could learn from their example.

Many other countries, both in Europe and elsewhere in the world, are turning to renewable energy to reduce their dependence on environmentally damaging and politically unstable fossil fuels. The European Renewable Energy Council suggests that we might obtain half our global energy supply by the middle of this century. In this chapter, we'll look at what our options are for finding environmentally and socially sustainable ways to meet our energy needs.

FIGURE 20.1 A 19,000 m² array of solar water heaters provides space heating for the town of Marstal on Ærø Island.

20.1 CONSERVATION

As the previous chapter and the opening story of this chapter suggest, we urgently need to move toward sustainable, environmentally friendly, affordable, politically progressive energy sources for a number of reasons. One of the easiest ways to avoid energy shortages and to relieve environmental and health effects of our current energy technologies is simply to use less. We have already seen the benefits of conservation. Energy consumption rose rapidly in the United States in the 1960s, but the price shocks of the 1970s brought energy use down sharply (fig. 20.2).

Although economic growth resumed in the 1980s and 1990s, conservation kept energy consumption relatively constant.

There are many ways to save energy

Much of the energy we consume is wasted. This statement isn't a simple admonishment to turn off lights and turn down furnace thermostats in winter; it's a technological challenge. Our ways of using energy are so inefficient that most potential energy in fuel is lost as waste heat, becoming a form of environmental pollution. Of the energy we do extract from primary resources, however, much is

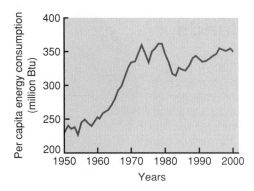

FIGURE 20.2 Per capita energy consumption in the United States rose rapidly in the 1960s. Price shocks in the 1970s encouraged conservation. Although GDP continued to grow in the 1980s and 1990s, higher efficiency kept per capita consumption relatively constant.
Source: U.S. Department of Energy.

FIGURE 20.3 Earth-sheltered homes take advantage of the stable temperatures and insulating qualities of the earth. This house has south-facing windows for maximum solar gain, and high clerestory windows that give light to the back of the house as well as summer ventilation.

used for frankly trivial or extravagant purposes. As chapter 19 shows, several European countries have higher standards of living than the United States, and yet use 30 to 50 percent less energy.

Many conservation techniques are relatively simple and highly cost effective. Compact fluorescent bulbs, for example, produce four times as much light as an incandescent bulb of the same wattage, and last up to ten times as long. Although they cost more initially, total lifetime savings can be $30 to $50 per fluorescent bulb.

Light-emitting diodes (LEDs) also are even more efficient, consuming 90 percent less energy and lasting hundreds of times as long as ordinary lightbulbs. They can produce millions of colors and be adjusted in brightness to suit ambient conditions. They are being used now in everything from flashlights and Christmas lights, to advertising signs, brake lights, exit signs, and street lights. New York city has replaced 11,000 traffic lights with LEDs. It also replaced 180,000 old refrigerators with new energy-saving models.

Many improvements in domestic energy efficiency have occurred in the past decade. Today's average new home uses one-half the fuel required in a house built in 1974, but much more can be done. Household energy losses can be reduced even further by better insulation, double or triple glazing of windows, thermally efficient curtains or window coverings, and by sealing cracks and loose joints. Reducing air infiltration is usually the cheapest, quickest, and most effective way of saving energy because it is the largest source of losses in a typical house. It doesn't take much skill or investment to caulk around doors, windows, foundation joints, electrical outlets, and other sources of air leakage.

According to new national standards, all new washing machines have to use 35 percent less water than older models. This makes them a little more expensive, but will pay back in seven years. It also cuts water use in the United States by 40 trillion liters (10.5 trillion gallons) per year and saves more electricity every year than is used to light all the homes in the United States. Air conditioners also are required to be about 20 percent more efficient than previous models.

For even greater savings, new houses can be built with extra thick superinsulated walls, air-to-air heat exchangers to warm incoming air, and even double-walled sections that create a "house within a house." The R-2000 program in Canada details how energy conservation can be built into homes. Special double-glazed windows that have internal reflective coatings and that are filled with an inert gas (argon or xenon) have an insulation factor of R11, the same as a standard 4-inch thick insulated wall or ten times as efficient as a single-pane window. Superinsulated houses now being built in Sweden require 90 percent less energy for heating and cooling than the average American home.

Orienting homes so that living spaces have passive solar gain in the winter and are shaded by trees or roof overhang in the summer also helps conserve energy. Earth-sheltered homes built into the south-facing side of a slope or protected on three sides by an earth berm are exceptionally efficient energy savers because they maintain relatively constant subsurface temperatures (fig. 20.3). Sod roofs provide good insulation, prevent rain runoff, and last longer than asphalt shingles. Because they are heavier, however, they need stronger supports.

Straw-bale construction offers both high insulating qualities and a renewable, inexpensive building material that can be assembled by amateurs (fig. 20.4). This isn't a new technique. Settlers on the Great Plains built straw-bale houses a century ago because they didn't have wood. Some of those houses are still standing. The bales are strong and will support the roof without any additional timber framing. They must be thoroughly water-proofed, however, with stucco, adobe, or plaster both inside and out so the straw doesn't decay. It's also important to seal them so mice and other vermin can't take up residence. The thick walls are terrific sound insulators as well as highly energy efficient. The cost can be less than a conventionally built home.

One of the most direct and immediate ways that individuals can save energy is to turn off appliances. Few of us realize how much electricity is used by appliances in a standby mode. You

FIGURE 20.4 Carolyn Roberts and her sons build a straw-bale house near Tucson, Arizona.

may think you've turned off your TV, DVD player, cable box, or printer, but they're really continuing to draw power in an "instant-on" mode. For the average home, standby appliances can represent up to 25 percent of the monthly electric bill. Home office equipment including computers, printers, cable modems, copiers, etc., usually are the biggest energy consumers (fig. 20.5). Putting your computer to sleep saves about 90 percent of the energy it uses when fully on, but turning it completely off is even better.

Industrial energy savings are another important part of our national energy budget. More efficient electric motors and pumps, new sensors and control devices, advanced heat-recovery systems, and material recycling have reduced industrial energy requirements significantly. In the early 1980s, U.S. businesses saved $160 billion per year through conservation. When oil prices collapsed, however, many businesses returned to wasteful ways.

Energy efficiency is a measure of energy produced compared to energy consumed. Table 20.1 shows the typical energy efficiencies of some power sources. Thermal-conversion machines, such as steam turbines in coal-fired or nuclear power plants, can turn no more than 40 percent of the energy in their primary fuel

TABLE 20.1

Typical Net Efficiencies of Some Power Sources

Electric Power Plants	Yield (Percent)
Hydroelectric (best case)	90
Co-generation	80
Fuel cell (hydrogen)	80
IGCC	45
Coal-fired generator	38
Oil-burning generator	38
Nuclear generator	30
Photovoltaic solar	15

Source: U.S. Department of Energy.

into electricity or mechanical power because of the need to reject waste heat. Does this mean that we can never increase the efficiency of fossil fuel use? No. Some waste heat can be recaptured and used for space heating, raising the net yield to 80 or 90 percent. The integrated gasification combined cycle (IGCC) process described in chapter 19 is an example of capture of waste heat. In another kind of process, fuel cells convert the chemical energy of a fuel directly into electricity without an intermediate combustion cycle. Since this process is not limited by waste heat elimination, its efficiencies can approach 80 percent with such fuel as hydrogen gas or methane. We'll discuss the special case of biofuel efficiency later in this chapter.

Transportation could be far more efficient

One of the areas in which most of us can accomplish the greatest energy conservation is in our transportation choices. You may not be able to build an energy-efficient house or persuade your utility company to switch from coal or nuclear to solar energy, but you can decide every day how you travel to school, to work, or for shopping or entertainment. Automobiles and light trucks account for 40 percent of the U.S. oil consumption and produce one-fifth of its carbon dioxide emissions. According the U.S. EPA, raising the average fuel efficiency of the passenger fleet by 3 miles per gallon (approx. 1.4 l/100 km), would save American consumers about $25 billion a year in fuel costs, reduce carbon dioxide emissions by 140 million metric tons per year, and save more oil than the maximum expected production from Alaska's Arctic National Wildlife Refuge.

The Bureau of Transportation Statistics reports that there are now more vehicles in the United States (214 million) than licensed drivers (190 million). More importantly, those vehicles are used for an average of 1 billion trips per day. Many of us drive now for errands or short shopping trips that might

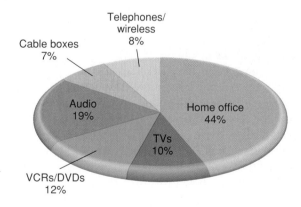

FIGURE 20.5 Typical standby energy consumption by household electrical appliances.
Source: U.S. Department of Energy.

have previously been made on foot. Some of that is due to the design of our cities (chapter 22). Suburban subdivisions have replaced compact downtown centers in most cities. Shopping areas are surrounded by busy streets and vast parking lots that are highly pedestrian unfriendly. But sometimes we use fuel inefficiently simply because we haven't thought about alternatives. The Census Bureau reports that three-quarters of all workers commute alone in private vehicles. Less than 5 percent use public transportation or carpool, and a mere 0.38 percent walk or travel by bicycle.

In response to the 1970s oil price shocks, automobile gas-mileage averages in the United States more than doubled from 13.3 mpg in 1973 to 25.9 mpg in 1988. Unfortunately, falling fuel prices of the 1990s discouraged further conservation. By 2006, the average fuel economy of America's passenger fleet was only 22.1 mpg miles a gallon. Most of this decrease was due to the popularity of SUVs and light trucks, which now account for half of all passenger vehicle sales in the United States. According to the Environmental Protection Agency (EPA), in 2006, SUVs averaged 18.5 miles per gallon (mpg), and pickups averaged 17 mpg, while cars averaged 24.6 mpg. Conservationists argue that efficiency standards should be raised to 44 mpg for cars and 33 mpg for SUVs and light trucks.

What can you do if you want to be environmentally responsible? The cheapest, least environmentally damaging, and healthiest alternative for short trips is walking. You need to get some exercise every day, why not make walking part of it? Next, in terms of minimal expense and environmental impact, is an ordinary bicycle. For trips less than 2 km, it's often quicker to go by bicycle than to find a parking space for your car. While many cities have downgraded their mass transit systems, you might be surprised at the places you can go with this option.

If you're only making short, local trips, why not consider one of the high-efficiency mini cars? The Daimlerchrysler "smart car," for example, has been available in Europe for several years and has now been approved for sale in the United States (fig. 20.6). They get 60 mpg and produce far less pollution than the average full-size car. Easy to maneuver in crowded city streets, two or three of these mini-autos can be parked head-on in a standard parking space.

You probably already know that **hybrid gasoline-electric engines** offer the best fuel economy and lowest emissions of any currently available vehicles. During most city driving, they depend mainly on quiet, emission-free, battery-powered electric motors. A small gasoline engine kicks in to help accelerate or when the batteries need recharging. This extends their range compared to pure electric vehicles. In 2007, the Toyota Prius had the highest mileage rating of any automobile sold in America: 60 mpg (25 km/l) in city driving and 51 mpg (22 km/l) on the highway. Many automakers are now offering hybrid models. Ford claims that half their vehicles will have this option in a few years. You should be aware that some so-called "mild hybrids" only use the electrical generator and battery pack to run accessories, such as video players and computers, not to enhance mileage.

FIGURE 20.6 High-efficiency "smart" cars have been available for many years in Europe. Getting the equivalent of 60 mpg, they produce far less pollution than a typical American car. They are easy to maneuver in crowded city streets, and two can park in a standard parking space.

An even greater savings can be achieved by **plug-in hybrids.** Recharging the batteries from ordinary household current at night can allow these vehicles to travel up to 64 km (40 mi) on the electric motor alone. Since most Americans only drive about 30 km per day, they'd rarely have to buy any gasoline. In most places, electricity costs the equivalent of about 50 cents per gallon. This means that we'll be generating more electricity, but it's easier to capture pollutants and greenhouse gases at a single, stationary power plant than from thousands of individual, mobile vehicles. You can already buy after-market kits to convert an ordinary hybrid into a plug-in, but auto manufacturers threaten to void your warranty if you do so. Several automakers promise to have plug-ins on the market soon.

Diesels already make up about half the autos sold in Europe because of their superior efficiency. A light-weight, four-passenger, diesel roadster that gets up to 150 mpg (62.5 km/l) is now being sold in Europe for about 11,000 euros. Most Americans think of diesels as noisy, smoke-belching, truck engines, but recent advances have made them much cleaner and quieter than they were a generation ago. Ultra low-sulfur diesel fuel and effective tailpipe emission controls could make these engines nearly as clean and energy-efficient as hybrids. Perhaps best of all would be to have flex-fuel or diesel plug-in hybrids that could burn ethanol or biodiesel when they need fuel. That could make us entirely independent from imported oil.

Both the United States and the European Union have announced plans to spend billions of dollars on research and development of hydrogen fuel-cell-powered vehicles. Using hydrogen gas for fuel, these vehicles would produce water as their only waste product. We'll discuss how fuel cells work in more detail later in this chapter. Although prototype fuel cell vehicles are already being tested in several places, even the most optimistic predictions are that it will take at least 20 years for this technology to be mass produced at a reasonable cost. Although hydrogen fuel could be produced with electricity from remote wind or solar facilities, providing a convenient and inexpensive way to get surplus energy to market, most hydrogen currently is created from natural gas, making it no cleaner or more efficient than simply burning the gas directly. While not calling for an end to fuel cell research, conservation groups are urging the government not to abandon other useful technologies, such as hybrid engines and conventional pollution control, while waiting for fuel cells.

Think About It

What barriers do you see to walking, biking, or mass transit in your home town? How could cities become more friendly to sustainable transportation? Why not write a letter to your city leaders or the editor of your newspaper describing your ideas?

Cogeneration produces both electricity and heat

One of the fastest growing sources of new energy is **cogeneration,** the simultaneous production of both electricity and steam or hot water in the same plant. By producing two kinds of useful energy in the same facility, the net energy yield from the primary fuel is increased from 30–35 percent to 80–90 percent. In 1900, half the electricity generated in the United States came from plants that also provided industrial steam or district heating. As power plants became larger, dirtier, and less acceptable as neighbors, they were forced to move away from their customers. Waste heat from the turbine generators became an unwanted pollutant to be disposed of in the environment. Furthermore, long transmission lines, which are unsightly and lose up to 20 percent of the electricity they carry, became necessary.

By the 1970s, cogeneration had fallen to less than 5 percent of our power supplies, but interest in this technology is being renewed. The capacity for cogeneration more than doubled in the 1980s to about 30,000 megawatts (MW). District heating systems are being rejuvenated, and plants that burn municipal wastes are being studied. New combined-cycle coal-gasification plants or "mini-nukes" (chapter 19) offer high efficiency and clean operation that may be compatible with urban locations. Small neighborhood- or apartment building-sized power-generating units are being built that burn methane (from biomass digestion), natural gas, diesel fuel, or coal (fig. 20.7). The Fiat Motor Company makes a small generator for about $10,000 that produces enough electricity and heat for four or five energy-efficient houses. These units are especially valuable for facilities like hospitals or computer centers that can't afford power outages.

Although you may not be buying a new house or car for a few years, and you probably don't have much influence over industrial policy or utility operation, there are things that all of us can do to save energy every day (What Can You Do? p. 453).

FIGURE 20.7 A technician adjusts a gas microturbine that produces on-site heat and electricity for businesses, industry, or multiple housing units.

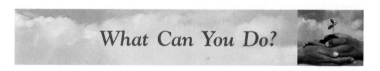

What Can You Do?

Some Things You Can Do to Save Energy

1. Drive less: make fewer trips, use telecommunications and mail instead of going places in person.
2. Use public transportation, walk, or ride a bicycle.
3. Use stairs instead of elevators.
4. Join a car pool or drive a smaller, more efficient car; reduce speeds.
5. Insulate your house or add more insulation to the existing amount.
6. Turn thermostats down in the winter and up in the summer.
7. Weatherstrip and caulk around windows and doors.
8. Add storm windows or plastic sheets over windows.
9. Create a windbreak on the north side of your house; plant deciduous trees or vines on the south side.
10. During the winter, close windows and drapes at night; during summer days, close windows and drapes if using air conditioning.
11. Turn off lights, television sets, and computers when not in use.
12. Stop faucet leaks, especially hot water.
13. Take shorter, cooler showers; install water-saving faucets and showerheads.
14. Recycle glass, metals, and paper; compost organic wastes.
15. Eat locally grown food in season.
16. Buy locally made, long-lasting materials.

20.2 TAPPING SOLAR ENERGY

The sun serves as a giant nuclear furnace in space, constantly bathing our planet with a free energy supply. Solar heat drives winds and the hydrologic cycle. All biomass, as well as fossil fuels and our food (both of which are derived from biomass), results from conversion of light energy (photons) into chemical bond energy by photosynthetic bacteria, algae, and plants. The average amount of solar energy arriving at the top of the atmosphere is 1,330 watts per square meter. About half of this energy is absorbed or reflected by the atmosphere (more at high latitudes than at the equator), but the amount reaching the earth's surface is some 10,000 times all the commercial energy used each year. However, this tremendous infusion of energy comes in a form that, until this century, has been too diffuse and low in intensity to be used except for environmental heating and photosynthesis. But if we could devise cost-effective ways to use this vast power source, we would never again have to burn fossil fuels. Figure 20.8 shows solar energy levels over the United States for a typical summer and winter day.

Solar collectors can be passive or active

Our simplest and oldest use of solar energy is **passive heat absorption,** using natural materials or absorptive structures with no moving parts to simply gather and hold heat. For thousands of years, people have built thick-walled stone and adobe dwellings that slowly collect heat during the day and gradually release

FIGURE 20.9 Taos Pueblo in northern New Mexico uses adobe construction to keep warm at night and cool during the day.

that heat at night (fig. 20.9). After cooling at night, these massive building materials maintain a comfortable daytime temperature within the house, even as they absorb external warmth.

A modern adaptation of this principle is a glass-walled "sunspace" or greenhouse on the south side of a building (fig. 20.10). Incorporating massive energy-storing materials, such as brick walls, stone floors, or barrels of heat-absorbing water into buildings also collects heat to be released slowly at night. An interior, heat-absorbing wall called a Trombe wall is an effective passive

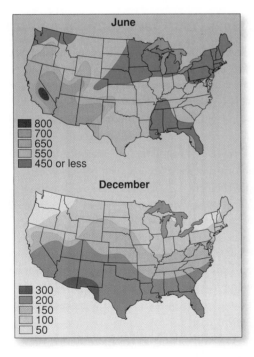

FIGURE 20.8 Average daily solar radiation in the United States in June and December. One langley, the unit for solar radiation, equals 1 cal/cm² of earth surface (3.69 Btu/ft²).
Source: Data from National Weather Bureau, U.S. Department of Commerce.

FIGURE 20.10 The Adam Joseph Lewis Center for Environmental Studies at Oberlin College is designed to be self-sustaining even in northern Ohio's cool, cloudy climate. Large, south-facing windows let in sunlight, while 370 m² of solar panels on the roof generate electricity. A constructed wetland outside and a living machine inside (see fig. 18.27) purify wastewater.

heat collector. Some Trombe walls are built of glass blocks enclosing a water-filled space or water-filled circulation tubes so heat from solar rays can be absorbed and stored, while light passes through to inside rooms.

Active solar systems generally pump a heat-absorbing, fluid medium (air, water, or an antifreeze solution) through a relatively small collector, rather than passively collecting heat in a stationary medium like masonry. Active collectors can be located adjacent to or on top of buildings rather than being built into the structure. Because they are relatively small and structurally independent, active systems can be retrofitted to existing buildings.

A flat black surface sealed with a double layer of glass makes a good solar collector. A fan circulates air over the hot surface and into the house through ductwork of the type used in standard forced-air heating. Alternatively, water can be pumped through the collector to pick up heat for space heating or to provide hot water. Water heating consumes 15 percent of the United States' domestic energy budget, so savings in this area alone can be significant. A simple flat panel with about 5 m^2 of surface can reach 95°C (200°F) and can provide enough hot water for an average family of four almost anywhere in the United States. In California, 650,000 homes now heat water with solar collectors. In Greece, Italy, Israel, and other countries where fuels are more expensive, up to 70 percent of domestic hot water comes from solar collectors. In Europe, municipal solar systems provide district heating for whole cities.

Storing solar energy is problematic

Sunshine doesn't reach us all the time, of course. How can solar energy be stored for times when it is needed? There are a number of options. In a climate where sunless days are rare and seasonal variations are minimal, a small, insulated water tank is a good solar energy storage system. For areas where clouds block the sun for days at a time or where energy must be stored for winter use, a large, insulated bin containing a heat-storing mass, such as stone, water, or clay, provides good solar energy storage. During the summer months, a fan blows the heated air from the collector into the storage medium. In the winter, a similar fan at the opposite end of the bin blows the warm air into the house. During the summer, the storage mass is cooler than the outside air, and it helps cool the house by absorbing heat. During the winter, it is warmer and acts as a heat source by radiating stored heat. In many areas, six or seven months' worth of thermal energy can be stored in 10,000 gallons of water or 40 tons of gravel, about the amount of water in a very small swimming pool or the gravel in two average-sized dump trucks.

20.3 HIGH-TEMPERATURE SOLAR ENERGY

Parabolic mirrors are curved reflecting surfaces that collect light and focus it into a concentrated point. There are two ways to use mirrors to collect solar energy to generate high temperatures. One

FIGURE 20.11 Parabolic mirrors focus sunlight on steam-generating tubes at this power plant in the California desert.

technique uses long curved mirrors focused on a central tube, containing a heat-absorbing fluid (fig. 20.11). Fluid flowing through the tubes reaches much higher temperatures than possible in a basic flat panel collector.

Another high-temperature system uses thousands of smaller mirrors arranged in concentric rings around a tall central tower. The mirrors, driven by electric motors, track the sun and focus its light on a heat absorber at the top of the "power tower" where molten salt is heated to temperatures as high as 500°C (1,000°F), which then drives a steam-turbine electric generator.

Under optimum conditions, a 50 ha (130 acres) mirror array should be able to generate 100 MW of clean, renewable power. The only power tower in the United States is Southern California Edison's Solar II plant in the Mojave Desert east of Los Angeles. Its 2,000 mirrors focused on a 100 m (300 ft) tall tower generates 10 MW or enough electricity for 5,000 homes at an operating cost far below that of nuclear power or oil. We haven't had enough experience with these facilities to know how reliable the mirrors, motors, heat absorbers, and other equipment will be over the long run.

If the entire U.S. electrical output came from such central tower solar steam generators, 60,000 km^2 of collectors would be needed. This is an area about half the size of South Dakota. It is less land, however, than would be strip mined in a 30-year period if all our energy came from coal or uranium. In contrast with windmill farms, which can be used for grazing or farming while also producing energy, mirror arrays need to be carefully protected and are not compatible with other land uses.

Simple solar cookers can save energy

Parabolic mirrors have been tested for home cooking in tropical countries where sunshine is plentiful and other fuels are scarce. They produce such high temperatures and intense light that they are dangerous, however. A much cheaper, simpler, and safer

FIGURE 20.12 A simple box of wood or cardboard, plastic, and foil can help reduce tropical deforestation, improve women's lives, and avoid health risks from smoky fires in developing countries. These inexpensive solar cookers could revolutionize energy use in developing tropical countries.

alternative is the solar box cooker (fig. 20.12). An insulated box costing only a few dollars, with a black interior and a glass or clear plastic lid, serves as a passive solar collector. Several pots can be placed inside at the same time. Temperatures only reach about 120°C (250°F) so cooking takes longer than an ordinary oven. Fuel is free, however, and the family saves hours each day usually spent hunting for firewood or dung. These solar ovens help reduce tropical forest destruction and reduce the adverse health effects of smoky cooking fires.

Utilities are promoting renewable energy

Energy policies in some states include measures to encourage conservation and alternative energy sources. Among these are: (1) "distributional surcharges" in which a small per kWh charge is levied on all utility customers to help renewable energy finance research and development, (2) "renewables portfolio" standards to require power suppliers to obtain a minimum percentage of their energy from sustainable sources, and (3) **green pricing** that allows utilities to profit from conservation programs and charge premium prices for energy from renewable sources. Perhaps your state has some or all of these in place.

Iowa, for example, has a Revolving Loan Fund supported by a surcharge on investor-owned gas and electric utilities. This fund provides low-interest loans for renewable energy and conservation. Many utilities now offer renewable energy options. You agree to pay a couple of dollars extra on your monthly bill, and they promise to use the money to build or buy renewable energy. Buying a 100 kW "block" of wind power provides the same environmental benefits as planting a half acre of trees or not driving an automobile 4,000 km (2,500 mi) per year. Not all green pricing plans are as straightforward as this, however. Some utilities collect the premium rates for facilities that

already exist or for energy sources, such as hydropower projects, that are technically "renewable" but still have adverse environmental effects.

Interestingly, some nonutility companies are investing in sustainable energy. BP, the company formerly known as British Petroleum, now says its initials stand for "Beyond Petroleum." It is investing in solar and other renewables. The company believes that the threat of global climate change requires us to search for new types of energy. Similarly, two European insurance companies, concerned about potential losses from storms and rising sea levels caused by global warming, are investing $5 million in Sunlight Power, a U.S. company that makes and services solar power systems for remote regions of developing countries where electric service is unavailable.

Photovoltaic cells capture solar energy

The photovoltaic cell offers an exciting potential for capturing solar energy in a way that will provide clean, versatile, renewable energy. This simple device has no moving parts, negligible maintenance costs, produces no pollution, and has a lifetime equal to that of a conventional fossil fuel or nuclear power plant.

Photovoltaic cells capture solar energy and convert it directly to electrical current by separating electrons from their parent atoms and accelerating them across a one-way electrostatic barrier formed by the junction between two different types of semiconductor material (fig. 20.13). The photovoltaic effect, which is the basis of these devices, was first observed in 1839 by French physicist

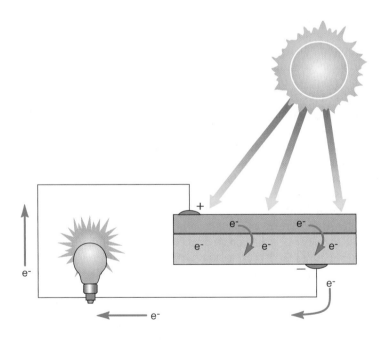

FIGURE 20.13 The operation of a photovoltaic cell. Boron impurities incorporated into the upper silicon crystal layers cause electrons (e-) to be released when solar radiation hits the cell. The released electrons move into the lower layer of the cell, thus creating a shortage of electrons, or a positive charge, in the upper layer and an oversupply of electrons, or negative charge, in the lower layer. The difference in charge creates an electric current in a wire connecting the two layers.

Alexandre-Edmond Becquerel, who also discovered radioactivity. His discovery didn't lead to any useful applications until 1954, when researchers at Bell Laboratories in New Jersey learned how to carefully introduce impurities into single crystals of silicon.

These handcrafted single-crystal cells were much too expensive for any practical use until the advent of the U.S. space program. In 1958, when *Vanguard I* went into orbit, its radio was powered by six palm-sized photovoltaic cells that cost $2,000 per peak watt of output, more than 2,000 times as much as conventional energy at the time. Since then, prices have fallen dramatically. In 1970, they cost $100 per watt; in 2007 they were less than $2.50 per watt. This makes solar energy cost-competitive with other sources in remote areas (more than 1 km from a power line).

Think About It

The 2005 U.S. Energy Bill had more than $12 billion in subsidies for the oil, coal, gas, and nuclear industries, but only one-sixth that much for renewable energy. Where might we be if that ratio had been reversed?

FIGURE 20.14 Roof-mounted solar panels (shiny area) can generate enough electricity for a house full of efficient appliances. On sunny days, this array can produce a surplus to sell back to the utility company, making it even more cost efficient.

During the last 25 years, the efficiency of energy capture by photovoltaic cells has increased from less than 1 percent of incident light to more than 15 percent under field conditions and over 75 percent in the laboratory. Promising experiments are under way using exotic metal alloys, such as gallium arsenide, and semiconducting polymers of polyvinyl alcohol, which are more efficient in energy conversion than silicon crystals. Photovoltaic prices are now dropping about 7 percent per year. When they reach $1 per watt (perhaps by 2020) their electricity should be competative with nuclear or coal-fired plants.

One of the most promising developments in photovoltaic cell technology in recent years is the invention of **amorphous silicon collectors.** First described in 1968 by Stanford Ovshinky, a self-taught inventor from Detroit, these noncrystalline silicon semiconductors can be made into lightweight, paper-thin sheets that require much less material than conventional photovoltaic cells. They also are vastly cheaper to manufacture and can be made in a variety of shapes and sizes, permitting ingenious applications. Roof tiles with photovoltaic collectors layered on their surface already are available (fig. 20.14). Even flexible films can be coated with amorphous silicon collectors. Silicon collectors already are providing power to places where conventional power is unavailable, such as lighthouses, mountaintop microwave repeater stations, villages on remote islands, and ranches in the Australian outback.

You probably already use amorphous silicon photovoltaic cells. They are being built into light-powered calculators, watches, toys, photosensitive switches, and a variety of other consumer products. Japanese electronic companies presently lead in this field, having foreseen the opportunity for developing a market for photovoltaic cells. This market is already more than $100 million per year. Japanese companies now have home-roof arrays capable of providing all the electricity needed for a typical home at prices in some areas

competitive with power purchased from a utility. And Shanghai, China, recently announced a plan to install photovoltaic collectors on 100,000 roofs. This is expected to generate 430 million kWh annually and replace 20,000 tons of coal per year.

The world market for solar energy is expected to grow rapidly in the near future, especially in remote places where conventional power isn't available. At least 2 billion people around the world now have no access to electricity. Most would like to have a modern power source if it were affordable. They may be able to enjoy the benefits of electrical power without the whole complex of power plants, transmission lines, air pollution, and utility companies.

Think about how solar power could affect your future energy independence. Imagine the benefits of being able to build a house anywhere and having a cheap, reliable, clean, quiet source of energy with no moving parts to wear out, no fuel to purchase, and little equipment to maintain. You could have all the energy you need without commercial utility wires or monthly energy bills. Coupled with modern telecommunications and information technology, an independent energy source would make it possible to live in the countryside and yet have many of the employment and entertainment opportunities and modern conveniences available in a metropolitan area.

Electrical energy is difficult and expensive to store

Storage is a problem for photovoltaic generation as well as other sources of electric power. Traditional lead-acid batteries are heavy and have low energy densities; that is, they can store only moderate amounts of energy per unit mass or volume. Acid from batteries is corrosive and lead from smelters or battery manufacturing is a serious health hazard for workers who handle these materials. A

FIGURE 20.15 A large battery array, together with inverters and regulators, store photovoltaic energy on the Manzanita Indian Reservation in California.

typical lead-acid battery array sufficient to store several days of electricity for an average home would cost about $5,000 and weigh 3 or 4 tons (fig. 20.15). Still, some communities are encouraging use of electric vehicles because they produce zero emissions.

Other types of batteries also have drawbacks. Metal-gas batteries, such as the zinc-chloride cell, use inexpensive materials and have relatively high-energy densities, but have shorter lives than other types. Sodium-sulfur batteries have considerable potential for large-scale storage. They store twice as much energy in half as much weight as lead-acid batteries. They require an operating temperature of about 300°C (572°F) and are expensive to manufacture. Alkali-metal batteries have a high storage capacity but are even more expensive. Lithium-ion batteries have very long lives and store more energy than other types, but are the most expensive of all. Recent advances in thin-film technology may hold promise for future energy storage but often require rare, toxic elements that are both costly and dangerous.

Another strategy is to store energy in a form that can be turned back into electricity when needed. Pumped-hydro storage involves pumping water to an elevated reservoir at times when excess electricity is available. The water is released to flow back down through turbine generators when extra energy is needed. Using a similar principle, pressurized air can be pumped into such reservoirs as natural caves, depleted oil and gas fields, abandoned mines, or special tanks. An Alabama power company

currently uses off-peak electricity to pump air at night into a deep salt mine. By day, the air flows back to the surface through turbines, driving a generator that produces electricity. Cool night air is heated to 1,600°F by compression plus geothermal energy, increasing pressure and energy yield.

An even better way to use surplus electricity may be electrolytic decomposition of water to H_2 and O_2. These gases can be liquefied (like natural gas) at −252°C (−423°F), making them easier to store and ship than most forms of energy. They are highly explosive, however, and must be handled with great care. They can be burned in internal combustion engines, producing mechanical energy, or they can be used to power fuel cells, which produce more electrical energy. There is a concern that if hydrogen escapes, it could destroy stratospheric ozone.

Flywheels are the subject of current experimentation for energy storage. Massive, high-speed flywheels, spinning in a nearly friction-free environment, store large amounts of mechanical energy in a small area. This energy is convertible to electrical energy. It is difficult, however, to find materials strong enough to hold together reliably when spinning at high speed. Flywheels have a disconcerting tendency to fail explosively and unexpectedly, sending shrapnel flying in all directions. Still, this might represent a useful technology if these problems can be overcome.

20.4 FUEL CELLS

Rather than store and transport energy, another alternative would be to generate it locally, on demand. **Fuel cells** are devices that use ongoing electrochemical reactions to produce an electric current. They are very similar to batteries except that rather than recharging them with an electrical current, you add more fuel for the chemical reaction.

Fuel cells are not new; the basic concept was recognized in 1839 by William Grove, who was studying the electrolysis of water. He suggested that rather than use electricity to break apart water and produce hydrogen and oxygen gases, it should be possible to reverse the process by joining oxygen and hydrogen to produce water and electricity. The term "fuel cell" was coined in 1889 by Ludwig Mond and Charles Langer, who built the first practical device using a platinum catalyst to produce electricity from air and coal gas. The concept languished in obscurity until the 1950s when the U.S. National Aeronautics and Space Administration (NASA) was searching for a power source for spacecraft. Research funded by NASA eventually led to development of fuel cells that now provide both electricity and drinkable water on every space shuttle flight. The characteristics that make fuel cells ideal for space exploration—small size, high efficiency, low emissions, net water production, no moving parts, and high reliability—also make them attractive for a number of other applications.

All fuel cells have similar components

All fuel cells consist of a positive electrode (the cathode) and a negative electrode (the anode) separated by an electrolyte, a material that allows the passage of charged atoms, called ions, but is

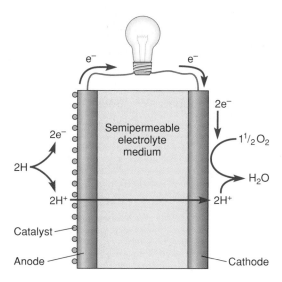

FIGURE 20.16 Fuel cell operation. Electrons are removed from hydrogen atoms at the anode to produce hydrogen ions (protons) that migrate through a semipermeable electrolyte medium to the cathode, where they reunite with electrons from an external circuit and oxygen atoms to make water. Electrons flowing through the circuit connecting the electrodes create useful electrical current.

impermeable to electrons (fig. 20.16). In the most common systems, hydrogen or a hydrogen-containing fuel is passed over the anode while oxygen is passed over the cathode. At the anode, a reactive catalyst, such as platinum, strips an electron from each hydrogen atom, creating a positively charged hydrogen ion (a proton). The hydrogen ion can migrate through the electrolyte to the cathode, but the electron is excluded. Electrons pass through an external circuit, and the electrical current generated by their passage can be used to do useful work. At the cathode, the electrons and protons are reunited and combined with oxygen to make water.

The fuel cell provides direct-current electricity as long as it is supplied with hydrogen and oxygen. For most uses, oxygen is provided by ambient air. Hydrogen can be supplied as a pure gas, but storing hydrogen gas is difficult and dangerous because of its volume and explosive nature. Liquid hydrogen takes far less space than the gas, but must be kept below $-250°C$ ($-400°F$), not a trivial task for most mobile applications. The alternative is a device called a **reformer** or converter that strips hydrogen from fuels such as natural gas, methanol, ammonia, gasoline, ethanol, or even vegetable oil. Many of these fuels can be derived from sustainable biomass crops. Even methane effluents from landfills and wastewater treatment plants can be used as a fuel source. Where a fuel cell can be hooked permanently to a gas line, hydrogen can be provided by solar, wind, or geothermal facilities that use electricity to hydrolyze water.

A fuel cell run on pure oxygen and hydrogen produces no waste products except drinkable water and radiant heat. When a reformer is coupled to the fuel cell, some pollutants are released (most commonly carbon dioxide), but the levels are typically far less than conventional fossil fuel combustion in a power plant or automobile engine. Although the theoretical efficiency of electrical generation of a fuel cell can be as high as 70 percent, the actual yield is closer to 40 or 45 percent. This is about the same as an integrated gasification combined cycle (IGCC) plant (chapter 19). On the other hand, the quiet, clean operation and variable size of fuel cells make them useful in buildings where waste heat can be captured for water heating or space heating. A new 45-story office building at 4 Times Square, for example, has two 200-kilowatt fuel cells on its fourth floor that provide both electricity and heat. This same building has photovoltaic panels on its façade, natural lighting, fresh air intakes to reduce air conditioning, and a number of other energy conservation features.

U.S. automakers have focused most of their efforts to improve efficiency of fuel cells. While this would reduce pollution, eliminate our dependence on imported oil, and make a good use for wind or solar energy, critics claim that it will take 20 years or more to develop automotive fuel cells and build the necessary infrastructure. We should concentrate, instead, on hybrid engines, they say. Iceland, with no fossil fuels, but abundant geothermal energy, is determined to be the world's first hydrogen-based economy. They have one hydrogen filling station and a fleet of fuel cell buses.

The current from a fuel cell is proportional to the size (area) of the electrodes, while the voltage is limited to about 1.23 volts per cell. A number of cells can be stacked together until the desired power level is reached. A fuel cell stack that provides almost all of the electricity needed by a typical home (along with hot water and space heating) would be about the size of a refrigerator. A 200 kilowatt unit fills a medium-size room and provides enough energy for 20 houses or a small factory (fig. 20.17). Tiny fuel cells running on methanol may soon be used in cell phones, pagers, toys, computers, videocameras, and other appliances now run by batteries. Rather than buy new batteries or spend hours

FIGURE 20.17 The Long Island Power Authority has installed 75 stationary fuel cells to provide reliable backup power.

TABLE 20.2

Fuel Cell Types

	Proton Electrolyte Membrane	Phosphoric Acid	Molten Carbonate	Solid Oxide
Electrolyte	Semipermeable organic polymer	Phosphoric acid	Liquid carbonate	Solid-oxide ceramic
Charge carrier	H^+	H^+	$CO_3^=$	$O^=$
Catalyst	Platinum	Platinum	Nickel	Perovskites (calcium titanate)
Operating temperature, °C	80	200	650	1,000
Cell material	Carbon or metal-based	Graphite-based	Stainless steel	Ceramic
Efficiency (percent)	Less than 40	40 to 50	50 to 60	More than 60
Heat cogeneration	None	Low	High	High
Status	Demonstration systems	Commercially available	Demonstration systems	Under development

Source: Alan C. Lloyd, 1999.

recharging spent ones, you might just add an eyedropper of methanol every few weeks to keep your gadgets operating.

Several different electrolytes can be used in fuel cells

Each fuel cell type has advantages and disadvantages (table 20.2). The design being developed for use in cars, buses, and trucks is called a proton exchange membrane (PEM). The membrane is a thin semipermeable layer of an organic polymer containing sulfonic acid groups that facilitate passage of hydrogen ions but block electrons and oxygen. The surface of the membrane is dusted with tiny particles of platinum catalyst. These cells have the advantage of being lightweight and operating at a relatively low temperature (80°C or 176°F). The fuel efficiency of PEM systems is typically less than 40 percent. Buses equipped with PEM stacks have been demonstrated in Chicago, Miami, and Vancouver, BC. DaimlerChrysler has a 5-passenger NECAR-4. Its 500 kg (1,100 lb) PEM stack costs $30,000 (compared to about $3,000 for a conventional gasoline engine) and produces performance comparable to most passenger cars, but takes up considerable interior room.

For stationary electrical generation, the most common fuel cell design uses phosphoric acid immobilized in a porous ceramic matrix as the electrolyte. Because this system operates at higher temperatures than PEM cells, less platinum is needed for the catalyst. It has a higher efficiency, 40 to 50 percent, but is heavier and larger than PEM cells. It also is less sensitive to carbon dioxide contamination than other designs. Hundreds of 200 kilovolt phosphoric acid fuel cells have been installed around the world. Some have run for decades. They supply dependable electricity in remote locations without the spikes and sags and risk of interruption common in utility grids. The largest fuel cell ever built was an 11 MW unit in Japan, that provides enough electricity for a small town.

A Central Park police station in New York City has a 200-kilowatt phosphoric acid fuel cell. The station is located in the middle of the park, so bringing in new electric lines would have cost $1.2 million and would have disrupted traffic and park use for months. A diesel generator was ruled out as too noisy and polluting. Solar photovoltaic panels were thought to be too obtrusive. A small, silent fuel cell provided just the right solution.

Carbonate fuel cells use an inexpensive nickel catalyst and operate at 650°C (1,200°F). The electrode is a very hot (thus the name molten carbonate) solution trapped in a porous ceramic. The charge carrier is carbonate ion, which is formed at the cathode where oxygen and carbon dioxide react in the presence of a nickel oxide catalyst. Migrating through the electrolyte, the carbonate ion reacts at the anode with hydrogen and carbon monoxide to release electrons. The high operating temperature of this design means that it can reform fuels internally and ionize hydrogen without expensive catalysts. Heat cogeneration is very good, but the high temperature makes these units more difficult to operate. It takes hours for carbonate fuel cells to get up to operating temperature, so they aren't suitable for short-term, quick-response uses.

The least developed fuel cell design is called solid oxide, because it uses a coated zirconium ceramic as an electrolyte. Oxygen ions formed by the titanium catalyst carry the charge across the electrolyte. Operating temperatures are 1,000°C (1,800°F). They have the highest fuel efficiency of any current design, but mass production of components has not yet been mastered, and these cells are still in the experimental stage.

20.5 Energy From Biomass

Photosynthetic organisms have been collecting and storing the sun's energy for more than 2 billion years. Plants capture about 0.1 percent of all solar energy that reaches the earth's surface. That kinetic energy is transformed, via photosynthesis, into chemical bonds in organic molecules (chapter 3). A little more than half of the energy that plants collect is spent in such metabolic activities as pumping water and ions, mechanical movement, maintenance of cells and tissues, and reproduction; the rest is stored in biomass.

FIGURE 20.18 Firewood is an important resource in this Siberian village near Lake Baikal.

The magnitude of this resource is difficult to measure. Most experts estimate useful biomass production at 15 to 20 times the amount we currently get from all commercial energy sources. It would be ridiculous to consider consuming all green plants as fuel, but biomass has the potential to become a prime source of energy. It has many advantages over nuclear and fossil fuels because of its renewability and easy accessibility. Biomass resources used as fuel include wood, wood chips, bark, branches, leaves, starchy roots, and other plant and animal materials.

We can burn biomass

Wood fires have been a primary source of heating and cooking for thousands of years. As recently as 1850, wood supplied 90 percent of the fuel used in the United States. Wood now provides less than 1 percent of the energy in the United States, but in many of the poorer countries of the world, wood and other biomass fuels provide up to 95 percent of all energy used (fig. 20.18). The 1,500 million cubic meters of fuelwood collected in the world each year is about half of all wood harvested.

In northern industrialized countries, wood burning has increased since 1975 in an effort to avoid rising oil, coal, and gas prices. Most of these northern areas have adequate wood supplies to meet demands at current levels, but problems associated with wood burning may limit further expansion of this use. Inefficient and incomplete burning of wood in fireplaces and stoves produces smoke laden with fine ash and soot and hazardous amounts of carbon monoxide (CO) and hydrocarbons. In valleys where inversion layers trap air pollutants, the effluent from wood fires can present a major source of air quality degradation and health risk. Polycyclic aromatic compounds produced by burning are especially worrisome because they are carcinogenic (cancer-causing).

In Oregon's Willamette Valley or in the Colorado Rockies, where woodstoves are popular and topography concentrates contaminants, as much as 80 percent of air pollution on winter days is attributable to wood fires. Several resort towns, such as Vail, Aspen, and Telluride, Colorado, have banned installation of new

FIGURE 20.19 This district heating and cooling plant supplies steam heat, air conditioning, and electricity to three-quarters of downtown St. Paul, Minnesota, at about twice the efficiency of a remote power plant. It burns mainly urban tree trimmings and, thus, is carbon neutral.

woodstoves and fireplaces because of high pollution levels. Oregon, Colorado, and Vermont now have emission standards for new woodstoves. The Environmental Protection Agency ranks wood burners high on a list of health risks to the general population, and standards are being considered to regulate the use of woodstoves nationwide and to encourage homeowners to switch to low-emission models.

Highly efficient and clean-burning woodstoves are available but expensive. Running exhaust gases through heat exchangers recaptures heat that would escape through the chimney, but if flue temperatures drop too low, flammable, tarlike creosote deposits can build up, increasing fire risk. A better approach is to design combustion chambers that use fuel efficiently, capturing heat without producing waste products. Brick-lined fireboxes with afterburner chambers to burn gaseous hydrocarbons do an excellent job. Catalytic combusters, similar to the catalytic converters on cars, also are placed inside stovepipes. They burn carbon monoxide and hydrocarbons to clean emissions and to recapture heat that otherwise would have escaped out the chimney in the chemical bonds of these molecules.

Wood chips, sawdust, wood residue, and other plant materials are being used in some places in the United States and Europe as a substitute for coal and oil in industrial boilers. St. Paul, Minnesota, for example has a district heating plant that supplies hot water to heat 25 million ft^2 of office and living space in 75 percent of all downtown buildings (fig. 20.19). This facility

FIGURE 20.20 A charcoal market in Ghana. Firewood and charcoal provide the main fuel for billions of people. Forest destruction results in wildlife extinction, erosion, and water loss.

burns 275,000 tons of wood per year (mostly from urban tree trimming) but has a flexible boiler that can burn a wide variety of fuels. It generates 25 MW of electricity and has a net efficiency of about 75 percent, or twice that of a typical coal-fired power plant.

Pollution-control equipment is easier to install and maintain in a central power plant like this than in individual home units.

Fuelwood is in short supply in many less-developed countries

Two billion people—about 40 percent of the total world population—depend on firewood and charcoal as their primary energy source. Of these people, three-quarters (1.5 billion) do not have an adequate, affordable supply. Most of these people are in the less-developed countries where they face a daily struggle to find enough fuel to warm their homes and cook their food. The problem is intensifying because rapidly growing populations in many developing countries create increasing demands for firewood and charcoal from a diminishing supply.

As firewood becomes increasingly scarce, women and children, who do most of the domestic labor in many cultures, spend more and more hours searching for fuel (fig. 20.20). In some places, it now takes eight hours, or more, just to walk to the nearest fuelwood supply and even longer to walk back with a load of sticks and branches that will only last a few days.

For people who live in cities, the opportunity to scavenge firewood is generally nonexistent and fuel must be bought from merchants. This can be ruinously expensive. In Addis Ababa, Ethiopia, 25 percent of household income is spent on wood for cooking fires. A circle of deforestation has spread more than 160 km (100 mi) around some major cities in India, and firewood costs up to ten times the price paid in smaller towns.

The poorest countries such as Ethiopia, Bhutan, Burundi, and Bangladesh depend on biomass for 90 percent of their energy. Often, the harvest is sustainable, consisting of deadwood, branches, trimmings, and shrubs. In Pakistan, for example, some

4.4 million tons of twigs and branches and 7.7 million tons of shrubs and crop residue are consumed as fuel each year with destruction of very few living trees.

In countries where fuel is scarce, however, desperate people often chop down anything that will burn. In Haiti, for instance, more than 90 percent of the once-forested land has been almost completely denuded and people cut down even valuable fruit trees to make charcoal they can sell in the marketplace. It is estimated that the 1,700 million tons of fuelwood now harvested each year globally is at least 500 million tons less than is needed. By 2025, if current trends continue, the worldwide demand for fuelwood is expected to be about twice current harvest rates while supplies will not have expanded much beyond current levels. Some places will be much worse than this average. In some African countries such as Mauritania, Rwanda, and the Sudan, firewood demand already is ten times the sustainable yield. Reforestation projects, agroforestry, community woodlots, and inexpensive, efficient, locally produced woodstoves could help alleviate expected fuelwood shortages in many places.

Dung and methane provide power

Where wood and other fuels are in short supply, people often dry and burn animal manure. This may seem like a logical use of waste biomass, but it can intensify food shortages in poorer countries. Not putting this manure back on the land as fertilizer reduces crop production and food supplies. In India, for example, where fuelwood supplies have been chronically short for many years, a limited manure supply must fertilize crops and provide household fuel. Cows in India produce more than 800 million tons of dung per year, more than half of which is dried and burned in cooking fires. If that dung were applied to fields as fertilizer, it could boost crop production of edible grains by 20 million tons per year, enough to feed about 40 million people.

When cow dung is burned in open fires, more than 90 percent of the potential heat and most of the nutrients are lost. Compare that to the efficiency of using dung to produce methane gas, an excellent fuel. In the 1950s, simple, cheap methane digesters were designed for villages and homes, but they were not widely used. Now, however, 6 million Chinese households use biogas for cooking and lighting. Two large municipal facilities in Nanyang will soon provide fuel for more than 20,000 families. Perhaps other countries will follow China's lead.

Methane gas is the main component of natural gas. It is produced by anaerobic decomposition (digestion by anaerobic bacteria) of any moist organic material. Many people are familiar with the fact that swamp gas is explosive. Swamps are simply large methane digesters, basins of wet plant and animal wastes sealed from the air by a layer of water. Under these conditions, organic materials are decomposed by anaerobic (oxygen-free) rather than aerobic (oxygen-using) bacteria, producing flammable gases instead of carbon dioxide. This same process may be reproduced artificially by placing organic wastes in a container and providing warmth and water

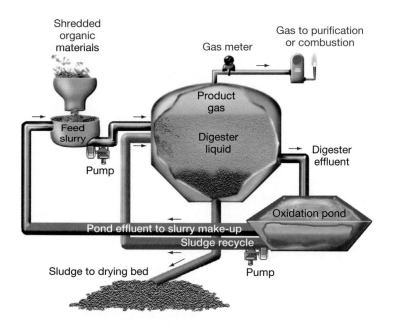

FIGURE 20.21 Continuous unit for converting organic material to methane by anaerobic fermentation. One kilogram of dry organic matter will produce 1–1.5 m^3 of methane, or 2,500–3,600 million calories per metric ton.

FIGURE 20.22 Harvesting marsh reeds (*Phragmites* sp.) in Sweden as a source of biomass fuel. In some places, biomass from wood chips, animal manure, food-processing wastes, peat, marsh plants, shrubs, and other kinds of organic material make a valuable contribution to energy supplies. Care must be taken, however, to avoid environmental damage in sensitive areas.

(fig. 20.21). Bacteria are ubiquitous enough to start the culture spontaneously.

Burning methane produced from manure provides more heat than burning the dung itself, and the sludge left over from bacterial digestion is a rich fertilizer, containing healthy bacteria as well as most of the nutrients originally in the dung. Whether the manure is of livestock or human origin, airtight digestion also eliminates some health hazards associated with direct use of dung, such as exposure to fecal pathogens and parasites.

How feasible is methane—from manure or from municipal sewage—as a fuel resource in developed countries? Methane is a clean fuel that burns efficiently. Any kind of organic waste material: livestock manure, kitchen and garden scraps, and even municipal garbage and sewage can be used to generate gas. In fact, municipal landfills are active sites of methane production, contributing as much as 20 percent of the annual output of methane to the atmosphere. This is a waste of a valuable resource and a threat to the environment because methane absorbs infrared radiation and contributes to the greenhouse effect (chapter 15). About 300 landfills in the United States currently burn methane and generate enough electricity together for a million homes. Another 600 landfills have been identified as potential sources for methane development. Hydrologists worry, however, that water will be pumped into landfills to stimulate fermentation, thus increasing the potential for groundwater contamination.

Cattle feedlots and chicken farms in the United States are a tremendous potential fuel source. Collectible crop residues and feedlot wastes each year contain 4.8 billion gigajoules (4.6 quadrillion BTUs) of energy, more than all the nation's farmers use. The Haubenschild farm in central Minnesota, for

instance, uses manure from 850 Holsteins to generate all the power needed for their dairy operation and still have enough excess electricity for an additional 80 homes. In January 2001, the farm saved 35 tons of coal, 1,200 gallons of propane, and made $4,380 from electric sales.

Municipal sewage treatment plants also routinely use anaerobic digestion as a part of their treatment process, and many facilities collect the methane they produce and use it to generate heat or electricity for their operations. Although this technology is well-developed, its utilization could be much more widespread.

Biofuels could replace some oil-based energy

As the opening story of this chapter shows, biomass can make a substantial contribution to renewable energy supplies. The islands of Samsø and Ærø get about half their space heating from biomass, both from straw and other crop wastes and from energy crops, such as reeds and elephant grass growing on land unsuitable for crops (fig. 20.22). As we discussed earlier, burning these crops in a industrial boiler for district heating makes it easier to install and maintain pollution-control equipment than in individual stoves. Most plant material has low sulfur, so it doesn't contribute to acid rain. And because it burns at a lower temperature than coal, it doesn't create as much nitrogen oxides. Of course, these crops are carbon neutral—that is, they absorb as much CO_2 in growing as they emit when burned. Wood chips and straw make great fuel for a boiler, but what about fuel for vehicles? You don't want to drive around with a hay bale in the backseat.

Biofuels, ethanol or methanol made from plant materials, or diesel fuel made from vegetable oils or animal fat, could meet much of our transportation needs. Alcohols can be burned directly in engines adapted to use these fuels, or they can be mixed with gasoline to be used in any normal engine. Ethanol makes gasoline burn cleaner and most states now require that 5 to 10 percent of an oxygenated additive, such as ethanol, be added to gasoline. Most ethanol is now made from grain, but it can also be made from wood chips, straw, or any **cellulosic** material (composed largely of cellulose).

Ethanol could be a solution to grain surpluses and bring a higher price for farmers than the food market offers. As one Midwestern politician pointed out, "The WTO is putting pressure on Europe and America to reduce farm subsidies, but there's no rule that forbids subsidies for energy crops." There has been much debate about whether ethanol represents a net energy gain (Exploring Science p. 465). Modern state-of-the-art ethanol refineries claim at least a 35 percent gain in ethanol made from corn (maize) over the energy content of the inputs. Removing high-value nutraceuticals, such as L-ribose and levoglucoseon before fermentation makes ethanol production much more profitable. Ethanol currently can be made for about $1 per gal (3.8l) in the United States, and about half that amount in Brazil. Local residents sometimes complain, however, about odors from ethanol plants.

The United States already has about 5 million **flex-fuel vehicles** that can burn variable mixtures of ethanol and gasoline, but few of them currently use anything beyond the standard 5 or 10 percent ethanol blend. Owners either don't know that they could burn up to 85 percent ethanol or they don't have access to biofuel. Vehicle manufacturers continue to make these engines because a special clause in the U.S. Energy Bill allows them to sell more gas guzzlers if they also produce flex-fuel vehicles. The ironic effect is that the existence of these alternative fuel vehicles actually results in more oil consumption.

Brazil is the current world leader in crop-based ethanol production. In 2006, Brazil refined some 18 billion l (4.7 billion gal) of ethanol, mostly from sugarcane waste called bagasse). By 2012, Brazil expects to double this output and to be exporting billions of liters of biofuels to other countries. Almost all new vehicles in Brazil now have flex-fuel engines. Most motorists mix their own combination (about 80 percent ethanol is most common) right at the pump (fig. 20.23). They are motivated to do so, because ethanol is about 40 percent cheaper than gasoline. Rapidly increasing soy and corn production in Brazil (chapter 7) could also provide vegetable oils for diesel fuel. Brazil gets about three times as much ethanol from a ton of sugarcane than the U.S. gets from a ton of shelled corn.

Ethanol production is growing explosively in the United States. Currently 101 biorefineries are operating and 39 more are under construction. Some people think we may have too many of these factories. In 2006, the U.S. produced 17.4 billion l of ethanol (4.6 billion gal), mostly from shelled corn. Like Brazil, the U.S. hopes to double this yield by 2012. Most American states now require small amounts of ethanol to improve air quality. Some states mandate higher ethanol content as a way of reducing oil dependence and supporting farmers. The highest of

FIGURE 20.23 Alcohol, mostly from sugarcane waste, is cheaper than either gasoline or diesel fuel in Brazil. All passenger vehicles in Brazil are required to use at least 20 percent ethanol.

these is in Minnesota which requires 20 percent ethanol in gasoline, the highest concentration considered feasible without modifying ordinary engines.

Should we use food for fuel?

Many Midwestern states see ethanol as a bonanza for farmers, but others doubt the wisdom of using food for fuel. Already, the 2 billion bushels of corn used per year for fuel production in the U.S. has increased prices by about 50 percent. This is great for farmers, but it has driven up the price of tortillas in Mexico (since NAFTA, the Mexican maize market has been dominated by the U.S.), a burden for poor people there. Most soy and corn grown in America are used for animal feed, so diverting a significant portion of the crop to fuel is driving up meat, milk, and egg prices. On the other hand, as much as 80 percent of the starting grain weight can be recovered as high-protein animal feed after fermentation, so ethanol production may not affect meat prices very much in the long run. Critics point out that even if we convert all our soy and corn crop to fuel, it would replace only about 12 percent of our current gasoline consumption. Increasing fuel economy standards by 12 percent would reduce oil consumption just as much, and would save taxpayers about $10 billion per year in ethanol subsidies.

Energy crops, such as switch grass, cattails, and hybrid poplar, could be grown specifically as an energy source on marginal

Biofuels (alcohol refined from plant material or diesel fuel made from vegetable oils or animal fats) are thought by many people to be the answer to both our farm crisis and our fuel needs. But do these crops represent a net energy gain? Or does it take more fossil fuel energy to grow, harvest, and process crop-based biofuels than you get back in the finished product? Some researchers calculate that crops can produce a net energy gain, while others insist that we can't even break even in converting biomass to fuel.

David Pimental from Cornell University, for example, is one of the most vocal critics of biofuels. He calculates that it takes 29 percent more energy to refine ethanol from corn than it yields. Soy-based biodiesel is equally inefficient, he maintains, and cellulose-based biofuels are even worse. Cellulosic crops take at least 50 percent more energy than they produce as ethanol, according to his calculations.

Among others who dispute these calculations are Bruce Dale of Michigan State University and John Sheehan from the U.S. National Renewable Energy Laboratory, who argue that biofuels produced by modern techniques represent a positive energy return. Dramatic improvements in farming productivity, they believe, coupled with much greater efficiency in ethanol fermentation now yield about 35 percent more energy in ethanol from corn than in the inputs, and cellulosic crops would be at least four times better.

A valuable addition to this debate comes from the work of ecologist David Tilman and his colleagues at the Cedar Creek Natural

History Area in Minnesota. This group grew experimental plots containing 1, 2, 4, 8, or 16 perennial herbaceous grassland species on degraded farmland for ten years. Each of the 152 plots contained a random mixture and number of species. All plots were unfertilized and irrigated only during establishment. Aboveground biomass was collected every fall, from each plot. Energy yields were calculated from dry biomass. Using numbers from other sources, the authors calculate energy outputs from various fuel derivation methods.

The most diverse plots (16 species), which they call low-input high-diversity grassland mixtures (LIHD), had the highest biomass yields (68.1 GJ/ha/yr) in this experiment. The fossil fuel energy needed to grow, harvest, and transport this biomass to a biofuel production facility is calculated to be 4.0 GJ/ha/yr. If the biomass were burned directly in an electric generating station, Tilman calculates, it would yield about 22 GJ/ha/yr, or about

5.5 times as much energy as the inputs. Cellulosic ethanol production would have about the same net energy yield. Production of biomass synfuel in an integrated gasification combined cycle (IGCC) facility similar to that described in chapter 19 would have the best net energy ratio (8.1) of any biofuel. These polycultures of native species can be grown on marginal farmland, so they don't need to displace food crops. Another great advantage is that prairie plants store carbon in their roots. While monocultures didn't have a significant net carbon storage, the most diverse mixtures have a net carbon storage of 4.4 Mg/ha/yr. Thus, they take up more carbon than is released during fuel production.

Corn-based ethanol produces more fuel energy per hectare than do LIHD mixtures, but it also requires far higher inputs for crop and fuel production. Thus, the net energy ratio for corn-based ethanol is far lower than that for native prairie species.

Biomass Fuel Efficiency

Fuel	Inputs (GJ/ha)	Outputs (GJ/ha)	Net Energy Ratio
Corn ethanol	75.0	93.8	1.2
Soy ethanol	15.0	28.9	1.9
Cellulosic electricity	4.0	22.0	5.5
Cellulosic ethanol	4.0	21.8	5.4
Cellulosic synfuel	4.0	32.4	8.1

Source: Tilman, et al. 2006. *Science* 314:1598.

land (fig. 20.24). These perennial crops require less cultivation (and therefore result in less erosion) than annual row crops. They may also reduce fertilizer and pesticide use. In his study of biodiversity and productivity in native prairie species, ecologist David Tilman found that experimental plots with the highest diversity (16 species) accumulated more than twice as much aboveground biomass as single species plots. Furthermore, these high-diversity plots also store 160 percent more carbon in the soil than do single species. These native plants are adapted to low-nitrogen soil and dry conditions, so they don't need irrigation, fertilizer, or pesticides. Tilman calls these mixed polycultures of perennial native species **low-input high-diversity biofuels,** and suggests they could be a good way to use marginal land that isn't suitable for crop production.

Cellulosic ethanol has a much better net yield than corn or soy fermentation. In the laboratory, up to 8.1 units of energy are produced from cellulosic crops for every unit invested. The biggest problem with fermenting cellulosic crops is the insoluble lignin that holds cells together. It's hoped that bacteria can be engineered to solubilize the lignin. If it could be removed, it would be a valuable by-product. So far, there are no commercial-scale cellulosic ethanol plants in North America, but the Department of Energy announced grants in 2007 totaling $385 million for six biorefinery projects. These pilot plants will use a variety of feedstocks including rice and wheat straw, milo stubble, switchgrass, organic waste from landfills, corn cobs and stalks, and woody crops.

Water is another worry about the sustainability of biofuels. Currently, it takes 3 to 6 liters of water for each liter of ethanol

FIGURE 20.24 Some of these 3-year-old hybrid poplars are 6 m tall. If they continue to grow at this rate, and cellulosic ethanol production becomes cost-effective, they may make a good energy crop for marginal land.

produced. In many plains states where grain is grown, there isn't enough water for both agriculture and fuel production. In some areas, plans for new processing facilities are being scaled back because of water shortages. So, although biofuels could make a useful contribution to our fuel supplies, especially in the interim between the age of fossil fuels and whatever comes next, there are concerns about the social and environmental impacts of a sudden switch to these alternate energy sources.

Just about anything organic, from turkey entrails to cow dung or sugarcane waste can be used to make biodiesel, and with oil over $70 a barrel in recent years, just about everything is. Factories have been set up next to meat-processing plants to convert waste material into biodiesel. They work well, but aren't currently economical given U.S. policies that favor fossil fuels. These plants are much more profitable in Europe, where waste disposal and carbon emission limits are more strict. Many people are converting diesel engines to burn discarded restaurant deep frying fat. They say that these "greasel" engines leave a pleasant whiff of french fries or egg rolls as they roll along.

A much greater potential for diesel fuel is from vegetable oil, which can be burned in most diesel engines without any processing except filtering. The European Union currently consumes about 4 million metric tons (1 billion gal) of biodiesel fuel, mostly made from rapeseed oil (called canola oil in North America). Plans are to triple this production by 2012 to 10 percent of all fuel consumed in Europe. Planners expect this move to create 250,000 jobs, reduce oil imports, improve air quality, save consumers billions of Euros, and reduce CO_2 emissions by 200 million tons per year. Already, however, this increased use of vegetable oils has reduced the amount available for human consumption. Tropical countries, such as Indonesia and Malaysia, hope that biodiesel demand will stimulate palm oil exports which have fallen out of favor for human diets. Conservationists, however, worry that increasing oil palm plantations will lead to more tropical forest destruction.

20.6 ENERGY FROM THE EARTH'S FORCES

The winds, waves, tides, ocean thermal gradients, and geothermal areas are renewable energy sources. Although available only in selected locations, these sources could make valuable contributions to our total energy supply.

Falling water has been used as an energy source since ancient times

The invention of water turbines in the nineteenth century greatly increased the efficiency of hydropower dams. By 1925, falling water generated 40 percent of the world's electric power. Since then, hydroelectric production capacity has grown 15-fold, but fossil fuel use has risen so rapidly that water power is now only 20 percent of total electrical generation. Still, many countries produce most of their electricity from falling water (fig. 20.25). Norway, for instance, depends on hydropower for 99 percent of its electricity. Currently, total world hydroelectric production is about 3,000 terrawatt hours (10^{12} Whr). Six countries, Canada, Brazil, the United States, China, Russia, and Norway, account for more than half that total. Approximately two-thirds of the economically feasible potential remains to be developed. Untapped hydro resources are still abundant in Latin America, Central Africa, India, and China.

Much of the hydropower development in recent years has been in enormous dams. There is a certain efficiency of scale in

FIGURE 20.25 Hydropower dams produce clean renewable energy but can be socially and ecologically damaging.

giant dams, and they bring pride and prestige to the countries that build them, but, as we discussed in chapter 17, they can have unwanted social and environmental effects. The largest hydroelectric dam in the world at present is the Three Gorges Dam on China's Yangtze River, which spans 2 km and will be 185 m (600 ft) tall when completed in 2009. Designed to generate 25,000 MW of power, this dam should produce as much energy as 25 large nuclear power plants when completed. The lake that it is creating already has displaced at least 1.5 million people and submerged 5,000 archaeological sites.

There are other problems with big dams, besides human displacement, ecosystem destruction, and wildlife losses. Dam failure can cause catastrophic floods and thousands of deaths. Sedimentation often fills reservoirs rapidly and reduces the usefulness of the dam for either irrigation or hydropower. In China, the Sanmenxia Reservoir silted up in only two years, and the Laoying Reservoir filled with sediment before the dam was even finished.

Rotting vegetation in artificial impoundments can have disastrous effects on water quality. When Lake Brokopondo in Suriname flooded a large region of uncut rainforest, underwater decomposition of the submerged vegetation produced hydrogen sulfide that killed fish and drove out villagers over a wide area. Acidified water from this reservoir ruined the turbine blades, making the dam useless for power generation. A recent study of one reservoir in Brazil suggested that decaying vegetation produced more greenhouse gases (carbon dioxide and methane) than would have come from generating an equal amount of energy by burning fossil fuels.

Floating water hyacinths (rare on free-flowing rivers) have already spread over reservoir surfaces behind the Tucurui Dam on the Amazon River in Brazil, impeding navigation and fouling machinery. Herbicides sprayed to remove aquatic vegetation have contaminated water supplies. Herbicides used to remove forests before dam gates closed caused similar pollution problems. Schistosomiasis, caused by parasitic flatworms called blood flukes (chapter 8), is transmitted to humans by snails that thrive in slow-moving, weedy tropical waters behind these dams. It is thought that 14 million Brazilians suffer from this debilitating disease.

As mentioned before, dams displace indigenous people. The Narmada Valley project in India will drown 150,000 ha of tropical forest and displace 1.5 million people, mostly tribal minorities and low-caste hill people. The Akosombo Dam built on the Volta River in Ghana nearly 20 years ago displaced 78,000 people from 700 towns. Few of these people ever found another place to settle, and those still living remain in refugee camps and temporary shelters. The Cree First Nations people of Canada also have been displaced by massive hydroelectric projects, most notably the James Bay project in Quebec and the Churchill/Nelson River diversion in Manitoba.

In tropical climates, large reservoirs often suffer enormous water losses. Lake Nasser, formed by the Aswan High Dam in Egypt, loses 15 billion m³ each year to evaporation and seepage. Unlined canals lose another 1.5 billion m³. Together, these losses represent one-half of the Nile River flow, or enough water to irrigate 2 million ha of land. The silt trapped by the Aswan Dam formerly fertilized farmland during seasonal flooding and pro-

vided nutrients that supported a rich fishery in the Delta region. Farmers now must buy expensive chemical fertilizers, and the fish catch has dropped almost to zero. As in South America, schistosomiasis is an increasingly serious problem.

If big dams—our traditional approach to hydropower—have so many problems, how can we continue to exploit the great potential of hydropower? Fortunately, there is an alternative to gigantic dams and destructive impoundment reservoirs. Small-scale, **low-head hydropower** technology can extract energy from small headwater dams that cause much less damage than larger projects. Some modern, high-efficiency turbines can even operate on **run-of-the-river flow.** Submerged directly in the stream and small enough not to impede navigation in most cases, these turbines don't require a dam or diversion structure and can generate useful power with a current of only a few kilometers per hour. They also cause minimal environmental damage and don't interfere with fish movements, including spawning migration. **Micro-hydro generators** operate on similar principles but are small enough to provide economical power for a single home. If you live close to a small stream or river that runs year-round and you have sufficient water pressure and flow, hydropower is probably a cheaper source of electricity for you than solar or wind power (fig. 20.26).

Small-scale hydropower systems also can cause abuses of water resources. The Public Utility Regulatory Policies Act of 1978 included economic incentives to encourage small-scale energy projects. As a result, thousands of applications were made to dam or divert small streams in the United States. Many of these projects have little merit. All too often, fish populations,

FIGURE 20.26 Solar collectors capture power only when the sun shines, but hydropower is available 24 hours a day. Small turbines such as this one can generate enough power for a single-family house with only 15 m (50 ft) of head and 200 l (50 g) per minute flow. The turbine can have up to four nozzles to handle greater water flow and generate more power.

aquatic habitat, recreational opportunities, and the scenic beauty of free-flowing streams and rivers are destroyed primarily to provide tax benefits for wealthy investors.

Wind energy is our fastest growing renewable source

Wind power played a crucial role in the settling of the American West, much of which has abundant underground aquifers, but little surface water. The strong, steady winds blowing across the prairies provided the energy to pump water that allowed ranchers and farmers to settle the land. By the end of the nineteenth century, nearly every farm or ranch west of the Mississippi River had at least one windmill, and the manufacture, installation, and repair of windmills was a major industry. The Rural Electrification Act of 1935 brought many benefits to rural America, but it effectively killed wind power development, and shifted electrical generation to large dams and fossil fuel-burning power plants. It's interesting to speculate what the course of history might have been if we had not spent trillions of dollars on fossil fuels and nuclear power, but instead had invested that money on small-scale, renewable energy systems.

The oil price shocks of the 1970s spurred a renewed interest in wind power. In the 1980s, the United States was a world leader in wind technology, and California hosted 90 percent of all wind power generators in the world. Poor management, technical flaws, and overdependence on subsidies, however, led to bankruptcy of many of the most important companies of that era, including Kenetech, once the world's largest manufacturer of wind generators. Now European companies dominate this (U.S.) $1 billion per year market.

Modern wind machines are far different from those employed a generation ago (fig. 20.27) The largest wind turbines now being built have towers up to 150 m tall with 62 m long blades that reach as high as a 45-story building. Each can generate 5 MW of electricity, or enough for 5,000 typical American homes. Out of commission for maintenance only about three days per year, many can produce power 90 percent of the time. Theoretically up to 60 percent efficient, modern windmills typically produce about 35 percent of peak capacity under field conditions. Currently, wind farms are the cheapest source of *new* power generation, costing as little as 3 cents/kWh compared to 4 to 5 cents/kWh for coal and five times that much for nuclear fuel. As table 20.3 shows, when the land consumed by mining is taken into account, wind power takes about one-third as much area and creates about five times as many jobs to create the same amount of electrical energy as coal.

Wind could meet all our energy needs

As this chapter's opening story shows, wind power offers an enormous potential for renewable energy. The World Meteorological Organization estimates that 80 million MW of wind power could be developed economically worldwide. This would be five times the total current global electrical generating capacity. Wind has a number of advantages over most other power sources. Wind farms have much shorter planning and construction times than

FIGURE 20.27 Renewable energy sources, such as wind, solar energy, geothermal power, and biomass crops, could eliminate our dependence on fossil fuels and prevent extreme global climate change, if we act quickly.

fossil fuel or nuclear power plants. Wind generators are modular (more turbines can be added if loads grow) and they have no fuel costs or air emissions. In the past decade, total wind generating capacity has increased 15-fold making it the fastest growing energy source in the world. With 75,000 MW of installed capacity in 2007, wind power is making a valuable contribution to reducing global warming. Wind does have limitations, however. Like solar energy, it is an intermittent source. Furthermore, not every place has strong enough or steady enough wind to make this an economical resource. Although modern windmills are more efficient than those of a few years ago, it takes a wind velocity between 7 m per second (16 mph) and 27 m per second (60 mph) to generate useful amounts of electricity.

In places like the Netherlands, which has winds above 16 mph an average of 245 days per year, windmills have long been recognized as a valuable energy source for pumping water and

TABLE 20.3

Jobs and Land Required for Alternative Energy Sources

Technology	Land Use (m² per Gigawatt-Hour for 30 Years)	Jobs (per Terawatt-Hour per Year)
Coal	3,642	116
Photovoltaic	3,237	175
Solar thermal	3,561	248
Wind	1,335	542

Source: Lester R. Brown, 1991.

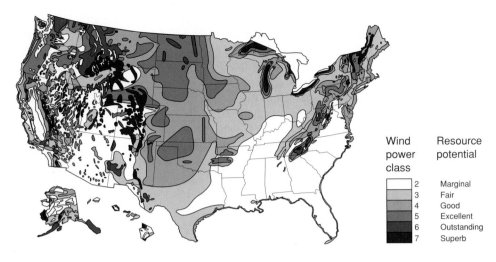

FIGURE 20.28 United States wind resource map. Mountain ranges and areas of the High Plains have the highest wind potential, but much of the country has a fair to good wind supply. **Source:** Data from U.S. Department of Energy.

Wind power class	Resource potential
2	Marginal
3	Fair
4	Good
5	Excellent
6	Outstanding
7	Superb

grinding grain. Germany, with 20,600 MW of installed capacity, now gets one-third of its electricity from wind power, and is now the world leader in this technology. Spain and the United States are tied for second with 11,600 MW each. India is fourth, with 6,300 MW. China has half as much installed capacity as India, but has potential for vastly more. The World Energy Council predicts that wind generating capacity could grow another tenfold by 2020.

While Europe, with its high population density, is focusing most attention to offshore wind farms, the bulk of North America's wind potential is situated across the Great Plains. Compared to offshore installations, which are costly because of the need to operate in deep water and withstand storms and waves, wind tower construction on land is relatively simple and cheap. There also is growing demand for wind projects from farmers, ranchers, and rural communities because of the economic benefits that wind energy brings. One thousand megawatts of wind power (equivalent to one large nuclear or fossil fuel plant) can create more than 3,000 permanent jobs, while paying about $4 million in rent to landowners and $3.6 million in tax payments to local governments. Seven Midwestern states—North Dakota, Texas, Kansas, South Dakota, Montana, Iowa, and Minnesota—all now have at least 1,000 MW of wind power, but wind-energy experts estimate this is less than 1 percent of their ultimate potential (fig. 20.28).

With each tower taking only about a 0.1 ha (0.25 acre) of cropland, farmers find that they can continue to cultivate 90 percent of their land while getting $2,000 or more in annual rent for each wind machine. An even better return results if the landowner builds and operates the wind generator, selling the electricity to the local utility. Annual profits can be as much as $100,000 per turbine, a wonderful bonus for use of 10 percent of your land. Cooperatives are springing up to help landowners finance, build, and operate their own wind generators. About 20 Native American tribes, for example, have formed a coalition to study wind power. Together, their reservations (which were sited in the windiest, least productive parts of the Great Plains) could generate at least 350,000 MW of electrical power, equivalent to about half of the current total U.S. installed capacity.

There are problems with wind energy. In some places, high bird mortality has been reported around wind farms. This seems to be particularly true in California, where rows of generators were placed at the summit of mountain passes where wind velocities are high but where migrating birds and bats are likely to fly into rotating blades. New generator designs and more careful tower placement seems to have reduced this problem in most areas. Some people object to the sight of large machines looming on the horizon. To others, however, windmills offer a welcome alternative to nuclear or fossil fuel-burning plants.

Many of the places most appropriate for wind development are far from the urban centers where power is needed (a major reason why people didn't settle in extremely windy places, is the wind). This means that some method is needed to transfer wind-generated power to the market. Currently, the only way to do that is high-voltage power lines (fig. 20.29). There may be more resistance to building thousands of kilometers of new power lines than to the wind farms themselves. An attractive alternative is to use the electricity from wind power to split water into hydrogen and oxygen gas. The gas could then be pumped through underground gas pipes to the city, where it could be used in fuel cells.

FIGURE 20.29 Dependence on wind or solar energy may require a vastly increased network of high-voltage power lines, many of which will cross places where people treasure now-unspoiled vistas.

Small windmills can generate enough electricity for a single home or farm, and are generally the least expensive form of renewable energy, if the winds are strong and steady enough in your area. The problem is how to store energy for windless days. Buying enough batteries to provide power for several days can be exorbitantly expensive, not to mention bulky and difficult to maintain. If you're able to hook up to the utility grid, the best solution to this problem is what's called reverse metering, in which you sell electricity back to your utility when you have a surplus and then draw from them when you have a shortage. If your system is sized right, it should pay for itself in about five years if wind speeds in your area are favorable.

Geothermal heat, tides, and waves could be valuable resources

The earth's internal temperature can provide a useful source of energy in some places. High-pressure, high-temperature steam fields exist below the earth's surface. Around the edges of continental plates or where the earth's crust overlays magma (molten rock) pools close to the surface, this **geothermal energy** is expressed in the form of hot springs, geysers, and fumaroles. Yellowstone National Park is the largest geothermal region in the United States. Iceland, Japan, and New Zealand also have high concentrations of geothermal springs and vents. Depending on the shape, heat content, and access to groundwater, these sources produce wet steam, dry steam, or hot water.

While few places have geothermal steam, the earth's warmth can help reduce energy costs nearly everywhere. Pumping water through buried pipes can extract enough heat so that a heat pump will operate more efficiently. Similarly, the relatively uniform temperature of the ground can be used to augment air conditioning in the summer (fig. 20.30). This can cut home heating costs by half in many areas, and pay for itself in five years.

Engineers are now exploring deep wells for community geothermal systems. Drilling 2,000 m (6,000 ft) in the American West gets you into rocks above 100°C. Fracturing them to expose more surface area, and pumping water in can produce enough steam to run an electrical generator at a cost significantly lower than conventional fossil fuel or nuclear power. The well is no more expensive than most oil wells, and the resource is never exhausted. Currently, about 60 new geothermal energy projects are being developed in the United States. This source could provide a significant energy supply eventually.

Ocean tides and waves contain enormous amounts of energy that can be harnessed to do useful work. A tidal station works like a hydropower dam, with its turbines spinning as the tide flows through them. A high-tide/low-tide differential of several meters is required to spin the turbines. Unfortunately, variable

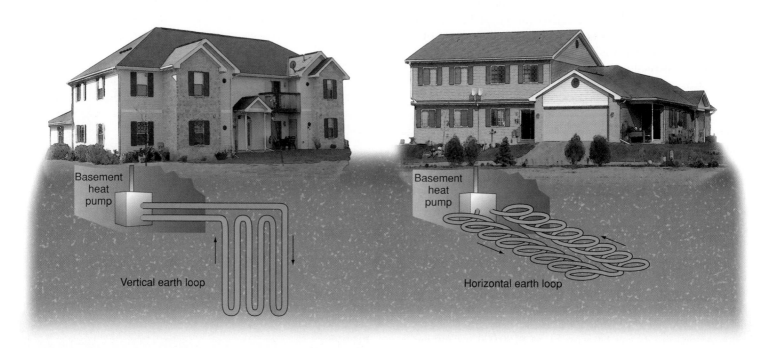

FIGURE 20.30 Geothermal energy can cut heating and cooling costs by half in many areas. In summer (shown here), warm water is pumped through buried tubing (earth loops) where it is cooled by constant underground temperatures. In winter, the system reverses and the relatively warm soil helps heat the house. Where space is limited (*left*), earth loops can be vertical. If more space is available (*right*) the tubing can be laid in shallow horizontal trenches. A heat exchanger concentrates heat, so water entering the house is far warmer than the ground temperature.

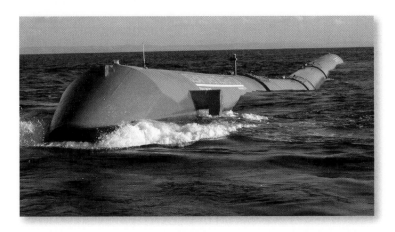

FIGURE 20.31 The Pelamis wave converter (named after a sea snake) is a 125 m long and 3.5 m diameter tube-hinged, so it undulates as ocean swells pass along it. This motion drives pistons that turn electrical generators. Energy experts calculate that capturing just 1 to 2 percent of global wave power could supply at least 16 percent of the world's electrical demand.

tidal periods often cause problems in integrating this energy source into the electric utility grid. Nevertheless, demand has kept some plants running for many decades.

Ocean wave energy can easily be seen and felt on any seashore. The energy that waves expend as millions of tons of water are picked up and hurled against the land, over and over, day after day, can far exceed the combined energy budget for both insolation (solar energy) and wind power in localized areas. Captured and turned into useful forms, that energy could make a substantial contribution to meeting local energy needs.

Dutch researchers estimate that 20,000 km of ocean coastline are suitable for harnessing wave power. Among the best places in the world for doing this are the west coasts of Scotland, Canada, the United States (including Hawaii), South Africa, and Australia. Wave energy specialists rate these areas at 40 to 70 kW per meter of shoreline. Altogether, it's calculated, if the technologies being studied today become widely used, wave power could amount to as much as 16 percent of the world's current electrical output.

Some of the designs being explored include oscillating water columns that push or pull air through a turbine, and a variety of floating buoys, barges, and cylinders that bob up and down as waves pass, using a generator to convert mechanical motion into electricity. It's difficult to design a mechanism that can survive the worst storms.

An interesting new development in this field is the Pelamis wave-power generator developed by the Scottish start-up company Ocean Power Delivery (fig. 20.31). The first application of this technology is now being built 5 km off the coast of Portugal.

It will use three units capable of producing some 2.25 MW of electricity, or enough to supply 1,500 Portuguese households. If preliminary trials go well, plans are to add 40 more units in a year or two. Each of the units consists of four cylindrical steel sections linked by hinged joints. Anchored to the seafloor at its nose, the snakelike machine points into the waves and undulates up and down and side to side as swells move along its 125 m length. This motion pumps fluid to hydraulic motors that drive electrical generators to produce power, which is carried to shore by underwater cables.

Pelamis's inventor, Richard Yemm, says that survivability is the most important feature of a wave-power device. Being offshore, the Pelamis isn't exposed to the pounding breakers that destroy shore-based wave-power devices. If waves get too steep, the Pelamis simply dives under them, much as a surfer dives under a breaker. These wave converters lie flat in the water and are positioned far offshore, so they are unlikely to stir up as much opposition as do the tall towers of wind generators.

Ocean thermal electric conversion might be useful

Temperature differentials between upper and lower layers of the ocean's water also are a potential source of renewable energy. In a closed-cycle **ocean thermal electric conversion (OTEC)** system, heat from sun-warmed upper ocean layers is used to evaporate a working fluid, such as ammonia or Freon, which has a low boiling point. The pressure of the gas produced is high enough to spin turbines to generate electricity. Cold water then is pumped from the ocean depths to condense the gas.

As long as a temperature difference of about 20°C (36°F) exists between the warm upper layers and cooling water, useful amounts of net power can, in principle, be generated with one of these systems. This differential corresponds, generally, to a depth of about 1,000 m in tropical seas. The places where this much temperature difference is likely to be found close to shore are islands that are the tops of volcanic seamounts, such as Hawaii, or the edges of continental plates along subduction zones (chapter 14) where deep trenches lie just offshore. The west coast of Africa, the south coast of Java, and a number of South Pacific islands, such as Tahiti, have usable temperature differentials for OTEC power.

Although their temperature differentials aren't as great as the ocean, deep lakes can have very cold bottom water. Ithaca, New York, has recently built a system to pump cold water out of Lake Cayuga to provide natural air conditioning during the summer. Cold water discharge from a Hawaiian OTEC system has been used to cool the soil used to grow cool-weather crops such as strawberries.

CONCLUSION

None of the renewable energy sources discussed in this chapter are likely to completely replace fossil fuels and nuclear power in the near future. They could, however, make a substantial collective contribution toward providing us with the conveniences we crave in a sustainable, environmentally friendly manner. They could also make us energy independent and balance our international payment deficit.

The World Energy Council projects that renewables could provide about 40 percent of world cumulative energy consumption under an idealized "ecological" scenario assuming that political leaders take global warming seriously and pass taxes to encourage conservation and protect the environment (fig. 20.32). This scenario also envisions measures to shift wealth from the north to south, and to enhance economic equity. By the end of the twenty-first century, renewable sources could provide all our energy needs if we take the necessary steps to make this happen.

Rising fuel prices and increasing dependence on imported oil have prompted demands for a U.S. energy policy. Environmentalists point to the dangers of air pollution, global climate change, and other environmental problems associated with burning of fossil fuels. Businesses stress the importance of a reliable energy supply for economic growth. While both call for a new policy, they disagree on what it should contain. Conservatives tend to favor increasing production and easing regulations on power plant operation and transmission-line siting, rather than limiting demand. Progressives, on the other hand, prefer conservation measures such as forcing automakers to increase average fuel

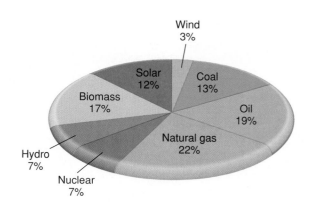

FIGURE 20.32 Idealized "ecological" scenario for cumulative world energy consumption, 2000 to 2100.
Source: World Energy Council, 2002.

efficiency of cars and light trucks and providing heating bill assistance for low-income households.

What path do you think we should take to achieve an ideal energy future?

REVIEWING LEARNING OUTCOMES

By now you should be able to explain the following points:

20.1 Remember that conservation can help us meet our energy needs.

- There are many ways to save energy.
- Transportation could be far more efficient.
- Cogeneration produces both electricity and heat.

20.2 Explain how we could tap solar energy.

- Solar collectors can be passive or active.
- Storing solar energy is problematic.

20.3 Discuss high-temperature solar energy.

- Simple solar cookers can save energy.
- Utilities are promoting renewable energy.
- Photovoltaic cells capture solar energy.
- Electrical energy is difficult and expensive to store.

20.4 Grasp the potential of fuel cells.

- All fuel cells have similar components.
- Several different electrolytes can be used in fuel cells.

20.5 Explain how we get energy from biomass.

- We can burn biomass.
- Fuelwood is in short supply in many less-developed countries.
- Dung and methane provide power.
- Biofuels could replace some oil-based energy.
- Should we use food for fuel?

20.6 Investigate energy from the earth's forces.

- Falling water has been used as an energy source since ancient times.
- Wind energy is our fastest growing renewable source.
- Wind could meet all our energy needs.
- Geothermal heat, tides, and waves could be valuable resources.
- Ocean thermal electric conversion might be useful.

PRACTICE QUIZ

1. Describe five ways that we could conserve energy individually or collectively.

2. Explain the principle of net energy yield. Give some examples.

3. What is the difference between active and passive solar energy?

4. How do photovoltaic cells generate electricity?

5. What is a fuel cell and how does it work?

6. Describe some problems with wood burning in both industrialized nations and developing nations.

7. How is methane made? Give an example of a useful methane source.

8. What are some advantages and disadvantages of large hydroelectric dams?

9. What are some examples of biomass fuel other than wood?

10. Describe how tidal power or ocean wave power generate electricity.

CRITICAL THINKING AND DISCUSSION QUESTIONS

1. What alternative energy sources are most useful in your region and climate? Why?

2. What can you do to conserve energy where you live? In personal habits? In your home, dormitory, or workplace?

3. Do you think building wind farms in remote places, parks, or scenic wilderness areas would be damaging or unsightly?

4. If you were the energy czar of your state, where would you invest your budget?

5. What could (or should) we do to help developing countries move toward energy conservation and renewable energy sources? How can we ask them to conserve when we live so wastefully?

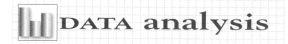 **DATA analysis**   **Energy Calculations**

Most college students either already own or are likely to buy an automobile and a computer sometime soon. How do these items compare in energy usage? Suppose that you were debating between a high-mileage car, such as the Honda Insight, or a sport utility vehicle, such as a Ford Excursion. How do the energy requirements of these two purchases measure up? To put it another way, how long could you run a computer on the energy you would save by buying an Insight rather than an Excursion?

Here are some numbers you need to know. The Insight gets about 75 mpg, while the Excursion gets about 12 mpg. A typical American drives about 15,000 mi per year. A gallon of regular, unleaded gasoline contains about 115,000 Btu on average. Most computers use about 100 watts of electricity. One kilowatt-hour (kWh) = 3,413 Btu.

1. How much energy does the computer use if it is left on continuously? (You really should turn it off at night or when it isn't in use, but we'll simplify the calculations.)

    100 watt/h × 24 h/day × 365 days/yr = _____ kWh/yr

2. How much gasoline would you save in an Insight, compared with an Excursion?

    a. Excursion:

    15,000 mi/yr ÷ 12 mpg   = _____ gal/yr

    b. Insight:

    15,000 mi/yr ÷ 75 mpg   = _____ gal/yr

    c. Gasoline savings $(a - b)$ = _____ gal/yr

    d. Energy savings:

    (gal × 115,000 Btu)   = _____ Btu/yr

    e. Converting Btu to kWh:

    (Btu × 0.00029 Btu/kWh) = _____ kWh/yr saved

3. How long would the energy saved run your computer? kWh/yr saved by Insight ÷ kWh/yr consumed by computer = _____

For Additional Help in Studying T [hand written overlay] ... hhe.com/cunningham10e. You will find additional practice quizzes and case s ... s for Google Earth™ mapping, and an extensive reading list, all of which will h ...

Handwritten notes:

1) 876

2) a) 1250
   b) 200
   c) 1050
   d) 120 million
   e) 35,000

3) 40 yrs.

Fresh Kills, the largest landfill in the United States,
closed in 2001 for lack of space.

C H A P T E R

21

Solid, Toxic, and Hazardous Waste

*We have no knowledge, so we have stuff; but stuff without
knowledge is never enough.*

—Greg Brown—

LEARNING OUTCOMES

After studying this chapter, you should be able to:

21.1 Identify the components of solid waste.

21.2 Describe how wastes have been—and are
being—disposed of or treated.

21.3 Identify how we might shrink the waste stream.

21.4 Investigate hazardous and toxic wastes.

Case Study The New Alchemy: Creating Gold from Garbage

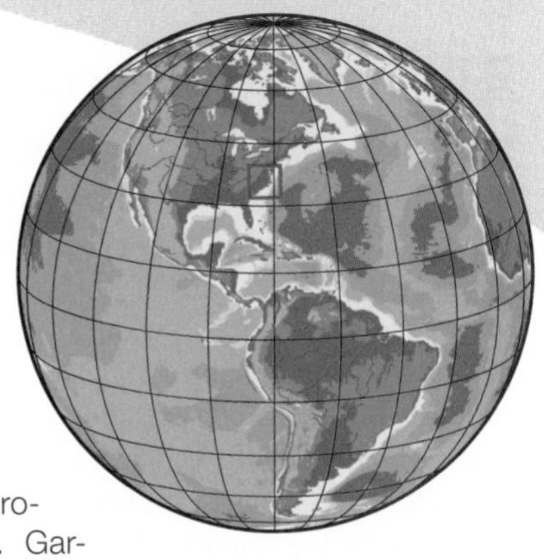

Most people think of recycling in terms of newspapers, plastic bottles, and other household goods. Your daily household recycling is the bedrock of recycling programs, but another growing and exciting area of recycling is done at commercial and industrial scales. The 230 million tons of garbage the United States produces each year includes some knotty problems: old furniture and carpeting, appliances and computers, painted wood, food waste. It's no wonder that we've simply dumped it all in landfills as long as we could. But landfill space is diminishing rapidly (fig. 21.1).

Incinerators are a common alternative, but they are expensive to build and operate, and they can produce dangerous air contaminants, including dioxins from burned plastics, and heavy metals.

One of our largest sources of waste is construction and demolition debris—the rubble left over when a building is torn down, remodeled, or built. Construction and demolition account for over 140 million tons of waste per year, about 1.5 kg per person per day—on top of the 230 million tons per year of municipal solid waste. All this mixed debris is normally trucked to landfills, but alternatives have emerged in recent years.

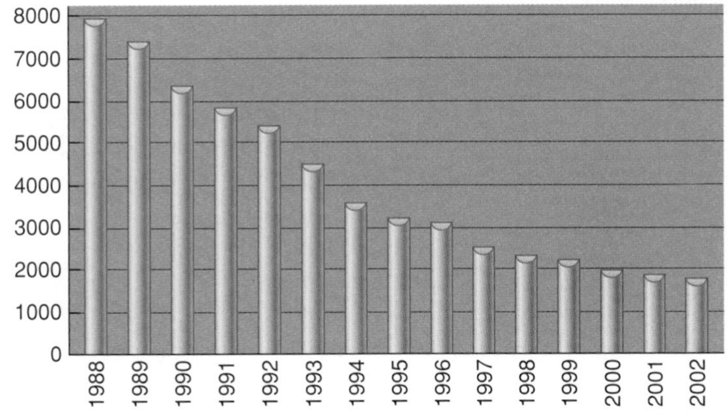

A slowly growing number of cities and companies are sending their waste, including construction debris, to commercial recyclers. One example is Taylor Recycling, based in Montgomery, New York. Starting as a tree removal business, Taylor has expanded to construction and demolition waste and now operates in four states. The company recycles and sells 97 percent of the mixed debris it receives, well above the industry average of 30 to 50 percent. Trees are ground and converted to mulch for landscaping. Dirt from stumps is screened and sold as clean garden soil. Mixed materials are sorted into recyclable glass, metals, and plastics. Construction debris is sorted and ground: broken drywall is ground to fresh gypsum, which is sold to drywall producers; wood is composted or burned; bricks are crushed for fill and construction material. Organic waste that can't be separated, such as food-soaked paper, is sent to a gasifier. The gasifier is like an enclosed, oxygen-free pressure cooker, which converts biomass to natural gas. The gas runs electric generators for the plant, and any extra gas can be sold. Waste heat warms the recycling facility. The 3 percent of incoming waste that doesn't get recycled is mainly mixed plastics, which are currently landfilled.

From their base outside of New York City, recycling is clearly a good idea. New York has used up most of its landfill space and now ships garbage to Virginia, Ohio, Pennsylvania, and South Carolina. Fuel and trucking costs alone drive up disposal costs, and with landfill capacity shrinking, tipping fees are climbing.

According to Jim Taylor, the 1,000 garbage trucks leaving New York City each day travel an average of 300 miles round-trip, at less than 4 miles per gallon of fuel.

The story of garbage processing is changing globally. Garbage has long been one of the United States' largest exports, but increasingly the country is exporting sorted recycled materials, as well. Chinese manufacturers are finding valuable material sources in American waste. In Western Europe, where environmental regulation and landfill space are both tight, recycling, composting, and conversion of biomass to gas are booming businesses. The Swiss company Kompogas, one of many companies processing garbage, ferments organic waste in giant tanks, producing methane, compost, and fertilizer.

New technologies are providing innovative strategies for recycling waste. One method that is getting attention is Changing World Technologies' thermal depolymerization process ("thermal" means it involves heating; "depolymerization" essentially means breaking down organic molecules) that converts almost any kind of organic waste into clean, usable diesel oil. Animal by-products, sewage sludge, shredded tires, and other waste can be converted with 85 percent energy efficiency. Pilot plants have been built for specialized sources, such as a turkey processing plant in Colorado, but the technology could accommodate mixed sources as well.

Recycling is a rapidly growing industry because it makes money coming and going. Recyclers are paid to haul away waste, which they turn into marketable products. The business is also exciting because these companies, like Taylor Recycling, see the huge social and economic benefits of environmental solutions. Often when we discuss environmental problems, businesses are part of the problem. But these examples show that business owners can be just as excited as anybody about environmental quality. With a good business model, being green can be very rewarding.

Garbage disposal may be one of the most exciting stories in environmental science, because it shows so much promise and innovation. So far, the examples discussed here make up a minority of our waste management strategies, but they are expanding. In this chapter we'll look at our other waste management methods, the composition of our waste, and some of the differences between solid waste and hazardous waste.

FIGURE 21.1 The number of landfills in the United States has declined steadily as space has become scarce and safety rules have increased. **Data Source:** U.S. EPA.

21.1 SOLID WASTE

Waste is everyone's business. We all produce wastes in nearly everything we do. According to the Environmental Protection Agency, the United States produces 11 billion tons of solid waste each year. About half of that amount consists of agricultural waste, such as crop residues and animal manure, which are generally recycled into the soil on the farms where they are produced. They represent a valuable resource as ground cover to reduce erosion and fertilizer to nourish new crops, but they also constitute the single largest source of nonpoint air and water pollution in the country. More than one-third of all solid wastes are mine tailings, overburden from strip mines, smelter slag, and other residues produced by mining and primary metal processing. Road and building construction debris is another major component of solid waste. Much of this material is stored in or near its source of production and isn't mixed with other kinds of wastes. Improper disposal practices, however, can result in serious and widespread pollution.

Industrial waste—other than mining and mineral production—amounts to some 400 million metric tons per year in the United States. Most of this material is recycled, converted to other forms, destroyed, or disposed of in private landfills or deep injection wells. About 60 million metric tons of industrial waste falls in a special category of hazardous and toxic waste, which we will discuss later in this chapter.

Municipal waste—a combination of household and commercial refuse—amounts to more than 200 million metric tons per year in the United States (fig. 21.2). That's approximately two-thirds of a ton for each man, woman, and child every year—twice as much per capita as Europe or Japan, and five to ten times as much as most developing countries.

The waste stream is everything we throw away

Does it surprise you to learn that you generate that much garbage? Think for a moment about how much we discard every year. There are organic materials, such as yard and garden wastes, food wastes, and sewage sludge from treatment plants; junked cars; worn out furniture; and consumer products of all types. Newspapers, magazines, advertisements, and office refuse make paper one of our major wastes (fig. 21.3). In spite of recent progress in recycling,

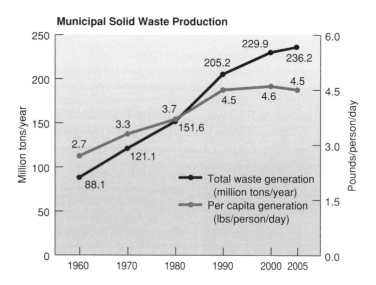

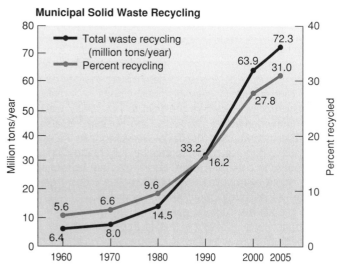

FIGURE 21.2 Bad news and good news in solid waste production. Per capita waste has risen steadily to more than 2 kg per person per day. Recycling rates are also rising, however. Recycling data include composting.
Source: Data from Environmental Protection Agency, 2006.

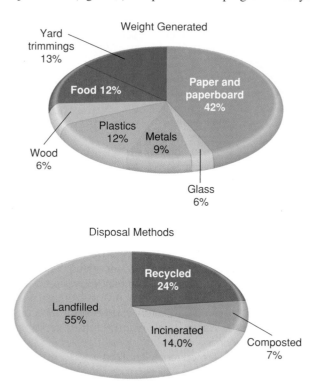

FIGURE 21.3 Composition of municipal solid waste in the United States by weight, before recycling, and disposal methods.
Source: Data from U.S. Environmental Protection Agency Office of Solid Waste Management, 2006.

many of the 200 *billion* metal, glass, and plastic food and beverage containers used every year in the United States end up in the trash. Wood, concrete, bricks, and glass come from construction and demolition sites, dust and rubble from landscaping and road building. All of this varied and voluminous waste has to arrive at a final resting place somewhere.

The **waste stream** is a term that describes the steady flow of varied wastes that we all produce, from domestic garbage and yard wastes to industrial, commercial, and construction refuse. Many of the materials in our waste stream would be valuable resources if they were not mixed with other garbage. Unfortunately, our collecting and dumping processes mix and crush everything together, making separation an expensive and sometimes impossible task. In a dump or incinerator, much of the value of recyclable materials is lost.

Another problem with refuse mixing is that hazardous materials in the waste stream get dispersed through thousands of tons of miscellaneous garbage. This mixing makes the disposal or burning of what might have been rather innocuous stuff a difficult, expensive, and risky business. Spray paint cans, pesticides, batteries (zinc, lead, or mercury), cleaning solvents, smoke detectors containing radioactive material, and plastics that produce dioxins and PCBs when burned are mixed willy-nilly with paper, table scraps, and other nontoxic materials. The best thing to do with household toxic and hazardous materials is to separate them for safe disposal or recycling, as we will see later in this chapter.

FIGURE 21.4 Trash disposal has become a crisis in the developing world, where people have adopted cheap plastic goods and packaging but lack good recycling or disposal options.

Think About It

Figure 21.2 shows a continuing increase in waste production per capita. What is the percentage increase per capita from 1960 to 2000? (*Hint:* calculate (4.6 − 2.7) ÷ 2.7.) What might account for this increase? Is there a relationship between waste production and our quality of life?

21.2 WASTE DISPOSAL METHODS

Where are our wastes going now? In this section, we will examine some historic methods of waste disposal as well as some future options. Notice that our presentation begins with the least desirable—but most commonly used—measures and proceeds to discuss some preferable options. Keep in mind as you read this that modern waste management reverses this order and stresses the "three R's" of reduction, reuse, and recycling before destruction or, finally, secure storage of wastes.

Open dumps release hazardous materials into air and water

For many people, the way to dispose of waste is to simply drop it someplace. Open, unregulated dumps are still the predominant method of waste disposal in most developing countries (fig. 21.4). The giant developing-world megacities have enormous garbage problems. Mexico City, one of the largest cities in the world, generates some 10,000 tons of trash *each day*. Until recently, most of this torrent of waste was left in giant piles, exposed to the wind and rain, as well as rats, flies, and other vermin. Manila, in the Philippines, generates a similar amount of waste, half of which goes to a giant, constantly smoldering dump called "Smoky Mountain." Over 20,000 people live and work on this mountain of refuse, scavenging for recyclable items or edible food scraps. In July 2000, torrential rains spawned by Typhoon "Kai Tak" caused part of the mountain to collapse, burying at least 215 people. The government would like to close these dumps, but how will the residents be housed and fed? Where else will the city put its garbage?

Most developed countries forbid open dumping, at least in metropolitan areas, but illegal dumping is still a problem. You have undoubtedly seen trash accumulating along roadsides and in vacant, weedy lots in the poorer sections of cities. Is this just a question of aesthetics? Consider the problem of waste oil and solvents. An estimated 200 million liters of waste motor oil are poured into the sewers or allowed to soak into the ground every year in the United States. This is about five times as much as was spilled by the *Exxon Valdez* in Alaska in 1989! No one knows the volume of solvents and other chemicals disposed of by similar methods.

Increasingly, these toxic chemicals are showing up in the groundwater supplies on which nearly half the people in America depend for drinking (chapter 18). An alarmingly small amount of oil or other solvents can pollute large quantities of drinking or irrigation water. One liter of gasoline, for instance, could theoretically make a million liters of water undrinkable. Open dumps are usually illegal in most developed countries today. They remain common, however, in developing countries.

FIGURE 21.5 Dumping of trash at sea is a global problem. Even on the most remote islands, beaches are covered with plastic flotsam and jetsam.

Ocean dumping is nearly uncontrollable

The oceans are vast, but not so large that we can continue to treat them as carelessly as has been our habit. Every year some 25,000 metric tons (55 million lbs) of packaging, including half a million bottles, cans, and plastic containers, are dumped at sea. Beaches, even in remote regions, are littered with the nondegradable flotsam and jetsam of industrial society (fig. 21.5). About 150,000 tons (330 million lbs) of fishing gear—including more than 1,000 km (660 mi) of nets—are lost or discarded at sea each year. Environmental groups estimate that 50,000 northern fur seals are entangled in this refuse and drown or starve to death every year in the North Pacific alone (see fig. 18.20).

You may have seen a recent television special in which Jean-Michel Cousteau led an expedition to Kure Atoll, the most remote of the Northwest Hawaiian Island chain (Kure is about 4,800 km from Honolulu or twice that far from the U.S. mainland). Underwater, this is one of the last pristine large-scale coral reef systems in the Pacific. Onshore, unfortunately, the scene is very different. This chain of islands in the newest U.S. national monument is contaminated with a shocking amount of refuse. Where does it all come from? No one lives on these tiny islets. The answer is:

from as far away as California, Mexico, and Japan. A giant circular ocean current sweeps up flotsam and jetsam into an area, about the size of Texas, called the eastern garbage patch. By some estimates more than 3 million tons of garbage circulate within this huge floating dump. Unfortunately, Hawaii lies in the path of these currents and even remote beaches accumulate trash that threatens endemic wildlife.

We often export waste to countries ill-equipped to handle it

Although most industrialized nations agreed to stop shipping hazardous and toxic waste to less-developed countries in 1989, the practice still continues. In 2006, for example, 400 tons of toxic waste were illegally dumped at 14 open dumps in Abidjan, the capital of the Ivory Coast. The black sludge—petroleum wastes containing hydrogen sulfide and volatile hydrocarbons—killed ten people and injured many others. At least 100,000 city residents sought medical treatment for vomiting, stomach pains, nausea, breathing difficulties, nosebleeds, and headaches. The sludge—which had been refused entry at European ports—was transported by an Amsterdam-based multinational company on a Panamanian-registered ship and handed over to an Ivorian firm (thought to be connected to corrupt government officials) to be dumped in the Ivory Coast. The Dutch company agreed to clean up the waste and pay the equivalent of (U.S.) $198 million to settle claims.

One of the greatest sources of toxic material currently going to developing countries is outdated electronic devices. There are at least 2 billion television sets and personal computers in use globally. Televisions often are discarded after only about five years, while computers, play-stations, cellular telephones, and other electronics become obsolete even faster. As many as 600 million computers are in use in the United States (twice as many as there are residents), and most will be discarded in the next few years. Only about 10 percent of the components are currently recycled. These computers contain at least 2.5 billion kg of lead (as well as mercury, gallium, germanium, nickel, palladium, beryllium, selenium, arsenic), and valuable metals, such as gold, silver, copper, and steel.

About 80 percent of **electronic waste** (or **e-waste**) is shipped overseas, mostly to China and other developing countries in Asia and Africa. There, villagers, including young children, break it apart to retrieve valuable metals. Often, this scrap recovery is done under primitive conditions where workers have little or no protective gear (fig. 21.6) and residue goes into open dumps. Health risks in this work are severe, especially for growing children. Soil, groundwater, and surface water contamination at these sites has been found to be as much as 200 times the World Health Organization's standards. An estimated 100,000 workers handle e-waste in China alone. The Basel Action Network, an international network of activists seeking to prevent the globalization of the toxic chemical trade, tracks international e-waste shipments and working conditions. The organization is named after the Swiss town where the agreement to ban international shipping of hazardous wastes was reached.

FIGURE 21.6 A Chinese woman smashes a cathode ray tube from a computer monitor in order to remove valuable metals. This kind of unprotected demanufacturing is highly hazardous to both workers and the environment.

Most of the world's obsolete ships are now dismantled and recycled in poor countries. The work is dangerous, and old ships often are full of toxic and hazardous materials, such as oil, diesel fuel, asbestos, and heavy metals. On India's Anlang Beach, for example, more than 40,000 workers tear apart outdated vessels using crowbars, cutting torches, and even their bare hands. Metal is dragged away and sold for recycling. Organic waste is often simply burned on the beach, where ashes and oily residue wash back into the water.

Think About It

Ocean dumping of both solid waste and hazardous waste is a chronic problem. Suppose you were a captain or a sailor on an ocean-going ship. What factors might influence your decision to dump waste oil, garbage, or occasional litter overboard? (Money? time? legal considerations about your cargo or waste?) Whose responsibility is ocean dumping? What steps could the international community take to reduce it?

Landfills receive most of our waste

Over the past 50 years most American and European cities have recognized the health and environmental hazards of open dumps. Increasingly, cities have turned to **sanitary landfills,** where solid waste disposal is regulated and controlled. To decrease smells and litter and to discourage insect and rodent populations, landfill operators are required to compact the refuse and cover it every day with a layer of dirt (fig. 21.7). This method helps control pollution, but the dirt fill also takes up as much as 20 percent of landfill space. Since 1994, all operating landfills in the United States have been required to control such hazardous substances as oil, chemical compounds, toxic metals, and contaminated rainwater that seeps through piles of waste. An impermeable clay and/or plastic lining underlies and encloses the storage area. Drainage systems are installed in and around the liner to catch drainage and to help monitor chemicals that may be leaking. Modern municipal solid-waste landfills now have many of the safeguards of hazardous waste repositories described later in this chapter.

Fresh kills Landfill on Staten Island, New York, was the world's largest (see photo p. 474). It officially closed in 2001, but then reopened to receive debris from the World Trade Center. It's named for the Fresh Kills River and estuary, which it spans.

More careful attention is now paid to the siting of new landfills. Sites located on highly permeable or faulted rock formations are passed over in favor of sites with less leaky geologic foundations. Landfills are being built away from rivers, lakes, floodplains, and aquifer recharge zones rather than near them, as was often done in the past. More care is being given to a landfill's long-term effects so that costly cleanups and rehabilitation can be avoided.

Historically, landfills have been a convenient and relatively inexpensive waste-disposal option in most places, but this situation is changing rapidly. Rising land prices and shipping

Compacted waste filling trench   Original terrain

Daily 6-inch earth cover

FIGURE 21.7 In a sanitary landfill, trash and garbage are crushed and covered each day to prevent accumulation of vermin and spread of disease. A waterproof lining is now required to prevent leaching of chemicals into underground aquifers.

costs, as well as increasingly demanding landfill construction and maintenance requirements, are making this a more expensive disposal method. The cost of disposing a ton of solid waste in Philadelphia went from $20 in 1980 to more than $100 in 1990. Union County, New York, experienced an even steeper price rise. In 1987, it paid $70 to get rid of a ton of waste; a year later, that same ton cost $420, or about $10 for a typical garbage bag. In the past decades, costs have continued to rise steadily, though not as sharply. The United States now spends about $10 billion per year to dispose of trash. A decade from now, it may cost Americans $100 billion per year to dispose of their garbage.

Suitable places for waste disposal are becoming scarce in many areas. Other uses compete for open space. Citizens have become more concerned and vocal about health hazards, as well as aesthetics. It is difficult to find a neighborhood or community willing to accept a new landfill. Since 1984, when stricter financial and environmental protection requirements for landfills took effect, more than 1,200 of the 1,500 existing landfills in the United States have closed. Many major cities are running out of local landfill space. They export their trash, at enormous expense, to neighboring communities and even other states. More than half the solid waste from New Jersey goes out of state, some of it up to 800 km (500 mi) away.

A positive trend in landfill management is methane recovery. Methane, or natural gas, is a natural product of decomposing garbage deep in a landfill. It is also an important "greenhouse gas." Normally methane seeps up to the landfill surface and escapes. At 300 U.S. landfills, the methane is being collected and burned. Cumulatively, these landfills could provide enough electricity for a city of a million people. Three times as many landfills could be recovering methane. Tax incentives could be developed to encourage this kind of resource recovery.

Incineration produces energy but causes pollution

Landfilling is still the disposal method for the majority of municipal waste in the United States (fig. 21.8). Faced with growing piles of garbage and a lack of available landfills at any price, however, public officials are investigating other disposal methods. The method to which they frequently turn is burning. Another term commonly used for this technology is **energy recovery,** or waste-to-energy, because the heat derived from incinerated refuse is a useful resource. Burning garbage can produce steam used directly for heating buildings or generating electricity. Internationally, well over 1,000 waste-to-energy plants in Brazil, Japan, and western Europe generate much-needed energy while also reducing the amount that needs to be landfilled. In the United States, more than 110 waste incinerators burn 45,000 tons of garbage daily. Some of these are simple incinerators; others produce steam and/or electricity.

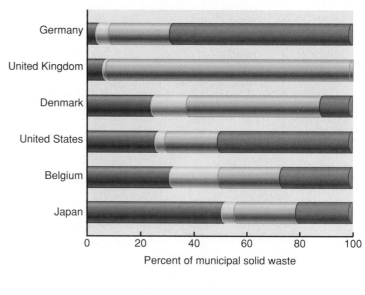

FIGURE 21.8 Percentage of municipal solid waste recycled, composted, incinerated, and landfilled in selected developed countries. **Source:** Eurostat, UNEP, 2003.

Types of Incinerators

Municipal incinerators are specially designed burning plants capable of burning thousands of tons of waste per day. In some plants, refuse is sorted as it comes in to remove unburnable or recyclable materials before combustion. This is called **refuse-derived fuel** because the enriched burnable fraction has a higher energy content than the raw trash. Another approach, called **mass burn,** is to dump everything smaller than sofas and refrigerators into a giant furnace and burn as much as possible (fig. 21.9).

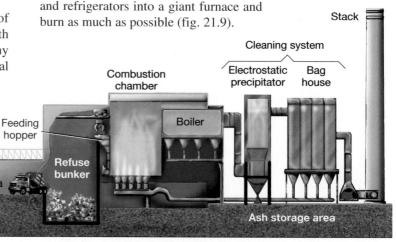

FIGURE 21.9 A diagram of a municipal "mass burn" garbage incinerator. Steam produced in the boiler can be used to generate electricity or to heat nearby buildings.

This technique avoids the expensive and unpleasant job of sorting through the garbage for nonburnable materials, but it often causes greater problems with air pollution and corrosion of burner grates and chimneys.

In either case, residual ash and unburnable residues representing 10 to 20 percent of the original volume are usually taken to a landfill for disposal. Because the volume of burned garbage is reduced by 80 to 90 percent, disposal is a smaller task. However, the residual ash usually contains a variety of toxic components that make it an environmental hazard if not disposed of properly. Ironically, one worry about incinerators is whether enough garbage will be available to feed them. Some communities in which recycling has been really successful have had to buy garbage from neighbors to meet contractual obligations to waste-to-energy facilities. In other places, fears that this might happen have discouraged recycling efforts.

Incinerator Cost and Safety

The cost-effectiveness of garbage incinerators is the subject of heated debates. Initial construction costs are high—usually between $100 million and $300 million for a typical municipal facility. Tipping fees at an incinerator, the fee charged to haulers for each ton of garbage dumped, are often much higher than those at a landfill. As landfill space near metropolitan areas becomes more scarce and more expensive, however, landfill rates are certain to rise. It may pay in the long run to incinerate refuse so that the lifetime of existing landfills will be extended.

Environmental safety of incinerators is another point of concern. The EPA has found alarmingly high levels of dioxins, furans, lead, and cadmium in incinerator ash. These toxic materials were more concentrated in the fly ash (lighter, airborne particles capable of penetrating deep into the lungs) than in heavy bottom ash. Dioxin levels can be as high as 780 parts per billion. One part per billion of TCDD, the most toxic dioxin, is considered a health concern. All of the incinerators studied exceeded cadmium standards, and 80 percent exceeded lead standards. Proponents of incineration argue that if they are run properly and equipped with appropriate pollution-control devices, incinerators are safe to the general public. Opponents counter that neither public officials nor pollution-control equipment can be trusted to keep the air clean. They argue that recycling and source reduction efforts are better ways to deal with waste problems.

The EPA, which generally supports incineration, acknowledges the health threat of incinerator emissions but holds that the danger is very slight. The EPA estimates that dioxin emissions from a typical municipal incinerator may cause one death per million people in 70 years of operation. Critics of incineration claim that a more accurate estimate is 250 deaths per million in 70 years.

One way to reduce these dangerous emissions is to remove batteries containing heavy metals and plastics containing chlorine before wastes are burned. Bremen, West Germany, is one of several European cities now trying to control dioxin emissions by keeping all plastics out of incinerator waste. Bremen is requiring

households to separate plastics from other garbage. This is expected to eliminate nearly all dioxins and other combustion by-products and prevent the expense of installing costly pollution-control equipment that otherwise would be necessary to keep the burners operating. Several cities have initiated a recycling program for the small "button" batteries used in hearing aids, watches, and calculators in an attempt to lower mercury emissions from its incinerator.

21.3 SHRINKING THE WASTE STREAM

Having less waste to discard is obviously better than struggling with disposal methods, all of which have disadvantages and drawbacks. In this section we will explore some of our options for recycling, reuse, and reduction of the wastes we produce.

Recycling captures resources from garbage

The term *recycling* has two meanings in common usage. Sometimes we say we are *recycling* when we really are *reusing* something, such as refillable beverage containers. In terms of solid waste management, however, **recycling** is the reprocessing of discarded materials into new, useful products (fig. 21.10). Some recycling processes reuse materials for the same purposes; for instance, old aluminum cans and glass bottles are usually melted and recast into new cans and bottles. Other recycling processes turn old materials into entirely new products. Old tires, for instance, are shredded and turned into rubberized road surfacing. Newspapers become cellulose insulation, kitchen wastes become a valuable soil amendment, and steel cans become new automobiles and construction materials.

FIGURE 21.10 Trucks with multiple compartments pick up residential recyclables at curbside, greatly reducing the amount of waste that needs to be buried or burned. For many materials, however, collection costs are too high and markets are lacking for recycling to be profitable.

What Do You Think?

Environmental Justice

Who do you suppose lives closest to toxic waste dumps, Superfund sites, or other polluted areas in your city or county? If you answered poor people and minorities, you are probably right. Everyday experiences tell us that minority neighborhoods are much more likely to have high pollution levels and unpopular industrial facilities such as toxic waste dumps, landfills, smelters, refineries, and incinerators than are middle- or upper-class, white neighborhoods.

One of the first systematic studies showing this inequitable distribution of environmental hazards based on race in the United States was conducted by Robert D. Bullard in 1978. Asked for help by a predominantly black community in Houston that was slated for a waste incinerator, Bullard discovered that all five of the city's existing landfills and six of eight incinerators were located in African-American neighborhoods. In a book entitled *Dumping on Dixie,* Bullard showed that this pattern of risk exposure in minority communities is common throughout the United States (fig. 1).

In 1987, the Commission for Racial Justice of the United Church of Christ published an extensive study of environmental racism. Its conclusion was that race is the most significant variable in determining the location of toxic waste sites in the United States. Among the findings of this study are:

- three of the five largest commercial hazardous waste landfills accounting for about 40 percent of all hazardous waste disposal in the United States are located in predominantly black or Hispanic communities.

- 60 percent of African Americans and Latinos and nearly half of all Asians, Pacific Islanders, and Native Americans live in communities with uncontrolled toxic waste sites.

- The average percentage of the population made up by minorities in communities without a hazardous waste facility is 12 percent. By contrast, communities with one hazardous waste facility have, on average, twice as high (24 percent) a minority population, while those with two or more such facilities average three times as high a minority population (38 percent) as those without one.

- The "dirtiest" or most polluted zip codes in California are in riot-torn South Central Los Angeles where the population is predominantly African American or Latino. Three-quarters of all blacks and half of all Hispanics in Los Angeles live in these polluted areas, while only one-third of all whites live there.

Race is claimed to be the strongest determinant of who is exposed to environmental hazards. Where whites can often "vote with their feet" and move out of polluted and dangerous neighborhoods, minorities are restricted by color barriers and prejudice to less desirable locations. In some areas, though, class or income also are associated with environmental hazards. The difference between *environmental racism* and other kinds of *environmental injustice* can be hard to define. Economic opportunity is often closely tied to race and cultural background in the United States.

Racial inequities also are revealed in the way the government cleans up toxic waste sites and punishes polluters (fig. 2). White communities see faster responses and get better results once toxic wastes are discovered than do minority communities. Penalties assessed against polluters of white communities average six times higher than those against polluters of minority communities. Cleanup is more thorough in white communities as well. Most toxic wastes in white communities are removed or destroyed. By contrast, waste sites in minority neighborhoods are generally only "contained" by putting a cap over them, leaving contaminants in place to potentially resurface or leak into groundwater at a later date. The growing environmental justice movement works to combine civil rights and social justice with environmental concerns to call for a decent, livable environment and equal environmental protection for everyone.

Ethical Considerations

What are the ethical considerations in waste disposal? Does everyone have a right to live in a clean environment or only a right to buy one if they can afford it? What would be a fair way to distribute the risks of toxic wastes? If you had to choose between an incinerator, a secure landfill, or a composting facility for your neighborhood, which would you take?

FIGURE 1 Native Americans protest toxic waste dumping on tribal lands.

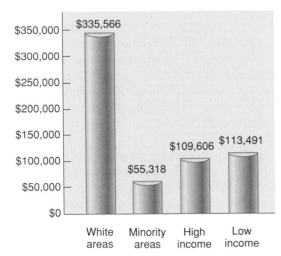

FIGURE 2 Hazardous waste law enforcement. The average fines or penalties per site for violation of the Resource Conservation and Recovery Act vary dramatically with racial composition of the communities where waste was dumped.
Source: M. Lavelle and M. Coyle, *The National Law Journal,* Vol. 15: 52–56, No. 3, September 21, 1992.

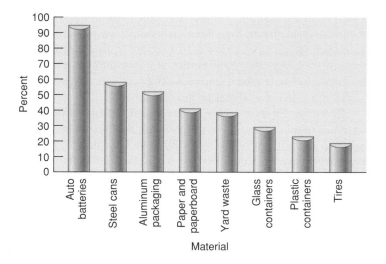

FIGURE 21.11 Recycling rates of selected materials in the United States.
Source: Environmental Protection Agency, 2003.

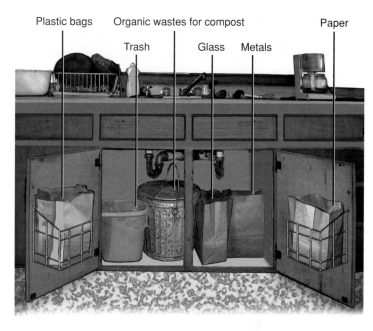

FIGURE 21.12 Source separation in the kitchen—the first step in a strong recycling program. One benefit of recycling is that it reminds us of our responsibility for waste management.

The high value of aluminum scrap ($2,500 per ton in 2007) has spurred a large percentage of aluminum recycling nearly everywhere (fig. 21.11). About two-thirds of all aluminum beverage cans are now recycled; up from only 15 percent in 1970. Aluminum recycling is so rapid that half of the cans now on grocery shelves will be made into another can within two months. Copper has been so valuable recently that thieves have been stripping copper pipes out of empty houses. Gas leaks have caused explosions.

These wild fluctuations in commodity prices are a big problem for recyclers. Newsprint, for example, which jumped to $160 a ton in 1995, dropped to $30 per ton in 2000 then climbed to $95 per ton in 2005. One day, it's so valuable that people are stealing it off the curb; the next day it's literally down in the dumps. It's hard to build a recycling program when you can't count on a stable price for your product.

Recycling saves money, materials, energy, and space

Recycling is usually a better alternative to either dumping or burning wastes. It saves money, energy, raw materials, and land space, while also reducing pollution. Recycling also encourages individual awareness and responsibility for the refuse produced (fig. 21.12). Some recycling facilities now have mechanical sorting machines so that homeowners don't have to separate recyclables into different categories. Everything can be placed in a single container.

Curbside pickup of recyclables costs around $35 per ton, as opposed to the $80 paid to dispose of them at an average metropolitan landfill. Many recycling programs cover their own expenses with materials sales and may even bring revenue to the community. Landfills continue

to dominate American waste disposal but recycling (including composting) has quadrupled since 1980 (fig. 21.13).

Another benefit of recycling is that it could cut our waste volumes drastically. Philadelphia is investing in neighborhood collection centers that will recycle 600 tons a day, enough to eliminate the need for a previously planned, high-priced incinerator. New York City closed its last remaining landfill, Fresh Kills, in 2001. The city now exports its 11,000 tons per day

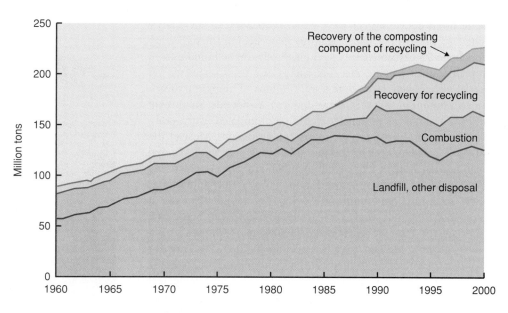

FIGURE 21.13 Disposal of municipal solid waste from 1960 to 2000. Landfills remain the dominant destination, but recycling and composting are increasing.
Source: Environmental Protection Agency, 2003.

of waste by truck, train, and barge, to New Jersey, Pennsylvania, Virginia, South Carolina, and Ohio. New York has set ambitious recycling goals of 50 percent waste reduction, but still the city recycles less than 30 percent of its household and office waste. In contrast, Minneapolis and Seattle recycle nearly 60 percent of domestic waste, Los Angeles and Chicago over 40 percent. In 2002, New York Mayor Michael Bloomberg raised a national outcry by canceling most of the city's recycling program. He argued that the program didn't pay for itself and the money should be spent to balance the city's budget. A year later, Bloomberg relented after realizing that it cost more to ship garbage than to recycle. Recycling was reinstated for nearly all recyclable materials.

Japan is probably the world's leader in recycling (see fig. 21.8). Short of land for landfills, Japan recycles about half its municipal waste and incinerates about 20 percent. The country has begun a push to increase recycling, because incineration costs almost as much. Some communities have raised recycling rates to 80 percent, and others aim to reduce waste altogether by 2020. This level of recycling is most successful when waste is well sorted. In Yokohama, a city of 3.5 million, there are now 10 categories of recyclables, including used clothing and sorted plastics. Some communities have 30 or 40 categories for sorting recyclables.

Recycling lowers our demands for raw resources. In the United States, we cut down 2 million trees every day to produce newsprint and paper products, a heavy drain on our forests. Recycling the print run of a single Sunday issue of the *New York Times* would spare 75,000 trees. Every piece of plastic we make reduces the reserves supply of petroleum and makes us more dependent on foreign oil. Recycling 1 ton of aluminum saves 4 tons of bauxite (aluminum ore) and 700 kg (1,540 lb) of petroleum coke and pitch, as well as keeping 35 kg (77 lb) of aluminum fluoride out of the air.

Recycling also reduces energy consumption and air pollution. Plastic bottle recycling can save 50 to 60 percent of the energy needed to make new ones. Making new steel from old scrap offers up to 75 percent energy savings. Producing aluminum from scrap instead of bauxite ore cuts energy use by 95 percent, yet we still throw away more than a million tons of aluminum every year. If aluminum recovery were doubled worldwide, more than a million tons of air pollutants would be eliminated every year.

Another problem in recycling is contamination. Most of the 24 billion plastic soft drink bottles sold every year in the United States are made of PET (polyethylene terephthalate), which can be melted and remanufactured into carpet, fleece clothing, plastic-strapping, and nonfood packaging. However, even a smidgen of vinyl—a single PVC (polyvinyl chloride) bottle in a truckload, for example—can make PET useless. Although most bottles are now marked with a recycling number, it's hard for consumers to remember which is which. A looming worry is the prospect of single-use, plastic beer bottles. Already being test marketed, these bottles are made of PET but are amber colored to block sunlight and have a special chemical coating to keep out oxygen, which would ruin the beer. The special color, interior coating, and vinyl cap lining will make these bottles incompatible

with regular PET, and it will probably cost more to remove them from the waste stream than the reclaimed plastic is worth. Plastic recycling already is down 50 percent from a decade ago because so many soft drink bottles are sold and consumed on the go, and never make it into recycling bins. Throw-away beer bottles are a looming threat to this industry.

Reducing litter is an important benefit of recycling. Ever since disposable paper, glass, metal, foam, and plastic packaging began to accompany nearly everything we buy, these discarded wrappings have collected on our roadsides and in our lakes, rivers, and oceans. Without incentives to properly dispose of beverage cans, bottles, and papers, it often seems easier to just toss them aside when we have finished using them. Litter is a costly as well as unsightly problem. The United States pays an estimated 32 cents for each piece of litter picked up by crews along state highways, which adds up to $500 million every year. "Bottle-bills" requiring deposits on bottles and cans have reduced littering in many states.

Our present public policies often tend to favor extraction of new raw materials. Energy, water, and raw materials are often sold to industries below their real cost to create jobs and stimulate the economy. For instance, in 1999, a pound of recycled clear PET, the material in most soft drink bottles, sold for about 40¢. By contrast, a pound of off-grade, virgin PET cost 25¢. Setting the prices of natural resources at their real cost would tend to encourage efficiency and recycling. State, local, and national statutes requiring government agencies to purchase a minimum amount of recycled material have helped create a market for used materials. Each of us can play a role in creating markets, as well. If we buy things made from recycled materials—or ask for them if they aren't available—we will help make it possible for recycling programs to succeed (fig. 21.14).

FIGURE 21.14 Commercial-scale recycling recovers and markets resources on a large scale. Consumers can help build markets for recycled goods.

Commercial-scale recycling and composting is an area of innovation

Recycling household waste is the bedrock of recycling programs, but large-scale recycling is growing rapidly. The most common large-scale recycling is **composting** municipal yard waste and tree trimmings. Composting allows natural aerobic (oxygen-rich) decomposition to reduce organic debris to a nutrient-rich soil amendment. Many people compost yard and garden waste in their backyards. Increasingly, cities and towns are providing compost facilities in order to save landfill space. Organic debris such as yard waste makes up 12 percent of our waste stream, and almost half of organic waste is composted (see fig. 21.3).

While compost is a useful material, its market value is low. Many new and exciting technologies are emerging that create still more marketable products, such as energy, from garbage. The Swiss company Kompogas, for example, ferments organic waste in giant tanks, producing natural gas (methane), compost, and fertilizer. The company makes money on both ends, by collecting waste and selling energy and fertilizer.

Every year thousands of tons of debris from building sites and demolition heads to landfills, but recycling facilities are collecting, sorting, and reselling increasing portions of this debris. One recycling facility in Newburgh, New York, recycles over 95 percent of the mixed wood, brick, concrete, metal scrap, and wallboard it receives each day. After sorting and separating, these materials are sold as mulch, crushed stone, gypsum, and recyclable metal and paper. The same company sorted and recycled over 500,000 tons of debris from the World Trade Center towers in just 9 months.

About 6 billion tons of animal wastes are produced from feedlots and processing plants each year in the United States, and these are especially difficult to process because they carry noxious odors and diseases. Industry produces another 5 billion tons per year of plastics, tires, waste oil, asphalt and other organic debris. A new technique called a thermal conversion process (TCP) has attracted much attention since 2003, when articles about it appeared in *Scientific American* and the *MIT Review*. This method essentially pressure-cooks animal manure, plastics, paper-processing waste, and even tires and urban sewage sludge. Extreme heat and pressure reduce organic molecules to simple hydrocarbons—oil, gasoline, and natural gas. An experimental plant in Missouri began commercial fuel production in 2004.

Although landfills and incinerators dwarf these new recycling technologies, recycling is likely to grow as land values and fuel prices continue to rise.

Demanufacturing is necessary for appliances and e-waste

Demanufacturing is the disassembly and recycling of obsolete products, such as TV sets, computers, refrigerators, and air conditioners. As we mentioned earlier, electronics and appliances are among the fastest-growing components of the global waste stream.

Americans throw away about 54 million household appliances, such as stoves and refrigerators, 12 million computers, and uncounted cell phones each year. Most office computers are used only 3 years; televisions last 5 years or so; refrigerators last longer, an average of 12 years. In the United States, an estimated 300 million computers await disposal in storage rooms and garages.

Demanufacturing is key to reducing the environmental costs of e-waste and appliances. A single personal computer can contain 700 different chemical compounds, including toxic materials (mercury, lead, and gallium), and valuable metals (gold, silver, copper), as well as brominated fire retardants and plastics. A typical personal computer has about $6 worth of gold, $5 worth of copper, and $1 of silver. Approximately 40 percent of lead entering U.S. landfills, and 70 percent of heavy metals, comes from e-waste. Batteries and switches in toys and electronics make up another 10 to 20 percent of heavy metals in our waste stream. These contaminants can enter groundwater if computers are landfilled, or the air if they are incinerated. When collected, these materials can become a valuable resource—and an alternative to newly mined materials.

To reduce these environmental hazards, the European Union now requires cradle-to-grave responsibility for electronic products. Manufacturers now have to accept used products or fund independent collectors. An extra $20 (less than one percent of the price of most computers) is added to the purchase price to pay for collection and demanufacturing. Manufacturers selling computers, televisions, refrigerators, and other appliances in Europe must also phase out many of the toxic compounds used in production. Japan is rapidly adopting European environmental standards, and some U.S. companies are following suit, in order to maintain their international markets. In the United States, at least 29 states have passed, or are considering, legislation to control disposal of appliances and computers, in order to protect groundwater and air quality.

Reuse is even more efficient than recycling

Even better than recycling or composting is cleaning and reusing materials in their present form, thus saving the cost and energy of remaking them into something else. We do this already with some specialized items. Auto parts are regularly sold from junkyards, especially for older car models. In some areas, stained glass windows, brass fittings, fine woodwork, and bricks salvaged from old houses bring high prices. Some communities sort and reuse a variety of materials received in their dumps (fig. 21.15).

In many cities, glass and plastic bottles are routinely returned to beverage producers for washing and refilling. The reusable, refillable bottle is the most efficient beverage container we have. This is better for the environment than remelting and more profitable for local communities. A reusable glass container makes an average of 15 round-trips between factory and customer before it becomes so scratched and chipped that it has to be recycled. Reusable containers also favor local bottling companies and help preserve regional differences.

FIGURE 21.15 Reusing discarded products is a creative and efficient way to reduce wastes. This recycling center in Berkeley, California, is a valuable source of used building supplies and a money saver for the whole community.

FIGURE 21.16 How much more do we need? Where will we put what we already have?
Source: Reprinted with special permission of Universal Press Syndicate.

Reducing waste is often the cheapest option

South Africa's effort to reduce the consumption of plastic bags (opening case study) demonstrates the multiple benefits of producing less waste. Excess packaging of food and consumer products is one of our greatest sources of unnecessary waste. Paper, plastic, glass, and metal packaging material make up 50 percent of our domestic trash by volume. Much of that packaging is primarily for marketing and has little to do with product protection (fig. 21.16). Manufacturers and retailers might be persuaded to reduce these wasteful practices if consumers ask for products without excess packaging. Canada's National Packaging Protocol (NPP) recommends that packaging minimize depletion of virgin resources and production of toxins in manufacturing. The preferred hierarchy is (1) no packaging, (2) minimal packaging, (3) reusable packaging, and (4) recyclable packaging.

Where disposable packaging is necessary, we still can reduce the volume of waste in our landfills by using materials that are compostable or degradable. **Photodegradable plastics** break down when exposed to ultraviolet radiation. **Biodegradable plastics** incorporate such materials as cornstarch that can be decomposed by microorganisms. These degradable plastics often don't decompose completely; they only break down to small particles that remain in the environment. In doing so, they can release toxic chemicals into the environment. And in modern, lined landfills they don't decompose at all. Furthermore, they make recycling less feasible and may lead people to believe that littering is okay.

Most of our attention in waste management focuses on recycling. But slowing the consumption of throw-away products is by far the most effective way to save energy, materials, and money. The 3R waste hierarchy—reduce, reuse, recycle—lists the most important strategy first. Industries are increasingly finding that reducing saves money. Soft drink makers use less aluminum per can than they did 20 years ago, and plastic bottles use less plastic. 3M has saved over $500 million in the past

Since the advent of cheap, lightweight, disposable food and beverage containers, many small, local breweries, canneries, and bottling companies have been forced out of business by huge national conglomerates. These big companies can afford to ship food and beverages great distances as long as it is a one-way trip. If they had to collect their containers and reuse them, canning and bottling factories serving large regions would be uneconomical. Consequently, the national companies favor recycling rather than refilling because they prefer fewer, larger plants and don't want to be responsible for collecting and reusing containers. In some circumstances, life-cycle assessment shows that washing and decontaminating containers takes as much energy and produces as much air and water pollution as manufacturing new ones.

In less affluent nations, reuse of all sorts of manufactured goods is an established tradition. Where most manufactured products are expensive and labor is cheap, it pays to salvage, clean, and repair products. Cairo, Manila, Mexico City, and many other cities have large populations of poor people who make a living by scavenging. Entire ethnic populations may survive on scavenging, sorting, and reprocessing scraps from city dumps.

What Can You Do?

Reducing Waste

1. Buy foods that come with less packaging; shop at farmers' markets or co-ops, using your own containers.

2. Take your own washable refillable beverage container to meetings or convenience stores.

3. When you have a choice at the grocery store between plastic, glass, or metal containers for the same food, buy the reusable or easier-to-recycle glass or metal.

4. When buying plastic products, pay a few cents extra for environmentally degradable varieties.

5. Separate your cans, bottles, papers, and plastics for recycling.

6. Wash and reuse bottles, aluminum foil, plastic bags, etc., for your personal use.

7. Compost yard and garden wastes, leaves, and grass clippings.

8. Write to your senators and representatives and urge them to vote for container deposits, recycling, and safe incinerators or landfills.

Source: Minnesota Pollution Control Agency.

FIGURE 21.17 According to the U.S. Environmental Protection Agency, industries produce about one ton of hazardous waste per year for every person in the United States. Responsible handling and disposal is essential.

30 years by reducing its use of raw materials, reusing waste products, and increasing efficiency. Individual action is essential, too (What Can You Do? p. 487).

In 2007, the European Union adopted new regulations that aim to reduce both landfills and waste incineration. For the first time, the waste hierarchy—prevention, reuse, recycling, then disposal only as a last resort—is formalized in law. By 2020, half of all E.U. municipal solid waste and 70 percent of all construction waste is expected to be reused or recycled as a result of this law. No recyclable waste will be allowed in landfills. This law also establishes the "polluter pays" principle (those who create pollution should pay for it), and the "proximity principle," which says that waste should be treated in the nearest appropriate facility to the site at which it was produced. Mixing of toxic waste is also forbidden, making reuse and reprocessing easier.

21.4 HAZARDOUS AND TOXIC WASTES

The most dangerous aspect of the waste stream we have described is that it often contains highly toxic and hazardous materials that are injurious to both human health and environmental quality. We now produce and use a vast array of flammable, explosive, caustic, acidic, and highly toxic chemical substances for industrial, agricultural, and domestic purposes (fig. 21.17). According to the EPA, industries in the United States generate about 265 million metric tons of *officially* classified hazardous wastes each year, slightly more than 1 ton for

each person in the country. In addition, considerably more toxic and hazardous waste material is generated by industries or processes not regulated by the EPA. Shockingly, at least 40 million metric tons (22 billion lbs) of toxic and hazardous wastes are released into the air, water, and land in the United States each year. The biggest source of these toxins are the chemical and petroleum industries (fig. 21.18).

Hazardous waste must be recycled, contained, or detoxified

Legally, a **hazardous waste** is any discarded material, liquid or solid, that contains substances known to be (1) fatal to humans or laboratory animals in low doses, (2) toxic, carcinogenic, mutagenic, or teratogenic to humans or other life-forms, (3) ignitable with a flash point less than 60°C, (4) corrosive, or (5) explosive or highly reactive (undergoes violent chemical reactions either by itself or when mixed with other materials). Notice that this definition includes both toxic and hazardous materials as defined in chapter 8. Certain compounds are exempt from regulation as hazardous waste

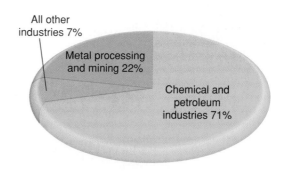

FIGURE 21.18 Producers of hazardous wastes in the United States. **Source:** Data from the U.S. EPA, 2002.

if they are accumulated in less than 1 kg (2.2 lb) of commercial chemicals or 100 kg of contaminated soil, water, or debris. Even larger amounts (up to 1,000 kg) are exempt when stored at an approved waste treatment facility for the purpose of being beneficially used, recycled, reclaimed, detoxified, or destroyed.

Most hazardous waste is recycled, converted to nonhazardous forms, stored, or otherwise disposed of on-site by the generators—chemical companies, petroleum refiners, and other large industrial facilities—so that it doesn't become a public problem. Still, the hazardous waste that does enter the waste stream or the environment represents a serious environmental problem. And orphan wastes left behind by abandoned industries remain a serious threat to both environmental quality and human health. For years, little attention was paid to this material. Wastes stored on private property, buried, or allowed to soak into the ground were considered of little concern to the public. An estimated 5 billion metric tons of highly poisonous chemicals were improperly disposed of in the United States between 1950 and 1975 before regulatory controls became more stringent.

Think About It

Hazardous waste is often poorly managed because it is invisible to the public. What steps do we take to make it invisible? Should the public be more involved in, or take more responsibility for, hazardous waste management? If most waste is produced by the chemical and petroleum industries (fig. 21.18), is there any way that you and your friends or family might help control hazardous waste production?

Federal Legislation

Two important federal laws regulate hazardous waste management and disposal in the United States. The Resource Conservation and Recovery Act (RCRA, pronounced "rickra") of 1976 is a comprehensive program that requires rigorous testing and management of toxic and hazardous substances. A complex set of rules require generators, shippers, users, and disposers of these materials to keep meticulous account of everything they handle and what happens to it from generation (cradle) to ultimate disposal (grave) (fig. 21.19).

The Comprehensive Environmental Response, Compensation and Liability Act (CERCLA or Superfund Act), passed in 1980 and modified in 1984 by the Superfund Amendments and Reauthorization Act (SARA), is aimed at rapid containment, cleanup, or remediation of abandoned toxic waste sites. This statute authorizes the Environmental Protection Agency to undertake emergency actions when a threat exists that toxic material will leak into the environment. The agency is empowered to bring suit for the recovery of its costs from potentially responsible parties such as site owners, operators, waste generators, or transporters.

SARA also established (under title III) community right to know and state emergency response plans that give citizens access to information about what is present in their communities. One of the most useful tools in this respect is the **Toxic Release Inventory,** which requires 20,000 manufacturing facilities to report annually on releases of more than 300 toxic materials. You can find specific information there about what is in your neighborhood.

The government does not have to prove that anyone violated a law or what role they played in a Superfund site. Rather, liability under CERCLA is "strict, joint, and several," meaning that anyone associated with a site can be held responsible for the entire cost of cleaning it up no matter

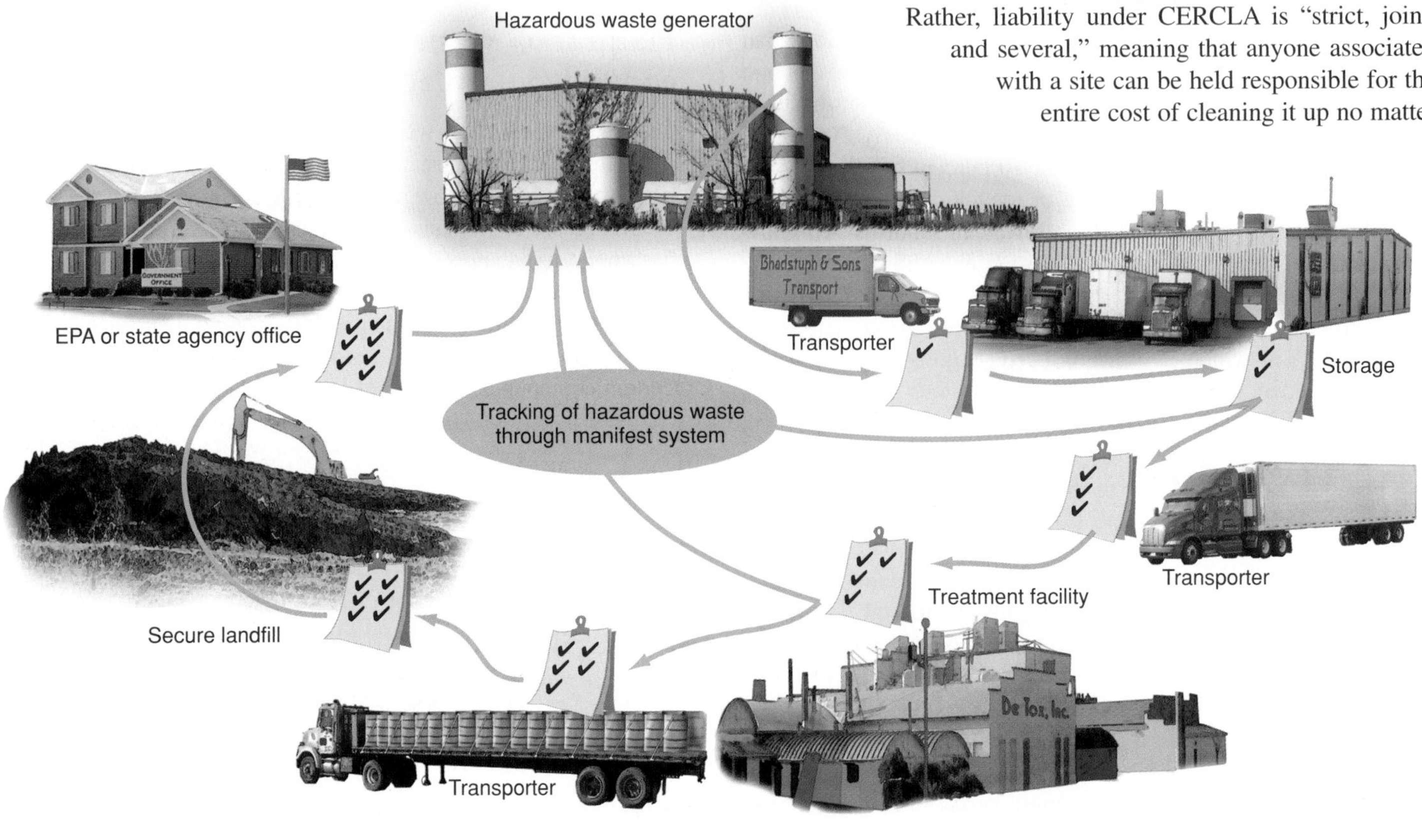

Hazardous waste generator

EPA or state agency office

Transporter

Storage

Tracking of hazardous waste through manifest system

Transporter

Secure landfill

Treatment facility

Transporter

FIGURE 21.19 Toxic and hazardous wastes must be tracked from "cradle to grave" by detailed shipping manifests.

how much of the mess they made. In some cases, property owners have been assessed millions of dollars for removal of wastes left there years earlier by previous owners. This strict liability has been a headache for the real estate and insurance businesses.

CERCLA was amended in 1995 to make some of its provisions less onerous. In cases where treatment is unavailable or too costly and it is likely that a less-costly remedy will become available within a reasonable time, interim containment is now allowed. The EPA also now has the discretion to set site-specific cleanup levels rather than adhere to rigid national standards.

Superfund sites are those listed for federal cleanup

The EPA estimates that there are at least 36,000 seriously contaminated sites in the United States. The General Accounting Office (GAO) places the number much higher, perhaps more than 400,000 when all are identified. By 2007, some 1,680 sites had been placed on the National Priority List (NPL) for cleanup with financing from the federal Superfund program. The **Superfund** is a revolving pool designed to (1) provide an immediate response to emergency situations that pose imminent hazards, and (2) to clean up or remediate abandoned or inactive sites. Without this fund, sites would languish for years or decades while the courts decided who was responsible to pay for the cleanup. Originally a $1.6 billion pool, the fund peaked at $3.6 billion. From its inception, the fund was financed by taxes on producers of toxic and hazardous wastes. Industries opposed this "polluter pays" tax, because current manufacturers are often not the ones responsible for the original contamination. In 1995, Congress agreed to let the tax expire. Since then the Superfund has dwindled, and the public has picked up an increasing share of the bill. In the 1980s the public covered less than 20 percent of the Superfund. Now, public funds have to pick up the entire cost of toxic waste cleanup.

Total costs for hazardous waste cleanup in the United States are estimated between $370 billion and $1.7 trillion, depending on how clean sites must be and what methods are used. For years, Superfund money was spent mostly on lawyers and consultants, and cleanup efforts were often bogged down in disputes over liability and best cleanup methods. During the 1990s, however, progress improved substantially, with a combination of rule adjustments and administrative commitment to cleanup. From 1993 to 2000, the number of completed NPL cleanups jumped from 155 to 757, almost half the list's 1,680 sites (fig. 21.20). Since 2000, progress has slowed again, due to underfunding and a lower priority in the federal government.

What qualifies a site for the NPL? These sites are considered to be especially hazardous to human health and environmental quality because they are known to be leaking or have a potential for leaking supertoxic, carcinogenic, teratogenic, or mutagenic materials (chapter 8). The ten substances of greatest concern or most commonly detected at Superfund sites are lead, trichloroethylene, toluene, benzene, PCBs, chloroform, phenol, arsenic, cadmium, and chromium. These and other hazardous or toxic materials are known to have contaminated groundwater at 75 percent of the sites now on the NPL. In addition, 56 percent of these

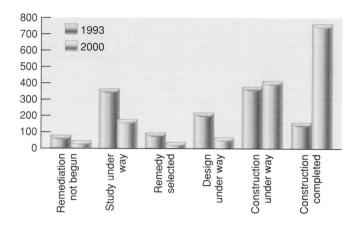

FIGURE 21.20 Progress on Superfund National Priority List (NPL) sites. After years of little progress, the number of completed sites jumped from 155 in 1993 to 757 in 2000. Over 90 percent of the 1,500 NPL sites are under construction or completed.
Source: Environmental Protection Agency, 2001.

sites have contaminated surface waters, and airborne materials are found at 20 percent of the sites.

Where are these thousands of hazardous waste sites, and how did they get contaminated? Old industrial facilities such as smelters, mills, petroleum refineries, and chemical manufacturing plants are highly likely to have been sources of toxic wastes. Regions of the country with high concentrations of aging factories such as the "rust belt" around the Great Lakes or the Gulf Coast petrochemical centers have large numbers of Superfund sites (fig. 21.21). Mining districts also are prime sources of toxic and hazardous waste. Within cities, factories and places such as railroad yards, bus repair barns, and filling stations where solvents, gasoline, oil, and other petrochemicals were spilled or dumped on the ground often are highly contaminated.

Some of the most infamous toxic waste sites were old dumps where many different materials were mixed together

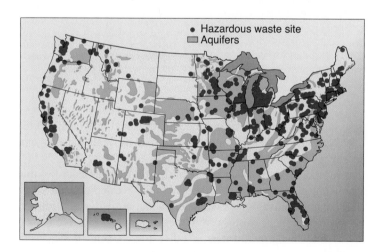

FIGURE 21.21 Some of the hazardous waste sites on the EPA priority cleanup list. Sites located on aquifer recharge zones represent an especially serious threat. Once groundwater is contaminated, cleanup is difficult and expensive. In some cases, it may not be possible.
Source: Environmental Protection Agency.

Getting contaminants out of soil and groundwater is one of the most widespread and persistent problems in waste cleanup. Once leaked into the ground, solvents, metals, radioactive elements, and other contaminants are dispersed and difficult to collect and treat. The main method of cleaning up contaminated soil is to dig it up, then decontaminate it or haul it away and store it in a landfill in perpetuity. At a single site, thousands of tons of tainted dirt and rock may require incineration or other treatment. Cleaning up contaminated groundwater usually entails pumping vast amounts of water out of the ground—hopefully extracting the contaminated water faster than it can spread through the water table or aquifer. In the United States alone, there are tens of thousands of contaminated sites on factories, farms, gas stations, military facilities, sewage treatment plants, landfills, chemical warehouses, and other types of facilities. Cleaning up these sites is expected to cost at least $700 billion.

Recently, a number of promising alternatives have been developed using plants, fungi, and bacteria to clean up our messes. *Phytoremediation* (remediation, or cleanup, using plants) can include a variety of strategies for absorbing, extracting, or neutralizing toxic compounds. Certain types of mustards and sunflowers can extract lead, arsenic, zinc, and other metals (*phytoextraction*). Poplar trees can absorb and break down toxic organic chemicals (*phytodegradation*). Reeds and other water-loving plants can filter water tainted with

sewage, metals, or other contaminants. Natural bacteria in groundwater, when provided with plenty of oxygen, can neutralize contaminants in aquifers, minimizing or even eliminating the need to extract and treat water deep in the ground. Radioactive strontium and cesium have been extracted from soil near the Chernobyl nuclear power plant using common sunflowers.

How do the plants, bacteria, and fungi do all this? Many of the biophysical details are poorly understood, but in general, plant roots are designed to efficiently extract nutrients, water, and minerals from soil and groundwater. The mechanisms involved may aid extraction of metallic and organic contaminants. Some plants also use toxic elements as a defense against herbivores—locoweed, for example, selectively absorbs elements such as selenium, concentrating toxic levels in its leaves. Absorption can be extremely seffective. Braken fern growing in Florida was

found to contain arsenic at concentrations more than 200 times higher than the soil in which it was growing.

Genetically modified plants are also being developed to process toxins. Poplars have been grown with a gene borrowed from bacteria that transform a toxic compound of mercury into a safer form. In another experiment, a gene for producing mammalian liver enzymes, which specialize in breaking down toxic organic compounds, was inserted into tobacco plants. The plants succeeded in producing the liver enzymes and breaking down toxins absorbed through their roots.

These remediation methods are not without risks. As plants take up toxins, insects could consume leaves, allowing contaminants to enter the food web. Some absorbed contaminants are volatilized, or emitted in gaseous form, through pores in plant leaves. Once toxic contaminants are absorbed into plants, the plants themselves are usually toxic and must be landfilled. But the cost of phytoremediation can be less than half the cost of landfilling or treating toxic soil, and the volume of plant material requiring secure storage ends up being a fraction of a percent of the volume of the contaminated dirt.

Cleaning up hazardous and toxic waste sites will be a big business for the foreseeable future, both in the United States and around the world. Innovations such as phytoremediation offer promising prospects for business growth as well as for environmental health and saving taxpayers' money.

Plants can absorb, concentrate, and even decompose toxic contaminants in soil and groundwater.

indiscriminately. For instance, Love Canal in Niagara Falls, New York, was an open dump used by both the city and nearby chemical factories as a disposal site. More than 20,000 tons of toxic chemical waste was buried under what later became a housing development. Another infamous example occurred in Hardeman County, Tennessee, where about a quarter of a million barrels of chemical waste were buried in shallow pits that leaked toxins into the groundwater. In other sites, liquid wastes were pumped into open lagoons or abandoned in warehouses.

Studies of who lives closest to Superfund and toxic release inventory sites reveal that minorities often are overrepresented in these neighborhoods. Charges of environmental racism have

been made, but this is difficult to show conclusively (What Do You Think? p. 482).

Brownfields present both liability and opportunity

Among the biggest problems in cleaning up hazardous waste sites are questions of liability and the degree of purity required. In many cities, these problems have created large areas of contaminated properties known as **brownfields** that have been abandoned or are not being used up to their potential because of real or suspected pollution. Up to one-third of all commercial and industrial sites in the urban core of many big cities fall in this category. In heavy industrial corridors the percentage typically is higher.

For years, no one was interested in redeveloping brownfields because of liability risks. Who would buy a property knowing that they might be forced to spend years in litigation and negotiations and be forced to pay millions of dollars for pollution they didn't create? Even if a site has been cleaned to current standards, there is a worry that additional pollution might be found in the future or that more stringent standards might be applied.

In many cases, property owners complain that unreasonably high levels of purity are demanded in remediation programs. Consider the case of Columbia, Mississippi. For many years a 35 ha (81 acre) site in Columbia was used for turpentine and pine tar manufacturing. Soil tests showed concentrations of phenols and other toxic organic compounds exceeding federal safety standards. The site was added to the Superfund NPL and remediation was ordered. Some experts recommended that the best solution was to simply cover the surface with clean soil and enclose the property with a fence to keep people out. The total costs would have been about $1 million.

Instead, the EPA ordered Reichhold Chemical, the last known property owner, to excavate more than 12,500 tons of soil and haul it to a commercial hazardous waste dump in Louisiana at a cost of some $4 million. The intention is to make the site safe enough to be used for any purpose, including housing—even though no one has proposed building anything there. According to the EPA, the dirt must be clean enough for children to play in it—even eat it every day for 70 years—without risk.

Similarly, in places where contaminants have seeped into groundwater, the EPA generally demands that cleanup be carried to drinking water standards. Many critics believe that these pristine standards are unreasonable. Former Congressman Jim Florio, a principal author of the original Superfund Act, says, "It doesn't make any sense to clean up a rail yard in downtown Newark so it can be used as a drinking water reservoir." Depending on where the site is, what else is around it, and what its intended uses are, much less stringent standards may be perfectly acceptable.

Recognizing that reusing contaminated properties can play a significant role in rebuilding old cities, creating jobs, increasing the tax base, and preventing needless destruction of open space at urban margins, programs have been established at both federal and state levels to encourage brownfield recycling. Adjusting purity standards according to planned uses and providing liability protection for nonresponsible parties gives developers and future purchasers confidence that they won't be unpleasantly surprised in the future with further cleanup costs. In some communities, former brownfields are being turned into "eco-industrial parks" that feature environmentally friendly businesses and bring in much needed jobs to inner-city neighborhoods.

Hazardous waste storage must be safe

What shall we do with toxic and hazardous wastes? In our homes, we can reduce waste generation and choose less toxic materials. Buy only what you need for the job at hand. Use up the last little bit or share leftovers with a friend or neighbor. Many common materials that you probably already have make excellent alternatives to commercial products (What Can You Do? p. 491). Dispose of unneeded materials responsibly (table 21.1).

Produce Less Waste

As with other wastes, the safest and least expensive way to avoid hazardous waste problems is to avoid creating the wastes in the

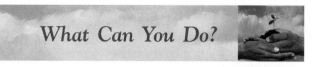

What Can You Do?

Alternatives to Hazardous Household Chemicals

Chrome cleaner: Use vinegar and nonmetallic scouring pad.

Copper cleaner: Rub with lemon juice and salt mixture.

Floor cleaner: Mop linoleum floors with 1 cup vinegar mixed with 2 gallons of water. Polish with club soda.

Brass polish: Use Worcestershire sauce.

Silver polish: Rub with toothpaste on a soft cloth.

Furniture polish: Rub in olive, almond, or lemon oil.

Ceramic tile cleaner: Mix 1/4 cup baking soda, 1/2 cup white vinegar, and 1 cup ammonia in 1 gallon warm water (good general purpose cleaner).

Drain opener: Use plunger or plumber's snake, pour boiling water down drain.

Upholstery cleaner: Clean stains with club soda.

Carpet shampoo: Mix 1/2 cup liquid detergent in 1 pint hot water. Whip into stiff foam with mixer. Apply to carpet with damp sponge. Rinse with 1 cup vinegar in 1 gal water. Don't soak carpet—it may mildew.

Window cleaner: Mix 1/3 cup ammonia, 1/4 cup white vinegar in 1 quart warm water. Spray on window. Wipe with soft cloth.

Spot remover: For butter, coffee, gravy, or chocolate stains: Sponge up or scrape off as much as possible immediately. Dab with cloth dampened with a solution of 1 teaspoon white vinegar in 1 quart cold water.

Toilet cleaner: Pour 1/2 cup liquid chlorine bleach into toilet bowl. Let stand for 30 minutes, scrub with brush, flush.

Pest control: Spray plants with soap-and-water solution (3 tablespoons soap per gallon water) for aphids, mealybugs, mites, and whiteflies. Interplant with pest repellent plants such as marigolds, coriander, thyme, yarrow, rue, and tansy. Introduce natural predators such as ladybugs or lacewings.

Indoor pests: Grind or blend 1 garlic clove and 1 onion. Add 1 tablespoon cayenne pepper and 1 quart water. Add 1 tablespoon liquid soap.

Moths: Use cedar chips or bay leaves.

Ants: Find where they are entering house, spread cream of tartar, cinnamon, red chili pepper, or perfume to block trail.

Fleas: Vacuum area, mix brewer's yeast with pet food.

Mosquitoes: Brewer's yeast tablets taken daily repel mosquitoes.

Note: test cleaners in small, inconspicuous area before using.

TABLE 21.1

How Should You Dispose of Household Hazardous Waste?

Flush to sewer system (drain or toilet)	Cleaning agents with ammonia or bleach, disinfectants, glass cleaner, toilet cleaner
Put dried solids in household trash	Cosmetics, putty, grout, caulking, empty solvent containers, water-based glue, fertilizer (without weed killer)
Save and deliver to a waste collection center	*Solvents:* cleaning agents (drain cleaner, floor wax-stripper, furniture polish, metal cleaner, oven cleaner), paint thinner and other solvents, glue with solvents, varnish, nail polish remover
	Metals: mercury thermometers, button batteries, NiCad batteries, auto batteries, paints with lead or mercury, fluorescent light bulbs/tubes/ballasts, electronics and appliances
	Poisons: bug spray, pesticides, weed killers, rat poison, insect poison, mothballs
	Other chemicals: antifreeze, gasoline, fuel oil, brake fluid, transmission fluid, paint, rust remover, hairspray, photo chemicals

Source: EPA, 2005.

Convert to Less Hazardous Substances

Several processes are available to make hazardous materials less toxic. *Physical treatments* tie up or isolate substances. Charcoal or resin filters absorb toxins. Distillation separates hazardous components from aqueous solutions. Precipitation and immobilization in ceramics, glass, or cement isolate toxins from the environment so that they become essentially nonhazardous. One of the few ways to dispose of metals and radioactive substances is to fuse them in silica at high temperatures to make a stable, impermeable glass that is suitable for long-term storage.

Incineration is a quick way to dispose of many kinds of hazardous waste. Incineration is not necessarily cheap—nor always clean—unless it is done correctly. Wastes must be heated to over 1,000°C (2,000°F) for a sufficient period of time to complete destruction. The ash resulting from thorough incineration is reduced in volume up to 90 percent and often is safer to store in a landfill or other disposal site than the original wastes. Nevertheless, incineration remains a highly controversial topic (fig. 21.22).

Several sophisticated features of modern incinerators improve their effectiveness. Liquid injection nozzles atomize liquids and mix air into the wastes so they burn thoroughly. Fluidized bed burners pump air from the bottom up through burning solid waste as it travels on a metal chain grate through the furnace. The air velocity is sufficient to keep the burning waste partially suspended. Plenty of oxygen is available, and burning is quick and complete. Afterburners add to the completeness of burning by igniting gaseous hydrocarbons not consumed in the incinerator. Scrubbers and precipitators remove minerals, particulates, and other pollutants from the stack gases.

Chemical processing can transform materials so they become nontoxic. Included in this category are neutralization, removal of metals or halogens (chlorine, bromine, etc.), and oxidation. The

first place. Manufacturing processes can be modified to reduce or eliminate waste production. In Minnesota, the 3M Company reformulated products and redesigned manufacturing processes to eliminate more than 140,000 metric tons of solid and hazardous wastes, 4 billion l (1 billion gal) of wastewater, and 80,000 metric tons of air pollution each year. They frequently found that these new processes not only spared the environment but also saved money by using less energy and fewer raw materials.

Recycling and reusing materials also eliminates hazardous wastes and pollution. Many waste products of one process or industry are valuable commodities in another. Already, about 10 percent of the wastes that would otherwise enter the waste stream in the United States are sent to surplus material exchanges where they are sold as raw materials for use by other industries. This figure could probably be raised substantially with better waste management. In Europe, at least one-third of all industrial wastes are exchanged through clearinghouses where beneficial uses are found. This represents a double savings: The generator doesn't have to pay for disposal, and the recipient pays little, if anything, for raw materials.

FIGURE 21.22 Actor Martin Sheen joins local activists in a protest in East Liverpool, Ohio, site of the largest hazardous waste incinerator in the United States. About 1,000 people marched to the plant to pray, sing, and express their opposition. Involving celebrities draws attention to your cause. A peaceful, well-planned rally builds support and acceptance in the broader community.

Sunohio Corporation of Canton, Ohio, for instance, has developed a process called PCBx in which chlorine in such molecules as PCBs is replaced with other ions that render the compounds less toxic. A portable unit can be moved to the location of the hazardous waste, eliminating the need for shipping them.

Biological waste treatment or **bioremediation** taps the great capacity of microorganisms to absorb, accumulate, and detoxify a variety of toxic compounds. Bacteria in activated sludge basins, aquatic plants (such as water hyacinths or cattails), soil microorganisms, and other species remove toxic materials and purify effluents. Recent experiments have produced bacteria that can decontaminate organic waste metals by converting them to harmless substances. Biotechnology offers exciting possibilities for finding or creating organisms to eliminate specific kinds of hazardous or toxic wastes. By using a combination of classic genetic selection techniques and high-technology gene-transfer techniques, for instance, scientists have recently been able to generate bacterial strains that are highly successful at metabolizing PCBs. There are concerns about releasing such exotic organisms into the environment, however (chapter 11). It may be better to keep these organisms contained in enclosed reaction vessels and feed contaminated material to them under controlled conditions.

Store Permanently

Inevitably, there will be some materials that we can't destroy, make into something else, or otherwise cause to vanish. We will have to store them out of harm's way. There are differing opinions about how best to do this.

Retrievable Storage. Dumping wastes in the ocean or burying them in the ground generally means that we have lost control of them. If we learn later that our disposal technique was a mistake, it is difficult, if not impossible, to go back and recover the wastes. For many supertoxic materials, the best way to store them may be in **permanent retrievable storage.** This means placing waste storage containers in a secure building, salt mine, or bedrock cavern where they can be inspected periodically and retrieved, if necessary, for repacking or for transfer if a better means of disposal is developed. This technique is more expensive than burial in a landfill because the storage area must be guarded and monitored continuously to prevent leakage, vandalism, or other dispersal of toxic materials. Remedial measures are much cheaper with this technique, however, and it may be the best system in the long run.

Secure Landfills. One of the most popular solutions for hazardous waste disposal has been landfilling. Although, as we saw earlier in this chapter, many such landfills have been environmental disasters, newer techniques make it possible to create safe, modern **secure landfills** that are acceptable for disposing of many hazardous wastes. The first line of defense in a secure landfill is a thick bottom cushion of compacted clay that surrounds the pit like a bathtub (fig. 21.23). Moist clay is flexible and resists cracking if the ground shifts. It is impermeable to groundwater and will safely contain wastes. A layer of gravel is

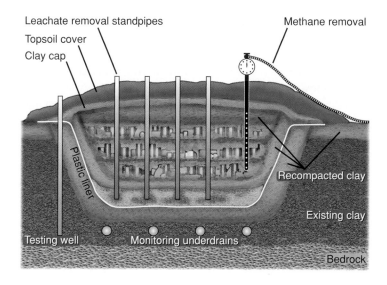

FIGURE 21.23 A secure landfill for toxic waste. A thick plastic liner and two or more layers of impervious compacted clay enclose the landfill. A gravel bed between the clay layers collects any leachate, which can then be pumped out and treated. Well samples are tested for escaping contaminants and methane is collected for combustion.

spread over the clay liner and perforated drain pipes are laid in a grid to collect any seepage that escapes from the stored material. A thick polyethylene liner, protected from punctures by soft padding materials, covers the gravel bed. A layer of soil or absorbent sand cushions the inner liner and the wastes are packed in drums, which then are placed into the pit, separated into small units by thick berms of soil or packing material.

When the landfill has reached its maximum capacity, a cover much like the bottom sandwich of clay, plastic, and soil—in that order—caps the site. Vegetation stabilizes the surface and improves its appearance. Sump pumps collect any liquids that filter through the landfill, either from rainwater or leaking drums. This leachate is treated and purified before being released. Monitoring wells check groundwater around the site to ensure that no toxins have escaped.

Most landfills are buried below ground level to be less conspicuous; however, in areas where the groundwater table is close to the surface, it is safer to build above-ground storage. The same protective construction techniques are used as in a buried pit. An advantage to such a facility is that leakage is easier to monitor because the bottom is at ground level.

Transportation of hazardous wastes to disposal sites is of concern because of the risk of accidents. Emergency preparedness officials conclude that the greatest risk in most urban areas is not nuclear war or natural disaster but crashes involving trucks or trains carrying hazardous chemicals through densely packed urban corridors. Another worry is who will bear financial responsibility for abandoned waste sites. The material remains toxic long after the businesses that created it are gone. As is the case with nuclear wastes (chapter 19), we may need new institutions for perpetual care of these wastes.

CONCLUSION

In many traditional societies, people reuse nearly everything because they can't afford to discard useful resources. Modern society, however, produces a prodigious amount of waste. Government policies and economies of scale make it cheaper and more convenient to extract virgin raw materials to make new consumer products rather than to reuse or recycle items that still have useful life. We're now beginning to recognize the impacts of this wasteful lifestyle. We see the problems associated with waste disposal as well as the impacts of energy and material resource extraction. The increasing toxicity of modern products makes waste reduction even more urgent. The mantra of reduction, reuse, and recycle is becoming more widely accepted.

There are increasing opportunities to exchange materials with others who can use them, or to recycle them into other products.

A big market for used construction supplies and surplus chemicals allows salvage of stuff that would otherwise go to landfills. Vehicles, electronics, and other complex products are demanufactured to reclaim valuable metals. Paint, used carpet, food and beverage containers, and many other unwanted consumer products are transformed into new merchandise. Organic matter can be composted into beneficial soil amendments. Some pioneers in sustainability find they can live comfortably while producing no waste at all if they practice reduction, reuse, and recycling faithfully.

How much waste do you produce, and where does it go after you toss it into the garbage can? What can you do to reduce your personal waste flow? Is recycling and reuse widely accepted in your community? If not, what could you do to change attitudes toward trash?

REVIEWING LEARNING OUTCOMES

By now you should be able to explain the following points:

21.1 Identify the components of solid waste.
- The waste stream is everything we throw away.

21.2 Describe how wastes have been—and are being—disposed of or treated.
- Open dumps release hazardous materials into air and water.
- Ocean dumping is nearly uncontrollable.
- We often export waste to countries ill-equipped to handle it.
- Landfills receive most of our waste.
- Incineration produces energy but causes pollution.

21.3 Identify how we might shrink the waste stream.
- Recycling captures resources from garbage.
- Recycling saves money, materials, energy, and space.

- Commercial-scale recycling and composting is an area of innovation.
- Demanufacturing is necessary for appliances and e-waste.
- Reuse is even more efficient than recycling.
- Reducing waste is often the cheapest option.

21.4 Investigate hazardous and toxic wastes.
- Hazardous waste must be recycled, contained, or detoxified.
- Superfund sites are those listed for federal cleanup.
- Brownfields present both liability and opportunity.
- Hazardous waste storage must be safe.

PRACTICE QUIZ

1. What are solid wastes and hazardous wastes? What is the difference between them?
2. Describe the difference between an open dump, a sanitary landfill, and a modern, secure, hazardous waste disposal site.
3. Why are landfill sites becoming limited around most major urban centers in the United States? What steps are being taken to solve this problem?
4. Describe some concerns about waste incineration.
5. List some benefits and drawbacks of recycling wastes. What are the major types of materials recycled from municipal waste and how are they used?
6. What is composting, and how does it fit into solid waste disposal?
7. Describe some ways that we can reduce the waste stream to avoid or reduce disposal problems.
8. List ten toxic substances in your home and how you would dispose of them.
9. What are brownfields and why do cities want to redevelop them?
10. What societal problems are associated with waste disposal? Why do people object to waste handling in their neighborhoods?

CRITICAL THINKING AND DISCUSSION QUESTIONS

1. A toxic waste disposal site has been proposed for the Pine Ridge Indian Reservation in South Dakota. Many tribal members oppose this plan, but some favor it because of the jobs and income it will bring to an area with 70 percent unemployment. If local people choose immediate survival over long-term health, should we object or intervene?

2. There is often a tension between getting your personal life in order and working for larger structural changes in society. Evaluate the trade-offs between spending time and energy sorting recyclables at home compared to working in the public arena on a bill to ban excess packaging.

3. Should industry officials be held responsible for dumping chemicals that were legal when they did it but are now known to be extremely dangerous? At what point can we argue that they *should* have known about the hazards involved?

4. Look at the discussion of recycling or incineration presented in this chapter. List the premises (implicit or explicit) that underlie the presentation as well as the conclusions (stated or not) that seem to be drawn from them. Do the conclusions necessarily follow from these premises?

5. The Netherlands incinerates much of its toxic waste at sea by a shipborne incinerator. Would you support this as a way to dispose of our wastes as well? What are the critical considerations for or against this approach?

 **DATA analysis** How Much Waste Do You Produce, and How Much Could You Recycle?

As people become aware of waste disposal problems in their communities, more people are recycling more materials. Some things are easy to recycle, such as newsprint, office paper, or aluminum drink cans. Other things are harder to classify. Most of us give up pretty quickly and throw things in the trash if we have to think too hard about how to recycle them.

1. Take a poll to find out how many people in your class know how to recycle the items in the table shown here. Once you have taken your poll, convert the numbers to percentages: divide the number who know how to recycle each item by the number of students in your class, and then multiply by 100.

2. Now find someone on your campus who works on waste management. This might be someone in your university/ college administration, or it might be someone who actually empties trash containers. (You might get more interesting and straightforward answers from the latter.) Ask the following questions: (1) Can this person fill in the items your class didn't know about? (2) Is there a college/ university policy about recycling? What are some of the points on that policy? (3) How much does the college spend each year on waste disposal? How many tuition payments does that total? (4) What are the biggest parts of the waste stream? (5) Does the school have a plan for reducing that largest component?

Item	Percentage Who Know How to Recycle
Newspapers	
Paperboard (cereal boxes)	
Cardboard boxes	
Cardboard boxes with tape	
Plastic drink bottles	
Other plastic bottles	
Styrofoam food containers	
Food waste	
Plastic shopping bags	
Plastic packaging materials	
Furniture	
Last year's course books	
Left-over paint	

For Additional Help in Studying This Chapter, please visit our website at www.mhhe.com/cunningham10e. You will find additional practice quizzes and case studies, flashcards, regional examples, place markers for Google Earth™ mapping, and an extensive reading list, all of which will help you learn environmental science.

Like mega cities in many developing countries, New Delhi, India suffers from traffic congestion and air pollution.

CHAPTER

Urbanization and Sustainable Cities

22

What kind of world do you want to live in? Demand that your
teachers teach you what you need to know to build it.

—Peter Kropotkin—

LEARNING OUTCOMES

After studying this chapter, you should be able to:

22.1 Define urbanization.

22.2 Describe why cities grow.

22.3 Understand urban challenges in the developing world.

22.4 Identify urban challenges in the developed world.

22.5 Explain smart growth.

Case Study   Curitiba: A Model Sustainable City

A few decades ago, Curitiba, Brazil, like many cities in the developing world, faced rapid population growth, air pollution, congested streets, and inadequate waste disposal systems (see photo opposite page). In 1969, Jamie Lerner, a landscape architect and former student activist, ran for mayor on a platform of urban renewal, social equity, and environmental protection. For the next 30 years, first as mayor, and then governor of Parana State, Lerner instituted social reforms and urban planning that have made the city an outstanding model of sustainable urban development. Today, with a population of about 1.8 million, Curitiba has an international reputation for progressive social programs, environmental protection, cleanliness, and livability. An international conference of mayors and city planners called Curitiba the most innovative city in the world.

The master plan instituted by Lerner called for rational municipal zoning and land-use regulations—rare provisions for a city in the developing world. An extensive network of parks and open space protected from future development extends throughout the city and provides access to nature as well as recreation opportunities. It also helps protect the watershed and prevent floods that once afflicted the city. Tanguá Park, for example, Curitiba's newest, is built on land once destined to be a garbage dump. The dump was moved elsewhere, however, and the park now encompasses about 250 ha of open space, including a rare stand of Brazilian pine (*Araucaria agustafolia*), the symbol of Parana State. Together, the many parks, bicycle trails, city gardens, and public riverbanks make the city a beautiful and pleasant place to live. Curitiba has more open space per capita than most American cities.

The heart of Curitiba's environmental and social plan is education for everyone. The first federal university in Brazil was built here in 1912. Free municipal schools give the city the highest literacy rate in the country. Environmental education reaches to everyone as well, School children study ecology along with Portuguese and math. With the help of children, who encourage their parents, the city has instituted a complex program that separates organic waste, trash, plastic, glass, and metal for recycling. Everything reusable is salvaged from the waste stream and sold as raw material to local industries. The city calculates that 1,200 trees per day are saved by paper recycled in this program. "Imagine if the whole of Brazil did this," Lerner exclaims. "We could save 26 million trees per year!" More than 70 percent of the city residents now participate in recycling.

Environmental protection and resource conservation also contribute to social welfare programs. Low-income Curitibanos are employed to pick up and sort recyclables, as well as to work in civic gardens and a city-run farm. Along with the many street sweepers in their distinctive orange suits, these programs provide jobs and pride while also keeping the city clean and beautiful. More than 1,100 municipal facilities provide services to the entire population. Free day care is available to all working mothers with children under 6 years old. Eldercare programs provide psychological services as well as food and housing to seniors. New job-training and small-business incubators run by the city help Curitibanos learn technical skills and launch new enterprises. The city has also built a technology park to attract new-economy businesses.

One of the crises that originally motivated Mayor Lerner was preservation of the historic central business district of his city. Traffic choked narrow streets and air pollution drove wealthy residents and businesses out of the central city, leaving beautiful old buildings to decay and eventually be torn down. In one of his boldest moves, Lerner banned most vehicles from the central business district and turned the streets into pedestrian malls. Rather than depend on private automobiles to move people around, Curitiba built a remarkably successful and economical bus-based rapid transit system. More than 340 feeder routes throughout the metropolitan area link to high-speed articulated buses that travel on dedicated busways. Everyone in the city lives within walking distance of frequent, economical, rapid public transportation. Today more than three-quarters of all trips made each day in the city utilize this innovative system, a percentage beyond the fondest imagination of most American city planners.

Benches, fountains, landscaping, distinctive lighting, and other amenities along with European-style shops and sidewalk cafes now make Curitiba's central district a pleasant place to stroll, shop, or simply gather with friends (fig. 22.1). Historic buildings have been refurbished, and many have been recycled for new purposes, Throughout the city center, walking streets are paved with characteristic black and white ceramic mosaics that draw on the city's colonial past.

Curitiba illustrates a number of ways that we can live sustainably with our environment and with each other. In this chapter, we'll look at other aspects of city planning and urban environments as well as some principles of ecological economics that help us understand the nature of resources and the choices we face both as individuals and communities.

FIGURE 22.1 Much of Curitiba's city center has been converted to pedestrian shopping streets, while historic buildings have been converted to new uses. Black and white tile mosaics are featured throughout the city.

22.1 URBANIZATION

For most of human history, the vast majority of people have lived in rural areas where they engaged in hunting and gathering, farming, fishing, or other natural-resource based occupations. Since the beginning of the Industrial Revolution about 300 years ago, however, cities have grown rapidly in both size and power (fig. 22.2). Now, for the first time ever, more people live in urban areas than in the country. In 1950, only 38 percent of the world population lived in cities (table 22.1). By 2030, that percentage is expected to nearly double. This means that over the next three decades about 3 billion people will crowd into cities. Some areas—Europe, North America, and Latin America—are already highly urbanized. Only Africa and Asia are below 45 percent urbanized.

Demographers predict that 90 percent of the human population growth in this century will occur in developing countries, and that almost all of that growth will occur in cities (fig. 22.3). Already huge **urban agglomerations** (mergers of multiple municipalities) are forming throughout the world. Some of these

TABLE 22.1

Urban Share of Total Population (Percentage)

	1950	2000	2030*
Africa	18.4	40.6	57.0
Asia	19.3	43.8	59.3
Europe	56.0	75.0	81.5
Latin America	40.0	70.3	79.7
North America	63.9	77.4	84.5
Oceania	32.0	49.5	60.7
World	38.3	59.4	70.5

*Projected

Source: United Nations Population Division, 2004.

megacities (urban areas with populations over 10 million people) are already truly enormous, some claiming up to 30 million people. Can cities this size—especially in poorer countries—supply all the public services necessary to sustain a civilized life?

If we are to learn to live sustainably—that is, to depend on renewable resources while also protecting environmental quality, biodiversity, and the ecological services on which all life depends—that challenge will have to be met primarily in the cities of the world, where most people will live. Many of us dream of moving to a secluded hideaway in the country, where we could grow our own food, chop wood, and carry water. But it probably isn't possible for 6.5 billion people (let alone the 8 to 9 billion expected by the end of this century) to live rural, subsistence lifestyles. Learning to live together in cities is probably the only way we'll survive.

Cities can be engines of economic progress and social reform. Some of the greatest promise for innovation comes from cities like Curitiba, where innovative leaders can focus knowledge and resources on common problems. Cities can be efficient places to live, where mass transportation can move people around and goods

FIGURE 22.2 In less than 20 years, Shanghai, China, has built Pudong, a new city of 1.5 million residents and 500 skyscrapers on former marshy farmland across the Huang Pu River from the historic city center. This kind of rapid urban growth is occurring in many developing countries.

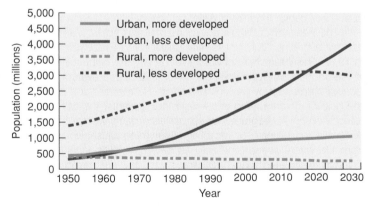

FIGURE 22.3 Growth of urban and rural populations in more-developed regions and in less-developed regions.
Source: United Nations Population Division. *World Urbanization Prospects,* 2004.

FIGURE 22.4 Since their earliest origins, cities have been centers of education, religion, commerce, politics, and culture. Unfortunately, they have also been sources of pollution, crowding, disease, and misery.

FIGURE 22.5 This village in Chiapas, Mexico, is closely tied to the land through culture, economics, and family relationships. While the timeless pattern of life here gives a great sense of identity, it can also be stifling and repressive.

and services are more readily available than in the country. Concentrating people in urban areas leaves open space available for farming and biodiversity. But cities can also be dumping grounds for poverty, pollution, and unwanted members of society. Providing food, housing, transportation, jobs, clean water, and sanitation to the 2 or 3 billion new urban residents expected to crowd into cities—especially those in the developing world—in this century may be one of the preeminent challenges of this century.

As the case study of Curitiba shows, there is much we can do to make our cities more livable. It's especially encouraging that a Brazilian city, where per capita income is only one-fifth that of the United States, can make so much progress. But what of countries where personal wealth is only one-tenth that of Brazil? What hope is there for them?

Cities have specialized functions as well as large populations

Since their earliest origins, cities have been centers of education, religion, commerce, record keeping, communication, and political power. As cradles of civilization, cities have influenced culture and society far beyond their proportion of the total population (fig. 22.4). Until about 1900, only a small percentage of the world's people lived permanently in urban areas, and even the greatest cities of antiquity were small by modern standards. The vast majority of humanity has always lived in rural areas where they subsisted on natural resources—farming, fishing, hunting, timber harvesting, animal herding, or mining.

Just what makes up an urban area or a city? Definitions differ. The U.S. Census Bureau considers any incorporated community to be a city, regardless of size, and defines any city with more than 2,500 residents as urban. More meaningful definitions are based on *functions*. In a **rural area,** most residents depend on agriculture or other ways of harvesting natural resources for their livelihood. In an **urban area,** by contrast, a majority of the people are not directly dependent on natural resource-based occupations.

A **village** is a collection of rural households linked by culture, custom, family ties, and association with the land (fig. 22.5). A **city,** by contrast, is a differentiated community with a population

and resource base large enough to allow residents to specialize in arts, crafts, services, or professions rather than natural resource-based occupations. While the rural village often has a sense of security and connection, it also can be stifling. A city offers more freedom to experiment, to be upwardly mobile, and to break from restrictive traditions, but it can be harsh and impersonal.

Beyond about 10 million inhabitants, an urban area is considered a supercity or **megacity.** Megacities in many parts of the world have grown to enormous size. Chongqing, China, having annexed a large part of Sichuan province and about 30 million people, claims to be the biggest city in the world. In the United States, urban areas between Boston and Washington, D.C., have merged into a nearly continuous megacity (sometimes called Bos-Wash) containing about 35 million people. The Tokyo-Yokohama-Osaka-Kobe corridor contains nearly 50 million people. Because these agglomerations have expanded beyond what we normally think of as a city, some geographers prefer to think of urbanized **core regions** that dominate the social, political, and economic life of most countries (fig. 22.6).

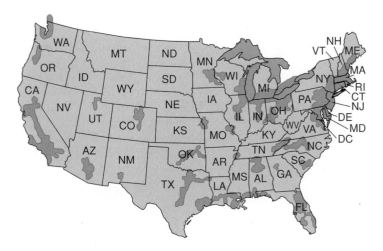

FIGURE 22.6 Urban core agglomerations (*lavender areas*) are forming megalopolises in many areas. While open space remains in these areas, the flow of information, capital, labor, goods, and services links each into an interacting system.
Source: U.S. Census Bureau.

Large cities are expanding rapidly

You can already see the dramatic shift in size and location of big cities. In 1900 only 13 cities in the world had populations over 1 million (table 22.2). All of those cities, except Tokyo and Peking were in Europe or North America. London was the only city in the world with more than 5 million residents. By 2007, there were at least 300 cities—100 of them in China alone—with more than 1 million residents. Of the 13 largest of these metropolitan areas, none are in Europe. Only New York City and Los Angeles are in a developed country. By 2025, it's expected that at least 93 cities will have populations over 5 million, and three-fourths of those cities will be in developing countries (fig. 22.7). In just the next 25 years, Mumbai, India; Delhi, India; Karachi, Pakistan; Manila, Philippines; and Jakarta, Indonesia all are expected to grow by at least 50 percent.

China represents the largest demographic shift in human history. Since the end of Chinese collectivized farming and factory work in 1986, around 250 million people have moved from rural areas to cities. And in the next 25 years an equal number is expected to join this vast exodus. In addition to expanding existing cities, China plans to build 400 new urban centers with populations of at least 500,000 over the next 20 years. Already at least half of the concrete and one-third of the steel used in construction around the world each year is consumed in China.

TABLE 22.2

The World's Largest Urban Areas (Populations in Millions)

1900		2015**	
London, England	6.6	Tokyo, Japan	31.0
New York, USA	4.2	New York, USA	29.9
Paris, France	3.3	Mexico City	21.0
Berlin, Germany	2.4	Seoul, Korea	19.8
Chicago, USA	1.7	São Paulo, Brazil	18.5
Vienna, Austria	1.6	Osaka, Japan	17.6
Tokyo, Japan	1.5	Jakarta, Indonesia	17.4
St. Petersburg, Russia	1.4	Delhi, India	16.7
Philadelphia, USA	1.4	Los Angeles, USA	16.6
Manchester, England	1.3	Beijing, China	16.0
Birmingham, England	1.2	Cairo, Egypt	15.5
Moscow, Russia	1.1	Manila, Philippines	13.5
Peking, China*	1.1	Buenos Aires, Brazil	12.9

*Now spelled Beijing.
**Projected.

Source: T. Chandler, Three Thousand Years of Urban Growth, 1974, Academic Press and World Gazetter, 2003.

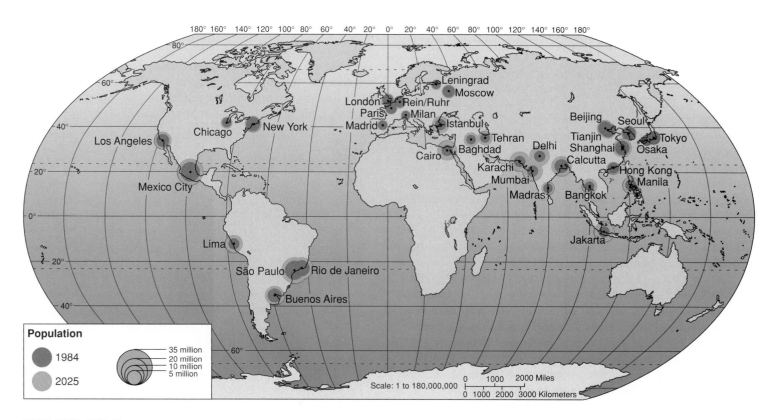

FIGURE 22.7 By 2025, at least 400 cities will have populations of 1 million or more, and 93 supercities will have populations above 5 million. Three-fourths of the world's largest cities will be in developing countries that already have trouble housing, feeding, and employing their people.

Consider Shanghai, for example. In 1985, the city had a population of about 10 million. It's now about 19 million—including at least 4 million migrant laborers. In the past decade, Shanghai has built 4,000 skyscrapers (buildings with more than 25 floors). The city already has twice as many tall buildings as Manhattan, and proposals have been made for 1,000 more. The problem is that most of this growth has taken place in a swampy area called Pudong, across the Huang Pu River from the historic city center (see fig. 22.2). Pudong is now sinking about 1.5 cm per year due to groundwater drainage and the weight of so many buildings.

Other Chinese cities have plans for similar massive building projects to revitalize blighted urban areas. Harbin, an urban complex of about 9 million people and the capital of Heilongjiang Province, for example, recently announced plans to relocate the entire city across the Songhua River on 740 km² (285 mi²) of former farmland. Residents hope these new towns will be both more livable for their residents and more ecologically sustainable than the old cities they're replacing. In 2005, the Chinese government signed a long-term contract with a British engineering firm to build at least five "eco-cities," each the size of a large Western capital. Plans calls for these cities to be self-sufficient in energy, water, and most food products, with the aim of zero emissions of greenhouse gases from transportation.

22.2 WHY DO CITIES GROW?

Urban populations grow in two ways: by natural increase (more births than deaths) and by immigration. Natural increase is fueled by improved food supplies, better sanitation, and advances in medical care that reduce death rates and cause populations to grow both within cities and in the rural areas around them (chapter 7). In Latin America and East Asia, natural increase is responsible for two-thirds of urban population growth. In Africa and West Asia, immigration is the largest source of urban growth. Immigration to cities can be caused both by **push factors** that force people out of the country and by **pull factors** that draw them into the city.

Immigration is driven by push and pull factors

People migrate to cities for many reasons. In some areas, the countryside is overpopulated and simply can't support more people. The "surplus" population is forced to migrate to cities in search of jobs, food, and housing. Not all rural-to-urban shifts are caused by overcrowding in the country, however. In some places, economic forces or political, racial, or religious conflicts drive people out of their homes. The countryside may actually be depopulated by such demographic shifts. The United Nations estimated that in 2002 at least 19.8 million people fled their native country and that about another 20 million were internal refugees within their own country, displaced by political, economic, or social instability. Many of these refugees end up in the already overcrowded megacities of the developing world.

Land tenure patterns and changes in agriculture also play a role in pushing people into cities. The same pattern of agricultural mechanization that made farm labor largely obsolete in the United States early in this century is spreading now to developing countries. Furthermore, where land ownership is concentrated in the hands of a wealthy elite, subsistence farmers are often forced off the land so it can be converted to grazing lands or monoculture cash crops. Speculators and absentee landlords let good farmland sit idle that otherwise might house and feed rural families.

Even in the largest and most hectic cities, many people are there by choice, attracted by the excitement, vitality, and opportunity to meet others like themselves. Cities offer jobs, housing, entertainment, and freedom from the constraints of village traditions. Possibilities exist in the city for upward social mobility, prestige, and power not ordinarily available in the country. Cities support specialization in arts, crafts, and professions for which markets don't exist elsewhere.

Modern communications also draw people to cities by broadcasting images of luxury and opportunity. An estimated 90 percent of the people in Egypt, for instance, have access to a television set. The immediacy of television makes city life seem more familiar and attainable than ever before. We generally assume that beggars and homeless people on the streets of developing nations' teeming cities have no other choice of where to live, but many of these people want to be in the city. In spite of what appears to be dismal conditions, living in the city may be preferable to what the country had to offer.

Government policies can drive urban growth

Government policies often favor urban over rural areas in ways that both push and pull people into the cities. Developing countries commonly spend most of their budgets on improving urban areas (especially around the capital city where leaders live), even though only a small percentage of the population lives there or benefits directly from the investment. This gives the major cities a virtual monopoly on new jobs, housing, education, and opportunities, all of which bring in rural people searching for a better life. In Peru, for example, Lima accounts for 20 percent of the country's population, but has 50 percent of the national wealth, 60 percent of the manufacturing, 65 percent of the retail trade, 73 percent of the industrial wages, and 90 percent of all banking in the country. Similar statistics pertain to São Paulo, Mexico City, Manila, Cairo, Lagos, Bogotá, and a host of other cities.

Governments often manipulate exchange rates and food prices for the benefit of more politically powerful urban populations but at the expense of rural people. Importing lower-priced food pleases city residents, but local farmers then find it uneconomical to grow crops. As a result, an increased number of people leave rural areas to become part of a large urban workforce, keeping wages down and industrial production high. Zambia, for instance, sets maize prices below the cost of local production to discourage farming and to maintain a large pool of workers for the mines. Keeping the currency exchange rate high stimulates export trade but makes it difficult for small farmers to buy the fuels, machinery, fertilizers, and seeds that they need. This depresses rural employment and rural income while stimulating the urban economy. The effect is to transfer wealth from the country to the city.

22.3 URBAN CHALLENGES IN THE DEVELOPING WORLD

Large cities in both developed and developing countries face similar challenges in accommodating the needs and by-products of dense populations. The problems are most intense, however, in rapidly growing cities of developing nations.

As figure 22.7 shows, most human population growth in the next century is expected to occur in the developing world, mainly in Africa, Asia, and South America. Almost all of that growth will occur in cities—especially the largest cities—which already have trouble supplying food, water, housing, jobs, and basic services for their residents. The unplanned and uncontrollable growth of those cities causes tragic urban environmental problems. Consider as you study urban conditions what responsibilities we in the richer countries have to help others who are less fortunate and how we might do so.

> ### Think About It
>
> How many of the large cities shown in figure 22.7 are in developing countries? What are some differences between large cities in wealthy countries and those in less-wealthy countries? If you were a farmer in India or China, what would encourage you to move to one of these cities?

Traffic congestion and air quality are growing problems

A first-time visitor to a supercity—particularly in a less-developed country—is often overwhelmed by the immense crush of pedestrians and vehicles of all sorts that clog the streets. The noise, congestion, and confusion of traffic make it seem suicidal to venture onto the street. Jakarta, for instance, is one of the most congested cities in the world (fig. 22.8). Traffic is chaotic almost all the time. People commonly spend three or four hours each way commuting to work from outlying areas. Bangkok also has monumental traffic problems. The average resident spends the equivalent of 44 days a year sitting in traffic jams. About 20 percent of all fuel is consumed by vehicles standing still. Hours of work lost each year are worth at least $3 billion.

Traffic congestion is expected to worsen in many developing countries as the number of vehicles increases but road construction fails to keep pace. Figure 22.9 shows trends between 1980 and 2000 in selected countries. All this traffic, much of it involving old, poorly maintained vehicles, combines with smoky factories, and use of wood or coal fires for cooking and heating to create a thick pall of air pollution in the world's supercities.

FIGURE 22.8 Motorized rickshaws, motor scooters, bicycles, street vendors, and pedestrians all vie for space on the crowded streets of Jakarta. But in spite of the difficulties of living here people work hard and have hope for the future.

Lenient pollution laws, corrupt officials, inadequate testing equipment, ignorance about the sources and effects of pollution, and lack of funds to correct dangerous situations usually exacerbate the problem.

What is its human toll? An estimated 60 percent of Kolkut's residents are thought to suffer from respiratory diseases linked to air pollution. Lung cancer mortality in Shanghai is reported to be four to seven times higher than rates in the countryside. There have been some encouraging success stories, however. As we saw

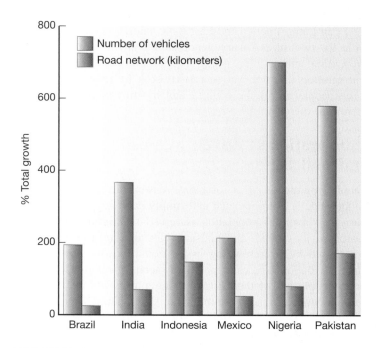

FIGURE 22.9 Transport growth in selected developing countries, 1980–2000.
Source: Earth Trends, 2006.

in chapter 16, air pollution in Delhi, India, decreased dramatically after vehicles were required to install emission controls and use cleaner fuels.

Insufficient sewage treatment causes water pollution

Few cities in developing countries can afford to build modern waste treatment systems for their rapidly growing populations. The World Bank estimates that only 35 percent of urban residents in developing countries have satisfactory sanitation services. The situation is especially desperate in Latin America, where only 2 percent of urban sewage receives any treatment. In Egypt, Cairo's sewer system was built about 50 years ago to serve a population of 2 million people. It is now being overwhelmed by more than 10 million people. Less than one-tenth of India's 3,000 towns and cities have even partial sewage systems and water treatment facilities. Some 150 million of India's urban residents lack access to sanitary sewer systems.

Figure 22.10 shows one of many tidal canals that crisscross Jakarta, and serve as the sewage disposal system for as much as half the 10 million city residents. In 2007, unusually heavy rain backed up these canals and flooded about half the city. Health officials braced for disease epidemics.

FIGURE 22.10 This tidal canal in Jakarta serves as an open sewer. By some estimates, about half of the 10 million residents of this city have no access to modern sanitation systems.

Some 400 million people, or about one-third of the population, in developing world cities do not have safe drinking water, according to the World Bank. Although city dwellers are somewhat more likely than rural people to have clean water, this still represents a large problem. Where people have to buy water from merchants, it often costs 100 times as much as piped city water and may not be any safer to drink. Many rivers and streams in developing countries are little more than open sewers, and yet they are all that poor people have for washing clothes, bathing, cooking, and—in the worst cases—for drinking. Diarrhea, dysentery, typhoid, and cholera are widespread diseases in these countries, and infant mortality is tragically high (chapter 8).

Many cities lack adequate housing

The United Nations estimates that at least 1 billion people—15 percent of the world's population—live in crowded, unsanitary slums of the central cities and in the vast shantytowns and squatter settlements that ring the outskirts of most developing world cities. Around 100 million people have no home at all. In Mumbai, India, for example, it is thought that half a million people sleep on the streets, sidewalks, and traffic circles because they can find no other place to live. In Brazil, perhaps 1 million "street kids" who have run away from home or been abandoned by their parents live however and wherever they can. This is surely a symptom of a tragic failure of social systems.

Slums are generally legal but inadequate multifamily tenements or rooming houses, either custom built to rent to poor people or converted from some other use. The chals of Mumbai, India, for example, are high-rise tenements built in the 1950s to house immigrant workers. Never very safe or sturdy, these dingy, airless buildings are already crumbling and often collapse without warning. Eighty-four percent of the families in these tenements live in a single room; half of those families consist of six or more people. Typically, they have less than 2 square meters of floor space per person and only one or two beds for the whole family. They may share kitchen and bathroom facilities down the hall with 50 to 75 other people. Even more crowded are the rooming houses for mill workers where up to 25 men sleep in a single room only 7 meters square. Because of this crowding, household accidents are a common cause of injuries and deaths in developing world cities, especially to children. Charcoal braziers or kerosene stoves used in crowded homes are a routine source of fires and injuries. With no place to store dangerous objects beyond the reach of children, accidental poisonings and other mishaps are a constant hazard.

Shantytowns are settlements created when people move onto undeveloped lands and build their own houses. Shacks are built of corrugated metal, discarded packing crates, brush, plastic sheets, or whatever building materials people can scavenge. Some shantytowns are simply illegal subdivisions where the landowner rents land without city approval. Others are spontaneous or popular settlements or **squatter towns** where people occupy land without the owner's permission. Sometimes this occupation

FIGURE 22.11 Homeless people have built shacks along this busy railroad track in Jakarta. It's a dangerous place to live, with many trains per day using the tracks, but for the urban poor, there are few other choices.

FIGURE 22.12 Plastic waste has created mountains of garbage, especially in developing countries with insufficient waste management systems.

involves thousands of people who move onto unused land in a highly organized, overnight land invasion, building huts and laying out streets, markets, and schools before authorities can root them out. In other cases, shantytowns just gradually "happen."

Called *barriads, barrios, favelas,* or *turgios* in Latin America, *bidonvillas* in Africa, or *bustees* in India, shantytowns surround every megacity in the developing world (fig. 22.11). They are not an exclusive feature of poor countries, however. Some 250,000 immigrants and impoverished citizens live in the *colonias* along the southern Rio Grande in Texas. Only 2 percent have access to adequate sanitation. Many live in conditions as awful as you would see in any developing world city.

About three-quarters of the residents of Addis Ababa, Ethiopia, or Luanda, Angola, live in squalid refugee camps. Two-thirds of the population of Calcutta live in unplanned squatter settlements and nearly half of the 20 million people in Mexico City live in uncontrolled, unauthorized shantytowns. Many governments try to clean out illegal settlements by bulldozing the huts and sending riot police to drive out the settlers, but the people either move back in or relocate to another shantytown.

These popular but unauthorized settlements usually lack sewers, clean water supplies, electricity, and roads. Often the land on which they are built was not previously used because it is unsafe or unsuitable for habitation. In Bhopal, India, and Mexico City, for example, squatter settlements were built next to deadly industrial sites. In Rio de Janeiro, La Paz (Bolivia), Guatemala City, and Caracas (Venezuela), they are perched on landslide-prone hills. In Lima (Peru), Khartoum (Sudan), and Nouakchott (Mauritania), shantytowns have spread onto sandy deserts. In Manila, thousands of people live in huts built on towering mounds of garbage and burning industrial waste in city dumps (fig. 22.12).

As desperate and inhumane as conditions are in these slums and shantytowns, many people do more than merely survive there. They keep themselves clean, raise families, educate their children, find jobs, and save a little money to send home to their parents. They learn to live in a dangerous, confusing, and rapidly changing world and have hope for the future. The people have parties; they sing and laugh and cry. They are amazingly adaptable and resilient. In many ways, their lives are no worse than those in the early industrial cities of Europe and America a century ago. Perhaps continuing development will bring better conditions to cities of the developing world as it has for many in the developed world.

22.4 URBAN CHALLENGES IN THE DEVELOPED WORLD

For the most part, the rapid growth of central cities that accompanied industrialization in nineteenth- and early twentieth-century Europe and North America has now slowed or even reversed. London, for instance, once the most populous city in the world, has lost nearly 2 million people, dropping from its high of 8.6 million in 1939 to about 6.7 million now. While the greater metropolitan area surrounding London has been expanding to about 10 million inhabitants, the city itself is now only the twelfth largest city in the world.

Many of the worst urban environmental problems of the more developed countries have been substantially reduced in recent years. Minority groups in inner cities, however, remain vulnerable to legacies of environmental degradation in industrial cities (What Do You Think? p. 505).

For most urban residents developed countries, in recent decades. Improved sanitation and medical care have reduced or totally eliminated many of the communicable diseases that once afflicted urban residents. Air and water quality have improved dramatically as heavy industry such as steel smelting and chemical manufacturing have moved to developing countries. In consumer and information economies, workers no longer need to be

What Do You Think?

People for Community Recovery

The Lake Calumet Industrial District on Chicago's far South Side is an environmental disaster area. A heavily industrialized center of steel mills, oil refineries, railroad yards, coke ovens, factories, and waste disposal facilities, much of the site is now a marshy wasteland of landfills, toxic waste lagoons, and slag dumps, around a system of artificial ship channels.

At the southwest corner of this degraded district sits Altgeld Gardens, a low-income public housing project built in the late 1940s by the Chicago Housing Authority. The 2,000 units of "The Gardens" or "The Projects," as they are called by the largely minority residents, are low-rise rowhouses, many of which are vacant or in poor repair. But residents of Altgeld Gardens are doing something about their neighborhood. People for Community Recovery (PCR) is a grassroots citizen's group

organized to work for a clean environment, better schools, decent housing, and job opportunities for the Lake Calumet neighborhood.

PCR was founded in 1982 by Mrs. Hazel Johnson, an Altgeld Gardens resident whose husband died from cancer that may have been pollution-related. PCR has worked to clean up more than two dozen waste sites and contaminated properties in their immediate vicinity. Often this means challenging authorities to follow established rules and enforce existing statutes. Public protests, leafleting, and community meetings have been effective in public education about the dangers of toxic wastes and have helped gain public support for cleanup projects.

PCR's efforts successfully blocked construction of new garbage and hazardous waste landfills, transfer stations, and incinerators in the Lake Calumet district. Pollution prevention programs have been established at plants still in operation. And PCR helped set up a community monitoring program to stop illegal dumping and to review toxic inventory data from local companies.

Education is an important priority for PCR. An environmental education center administered by community members organizes workshops, seminars, fact sheets, and outreach for citizens and local businesses. A public health education and screening program has been set up to improve community health. Partnerships have been established with nearby Chicago State University to provide technical assistance and training in environmental issues.

PCR also works on economic development. Environmentally responsible products and services are now available to residents. Jobs that are being created as Green businesses are brought into the community. Wherever possible, local people and minority contractors from the area are hired to clean up waste sites and restore abandoned buildings. Job training for youth and adults as well as retraining for displaced workers is a high priority. Funding for these projects has come from fines levied on companies for illegal dumping.

PCR and Mrs. Johnson have received many awards for their fight against environmental racism and despair. In 1992, PCR was the recipient of the President's Environmental and Conservation Challenge Award. PCR is the only African-American grassroots organization in the country to receive this prestigious award.

Although Altgeld Gardens is far from clean, much progress has been made. Perhaps the most important accomplishment is community education and empowerment. Residents have learned how and why they need to work together to improve their living conditions. Could these same lessons be useful in your city or community? What could you do to help improve urban environments where you live?

The Calumet industrial district in South Chicago.

---

concentrated in central cities. They can live and work in dispersed sites. Automobiles now make it possible for much of the working class to enjoy amenities such as single-family homes, yards, and access to recreation that once were available only to the elite.

In the United States, old, dense manufacturing cities such as Philadelphia and Detroit have lost population as industry has moved to developing countries. In a major demographic shift,

both businesses and workers have moved west and south. Some of the most rapidly growing metropolitan areas like Phoenix, Arizona; Boulder, Colorado; Austin, Texas; and San Jose, California, are centers for high-technology companies located in landscaped suburban office parks. These cities often lack a recognizable downtown, being organized instead around low-density housing developments, national-chain shopping malls, and extensive freeway networks. For many high-tech companies, being located near industrial centers and shipping is less important than a good climate, ready access to air travel, and amenities such as natural beauty and open space.

Urban sprawl consumes land and resources

While the move to suburbs and rural areas has brought many benefits to the average citizen, it also has caused numerous urban problems. Cities that once were compact now spread over the landscape, consuming open space and wasting resources. This pattern of urban growth is known as **sprawl.** While there is no universally accepted definition of the term, sprawl generally includes the characteristics outlined in table 22.3. As former Maryland Governor Parris N. Glendening said, "In its path, sprawl consumes thousands of acres of forests and farmland, woodlands and wetlands. It requires government to spend millions extra to build new schools, streets, and water and sewer lines." And Christine Todd Whitman, former New Jersey governor and head of the Environmental Protection Agency, said, "Sprawl eats up our open space. It creates traffic jams that boggle the mind and pollute the air. Sprawl can make one feel downright claustrophobic about our future."

In most American metropolitan areas, the bulk of new housing is in large, tract developments that leapfrog out beyond the edge of the city in a search for inexpensive rural land with few restrictions on land use or building practices (fig. 22.13). The U.S. Department of Housing and Urban Development estimates that urban sprawl consumes some 200,000 ha (roughly

FIGURE 22.13 Huge houses on sprawling lots consume land, alienate us from our neighbors, and make us ever more dependent on automobiles. They also require a lot of lawn mowing!

500,000 acres) of farmland each year. Because cities often are located in fertile river valleys or shorelines, much of that land would be especially valuable for producing crops for local consumption. But with planning authority divided among many small, local jurisdictions, metropolitan areas have no way to regulate growth or provide for rational, efficient resource use. Small towns and township or county officials generally welcome this growth because it profits local landowners and business people. Although the initial price of tract homes often is less than comparable urban property, there are external costs in the form of new roads, sewers, water mains, power lines, schools, and shopping centers and other extra infrastructure required by this low-density development.

Landowners, builders, real estate agents, and others who profit from this crazy-quilt development pattern generally claim that growth benefits the suburbs in which it occurs. They promise that adding additional residents will lower the average taxes for everyone, but in fact, the opposite often is true. In a study titled *Better Not Bigger*, author Eben Fodor analyzed the costs of medium-density and low-density housing. In suburban Washington, D.C., for instance, each new house on a quarter acre (0.1 ha) lot cost $700 more than it paid in taxes. A typical new house on a 5 acre (2 ha) lot, however, cost $2,200 more than it paid in taxes because of higher expenses for infrastructure and services. Ironically, people who move out to rural areas to escape from urban problems such as congestion, crime, and pollution often find that they have simply brought those problems with them. A neighborhood that seemed tranquil and remote when they first moved in, soon becomes just as crowded, noisy, and difficult as the city they left behind as more people join them in their rural retreat.

In a study of 58 large American urban areas, author and former mayor of Albuquerque, David Rusk, found that between 1950 and 1990, populations grew 80 percent, while land area grew 305 percent. In Atlanta, Georgia, the population grew 32 percent between 1990 and 2000, while the total metropolitan area

---

TABLE 22.3

Characteristics of Urban Sprawl

1. Unlimited outward extension.
2. Low-density residential and commercial development.
3. Leapfrog development that consumes farmland and natural areas.
4. Fragmentation of power among many small units of government.
5. Dominance of freeways and private automobiles.
6. No centralized planning or control of land uses.
7. Widespread strip malls and "big-box" shopping centers.
8. Great fiscal disparities among localities.
9. Reliance on deteriorating older neighborhoods for low-income housing.
10. Decaying city centers as new development occurs in previously rural areas.

Source: Excerpt from speech by Anthony Downs at the CTS Transportation Research Conference, as appeared on Website by Planners Web, Burlington, VT, 2001.

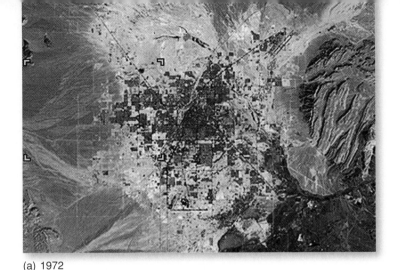

(a) 1972

(b) 1992

FIGURE 22.14 Satellite images of Las Vegas, Nevada, in 1972 (a) and 1992 (b) show how the metropolitan area has grown over two decades.

increased by 300 percent. The city is now more than 175 km across. Atlanta loses an estimated $6 million to traffic delays every day. By far the fastest growing metropolitan region in the United States is Las Vegas, Nevada, which doubled its population but quadrupled its size in the 1990s (fig. 22.14*a* and *b*).

Expanding suburbs force long commutes

The U.S. Interstate Highway System was the largest construction project in human history. Originally justified as necessary for national defense, it was really a huge subsidy for the oil, rubber, automobile, and construction industries. Its 72,000 km (45,000 mi) of freeways probably did more than anything to encourage sprawl and change America into an auto-centered society.

Because many Americans live far from where they work, shop, and recreate, they consider it essential to own a private automobile. The average U.S. driver spends about 443 hours per year behind a steering wheel. This means that for most people, the equivalent of one full 8-hour day per week is spent sitting in an automobile. Of the 5.8 billion barrels of oil consumed each year in the United States (60 percent of which is imported), about two-thirds is burned in cars and trucks. As chapter 16 shows, about two-thirds of all carbon monoxide, one-third of all nitrogen oxides, and one-quarter of all volatile organic compounds emitted each year from human-caused sources in the United States are released by automobiles, trucks, and buses.

Building the roads, parking lots, filling stations, and other facilities needed for an automobile-centered society takes a vast amount of space and resources. In some metropolitan areas it is estimated that one-third of all land is devoted to the automobile. To make it easier for suburban residents to get from their homes to jobs and shopping, we provide an amazing network of freeways and highways. At a cost of several trillion dollars to build, the interstate highway system was designed to allow us to drive at high speeds from source to destination without ever having to stop. As more and more drivers clog the highways, however, the reality is far different. In Los Angeles, for example, which has the worst congestion in the United States, the average speed in 1982 was 58 mph (93 km/hr), and the average driver spent less than 4 hours per year in traffic jams. In 2000, the average speed in Los Angeles was only 35.6 mph (57.3 km/hr), and the average driver spent 82 hours per year waiting for traffic. Although new automobiles are much more efficient and cleaner operating than those of a few decades ago, the fact that we drive so much farther today and spend so much more time idling in stalled traffic means that we burn more fuel and produce more pollution than ever before.

Altogether, it is estimated that traffic congestion costs the United States $78 billion per year in wasted time and fuel. Some people argue that the existence of traffic jams in cities shows that more freeways are needed. Often, however, building more traffic lanes simply encourages more people to drive farther than before. Rather than ease congestion and save fuel, more freeways can exacerbate the problem (fig. 22.15).

Sprawl impoverishes central cities from which residents and businesses have fled. With a reduced tax base and fewer civic leaders living or working in downtown areas, the city is unable to maintain its infrastructure. Streets, parks, schools, and civic buildings fall into disrepair at the same time that these facilities are being built at great expense in new suburbs. The poor who are left behind when the upper and middle classes abandon the city center often can't find jobs where they live and have no way to commute to the suburbs where jobs are now located. About one-third of Americans are too young, too old, or too poor to drive. For these people, car-oriented development causes isolation and makes daily tasks like grocery shopping very difficult. Parents, especially mothers, spend long hours transporting young children. Teenagers and aging grandparents are forced to drive, often presenting a hazard on public roads.

Sprawl also is bad for your health. By encouraging driving and discouraging walking, sprawl promotes a sedentary lifestyle that contributes to heart attacks and diabetes, among other problems. In Atlanta, for example, the lowest-density suburbs tend to have significantly higher rates of overweight residents than the highest-density neighborhoods.

FIGURE 22.15 Building new freeways to reduce congestion is like trying to solve a weight problem by buying bigger clothes.

Finally, sprawl fosters uniformity and alienation from local history and natural environment. Housing developments often are based on only a few standard housing styles, while shopping centers and strip malls everywhere feature the same national chains. You could drive off the freeway in the outskirts of almost any big city in America and see exactly the same brands of fast-food restaurants, motels, stores, filling stations, and big-box shopping centers.

Think About It

Who benefits most from urban sprawl, and who benefits least? In what ways do you benefit and suffer from sprawl? Do home buyers initiate the process of urban expansion, or do developers? What conditions help make this process so persistent?

Mass-transit could make our cities more livable

As the opening Case Study for this chapter shows, Curitiba provides an excellent example of successful mass transit and environmentally sound, socially just, and economically sustainable urban planning. Curitiba's bus rapid transit system focuses around five transportation corridors radiating from the city center

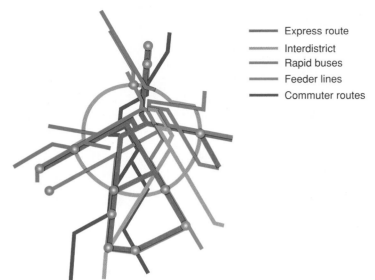

— Express route
— Interdistrict
— Rapid buses
— Feeder lines
— Commuter routes

FIGURE 22.16 Curitiba's transit system has more than 340 interlinked bus routes. Neighborhood feeder lines connect to high-volume, express buses running on dedicated busways. Interdistrict routes carry passengers between suburbs, while specialized commuter routes carry workers directly from distant neighborhoods to the city center. Altogether this system moves more than 1.9 million people per day, or more than 75 percent of all personal transportation in the city.

(fig. 22.16). Within these corridors, high-speed, articulated buses, each of which can carry 270 passengers, travel on dedicated roadways closed to all other vehicles. These bus-trains make limited stops at transfer terminals that link to 340 feeder routes extending throughout the city. Everyone in the city is within walking distance of a bus stop that has frequent, convenient, affordable service.

Each bus station along the dedicated busway is made up of tubular structures elevated to be at the same height as the bus floor, providing access for disabled persons and making entry and exit quicker and easier for everyone (fig. 22.17). Passengers pay as they

FIGURE 22.17 High-speed, bi-articulated buses travel on dedicated transit ways in Curitiba and make limited stops at elevated tubular bus terminals that allow many passengers to disembark and reload during 60-second stops.

go through a turnstile to enter the terminal. When a bus pulls up, it opens multiple doors through which passengers can enter or exit. Dozens of passengers might disembark and an equal number get on the bus during the 60-second stop, making a trip over many kilometers remarkably fast. The system charges a single fee for a trip in one direction, regardless of the number of transfers involved. This makes the system equitable for those who can't afford to live in expensive neighborhoods close to the city center.

Curitiba's buses make more than 21,000 trips per day, traveling more than 440,000 km (275,000 mi) and carrying 1.9 million passengers, or about three-quarters of all personal trips within the city. One of the best things about this system is its economy. Working with existing roadways for the most part, the city was able to construct this system for one-tenth the cost of a light rail system or freeway system, and one-hundredth the cost of a subway. The success of bus rapid transit has allowed Curitiba to remain relatively compact and to avoid the sprawl engendered by an American-style freeway system.

22.5 SMART GROWTH

Smart growth is a term that describes such strategies for well-planned developments that make efficient and effective use of land resources and existing infrastructure. An alternative to haphazard, poorly planned sprawling developments, smart growth involves thinking ahead to develop pleasant neighborhoods while minimizing the wasteful use of space and tax dollars for new roads and extended sewer and water lines.

Smart growth aims to make land-use planning democratic. Public discussions allow communities to guide planners. Mixing land uses, rather than zoning exclusive residential areas far separate from commercial areas, makes living in neighborhoods more enjoyable. By planning a range of housing styles and costs, smart growth allows people of all income levels, including young families and aging grandparents, to find housing they can afford. Open communication between planners and the community helps make urban expansion fair, predictable, and cost-effective.

Smart growth approaches acknowledge that urban growth is inevitable; the aim is to direct growth, to make pleasant spaces for us to live, and to preserve some accessible, natural spaces for all to enjoy (table 22.4). It strives to promote the safety, livability, and revitalization of existing urban and rural communities.

Smart growth protects environmental quality. It attempts to reduce traffic and to conserve farmlands, wetlands, and open space. This may mean restricting land use, but it also means finding economically sound ways to reuse polluted industrial areas within the city (fig. 22.18). As cities grow and transportation and communications enable communities to interact more, the need for regional planning becomes both more possible and more pressing. Community and business leaders need to make decisions based on a clear understanding of regional growth needs and how infrastructure can be built most efficiently and for the greatest good.

TABLE 22.4

Goals for Smart Growth

1. Create a positive self-image for the community.
2. Make the downtown vital and livable.
3. Alleviate substandard housing.
4. Solve problems with air, water, toxic waste, and noise pollution.
5. Improve communication between groups.
6. Improve community member access to the arts.

Source: Vision 2000, Chattanooga, TN.

One of the best examples of successful urban land-use planning in the United States is Portland, Oregon, which has rigorously enforced a boundary on its outward expansion, requiring, instead, that development be focused on in-filling unused space within the city limits. Because of its many urban amenities, Portland is considered one of the best cities in America. Between 1970 and 1990, the Portland population grew by 50 percent but its total land area grew only 2 percent. During this time, Portland property taxes decreased 29 percent and vehicle miles traveled increased only 2 percent. By contrast, Atlanta, which had similar population growth, experienced an explosion of urban sprawl that increased its land area three-fold, drove up property taxes 22 percent, and increased traffic miles by 17 percent. A result of this expanding traffic and increasing congestion was that

FIGURE 22.18 Many cities have large amounts of unused open space that could be used to grow food. Residents often need help decontaminating soil and gaining access to the land.

Atlanta's air pollution increased by 5 percent, while Portland's, which has one of the best public transit systems in the nation, decreased by 86 percent.

Garden cities and new towns were early examples of smart growth

The twentieth century saw numerous experiments in building **new towns** for society at large that try to combine the best features of the rural village and the modern city. One of the most influential of all urban planners was Ebenezer Howard (1850–1929), who not only wrote about ideal urban environments but also built real cities to test his theories. In *Garden Cities of Tomorrow,* written in 1898, Howard proposed that the congestion of London could be relieved by moving whole neighborhoods to **garden cities** separated from the central city by a greenbelt of forests and fields.

In the early 1900s, Howard worked with architect Raymond Unwin to build two garden cities outside of London, Letchworth and Welwyn Garden. Interurban rail transportation provided access to these cities. Houses were clustered in "superblocks" surrounded by parks, gardens, and sports grounds. Streets were curved. Safe and convenient walking paths and overpasses protected pedestrians from traffic. Businesses and industries were screened from housing areas by vegetation. Each city was limited to about 30,000 people to facilitate social interaction. Housing and jobs were designed to create a mix of different kinds of people and to integrate work, social activities, and civic life. Trees and natural amenities were carefully preserved and the towns were laid out to maximize social interactions and healthful living. Care was taken to meet residents' psychological needs for security, identity, and stimulation.

Letchworth and Welwyn Garden each have 70 to 100 people per acre. This is a true urban density, about the same as New York City in the early 1800s and five times as many people as most suburbs today. By planning the ultimate size in advance and choosing the optimum locations for housing, shopping centers, industry, transportation, and recreation, Howard believed he could create a hospitable and satisfying urban setting while protecting open space and the natural environment. He intended to create parklike surroundings that would preserve small-town values and encourage community spirit in neighborhoods.

Planned communities also have been built in the United States following the theories of Ebenezer Howard, but most plans have been based on personal automobiles rather than public transit. Radburn, New Jersey, was designed in the 1920s, and two highly regarded new towns of the 1960s are Reston, Virginia, and Columbia, Maryland. More recent examples, such as Seaside in northern Florida, represent a modern movement in new towns known as "new urbanism."

New urbanism advanced the ideas of smart growth

New towns and garden cities included many important ideas, but they still left cities behind. Rather than abandon the cultural history and infrastructure investment in existing cities, a group of architects and urban planners is attempting to redesign metropolitan areas to make them more appealing, efficient, and livable. In the United States, Andres Duany, Peter Calthorpe, and others have led this movement and promoted the term "new urbanism" to describe it. Sometimes called a neo-traditionalist approach, their designs attempt to recapture a small-town neighborhood feel in new developments. The goal of new urbanism has been to rekindle Americans' enthusiasm for cities. New urbanist architects do this by building charming, integrated, walkable developments. Sidewalks, porches, and small front yards encourage people to get outside and be sociable. A mix of apartments, townhouses, and detached houses in a variety of price ranges ensures that neighborhoods will include a diversity of ages and income levels. Some design principles of this movement include:

- Limit city size or organize them in modules of 30,000 to 50,000 people, large enough to be a complete city but small enough to be a community. A greenbelt of agricultural and recreational land around the city limits growth while promoting efficient land use. By careful planning and cooperation with neighboring regions, a city of 50,000 people can have real urban amenities such as museums, performing arts centers, schools, hospitals, etc.

- Determine in advance where development will take place. Such planning protects property values and prevents chaotic development in which the lowest uses drive out the better ones. It also recognizes historical and cultural values, agricultural resources, and such ecological factors as impact on wetlands, soil types, groundwater replenishment and protection, and preservation of aesthetically and ecologically valuable sites.

- Locate everyday shopping and services so people can meet daily needs with greater convenience, less stress, less automobile dependency, and less use of time and energy. Provide accessible, sociable public spaces (fig. 22.19).

- Increase jobs in the community by locating offices, light industry, and commercial centers in or near suburbs, or by enabling work at home via computer terminals. These alternatives save commuting time and energy and provide local jobs.

- Encourage walking or the use of small, low-speed, energy-efficient vehicles (microcars, motorized tricycles, bicycles, etc.) for many local trips now performed by full-size automobiles.

- Promote more diverse, flexible housing as alternatives to conventional, detached single-family houses. "In-fill" building between existing houses saves energy, reduces land costs, and might help provide a variety of living arrangements. Allowing owners to turn unused rooms into rental units provides space for those who can't afford a house and brings income to retired people who don't need a whole house themselves.

- Create housing "superblocks" that use space more efficiently and foster a sense of security and community. Widen peripheral arterial streets and provide pedestrian overpasses so traffic flows smoothly around residential areas; narrow streets within

FIGURE 22.19 This walking street in Queenstown, New Zealand, provides opportunities for shopping, dining, and socializing in a pleasant outdoor setting.

FIGURE 22.20 This award-winning green roof on the Chicago City Hall is functional as well as beautiful. It reduces rain runoff by about 50 percent, and keeps the surface as much as 30°F cooler than a conventional roof on hot summer days.

blocks, to slow traffic so children can play more safely. The land released from streets can be used for gardens, linear parks, playgrounds, and other public areas that will foster community spirit and encourage people to get out and walk.

Green urbanism promotes sustainable cities

While new urbanism has promoted livable neighborhoods and raised interest in cities, critics point out that green urbanist developments, like garden cities and new towns, have often been **greenfield developments,** projects built on previously undeveloped farmlands or forests on the outskirts of large cities. In addition to contributing to sprawl, developments built on greenfields still require most residents to commute to work by private car, which undermines efforts to reduce car dependence. Goals of mixed-income neighborhoods also fall short, because the architect-designed houses rarely fall into middle- or low-income price ranges.

"Green urbanism" is a term that describes many strategies to redevelop existing cities to promote ecologically sound practices. Many green urbanist ideas are demonstrated in the BedZED project in London, England (What Do You Think? p. 573).

European cities have been especially innovative in green planning. Stockholm, Sweden, has expanded by building small satellite suburbs linked to the central city by commuter rails and by bicycle routes that pass through a network of green spaces that reach far into the city. Copenhagen, Denmark, has rebuilt most of its transportation infrastructure since the 1960s, including more than 300 km of well-marked bike lanes and separated bike trails. Thirty percent of all trips through central Copenhagen are made using public transportation, and 14 percent of the trips are made by bicycle.

Green building strategies are encouraged in many European cities. Many German cities now require that half of all new development must be vegetated. An increasingly popular strategy to meet this rule is "green roofs"—with growing grass or other vegetation (fig. 22.20). Green roofs absorb up to 70 percent of rain water, provide bird and butterfly habitat, insulate homes, and, contrary to old mythology, are structurally sturdy and long-lasting.

These are some common principles of green urbanist planning:

- When building new structures, focus on in-fill development—filling in the inner city so as to help preserve green space in and around cities. Where possible, focus on **brownfield developments,** building on abandoned, reclaimed industrial sites. Brownfields have been eyesores and environmental liabilities in cities for decades, but as urban growth proceeds, they are becoming an increasingly valuable land resource.

- Build high-density, attractive, low-rise, mixed-income housing near the center of cities or near public transportation routes (fig. 22.21a). Densely packed housing saves energy as well as reducing infrastructure costs per person.

- Provide incentives for alternative transportation, such as reserved parking for shared cars (fig. 22.21b) or bicycle routes and bicycle parking spaces. Figure 22.21c shows an 8,000-bicycle parking garage at the train station in Leiden, the Netherlands. An 8,000-car parking garage at the station would cut out the heart of the city. Discourage car use by minimizing the amount of space devoted to driving and parking cars, or by charging for parking space, once realistic alternatives are available.

(a)

(b)

(c)

FIGURE 22.21 Green urbanism includes (a) concentrated, low-rise housing, (b) car-sharing clubs that receive special parking allowances, and (c) alternative transportation methods. These examples are from Amsterdam and Leiden, the Netherlands.

- Encourage ecological building techniques, including green roofs, passive solar energy use, water conservation systems, solar water heating, wind turbines, and appliances that conserve water and electricity.

- Encourage co-housing—groups of households clustered around a common green space that share child care, gardening, maintenance, and other activities. Co-housing can reduce consumption of space, resources, and time while supporting a sense of community.

- Provide facilities for recycling organic waste, building materials, appliances, and plastics, as well as metals, glass, and paper.

- Invite public participation in decision making. Emphasize local history, culture, and environment to create a sense of community and identity. Coordinate regional planning through metropolitan boards that cooperate with but do not supplant local governments.

Think About It

List ten aspects of a city you know that are environmentally or socially unsustainable. Choose one and propose a solution to fix it. Compare notes with colleagues in your class. Did you come up with the same lists? The same solutions?

Open space design preserves landscapes

Traditional suburban development typically divides land into a checkerboard layout of nearly identical 1 to 5 ha parcels with no designated open space (fig. 22.22, *top*). The result is a sterile landscape consisting entirely of house lots and streets. This style of development, which is permitted—or even required—by local zoning and ordinances, consumes agricultural land and fragments wildlife habitat. Many of the characteristics that people move to the country to find—space, opportunities for outdoor recreation, access to wild nature, a rural ambience—are destroyed by dividing every acre into lots that are "too large to mow but too small to plow."

An interesting alternative known as **conservation development,** cluster housing, or open space zoning preserves at least half of a subdivision as natural areas, farmland, or other forms

FIGURE 22.22 Conventional subdivision (*top*) and an open space plan (*bottom*). Although both plans provide 36 home sites, the conventional development allows for no open space. Cluster housing on smaller lots in the open space design preserves at least half the area as woods, prairie, wetlands, farms, or other conservation lands, while providing residents with more attractive vistas and recreational opportunities than a checkerboard development.

What Do You Think?

The Architecture of Hope

How sustainable and self-sufficient can urban areas be? An exciting experiment in minimal impact in London gives us an image of what our future may be. BedZED, short for the Beddington Zero Energy Development is an integrated urban project built on the grounds of an old sewage plant in South London. BedZED's green strategies begin with recycling the ground on which it stands. Designed by architect Bill Dunster and his colleagues, the complex demonstrates dozens of energy-saving and water-saving ideas. BedZED has been occupied, and winning awards, since it was completed in 2003.

Most of BedZED's innovations involve the clever combination of simple, even conventional ideas. Expansive, south-facing, triple-glazed windows provide abundant light, minimize the use of electric lamps, and provide passive solar heat in the winter. Thick, superinsulated walls keep interiors warm in winter and cool in summer. Rotating "wind cowls" on roofs turn to catch fresh breezes, which cool spaces in summer. In winter, heat exchangers use the heat of stale, outgoing air to warm fresh, incoming air, Energy used in space heating is nearly eliminated. Building materials are recycled, reclaimed, or renewable, which reduces the "embodied energy" invested in producing and transporting them.

BedZED does use energy, but the complex generates its own heat and electricity with a small, on-site, superefficient plant that uses local tree trimmings for fuel. Thus BedZED uses no *fossil* fuels, and it is "carbon-neutral" because the carbon dioxide released by burning wood was recently captured from the air by trees. In addition, photovoltaic cells on roofs provide enough free energy to power 40 solar cars. Fuel bills for BedZED residents can be as little as 10 percent of what other Londoners pay for similar-sized homes.

Water-efficient appliances and toilets reduce water use. Rainwater collection systems provide "green water" for watering gardens, flushing toilets, and other nonconsumptive uses. Reed-bed filtration systems purify used water without chemicals. Water meters allow residents to see how much water they use. Just knowing about consumption rates helps encourage conservation. Residents use about half as much water per person as other Londoners.

BedZED residents can save money and time by not using, or even owning a car. Office space is available on-site, so some residents can work where they live, and the commuter rail station is just a ten-minute walk away. The site is also linked to bicycle trails that facilitate bicycle commuting. Car pools and rent-by-the-hour auto memberships allow many residents to avoid owning (and parking) a vehicle altogether.

Building interiors are flooded with natural light, ceilings are high, and most residences have rooftop gardens. Community events and common spaces encourage humane, healthy lifestyles and community ties.

South-facing windows heat homes, and colorful, rotating "wind cowls" ventilate rooms at BedZED, an ecological housing complex in South London, U.K.

Child-care services, shops, entertainment, and sports facilities are built into the project. The approximately 100 housing units are designed for a range of income levels ensuring a racially, ethnically, and age-diverse community. Prices are lower than many similar-sized London homes, and few in this price range or inner-city location have abundant sunlight or gardens.

Similar projects are being built across Europe and even in some developing countries, such as China. Architect Dunster says that BedZED-like developments on cleaned-up brownfields could provide all the 3 million homes that the U.K. expects to need in the next decade with no sacrifice of open space. And as green building techniques, designs, and materials become standard, he argues, they will cost no more than conventional, energy-wasting structures.

What do you think? Would you enjoy living in a dense, urban setting such as BedZED? Would it involve a lower standard of living than you now have? How much would it be worth to avoid spending 8 to 10 hours per week not fighting bumper-to-bumper traffic while commuting to school or work? The average cost of owning and driving a car in the United States is about $9,000 per year. What might you do with that money if owning a vehicle were unnecessary? Try to imagine what urban life might be like if most private automobiles were to vanish. If you live in a typical American city, how much time do you have to enjoy the open space that the suburbs and freeways were supposed to provide? Perhaps, most importantly, how will we provide enough water, energy, and space for the 3 billion people expected to crowd into cities worldwide over the next few decades if we don't adopt some of the sustainable practices and approaches represented by BedZED?

---

of open space. Among the leaders in this design movement are landscape architects Ian McHarg, Frederick Steiner, and Randall Arendt. They have shown that people who move to the country don't necessarily want to own a vast acreage or to live miles from the nearest neighbor; what they most desire is long views across an interesting landscape, an opportunity to see wildlife, and access to walking paths through woods or across wildflower meadows.

By carefully clustering houses on smaller lots, a conservation subdivision can provide the same number of buildable lots as a conventional subdivision and still preserve 50 to 70 percent of the land as open space (fig. 22.22, *bottom*). This not only reduces development costs (less distance to build roads, lay telephone lines, sewers, power cables, etc.) but also helps foster a greater sense of community among new residents. Walking paths and recreation areas get people out of their houses to meet their neighbors. Home owners have smaller lots to care for and yet everyone has an attractive vista and a feeling of spaciousness.

An award-winning example of cluster development is Jackson Meadow, near Stillwater, Minnesota (fig. 22.23). The 64 single-family, custom-designed houses are gathered on just one-third of the project's 336 acres. Two hundred acres are reserved for recreation and scenery. Because the houses are clustered, the developer was able to share one central well and pump house between them, instead of drilling 64 separate wells. In most remote developments of this size, wastewater from these homes would be treated in 64 separate, underground septic systems and leach fields. Here, wastewater is collected and drained into a constructed wetland septic system, where natural bacteria treat water. Water treatment in constructed wetlands is clean, safe, chemical-free, and odorless when it is designed correctly.

Urban habitat can make a significant contribution toward saving biodiversity. In a ground-breaking series of habitat conservation plans triggered by the need to protect the endangered California gnatcatcher, some 85,000 ha (210,000 acres) of coastal scrub near San Diego was protected as open space within the

FIGURE 22.23 Jackson Meadows, an award-winning cluster development near Stillwater, Minnesota, groups houses at sociable distances and preserves surrounding open space for walking, gardening, and scenic views from all houses.

rapidly expanding urban area. This is an area larger than Yosemite Valley, and will benefit many other species as well as humans.

CONCLUSION

What can be done to improve conditions in cities? Curitiba, Brazil, is an outstanding example of green design to improve transportation, protect central cities, and create a sense of civic pride. Other cities have far to go, however, before they reach this standard. Among the immediate needs are housing, clean water, sanitation, food, education, health care, and basic transportation for their residents. The World Bank estimates that interventions to improve living conditions in urban households in the developing world could average the annual loss of almost 80 million "disability-free" years of life. This is about twice the feasible benefit estimated from all other environmental programs studied by the World Bank.

Many planners argue that social justice and sustainable economic development are answers to the urban problems we have discussed in this chapter. If people have the opportunity and money to buy

better housing, adequate food, clean water, sanitation, and other things they need for a decent life, they will do so. Democracy, security, and improved economic conditions help in slowing population growth and reducing rural-to-city movement. An even more important measure of progress may be institution of a social welfare safety net guaranteeing that old or sick people will not be abandoned and alone.

Some countries have accomplished these goals even without industrialization and high incomes. Sri Lanka, for instance, has lessened the disparity between the core and periphery of the country. Giving all people equal access to food, shelter, education, and health care eliminates many incentives for interregional migration. Both population growth and city growth have been stabilized, even though the per capita income is only $800 per year. What do you think; could we help other countries do something similar?

REVIEWING LEARNING OUTCOMES

By now you should be able to explain the following points:

22.1 Define urbanization.

- Cities have specialized functions as well as large populations.
- Large cities are expanding rapidly.

22.2 Describe why cities grow.

- Immigration is driven by push and pull factors.
- Government policies can drive urban growth.

22.3 Understand urban challenges in the developing world.

- Traffic congestion and air quality are growing problems.
- Insufficient sewage treatment causes water pollution.
- Many cities lack adequate housing.

22.4 Identify urban challenges in the developed world.

- Urban sprawl consumes land and resources.
- Expanding suburbs force long commutes.
- Mass-transit could make our cities more livable.

22.5 Explain smart growth.

- Garden cities and new towns were early examples of smart growth.
- New urbanism advanced the ideas of smart growth.
- Green urbanism promotes sustainable cities.
- Open space design preserves landscapes.

PRACTICE QUIZ

1. What is the difference between a city and a village and between rural and urban?

2. How many people now live in cities, and how many live in rural areas worldwide?

3. What changes in urbanization are predicted to occur in the next 30 years, and where will that change occur?

4. From memory, list five of the world's largest cities. Check your list against table 22.2. How many were among the largest in 1900?

5. Describe the current conditions in a typical megacity of the developing world. What forces contribute to its growth?

6. Describe the difference between slums and shantytowns.

7. Why are urban areas in U.S. cities decaying?

8. How has transportation affected the development of cities? What have been the benefits and disadvantages of freeways?

9. Describe some ways that American cities and suburbs could be redesigned to be more ecologically sound, socially just, and culturally amenable.

10. Explain the difference between greenfield and brownfield development. Why is brownfield development becoming popular?

CRITICAL THINKING AND DISCUSSION QUESTIONS

1. Picture yourself living in a rural village or a developing world city. What aspects of life there would you enjoy? What would be the most difficult for you to accept?

2. Why would people move to one of the megacities of the developing world if conditions are so difficult there?

3. A city could be considered an ecosystem. Using what you learned in chapters 3 and 4, describe the structure and function of a city in ecological terms.

4. Look at the major urban area(s) in your state. Why were they built where they are? Are those features now a benefit or drawback?

5. Weigh the costs and benefits of automobiles in modern American life. Is there a way to have the freedom and convenience of a private automobile without its negative aspects?

6. Boulder, Colorado, has been a leader in controlling urban growth. One consequence is that the city has stayed small and charming, so housing prices have skyrocketed and poor people have been driven out. If you lived in Boulder, what solutions might you suggest? What do you think is an optimum city size?

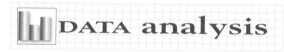 **DATA analysis** Using a Logarithmic Scale

We've often used very large numbers in this book. Millions of people suffer from common diseases. Hundreds of millions are moving from the country to the city. Billions of people will probably be added to the world population in the next half century. Cities that didn't exist a few decades ago now have millions of residents. How can we plot such rapid growth and such huge numbers? If you use ordinary graph paper, making a scale that goes to millions or billions will run off the edge of the page unless you make the units very large.

Figure 1, for example, shows the growth of Mumbai, India, over the past 150 years plotted with an **arithmetic scale** (showing constant intervals) for the Y-axis. It looks as if there is very little growth in the first third of this series and then explosive growth during the last few decades, yet we know that the *rate* of growth was actually greater at the beginning than at the end of this time. How could we display this differently? One way to make the graph easier to interpret is to use a **logarithmic scale.** A logarithmic scale, or "log scale," progresses by factor of 10. So the Y-axis would be numbered 0, 1, 10, 100, 1,000. . . . The effect on a graph is to spread out the smaller values and compress the larger values. In figure 2, the same data are plotted using a

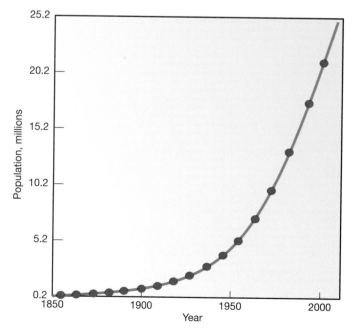

FIGURE 1 The growth of Mumbai.

log scale for the Y-axis, which makes it much easier to see what happened throughout this time period.

1. Do these two graphing techniques give you a different impression of what's happening in Mumbai?

2. How might researchers use one or the other of these scales to convey a particular message or illustrate details in a specific part of the growth curve?

3. Approximately how many people lived in Mumbai in 1850?

4. How many lived there in 2000?

5. When did growth of Mumbai begin to slow?

6. What percentage did the population increase between 1850 and 2000?

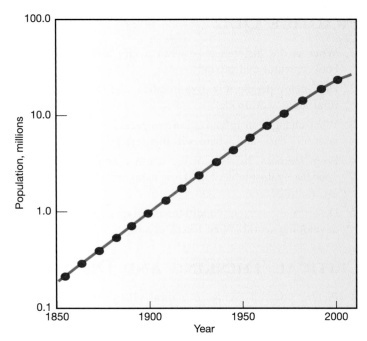

FIGURE 2 The growth of Mumbai.

For Additional Help in Studying This Chapter, please visit our website at www.mhhe.com/cunningham10e. You will find additional practice quizzes and case studies, flashcards, regional examples, place markers for Google Earth™ mapping, and an extensive reading list, all of which will help you learn environmental science.

Fishing once provided a livelihood for most residents of this small Norwegian village, but heavily-subsidized deep-sea trawlers and floating fish processing factories from far-distant countries have now depleted the resource. The economic choices we make often have unintended consequences.

C H A P T E R

23

Ecological Economics

Unleashing the energy and creativity in each human being is the answer to poverty.

—*Muhammad Yunus*—

LEARNING OUTCOMES

After studying this chapter, you should be able to:

23.1 Analyze economic worldviews.

23.2 Scrutinize population, technology, and scarcity.

23.3 Investigate natural resource accounting.

23.4 Summarize how market mechanisms can reduce pollution.

23.5 Study trade, development, and jobs.

23.6 Evaluate green business.

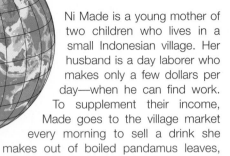

Case Study Loans That Change Lives

Ni Made is a young mother of two children who lives in a small Indonesian village. Her husband is a day laborer who makes only a few dollars per day—when he can find work. To supplement their income, Made goes to the village market every morning to sell a drink she makes out of boiled pandamus leaves, coconut milk, and pink tapioca (fig. 23.1). A small loan would allow her to rent a covered stall during the rainy season and to offer other foods for sale. The extra money she could make could change her life. But traditional banks consider Made too risky to lend to, and the amounts she needs too small to bother with.

Around the world, billions of poor people find themselves in the same position as Made; they're eager to work to build a better life for themselves and their families, but lack resources to succeed. Now, however, a financial revolution is sweeping around the world. Small loans are becoming available to the poorest of the poor. This new approach was invented by Dr. Muhammad Yunus, professor of rural economics at Chittagong University in Bangladesh. Talking to a woman who wove bamboo mats in a village near his university, Dr. Yunus learned that she had to borrow the few taka she needed each day to buy bamboo and twine. The interest rate charged by the village moneylenders consumed nearly all her profits. Always living on the edge, this woman, and many others like her, couldn't climb out of poverty.

To break this predatory cycle, Dr. Yunus gave the woman and several of her neighbors small loans totaling about 1,000 taka (about $20). To his surprise, the money was paid back quickly and in full. So he offered similar amounts to other villagers with similar results. In 1983, Dr. Yunus started the Grameen (village) Bank to show that "given the support of financial capital, however small, the poor are fully capable of improving their lives." His experiment has been tremendously successful. By 2007, the Grameen Bank had almost 7 million customers, 97 percent of them women. It had loaned more than $6 billion with 98 percent repayment, nearly twice the collection rate of commercial Bangladesh banks.

The Grameen Bank provides credit to poor people in rural Bangladesh without the need for collateral. It depends, instead, on mutual trust, accountability, participation, and creativity of the borrowers themselves. Microcredit is now being offered by hundreds of organizations in 43 other countries. Institutions from the World Bank to religious charities make small loans to worthy entrepreneurs. Wouldn't you like to be part of this movement? Well, now you can. You don't have to own a bank to help someone in need.

A brilliant way to connect entrepreneurs in developing countries with lenders in wealthy countries is offered by Kiva, a San Francisco-based technology start-up. The idea for Kiva, which means unity or cooperation in Swahili, came from Matt and Jessica Flannery. Jessica had worked in East Africa with the Village Enterprise Fund, a California nonprofit that provides training, capital, and mentoring to small businesses in developing countries. Jessica and Matt wanted to help some of the people she had met, but they weren't wealthy enough to get into microfinancing on their own. Joining with four other young people with technology experience, they created Kiva, which uses the power of the Internet to help the poor.

Kiva partners with about a dozen development nonprofits with staff in developing countries. The partners identify hardworking entrepreneurs who deserve help. They then post a photo and brief introduction to each one on the Kiva web page. You can browse the collection to find someone whose story touches you. The minimum loan is generally $25. Your loan is bundled with that of others until it reaches the amount needed by the borrower. You make your loan using your credit card (through PayPal, so it's safe and easy). The loan is generally repaid within 12 to 18 months (although without interest). At that point, you can either withdraw the money, or use it to make another loan.

The in-country staff keeps track of the people you're supporting and monitors their progress, so you can be confident that your money will be well used. In less than a year, Kiva has raised more than $2 million and helped thousands of people like Made change their lives.

FIGURE 23.1 A small amount of seed money would allow this young mother to expand her business and help provide for her family.

Case Study *continued*

Loan requests often are on their web page for only a few hours before being filled. Wouldn't you like to take part in this innovative person-to-person human development project? Check out Kiva.org.

In this chapter, we'll look further at both microlending and conventional financing for human development. We'll also look at the role of natural resources in national economies, and how ecological economics is bringing ecological insights into economic analysis. We'll examine cost-benefit analysis as well as other measures of human well-being and genuine progress. Finally, we'll look at how market mechanisms can help us solve environmental problems, and how businesses can contribute to sustainability.

23.1 ECONOMIC WORLDVIEWS

Economy is the management of resources to meet our needs in the most efficient manner possible. It typically deals with choices and alternatives. Because we don't have unlimited ability to produce, distribute, or consume goods and services, economists ask, "Should we make bread or bullets? What are the trade-offs between them, and which would result in the greatest benefits?" (fig. 23.2). Interestingly, ecology and economy are derived from the same root words and concerns. *Oikos* (ecos) is the Greek word for household. Economics is the *nomos,* or counting, of the household resources. Ecology is the *logos,* or logic of how the household works. In both disciplines, the household is expanded to include the whole world. As you will read in this chapter, economics provides important tools for understanding and managing resources. Economics has also evolved as our understanding of resources has changed.

FIGURE 23.2 Bread or bullets? What are the costs and benefits of each? And what are the trade-offs between them?

Can development be sustainable?

By now it is clear that security and living standards for the world's poorest people are inextricably linked to environmental protection. One of the most important questions in environmental science is how we can continue improvements in human welfare within the limits of the earth's natural resources. *Development* means improving people's lives. *Sustainability* means living on the earth's renewable resources without damaging the ecological processes that support us all (table 23.1). **Sustainable development** is an effort to marry these two ideas. A popular definition describes this goal as "meeting the needs of the present without compromising the ability of future generations to meet their own needs."

But is this possible? As you've learned elsewhere in this book, many people argue that our present population and economic levels are exhausting the world's resources. There's no way, they insist, that more people can live at a higher standard without irreversibly degrading our environment. Others claim that there's enough for everyone if we just share equitably and live modestly. Let's look a little deeper into this important debate.

Our definitions of resources shape how we use them

Capital is any form of wealth available for use in the production of more wealth. Economists distinguish between *natural capital* (goods and services provided by nature), *human* or *cultural capital* (knowledge, experience, and human enterprise), and *manufactured* or *built capital* (tools, infrastructure, and technology). Social scientists would add *social capital*

TABLE 23.1

Goals for Sustainable Natural Resource Use

- Harvest rates for renewable resources (those like organisms that regrow or those like fresh water that are replenished by natural processes) should not exceed regeneration rates.
- Waste emissions should not exceed the ability of nature to assimilate or recycle those wastes.
- Nonrenewable resources (such as minerals) may be exploited by humans, but only at rates equal to the creation of renewable substitutes.

FIGURE 23.3 Nonrenewable resources, such as the oil from this forest of derricks in Huntington Beach, California, are irreplaceable. Once they're exhausted (as this oil field was half a century ago) they will never be restored on a human time scale.

FIGURE 23.4 Nature provides essential ecological services, such as the biological productivity, water storage and purification, and biodiversity protection in this freshwater marsh and its surrounding forest. Ironically, while biological resources are infinitely renewable, if they're damaged by our actions they may be lost forever.

(shared values, trust, cooperative spirit, and community organization) to this list. A **resource** is anything with potential use in creating wealth or giving satisfaction. Natural resources can be either renewable or nonrenewable, as well as tangible or intangible. Although we generally define these terms from a human perspective, many resources we use are important to other species as well.

In general, **nonrenewable resources** are the earth's geological endowment: the minerals, fossil fuels, and other materials present in fixed amounts in the environment (fig. 23.3). Although many of these resources are renewed or recycled by geological or ecological processes, the time scales to do so are so long by human standards that the resource will be gone once present supplies are exhausted. Predictions abound that we are in imminent danger of running out of one or another of these exhaustible resources. The actual available supplies of many commodities—such as metals—can be effectively extended, however, by more efficient use, recycling, substitution of one material for another, or better extraction from dilute or remote sources.

Renewable resources are things that can be replenished or replaced. They include sunlight—our ultimate energy source—and the biological organisms and biogeochemical cycles that provide essential ecological services (fig. 23.4 and table 23.2).

Because biological organisms and ecological processes are self-renewing, we often can harvest surplus organisms or take advantage of ecological services without diminishing future availability, if we do so carefully. Unfortunately, our stewardship of these resources often is less than ideal. Even once vast biological populations such as passenger pigeons, American bison (buffalo), and Atlantic cod, for instance, were exhausted by overharvesting in only a few years. Similarly, we are now upsetting climatic systems with potentially disastrous results (see chapter 15). Mismanagement of renewable resources often makes them more ephemeral and limited than fixed geological resources.

Abstract or **intangible resources** include open space, beauty, serenity, wisdom, diversity, and satisfaction (fig. 23.5). Paradoxically, these resources can be both infinite *and* exhaustible. There is no upper limit to the amount of beauty, knowledge, or compassion that can exist in the world, yet they can be easily destroyed. A single piece of trash can ruin a beautiful vista, or a single cruel remark can spoil an otherwise perfect day. On the other hand, unlike tangible resources that usually are reduced by use or sharing, intangible resources often are increased by use and multiplied by being shared. Nonmaterial assets can be important economically. Information management and tourism—both based on intangible resources—have become two of the largest and most powerful industries in the world.

TABLE 23.2

Important Ecological Services

1. *Regulate* global energy balance; chemical composition of the atmosphere and oceans; local and global climate; water catchment and groundwater recharge; production, storage, and recycling of organic and inorganic materials; maintenance of biological diversity.

2. *Provide* space and suitable substrates for human habitation, crop cultivation, energy conversion, recreation, and nature protection.

3. *Produce* oxygen, fresh water, food, medicine, fuel, fodder, fertilizer, building materials, and industrial inputs.

4. *Supply* aesthetic, spiritual, historic, cultural, artistic, scientific, and educational opportunities and information.

Source: R. S. de Groot, *Investing in Natural Capital,* 1994.

FIGURE 23.5 Scenic beauty, solitude, and relatively untouched nature in this Colorado wilderness are treasured by many people, yet hard to evaluate in economic terms.

FIGURE 23.6 Informal markets such as this one in Bali, Indonesia, may be the purest example of willing sellers and buyers setting prices based on supply and demand.

Classical economics examines supply and demand

Classical economics originally was a branch of moral philosophy concerned with how individual interest and values intersect with larger social goals. The founder of modern Western economics, Adam Smith (1723–1790), was an ethicist concerned with individual freedom of choice. Smith's landmark book *Inquiry into the Nature and Causes of the Wealth of Nations,* published in 1776, argued that

> Every individual endeavors to employ his capital so that its produce may be of the greatest value. He generally neither intends to promote the public interest, nor knows how much he is promoting it. He intends only his own security, only his own gain. And he is in this led by an *invisible hand* to promote an end which was no part of his intention. By pursuing his own interests he frequently promotes that of society more effectually than when he really intends to.

This statement often is taken as justification for the capitalist system, in which market competition between willing sellers and buyers is believed to bring about the greatest efficiency of resource use, the optimum balance between price and quality, and to be critical for preserving individual liberty (fig. 23.6). As we will discuss later in this chapter, however, *laissez faire* market systems rarely incorporate factors such as social or environmental costs.

David Ricardo (1772–1823), an important contemporary of Smith, introduced a better understanding of the relation between supply and demand in economics. **Demand** is the amount of a product or service that consumers are willing and able to buy at various possible prices, assuming they are free to express their preferences. **Supply** is the quantity of that product being offered for sale at various prices, other things being equal. Classical

economics proposes that there is a direct, inverse relationship between supply and demand (fig. 23.7). As prices rise, the supply increases and demand falls. The reverse holds true as price decreases. The difference between the cost of production and the price buyers are willing to pay, Ricardo called "rent." Today we call it profit.

In a free market of independent and intelligent buyers and sellers, supply and demand should come into a **market equilibrium,** represented in figure 23.7 by the intersection of the two curves. In real life, prices are not determined strictly by total supply and demand as much as what economists called **marginal costs and benefits.** Sellers ask themselves, "What would it cost to produce one more unit of this product or service? Suppose I

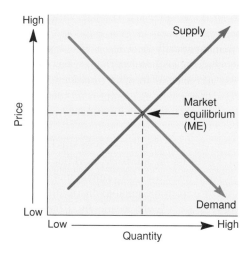

FIGURE 23.7 Classic supply/demand curves. When price is low, supply is low and demand is high. As prices rise, supply increases but demand falls. Market equilibrium is the price at which supply and demand are equal.

add one more worker or buy an extra supply of raw materials, how much profit could I make?" Buyers ask themselves similar questions, "How much would I benefit and what would it cost if I bought one more widget?" If both buyer and seller find the marginal costs and benefits attractive, a sale is made.

There are exceptions, however, to this theory of supply and demand. Consumers will buy some things regardless of cost. Raising the price of cigarettes, for instance, doesn't necessarily reduce demand. We call this price inelasticity. Other items have **price elasticity;** that is, they follow supply/demand curves exactly. When price goes up, demand falls and vice versa.

John Stuart Mill (1806–1873), another important classical philosopher/economist, believed that perpetual growth in material well-being is neither possible nor desirable. Economies naturally mature to a steady state, he believed, leaving people free to pursue nonmaterialistic goals. He didn't regard this equilibrium state to be necessarily one of stagnation or poverty. In *Principles of Political Economy,* he wrote, "It is scarcely necessary to remark that a stationary condition of capital and population implies no stationary state of human improvement. There would be as much scope as ever for all kinds of mental culture, and moral and social progress; as much room for improving the art of living, and much more likelihood of its being improved when minds cease to be engrossed by the art of getting on." This view has much in common with the concepts of a steady-state economy and sustainable development that we will examine later in this chapter.

Neoclassical economics emphasizes growth

Toward the end of the nineteenth century, the field of economics divided into two broad camps. **Political economy** continued the tradition of moral philosophy and concerned itself with social structures, value systems, and relationships among the classes. This group included reformers such as Karl Marx and E. F. Schumacher, along with many socialists, anarchists, and utopians. The other camp, called **neoclassical economics,** adapted principles of modern science to economic analysis. The late Milton Friedman was a leader in free-market, neoclassical economics. This approach strives to be mathematically rigorous, noncontextual, abstract, and predictive. Neoclassical economists claim to be objective and value free, leaving social concerns to other disciplines. Like their classical predecessors, they retain an emphasis on scarcity and the interaction of supply and demand in determining prices and resource allocation (fig. 23.8).

Continued economic growth is considered to be both necessary and desirable in the neoclassical worldview. Growth is seen as the only way to maintain full employment and avoid class conflict arising from inequitable distribution of wealth. Natural resources are viewed as merely a factor of production rather than a critical supply of materials, services, and waste sinks by neoclassical economics. Because factors of production are thought to be interchangeable and substitutable, materials and services provided by the environment are not considered indispensable. As one resource becomes scarce, neoclassical economists believe, substitutes will be found.

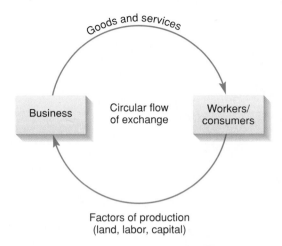

FIGURE 23.8 The neoclassical model of the economy focuses on the flow of goods, services, and factors of production (land, labor, capital) between business and individual workers and consumers. The social and environmental consequences of these relationships are irrelevant in this view.

Ecological economics incorporates principles of ecology

Classical and neoclassical economics have usually focused on human resources, such as buildings, roads, or labor. Natural systems are essential for economic productivity; rivers absorb wastewater, bacteria decompose waste, and winds carry away smoke that would otherwise debilitate workers, but classical and neoclassical economics treat these natural services as external to the costs of production. For example, energy producers have calculated profits against the price of coal, the costs of labor, investments in new buildings, and furnaces, but they have not accounted for the climate's absorption of carbon dioxide, sulfur dioxide, or mercury, nor have they accounted for the public health costs associated with polluted air. These have been **external costs,** or costs outside the accounting system. **Internal costs,** in contrast, are expenses considered to be the normal cost of doing business.

Often resources are externalized (treated as external costs) because they're free. Air or sunlight, for example, are provided without charge by nature and, thus, are usually excluded from accounting systems. *Natural resource economics* assigns a value to some of nature's resources, such as clean water, forests, or even biodiversity. Like neoclassical economics, however, natural resource economics assumes that natural services are abundant, and thus cheap, while manufactured capital is limited, and thus expensive (fig. 23.9).

Ecological economics focuses on the value of natural services and tries to include those services into price calculations. More than natural resource economics, ecological economics assumes that natural resources are limited and valuable, while manufactured capital is abundant (fig. 23.10).

This is because ecological economics focuses explicitly on ecological concepts such as systems (chapter 2), thermodynamics, and material cycles (chapter 3). In a system, all components are interdependent. Disrupting one component (such as climate

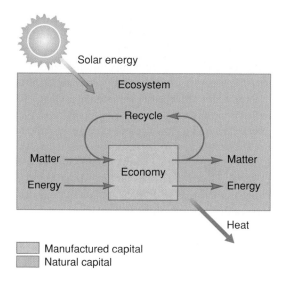

FIGURE 23.9 In natural resource economics, the economy is seen as an open subset of the ecosystem. Matter and energy are consumed in the form of raw resources and rejected as waste. Material is recycled by ecological processes, while heat is ejected back to space. Manufactured or "built" capital is regarded as scarce—and therefore valuable—while natural capital is regarded as plentiful and, therefore, cheap.

conditions) risks destabilizing other components (such as agricultural production) in unpredictable and possibly catastrophic ways. Thermodynamics and material cycles teach that energy and materials cycle through systems, constantly being reused; one organism's waste is another's nutrient or energy source. This perspective allows us to see environmental resources, and other species, as limited in supply, valuable, and often fragile.

Systems analysis also raises the concern of carrying capacities for human populations and questions the idea of unlimited economic growth. Many ecological economists, such as Herman

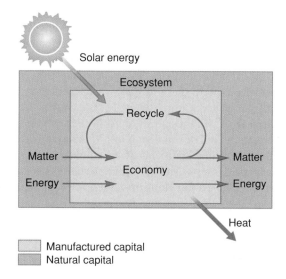

FIGURE 23.10 An ecological economic view reverses ideas of scarcity. Because material recycling is considered to be integral to the economy, manufactured or built capital is seen to be large compared to limited supplies of natural capital.

Daly of the University of Maryland, promote a **steady-state economy,** characterized by low human birth rates and death rates, the use of renewable energy sources, material recycling, and an emphasis on durability, efficiency, and stability. "Throughput," the volume of materials and energy consumed and of waste produced, should be minimized. This model contrasts sharply with the neoclassical emphasis on growth based on ever-increasing consumption and waste production. The steady-state idea reflects the notion of an ecosystem in equilibrium, or a population below its carrying capacity, where overall conditions remain stable and catastrophic conditions are unlikely to develop.

Ecological economics tries to make producers account for social costs, as well as environmental costs. For example, if a power plant releases soot from its smokestack, populations downwind suffer from air pollution. Suppose the population living downwind incurs health expenses, lost work days, and illness as a result. Is it the power plant's responsibility to pay for these health expenses and lost wages? Traditional economics has said no, that society should bear responsibility. Ecological economists say yes, these costs should be considered part of the cost of power production, and the added expenses should be included in the price of electricity. In general, the cost of cleaning up a power plant usually is lower than the cost of health care and lost productivity in a population. But calculating who should pay, and how much, requires that accountants internalize these external costs.

In economic terms, the extra costs of illness and lost work days, as well as other costs associated with pollution, are market inefficiencies; they represent inefficient overall use of resources (money, time, energy, materials) because of incomplete accounting of costs and benefits.

Ecological economics also questions the basic economic assumption that all goods can be compared according to their monetary value, so that all goods, including bread and bullets, or guns and butter, are essentially interchangeable and can be compared on the same monetary scale. Ecological economists propose that some aspects of nature are irreplaceable and essential, such as beauty, space, wild animals, or quiet. If natural amenities are essential, then it follows that human population growth and resource consumption must have limits.

Ecological economists have developed methods to measure well-being, such as the Genuine Progress Index, discussed later in this chapter. These measurements allow us to assess growth in terms beyond simply the amount of money changing hands.

Think About It

What do you suppose are the internal costs (those calculated into the price) when you buy a gallon of gas? What might be some external costs? How many of those costs are geographically expressed where you live? Are there any costs that might be paid for through your taxes? Are there any that are not paid for by your taxes? (*Hint:* Think of the conditions necessary to get oil safely delivered to your gas station. How are those conditions created or maintained?)

FIGURE 23.11 Adding more cattle to the Brazilian Cerrado (savanna) increases profits for individual ranchers, but is bad for biodiversity and environmental quality.

Communal property resources are a classic problem in ecological economics

In 1968, biologist Garret Hardin wrote a widely quoted article entitled **"The Tragedy of the Commons"** in which he argued that any commonly held resource inevitably is degraded or destroyed because the narrow self-interests of individuals tend to outweigh public interests. Hardin offered as a metaphor the common woodlands and pastures held by most colonial New England villages. In deciding how many cattle to put on the commons, Hardin explained, each villager would attempt to maximize his or her own personal gain. Adding one more cow to the commons could mean a substantially increased income for an individual farmer. The damage done by overgrazing, however, would be shared among all the farmers (fig. 23.11). Hardin concluded that the only solution would be either to give coercive power to the government or to privatize the resource.

Hardin intended this dilemma, known in economics as the "free-rider" problem, to warn about human overpopulation and resource availability. Other authors have used his metaphor to explain such diverse problems as African famines, air pollution, fisheries declines, and urban crime. What Hardin was really describing, however, was an **open access system** in which there are no rules to manage resource use. In fact, many communal resources have been successfully managed for centuries by cooperative arrangements among users. Some examples include Native American management of wild rice beds and hunting grounds; Swiss village-owned mountain forests and pastures; Maine lobster fisheries; communal irrigation systems in Spain, Bali, and Laos; and nearshore fisheries almost everywhere in the world. A large body of literature in economics and social sciences describes how these cooperative systems work. Among the features shared by **communal resource management systems** are: (1) community members have lived on the land or used the resource for a long time and

anticipate that their children and grandchildren will as well, thus giving them a strong interest in sustaining the resource and maintaining bonds with their neighbors; (2) the resource has clearly defined boundaries; (3) the community group size is known and enforced; (4) the resource is relatively scarce and highly variable so that the community is forced to be interdependent; (5) management strategies appropriate for local conditions have evolved over time and are collectively enforced; that is, those affected by the rules have a say in them; (6) the resource and its use are actively monitored, discouraging anyone from cheating or taking too much; (7) conflict resolution mechanisms reduce discord; and (8) incentives encourage compliance with rules, while sanctions for noncompliance keep community members in line.

Rather than being the only workable solution to problems in common pool resources, privatization and increasing external controls often prove to be disastrous. In places where small villages have owned and operated local jointly held forests and fishing grounds for generations, nationalization and commodification of resources generally have led to rapid destruction of both society and ecosystems. Where communal systems once enforced restraint over harvesting, privatization encouraged narrow self-interest and allowed outsiders to take advantage of the weakest members of the community.

A tragic example is the forced privatization of Indian reservations in the United States. Failing to recognize or value local knowledge and forcing local people to participate in a market economy allowed outsiders to disenfranchise native people and resulted in disastrous resource exploitation. Learning to distinguish between open access systems and communal property regimes is important in understanding how best to manage natural resources.

23.2 POPULATION, TECHNOLOGY, AND SCARCITY

Are we about to run out of essential natural resources? It stands to reason that if we consume a fixed supply of nonrenewable resources at a constant rate, eventually we'll use up all the economically recoverable reserves. There are many warnings in the environmental literature that our extravagant depletion of nonrenewable resources sooner or later will result in catastrophe. The dismal prospect of Malthusian diminishing returns and a life of misery, starvation, and social decay inspire many environmentalists to call for an immediate change to voluntary simplicity and lower consumption rates. Models for exploitation rates of nonrenewable resources—called Hubbert curves, after Stanley Hubbert who developed them—often closely match historic experience for natural resource depletion (fig. 23.12).

Many economists, however, contend that neither supply/demand relationships nor economically recoverable reserves are rigidly fixed. Human ingenuity and enterprise often allow us to respond to scarcity in ways that postpone or alleviate the dire effects predicted by modern Jeremiahs. In the next section we will look at some of the arguments for and against limits to growth of the global economy.

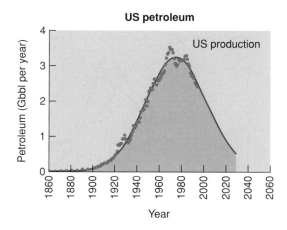

FIGURE 23.12 United States petroleum production, 1860 to 2000. Dots indicate actual production. The bell-shaped curve is a theoretical Hubbert curve for a nonrenewable resource. The shaded area under the curve, representing 220 Gbbl (Gbbl = Gigabarrels or billions of standard 42-gallon barrels), is an estimate of the total economically recoverable resource.

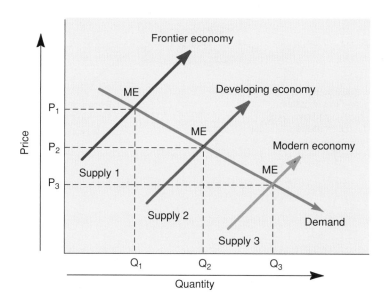

FIGURE 23.13 Supply and demand curves at three different stages of economic development. At each stage there is a market equilibrium point at which supply and demand are in balance. As the economy becomes more efficient, the equilibrium shifts so there is a larger quantity available at a lower price than before. (P = price, Q = quantity, ME = market equilibrium)

Scarcity can lead to innovation

In a pioneer or frontier economy, methods for harvesting resources and turning them into useful goods and services tend to be inefficient and wasteful. This may not matter, however, if the supply of resources exceeds the demand for them. The loggers, for example, who swarmed across the Great Lakes Forest at the beginning of the twentieth century, wasted a vast amount of wood. Their inefficiency didn't seem important, however, because the supply of trees appeared infinite, while labor, capital, and means for getting lumber to markets was scarce. As markets and societies develop, however, better technology and more efficient systems allow people to create the same amount of goods and services using far fewer resources. We now produce hundreds of times the crop yield with less labor from the same land that pioneers farmed.

This increasing technological efficiency can dramatically shift supply and demand relationships (fig. 23.13). As technology makes goods and services cheaper to produce, the quantity available at a given price can increase greatly. The market equilibrium or the point at which supply and demand equilibrate will shift to lower prices and higher quantities as a market matures.

Scarcity can actually serve as a catalyst for innovation and change (fig. 23.14). As materials become more expensive and difficult to obtain, it becomes cost-effective to discover new supplies or to use available ones more efficiently. The net effect is as if a new supply of resources has been created or discovered. Several important factors play a role in this cycle of technological development:

- Technical inventions can increase efficiency of extraction, processing, use, and recovery of materials.

- Substitution of new materials or commodities for scarce ones can extend existing supplies or create new ones. For instance, substitution of aluminum for copper, concrete for structural

steel, grain for meat, and synthetic fibers for natural ones all remove certain limits to growth.

- Trade makes remote supplies of resources available and may also bring unintended benefits in information exchange and cultural awakening.

- Discovery of new reserves through better exploration techniques, more investment, and looking in new areas becomes rewarding as supplies become limited and prices rise.

- Recycling becomes feasible and accepted as resources become more valuable. Recycling now provides about 37 percent of the iron and lead, 20 percent of the copper, 10 percent of the aluminum, and 60 percent of the antimony that is consumed each year in the United States.

Carrying capacity is not necessarily fixed

Despite repeated warnings that rapidly growing populations and increasing affluence are bound to exhaust natural resources and result in rapid price increases, technological developments of the sort described earlier have resulted in price decreases for most raw materials over the last hundred years. Consider copper for example. Twenty years ago worries about impending shortages led the United States to buy copper and store it in strategic stockpiles. Estimates of the amount of this important metal needed for electric motors, telephone lines, transoceanic cables, and other uses essential for industrialized society far exceeded known reserves. It looked as if severe shortages and astronomical price increases were inevitable. But then aluminum power lines, satellites, fiber optics, integrated circuits, microwave transmission, and other inventions greatly diminished the need for copper. Although prices

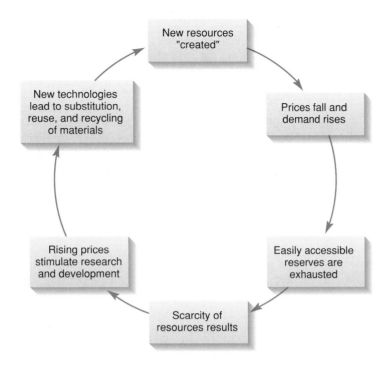

FIGURE 23.14 Scarcity/development cycle. Paradoxically, resource use and depletion of reserves can stimulate research and development, the substitution of new materials, and the effective creation of new resources.

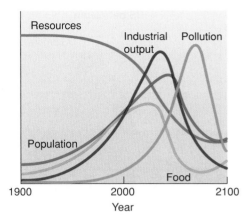

FIGURE 23.15 A run of one of the world models in *Limits to Growth*. This model assumes business-as-usual for as long as possible until Malthusian limits cause industrial society to crash. Notice that pollution continues to increase well after industrial output, food supplies, and population have all plummeted.

Economic models compare growth scenarios

In the early 1970s, an influential study of resource limitations was funded by the Club of Rome, an organization of wealthy business owners and influential politicians. The study was carried out by a team of scientists from the Massachusetts Institute of Technology headed by the late Donnela Meadows. The results of this study were published in the 1972 book *Limits to Growth*. A complex computer model of the world economy was used to examine various scenarios of different resource depletion rates, growing population, pollution, and industrial output. Given the Malthusian assumptions built into this model, catastrophic social and environmental collapse seemed inescapable.

Figure 23.15 shows one example of the world model. Food supplies and industrial output rise as population grows and resources are consumed. Once past the carrying capacity of the environment, however, a crash occurs as population, food production, and industrial output all decline precipitously. Pollution continues to grow as society decays and people die, but, eventually, it also falls. Notice the similarity between this set of curves and the "boom and bust" population cycles described in chapter 6.

Many economists criticized these results because they discount technological development and factors that might mitigate the effects of scarcity. In 1992, the Meadows group published updated computer models in *Beyond the Limits* that include technological progress, pollution abatement, population stabilization, and new public policies that work for a sustainable future. If we adopt these changes sooner rather than later, the computer shows an outcome like that in figure 23.16, in which all factors stabilize sometime in this century at an improved standard of living for everyone. Of course neither of these computer models shows what will happen, only what some possible outcomes *might* be, depending on the choices we make.

are highly variable because of world politics and trade policies, the general trend for most materials has been downward in this century. It is as if the carrying capacity of our natural resources—at least in terms of copper—has been increased.

Economists generally believe that this pattern of substitutability and technological development is likely to continue into the future. Ecologists generally disagree. There are bound to be limits, they argue, to how many people our environment can support. An interesting example of this debate occurred in 1980. Ecologist Paul Ehrlich bet economist Julian Simon that increasing human populations and growing levels of material consumption would inevitably lead to price increases for natural resources. They chose a package of five metals—chrome, copper, nickel, tin, and tungsten—priced at the time at $1,000. If, in ten years, the combined price (corrected for inflation) was higher than $1,000, Simon would pay the difference. If the combined price had fallen, Ehrlich would pay. In 1990 Ehrlich sent Simon a check for $576.07; the price for these five metals had fallen 47.6 percent.

Does this prove that resource abundance will continue indefinitely? Hardly. Ehrlich claims that the timing and set of commodities chosen simply were the wrong ones. The fact that we haven't yet run out of raw materials doesn't mean that it will never happen. Many ecological economists now believe that some nonmarket resources such as ecological processes may be more irreplaceable than tangible commodities like metals. What do you think? Are we approaching the limits of nature to support more humans and more consumption? Which resources, if any, do you think are most likely to be limiting in the future?

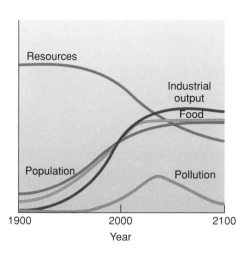

FIGURE 23.16 A run of the world model from *Beyond the Limits.* This model assumes that population and consumption are curbed, new technologies are introduced, and sustainable environmental policies are embraced immediately, rather than after resources are exhausted.

Why not conserve resources?

Even if large supplies of resources are available or the technological advances to mitigate scarcity exist, wouldn't it be better to reduce our use of natural resources so they will last as long as possible? Will anything be lost if we're frugal now and leave more to be used by future generations? Many economists would argue that resources are merely a means to an end rather than an end in themselves. They have to be used to have value.

If you bury your money in a jar in the backyard, it will last a long time but may not be worth much when you dig it up. If you invest it productively, you may have much more in the future than you do now. Furthermore, a window of opportunity for investment may be open now but not later. Suppose that 300 years ago our ancestors had decided that the Industrial Revolution should not be allowed to continue because it required too much resource consumption. There certainly would be more easily accessible resources available now, but would we be better off than we are? What do you think? Where is the proper balance between using our resources now or saving them for the future?

23.3 NATURAL RESOURCE ACCOUNTING

How can we determine the value of environmental goods and services? Some of the most crucial natural resources are not represented by monetary prices in the marketplace. Certain resource allocation decisions are political or social. Others are simply ignored. Groundwater, sunlight, clean air, biological diversity, and other assets that we all share in common often are treated as public goods that anyone can use freely. Until they are transformed by human activities, natural resources are regarded as having little value. Ecological economics calls for recognition of the real value of those resources in calculating economic progress. In this section we'll look at some suggestions for how we could do this.

Gross national product is our dominant growth measure

The most common way to measure a nation's output is **gross national product (GNP).** GNP can be calculated in two ways. One is the money flow from households to businesses in the form of goods and services purchased (see fig. 23.8). The other is to add up all the costs of production in the form of wages, rent, interest, taxes, and profit. In either case, a subtraction is made for capital depreciation, the wear and tear on machines, vehicles, and buildings used in production. Some economists prefer **gross domestic product (GDP),** which includes only the economic activity within national boundaries. Thus the vehicles made and sold by Ford in Europe don't count in GDP.

Both GNP and GDP are criticized as a measure of real progress or well-being because it doesn't attempt to distinguish between economic activities that are beneficial or harmful. A huge oil spill that pollutes beaches and kills wildlife, for example, shows up as a positive addition to GNP because of the business generated by cleanup efforts.

Ecological economists also criticize GNP because it doesn't account for natural resources used up or ecosystems damaged by economic activities. Robert Repeto of the World Resources Institute estimates that soil erosion in Indonesia, for instance, reduces the value of crop production about 40 percent per year. If natural capital were taken into account, Indonesian GNP would be reduced by at least 20 percent annually.

Similarly, Costa Rica experienced impressive increases in timber, beef, and banana production between 1970 and 1990. But decreased natural capital during this period represented by soil erosion, forest destruction, biodiversity losses, and accelerated water runoff add up to at least $4 billion or about 25 percent of annual GNP.

Alternate measures account for well-being

A number of systems have been proposed as alternatives to GNP that reflect genuine progress and social welfare. In their 1989 book, Herman Daly and John Cobb proposed a **genuine progress index (GPI)** that takes into account real per capita income, quality of life, distributional equity, natural resource depletion, environmental damage, and the value of unpaid labor. They point out that while per capita GNP in the United States nearly doubled between 1970 and 2000, per capita GPI increased only 4 percent (fig. 23.17). Some social service organizations would add to this index the costs of social breakdown and crime, which would decrease real progress even further over this time span.

A newer measure is the **Environmental Performance Index (EPI)** created by researchers at Yale and Columbia Universities to evaluate national sustainability and progress toward achievement of the United Nations Millennium Development Goals. The EPI is based on sixteen indicators tracked in six categories: environmental health, air quality, water resources, productive natural resources, biodiversity and habitat, and sustainable energy. The top-ranked countries—New Zealand, Sweden, Finland, the Czech

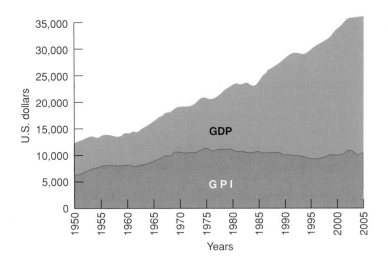

FIGURE 23.17 Although per capita GDP in the United States nearly doubled between 1970 and 2000 in inflation-adjusted dollars, a genuine progress index that takes into account natural resource depletion, environmental damage, and options for future generations hardly increased at all. **Source:** Data from *Redefining Progress,* 2006.

FIGURE 23.18 Raising agricultural productivity and rural incomes are high priorities of the UN Millennium Development Goals.

Republic, and the United Kingdom—all commit significant resources and effort to environmental protection. In 2006, the United States ranked 28th in the EPI, or lower than Malaysia, Costa Rica, Columbia, and Chile, all of which have between 6 and 15 times lower GDP than the U.S. See Data Analysis (p. 539) for a graph of human development index (HDI) versus EPI.

The United Nations Development Program (UNDP) uses a benchmark called the human development index (HDI) to track social progress. HDI incorporates life expectancy, educational attainment, and standard of living as critical measures of development. Gender issues are accounted for in the gender development index (GDI), which is simply HDI adjusted or discounted for inequality or achievement between men and women.

In its annual Human Development Report, the UNDP compares country-by-country progress. As you might expect, the highest development levels are generally found in North America, Europe, and Japan. In 2006, Norway ranked first in the world in both HDI and GDI. The United States ranked eighth while Canada was sixth. The 25 countries with the lowest HDI in 2006 were all in Africa. Haiti ranks the lowest in the Western Hemisphere.

Although poverty remains widespread in many places, encouraging news also can be found in development statistics. Poverty has fallen more in the past 50 years, the UNDP reports, than in the previous 500 years. Child death rates in developing countries as a whole have been more than halved. Average life expectancy has increased by 30 percent while malnutrition rates have declined by almost a third. The proportion of children who lack primary school has fallen from more than half to less than a quarter. And the share of rural families without access to safe water has fallen from nine-tenths to about one-quarter.

Some of the greatest progress has been made in Asia. China and a dozen other countries with populations that add up to more than 1.6 billion, have decreased the proportion of their people living below the poverty line by half. Still, in the 1990s the number of people with incomes less than $1 per day increased by almost 100 million to 1.3 billion—and the number appears to be growing in every region except Southeast Asia and the Pacific. Even in industrial countries, more than 100 million people live below the poverty line and 37 million are chronically unemployed.

Economic growth can be a powerful means of reducing poverty, but its benefits are not automatic. The GNP of Honduras, for instance, grew 2 percent per year in the 1980s and yet poverty doubled. To combat poverty, the UNDP calls for "pro-poor growth" designed to spread benefits to everyone. Specifically, some key elements of this policy would be to: (1) create jobs that pay a living wage, (2) lessen inequality, (3) encourage small-scale agriculture, microenterprises, and the informal sector, (4) foster technological progress, (5) reverse environmental decline in marginal regions, (6) speed demographic transitions, and (7) provide education for all. Since about three-quarters of the world's poorest people live in rural areas, raising agricultural productivity and incomes is a high priority for these actions (fig. 23.18).

New approaches incorporate nonmarket values

New tools and new approaches are needed to represent nature in national accounting and human development. Among the natural resource characteristics that ecological economists suggest be taken into account include:

- Use values: the price we pay to use or consume a resource
- Option value: preserving options for the future
- Existence value: those things we like to know still exist even though we may never use or even see them
- Aesthetic values: aspects we appreciate for their beauty
- Cultural values: factors important for cultural identity
- Scientific and educational values: information or experience-rich aspects of nature

How can we measure these values of natural resources and ecological services when they are not represented in market systems? Ecological economists often have to resort to "shadow pricing" or other indirect valuation methods for natural resources. For instance, what is the worth of a day of canoeing on a wild river? We might measure opportunity costs such as how much we pay to get to the river or to rent a canoe. The direct out-of-pocket costs might represent only a small portion, however, of what it is really worth to participants. Another approach is contingent valuation in which potential resource users are asked, "How much would you be willing to pay for this experience?" or "What price would you be willing to accept to sell your access or forego this opportunity?" These approaches are controversial because people may report what they think they ought to pay rather than what they would really pay for these activities.

Several ecological economists have attempted to put a price on the goods and services provided by natural ecosystems. Although many ecological processes have no direct market value, we can estimate replacement costs, contingent values, shadow prices, and other methods of indirect assessment to determine a rough value. For instance, we now dispose of much of our wastes by letting nature detoxify them. How much would it cost if we had to do this ourselves? Ecological economists look at everything from recreational beaches to forest lumber to hidden services such as the ocean's regulation of atmospheric carbon dioxide to pollination of crops by insects.

The estimated annual value of all ecological goods and services provided by nature range from $16 trillion to $54 trillion, with a median worth of $33 trillion, or about three-fourths the combined annual GNPs of all countries in the world (table 23.3). These estimates are probably understated because they omit ecosystem services from several biomes, such as deserts and tundra, that are poorly understood in terms of their economic contributions. The most valuable ecosystems in terms of biological processes are wetlands and coastal estuaries because of their high level of biodiversity and their central role in many biogeochemical cycles.

In 2003, economists from Cambridge University (U.K.) estimated that protecting a series of nature reserves representing samples of all major biomes would cost (U.S.) $45 billion per year, but would preserve ecological services worth between $4.4 trillion to $5.2 trillion annually.

Cost-benefit analysis aims to optimize resource use

One way to evaluate public projects is to analyze the costs and benefits they generate in a **cost-benefit analysis (CBA).** This process attempts to assign values to resources as well as to social and environmental effects of carrying out or not carrying out a given undertaking. It tries to find the optimal efficiency point at which the marginal cost of pollution control equals the marginal benefits (fig. 23.19).

CBA is one of the main conceptual frameworks of resource economics and is used by decision makers around the world as a way of justifying the building of dams, roads, and airports, as well

TABLE 23.3

Estimated Annual Value of Ecological Services

Ecosystem Services	Value (Trillion $U.S.)
Soil formation	17.1
Recreation	3.0
Nutrient cycling	2.3
Water regulation and supply	2.3
Climate regulation (temperature and precipitation)	1.8
Habitat	1.4
Flood and storm protection	1.1
Food and raw materials production	0.8
Genetic resources	0.8
Atmospheric gas balance	0.7
Pollination	0.4
All other services	1.6
Total value of ecosystem services	**33.3**

Source: Adapted from R. Costanza et al., "The Value of the World's Ecosystem Services and Natural Capital," *Nature,* Vol. 387 (1997).

as in considering what to do about biodiversity loss, air pollution, and global climate change. Deeply entrenched in bureaucratic practice and administrative culture, this technique has become much more widespread in American public affairs since the Reagan administration's executive orders in the 1980s calling for the application of CBA to all regulatory decisions and legislative proposals. Many conservatives see CBA as a way of eliminating what they consider to be unnecessary and burdensome requirements to protect clean air, clear water, human health, or biodiversity. They would like to add a requirement that all regulations be shown to be cost-effective.

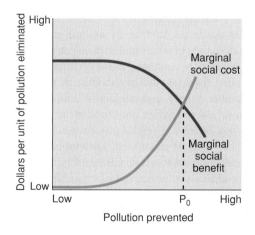

FIGURE 23.19 To achieve maximum economic efficiency, regulations should require pollution prevention up to the optimum point (P_0) at which the costs of eliminating pollution just equal the social benefits of doing so.

FIGURE 23.20 What is the value of solitude or beauty? How would you assign costs and benefits to a scene such as this?

The first step in CBA is to identify who or what might be affected by a particular plan. What are the potential outcomes and results? What alternative actions might be considered? After identifying and quantifying all the contingent factors, an attempt is made to assign monetary costs and benefits to each component. Usually the direct expenses of a project are easy to ascertain. How much will you have to pay for land, materials, and labor? The monetary worth of lost opportunities—to swim or fish in a river, or to see birds in a forest—on the other hand, are much harder to appraise. How would you put a price on good health or a long life? It's also important to ask who will bear the costs and who will reap the benefits of any proposal. Are there consequences that cannot be given a monetary price? What is a bug or a bird worth, for instance, or the opportunity for solitude or inspiration (fig. 23.20)? Eventually, the decision maker compares all the costs and benefits to see whether the project is justified or whether some alternative action might bring more benefits at less cost.

At the same time that some environmentalists are using CBA as a way of obtaining recognition for environmental issues within the terms of mainstream discourse, others are challenging this method of resource accounting as amoral and deeply flawed. Grassroots opponents of roads and hydroelectric dams around the world have repeatedly contested the ways that CBA values land, forests, streams, fisheries and livelihoods, as well as its reliance on unaccountable experts and neglect of equity issues. Ordinary people often refuse to answer questions about how much money they would pay to save a wilderness or how much they would accept to allow it to be destroyed. Developing world delegates to the Intergovernmental Panel on Climate Change angrily rejected a cost-benefit analysis of policy options regarding global warming that assigned a higher value to lives of people in industrialized countries compared to those in less-developed nations.

A study by the Economic Policy Institute of Washington, D.C., found that costs for complying with environmental regulations are almost always less than industry and even governments estimate they will be. For example, electric utilities in the United States claimed that it would cost $4 to $5 billion to meet the 1990 Clean Air Act. But by 1996, utilities were actually saving $150 million per year. Similarly, when CFCs were banned, automobile manufacturers protested it would add $1,200 to the cost of each new car. The actual cost was about $40.

Some other criticisms of CBA include its absence of standards, inadequate attention to alternatives, and the placing of monetary values on intangible and diffuse or distant costs and benefits. Who judges how costs and benefits will be estimated? How can we compare things as different as the economic gain from cheap power with loss of biodiversity or the beauty of a free-flowing river? Critics claim that placing monetary values on everything could lead to a belief that only money and profits count and that any behavior is acceptable as long as you can pay for it. Sometimes speculative or even hypothetical results are given specific numerical values in CBA and then treated as if they are hard facts. Risk-assessment techniques (see chapter 8) may be more appropriate for comparing uncertainties.

23.4 MARKET MECHANISMS CAN REDUCE POLLUTION

We are becoming increasingly aware that our environment and economy are mutually interconnected. Natural resources and ecological services are essential for a healthy economy, and a vigorous economy can provide the means to solve environmental problems. In this section, we'll explore some of these links.

Using market forces

As you've learned from previous chapters in this book, most scientists regard global climate change as the most serious environmental problem we face currently. In 2006, the business world got a harsh warning about this problem from British economist, Sir Nicolas Stern. Commissioned by the British treasury department to assess the threat of global warming, Sir Nicolas, who formerly was chief economist at the World Bank, issued a 700-page study that concluded that if we don't act to control greenhouse gases, the damage caused by climate change could be equivalent to losing as much as 20 percent of the global GDP every year. This could have an impact on our lives and environment greater than the worldwide depression or the great wars of the twentieth century.

We have many options for combating climate change, but economists believe that **market forces** can reduce pollution more efficiently than rigid rules and regulations. Assessing a tax, for example, on each ton of carbon emitted could have the desired effect of reducing greenhouse gases and controlling climate change, but could still allow industry to search for the most cost-effective ways to achieve these goals. It also creates a continuing incentive to search for better ways to reduce emissions. The more you reduce your discharges, the more you save.

The cost of climate control won't be cheap. Sir Nicolas calculates that it will take about $500 billion per year (1 percent of global GDP) to avoid the worst impacts of climate change if we

FIGURE 23.21 Markets for low-carbon energy could be worth $500 billion per year by 2050, and could create millions of high-paying jobs.

act now. Half a trillion dollars is a lot of money, but it's a bargain compared to his estimates of $10 trillion in annual losses and costs of climate change in 50 years if we don't do something. And the longer we wait, the more expensive carbon reduction and adaptation are going to be.

On the other hand, reducing greenhouse gas emissions and adapting to climate change will create significant business opportunities, as new markets are created in low-carbon energy technologies and services (fig. 23.21). These markets could create millions of jobs and be worth hundreds of billions of dollars every year. Already, Europe has more than 5 million jobs in renewable energy, and the annual savings from solar, wind, and hydro power are saving the European Union about $10 billion per year in avoided oil and natural gas imports. Being leaders in the fields of renewable energy and carbon reduction gives pioneering countries a tremendous business advantage in the global marketplace. Markets for low-carbon energy could be worth $500 billion per year by 2050, according to the Stern report.

Is emissions trading the answer?

They Kyoto Protocol, which was negotiated in 1997, and has been ratified by every industrialized nation in the world except the United States and Australia, sets up a mechanism called **emissions trading** to control greenhouse gases. This is also called a **cap-and-trade** approach. The first step is to mandate upper limits (the cap), on how much each country, sector, or specific industry is allowed to emit. Companies that can cut pollution by more than they're required can sell the **credit** to other companies that have more difficulty meeting their mandated levels. Suppose you've just built a state-of-the-art integrated gasification combined cycle (IGCC) power plant (chapter 19) that allows you to capture and store CO_2 for about $20 per ton. You're allowed to emit a thousand tons of CO_2 per day, but you could easily cut your releases below that amount.

Suppose, further, that your neighboring utility has a dirty, old coal-fired power plant for which it would cost $60 per ton to keep CO_2 out of the smokestack. You might strike a deal with your neighbor. For $40 per ton, you'll reduce your emissions enough to meet his mandated cap. You make $20 per ton, and your neighbor saves $20 per ton. Both benefit. But if your neighbor can find an even cheaper way to **offset** his carbon emissions, he's free to do so. This creates an incentive to continually search for ever more cost-effective ways to reduce emissions.

Here's another example of carbon trading that might have even greater application to your life. Suppose you drive an old car that doesn't get very good mileage. You may feel guilty about the CO_2 you're emitting, but perhaps you can't afford to buy a new, more efficient vehicle. There are several organizations that will now sell an offset to you. For about $20 per ton (or about $100 per year) for the average American car, they'll plant trees, build a windmill, or provide solar lights to a village in a developing country to compensate for your emissions. You can take pride in being **carbon-neutral** at a far lower price than buying a new automobile.

Sulfur trading offers a good model

The 1990 U.S. Clean Air Act created one of the first market-based systems for reducing air pollution. It mandated a decrease in acid-rain-causing sulfur dioxide (SO_2) from power plants and other industrial facilities. A SO_2 targeted reduction was set at 10 million tons per year, leaving it to industry to find the most efficient way to do this. The government expected that meeting this goal would cost companies up to $15 billion per year, but the actual cost has been less than one-tenth of that. Prices on the sulfur exchange have varied from $60 to $800 per ton depending on the availability and price of new technology, but most observers agree that the market has found much more cost-effective ways to achieve the desired goal than rigid rules would have created.

This program is regarded as a shining example of the benefits of market-based approaches. There are complaints, however, that while nationwide emissions have come down, "hot spots" remain where local utilities have paid for credits rather than install pollution abatement equipment. If you're living in one of these hot spots and continuing to breathe polluted air, it's not much comfort to know that nationwide average air quality has improved. Currently, credits and allowances of more than 30 different air pollutants are traded in international markets.

Carbon trading is already at work

Climate change is revolutionizing global economics. In 2006, approximately (U.S.) $28 billion worth of climate credits, equivalent to 1 billion tons of CO_2, traded hands on international markets. It's expected that this market will expand by nearly five-fold in 2008. By far the most active market currently is the Amsterdam-based European Climate Exchange. The United States has a climate market in Chicago, but at this point, participation is only voluntary because the U.S. doesn't have mandatory emissions limits, and carbon credits are selling for only about one-tenth the price they are in Europe.

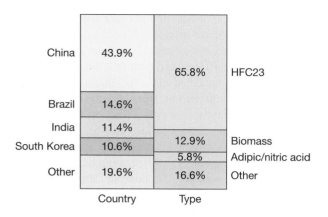

FIGURE 23.22 Worldwide emissions reductions payments by country and type. Currently, four countries are collecting 80 percent of all proceeds from emissions trading, and two-thirds of those payments are going for relatively cheap HFC23 incineration. Is this fair?
Source: United Nations, 2007.

In 2006, more than 80 percent of the international emissions payments went to just four countries, and nearly two-thirds of those payments were for reductions of the refrigerant HFC-23 (fig. 23.22). Most entrepreneurs are uninterested in deals less than about $250,000. Smaller projects just aren't worth the time and expense of setting them up. In one of the biggest deals so far, a consortium of British bankers signed a contract to finance an incinerator on a large chemical factory in Quzhou, in China's Zhejiang Province. The incinerator will destroy hydrofluorocarbon (HFC-23) that previously had been vented into the air. This has a double benefit: HFC-23 destroys stratospheric ozone, and it also is a potent greenhouse gas (approximately 11,700 times as powerful as CO_2). The $500 million deal will remove the climate-changing equivalent of the CO_2 emitted by 1 million typical American cars each driven 20,000 km per year. But the incinerator will cost only $5 million—a windfall profit to be split between the bankers and the factory owners.

There's a paradox in this deal. HFC production in China and India is soaring because a growing middle class fuels a demand for refrigerators and air conditioners. The huge payments flowing into these countries under the Kyoto Protocol are helping their economies grow and increasing middle-class affluence, and thus creating more demand for refrigerators and air conditioners. Furthermore, air conditioners using this refrigerant are much less energy efficient than newer models, so their increasing numbers are driving the demand for electricity, which currently is mostly provided by coal-fired power plants.

Critics of our current emissions markets point out that this mechanism was originally intended to encourage the spread of renewable energy and nonpolluting technology to developing countries in places such as sub-Saharan Africa. It was envisioned as a way to spread solar panels, windmills, tree farms, and other technologies that would provide climate control and also speed development of the poorest people. Instead, marketing emission credits, so far, is benefiting primarily bankers, consultants, and factory owners and is leading to short-term fixes rather than fundamental, long-term solutions.

23.5 TRADE, DEVELOPMENT, AND JOBS

Trade can be a powerful tool in making resources available and raising standards of living. Think of the things you now enjoy that might not be available if you had to live exclusively on the resources available in your immediate neighborhood. Too often, the poorest, least powerful people suffer in this global marketplace. To balance out these inequities, nations can deliberately invest in economic development projects. In this section, we'll look at some aspects of trade, development, business, and jobs that have impacts on our environment and welfare.

International trade brings benefits but also intensifies inequities

The banking and trading systems that regulate credit, currency exchange, shipping rates, and commodity prices were set up by the richer and more powerful nations in their own self-interest. The General Agreement on Tariffs and Trade (GATT) and World Trade Organization (WTO) agreements, for example, negotiated primarily between the largest industrial nations, regulate 90 percent of all international trade.

These systems tend to keep the less-developed countries in a perpetual role of resource suppliers to the more-developed countries. The producers of raw materials, such as mineral ores or agricultural products, get very little of the income generated from international trade (fig. 23.23).

Policies of the WTO and the IMF have provoked criticism and resistance in many countries. As a prerequisite for international development loans, the IMF frequently requires debtor nations to adopt harsh "structural adjustment" plans that slash welfare programs and impose cruel hardships on poor people. The WTO has issued numerous rulings that favor international

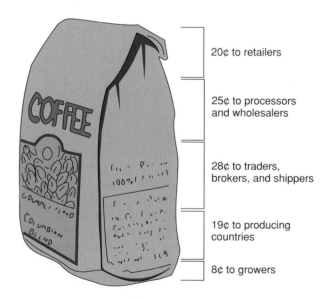

20¢ to retailers

25¢ to processors and wholesalers

28¢ to traders, brokers, and shippers

19¢ to producing countries

8¢ to growers

FIGURE 23.23 What do we really pay for when we purchase a dollar's worth of coffee?

trade over pollution prevention or protection of endangered species. Trade conventions such as the North American Free Trade Agreement (NAFTA) have been accused of encouraging a "race to the bottom" in which companies can play one country against another and move across borders to find the most lax labor and environmental protection standards.

Aid often doesn't help the people who need it

No single institution has more influence on financing and policies of developing countries than the World Bank. Of some $25 billion loaned each year for development projects by international agencies, about two-thirds comes from the World Bank. Founded in 1945 to fund reconstruction of Europe and Japan, the World Bank shifted its emphasis to aid developing countries in the 1950s. Many of its projects have had adverse environmental and social effects, however. Its loans often go to corrupt governments and fund ventures such as nuclear power plants, huge dams, and giant water diversion schemes. Former U.S. Treasury Secretary Paul O'Neill said that these loans have driven poor countries "into a ditch" by loading them with unpayable debt. He said that funds should not be loans, but rather grants to fight poverty.

Microlending helps the poorest of the poor

As the opening case study for this chapter shows, tiny loans can change the lives of the poorest of the poor. The Grameen Bank, founded by Dr. Muhammad Yunus, has assisted millions of people—most of them low-status women who have no other way to borrow money at reasonable interest rates. This model is now being used by hundreds of other development agencies around the world (fig. 23.24). Even in the United States, organizations assist microenterprises with loans, grants, and training. The Women's Self-Employment Project in Chicago, for instance, teaches job skills to single mothers in housing projects. Similarly, "tribal circle" banks on Native American reservations successfully finance microscale economic development ventures.

One of the most important innovations of the Grameen Bank is that borrowers take out loans in small groups. Everyone in the group is responsible for each other's performance. The group not only guarantees loan repayment, it helps businesses succeed by offering support, encouragement, and advice. Where banks depend on the threat of foreclosure and a low credit rating to ensure debt repayment, the Grameen Bank has something at least as powerful for poor villagers—the threat of letting down your neighbors and relatives. Becoming a member of a Grameen group also requires participation in a savings program that fosters self-reliance and fiscal management.

Microlending is built on a series of small loans. You start with a very small amount: the equivalent of only a few dollars. When that loan is repaid, you can take out another one. This creates an ongoing relationship between the borrowers and lenders. Success builds on success. And as each loan is granted, it allows the lender to learn about other aspects of the borrower's life, and to help in ways other than merely providing money. The process of running

FIGURE 23.24 With a loan of only a few dollars, this Chinese coal deliverer could buy his own cart and more than double his daily income. If you could make a tiny loan that would change his life, wouldn't you do it?

a successful business and repaying each installment of the loan on time transforms many individuals. Women, who previously had little power or self-esteem, are empowered with a sense of pride and accomplishment. Dr. Yunus discovered that money going to families through women helped the families much more than the same amount of money in men's salaries.

The most recent venture for the Grameen Bank is providing mobile phone service to rural villages. Supplying mobile phones to poor women not only allows them to communicate, it provides another business opportunity. They rent out their phone to neighbors, giving the owner additional income, and linking the whole village to the outside world. Suddenly, people who had no access to communication can talk with their relatives, order supplies from the city, check on prices at the regional market, and decide when and where to sell their goods and services. This is a great example of "bottom-up development." Founded in 1996, Grameen Phone now has 2.5 million subscribers and is Bangladesh's largest mobile phone company.

In 2006, Dr. Yunus received the Nobel Peace Prize for his groundbreaking work in helping the poor. He has formed a political party called "Nagorik Shakti," (Citizen Power), and is running for national office. Changing the lives of the poor continues to be his highest priority.

23.6 GREEN BUSINESS

During the first Industrial Revolution 200 years ago, raw materials such as lumber, minerals, and clean water seemed inexhaustible, while nature was regarded as something to be tamed and civilized. The "solution to pollution is dilution," was a common assumption in disposal of unwanted wastes. Today business leaders are increasingly discovering that they can save money by greening up their business practices. Fuel efficiency saves money and cuts greenhouse gases. Conversion to less-hazardous chemicals can cut disposal costs, health costs, and production costs. In addition, these companies win public praise and new customers

Eco-efficient Business Practices

In 1994, in response to customers' concerns about health problems caused by chemical fumes from new carpeting, wall coverings, and other building materials, Ray Anderson, founder and CEO of Interface, Inc., a billion-dollar-a-year interior furnishing company, decided to review company environmental policy. What he found was that the company really didn't have an environmental vision other than to obey all relevant laws and comply with regulations. He also learned that carpeting—of which Interface was the world's third-largest manufacturer—is one of the highest volume and longest lasting components in landfills. A typical carpet is made of nylon embedded in fiberglass and polyvinyl chloride. After a useful life of about 12 years, most carpeting is ripped up and discarded. Every year, more than 770 million m^2 (920 million yd^2) of carpet weighing 1.6 billion kg (3.5 billion lbs) ends up in U.S. landfills. The only recycling that most manufacturers do is to shave off some of the nylon for remanufacture. Everything else is buried in the ground where it will last at least 20,000 years.

At about the same time that Interface was undergoing its environmental audit, Anderson was given a copy of Paul Hawken's book *The Ecology of Commerce.* Reading it, he said, was like "a spear through the chest." He vowed to turn his company around, to make its goal sustainability instead of simply maximizing profits. Rather than sell materials, Interface would focus on selling service. The key is what Anderson calls an "evergreen lease." First of all, the carpet is designed to be completely recyclable. Where most flooring companies merely sell carpet, Interface offers to lease carpets to customers. As carpet tiles wear out, old ones are removed and replaced as part of the lease. The customer pays no installation or removal charges, only a monthly fee for constantly fresh-looking and functional carpeting. Everything in old carpet is used to make new product. Only after many reincarnations as carpet, are materials finally sent to the landfill.

Dramatic changes have been made at Interface's 26 factories. Toxic air emissions have been nearly eliminated by changing manufacturing processes and substituting nontoxic materials for more dangerous ones. Solar power and methane from a landfill are replacing fossil fuel use. Interface may be the first carbon-neutral manufacturing company in America. Less waste is produced as more material is recycled and products are designed for eco-efficiency. The total savings from pollution prevention and recycling in 2007 was $150 million.

Not only has Interface continued to be an industry leader, it was named one of the "100 Best Companies to Work For in America" by *Fortune* magazine. Ray Anderson has become a popular speaker on the

Ray Anderson.
Courtesy Ray Anderson, Interface, Inc.

topic of eco-efficiency and clean production. He cochairs the President's Council on Sustainable Development, was named Entrepreneur of the Year by Ernest & Young, and was the Georgia Conservancy's Conservationist of the Year in 1998. Anderson's book, *Mid-Course Correction: Toward a Sustainable Enterprise,* published in 1999 by Chelsea Green, has won critical acclaim.

Transforming an industry as large as interior furnishing has not been an overnight success. "Like aircraft carriers," Anderson says, "big businesses don't turn on a dime." Still, he has shown that the principles of sustainability and financial success can coexist and can lead to a new prosperity that includes both environmental and human dividends. His motto, that we should "put back more than we take and do good to the Earth, not just no harm," has become a vision for a new industrial revolution that now is reaching many companies beyond his own.

Ethical Considerations

What responsibilities do businesses have to protect the environment or save resources beyond the legal liabilities spelled out in the law? None whatever, according to conservative economist Milton Friedman. In fact, Friedman argues, it would be unethical for corporate leaders to consider anything other than maximizing profits. To spend time or resources doing anything other than making profits and increasing the value of the company is a betrayal of their duty. What do you think? Should social justice, sustainability, or environmental protection be issues of concern to corporations?

by demonstrating an interest in our shared environment. By conserving resources, they also help ensure the long-term survival at their own corporations.

Known by a variety of names, including eco-efficiency, clean production, pollution prevention, industrial ecology, natural capitalism, restorative technology, the natural step, environmentally preferable products, design for the environment, and the next industrial revolution, this movement has had some remarkable successes and presents an encouraging pathway for how we might achieve both environmental protection and social welfare.

Some of the leaders in this new approach to business include Paul Hawken, William McDonough, Ray Anderson, Amory Lovins, David Crockett, and John and Nancy Todd.

Operating in a socially responsible manner consistent with the principles of sustainable development and environmental protection, they have shown, can be good for employee morale, public relations, and the bottom line simultaneously. Environmentally conscious or "green" companies such as the Body Shop, Patagonia, Aveda, Malden Mills, Johnson and Johnson, and Interface, Inc. (What Do You Think? p. 534) consistently earn high marks from community and environmental groups. Conserving resources, reducing pollution, and treating employees and customers fairly may cost a little more initially, but can save money and build a loyal following in the long run.

New business models follow concepts of ecology

Paul Hawken's 1993 book, *The Ecology of Commerce,* was a seminal influence in convincing many people to reexamine the role of business and economics in environmental and social welfare. Basing his model for a new industrial revolution on the principles of ecology, Hawken points out that almost nothing is discarded or unused in nature. The wastes from one organism become the food of another. Industrial processes, he argues, should be designed on a similar principle (table 23.4). Rather than a linear pattern in which we try to maximize the throughput of material and minimize labor, products and processes should be designed to

- be energy efficient;
- use renewable materials;
- be durable and reusable or easily dismantled for repair and remanufacture, nonpolluting throughout their entire life cycle;
- provide meaningful and sustainable livelihoods for as many people as possible;
- protect biological and social diversity;
- use minimum and appropriate packaging made of reusable or recyclable materials.

We can do all this and at the same time increase profits, reduce taxes, shrink government, increase social spending, and restore our environment, Hawken claims. Recently, Hawken has served as chairperson for The Natural Step in America, a

TABLE 23.4

Goals for an Eco-Efficient Economy

- Introduce no hazardous materials into the air, water, or soil.
- Measure prosperity by how much natural capital we can accrue in productive ways.
- Measure productivity by how many people are gainfully and meaningfully employed.
- Measure progress by how many buildings have no smokestacks or dangerous effluents.
- Make the thousands of complex governmental rules unnecessary that now regulate toxic or hazardous materials.
- Produce nothing that will require constant vigilance from future generations.
- Celebrate the abundance of biological and cultural diversity.
- Live on renewable solar income rather than fossil fuels.

TABLE 23.5

The Natural Step: System Conditions for Sustainability

1. Minerals and metals from the earth's crust must not systematically increase in nature.
2. Materials produced by human society must not systematically increase in nature.
3. The physical basis for biological productivity must not be systematically diminished.
4. The use of resources must be efficient and just with respect to meeting human needs.

movement started in Sweden by Dr. K. H. Robert, a physician concerned about the increase in environmentally related cancers. Through a consensus process, a group of 50 leading scientists endorsed a description of the living systems on which our economy and lives depend. More than 60 major European corporations and 55 municipalities have incorporated sustainability principles (table 23.5) into their operations.

Another approach to corporate responsibility is called the **triple bottom line.** Rather than reporting only net profits as a measure of success, ethically sensitive corporations include environmental effects and social justice programs as indications of genuine progress.

Corporations committed to eco-efficiency and clean production include such big names as Monsanto, 3M, DuPont, Duracell, and Johnson and Johnson. Following the famous three Rs—reduce, reuse, recycle—these firms have saved money and gotten welcome publicity. Savings can be substantial. Slashing energy use and redesigning production to use less raw material and to produce less waste is reported to have saved DuPont $3 billion over the past decade, while also reducing greenhouse emissions 72 percent.

Think About It

Most designs for environmental efficiency involve relatively simple rethinking of production or materials. Many of us might be able to save money, time, or other resources in our own lives just by thinking ahead. Think about your own daily life: Could you use new strategies to reduce consumption or waste in recreational activities, cooking, or shopping? In transportation? In housing choices?

Efficiency starts with design of products and processes

Our current manufacturing system often is incredibly wasteful. On average, for every truckload of products delivered in the United States, 32 truckloads of waste are produced along the way. The automobile is a typical example. Industrial ecologist, Amory Lovins, calculates that for every 100 gallons (380 l) of gasoline burned in your car engine, only one percent (1 gal or 3.8 l)

TABLE 23.6

McDonough Design Principles

Inspired by the way living systems actually work, Bill McDonough offers three simple principles for redesigning processes and products:

1. *Waste equals food.* This principle encourages elimination of the concept of waste in industrial design. Every process should be designed so that the products themselves, as well as leftover chemicals, materials, and effluents, can become "food" for other processes.
2. *Rely on current solar income.* This principle has two benefits: First, it diminishes, and may eventually eliminate, our reliance on hydrocarbon fuels. Second, it means designing systems that sip energy rather than gulping it down.
3. *Respect diversity.* Evaluate every design for its impact on plant, animal, and human life. What effects do products and processes have on identity, independence, and integrity of humans and natural systems? Every project should respect the regional, cultural, and material uniqueness of its particular place.

actually moves passengers. All the rest is used to move the vehicle itself. The wastes produced—carbon dioxide, nitrogen oxides, unburned hydrocarbons, rubber dust, heat—are spread through the environment where they pollute air, water, and soil.

Architect William McDonough urges us to rethink design approaches (table 23.6). In the first place, he says, we should question whether the product is really needed. Could we provide the same service in a more eco-efficient manner? According to McDonough, products should be divided into three categories:

1. *Consumables* are products like food, natural fabrics, or paper that can harmlessly go back to the soil as compost.
2. *Service products* are durables such as cars, TVs, and refrigerators. These products should be leased to the customer to provide their intended service, but would always belong to the manufacturer. Eventually they would be returned to the maker, who would be responsible for recycling or remanufacturing the product. Knowing that they will have to dismantle the product at the end of its life will encourage manufacturers to design for easy disassembly and repair.
3. *Unmarketables* are compounds like radioactive isotopes, persistent toxins, and bioaccumulative chemicals. Ideally, no one would make or use these products. But because eliminating their use will take time, McDonough suggests that in the mean time these materials should belong to the manufacturer and be molecularly tagged with the maker's mark. If they are discovered to be discarded illegally, the manufacturer would be held liable.

Following these principles, McDonough Bungart Design Chemistry has created nontoxic, easily recyclable materials to use in buildings and for consumer goods. Among some important and

innovative "green office" projects designed by the McDonough and Partners architectural firm are the Environmental Defense Fund headquarters in New York City, the Environmental Studies Center at Oberlin College in Ohio (see fig. 20.10), the European Headquarters for Nike in Hilversum, the Netherlands, and the Gap Corporate Offices in San Bruno, California (fig. 23.25). Intended to promote employee well-being and productivity as well as eco-efficiency, the Gap building has high ceilings, abundant skylights, windows that open, a full-service fitness center (including pool), and a landscaped atrium for each office bay that brings the outside in. The roof is covered with native grasses. Warm interior tones and natural wood surfaces (all wood used in the building was harvested by certified sustainable methods) give a friendly feeling. Paints, adhesives, and floor coverings are low toxicity and the building is one-third more energy efficient than strict California laws require. A pleasant place to work, the offices help recruit top employees and improve both effectiveness and retention. As for the bottom line, Gap, Inc. estimates that the increased energy and operational efficiency will have a four- to eight-year payback.

Green consumerism gives the public a voice

Consumer choice can play an important role in persuading businesses to produce eco-friendly goods and services (What Can You Do? p. 537). Increasing interest in environmental and social sustainability has caused an explosive growth of green products. The National Green Pages published by Co-Op America currently lists more than 2,000 green companies. You can find ecotravel agencies, telephone companies that donate profits to environmental groups, entrepreneurs selling organic foods, shade-grown coffee, straw-bale houses, geodesic-dome kits, paint thinner made from orange peels, sandals made from recycled auto tires, earthworms for composting, and a plethora of hemp products

FIGURE 23.25 The award-winning Gap, Inc. corporate offices in San Bruno, California, demonstrate some of the best features of environmental design. A roof covered with native grasses provides insulation and reduces runoff. Natural lighting, an open design, and careful relation to its surroundings all make this a good place to work.

including burgers, ale, clothing, shoes, rugs, balm, shampoo, and insect repellent. Although these eco-entrepreneurs represent a tiny sliver of the $7 trillion per year U.S. economy, they often serve as pioneers in developing new technologies and offering innovative services.

In some industries eco-entrepreneurs have found profitable niches (organic, naturally colored clothing, for example) within a larger market. In other cases, once a consumer demand has built up, major companies add green products or services to their inventory. Natural foods, for instance, have grown from the domain of a few funky, local co-ops to a $7 billion market segment. Most supermarket chains now carry some organic food choices. Similarly, natural-care health and beauty products are now more than 10 percent of a $33 billion industry. By supporting these products, you can ensure that they will continue to be available and, perhaps, even help expand their penetration into the market.

Environmental protection creates jobs

For years business leaders and politicians have portrayed environmental protection and jobs as mutually exclusive. Pollution control, protection of natural areas and endangered species, limits on use of nonrenewable resources, they claim, will strangle the economy and throw people out of work. Ecological economists dispute this claim, however. Their studies show that only 0.1 percent of all large-scale layoffs in the United States in recent years were due to government regulations (fig. 23.26). Environmental protection, they argue, is not only necessary for a healthy economic system, it actually creates jobs and stimulates business.

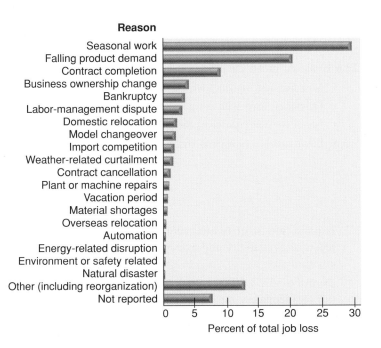

FIGURE 23.26 Although opponents of environmental regulation often claim that protecting the environment costs jobs, studies by economist E. S. Goodstein show that only 0.1 percent of all large-scale layoffs in the United States were the result of environmental laws.
Source: E. S. Goodstein, Economic Policy Institute, Washington, D.C.

Personally Responsible Consumerism

There are many things that each of us can do to lower our ecological impacts and support green businesses through responsible consumerism and ecological economics.

- Practice living simply. Ask yourself if you really need more material goods to make your life happy and fulfilled.

- Rent, borrow, or barter when you can. Do you really need to have your own personal supply of tools, machines, and other equipment that you use, at most, once per year?

- Recycle or reuse building materials: doors, windows, cabinets, appliances. Shop at salvage yards, thrift stores, yard sales, or other sources of used clothes, dishes, appliances, etc.

- Consult the National Green Pages from Co-Op America for a list of eco-friendly businesses. Write one letter each month to a company from which you buy goods or services and ask them what they are doing about environmental protection and human rights.

- Buy green products. Look for efficient, high quality materials that will last and that are produced in the most environmentally friendly manner possible. Subscribe to clean energy programs if they are available in your area. Contact your local utility and ask that they provide this option if they don't now.

- Buy products in bulk if you can or look for the least amount of packaging. Choose locally grown or locally made products made under humane conditions by workers who receive a fair wage.

- Think about the total life-cycle costs of the things you buy, including environmental impacts, service charges, energy use, and disposal costs as well as initial purchase price.

- Stop junk mail. Demand that your name be removed from mass-mailing lists.

- Invest in socially and environmentally responsible mutual funds or green businesses when you have money for investment.

Recycling, for instance, makes more new jobs than extracting virgin raw materials. This doesn't necessarily mean that recycled goods are more expensive than those from virgin resources. We're simply substituting labor in the recycling center for energy and huge machines used to extract new materials in remote places.

Japan, already a leader in efficiency and environmental technology, has recognized the multibillion dollar economic potential of green business. The Japanese government is investing (U.S.) $4 billion per year on research and development that targets seven areas, ranging from utilitarian projects such as biodegradable plastics and heat-pump refrigerants to exotic schemes such as carbon-dioxide-fixing algae and hydrogen-producing microbes.

A World Wildlife Fund study predicts that over the next decade, energy conservation, renewable energy sources, and other environmental protection programs could result in 2 million new jobs in the United States if new technologies are embraced and promoted. The net economic impact could be as much as $10 billion per year.

CONCLUSION

At the 1972 Stockholm Conference on the Human Environment, Indira Gandhi, then prime minister of India, stated that "Poverty is the greatest polluter of them all." What she meant was that the world's poorest people are too often both the victims and agents of environmental degradation. They are forced to meet short-term survival needs at the cost of long-term sustainability. But "charity is not an answer to poverty," Dr. Muhammad Yunus says, "It only helps poverty to continue. It creates dependency and takes away individual's initiative . . . Poverty isn't created by the poor, it's created by the institutions and policies that surround them . . . All we need to do is to make appropriate changes in the institutions and policies, and/or create new ones." The microcredit revolution he started may be the key for breaking the cycle of poverty and changing the lives of the poor.

Emissions trading could also be a way to aid poor countries. It could encourage the spread of renewable energy and nonpolluting technology to sub-Saharan Africa, Asia, and Latin America. Many of the countries where biodiversity is highest, and programs to stop deforestation, replant trees, and install solar, wind, and biomass generators can do the most good, are also poor countries most in need of development aid. Incorporating ecological knowledge into our economic policies could help us value nature as well as assist in meeting the UN Millennium goals of eradicating extreme poverty and hunger while also ensuring environmental sustainability.

REVIEWING LEARNING OUTCOMES

By now you should be able to explain the following points:

23.1 Analyze economic worldviews.
- Can development be sustainable?
- Our definitions of resources shape how we use them.
- Classical economics examines supply and demand.
- Neoclassical economics emphasizes growth.
- Ecological economics incorporates principles of ecology.
- Communal property resources are a classic problem in ecological economics.

23.2 Scrutinize population, technology, and scarcity.
- Scarcity can lead to innovation.
- Carrying capacity is not necessarily fixed.
- Economic models compare growth scenarios.
- Why not conserve resources?

23.3 Investigate natural resource accounting.
- Gross national product is our dominant growth measure.
- Alternate measures account for well-being.

- New approaches incorporate nonmarket values.
- Cost-benefit analysis aims to optimize resource use.

23.4 Summarize how market mechanisms can reduce pollution.
- Using market forces.
- Is emissions trading the answer?
- Sulfur trading offers a good model.
- Carbon trading is already at work.

23.5 Study trade, development, and jobs.
- International trade brings benefits but also intensifies inequities.
- Aid often doesn't help the people who need it.
- Microlending helps the poorest of the poor.

23.6 Evaluate green business.
- New business models follow concepts of ecology.
- Efficiency starts with design of products and processes.
- Green consumerism gives the public a voice.
- Environmental protection creates jobs.

PRACTICE QUIZ

1. Define *economics* and distinguish between classical, neoclassical, and ecological economics.
2. Define *resources* and give some examples of renewable, nonrenewable, and intangible resources.
3. List three economic categories of resources and describe the differences among them.
4. Describe the relationship between supply and demand.
5. Identify some important ecological services on which our economy depends.
6. Describe how cost-benefit ratios are determined and how they are used in natural resource management.
7. Explain how scarcity and technological progress can extend resource availability and extend the carrying capacity of the environment.
8. Describe how GNP is calculated and explain why this may fail to adequately measure human welfare and environmental quality. Discuss some alternative measures of national progress.
9. What is microlending, and what are its benefits?
10. List some of the characteristics of an eco-efficient economic system.

CRITICAL THINKING AND DISCUSSION QUESTIONS

1. When the ecologist warns that we are using up irreplaceable natural resources and the economist rejoins that ingenuity and enterprise will find substitutes for most resources, what underlying premises and definitions shape their arguments?

2. How can intangible resources be infinite and exhaustible at the same time? Isn't this a contradiction in terms? Can you find other similar paradoxes in this chapter?

3. What would be the effect on the developing countries of the world if we were to change to a steady-state economic system? How could we achieve a just distribution of resource benefits while still protecting environmental quality and future resource use?

4. Resource use policies bring up questions of intergenerational justice. Suppose you were asked: "What has posterity ever done for me?" How would you answer?

5. If you were doing a cost-benefit study, how would you assign a value to the opportunity for good health or the existence of rare and endangered species in faraway places? Is there a danger or cost in simply saying some things are immeasurable and priceless and therefore off limits to discussion?

6. If natural capitalism or eco-efficiency has been so good for some entrepreneurs, why haven't all businesses moved in this direction?

 **DATA analysis**   **Evaluating Human Development**

The human development index (HDI) is a measure created by the United Nations Development Programme to track social progress. HDI incorporates life expectancy, adult literacy, children's education, and standard of living indicators to measure human development. The 2006 report draws on statistics from 175 countries. While there has been encouraging progress in most world regions, the index shows that widening inequality is taking a toll on global human development.

The graph shows trends in the HDI by world region. Study this graph carefully, and answer the following questions: (*Hint:* you may have to use a search engine to find some answers.)

1. Which region has the highest HDI rating?

2. What does OECD stand for?

3. Which region has made the greatest progress over the past 30 years, and how much has its HDI increased?

4. Which region has shown the least progress in human development?

5. What historic events could explain the reduction in Europe and the CIS between 1990 and 1995?

6. How much lower is the HDI ranking of sub-Saharan Africa from the OECD?

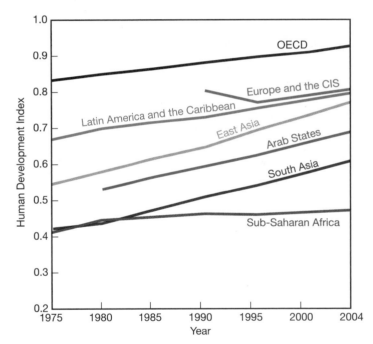

Trends in human development, 1975–2004.
Source: United Nations Development Programme, 2006.

For Additional Help in Studying This Chapter, please visit our website at www.mhhe.com/cunningham10e. You will find additional practice quizzes and case studies, flashcards, regional examples, place markers for Google Earth™ mapping, and an extensive reading list, all of which will help you learn environmental science.

The ability of ordinary citizens to petition their government
and to participate in public policy formation is a
hallmark of democracy.

C H A P T E R

24

Environmental Policy, Law, and Planning

The power to command frequently causes failure to think.

—Barbara Tuchman—

LEARNING OUTCOMES

After studying this chapter, you should be able to:

24.1 Define environmental policy.

24.2 Discuss environmental law.

24.3 Describe international treaties and conventions.

24.4 Summarize dispute resolution and planning.

Case Study The Snail Darter versus Tellico Dam

In the 1970s, a paperclip-sized fish in Tennessee made national headlines and helped establish national environmental policy. The controversy arose because construction of the Tellico Dam on the Little Tennessee River threatened to eradicate the only known population of the snail darter (*Percina tanasi*), a tiny member of the perch family (fig. 24.1). The dam, which was being built by the Tennessee Valley Authority (TVA), was intended to promote economic development, control floods, and provide recreation. Opponents had fought against this project for years, claiming that the dam wasn't needed, cost too much, offered insufficient benefits, and would destroy forests, wildlife habitat, family farms, Native American archeological sites, and already rare free-flowing sections of river in the area. None of these arguments dissuaded the TVA, however, which had already built more than 65 dams and impounded 4,000 km (2,500 mi) of rivers in the Tennessee and adjacent watersheds.

The Tellico Dam was roughly 85 percent complete (at a cost of about $80 million) in 1973 when an aquatic biologist discovered a new variety of darters living in shallow shoals at Coytee Springs, just upstream from the dam site. As far as biologists knew at the time, the total population of only a few hundred animals confined to the clear, cool, rapidly flowing currents and rocky habitat of this particular stretch of river were the only living examples of their species. Their absence from nearby deeper sections of the river suggested that if the dam were completed, the snail darters probably wouldn't survive in the deep, still reservoir the dam would create.

FIGURE 24.1 The tiny snail darter almost blocked construction of the $100 million Tellico Dam on the Little Tennessee River. The Endangered Species Act, which would have protected this rare and obscure species, is arguably our most controversial environmental law.

Shortly before discovery of the snail darter, Congress had passed—and President Nixon had signed—a revision of the Endangered Species Act (ESA) directing that federal agencies "shall insure that actions do not jeopardize the continued existence of endangered species or result in the destruction or modification of habitat . . ." This new version of the ESA also allowed citizen intervention on behalf of endangered species. Anyone can petition the Fish and Wildlife Service to study a species for listing as threatened or endangered. If scientific evidence shows that the species or its habitat are threatened, ESA protection takes effect.

While the snail darter was being studied for endangered species status, the TVA was ordered to suspend dam construction. In 1975, the snail darter was declared an endangered species. A landmark series of court cases ensued, challenging the legality of halting such an expensive project to protect a fish with "no economic or recreational value." Howard Baker, the senior senator from Tennessee said, "We didn't intend this Act to protect cold slimy things, but rather warm fuzzy things like eagles and polar bears." This case

eventually made its way to the U.S. Supreme Court, which ruled in 1978 that the "plain intent" of the law is to save species from extinction "whatever the cost."

In response, Congress amended the ESA to create a committee known as the "God Squad," which can exclude species from protection if it deems that economic benefits outweigh species value. To the dismay of its sponsors, this committee agreed with the courts that protection of the snail darter should take precedence over Tellico Dam. The Tennessee congressional delegation responded by slipping a rider into an appropriations bill exempting the Tellico Dam from the ESA. The TVA then completed the dam, destroying all the then-known snail darter habitat.

As it turns out, several other small, isolated populations of snail darters have been discovered elsewhere in the Tennessee River since the dam was finished, and their status was downgraded in 1980 from endangered to threatened. The Tellico Dam is now 25 years old, but most of the economic benefits its promoters claimed have yet to be fulfilled. The "God Squad" has very rarely been called on to exclude a species from ESA protection. Still, this case was an important test of the ESA. It raised questions about whether our policy should be to protect all species or only warm, fuzzy, appealing ones.

Do human needs and economic interests take precedence over preservation of biodiversity? How much sacrifice are we willing to make to protect economically insignificant organisms?

The ESA continues to be one of the most contentious environmental policies in the United States. Conservatives complain that it tramples on property rights and impedes businesses. Conservationists regard it as the most powerful tool at their disposal to protect environmental quality. Both sides call for revisions of this important act, but their intents are diametrically opposed. Property rights advocates want to weaken or eliminate the ESA altogether, while conservationists wants it to focus on endangered ecosystems and biological communities rather than single species.

This case study illustrates the interactions between the legislative, judicial, and administrative branches of government, and the struggles to form and implement environmental policies in the face of competing interests and values. In this chapter, we'll look at how policy is created, how laws are established and tested, and how regulatory agencies implement those policies and laws.

24.1 ENVIRONMENTAL POLICY

The term "policy" is used in many different ways to indicate both formal and informal decisions or intentions at a personal, community, national, or international level. You might have an informal policy never to accept telemarketing calls; your church may have an open-door policy for visitors; and many countries have a policy not to negotiate with terrorists. At the same time, the U.S. Clean Air Act is a formal statement of national policy on acceptable air quality, while the U.N. Convention on Global Climate Change represents the official intentions of many nations to curb greenhouse gases. Interestingly, policy can describe both an actual document as well as a contractual agreement, such as when you buy insurance.

At its core, then, **policy** is a plan or statement of intentions—either written or stated—about a course of action or inaction intended to accomplish some end. Some political scientists limit the term public policy to the principles, laws, executive orders, codes, or goals established by some government body or institution. For the purposes of this chapter, **environmental policy** will be taken as those official rules and regulations concerning the environment that are adopted, implemented, and enforced by some governmental agency as well as general public opinion about environmental issues (fig. 24.2).

How is policy created?

The policies we establish depend to a great extent on the system within which they operate. For many of us, the ideal political system is one that is open, honest, transparent, reaches the best possible decisions to maximize benefits to everyone. In a pluralistic, democratic society, we aim to give everyone an equal voice in policy making. Ideally, many separate interests put forward their solutions to public problems that are discussed, debated, and evaluated fairly and equally. Facts and access are open to everyone. Policy choices are made democratically but compassionately;

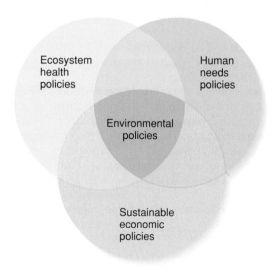

FIGURE 24.2 The best environmental policies incorporate economic, ecological, and social/cultural considerations.

implementation is reasonable, fair, and productive. Unfortunately, this isn't always the way our system works. Can you think of some alternate explanations of how we create policy?

Politics as Power

According to some observers, politics is really the struggle among often unequal, competing interest groups as they strive to shape public policy to suit their own agendas. The political system, in this view, manages group conflict by (1) establishing rules to ensure civil competition, (2) encouraging compromises and balancing interests to the extent possible, (3) codifying compromises as public policy, and (4) enforcing laws and rules based on that policy. Where this form of power politics is operational, it often results in a tyranny of a powerful elite over the impotent masses. Those elites manipulate public opinion and give up only as much power and wealth as necessary to maintain overall control.

However, while self-interest and power politics clearly are important forces in American public life, they can't account for all of the civil rights and environmental movements, the war on poverty, and public interest that characterized much of the 1960s and 1970s. The force of ideals, values, and altruism sometimes carry the day even in our winner-take-all system. We seem to go through periods of public spirit, optimism, and openness to change every few decades that give our public life a sense of generosity and good will.

Rational Choice

Another model for public decision making is **rational choice** and science-based management. In this utilitarian approach, no policy should have greater total costs than benefits. In choosing between policy alternatives, we should always prefer those with the greatest cumulative welfare and the least negative impacts. Professional administrators would weigh various options and make an objective, methodical decision that would bring maximum social gain. As chapters 2 and 23 illustrate, there are many arguments against simply applying utilitarian, cost-benefit approaches in public decision making. For example:

- Many conflicting values and needs cannot be compared because they aren't comparable or we don't have perfect information.

- There are few generally agreed-upon broad societal goals but rather benefits to specific groups and individuals, many of which are in conflict.

- Policymakers generally are not motivated to make decisions on the basis of societal goals, but rather to maximize their own rewards: power, status, money, or reelection.

- Large investments in existing programs and policies create "path dependence" and "sunken costs" that prevent policymakers from considering good alternatives foreclosed by previous decisions.

- Uncertainty about consequences of various policy options compels decision makers to stick as closely as possible to previous policies to reduce the likelihood of adverse, calamitous, unanticipated consequences.

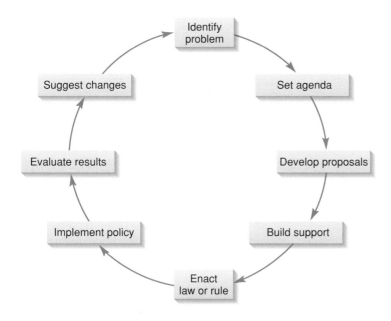

FIGURE 24.3 The policy cycle.

- Policymakers, even if well-meaning, don't have sufficient intelligence or adequate data or models to calculate accurate costs and benefits when large numbers of diverse political, social, economic, and cultural values are at stake.

- The segmented nature of policy making in large bureaucracies makes it difficult to coordinate decision making.

Policy formation follows predictable steps

How do policy issues and options make their way onto the stage of public debate? In this section, we will look at the **policy cycle** by which problems are identified and acted upon in the public arena (fig. 24.3). The first stage in this process is problem identification. Sometimes the government identifies issues for groups that have no voice or don't recognize problems themselves. In other cases, the public identifies a problem such as loss of biodiversity or health effects exposure to toxic waste and demands redress by the government. In either case, proponents describe the issue—either privately or publicly— and characterize the risks and benefits of their preferred course of action. Seizing the initiative in issue identification often allows leaders to define terms, set the agenda, organize stake- holders, choose tactics, aggregate related issues, and legitimate (or de-legitimate) issues and actors. It can be a great advantage to define the terms or choose the location of a debate. Next, stakeholders develop proposals for preferred policy options, often in the form of legislative proposals or administrative rules. Proponents build support for their position through media campaigns, public education, and personal lobbying of decision makers. By following the legislative or administrative process through its many steps, interest groups ensure that their proposals finally get enacted into law or established as a rule or regulation.

The next step is implementation. Ideally, government agen- cies will faithfully carry out policy directives as they organize bureaucracies, provide services, and enforce rules and regula- tions, but often it takes continued monitoring to make sure the system works as it should. Evaluating the results of policy deci- sions is as important as establishing them in the first place. Mea- suring impacts on target and nontarget populations shows us whether the intended goals, principles, and course of action are being attained. Finally, suggested changes or adjustments are considered that will make the policy fairer or more effective.

There generally are two different routes by which this cycle is carried out. Special economic interest groups such as industry associations, labor unions, or wealthy and powerful individuals don't need (or often want) much public attention or support for their policy initiatives. They generally carry out the steps of issue identification, agenda setting, and proposal development privately because they can influence legislative or administrative processes directly through their contacts with decision makers. Public interest groups, on the other hand, often lack direct access to corridors of power and need to rally broad general support to legitimate their proposals. An important method for getting their interests on the table is to attract media attention. Organizing a dramatic protest or media event can generate a lot of free publicity. Announcing some dire threat or sensational claim is a good way to gain attention. The problem is that it takes ever increasing levels of hysteria and hyperbole to get yourself heard in the flood of shocking news with which we are bombarded every day. Ironically, many groups that bemoan the overload of rhetorical overstatement that engulfs us, contribute to it in order to be heard above the din.

Is a clean, healthy environment a basic human right?

A long history in international law argues that we all have an inalienable right to a safe, sustainable environment (fig. 24.4). The 1982 World Charter for Nature, for example, asserts that

FIGURE 24.4 Do we have a basic human right to a clean environ- ment? Who's responsible for keeping our environment safe and healthful?

"man's needs can be met only by ensuring the proper functioning of natural systems," and that it is "an essential human right to means of redress when the human environment has suffered damage or degradation." The 1987 World Commission on Environment and Development (famous for defining sustainable development as meeting the needs of the present without compromising the ability of future generations to meet their own needs) went further in stating, "All human beings have the fundamental right to an environment adequate for their health and well-being."

Of the 194 nations in the world, 109 now have constitutional provisions for protection of the environment and natural resources. One hundred of them specifically recognize the right to a clean and healthy environment and/or the state's obligation to prevent environmental harm. The 1988 American Convention on Human Rights expressly declares, "everyone has a right to live in a healthy environment and to have access to basic public services" and that signatory parties "shall promote the protection, preservation, and improvement of the environment." In spite of having agreed to this Convention, however, the United States has not accepted environmental protection as a fundamental human right. When the National Environmental Policy Act (NEPA) was introduced in Congress in 1969, supporters argued that it should be a constitutional amendment. Disagreements about what clean and healthy mean, along with worries about how much it might impede commerce and industry to make such a guarantee, limited NEPA to a statute with more restricted, but still important powers than a constitutional amendment. A statute, however, is much easier to ignore or overturn than a constitutional provision. What do you think? Does everyone have an inherent right to a clean, healthy environment? How would you define these terms?

24.2 ENVIRONMENTAL LAW

Laws are rules set by authority, society or custom. Church laws, social mores, administrative regulations, and a variety of other codes of behavior can be considered laws if they are backed by some enforcement power. Government laws are established by federal, state, or local legislative bodies or administrative agencies. **Environmental law** constitutes a special body of official rules, decisions, and actions concerning environmental quality, natural resources, and ecological sustainability. Each branch of government plays a role in establishing the rules of law. **Statute law** consists of formal documents or decrees enacted by the legislative branch of government declaring, commanding, or prohibiting something. It represents the formal will of the legislature. **Case law** is derived from court decisions in both civil and criminal cases. **Administrative law** rises from executive orders, administrative rules and regulations, and enforcement decisions in which statutes passed by the legislature are interpreted in specific applications and individual cases. Because every country has different legislative and legal processes, this chapter will focus primarily on the U.S. system in the interest of simplicity and space.

A brief environmental history

Before looking at how each of these branches of government functions it might be useful to review how the underlying attitudes toward our environment and the policies that underlie our environmental laws have changed over the past century. For most of its history, U.S. environmental policy has had a laissez-faire or hands-off attitude toward business and private property. Pollution and environmental degradation were regarded as the unfortunate but necessary cost of doing business. If you didn't like the smell of a tannery or the sight of a waste dump, you were free to go somewhere else. While there were some early laws forbidding gross interference with another person's property or rights—the Rivers and Harbors Act of 1899, for example, made it illegal to dump so much refuse in waterways that navigation was blocked—in general, everyone was free to do whatever they wanted on their own property. People were either unaware of, or didn't pay much attention to, the fact that pollutants can move through the air, soil, and water to endanger people distant from the source.

The emergence of the modern environmental movement in the 1960s and 1970s marked a dramatic turning point in our understanding of the dangerous consequences of pollution and our demands to be protected from it. Rachel Carson's *Silent Spring* (1962) and Barry Commoner's *Closing Circle* (1971) alerted the public to the ecological and health risks of pesticides, hazardous wastes, and toxic industrial effluents. Public activism in the civil rights and antiwar movements was carried over to environmental protests and demands for environmental protection. Emergence of new media—especially television—provided access to environmental news and made events in faraway places seem immediate and important.

The 1969 blowout of an oil well in the Santa Barbara Channel just off the coast of southern California is a good example of how the convergence of actors, events, timing, and media attention can shape public opinion and influence the policy cycle. For many weeks, black, gooey crude oil washed up onto beautiful southern California beaches. The oil spill made a perfect story for TV. The continuing saga was ideal for nightly updates. The setting guaranteed good photos and was readily recognized by the viewing public as an important place (fig. 24.5). National news networks had just developed the capacity for live satellite feeds and were hungry to use their new technology. Los Angeles was one of few locations with reliable uplinks and Santa Barbara was close enough for a film crew to go out every day to get some good footage and be back in the studio in time for the five o'clock news.

Because the story was ongoing, it fit well in the 30-second spots characteristic of TV news. The audience was familiar with both the issue and the images that described it. Reporters didn't have to spend precious seconds explaining what was happening, but could just give a sporting-event-like update on which side was winning today. A policy debate in Congress might be more important, but is too complex to explain in a few seconds and doesn't provide exciting visuals. The wealthy residents of Santa Barbara were media-savvy, and had the influence and contacts to publicize

FIGURE 24.5 Beach cleanup efforts after the Santa Barbara oil spill in 1969 made excellent media material and had an important role in U.S. environmental policy.

the oil spill. While some of the cleanup efforts were not very effective, they made great visual footage. Attractive young people, smudged with oil, trying vainly to sweep gooey crud off a beautiful beach made ideal TV footage. Although the Santa Barbara oil spill wasn't nearly as big as others around the world, it played an important role in mobilizing public opinion and was an major factor in passage of the 1972 U.S. Clean Water Act.

As a result of awakened public concern about environmental issues, more than 27 major federal laws for environmental protection and hundreds of administrative regulations were established in the United States in the environmental decade of the 1970s. One of the most important of these is the National Environmental Policy Act (NEPA), which requires environmental impact statements for all major federal projects. Because of its power, NEPA has been the target of criticism and repeated attacks (What Do You Think? p. 546).

The statutes, case law, rules, precedents, and agencies resulting from that period created the foundation on which much of our current environmental protection rests. In the initial phase of this environmental revolution, the main focus was on direct regulation and litigation to force malefactors to obey the law. In recent years, attention has shifted from end-of-the-pipe command and control to pollution prevention and collaborative methods that can provide win/win solutions for all stakeholders. We'll look at some of those alternative dispute resolution approaches later in this chapter. But next, let's look more closely at how environmental law is established and administered.

Statutory law: The legislative branch

Establishing laws at either the state or federal level is one of the most important ways of protecting our environment. Many environmental groups spend a good deal of their time and resources trying to influence the legislative process. In this section we'll look at how that system works.

Federal laws (statutes) are enacted by Congress and must be signed by the president. They originate as legislative proposals called bills, which are usually drafted by the congressional staff, often in consultation with representatives of various interest groups. Thousands of bills are introduced every year in Congress. Some are very narrow, providing funds to build a specific section of road or to help a particular person, for instance. Others are extremely broad, perhaps overhauling the social security system or changing the entire tax code. Similarly, environmental legislation might deal with a very specific local problem or a national or international issue. Often a number of competing bills on a single issue may be introduced as proponents from different sides attempt to incorporate their views into law. A bill may have a single sponsor if it is the pet project of a particular legislator, or it may have 100 or more coauthors if it is an issue of national importance.

A Convoluted Path

After introduction, each bill is referred to a committee or subcommittee with jurisdiction over the issue for hearings and debate. Most hearings take place in Washington, but if the bill is controversial or legislators want to attract publicity for themselves or the issue, they may conduct field hearings closer to the site of the controversy. The public often has an opportunity to give testimony at field hearings (fig. 24.6). Although it's not likely that you will change the opinions of many legislators no matter how fervent or cogent your testimony, these events can be a good place to gain attention and educate the public about a topic. Hearings and debates also build a record of legislative intent that can be valuable in later interpretation and implementation of laws by courts and administrative agencies.

If a bill has sufficient support within the subcommittee, its language will be "marked up" or revised and modified to be more widely acceptable and to improve its chances of passage. At this stage several competing bills might be combined into a single

FIGURE 24.6 Citizens line up to testify at a legislative hearing. By getting involved in the legislative process, you can be informed and have an impact on governmental policy.

What Do You Think?

Every major federal project in the United States must be preceded by an Environmental Impact Statement.

Does NEPA Need an Overhaul?

Signed into law by President Nixon in 1970, the **National Environmental Policy Act (NEPA)** is the cornerstone of U.S. environmental policy. Conservationists see this act as a powerful tool for environmental protection, but commercial interests blame it for gridlock and consider it an impediment to business. President Bush has created a task force to study streamlining NEPA and make it less burdensome for industry. Conservationists worry this is really an effort to weaken the law and return to laissez-faire resource management.

NEPA does three important things: (1) it authorizes the Council on Environmental Quality (CEQ), the oversight board for general environmental conditions; (2) it directs federal agencies to take environmental consequences into account in decision making; and (3) it requires an **environmental impact statement (EIS)** be published for every major federal project likely to have an important impact on environmental quality. NEPA doesn't forbid environmentally destructive activities if they comply otherwise with relevant laws, but it demands that agencies admit publically what they plan to do. Once embarrassing information is revealed, however, few agencies will bulldoze ahead, ignoring public opinion. And an EIS can provide valuable information about government actions to public interest groups that wouldn't otherwise have access to these resources.

What kinds of projects require an EIS? The activity must be federal and it must be major, with a significant environmental impact. Evaluations are always subjective as to whether specific activities meet these characteristics. Each case is unique and depends on context, geography, the balance of beneficial versus harmful effects, and whether any areas of special cultural, scientific, or historical importance might be affected. A complete EIS for a project is usually time-consuming and costly. The final document is often hundreds of pages long and generally takes six to nine months to prepare. Sometimes just requesting an EIS is enough to sideline a questionable project. In other cases, the EIS process gives adversaries time to rally public opposition and information with which to criticize what's being proposed. If agencies don't agree to prepare an EIS voluntarily, citizens can petition the courts to force them to do so.

Every EIS must contain the following elements: (1) purpose and need for the project, (2) alternatives to the proposed action (including taking no action), (3) a statement of positive and negative environmental impacts of the proposed activities. In addition, an EIS should make clear the relationship between short-term resources and long-term productivity, as well as any irreversible commitment of resources resulting from project implementation.

Among the areas in which lawmakers have tried recently to ignore or limit NEPA include forest policy, energy exploration, and marine wildlife protection. The "Healthy Forest Initiative," for example, called

for bypassing EIS reviews for logging or thinning projects, and prohibited citizen appeals of forest management plans (chapter 12). Similarly, when the Bureau of Land Management proposed 77,000 coal-bed methane wells in Wyoming and Montana, promoters claimed that water pollution and aquifer depletion associated with this technology didn't require environmental review (chapter 19). And in the 2005 Energy Bill, Congress inserted a clause that exempts energy companies from NEPA requirements in a number of situations, with the aim of speeding energy development on federal land.

Interestingly, at the same time that the United States seems intent on weakening NEPA, a new Chinese law that requires environmental impact statements for major projects was used recently to oppose a series of dams on the Nu River in southwestern China. The Nu, which was one of only two free-flowing rivers in China, had been targeted for 13 large hydropower dams in spite of the fact that UNESCO designated it a World Heritage Site. The area has been described as the most biologically diverse temperate ecosystem in the world. At first, it appeared that this huge development would go through without any public scrutiny. This new law came into force during dam planning, however, and the spotlight it cast on environmental and social costs appears to have doomed the project. In autocratic governments, environmental protests are often the only form of government criticism allowed, and often are the first step in spreading democracy. This was true in the former Soviet Union, as well as in China.

To be informed environmental citizens, we all need to know something about how policies and laws like NEPA are created and applied. What do you think? Is NEPA an essential safeguard for our environmental quality, or merely an antiquated impediment for business and personal property rights? What information would you need to judge the effectiveness of this important law?

---

compromise version. If the compromise bill garners sufficient support, it is forwarded to the full committee for more hearings, debate, and a vote. If it fails in the full committee, the bill is sent back to the subcommittee for more work and further compromise. A bill that succeeds in the full committee is reported to the full House or Senate for a floor debate. Often opponents of a bill will attempt to amend it during each of these stages to lessen its impact or to make it so

unpalatable that even the original authors can no longer support it. Some amendments add completely unrelated material or even reverse the intent of a bill. As bills move through this convoluted pathway, interested parties can follow their progress in the *Congressional Quarterly Weekly,* a publication both in print and online that keeps track of proposed legislation. Many environmental groups also maintain websites with up-to-date information on events in Congress.

TABLE 24.1

Major U.S. Environmental Laws

Legislation	Provisions
Wilderness Act of 1964	Established the national wilderness preservation system.
National Environmental Policy Act of 1969	Declared national environmental policy, required Environmental Impact Statements, created Council on Environmental Quality.
Clean Air Act of 1970	Established national primary and secondary air quality standards. Required states to develop implementation plans. Major amendments in 1977 and 1990.
Clean Water Act of 1972	Set national water quality goals and created pollutant discharge permits. Major amendments in 1977 and 1996.
Federal Pesticides Control Act of 1972	Required registration of all pesticides in U.S. commerce. Major modifications in 1996.
Marine Protection Act of 1972	Regulated dumping of waste into oceans and coastal waters.
Coastal Zone Management Act of 1972	Provided funds for state planning and management of coastal areas.
Endangered Species Act of 1973	Protected threatened and endangered species, directed FWS to prepare recovery plans.
Safe Drinking Water Act of 1974	Set standards for safety of public drinking-water supplies and to safeguard groundwater. Major changes made in 1986 and 1996.
Toxic Substances Control Act of 1976	Authorized EPA to ban or regulate chemicals deemed a risk to health or the environment.
Federal Land Policy and Management Act of 1976	Charged the BLM with long-term management of public lands. Ended homesteading and most sales of public lands.
Resource Conservation and Recovery Act of 1976	Regulated hazardous waste storage, treatment, transportation, and disposal. Major amendments in 1984.
National Forest Management Act of 1976	Gave statutory permanence to national forests. Directed USFS to manage forests for "multiple use."
Surface Mining Control and Reclamation Act of 1977	Limited strip mining on farmland and steep slopes. Required restoration of land to original contours.
Alaska National Interest Lands Act of 1980	Protected 40 million ha (100 million acres) of parks, wilderness, and wildlife refuges.
Comprehensive Environmental Response, Compensation and Liability Act of 1980	Created $1.6 billion "Superfund" for emergency response, spill prevention, and site remediation for toxic wastes. Established liability for cleanup costs.
Superfund Amendments and Reauthorization Act of 1994	Increased Superfund to $8.5 billion. Shares responsibility for cleanup among potentially responsible parties. Emphasizes remediation and public "right to know."

Source: N. Vig and M. Kraft, *Environmental Policy in the 1990s*, 3rd Congressional Quarterly Press.

By the time an issue has passed through both the House and Senate, the versions approved by the two bodies are likely to be different. They go then to conference committee to iron out any differences between them. After going back to the House and Senate for confirmation, the final bill goes to the president, who may either sign it into law or veto it. If the president vetoes the bill, it may still become law if two-thirds of the House and Senate vote to override the veto. If the president takes no action within ten days of receiving a bill from Congress, the bill becomes law without his signature. One exception to this procedure is that if Congress adjourns before the ten-day period elapses, the bill does not become law. The president, by doing nothing, is said to have exercised a "pocket veto."

The Thomas website, maintained by Congress, also has current information about the progress of legislation. You can find out how your senator and representative voted on critical environmental issues by consulting The League of Conservation Voters, which ranks each member of Congress on their voting record. The Defenders of Wildlife maintains a daily e-mail environmental news service called Greenwire that has up-to-date information about what's happening in Washington. Table 24.1 lists some of the most important recent federal environmental legislation.

Legislative Riders

There are two types of legislation: authorizing bills become law, while appropriation bills provide funds for federal agencies and programs. Appropriation bills can have language attached expressing the intent of Congress, but, in theory, at least, are not supposed to make policy, merely fund existing plans and projects. Legislators who can't muster enough votes to pass pet projects through regular channels often will try to add authorizing amendments called **riders** into completely unrelated funding bills. Even if they oppose the riders, other members of Congress have a difficult time voting against an appropriation package for disaster relief or to fund programs that benefit their districts. Often this happens in conference committee because when the conference report goes back to the House and Senate, the vote is either to accept or reject with no opportunity to debate or amend further.

Starting with the 104th Congress, antienvironmental forces began using this tactic to roll back environmental protections and gain access to natural resources. Environmental groups were outraged, for instance, when riders were attached to 1996 supplemental spending bills that put a moratorium on listing additional species under the Endangered Species Act and exempted "salvage" logging on public lands from environmental laws. In subsequent years, numerous antienvironmental riders have been attached to appropriation bills. The 2004 Omnibus spending bill, for example, included numerous special-interest amendments to prevent administrative appeals and judicial reviews of environmentally destructive government policies, allow increased logging and road building in Alaska's Tongass National Forest, cut funding for land conservation, weaken national organic labeling standards, and expand forest-thinning projects. Generally, riders are tacked onto completely unrelated bills that legislators will have difficulty voting against. A rider to eliminate critical habitat for endangered species, for example, was hung on a veteran's health care bill. Congressional leaders pledged to end this practice, but little has been done so far to stop it.

Lobbying

Groups or individuals with an interest in pending legislation can often cause a great deal of influence by **lobbying,** or using personal contacts, public pressure, and political action to persuade legislators to vote or act in their favor. The term derives from the habit of partisans and professional lobbyists to lurk in the hallways and lobbies of Congress hoping to snare a passing legislator to urge them to vote in a specific way. We also use the same term to describe efforts to influence administrative agencies.

Most major environmental organizations maintain offices in Washington from which they monitor legislative and administrative programs and policies. Hundreds of professional environmental lobbyists and volunteer activists attend hearings, meet with legislators and agency personnel, draft proposed legislation and administrative rules, and attempt in a variety of ways to shape the national environmental agenda. They join thousands of other amateur and professional lobbyists representing industry and business organizations, workers, property owners, religious groups, ethnic associations, and just about every other kind of interest group that you can imagine. Walking the halls of Congress or those of the House or Senate office buildings, you see an amazing mixture of people attempting to be heard. It's fascinating to be part of this process.

FIGURE 24.7 Making a ruckus on behalf of environmental protection can attract attention to your cause.

In a survey of professional lobbyists, a majority agreed that personal contacts were the most effective way to influence decision makers. Undoubtedly the best way to make contact is through personal friends of a legislator or someone to whom they owe political allegiance. Having a famous movie star, a person of great power or wealth, or some other celebrity represent your group also can help open doors for your ideas. Your own senator or representative is more likely to be interested in your views than someone with whom you have no connection. They place a high priority on responding to their own constituents who can vote them in (or out) of office. But even if you aren't rich or illustrious or politically connected, you can often get a fair hearing from legislative staff—if not their bosses—if you have a persuasive case concerning an important topic.

What can those of us do to have an impact on legislation who can't afford to go to Washington to be directly involved? Getting involved in local election campaigns can greatly increase your access to legislators. Writing letters or making telephone calls also are highly effective ways to get your message across. You'd be surprised at how few letters or calls legislators receive even on important national issues. Your voice can have an important impact. How to write an effective letter is described further in chapter 25. All legislators now have e-mail addresses, although it isn't clear how much weight this form of communication carries.

Getting media attention can sway the opinions of decision makers. Organizing protests, marches, demonstrations, street theater, or other kinds of public events can call attention to your issue (fig. 24.7). Public education campaigns, press conferences, TV ads, and a host of other activities can be helpful. Tax-exempt (503c) organizations can't lobby directly or engage in politics, but there is a murky line between educational ads and outright campaigning and lobbying. Joining together with other like-minded groups can greatly increase your clout and ability to get things done. It's hard for a single individual or even a small group to have much impact, but if you can organize a mass movement, you may be very effective.

Case law: The judicial branch

Over the past 30 years, appeals to the judicial system have often been the most effective ways for seeking redress for environmental damage and forcing changes in how things are done. Activist judges and sympathetic juries in both federal and state court systems have been willing to take a stand where legislatures have been too timid or conservative to do so. Many groups spend a great deal of their time and energy bringing lawsuits that will shape environmental policy. The Environmental Defense Fund, for example, operates primarily in this arena. In the early days of the organization, their motto was, "Sue the bastards, that will get their attention." Even if you're not interested in environmental issues, it's worthwhile knowing something about how this system works. You may find yourself in court someday.

The judicial branch of government establishes environmental law by ruling on the constitutionality of statutes and interpreting their meaning. We describe the body of legal opinions built up by many court cases as case law. Often legislation is written in vague and general terms so as to make it widely enough accepted to gain passage. Congress, especially in the environmental area, often leaves it to the courts to "fill in the gaps." As one senator said when Congress was about to pass the Superfund legislation, "All we know is that the American people want these hazardous waste sites cleaned up … Let the courts worry about the details." When trying to interpret a law, the courts depend on the legislative record from hearings and debates to determine congressional intent. What was a particular statute meant to do by those who wrote and passed it?

The Court System

The United States is divided into 96 federal court districts, each of which has at least one trial court. Over these district courts are the circuit court of appeals, which hears disputes arising from questions about procedural issues and interpretations of the law in district courts. There are 12 geographic regions for the appeals courts. The federal courts have jurisdiction over federal criminal prosecutions, claims against the federal government, claims arising under federal statutes or treaties, and cases in which defendants or plaintiffs come from two or more states. The residence of the defendant or location of the property in dispute usually determines the venue, or the court in which each case is heard. Each state has its own courts that generally parallel the federal system. These courts have jurisdiction over cases arising from state laws. The U.S. Supreme Court is the court of last resort for appeals for both federal and state court systems.

A trial court judge presides over trials, rules on motions made by attorneys, and decides questions of law, such as what evidence is admissible, and what law applies to the case (fig. 24.8). The judge controls the pace of the proceedings and

FIGURE 24.8 In a trial court, the judge presides over trials, rules on motions, and interprets the law.

maintains decorum in the court room. Although Thomas Jefferson said, "Ours is a government of laws, not men," in reality, the judge has tremendous power over the outcome of a trial. Certain courts earn a reputation of being pro- or antienvironmental. Litigants always try to know something about the judge's ideology before bringing a case to trial.

The first judge to hear a case arising from a particular statute or situation has the greatest latitude to interpret the law and set a **precedent** to be used as an example in subsequent trials. These decisions are binding, however, only on those courts on a lower level and in the same system. Precedents from the California courts, for instance, are not binding in Arizona courts. Furthermore, if a judge *distinguishes* a case—determines it is different from other cases—she or he is not obliged to adhere to prior precedents. Or a judge can always simply overturn a clearly applicable precedent as a matter of "correcting the law" where technology or changing community values has made prior decisions outdated.

Legal Thresholds

Before a trial can start, the litigants must satisfy certain threshold requirements. The first of these is **standing,** or whether the participants have a right to stand before the bar and be heard. The main criteria for standing is a valid interest in the case. Plaintiffs must show that they are materially affected by the situation they petition the court to redress. This is an important point in environmental cases. Groups or individuals often want to sue a person or corporation for degrading the environment. But unless they can show that they personally suffer from the degradation, courts are likely to deny standing.

In a landmark 1969 case, for example, the Sierra Club challenged a decision of the Forest Service and the Department of the Interior to lease public land in California to Walt Disney Enterprises for a ski resort. The land in question was a beautiful valley that cut into the southern boundary of Sequoia National

Park (see fig. 2.11). Building a road into the valley would have necessitated cutting down a grove of giant redwood trees within the park. The Sierra Club argued that it should be granted standing in the case to represent the trees, animals, rocks, and mountains that couldn't defend their own interests in court. After all, the club pointed out corporations—such as Disney Enterprises—are treated as persons and represented by attorneys in the courts. Why not grant trees the same rights? The case went all the way to the Supreme Court, which ruled that the Sierra Club failed to show that it or any of its members would be materially affected by the development.

In addition to standing, plaintiffs must show that their case represents a "live" legal dispute that is likely to result in a final and meaningful judgment, and that there is a present controversy for which a decision is needed. In other words, you can't bring a suit over a hypothetical situation or one that is no longer cogent. For example, suppose you want to stop a development project that would destroy endangered species habitat. If the defendant can show that the species couldn't inhabit the habitat (if it were already extinct, for instance), the case is moot (of no practical importance) and the courts will refuse to hear it.

Criminal law derives from those federal and state statutes that prohibit wrongs against the state or society, such as arson, rape, murder, and robbery. Serious crimes, like murder or rape, that are punishable by long-term incarceration or heavy fines are called felonies. Lesser crimes, such as shoplifting or vandalism, that result in smaller fines or shorter sentences in a county or city jail are labeled misdemeanors. Definitions vary from state to state. What may be a felony crime in one state may be only a misdemeanor in another.

A criminal case is always initiated by a government prosecutor. Guilt or innocence of the defendant is determined by a jury of peers, but the sentence is imposed, often in consultation with the jury, by the judge. The judge is responsible for keeping order in the hearings and for determining points of law. The jury acts as a fact-finding body that weighs the truth and reliability of the witnesses and evidence.

Violation of many environmental statutes constitutes criminal offenses. In 1975, the U.S. Supreme Court ruled that corporate officers can be held criminally liable for violations of environmental laws if they were grossly negligent, or the illegal actions can be considered willful and knowing violations. In 1982, the EPA created an Office for Criminal Investigation. Under Bill Clinton, prosecutions for environmental crimes rose to nearly 600 per year. They fell by 75 percent under George W. Bush, however.

Nevertheless, the American public reports that they want environmental protection (fig. 24.9)

Deliberate, egregious pollution cases can lead to criminal prosecution. The president of a Colorado company for example, was sentenced to 14 years in prison for knowingly dumping chlorinated solvents that contaminated the water table. The company itself was fined nearly $1 million and put on probation for ten years.

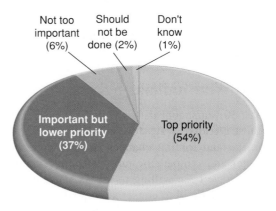

FIGURE 24.9 More than half the respondents in a Princeton University survey in 2000 said that environmental protection should be a top priority for the president and congress.

> ### Think About It
>
> How much environmental protection do we need? Were we being overprotected during the Clinton administration, or is our environment so clean and safe now that we don't need enforcement?

Too often international environmental crimes go unpunished because of jurisdictional complications. The European Union has called for a Global Environmental Crime Intelligence Unit, much like the renowned Interpol, to investigate illegal logging, waste dumping, and other transborder crimes.

Civil law is defined as a body of laws regulating relations between individuals or between individuals and corporations. Issues such as property rights, and personal dignity and freedom are protected by civil law. In some cases, legislative statutes, such as the Civil Rights Act, establish specific aspects of civil law. In other cases, where no particular statute exists, custom and a body of previous court decisions, collectively called **common law,** establish precedents that constitute a working definition of individual rights and responsibilities. Cases that seek compensation for damages, such as those caused by pollution, are called **tort law** (tort derives from the Middle English word for injury). This kind of civil action is usually initiated by the attorney representing the injured party (the plaintiff). The defendant in a civil case has a right to be tried by a jury, but in highly technical issues, this right often is waived and the case is heard only by the judge. Being found guilty of a civil offense can result in financial penalties but not jail time.

In contrast to a criminal case where the burden of proof lies with the prosecution, and defendants are considered innocent until proven guilty, civil cases can be decided on a "preponderance of evidence." This makes civil cases considerably easier to win than criminal cases where the evidence is ambiguous. A number of mitigating factors also are taken into account in determining guilt and assigning penalties in civil cases. Culpability is based on whether the defendant could reasonably have anticipated and avoided the offense. A "good faith effort" to comply or solve the problem can be a factor. The compliance history is important. Is this a first offense or a habitual repeater? Finally, is there evidence of economic benefit to the perpetrator? That is, did the violator gain personally from the action? If so, it is more likely that willful intent was involved.

Most people consider being convicted of a criminal offense much more serious than losing a civil case, because the former can lead to incarceration while the latter only costs money. Civil judgments can be costly, however. A group of Alaskan fishermen won $5 billion from the Exxon oil company for damages caused by the 1989 *Exxon Valdez* oil spill. Civil cases can be brought in both state and federal court. In 2000, the Koch oil company, one of the largest pipeline and refinery operators in the United States, agreed to pay $35 million in fines and penalties to state and federal authorities for negligence in more than 300 oil spills in Texas, Oklahoma, Kansas, Alabama, Louisiana, and Missouri between 1990 and 1997. Koch also agreed to spend more than $1 billion on cleanup and improved operations.

Sometimes the purpose of a civil suit is to seek an injunction or some other form of equitable relief from the actions of an individual, a corporation, or a governmental agency. You might ask the courts, for example, to order the government to cease and desist from activities that are in violation of either the spirit or the letter of the law. This sort of civil action is heard only by a judge; no jury is present. Environmental groups have been very successful in asking courts to stop logging and mining operations, to enforce implementation of the endangered species act, to require agencies to enforce air and water pollution laws, and a host of other efforts to protect the environment and conserve natural resources. Often, rather than sue a corporation directly for environmental damage, it is more effective to sue the government for not enforcing laws that would have prevented the damage. A big corporation with deep pockets can afford squadrons of lawyers and may have the resources and incentive to tie up litigation for years with motions and counter suits. Federal or state agencies may be more inclined to agree that you are right and to be willing to settle the matter quickly.

Adversarial Approaches and SLAPP Suits

The American legal system is adversarial, pitting one side against the other in an effort to distinguish right from wrong, or innocence from guilt. In a trial, each side tries to make the strongest possible arguments for its position, and to point out the faults in the opponent's case. The jury, as neutral fact-finder, hears arguments from both sides and makes an objective decision. We believe that this approach gives the best chance for truth to be discovered. It is time-consuming and costly, however, and doesn't always result in justice. Everyone has heard of cases where it seems perfectly obvious that the defendant is guilty, and yet they get off because certain evidence is inadmissible, or they simply have lawyers who can dazzle or confuse the jury with deceptive arguments. This system promotes strife and confrontation. It doesn't encourage compromise and often leaves lasting hatreds that make future cooperation impossible.

Because defending a lawsuit is so expensive, the mere threat of litigation can be a chilling deterrent. Increasingly, environmental activists are being harassed with **strategic lawsuits against public participation (SLAPP).** Citizens who criticize businesses that pollute or government agencies that are derelict in their duty to protect the environment are often sued in retaliation. While most of these preemptive strikes are groundless and ultimately dismissed, defending yourself against them can be exorbitantly expensive and take up time that might have been spent working on the original issue. Public interest groups and individual activists—many of whom have little money to defend themselves—often are intimidated from taking on polluters. For example, a West Virginia farmer wrote an article about a coal company's pollution of the Buckhannon River. The company sued him for $200,000 for defamation. Similarly, citizen groups fighting a proposed incinerator in upstate New York were sued for $1.5 million by their own county governments. A Texas woman called a nearby landfill a dump—and her husband was named in a $5 million suit for failing to "control his wife." Of course these suits also are expensive for the company or agency that initiates them, but they may be far cheaper than paying a fine or scrapping a big project.

Administrative law: The executive branch

More than 100 federal agencies and thousands of state and local boards and commissions have environmental oversight. They usually have power to set rules, adjudicate disputes, and investigate misconduct. Federal agencies often delegate power to a matching state agency in order to decentralize authority. The enabling legislation to create each agency is called an "organic" act because it establishes a basic unit of governmental organization. In the federal government, most executive agencies come under the jurisdiction of cabinet-level departments such as Agriculture, Interior, or Justice (fig. 24.10).

Agency rule-making and standard-setting can be either formal or informal. In an informal case, notice and background for proposed rules are published in the Federal Register. Opportunities for all interested parties to submit comments are provided. This is often an important avenue for environmentally concerned citizens and public interest groups to have an impact on environmental policy. In formal rule making, a public hearing is held with witnesses and testimony much like a civil trial. Witnesses can be cross-examined. A complete transcript is made and final findings are published in the public record. It is generally more difficult for individuals to intervene in a formal hearing, although sometimes there is an opportunity to submit written comments. Rule-making is often a complex, highly technical process that is difficult for citizen groups to understand and monitor. The proceedings are usually less dramatic and colorful than criminal trials, and yet can be very important for environmental protection.

Executive orders also can be powerful agents for change. In 1994, for instance, President Bill Clinton issued Executive Order 12898 requiring all federal agencies to collect data on effects of pollution on minorities, and to develop strategies to

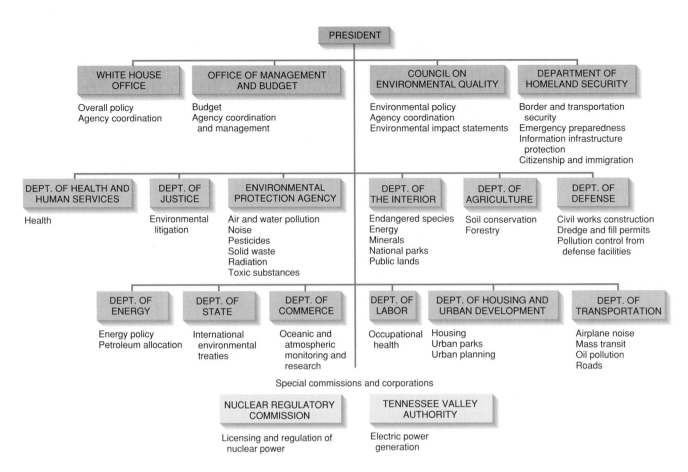

FIGURE 24.10 Major agencies of the Executive Branch of the U.S. federal government with responsibility for resource management and environmental protection.
Source: U.S. General Accounting Office.

promote environmental justice. During his two terms in office, Clinton used the Antiquities Act to establish 22 new national monuments (fig. 24.11). In addition, he expanded dozens of existing national parks and wildlife refuges. Altogether, Clinton ordered protection for about 36 million ha (90 million acres) of nature preserves, the largest of which was the Pacific Ocean reserve composed of 34 million ha of ocean and coral reefs northwest of Hawaii. In addition, the U.S. Forest Service, under Clinton appointee Mike Dombeck, ordered a moratorium on road building and logging on nearly 24 million ha of *de facto* forest wilderness.

Rules and policies made by executive decree in one administration can be quickly undone in the next one. In his first day in office, President George W. Bush ordered all federal agencies to suspend or ignore more than 60 rules and regulations from the Clinton administration. In addition, Mr. Bush called for a sweeping overhaul of environmental laws to ease restrictions on businesses and to speed decisions on development projects. His supporters regard these policies as merely restoring reason and balance to government; critics saw it as a radical ideological campaign to roll back environmental protections and social progress. Because most of this agenda was pursued through

agency regulations and executive orders, most Americans, distracted by terrorism and lingering wars in Afghanistan and Iraq, were unaware of the magnitude and implications of this abrupt policy shift.

Regulatory Agencies

The EPA is the primary agency with responsibility for protecting environmental quality. Created in 1970, at the same time as NEPA, the EPA is a cabinet-level department, with more than 18,000 employees and ten regional offices. Often in conflict with Congress, other agencies of the Executive Branch, and environmental groups, the EPA has to balance many competing interests and conflicting opinions. Greatly influenced by politics, the agency changes dramatically depending on which party is in power and what attitudes toward the environment prevail at any given time. Under the Nixon and Carter administrations, the EPA grew rapidly and enforced air and water quality standards vigorously. It declined sharply during the Reagan administration, then recovered under Bill Clinton.

The Departments of the Interior and Agriculture are to natural resources what the EPA is to pollution. Interior is home to the National Park Service, which is responsible for more than

FIGURE 24.11 In his two terms in office, President Bill Clinton created 22 new national monuments, including the U.S. Virgin Islands Coral Reef Monument shown here.

FIGURE 24.12 Smokey Bear symbolizes the Forest Service's role in extinguishing forest fires.

376 national parks, monuments, historic sites, and recreational areas. It also houses the Bureau of Land Management (BLM), which administers some 140 million ha of land, mostly in the western United States. In addition, Interior is home to the U.S. Fish and Wildlife Service, which operates more than 500 national wildlife refuges and administers endangered species protection.

The Department of Agriculture is home to the U.S. Forest Service, which manages about 175 national forests and grasslands, totaling some 78 million ha. With 39,000 employees, the Forest Service is nearly twice as large as the EPA (fig. 24.12). The Department of Labor houses the Occupational Safety and Health Agency (OSHA), which oversees workplace safety. Research that forms the basis for OSHA standards is carried out by the National Institute for Occupational Safety and Health (NIOSH). In addition, several independent agencies that are not tied to any specific department also play a role in environmental protection and public health. The Consumer Products Safety Commission passes and enforces regulations to protect consumers, and the Food and Drug Administration is responsible for the purity and wholesomeness of food and drugs.

All of these agencies have a tendency to be "captured" by the industries they are supposed to be regulating. Many of the people with expertise to regulate specific areas came from the industry or sector of society that their agency oversees. Furthermore, the people they work most closely with and often develop friendships with are those they are supposed to watch. And when they leave the agency to return to private life—as many do when the administration changes—they are likely to go back to the same industry or sector where their experience and expertise lies. The effect is often what's called a "revolving door," where workers move back and forth between industry and government. As a result regulators often become overly sympathetic with and protective of the industry they should be overseeing.

A short clause slipped quietly into the 2006 energy bill threatens all these agencies. The clause allows the president to appoint a "Sunset Commission" that could terminate any federal program or agency judged ineffective or "not producing results." While this text had not become law at the time of this writing, it expresses the desire of some to greatly reduce the size and scope of the federal government. What do you think? Which of the agencies in figure 24.10 contribute to your quality of life? Which would you like to see eliminated?

Administrative Courts

Administrative regulations and rule-making procedures often are objected to by affected parties. Over the past decade, 80 percent of the rules made by the EPA, for example, were challenged in court. It is very unusual for the courts to overturn agency rules, but the challenge buys time for regulated corporations to continue business as usual for a while longer. **Administrative courts** hear challenges to agency rules and regulations. An administrative judge can consider both the validity of the rule and its application to a specific case. If the parties dispute the judge's findings, they can appeal to a district court. The courts rarely overturn agency rules unless (1) the enabling act is too vague or unconstitutional, (2) the agency has gone beyond the scope of power granted by the legislature, or (3) the agency didn't follow proper procedures.

Administrative courts also hear enforcement cases where an individual or corporation has violated an agency rule or standard. Suppose, for instance, that a factory is found to be exceeding allowable air pollution emissions. After an investigation, a complaint is filed with an administrative judge. A hearing is scheduled and the judge listens to both sides of the case and issues an opinion, which is usually a recommendation to the head of the responsible agency for a penalty or remediation action. This decision also can be appealed to the circuit court of appeals.

The rules for evidence are usually less strict in an administrative case than in a criminal action. The administrative judge acts as both fact finder and decision maker. There is no jury. The judge can question witnesses and ask for additional evidence. The emphasis is on finding the truth rather than sticking to strict rules of procedure. Administrative courts often recommend relatively small penalties as a way of encouraging early settlement. In one case, a company charged with a violation of EPA standards was assessed a $500 fine. The company appealed the case to the federal district court, which imposed a $10,000 penalty.

24.3 INTERNATIONAL TREATIES AND CONVENTIONS

As recognition of the interconnections in our global environment has advanced, the willingness of nations to enter into protective covenants and treaties has grown concomitantly. Table 24.2 lists some major international treaties and conventions, while figure 24.13 shows the number of participating parties in them. Note that the earliest of these conventions has no nations as participants; they were negotiated entirely by panels of experts. Not only the number of parties taking part in these negotiations has grown, but the rate at which parties are signing on and the speed at which agreements take force also have increased rapidly. The Convention on International Trade in Endangered Species (CITES), for example, was not enforced until 14 years after ratification, but the Convention on Biological Diversity was enforceable after just one year, and had 160 signatories only four years after introduction. Over the past 25 years, more than 170 treaties and conventions have been negotiated to protect our global environment. Designed to regulate activities ranging from intercontinental shipping of hazardous waste, to deforestation, overfishing, trade in endangered species, global warming, and wetland protection, these agreements theoretically cover almost every aspect of human impacts on the environment.

Unfortunately, many of these environmental treaties constitute little more than vague, good intentions. In spite of the fact that we often call them laws, there is no body that can legislate or enforce international environmental protection. The United Nations and a variety of regional organizations bring stakeholders together to negotiate solutions to a variety of problems but the agreed-upon solutions generally rely on moral persuasion and public embarrassment for compliance. Most nations are unwilling to give up sovereignty. There is an international court, but it has no enforcement power. Nevertheless, there are creative ways to strengthen international environmental protection.

One of the principal problems with most international agreements is the tradition that they must be by unanimous consent. A single recalcitrant nation effectively has veto power over the wishes of the vast majority. For instance, more than 100 countries at the UN Conference on Environment and Development (UNCED), held in Rio de Janeiro in 1992, agreed to restrictions on the release of greenhouse gases. At the insistence of U.S. negotiators, however, the climate convention was reworded so that it only *urged*—but did not require—nations to stabilize their emissions.

New approaches can make treaties effective

As a way of avoiding these problems, some treaties incorporate innovative voting mechanisms. When a consensus cannot be reached, they allow a qualified majority to add stronger measures in the form of amendments that do not need ratification. All members are legally bound to the whole document unless they expressly object. This approach was used in the Montreal Protocol, passed

---

TABLE 24.2

Some Important International Treaties

CBD: Convention on Biological Diversity 1992 (1993)

CITES: Convention on International Trade on Endangered Species of Wild Fauna and Flora 1973 (1987)

CMS: Convention on the Conservation of Migratory Species of Wild Animals 1979 (1983)

Basel: Basel Convention on the Transboundary Movements of Hazardous Wastes and their Disposal 1989 (1992)

Ozone: Vienna Convention for the Protection of the Ozone Layer and Montreal Protocol on Substances that Deplete the Ozone Layer 1985 (1988)

UNFCCC: United Nations Framework Convention on Climate Change 1992 (1994)

CCD: United Nations Convention to Combat Desertification in those Countries Experiencing Serious Drought and/or Desertification, Particularly in Africa 1994 (1996)

Ramsar: Convention on Wetlands of International Importance especially as Waterfowl Habitat 1971 (1975)

Heritage: Convention Concerning the Protection of the World Cultural and Natural Heritage 1972 (1975)

UNCLOS: United Nations Convention on the Law of the Sea 1982 (1994)

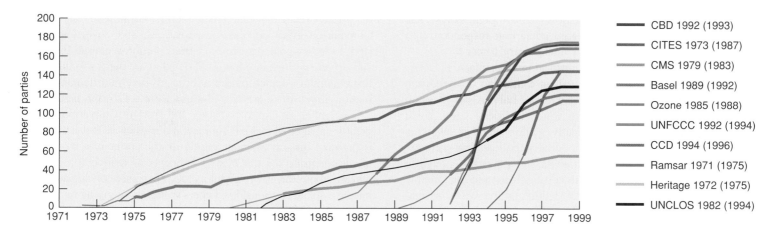

FIGURE 24.13 Number of participating parties in some major international environmental treaties. The thick portion of each line shows when the agreement went into effect (date in parentheses). See table 24.2 for complete treaty names.
Source: United Nations Environment Programme from Global Environment Outlook 2000.

in 1987 to halt the destruction of stratospheric ozone by chlorofluorocarbons (CFCs). The agreement allowed a vote of two-thirds of the 140 participating nations to amend the protocol. Although initially the protocol called for only a 50 percent reduction in CFC production, subsequent research showed that ozone was being depleted faster than previously thought. The protocol was strengthened by amendment to an outright ban on CFC production in spite of the objection of a few countries.

Where strong accords with meaningful sanctions cannot be passed, sometimes the pressure of world opinion generated by revealing the sources of pollution can be effective. NGOs and others can use this information to expose violators. For example, the environmental group Greenpeace discovered monitoring data in 1990 showing that Britain was disposing of coal ash in the North Sea. Although not explicitly forbidden by the Oslo Convention on ocean dumping, this evidence proved to be an embarrassment, and the practice was halted.

Trade sanctions can be an effective tool to compel compliance with international treaties. The Montreal Protocol, for example, bound signatory nations not to purchase CFCs or products made using them from countries that refused to ratify the treaty. Because many products employed CFCs in their manufacture, this stipulation proved to be very effective. On the other hand, trade agreements also can work against environmental protection. The World Trade Organization was established to make international trade more fair and to encourage development. It has been used, however, to subvert national environmental laws. In a ruling in 1998, the WTO forbid the United States from restricting imports of shrimp from Thailand, Malaysia, India, and Pakistan that were caught with nets that trap endangered sea turtles. Similarly, under provisions of the North American Free Trade Agreement, the Methanex Corporation of Canada filed a $1 billion claim against California for banning MTBE (a suspected carcinogen) in gasoline. Methanex claims that their sales of methanol (an ingredient of MTBE) will be harmed by this ban.

International governance has been controversial

Increasingly, private citizens and nongovernmental organizations collect to protest policies of the far-reaching institutions like the World Bank, the International Monetary Fund, and the World Trade Organization (WTO). In 1999, for example, more than 50,000 people gathered in Seattle, Washington, to demonstrate their opposition to the WTO. Including farmers, workers, animal rights activists, environmentalists, and indigenous peoples, the protestors represented a wide range of complaints about how international politics impact their lives (fig. 24.14). Although most demonstrators were peaceful, a small group of anarchists

FIGURE 24.14 Protestors demonstrate against monetary policies and international institutions that threaten livelihoods and environmental quality.

and vandals set fires, broke windows, and looted stores. The police, unprepared for such a large gathering, responded with what many observers regarded as unnecessary force.

An even more violent demonstration took place in 2001, when 100,000 protestors converged on a meeting of the Group of Eight Industrialized Nations in Genoa, Italy. Again, a small group of radicals started a riot in which one person was killed and hundreds were injured. Ironically, while many demonstrators called for a return to insular, nationalist policies, others were protesting because global institutions aren't powerful enough to protect them from transnational corporations and inequitable financial arrangements. The 2003 meeting of the WTO in Cancun, Mexico, for example, collapsed when delegates from developing nations insisted that the $300 billion in subsidies paid every year to the world's wealthiest farmers undermined the livelihoods of millions of poor farmers around the world.

Will globalization bring better environmental governance?

The rapid pace of **globalization**—the revolution in communications, transportation, finances and commerce that has brought about increasing interdependence of national economies—offers both opportunities and challenges to environmental management. Increasingly, we recognize that international cooperation is essential for conserving resources and maintaining a healthy environment. International commissions and conventions are paying ever more attention to how good environmental decisions are made. Do democratic rights and civil liberties contribute to better environmental management? Should local citizens or advocacy groups have the right to appeal a decision they believe harms an ecosystem or is unfair? What is the best way to fight corruption among those who manage our forests, water, parks, and mineral resources? These are all questions about how we make environmental decisions and who makes them—a process called **environmental governance.**

One of the strongest arguments for encouraging better governance is that it requires us to focus on the social dimensions of natural resource use and ecosystem management, in addition to the technical details of how to manage. This includes how we value ecosystems, how we form environmental policies, how we negotiate trade-offs between conflicting uses or values, and, finally, how we make sure the costs and benefits of our decisions—including impacts on the poor—are equitably shared. A basic principle of governance is that how we decide and who gets to decide often determine what we decide.

The Aarhus Convention of 1988 specifically addresses these issues. First negotiated as a regional agreement among European countries, this document has now been ratified by 40 nations in Europe and Asia, and is open to signature by all nations of the world. In addition to recognizing the basic right of every person of present and future generations to a healthy environment, the convention also specifies how authorities at all levels will provide fair and transparent decision-making processes, access to information, and right to redress of grievances. Individuals don't need to prove legal standing to request information or comment on official decisions that affect their environment. The Aarhus Convention gives citizens, organizations, and governments the right to investigate and seek to curtail pollution caused by public and private entities in other countries that are parties to the treaty. For example, a Dutch public interest group could demand information about air or water emissions from a German factory.

Adopting and implementing the Aarhus Convention could greatly enhance global environmental governance. But while there is growing interest in endorsing the Aarhus principles worldwide, many countries see the treaty's concepts of democratic decision making about the environment as too progressive or threatening to business social relationships. Others worry about giving up national sovereignty and independence. What do you think? Would you support these principles? How would you advise your legislators to vote if the convention is introduced in your country?

24.4 DISPUTE RESOLUTION AND PLANNING

The adversarial approach of our current legal system often fails to find good solutions for many complex environmental problems. Identifying an enemy and punishing him for transgressions seems more important to us than finding win/win compromises. Gridlocks occur in which conflicts between adversaries breed mutual suspicion and decision paralysis. The result is continuing ecosystem deterioration, economic stagnation, and growing incivility and confrontation. The complexity of many environmental problems arises from the fact that they are not purely ecological, economic, or social, but a combination of all three. They require an understanding of the interrelations between nature and people. Are there ways to break these logjams and find creative solutions? In this section we will look at some new developments in mediation, dispute resolution, and alternative procedures for environmental decision making.

Wicked problems don't have simple answers

Rational choice theories of planning and decision making assume that if we just collect more data, buy faster computers to crunch numbers, build more complex models, and spend more money, any problem should be resolvable. More information, it's assumed, will lead automatically to better management. This "bigger hammer" approach may be effective in problems that are difficult but relatively straightforward. Increasingly, however, we have come to recognize that many of the most important problems we face don't fit this pattern. Questions like what ecosystem health means, or how clean is clean, don't have simple right or wrong answers. They depend on your worldview and how you define these terms. Different people come to different conclusions even if they share the same information.

Environmental scientists describe problems with no simple right or wrong answers as being **wicked problems,** not in the sense of having malicious intent, but rather as obstinate or intractable.

These problems often are nested within other sets of interlocking issues. The definition of both the problem and its solutions differ for various stakeholders. There are no value-free, objective answers for these dilemmas, only choices that are better or worse depending on your viewpoint. Wicked problems are important and have serious consequences, but also are complex and have a poor match between who bears the costs and who bears the benefits on any proposed solution. They usually can't be solved by simple rules and regulations, more scientific research, or appeals to ethics. Often the best solution comes from community-based planning and consensus building. Inherent uncertainty gives these questions no clear end point. You cannot know when all possible solutions have been explored.

Recent advancement in understanding how ecological systems work gives us some insight into many wicked problems. Like biological organisms, social problems often change and evolve over time. Their history unfolds in complex ways, depending on chance interactions and unpredictable events. Like ecological systems, there may never be a stable equilibrium in many environmental issues. Each involves an assemblage of issues and actors that are unique in time and place. They can't be standardized. There are no good precedents from previous experience. Their solutions are unique, and what may work today, may not be applicable tomorrow. How can we learn to cope with such uncertainty?

Adaptive management introduces science to planning

One promising approach to solving wicked environmental problems comes from the work of ecologists C. S. Holling and Lance Gunderson, and planners Steven Light and Kai Lee, among others. Starting with the observation that human understanding of nature is imperfect, this group believes that all human interactions with nature should be experimental. They suggest that environmental policies should incorporate **adaptive management,** or "learning by doing," designed from the outset to test clearly formulated hypotheses about the ecological, social, and economic impacts of the actions being undertaken (fig. 24.15). Rather than assume that what seemed the best initial policy option will always remain so, we need to carefully monitor how conditions are changing and what effects we are having on both target and nontarget elements of the system. If our policy succeeds, the hypothesis is affirmed. But if the policy fails, an adaptive design still permits learning, so that future decisions can proceed from a better base of understanding. The goal of adaptive management and experimental design is to enable us to live with the unexpected. They aim to yield understanding as much as to produce answers or solutions (table 24.3). This approach to natural resources is similar to—but more explicitly experimental than—ecosystem management.

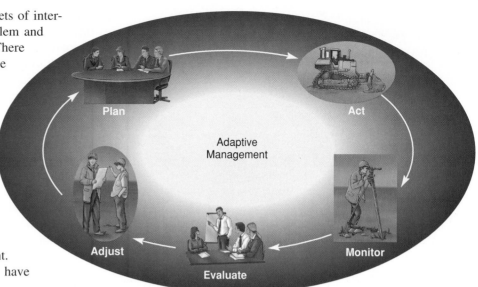

FIGURE 24.15 Adaptive management recognizes that we need to treat management plans for ecosystem as a scientific experiment in which we monitor, evaluate, and adjust our policies to fit changing conditions and knowledge.

Resilience is important in ecosystems and institutions

The great economist Joseph Schumpeter described "waves of creative destruction" that transform economic systems. Another insight from Holling and his collaborators is that similar cycles of destructive creation operate in both ecological systems and in

---

TABLE 24.3

Institutional Conditions for Adaptive Management

1. There is a mandate to take action in the face of uncertainty.
2. Decision makers are aware they are experimenting anyway.
3. Decision makers care about improving outcomes over biological time scales.
4. Preservation of pristine environments is no longer an option, and human intervention cannot produce desired outcomes predictably.
5. Resources are sufficient to measure ecosystem-scale behavior.
6. Theory, models, and field methods are available to estimate and infer ecosystem-scale behavior.
7. Hypotheses can be formulated.
8. Organizational culture encourages learning from experience.
9. There is sufficient stability to measure long-term outcomes; institutional patience is essential.

Source: Kai N. Lee, *Compass and Gyroscope*, 1993. Copyright © 1993 Island Press. Reprinted by permission of Alexander Hoyt & Associates.

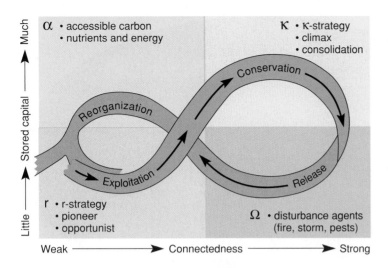

α • accessible carbon
 • nutrients and energy

κ • κ-strategy
 • climax
 • consolidation

Conservation

Reorganization

Exploitation

Release

r • r-strategy
 • pioneer
 • opportunist

Ω • disturbance agents
 (fire, storm, pests)

Much ← Stored capital → Little

Weak ————→ Connectedness ————→ Strong

FIGURE 24.16 The creative-destruction cycle. Resilience, or the ability to reorganize and recover from disturbance, is the most important characteristic of both natural and human systems.

policy institutions (fig. 24.16). This is a familiar process that occurs in secondary succession (chapter 4). The release phase of the cycle occurs when factors such as fires, storms, or pests disturb a biological community, mobilizing nutrients and making space available for new growth. During the reorganization phase, pioneer and opportunist species colonize the new habitat. These species grow rapidly on the accessible carbon, nutrient, and energy sources during the exploitation stage. As the community matures, both the stored capital and connectedness increase until the ecosystem reaches a stage at which the system is poised for some new disturbance that starts the cycle again.

The most important characteristic of natural systems is their **resilience,** or ability to recover from disturbance. This doesn't imply that the ecosystem always returns to the exact condition it was in before the disturbance. It may have a new assemblage of species, or different set of physical conditions, but if it is resilient, the system has the ability to reorganize itself in creative and constructive ways. "Environmental quality is not achieved by attempting to eliminate change or surprises," Holling observed. The goal, instead, is resilience in the face of surprise. Surprise can be counted on. Resilience comes from adaptation to stress, from survival of the fittest in a turbulent environment.

In studying a variety of natural resource management regimes, Holling and others observed that human institutions also follow a similar pattern. In studying a variety of natural resource management issues ranging from restoration of the Florida Everglades, to control of spruce budworm in New England forests, to cattle grazing in South Africa, to protection of salmon in the Pacific Northwest, they observed that every attempt to manage ecological variables one factor at a time inexorably leads to less resilient ecosystems, more rigid management systems, and more dependent societies. Initial success sets the conditions for eventual collapse. Take the example of forest fire suppression. For 70 years, the U.S. Forest Service has had a very effective

policy of putting out all forest fires. The result has been that flammable debris has built up in the forests so that major conflagrations are now inevitable. During this time, however, people have felt safe moving to the borders of the forests and now there is a large population with a huge investment in property that needs to be protected from fire. Furthermore, a big bureaucracy has built up whose *raison d'etre* is to fight fires. It takes more and more money to forestall a calamity that becomes increasingly likely because of our efforts to prevent it.

What happens in each of these cases is that our goal to control variability in ecological systems leads us to a narrow purpose and to focus exclusively on solving a single problem. But elements of the system change gradually as a consequence of our management success in ways that we did not anticipate. As more homogenous ecosystems develop over a landscape scale, resilience decreases, and it becomes more likely that the system will flip suddenly into a new regime. What can we do to avoid this trap? Table 24.4 suggests some important lessons for ecological managers.

The precautionary principle urges institutional caution

One response to the uncertainty of wicked problems and chaotic, nonlinear, discontinuous systems is to plan a margin of safety for error or surprises. Drawing on studies of ecological systems, many conservation biologists advocate a **precautionary principle** that says that when an activity raises threats of harm to

TABLE 24.4

Planning for Resilience

1. Interdisciplinary, integrated modes of inquiry are needed for adaptive management of wicked problems.

2. We must recognize that these problems are fundamentally nonlinear and that we need nonlinear approaches to them.

3. Interactions between slow ecological processes such as global climate change or soil erosion in the American cornbelt are difficult to study together with the fast processes that bring creative destruction such as potential collapse of Antarctic ice sheets or appearance of a dead zone in the Gulf of Mexico, but we need to look for connections.

4. The spatial and temporal scales of our concerns are widening. We now need to consider global connections and problems in our planning.

5. Both ecological and social systems are evolutionary and are not amenable to simple solutions based on knowledge of small parts of the whole or on assumptions of constancy or stability of fundamental relationships.

6. We need adaptive management policies that focus on building resilience and the capacity of renewal in both ecosystems and human institutions.

Source: L. Gunderson, C. Holling, and S. Light, *Barriers and Bridges to the Renewal of Ecosystems and Institutions,* 1995, Columbia University Press.

human health or the environment, precautionary measures should be taken even if some cause and effect relationships are not fully established scientifically. At a meeting at the Wingspread Center in 1998, an international group of scientists, government officials, lawyers, and grassroots environmental activists agreed on four basic tenets of precautionary action:

- People have a duty to take anticipatory steps to prevent harm. If you have a reasonable suspicion that something bad might be going to happen, you have an obligation to try to stop it.

- The burden of proof of carelessness of a new technology, process, activity, or chemical lies with the proponents, not with the general public.

- Before using a new technology, process, or chemical, or starting a new activity, people have an obligation to examine a full range of alternatives, including the alternative of not using it.

- Decisions applying the precautionary principle must be open, informed, and democratic, and must include the affected parties.

The European Union has adopted this principle as the basis of its environmental policy, but American opponents of this approach claim that it could prevent the country from doing anything productive or innovative. Many American firms that do business in Europe—virtually all of the largest corporations are in this category—are having to change their manufacturing processes to adapt to more stringent E.U. standards. For example, lead, mercury, and other hazardous materials must be eliminated from electronics, toys, cosmetics, clothing, and a variety of other consumer products. A proposal currently being debated by the E.U. would require testing of thousands of chemicals, cost industry billions of dollars, and lead to many more products and compounds being pulled off the market. What do you think? Is this just common sense or an invitation to decision paralysis?

Arbitration and mediation can help settle disputes

Another set of alternatives to the adversarial nature of litigation and administrative challenges is the growing field of dispute resolution. Increasingly used to avoid the time, expense, and winner-take-all confrontation inherent in tort law, these techniques encourage compromise and workable solutions with which everyone can live.

Arbitration is a formal process of dispute resolution somewhat like a trial. There are stringent rules of evidence, cross-examination of witnesses, and the process results in a legally binding decision. The arbitrator takes a more active role than a judge, however, and is not as constrained by precedent. The arbitrator is more interested in resolving the dispute rather than strict application or interpretation of the law. Arbitrators must have formal training and be certified by the Federal Mediation and Conciliation Service or the American Arbitration Association. Arbitration is usually an attractive prospect if you don't think you could win a formal lawsuit, but why would anyone agree to arbitration if they think they can win the whole enchilada in court? They might take this route

FIGURE 24.17 Mediation encourages stakeholders to discuss issues and try to find a workable compromise.

just to avoid disagreeable surprises. Juries can be fickle. Furthermore, they might want to get an unpleasant process over with sooner rather than later. In addition, arbitration often is written into contracts so that the disputants have no choice over the matter.

There are disadvantages to arbitration. It doesn't create a legally binding precedent, something that often is the main motivation for a lawsuit. There is less opportunity to appeal if you don't like the decision you get. There also is less protection from self-incrimination, false witnesses, or evidence you didn't expect. You don't generate nearly as much publicity because the proceedings and record are not public. For some litigants, the publicity generated by a trial is more valuable than the settlement itself. Finally, you are less likely to win the whole thing. Some sort of compromise is the most likely outcome.

Mediation is a process in which disputants are encouraged to sit down and talk to see if they can come up with a solution by themselves (fig. 24.17). The mediator makes no final decision but is simply a facilitator of communication. This process is especially useful in complex issues where there are multiple stakeholders with different interests, as is often the case in environmental controversies. For example, a mediation was attempted to work out policy and management disputes about the Boundary Waters Canoe Area Wilderness in northern Minnesota. Local property owners, resort operators, anglers who wanted to use motorboats on wilderness lakes, wildlife protection groups, wilderness advocates, and the U.S. Forest Service all were represented around the table. Each group had its own agenda, although they all quickly coalesced into pro-motor and anti-motor factions. Interestingly, although we usually think of environmentalists as being relatively disadvantaged in these issues, in this case, the local folks felt they were unfairly outgunned by the high-powered lawyers who represented some of the pro-wilderness and wildlife groups. Although the participants agreed on many points and it seemed as if the mediation was about to come up with a workable compromise, at the last minute everything fell apart and the groups refused to agree about anything. Each side accused the other of intransigence and bad will, and the issues they disagreed upon will likely end up in court.

This example illustrates both the promises and perils of mediation. It can be quicker and cheaper than court battles. It can lead to compromise and understanding that will lead to further cooperation, and that will solve problems faster than endless appeals. And it may find creative solutions that satisfy multiple parties and interests. On the other hand no one can be forced to mediate or to do so in good faith. Rancorous participants can tie up the process in long, pointless arguments that only make others more angry. Ultimately, there can be a tyranny of the minority. A single person can veto an agreement that everyone else wants. Furthermore, mediation represses or denies certain irreconcilable structural conflicts, giving the impression of equality between disputants when none really exists. Unequal negotiating skills of the participants can lead to unfair outcomes and even more rancor and paranoia than before the mediation was attempted. As is the case in arbitration, mediation doesn't generate the publicity and complete victory that some groups may desire.

FIGURE 24.18 The Bay of Fundy has the greatest tidal range in the world. It is the site of innovative community-based environmental planning process.

Community-based planning can help solve environmental problems

Over the past several decades, natural resource managers have come to recognize the value of holistic, adaptive, multiuse, multivalue approach to planning. Involving all stakeholders and interest groups early in the planning process can help avoid the "train wrecks" in which adversaries become entrenched in nonnegotiable positions. Working with local communities can tap into traditional knowledge and gain acceptance for management plans that finally emerge from policy planning. This approach is especially important in nonlinear, nonequilibrium systems and wicked problems. Among the more important reasons to use collaborative approaches are:

- The way wicked problems are formulated depends on your worldview. Incorporating a variety of perspectives early in the process is more likely to lead to the development of acceptable solutions in the end.

- People have more commitment to plans they have helped develop. The first stage is therefore to identify those involved and to engage them in the process.

- There is truth in the old adage that "two heads are better than one." Involving multiple stakeholders and multiple sources of information enriches the process.

- Community-based planning provides access to situation-specific information and experience that can often only be obtained by active involvement of local residents.

- Participation is an important management tool. Project-threatening resistance on the part of certain stakeholders can be minimized by inviting active cooperation of all stakeholders throughout the planning process.

- The knowledge and understanding needed by those who will carry out subsequent phases of a project can only be gained through active participation.

A good example of community-based planning can be seen in the Atlantic Coastal Action Programme (ACAP) in eastern Canada. The purpose of this project is to develop blueprints for the restoration and maintenance of environmentally degraded harbors and estuaries in ways that are both biologically and socially sustainable. Officially established under Canada's Green Plan and supported by Environment Canada, this program created 13 community groups, some rural and some urban, with membership in each dominated by local residents. Federal and provincial government agencies are represented primarily as nonvoting observers and resource people. Each community group is provided with core funding for full-time staff who operate an office in the community and facilitate meetings.

Four of the 13 ACAP sites are in the Bay of Fundy, an important and unique estuary lying between New Brunswick and Nova Scotia. Approximately 270 km long, and with an area of more than 12,000 km^2, the bay, together with the nearby Georges Bank and the Gulf of Maine once formed one of the richest fisheries in the world. With the world's highest recorded tidal range (up to 16 m at maximum spring tide), the bay still sustains a great variety of fishery and wildlife resources, and provides habitat for a number of rare or endangered species. Now home to more than 1 million people, the coastal region is an important agricultural, lumbering, and paper-producing region (fig. 24.18).

Since European settlement began in 1604, the Bay of Fundy region has experienced great changes in population growth, resource use, and human-induced ecosystem change. More than 80 percent of the salt marshes present in 1604 have been eliminated or degraded. Pollution and sediment damage harbors and biological communities. Overfishing and introduction of exotic species have resulted in endemic species declines. The collapse of cod, halibut, and haddock fishing has had devastating economic effects on the regional economy and the livelihoods of local residents. Aquaculture is now a more valuable activity than all wild fisheries.

To cope with these complex, intertwined social and biological problems, ACAP is bringing together different stakeholders from around the bay to create comprehensive plans for ecological, economic, and social sustainability. Through citizen monitoring and adaptive management, the community builds social capital (knowledge, cooperative spirit, trust, optimism, working relations), develops a sense of ownership in the planning process, and eliminates some of the fears and sectorial rivalry that often divides local groups, outsiders, and government agents.

On the other hand, giving a greater voice and increased power to local communities could simply result in the foxes guarding the hen house. How would you balance the general public interest with those of local stakeholders?

Some nations have developed green plans

Several national governments have undertaken integrated environmental planning that incorporates community round-tables for vision development. Canada, New Zealand, Sweden, and Denmark all have so-called **green plans** or comprehensive, long-range national environmental strategies. The best of these plans weave together complex systems, such as water, air, soil, and energy, and mesh them with human factors such as economics, health, and carrying capacity. Perhaps the most thorough and well-thought-out green plan in the world is that of the Netherlands.

Developed in the 1980s through a complex process involving the public, industry, and government, the 400-page Dutch plan contains 223 policy changes aimed at reducing pollution and establishing economic stability. Three important mechanisms have been adopted for achieving these goals: integrated life-cycle management, energy conservation, and improved product quality. These measures should make consumer goods last longer and be more easily recycled or safely disposed of when no longer needed. For example, auto manufacturers are now required to design cars so they can be repaired or recycled rather than being discarded.

Among the guiding principles of the Dutch green plan are: (1) the "stand-still" principle that says environmental quality will not deteriorate, (2) abatement at the source rather than cleaning up afterward, (3) the "polluter pays" principle that says users of a resource pay for negative effects of that use, (4) prevention of unnecessary pollution, (5) application of the best practicable means for pollution control, (6) carefully controlled waste disposal, and (7) motivating people to behave responsibly.

The Netherlands have invested billions of guilders in implementing this comprehensive plan. Some striking successes

FIGURE 24.19 Under the Dutch green plan, 250,000 ha (600,000 acres) of drained agricultural land are being restored to wetland and 40,000 ha (99,000 acres) are being replanted as woodland.

already have been accomplished. Between 1980 and 1990, emissions of sulfur dioxide, nitrogen oxides, ammonia, and volatile organic compounds were reduced 30 percent. By 1995, pesticide use had been reduced 25 percent from 1988 levels, and chlorofluorocarbon use had been virtually eliminated. By 1998, industrial wastewater discharge into the Rhine River was 70 percent less than a decade earlier. Some 250,000 ha (more than 600,000 acres) of former wetlands that had been drained for agriculture are being restored as nature preserves and 40,000 ha (99,000 acres) of forest are being replanted. This is remarkably generous and foresighted in such a small, densely populated country, but the Dutch have come to realize they cannot live without nature (fig. 24.19).

Not all goals have been met so far. Planned reductions in CO_2 emissions failed to materialize when cheap fuel prices encouraged fuel-inefficient cars. Currently a carbon tax is being considered. A sudden population increase caused by immigration from developing countries and Eastern Europe also complicates plan implementation, but the basic framework of the Dutch plan has much to recommend it, nevertheless. Other countries would be more sustainable and less environmentally destructive if they were to adopt a similar plan.

CONCLUSION

Otto von Bismark once said, "Laws are like sausages, it is better not to see them being made." Still, if you hope to improve your environmental quality, it's helpful to understand how both policy and laws are made and enforced. Laws, such as the Endangered Species Act have been among the most effective tools that conservationists have had to protect biodiversity and habitat. As the opening case study for this chapter shows, there's a constant struggle between those who want to strengthen environmental laws, and those who want to reduce or eliminate them.

You may think that ordinary individuals have little opportunity to influence either policy or law, but you might be surprised at how much impact you can have if you get involved. Probably the best way to participate in environmental policy formation or passage of environmental laws is to join an organization, such as one of the national conservation organizations. Being part of a group amplifies your influence. But even as an individual, you can make an impression. Write to or call your legislator. They do pay attention to constituents. Participate in public planning sessions by agencies. Make a statement on behalf of your favorite cause. Sign a petition. Send in a comment on proposed actions. You can also influence both policymakers and your fellow citizens by writing a letter to the editor of your local newspaper. In chapter 25, we'll have more information on ways that individuals and groups are using the principles you've learned in this chapter about policy and law to bring about positive change.

REVIEWING LEARNING OUTCOMES

By now you should be able to explain the following points:

24.1 Define environmental policy.
- How is policy created?
- Policy formation follows predictable steps.
- Is a clean, healthy environment a basic human right?

24.2 Discuss environmental law.
- A brief environmental history.
- Statutory law: The legislative branch.
- Case law: The judicial branch.
- Administrative law: The executive branch.

24.3 Describe international treaties and conventions
- New approaches can make treaties effective.
- International governance has been controversial.
- Will globalization bring better environmental governance?

24.4 Summarize dispute resolution and planning.
- Wicked problems don't have simple answers.
- Adaptive management introduces science to planning.
- Resilience is important in ecosystems and institutions.
- The precautionary principle urges institutional caution.
- Arbitration and mediation can help settle disputes.
- Community-based planning can help solve environmental problems.
- Some nations have developed green plans.

PRACTICE QUIZ

1. What is the policy cycle, and how does it work?
2. Describe the path of a bill through Congress. When are riders and amendments attached?
3. What are the differences and similarities between civil, criminal, and administrative law?
4. List some of the major U.S. environmental laws of the past 30 years.
5. Why have some international environmental treaties and conventions been effective while most have not? Describe two such treaties.
6. Define *globalization* and describe how it impacts environmental quality.
7. What are wicked problems? Why are they difficult?
8. What is resilience? Why is it important?
9. What is collaborative, community-based planning?
10. What is unique about the Dutch green plan?

CRITICAL THINKING AND DISCUSSION QUESTIONS

1. In your opinion, how much environmental protection is too much? Think of a practical example in which some stakeholders may feel oppressed by government regulations. How would you justify or criticize these regulations?

2. Which is the most important step in the policy cycle? If you were leader of a major environmental group, where would you put your efforts in establishing policy?

3. Do you believe that trees, wild animals, rocks, or mountains should have legal rights and standing in the courts? Why or why not? Are there partial rights or some other form of protection you would favor for nature?

4. It's sometimes difficult to determine whether a lawsuit is retaliatory or based on valid reason. How would you define a SLAPP suit, and differentiate it from a legitimate case?

5. Create a list of arguments for and against an international body with power to enforce global environmental laws. Can you see a way to create a body that could satisfy both reasons for and against this power?

6. Take a current environmental problem. If you were an environmental leader trying to resolve this problem, would you choose litigation, arbitration, or mediation? What are your reasons for favoring or rejecting each one?

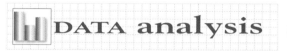 **DATA analysis** **Scatter Plots and Regression Analysis**

Often environmental scientists want to know if one factor causes another to change. For example, does improving human well-being require a decline in environmental quality? This trade-off is often an assumption in political and economic debates, in which people insist that we must sacrifice our environment if we want better jobs. But how can you know whether this assumption is true?

One way to test this assumption is to plot those two factors for a number of places and observe what the relationship really is. For example, the United Nations calculates the human development index (HDI), while researchers from Columbia and Yale Universities have calculated environmental performance index (EPI) for most countries in the world (see chapter 23 for more discussion of these measures). Each of these indices is simply a number boiled down to represent many useful measures, such as life expectancy, access to education, availability of safe drinking water, sustainability, and effective governance.

We can make a graph, with HDI on one axis and EPI on the other. On this graph, we can plot a point for each country. This graph is called a scatter plot. In figure 1, the cloud of points define an upward trend from left to right. This scatter plot helps you visually interpret the relationship between these two variables.

To describe the relationship further, you can draw a "regression" line through the points (the straight line through the cloud of dots in figure 1). We've created some hypothetical graphs in figure 2 to help you understand this concept. A computer can find the line that fits the dots best—that is, the line that minimizes the vertical distance to all the points (black dashed vertical lines in figure 2a). If the sum of all these distances is small (compared to a flat line drawn at the average Y value, figure 2b), you know the line describes the data set well. Also, if the line is steep, you know there is a strong relationship between the variables. If the line is nearly flat, there is little or no relationship.

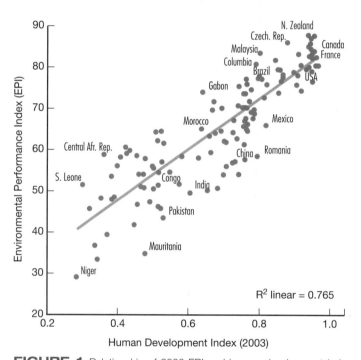

FIGURE 1 Relationship of 2006 EPI and human development index.

How *well* does HDI explain EPI? We use an R^2 value to answer this. R^2 compares the sum of black dashed lines in figure 2a to the sum of dashed lines in figure 2b. If $R^2 = 1$, then 100 percent of variation in the Y variable is explained by the X variable. For complicated issues such as development and environment, an R^2 of 0.7 or more is often considered very strong.

These ideas and statistics are very important concepts in environmental science. Think about them next time you see a graph—especially in a policy or economics debate.

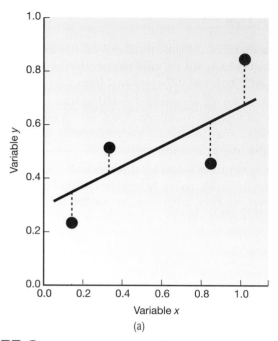

(a)

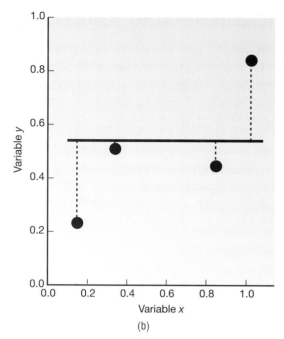

(b)

FIGURE 2 A regression line is defined to minimize the distance between the line and all points. A horizontal line can be drawn for the same points at the average Y value (b). A regression coefficient (R^2) describes the strength of the relationship by comparing the cumulative lenghts of dashed lines in (a) and (b).

Study figure 1 and answer the following questions:

1. In your own words, what relationship between environmental performance index and the human development index is shown in this graph?

2. What are the highest and lowest environmental performance and human development rankings shown, and which countries have those rankings?

3. List several factors that might describe differences between countries at the top and bottom of this curve. Discuss with your colleagues if necessary.

4. Identify one or two countries with the greatest variance (i.e., distance from the regression line). Do they have unusually low environmental conditions or human development conditions?

5. Some countries, such as Sierra Leone and India, have very similar environmental performance index, but very different human development index ratings. How would you interpret this information?

6. Explain these terms: scatter plot, regression line, and R^2.

For Additional Help in Studying This Chapter, please visit our website at www.mhhe.com/cunningham10e. You will find additional practice quizzes and case studies, flashcards, regional examples, place markers for Google Earth™ mapping, and an extensive reading list, all of which will help you learn environmental science.

Ecotourism and whale watching provide jobs for local people and help protect the Laguna San Ignacio.

25

What Then Shall We Do?

*You must be the change you wish to see
in the world.*
—Mahatma Gandhi—

LEARNING OUTCOMES

After studying this chapter, you should be able to:

16.1 Explain how we can make a difference.

16.2 Summarize environmental education.

16.3 Evaluate what individuals can do.

16.4 Review how we can work together.

16.5 Investigate campus greening.

16.6 Define the challenge of sustainability.

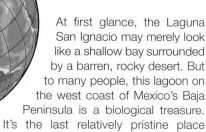

Case Study Saving a Gray Whale Nursery

At first glance, the Laguna San Ignacio may merely look like a shallow bay surrounded by a barren, rocky desert. But to many people, this lagoon on the west coast of Mexico's Baja Peninsula is a biological treasure. It's the last relatively pristine place where gray whales congregate each winter to mate, give birth, and nurse their calves. Pacific gray whales (*Eschrichtius robustus*) make a round trip of about 16,000 km (10,000 mi) every year between their summer feeding grounds north of the Arctic Circle and the Baja (fig. 25.1). The warm, salty water of bays like San Ignacio give calves extra buoyancy that helps them swim and nurse, while also sheltering them from predators and winter storms.

In the nineteenth century, a whaling captain named Scammon discovered the winter calving areas in Baja. The enclosed bays that once protected the whales became killing grounds. In a short time, the Pacific population was reduced from an estimated 25,000 animals to only a few thousand. Whaling bans have allowed the species to rebound to nearly its prehunting population, a great success story in endangered species protection. In 1994, Pacific gray whales were removed from the U.S. endangered species list.

In 1954, the same year that Mexico banned commercial whaling, a sea salt extraction facility was built in Guerro Negro bay (formerly Scammon's Lagoon) and the nearby Ojo de Liebre just north of Laguna San Ignacio. These saltworks, which are now operated by Expotadora de Sal and jointly owned by the Mitsubishi Company and the Mexican government, are the largest in the world, producing 6.5 million metric tons of salt per year. Concern about the effects of this huge industrial development on both the whales and the surrounding desert caused Mexican President Miguel de La Madrid to establish the Vizcaino Biosphere Reserve in 1988, including all three lagoons plus 2.4 million ha (6 million acres) of surrounding desert.

In 1994, however, Expotadora de Sal announced intentions to build an even bigger saltworks at Laguna San Ignacio, Plans called for 300 km² (116 mi²) of salt evaporation ponds carved out of the shoreline and filled by diesel engines that would pump 23,000 l (nearly 6,000 gal) of seawater per second. A 1.6 km (1 mi) long concrete pier built across the lagoon would transport the salt to an offshore loading area that would fill more than 120 salt tankers per year. The threat to whale survival from this immense operation was evident.

One of Mexico's leading environmental groups, el Grupo de los Cien (the Group of 100) started a campaign to stop this huge industrial development. They joined with other nongovernmental groups (NGOs), including the Natural Resources Defense Council (NRDC) and

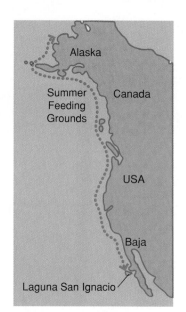

FIGURE 25.1 Gray whale migration route from Alaska to Baja, California.

the International Fund for Animal Welfare (IFAW), to raise public awareness and to lobby the Mexican government. The campaign took a number of different approaches. One of these was to organize whale-watching trips featuring movie stars, such as Glenn Close and Pierce Brosnan, to gain attention and educate the public about the issue. Newspaper ads and magazine articles criticized the industrialization of San Ignacio. One of these, entitled "An Unacceptable Risk," presented the scientific value of the lagoon and was signed by 33 of the world's most famous scientists, including several Nobel laureates.

Environmentalists also lobbied Mitsubishi directly, threatening to boycott their cars, TVs, electronics, and other products. A 1998 UN World Heritage Conference in Kyoto, Japan, provided an excellent opportunity to meet face-to-face with company leaders. Activists said to Hajime Koga, manager of Mitsubishi's "Salt Team," "You would never contemplate such a project in a World Heritage site in Japan. Why would you destroy one in another country?" The company was amazed to receive more than 1 million petitions, letters, and emails from all over the world, criticizing their expanded saltworks. Although the environmental NGOs weren't successful in obtaining "In Danger" designation for the biosphere reserve at the conference in Kyoto, they did get this classification at the next meeting of the World Heritage Committee in Marrakech, Morocco, in 2000.

In 2002, Expotadora de Sal announced that it was abandoning plans for Laguna San Ignacio. Mitsubishi said it was the first time in its history that it had changed its policy because of environmental concerns. Simply blocking development isn't enough; long-term solutions need to be economically sustainable as well as scientifically sound and socially just. The 35,000 Mexicans who live within the biosphere reserve need to make a living. In 2005, local residents and environmental NGOs signed an agreement to preserve 50,000 ha (124,000 acres) of land around Laguna San Ignacio. The Ejido Luis Echeverria, a land cooperative, which owns the land, will limit development in exchange for a $25,000 annual payment to be used for low-impact projects, such as ecotourism and whale watching. Eventually, conservationists hope to reach a similar agreement with five other *ejidos* to extend protection to 4,000 km² (1 million acres) of the Vizcaino Biosphere Reserve. This will cost about $10 million.

This case study demonstrates some of the steps in influencing public policy. First, you have to gather information and understand the science that informs your issue. You should recognize how, and by whom, policy is made. You must evaluate which of the many techniques for educating the public and shaping opinion can be effective. And you need to learn how to work with other groups, and to reason with those whose opinions you hope to sway. In this chapter, we'll study how individuals and groups can work to affect this process.

25.1 MAKING A DIFFERENCE

Throughout this book you have read about environmental problems, from climate change to biodiversity to energy policy debates. Biodiversity is disappearing at the fastest rate ever known; major ocean fisheries have collapsed; within 50 years, it is expected that two-thirds of countries will experience water shortages, and 3 billion people may live in slums. You have also seen that, as we have come to understand these problems, many exciting innovations have been developed to deal with them. New irrigation methods reduce agricultural water use; bioremediation provides inexpensive methods to treat hazardous waste; new energy sources, including wind, solar, and even pressure-cooked garbage, offer strategies for weening our society from its dependence on oil and gas. Growth of green consumerism has developed markets for recycled materials, low-energy appliances, and organic foods. Population growth continues, but its rate has plummeted from a generation ago.

Stewardship for our shared resources is increasingly understood to be everybody's business. The environmental justice movement (chapter 23) has shown that minority groups and the poor frequently suffer more from pollution than wealthy or white people. African Americans, Latinos, and other minority groups have a clear interest in pursuing environmental solutions. Religious groups are voicing new concerns about preserving our environment (chapter 2). Farmers are seeking ways to save soil and water resources (chapter 9). Loggers are learning about sustainable harvest methods (chapter 12). Business leaders are discovering new ways to do well by doing good work for society and the environment (chapter 21). These changes are exciting, though many challenges remain.

Whatever your skills and interests, you can contribute to understanding and protecting our common environment. If you enjoy science, there are many disciplines that contribute to environmental science. As you know by now, biology, chemistry, geology, ecology, climatology, geography, demography, and other sciences all provide essential ideas and data to environmental science. Environmental scientists usually focus on one of these disciplines, but their work also serves the others. An environmental chemist, for example, might study contaminants in a stream system, and this work might help an aquatic ecologist understand changes in a stream's food web.

You can also help seek environmental solutions if you prefer writing, art, working with children, history, politics, economics, or other areas of study. As you have read, environmental science depends on communication, education, good policies, and economics as well as on science.

In this chapter, we will discuss some of the steps you can take to help find solutions to environmental problems. You have already taken the most important step, educating yourself. When you understand how environmental systems function—from nutrient cycles and energy flows to ecosystems, climate systems, population dynamics, agriculture, and economies—you can develop well-informed opinions and help find useful answers (fig. 25.2).

FIGURE 25.2 What lives in a tide pool? Learning to appreciate the beauty, richness, and diversity of the natural world is important if we are to protect it.

25.2 ENVIRONMENTAL EDUCATION

In 1990 Congress recognized the importance of environmental education by passing the National Environmental Education Act. The act established two broad goals: (1) to improve understanding among the general public of the natural and built environment and the relationships between humans and their environment, including global aspects of environmental problems, and (2) to encourage postsecondary students to pursue careers related to the environment. Specific objectives proposed to meet these goals include developing an awareness and appreciation of our natural and social/cultural environment, knowledge of basic ecological concepts, acquaintance with a broad range of current environmental issues, and experience in using investigative, critical-thinking, and problem-solving skills in solving environmental problems (fig. 25.3). Several states, including Arizona, Florida,

FIGURE 25.3 Environmental education helps develop awareness and appreciation of ecological systems and how they work.

TABLE 25.1

Outcomes from Environmental Education

The natural context: An environmentally educated person understands the scientific concepts and facts that underlie environmental issues and the interrelationships that shape nature.

The social context: An environmentally educated person understands how human society is influencing the environment, as well as the economic, legal, and political mechanisms that provide avenues for addressing issues and situations.

The valuing context: An environmentally educated person explores his or her values in relation to environmental issues; from an understanding of the natural and social contexts, the person decides whether to keep or change those values.

The action context: An environmentally educated person becomes involved in activities to improve, maintain, or restore natural resources and environmental quality for all.

Source: *A Greenprint for Minnesota,* Minnesota Office of Environmental Education, 1993.

TABLE 25.2

The Environmental Scientist's Bookshelf

What are some of the most influential and popular environmental books? In a survey of environmental experts and leaders around the world, the top 12 best books on nature and the environment were:

A Sand County Almanac by Aldo Leopold (100)[1]
Silent Spring by Rachel Carson (81)
State of the World by Lester Brown and the Worldwatch Institute (31)
The Population Bomb by Paul Ehrlich (28)
Walden by Henry David Thoreau (28)
Wilderness and the American Mind by Roderick Nash (21)
Small Is Beautiful: Economics as if People Mattered by E. F. Schumacher (21)
Desert Solitaire: A Season in the Wilderness by Edward Abbey (20)
The Closing Circle: Nature, Man, and Technology by Barry Commoner (18)
The Limits to Growth: A Report for the Club of Rome's Project on the Predicament of Mankind by Donella H. Meadows, et al. (17)
The Unsettling of America: Culture and Agriculture by Wendell Berry (16)
Man and Nature by George Perkins Marsh (16)

[1]Indicates number of votes for each book. Because the preponderance of respondents were from the United States (82 percent), American books are probably overrepresented.

From Robert Merideth, *The Environmentalist's Bookshelf: A Guide to the Best Books,* 1993, by G. K. Hall, an imprint of Macmillan, Inc. Reprinted by permission.

Maryland, Minnesota, Pennsylvania, and Wisconsin, have successfully incorporated these goals and objectives into their curricula (table 25.1).

A number of organizations have been established to teach ecology and environmental ethics to elementary and secondary school students, as well as to get them involved in active projects to clean up their local community. Groups such as Kids Saving the Earth or Eco-Kids Corps are an important way to reach this vital audience. Family education results from these efforts as well. In a World Wildlife Fund survey, 63 percent of young people said they "lobby" their parents about recycling and buying environmentally responsible products.

Environmental literacy means understanding our environment

Speaking in support of the National Environmental Education Act, former Environmental Protection Agency administrator William K. Reilly called for broad **environmental literacy** in which every citizen is fluent in the principles of ecology and has a "working knowledge of the basic grammar and underlying syntax of environmental wisdom." Environmental literacy, according to Reilly can help establish a stewardship ethic—a sense of duty to care for and manage wisely our natural endowment and our productive resources for the long haul. "Environmental education," he says, "boils down to one profoundly important imperative: preparing ourselves for life in the next century. When the twenty-first century rolls around, it will not be enough for a few specialists to know what is going on while the rest of us wander about in ignorance."

You have made a great start toward learning about your environment by reading this book and taking a class in environmental science. Pursuing your own environmental literacy is a life-long process. Some of the most influential environmental books of all time examine environmental problems and suggest solutions (table 25.2). To this list we'd add some personal favorites: *The Singing Wilderness* by Sigurd F. Olson, *My First Summer in the Sierra* by John Muir, and *Encounters with the Archdruid* by John McPhee.

Citizen science encourages everyone to participate

While university classes often tend to be theoretical and abstract, many students are discovering they can make authentic contributions to scientific knowledge through active learning and undergraduate research programs. Internships in agencies or environmental organizations are one way of doing this. Another is to get involved in organized **citizen science** projects in which ordinary people join with established scientists to answer real scientific questions. Community-based research was pioneered in the Netherlands, where several dozen research centers now study environmental issues ranging from water quality in the Rhine River, cancer rates by geographic area, and substitutes for harmful organic solvents. In each project, students and neighborhood groups team with scientists and university personnel to collect data. Their results have been incorporated into official government policies.

Similar research opportunities exist in the United States and Canada. The Audubon Christmas Bird Count is a good example (Exploring Science p. 570). Earthwatch offers a much smaller but more intense opportunity to take part in research. Every year hundreds of Earthwatch projects each field a team of a dozen or so volunteers who spend a week or two working on issues ranging from loon nesting behavior to archaeological digs. The American River Watch organizes teams of students to measure water quality. You might be able to get academic credit as well as helpful practical experience in one of these research experiences.

FIGURE 25.4 Many interesting, well-paid jobs are opening up in environmental fields. Here an environmental technician takes a sample from a monitoring well for chemical analysis.

Environmental careers range from engineering to education

The need for both environmental educators and environmental professionals opens up many job opportunities in environmental fields. The World Wildlife Fund estimates, for example, that 750,000 new jobs will be created over the next decade in the renewable energy field alone. Scientists are needed to understand the natural world and the effects of human activity on the environment. Lawyers and other specialists are needed to develop government and industry policy, laws, and regulations to protect the environment. Engineers are needed to develop technologies and products to clean up pollution and to prevent its production in the first place. Economists, geographers, and social scientists are needed to evaluate the costs of pollution and resource depletion and to develop solutions that are socially, culturally, politically, and economically appropriate for different parts of the world. In addition, business will be looking for a new class of environmentally literate and responsible leaders who appreciate how products sold and services rendered affect our environment.

Trained people are essential in these professions at every level, from technical and clerical support staff to top managers. Perhaps the biggest national demand over the next few years will be for environmental educators to help train an environmentally literate populace. We urgently need many more teachers at every level who are trained in environmental education. Outdoor activities and natural sciences are important components of this mission, but environmental topics such as responsible consumerism, waste disposal, and respect for nature can and should be incorporated into reading, writing, arithmetic, and every other part of education.

Green business and technology are growing fast

Can environmental protection and resource conservation—a so-called green perspective—be a strategic advantage in business? Many companies think so. An increasing number are jumping on the environmental bandwagon, and most large corporations now have an environmental department. A few are beginning to explore integrated programs to design products and manufacturing processes to minimize environmental impacts. Often called "design for the environment," this approach is intended to avoid problems at the beginning rather than deal with them later on a case-by-case basis. In the long run, executives believe this will save money and make their businesses more competitive in future markets. The alternative is to face increasing pollution control and waste disposal costs—now estimated to be more than $100 billion per year for all American businesses—as well as to be tied up in expensive litigation and administrative proceedings.

The market for pollution-control technology and know-how is also expected to be huge. Many companies are positioning themselves to cash in on this enormous market. Germany and Japan appear to be ahead of America in the pollution-control field because they have had more stringent laws for many years, giving them more experience in reducing effluents.

The rush to "green up" business is good news for those looking for jobs in environmentally related fields, which are predicted to be among the fastest growing areas of employment during the next few years. The federal government alone projects a need to hire some 10,000 people per year in a variety of environmental disciplines (fig. 25.4). How can you prepare yourself to enter this market? The best bet is to get some technical training: Environmental engineering, analytical chemistry, microbiology, ecology, limnology, groundwater hydrology, or computer science all have great potential. Currently, a chemical engineer with a graduate degree and some experience in an environmental field can practically name his or her salary. Some other very good possibilities are environmental law and business administration, both rapidly expanding fields.

For those who aren't inclined toward technical fields, there are many opportunities for environmental careers. A good liberal arts education will help you develop skills such as communication, critical thinking, balance, vision, flexibility, and caring that should serve you well. Large companies need a wide variety of people; small companies need a few people who can do many things well. There are many opportunities for planners (chapter 22), health professionals (chapter 8), writers, teachers, and policymakers.

Every Christmas since 1900, dedicated volunteers have counted and recorded all the birds they can find within their team's designated study site (fig. 1). This effort has become the largest, longest-running, citizen-science project in the world. For the 100th count, nearly 50,000 participants in about 1,800 teams observed 58 million birds belonging to 2,309 species. Although about 70 percent of the counts in 2000 were made in the United States or Canada, 650 teams in the Caribbean, Pacific Islands, and Central and South America also participated. Participants enter their bird counts on standardized data sheets, or submit their observations over the Internet. Compiled data can be viewed and investigated online, almost as soon as they are submitted.

Frank Chapman, the editor of *Bird-Lore* magazine and an officer in the newly formed Audubon Society, started the Christmas Bird Count in 1900. For years, hunters had gathered on Christmas Day for a competitive hunt, often killing hundreds of birds and mammals as teams tried to outshoot each other. Chapman suggested an alternative contest: to see which team could observe and identify the most birds, and the most species, in a day. The competition has grown and spread. In the 100th annual count, the winning team was in Monte Verde, Costa Rica, with an amazing 343 species tallied in a single day.

The tens of thousands of bird-watchers participating in the count gather vastly more information about the abundance and distribution of birds than biologists could gather alone. These data provide important information for scientific research on bird migrations, populations, and habitat change. Now that the entire record for a century of bird data is available on the BirdSource website (www.birdsource.org), both professional ornithologists and amateur

FIGURE 1 Citizen-science projects, such as the Christmas Bird Count, encourage people to help study their local environment.

bird-watchers can study the geographical distribution of a single species over time, or they can examine how all species vary at a single site through the years. Those concerned about changing climate can look for variation in long-term distribution of species. Climatologists can analyze the effects of weather patterns such as El Niño or La Niña on where birds occur.

One of the most intriguing phenomena revealed by this continent-wide data collection is irruptive behavior: that is, appearance of massive numbers of a particular species in a given area in one year, and then their move to other places in subsequent years following weather patterns, food availability, and other factors.

In 2005 the 105th Christmas Bird Count, collected data on nearly 70 million birds from 2,019 volunteer groups. This citizen-science effort has produced a rich, geographically broad data set far larger than any that could be produced by professional scientists (fig. 2). Following the success of the Christmas Bird Count, other citizen-science projects have been initiated. Project Feeder Watch, which began in the 1970s, has more than 15,000 participants, from schoolchildren and backyard bird-watchers to dedicated birders. The Great Backyard Bird Count of 2005 collected records on over 600 species and more than 6 million individual birds. In other areas, farmers have been enlisted to monitor pasture and stream health; volunteers monitor water quality in local streams and rivers; and nature reserves solicit volunteers to help gather ecological data. You can learn more about your local environment, and contribute to scientific research, by participating in a citizen-science project. Contact your local Audubon chapter or your state's department of natural resources to find out what you can do.

How does counting birds contribute to sustainability? Citizenscience projects are one way individuals can learn more about the scientific process, become familiar with their local environment, and become more interested in community issues. In this chapter, we'll look at other ways individuals and groups can help protect nature and move toward a sustainable society.

Black-capped chickadee
Christmas Bird Count
2002-03

> 500
100–500
50–100
10–50
< 10
0

FIGURE 2 Volunteer data collection can produce a huge, valuable data set. Christmas Bird Count data, such as this map, are available online. Data from Audubon Society.

25.3 WHAT CAN INDIVIDUALS DO?

Some prime reasons for our destructive impacts on the earth are our consumption of resources and disposal of wastes. Technology has made consumer goods and services cheap and readily available in the richer countries of the world. As you already know, we in the industrialized world use resources at a rate out of proportion to our percentage of the population. If everyone in the world were to attempt to live at our level of consumption, given current methods of production, the results would surely be disastrous. In this section we will look at some options for consuming less and reducing our environmental impacts. Perhaps no other issue in this book represents so clear an ethical question as the topic of responsible consumerism.

How much is enough?

A century ago, economist and social critic, Thorstein Veblen, in his book, *The Theory of the Leisure Class,* coined the term **conspicuous consumption** to describe buying things we don't want or need just to impress others. How much more shocked he would be to see current trends. The average American now consumes twice as many goods and services as in 1950. The average house is now more than twice as big as it was 50 years ago, even though the typical family has half as many people. We need more space to hold all the stuff we buy. Shopping has become the way many people define themselves (fig. 25.5). As Marx predicted, everything has become commodified; getting and spending have eclipsed family, ethnicity, even religion as the defining matrix of our lives. But the futility and irrelevance of much American consumerism leaves a psychological void. Once we possess things, we find they don't make us young, beautiful, smart, and interesting as they promised. With so much attention on earning and spending money, we don't have time to have real friends, to cook real food, to have creative hobbies, or to do work that makes us feel we have accomplished something with our lives. Some social critics call this drive to possess stuff "affluenza."

A growing number of people find themselves stuck in a vicious circle: They work frantically at a job they hate, to buy things they don't need, so they can save time to work even longer hours. Seeking a measure of balance in their lives, some opt out of the rat race and adopt simpler, less-consumptive lifestyles. As Thoreau wrote in *Walden,* "Our life is frittered away by detail . . . simplify, simplify."

We can choose to reduce our environmental impact

Marketers and trend spotters have noticed that an increasing number of people in affluent countries are becoming concerned about the effects of pollution and social inequity. There's a name for consumers who worry about the environment, want products to be produced in a fair, sustainable way, and use purchasing power to express their values. They're called Lohas, an acronym for "lifestyles of health and sustainability." Encompassing things like organic food, energy-efficient appliances and automobiles, natural home care and health products, active vacations and eco-tourism, the total market for this group represents $230 billion per year, according to Natural Business Communications, a company in Colorado that publishes *The Lohas Journal* and is credited with coining the term. Altogether, 68 million Americans—about one-third of the adult population—qualify as Lohas, consumers who take environmental and social issues into account when they make purchases. Ninety percent of this group say they prefer to make purchases from companies that share their values, and many say they are willing to pay a premium for products and services they consider healthier for themselves, their families, society, and the environment. Merchants flock to annual Lohas business conferences to learn how to tap into this important market.

Another name for people who are deeply concerned about nature and want to be involved in creating a new and better way of life is **cultural creatives,** a term introduced by Paul Ray and Sherry Ruth Anderson in a book by the same title. Ray and Anderson describe this group as socially conscious, involved in improving communities, and willing to translate values into action. They are strongly aware of environmental problems and want to do something to remedy them. Most cultural creatives place great importance on helping other people, care intensely about psychological or spiritual development, and volunteer for

FIGURE 25.5 Is this our highest purpose?

one or more good causes. They dislike the modern emphasis on wealth, consumerism and power, and enjoy learning about new places and people and alternative ways of life.

Neither cultural creatives nor Lohas are defined by particular demographic characteristics. They work at all sorts of jobs and occupy every economic level. The majority are mainstream in their religious beliefs and are no more liberal or conservative than the U.S. average. One important trait is that about two-thirds of them are women, and many of the values most important to them—relationships, family life, children, education, and responsibility—are traditionally thought of as women's issues. Because women now do a majority of family shopping as well as hold more than half of all personal wealth in America, both businesses and non-profits are beginning to pay attention to these concerns.

Recognizing that making people feel guilty about their lifestyles and purchasing habits isn't working, many organizations are now attempting to find ways to make sustainable living something consumers will adopt willingly. The goal is economically, socially, and environmentally viable solutions that allow people to enjoy a good quality of life while consuming fewer natural resources and polluting less. A good example of this approach is a British automaker that provides a mountain bike with every car it sells, urging buyers to use the bike for short journeys. Another example cited by UNEP is European detergent makers who encourage customers to switch to low-temperature washing liquids and powders, not just to save energy but because it's good for their clothes.

Although each of our individual choices may make a small impact, collectively they can be important. The What Can You Do? box on p. 572 offers some suggestions for reducing waste and pollution.

"Green washing" can mislead consumers

Although many people report they prefer to buy products and packaging that are socially and ecologically sustainable, there is a wide gap between what consumers say in surveys about purchasing habits and the actual sales data. Part of the problem is accessibility and affordability. In many areas, green products either aren't available or are so expensive that those on limited incomes (as many living in voluntary simplicity are) can't afford them. Although businesses are beginning to recognize the size and importance of the market for "green" merchandise, the variety of choices and the economies of scale haven't yet made them as accessible as we would like.

Another problem is that businesses, eager to cash in on this premium market, offer a welter of confusing and often misleading claims about the sustainability of their offerings. Consumers must be wary to avoid "green scams" that sound great but are actually only overpriced standard items. Many terms used in advertising are vague and have little meaning. For example:

- "Nontoxic" suggests that a product has no harmful effects on humans. Since there is no legal definition of the term, however, it can have many meanings. How nontoxic is the product? And to whom? Substances not poisonous to humans can be harmful to other organisms.

What Can You Do?

Reducing Your Impact

Purchase Less

Ask yourself whether you really need more stuff.

Avoid buying things you don't need or won't use.

Use items as long as possible (and don't replace them just because a new product becomes available).

Use the library instead of purchasing books you read.

Make gifts from materials already on hand, or give nonmaterial gifts.

Reduce Excess Packaging

Carry reusable bags when shopping and refuse bags for small purchases.

Buy items in bulk or with minimal packaging; avoid single-serving foods.

Choose packaging that can be recycled or reused.

Avoid Disposable Items

Use cloth napkins, handkerchiefs, and towels.

Bring a washable cup to meetings; use washable plates and utensils rather than single-use items.

Buy pens, razors, flashlights, and cameras with replaceable parts.

Choose items built to last and have them repaired; you will save materials and energy while providing jobs in your community.

Conserve Energy

Walk, bicycle, or use public transportation.

Turn off (or avoid turning on) lights, water, heat, and air conditioning when possible.

Put up clotheslines or racks in the backyard, carport, or basement to avoid using a clothes dryer.

Carpool and combine trips to reduce car mileage.

Save Water

Water lawns and gardens only when necessary.

Use water-saving devices and fewer flushes with toilets.

Don't leave water running when washing hands, food, dishes, and teeth.

Based on material by Karen Oberhauser, Bell Museum Imprint, University of Minnesota, 1992. Used by permission.

- "Biodegradable," "recyclable," "reusable," or "compostable" may be technically correct but not signify much. Almost everything will biodegrade *eventually*, but it may take thousands of years. Similarly, almost anything is potentially recyclable or reusable; the real question is whether there are programs to do so in your community. If the only recycling or composting program for a particular material is half a continent away, this claim has little value.

- "Natural" is another vague and often misused term. Many natural ingredients—lead or arsenic, for instance—are highly toxic. Synthetic materials are not necessarily more dangerous or environmentally damaging than those created by nature.

- "Organic" can connote different things in different places. There are loopholes in standards so that many synthetic chemicals can be included in "organics." On items such as shampoos and skin-care products, "organic" may have no significance at all. Most detergents and oils are organic chemicals, whether they are synthesized in a laboratory or found in nature. Few of these products are likely to have pesticide residues anyway.

- "Environmentally friendly," "environmentally safe," and "won't harm the ozone layer" are often empty claims. Since there are no standards to define these terms, anyone can use them. How much energy and nonrenewable material are used in manufacture, shipping, or use of the product? How much waste is generated, and how will the item be disposed of when it is no longer functional? One product may well be more environmentally benign than another, but be careful who makes this claim.

Certification identifies low-impact products

Products that claim to be environmentally friendly are being introduced at 20 times the normal rate for consumer goods. To help consumers make informed choices, several national programs have been set up to independently and scientifically analyze environmental impacts of major products. Germany's Blue Angel, begun in 1978, is the oldest of these programs. Endorsement is highly sought after by producers since environmentally conscious shoppers have shown that they are willing to pay more for products they know have minimum environmental impacts. To date, more than 2,000 products display the Blue Angel symbol. They range from recycled paper products, energy-efficient appliances, and phosphate-free detergents to refillable dispensers.

Similar programs are being proposed in every Western European country as well as in Japan and North America. Some are autonomous, nongovernmental efforts like the United States' Green Seal program (managed by the Alliance for Social Responsibility in New York). Others are quasigovernmental institutions such as the Canadian Environmental Choice programs.

The best of these organizations attempt "cradle-to-grave" **life-cycle analysis** (fig. 25.6) that evaluates material and energy inputs and outputs at each stage of manufacture, use, and disposal of the product. While you need to consider your own situation in making choices, the information supplied by these independent agencies is generally more reliable than self-made claims from merchandisers.

Green consumerism has limits

To quote Kermit the Frog, "It's not easy being green." Even with the help of endorsement programs, doing the right thing from an environmental perspective may not be obvious. Often we are faced with complicated choices. Do the social benefits

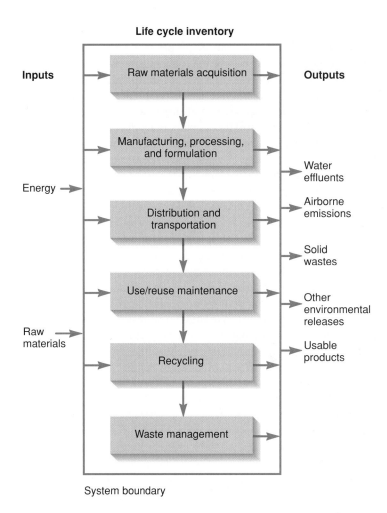

FIGURE 25.6 At each stage in its life cycle, a product receives inputs of materials and energy and produces outputs of materials or energy that move to subsequent phases and wastes that are released into the environment.

of buying rainforest nuts justify the energy expended in transporting them here, or would it be better to eat only locally grown products? In switching from Freon propellants to hydrocarbons, we spare the stratospheric ozone but increase hydrocarbon-caused smog. By choosing reusable diapers over disposable ones, we decrease the amount of material going to the landfill, but we also increase water pollution, energy consumption, and pesticide use (cotton is one of the most pesticide-intensive crops grown in the United States).

When the grocery store clerk asks you, "Paper or plastic?" you probably choose paper and feel environmentally virtuous, right? Everyone knows that plastic is made by synthetic chemical processes from nonrenewable petroleum or natural gas. Paper from naturally growing trees is a better environmental choice, isn't it? Well, not necessarily. In the first place, paper making consumes water and causes much more water pollution than does plastic manufacturing. Paper mills also release air pollutants, including foul-smelling sulfides and captans as well as highly toxic dioxins.

Furthermore, the brown paper bags used in most supermarkets are made primarily from virgin paper. Recycled fibers aren't strong enough for the weight they must carry. Growing, harvesting, and transporting logs from agroforestry plantations can be as environmentally disruptive as oil production. It takes a great deal of energy to pulp wood and dry newly made paper. Paper is also heavier and bulkier to ship than plastic. Although the polyethylene used to make a plastic bag contains many calories, in the end, paper bags are generally more energy-intensive to produce and market than plastic ones.

If both paper and plastic go to a landfill in your community, the plastic bag takes up less space. It doesn't decompose in the landfill, but neither does the paper in an air-tight, water-tight landfill. If paper is recycled but plastic is not, then the paper bag may be the better choice. If you are lucky enough to have both paper and plastic recycling, the plastic bag is probably a better choice since it recycles more easily and produces less pollution in the process. The best choice of all is to bring your own reusable cloth bag.

Complicated, isn't it? We often must make decisions without complete information, but it's important to make the best choices we can. Don't assume that your neighbors are wrong if they reach conclusions different from yours. They may have valid considerations of which you are unaware. The truth is that simple black and white answers often don't exist.

Taking personal responsibility for your environmental impact can have many benefits. Recycling, buying green products, and other environmental actions not only set good examples for your friends and neighbors, they also strengthen your sense of involvement and commitment in valuable ways. There are limits, however, to how much we can do individually through our buying habits and personal actions to bring about the fundamental changes needed to save the earth. Green consumerism generally can do little about larger issues of global equity, chronic poverty, oppression, and the suffering of millions of people in the developing world. There is a danger that exclusive focus on such problems as whether to choose paper or plastic bags, or to sort recyclables for which there are no markets, will divert our attention from the greater need to change basic institutions.

25.4 HOW CAN WE WORK TOGETHER?

While a few exceptional individuals can be effective working alone to bring about change, most of us find it more productive and more satisfying to work with others.

Collective action multiplies individual power (fig. 25.7). You get encouragement and useful information from meeting regularly with others who share your interests. It's easy to get discouraged by the slow pace of change; having a support group helps maintain your enthusiasm. You should realize, however, that there is a broad spectrum of environmental and social action groups. Some will suit your particular interests, preferences, or beliefs more than others. In this section, we will look at some environmental organizations as well as options for getting involved.

FIGURE 25.7 Working together with others can give you energy, inspiration, and a sense of accomplishment.

As the opening case study for this chapter shows, individuals and organizations can bring about major changes in governmental and corporate policy. Mitsubishi had already invested a great deal of money planning for the salt works at Laguna San Ignacio. Their potential profits could have been millions of dollars per year. But the persuasive moral arguments of environmental groups, plus the risk of international embarrassment made them rethink their position. This example shows the power of persistence and organization.

National organizations are influential but sometimes complacent

Among the oldest, largest, and most influential environmental groups in the United States are the National Wildlife Federation, the World Wildlife Fund, the Audubon Society, the Sierra Club, the Izaak Walton League, Friends of the Earth, Greenpeace, Ducks Unlimited, the Natural Resources Defense Council, and The Wilderness Society. Sometimes known as the "group of 10," these organizations are criticized by radical environmentalists for their tendency to compromise and cooperate with the establishment. Although many of these groups were militant—even extremist—in their formative stages, they now tend to be more staid and conservative. Members are mostly passive and know little about the inner workings of the organization, joining as much for publications or social aspects as for their stands on environmental issues. Collectively, these groups grew rapidly during the 1980s (fig. 25.8), but many of their new members had little contact with them beyond making a one-time donation.

Lack of progress over the past decade in important areas, such as global climate change, despite millions of dollars spent by major environmental organizations has led some critics to discuss the "death of environmentalism." They charge that professional staff have become more concerned about protecting their jobs and access to corridors of power in Washington than in bringing about change. Some observers argue that we need to abandon established structures and ways of thinking to come up with new approaches and new coalitions to protect our environment.

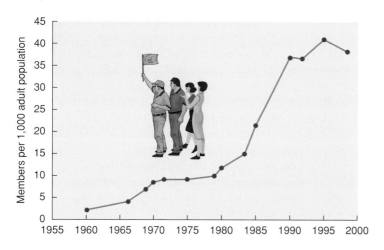

FIGURE 25.8 Growth of national environmental organizations in the United States.

Still, the established groups are powerful and important forces in environmental protection. Their mass membership, large professional staffs, and long history give them a degree of respectability and influence not found in newer, smaller groups. The Sierra Club, for instance, with about half a million members and chapters in almost every state, has a national staff of about 400, an annual budget over $20 million, and 20 full-time professional lobbyists in Washington, D.C. These national groups have become a potent force in Congress, especially when they band together to pass specific legislation, such as the Alaska National Interest Lands Act or the Clean Air Act.

In a survey that asked congressional staff and officials of government agencies to rate the effectiveness of groups that attempt to influence federal policy on pollution control, the top five were national environmental organizations. In spite of their large budgets and important connections, the American Petroleum Institute, the Chemical Manufacturers Association, and the Edison Electric Institute ranked far behind these environmental groups in terms of influence.

Although much of the focus of the big environmental groups is in Washington, Audubon, Sierra Club, and Izaak Walton have local chapters, outings, and conservation projects. This can be a good way to get involved. Go to some meetings, volunteer, offer to help. You may have to start out stuffing envelopes or some other unglamorous job, but if you persevere, you may have a chance to do something important and fun. It's a good way to learn and meet people.

Some environmental groups, such as the Environmental Defense Fund (EDF), The Nature Conservancy (TNC), the National Resources Defense Council (NRDC), and the Wilderness Society (WS), have limited contact with ordinary members except through their publications. They depend on a professional staff to carry out the goals of the organization through litigation (EDF and NRDC), land acquisition (TNC), or lobbying (WS). Although not often in the public eye, these groups can be very effective because of their unique focus. TNC buys land of high ecological value that is threatened by development. With more than 3,200 employees and assets around $3 billion, TNC manages 7 million acres in

FIGURE 25.9 The Nature Conservancy buys land with high biodiversity or unique natural values to protect it from misuse and development.

what it describes as the world's largest private sanctuary system (fig. 25.9). Still, the Conservancy is controversial for some of its management decisions, such as gas and oil drilling in some reserves, and including executives from some questionable companies on its governing board and advisory council. The Conservancy replies that it is trying to work with these companies to bring about change rather than just criticize them.

Radical groups capture attention and broaden the agenda

A striking contrast to the mainline conservation organizations are the direct action groups, such as Earth First!, Sea Shepherd, and a few other groups that form either the "cutting edge" or the "radical fringe" of the environmental movement, depending on your outlook. Often associated with the deep ecology philosophy and bioregional ecological perspective, the strongest concerns of these militant environmentalists tend to be animal rights and protection of wild nature. Their main tactics are civil disobedience and attention-grabbing actions, such as picketing, protest marches, road blockades, and other demonstrations. Some of these actions are humorous and lighthearted, such as street theater that gets a point across in a nonthreatening way (fig. 25.10). Many of these techniques are borrowed from the civil rights movement and Mahatma Gandhi's nonviolent civil disobedience. While often more innovative than the mainstream organizations, pioneering new issues and new approaches, the tactics of these groups can be controversial.

Greenpeace, for example, is notorious (or famous, depending on your perspective) for attention-grabbing actions, such as draping protest signs from buildings, bridges, and other tall structures,

FIGURE 25.10 Street theater can be a humorous, yet effective, way to convey a point in a nonthreatening way. Confrontational tactics get attention, but they may alienate those who might be your allies and harden your opposition.

or pursuing whaling vessels in small rubber runabouts. Some people regard these tactics as meaningless stunts that contribute little to a constructive dialog on how to solve real problems. Others see them as useful tools in gaining public attention to serious problems. Is civil disobedience dangerous and counterproductive, or is it brave and constructive? Remember that many major social movements—from the slavery abolition, labor rights, and women's suffrage movements of the eighteenth and nineteenth centuries to civil rights and anti-war movements of the twentieth century gained much of their momentum from mass demonstrations and protests that many contemporaries regarded as unjustified and inexcusable. Nevertheless, these actions resulted in crucial social change.

How far you can go in disobeying rules and customs to influence public opinion and change public policy remains a difficult question. Is it better to try to overturn society or to work for progressive change within existing political, economic, and social systems? Is it more important to work for personal perfection or collective improvement? There may be no single answer to these questions: it's good to have people working in many different ways to find solutions.

International nongovernmental organizations mobilize many people

As the opening Case Study in this chapter shows, international **nongovernmental organizations (NGOs)** can be vital in the struggle to protect areas of outstanding biological value. Without this help, local groups could never mobilize the public interest or financial support for projects, such as saving Laguna San Ignacio.

The rise in international NGOs in recent years has been phenomenal. At the Stockholm Conference in 1972, only a handful of environmental groups attended, almost all from fully developed countries. Twenty years later, at the Rio Earth Summit, more than 30,000 individuals representing several thousand environmental groups, many from developing countries, held a global Ecoforum to debate issues and form alliances for a better world.

Some NGOs are located primarily in the more highly developed countries of the north and work mainly on local issues. Others are headquartered in the north but focus their attention on the problems of developing countries in the south. Still others are truly global, with active groups in many different countries. A few are highly professional, combining private individuals with representatives of government agencies on quasi-government boards or standing committees with considerable power. Others are on the fringes of society, sometimes literally voices crying in the wilderness. Many work for political change, more specialize in gathering and disseminating information, and some undertake direct action to protect a specific resource.

Public education and consciousness-raising using protest marches, demonstrations, civil disobedience, and other participatory public actions and media events are generally important tactics for these groups. Greenpeace, for instance, carries out well-publicized confrontations with whalers, seal hunters, toxic waste dumpers, and others who threaten very specific and visible resources. Greenpeace may well be the largest environmental organization in the world, claiming some 2.5 million contributing members.

In contrast to these highly visible groups, others choose to work behind the scenes, but their impact may be equally important. Conservation International has been a leader in debt-for-nature swaps to protect areas particularly rich in biodiversity. It also has some interesting initiatives in economic development, seeking products made by local people that will provide livelihoods along with environmental protection (fig. 25.11).

FIGURE 25.11 International conservation groups often initiate economic development projects that provide local alternative to natural resource destruction.

25.5 CAMPUS GREENING

Colleges and universities can be powerful catalysts for change. Across North America, and around the world, students and faculty are studying sustainability and carrying out practical experiments in sustainable living and ecological restoration.

Organizations for secondary and college students often are among our most active and effective groups for environmental change. The largest student environmental group in North America is the **Student Environmental Action Coalition (SEAC).** Formed in 1988 by students at the University of North Carolina at Chapel Hill, SEAC has grown rapidly to more than 30,000 members in some 500 campus environmental groups. SEAC is both an umbrella organization and grassroots network that functions as an information clearinghouse and a training center for student leaders. Member groups undertake a diverse spectrum of activities ranging from politically neutral recycling promotion to confrontational protests of government or industrial projects. National conferences bring together thousands of activists who share tactics and inspiration while also having fun. If there isn't a group on your campus, why not look into organizing one?

Another important student organizing group is the network of Public Interest Research Groups active on most campuses in the United States. While not focused exclusively on the environment, the PIRGs usually include environmental issues in their priorities for research. By becoming active, you could probably introduce environmental concerns to your local group if they are not already working on problems of importance to you. Remember that you are not alone. Others share your concerns and want to work with you to bring about change; you just have to find them. There is power in working together.

Environmental leadership can be learned

One of the most important skills you are likely to lean in SEAC or other groups committed to social change is how to organize. This is a dynamic process in which you must adapt the realities of your circumstances and the goals of your group, but there are some basic principles that apply to most situations (table 25.3). Using communications media to get your message out is an important part of the modern environmental movement. Table 25.4 suggests some important considerations in planning a media campaign.

It's probably not a surprise to anyone that the Internet is changing our world. You may not have thought, however, about how it can effect an environmental crusade. In 2007, author Bill McKibben worked with a small group of recent college graduates to organize the "Step it Up" campaign to demand national action to combat global climate change. They realized that they didn't have the financial or organizational muscle to mount a conventional campaign, so they turned to the Internet. Reaching other students through blogs and online journals, such as Grist, they built a huge, grassroots environmental protest movement.

With hardly any attention from the conventional press, they organized nearly 1,500 events involving thousands of individuals across the United States. Instead of a massive march on Washington

TABLE 25.3
Organizing an Environmental Campaign

1. What do you want to change? Are your goals realistic, given the time and resources you have available?
2. What and who will be needed to get the job done? What resources do you have now and how can you get more?
3. Who are the stakeholders in this issue? Who are your allies and constituents? How can you make contact with them?
4. How will your group make decisions and set priorities? Will you operate by consensus, majority vote, or informal agreement?
5. Have others already worked on this issue? What successes or failures did they have? Can you learn from their experience?
6. Who has the power to give you what you want or to solve the problem? Which individuals, organizations, corporations, or elected officials should be targeted by your campaign?
7. What tactics will be effective? Using the wrong tactics can alienate people and be worse than taking no action at all.
8. Are there social, cultural, or economic factors that should be recognized in this situation? Will the way you dress, talk, or behave offend or alienate your intended audience? Is it important to change your appearance or tactics to gain support?
9. How will you know when you have succeeded? How will you evaluate the possible outcomes?
10. What will you do when the battle is over? Is yours a single-issue organization, or will you want to maintain the interest, momentum, and network you have established?

Source: Based on material from "Grassroots Organizing for Everyone" by Claire Greensfelder and Mike Roselle from *Call to Action,* 1990 Sierra Book Club Books.

(and the time, expense, and greenhouse gas emissions required to get huge numbers of people to a single location), they encouraged small groups to gather in their local communities to hike, bike, climb, walk, swim, kayak, canoe, or simply sit or stand with banners proclaiming a commitment to action (fig. 25.12). Calling this electronic environmentalism, they showed how it's now possible to link many local systems into a virtual network. Another important example of this is the dramatic change in political campaigns in recent years, ranging from fundraising by web organizations, such as MoveOn.org, or the rapid spread of information through blogs, YouTube, and MySpace.

Schools can be environmental leaders

Colleges and universities can be sources of information and experimentation in sustainable living. They have knowledge and expertise to figure out how to do new things, and they have students who have the energy and enthusiasm to do much of the research, and for whom that discovery will be a valuable learning experience. At

FIGURE 25.12 Protests, marches, and public demonstrations can be an effective way to get your message out and to influence legislators.

more than 100 universities and colleges across America, graduating students have taken a pledge that reads:

> "I pledge to explore and take into account the social and environmental consequences of any job I consider and will try to improve these aspects of any organization for which I work."

Could you introduce something similar at your school?

Campuses often have building projects that can be models for sustainability research and development. More than 110 colleges have built, or are building structures certified by the U.S. Green Building Council. Some recent examples of prize-winning sustainable design can be found at Stanford University, Oberlin College in Ohio, and the University of California at Santa Barbara. Stanford's Jasper Ridge building will provide classroom, laboratory, and office space for its biological research station. Stanford students worked with the administration to develop *Guidelines for Sustainable Buildings,* a booklet that covers everything from energy-efficient lighting to native landscaping. With 275 photovoltaic panels to catch sunlight, there should be no need to buy electricity for the building. In fact, it's expected that surplus energy will be sold back to local utility companies to help pay for building operation.

Oberlin's Environmental Studies Center, designed by architect Bill McDonough, features 370 m^2 of photovoltaic panels on its roof, a geothermal well to help heat and cool the building, large south-facing windows for passive solar gain, and a "living machine" for water treatment, including plant-filled tanks in an indoor solarium and a constructed wetland outside (see figs. 18.27 and 20.10).

UCSB's Bren School of Environmental Science and Management looks deceptively institutional but claims to be the most environmentally state-of-the-art structure of its kind in the United States (fig. 25.13). It wasn't originally intended to be a particularly green building, but planners found that some simple features like having large windows that harvest natural light and open to let ocean breezes cool the interior make the building both more functional and more appealing. Motion detectors control light levels and sensors monitor and refresh the air when there is too much CO_2 putting students to sleep. More than 30 percent of interior materials are recycled. Solar panels supply 10 percent of

many colleges and universities, students have undertaken campus audits to examine water and energy use. waste production and disposal, paper consumption, recycling, buying locally produced food, and many other examples of sustainable resource consumption. At

FIGURE 25.13 The University of California at Santa Barbara claims its new Bren School of Environmental Science and Management is the most environmentally friendly building of its kind in the United States.

the electricity, and the building exceeds federal efficiency standards by 30 percent. "The overriding and very powerful message is it really doesn't cost any more to do these things," says Dennis Aigner, dean of Bren School.

These facilities can become important educational experiences. At Carnegie Mellon University in Pittsburg, students helped design a green roof for Hamershlag Hall. They now monitor how the living roof is reducing storm water drainage and improving water quality. A kiosk inside the dorm shows daily energy use and compares it to long-term averages. Classrooms within the dorm offer environmental science classes in which students can see sustainability in action. Green dorms are popular with students. They appreciate natural lighting, clean air, lack of allergens in building materials, and other features of LEED-certified buildings. One of the largest green dorms in the country is at the University of South Carolina, where more than 100 students are on a waiting list for a room.

A recent study by the Sustainable Endowments Institute evaluated more than 100 of the leading colleges and universities in the United States on their green building policies, food and recycling programs, climate change impacts, and energy consumption. The report card ranked Dartmouth, Harvard, Stanford, and Williams as the top of the "A list" of 23 greenest campuses. Berea College in Kentucky got special commendation as a small school with a strong commitment to sustainability. It's "ecovillage" has a student-designed house that produces its own electricitiy and treats waste water in a living system. The college has a full-time sustainability coordinator to provide support to campus programs, community outreach, and teaching. Some other campuses with academic programs in sustainability include Arizona State in Tempe, and Northern Arizona University in Flagstaff.

Your campus can reduce energy consumption

The Campus Climate Challenge, recently launched by a coalition of nonprofit groups, seeks to engage students, faculty, and staff at 500 college campuses in the United States and Canada in a long-term campaign to eliminate global warming pollution. Many campuses have invested in clean energy, set strict green building standards for new construction, purchased fuel-efficient vehicles,

and adopted other policies to save energy and reduce their greenhouse gas emissions. Some examples include Concordia University in Austin, Texas, the first college or university in the country to purchase all of its energy from renewable sources. The 5.5 million kilowatt-hours of "green power" it uses each year will eliminate about 8 million pounds of CO_2 emissions annually, the equivalent of planting 1,000 acres of trees or taking 700 cars off the roads. Emory University in Atlanta, Georgia, is a leader in green building standards, with 11 buildings that are or could become LEED certified. Emory's Whitehead Biomedical Research Building was the first facility in the Southeast to be LEED certified. Like a number of other colleges, Carleton College in Northfield, Minnesota, has built its own windmill, which is expected to provide about 40 percent of the school's electrical needs. The $1.8 million wind turbine is expected to pay for itself in about ten years. The Campus Climate Challenge website at http://www.energyaction.net contains valuable resources, including strategies and case studies, an energy action packet, a campus organizing guide, and more.

At many schools, students have persuaded the administration to buy locally produced food and to provide organic, vegetarian, and fair trade options in campus cafeterias. This not only benefits your health and the environment, but can also serve as a powerful teaching tool and everyday reminder that individuals can make a difference. Could you do something similar at your school? See the Data Analysis box at the end of this chapter for other suggestions.

25.6 SUSTAINABILITY IS A GLOBAL CHALLENGE

As the developing countries of the world become more affluent, they are adopting many of the wasteful and destructive lifestyle patterns of the richer countries. Automobile production in China, for example, is increasing at about 19 percent per year, or doubling every 3.7 years. By 2030 there could be nearly as many automobiles in China than the United States. What will be the effect on air quality, world fossil fuel supplies, and global climate if that growth rate continues? Already, two-thirds of the children in Shenzhen, China's wealthiest province, suffer from lead poisoning, probably caused by use of leaded gasoline. And, as chapter 8 points out, diseases associated with affluent lifestyles—such as obesity, diabetes, heart attacks, depression, and traffic accidents—are becoming the leading causes of morbidity and mortality worldwide.

We would all benefit by helping developing countries access more efficient, less-polluting technologies. Education, democracy, and access to information are essential for sustainability (fig. 25.14). It is in our best interest to help finance protection of our common future in some equitable way. Maurice Strong, chair of the Earth Charter Council, estimates that development aid from the richer countries should be some $150 billion per year, while internal investments in environmental protection by developing countries will need to be about twice that amount. Many scholars and social activists believe that poverty is at the core of many of the world's most serious human problems: hunger, child deaths, migrations, insurrections, and environmental

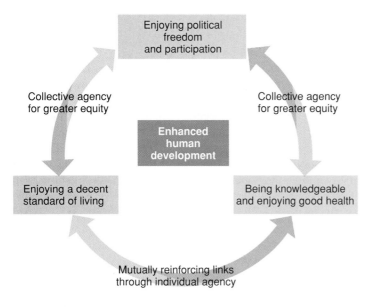

FIGURE 25.14 Human development, democracy, and education are mutually reinforcing.
Source: UN, 2002.

FIGURE 25.15 A model for integrating ecosystem health, human needs, and sustainable economic growth.

degradation. One way to alleviate poverty is to foster economic growth so there can be a bigger share for everyone.

Strong economic growth already is occurring in many places. The World Bank projects that if current trends continue, economic output in developing countries will rise by 4 to 5 percent per year in the next 40 years. Economies of industrialized countries are expected to grow more slowly but could still triple over that period. Altogether, the total world output could be quadruple what it is today.

That growth could provide funds to clean up environmental damage caused by earlier, wasteful technologies and misguided environmental policies. It is estimated to cost S350 billion per year to control population growth, develop renewable energy sources, stop soil erosion, protect ecosystems, and provide a decent standard of living for the world's poor. This is a great deal of money, but it is small compared to over $1 trillion per year spent on wars and military equipment.

While growth simply implies an increase in size, number, or rate of something, development, in economic terms, means a real increase in average welfare or well-being. **Sustainable development** based on the use of renewable resources in harmony with ecological systems is an attractive compromise to the extremes of no growth versus unlimited growth (fig. 25.15). Perhaps the best definition of this goal is that of the World Commission on Environment and Development, which defined sustainable development in *Our Common Future* as "meeting the needs of the present without compromising the ability of future generations to meet their own needs." Some goals of sustainable development include:

- A demographic transition to a stable world population of low birth and death rates.

- An energy transition to high efficiency in production and use, coupled with increasing reliance on renewable resources.

- A resource transition to reliance on nature's "income" without depleting its "capital."

- An economic transition to sustainable development and a broader sharing of its benefits.

- A political transition to global negotiation grounded in complementary interests between North and South, East and West.

- An ethical or spiritual transition to attitudes that do not separate us from nature or each other.

Notice that these goals don't apply just to developing countries. It's equally important that those of us in the richer countries adopt these targets as well. Supporting our current lifestyles is much more resource intensive and has a much greater impact on our environment than the billions of people in poorer countries. Many environmental scientists prefer to simply use the term **sustainability** to describe the search for ways of living more lightly on the earth because it can include residents of both the developed and developing world.

In 2000, United Nations Secretary-General Kofi Annan called for a **millennium assessment** of the consequences of ecosystem change on human well-being as well as the scientific basis for actions to enhance the conservation and sustainable use of those systems. More than 1,360 experts from around the world worked on technical reports about the conditions and trends of ecosystems, scenarios for the future, and possible responses.

TABLE 25.5

Millennium Development Goals

Goals	Specific Objectives
1. Eradicate extreme poverty and hunger.	1a. Reduce by half the proportion of people living on less than a dollar a day. 1b. Reduce by half the proportion of people who suffer from hunger.
2. Achieve universal primary education.	2a. Ensure that all boys and girls complete a full course of primary schooling.
3. Promote gender equality and empower women.	3a. Eliminate gender disparity in primary and secondary education by 2015.
4. Reduce child mortality.	4a. Reduce by two-thirds the mortality rate among children under five.
5. Improve maternal health.	5a. Reduce by three-quarters the maternal mortality ratio.
6. Combat HIV/AIDS, malaria, and other diseases.	6a. Halt and begin to reverse the spread of HIV/AIDS. 6b. Halt and begin to reverse the spread of malaria and other major diseases.
7. Ensure environmental sustainability.	7a. Integrate the principles of sustainable development into policies and programs; reverse the loss of environmental resources. 7b. Reduce by half the proportion of people without sustainable access to safe drinking water. 7c. Achieve significant improvement, in the lives of 100 million slum dwellers by 2020.
8. Develop a global partnership for development.	8a. Develop further an open trading and financial system that is rule-based, predictable, and nondiscriminatory, including a commitment to good governance, development, and poverty reduction. 8b. Address the least-developed countries' special needs. This includes tariff- and quota-free access for their exports; enhanced debt relief for heavily indebted poor countries.

The findings from the millennium assessment serve as a good summary for this book. Among the key conclusions are:

- All of us depend on nature and ecosystem services to provide the conditions for a decent, healthy, and secure life.

- We have made unprecedented changes to ecosystems in recent decades to meet growing demands for food, fresh water, fiber, and energy.

- These changes have helped improve the lives of billions, but at the same time they weakened nature's ability to deliver other key services, such as purification of air and water, protection from disasters, and the provision of medicine.

- Among the outstanding problems we face are the dire state of many of the world's fish stocks, the intense vulnerability of the 2 billion people living in dry regions, and the growing threat to ecosystems from climate change and pollution.

- Human actions have taken the planet to the edge of a massive wave of species extinctions, further threatening our own well-being.

- The loss of services derived from ecosystems is a significant barrier to reducing poverty, hunger, and disease.

- The pressures on ecosystems will increase globally unless human attitudes and actions change.

- Measures to conserve natural resources are more likely to succeed if local communities are given ownership of them, share the benefits, and are involved in decisions.

- Even today's technology and knowledge can reduce considerably the human impact on ecosystems. They are unlikely to be deployed fully, however, until ecosystem services cease to be perceived as free and limitless.

- Better protection of natural assets will require coordinated efforts across all sections of governments, businesses, and international institutions.

As a result of this assessment, the United Nations has developed a set of goals and objectives for sustainable development (table 25.5). From what you've learned in this book, how do you think we could work—individually and collectively—to accomplish these goals?

CONCLUSION

All through this book you've seen evidence of environmental degradation and resource depletion, but there are also many cases in which individuals and organizations are finding ways to stop pollution, use renewable rather than irreplaceable resources, and even restore biodiversity and habitat. Sometimes all it takes is the catalyst of a pilot project to show people how things can be done differently to change attitudes and habits. In this chapter, you've learned some practical approaches to living more lightly on the world individually as well as working collectively to create a better world.

Public attention to issues in the United States seems to run in cycles. Concern builds about some set of problems, and people are willing to take action to find solutions, but then interest wanes and other topics come to the forefront. For the past decade, the American public has consistently said that the environment is very important, and that government should pay more attention to environmental quality. Nevertheless, people haven't shown this concern for the environment to be a very high priority, either in personal behavior or in how they vote.

Recently, however, the whole world seems to have reached a tipping point. Countries, cities, companies, and campuses all are vying to be the most green. This may be a very good time to work on social change and sustainable living. We hope that you'll find the information in this chapter helpful. As the famous anthropologist Margaret Mead said, "Never doubt that a small group of thoughtful, committed people can change the world. Indeed, it is the only thing that ever has."

REVIEWING LEARNING OUTCOMES

By now you should be able to explain the following points:

25.1 Explain how we can make a difference.

25.2 Summarize environmental education.
- Environmental literacy means understanding our environment.
- Citizen science encourages everyone to participate.
- Environmental careers range from engineering to education.
- Green business and technology are growing fast.

25.3 Evaluate what individuals can do.
- How much is enough?
- We can choose to reduce our environmental impact.
- "Green washing" can mislead consumers.
- Certification identifies low-impact products.
- Green consumerism has limits.

25.4 Review how we can work together.
- National organizations are influential but sometimes complacent.
- Radical groups capture attention and broaden the agenda.
- International nongovernmental organizations mobilize many people.

25.5 Investigate campus greening.
- Environmental leadership can be learned.
- Schools can be environmental leaders.
- Your campus can reduce energy consumption.

25.6 Define the challenge of sustainability.

PRACTICE QUIZ

1. Describe four major contexts for outcomes from environmental education.
2. Define *conspicuous consumption*.
3. Describe Lohas and cultural creatives.
4. Give two examples of green washing.
5. List five things that you can do to reduce your environmental impact.
6. List six stages in the Life Cycle Inventory at which we can analyze material and energy balances of products.
7. Identify the ten biggest environmental organizations.
8. List six goals of sustainable development.
9. Identify two key messages from the UN millennium assessment that you believe are most important for environmental science.
10. Identify two goals or objectives from the UN millennium goals that you believe are most important for environmental science.

CRITICAL THINKING AND DISCUSSION QUESTIONS

1. What lessons do you derive from the case study about protecting Laguna San Ignacio. If you were interested in protecting habitat and resources somewhere else in the world, which of the tactics used in this effort might you use for your campaign?

2. Reflect on how you learned about environmental issues. What have been the most important formative experiences or persuasive arguments in shaping your own attitudes. If

you were designing an environmental education program for youth, what elements would you include?

3. How might it change your life if you were to minimize your consumption of materials and resources? Which aspects could you give up, and what is absolutely essential to your happiness and well-being. Does your list differ from that of your friends and classmates?

4. Have you ever been involved in charitable or environmental work? What were the best and worst aspects of that experience? If you haven't yet done anything of this sort, what activities seem appealing and worthwhile to you?

5. What green activities are now occurring at your school? How might you get involved?

6. In the practice quiz, we asked you to identify two key messages from the millennium assessment and two goals and objectives that you believe are most important for environmental science. Why did you choose these messages and goals? How might we accomplish them?

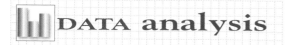 **DATA analysis**   **Campus Environmental Audit**

How sustainable is your school? What could you, your fellow students, the faculty, staff, and administration do to make your campus more environmentally friendly? Perhaps you and your classmates could carry out an environmental audit of your school. Some of the following items are things you could observe for yourself; other information you'd need to get from the campus administrators.

1. *Energy* How much total energy does your campus use each year? Is any of it from renewable sources? How does your school energy use compare to that of a city with the same population? Could you switch to renewable sources? How much would that cost? How long would the payback time be for various renewable sources? Is there a campus policy about energy conservation? What would it take to launch a campaign for using resources efficiently?

2. *Buildings.* Are any campus buildings now LEED certified? Do any campus buildings now have compact fluorescent bulbs, high-efficiency fans, or other energy-saving devices? Do you have single-pane or double-pane windows? Are lights turned off when rooms aren't in use? At what temperatures (winter and summer) are classrooms, offices, and dorms maintained? Who makes this decision? Could you open a window in hot weather? Are new buildings being planned? Will they be LEED certified? If not, why?

3. *Transportation.* Does your school own any fuel-efficient vehicles (hybrids or other high-mileage models)? If you were making a presentation to an administrator to encourage him or her to purchase efficient vehicles, what arguments would you use? How many students commute to campus? Are they encouraged to carpool or use public

transportation? How might you promote efficient transportation? How much total space on your campus is devoted to parking? What's the cost per vehicle to build and maintain parking? How else might that money be spent to facilitate efficient transportation? Where does runoff from parking lots and streets go? What are the environmental impacts of this storm runoff?

4. *Water use.* What's the source of your drinking water? How much does your campus use? Where does wastewater go? How many toilets are on the campus? How much water does each use for every flush? How much would it cost to change to low-flow appliances? How much would it save in terms of water use and cost?

5. *Food.* What's the source of food served in campus dining rooms? Is any of it locally grown or organic? How much junk food is consumed annually? What are the barriers to buying locally grown, fair-trade, organic, free-range food? Does the campus grow any of its own food? Would that be possible?

6. *Ecosystem restoration.* Are there opportunities for reforestation, stream restoration, wetland improvements, or other ecological repair projects on your campus. What percentage of the vegetation on campus is native? What might be the benefits of replacing non-native species with indigenous varieties? Have gardeners considered planting species that provide food and shelter for wildlife?

What other aspects of your campus life could you study to improve sustainability? How could you organize a group project to promote beneficial changes in your school's environmental impacts?

Glossary

A

abiotic Nonliving.

abundance The number or amount of something.

acid precipitation Acidic rain, snow, or dry particles deposited from the air due to increased acids released by anthropogenic or natural resources.

acids Substances that release hydrogen ions (protons) in water.

active learner Someone who understands and remembers best by doing things physically.

active solar system A mechanical system that actively collects, concentrates, and stores solar energy.

acute effects Sudden, severe effects.

acute poverty Insufficient income or access to resources needed to provide the basic necessities for life such as food, shelter, sanitation, clean water, medical care, and education.

adaptation The acquisition of traits that allow a species to survive and thrive in its environment.

adaptive management A management plan designed from the outset to "learn by doing," and to actively test hypotheses and adjust treatments as new information becomes available.

administrative courts Courts that hear enforcement cases for agencies or consider appeals to agency rules.

administrative law Executive orders, administrative rules and regulations, and enforcement decisions by administrative agencies and special administrative courts.

aerobic Living or occurring only in the presence of oxygen.

aerosols Minute particles or liquid droplets suspended in the air.

aesthetic degradation Changes in environmental quality that offend our aesthetic senses.

agroecology Practices, such as low-input farming, community-supported agriculture, and agroecology that use the principles of ecology to show how we can live more harmoniously with the land while still producing the food and fiber we need.

albedo A description of a surface's reflective properties.

allergens Substances that activate the immune system.

allopatric speciation Species that arise from a common ancestor due to geographic barriers that cause reproductive isolation.

ambient air The air immediately around us.

amino acid An organic compound containing an amino group and a carboxyl group; amino acids are the units or building blocks that make peptide and protein molecules.

amorphous silicon collectors Photovoltaic cells that collect solar energy and convert it to electricity using noncrystalline (randomly arranged) thin films of silicon.

anaerobic respiration The incomplete intracellular breakdown of sugar or other organic compounds in the absence of oxygen that releases some energy and produces organic acids and/or alcohol.

analytical thinking Asks, how can I break this problem down into its constituent parts?

anemia Low levels of hemoglobin due to iron deficiency or lack of red blood cells.

annual A plant that lives for a single growing season.

anthropocentric The belief that humans hold a special place in nature; being centered primarily on humans and human affairs.

antigens Chemical compounds to which antibodies bind.

appropriate technology Technology that can be made at an affordable price by ordinary people using local materials to do useful work in ways that do the least possible harm to both human society and the environment.

aquaculture Growing aquatic species in net pens or tanks.

aquifers Porous, water-bearing layers of sand, gravel, and rock below the earth's surface; reservoirs for groundwater.

arbitration A formal process of dispute resolution in which there are stringent rules of evidence, cross-examination of witnesses, and a legally binding decision made by the arbitrator that all parties must obey.

arithmetic scale One that uses ordinary numbers as units in a linear sequence.

artesian well The result of a pressurized aquifer intersecting the surface or being penetrated by a pipe or conduit, from which water gushes without being pumped; also called a spring.

asthma A distressing disease characterized by shortness of breath, wheezing, and bronchial muscle spasms.

atmospheric deposition Sedimentation of solids, liquids, or gaseous materials from the air.

atom The smallest unit of matter that has the characteristics of an element; consists of three main types of subatomic particles: protons, neutrons, and electrons.

atomic number The characteristic number of protons per atom of an element. Used as an identifying attribute.

autotroph An organism that synthesizes food molecules from inorganic molecules by using an external energy source, such as light energy.

B

barrier islands Low, narrow, sandy islands that form offshore from a coastline.

bases Substances that bond readily with hydrogen ions.

BAT *See* best available, economically achievable technology.

Batesian mimicry Evolution by one species to resemble the coloration, body shape, or behavior of another species that is protected from predators by a venomous stinger, bad taste, or some other defensive adaptation.

benthic The bottom of a sea or lake.

best available, economically achievable technology (BAT) The best pollution control available.

best practicable control technology (BPT) The best technology for pollution control available at reasonable cost and operable under normal conditions.

beta particles High-energy electrons released by radioactive decay.

bill A piece of legislation introduced in Congress and intended to become law.

binomials Two part names (genus and species, usually in Latin) invented by Carl Linneaus to show taxonomic relationships.

bioaccumulation The selective absorption and concentration of molecules by cells.

biocentric preservation A philosophy that emphasizes the fundamental right of living organisms to exist and to pursue their own goods.

biocentrism The belief that all creatures have rights and values; being centered on nature rather than humans.

biochemical oxygen demand A standard test of water pollution measured by the amount of dissolved oxygen consumed by aquatic organisms over a given period.

biocide A broad-spectrum poison that kills a wide range of organisms.

biodegradable plastics Plastics that can be decomposed by microorganisms.

biodiversity The genetic, species, and ecological diversity of the organisms in a given area.

biodiversity hot spots Areas with exceptionally high numbers of endemic species.

biofuel Fuels such as ethanol, methanol, or vegetable oils from crops.

biogeochemical cycles Movement of matter within or between ecosystems; caused by living organisms, geological forces, or chemical reactions. The cycling of nitrogen, carbon, sulfur, oxygen, phosphorus, and water are examples.

biogeographical area An entire self-contained natural ecosystem and its associated land, water, air, and wildlife resources.

biological community The populations of plants, animals, and microorganisms living and interacting in a certain area at a given time.

biological controls Use of natural predators, pathogens, or competitors to regulate pest populations.

biological or biotic factors Organisms and products of organisms that are part of the environment and potentially affect the life of other organisms.

biological oxygen demand (BOD) A standard test for measuring the amount of dissolved oxygen utilized by aquatic microorganisms.

biological pests Organisms that reduce the availability, quality, or value of resources useful to humans.

biological resources The earth's organisms.

biomagnification Increase in concentration of certain stable chemicals (for example, heavy metals or fat-soluble pesticides) in successively higher trophic levels of a food chain or web.

biomass The total mass or weight of all the living organisms in a given population or area.

biomass fuel Organic material produced by plants, animals, or microorganisms that can be burned directly as a heat source or converted into a gaseous or liquid fuel.

biomass pyramid A metaphor or diagram that explains the relationship between the amounts of biomass at different trophic levels.

biome A broad, regional type of ecosystem characterized by distinctive climate and soil conditions and a distinctive kind of biological community adapted to those conditions.

bioremediation Use of biological organisms to remove or detoxify pollutants from a contaminated area.

biosphere The zone of air, land, and water at the surface of the earth that is occupied by organisms.

biosphere reserves World heritage sites identified by the IUCN as worthy for national park or wildlife refuge status because of high biological diversity or unique ecological features.

biota All organisms in a given area.

biotic Pertaining to life; environmental factors created by living organisms.

biotic potential The maximum reproductive rate of an organism, given unlimited resources and ideal environmental conditions. Compare with environmental resistance.

birth control Any method used to reduce births, including abstinence, delayed marriage, contraception; devices or medication that prevent implantation of fertilized zygotes, and induced abortions.

black lung disease Inflammation and fibrosis caused by accumulation of coal dust in the lungs or airways. *See* respiratory fibrotic agents.

blind experiments Those in which those carrying out the experiment don't know until after data has been gathered and analyzed which was the experimental treatment and which was the control.

blue revolution New techniques of fish farming that may contribute as much to human nutrition as miracle cereal grains but also may create social and environmental problems.

body burden The sum total of all persistent toxins in our body that we accumulate from our air, water, diet, and surroundings.

bog An area of waterlogged soil that tends to be peaty; fed mainly by precipitation; low productivity; some bogs are acidic.

boom-and-bust cycles Population cycles characterized by repeated overshoot of the carrying capacity of the environment followed by population crashes.

boreal forest A broad band of mixed coniferous and deciduous trees that stretches across northern North America (and also Europe and Asia); its northernmost edge, the taiga, intergrades with the arctic tundra.

BPT *See* best practical control technology.

breeder reactor A nuclear reactor that produces fuel by bombarding isotopes of uranium and thorium with high-energy neutrons that convert inert atoms to fissionable ones.

bronchitis A persistent inflammation of bronchi and bronchioles (large and small airways in the lungs).

brownfield development Building on abandoned or reclaimed polluted industrial sites.

brownfields Abandoned or underused urban areas in which redevelopment is blocked by liability or financing issues related to toxic contamination.

buffalo commons A large open area proposed for the Great Plains in which wildlife and native people could live as they once did without interference by industrialized society.

C

cancer Invasive, out-of-control cell growth that results in malignant tumors.

cap-and-trade An approach to controlling pollution by mandating upper limits (the cap), on how much each country, sector, or specific industry is allowed to emit. Companies that can cut pollution by more than they're required can sell the credit to other companies that have more difficulty meeting their mandated levels.

capital Any form of wealth, resources, or knowledge available for use in the production of more wealth.

captive breeding Raising plants or animals in zoos or other controlled conditions to produce stock for subsequent release into the wild.

carbamates Urethanes such as carbaryl, aldicarb, etc. that are used as pesticides.

carbohydrate An organic compound consisting of a ring or chain of carbon atoms with hydrogen and oxygen attached; examples are sugars, starches, cellulose, and glycogen.

carbon cycle The circulation and reutilization of carbon atoms, especially via the processes of photosynthesis and respiration.

carbon neutral A system or process that doesn't release more carbon to the atmosphere than it consumes.

carbon management Storing CO_2 or using it in ways that prevent its release into the air.

carbon monoxide (CO) Colorless, odorless, nonirritating but highly toxic gas produced by incomplete combustion of fuel, incineration of biomass or solid waste, or partially anaerobic decomposition of organic material.

carbon sequestration Storing carbon (usually in the form of CO_2) in geological formations or at the bottom of the ocean.

carbon sink Places of carbon accumulation, such as in large forests (organic compounds) or ocean sediments (calcium carbonate); carbon is thus removed from the carbon cycle for moderately long to very long periods of time.

carbon source Originating point of carbon that reenters the carbon cycle; cellular respiration and combustion.

carcinogens Substances that cause cancer.

carnivores Organisms that mainly prey upon animals.

carrying capacity The maximum number of individuals of any species that can be supported by a particular ecosystem on a long-term basis.

case law Precedents from both civil and criminal court cases.

cash crops Crops that are sold rather than consumed or bartered.

catastrophic systems Dynamic systems that jump abruptly from one seemingly steady state to another without any intermediate stages.

cell Minute biological compartments within which the processes of life are carried out.

cellular respiration The process in which a cell breaks down sugar or other organic compounds to release energy used for cellular work; may be anaerobic or aerobic, depending on the availability of oxygen.

cellulosic Material composed primarily of cellulose.

chain reaction A self-sustaining reaction in which the fission of nuclei produces subatomic particles that cause the fission of other nuclei.

chaotic systems Systems that exhibit variability, which may not be necessarily random, yet whose complex patterns are not discernible over a normal human time scale.

chaparral Thick, dense, thorny evergreen scrub found in Mediterranean climates.

chemical bond The force that holds atoms together in molecules and compounds.

chemical energy Potential energy stored in chemical bonds of molecules.

chemosynthesis The process in which inorganic chemicals, such as hydrogen sulfide (HS) or hydrogen gas (H^2), serve as an energy source for synthesis of organic molecules.

chlorinated hydrocarbons Hydrocarbon molecules to which chlorine atoms are attached.

chlorofluorocarbons (CFCs) Chemical compounds with a carbon skeleton and one or more attached chlorine and fluorine atoms. Commonly used as refrigerants, solvents, fire retardants, and blowing agents.

chloroplasts Chlorophyll-containing organelles in eukaryotic organisms; sites of photosynthesis.

chronic effects Long-lasting results of exposure to a toxin; can be a permanent change caused by a single, acute exposure or a continuous, low-level exposure.

chronic food shortages Long-term undernutrition and malnutrition; usually caused by people's lack of money to buy food or lack of opportunity to grow it themselves.

chronic obstructive lung disease Irreversible damage to the linings of the lungs caused by irritants.

chronically undernourished Those people whose diet doesn't provide the 2,200 kcal per day, on average, considered necessary for a healthy productive life.

citizen science Projects in which trained volunteers work with scientific researchers to answer real-world questions.

city A differentiated community with a sufficient population and resource base to allow residents to specialize in arts, crafts, services, and professional occupations.

civil law A body of laws regulating relations between individuals or between individuals and corporations concerning property rights, personal dignity and freedom, and personal injury.

classical economics Modern, western economic theories of the effects of resource scarcity, monetary policy, and competition on supply and demand of goods and services in the marketplace. This is the basis for the capitalist market system.

clear-cut Cutting every tree in a given area, regardless of species or size; an appropriate harvest

method for some species; can be destructive if not carefully controlled.

climate A description of the long-term pattern of weather in a particular area.

climax community A relatively stable, long-lasting community reached in a successional series; usually determined by climate and soil type.

closed canopy A forest where tree crowns spread over 20 percent of the ground; has the potential for commercial timber harvests.

cloud forests High mountain forests where temperatures are uniformly cool and fog or mist keeps vegetation wet all the time.

coal gasification The heating and partial combustion of coal to release volatile gases, such as methane and carbon monoxide; after pollutants are washed out, these gases become efficient, clean-burning fuel.

coal-to-liquid (CTL) technology Turning coal into liquid fuel.

coal washing Coal technology that involves crushing coal and washing out soluble sulfur compounds with water or other solvents.

Coastal Zone Management Act Legislation of 1972 that gave federal money to 30 seacoast and Great Lakes states for development and restoration projects.

co-composting Microbial decomposition of organic materials in solid waste into useful soil additives and fertilizer; often, extra organic material in the form of sewer sludge, animal manure, leaves, and grass clippings are added to solid waste to speed the process and make the product more useful.

coevolution The process in which species exert selective pressure on each other and gradually evolve new features or behaviors as a result of those pressures.

cogeneration The simultaneous production of electricity and steam or hot water in the same plant.

cold front A moving boundary of cooler air displacing warmer air.

coliform bacteria Bacteria that live in the intestines (including the colon) of humans and other animals; used as a measure of the presence of feces in water or soil.

commensalism A symbiotic relationship in which one member is benefited and the second is neither harmed nor benefited.

common law The body of court decisions that constitute a working definition of individual rights and responsibilities where no formal statutes define these issues.

communal resource management systems Resources managed by a community for long-term sustainability.

community-supported agriculture (CSA) A program in which you make an annual contribution to a local farm in return for weekly deliveries of a "share" of whatever the farm produces.

competitive exclusion A theory that no two populations of different species will occupy the same niche and compete for exactly the same resources in the same habitat for very long.

complexity (ecological) The number of species at each trophic level and the number of trophic levels in a community.

composting The biological degradation of organic material under aerobic (oxygen-rich) conditions to produce compost, a nutrient-rich soil amendment and conditioner.

compound A molecule made up of two or more kinds of atoms held together by chemical bonds.

concentrated animal feeding operations Facilities in which large numbers of animals spend most or all of their life in confinement.

concept map A two-dimensional representation of the relationship between key ideas. A flow chart or graph of ideas.

conclusion A statement that follows logically from a set of premises.

condensation The aggregation of water molecules from vapor to liquid or solid when the saturation concentration is exceeded.

condensation nuclei Tiny particles that float in the air and facilitate the condensation process.

confidence limits A statistical measure of the quality of data that tells you how close the sample's average probably is to the average for the entire population of that species.

conifers Needle-bearing trees that produce seeds in cones.

conservation development Consideration of landscape history, human culture, topography, and ecological values in subdivision design. Using cluster housing, zoning, covenants, and other design features, at least half of a subdivision can be preserved as open space, farmland, or natural areas.

conservation medicine A medical field that attempts to understand how environmental changes threaten our own health as well as that of the natural communities on which we depend for ecological services.

conservation of matter In any chemical reaction, matter changes form; it is neither created nor destroyed.

conspicuous consumption A term coined by economist and social critic Thorstein Veblen to describe buying things we don't want or need to impress others.

consumer An organism that obtains energy and nutrients by feeding on other organisms or their remains. *See also* heterotroph.

consumption The fraction of withdrawn water that is lost in transmission or that is evaporated, absorbed, chemically transformed, or otherwise made unavailable for other purposes as a result of human use.

contour plowing Plowing along hill contours; reduces erosion.

control rods Neutron-absorbing material inserted into spaces between fuel assemblies in nuclear reactors to regulate fission reaction.

controlled studies Those in which comparisons are made between experimental and control populations that are identical (as far as possible) in every factor except the one variable being studied.

convection currents Rising or sinking air currents that stir the atmosphere and transport heat from one area to another. Convection currents also occur in water; *see* spring overturn.

conventional pollutants The seven major pollutants (sulfur dioxide, carbon monoxide, particulates, hydrocarbons, nitrogen oxides, photochemical oxidants, and lead) identified and regulated by the U.S. Clean Air Act.

cool deserts Deserts such as the American Great Basin characterized by cold winters and sagebrush.

coral bleaching Whitening of corals caused by expulsion of symbiotic algae—often resulting from high water temperatures, pollution, or disease.

coral reefs Prominent oceanic features composed of hard, limy skeletons produced by coral animals; usually formed along edges of shallow, submerged ocean banks or along shelves in warm, shallow, tropical seas.

core The dense, intensely hot mass of molten metal, mostly iron and nickel, thousands of kilometers in diameter at the earth's center.

core habitat Essential habitat for a species.

core region The primary industrial region of a country; usually located around the capital or largest port; has both the greatest population density and the greatest economic activity of the country.

Coriolis effect The influence of friction and drag on air layers near the earth; deflects air currents to the direction of the earth's rotation.

cornucopian fallacy The belief that nature is limitless in its abundance and that perpetual growth is not only possible but essential.

corridor A strip of natural habitat that connects two adjacent nature preserves to allow migration of organisms from one place to another.

cost-benefit analysis (CBA) An evaluation of large-scale public projects by comparing the costs and benefits that accrue from them.

cover crops Plants, such as rye, alfalfa, or clover, that can be planted immediately after harvest to hold and protect the soil.

creative thinking Asks, how could I do this differently?

credit An amount of pollution a company is allowed to sell when they reduce emissions below their allowed cap. *See* cap-and-trade.

criminal law A body of court decisions based on federal and state statutes concerning wrongs against persons or society.

criteria pollutants *See* conventional pollutants.

critical factor The single environmental factor closest to a tolerance limit for a given species at a given time. *See* limiting factors.

critical thinking An ability to evaluate information and opinions in a systematic, purposeful, efficient manner.

croplands Lands used to grow crops.

crude birth rate The number of births in a year divided by the midyear population.

crude death rate The number of deaths per thousand persons in a given year; also called crude mortality rate.

crust The cool, lightweight, outermost layer of the earth's surface that floats on the soft, pliable underlying layers; similar to the "skin" on a bowl of warm pudding.

cultural creatives People who are socially conscious, involve in improving communities and willing to translate values into action.

cultural eutrophication An increase in biological productivity and ecosystem succession caused by human activities.

D

debt-for-nature swap Forgiveness of international debt in exchange for nature protection in developing countries.

deciduous Trees and shrubs that shed their leaves at the end of the growing season.

decline spiral A catastrophic deterioration of a species, community, or whole ecosystem; accelerates as functions are disrupted or lost in a downward cascade.

decomposers Fungi and bacteria that break complex organic material into smaller molecules.

deductive reasoning Deriving testable predictions about specific cases from general principles.

deep ecology A philosophy that calls for a profound shift in our attitudes and behavior toward nature.

deforestation Forest removal.

degradation (of water resource) Deterioration in water quality due to contamination or pollution; makes water unsuitable for other desirable purposes.

Delaney Clause A controversial amendment to the Federal Food, Drug, and Cosmetic Act, added in 1958, prohibiting the addition of any known cancer-causing agent to processed foods, drugs, or cosmetics.

delta Fan-shaped sediment deposit found at the mouth of a river.

demand The amount of a product that consumers are willing and able to buy at various possible prices, assuming they are free to express their preferences.

demanufacturing Disassembly of products so components can be reused or recycled.

demographic bottleneck A population founded when just a few members of a species survive a catastrophic event or colonize new habitat geographically isolated from other members of the same species.

demographic transition A pattern of falling death rates and birthrates in response to improved living conditions; could be reversed in deteriorating conditions.

demography Vital statistics about people: births, marriages, deaths, etc.; the statistical study of human populations relating to growth rate, age structure, geographic distribution, etc., and their effects on social, economic, and environmental conditions.

denitrifying bacteria Free-living soil bacteria that converts nitrates to gaseous nitrogen and nitrous oxide.

density-dependent Factors affecting population growth that change as population size changes.

dependency ratio The number of nonworking members compared to working members for a given population.

dependent (response) variable A variable that is affected by the condition being altered in a manipulative experiment.

desalinization (or desalination) Removal of salt from water by distillation, freezing, or ultrafiltration.

desert A type of biome characterized by low moisture levels and infrequent and unpredictable precipitation. Daily and seasonal temperatures fluctuate widely.

desertification Conversion of productive lands to desert.

detritivore Organisms that consume organic litter, debris, and dung.

dew point The temperature at which condensation occurs for a given concentration of water vapor in the air.

dieback A sudden population decline; also called a population crash.

diminishing returns A condition in which unrestrained population growth causes the standard of living to decrease to a subsistence level where poverty, misery, vice, and starvation makes life permanently drab and miserable. This dreary prophecy has led economics to be called "the dismal science."

disability-adjusted life years (DALY) A measure of premature deaths and losses due to illnesses and disabilities in a population.

discharge The amount of water that passes a fixed point in a given amount of time; usually expressed as liters or cubic feet of water per second.

disclimax community *See* equilibrium community.

discount rates The difference between present value and future value of a resource. Generally equivalent to an interest rate.

disease A deleterious change in the body's condition in response to destabilizing factors, such as nutrition, chemicals, or biological agents.

dissolved oxygen (DO) content Amount of oxygen dissolved in a given volume of water at a given temperature and atmospheric pressure; usually expressed in parts per million (ppm).

disturbance-adapted species Species that depend on disturbances to succeed.

disturbances Periodic, destructive events such as fire or floods; changes in an ecosystem that affect (positively or negatively) the organisms living there.

diversity (species diversity, biological diversity) The number of species present in a community (species richness), as well as the relative abundance of each species.

DNA (deoxyribonucleic acid) A giant molecule composed of millions or billions of nucleotides (sugars and bases called purines and pyramidines held together by phosphate bonds) that form a double helix and store genetic information in all living cells.

dominant plants Those plant species in a community that provide a food base for most of the community; they usually take up the most space and have the largest biomass.

double-blind design One in which neither the experimenter nor the subjects know until after data has been gathered and analyzed which was the experimental treatment and which was the control.

downbursts Sudden, very strong, downdrafts of cold air associated with an advancing storm front.

drip irrigation Uses pipe or tubing perforated with very small holes to deliver water one drop at a time directly to the soil around each plant.

dry alkali injection Spraying dry sodium bicarbonate into flue gas to absorb and neutralize acidic sulfur compounds.

E

Earth Charter A set of principles for sustainable development, environmental protection, and social justice developed by a council appointed by the United Nations.

earthquakes Sudden, violent movement of the earth's crust.

ecocentric (ecologically centered) A philosophy that claims moral values and rights for both organisms and ecological systems and processes.

ecofeminism A pluralistic, nonhierarchical, relationship-oriented philosophy that suggests how humans could reconceive themselves and their relationships to nature in nondominating ways as an alternative to patriarchal systems of domination.

ecojustice Justice in the social order and integrity in the natural order.

ecological development A gradual process of environmental modification by organisms.

ecological diseases Emergent diseases (new or rarely seen diseases) that cause devastating epidemics among wildlife and domestic animals.

ecological economics A relatively new field that brings the insights of ecology to economic analysis.

ecological equivalents Different species that occupy similar ecological niches in similar ecosystems in different parts of the world.

ecological footprint A measure that computes the demands placed on nature by individuals and nations.

ecological niche The functional role and position of a species (population) within a community or ecosystem, including what resources it uses, how and when it uses the resources, and how it interacts with other populations.

ecological succession The process by which organisms occupy a site and gradually change environmental conditions so that other species can replace the original inhabitants.

ecology The scientific study of relationships between organisms and their environment. It is concerned with the life histories, distribution, and behavior of individual species as well as the structure and function of natural systems at the level of populations, communities, and ecosystems.

economic development A rise in real income *per person;* usually associated with new technology that increases productivity or resources.

economic growth An increase in the total wealth of a nation; if population grows faster than the economy, there may be real economic growth, but the share per person may decline.

economic thresholds In pest management, the point at which the cost of pest damage exceeds the costs of pest control.

ecosystem A specific biological community and its physical environment interacting in an exchange of matter and energy.

ecosystem management An integration of ecological, economic, and social goals in a unified systems approach to resource management.

ecosystem restoration To reinstate an entire community of organisms to as near its natural condition as possible.

ecotone A boundary between two types of ecological communities.

ecotourism A combination of adventure travel, cultural exploration, and nature appreciation in wild settings.

edge effects A change in species composition, physical conditions, or other ecological factors at the boundary between two ecosystems.

effluent sewerage A low-cost alternative sewage treatment for cities in poor countries that combines some features of septic systems and centralized municipal treatment systems.

electron A negatively charged subatomic particle that orbits around the nucleus of an atom.

electronic waste *See* e-waste.

electrostatic precipitators The most common particulate controls in power plants; fly ash particles pick up an electrostatic surface charge as they pass between large electrodes in the effluent stream, causing particles to migrate to the oppositely charged plate.

element A molecule composed of one kind of atom; cannot be broken into simpler units by chemical reactions.

El Niño A climatic change marked by shifting of a large warm water pool from the western Pacific Ocean towards the east. Wind direction and precipitation patterns are changed over much of the Pacific and perhaps around the world.

emergent diseases A new disease or one that has been absent for at least 20 years.

emergent properties Characteristics of whole, functioning systems that are quantitatively or qualitatively greater than the sum of the systems' parts.

emigration The movement of members from a population.

emission standards Regulations for restricting the amounts of air pollutants that can be released from specific point sources.

emissions trading Programs in which companies that have cut pollution by more than they're required to can sell "credits" to other companies that still exceed allowed levels.

endangered species A species considered to be in imminent danger of extinction.

endemism A state in which species are restricted to a single region.

endocrine disrupters Chemicals that disrupt normal hormone functions.

energy The capacity to do work (that is, to change the physical state or motion of an object).

energy crops Crops that can be used to make ethanol or diesel fuel.

energy efficiency A measure of energy produced compared to energy consumed.

energy pyramid A representation of the loss of useful energy at each step in a food chain.

energy recovery Incineration of solid waste to produce useful energy.

entropy Disorder in a system.

environment The circumstances or conditions that surround an organism or group of organisms as well as the complex of social or cultural conditions that affect an individual or community.

environmental economics *See* ecological economics.

environmental ethics A search for moral values and ethical principles in human relations with the natural world.

environmental governance Rules and regulations that govern our impacts on the environment and natural resources.

environmental health The science of external factors that cause disease, including elements of the natural, social, cultural and technological worlds in which we live.

environmental hormones Chemical pollutants that substitute for, or interfere with, naturally occurring hormones in our bodies; these chemicals may trigger reproductive failure, developmental abnormalities, or tumor promotion.

environmental impact statement (EIS) An analysis, required by provisions in the National Environmental Policy Act of 1970, of the effects of any major program a federal agency plans to undertake.

environmental indicators Organisms or physical factors that serve as a gauge for environmental changes. More specifically, organisms with these characteristics are called bioindicators.

Environmental Performance Index (EPI) A measure that evaluates national sustainability and progress toward achievement of the United Nations Millennium Development Goals.

environmentalism Active participation in attempts to solve environmental pollution and resource problems.

environmental justice A recognition that access to a clean, healthy environment is a fundamental right of all human beings.

environmental law The special body of official rules, decisions, and actions concerning environmental quality, natural resources, and ecological sustainability.

environmental literacy Fluency in the principles of ecology that gives us a working knowledge of the basic grammar and underlying syntax of environmental wisdom.

environmental policy The official rules or regulations concerning the environment adopted, implemented, and enforced by some governmental agency.

environmental racism Decisions that restrict certain people or groups of people to polluted or degraded environments on the basis of race.

environmental resistance All the limiting factors that tend to reduce population growth rates and set the maximum allowable population size or carrying capacity of an ecosystem.

environmental resources Anything an organism needs that can be taken from the environment.

environmental science The systematic, scientific study of our environment as well as our role in it.

enzymes Molecules, usually proteins or nucleic acids, that act as catalysts in biochemical reactions.

epidemiology The study of the distribution and causes of disease and injuries in human populations.

epiphyte A plant that grows on a substrate other than the soil, such as the surface of another organism.

equilibrium community Also called a **disclimax community;** a community subject to periodic disruptions, usually by fire, that prevent it from reaching a climax stage.

estuary A bay or drowned valley where a river empties into the sea.

ethics A branch of philosophy concerned with right and wrong.

eukaryotic cell A cell containing a membrane-bounded nucleus and membrane-bounded organelles.

eutrophic Rivers and lakes rich in organisms and organic material (*eu* = truly; *trophic* = nutritious).

evaporation The process in which a liquid is changed to vapor (gas phase).

evolution A theory that explains how random changes in genetic material and competition for scarce resources cause species to change gradually.

e-waste Discarded electronic equipment such as computers, cell phones, television sets, etc.

exhaustible resources Generally considered the earth's geologic endowment: minerals, nonmineral resources, fossil fuels, and other materials present in fixed amounts in the environment.

existence value The importance we place on just knowing that a particular species or a specific organism exists.

exotic organisms Alien species introduced by human agency into biological communities where they would not naturally occur.

exponential growth Growth at a constant rate of increase per unit of time; can be expressed as a constant fraction or exponent. *See* geometric growth.

external costs Expenses, monetary or otherwise, borne by someone other than the individuals or groups who use a resource.

extinction The irrevocable elimination of species; can be a normal process of the natural world as species out-compete or kill off others or as environmental conditions change.

extirpate To destroy totally; extinction caused by direct human action, such as hunting, trapping, etc.

extreme poverty Living on less than $1 (U.S.) per day.

F

family planning Controlling reproduction; planning the timing of birth and having as many babies as are wanted and can be supported.

famines Acute food shortages characterized by large-scale loss of life, social disruption, and economic chaos.

fauna All of the animals present in a given region.

fecundity The physical ability to reproduce.

fen An area of waterlogged soil that tends to be peaty; fed mainly by upwelling water; low productivity.

feral A domestic animal that has taken up a wild existence.

fermentation (alcoholic) A type of anaerobic respiration that yields carbon dioxide and alcohol.

fertility Measurement of actual number of offspring produced through sexual reproduction; usually described in terms of number of offspring of females, since paternity can be difficult to determine.

fetal alcohol syndrome A tragic set of permanent physical and mental and behavioral birth defects that result when mothers drink alcohol during pregnancy.

fibrosis The general name for accumulation of scar tissue in the lung.

filters A porous mesh of cotton cloth, spun glass fibers, or asbestos-cellulose that allows air or liquid to pass through but holds back solid particles.

fire-climax community An equilibrium community maintained by periodic fires; examples include grasslands, chaparral shrubland, and some pine forests.

first law of thermodynamics States that energy is *conserved;* that is, it is neither created nor destroyed under normal conditions.

flex-fuel vehicles Vehicles that can burn variable mixtures of gasoline and ethanol.

floodplains Low lands along riverbanks, lakes, and coastlines subjected to periodic inundation.

flora All of the plants present in a given region.

flue-gas scrubbing Treating combustion exhaust gases with chemical agents to remove pollutants. Spraying crushed limestone and water into the exhaust gas stream to remove sulfur is a common scrubbing technique.

food aid Financial assistance intended to boost less-developed countries' standards of living.

food chain A linked feeding series; in an ecosystem, the sequence of organisms through which energy and materials are transferred, in the form of food, from one trophic level to another.

food security The ability of individuals to obtain sufficient food on a day-to-day basis.

food surpluses Excess food supplies.

food web A complex, interlocking series of individual food chains in an ecosystem.

forest management Scientific planning and administration of forest resources for sustainable harvest, multiple use, regeneration, and maintenance of a healthy biological community.

fossil fuels Petroleum, natural gas, and coal created by geological forces from organic wastes and dead bodies of formerly living biological organisms.

founder effect The effect on a population founded when just a few members of a species survive a catastrophic event or colonize new habitat geographically isolated from other members of the same species.

freezing condensation A process that occurs in the clouds when ice crystals trap water vapor. As the ice crystals become larger and heavier, they begin to fall as rain or snow.

fresh water Water other than seawater; covers only about 2 percent of earth's surface, including streams, rivers, lakes, ponds, and water associated with several kinds of wetlands.

freshwater ecosystems Ecosystems in which the fresh (nonsalty) water of streams, rivers, ponds, or lakes plays a defining role.

front The boundary between two air masses of different temperature and density.

fuel assembly A bundle of hollow metal rods containing uranium oxide pellets; used to fuel a nuclear reactor.

fuel cells Mechanical devices that use hydrogen or hydrogen-containing fuel such as methane to produce an electric current. Fuel cells are clean, quiet, and highly efficient sources of electricity.

fuel-switching A change from one fuel to another.

fuelwood Branches, twigs, logs, wood chips, and other wood products harvested for use as fuel.

fugitive emissions Substances that enter the air without going through a smokestack, such as dust from soil erosion, strip mining, rock crushing, construction, and building demolition.

fumigants Toxic gases such as methyl bromine that are used to kill pests.

fungi One of the five kingdom classifications; consists of nonphotosynthetic, eukaryotic organisms with cell walls, filamentous bodies, and absorptive nutrition.

fungicide A chemical that kills fungi.

G

Gaia hypothesis A theory that the living organisms of the biosphere form a single, complex interacting system that creates and maintains a habitable Earth; named after Gaia, the Greek "Earth mother" goddess.

gamma rays Very short wavelength forms of the electromagnetic spectrum.

gap analysis A biogeographical technique of mapping biological diversity and endemic species to find gaps between protected areas that leave endangered habitats vulnerable to disruption.

garden city A new town with special emphasis on landscaping and rural ambience.

gasohol A mixture of gasoline and ethanol.

gene A unit of heredity; a segment of DNA nucleus of the cell that contains information for the synthesis of a specific protein, such as an enzyme.

gene banks Storage for seed varieties for future breeding experiments.

general fertility rate Crude birthrate multiplied by the percentage of reproductive age women.

genetic assimilation The disappearance of a species as its genes are diluted through crossbreeding with a closely related species.

genetic drift The gradual changes in gene frequencies in a population due to random events.

genetic engineering Laboratory manipulation of genetic material using molecular biology techniques to create desired characteristics in organisms.

genetically modified organisms (GMOs) Organisms whose genetic code has been altered by artificial means such as interspecies gene transfer.

genuine progress index (GPI) An alternative to GNP or GDP for economic accounting that measures real progress in quality of life and sustainability.

geographic information systems (GIS) Spatial data, such as boundaries or road networks and computer software to display and analyze those data.

geographic isolation *See* allopatric speciation.

geometric growth Growth that follows a geometric pattern of increase, such as 2, 4, 8, 16, etc. *See* exponential growth.

geothermal energy Energy drawn from the internal heat of the earth, either through geysers, fumaroles, hot springs, or other natural geothermal features, or through deep wells that pump heated groundwater.

germ plasm Genetic material that may be preserved for future agricultural, commercial, and ecological values (plant seeds or parts or animal eggs, sperm, and embryos).

global environmentalism A concern for, and action to help solve, global environmental problems.

globalization The revolution in communications, transportation, finances and commerce that has brought about increasing inter-dependence of national economies.

grasslands A biome dominated by grasses and associated herbaceous plants.

greenfield development Housing projects built on previously undeveloped farmlands or forests on the outskirts of large cities.

greenhouse effect Gases in the atmosphere are transparent to visible light but absorb infrared (heat) waves that are reradiated from the earth's surface.

green plans Integrated national environmental plans for reducing pollution and resource consumption while achieving sustainable development and environmental restoration.

green political parties Political organizations based on environmental protection, participatory democracy, grassroots organization, and sustainable development.

green pricing Setting prices to encourage conservation or renewable energy. Plans that invite customers to pay a premium for energy from renewable sources.

green revolution Dramatically increased agricultural production brought about by "miracle" strains of grain; usually requires high inputs of water, plant nutrients, and pesticides.

gross domestic product (GDP) The total economic activity within national boundaries.

gross national product (GNP) The sum total of all goods and services produced in a national economy. Gross domestic product (GDP) is used to distinguish economic activity within a country from that of off-shore corporations.

groundwater Water held in gravel deposits or porous rock below the earth's surface; does not include water or crystallization held by chemical bonds in rocks or moisture in upper soil layers.

gully erosion Removal of layers of soil, creating channels or ravines too large to be removed by normal tillage operations.

H

habitat The place or set of environmental conditions in which a particular organism lives.

habitat conservation plans Agreements under which property owners are allowed to harvest resources or develop land as long as habitat is conserved or replaced in ways that benefit resident endangered or threatened species in the long run. Some incidental "taking" or loss of endangered species is generally allowed in such plans.

Hadley cells Circulation patterns of atmospheric convection currents as they sink and rise in several intermediate bands.

hazardous Describes chemicals that are dangerous, including flammables, explosives, irritants, sensitizers, acids, and caustics; may be relatively harmless in diluted concentrations.

hazardous air pollutants (HAPs) Especially dangerous air pollutants including carcinogens, neurotoxins, mutagens, teratogens, endocrine system disrupters and other highly toxic compounds.

hazardous waste Any discarded material containing substances known to be toxic, mutagenic, carcinogenic, or teratogenic to humans or other life-forms; ignitable, corrosive, explosive, or highly reactive alone or with other materials.

health A state of physical and emotional well-being; the absence of disease or ailment.

heap-leach extraction A technique for separating gold from extremely low-grade ores. Crushed ore is piled in huge heaps and sprayed with a dilute alkaline-cyanide solution, which percolates through the pile to extract the gold, which is separated from the effluent in a processing plant. This process has a high potential for water pollution.

heat A form of energy transferred from one body to another because of a difference in temperatures.

heat capacity The amount of heat energy that must be added or subtracted to change the temperature of a body; water has a high heat capacity.

heat of vaporization The amount of heat energy required to convert water from a liquid to a gas.

herbicide A chemical that kills plants.

herbivore An organism that eats only plants.

heterotroph An organism that is incapable of synthesizing its own food and, therefore, must feed upon organic compounds produced by other organisms.

high-level waste repository A place where intensely radioactive wastes can be buried and remain unexposed to groundwater and earthquakes for tens of thousands of years.

high-quality energy Intense, concentrated, and high-temperature energy that is considered high-quality because of its usefulness in carrying out work.

HIPPO Habitat destruction, Invasive species, Pollution, Population (human), and Overharvesting, the leading causes of extinction.

holistic science The study of entire, integrated systems rather than isolated parts. Often takes a descriptive or interpretive approach.

homeostasis Maintaining a dynamic, steady state in a living system through opposing, compensating adjustments.

Homestead Act Legislation passed in 1862 allowing any citizen or applicant for citizenship over 21 years old and head of a family to acquire 160 acres of public land by living on it and cultivating it for five years.

host organism An organism that provides lodging for a parasite.

hot desert Deserts of the American Southwest and Mexico; characterized by extreme summer heat and cacti.

human development index (HDI) A measure of quality of life using life expectancy, child survival, adult literacy, childhood education, gender equity and access to clan water and sanitation as well as income.

human ecology The study of the interactions of humans with the environment.

human resources Human wisdom, experience, skill, labor, and enterprise.

humus Sticky, brown, insoluble residue from the bodies of dead plants and animals; gives soil its structure, coating mineral particles and holding them together; serves as a major source of plant nutrients.

hurricanes Large cyclonic oceanic storms with heavy rain and winds exceeding 119 km/hr (74 mph).

hybrid gasoline-electric engines A small gasoline engine generates electricity that is stored in batteries and powers electric motors that drive vehicle wheels.

hybrid gasoline-electric vehicles Automobiles that run on electric power and a small gasoline or diesel engine.

hydrologic cycle The natural process by which water is purified and made fresh through evaporation and precipitation. This cycle provides all the freshwater available for biological life.

hypothesis A provisional explanation that can be tested scientifically.

I

igneous rocks Crystalline minerals solidified from molten magma from deep in the earth's interior; basalt, rhyolite, andesite, lava, and granite are examples.

inbreeding depression In a small population, an accumulation of harmful genetic traits (through random mutations and natural selection) that lowers viability and reproductive success of enough individuals to affect the whole population.

independent variable A factor not affected by the condition being altered in a manipulative experiment.

indicator species Those whose critical tolerance limits can be used to judge environmental conditions.

inductive reasoning Inferring general principles from specific examples.

industrial revolution Advances in science and technology that have given us power to understand and change our world.

industrial timber Trees used for lumber, plywood, veneer, particleboard, chipboard, and paper; also called roundwood.

inertial confinement A nuclear fusion process in which a small pellet of nuclear fuel is bombarded with extremely high-intensity laser light.

infiltration The process of water percolation into the soil and pores and hollows of permeable rocks.

informal economy Small-scale family businesses in temporary locations outside the control of normal regulatory agencies.

inherent value Ethical values or rights that exist as an intrinsic or essential characteristic of a particular thing or class of things simply by the fact of their existence.

inholdings Private lands within public parks, forests, or wildlife refuges.

inorganic pesticides Inorganic chemicals such as metals, acids, or bases used as pesticides.

insecticide A chemical that kills insects.

insolation Incoming solar radiation.

instrumental value Value or worth of objects that satisfy the needs and wants of moral agents. Objects that can be used as a means to some desirable end.

intangible resources Factors such as open space, beauty, serenity, wisdom, diversity, and satisfaction that cannot be grasped or contained. Ironically, these resources can be both infinite and exhaustible.

integrated gasification combined cycle (IGCC) A power plant that heats fuel (usually coal, but could be biomass or other sources) to high temperatures and pressures in the presence of 96 percent oxygen. Hydrogen is separated from hydrocarbons and separated from CO_2 and other contaminants. The hydrogen is burned in a gas turbine and surplus heat drives a steam turbine, both of which generate electricity.

integrated pest management (IPM) An ecologically based pest-control strategy that relies on natural mortality factors, such as natural enemies, weather, cultural control methods, and carefully applied doses of pesticides.

Intergovernmental Panel on Climate Change (IPCC) An international organization formed to assess global climate change and its impacts. The IPCC is concerned with social, economic, and environmental impacts of climate change, and it was established by the United Nations Environment Program and the World Meteorological Organization.

internal costs The expenses, monetary or otherwise, borne by those who use a resource.

interplanting The system of planting two or more crops, either mixed together or in alternating rows, in the same field; protects the soil and makes more efficient use of the land.

interpretive science Explanation based on observation and description of entire objects or systems rather than isolated parts.

interspecific competition In a community, competition for resources between members of *different* species.

intraspecific competition In a community, competition for resources among members of the *same* species.

invasive species Organisms that thrive in new territory where they are free of predators, diseases or resource limitations that may have controlled their population in their native habitat.

ionizing radiation High-energy electromagnetic radiation or energetic subatomic particles released by nuclear decay.

ionosphere The lower part of the thermosphere.

ions Electrically charged atoms that have gained or lost electrons.

I-PAT formula A formula that says our environmental impacts (I) are the product of our population size (P) times affluence (A) and the technology (T) used to produce the goods and services we consume.

island biogeography The study of rates of colonization and extinction of species on islands or other isolated areas based on size, shape, and distance from other inhabited regions.

isotopes Forms of a single element that differ in atomic mass due to a different number of neutrons in the nucleus.

J

J curve A growth curve that depicts exponential growth; called a J curve because of its shape.

jet streams Powerful winds or currents of air that circulate in shifting flows; similar to oceanic currents in extent and effect on climate.

joule A unit of energy. One joule is the energy expended in 1 second by a current of 1 amp flowing through a resistance of 1 ohm.

K

k-selected species Species that reproduce more slowly, occupy higher trophic levels, have fewer offspring, longer life-spans, and greater intrinsic control of population growth than *r*-selected species.

keystone species A species whose impacts on its community or ecosystem are much larger and more influential than would be expected from mere abundance.

kinetic energy Energy contained in moving objects such as a rock rolling down a hill, the wind blowing through the trees, or water flowing over a dam.

known resources Those that have been located but are not completely mapped but, nevertheless, are likely to become economical in the foreseeable future.

kwashiorkor A widespread human protein deficiency disease resulting from a starchy diet low in protein and essential amino acids.

Kyoto Protocol An international agreement to reduce greenhouse gas emissions.

L

La Niña The part of a large-scale oscillation in the Pacific (and, perhaps, other oceans) in which trade winds hold warm surface waters in the western part of the basin and cause upwelling of cold, nutrient-rich, deep water in the eastern part of the ocean.

landfills Land disposal sites for solid waste; operators compact refuse and cover it with a layer of dirt to minimize rodent and insect infestation, wind-blown debris, and leaching by rain.

land reform Democratic redistribution of landownership to recognize the rights of those who actually work the land to a fair share of the products of their labor.

landscape ecology The study of the reciprocal effects of spatial pattern on ecological processes. A study of the ways in which landscape history shapes the features of the land and the organisms that inhabit it as well as our reaction to, and interpretation of, the land.

landslide The sudden fall of rock and earth from a hill or cliff. Often triggered by an earthquake or heavy rain.

latent heat Stored energy in a form that is not sensible (detectable by ordinary senses).

LD50 A chemical dose lethal to 50 percent of a test population.

less-developed countries (LDC) Nonindustrialized nations characterized by low per capita income, high birthrates and death rates, high population growth rates, and low levels of technological development.

life-cycle analysis Evaluation of material and energy inputs and outputs at each stage of manufacture, use, and disposal of a product.

life expectancy The average age that a newborn infant can expect to attain in a particular time and place.

life span The longest period of life reached by a type of organism.

limiting factors Chemical or physical factors that limit the existence, growth, abundance, or distribution of an organism.

lipid A nonpolar organic compound that is insoluble in water but soluble in solvents, such as alcohol and ether; includes fats, oils, steroids, phospholipids, and carotenoids.

liquid metal fast breeder A nuclear power plant that converts uranium 238 to plutonium 239; thus, it creates more nuclear fuel than it consumes. Because of the extreme heat and density of its core, the breeder uses liquid sodium as a coolant.

lobbying Using personal contacts, public pressure, or political action to persuade legislators to vote in a particular manner.

locavore Someone who eats locally grown, seasonal food.

logarithmic scale One that uses logarithms as units in a sequence that progresses by a factor of 10. That is, each subsequent increment on the scale is 10 times the one that precedes it.

logical learner Someone who understands and remembers best by thinking through a topic and finding logical reasons for statements.

logical thinking Asks, can the rules of logic help understand this?

logistic growth Growth rates regulated by internal and external factors that establish an equilibrium with environmental resources.

longevity The length or duration of life; compare to survivorship.

low-head hydropower Small-scale hydro technology that can extract energy from small headwater dams; causes much less ecological damage.

low-input high-diversity biofuels Mixed polycultures of perennial native species that don't require minimal amounts of cultivation, fertilizer, irrigation, or pesticides when grown as energy crops.

low-quality energy Diffuse, dispersed energy at a low temperature that is difficult to gather and use for productive purposes.

LULUs *Locally Unwanted Land Uses* such as toxic waste dumps, incinerators, smelters, airports, freeways, and other sources of environmental, economic, or social degradation.

M

magma Molten rock from deep in the earth's interior; called lava when it spews from volcanic vents.

magnetic confinement A technique for enclosing a nuclear fusion reaction in a powerful magnetic field inside a vacuum chamber.

malignant tumor A mass of cancerous cells that have left their site of origin, migrated through the body, invaded normal tissues, and are growing out of control.

malnourishment A nutritional imbalance caused by lack of specific dietary components or inability to absorb or utilize essential nutrients.

Man and Biosphere (MAB) program A design for nature preserves that divides protected areas into zones with different purposes. A highly protected core is surrounded by a buffer zone and peripheral regions in which multiple-use resource harvesting is permitted.

mangroves Trees from a number of genera that live in salt water.

manipulative experiment One in which some conditions are deliberately altered while others are held constant to study cause-and-effect relationships.

mantle A hot, pliable layer of rock that surrounds the earth's core and underlies the cool, outer crust.

marasmus A widespread human protein deficiency disease caused by a diet low in calories and protein or imbalanced in essential amino acids.

marginal costs and benefits The costs and benefits of producing one additional unit of a good or service.

marine Living in or pertaining to the sea.

market equilibrium The dynamic balance between supply and demand under a given set of conditions in a "free" market (one with no monopolies or government interventions).

market forces Depending on capitalist market systems to achieve national goals.

marsh Wetland without trees; in North America, this type of land is characterized by cattails and rushes.

mass burn Incineration of unsorted solid waste.

mass wasting Mass movement of geologic materials downhill caused by rockslides, avalanches, or simple slumping.

matter Anything that takes up space and has mass.

mean An average.

mediation An informal dispute resolution process in which parties are encouraged to discuss issues openly but in which all decisions are reached by consensus and any participant can withdraw at any time.

Mediterranean climate areas Specialized landscapes with warm, dry summers; cool, wet winters; many unique plant and animal adaptations; and many levels of endemism.

megacity *See* megalopolis.

megalopolis Also known as a megacity or supercity; megalopolis indicates an urban area with more than 10 million inhabitants.

megawatt (MW) Unit of electrical power equal to 1,000 kilowatts or 1 million watts.

mesosphere The atmospheric layer above the stratosphere and below the thermosphere; the middle layer; temperatures are usually very low.

metabolism All the energy and matter exchanges that occur within a living cell or organism; collectively, the life processes.

metamorphic rock Igneous and sedimentary rocks modified by heat, pressure, and chemical reactions.

metapopulation A collection of populations that have regular or intermittent gene flow between geographically separate units.

methane hydrate Small bubbles or individual molecules of methane (natural gas) trapped in a crystalline matrix of frozen water.

micorrhizal symbiosis An association between the roots of most plant species and certain fungi. The plant provides organic compounds to the fungus, while the fungus provides water and nutrients to the plant.

microbial agents Or biological controls, are beneficial microbes (bacteria, fungi) that can be used to suppress or control pests.

micro-hydro generators Small power generators that can be used in low-level rivers to provide economical power for four to six homes, freeing them from dependence on large utilities and foreign energy supplies.

mid-ocean ridges Mountain ranges on the ocean floor created where molten magma is forced up through cracks in the planet's crust.

Milankovitch cycles Periodic variations in tilt, eccentricity, and wobble in the earth's orbit; Milutin Milankovitch suggested that it is responsible for cyclic weather changes.

millennium assessment A set of ambitious environmental and human development goals established by the United Nations in 2000.

milpa agriculture An ancient farming system in which small patches of tropical forests are cleared and perennial polyculture agriculture practiced and is then followed by many years of fallow to restore the soil; also called **swidden agriculture.**

mineral A naturally occurring, inorganic, crystalline solid with definite chemical composition and characteristic physical properties.

minimum viable population size The number of individuals needed for long-term survival of rare and endangered species.

mitigation Repairing or rehabilitating a damaged ecosystem or compensating for damage by providing a substitute or replacement area.

mixed perennial polyculture Growing a mixture of different perennial crop species (where the same plant persists for more than one year) together in the same plot.

models Simple representations of more complex systems.

molecule A combination of two or more atoms.

monitored, retrievable storage Holding wastes in underground mines or secure surface facilities such as dry casks where they can be watched and repackaged, if necessary.

monoculture agroforestry Intensive planting of a single species; an efficient wood production approach, but one that encourages pests and disease infestations and conflicts with wildlife habitat or recreation uses.

monsoon A seasonal reversal of wind patterns caused by the different heating and cooling rates of the oceans and continents.

montane coniferous forests Coniferous forests of the mountains consisting of belts of different forest communities along an altitudinal gradient.

moral agents Beings capable of making distinctions between right and wrong and acting accordingly. Those whom we hold responsible for their actions.

moral extensionism Expansion of our understanding of inherent value or rights to persons, organisms, or things that might not be considered worthy of value or rights under some ethical philosophies.

moral subjects Beings that are not capable of distinguishing between right or wrong or that are not able to act on moral principles and yet are capable of being wronged by others.

morals A set of ethical principles that guide our actions and relationships.

morbidity Illness or disease.

more-developed countries (MDC) Industrialized nations characterized by high per capita incomes, low birth and death rates, low population growth rates, and high levels of industrialization and urbanization.

mortality Death rate in a population; the probability of dying.

Müellerian mimicry Evolution of two species, both of which are unpalatable and, have poisonous stingers or some other defense mechanism, to resemble each other.

mulch Protective ground cover, including both natural products and synthetic materials that protect the soil, save water, and prevent weed growth.

multiple use Many uses that occur simultaneously; used in forest management; limited to mutually compatible uses.

mutagens Agents, such as chemicals or radiation, that damage or alter genetic material (DNA) in cells.

mutation A change, either spontaneous or by external factors, in the genetic material of a cell; mutations in the gametes (sex cells) can be inherited by future generations of organisms.

mutualism A symbiotic relationship between individuals of two different species in which both species benefit from the association.

N

NAAQS *National Ambient Air Quality Standard*; federal standards specifying the maximum allowable levels (averaged over specific time periods) for regulated pollutants in ambient (outdoor) air.

natality The production of new individuals by birth, hatching, germination, or cloning.

natural experiment A study of events that have already happened.

natural history The study of where and how organisms carry out their life cycles.

natural increase Crude death rate subtracted from crude birthrate.

natural organic pesticides "Botanicals" or organic compounds naturally occurring in plants, animals or microbes that serve as pesticides.

natural resources Goods and services supplied by the environment.

natural selection The mechanism for evolutionary change in which environmental pressures cause certain genetic combinations in a population to become more abundant; genetic combinations best adapted for present environmental conditions tend to become predominant.

negative feedback loop A situation in which a factor or condition causes changes that reduce that factor or condition.

neoclassical economics A branch of economics that attempts to apply the principles of modern science to economic analysis in a mathematically rigorous, noncontextual, abstract, predictive manner.

neo-Luddites People who reject technology as the cause of environmental degradation and social disruption. Named after the followers of Ned Ludd who tried to turn back the Industrial Revolution in England.

neo-Malthusian A belief that the world is characterized by scarcity and competition in which too many people fight for too few resources. Named for Thomas Malthus, who predicted a dismal cycle of misery, vice, and starvation as a result of human overpopulation.

NEPA NEPA, the National Environmental Policy Act, is the cornerstone of U.S. environmental policy. It authorizes the Council on Environmental Quality, directs federal agencies to take environmental consequences into account when making decisions, and requires an environmental impact statement for every major federal project likely to have adverse environmental effects.

net energy yield Total useful energy produced during the lifetime of an entire energy system minus the energy used, lost, or wasted in making useful energy available.

neurotoxins Toxic substances, such as lead or mercury, that specifically poison nerve cells.

neutron A subatomic particle, found in the nucleus of the atom, that has no electromagnetic charge.

new towns Experimental urban environments that seek to combine the best features of the rural village and the modern city.

nihilists Those who believe the world has no meaning or purpose other than a dark, cruel, unceasing struggle for power and existence.

NIMBY *Not In My BackYard*: the rallying cry of those opposed to LULUs.

nitrate-forming bacteria Bacteria that convert nitrites into compounds that can be used by green plants to build proteins.

nitrite-forming bacteria Bacteria that combine ammonia with oxygen to form nitrites.

nitrogen cycle The circulation and reutilization of nitrogen in both inorganic and organic phases.

nitrogen-fixing bacteria Bacteria that convert nitrogen from the atmosphere or soil solution into ammonia that can then be converted to plant nutrients by nitrite- and nitrate-forming bacteria.

nitrogen oxides Highly reactive gases formed when nitrogen in fuel or combustion air is heated to over 650°C (1,200°F) in the presence of oxygen or when bacteria in soil or water oxidize nitrogen-containing compounds.

noncriteria pollutants *See* unconventional air pollutants.

nongovernmental organizations (NGOs) A term referring collectively to pressure and research groups, advisory agencies, political parties, professional societies, and other groups concerned about environmental quality, resource use, and many other issues.

nonpoint sources Scattered, diffuse sources of pollutants, such as runoff from farm fields, golf courses, construction sites, etc.

nonrenewable resources Minerals, fossil fuels, and other materials present in essentially fixed amounts (within human time scales) in our environment.

nuclear fission The radioactive decay process in which isotopes split apart to create two smaller atoms.

nuclear fusion A process in which two smaller atomic nuclei fuse into one larger nucleus and release energy; the source of power in a hydrogen bomb.

nucleic acids Large organic molecules made of nucleotides that function in the transmission of hereditary traits, in protein synthesis, and in control of cellular activities.

nucleus The center of the atom; occupied by protons and neutrons. In cells, the organelle that contains the chromosomes (DNA).

nuées ardentes Deadly, denser-than-air mixtures of hot gases and ash ejected from volcanoes.

numbers pyramid A diagram showing the relative population sizes at each trophic level in an ecosystem; usually corresponds to the biomass pyramid.

O

obese Generally considered to be a body mass greater than 30 kg/m^2, or roughly 30 pounds above normal for an average person.

ocean shorelines Rocky coasts and sandy beaches along the oceans; support rich, stratified communities.

ocean thermal electric conversion (OTEC) Energy derived from temperature differentials between warm ocean surface waters and cold deep waters. This differential can be used to drive turbines attached to electric generators.

oceanic islands Islands in the ocean; formed by breaking away from a continental landmass, volcanic action, coral formation, or a combination of sources; support distinctive communities.

offset allowances A controversial component of air quality regulations that allows a polluter to avoid installation of control equipment on one source with an "offsetting" pollution reduction at another source.

oil shale A fine-grained sedimentary rock rich in solid organic material called kerogen. When heated, the kerogen liquefies to produce a fluid petroleum fuel.

old-growth forests Forests free from disturbance for long enough (generally 150 to 200 years) to have mature trees, physical conditions, species diversity, and other characteristics of equilibrium ecosystems.

oligotrophic Condition of rivers and lakes that have clear water and low biological productivity (*oligo* = little; *trophic* = nutrition); are usually clear, cold, infertile headwater lakes and streams.

omnivore An organism that eats both plants and animals.

open access system A commonly held resource for which there are no management rules.

open canopy A forest where tree crowns cover less than 20 percent of the ground; also called woodland.

open range Unfenced, natural grazing lands; includes woodland as well as grassland.

open system A system that exchanges energy and matter with its environment.

optimum The most favorable condition in regard to an environmental factor.

orbital The space or path in which an electron orbits the nucleus of an atom.

organic compounds Complex molecules organized around skeletons of carbon atoms arranged in rings or chains; includes biomolecules, molecules synthesized by living organisms.

organophosphates Organic molecules to which phosphate group(s) are attached.

overburden Overlying layers of noncommercial sediments that must be removed to reach a mineral or coal deposit.

overgrazing Allowing livestock to eat so much forage that it damages the ecological health of the habitat.

overharvesting Harvesting so much of a resource that it threatens its existence.

overnutrition Receiving too many calories.

overshoot The extent to which a population exceeds the carrying capacity of its environment.

oxygen cycle The circulation and reutilization of oxygen in the biosphere.

oxygen sag Oxygen decline downstream from a pollution source that introduces materials with high biological oxygen demands.

ozone A highly reactive molecule containing three oxygen atoms; a dangerous pollutant in ambient air. In the stratosphere, however, ozone forms an ultraviolet absorbing shield that protects us from mutagenic radiation.

P

Pacific Decadal Oscillation (PDO) A large pool of warm water that moves north and south in the Pacific Ocean every 30 years or so and has large effects on North America's climate.

parabolic mirrors Curved mirrors that focus light from a large area onto a single, central point, thereby concentrating solar energy and producing high temperatures.

paradigm A model that provides a framework for interpreting observations.

parasite An organism that lives in or on another organism, deriving nourishment at the expense of its host, usually without killing it.

parsimony If two explanations appear equally plausible, choose the simpler one.

particulate material Atmospheric aerosols, such as dust, ash, soot, lint, smoke, pollen, spores, algal cells, and other suspended materials; originally applied only to solid particles but now extended to droplets of liquid.

parts per billion (ppb) Number of parts of a chemical found in 1 billion parts of a particular gas, liquid, or solid mixture.

parts per million (ppm) Number of parts of a chemical found in 1 million parts of a particular gas, liquid, or solid mixture.

parts per trillion (ppt) Number of parts of a chemical found in 1 trillion (10^{12}) parts of a particular gas, liquid, or solid mixture.

passive heat absorption The use of natural materials or absorptive structures without moving parts to gather and hold heat; the simplest and oldest use of solar energy.

pastoralists People who make a living by herding domestic livestock.

pasture Grazing lands suitable for domestic livestock.

patchiness Within a larger ecosystem, the presence of smaller areas that differ in some physical conditions and thus support somewhat different communities; a diversity-promoting phenomenon.

pathogen An organism that produces disease in a host organism, disease being an alteration of one or more metabolic functions in response to the presence of the organism.

peat Deposits of moist, acidic, semidecayed organic matter.

pelagic Zones in the vertical water column of a water body.

pellagra Lassitude, torpor, dermatitis, diarrhea, dementia, and death brought about by a diet deficient in tryptophan and niacin.

peptides Two or more amino acids linked by a peptide bond.

perennial species Plants that grow for more than two years.

permafrost A permanently frozen layer of soil that underlies the arctic tundra.

permanent retrievable storage Placing waste storage containers in a secure building, salt mine, or bedrock cavern where they can be inspected periodically and retrieved, if necessary.

persistent organic pollutants (POPs) Chemical compounds that persist in the environment and retain biological activity for long times.

pest Any organism that reduces the availability, quality, or value of a useful resource.

pesticide Any chemical that kills, controls, drives away, or modifies the behavior of a pest.

pesticide treadmill A need for constantly increasing doses or new pesticides to prevent pest resurgence.

pest resurgence Rebound of pest populations due to acquired resistance to chemicals and nonspecific destruction of natural predators and competitors by broadscale pesticides.

pH A value that indicates the acidity or alkalinity of a solution on a scale of 0 to 14, based on the proportion of H^+ ions present.

phosphorus cycle The movement of phosphorus atoms from rocks through the biosphere and hydrosphere and back to rocks.

photochemical oxidants Products of secondary atmospheric reactions. *See* smog.

photodegradable plastics Plastics that break down when exposed to sunlight or to a specific wavelength of light.

photosynthesis The biochemical process by which green plants and some bacteria capture light energy and use it to produce chemical bonds. Carbon dioxide and water are consumed while oxygen and simple sugars are produced.

photosynthetic efficiency The percentage of available light captured by plants and used to make useful products.

photovoltaic cell An energy-conversion device that captures solar energy and directly converts it to electrical current.

physical or abiotic factors Nonliving factors, such as temperature, light, water, minerals, and climate, that influence an organism.

phytoplankton Microscopic, free-floating, autotrophic organisms that function as producers in aquatic ecosystems.

pioneer species In primary succession on a terrestrial site, the plants, lichens, and microbes that first colonize the site.

plankton Primarily microscopic organisms that occupy the upper water layers in both freshwater and marine ecosystems.

plasma A hot, electrically neutral gas of ions and free electrons.

plug-in hybrids Vehicles with hybrid gasoline-electric engines adapted with a larger battery array (enough to propel the vehicle for 50 km or so on the batteries alone) and a plug-in to recharge the batteries from a standard electric outlet.

poachers Those who hunt wildlife illegally.

point sources Specific locations of highly concentrated pollution discharge, such as factories, power plants, sewage treatment plants, underground coal mines, and oil wells.

policy A societal plan or statement of intentions intended to accomplish some social good.

policy cycle The process by which problems are identified and acted upon in the public arena.

political economy The branch of economics concerned with modes of production, distribution of benefits, social institutions, and class relationships.

pollution To make foul, unclean, dirty; any physical, chemical, or biological change that adversely affects the health, survival, or activities of living organisms or that alters the environment in undesirable ways.

pollution charges Fees assessed per unit of pollution based on the "polluter pays" principle.

polycentric complex Cities with several urban cores surrounding a once dominant central core.

population A group of individuals of the same species occupying a given area.

population crash A sudden population decline caused by predation, waste accumulation, or resource depletion; also called a dieback.

population explosion Growth of a population at exponential rates to a size that exceeds environmental carrying capacity; usually followed by a population crash.

population momentum A potential for increased population growth as young members reach reproductive age.

positive feedback loop A situation in which a factor or condition causes changes that further enhance that factor or condition.

postmaterialist values A philosophy that emphasizes quality of life over acquisition of material goods.

post-modernism A philosophy that rejects the optimism and universal claims of modern positivism.

potential energy Stored energy that is latent but available for use. A rock poised at the top of a hill or water stored behind a dam are examples of potential energy.

power The rate of energy delivery; measured in horsepower or watts.

precautionary principle The decision to leave a margin of safety for unexpected developments.

precedent An act or decision that can be used as an example in dealing with subsequent similar situations.

precycling Making environmentally sound decisions at the store and reducing waste before we buy.

predation The act of feeding by a predator.

predator An organism that feeds directly on other organisms in order to survive; live-feeders, such as herbivores and carnivores.

predator-mediated competition A situation in which predation reduces prey populations and gives an advantage to competitors that might not otherwise be successful.

premises Introductory statements that set up or define a problem. Those things taken as given.

prevention of significant deterioration A clause of the Clean Air Act that prevents degradation of existing clean air; opposed by industry as an unnecessary barrier to development.

price elasticity A situation in which supply and demand of a commodity respond to price.

primary pollutants Chemicals released directly into the air in a harmful form.

primary productivity Synthesis of organic materials (biomass) by green plants using the energy captured in photosynthesis.

primary standards Regulations of the 1970 Clean Air Act; intended to protect human health.

primary succession An ecological succession that begins in an area where no biotic community previously existed.

primary treatment A process that removes solids from sewage before it is discharged or treated further.

principle of competitive exclusion A result of natural selection whereby two similar species in a community occupy different ecological niches, thereby reducing competition for food.

producer An organism that synthesizes food molecules from inorganic compounds by using an external energy source; most producers are photosynthetic.

production frontier The maximum output of two competing commodities at different levels of production.

productivity The synthesis of new organic material. That done by green plants using solar energy is called primary productivity.

prokaryotic Cells that do not have a membrane-bounded nucleus or membrane-bounded organelles.

promoters Agents that are not carcinogenic but that assist in the progression and spread of tumors; sometimes called cocarcinogens.

pronatalist pressures Influences that encourage people to have children.

proteins Chains of amino acids linked by peptide bonds.

proton A positively charged subatomic particle found in the nucleus of an atom.

proven reserves *See* proven resources.

proven resources Those that have been thoroughly mapped and are economical to recover at current prices with available technology.

public trust A doctrine obligating the government to maintain public lands in a natural state as guardians of the public interest.

pull factors (in urbanization) Conditions that draw people from the country into the city.

push factors (in urbanization) Conditions that force people out of the country and into the city.

R

***r*-selected species** Species that tend to have rapid reproduction and high offspring mortality. They frequently overshoot carrying capacity of their environment and display boom and bust cycles. They lack intrinsic population controls and tend to occupy lower trophic levels in food webs than *k*-selected species.

radiative evolution Divergence from a common ancestor into two or more new species.

radioactive An unstable isotope that decays spontaneously and releases subatomic particles or units of energy.

radioactive decay A change in the nuclei of radioactive isotopes that spontaneously emit high-energy electromagnetic radiation and/or subatomic particles while gradually changing into another isotope or different element.

radionucleides Isotopes that exhibit radioactive decay.

rainforest A forest with high humidity, constant temperature, and abundant rainfall (generally over 380 cm [150 in] per year); can be tropical or temperate.

rain shadow Dry area on the downwind side of a mountain.

rangeland Grasslands and open woodlands suitable for livestock grazing.

rational choice Public decision making based on reason, logic, and science-based management.

recharge zone Area where water infiltrates into an aquifer.

reclamation Chemical, biological, or physical cleanup and reconstruction of severely contaminated or degraded sites to return them to something like their original topography and vegetation.

recoverable resources Those accessible with current technology but not economical under current conditions.

re-creation Construction of an entirely new biological community to replace one that has been destroyed on that or another site.

recycling Reprocessing of discarded materials into new, useful products; not the same as reuse of materials for their original purpose, but the terms are often used interchangeably.

red tide A population explosion or bloom of minute, single-celled marine organisms called dinoflagellates. Billions of these cells can accumulate in protected bays where the toxins they contain can poison other marine life.

reduced tillage systems Systems, such as minimum till, conserve-till, and no-till, that preserve soil, save energy and water, and increase crop yields.

reflective thinking Asks, what does this all mean?

reformer A device that strips hydrogen from fuels such as natural gas, methanol, ammonia, gasoline, or vegetable oil so they can be used in a fuel cell.

refuse-derived fuel Processing of solid waste to remove metal, glass, and other unburnable materials; organic residue is shredded, formed into pellets, and dried to make fuel for power plants.

regenerative farming Farming techniques and land stewardship that restore the health and productivity of the soil by rotating crops, planting ground cover, protecting the surface with crop residue, and reducing synthetic chemical inputs and mechanical compaction.

regolith The parent material between bedrock and subsoil (sometimes called the C horizon) that is made of weathered rock fragments and very little organic material and from which soil is formed.

regulations Rules established by administrative agencies; regulations can be more important than statutory law in the day-to-day management of resources.

rehabilitate land A utilitarian program to make an area useful to humans.

rehabilitation To rebuild elements of structure or function in an ecological system without necessarily achieving complete restoration to its original condition.

relative humidity At any given temperature, a comparison of the actual water content of the air with the amount of water that could be held at saturation.

relativists Those who believe moral principles are always dependent on the particular situation.

remediation Cleaning up chemical contaminants from a polluted area.

renewable resources Resources normally replaced or replenished by natural processes; resources not depleted by moderate use; examples include solar energy, biological resources such as forests and fisheries, biological organisms, and some biogeochemical cycles.

renewable water supplies Annual freshwater surface runoff plus annual infiltration into underground freshwater aquifers that are accessible for human use.

replication Repeating studies or tests to verify reliability.

reproducibility Making an observation or obtaining a particular result more than once.

reproductive isolation Barriers (geographical, behavioral, or biological) that prevent gene flow between members of a species.

residence time The length of time a component, such as an individual water molecule, spends in a particular compartment or location before it moves on through a particular process or cycle.

resilience The ability of a community or ecosystem to recover from disturbances.

resistance (inertia) The ability of a community to resist being changed by potentially disruptive events.

resource In economic terms, anything with potential use in creating wealth or giving satisfaction.

resource partitioning In a biological community, various populations sharing environmental resources through specialization, thereby reducing direct competition. *See also* ecological niche.

resource scarcity A shortage or deficit in some resource.

restoration To bring something back to a former condition. Ecological restoration involves active manipulation of nature to re-create conditions that existed before human disturbance.

restoration ecology Seeks to repair or reconstruct ecosystems damaged by human actions.

riders Amendments attached to bills in conference committee, often completely unrelated to the bill to which they are added.

rill erosion The removing of thin layers of soil as little rivulets of running water gather and cut small channels in the soil.

risk Probability that something undesirable will happen as a consequence of exposure to a hazard.

risk assessment Evaluation of the short-term and long-term risks associated with a particular activity or hazard; usually compared to benefits in a cost-benefit analysis.

RNA Ribonucleic acid; nucleic acid used for transcription and translation of the genetic code found on DNA molecules.

rock A solid, cohesive, aggregate of one or more crystalline minerals.

rock cycle The process whereby rocks are broken down by chemical and physical forces; sediments are moved by wind, water, and gravity, sedimented and reformed into rock, and then crushed, folded, melted, and recrystallized into new forms.

rotational grazing Confining animals to a small area for a short time (often only a day or two) before shifting them to a new location.

ruminant animals Cud-chewing animals, such as cattle, sheep, goats, and buffalo, with multichambered stomachs in which cellulose is digested with the aid of bacteria.

runoff The excess of precipitation over evaporation; the main source of surface water and, in broad terms, the water available for human use.

run-of-the-river flow Ordinary river flow not accelerated by dams, flumes, etc. Some small, modern, high-efficiency turbines can generate useful power with run-of-the-river flow or with a current of only a few kilometers per hour.

rural area An area in which most residents depend on agriculture or the harvesting of natural resources for their livelihood.

S

S curve A curve that depicts logistic growth; called an S curve because of its shape.

salinity Amount of dissolved salts (especially sodium chloride) in a given volume of water.

salinization A process in which mineral salts accumulate in the soil, killing plants; occurs when soils in dry climates are irrigated profusely.

salt marsh Shallow wetlands along coastlines that are flooded regularly or occasionally with seawater.

saltwater intrusion Movement of saltwater into freshwater aquifers in coastal areas where groundwater is withdrawn faster than it is replenished.

salvage logging Harvesting timber killed by fire, disease, or windthrow.

sample To analyze a small but representative portion of a population to estimate the characteristics of the entire class.

sanitary landfills A landfill in which garbage and municipal waste is buried every day under enough soil or fill to eliminate odors, vermin, and litter.

saturation point The maximum concentration of water vapor the air can hold at a given temperature.

savannas Open grasslands with sparse tree cover.

scavenger An organism that feeds on the dead bodies of other organisms.

science A process for producing knowledge methodically and logically.

scientific consensus A general agreement among informed scholars.

scientific method A systematic, precise, objective study of a problem. Generally this requires observation, hypothesis development and testing, data gathering, and interpretation.

scientific theory An explanation supported by many tests and accepted by a general consensus of scientists.

secondary pollutants Chemicals modified to a hazardous form after entering the air or that are formed by chemical reactions as components of the air mix and interact.

secondary recovery technique Pumping pressurized gas, steam, or chemical-containing water into a well to squeeze more oil from a reservoir.

secondary standards Regulations of the 1972 Clean Air Act intended to protect materials, crops, visibility, climate, and personal comfort.

secondary succession Succession on a site where an existing community has been disrupted.

secondary treatment Bacterial decomposition of suspended particulates and dissolved organic compounds that remain after primary sewage treatment.

second law of thermodynamics States that, with each successive energy transfer or transformation in a system, less energy is available to do work.

secure landfill A solid waste disposal site lined and capped with an impermeable barrier to prevent leakage or leaching. Drain tiles, sampling wells, and vent systems provide monitoring and pollution control.

sedimentary rock Deposited material that remains in place long enough or is covered with enough material to compact into stone; examples include shale, sandstone, breccia, and conglomerates.

sedimentation The deposition of organic materials or minerals by chemical, physical, or biological processes.

selection pressures Factors in the environment that favor successful reproduction of individuals possessing heritable traits and that reduce viability and fertility of those individuals not possessing those traits.

selective cutting Harvesting only mature trees of certain species and size; usually more expensive than clear-cutting, but it is less disruptive for wildlife and often better for forest regeneration.

seriously undernourished Those who receive less than 80 percent of their minimum daily caloric requirements.

shantytowns Settlements created when people move onto undeveloped lands and build their own shelter with cheap or discarded materials; some are simply illegal subdivisions where a landowner rents land without city approval; others are land invasions.

sheet erosion Peeling off thin layers of soil from the land surface; accomplished primarily by wind and water.

shelterwood harvesting A harvest method in which mature trees are removed in a series of linear cuts leaving enough trees in unharvested strips to protect from wind storms or other disruptive factors.

sick building syndrome Headaches, allergies, chronic fatigue and other symptoms caused by poorly vented indoor air contaminated by pathogens or toxins.

significant numbers Meaningful numbers whose accuracy can be verified.

sinkholes A large surface crater caused by the collapse of an underground channel or cavern; often triggered by groundwater withdrawal.

sludge Semisolid mixture of organic and inorganic materials that settles out of wastewater at a sewage treatment plant.

slums Legal but inadequate multifamily tenements or rooming houses; some are custom built for rent to poor people, others are converted from some other use.

smart growth Efficient use of land resources and existing urban infrastructure.

smelting Heating ores to extract metals.

smog The term used to describe the combination of smoke and fog in the stagnant air of London; now often applied to photochemical pollution products or urban air pollution of any kind.

social justice Equitable access to resources and the benefits derived from them; a system that recognizes inalienable rights and adheres to what is fair, honest, and moral.

soil A complex mixture of weathered mineral materials from rocks, partially decomposed organic molecules, and a host of living organisms.

soil horizons Horizontal layers that reveal a soil's history, characteristics, and usefulness.

soil profile All the vertical layers or horizons that make up a soil in a particular place.

southern pine forest United States coniferous forest ecosystem characterized by a warm, moist climate.

speciation The generation of new species.

species A population of morphologically similar organisms that can reproduce sexually among themselves but that cannot produce fertile offspring when mated with other organisms.

species diversity The number and relative abundance of species present in a community.

species recovery plan A plan for restoration of an endangered species through protection, habitat management, captive breeding, disease control, or other techniques that increase populations and encourage survival.

sprawl Unlimited outward extension of city boundaries that lowers population density, consumes open space, generates freeway congestion, and causes decay in central cities.

spring overturn Springtime lake phenomenon that occurs when the surface ice melts and the surface water temperature warms to its greatest density at 4°C and then sinks, creating a convection current that displaces nutrient-rich bottom waters.

squatter towns Shantytowns that occupy land without owner's permission; some are highly organized movements in defiance of authorities; others grow gradually.

stability In ecological terms, a dynamic equilibrium among the physical and biological factors in an ecosystem or a community; relative homeostasis.

stable runoff The fraction of water available year-round; usually more important than total runoff when determining human uses.

Standard Metropolitan Statistical Area (SMSA) An urbanized region with at least 100,000 inhabitants with strong economic and social ties to a central city of at least 50,000 people.

standing The right to take part in legal proceedings.

statistics Numbers that let you evaluate and compare things.

statute law Formal documents or decrees enacted by the legislative branch of government.

statutory law Rules passed by a state or national legislature.

steady-state economy Characterized by low birth and death rates, use of renewable energy sources, recycling of materials, and emphasis on durability, efficiency, and stability.

stewardship A philosophy that holds that humans have a unique responsibility to manage, care for, and improve nature.

strategic lawsuits against public participation (SLAPP) Lawsuits that have no merit but are brought merely to intimidate and harass private citizens who act in the public interest.

strategic metals and minerals Materials a country cannot produce itself but that it uses for essential materials or processes.

stratosphere The zone in the atmosphere extending from the tropopause to about 50 km (30 mi) above the earth's surface; temperatures are stable or rise slightly with altitude; has very little water vapor but is rich in ozone.

stratospheric ozone The ozone (O_3) occurring in the stratosphere 10 to 50 km above the earth's surface.

stress-related diseases Diseases caused or accentuated by social stresses such as crowding.

strip cutting Harvesting trees in strips narrow enough to minimize edge effects and to allow natural regeneration of the forest.

strip farming Planting different kinds of crops in alternating strips along land contours; when one crop is harvested, the other crop remains to protect the soil and prevent water from running straight down a hill.

strip mining Removing surface layers over coal seams using giant, earth-moving equipment; creates a huge open-pit from which coal is scooped by enormous surface-operated machines and transported by trucks; an alternative to deep mines.

structure (in ecological terms) Patterns of organization, both spatial and functional, in a community.

Student Environmental Action Coalition (SEAC) A student-based environmental organization that is both an umbrella organization and a grassroots network to facilitate environmental action and education on college campuses.

subduction The process by which one tectonic plate is pushed down below another as plates crash into each other.

sublimation The process by which water can move between solid and gaseous states without ever becoming liquid.

subsidence A settling of the ground surface caused by the collapse of porous formations that result from withdrawal of large amounts of groundwater, oil, or other underground materials.

subsoil A layer of soil beneath the topsoil that has lower organic content and higher concentrations of fine mineral particles; often contains soluble compounds and clay particles carried down by percolating water.

sulfur cycle The chemical and physical reactions by which sulfur moves into or out of storage and through the environment.

sulfur dioxide A colorless, corrosive gas directly damaging to both plants and animals.

Superfund A fund established by Congress to pay for containment, cleanup, or remediation of abandoned toxic waste sites. The fund is financed by fees paid by toxic waste generators and by cost-recovery from cleanup projects.

supply The quantity of that product being offered for sale at various prices, other things being equal.

surface mining Some minerals are also mined from surface pits. See strip mining.

surface tension A condition in which the water surface meets the air and acts like an elastic skin.

survivorship The percentage of a population reaching a given age or the proportion of the maximum life span of the species reached by any individual.

sustainability Living within the bounds of nature based on renewable resources used in ways that don't deplete nonrenewable resources, harm essential ecological services, or limit the ability of future generations to meet their own needs.

sustainable agriculture An ecologically sound, economically viable, socially just, and humane agricultural system. Stewardship, soil conservation, and integrated pest management are essential for sustainability.

sustainable development A real increase in well-being and standard of life for the average person that can be maintained over the long-term without degrading the environment or compromising the ability of future generations to meet their own needs.

sustained yield Utilization of a renewable resource at a rate that does not impair or damage its ability to be fully renewed on a long-term basis.

swamp Wetland with trees, such as the extensive swamp forests of the southern United States.

swidden agriculture *See* milpa agriculture.

symbiosis The intimate living together of members of two different species; includes **mutualism, commensalism,** and, in some classifications, **parasitism.**

sympatric speciation Species that arise from a common ancestor due to biological or behavioral barriers that cause reproductive isolation even though the organisms live in the same place.

synergism An interaction in which one substance exacerbates the effects of another. The sum of the interaction is greater than the parts.

synergistic effects When an injury caused by exposure to two environmental factors together is greater than the sum of exposure to each factor individually.

systemic A condition or process that affects the whole body; many metabolic poisons are systemic.

systems Networks of interactions among many interdependent factors.

T

taiga The northernmost edge of the boreal forest, including species-poor woodland and peat deposits; intergrading with the arctic tundra.

tailings Mining waste left after mechanical or chemical separation of minerals from crushed ore.

taking Unconstitutional confiscation of private property.

tar sands Sand deposits containing petroleum or tar.

technological optimists Those who believe that technology and human enterprise will find cures for all our problems. Also called **Promethean environmentalism.**

tectonic plates Huge blocks of the earth's crust that slide around slowly, pulling apart to open new ocean basins or crashing ponderously into each other to create new, larger landmasses.

temperate rainforest The cool, dense, rainy forest of the northern Pacific coast; enshrouded in fog much of the time; dominated by large conifers.

temperature A measure of the speed of motion of a typical atom or molecule in a substance.

temperature inversions A stable layer of warm air overlays cooler air, trapping pollutants near ground level.

teratogens Chemicals or other factors that specifically cause abnormalities during embryonic growth and development.

terracing Shaping the land to create level shelves of earth to hold water and soil; requires extensive hand labor or expensive machinery, but it enables farmers to farm very steep hillsides.

territoriality An intense form of intraspecific competition in which organisms define an area surrounding their home site or nesting site and defend it, primarily against other members of their own species.

tertiary treatment The removal of inorganic minerals and plant nutrients after primary and secondary treatment of sewage.

thermal plume A plume of hot water discharged into a stream or lake by a heat source, such as a power plant.

thermocline In water, a distinctive temperature transition zone that separates an upper layer that is mixed by the wind (the epilimnion) and a colder, deep layer that is not mixed (the hypolimnion).

thermodynamics A branch of physics that deals with transfers and conversions of energy.

thermodynamics, first law Energy can be transformed and transferred, but cannot be destroyed or created.

thermodynamics, second law With each successive energy transfer or transformation, less energy is available to do work.

thermosphere The highest atmospheric zone; a region of hot, dilute gases above the mesosphere extending out to about 1,600 km (1,000 mi) from the earth's surface.

Third World Less-developed countries that are not capitalistic and industrialized (First World) or centrally-planned socialist economies (Second World); not intended to be derogatory.

threatened species While still abundant in parts of its territorial range, this species has declined significantly in total numbers and may be on the verge of extinction in certain regions or localities.

tidal station A dam built across a narrow bay or estuary traps tide water flowing both in and out of the bay. Water flowing through the dam spins turbines attached to electric generators.

tide pool Depressions in a rocky shoreline that are flooded at high tide but cut off from the ocean at low tide.

timberline In mountains, the highest-altitude edge of forest that marks the beginning of the treeless alpine tundra.

tolerance limits *See* limiting factors.

topsoil The first true layer of soil; layer in which organic material is mixed with mineral particles; thickness ranges from a meter or more under virgin prairie to zero in some deserts.

tornado A violent storm characterized by strong swirling winds and updrafts; tornadoes form when a strong cold front pushes under a warm, moist air mass over the land.

tort law Court cases that seek compensation for damages.

total fertility rate The number of children born to an average woman in a population during her entire reproductive life.

total growth rate The net rate of population growth resulting from births, deaths, immigration, and emigration.

total maximum daily loads (TMDL) The amount of particular pollutant that a water body can receive from both point and nonpoint sources and still meet water quality standards.

toxic colonialism Shipping toxic wastes to a weaker or poorer nation.

Toxic Release Inventory A program created by the Superfund Amendments and Reauthorization Act of 1984 that requires manufacturing facilities and waste handling and disposal sites to report annually on releases of more than 300 toxic materials.

toxins Poisonous chemicals that react with specific cellular components to kill cells or to alter growth or development in undesirable ways; often harmful, even in dilute concentrations.

tradable permits Pollution quotas or variances that can be bought or sold.

tragedy of the commons An inexorable process of degradation of communal resources due to selfish self-interest of "free riders" who use or destroy more than their fair share of common property. *See* open access system.

transitional zone A zone in which populations from two or more adjacent communities meet and overlap.

transpiration The evaporation of water from plant surfaces, especially through stomates.

triple bottom line Corporate accounting that reports social and environmental costs and benefits as well as merely economic ones.

trophic level Step in the movement of energy through an ecosystem; an organism's feeding status in an ecosystem.

tropical rainforests Forests in which rainfall is abundant—more than 200 cm (80 in) per year—and temperatures are warm to hot year-round.

tropical seasonal forest Semievergreen or partly deciduous forests tending toward open woodlands and grassy savannas dotted with scattered, drought-resistant tree species; distinct wet and dry seasons, hot year-round.

tropopause The boundary between the troposphere and the stratosphere.

troposphere The layer of air nearest to the earth's surface; both temperature and pressure usually decrease with increasing altitude.

tsunami Giant seismic sea swells that move rapidly from the center of an earthquake; they can be 10 to 20 meters high when they reach shorelines hundreds or even thousands of kilometers from their source.

tundra Treeless arctic or alpine biome characterized by cold, harsh winters, a short growing season, and potential for frost any month of the year; vegetation includes low-growing perennial plants, mosses, and lichens.

U

unconventional air pollutants Toxic or hazardous substances, such as asbestos, benzene, beryllium, mercury, polychlorinated biphenyls, and vinyl chloride, not listed in the original Clean Air Act because they were not released in large quantities; also called noncriteria pollutants.

unconventional oil Resources such as shale oil and tar sands that can be liquefied and used like oil.

undernourished Those who receive less than 90 percent of the minimum dietary intake over a long-term time period; they lack energy for an active, productive life and are more susceptible to infectious diseases.

undiscovered resources Speculative or inferred resources or those that we haven't even thought about.

universalists Those who believe that some fundamental ethical principles are universal and unchanging. In this vision, these principles are valid regardless of the context or situation.

upwelling   Convection currents within a body of water that carry nutrients from bottom sediments toward the surface.

urban agglomerations   An aggregation of many cities into a large metropolitan area.

urban area   An area in which a majority of the people are not directly dependent on natural resource-based occupations.

urbanization   An increasing concentration of the population in cities and a transformation of land use to an urban pattern of organization.

utilitarian conservation   A philosophy that resources should be used for the greatest good for the greatest number for the longest time.

utilitarianism   *See* utilitarian conservation.

utilitarians   Those who hold that an action is right that produces the greatest good for the greatest number of people.

V

values   An estimation of the worth of things; a set of ethical beliefs and preferences that determine our sense of right and wrong.

verbal learner   Someone who understands and remembers best by listening to the spoken word.

vertical stratification   The vertical distribution of specific subcommunities within a community.

vertical zonation   Terrestrial vegetation zones determined by altitude.

village   A collection of rural households linked by culture, custom, and association with the land.

visible light   A portion of the electromagnetic spectrum that includes the wavelengths used for photosynthesis.

visual learner   Someone who understands and remembers best by reading, or looking at pictures and diagrams.

vitamins   Organic molecules essential for life that we cannot make for ourselves; we must get them from our diet; they act as enzyme cofactors.

volatile organic compounds (VOCs)   Organic chemicals that evaporate readily and exist as gases in the air.

volcanoes   Vents in the earth's surface through which gases, ash, or molten lava are ejected. Also a mountain formed by this ejecta.

voluntary simplicity   Deliberately choosing to live at a lower level of consumption as a matter of personal and environmental health.

vulnerable species   Naturally rare organisms or species whose numbers have been so reduced by human activities that they are susceptible to actions that could push them into threatened or endangered status.

W

warm front   A long, wedge-shaped boundary caused when a warmer advancing air mass slides over neighboring cooler air parcels.

waste stream   The steady flow of varied wastes, from domestic garbage and yard wastes to industrial, commercial, and construction refuse.

water cycle   The recycling and reutilization of water on earth, including atmospheric, surface, and underground phases and biological and nonbiological components.

water droplet coalescence   A mechanism of condensation that occurs in clouds too warm for ice crystal formation.

water scarcity   Annual available freshwater supplies less than 1,000 m^3 per person.

water stress   A situation when residents of a country don't have enough accessible, high-quality water to meet their everyday needs.

water table   The top layer of the zone of saturation; undulates according to the surface topography and subsurface structure.

waterlogging   Water saturation of soil that fills all air spaces and causes plant roots to die from lack of oxygen; a result of overirrigation.

watershed   The land surface and groundwater aquifers drained by a particular river system.

weather   Description of the physical conditions of the atmosphere (moisture, temperature, pressure, and wind).

weathering   Changes in rocks brought about by exposure to air, water, changing temperatures, and reactive chemical agents.

wetland mitigation   Replacing a wetland damaged by development (roads, buildings, etc.) with a new or refurbished wetland.

wetlands   Ecosystems of several types in which rooted vegetation is surrounded by standing water during part of the year. *See also* swamp, marsh, bog, fen.

wicked problems   Problems with no simple right or wrong answer where there is no single, generally agreed-on definition of or solution for the particular issue.

wilderness   An area of undeveloped land affected primarily by the forces of nature; an area where humans are visitors who do not remain.

Wilderness Act   Legislation of 1964 recognizing that leaving land in its natural state may be the highest and best use of some areas.

wildlife   Plants, animals, and microbes that live independently of humans; plants, animals, and microbes that are not domesticated.

wildlife refuges   Areas set aside to shelter, feed, and protect wildlife; due to political and economic pressures, refuges often allow hunting, trapping,

mineral exploitation, and other activities that threaten wildlife.

windbreak   Rows of trees or shrubs planted to block wind flow, reduce soil erosion, and protect sensitive crops from high winds.

wind farms   Large numbers of windmills concentrated in a single area; usually owned by a utility or large-scale energy producer.

wise use groups   A coalition of ranchers, loggers, miners, industrialists, hunters, off-road vehicle users, land developers, and others who call for unrestricted access to natural resources and public lands.

withdrawal   A description of the total amount of water taken from a lake, river, or aquifer.

woodland   A forest where tree crowns cover less than 20 percent of the ground; also called open canopy.

work   The application of force through a distance; requires energy input.

world conservation strategy   A proposal for maintaining essential ecological processes, preserving genetic diversity, and ensuring that utilization of species and ecosystems is sustainable.

World Trade Organization (WTO)   An association of 135 nations that meet to regulate international trade.

worldviews   Sets of basic beliefs, images, and understandings that shape how we see the world around us.

X

X ray   Very short wavelength in the electromagnetic spectrum; can penetrate soft tissue; although it is useful in medical diagnosis, it also damages tissue and causes mutations.

Y

yellowcake   The concentrate of 70 to 90 percent uranium oxide extracted from crushed ore.

Z

zero population growth (ZPG)   The number of births at which people are just replacing themselves; also called the replacement level of fertility.

zone of aeration   Upper soil layers that hold both air and water.

zone of leaching   The layer of soil just beneath the topsoil where water percolates, removing soluble nutrients that accumulate in the subsoil.

zone of saturation   Lower soil layers where all spaces are filled with water.

Credits

PHOTOGRAPHS

Chapter L

Opener: p. 1: David L. Hansen, University of Minnesota Agricultural Experiment Station; p. 2: © The McGraw-Hill Companies, Inc./Barry Barker, photographer; p. 4: © Vol. 198/Corbis; p. 6: © PhotoDisc.

Chapter 1

Opener: © China Photos/Getty Images; 1.1: Courtesy of Dr. Delbert Swanson; 1.3a: Courtesy of the Bancroft Library. University of California, Berkeley; 1.3b: Courtesy of Grey Towers National Historic Landmark; 1.3c: © Bettmann/Corbis; 1.3d: © AP/Wide World Photos; 1.4: © William P. Cunningham; 1.5a: © AP/Wide World Photos; 1.5b: © Tom Turney/The Brower Fund/Earth Island Institute; 1.5c: With permission of the University Archives, Columbia University in the City of New York and Columbia Library; 1.5d: © AP/Wide World Photos; 1.6: © Earth Imaging/Stone/Getty; 1.7: © Vol. 6/Corbis Royalty Free; 1.8: © Corbis Royalty Free; 1.10: © Vol. 244/Corbis Royalty Free; 1.12: © Norbert Schiller/The Image Works; 1.14: © Justin Guariglia/Corbis; 1.17: © The McGraw-Hill Companies/Barry Barker, photographer; 1.18: © Stone/Getty Images; 1.19: © The McGraw-Hill Companies, Inc./Barry Barker, photographer.

Chapter 2

Opener: © 2005 Gary Braasch/World View of Global Warming; 2.2 & 2.3: David L. Hansen, University of Minnesota Agricultural Experiment Station; 2.4: © William P. Cunningham; 2.7: © The McGraw-Hill Companies, Inc./John A. Karachewski, photographer; 2.9: © Mary Ann Cunningham; 2.11: © Carr Clifton; 2.12: © William P. Cunningham; 2.13: © Sam Kittner; 2.14: © William P. Cunningham.

Chapter 3

Opener: © John Warden/Stone/Getty Images; p. 55: © G.I. Bernard/Animals Animals Earth Scenes; 3.8: © William P. Cunningham; 3.9: NOAA; p. 68: The SeaWIFS Project, NASA/Goddard Space Flight Center and ORBIMAGE; 3.22: © David Dennis/Tom Stack & Associates.

Chapter 4

Opener: © Galen Rowell/Corbis; 4.1: Portrait of Charles Darwin, 1840 by George Richmond (1809–96). Down House, Downe, Kent, UK/Bridgeman Art Library; 4.2: © Vol. 6/Corbis Royalty Free; 4.3: © William P. Cunningham; 4.5: © Digital Vision Ltd./Royalty Free; 4.6 & 4.13: © William P. Cunningham; 4.14: © Ray Coleman/Photo Researchers; 4.15: © D. P. Wilson/Photo Researchers; 4.16: © Michael Fogdon/Animals Animals Earth Scenes; 4.17a & b: © Edward S. Ross; 4.18: © Tom Finkle; 4.19a: © William P. Cunningham; 4.19b: PhotoDisc Royalty Free; 4.19c: © William P. Cunningham; 4.20 © Vol. 6 PhotoDisc/Getty Images; 4.22a: © Corbis Royalty Free; 4.22b: © Eric and David Hosking/Corbis; 4.22c: © Fred Bavindam/Peter Arnold; p. 91: © L. David Mech; 4.24: © Vol. 262/Corbis Royalty Free; 4.27 & 4.28: © William P. Cunningham; 4.29: © Gerals Lacz/Peter Arnold.

Chapter 5

Opener 5: © William P. Cunningham; 5.1: © Vol. DV13/Getty Images; 5.5: © Vol. 60/PhotoDisc; 5.6: © Vol. 6/Corbis; 5.7: © William P. Cunningham; 5.8: © Mary Ann Cunningham; 5.9: © William P. Cunningham; 5.10: © Vol. 90/Corbis; 5.11: © William P. Cunningham; 5.12: © Vol. 74/PhotoDisc; 5.13: Courtesy of SeaWiFS/NASA; 5.15: NOAA; 5.16a: © Vol. 89/PhotoDisc; 5.16b: © Robert Garvey/Corbis; 5.16c: © Andrew Martinez/Photo Researchers; 5.16d: © Pat O'Hara/Corbis; 5.17: USGS; 5.18: © Stephen Rose; 5.21a, b: © Vol. 16/PhotoDisc; 5.21c: © William P. Cunningham.

Chapter 6

Opener: © Tom Stewart/Corbis; 6.2: Courtesy of National Museum of Natural History, Smithsonian Institution; p. 121: © Vol. 72/Corbis; 6.7a: © Digital Vision/Getty Images Royalty Free; 6.7b: © Stockbyte Royalty Free; 6.7c: © PhotoDisc; 6.7d: © Corbis Royalty Free; 6.9: © Art Wolfe/Stone/Getty Images; 6.13: © The McGraw-Hill Companies, Inc./Barry Barker, photographer.

Chapter 7

Opener: © The McGraw-Hill Companies, Inc./Barry Barker, photographer; 7.1: © Apichart Weerawong/AP/Wide World Photos; 7.2: Courtesy of John Cunningham; 7.9: © Alain Le Grasmeur/Corbis; 7.13a: © William P. Cunningham; 7.13b: © Vol. 17/PhotoDisc/Getty Images.

Chapter 8

Opener 8: © The Carter Center/Emily Staub; 8.1: Courtesy of Donald R. Hopkins; 8.4: © William P. Cunningham; 8.5a: © Vol. 40/Corbis; 8.5b: © Image Source/Getty Royalty Free; 8.5c: Courtesy of Stanley Erlandsen, University of Minnesota; 8.8: © Vol. 72/Corbis; 8.9: © Vol. 53/Corbis; 8.19: © The McGraw-Hill Companies, Inc./Sharon Farmer, photographer.

Chapter 9

Opener 9: © John Maier/The New York Times; 9.5: © Norbert Schiller/The Image Works; 9.6: © Lester Bergman/Corbis; 9.7: © Scott Daniel Peterson; 9.10 & 9.11a,c: © William P. Cunningham; 9.12: Courtesy U.S. Department of Agriculture Conservation Services; 9.13 & 9.18: © FAO Photo/R. Faidutti; 9.20a: Photo by Lynn Betts, courtesy of USDA Natural Resources Conservation Service; 9.20b: Photo by Jeff Vanuga, courtesy of USDA Natural Resources Conservation Center; 9.20c: © Corbis Royalty Free; 9.21: © William P. Cunningham; 9.22: © Vol. 13/Corbis; 9.23-9.25, p. 200: © William P. Cunningham; 9.29: Photo by Lynn Betts, courtesy of USDA Natural Resources Conservation Service; 9.30: © William P. Cunningham; 9.31: Courtesy of David Hansen, College of Agriculture, University of Minnesota; 9.32: © Tom Sweeney/Minneapolis Star Tribune.

Chapter 10

Opener: Courtesy of Scott Bauer, U.S. Department of Agriculture; 10.1: © William P. Cunningham; 10.2: © Bettmann/Corbis; 10.5: © William P. Cunningham; 10.6: © Corbis Royalty Free; 10.7: © Stephen Dalton/Photo Researchers; 10.8: © SPL/Photo Researchers; 10.9: © Joe Munroe/Photo Researchers; p. 214: © Howard K. Suzuki/Aquatic Life Structures; 10.11: © Norm Thomas/Photo Researchers; 10.12: Photo by Tim McCabe, courtesy of USDA Natural Resources Conservation Service; 10.15: © William P. Cunningham; 10.16: © FAO Photo/W. Ciesla; 10.18: © W.C. Wood/Tanimura and Antle, Inc.; p. 221: Courtesy of Bill Wilcke; 10.21 & 10.22: © William P. Cunningham.

Chapter 11

Opener: Courtesy of David Tilman; 11.1: Courtesy of David Tilman; 11.2: © Vol. 112/PhotoDisc; 11.4: © Vol. 244/PhotoDisc; 11.6–11.8 & 11.11: © William P. Cunningham; 11.12: © Corbis Royalty Free; 11.13: Courtesy of Bell Museum, University of MN. Photo taken by Mary Ann Cunningham; 11.15a: Courtesy of Dr. Mitchell Eaton; 11.15b: © William P. Cunningham; 11.15c: © Lynn Funkhouser/Peter Arnold; 11.16: © Vol. 6/Photo Disc; 11.17: © Tom McHugh/Photo Researchers; p. 244 (top): © Digital Vision Royalty Free; p. 244 (bottom): © Mary Ann Cunningham; 11.20 & 11.21: © William P. Cunningham; 11.22: Courtesy of Dr. Ronald Tilson, Minnesota Zoological Garden.

Chapter 12

Opener: © Ron Thiele; 12.5: © William P. Cunningham; 12.6: © Digital Vision/Getty Images; 12.8a-c: Courtesy United Nations Environment Programme; 12.9: © Steven P. Lynch; 12.10: © Greg Vaughn/Tom Stack & Associates; 12.11: © Gary Braasch/Stone/Getty Images; 12.12: © William P. Cunningham; 12.13: U.S. Forest Service; 12.14: © William P. Cunningham; 12.16: © Mary Ann Cunningham; 12.17: Courtesy David L. Hansen, University of Minnesota, Agriculture Experiment Station; 12.18 & 12.20: © William P. Cunningham; 12.22: Courtesy American Heritage Center, University of Wyoming; 12.23: © William P. Cunningham; 12.24: © Richard Hamilton Smith/Corbis; 12.25: © Corbis/Royalty Free; 12.26: © William P. Cunningham; 12.30: Courtesy of R.O. Bierregaard.

Chapter 13

Opener: © Mary Ann Cunningham; 13.3, 13.4, 13.5b,c & 13.6: © William P. Cunningham; 13.7: © Mark Edwards/Peter Arnold; 13.8: © Vol. 5/PhotoDisc/Getty Images; 13.9, 13.10 & 13.12: © William P. Cunningham; 13.14a: Courtesy of Wisconsin-Madison Archives; 13.14b: Courtesy University of Wisconsin Arboretum; 13.15: Courtesy Minnesota Department of Natural Resources; 13.16 & 13.17 © William P. Cunningham; 13.19: © Tom Bean/Corbis; 13.20: Courtesy U.S. Army Corps of Engineers; 13.21: © Chris Harris; 13.22: Courtesy U.S. Army Corps of Engineers; 13.23: © Jeff Greenberg/Photo Researchers; p. 293: © Mary Ann Cunningham; 13.25: Courtesy of Sout Florida Water Management District; 13.27: Courtesy NOAA; 13.28: Photo by Don Poggensee, USDA Natural Resources Conservation Service; 13.29: © William P. Cunningham; 13.30: Courtesy Federal Interagency Stream Corridor Restoration Handbook, NRCS, USDA; 13.22a,b, 13.32a,c & 13.34: © William P. Cunningham.

Chapter 14

Opener: © William P. Cunningham; 14.7: © Corbis Royalty Free; 14.9: Courtesy of David McGeary; 14.10: © PhotoDisc Royalty Free; p. 309: © Bryan F. Peterson; 14.12: © James P. Blair/National Geographic Image Collection; 14.13: © AP Photo/Bob Bird; 14.15: © Joseph Nettis/Photo Researchers; 14.16a: © Vol. 34/PhotoDisc/Getty; 14.16b,c: © Corbis Royalty Free; 14.17: Charles Mueller, U.S. Geological Survey; p. 317: Department of Energy; 14.20: Chris G. Newhall/U.S. Geological Survey; 14.21: © Los Angeles Times, photo by Al Seib.

Chapter 15

Opener & 15.1: © William P. Cunningham; 15.8: © Vic Engelbert/ Photo Researchers; 15.10a: Images produced by Hal Pierce, Laboratory for Atmosphere, NASA, Goddard Space Flight Center/NOAA; 15.10b: © Marc Serota/Corbis; 15.10c: © William P. Cunningham; 15.11: Courtesy Candace Kohl, University of California, San Diego; 15.17: © Vol. 244/Corbis; 15.19a, b: Photographer Lisa McKeon, courtesy of Glacier National Park Archives; 15.20: © AP/Wide World Photos; p. 340: © William P. Cunningham; 15.23: © Stonehaven CCS Canada, Inc.; 15.24: © Vol. 23/PhotoDisc/Getty.

Chapter 16

Opener: © Corbis Royalty Free; 16.2a,b: © William P. Cunningham; 16.3: U.S. Geological Survey; 16.4 & 16.7: © William P. Cunningham; 16.9: Image courtesy of Norman Kuring, SeaWIFS Project; 16.14: © Guang Hui Xie/The Image Bank/Getty Images; 16.18: NASA; 16.20: © McGraw-Hill Companies, Inc., John Thoeming, photographer; 16.22–16.24: © William P. Cunningham; 16.25: © Vol. 44/PhotoDisc; 16.27: © William P. Cunningham; 16.28: © John Cunningham/Visuals Unlimited; 16.30: © Corbis Royalty Free; 16.32: © vario images GmbH & Co.KG/Alamy; 16.33: © William P. Cunningham.

Chapter 17

Opener, 17.2 & 17.3: © William P. Cunningham; 17.7: NASA; 17.9: © William P. Cunningham; 17.12: © Ray Ellis/Photo Researchers; 17.15a, c, & d: EROS Data Center, USGS; 17.15b: NASA; 17.16a: Photo by Jeff Vanuga, USDA Natural Resources Conservation Service; 17.16b: Photo by Tim McCabe, USDA Natural Resources Conservation Service; 17.16c: Photo by Lynn Betts, USDA Natural Resources Conservation Service; 17.18: Courtesy NREL/PIX; 17.20: Courtesy of USDA, NRCS, photo by Lynn Betts; 17.21: © William P. Cunningham; 17.22: Courtesy of Tim McCabe, Soil Conservation Services, USDA; 17.23: © William P. Cunningham; 17.24: © Josh Baker/Colorado Whitewater Photography; 17.25: © John Gerlach/Visuals Unlimited; 17.28 & 17.29: © William P. Cunningham.

Chapter 18

Opener: © Peter Menzel Photography; 18.2: © Simon Fraser/SPL/Photo Researchers; 18.3: © Joe McDonald/Animals Animals Earth Scenes; 18.4: © Dr. Parvinder Sethi; 18.6: © Roger A. Clark/Photo Researchers; 18.8: © William P. Cunningham; 18.10: © Rob and Melissa Simpson/Valan Photos; 18.12: © Lawrence Lowry/Photo Researchers; 18.15: © William P. Cunningham; 18.16: © Les Stone/Sygma/Corbis; 18.19: © Steve Austin/Papilo/Corbis; 18.20: Courtesy of Joe Lucas/Marine Entanglement Research Program/National Marine Fisheries Service NOAA; 18.21: © McGraw-Hill Companies, Inc., Bob Coyle, photographer; 18.25: Courtesy FAO/9717/I. de Borhegyi; 18.27: Courtesy of National Renewable Energy Laboratory/NREL/PIX.

Chapter 19

Opener: © George Steinmetz/Corbis; 19.2: © Vol. 160/Corbis; 19.4: © Owen Franken/Stock Boston; 19.8: © Vol. 39/PhotoDisc; p. 403: © William P. Cunningham; 19.13: Courtesy of Office of Response and Restoration, NOAA; p. 434: © Mary Ann Cunningham; 19.14 © William P. Cunningham; 19.15 © Regis Martin/The New York Times; p. 437: © Peter Aengst/Wilderness Society/Lighthawk; 19.17: © Corbis Royalty Free; 19.24: Courtesy Office of Civilian Radioactive Waste Management, DOE; 19.26: Courtesy of Department of Energy.

Chapter 20

Opener: © Mary Ann Cunningham; 20.1: Courtesy of Tom Finkle; 20.3: Courtesy of National Renewable Energy Laboratory/NREL/PIX; 20.4: Courtesy of Carolyn Roberts/A House of Straw; 20.6: Courtesy Tom Finkle; 20.7: Courtesy of Capstone Micro Turbines; 20.9: © William P. Cunningham; 20.10: Courtesy of National Renewable Energy Laboratory/NREL/PIX; 20.11: © McGraw-Hill Companies, Inc., Doug Sherman, photographer; 20.12: © William P. Cunningham; 20.14: Courtesy of National Renewable Energy Laboratory/NREL/PIX; 20.15: Courtesy Warren Gretz, National Renewable Energy Laboratory, DOE; 20.17: Courtesy of Long Island Power Authority; 20.18 & 20.19: © William P. Cunningham; 20.20: Courtesy Warren Gretz, National Renewable Energy Laboratory, DOE; 20.22: Courtesy of Dr. Douglas Pratt; 20.23 & 20.24: William P. Cunningham; 20.25: © The McGraw-Hill Companies, Inc./Barry Barker, photographer; 20.26: Courtesy of Burkhardt Turbines; 20.27: National Renewable Energy Lab; 20.29: © William P. Cunningham; 20.31: © Ocean Power Delivery Ltd.

Chapter 21

Opener: © Ray Pforther/Peter Arnold; 21.4: © William P. Cunningham; 21.5: © Vol. 31/PhotoDisc; 21.6: © Basel Action Network 2006; 21.10: © William P. Cunningham; p. 482: © Barbara Gauntt; 21.14: David L. Hansen, University of Minnesota Agricultural Experiment Station; 21.15: Courtesy of Urban Ore, Inc., Berkeley, California; 21.17: © Michael Greenlar/The Image Works; 21.22: © Pier van Lier.

Chapter 22

Opener: © Reuters/Corbis; 22.1: © William P. Cunningham; 22.2 & 22.4: © Corbis Royalty Free; 22.5, 22.8 & 22.10: © William P. Cunningham; 22.11: Courtesy Dr. Helga Leitner; 22.12: © Fred McConnaughey/Photo Researchers; 22.13: © Regents of the University of Minnesota. All Rights Reserved. Used with permission of the Metropolitan Design Center; 22.14a, b: U.S. Geological Survey; 22.15, 22.17, 22.18 & 22.19: © William P. Cunningham; 22.20: © Roofscapes, Inc. Used by permission; all rights reserved; 22.21a-c: © Mary Ann Cunningham; p. 513: © Bill Dunster; 22.23: © William P. Cunningham.

Chapter 23

Opener, 23.1–23.6 & 23.11: © William P. Cunningham; 23.18: © McGraw-Hill Companies, Inc./Barry Barker, photographer; 23.20: © William P. Cunningham; 23.21: © Mary Ann Cunningham; 23.24: © William P. Cunningham; p. 534: Courtesy Ray Anderson, Interface, Inc.; 23.25: © Mark Luthringer.

Chapter 24

Opener: © William P. Cunningham; 24.1: U.S. Fish & Wildlife Service; 24.4: © Vol. 25/PhotoDisc; 24.5: © Esther Henderson/Photo Researchers; 24.6: © Bob Daemmrich/The Image Works; p. 546: © Vol. 16/Corbis; 26.7: © William P. Cunningham; 24.8: © Vol. 25/PhotoDisc; 24.11 & 24.12: © William P. Cunningham; 24.14: Courtesy of Tom Finkle; 24.17: © John Riley/Stone/Getty Images; 24.18: © Phil Degginger/Animals Animals Earth Scenes; 24.19: © David L. Brown/Tom Stack & Associates.

Chapter 25

Opener: © Robert Sobol; 25.2–25.4, p. 570 & 25.7: © William P. Cunningham; 25.9: David L. Hansen, University of Minnesota Agricultural Experiment Station; 25.10–25.13: © William P. Cunningham.

LINE ART/TEXT CREDITS

Chapter L

L Table 1.2, p. 4: Source: Dr. Melvin Northrup, Grand Valley State University; L Table 1.4, p. 8: Source: Courtesy of Karen Warren, Philosophy Department, Macaelester College, St. Paul, MN.

Chapter 1

Table 1.1, p. 25: Source: UNDP Human Development Index 2006; Figure 1.9, p. 21: Source: NASA, 2002; Figure 1.11, p. 24: Source: World Bank, 2000; Figure 1.16, p. 28: Graph from World Energy Assessment, UNDP 2000, Figure 3.10, p. 95; Figure 1.20, p. 30: Source: Norman Myers, Conservation International and Cultural Survival Inc., 2002; TA1.1, p. 23: From LIVING PLANET, 2004, p. 10. Copyright © 2004 World Wide Fund for Nature (Formerly World Wildlife Fund). All rights reserved; TA1.2, p. 23: From LIVING PLANET, 2004, p. 10. Copyright © 2004 World Wide Fund for Nature (Formerly World Wildlife Fund). All rights reserved.

Chapter 2

TA2.1, p. 38: Source: Data from U.S. Environmental Protection Agency.

Chapter 3

Figure 3.1, p. 52: From "Salmon and alder as nitrogen sources to riparian forests in a boreal Alaskan watershed" by Robert Naiman and James Helfield in OECOLOGIA (2002) 133:573–582, Figure 3, p. 576. Copyright © 2002. Reprinted by permission of Springer Science and Business Media and the authors, Robert Naiman and James Helfield; Figure 3.10, p. 60: Source: Principles of Ecology: Matter, Energy and Life; Figure 3.16, p. 64: Source: Howard T. Odum, "Trophic Structure and Productivity of Silver Springs, Florida," in Ecological Monographs, 27:55–112, 1957, Ecological Society of America; Figure 3.20, p. 67: Source: Intergovernmental Panel for Climate Change (IPCC), 2002; Figure 3.21, p. 69: Source: Galloway and Cowling, 2002.

Chapter 4

Figure 4.8, p. 79: Source: Original observation by R.H. MacArthur (1958); Figure 4.25, p. 92: Verner, Jared. WILDLIFE 2000. © 1986. Reprinted by permission of The University of Wisconsin Press; TA4.3, p. 97: Source: Gause, Georgyi Frantsevitch 1934. THE STRUGGLE FOR EXISTENCE; TA4.4, p. 97: Source: Gause, Georgyi Frantsevitch 1934. THE STRUGGLE FOR EXISTENCE.

Chapter 5

Table 5.1, p. 112: Source: Hahhan, Lee, et al., "Human Disturbance and Natural Habitat: A Biome Level Analysis of a Global Data et," in Biodiversity and Conservation, 1995, Vol. 4: 128–155; Figure 5.2, p. 100: From COMMUNITIES AND ECOSYSTEMS, 2nd Edition by Robert C. Whittaker. Copyright © 1975. Reprinted by permission of Pearson Education, Inc., Upper Saddle River, NJ; Figure 5.3, p. 101: Source: WWF Ecoregions; Figure 5.22, p. 112: Source: United Nations Environment Programme, Global Environmental Outlook, 1997.

Chapter 6

Figure 6.1, p. 117: Source: World Resources Institute, 2000; Figure 6.5, p. 122: Source: National Vital Statistics Reports 51(3): 2002; Figure 6.8, p. 125: Source: D.A. MacLulich, Fluctuations In the Numbers of the Varying Hare (Lepus americus), Toronto: University of Toronto Press, 1937, reprinted 1974; Figure 6.10, p. 126: Source: Baesd on MacArthur and Wilson, The Theory of Island Biogeography, 1967, Princeton University Press; Figure 6.11, p. 127: Source: H.L. Jones and J. Diamond, "Short-term-base Studies of Turnover in Breeding Bird Populations on the California Coast Island," in Condor, Vol. 78: 526–549, 1976.

Chapter 7

Table 7.1, p. 134: Source: United Nations Population Division; Table 7.2, p. 137: Source: Population Reference Bureau 2006; Table 7.3, p. 138: Source: Population Reference Bureau 2006; Table 7.4, p. 141: Source: Population Reference Bureau 2006; Table 7.5, p. 149: Source: U.S. Food and Drug Administration, BIRTH CONTROL GUIDE, 2003 Revision; Figure 7.3, p. 134: Source: Figure redrawn with permission from Population Bulletin, Vol. 18, no. 18, 1985. Population Reference Bureau; TA7.3, p. 153: Source: U.S. Census Bureau; TA7.4, p. 153: Source: U.S. Census Bureau; Figure 7.5, p. 136: Source: UN Population Division, 2005; Figure 7.6, p. 137: Source: UN Population Division, 2006; Figure 7.7, p. 138: Source: World Bank, 2000; Figure 7.8, p. 139: Source: UN, ESA/WP, 180, World Population Prospects, 2002 Revision, February 2003, p. 17; Figure 7.10, p. 142: Source: The World Bank, World Development Indicators 1997, the World Bank, Washington DC, 1997; Figure 7.11, p. 142: Source: U.S. Census Bureau, 2006; Figure 7.14, p. 144: Sources: Population Reference Bureau and U.S. Bureau of the Census; Figure 7.15, p. 146: Source: http://www.uwmc.uwc.edu/geography/Demotrans/demtran.htm. Reprinted by permission of Keith Montgomery; Figure 7.17, p. 148: Source: Worldwatch Institute 2003; Figure 7.18, p. 149: Source: UN Population Division, 2004; Figure 7.19, p. 150: Source: Population Reference Bureau, 2003; Figure 7.20, p. 150: Data Source: U.N. Population Division.

Chapter 8

Table 8.1, p. 157: Source: World Health Organization, 2002; Table 8.2, p. 162: Source: U.S. EPA, 2003; Table 8.3, p. 165: Source: U.S. Department of Health and Human Services, 1995; Table 8.4, p. 174: Source: U.S. National Safety Council, 2003; Table 8.5, p. 175: Source: Environmental Protection Agency; Figure 8.2, p. 156: Source: WHO, 2002; Figure 8.3, p. 156: Data from U.N. Population Division 2006; Figure 8.6, p. 159: Source: Data from U.S. Centers for Disease Control and Prevention; Figure 8.7, p. 159: Source: CDC and USGS; Figure 8.10, p. 161: "How Microbes Acquire Antibiotic Resistance" from TIME, September 12, 1994. Copyright © 1994 TIME Inc.; TA8.2, p. 167; Pickle, J.L., et al ENV HEALTH PERSPECTIVES 1998, 106:745–50. Fata Residential Injuries among U.S. Children, 1985 to 1997 from www.centerforhealthyhousing.org; TA8.3, p. 167: Pickle, J.L., et al ENV HEALTH PERSPECTIVES 1998, 106:745–50. Fata Residential Injuries among U.S. Children, 1985 to 1997 from www.centerforhealthyhousing.org; Figure 8.20, p. 174: Source: D.E. Patton, "USEPA's Framework for Ecological Risk Assessment" in Human Ecological Risk Assessment,

Vol. 1, No. 4; Figure 8.21, p. 175: Reprinted with permission of the Star-Tribune, Minneapolis-St. Paul; TA8.4, p. 177: Source: Data from Slovic, Paul. 1987. Perception of Risk, SCIENCE, 236(4799):286–290.

Chapter 9

Table 9.1, p. 184: Source: Food and Agriculture Organization (FAO), 2006; Table 9.2, p. 202: Source: Based on 14 years' data From Missouri Experiment Station, Columbia, MO; Figure 9.2, p. 181: FIVIMS Hunger Map from Food and Agriculture Organization of the United Nations website; Figure 9.3, p. 181: Source: Food and Agriculture Organization (FAO), 2002; Figure 9.4, p. 181: Source: Food and Agriculture Organization (FAO), 2002; Figure 9.8, p. 183: Source: Worldwatch Institute 2001; Figure 9.9, p. 184: (a) Data from Willett and Stampfer, 2002, (b) USDA, 2005; Figure 9.11, p. 185: (c) Data from USDA, 2005; Figure 9.15, p. 189: Source: Soil Conservation Service; Figure 9.19, p. 192: Source: Data from USDA; Figure 9.28, p. 198: Source: Data from Information Systems for Biotechnology, Virginia Tech University; TA9.3, p. 206: Source: World Wide Fund for Nature (WWF), 2000.

Chapter 10

Chapter 10, What can you do?, p. 218: Source: Citizen's Guide to Pest Control and Pesticide Safety, EPA 730-K-95-001; Table 10.1, p. 223: Source: Environmental Working Group, 2002; Figure 10.3, p. 210: Source: U.S. EPA 2000; Figure 10.4, p. 210: Source: U.S. EPA 2002; Figure 10.10, p. 213: Source: Worldwatch Institute, 2003; Figure 10.13, p. 217: Source: From E.A. Guillette, et al., "An Anthropological Approach to the Evaluation of Preschool Children Exposed to Pesticides in Mexico" in Environmental Health Perspective, 106(6):347–353, 1998. U.S. Dept. of Health and Human Services; Figure 10.19, p. 220: Source: Tolba, et al., World Environment, 1972–1992, p. 307, Chapman & Hall, 1992 United Nations Environment Programme; Figure 10.20, p. 222: Source: National Academy of Sciences; TA10.4, p. 227: Data source: C.J. Olson Market Research Inc. for Minnesota Department of Agriculture. QUANTITATIVE RESEARCH REGARDING PEST MANAGEMENT IN MINNESOTA K-12 SCHOOLS, 1999.

Chapter 11

Figure 11.5, p. 232: Source: Conservation International; Table 11.1, p. 231: Source: IUCN 2006; Table 11.3, p. 235: Source: W.W. Gibbs, 2001.

Chapter 12

Table 12.1, p. 264: Source: Data from USFS, 2002; Table 12.2, p. 268: Source: Data from World Conservation Union, 1990; Figure 12.2, p. 254: Source: Food and Agriculture Organization (FAO), 2002; Figure 12.3, p. 255: Source: UN Food and Agricultural Organization (FAO); Figure 12.4, p. 255: Source: United Nations Food and Agriculture Organization; Figure 12.7, p. 257: Source: Data from Food and Agriculture Organization of the United Nations, 2002; TA12.1, p. 259: Source: Wildlife Conservation Society; TA12.2, p. 259: Source: Wildlife Conservation Society; TA12.3, p. 259: Source: Wildlife Conservation Society; Figure 12.15, p. 266: Source: World Resource Institute, 2004; Figure 12.19, p. 267: Source: UN World Commission on Protected Areas; Figure 12.21, p. 269: Data from J. Hoekstra, et al, 2005.

Chapter 13

Figure 13.1, p. 278: Source: USGS; Figure 13.2, p. 279: Data Source: Walker and Moral, 2003; Figure 13.11, p. 286: Data from S. Packard and J. Balaban, 1994; Figure 13.18, p. 290: Source: http://www.peaceparks.org.

Chapter 14

Table 14.3, p. 313: Source: Blacksmith Institute 2006; Table 14.4, p. 314: Source: E.T. Hayes, Implications of Materials Processing, 1997; Table 14.5, p. 315: Source: The Disaster Center 2005; Table 14.6, p. 316: Source: B. Gutenberg in Earth by F. Pres and R. Siever, 1978, W.H. Freeman & Company; Figure 14.3, p. 305: Source: U.S. Department of the Interior, U.S. Geological Survey; Figure 14.14, p. 313: Source: George Laycock, Audubon Magazine, Vol. 91(7), July 1989, p. 77; Figure 14.19, p. 318: Source: National Oceanic and Atmospheric Administration, from "U.S. Announces Plans for Improved Tsunami Warning and Detection System."

Chapter 15

Figure 15.2, p. 324: Source: Courtesy of Dr. William Culver, St. Petersburg Junior College; Figure 15.12, p. 332: Source: Data from United Nations Environment Programme; Figure 15.15, p. 334: Source: Data from NASA Earth Observatory, 2002; TA15.3, p. 346: Data from Peter Reich; TA15.3, p. 346: Climate Change 2007: The Physical Science Basis Summary for Policymakers. Reprinted by permission of Intergovenmental Panel on Climate Change; Figure 15.21, p. 339: Climate Change 2007: The Physical Science Basis Summary for Policymakers. Reprinted by permission of Intergovenmental Panel on Climate Change; Figure 15.22, p. 342: Data from Socolow, et al, ENVIRONMENT, Vol. 46, No. 10, p. 11, 2004.

Chapter 16

Table 16.1, p. 351: Source: UNEP, 1999; Figure 16.1, p. 348: Source: EPA, 1998; Figure 16.5, p. 351: Source: UNEP, 1999; Figure 16.6, p. 352: Source: UNEP, 1999; Figure 16.8, p. 352: Source: UNEP, 1999; Figure 16.11, p. 355: Source: UNEP, 1999; Figure 16.12, p. 355: Source: Data from B.J. Finlayson-Pitts and J.N. Pitts, Atmospheric Chemistry, 1986, John Wiley & Sons, Inc.; Figure 16.13, p. 356: Source: Environmental Defense Fund based on EPA data, 2003; Figure 16.26, p. 365: Source: National Atmospheric Deposition Program/National Trends Network, 2000. http://nadp.sws. uiuc.edu; Figure 16.31, p. 369: Source: Environmental Protection Agency, 2002.

Chapter 17

Table 17.1, p. 377: Data from U.S. Geological Survey; Table 17.3, p. 380: Source: World Resources Institute; Figure 17.8, p. 379: Source: U.S. Geological Survey; Figure 17.13, p. 382: Source: Data from U.S. Department of Interior; Figure 17.14, p. 382: Source: UNEP, 2002; Figure 17.17, p. 384: Source: EPA, 2004; Figure 17.19, p. 386: Climate Change 2007: The Physical Science Basis Summary for Policymakers. Reprinted by permission of Intergovenmental Panel on Climate Change; Figure 17.30, p. 394: Source: World Bank estimates based on research paper by Dennis Anderson and William Cavendish, "Efficiency and Substitution in Pollution Abatement: Simulation Studies in Three Sectors;" TA17.3, p. 396: Freshwater Stress and Scarcity in Africa by 2025, VITAL WATER GRAPHICS, reproduced with kind permission from the United Nations Environment Programme (www.unep.org/vitalwater/management.htm).

Chapter 18

Figure 18.5, p. 401: Source: UNESCO, 2002; Figure 18.14, p. 409: Source: USDA and Natural Resources Conservation Service; Figure 18.18, p. 412: Source: EPA Safe Drinking Water Information System 2001; Figure 18.24, p. 417: © 2001 Sidney Harris, ScienceCartoonsPlus.com. Reprinted by permission; Figure 18.26, p. 418: Source: World Bank estimates based on research paper by Dennis Anderson and William Cavendish, "Efficiency and Substitution in Pollution Abatement: Simulation Studies in Three Sectors;" Figure 18.28, p. 421: Source: B. Cohen and E. Bondoc, "Weed Killers by the Glass," Environmental Working Group 1995.

Chapter 19

Figure 19.3, p. 426: Source: Data from British Petroleum, 2006; Figure 19.11, p. 432: Source: Jean Laherrere, www.hubbertpeak.org; Figure 19.12, p. 433: Source: British Petroleum, 2006; Figure 19.16, p. 436: Source: British Petroleum, 2006.

Chapter 20

Table 20.1, p. 451: Source: U.S. Department of Energy; Table 20.2, p. 460: Source: Alan C. Lloyd, 1999; Table 20.3, p. 468: Source: Lester R. Brown, 1991; Figure 20.2, p. 450: Source: U.S. Department of Energy; Figure 20.5, p. 451: Source: U.S. Department of Energy; Figure 20.8, p. 454: Source: National Weather Bureau, U.S. Department of Commerce; p. 465: Tilman et al, 2006. Science 314:1596; Figure 20.28, p. 469: Source: Data from U.S. Department of Energy; Figure 20.32, p. 472: Source: World Energy Council, 2002.

Chapter 21

What can you do? Page 487: Source: Minnesota Pollution Control Agency; Table 21.1, p. 492: Source: EPA,

2005; Figure 21.1, p. 475: Data Source: U.S. EPA; Figure 21.2, p. 476: Source: Data from Environmental Protection Agency, 2006; Figure 21.3, p. 476: Source: Data from U.S. Environmental Protection Agency Office of Solid Waste Management, 2006; Figure 21.8, p. 480: Source: Eurostat, UNEP, 2003; TA21.3, p. 482: Source: M. Lavelle and M. Coyle, The National Law Journal, Vol. 15:52–56, No. 3, September 21, 1992; Figure 21.11, p. 483: Source: Environmental Protection Agency, 2003; Figure 21.13, p. 483: Source: Environmental Protection Agency, 2003; Figure 21.18, p. 487: Source: Data from the U.S. EPA, 2002; Figure 21.20, p. 489: Source: Environmental Protection Agency, 2001; Figure 21.21, p. 489: Source: Environmental Protection Agency.

Chapter 22

Table 22.1, p. 498: Source: United Nations Population Division, 2004; Table 22.2, p. 500: Source: T. Chandler, Three Thousand Years of Urban Growth, 1974, Academic Press and World Gazetter, 2003; Table 22.3, p. 506: Source: Excerpt from speech by Anthony Downs at the CTS Transportation Research Conference, as appeared on Website by Planners Web, Burlington, VT 2001; Table 22.4, p. 509: Source: Vision 2000, Chattanooga, TN; Figure 22.3, p. 498: Source: United Nations Population Division, World Urbanization Prospects, 2004; Figure 22.6, p. 499: Source: U.S. Census Bureau; Figure 22.9, p. 499: Source: EarthTrends, 2006.

Chapter 23

Table 23.2, p. 520: Source: R.S. de Groot, Investing in Natural Capital, 1994; Table 23.3, p. 529: Source: Adapted from R. Costanza et al, "The Value of the World's Ecosystem Services and Natural Capital," Nature, Vol. 387 (1997); Figure 23.9, p. 523: Source: T. Prugh, National Capital and Human Economic Survival 1995, ISEE; Figure 23.10, p. 523: Source: Herman Daly in A.M. Jansson, et al, Investing in Natural Capital, ISEE; Figure 23.17, p. 528: Source: Redefining Progress, 2006; Figure 23.22, p. 532: Source: United Nations, 2007; Figure 23.26, p. 537: Source: E.S. Goodstein, Economic Policy Institute, Washington, DC; TA23.3, p. 539: Source: United Nations Development Programme, 2006.

Chapter 24

Table 24.1, p. 547: Source: N. Vig and M. Kraft, Environmental Policy in the 1990s, 3rd Congressional Quarterly Press; Table 24.3, p. 557: Kai N. Lee, Compass and Gyroscope, 1993, Coypright © 1993 Island Press, Reprinted by permission of Alexander Hoyt & Associates; Table 24.4, p. 558: Source: L. Gunderson, C. Holling and S. Light, Barriers and Bridges to the Renewal of Ecosystems and Institutions, 1995, Columbia University Press; Figure 24.2, p. 542: Source: Modified from Ray Grizzle, Bioscience, Vol. 44(4), April 1994; Figure 24.10, p. 552: Source: U.S. General Accounting Office; Figure 24.13, p. 555: Source: United Nations Environment Programme from Global Environment Outlook 2000; Figure 24.16, p. 558: Source: Buzz Holling, Barriers & Bridges, edited by Gunderson, Holing and Light, 1995. Columbia University Press. Reprinted by permission of the authors; Data Analysis Figure 1, Graphic from PILOT 2006 EPI, Figure 12, p. 78. Reprinted by permission of Environmental Performance Index, Yale University.

Chapter 25

Table 25.1, p. 568: Source: A Greenprint for Minnesota, Minnesota Office of Environmental Education, 1993; Table 25.2, p. 568: From Robert Merideth, The Environmentalist's Bookshelf: A Guide to the Best Books, 1993 by G.K. Hall, an imprint of Macmillan, Inc. Reprinted by permission; What can you do? p. 572: Based on material by Karen Oberhauser, Bell Museum Imprint, University of Minnesota, 1992. Used by permission. Table 25.3, p. 577: Based on material from "Grassroots Organizing for Everyone" by Claire Greensfelder and Mike Roselle from Call to Action, 1990 Sierra Book Club Books; TA25.3, p. 570: Reprinted by permission of Audubon; Figure 25.5, p. 571: © 1990 Bruce von Alten; Figure 25.6, p. 573: Source: After Society of Environmental Toxicology and Chemistry, 1991; Figure 25.14, p. 580: Source: UN, 2002; Figure 25.15, p. 580: Modified from Raymond Grizzle and Christopher Barrett, personal communication.

animals. *See also* wildlife; *individual species*
    intrinsic or instrumental values and, 44–45
    introduced species, 95
    lab animals, toxicity testing on, 170–171
    moral value and, 44–45
    water quality and pet waste disposal, 409
Annan, Kofi, 194, 580
antagonistic reactions, 169
Antarctica, stratospheric ozone destruction, 360–361
antibiotics, resistance, 82, 161
antigens, 163
ants, symbiotic relationships in, 86
aquaculture, 186–187
aquifers, 379, 391. *See also* groundwater
Aral Sea (Kazakhstan, Uzebekistan), 383
arbitration, 559
architecture
    acid precipitation damage to buildings, 366
    controlling water pollution from construction, 414
Arctic National Wildlife Refuge (ANWR) (Alaska)
    debate over drilling for oil, 433–434, 435, 451
    and natural gas, 437
arctic tundra, 105, 113
Arendt, Randall, 513
arguments, unpacking, 10
aridosols, 191
Army Corps of Engineers, US
    dam construction, 292, 386, 388
    Everglades restoration, 294
    flood control efforts, Mississippi River, 278
    reclamation projects, 299
Arrhenius, Svante, 334
arsenic
    from coal burning, 430
    as water pollutant, 405
artesian wells, 379, 411
Asia. *See also individual countries*
    AIDS in, 137, 159
    air pollution, 359–360
    arable land use in East, 191
    deforestation in, 22
    electronic waste shipments to, 478
    fertility rates, 139
    land degradation in, 192
    malaria, 161
    night soil, collecting, 415
    poverty and, 25
    reforestation programs, 256
Asian Americans, environmental racism, 47
asphyxiants, 363
Atlantic Coastal Action Programme (ACAP), 560, 561
atmosphere
    composition, 323
    energy balance, 325
    four zones, 324–325
    structure, 323–325
    as water compartment, 381
atmospheric deposition, 400
atomic numbers, 53
atoms, 53
Atoms for Peace (speech), 437
atrazine, 169
    effects of exposure in frogs, 216
    as water pollutant, 406, 412, 421
Audubon, John James, 239
Audubon Society, 574, 575
    Christmas Bird Watch, 569, 570
aurora australis, 325
aurora borealis, 325
Australia
    drinking water in Queensland, 393
    marine reserve, 271
    prickly pear cactus, introduction of, 219
    surface-water quality in, 409

    water pricing and allocation policies, 394
    wealth in, 25
Austria, air pollution reduction, 370
automobiles. *See also* diesel engines
    flex-fuel vehicles, 464
    fuel cells, improving efficiency of, 459
    hybrid gasoline-electric engines, 452
    hydrogen fuel-cell-powered vehicles, 452–453
    improving efficiency of, 451–453
    plug-in hybrids, 452
    positive crankcase ventilation systems, 367
Avery, Dennis, 224
Azodrin, 213

Babbitt, Bruce, 388
bacteria, as water pollutants, 401–402
bald eagles, 242, 243
Baloney Detection Kit, 42
bamboo, 78
Bangladesh
    biomass for energy, dependence on, 462
    Grameen Bank in, 518, 533
    monsoon rains, 329, 360
barrier islands, 109–110
Bartholomew (patriarch), 46
Basel Action Network, 478
bases, 54–55, 405
Bates, H. W., 85
Batesian mimicry, 85
bathypelagic zone, 106
batteries
    alkali-metal, 458
    lead-acid, 457–458
    lithium-ion, 458
    metal-gas, 458
    sodium-sulfur, 458
Beagle, H.M.S., 75
Beatty, Mollie, 252
Becquerel, Alexandre-Edmond, 457
BedZED, ecological housing complex
    (United Kingdom), 511, 513
bees, honeybees, 208
beluga whales, 215
benthos
    freshwater, 110
    marine, 106
benzene, 356, 357
Bermuda, Nonsuch Island recovery project, 282–283
best available, economically achievable technology
    (BAT), 420
best practicable control technology (BPT), 419
Better Not Bigger (Fodor), 506
Beyond the Limits, 526
Bhutan, biomass for energy, dependence on, 462
bias, and science, 37–39
Bible, 46
Bill and Melinda Gates Foundation, 162, 197
Bingham, Sally, 34
binomials, 82
bioaccumulation, 168
biocentric preservation, 18
biochemical oxygen demand (BOD), 402
biocides, 209–210
biodegradable plastics, 486
biodiesel fuels, 466
biodiversity, 228–249
    aesthetic and cultural benefits of, 234–235
    benefits of, 233–235
    commercial products, loss of biodiversity from,
        240–241
    defined, 230
    and drugs, 233–234
    ecological benefits of, 234
    and economic stability, 229

    endangered species management, 241–243,
        245–247
    extinction (*see* extinction)
    and food, 233
    habitat destruction, loss of biodiversity
        through, 235–236
    hot spots, 232–233
    human population, loss of biodiversity from, 239
    invasive species, 236–238
    and medicines, 233–234
    molecular techniques in taxonomic relationships,
        230–231
    overharvesting, loss of biodiversity through, 239–240
    pollution, loss of diversity through, 239
    species (*see* species)
    threats to, 235–241
biofuels, 463–464, 465
biological communities, 87–95
    abundance, 87–88
    climax community, 92–93
    community structure, 89
    complexity and connectedness, 90
    defined, 62
    disturbances, 93–95
    ecological succession, 93
    edges and boundaries, 90, 92
    introduced species, 95
    productivity, 87
    resilience, 90
    stability, 90
biological controls, 211
biological pests, 208
biomagnification, 168
biomass
    accumulation in world ecosystems, 87, 88
    burning, 461–462
    defined, 62
    dung, as fuel, 462–463
    ecological pyramids and, 65–66
    energy from, 460–466
    fuel efficiency, 465
    fuelwood, 462
    methane (*see* methane gas)
biomes
    defined, 100
    freshwater ecosystems, 110–111
    human disturbance, 111–113
    marine ecosystems, 106–110
    terrestrial, 100–106
bioorganic compounds, 55
biopiracy, 233
bioremediation, 493
biosphere reserves, 272–273
biotic population factors, 124
biotic potential, 118
birds. *See also individual species*
    abundance and diversity in, 88
    Audubon Christmas Bird Watch, 569, 570
    bird colonies on Falkland Islands, 89
    Breeding Bird Survey, 91
    Galápagos species, 79, 81–82
    live specimens, importation of, 240–241
    Pelican Island, first national bird reservation, 282
birth control, 148–149
birth dearth, 145
birth rates, 22, 138, 144–145
Bismark, Otto von, 562
bison, American. *See* buffaloes
bisphenol A (BPA), 169
black blizzards, 353
black lung disease, 430
Blacksmith Institute, 313, 370
Blair, Tony, 22
blind experiments, 39

Bloomberg, Michael, 484
blue baby syndrome, 412
body burden, 169
Boettner, George, 221
bogs, 111
boiling water nuclear reactors (BWR), 440
Boisclair, Andre, 223
Bolivia, debt-for-nature swaps in, 258
Bonaparte, Napoleon, 344
Bonhoeffer, Dietrich, 33
boom-and-bust cycles, in populations, 119
boreal forests, 100, 104–105, 254–255
Borlaug, Norman, 197
Bos-Wash megacity, 499
botanical gardens, 247
bottom trawling, 236
bottom-up development, 533
boundaries, in biological communities, 90, 92
Brazil, 352
    agriculture in the Cerrado, 178, 179–180, 191
    air pollution controls in Cubatao, 370–371
    crop-based ethanol production, 464
    deforestation in, 257
    environmental protection in urban planning in
        Curitiba, 497, 508–509
    forest destruction, 185
    hydropower use in, 28, 466, 467
    landownership disputes, 185
    old-growth forests in, 255
    parks and preserves in, 268
    waste-to-energy plants, 480
    water usage, 382
breeder nuclear reactors, 441–442
Breeding Bird Survey, 91
breeding programs, 247–248, 281
Brewster, William, 18
British Antarctic Atmospheric Survey, 360
British Petroleum, 344, 456
Broecker, Wallace, 322, 328
bronchitis, 363
Bronx Zoo (New York), 247
Brosnan, Pierce, 566
Brower, David, 19
brown bears, 128, 242
Brown, Greg, 474
Brown, Lester, 146
Browner, Carol, 415
brownfield developments, 511
brownfields, 299, 490–491
barrier islands, 109
Brundtland, Gro Harlem, 20, 27
Bt crops, 199, 218
bubonic plague, 133, 315
Buddhism, 46
buffalo commons, 289
buffaloes, 244, 520
    bison introduction onto prairies, 288, 289,
        290–291
    hunted to near extinction, 239
buildings, acid precipitation damage to, 366
built capital, 519
bull thistle, 77
Bullard, Robert D., 482
Bureau of Land Management, 309, 546, 553
Bureau of Reclamation, 299, 386
Bureau of Transportation Statistics, 451
Burke, Edmund, 344
Burroughs, John, 116
Burundi, biomass for energy, dependence on, 462
Bush, George W., 424
    Clean Air Act, 368
    environmental rules and policies under, 550, 552
    forest policy, 261
    global warming issues, 323, 335

new deep-ocean assessment buoys, plan to
    deploy, 318
refusal to honor Kyoto Protocol, 341
renewable and alternatives fuels, 431
wetlands protection, 291
butterflies, use of Batesian mimicry, 85

Cabot, John, 117
calcium carbonate, 67
calcium sulfate (gypsum), 71
California
    air pollution reduction in, 369
    cat parasite killing sea otters, 160
    chaparral landscape, 103
    condors, 243, 246, 247
    cotton yields in 1990s, plunging of, 215
    dams, removal of aging, 388
    DDT by-product in pregnant women
        (Los Angeles), 215
    domestic water use, 382, 393
    Edison's Solar II plant, Mojave Desert, 455
    global warming, 323
    groundwater depletion, 392
    Headwaters Redwood Forest, 46
    herbicide spill into Sacramento River, 213
    honeybee shortage, 208
    oil spill in Santa Barbara Channel, 544–545
    pesticide exposure, young children and, 217
    pests resistance in, 213
    redwoods, 104
    San Onofre Nuclear Plant, 438, 444
    sea lions with tumors, 160
    sudden oak death syndrome, 160
    temperature inversions in Los Angeles, 358
    traffic congestion in Los Angeles, 507
    wastewater treatment in Arcata, 397, 398
    water diversion controversy, Los Angeles,
        390–391
    wetland disturbances in, 113
    wind power use, 468
Calment, Jeanne Louise, 141
Calthorpe, Peter, 510
Camel's Hump Mountain (Vermont), 365
Cameroon, family size, 144
Campus Climate Challenge, 579
campus greening, 577–579
Canada
    Atlantic Coastal Action Programme, 560
    carbon management program in, 343–344
    Churchill Nelson River diversion in
        Manitoba, 467
    energy consumption, 427
    flooding of ancestral First Nations lands, debate
        over, 387, 467
    Great Bear Rainforest, 253, 260, 268, 272–273
    green plans, 560, 561
    hydropower use in, 466
    insecticide spraying in Canadian forests, 213
    irrigation methods, 195
    James Bay hydroelectric project, debate over,
        28, 467
    lynx population, 124–125
    mercury contamination in, 355
    Montreal Protocol, 361–362, 555
    National Packaging Protocol (NPP), 486
    native pollinators in New Brunswick, spraying
        for, 208
    natural gas source, for US, 436
    nuclear reactors, 440
    oil production, 427–428
    old-growth forests in, 255, 260
    parks and preserves in, 268
    persistent pollutants and, 215
    Pest Control Products Act and Regulations, 222

pesticide regulation, 222, 223
placer mining, 311
R-2000 program, 450
rich and poor in, 25
slow-poke nuclear reactors, 441
tar sands, 435
Water Act (1970), 409
water usage, 382, 386
wood and paper pulp production in, 255
Canary Islands
    Cumbre Vieja volcano, 318
    insects introduced from, 86
cancers, 165
    cancer rate from HAPs, 356
    global cancer rates, 157
    saccharin and, 174
    smoking and lung cancer, 357, 363
    uranium mining and lung cancer, 438
cap-and-trade program, in Kyoto Protocol, 531
capital, in economics, 519–520
captive breeding and species survival plans,
    247–248
carbamates, 211
carbaryl, 164
carbohydrates, 55, 56
carbon
    atom, 53, 54
    carbon enrichment studies, 340
    forests as carbon sink, 254–255
    management, 343–344
    in organic compounds, 55–57
    trading, 531–532
carbon cycle, 67–68
carbon dioxide
    as air pollutant, 349, 350, 351, 369
    atmospheric, 324, 325
    carbon enrichment studies, 340
    from coal burning, 429, 430–431
    from deep sea organisms, 59
    emissions, 433
    emissions, reducing, 342
    in global climate change, 334–336
    in Kyoto Protocol, 341–342
    molecule, 54
    in photosynthesis, 60–61, 67
    rising concentrations, 22, 25, 26
    vehicle emissions, 451
carbon enrichment studies, 340
carbon management, 343–344
carbon monoxide
    as air pollutant, 351, 353, 357, 363
    from mining, 429
    from wood burning, 461
carbon neutral systems, 343
carbon oxides, as air pollutant, 353
carbon sequestration, 431
carbon sinks, 67
carbon tetrachloride, 357
carbonate fuel cells, 460
carcinogens, 165, 171, 172
careers, environmental, 569
carnivores, 63, 84, 168
carrying capacity, 118–119, 135–136, 525–526
Carson, Rachel, 18–19, 156, 208, 209, 544
case law, 544, 549–551
Cathedral, Grace, 34
Catlin, George, 289, 290
Catskill Mountains (New York), 413, 414
Cedar Creek Natural History (Minnesota), 228, 229
cells, 57
cellular respiration, 61, 67
cellulose-based biofuels, 465
cellulosic ethanol, 464, 465
Census Bureau, US, 25, 452, 499

core, earth's, 304
core habitat, 274
core regions, 499
Coriolis effect, 327, 328, 330
corn-based ethanol, 464, 465
corridors, of natural habitat, 273
cost-benefit analysis (CBA), 529–530
Costa Rica
    as biodiversity hot spot, 234
    conversion of forests to pasture land, data on, 235
    debt-for-nature swaps in, 258
    gross national product, 527
    Guanacaste National Park, rebuilding of, 258
    integrated pest management in, 220
cotton, 198, 215
Council on Environmental Quality, 366
court system, 549
courts, administrative, 553–554
Cousteau, Jean-Michel, 478
covalent bonds, 54
cover crops, 202
Cowles, Henry Chandler, 92
Creation Care Network, 34, 46
creative thinking, 8
credit, in emissions trading, 531
Cree people, 387, 467
Creutzfeldt-Jacob disease, 160
criminal law, 550
criteria pollutants, 350, 369
critical factors, 76
critical limits, 76–77
critical-thinking skills, 2, 8–10, 11, 41–42
Crockett, David, 534
CropLife America, 210, 212
crude birth rate, 138
crude death rates, 140–141
crust, earth's, 304
Crutzen, Paul, 22, 359, 362
Cuba
    amphibian species in, 126
    integrated pest management in, 220
    organic farming in, 221
cultural capital, 519
cultural creatives, 571
cultural eutrophication, 402–403
culture, environment and, 134–135
cyanide, used in metal processing, 313
cyclones, 330
cyclonic storms, 330–331
cyclonic winds, 327
cytosine, 56

Dai Qing, 20
Daily, Gretchen, 128, 230
dairy, dietary, 185–186
Dale, Bruce, 465
Daly, Herman, 523, 527
dams, 386–392
    controversy over, 387
    ecosystems, damage to, 387
    failures, 386–387, 467
    human populations, displacement of, 387, 467
    hydropower, 466–468
    impact on fish, 387–388
    removal, 389
    sedimentation, 388, 389
    snail darter/Tellico Dam controversy, 243, 541
    as solution to wetlands degradation, 292
    South-to-North Water Diversion Project (China), 390
Darwin, Charles, 75, 79, 117
DDT (dichlorodiphenyltrichloroethane), 22, 164
    banning, 216, 413
    in bioaccumulation and biomagnification, 168

as chlorinated hydrocarbons, 211
discovery of, 209
historical overview, 209
indiscriminate use of, 213, 215
persistence, 215
spill in Lake Apopka (Florida), 214
use in malaria-prone countries, 161, 212
as water pollutant, 406
dead zones, 186, 194, 403, 404
death rates, 137, 138, 140–141, 145–146
debt-for-nature swaps, 258, 576
deciduous forests, 102, 104
decomposer organisms, 63, 84
deductive reasoning, in science, 35–37
Deep-ocean Assessment and Reporting of Tsunami (DART) buoys, 318
deer, white-tail populations, debate over, 121
Defenders of Wildlife, 547
degraded water, 382
demand, in classical economics, 521
demanufacturing, 485
demographic bottlenecks, 127–128
demographic transition, 145–148
    defined, 145
    development, role of, 145–146
    optimistic view, 146
    pessimistic view, 146–147
    social justice, 147
    women's rights and, 148
demographics, 136–143
    defined, 136
Denmark
    green planning in Copenhagen, 511
    green plans, 561
    renewable energy islands in, 448, 449
    wind power in, 343, 448
density-dependent population factors, 120, 124–126
density-independent population factors, 124
deoxyribonucleic acid (DNA). See DNA (deoxyribonucleic acid)
dependency ratios, 142–143
dependent variables, 39
depression, 157
desalination, 392
Descartes, René, 44
desertification, 193, 194, 264
deserts, 102, 103, 113
detritivores, 63, 84
developing countries
    chronic hunger in, 180
    depression in, 157
    disease burden in, 161
    emissions markets in, 532
    international aid to, 533
    sanitation on, 503
    sewage treatment in, 503
    urbanization, 500, 502, 504
    water quality in, 410
development
    conservation development, 512–513
    greenfield developments, 511
    sustainable development, 27–30, 519
development aid, 29, 533
dew point, 375
diabetes, 157, 165
Diamond, Jared, 126
diarrhea, 157
dieldrin, 164
diesel engines, 653
    air pollution from, 353, 362, 370, 433
    efficiency of, 452
    vegetable oil for diesel fuel, 464, 466
diet affecting health, 165

Dioum, Baba, 3
dioxins, 164, 406, 481
disability-adjusted life years (DALYs), 156
discharge, rivers, 380
disease burden, 156–157, 161
diseases, infectious, 22, 157–160, 339
Disney Corporation, 45, 549–550
dispute resolution, 556–561
dissolved oxygen (DO) content, 402
disturbance-adapted species, 94
disturbances
    in biological communities, 93–95
    defined, 93
    human disturbance, 111–113
    in systems, 43
disulfide (pyrite), 71
diversity, in biological communities, 87–88, 89, 90
DNA (deoxyribonucleic acid), 56–57
    analysis, DNA, 230–231
    in genetic engineering, 197
    in natural selection, 76, 79, 80
    sequencing, benefits of, 230–231
Dombeck, Mike, 552
Donne, John, 48
double-blind experiments, 39
downbursts, 331
drip irrigation, 384
droughts, 258, 262, 337, 381, 385
drugs, biodiversity and, 233–234
Duany, Andres, 510
Duck Stamp Act (1934), 296
Ducks Unlimited, 574
duckweeds, 419
Dumping on Dixie (Bullard), 482
dung, 357, 462–463
Dunster, Bill, 513
Durning, Alan, 29–30, 272
Dursban, 222
dust, as air pollutant, 353–354, 358–359
dust plumes, 358

e-waste, 485
earth
    carbon cycle, 67–68
    composition, 304
    hydrologic cycle, 66–67
    layers, 304
    nitrogen cycle, 68–70
    phosphorus cycle, 70–71
    picture, 20
    planet Earth, 20, 304
    sulfur cycle, 71
    tectonic processes, 304–306
Earth Charter Council, 579
Earth Day, 19
Earth First, 575
earth-imaging satellites, 68
Earth Island Institute, 19
earthquakes
    cause, 305–306
    effects of, 316–318
Earthwatch, 569
eastern garbage patch, 478
Eaton, Frederic, 390
Ebola, 158, 160
eco-efficient business practices, 534
Eco-Kids Corps, 568
ecofeminism, 45
ecological diseases, 160
ecological diversity, 230
ecological economics, 522–523
ecological engineering, 419
ecological footprint, calculating, 23

overgrazing (*cont.*)
  indicators of, 7
  threats to US rangelands, 264–265
overharvesting, extinction and, 239–240
overshoots, 119
Ovshinky, Stanford, 457
oxidation, of atoms, 54
oxisols, 191
oxygen
  atom, 53
  molecule, 54
  in photosynthesis, 60–61
oxygen sag, 402
ozone
  as air pollutant, 356
  atmospheric, 324
  solar energy and, 60
  stratospheric, 360–361

Pacific Decadal Oscillation, 334
PacifiCorp, 388
Pakistan, fuelwood, scarcity of, 462
pandas, giant, 78
paper pulp, 255
parabolic mirrors, 455
paradigm shifts, 40–41
parasites, 84
parasitism, 86
parathion, 164
parks and preserves, 267–274
  ecotourism, 271–272
  historical overview, 267
  marine ecosystems, 271
  native people and nature protection, 272–273
  nature preserves, 267–269
  size and shape of, 273–274
particulate removal, air pollution control by, 367
particulates
  as air pollutant, 351, 353–354, 357, 369
  defined, 353
passenger pigeons, 239, 520
passive heat absorption, 454
pastoralists, 264
pasture, 264
  grasslands (*see* grasslands)
  rangelands, 264–265
pathogens, 84, 158
pebble-bed nuclear reactor, 440
pelagic zone, 106
Pelamis wave-power generator, 471
Pennsylvania
  Johnstown flood (1889), 386
  Three Mile Island accident (1979), 440, 444
People for Community Recovery (PCR), 505
perchlorate, 169
perennial species, 202
perfluorooctane sulfonate (PFOS), 169
perfluorooctanoic acid (PFOA), 169
permafrost melting, arctic, 339, 341
persistent organic pollutants (POPs), 82, 168–169, 216
Peru
  cholera in, 158
  pesticide resistant boll weevils in Canete Valley, 215
pest resurgence, 213
pesticide treadmill, 213
pesticides. *See also* toxins
  alternatives, 218–221
  behavioral changes, as alternative to, 218
  benefits, 211–212
  biological controls, 211, 218–219
  crop protection, 212
  defined, 208–209
  exposure, reducing, 222–225

  in groundwater, 412
  as health hazards, 212
  historical overview, 209
  as human health problem, 216–217
  inert ingredients, 222
  insect-borne diseases, controlling, 211–212
  integrated pest management, 219–221
  loss of toxicity, 168
  new pests, creation of, 215
  nontarget species, effects on, 213
  organic farming, 223–224
  persistence, 215–216
  problems, 209–210, 212–213, 215–217
  regulation of, 222–223
  resistance, 81, 82, 161, 213, 215
  spraying, 84
  types, 210–211
  as water pollutants, 406
pests
  defined, 208
  genetic modification of crops for controlling, 198–199
petroleum, 412
Pfiesteria piscicida, 403
pH, 55, 56, 365
Philippines
  as biodiversity hot spot, 232
  scavenging in Manila, 477
  Smoky Mountain open dump in Manila, 477
  wetland disturbances in, 113
Phillips, John, 296
phosphates, 409
phosphorus cycle, 70–71
photochemical oxidants, 355–356
photodegradable plastics, 486
photosynthesis
  carbon cycle, 67
  defined, 60, 62
  energy captured by, 60–61, 326
  remote sensing of, 68
photovoltaic cells, 456–457
phthalates, 169
phylogenetic species concept (PSC), 230
phytoplankton, 106
pigeons, passenger, 239, 520
Pimental, David, 135, 213, 465
Pinchot, Gifford, 17, 18, 281, 284
pioneer species, 93
placer mining, 311
Planning (journal), 289
plants
  air pollution, damage to plants from, 363–364
  clustering for protection, 89
  endangered species products, buying, 240
  endemic, 78
  in hydrologic cycle, 375
  indicator species, 77
  invasive species, 236–238, 280–281
  live specimens, importation of, 240
  as natural pesticides, 219
  nitrogen-fixing, 69
  species, disappearance of, 22
  tree planting, 284–285, 343
plastics, degradable, 486
Plato, 17
plug-in hybrid vehicles, 452
plutonium, 441–442, 444
point sources, of water pollution, 399
Poivre, Pierre, 17
polar bears
  concentrations of chlorinated compounds in, 215
  effects of climate change on, 336
  pesticides in, 360

policy
  defined, 543
  environmental (*see* environmental law)
political economy, 522
politics
  in environmental policy, 542
  and family planning, 150–151
polluter pays principle, 487, 489
pollution
  air (*see* air pollution)
  extinction rates accelerated due to, 239
  market-based mechanisms used to reduce, 530–532
  polluted wetlands sites, 298–299
  in start of environmental movement, 18–19
  water (*see* water pollution)
polybrominated diphenyl ethers (PBDE), 168
polychlorinated biphenyls (PCBs), 22, 163, 164, 360, 413
polyculture of fish and seafood, 187
polyethylene terephthalate (PETs), 484
polyvinyl chloride (PVCs), 484
ponds, 380
Popper, Frank and Debora, 289
Population and Community Development Association (Thailand), 132
Population Bomb, The (Ehrlich), 136
population crash, 119
population dynamics, 116–130
  biotic potential, 118, 124
  carrying capacity, 118–119
  density-dependent factors, 120, 124–126
  density-independent factors, 124
  emigration, 124
  exponential growth, 118
  fecundity, 122
  fertility, 122, 138–139, 148
  immigration, 124
  interspecific interactions, 124–125
  intraspecific interactions, 125
  life expectancy, 122
  life span, 122–123
  logistic growth, 119–120
  mortality, 122–123
  natality, 121
  r-selected species and K-selected species, 120
  stress and crowding, 125
  survivorship, 122–123
population momentum, 142
populations
  defined, 62
  growth rates, 21
  human (*see* human populations)
positive feedback loop, 43, 326
potential energy, 58
poverty
  extreme, 24, 25, 26, 28–29
  food security threat, 180
Powell, John Wesley, 289
power, 426
power towers, 455
prairies
  bison introduction, 289, 290–291
  fire for prairie restoration, 288–289
  restoring, 287–291
  shortgrass prairie, preserving, 289–290
precautionary principle, 558–559
precedents, 549
precipitation. *See also* rainfall
  average annual, 376
  in biome distribution, 100, 101
  uneven distribution, 376
  water cycle and, 375
precision, in science, 35

ultraviolet radiation, 59, 60, 324, 360–361
umbrella species, 243
underground mining, 311
undernourishment, 25, 180–181
unfunded mandates, 420–421
United Arab Emirates, population growth rates, 136
United Church of Christ, Commission on Racial
    Justice, 482
United Kingdom
    BedZED, ecological housing complex in, 511, 513
    garden cities outside London, 510
    MAGNOX nuclear reactor design, 440
    Windscale Plutonium Reactor, 440
United Nations
    air pollution, data on, 22, 362
    and carbon sequestration, 431
    clean drinking water and sanitation, data on, 385
    Conference of Biodiversity (Paris, 2005), 240
    Conference on Environment and Development
        (Rio de Janeiro, 1992), 554
    Conference on the Human Environment
        (Stockholm, 1972), 538, 576
    Convention on Biodiversity, 234
    Convention on Global Climate Change, 542
    corals reefs, data on, 108
    desertification, data on, 194
    Development Agency, 29
    Development Programme, 233, 528
    Earth Summit (1992), 341
    Educational, Scientific, and Cultural Organization,
        272, 546
    Environment Programme, 17, 22, 235, 360, 572
    Food and Agriculture Organization, 135, 180,
        182–183, 185, 186, 187, 191, 196, 254,
        256, 257
    food production, data on, 21
    High Commission on Refugees, 143
    International Conference on Population and
        Development (Cairo, 1994), 150
    Kyoto Protocol on Global Climate Change
        (Japan, 2005), 24
    lack of adequate housing, data on, 503
    millennium assessment, 580–581
    Millennium Development Project, 28–29, 275, 527
    population displacement, data on, 501
    Population Division, 22, 149
    Population Fund, 150
    sewage disposal, data on, 400
    tree planting campaign, 285
    World Food Summit (2003), 180
    World Heritage Conference (1998), 566
United States
    air pollution in, 356, 357, 358, 359, 362, 369
    automobiles, 451–452
    cancer in, 165
    coal mining in, 430
    concentrated animal feeding operations, 186
    consumption levels in, 23
    desalination in, 392
    development aid from, 29, 162
    domestic water use, 382
    droughts, cycle of, 381
    earthquakes, 316–317
    electrical energy sources, 444
    endangered species protection and recovery, 243
    energy consumption, 427–428, 449, 450
    environmental groups, most influential, 574
    environmental racism in, 482
    ethanol production in, 464
    farm subsidies in, 187–188
    flex-fuel vehicles in, 464
    genetically modified crops, use of, 198, 199
    geothermal springs and vents, 470

global warming, 336
gross domestic product, 162
groundwater, dependence on, 391–392
groundwater pollution, 412
hazardous wastes in, 487, 489
honeybee shortage, 208
hunting and fishing laws, 241–242
hydropower use in, 466
integrated pest management in Massachusetts, 220
International Conference on Population and
    Development, refusal to reaffirm, 150
irrigation methods, 195
irrigation provided by government, 384
and Kyoto Protocol on Global Climate Change, 341
landfills burning methane, 463
life expectancy, 122, 141
megalopolises, 499
mercury contamination in, 355
mine reclamation in, 430
National Ambient Air Quality Standards, 369
natural gas resources, 435–436
nuclear power, 437–438, 440, 441, 442, 444, 445
obesity in, 165, 183
old-growth forests in, 260
organic farming in, 223–224
overgrazing in, 264, 265
parks and preserves in, 268, 269–270
Pelican Island, first national bird reservation, 282
pesticide usage, 209, 210, 216, 217
population growth, 141, 142, 143
population politics, 150
population shift west and south, 504, 506–507
recycling in, 484
sediment accumulation in reservoirs, 388
synfuel projects, 431
trash disposal costs, 480
waste incinerators in, 480
waste production, 476, 477
water pollution in, 409, 411, 413
water pricing and allocation policies, 393–394
wealth in, 25
West Nile virus, 159
wetland disturbances in, 113
wildfires in, 262
wind energy use, 468, 469
wood and paper pulp production in, 255
wood, energy from, 461
unmarketables, 536
Unwin, Raymond, 510
urban agglomeration, 498
urban areas
    air quality, 502–503
    challenges, 504–509
    defined, 499
    garden cities, 510
    housing, lack of sufficient, 503–504
    mass-transit, 508–509
    shantytowns, 503
    slums, 503
    smart growth, 509–514
    squatter towns, 503–504
    traffic congestion, 502–503
    water use, 503
    world's largest (chart of), 500
urban runoff, 414
urban sprawl, 506–507
urbanization, 498–501
    brownfield developments, 511
    commute times, expanding, 507–508
    core regions, 499
    in developed world, 502–504
    governmental policies and, 501–502
    green urbanism, 511–512

greenfield developments, 511
new urbanism, 510–511
population shift toward, 498, 499, 500–501
pull factors, 501
push factors, 501
rate of growth, 498, 500–501
reasons for, 501–502
smart growth, 509–514
sprawl, 506–507
world map of, 500
utilitarian conservation, 18

Veblen, Thorstein, 571
Venezuela
    parks and preserves in, 268
    tar sands, 435
verbal learners, 5
Vermont
    emissions standards for woodstoves, 461
    nuclear power, use of, 444
    reforesting of, 282
vertical stratification, 106
vertical zonation, 100
vertisols, 191
Vietnam, deforestation in, 257
villages, defined, 499
vinblastine, 233
vincristine, 233
viral diseases, 212
Viravaidya, Mechai, 132
Virginia, Reston, as planned community, 510
visual learners, 5
vitamin A, deficiency, 183
volatile organic compounds (VOCs), 351, 355
volcanoes, 318–319, 333
    air pollution from, 349
    Cumbre Vieja (Canary Islands), 318
    Mount St. Helens (Washington, 1980), 318
    Mount Vesuvius (Italy), 318
    Nevado del Ruiz (Columbia, 1985), 318
    ring of fire, 318
    Tambora, Indonesia, 318–319
    tectonic processes and, 304–306
vulnerable species, 242

Wallace, Alfred, 85
Ward, Barbara, 16
warm fronts, 330
Warming, J. E. B., 92
Warren, Karen J., 8
Washington
    Mount St. Helens volcano (1980), 318
    Seattle demonstration against WTO, 555–556
Washington, D. C.
    cost of housing in, 506
    Mineral Policy Center, 312
    songbird population decline in, 91
waste disposal, 477–481
    exporting waste, 478–479
    hazardous wastes (see hazardous wastes)
    incineration (see incineration)
    landfills (see landfills)
    ocean dumping, 478
    open dumps, 477
waste-to-energy, 480
wastes
    demanufacturing, 485
    e-waste, 485
    hazardous, 487–493
    reducing, 486–487
    reusing, 485–486
    shrinking the waste stream, 481–487
    solid, 476–477